HANDBOOK OF
Carbohydrate Engineering

HANDBOOK OF

Carbohydrate
Engineering

Edited by
Kevin J. Yarema

CRC Press
Taylor & Francis Group
Boca Raton London New York

CRC Press is an imprint of the
Taylor & Francis Group, an **informa** business

A TAYLOR & FRANCIS BOOK

CRC Press
Taylor & Francis Group
6000 Broken Sound Parkway NW, Suite 300
Boca Raton, FL 33487-2742

First issued in paperback 2019

© 2005 by Taylor & Francis Group, LLC
CRC Press is an imprint of Taylor & Francis Group, an Informa business

No claim to original U.S. Government works

ISBN-13: 978-1-57444-472-8 (hbk)
ISBN-13: 978-0-367-39271-0 (pbk)

Library of Congress Card Number 2004063488

Library of Congress Cataloging-in-Publication Data

Handbook of carbohydrate engineering / edited by Kevin J. Yarema.
     p. cm.
  Includes bibliographical references and index.
  ISBN 1-57444-472-7 (alk. paper)
  1. Carbohydrates--Biotechnology--Handbooks, manuals, etc. 2. Carbohydrates--Research--Methodology--Handbooks, manuals, etc. I. Yarema, Kevin J.

TP248.65.P64H36 2005
660.6'3--dc22
                                                      2004063488

**Visit the Taylor & Francis Web site at
http://www.taylorandfrancis.com**

**and the CRC Press Web site at
http://www.crcpress.com**

Introduction

Zhonghui Sun and Kevin J. Yarema

The term "carbohydrate engineering," has many potential meanings. Strictly speaking, engineering refers to the process of improving a product continuously by standardizing preceding formulations and recombining them to make new and more sophisticated systems. In many respects, based on this definition, Mother Nature herself is the ultimate engineer as she is continually creating novel and increasingly complicated forms of life from sets of "standardized" basic building blocks. Carbohydrates are the perfect example of Nature's engineering efforts as a relatively small handful of basic monosaccharides (nine in humans) are assembled into vastly diverse assemblies of oligo- and polysaccharides that have thousands of structural and regulatory roles. The carbohydrate engineer, upon becoming acquainted with these versatile and incredibly useful molecules, is tempted to jump right in and get to work assisting Nature in designing even more carbohydrate-based structures and adapting them for yet additional applications. But to get started, a number of basic "nuts and bolts" level issues must first be addressed; specifically, one must know what *can* be done, and *how* to do it, by using current and emerging technologies. To address fully these issues and provide the researcher with the optimal set of tools, the term "carbohydrate engineering" is broadly interpreted in this volume.

To some, "carbohydrate engineering" might imply the application of rigorous engineering principles or a technology-based approach to the study or manipulation of carbohydrates. Indeed, engineering and technology have greatly benefited carbohydrate research by providing sophisticated instrumentation, such as that utilized in Chapters 4 to 6[1–3] of this book, which has been indispensable for the recent blossoming of the glycobiology field. Increasingly, "hard-core" engineering-type approaches are being used to complement more empirical, biology-based methods in the study of carbohydrates, as exemplified by the modeling methods outlined in Chapters 7[4] and 8.[5] For others, however, the term "engineering" primarily denotes practicality, especially when used in a biological context. To illustrate, the fields of "genetic engineering" and "tissue engineering" are both highly multi- and interdisciplinary endeavors that eagerly use tools, regardless of if they incorporate any traditional or rigorous engineering aspects, required to get the job done. More specifically, genetic engineering is a mix of microbiology, cell biology, synthetic organic chemistry, biochemistry, and genetics that greatly benefits from technology such as automated DNA synthesizers and sequencers. As an aside, advances in genetic engineering have provided the carbohydrate engineer with powerful tools to manipulate the genes involved in glycosylation. Indeed, many chapters in this book, listed in more detail below in Sections 3 and 4 of the Introduction, exploit recombinant DNA methods to synthesize, modify, and even degrade, complex carbohydrates. Similarly, tissue engineering (the field which seeks to create replacement organs and tissues for worn-out or diseased body parts) exploits and combines advances in cell and molecular biology, polymer chemistry, materials science, developmental biology, and carbohydrates research (especially in the use of polysaccharides as out-lined in Chapter 29.[6] This book takes a cue from these fields and broadly defines "carbohydrate engineering" as the study or manipulation of carbohydrates and their biosynthetic processes whatever methods best facilitate the development of practical applications for these complex and versatile molecules.

On a practical level, the carbohydrate engineer is interested in two basic questions; the first is "What use can I find for this carbohydrate?" and the second is "What methods are available to manipulate this carbohydrate?" To address the first issue, namely identifying research, biomedical, or industrial uses for carbohydrates, a basic knowledge of the chemical structure and biophysical properties of these complex molecules is a necessity. Carbohydrates, like proteins, DNAs, and RNAs, are biopolymers built from a basic set of "building blocks," in this case monosaccharides. Unlike proteins and nucleic acids, however, the biosynthesis of carbohydrates is not template-based; moreover they can be assembled into branching structures of greatly increased complexity compared with completely linear macromolecules. These two factors considerably complicate the study and manipulation of carbohydrates, which can be further classified as oligosaccharides and polysaccharides. In general, smaller, branching structures composed of approximately less than 20 ± 15 (definitions are fairly loose here) constituent monosaccharide units are referred to as oligosaccharides; these molecules often play regulatory or signaling roles. Polysaccharides are typically much larger, mostly linear structures that consist of dozens, or even hundreds, of monosaccharide units. A basic review of carbohydrate structures produced in mammalian cells is provided in Chapter 1[7] and specific carbohydrates, or classes of these molecules, are explored in much greater detail in subsequent chapters. Upon knowing the carbohydrates available, efforts continue to exploit these molecules for a host of practical purposes, a few of which are described in this book.

Once a specific application has been determined for a particular carbohydrate, the next question of "What methods are available to isolate or synthesize as well as purify and characterize this complex sugar?" becomes salient. In many cases, a basic understanding of the strategies that Nature uses to create carbohydrates offers valuable assistance in laboratory and biotechnological strategies to manufacture glycans and often Nature's methods can be adapted, sometimes directly and sometimes in roundabout, but highly ingenious, ways by the carbohydrate engineer. For instance, on many occasions, carbohydrates can be directly harvested from biological sources and on other occasions, cells can be employed as factories to produce glycoconjugates or polysaccharides. Cell culture-based production strategies, while they appear simple and lucrative at first because a cell is doing all — or at least much of — the work, nevertheless benefit from a detailed understanding of underlying biosynthetic processes. Metabolic engineers have learned from bitter experience — or in the case of carbohydrate engineers from bittersweet experience — that pathway bottlenecks afflict product quantity and quality. Accordingly, a thorough understanding of the intricacies of the metabolic networks responsible for carbohydrate biosynthesis is a valuable asset in identifying and confronting these problems.

An important benefit of, and indeed often, the primary motivation for the study of the basic glycosylation processes of a cell is to gain clues on how to intervene in diseases where carbohydrate metabolism has gone awry. Therefore the efforts of an engineer seeking to optimize the cellular production of carbohydrates are likely to overlap significantly with the biomedical researcher who studies the role of glycosylation in human disease. In fact, the general idea of an engineer as physician is rapidly gaining currency as evidenced by the number of universities that have established or expanded biomedical engineering departments in the past decade. This book fittingly includes a critical mass of chapters with a biomedical focus, as listed below in Section 6 of the Introduction. Coming back to the topic of practical methods to produce carbohydrates for "real-world" applications, and moving a step farther from Nature's cell-based production methods but still utilizing her tools, glycosylation enzymes are increasingly being used to assemble oligosaccharides with stereo- and regiospecificity that would have proved to be significant challenges to the synthetic chemist. Notwithstanding such challenges, purely synthetic means to produce complex carbohydrates are achieving even more impressive results and efforts are now in progress to make these methods accessible to the nonspecialist.

The definition of "carbohydrate engineering" used in this volume integrates aspects of basic biology, synthetic chemistry, enzymology, complex instrumentation, and sophisticated modeling toward the production of molecules intended for biomedical, research, or industrial applications. Clearly, it is beyond the scope of one volume to address each of these different subdisciplines in depth — indeed, if such an approach was attempted, this book would largely reproduce other specialized books currently available and exclusively devoted to any of the topics of the purification, the physical characterization, the chemical synthesis, or the biosynthetic production of carbohydrates. Instead, the goal of this book is to provide a "big picture" that gives an overview of the diverse tools available to today's carbohydrate engineer. At the same time, however, the goal of each chapter is also to provide detailed technical information designed to be of value to the specialist.

A quick perusal of the table of contents, or a more detailed reading of the chapter descriptions presented below (in Sections I–VII of the Introduction) provides a snapshot of the numerous tools available to the carbohydrate engineer. Moreover, as mentioned, carbohydrate engineering is a truly multi-disciplinary endeavor where techniques and approaches are mixed and matched as needed to increase greatly the diversity of goals capable of being pursued. For organizational purposes, the chapters are first sorted into five sections based on the *methodology* (as listed below). In addition, topics are also categorized based on the *application* (Section VI) or *classification* of sugar (Section VII). In many cases, a single chapter spans (and is discussed in) two or more of these areas and is therefore mentioned multiple times.

I. ISOLATION, PURIFICATION, AND STRUCTURAL CHARACTERIZATION OF CARBOHYDRATES

In general, the abilities to isolate, purify, and characterize complex carbohydrates are necessary prerequisites to the successful manipulation and exploitation of these complex molecules for any practical purpose. The enormous structural diversity of carbohydrates provides a significant obstacle to achieve these ends; for example, a single site of glycosylation on a protein may host one of several dozen potential glyco-forms.[8–11] or a single repeating disaccharide unit in a polysaccharide may appear in any one of up to 32 different manifestations after epimerization and sulfation reactions take place. This complexity makes it difficult to separate the many similar, but nevertheless different, forms of naturally occurring (or synthetic) carbohydrates and purify them to homogeneity. In some cases, the separation step can be sidestepped if it is possible to analyze or detect specific structures within an unpurified mixture of sugars as they occur *in situ*. For example, Bovin and co-workers[12] describe a "lectin-omics" characterization of cell surface-displayed carbohydrates in Chapter 2, where the objective is to recognize certain epitopes in complex mixtures using lectins or antibodies complementary to a subset of carbohydrate structures (e.g., α-2,6-linked sialic acid or the sialyl Lewis X tetrasaccharide). It can be a daunting, but absolutely crucial task, to use such biological probes to recognize subtle variations in sugar structure, such as the various *O*-acetylated forms of sialic acid recognized as binding sites by human parasites as discussed by Mandal and co-authors[13] in Chapter 3. In other situations, a subset of carbohydrate structures can be altered by metabolic engineering methods to allow the "tagging" and subsequent detection of certain epitopes as discussed by Hinderlich and co-authors[14] in Chapter 13 for sialic acid.

Many times the specialized situations discussed above do not provide an adequate solution to the carbohydrate engineer; instead, the complete physical characterization of a specific carbohydrate of interest is necessary. In these cases, the carbohydrate must be purified to (relative) homogeneity before structural analysis can proceed. In recent years, there has been rapid development of many separation techniques designed to purify complex carbohydrates that include specialized two-dimension (2D) gel electrophoretic[15–19] and HPLC techniques.[20–24] Upon purification, methods must be available to characterize fully an oligosaccharide that are capable of distinguishing between D and L sugars, sequence composition, anomericity, linking positions, and in the case of most oligosaccharides (with the exception of linear polysaccharides), branching structures. Nuclear magnetic resonance (NMR) provides a method for complete structural characterization of carbohydrates of modest size.[25–28] Unfortunately NMR, or complementary methods such as x-ray crystallography,[27,29] require relatively large amounts of material, which is a significant problem considering the trace amounts of many oligosaccharides available after (or even before) purification. Compounding the difficulty of analysis is that, unlike other biomacromolecules such as nucleic acids and proteins, the synthesis of carbohydrates is not template-based. As a consequence, PCR-type amplification methods do not apply, and structural analysis must proceed on the limited quantities of material that can be isolated. Mass spectrometry techniques, in part because of their capacity to analyze minute quantities, are becoming increasingly popular for carbohydrate characterization; these methods often employ instrumentation, such as capillary electrophoresis, which both isolates and characterizes the molecule under test.[30–35] A practical application of sophisticated analytical instrumentation has been to facilitate the diagnosis of congenital disorders of glycosylation (CDG), as discussed by Kościelak in Chapter 4.[1] Another "real-world" situation where carbohydrate characterization is important lies in human milk, where over 900 different oligosaccharides provide a "glycode-fense" for the newborn. Clearly, in order to reproduce these sugars commercially, they must first be identified; this task is described by Hueso and colleagues[2] in Chapter 5. To complement the oligosaccharide analysis described in these aforementioned chapters, several chapters also focus on polysaccharide characterization, as exemplified by the thorough discussion of glycosoaminoglycan (GAG) analysis by Sasisekharan and colleagues[3] in Chapter 6.

The conformational analysis of carbohydrates differs in several ways from that of other biomolecules. For instance, many glycans exhibit numerous conformations that coexist in solution at room or physiological temperatures and a conformational analysis of a carbohydrate must address both spatial and temporal properties. Molecular dynamic simulations are well suited for such condensed phase modeling. The interaction between a carbohydrate and a second biological macromolecule, either covalently linked or not, is capable of having a significant impact on either moiety; such interactions are also amenable to computational approaches. Methods to model carbohydrates have undergone rapid development during the past few years with advances in improving force fields, coupling of NMR measurements with molecular dynamics simulations, direct calculation of thermodynamic properties, application of quantum chemical methods on a large scale, and web access to modeling.[36–38] In this book, the basics of structural modeling of polysaccharides are discussed by Stortz in Chapter 7 by using the carrageenan family of compounds as an illustrative example.[4]

II. STUDY OF GLYCOSYLATION IN LIVING CELLS

Cells can be regarded as miniature factories with all the necessary metabolic machinery to produce complex carbohydrates similar to the way they are now exploited for the production of recombinant proteins. However, significant challenges remain before the production of complex sugars in cell culture or fermentation conditions becomes routine (although it should be noted that intense efforts are now in progess, many of which are surveyed in the current book as outlined in the next section). More specifically, although the basic outline of complex carbohydrate biosynthesis is known (as presented in Chapter 1[7]), both the mechanisms that (i) determine the precise carbohydrate structures and (ii) provide high product yields are still not fully explained. Each of these factors is now briefly discussed (below) to illustrate the types of challenges a carbohydrate engineer faces when contemplating the use of a biological system for the manufacture of a sugar-based product.

A continuing puzzle in the biosynthesis of complex carbohydrates is that different classes of glycoproteins, glycolipids, or GAGs are made by enzymes not always unique to each classification group making it unclear how the synthesis of specific structures is maintained. A partial answer lies in the existence of multiple isozymes for several of the transferases that differ in substrate specificity. One enzyme may serve multiple functions, multiple enzymes may serve overlapping functions, or multiple enzymes may form complexes and function together in the synthesis of these macromolecules. Another key factor, alluded to in the previous sentence, is that biosynthetic enzymes form assemblies confined to specific locations within the ER or Golgi. One example of such spatial localization is provided by the stepwise elaboration of N-linked glycoproteins by enzymes that habitate either the medial- or *trans*-Golgi compartments depending on the glycosidic linkage they process. In another instance, the production of one form of GAG over another is specified when Gal transferases do not interact with GlcA transferase I simply due to their location with cellular organelles.[39] Clearly, the presence of multienzymatic complexes is an important method of fine-tuning and controlling oligo- and polysaccharide synthesis,[40] as well as postsynthetic modifications; these issues are discussed in detail by Sugahara and colleagues[41] in Chapter 9. Unfortunately, the manipulation of enzyme complexes by traditional genetic engineering methods remains difficult and presents significant challenges to one seeking to produce a specific carbohydrate.

Another concern of the carbohydrate engineer is optimization of product output. Achieving high product yields when using a metabolic system can be a formidable task considering that the "glycosylation machinery" of a cell consists of hundreds of components (as discussed further in Chapters 1 and 8, amongst others). To use GAG synthesis as a specific example to illustrate these challenges, it was once thought that the transport of metabolic precursors into the ER/Golgi was the principal mechanism by which the rate of polysaccharide synthesis was controlled. However, in an experiment where cells were incubated with *p*-nitrophenyl-β-D-xylopyranoside, an initiator of GAG biosynthesis, a multifold increase in chondroitin synthesis was observed suggesting that there is an excess of processing enzymes in the pathway and additional regulatory mechanisms are in place.[42] The identification and overcoming of metabolic "bottlenecks" that interfere with, or limit, production of a carbohydrate-based product is an important objective of the carbohydrate engineer. Accordingly, metabolic flux considerations are the topic of chapters in this book that include the discussion of large-scale oligosaccharide production in bacteria by Koizumi[43] in Chapter 10 and fructan biosynthesis in plants by Cairns and Perret[44] in Chapter 11. On a theoretical level, computational metabolic modeling methods described by Murrell and co-authors[5] in Chapter 8 also hold promise in optimizing product output.

To further assist a carbohydrate engineer in his or her efforts to produce a sugar-based molecule by biological methods, several chapters of this book provide detailed descriptions of the biosynthesis of various types of carbohydrates over a wide range of different organisms. Beyond the basic description provided in Chapter 1,[7] Stanley and Patnaik,[45] in Chapter 12, discuss mutant "Lec" CHO cell lines that have been invaluable for dissecting the basics of mammalian glycosylation over the past three decades.[45] Similarly, the uncovering of the molecular defects responsible for congenital disorders of glycosylation, discussed by Kościelak[1] in Chapter 4, has shed considerable light on human glycosylation processes. The use of nonnatural monosaccharides as metabolic probes, described by Hinderlich and co-authors[14] in Chapter 13, has been valuable in uncovering details of sialic acid biosynthesis. Finally, an in-depth description of mammalian GAG biosynthesis is provided by Mizumoto and co-authors[41] in Chapter 9. Moving beyond mammalian glycosylation, organisms ranging from insect and plant cells to yeast and bacteria, each produce characteristic complements of carbohydrates that offer distinctive advantages that can be exploited in a wide range of carefully-tailored applications.

Insect cells and yeast have increasingly been seen as viable alternates for the production of glycosylated proteins and illustrate both the advantages and disadvantages mentioned above. An advantage to each system is that large scale and economical (compared to mammalian cell) culture conditions apply; a disadvantage is that oligosaccharides produced by both cell types have important structural differences from their human counterparts that threaten their therapeutic use. To overcome these difficulties, an in-depth understanding of the inherent glycosylation capabilities of both insect cells and yeast is needed and supplied in Chapters 14 and 15 by Viswanathan and Betenbaugh[46] and Contreras and co-workers,[47] respectively. Plants are also an important source of polysaccharides and, as mentioned above, an overview of the synthesis of plant wall carbohydrates is given by Cairns and Perret[44] in Chapter 11. Finally, bacteria are becoming increasingly important in carbohydrate production schemes, and chapters detailing the glycosylation capabilities of prokaryotes include a description of capsule biosynthesis by Taylor and Roberts[48] in Chapter 16[48] and the use of metabolic analogs as biochemical probes of cell wall synthesis by Sadamoto and Nishimura[49] (Chapter 17).

III. EXPLOITATION OF THE "GLYCOSYLATION MACHINERY" FOR THE PRODUCTION OF COMPLEX CARBOHYDRATES IN LIVING CELLS

The cloning and expression of an ever-growing number of biosynthetic enzymes involved in glycosylation has opened the door for the production of complex carbo-hydrates in cell culture (this section) or cell-free (see Section IV, below) systems. For example, a survey of the literature provides many examples of the use of the common bacterium *Escherichia coli* as a living factory to produce carbohydrates that include chitooligosaccharides,[50] hyaluron,[51] the α-gal xenotransplantation epitope,[52] or *N*-acetyllactosamine-tagged hexasaccharides.[53] In this book, several chapters are devoted to the important topic of bacterial production of polysaccharides. As mentioned above, efforts are in progress to exploit bacterial capsules for polysaccharide engineering as described by Taylor and Roberts[48] in Chapter 16, and new properties can be metabolically engineered into bacterial cell wall polysaccharides by the chemical-biology approach pioneered by Sadamoto and Nishimura[49] (Chapter 17). Bach and Gutnick[54] then describe a specific application for bacterially produced polysaccharides, namely the engineering of these molecules to have heavy metal binding abilities (Chapter 18). For applications such as wastewater cleanup, large amounts of polysaccharides are likely to be needed; consequently, efforts to produce them as economically and efficiently as possible assume paramount importance. This topic is addressed by Koizumi[43] in Chapter 10.

In addition to free-standing carbohydrates, the workhorse *E. coli* host, which has been widely used for the expression of nonglycosylated recombinant proteins, is now being considered for the production of glycoproteins. These efforts, long stymied by the lack of endogenous protein glycosylation in prokaryotes, are now approaching to fruition as clever *E. coli* expression systems capable of glycosylating proteins are coming online. In one example, myoglobin containing β-GlcNAc-serine at a defined position has been expressed in good yield and with high fidelity.[55] Recently, the surprising discovery that *Campylobacter jejuni* produces *N*-linked glycoproteins and gangliosides by glycosylation pathways remarkably similar to those found in eukaryotes has led to efforts to transfer this system to *E. coli*.[56,57] However, the ultimate production of a full repertoire of glycoproteins in these versatile microorganisms remains uncertain as carbohydrate-processing enzymes often need to be localized within the endoplasmic reticulum and Golgi apparatus, organelles lacking in bacteria, to ensure the fidelity of biosynthetic process for specific oligosaccharides. Instead, more appropriate eukaryotic expression systems that employ rodent, insect, or even yeast cells are under development. Glycosylation patterns, however, tend to deviate from those found in human cells even when rodent cells are used as hosts.[11] Insect and yeast cells diverge even further from human glycosylation as they lack the capability to install sialic acid. To overcome these limitations, efforts are in progress to "humanize" production of recombinant glycans; for example, the sialic acid pathway can be engineered *de novo* into insect cells[58–60] as described by Viswanathan and Betenbaugh[46] in Chapter 14. Similarly, promising efforts are in progress to employ various types of fungi for production of humanized *N*-linked glycans, as described by Contreras and co-authors[47] in Chapter 15.

The insect and yeast expression systems described in Chapters 14 and 15, as well as the plant and bacterial polysaccharide-producing hosts described in Chapters 11 and 18, illustrate the value of having a catalog of available glycosylation genes to complement deficiencies in any particular cell type. In general, the identification and cloning of the glycosyltransferases and glycosidases involved in the glycosylation pathways from many different organisms[61] will provide maximum versatility for the development of cell-based expression systems designed for the production of recombinant glycoproteins. The availability of multiple expression system options serves the carbohydrate engineer well as is epitomized by Morrison's discussion of engineered antibodies in Chapter 20,[62] whose structure, function, and pharmacokinetic properties are finely tuned by their exact glycosylation status, which in turn is determined by their expression host.

IV.   CELL-FREE ENZYMATIC METHODS FOR CARBOHYDRATE SYNTHESIS

Significant progress continues toward the use of glycosidases and glycosyltransferases as synthetic tools. The use of these enzymes for the preparation of oligosaccharides provides an interesting alternative to cell-based (see Section III, above) as well as traditional chemical synthesis (see Section V, below) methods. Enyzmatic methods offer the advantage of allowing regio- and stereoselectivity without the need for protecting groups and are becoming valuable tools for stereocontrolled oligosaccharide synthesis in cell-free systems.[62,63] Furthermore, the use of solid-phase techniques offers easy workup procedures and the prospect of automatability.[64] Studies on glycosyltransferases have defined reaction mechanisms and demonstrated that in many cases these enzymes can be coaxed to accept a range of natural and nonnatural substrates thereby increasing synthetic options — if not by the native enzyme, then by "engineered" enzymes, as discussed by Rossi and colleagues[65] in Chapter 21. Effective methodology for the enzymatic synthesis of defined glycoproteins has also been demonstrated.[66] Another twist is that, while not strictly a synthetic process, enzymes are proving valuable to the modification of common carbohydrates isolated from natural sources into higher-value products by using enzymes produced by Aspergillus as discussed by de Vries and colleagues[67] in Chapter 22. In other cases, a very similar objective is to degradatively alter a polysaccharide to either improve food quality or produce industrially valuable products such as ethanol; these topics are described in Chapter 23 by Eyzaguirre and co-authors,[68] who employ β-glucosidases to process cellubiose.

V. CHEMICAL SYNTHETIC METHODS

Organic chemists have developed even more sophisticated approaches to carbohydrate chemistry for over a century since 1894 when Emil Fisher first identified the 16 stereoisomers for the aldohexoses ($C_6H_{12}O_6$), of which D-glucose is the most well-known member. Building on this groundbreaking work, totally synthetic methods to assemble monosaccharides into larger structures of fully defined composition slowly developed for a century until the 1990s when efforts culminated in the production of molecules as complex as the adenocarcinoma antigen KH-1.[69] Semisynthetic methods that make use of organic chemistry to provide certain glycosidic bonds, and the use of recombinant enzymes, often combined with synthetic nucleotide sugar donors[70] to provide other bonds are growing in popularity. Going a step further, synthesis of saccharide structures linked to lipids,[71–73] proteins,[70,74–77] or both as exemplified by the total synthesis of GPI-anchored peptide[78] are being pursued. Carbohydrates have also been assembled into microarrays as a step in the development of high-throughput methods to elucidate recognition events between carbohydrates and proteins (or other interacting partners).[79–84]

Production of highly complex carbohydrate structures, while an amazing achievement, requires highly specialized organic synthetic skills and a large input of time and effort that puts this approach beyond the scope of the nonspecialist. Fortunately, recent directions in the synthesis of complex carbohydrates have shifted to the development of synthetic tools that are simple and easily accessible to the non-specialist and therefore stand to benefit greatly the carbohydrate engineer and glycobiologist. Specific steps in this direction include the use of combinatorial methods as outlined in References 85–87 and automated synthesis.[64,88–90] Another intriguing approach is the use of sets of building blocks for programed, sequential one-pot synthesis where the assembly of an oligosaccharide library has been achieved in a practical and efficient manner. In this work, a di- or trisaccharide was selectively formed without self-condensation and subsequently reacted *in situ* with an anomerically inactive glycoside (mono- or disaccharide) by using a thioglycoside (mono- or disaccharide) with one free hydroxyl group as acceptor and donor coupled with another fully protected thioglycoside. This strategy provides tri- or tetrasaccharides in high overall yield with the help of the anomeric reactivity values of thioglycosides. So far, this approach has been demonstrated for the rapid assembly of 33 linear or branched fully protected oligosaccharides using designed building blocks.[62,91–93]

The synthetic methods described in this book adhere to the rationale described above where the intent is to provide simple (relatively speaking) tools for the non-specialist. von Itzstein and colleagues[94] focus on synthetic approaches to sialic acid in Chapter 24, describing methods to synthesize a wide range of structural variants of this key sugar. Continuing with sialic acid, Hinderlich and co-workers[14] describe synthetic analogs of ManNAc used to manipulate metabolically sialic acid biosynthesis in mammalian cells in Chapter 13; Sadamoto and Nishimura[49] extend this "chemical biology" strategy to bacterial cell wall polysaccharides in Chapter 17. The ability to create molecules that are partly sugar and partly nonsugar is assuming paramount importance with the realization that the sugars are usually not just a superfluous decoration for a biomolecule, but are rather a critical determinant of structure and function. In this book, Schweizer[95] describes the combinatorial synthesis of sugar–amino acid hybrids in Chapter 25, where carbohydrate moieties are used to provide novel properties to amino acids. The intimate interface between sugar and peptide, and the resulting impact of the carbohydrate on protein structure and function, is revisited on a larger molecular scale in Morrison's discussion of antibody glycosylation in Chapter 20.[96] In addition to the handful of chapters whose primary focus is on chemical synthetic methods, chemistry is sprinkled throughout many additional chapters to a greater or lesser extent. One example of a topic presented with a significant chemical component is provided by Bach and Gutnick[54] in Chapter 18, where they describe the chemical modification of bacterially produced polysaccharides. Conversely, Chapter 1, which provides the chemical structures of basic monosaccharides and a few simple oligosaccharides, exemplifies many of the chapters that do not focus on chemistry *per se*, but recognize that a thorough understanding of the topic under discussion benefits from molecular level detail.

VI. APPLICATIONS OF CARBOHYDRATE ENGINEERING

A. INDUSTRIAL AND BIOTECHNOLOGICAL APPLICATIONS

1. Large-Scale Production Processes

A primary consideration when a carbohydrate engineer contemplates the development of sugar-based products is whether there are efficient and cost-effective methods available for the production of these often-times extremely expensive compounds. Cell-based production methods are attractive based on the rationale that a cell can seemingly effortlessly perform complex and incredibly challenging synthetic reactions. In this book, several chapters describe the development of expression systems based on insect cells (Chapter 14), yeast (Chapter 15), plant cells (Chapter 11), or bacteria (Chapters 10 and 16). Alternatively, cell-free enzyme-based systems are also showing promise as efficient production of increasing numbers of recombinant glycosylation-processing enzymes is becoming available. These enzymes are described throughout this book, including Chapters 1, 4, 8–10, 11–18, 22, and 23; an especially intriguing approach with industrial relevance is the synthesis of oligosaccharides by hyperthermophilic glycosidases as described by Rossi and colleagues[65] in Chapter 21.

2. Creation of New Products from Abundant Polysaccharides

The goal of the efforts described in Section VI(A)1, above, is the *de novo* production of an oligosaccharide or polysaccharide, which can be a technically difficult as well as costly proposition. An alternative approach to the development of carbohydrate-based products is to use polysaccharides that Nature provides in large quantities, but that have very little intrinsic value of their own, as raw materials. Two examples are chitin from shellfish and cellulobiose from plant cell walls. Just as Nature postsynthetically processes polysaccharides to provide them with an expanded repertoire of structures and functions, as discussed in Chapters 1,[7] 6,[3] 7,[4] and 9,[41] modification of abundant but low-value chitin- and cellulose-based polysaccharides can be accomplished by enzymatic methods to convert them into a variety of higher value products. These processes are discussed in detail by de Vries and co-workers, who modify plant cell wall polysaccharides using enzymes from Aspergillus as well as by Eyzaguirre and colleagues, who exploit β-glucosidases from filamentous fungi in Chapters 22[67] and 23,[68] respectively.

3. Examples of Industrial-Scale Exploitation of Carbohydrates — Environmental Remediation and Binding of Contaminants

A thorough listing of already-established and newly developing industrial uses for carbohydrates is beyond the scope of the Introduction (and this book); however, a flavor of their versatility can be sampled throughout this volume. Here, we will briefly mention just one application that spans industrial and environmental areas, illustrating how these two endeavors often at odds with each other can be reconciled via use of polysaccharides. Specifically, these molecules are showing remarkable utility for the removal of trace contaminants from both industrial and agricultural products, as well as from the environment. Different categories of polysaccharides can be exquisitely fine-tuned to selectively chelate certain metals even when they appear in miniscule amounts, this topic is discussed in depth by Bach and Gutnick[54] in Chapter 18. Polysaccharide-based bioremediation approaches are not limited to metals; for instance, No and co-authors[97] describe how chitosan can remove compounds ranging from suspended solids, dyes, pesticides, to toxicants from wastewater in Chapter 19.

B. BIOMEDICINE AND HUMAN HEALTH

1. Study of Human Disease and the Development of Diagnostic Tools

There is a growing realization that many, if not most, human diseases have a carbohydrate-related component; it has been a formidable challenge, however, merely to describe the molecular basis of carbohydrates in disease. Many examples illustrating recent advances towards determining the roles that sugars play in disease are provided in this book; these include Stanley and Patnaik's[45] description of glycosylation defects in mutated hamster cells that shed light on abnormalities found in human cancer in Chapter 12, Kościelak's[1] discussion of the molecular basis of CDGs in Chapter 4, and Poulsen's[98] overview of the effects of inulin-type fructans on large intestinal physiology and cancer in Chapter 26. Also, carbohydrates almost always supply a necessary interface between a pathogen and host; examples of infectious disease are given in Hueso's[2] discussion of milk oligosaccharides in Chapter 5, Bovin's[12] discussion of influenza virus in Chapter 2, and Mandal and co-workers'[13] description of *O*-acetyl sialic acids in parasitic disease in Chapter 3. In many cases, it is possible to extend laboratory methods used in these studies to provide clinical diagnosis by streamlining assays; the next step is to actually intervene therapeutically in these processes. While more challenging, this next step is already being taken in many cases. For example, Stanley and Paitnak's[45] cell lines have been instrumental in improving the pharmacokinetic properties of glucocerebrosidase used in enzyme replacement therapy for Gaucher's disease and Hueso's[2] catalogue of key oligosaccharides found in milk provides the first step towards their inclusion in commercial infant formulas. Similarly, both Hinderlich[14] and von Itzstein[94] are developing sialic acid analogs to modulate viral infection, as described in Chapters 13 and 24, respectively.

2. Therapeutic Products

As alluded to in the previous section, the field of glycobiology is currently undergoing a major transformation from being a largely descriptive field endeavoring to elucidate basic biology of glycosylation to become actively involved in developing actual methods to intervene in disease by correcting molecular defects. Indeed, the promise of "Glycomedicine" has been highlighted in recent special issues of high profile journals such as *Science*[99] and *Scientific American.*[100] Monosaccharide-based tools are being developed, as discussed by Hinderlich[14] in Chapter 13, by Sadamoto and Nishimura,[49] Chapter 17, by von Itzstein[94] in Chapter 24, and Schweizer[95] in Chapter 25. At the opposite extreme with respect to size, polysaccharides are finding a plethora of therapeutic uses as overviewed by Taylor and Roberts in Chapter 16. Detailed discussions of specific applications then provided by Pandey and Khuller,[101] who describe the uses of alginate as a drug delivery carrier in Chapter 27, by Liu and co-workers,[102] who cover the prospects for polysaccharide-mediated gene therapy/DNA delivery in Chapter 28, and by Hutmacher and colleagues[6] who give an overview of the many uses polysaccharides are finding in the rapidly growing field of tissue engineering in Chapter 29. Finally, on a slightly different topic, engineered antibodies, discussed by Morrison[96] in Chapter 20 provide an outstanding example of the importance of glycosylation in achieving the intended structure and function of a recombinant glycoprotein; such considerations give impetus to "humanized" production systems described in Chapters 14 and 15.

VII. TYPES OF CARBOHYDRATES — POLYSACCHARIDES, OLIGOSACCHARIDES, AND MONOSACCHARIDES

A. POLYSACCHARIDES

1. Biosynthesis, Postsynthetic Processing, and Characterization

Several chapters describe the biosynthesis of polysaccharides; these include Chapter 9 for glycosoaminoglycans,[41] Chapter 11 for fructan, and Chapters 16–18 for bacterial polysaccharides. Postsynthetic processing of polysaccharides is covered in Chapters 22 and 23. Detailed descriptions of the structural characterization and modeling of polysaccharide conformation are provided in Chapters 6 and 7.

2. Specific Classes of Polysaccharides

Several chapters focus on specific types of polysaccharides. For example, chitin and chitosan are the topic of Chapter 18 where methods for bacterial production of modified forms are discussed by Bach and Gutnick;[54] of Chapter 28 where their utility in DNA delivery are described by Liu and co-workers,[102] and in Chapter 19 where their role in wastewater cleanup is described by No and co-authors, and finally, in Chapter 29 where chitosans (as well as other polysaccharides) are considered as scaffolds for tissue engineering by Hutmacher and colleagues.[6] Fructans are discussed in two chapters, their biosynthesis in Chapter 11[44] and their effects on large intestinal physiology and cancer in Chapter 26.[98] Finally, uses of cellubiose after enzymatic modification are considered in depth in Chapter 23[68] and the prospects for alginate-mediated drug delivery are outlined in Chapter 27.[101]

B. OLIGOSACCHARIDES

1. Biosynthesis, Analysis and Modification

A brief overview of mammalian oligosaccharide biosynthesis is provided[7] in Chapter 1[7] and amplified[1,45] in Chapters 4[1] and 12.[45] Methods to detect, isolate, and analyze oligosaccharides are given in Chapters 2–4[1,12,13] and 8.[5] Methods to synthesize or modify oligosaccharides by metabolic, enzymatic, or chemical means are described in Chapters 10,[43] 12–15,[14,45–47] 21,[65] and 24.[94]

2. N-Linked Oligosaccharides

Of the various types of oligosaccharides (i.e., those found on glycolipids or O-linked to proteins), the largest group discussed in this book consists of the N-linked glycans that comprise the predominant class of mammalian oligosaccharides. Efforts to unravel their biosynthesis are described in Chapters 8 and 12, to characterize their composition in Chapters 2–5,[1,2,12,13] and to develop expression systems in Chapters 10,[43] 12–15,[14,45–47] and 20.[96]

C. MONOSACCHARIDES

1. Basic Building Blocks for the Assembly of Complex Carbohydrates

While the primary focus of this book is on oligo- and polysaccharides, the production of these larger carbohydrates begins with basic monosaccharide "building blocks" that cannot be overlooked by the carbohydrate engineer. The basic biosynthetic procedures needed to supply a cell with nucleotide sugar "building blocks" is outlined in Chapter 1[7] and amplified in Chapter 10[43] where large-scale production methods for these valuable compounds are described. As an alternative to biological supply of monosaccharides, chemical synthesis is possible, with the advantage of endowing these building blocks with novel physical and chemical properties reflected in the larger product. This type of "chemical-biology" approach is described in Chapters 14,[14] 17,[49] 24,[94] and 25.[95]

2. Sialic Acid

Similar to the 20 amino acids commonly found in proteins and four bases found in nucleic acids, nine monosaccharide building blocks are commonly used in mammalian cells (with several additional ones in other phyla). The most intriguing of these basic sugars is sialic acid; in one author's opinion, "Sialic acids are not only the most interesting molecules in the world, but also the most important."[103] Despite the possible hyperbole in this statement, sialic acid is featured prominently in the current book in accordance with this elevated status. This sugar gains at least passing mention in virtually all the chapters touching on oligosaccharides and is the primary focus of several chapters. First, its biosynthesis is used a model to apply metabolic modeling methods to glycosylation by Levchenko and co-authors[5] in Chapter 8. Hinderlich and co-workers[14] then discuss the biochemical engineering of sialic acid metabolism in Chapter 13 and Viswanathan and Betenbaugh,[46] as well as Contreras and co-workers[47] all describe efforts to create insect and yeast systems, respectively, capable of sialylation in Chapters 14 and 15. The role of sialic acid in the infection process of pathogens and parasites is discussed by Bovin and co-workers[12] in Chapter 2, and Mandal and co-authors[13] in Chapter 3. Finally, von Izstein and colleagues describe synthetic routes to sialic acid analogs designed for modulation of viral binding and selectin-mediated cell adhesion events in Chapter 24.[94]

REFERENCES

1.   Kościelak, J., Congenital disorders of glycosylation, *Handbook of Carbohydrate Engineering*, Yarema, K., (Ed.), Dekker/CRC Press, Boca Raton, 2005, chap. 4.
2.   Hueso, P., Martín-Sosa, S., and Martín, M.-J., Role of milk carbohydrates in preventing bacterial adhesion, *Handbook of Carbohydrate Engineering*, chap. 5.
3.   Pojasek, K., Raman, R., and Sasisekharan, R., Structural characterization of glycosaminoglycans, *Handbook of Carbohydrate Engineering*, chap.6.
4.   Stortz, C.A., Carrageenans: structural and conformational studies, *Handbook of Carbohydrate Engineering*, chap.7.
5.   Murrell, M.P., Yarema, K.J., and Levchenko, A., Computational modeling of glycosylation, *Handbook of Carbohydrate Engineering*, chap. 8.
6.   Hutmacher, D.W., Leong, D.T.W., and Chen, F., Polysaccharides in tissue engineering applications, *Handbook of Carbohydrate Engineering*, chap. 29.
7.   Chen, H., Wang, Z., Sun, Z., Kim, E.J., and Yarema, K.J., Mammalian glycosylation: An overview of carbohydrate biosynthesis, *Handbook of Carbohydrate Engineering*, chap. 1.
8.   Endo, T., Groth, D., Prusiner, S.B., and Kobata, A., Diversity of oligosaccharide structures linked to asparagines of the scrapie prion protein, *Biochem*istry, 28, 8380–8388, 1989.
9.   Rudd, P.M., Endo, T., Colominas, C., Groth, D., Wheeler, S.F., Harvey, D.J., Wormald, M.R., Serban, H., Prusiner, S.B., Kobata, A., and Dwek, R.A., Glycosylation differences between the normal and pathogenic prion protein isoforms, *Proc. Natl. Acad. Sci.* USA, 96, 13044–13049, 1999.
10. Gervais, A., Hammel, Y.A., Pelloux, S., Lepage, P., Baer, G., Carte, N., Sorokine, O., Strub, J.M., Koerner, R., Leize, E., and Van Dorsselaer, A., Glycosylation of human recombinant gonadotrophins: characterization and batch-to-batch consistency, *Glycobiology*, 13, 179–189, 2003.

11. Golhke, M., Mach, U., Nuck, R., Zimmermann-Kordmann, M., Grunow, D., Fieger, C., Volz, B., Tauber, R., Thomas, P., Debus, N., and Reutter, W., Carbohydrate structures of soluble human L-selectin recombinantly expressed in baby-hamster kidney cells, *Biotechnol. Appl. Biochem.*, 32, 41–51, 2000.

12. Rapoport, E.M., Mochalova, L.V., Gabius, H.-J., Romanova, J., and Bovin, N.V., Patterning of lectins of Vero and MDCK cells and influenza viruses, The search for additional virus/cell interactions, *Handbook of Carbohydrate Engineering,* chap. 2.

13. Chava, A.K., Mitali Chatterjee, M., and Mandal, C., *O*-Acetyl sialic acids in parasitic diseases, *Handbook of Carbohydrate Engineering,* chap. 3.

14. Hinderlich, S., Oetke, C., and Pawlita, M., Biochemical engineering of sialic acids, *Handbook of Carbohydrate Engineering,* chap. 13.

15. Kuster, B., Krogh, T.N., Mortz, E., and Harvey, D.J., Glycosylation analysis of gel-separated proteins, *Proteomics,* 1, 350–361, 2001.

16. Kishino, S. and Miyazaki, K., Separation methods for glycoprotein analysis and preparation, *J. Chromatogr. B. Biomed. Sci. Appl.,* 699, 371–381, 1997.

17. Taniguchi, N., Ekuni, A., Ko, J.H., Miyoshi, E., Ikeda, Y., Ihara, Y., Nishikawa, A., Honke, K., and Takahashi, M., A glycomic approach to the identification and characterization of glycoprotein function in cells transfected with glycosyltransferase genes, *Proteomics,* 1, 239–247, 2001.

18. Packer, N.H., Ball, M.S., and Devine, P.L., Glycoprotein detection of 2-D separated proteins, *Methods Mol. Biol.,* 112, 341–352, 1999.

19. Koch, G.L. and Smith, M.J., The analysis of glycoproteins in cells and tissues by two-dimensional polyacrylamide gel electrophoresis, *Electrophoresis,* 11, 213–219, 1990.

20. Rohrer, J.S., Analyzing sialic acids using high-performance anion-exchange chromatography with pulsed amperometric detection, *Anal. Biochem.,* 283, 3–9, 2000.

21. Stroop, C.J., Bush, C.A., Marple, R.L., and LaCourse, W.R., Carbohydrate analysis of bacterial polysaccharides by high-pH anion-exchange chromatography and online polarimetric determination of absolute configuration, *Anal. Biochem.,* 303, 176–185, 2002.

22. Campo, G.M, Campo, S., Ferlazzo, A.M., Vinci, R., and Calatroni, A., Improved high-performance liquid chromatographic method to estimate aminosugars and its application to glycosaminoglycan determination in plasma and serum, *J. Chromatogr. B. Biomed. Sci. Appl.,* 765, 151–160, 2001.

23. Cataldi, T.R., Campa, C., and De Benedetto, G.E., Carbohydrate analysis by high-performance anion-exchange chromatography with pulsed amperometric detection: the potential is still growing, *Fresenius J. Anal. Chem.,* 368, 739–758, 2000.

24. Hirabayashi, J. and Kasai, K.-I., Separation technologies for glycomics, *J. Chromatogr. B.,* 771, 67–87, 2002.

25. Jimenez-Barbero, J., Asensio, J.L., Canada, F.J., and Poveda, A., Free and protein-bound carbohydrate structures, *Curr. Opin. Struct. Biol.,* 9, 549–555, 1999.

26. Bush, C.A., Martin-Pastor, M., and Imberty, A., Structure and conformation of complex carbohydrates of glycoproteins, glycolipids, and bacterial polysaccharides, *Annu. Rev. Biophys. Struct.,* 28, 269–293, 1999.

27. Wormald, M.R., Petrescu, A.J., Pao, Y.-L., Glithero, A., Elliot, T., and Dwek, R.A., Conformational studies of oligosaccharides and glycopeptides: complementarity of NMR, X-ray crystallography, and molecular modelling, *Chem. Rev.,* 102, 371–386, 2002.

28. Martin-Pastor, M., Canales-Mayordomo, A., and Jimenez-Barbero, J., NMR experiments for the measurement of proton–proton and carbon–carbon residual dipolar couplings in uniformly labelled oligosaccharides, *J. Biomol. NMR.,* 26, 345–353, 2003.

29. Chandrasekaran, R., X-ray diffraction of food polysaccharides, *Adv. Food Nutr. Res.,* 42, 131–210, 1998.

30. Cooper, C.A., Gasteiger, E., and Packer, N.H., GlycoMod — A software tool for determining glycosylation compositions from mass spectrometric data, *Proteomics*, 1, 340–349, 2001.

31. Zaia, J., Mass spectrometry of oligosaccharides, *Mass Spectrom. Rev.*, 23, 161–227, 2004.

32. Sturiale, L., Naggi, A., and Torri, G., MALDI mass spectrometry as a tool for characterizing glycosaminoglycan oligosaccharides and their interaction with proteins, *Semin. Thromb. Hemost.*, 27, 465–472, 2001.

33. Harvey, D.J., Matrix-assisted laser desorption/ionization mass spectrometry of carbohydrates, *Mass Spectrom. Rev.*, 18, 349–450, 1999.

34. Que, A.H., Mechref, Y., Huang, Y., Taraszka, J.A., Clemmer, D.E., and Novotny, M.V., Coupling capillary electrochromatography with electrospray Fourier transform mass spectrometry for characterizing complex oligosaccharide pools, *Anal. Chem.*, 75, 1684–1690, 2003.

35. Robbe, C., Capon, C., Flahaut, C., and Michalski, J.-C., Microscale analysis of mucin-type *O*-glycans by a coordinated flurophore-assisted carbohydrate electrophoresis and mass spectrometry approach, *Electrophoresis*, 24, 611–621, 2003.

36. Dyekjaer, J.D. and Rasmussen, K., Recent trends in carbohydrate modeling, *Mini Rev. Med. Chem,*, 3, 713–717, 2003.

37. Kuttel, M., Brady, J.W., and Naidoo, K.J., Carbohydrate solution simulations: producing a force field with experimentally consistent primary alcohol rotational frequencies and populations, *J. Comput. Chem.*, 23, 1236–1243, 2002.

38. Woods, R.J., Computational carbohydrate chemistry: what theoretical methods can tell us, *Glycoconj. J,* 15, 209–216, 1998.

39. Sugumaran, G., Katsman, M., and Silbert, J.E., Subcellular co-localization and potential interaction of glucuronyltransferases with nascent proteochondroitin sulphate at Golgi sites of chondroitin synthesis, *Biochem. J.*, 329, 203–208, 1998.

40. McCormick, C., Duncan, G., Goutsos, K.T., and Tufaro, F., The putative tumor suppressors EXT1 and EXT2 form a stable complex that accumulates in the Golgi apparatus and catalyzes the synthesis of heparan sulfate, *Proc. Natl. Acad. Sci. USA*, 97, 668–673, 2000.

41. Mizumoto, S., Uyama, T., Mikami, T., Kitagawa, H., and Sugahara, K., Biosynthetic pathways for differential expression of functional chondroitin sulfate and heparan sulfate, *Handbook of Carbohydrate Engineering*, chap. 9.

42. Parry, G., Farson, D., Cullen, B., and Bissell, M.J., p-Nitrophenyl-β-D-xyloside modulates proteoglycan synthesis and secretory differentiation in mouse mammary epithelial cell cultures, *In Vitro Cell Dev. Biol.*, 24, 1217–1222, 1988.

43. Koizumi, S., Large-scale production of oligosaccharides using engineered bacteria, *Handbook of Carbohydrate Engineering*, chap. 10.

44. Cairns, A.J. and Perret, S.J., Fructan biosynthesis in genetically-modified plants, *Handbook of Carbohydrate Engineering*, chap. 11.

45. Stanley, P. and Patnaik, S.K., Chinese hamster ovary (CHO) glycosylation mutants for glycan engineering, *Handbook of Carbohydrate Engineering*, chap. 12.

46. Viswanathan, K. and Betenbaugh, M., Engineering sialic acid synthetic ability into insect cells: Identifying metabolic bottlenecks and devising strategies to overcome them, *Handbook of Carbohydrate Engineering*, chap. 14.

47. Callewaert, N., Vervecken, W., Geysens, S., and Contreras, R., *N*-glycan engineering in yeasts and fungi: progress towards human-like glycosylation, *Handbook of Carbohydrate Engineering*, chap. 15.

48. Taylor, C.M., and Roberts, I.S., Bacterial capsules: A route for polysaccharide engineering, *Handbook of Carbohydrate Engineering*, chap. 16.

49. Sadamoto, R. and Nishimura, S.-I., Chemo-biological approach to modification of the bacterial cell wall, *Handbook of Carbohydrate Engineering,* chap.17.

50. Samain, E., Drouillard, S., Heyraud, A., Driguez, H., and Geremia, R.A., Gram-scale synthesis of recombinant chitooligosaccharides in *Escherichia coli. Carbohydr. Res.*, 302, 35–42, 1997.

51. Hoshi, H., Nakagawa, H., Nishiguchi, S., Iwata, K., Niikura, K., Monde, K., and Nishimura, S., An engineered hyaluronan synthase: characterization for recombinant human hyaluronan synthase in *Escherichia coli, J. Biol. Chem.*, 279, 2341–2349, 2004.

52. Bettler, E., Imberty, A., Priem, B., Chazalet, V., Heyraud, A., Joziasse, D.H., and Geremia, R.A., Production of recombinant xenotransplantation antigen in *Escherichia coli, Biochem. Biophys. Res. Commun.*, 302, 620–624, 2003.

53. Bettler, E., Samain, E., Chazalet, V., Bosso, C., Heyraud, A., Joziasse, D.H., Wakarchuk, W.W., Imberty, A., and Geremia, A.R., The living factory: *in vivo* production of *N*-acetyl-lactosamine containing carbohydrates in *E. coli, Glycoconj. J.*, 16, 205–212, 1999.

54. Bach, H. and Gutnick, D.L., Engineering bacterial biopolymers for the biosorption of heavy metals, *Handbook of Carbohydrate Engineering,* chap. 18.

55. Zhang, Z., Gildersleeve, J., Yang, Y.Y., Xu, R., Loo, J.A., Uryu, S., Wong, C.H., and Schultz, P.G., A new strategy for the synthesis of glycoproteins, *Science*, 303, 371–373, 2004.

56. Kowarik, M., Wacker, M., and Aebi, M., Production of *N*-linked glycoproteins in *E. coli, Glycobiology*, 13, 851, 2003.

57. Blixt, O., Razi, N., Warnock, D., Gilbert, M., Paulson, J.C., and Wakarchuk, W.W., Efficient chemoenzymatic synthesis of ganglioside mimics GD3, GT3, GM2, GD2, GT2 and GA2, *Glycobiology*, 13, 894, 2003.

58. Jarvis, D.L., Developing baculovirus-insect cell expression systems for humanized recombinant glycoprotein production, *Virology*, 310, 1–7, 2003.

59. Joshi, L., Shuler, M.L., and Wood, H.A., Production of a sialylated *N*-linked glyco-protein in insect cells, *Biotechnol. Prog.*, 17, 822–827, 2001.

60. Lopez, M., Tetaert, D., Juliant, S., Gazon, M., Cerutti, M., Verbert, A., and Delannoy, P., *O*-glycosylation potential of lepidopteran insect cell lines, *Biochim. Biophys. Acta*, 1427, 49–61, 1999.

61. Meynial-Salles, I. and Combes, D., *In vitro* glycosylation of proteins: an enzymatic approach. *J. Biotechnol.*, 46, 1–14, 1996.

62. Koeller, K.M. and Wong, C.H., Complex carbohydrate synthesis tools for glycobiol-ogists: enzyme-based approach and programmable one-pot strategies, *Glycobiology*, 10, 1157–1169, 2000.

63. Ichikawa, Y., Look, G.C., and Wong, C.H., Enzyme-catalyzed oligosaccharide syn-thesis, *Anal. Biochem.*, 202, 215–238, 1992.

64. Tolborg, J.F., Petersen, L., Jensen, K.J., Mayer, C., Jakeman, D.L., Warren, R.A., and Withers, S.G., Solid-phase oligosaccharide and glycopeptide synthesis using gly-cosynthases, *J. Org. Chem.*, 67, 4143–4149, 2002.

65. Moracci, M., Cobucci-Ponzano, B., Perugino, G., Giordano, A., Trincone, A., and Rossi, M., Recent developments in the synthesis of oligosaccharides by hyperther-mophilic glycosidases, *Handbook of Carbohydrate Engineering,* chap. 21.

66. Watt, G.M., Lowden, P.A., and Flitsch, S.L., Enzyme-catalyzed formation of glyco-sidic linkages, *Curr. Opin. Struct. Biol.,* 7, 652–660, 1997.

67. de Vries, R.P., McCann, M.C., and Visser, J., Modification of plant cell wall polysac-charides using enzymes from *Aspergillus, Handbook of Carbohydrate Engineering,* chap. 22.

68. Eyzaguirre, J., M. Hidalgo, and A. Leschot, β-Glucosidases from filamentous fungi: properties, structure and applications, *Handbook of Carbohydrate Engineering,* chap. 23.

69.   Deshpande, P.P. and Danishefsky, S.J., Total synthesis of the potential anticancer vaccine KH-1 adenocarcinoma antigen, *Nature*, 387, 164–166, 1997.

70.   Bulter, T., Schumacher, T., Namdjou, D.J., Gutierrez Gallego, R., Clausen, H., and Elling, L., Chemoenzymatic synthesis of biotinylated nucleotide sugars as substrates for glycosyltransferases, *Chem. BioChem.*, 2, 884–894, 2001.

71.   Duclos Jr, R.I., The total synthesis of ganglioside GM3, *Carbohydr. Res.*, 328,489-507, 2000.

72.   Koeller, K.M. and Wong, C.H., Chemoenzymatic synthesis of sialyl-trimeric-Lewis x, *Chemistry*, 6, 1243–1251, 2000.

73.   Earle, M.A., Manku, S., Hultin, P.G., Li, H., and Palcic, M.M., Chemoenzymatic synthesis of a trimeric ganglioside GM3 analogue, *Carbohydr. Res.*, 301, 1–4, 1997.

74.   Fujita, K. and Takegawa, K., Chemoenzymatic synthesis of neoglycoproteins using transglycosylation with endo-β-N-acetylglucosaminidase A, *Biochem. Biophys. Res. Commun.*, 282, 678–682, 2001.

75.   Matsuo, I., Isomura, M., Miyazaki, T., Sakakibara, T., and Ajisaka, K., Chemoenzymatic synthesis of the branched oligosaccharides which correspond to the core structures of N-linked sugar chains. *Carbohydr. Res.*, 305, 401–413, 1997.

76.   Blixt, O., Allin, K., Pereira, L., Datta, A., and Paulson, J.C., Efficient chemoenzymatic synthesis of O-linked sialyl oligosaccharides, *J. Am. Chem. Soc.*, 124, 5739–5746, 2002.

77.   Marcaurelle, L.A. and Bertozzi, C.R., Recent advances in the chemical synthesis of mucin-like glycoproteins, *Glycobiology*, 12,69R-77R, 2002.

78.   Xue, J., Shao, N., and Guo, Z., First total sysnthzesis of GPI-anchored peptide, *J. Org. Chem.*, 68, 4020–4029, 2003.

79.   Fazio, F., Bryan, M.C., Blixt, O., Paulson, J.C., and Wong, C.H., Synthesis of sugar arrays in microtiter plate, *J. Am. Chem. Soc.*, 124, 14397–14402, 2002.

80.   Bryan, M.C., Plettenburg, O., Sears, P., Rabuka, D., Wacowich-Sgarbi, S., and Wong, C.H., Saccharide display on microtiter plates, *Chem. Biol.*, 9, 713–729, 2002.

81.   Feizi, T., Fazio, F., Chai, W., and Wong, C.H., Carbohydrate microarrays - a new set of technologies at the frontiers of glycomics, *Curr. Opin. Struct. Biol.*, 13, 637–645, 2003.

82.   Wang, D., Liu, S., Trummer, B.J., Deng, C., and Wang, A., Carbohydrate microarrays for the recognition of cross-reactive molecular markers of microbes and host cells, *Nat. Biotechnol.*, 20, 275–281, 2002.

83.   Love, K.R. and Seeberger, P.H., Carbohydrate arrays as tools for glycomics, *Angew Chem. Int. Ed. Engl.*, 41, 3583–3586, 2002.

84.   Park, S., Lee, M.R., Pyo, S.J., and Shin, I., Carbohydrate chips for studying high-throughput carbohydrate-protein interactions, *J. Am. Chem. Soc.*, 126, 4812–4819, 2004.

85.   Sears, P. and Wong, C.H., Toward automated synthesis of oligosaccharides and glycoproteins. *Science*, 291, 2344–2350, 2001.

86.   Grathwohl, M., Drinnan, N., Broadhurst, M., West, M.L., and Meutermans, W., Solid-phase oligosaccharide chemistry and its application to library synthesis, *Methods Enzymol.*, 369, 248–267, 2003.

87.   Kahne, D., Combinatorial approaches to carbohydrates, *Curr. Opin. Chem. Biol.*, 1, 130–135, 1997.

88.   Plante, O.J., Palmacci, E.R., and Seeberger, P.H., Automated synthesis of polysaccharides, *Methods Enzymol.*, 369, 235–248, 2003.

89.   Plante, O.J., Palmacci, E.R., and Seeberger, P.H., Development of an automated oligosaccharide synthesizer, *Adv. Carbohydr. Chem. Biochem.*, 58, 35–54, 2003.

90. Macmillan, D. and Daines, A.M., Recent developments in the synthesis and discovery of oligosaccharides and glycoconjugates for the treatment of disease, *Curr. Med. Chem.*, 10, 2733–2773, 2003.

91. Ye, X.S. and Wong, C.H., Anomeric reactivity-based one-pot oligosaccharide synthesis: a rapid route to oligosaccharide libraries. *J. Org. Chem.*, 65, 2410–2431, 2000.

92. Koeller, K.M. and Wong, C.H., Synthesis of complex carbohydrates and glycoconjugates: enzyme-based and programmable one-pot strategies, *Chem. Rev.*, 100, 4465–4494, 2000.

93. Mong, T.K., Lee, H.K., Duron, S.G., and Wong, C.H., Reactivity-based one-pot total synthesis of fucose GM1 oligosaccharide: a sialylated antigenic epitope of small-cell lung cancer, *Proc. Natl. Acad. Sci.* USA, 100, 797–802, 2003.

94. Wilson, J.C., Kiefel, M.J., and von Itzstein, M., Sialic acids and sialylmimetics: useful chemical probes of sialic acid-recognising proteins, *Handbook of Carbohydrate Engineering,* chap. 24.

95. Schweizer, F., Engineering carbohydrate scaffolds into the side chains of amino acids and use in combinatorial synthesis, *Handbook of Carbohydrate Engineering,* chap. 25.

96. Morrison, S.L., The role of glycosylation in engineered antibodies, *Handbook of Carbohydrate Engineering,* chap. 20.

97. No, H.K., Prinyawiwatkul, W., and Meyers, S.P., Treatment of wastewaters with the biopolymer chitosan, *Handbook of Carbohydrate Engineering,* Chapter 19.

98. Poulsen, M., Effects of short- and long-chained fructans on large intestinal physiology and development of pre-neoplastic lesions in rats, *Handbook of Carbohydrate Engineering,* Chapter 26.

99. Special Issue: Carbohydrates and Glycobiology. *Science*, 291, 2263–2503, 2001.

100. Special Issue: Sweet Medicines, *Sci. Am.*, 287, 40–47, 2002.

101. Pandey, R. and Khuller, G.K., Alginate as drug delivery carrier, *Handbook of Carbohydrate Engineering,* Chapter 27.

102. Liu, W.G., Lu, W.W., and Yao, K.D., Chitosan-based nonviral vectors for gene delivery, *Handbook of Carbohydrate Engineering,* chap. 28.

103. Vimr, E.R., Kalivoda, K.A., Deszo, E.L., and Steenbergen, S.M., Diversity of microbial sialic acid metabolism, *Microbiol. Mol. Biol. Rev.*, 68, 132–153, 2004.

Contributors

Horacio Bach
Division of Infectious Diseases
University of British Columbia
Vancouver, British Columbia
Canada

Michael J. Betenbaugh
Department of Biomolecular and Chemical Engineering
The Johns Hopkins University
Baltimore, Maryland, USA

N. V. Bovin
Shemyakin Institute of Bioorganic Chemistry
Russian Academy of Sciences
Moscow, Russia

Andrew J. Cairns
Plant Genetics and Breeding Department
Institute of Grassland and Environmental Research
Wales, UK

Nico Callewaert
Department of Molecular Biomedical Research
Institute for Microbiology
Swiss Federal Institute of Technology
Zürich, Switzerland

Mitali Chatterjee
Dr. BC Roy Postgraduate Institute of Basic Medical Sciences
Kolkata, India

Anil Kumar Chava
Immunobiology Division
Indian Institute of Chemical Biology
Kolkata, India

Fulin Chen
Division of Bioengineering
National University of Singapore
Singapore

Hao Chen
Department of Biomedical Engineering
The Johns Hopkins University
Baltimore, Maryland, USA

Beatrice Cobucci-Ponzano
Institute of Protein Biochemistry
Consiglio Nazionale delle Ricerche
Naples, Italy

Roland Contreras
Department of Molecular Biomedical Research
Ghent University and Flanders Interuniversity Institute for Biotechnology
Ghent-Zwijnaarde, Belgium

Ronald P. de Vries
Department of Microbiology
Utrecht University
Utrecht, The Netherlands

Jaime Eyzaguirre
Departamento de Ciencias Biológicas
Universidad Andrés Bello
Santiago, Chile

H.-J. Gabius
Institut für Physiologische Chemie
Ludwig-Maximilians-Universität
Munich, Germany

Steven Geysens
Department of Molecular Biomedical Research
Ghent University and Flanders Interuniversity Institute for Biotechnology
Ghent-Zwijnaarde, Belgium

Assunta Giordano
Istituto di Chimica Biomolecolare
Consiglio Nazionale delle Ricerche
Pozzuoli, Italy

David L. Gutnick
Department of Molecular Microbiology and Biotechnology
George S. Wise Faculty of Life Sciences
Tel Aviv University
Tel Aviv, Israel

Mauricio Hidalgo
Departamento de Genética Molecular y Microbiología
Santiago, Chile

Stephan Hinderlich
Charité Universitätsmedizin Berlin
Institut für Biochemie und Molekularbiologie
Berlin-Dahlem, Germany

Pablo Hueso
Departamento de Bioquímica y Biología Molecular
Universidad de Salamanca
Salamanca, Spain

Dietmar W. Hutmacher
Division of Bioengineering
Faculty of Engineering and Department of Orthopaedic Surgery
Faculty of Medicine
National University of Singapore
Singapore

Eun Jeong Kim
Department of Biomedical Engineering
The Johns Hopkins University
Baltimore, Maryland, USA

G. K. Khuller
Department of Biochemistry
Postgraduate Institute of Medical Education and Research
Chandigarh, India

Milton J. Kiefel
Institute for Glycomics
Griffith University
Queensland, Australia

Hiroshi Kitagawa
Department of Biochemistry
Kobe Pharmaceutical University
Kobe, Japan

Satoshi Koizumi
Tokyo Research Laboratories
Kyowa Hakko Kogyo Co., Ltd.
Machida, Tokyo, Japan

Jerzy Kościelak
Department of Biochemistry
Institute of Haematology and Blood Transfusion
Warsaw, Poland

David T.W. Leong
Department of Biological Science
National University of Singapore
Singapore

Andrés Leschot
Departamento de Genética Molecular y Microbiología
Santiago, Chile

Andre Levchenko
Department of Biomedical Engineering
The Johns Hopkins University
Baltimore, Maryland, USA

Wen Guang Liu
Research Institute of Polymeric Materials
Tianjin University
People's Republic of China

William W. Lu
Department of Orthopaedic Surgery
The University of Hong Kong
Hong Kong

Chitra Mandal
Immunobiology Division
Indian Institute of Chemical Biology
Kolkata, India

María-Jesús Martín
Departamento de Bioquímica y Biología Molecular
Universidad de Salamanca
Salamanca, Spain

Samuel Martín-Sosa
Departamento de Bioquímica y Biología Molecular
Universidad de Salamanca
Salamanca, Spain

Maureen C. McCann
Department of Biological Sciences
Purdue University
West Lafayette, Indiana, USA

Samuel P. Meyers
Department of Food Science
Louisiana State University
Baton Rouge, Louisiana, USA

Tadahisa Mikami
Department of Biochemistry
Kobe Pharmaceutical University
Kobe, Japan

Shuji Mizumoto
Department of Biochemistry
Kobe Pharmaceutical University
Kobe, Japan

L.V. Mochalova
Shemyakin Institute of Bioorganic Chemistry
Russian Academy of Sciences
Moscow, Russia

Marco Moracci
Institute of Protein Biochemistry
Consiglio Nazionale delle Ricerche
Naples, Italy

Sherie L. Morrison
Department of Microbiology, Immunology and Molecular Genetics
University of California
Los Angeles, California, USA

Michael P. Murrell
Department of Biological Engineering
Massachusetts Institute of Technology
Cambridge, Maryland, USA

Shin-Ichiro Nishimura
Division of Biological Sciences
Hokkaido University
Sapporo, Japan

Hong Kyoon No
Department of Food Science and Technology
Catholic University of Daegu
Hayang, South Korea

Cornelia Oetke
Research Program Infection and Cancer
German Cancer Research Center
Heidelberg, Germany

Rajesh Pandey
Department of Biochemistry
Postgraduate Institute of Medical Education and Research
Chandigarh, India

Santosh K. Patnaik
Department of Cell Biology
Albert Einstein College Medicine
New York, New York, USA

Michael Pawlita
Research Program for Infection and Cancer
German Cancer Research Center
Heidelberg, Germany

Sophie J. Perret
Plant Genetics and Breeding Department
Institute of Grassland and Environmental Research
Wales, UK

Giuseppe Perugino
Institute of Protein Biochemistry
Consiglio Nazionale delle Ricerche
Naples, Italy

Kevin Pojasek
Biological Engineering Division
Massachusetts Institute of Technology
Cambridge, Massachusetts, USA

Morten Poulsen
Department of Toxicology and Risk Assessment
Danish Institute for Food and Veterinary Research
Søborg, Denmark

Witoon Prinyawiwatkul
Department of Food Science
Louisiana State University
Baton Rouge, Louisiana, USA

Rahul Raman
Biological Engineering Division
Massachusetts Institute of Technology
Cambridge, Massachusetts, USA

E.M. Rapoport
Shemyakin Institute of Bioorganic Chemistry
Russian Academy of Sciences
Moscow, Russia

Ian S. Roberts
Faculty of Life Sciences
University of Manchester
Manchester, UK

J. Romanova
Institute of Applied Microbiology
University of Natural Resources and Applied Life Sciences
Vienna, Austria

Mosé Rossi
Institute of Protein Biochemistry
Consiglio Nazionale delle Ricerche
Naples, Italy

Reiko Sadamoto
Division of Biological Sciences
Hokkaido University
Sapporo, Japan

Ram Sasisekharan
Biological Engineering Division
Massachusetts Institute of Technology
Cambridge, Massachusetts, USA

Frank Schweizer
Department of Chemistry
University of Manitoba
Winnipeg, Manitoba, Canada

Pamela Stanley
Department of Cell Biology
Albert Einstein College Medicine
New York, New York, USA

Carlos A. Stortz
Departamento de Química Orgánica-CIHIDECAR
Universidad de Buenos Aires
Buenos Aires, Argentina

Kazuyuki Sugahara
Department of Biochemistry
Kobe Pharmaceutical University
Kobe, Japan

Zhonghui Sun
Department of Biomedical Engineering
The Johns Hopkins University
Baltimore, Maryland, USA

Clare M. Taylor
Faculty of Life Sciences
University of Manchester
Manchester, UK

Antonio Trincone
Istituto di Chimica Biomolecolare
Consiglio Nazionale delle Ricerche
Pozzuoli, Italy

Toru Uyama
Department of Biochemistry
Kobe Pharmaceutical University
Kobe, Japan

Wouter Vervecken
Department of Molecular Biomedical Research
Ghent University and Flanders Interuniversity Institute for Biotechnology
Ghent-Zwijnaarde, Belgium

Jaap Visser
Fungal Genetics and Technology Consultancy
Wageningen, The Netherlands

Karthik Viswanathan
Department of Biomolecular and Chemical Engineering
The Johns Hopkins University
Baltimore, Maryland, USA

Mark von Itzstein
Institute for Glycomics, Griffith University
Queensland, Australia

Zhiyun Wang
Department of Biomedical Engineering
The Johns Hopkins University
Baltimore, Maryland, USA

Jennifer C. Wilson
Institute for Glycomics
Griffith University
Queensland, Australia

Kang De Yao
Research Institute of Polymeric Materials
Tianjin University
People's Republic of China

Kevin J. Yarema
Department of Biomedical Engineering
The Johns Hopkins University
Baltimore, Maryland, USA

Table of Contents

1 Mammalian Glycosylation: An Overview of Carbohydrate Biosynthesis

Hao Chen, Zhiyun Wang, Zhonghui Sun, Eun Jeong Kim, and Kevin J. Yarema

CONTENTS

I. INTRODUCTION — AN OVERVIEW OF MAMMALIAN GLYCOSYLATION

This chapter provides a basic overview of mammalian glycosylation with a focus on carbohydrate biosynthesis in human cells. The twin objectives of this chapter are to

1

provide a brief synopsis of this topic to benefit the reader who is not a glycobiology expert as well as to serve as a framework for the subsequent carbohydrate engineering chapters presented in this book.* These two goals are complementary in many ways; for example, a fundamental knowledge of *what* carbohydrates are produced in nature provides a carbohydrate engineer with a catalog of materials available for use. Similarly, an understanding of *how* nature produces these complex molecules often provides hints that facilitate their laboratory production or industrial manufacture.

Glycosylation is an extremely complex process and even a cursory description of carbohydrate biosynthesis across a range of diverse organisms such as mammals, marine organisms, fungi, insects, plants, and bacteria is beyond the scope of a single chapter. Instead, the rudimentary overview of glycosylation provided here will be limited to human (and when specific information is not available for human, to other mammalian) cells. Of course, many carbohydrates produced by other organisms are of considerable interest to the carbohydrate engineer and therefore it is gratifying that glycosylation processes share many similarities across species, and even kingdom, boundaries, making the present discussion of human glycosylation broadly relevant. For instance, many of the basic monosaccharide "building blocks" as well as the enzymes that assembly them into complex carbohydrates are widely shared among all organisms. At the same time, important aspects of the biosynthetic process diverge between different species and kingdoms, for example, many organisms employ an expanded repertoire of monosaccharides not found in mammals. These additional monosaccharides endow non-mammalian carbohydrates with unusual and novel properties valuable for biomedical or industrial applications. In addition, enzymes from different species can have significantly different properties, as illustrated in several chapters in this volume.[1–8]

The glycosylation process is divided into five broad (and arbitrarily defined) steps for the purposes of discussion in this chapter (Figure 1.1). The first step involves the cellular uptake of basic monosaccharides, such as glucose or fructose, from extracellular sources such as the diet. In the second step, dietary sugars are diversified by conversion into additional monosaccharides: the most abundant monosaccharide, glucose, for example, can (in theory at least) be converted into any of the other monosaccharides used in glycosylation. The resulting nine monosaccharides are then transformed into high-energy nucleotide sugar donors. The next step entails the transport of the nucleotide sugar donors from the cytosol into the lumens of the endoplasmic reticulum (ER) or Golgi apparatus, the sites where the majority of glycoconjugate biosynthesis (Step 4) occurs. Once a complex carbohydrate is synthesized, postsynthetic processing often takes place to endow the glycoconjugate with an expanded repertoire of physical and chemical properties (Step 5). Finally, although not discussed in detail here, mature glycans are recycled via salvage pathways and their components are reused, providing an alternative source (to dietary intake) for a cell to obtain the basic molecular building blocks used in carbohydrate biosynthesis.

---

*A detailed overview of the chapters appearing in this book is provided in the Introduction. It should be noted, however, that they are also referenced throughout Chapter 1 when they provide additional elaboration of a specific aspect of glycosylation that is beyond the scope of the current discussion.

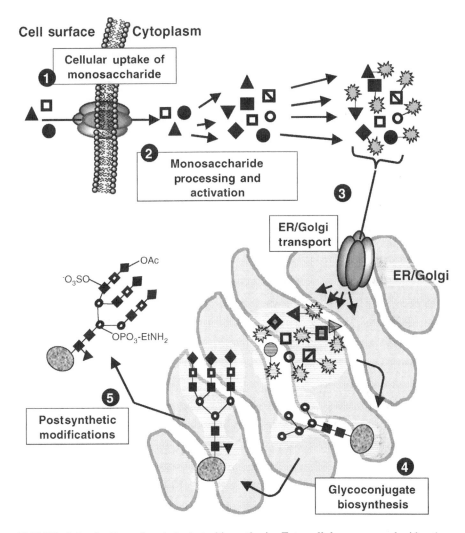

FIGURE 1.1 Outline of carbohydrate biosynthesis. Extracellular monosaccharides (see Figure 1.2 for chemical structures) are taken into cells (Step 1) by families of transporters (listed in Table 1.1) and are then diversified by a series of phosphorylation, epimerization, and acetylation reactions (Step 2 and Figure 1.3), then activated and converted into high-energy nucleotide sugar donors (see Figure 1.4). The sugar donors are either used in the cytoplasm or transported into the ER or Golgi (Step 3 and Table 1.3), where the majority of cellular glycoconjugate biosynthesis occurs (Step 4). Finally, co- or postsynthetic chemical modification of the carbohydrate fine-tunes structure and function (Step 5).

II. CARBOHYDRATE BIOSYNTHESIS

A. Cellular Intake of Sugars

Mammalian cells obtain monosaccharide "building blocks" needed for glycosylation from dietary sources, endogenous metabolism, and salvage pathways. The ultimate

sources of monosaccharides in mammals are dietary carbohydrates obtained as polysaccharides or oligosaccharides and broken down into simple carbohydrates (monosaccharides and disaccharides) by glycosidases found in the digestive system. The
major monosaccharides resulting from carbohydrate digestion are glucose, galactose,
and fructose. The most abundant dietary monosaccharide is glucose, which is the
hydrolysis product of starch (or cellulose when digested by symbiotic microbes).
Other simple monosaccharides are also found in food sources, for instance, fructose
and the disaccharide sucrose are found in many fruits, vegetables, and honey, and
galactose is plentiful in milk. The simple sugars can be absorbed into the blood stream
once they are in the monosaccharide form and taken up by cells through the action of
membrane transporters (Table 1.1). Inside a cell, the full set of monosaccharides

TABLE 1.1
Hexose (Monosaccharide) Transporters Found in the Plasma Membrane of Mammalian Cells

Type	Tissue	Sugar Transported (K_m in mM)	Function and Comments	Ref.
SGLT Family			AKA[a] the SLC5A family, specific member numbers are shown in the left column	10
SGLT-1 (A1)	Small intestine, kidney	Glucose (0.1–0.8) galactose	Major apical glucose transporter	14,15
SGLT-2 (A2)	Kidney	Glucose (1.6)	Glucose uptake	16,17
SGLT-3 (A4)	Small intestine, muscle		Serves as a glucose sensor, not a transporter	18,19
SGLT-4				
SGLT-5 (A10)			Identified through full-length sequencing matching to gi:22748760	20
SGLT-6 (A11)		*Myo*-inositol (0.12)	AKA, the sodium *myo*-inositolcotransporter 2 (SMIT2)	21,22
GLUT Family				10,13
GLUT-1 (subfamily I)	Erythrocytes, blood /brain barrier	Glucose (3–5) galactose mannose glucosamine	Primary glucose transporter of fetal	23,24
GLUT-2 (sfI)	Liver, small intestine, brain	Glucose (17) galactose (92) mannose (125) fructose (76) glucosamine (0.8)	High capacity and low affinity	24–26
GLUT-3 (sfI)	Neurons, placenta	Glucose (1–2) galactose mannose	Primary glucose transporter of neurons	27

(Continued)

TABLE 1.1 (Continued)

Type	Tissue	Sugar Transported (Km in mM)	Function and Comments	Ref.
		maltose xylose dehydroascorbate		
GLUT-4 (sfI)	Adipose tissue, skeletal muscle,	Glucose (5) glucosamine (3.9) dehydroascorbate	Insulin-stimulated glucose uptake	28
GLUT-5 (sfII)	Small intestine	Fructose (6) glucose (in rat)	Primarily fructose absorption	29,30
GLUT-6 (sfIII)	Adipose tissue, leukocytes, brain	Transports glucose (1–5) in liposomes	Retained in an intracellular compartment and not responsive to insulin; formerly designated as "GLUT-9"	31,32
GLUT-7 (sfII)	Liver	Fructose glucose (?)	Supposed to transport fructose based on homology to GLUT-5	12
GLUT-8 (sfIII)	Testis, blastocysts, brain, muscle, adipocytes	Glucose (2) galactose fructose	Retained in an intracellular compartment and not responsive to insulin; formerly designated as "GLUTX1"	31,33
GLUT-9 (sfII)	Liver, kidney, small intestine, placenta, lung, and leukocytes		Supposed to transport fructose based on homology to GLUT-5 although no sugar transport activity has been demonstrated to date	13,34
GLUT-10 (sfIII)	Heart, lung, brain, liver, skeletal muscle, pancreas, placenta, and kidney	2-DOG[b] (0.3) glucose galactose		35–37
GLUT-11 (sfII)	Heart, skeletal muscle	Glucose fructose	mRNA is detected in many tissues, GLUT-11 occurs in three alternative splice forms (a, b, and c)	38–40
GLUT-12 (sfIII)	Heart, skeletal muscle, brown adipose tissue, prostate, mammary gland		Substrate specificity and sugar transport activity is unknown	37,41,42
HMIT-1 (sfIII)	Brain	*Myo*-inositol (0.1)		43

[a] AKA = "also known as".

[b] 2-DOG = 2-deoxyglucose.

needed for glycoconjugate synthesis (Figure 1.2) can be derived from dietary sugars through a series of epimerization and acetylation reactions (Figure 1.3).

Absorption of monosaccharides into cells in the gastrointestinal system occurs across the brush-border and basolateral membranes of enterocytes and is mediated by both sodium-dependent and -independent membrane proteins. Cellular intake of simple sugars is a carrier-mediated process that exhibits enzyme-like characteristics including substrate specificity, stereospecificity, and saturation kinetics. Glucose transporters, for example, follow simple Michaelis–Menten saturation kinetics and behave like enzymes when D-glucose on one side of a membrane is considered to be the substrate and the product is considered to be D-glucose on the other side of the membrane. At least two types of monosaccharide transporters move monosaccharides from the intestinal lumen into the epithelial cell.[9] The first category, the SLC5A gene family, consists of up to six Na^+-monosaccharide energy-dependent co-transporters known as sodium–glucose transporters (SGLTs).[10,11] Hydrolysis of ATP by the action of the sodium-potassium ATPase provides energy to export three Na^+ ions in exchange for the import of two K^+ ions and maintains a low intracellular level of sodium. The

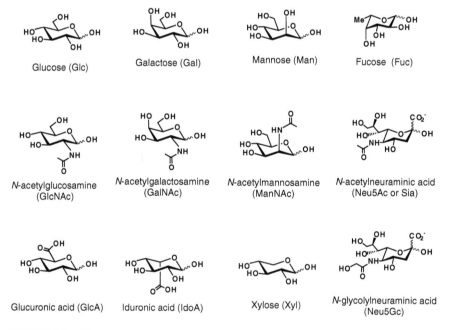

FIGURE 1.2   Chemical structures of common monosaccharides found in mammalian carbohydrates. The names, common abbreviations, and chemical structures of the nine monosaccharides used for mammalian glycoconjugate biosynthesis are shown along with ManNAc, an intermediate used for sialic acid biosynthesis (Figure 1.3), and IdoA, a postsynthetic epimer of GlcA found in GAGs (Figure 1.15). In addition, two forms of sialic acid are shown; Neu5Ac is the most common form of this sugar in humans while all other mammals also produce Neu5Gc. All of these monosaccharides occur in the D-conformation in mammals except for L-fucose.

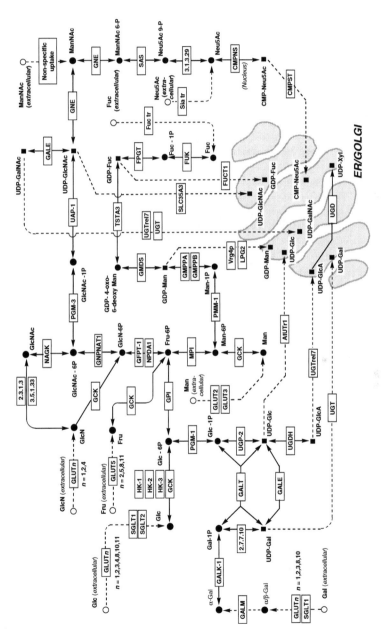

FIGURE 1.3 Outline of monosaccharide processing reactions. This diagram depicts transport steps (dashed lines) for the cellular intake of simple carbohydrates, the basic processing reactions used to produce monosaccharide diversity and manufacture nucleotide sugar donors (solid lines), and a further set of transport reactions where the sugar donors are taken into the ER or Golgi (dashed lines). Additional information about the transporters is provided in Tables 1.1 and 1.3 (for plasma membrane and ER or Golgi transporters, respectively) and Table 1.2 provides information on the processing enzymes.

TABLE 1.2
Enzymes Involved in Monosaccharide Processing and Carbohydrate Assembly

Designation	Name and Description	Ref.
2.3.1.3[a]	Glucosamine *N*-acetyltransferase	56
2.4.1.47[a]	*N*-acylsphingosine galactosyltransferases (AKA[b] GalCer synthase)	57
2.4.1.147[a]	Core 3 β-GlcNAc-transferase	GI:17384687
2.4.1.148[a]	Core 6 β-GalNAc-transferase B	58
2.4.1.152[a]	α-3-L-Fucosyltransferase	59
2.7.7.10[a]	Galactose-1-phosphate uridylyltransferase	60
3.1.3.29[a]	*N*-acylneuraminic acid 9-phosphate phosphatase	61
3.5.1.33[a]	*N*-acetylglucosamine deacetylase	62
A4GALT	α-1,4-galactosyltransferase; globotriaosylceramide/CD77 synthase	63,64
ABO	ABO blood group (transferase A, α-1-3-*N*-acetylgalactosaminyltransferase; transferase B, α-1,3-galactosyltransferase)	65,66
B3GALT3	GlcNAc β-1,3-galactosyltransferase 3 (AKA[b] globoside synthase; globotriaosylceramide 3-β-*N*-acetylgalactosaminyltransferase	67
B3GALT4	UDP-Gal:β-GlcNAc β-1,3-galactosyltransferase, polypeptide 4	68,69
B3GALT5	UDP-Gal:β-GlcNAc β-1,3-galactosyltransferase, polypeptide 5	68,70
B3GNT3	UDP-GlcNAc:β-Gal β-1,3-*N*-acetylglucosaminyltransferase 3	68,71,72
B3GNT5	UDP-GlcNAc:β-Gal β-1,3-*N*-acetylglucosaminyltransferase 5	73
B4GALT1	UDP-Gal:β-GlcNAc β-1,4-galactosyltransferase, polypeptide 1	
C1GALT1	Core 1 UDP-galactose:*N*-acetylgalactosamine-α-*R* β-1,3-galactosyltransferase	68,74
CMPNS	CMP-Neu5Ac synthetase	75
CSAH	Cyclic sialic acid hydrolase	76
CTBS	di-*N*-acetyl-Chitobiase	77,78
DDOST	Dolichyl-diphosphooligosaccharide-protein glycosyltransferase	79
EXT1	Exostoses (multiple) 1 (GAG biosynthesis)	80,81
EXT2	Exostoses (multiple) 2 (GAG biosynthesis)	82,83
EXTL1	Exostoses (multiple)-like 1 (GAG biosynthesis)	84,85
EXTL2	Exostoses (multiple)-like 2 (GAG biosynthesis)	86,87
EXTL3	Exostoses (multiple)-like 3 (GAG biosynthesis)	88
FPGT	Fucose-1-phosphate guanylyltransferase	89
FS	Forssman glycolipid synthetase	90,91
FUK	L-fucose kinase	92
FUT1	Fucosyltransferase 1	93,94
FUT2	Fucosyltransferase 2	94,95
FUT3	Fucosyltransferase 3	96,97
FUT5	Fucosyltransferase 5	96,98
FUT6	Fucosyltransferase 6	96,99
FUT8	Fucosyltransferase 8	100,101
GALE	UDP-Galactose-4-epimerase	102
GALK1	Galactokinase 1	103
GALM	Galactose mutarotase (aldose 1-epimerase),	104

(Continued)

TABLE 1.2 (Continued)

Designation	Name and Description	Ref.
GALNT6	UDP-N-acetyl-α-D-galactosamine:polypeptide N-acetylgalactosaminyltransferase 6	105
GALGT	UDP-N-acetyl-α-D-galactosamine:(N-acetylneuraminyl)-galactosylglucosylceramide N-acetylgalactosaminyltransferase	106
GALT	Galactose-1-phosphate uridylyltransferase	107
GCK	Glucokinase (AKA hexokinase 4)	108
GCNT1	Glucosaminyl (N-acetyl) transferase 1, core 2 (β-1,6-N-acetylglucosaminyltransferase)	109
GCS1	Glucosidase 1	110,111
GFPT-1	Glutamine-fructose-6-phosphate transaminase 1	112
Ggta1	β-D-galactosyl-1,4-N-acetyl-D-glucosaminide α-1,3-galactosyltransferase (mouse)	113,114
GMDS	GDP-mannose 4,6-dehydratase	115,116
GMPPA	GDP-mannose pyrophosphorylase A	20
GMPPB	GDP-mannose pyrophosphorylase B	117
GN6ST	N-acetylglucosamine-6-O-sulfotransferase	118–121
GNE	UDP-GlcNAc 2-Epimerase/ManNAc 6-kinase	122
GNPI	Glucosamine-6-phosphate deaminase 1	123
Gnpnat-1	Glucosamine-phosphate N-acetyltransferase 1 (mouse)	124
GNPNAT-1	Glucosamine-phosphate N-acetyltransferase 1 (human)	20
GPI	Glucose phosphate isomerase	125
HK-1	Hexokinase 1 isoform HKI-td; brain form hexokinase	126,127
HK-2	Hexokinase 2 (muscle)	128
HK-3	Hexokinase 3 ATP:D-hexose 6-phosphotransferase	129,130
LPG2	Leishmania GDP-mannose transporter	131
MAN1A2	Mannosidase, α, Class 1A, Member 2	132
MAN2A2	Mannosidase, α, Class 2A, member 2	133
MGAT1	Mannosyl(α-1,3-)-glycoprotein β-1,2-N-acetylglucosaminyltransferase I	134,135
MGAT2	Mannosyl(α-1,6-)-glycoprotein β-1,2-N-acetylglucosaminyltransferase II	136
MGAT3	Mannosyl(β-1,4-)-glycoprotein β-1,4-N-acetylglucosaminyltransferase	137
MGAT4B	Mannosyl(α-1,3-)-glycoprotein β-1,4-N-acetylglucosaminyltransferase, isoenzyme B	138
MGAT5	Mannosyl(α-1,6-)-glycoprotein β-1,6-N-acetylglucosaminyltransferase	139,140
MPI	Mannose phosphate isomerase	141
NAGK	N-acetylglucosamine kinase	142
PGM1	Phosphoglucomutase 1	143
PGM3	Phosphoglucomutase 3	144,145
PMM1	Phosphomannomutase 1	146
SAC	Sialic acid cyclase	76,147
SANAE	Sialic acid N-acetylesterase	76,148–150

(Continued)

TABLE 1.2 (Continued)

Designation	Name and Description	Ref.
SANAT	Sialic acid *N*-acetyltransferase	76,148–150
SIAT4A	Sialyltransferase 4A	151–153
SIAT4B	Sialyltransferase 4B	151,154
SIAT4C	Sialyltransferase 4C	151,155
SIAT6	Sialyltransferase 6	156
SIAT7A	Sialyltransferase 7A	157,158
SIAT7B	Sialyltransferase 7B	159,160
SIAT8A	Sialyltransferase 8A	161,162
SIAT8E	Sialyltransferase 8E	
SIAT9	Sialyltransferase 9	163
TSTA3	Tissue-specific transplantation antigen P35B	115
UAP1	UDP-*N*-GlcNAc Pyrophosphorylase 1	164
UDG	UDP-glucuronate decarboxylase 1 (UXS-1)	165
UGDH	UDP-glucose dehydrogenase	166
UGP-2	UDP-glucose pyrophosphorylase 2	20

[a]EC (Enzyme Commission) numbers are provided in cases where an enzyme activity for a specific transformation has been described but a specific protein responsible for this activity has not yet been cloned and characterized.
[b]AKA = "also known as."

sodium gradient between the intestinal lumen and the cytoplasm provides the driving force for active carbohydrate transport; emerging evidence indicates that water is co-transported along with Na$^+$ and sugar through SGLTs. The SLC2A gene family of 13 currently known monosaccharide transporters were previously known as the GLUT transporters. They provide a second general method of sugar uptake by mammalian cells where monosaccharides transverse a carbohydrate concentration gradient by a sodium-independent facilitated diffusion mechanism.[12,13]

The two families of monosaccharide transporters both prominently facilitate the intake of glucose into cells as would be expected considering their names. It is not surprising that the entry of glucose into a cell is mediated via multiple transporters and is a process of considerable complexity,[10] because this sugar is the main source of energy for mammalian cells. Various members of these "glucose" transporter families, however, transport additional monosaccharides in addition to, or instead of, glucose. Fructose transport, for example, is carried out primarily by GLUT-5; mannose transport also occurs efficiently and supplies up to 80% of the mannose used for glycoprotein synthesis in some cell types.[44] A fucose-specific transporter has been found in several types of mammalian cells[45] and various transporters also recognize galactose, glucosamine, and a variety of additional sugars as listed in Table 1.1. Monosaccharide transport remains an area of active investigation as efforts continue to specify the exact sugars transported by several of the GLUT transporters that currently remain unknown or unclear (Table 1.1). Conversely other sugars, exemplified by sialic acid, are efficiently taken into cells but the transporter involved[46] as well as the transport mechanism are not well characterized at a molecular level.[47]

TABLE 1.3
Nucleotide Transport in the ER and Golgi

Nucleotide Sugar Donor	Compartment[a] ER	Golgi	Human Gene	Transporter Name	Ref.
ATP				Rat Golgi membrane ATP transporter	183
CMP-Sia	− − −	+ + +	SLC35A1	CMPST; CMP-sialic acid Golgi transporter	184
GDP-Fuc	− − −	+ + + +	SLC35C1	FUCT1; GDP-fucose transporter 1	185,186
GDP-Man	− − −	+ + + +		Vrg4p; Yeast GDP-mannose Golgi transporter	187
PAPS	− − −	+ + + +	PAPST1	PAPS transporter	188,189
UDP-Gal	− − −	+ + + +	SLC35A2	UGT; UDP-galactose transporter	190–192
UDP-GalNAc	+ +	+ + + +	SLC35A2 SLC35D1	UGT (see UDP-Gal) UGTrel7 (see UDP-Gal/UDP-GlcA)	
UDP-Glc	− − −	+ + + +		AtUTr1; *Arabidopsis thaliana* UDP-galactose/UDP-glucose transporter	193
UDP-GlcA	+ + + +	+ + + +	SLC35D1	UGTrel7; UDP-glucuronic acid/ UDP-*N*-acetyl-galactosamine transporter	194
UDP-GlcNAc	+ +	+ + + +	SLC35A3	UDP-GlcNAc transporter	195
UDP-Xyl	+ +	+ + + +			

[a]The relative distribution of the nucleotide transporters in the ER and Golgi is indicated by the number of plus (+) signs. A minus (−) indicates that the transporter is not found in that compartment.

In addition to dietary sources, monosaccharides can be salvaged from glyco-conjugates degraded within the same or proximal cells, for example, 15 to 90% sialic acid is recycled for use in the production of new glycoproteins. In cell culture media, fetal bovine serum provides a rich source of salvaged sugars for glycosylation. The reutilization of salvaged sugars usually begins with their hydrolysis to monosaccharides by glycosidases that are active at the low pH found in lysosomes.[48] The monosaccharides exit the lysosome via transporters specific for acidic sugars such as sialic or glucuronic acids,[49,50] neutral monosaccharides including glucose, galactose, mannose, and fucose,[51,52] or aminosugars such as GlcNAc and GalNAc.[53] Upon egress from the lysosome, these sugars enter the cytoplasm where many of the enzymes that process monosaccharides and activate them to nucleotide sugar donors are located allowing them to be used in a subsequent round of oligosaccharide biosynthesis.[54]

B. INTRACELLULAR MONOSACCHARIDE PROCESSING

Typically, only glucose and fructose are obtained in large quantities from the diet whereas ten monosaccharides, nine of them produced in the cytosol, commonly occur in mammalian glycans (Figure 1.2). Smaller amounts of glucosamine, fructose, mannose, fucose, sialic acid, and galactose can also be obtained from food but each of these sugars, as well as the additional monosaccharides GlcNAc, GalNAc, xylose, and glucuronic acid (and its epimer iduronic acid) must be produced intracellularly by the metabolic network outlined in Figure 1.3 to provide adequate amounts of these sugars to supply glycoconjugate biosynthesis. Regardless of whether a monosaccharide is obtained from dietary sources, intracellular processing reactions or salvage pathways, the sugar must be activated and converted into its high-energy nucleotide sugar donor form for use in oligosaccharide biosynthesis. The nucleotide sugars are coupled to UDP for glucose, galactose, GlcNAc, GlcA, GalNAc, and xylose; to GDP for mannose and fucose; or CMP for sialic acids (Figure 1.4). The formation of glycosidic bonds that link monosaccharides to proteins or lipids, or to each other, is driven in the forward direction by the release of free energy provided when a monosaccharide is released from the nucleotide di- (or mono-) phosphate sugar form in glycosyltransferases-mediated reactions. After synthesis in the cytosol, nucleotide sugar donors are imported into the ER and Golgi by the action of the transporters outlined in Figure 1.3 and listed in Table 1.3.[55] The metabolism of each of the sugar is now discussed briefly.

1. *Glucose*: Glucose can meet several fates in a cell. Generally, it is phosphorylated in an ATP-dependent reaction to form Glc-6P upon entering a cell. Glc-6P can be converted into Glc-1P by phosphoglucomutase and then activated and used with the UDP-Glc nucleotide sugar donor, which can be further converted into UDP-GlcA or UDP-Gal. In an alternative reaction, Glc-6P can be isomerized to Fru-6P that can then be converted into Fru-1,6P$_2$ and used for energy production via the glycolysis pathway. Alternatively, Fru-6P can be converted into GlcN-6P; this intermediate can subsequently be used for the biosynthesis of GlcNAc, GalNAc, or sialic acid.

2. *Fructose*: Although glucose is the most abundant exogenously obtained monosaccharide, other sugars, predominantly fructose, are absorbed by cells in significant quantities. Fructose, whether obtained from exogenous sources or converted from glucose, is phosphorylated to form Fru-6P, a sugar primarily used as an energy source in the glycolysis pathway after further phosphorylation to Fru-1,6P$_2$. Alternatively, as mentioned above, Fru-6P can be converted into GlcN-6P and subsequent sugars. The relative distributions of flux between these different pathways is cell-type-dependent, for example, liver, kidney, and muscle cells each maintain a distinct level of Fru-6P.

3. *Galactose*: The conversion of galactose into UDP-Gal and UDP-Glc via the Leloir pathway,[167] which consists of the GALM, GALK-1, GALT, and GALE enzymes shown on the lower left area of Figure 1.3, is another (in addition to glycolysis) well-studied segment of the mammalian glycosylation process. Briefly, galactose is converted into UDP-Gal by first being phosphorylated to Gal-1P; Gal-1P is then activated in a reaction catalyzed by galactose-1-phosphate uridylyltransferase (GALT, Table 1.2) by acquisition of an uridyl group from UDP-Glc to form UDP-Gal and regenerate Glc-1P. Interestingly, the galactose moiety of UDP-Gal can then be epimerized to glucose in a reversible reaction catalyzed by UDP-galactose-4-epimerase (GALE). The conversion of UDP-Glc into UDP-Gal is essential for the synthesis of galactosyl residues in complex polysaccharides and glycoproteins if the amount of galactose in the diet is inadequate to meet these needs. In certain cases, UDP-Gal can be formed from Gal-1P in a UTP-dependent reaction catalyzed by galactose-1-phosphate uridylyltransferase (EC 2.7.7.10) activity.

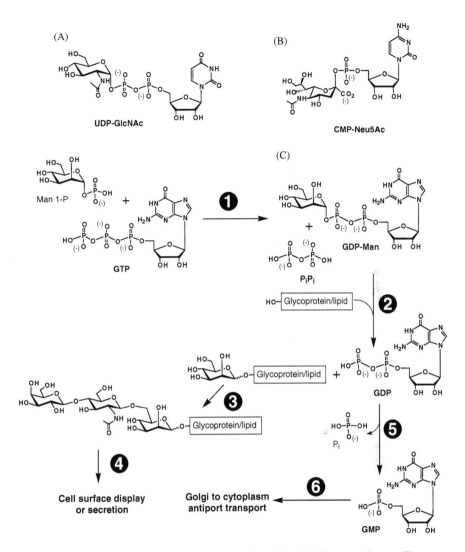

FIGURE 1.4 Structure, biosynthesis, and recycling of nucleotide sugar donors. The structure of UDP-linked nucleotide sugar donors is represented by UDP-GlcNAc (A), CMP-Neu5Ac shows the structure of CMP-linked sialic acids (B), and GDP-Man represents GDP-linked nucleotide sugar donors (C). The formation of a nucleotide sugar donor is illustrated by the condensation of GTP and Man-1P to form GDP-Man (Step 1); the mannose residue is added to a growing oligosaccharide chain of a glycoprotein or glycolipid (Step 2); additional sugar residues may then be added to the growing saccharide chain (Step 3) followed by its ultimate display on the cell surface or secretion (Step 4). In a parallel set of reactions, GDP produced during Step 2 is dephosphorylated to GMP (Step 5), which is transported out of the Golgi by an SLC35 family transporter (Step 6, see Table 1.3).

4. *N-acetylglucosamine (GlcNAc)*: In some organisms, GlcNAc can be derived directly from acetylation of glucosamine via glucosamine *N*-acetyltransferase activity (EC 2.3.1.3); GlcNAc then can be directly phosphorylated to form GlcNAc-6P.

GlcNAc-6P, the first committed intermediate in the cytoplasmic biosynthetic pathway leading to the formation of UDP-GlcNAc, can also be produced in a complex reaction involving ammonia transfer from L-glutamine to Fru-6P as well as sugar phosphate isomerization by glutamine-fructose-6-phosphate transaminase 1 (GFPT-1), followed by acetylation by glucosamine-phosphate N-acetyltransferase (GNPNAT-1). UDP-GlcNAc is an activated precursor of numerous macromolecules containing amino sugars, including chitin and mannoproteins in fungi, peptidoglycan and lipopolysaccharides in bacteria, and glycoproteins in mammals. UDP-GlcNAc can also be converted into UDP-GalNAc and used as a substrate for protein O-GlcNAc modification.[168,169]

5. *Glucuronic acid*: UDP-GlcA is synthesized directly from UDP-Glc by oxidation at the C-6-OH position, accompanied by the reduction of two equivalents of NAD$^+$. This acidic sugar is used primarily for glycosaminoglycan biosynthesis, and also for adding GlcA to N- and O-glycans as well as glycolipids. A large class of glucuronosyl transferases also occurs in cells; these enzymes utilize UDP-GlcA for glucuronidation of bile acids and for detoxification of xenobiotic compounds as the addition of GlcA increases the solubility of lipophilic molecules.

6. *Iduronic acid (L-IdoA)*: This sugar is the C-5 epimer of GlcA and, unlike all other monosaccharides occurring in complex carbohydrates produced in mammalian cells, IdoA is not derived from a sugar nucleotide donor but is formed by the C-5 epimerization of GlcA after its incorporation into the newly synthesized saccharide chain. IdoA primarily occurs in glycosaminoglycans (GAGs) polysaccharide chains, as discussed subsequently in this chapter, and has slightly different charge characteristics and conformational flexibility than GlcA; these features allow IdoA-containing polysaccharides to achieve multiple energetically favorable conformations.

7. *Xylose*: Biosynthesis of UDP-Xyl is catalyzed by UDP-GlcA decarboxylase (UGD) isozymes, all of which convert UDP-GlcA into UDP-Xyl[165] in reactions rendered essentially irreversible by the release of gaseous carbon dioxide. UDP-Xyl feedback inhibits the upstream enzymes UDP-glucose dehydrogenase, UDP-Glc pyrophosphorylase, as well as the UDP-GlcA decarboxylase involved in its own synthesis.[170]

8. *N-acetylgalactosamine (GalNAc)*: GalNAc forms the linkage between carbohydrate chains and a serine or threonine residue of the protein backbone in mucins and many of the other O-linked oligosaccharides found in mammals. GalNAc is also commonly found in many of the glycosphingolipids (GSLs) present in animal cell membranes and is a major component of proteoglycans such as chondroitin sulfates. GalNAc is installed in mammalian carbohydrates by the UDP-GalNAc nucleotide sugar donor, a compound derived almost entirely from the epimerization of UDP-GlcNAc instead of *de novo* synthesis from galactose intermediates.

9. *Mannose*: Eukaryotic cells contain mannose primarily in N-linked oligosaccharides and glycophosphatidylinositol (GPI) membrane anchors. The primary pathway for the production of activated mannose involves the phosphorylation of mannose to Man-6P by a glucokinase (GCK; see Figure 1.3 and Table 1.1); alternatively, Man-6P can be derived by the epimerization of Fru-6P by mannose phosphate isomerase (MPI). Man-6P can then be converted into Man-1P by phosphomannomutase 1 (PMM-1) and activated to form the nucleotide sugar donor GDP-Man by

GDP-mannose pyrophosphorylase. GDP-Man can be used directly for the formation of the lipid-linked oligosaccharide on the cytosolic face of the ER. Mannose provided by GDP-Man can also be transferred to dolichol phosphate to form Dol-P-Man in the ER membrane.[44]

10. *Fucose*: In one biosynthetic route, the formation of GDP-Fuc begins with GDP-Man and is catalyzed by the sequential action of two enzymes, GDP-mannose 4,6-dehydratase (GMDS), and the tissue-specific transplantation antigen P35B (TSTA3), to yield GDP-Fuc. Alternatively, the "salvage" pathway begins with cytosolic fucose derived from extracellular sources or from intracellular catabolic sources. In this case, the two enzymes L-fucose kinase (FUK) and fucose-1-phosphate guanylyltransferase (FPGT) work together to yield GDP-Fuc sequentially.

11. *Sialic acids*: Sialic acids are a family of unusual nine-carbon monosaccharides found at the nonreducing terminus of many glycoconjugates. The name "sialic acid" applies to a large family comprising over 50 natural forms of this sugar[171] as well as a growing number of nonnatural sialic acid analogs.[172–175] The sialic acid structure shown in Figure 1.2, also known as "Neu5Ac" (*N*-acetylneuraminic acid), is the most common form of this sugar found in humans. The biosynthetic pathway for sialic acid is more complicated compared with those of the other activated sugars; it begins with the production of ManNAc from UDP-GlcNAc, which is the first committed intermediate for sialic acid biosynthesis, by UDP-GlcNAc 2-epimerase/ManNAc 6-kinase (GNE). In the next step, sialic acids are derived by the condensation of ManNAc-6P (for Neu5Ac) or Man-6P (for KDN) with activated forms of pyruvate by sialic acid synthase (SAS). NeuAc 9-phosphatase (EC 3.1.3.29) and CMP-Neu5Ac synthetase (CMPNS) then sequentially produce the CMP-sialic acid nucleotide sugar donor. Unlike most monosaccharide processing reactions that occur entirely in the cytosol, the condensation of CTP with sialic acids to form CMP-sialic acid nucleotide sugar donors occurs in the nucleus with subsequent export of the activated precursor to the cytoplasm[122] and then further transport into the Golgi where sialylation reactions take place. Sialic acid is emerging as an incredibly important cell surface epitope, and several chapters in this book focus on this sugar; these include the description of the biochemical engineering of this pathway by Hinderlich and co-authors,[176] synthetic methods to develop sialic acid analogs and glycomimetics by Wilson and co-authors,[177] detection methods for sialoglycans that facilitate viral entry into cells[178] as well as *O*-acetylated forms of sialic acid implicated in parasitic infections.[179] Given the importance of this sugar, it is not surprising that intense efforts are in progress to develop expression systems for the production of glycoproteins with "humanized" patterns of sialic acid in both insect cells[180] and yeast.[3]

12. *N-acetylmannosamine (ManNAc)*: ManNAc is not found on the cell surface of mammalian cells, but is briefly discussed here because of its emerging importance as a tool for the biochemical engineering of sialic acid biosynthesis and cell surface presentation. This topic, discussed in greater detail in this book by Hinderlich and co-authors,[176] relies on the ability of nonnatural, exogenously supplied ManNAc analogs to intercept the sialic acid pathway in living cells and be converted into the corresponding nonnatural sialosides. By using this approach, the careful design of an analog allows the cell surface to be endowed with novel physical and chemical properties that fine-tune a host of cellular responses.[173–176]

C.   NUCLEOTIDE SUGAR TRANSPORT INTO THE ER AND GOLGI

Once nucleotide sugar donors are produced in the cytoplasm of a cell, they must be localized to the appropriate cellular sites where the biosynthesis of complex carbohydrates occurs. Upon being manufactured, these building blocks are correctly situated only for the production of the dolichol-linked 14-mer used for *N*-linked glycan biosynthesis on the cytosolic face of the ER. The biosynthesis, processing, and post-synthetic modification of the majority of secreted and membrane-bound proteins and lipids occur in the lumens of the ER and the Golgi apparatus.[181,182] Nucleotide sugars cannot freely diffuse across the membranes of these organelles making it necessary to translocate the sugar donors from the cytosol with the aid of membrane proteins. More specifically, the SLC35 nucleotide sugar transporter family, consisting of at least 17 members, is responsible for this task. In the past, it has been assumed that there is a specific transporter for each sugar but this simplified view is proving incorrect as evidence accumulates that certain transporters accept multiple substrates; conversely, sugars such as UDP-GlcNAc are transported by multiple transporters.

In vitro studies performed with Golgi vesicles derived from rat liver, mammary gland, and yeast cells have shown that the transport of nucleotide sugar donors is organelle-specific. UDP-Gal, for example, is transported solely into vesicles from the Golgi apparatus, whereas UDP-GlcA has a high rate of intake into both the ER and Golgi (Table 1.3). These compounds enter the ER or Golgi lumens through transporters with K_m values in the range of 1–10 μM; the transport of nucleotide sugar into these vesicles concentrates them relative to their cytoplasmic levels as lumen concentrations of often 100 μM or higher. Transport is neither inhibited nor stimulated by the presence of different ionophores in the medium and does not require ATP for energy, but is linked to the export of the corresponding nucleoside monophosphate. Specifically, UMP is exported upon the import of UDP-Gal, UDP-Glc, UDP-GlcA, UDP-GlcNAc, and UDP-GalNAc; GMP is exported during GDP-Man and GDP-Fuc import and CMP is exchanged for CMP-sialic acid. A wrinkle in this process is that UMP and GMP are not directly produced during glycosylation, instead, UDP and GDP are generated and enzymatic mono-dephosphorylation is required to generate UMP and GMP. As will be discussed in more detail below, the complete synthesis and processing of complex carbohydrates often involves post-synthetic modifications including epimerizations, phosphorylation, and sulfation. It is necessary, therefore, to also transport the raw materials for these reactions, namely ATP and PAPS, into the ER or Golgi (see Table 1.3).

D.   GLYCOCONJUGATE BIOSYNTHESIS: GLYCOPROTEINS, GLYCOLIPIDS, AND POLYSACCHARIDES

Glycosylation is the predominant post-translation modification of proteins and classification of glycoproteins is often done on the basis of the sugar linkage to the underlying protein. As shown in Figure 1.5, nature provides numerous options to achieve the linkage of a sugar to a peptide (and, as an aside, chemists can significantly expand these as demonstrated in Chapter 25.[196] In this chapter, we will focus on the most prevalent forms in mammals, specifically those attached to proteins by *O*- or *N*-linked glycosidic bonds. In addition to glycoproteins, the

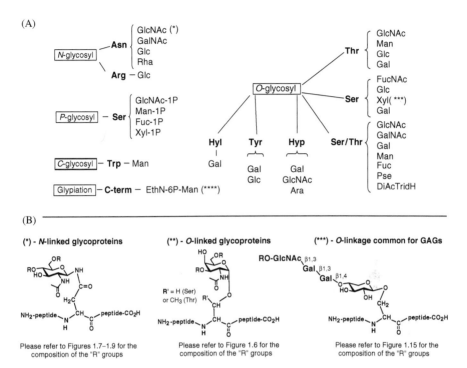

FIGURE 1.5 Naturally occurring carbohydrate–peptide linkages. A compilation of the carbohydrate linkages to peptides found in nature (as of the end of 2003) is given in (A); only a few of these structures commonly occur in mammals, however. The mammalian linkages, shown in more detail in (B), include oligosaccharides N-linked (*) and O-linked (**) to a typical glycoprotein. In addition, polysaccharides found in most GAGs are O-linked (***) to a serine via a xylose residue. Finally, glypiation of GPI membrane anchors (*****) is shown in more detail in Figure 1.14.

biosynthesis of glycolipids and GPI anchors will be discussed briefly, and an overview of mammalian polysaccharide synthesis will also be provided.

1. O-Linked Glycosylation

O-linked glycosylation proceeds by the stepwise addition of sugars directly onto a polypeptide chain. The first sugar residue of an O-glycan is usually GalNAc-linked to the hydroxyl group of serine or threonine (Figure 1.4) or, in the case of collagen, a single galactose or the disaccharide glucosylgalactose linked to a hydroxylysine residue. The first step, the transfer of a GalNAc residue from UDP-GalNAc to the hydroxyl group of Ser or Thr residue, is catalyzed by the GALNT6 GalNAc transferase (Table 1.2) to form a structure known as the T_n antigen (Figure 1.6). The T_n antigen is then translocated into the *trans*-Golgi, where the stepwise biosynthetic process continues with addition of galactose, GalNAc, or GlcNAc to the proximal GalNAc by specific glycosyltransferases to form a set of eight "core" structures

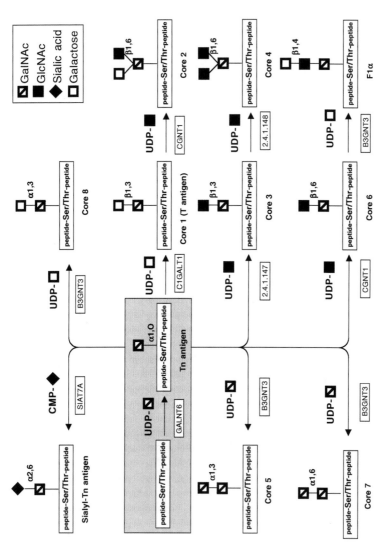

FIGURE 1.6 Outline of *O*-glycan biosynthesis showing the formation of "core" structures. The biosynthesis of *O*-linked glycoproteins begins with the formation of a common set of "core" structures (Cores 1 through 8); each core structure can be further elaborated to produce hundreds (and possibly thousands) of additional structures. Additional information on the enzymes that catalyze each step is provided in Table 1.2.

(Cores 1–8, Figure 1.6). Sialic acids are often added to the termini of the T_n antigen or other core structures; alternatively, the core structures can be elongated by the step-by-step construction of many additional oligosaccharide structures.

2. N-Linked Glycosylation

N-linked glycosylation, the biosynthetic process that produces the predominant form of carbohydrate found on most mammalian glycoproteins, is considerably more complex than O-linked glycan biosynthesis. (A much more detailed description of N-linked glycosylation is provided in this book by Callewaert and co-authors,[3] who are interested in exploiting these enzymes for production of glycoproteins in yeast, as well as by Kościelak who discusses N-linked glycosylation in depth[197] from the perspective of congenital human disease and by Stanley and Patnaik's discussion of "Lec" CHO cells.[198] These chapters nicely illustrate the necessity of a thorough basic understanding of the biosynthetic processes and render only a brief discussion necessary here.) N-linked glycosylation begins with the synthesis of a lipid-linked oligosaccharide (LLO) precursor dolichol-$GlcNAc_2Man_9Glc_3$ (Figure 1.7). Dolichol is a 75 to 95 carbon atom-long, highly hydrophobic, polyisoprenoid lipid that spans the membrane four or five times and is linked to the $GlcNAc_2Man_9Glc_3$ oligosaccharide by a pyrophosphoryl group. Synthesis of this precursor begins on the cytosolic face of the rough ER by the addition of two GlcNAc and five mannose residues to dolichol phosphate. A "flippase" then inverts the dolichol pyrophosphoryl oligosaccharide, orienting it to face the luminal face of the ER, and four additional mannose and three glucose units are added to the growing oligosaccharide.

After the synthesis of the LLO is completed, the oligosaccharide precursor is transferred *en bloc* from the dolichol to an asparagine (Asn) residue located within the Asn-X-Ser/Thr consensus sequence of amino acids, where X is any amino acid except proline, on the nascent polypeptide. This process is catalyzed by dolichyl-diphosphooligosaccharide-protein glycosyltransferase (DDOST[79])and, unlike the O-linked glycosylation process that occurs posttranslationally in the Golgi apparatus, N-linked glycosylation occurs co-translationally in the ER. The addition of the oligosaccharide to the peptide at an early stage of glycoprotein synthesis allows the carbohydrate to play a key role in the quality control and folding of a newly synthesized protein in a process, where the three glucose residues are trimmed by the sequential action of the GCS1[110,111] and chitobiase (CTBS[77,78]) glucosidases (Table 1.2) and iteratively reelaborated if necessary.[199] Finally, upon the successful folding of the protein, any remaining glucose residues are removed, and then the α-1, 2-linked mannose residues of the oligosaccharide are trimmed by Mannosidase 1 in the *cis*-Golgi to form the high-mannose oligosaccharide that serves as an intermediate for the production of complex and hybrid-type N-linked glycans,[200] as outlined in Figure 1.7 and Figure 1.8 and discussed briefly below.

a. Biosynthesis of the Complex Type Oligosaccharides
The intermediary N-Asn-linked $GlcNAc_2Man_5$ oligosaccharide is transported from the *cis*- to the medial-Golgi, where a GlcNAc residue is first added to the α-1,3-linked mannose by mannosyl (α-1,3)-glycoprotein β-1,2-N-acetylglucosaminyltransferase

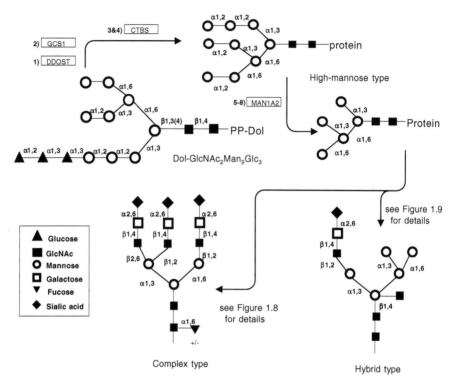

FIGURE 1.7 Outline of *N*-glycan biosynthesis. The Dol-GlcNac$_2$Man$_9$Glc$_3$ 14-mer forms the starting point for *N*-glycosylation. This structure is transferred *en bloc* to an asparagine residue within a consensus sequence by DDOST (Step 1) and then the three glucose residues on the lower branch (as drawn) are removed by the sequential action of GCS1 and CTBS (acting twice). MAN1A2 then removes the four α-1,2-linked mannose residues to generate the high-mannose structure that can be rebuilt to form complex type (Figure 1.8) or hybrid-type *N*-linked glycans (Figure 1.9).

(MGAT1[134,135]). Two additional mannose residues are removed by Mannosidase II (MAN2A2[133]) and then the α-6-mannoside-β-1,2-GlcNAc transferase II gene (MGAT2[136]) adds one GlcNAc residue. Fucosyltransferase FUT8 then may or may not add a fucose residue to the GlcNAc residue proximal to the protein.[100,101] At this stage, GlcNAc transferase IV (MGAT4[138]) has the option to add a GlcNAc residue to the α-1,3-linked mannose residue to form a triantennary oligosaccharide (as shown in Figure 1.8) and GlcNAc transferase V (MGAT5[201]) may further add a GlcNAc to the adjacent α-1,6-linked mannose to form a quaternary oligosaccharide (not shown). A galactose (or GalNAc) residue can then be added to the GlcNAc residues by a β-GlcNAc β-1,4-galactosyltransferase (such as B4GALT1[139,140]) and, finally, the branches can be capped with terminal sialic acid residues. It is important to note that the relatively simple oligosaccharide depicted in Figure 1.8 is shown for illustrative purposes only, and is only one of an incredibly diverse range of structures that have been characterized.

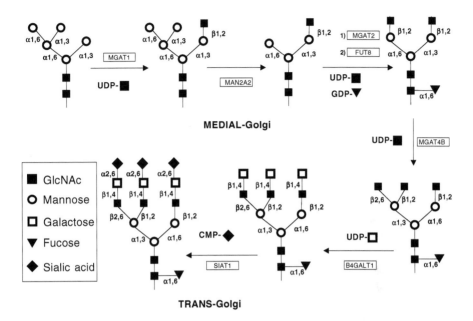

FIGURE 1.8 Synthesis of complex type *N*-linked oligosaccharides. After the addition of a β-1,2-linked GlcNAc residue, two additional mannose residues are trimmed from the high mannose structure. The resulting hexasaccharide is then reelaborated to form complex type *N*-linked oligosaccharides. The example of the final structure is shown for illustrative purposes only; in reality, a plethora of diverse structures exist as discussed in the text.

b. Hybrid Type N-Linked Oligosaccharides

Hybrid type *N*-glycans (Figure 1.9) are synthesized from the high-mannose oligosaccharide when GlcNAc transferase III (MGAT3[137]) adds a bisecting GlcNAc residue to the core mannose residue prior to the trimming of two residues by Mannosidase II in the medial-Golgi.[202,203] The structure is then translocated into the *trans*-Golgi where further elaboration of the structure, for example, by the addition of galactose and sialic acid as shown in Figure 1.9, occurs to form a "hybrid type" *N*-linked oligosaccharide.

3. Glycolipids

Glycolipids, molecules composed of a carbohydrate moiety linked to a hydrophobic aglycon, are ubiquitous although quantitatively minor components of biological membranes.[204,205] There are two main classes of glycolipids; glycoglycerolipids are common in bacteria and plants and glycosphingolipids (GSLs) comprise the majority of glycolipids in animals (Figure 1.10). This chapter will focus on mammalian GSLs and will provide a brief discussion of the structure and biosynthesis of basic structures. GSL biosynthesis begins with ceramide, which is the long-chain amino alcohol D-erythrosphingosine *N*-acylated with a fatty acid. Ceramide can be either glucosylated to form glucosylceramide (GlcCer) or galactosylated to form galactosylceramide

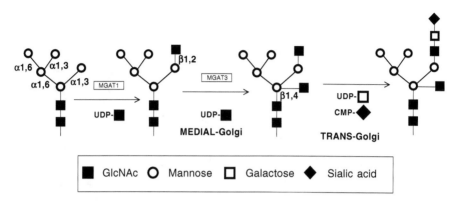

FIGURE 1.9 Synthesis of hybrid-type *N*-linked glycans. Unlike the biosynthesis of complex type *N*-linked oligosaccharides where two additional mannose residues are removed from the high mannose structure (Figure 1.8), hybrid-type *N*-linked oligosaccharides are built on the GlcNAc$_2$Man$_5$ heptasaccharide structure. Once again, the final structure is shown for only one of many possible examples of *N*-linked hybrid oligosaccharides.

(GalCer). After the addition of a galactose residue to GlcCer to form LacCer, LacCer subsequently serves as a core molecule for hundreds of GSLs that can be further divided into three main classes (Figure 1.10, top). First, the "blood group series" are derived from further elaboration of LacCer with a β-1,3-linked GlcNAc (the blood group GSLs can be subdivided into the lacto- and neo-lacto series, the latter is outlined in Figure 1.11). Next, the Globo Series is derived from the P^k antigen formed upon the addition of an α-1,4-linked galactose to LacCer (Figure 1.12). Finally, an outline of basic ganglioside structures (a ganglioside is a sialic acid-bearing GSL and includes some members of the lacto and globo structures) is given in Figure 1.13.[206] Unlike the huge diversity of GlcCer-derived GSLs, GalCer (Figure 1.10, bottom) is used to provide a relatively small number of GSLs.

4. GPI Membrane Anchors

GPI membrane anchors are found ubiquitously throughout eukaryotes and provide an elaborate mechanism for the attachment of proteins to the cell surface (Figure 1.14). The polypeptide chain of a GPI-anchored protein is attached through a C-terminal carboxyl group to a short carbohydrate sequence that is in turn attached to a phosphatidylinositol (PI) lipid structure.[207] The biosynthesis of GPI structures is described in review articles[208–210] and will be outlined only briefly here. In mammalian cells, the core GPI structure is synthesized on the luminal face of the ER from PI, UDP-GlcNAc, Dol-P-mannose and phosphoethanolamine. In the first step, GlcNAc is transferred to a phosphoinositol (PI) acceptor to form GlcNAc-PI. A fatty acid is then added to the inositol ring and three mannose residues are transferred to the elongating GPI core followed by the transfer of ethanolamine phosphates to the mannose residues. Proteins that are destined for GPI anchoring have both an N- and C-terminal signal peptide that allows the nascent peptide to transverse the ER and be transferred to the preformed GPI anchor in a transamidase reaction.[209] The core GPI anchor structure is conserved

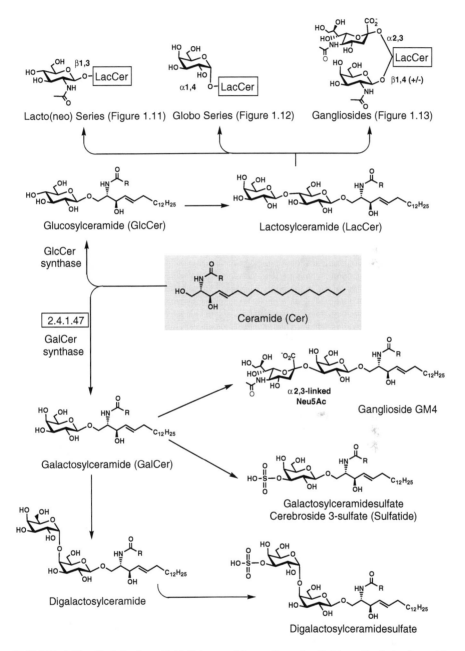

FIGURE 1.10 Carbohydrate–lipid linkages. Mammalian glycolipid synthesis begins with ceramide (shaded, center). Sequential addition of glucose and galactose residues (top pathway) produces LacCer, a compound used for hundreds of glycolipids that can be divided into three additional classes: the lacto(neo) series, the globo series, and gangliosides (each of these groups is presented in more detail in Figures 1.11, 1.12, and 1.13, respectively). By contrast, the initial addition of galactose to ceramide to produce GalCer (bottom pathway) leads to a relatively small number of final structures, including the ones shown in the bottom half of this figure.

Lacto series glycosphingolipid biosynthesis

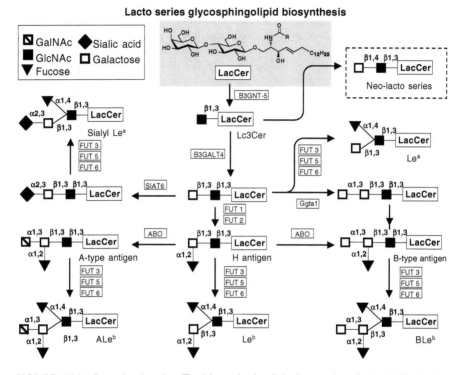

FIGURE 1.11 Lacto(neo) series. The biosynthesis of the lactoseries of glycolipids begins when LacCer (shaded, top center) is converted into Lc3Cer; Lc3Cer is further derivatized with a β-1,3-linked galactose followed by fucose, GlcNAc, and sialic acid to form the lactoseries of ganglioside (shown). Alternatively, after addition of the initial GlcNAc to LacCer, if a β-1,4- (instead of β-1,3-) linked galactose is then added to Lc3Cer, the resulting ceramide-disaccharide structure forms the core unit for Neo-lactoseries biosynthesis that contains structures such as the sialyl Lewis x tetrasaccharide (not shown).

among eukaryotes but considerable microheterogeneity exists;[209] in particular, parasitic organisms often have appendages that distinguish their GPI anchors from those of mammals and therefore offer attractive therapeutic targets.[211]

5. Polysaccharide Biosynthesis

GAGs are complex polysaccharides located in the extracellular matrix and attached covalently to cell surface proteoglycans. These polysaccharides, although based on repeating disaccharide units, are remarkably heterogeneous and show organism- as well as tissue-specific patterns of expression. In general, a mammalian GAG belongs to one of four primary groups: hyaluronic acid, keratan sulfate, heparan sulfate, and its more highly sulfated counterpart heparin, and chondroitin sulfate and its epimer dermatan sulfate (Figure 1.15). GAGs serve a variety of functions in the body that include modulating cell adhesion, differentiation, migration, as well as the regulation and the maintenance of proliferation. Each GAG has type-specific functions as

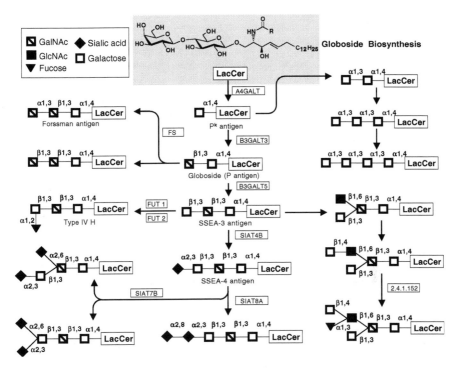

FIGURE 1.12   Globoside series. Lactosylceramide is converted into the P^k antigen by the addition of an α-1,4-linked galactose residue. After that, galactose as well as GalNAc, GlcNAc, fucose, and sialic acid residues are added to the P^k antigen to produce the globoside series of GSLs for which a few basic core structures are shown.

well. For instance, hyaluronic acid plays an important role in the lubrication and cushioning of joints and interacts with CD44 during cell migration.[212] Keratan sulfate is well known as a major constituent of the cornea, an important component of connective tissue, and has been identified in the brain.[213] Heparan sulfate-containing proteoglycans participate in angiogenesis, morphogenesis, regulation, and sequestration of growth factors and developmental signaling molecules; heparin expression is restricted to mast cells of the lymphatic system and historically has been used as a clinical anticoagulant. Chondroitin and dermatan sulfate, present in connective matrices and on cell membranes, are important to the formation of mammalian neural networks.

The basic biosynthetic processes for the production of mammalian polysaccharides are now known and have been discussed in detail in recent reviews[214–217] and in this book,[1,218] and will therefore only be presented briefly in this chapter. Similar to the smaller protein-linked oligosaccharides discussed above, polysaccharide synthesis is mediated by enzymes localized to the ER–Golgi complex and *trans*-Golgi network and uses sugar nucleotide precursors and 3′-phosphoadenosine-5′-phosphosulfate (PAPS) formed by enzymes in the cytosol and transported to the lumen of the ER or Golgi. Glycosaminoglycans (with the exceptions of hyaluronic acid and

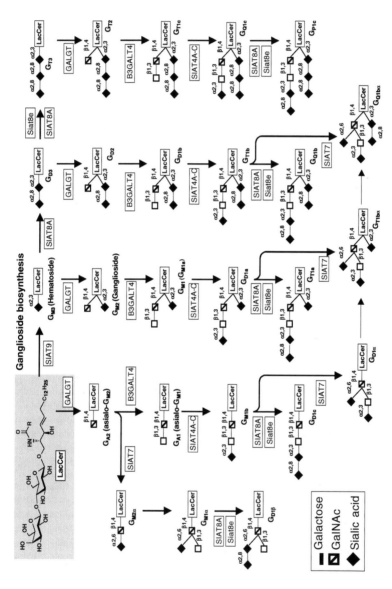

FIGURE 1.13 Ganglioside biosynthesis. Typically, ganglioside biosynthesis follows one of two routes starting from LacCer. An α-2,3-linked sialic acid residue can be added directly to LacCer by SIAT9 to produce a ganglioside, G_{M3}; alternatively, a β-2,3-linked GalNAc residue can be added before sialic acid residues are attached. Interestingly, certain gangliosides such as $G_{D1\alpha}$, $G_{T1b\alpha}$, and $G_{Q1b\alpha}$ (bottom row) can be derived from either route.

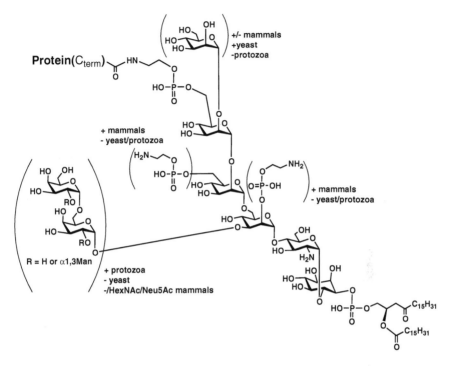

FIGURE 1.14 GPI membrane anchor structure. The core GPI structure is shown along with appendages that may, or may not, be present depending on phylogenetic origin.

certain forms of keratan sulfate that exist unattached to proteins) are covalently linked to proteoglycans via a serine residue and an *O*-linked tetrasaccharide attachment sequence (GlcA-Gal-Gal-Xyl). In the first step, xylose is transferred to the hydroxyl group of serine via UDP-xylose by xylosyltransferase[219,220] in the ER. Then, continuing into early Golgi compartments, Gal/GlcNAc and GlcA are attached by the sequential action of Gal transferases I and II, and GlcA transferase I. Next, alternating Gal/GlcNAc and GlcA sugars are added to the growing polymer chain; some of the transferases are bifunctional, or work in concert, to catalyze the addition of both subunits of the repeating disaccharide to ensure a greater degree of polymerization.

Finally, after assembly of the basic carbohydrate chain, GAGs are postsynthetically modified at various positions along their length to provide the final product by *N*-deacetylation of GlcNAc, *N*-sulfation at unsubstituted amino groups, C_5 epimerization of D-GlcA residues to form L-iduronic acid (IdoA), or multiple *O*-sulfation of sugars via various sulfotransferases.[221] Not all reactions are carried through to completion; therefore by "mixing and matching" any or all of these modifications a single disaccharide repeating unit can assume any one of up to 32 different structural forms resulting in the display of diverse heterogeneity in the final GAGs (Figure 1.15). Generally, postsynthetic modifications occur in "patches," enabling GAGs to interact with ligands

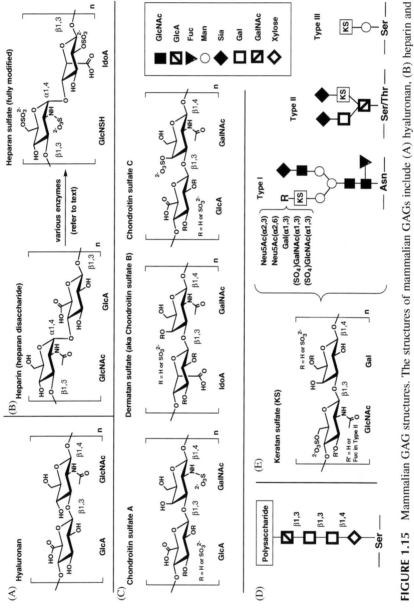

FIGURE 1.15 Mammalian GAG structures. The structures of mammalian GAGs include (A) hyaluronan, (B) heparin and fully modified heparan sulfate, and (C) chondroitin sulfate and dermatan sulfate. In general, GAGs are linked to proteins by the serine-*O*-xylose-linked tetrasaccharide shown in (D). An exception is keratan sulfate, which employs the three different linkages as shown in (E).

through both their primary saccharide sequences and the pattern of negative charges engendered by sulfation unique to each glycan.

a. Hyaluronan

Hyaluronan (HA, hyaluronic acid) is the simplest of the glycosaminoglycans in structure (Figure 1.15A) and exists unattached to any protein. It consists of repeating GlcA and GlcNAc residues and can achieve a size up to ~10^6 Da; this large molecular mass contributes to the ability of hyaluronan to resist compressive forces. The sulfated residues give the molecule an anionic character that allows it to bind water molecules to form hydrogels. These two features render hyaluronan an attractive molecule for use in tissue engineering applications. Three mammalian hyaluronic acid synthase genes have been cloned.[222,223] The first, HA synthase I, was found to stimulate HA production in mouse mammary carcinoma cells[224] and two other synthases were identified by homology to this first cloned synthase.[225] HA synthases possess seven transmembrane regions and a predicted cytoplasmic loop containing the catalytic glycosyltransferase sites. The relevant amino acids involved in this activity have been studied via mutagenesis in murine HA proteins.[226] Interestingly, studies have shown HA synthases to be capable of transglycosylation with both GlcNAc as well as GlcA. Each synthase is capable of *in vitro* synthesis of HA, yet they have different binding affinities for GlcA and GlcNAc, and catalyze the synthesis of HA strands of varying molecular weight.[227] Furthermore, genes encoding these proteins are controlled by distinct regulatory elements and their transcription is regulated with respect to spatial and temporal factors. Embryos that were homozygous recessive for the HA synthase II gene perished with severe cardiac and vascular abnormalities,[228] while overexpression of hyaluronan has been correlated with tumor growth and metastasis, suggesting that the development of methods to manipulate these carbohydrates *in vivo* could have profound clinical implications.

b. Heparan Sulfate and Heparin

Heparan sulfate and heparin chains are composed of the same alternating monosaccharide residues, GlcNAc and GlcA, found in hyaluronan; the β-1,4-linkage in hyaluronan, however, is replaced by an α-1,4-glycosidic bond (Figure1.15B). Alternating GlcNAc and GlcA sugars are added to the growing polymer chain by recently cloned tumor suppressor EXT family genes.[229] In particular, the EXT1 and EXT2 genes, previously linked to hereditary multiple exostoses (HME), a disease which leads to cartilage-capped tumors, exhibit GlcA as well as GlcNAc transferase activities. In addition to the EXT family, highly homologous EXT-like genes (EXTL1-3) have been isolated. EXTL-1 has been shown to catalyze α-1, 4-GlcNAc transfer for the synthesis of the repeating chain while EXTL-3 appears to harbor both GlcNAcT I and GlcNAcT II activities.[213] EXT genes have also been relatively well conserved between humans, *Drosophila melanogaster*, and *Caenorhabditis elegans*. Deficiencies in EXT biosynthetic enzymes impair heparan sulfate synthesis and lead to developmental defects in the latter two species.[230]

Similar to most glycosyltransferases, EXT1 and EXT2 are type II transmembrane proteins, with an N-terminal cytoplasmic tail, a single transmembrane domain, a stem, and a large luminally oriented globular domain with enzymatic activity. During heparan sulfate polymerization, EXT1 and EXT2 form a hetero-oligomeric complex that is relocalized to the Golgi.[231,232] The complex has significantly higher amounts of glycosyltransferase activity than seen with individual enzymes, and appears to be the active form of the enzymes *in vivo*. Heparan sulfate biosynthesis also requires galactosyltransferase-1, uronosyl 5-epimerase, uronosyl 2-*O*-sulfotransferase, and GlcNAc *N*-deacetylase/*N*-sulfotransferase[233] to create the final polysaccharide complete with the modifications shown in Figure 1.15B. Evidence suggests that many (or all) of these enzymes are in close physical proximity with each other because when pairs of these enzymes were studied, with one of the two relocalized to the ER by means of replacing their cytoplasmic tails with an ER retention signal, it was found that repositioning the epimerase led to a concomitant reposition of the 2-*O*-sulfotransferase. Uronosyl-5-epimerase and 2-*O*-sulfotransferase also interact *in vivo*, and physical association of these two proteins is important for both epimerase stability as well as Golgi translocation.[234]

c. Chondroitin Sulfate

Chondroitin sulfate shares several postsynthetic modifications with heparan sulfate but is based on a different disaccharide polymer unit. Instead of alternating GlcNAc and α-1,4-linked GlcA residues, GalNAc replaces GlcNAc, and GlcA is attached to this sugar via a β-1,4 glycosidic linkage (Figure 1.15C). The first hexosamine transfer is the crucial determinant of whether the heparin sulfate or chondroitin sulfate GAG is synthesized. The addition of α-GlcNAc initiates heparin or heparan sulfate synthesis, while the addition of a β-GalNAc prompts chondroitin sulfate synthesis. EXTL-2, which has *N*-acetylhexosaminyltransferase activity,[235] can transfer GalNAc or GlcNAc to the tetrasaccharide linkage region via an α-1,4 linkage and may be the enzyme that determines whether heparan sulfate or chondroitin sulfate chains are produced. Two decades ago it was thought that one enzyme was responsible for the addition of the initial GalNAc to the attachment region and a different enzyme, *N*-acetylgalactosaminyl transferase, played a role in the polymerization of chondroitin sulfate.[236] More recently, however, GalNAc transferase I has been cloned and found to add GalNAc both to the polymerizing chain as well as to the attachment region.[237] Additional complexity resulted from the discovery of chondroitin synthase; this enzyme bears similarity to the transferases involved in heparan sulfate synthesis and can catalyze both GlcA and GalNAc transfer.[238, 239] Sulfation along the length of the polysaccharide chain is carried out by various sulfotransferases. Sulfation of the 2- or 4-positions of GalNAc generates chondroitin sulfates A and C, respectively; in addition, GlcA may be sulfated at the 2- and 3-hydroxyl positions. Sulfation occurs while the polysaccharide chain is undergoing polymerization, and takes place in an all or none fashion over local regions of the chain; consequently, partially sulfated segments of chondroitin are rarely found. Termination of chondroitin sulfate chains can occur on unsulfated GlcA, GalNAc, GalNAc 4-sulfate, 6-sulfate, or 4,6-disulfate residues (GalNAc-6-transferases are now known).[240, 241] Chondroitin sulfate B, now more commonly referred to as dermatan sulfate, is formed when GlcA residues of chondroitin sulfate are epimerized at the C-5 position to form IdoA.

d. Keratan Sulfate

It is different from other mammalian GAGs in that it does not contain an urosolic acid (GlcA or IdoA) but instead consists of a galactose/GlcNAc repeating unit. Keratan sulfate (KS) is classified into three types based on the linkage that connects it to the underlying protein.[215, 242] Type I is abundant in the cornea and consists of repeating units of galactose and Glc-6-O-sulfate; this polysaccharide is attached to a mannose of an N-linked glycan (Figure 1.15E). Type II KS is found in cartilage and is exclusively linked to the protein aggrecan via a mucin "Core 2" structure (see Figure 1.7) and consists of galactose alternating with Gal-6-O-sulfate. The terminal ends of both Type I and Type II KS can be decorated with a sialic acid residue; in addition, Type I can hold a galactose or GlcNAc terminal residue. A third form of KS, found in the brain, is linked to protein via an unusual mannose residue 2-O-linked to serine.[213,215] The regulatory mechanisms involved in the initiation of KS from each of these attachment sequences remain to be well characterized,[215] but in each case are thought to involve the initial addition of galactose to the underlying sugar by β-1,4-galactosyltransferase (β4Gal-T1)[243,244] followed by sequential addition of GlcNAc and galactose. Subsequent addition of galactose is, at least in part, also catalyzed by β4Gal-T1 while the requisite GlcNAc-transferase has not yet been unambiguously identified.

E. POSTSYNTHETIC MODIFICATIONS FINE-TUNE CARBOHYDRATE STRUCTURE AND FUNCTION

Glycosylation has been described as nature's way of fine-tuning the structure and function of proteins and lipids. An excellent example of this phenomenon is provided in Chapter 20, where the role of glycosylation in specifying the properties of engineered antibodies is discussed.[245] Interestingly, carbohydrates themselves are also subject to post- (or co-) synthetic modifications that further increase their structural diversity and range of possible biological responses. Many of these modifications have already been discussed in this chapter: for instance, the carbohydrate backbone of GPI anchors (Figure 1.14) is decorated with phosphatidylethanolamine and human GAG polysaccharides are almost always subject to epimerization and sulfation reactions (Figure 1.15).

These modifications change the physical and structural properties of the host molecule and can also significantly alter the resulting biological behavior of the host cell. In one example, acetylation of sialic acid at the 9-O- or 7-O-position converts ganglioside GD3 from having potent apoptosis-inducing properties to being antiapoptotic (see References 246 and 247 and Chapter 8[248]). In a second example, shown in Figure 1.16, multiple posttranslational modifications control the activity of the sialyl Lewis X (sLe[X]) tetrasaccharide in the leukocyte homing process. sLe[X] has been shown to function as a ligand for all three (E-, L-, and P-) selectins when studied in *in vitro* assays but *in vivo* specificity is more stringent and is determined by several factors. First, the identity of the host molecule plays a role in specificity; sLe[X] is a ligand for P-selectin when attached to PSGL-1 or a ligand for L-selectin when presented on CD34 or GlyCAM-1, and growing evidence suggests that it is a ligand for E-selectin when it is part of a ganglioside.[249] Further discrimination is achieved by sulfation at the 6-position of the GlcNAc and galactose residues;

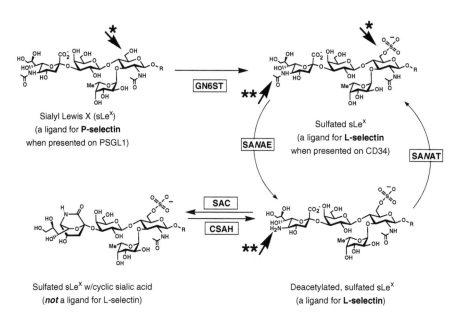

FIGURE 1.16 Postsynthetic modification of Sialyl Lewis X (sLe^X). Postsynthetic modification of carbohydrate structures controls their biological activity as exemplified by sLe^X and discussed in the text.

this modification is required for binding to L-selectin under physiological conditions.[118,120] Modification of sialic acid residue of sLe^X by stepwise deacetylation of the 5- position followed by cyclization of the sialic acid residue inhibits binding to L-selectin. It has been proposed that these two posttranslational modifications of sLe^X allow the immune system to control the recruitment of leukocytes at different rates in routine homing and inflammation. The negative feedback sialic acid cyclase system allows routine homing of leukocytes at a "slow and steady" rate, whereas massive accumulation of lymphocytes at sites of inflammatory lesions is mediated mainly by the interaction of nonsulfated sLe^X with P-selectin.[76,147]

The example discussed above where sialic acids act as a sensitive modulator of selectin interactions is just one of the many examples where this one sugar has an impact on important biological behavior; metabolic and chemical pursuits toward expanding nature's library of ~50 different forms of sialic acid [171] are described in Chapters 13 and 24.[176, 177] Of course, other monosaccharide residues besides sialic acid undergo postsynthetic modifications, especially when they appear in polysaccharides. As discussed above, as well as in Chapters 6 and 9,[1, 218] mammalian glycosoaminoglycans readily undergo epimerization and sulfation reactions. Polysaccharides from plants, marine organisms, and bacteria enjoy an expanded repertoire of modification reactions, becoming modified with acetyl, succinyl, pyruvylyl, and phosphoryl groups. This wealth of diversity can exquisitely fine-tune the properties of these molecules, opening the door for a wide range of biomedical, biotechnological, industrial, and environmental applications as described in many chapters in this book.[2, 4–8, 250–254]

REFERENCES

1. Mizumoto, S., Uyama, T., Mikami, T., Kitagawa, H., and Sugahara, K., Biosynthetic pathways for differential expression of functional chondroitin sulfate and heparan sulfate, in *Handbook of Carbohydrate Engineering*, Yarema, K., Ed., Dekker/CRC Press, Boca Raton, 2005, chap. 9.

2. Cairns, A.J. and Perret, S.J., Fructan biosynthesis in genetically-modified plants, in *Handbook of Carbohydrate Engineering*, Yarema, K., Ed., Dekker/CRC Press, Boca Raton, 2005, chap. 11.

3. Callewaert, N., Vervecken, W., Geysens, S., and Contreras, R., *N*-glycan engineering in yeasts and fungi: progress towards human-like glycosylation, in *Handbook of Carbohydrate Engineering*, Yarema, K., Ed., Dekker/CRC Press, Boca Raton, 2005, chap. 15.

4. Taylor, C.M. and Roberts, I.S., Bacterial capsules: a route for polysaccharide engineering, in *Handbook of Carbohydrate Engineering*, Yarema, K., Ed., Dekker/CRC Press, Boca Raton, 2005, chap. 16.

5. Bach, H. and Gutnick, D.L., Engineering bacterial biopolymers for the biosorption of heavy metals, in *Handbook of Carbohydrate Engineering*, Yarema, K., Ed., Dekker/CRC Press, Boca Raton, 2005, chap. 18.

6. Moracci, M., Cobucci-Ponzano, B., Perugino, G., Giordano, A., Trincone, A., and Rossi, M., Recent developments in the synthesis of oligosaccharides by hyperthermophilic glycosidases, in *Handbook of Carbohydrate Engineering*, Yarema, K., Ed., Dekker/CRC Press, Boca Raton, 2005, chap. 21.

7. de Vries, R.P., McCann, M.C., and Visser, J., Modification of plant cell wall polysaccharides using enzymes from *Aspergillus*, in *Handbook of Carbohydrate Engineering*, Yarema, K., Ed., Dekker/CRC Press, Boca Raton, 2005, chap. 22.

8. Eyzaguirre, J., Hidalgo, M., and Leschot, A., β-Glucosidases from filamentous fungi: properties, structure and applications, in *Handbook of Carbohydrate Engineering*, Yarema, K., Ed., Dekker/CRC Press, Boca Raton, 2005, chap. 23.

9. Bell, G.I., Burant, C.F., Takeda, J., and Gould, G.W., Structure and function of mammalian facilitative sugar transporters, *J. Biol. Chem.*, 268, 19161–19164, 1993.

10. Wood, I.S. and Trayhurn, P., Glucose transporters (GLUT and SGLT): expanded families of sugar transport proteins, *Br. J. Nutr.*, 89, 3–9, 2003.

11. Wright, E.M. and Turk, E., The sodium/glucose cotransport family SLC5, *Pflugers Arch.*, 447, 510–518, 2004.

12. Joost, H.G. and Thorens, B., The extended GLUT-family of sugar/polyol transport facilitators, nomenclature, sequence characteristics, and potential function of its novel members (review), *Mol. Membr. Biol.*, 18, 247–256, 2001.

13. Uldry, M. and Thorens, B., The SLC2 family of facilitated hexose and polyol transporters, *Pflugers Arch.*, 447, 480–489, 2004.

14. Turk, E., Klisak, I., Bacallao, R., Sparkes, R.S., and Wright, E.M., Assignment of the human Na+/glucose cotransporter gene SGLT1 to chromosome 22q13.1, *Genomics*, 17, 752–754, 1993.

15. Turk, E., Martin, M.G., and Wright, E.M., Structure of the human Na+/glucose cotransporter gene SGLT1, *J. Biol. Chem.*, 269, 15204–15209, 1994.

16. Wells, R.G., Pajor, A.M., Kanai, Y., Turk, E., Wright, E.M., and Hediger, M.A., Cloning of a human kidney cDNA with similarity to the sodium–glucose cotransporter, *Am. J. Physiol.*, 263, F459–F465, 1992.

17. Kanai, Y., Lee, W.-S., You, G., Brown, D., and Hediger, M.A., The human kidney low affinity Na(+)/glucose cotransporter SGLT2, delineation of the major renal reabsorptive mechanism for D-glucose, *J. Clin. Invest.*, 93, 397–404, 1994.

18. Diez-Sampedro, A., Hirayama, B.A., Osswald, C., Gorboulev, V., Baumgarten, K., Volk, C., Wright, E.M., and Koepsell, H., A glucose sensor hiding in a family of transporters, *Proc. Natl. Acad. Sci. USA*, 100, 11753–11758, 2003.

19. Jung, H., The sodium/substrate symporter family: structural and functional features, *FEBS Lett.*, 529, 73–77, 2002.

20. Strausberg, R.L., Feingold, E.A., Grouse, L.H. et al., Generation and initial analysis of more than 15,000 full-length human and mouse cDNA sequences, *Proc. Natl. Acad. Sci. USA*, 99, 16899–16908, 2002.

21. Coady, M.J., Wallendorff, B., Gagnon, D.G., and Lapointe, J.Y., Identification of a novel Na⁺/myo-inositol cotransporter, *J. Biol. Chem.*, 277, 35219–35224, 2002.

22. Roll, P., Massacrier, A.,,. Pereira, S., Robaglia-Schlupp, A., Cau, P., and Szepetowski, P., New human sodium/glucose cotransporter gene (KST1): identification, characterization, and mutation analysis in ICCA (infantile convulsions and choreoathetosis) and BFIC (benign familial infantile convulsions) families, *Gene*, 285, 141–148, 2002.

23. Mueckler, M., Caruso, C., Baldwin, S.A., Panico, M., Blench, I., Morris, H.R., Allard, W.J., Lienhard, G.E., and Lodish, H.F., Sequence and structure of a human glucose transporter, *Science*, 229, 941–945, 1985.

24. Uldry, M., Ibberson, M., Hosokawa, M., and Thorens, B., GLUT2 is a high affinity glucosamine transporter, *FEBS Lett.*, 524, 199–203, 2002.

25. Fukumoto, H., Seino, S., Imura, H., Seino, Y., Eddy, R.L., Fukushima, Y., Byers, M.G., Shows, T.B., and Bell, G.I., Sequence, tissue distribution, and chromosomal localization of mRNA encoding a human glucose transporter-like protein, *Proc. Natl. Acad. Sci. USA*, 85, 5434–5438, 1988.

26. Roncero, I., Alvarez, E., Chowen, J.A., Sanz, C., Rabano, A., Vazquez, P., and Blazquez, E., Expression of glucose transporter isoform GLUT-2 and glucokinase genes in human brain, *J. Neurochem.*, 88, 1203–1210, 2004.

27. Kayano, T., Fukumoto, H., Eddy, R.L., Fan, Y.S., Byers, M.G., Shows, T.B., and Bell, G.I., Evidence for a family of human glucose transporter-like proteins. Sequence and gene localization of a protein expressed in fetal skeletal muscle and other tissues, *J. Biol. Chem.*, 263, 15245–15248, 1988.

28. Fukumoto, H., Kayano, T., Buse, J.B., Edwards, Y.H., Pilch, P.F., Bell, G.I., and Seino, S., Cloning and characterization of the major insulin-responsive glucose transporter expressed in human skeletal muscle and other insulin-responsive tissues, *J. Biol. Chem.*, 264, 7776–7779, 1989.

29. Kayano, T., Burant, C.F., Fukumoto, H., Gould, G.W., Fan, Y.S., Eddy, R.L., Byers, M.G., Shows, T.B., Seino, S., and Bell, G.I., Human facilitative glucose transporters. Isolation, functional characterization, and gene localization of cDNAs encoding an isoform (GLUT5) expressed in small intestine, kidney, muscle, and adipose tissue and an unusual glucose transporter pseudogene-like sequence (GLUT6), *J. Biol. Chem.*, 265, 13276–13282, 1990.

30. Rand, E.B., Depaoli, A.M., Davidson, N.O., Bell, G.I., and Burant, C.F., Sequence, tissue distribution, and functional characterization of the rat fructose transporter GLUT5, *Am. J. Physiol.*, 264, G1169–G1176, 1993.

31. Lisinski, I., Schurmann, A., Joost, H.G., Cushman, S.W., and Al-Hasani, H., Targeting of GLUT6 (formerly GLUT9) and GLUT8 in rat adipose cells, *Biochem. J.*, 358, 517–522, 2001.

32. Doege, H., Bocianski, A., Joost, H.G., and Schurmann, A., Activity and genomic organization of human glucose transporter 9 (GLUT9), a novel member of the family of sugar-transport facilitators predominantly expressed in brain and leucocytes, *Biochem. J.*, 350, 771–776, 2000.

33. Doege, H., Schurmann, A., Bahrenberg, G., Brauers, A., and Joost, H.G., GLUT8, a novel member of the sugar transport facilitator family with glucose transport activity, *J. Biol. Chem.*, 275, 16275–16280, 2000.

34. Phay, J.E., Hussain, H.B., and Moley, J.F., Cloning and expression analysis of a novel member of the facilitative glucose transporter family, SLC2A9 (GLUT9), *Genomics*, 66, 217–220, 2000.

35. Dawson, P.A., Mychaleckyj, J.C., Fossey, S.C., Mihic, S.J., Craddock, A.L., and Bowden, D.W., Sequence and functional analysis of GLUT10, a glucose transporter in the Type 2 diabetes-linked region of chromosome 20q12–13.1, *Mol. Genet. Metab.*, 74, 186–199, 2001.

36. McVie-Wylie, A.J., Lamson, D.R., and Chen, Y.T., Molecular cloning of a novel member of the GLUT family of transporters, SLC2a10 (GLUT10), localized on chromosome 20q13.1, a candidate gene for NIDDM susceptibility, *Genomics*, 113–117, 2001.

37. Wood, I.S., Hunter, L., and Trayhurn, P., Expression of Class III facilitative glucose transporter genes (GLUT-10 and GLUT-12) in mouse and human adipose tissues, *Biochem. Biophys. Res. Commun.*, 308, 43–49, 2003.

38. Doege, H., Bocianski, A., Scheepers, A., Axer, H., Eckel, J., Joost, H.G., and Schurmann, A., Characterization of human glucose transporter (GLUT) 11 (encoded by SLC2A11), a novel sugar-transport facilitator specifically expressed in heart and skeletal muscle. *Biochem. J.*, 359, 443–449, 2001.

39. Wu, X., Li, W., Sharma, V., Godzik, A., and Freeze, H.H., Cloning and characterization of glucose transporter 112, a novel sugar transporter that is alternatively spliced in various tissues, *Mol. Genet. Metab.*, 76, 37–45, 2002.

40. Sasaki, T., Minoshima, S., Shiohama, A., Shintani, A., Shimuzu, A., Asakawa, S., Kawasaki, K., and Shimizu, N., Molecular cloning of a member of the facilitative glucose transporter gene family GLUT11 (SLC2A11) and identification of transcription variants, *Biochem. Biophys. Res. Commun.*, 289, 1218–1224, 2001.

41. Macheda, M.L., Williams, E.D., Best, J.D., Wlodek, M.E., and Rogers, S., Expression and localisation of GLUT1 and GLUT12 glucose transporters in the pregnant and lactating rat mammary gland, *Cell Tissue Res.*, 311, 91–97, 2003.

42. Rogers, S., Macheda, M.L., Docherty, S.E., Carty, M.D., Henderson, M.A., Soeller, W.C., Gibbs, E.M., James, D.E., and Best, J.D., Identification of a novel glucose transporter-like protein-GLUT-12, *Am. J. Physiol. Endocrinol. Metab.*, 282, E733–E738, 2002.

43. Uldry, M., Ibberson, M., Horisberger, J.D., Chatton, J.Y., Riederer, B.M., and Thorens, B., Identification of a mammalian H+-*myo*-inositol symporter expressed predominantly in the brain, *EMBO J.*, 20, 4467–4477, 2001.

44. Panneerselvam, K. and Freeze, H.H., Mannose enters mammalian cells using a specific transporter that is insensitive to glucose, *J. Biol. Chem.*, 271, 9417–9421, 1996.

45. Wiese, T.J., Dunlap, J.A., and Yorek, M.A., L-Fucose is accumulated via a specific transport system in eukaryotic cells, *J. Biol. Chem.*, 269, 22705–22711, 1994.

46. Bulai, T., Bratosin, D., Artenie, V., and Montreuil, J., Uptake of sialic acid by human erythrocyte. Characterization of a transport system, *Biochimie*, 85, 241–244, 2003.

47. Oetke, C., Hinderlich, S., Brossmer, R., Reutter, W., Pawlita, M., and Keppler, O.T., Evidence for efficient uptake and incorporation of sialic acid by eukaryotic cells, *Eur. J. Biochem.*, 268, 4553–4561, 2001.

48. Tettamanti, G., Bassi, R., Viani, P., and Riboni, L., Salvage pathways in glycosphingolipid metabolism, *Biochimie.*, 85, 423–437, 2003.

49. Havelaar, A.C., Mancini, G.M., Beerens, C.E., Souren, R.M., and Verheijen, F.W., Purification of the lysosomal sialic acid transporter. Functional characteristics of a monocarboxylate transporter, *J. Biol. Chem.*, 273, 34568–34574, 1998.

50. Mancini, G.M., de Jonge, H.R., Galjaard, H., and Verheijen, F.W., Characterization of a proton-driven carrier for sialic acid in the lysosomal membrane. Evidence for a group-specific transport system for acidic monosaccharides, *J. Biol. Chem.*, 264, 15247–15254, 1989.

51. Mancini, G.M., Beerens, C.E., and Verheijen, F.W., Glucose transport in lysosomal membrane vesicles. Kinetic demonstration of a carrier for neutral hexoses, *J. Biol. Chem.*, 265, 12380–12387, 1990.

52. Jonas, A.J., Conrad, P., and Jobe, H., Neutral-sugar transport by rat liver lysosomes, *Biochem. J.*, 272, 323–326, 1990.

53. Jonas, A.J., Speller, R.J., Conrad, P.B., and Dubinsky, W.P., Transport of N-acetyl-D-glucosamine and N-acetyl-D-galactosamine by rat liver lysosomes, *J. Biol. Chem.*, 264, 4953–4956, 1989.

54. Reichner, J.S., Helgemo, S.L., and Hart, G.W., Recycling cell surface glycoproteins undergo limited oligosaccharide reprocessing in LEC1 mutant Chinese hamster ovary cells, *Glycobiology*, 8, 1173–1182, 1998.

55. Hirschberg, C.B. and Snider, T., Topography of glycosylation in the rough endoplasmic reticulum and Golgi apparatus, *Annu. Rev. Biochem.*, 56, 63, 1987.

56. Chou, T.C. and Soodak, M., The acetylation of D-glucosamine by pigeon liver extracts, *J. Biol. Chem.*, 196, 105–109, 1952.

57. Fujino, Y. and Nakano, M. Enzymic synthesis of cerebroside from ceramide and uridine diphosphate galactose, *Biochem. J.*, 113, 573–575, 1969.

58. Brockhausen, I., Rachaman, E.S., Matta, K.L., and Schachter, H., The separation by liquid chromatography (under elevated pressure) of phenyl, benzyl, and O-nitrophenyl glycosides of oligosaccharides. Analysis of substrates and products for four N-acetyl-D-glucosaminyl-transferases involved in mucin synthesis, *Carbohydr. Res.*, 120, 3–16, 1983.

59. Johnson, P.H., Yates, A.D., and Watkins, W.M., Human salivary fucosyltransferase, evidence for two distinct α-3-L-fucosyltransferase activities one of which is associated with the Lewis blood Le gene, *Biochem. Biophys. Res. Commun.*, 100, 1611–1618, 1981.

60. Isselbacher, K.J., A mammalian uridinediphosphate galactose pyrophosphorylase, *J. Biol. Chem.*, 232, 429–444, 1958.

61. Jourdian, G.W., Swanson, A., Watson, D., and Roseman, S., N-acetylneuraminic (sialic) acid 9-phosphatase, *Methods Enzymol.*, 8, 205–208, 1966.

62. Roseman, S., Glucosamine metabolism, I. N-acetylglucosamine deacetylase, *J. Biol. Chem.*, 226, 115–123, 1957.

63. Kojima, Y., Fukumoto, S., Furukawa, K., Okajima, T., Wiels, J., Yokoyama, K., Suzuki, Y., Urano, T., Ohta, M., and Furukawa, K., Molecular cloning of globotriaosylceramide/CD77 synthase, a glycosyltransferase that initiates the synthesis of globo series glycosphingolipids, *J. Biol. Chem.*, 275, 15152–15156, 2000.

64. Steffensen, R., Carlier, K., Wiels, J., Levery, S.B., Stroud, M., Cedergren, B., Nilsson Sojka, B., Bennett, E.P., Jersild, C., and Clausen, H., Cloning and expression of the histo-blood group Pk UDP-galactose: Gal-4βG1cβ1-cer α1, 4-galactosyltransferase. Molecular genetic basis of the p phenotype, *J. Biol. Chem.*, 275, 16723–16729, 2000.

65. Yamamoto, F., Marken, J., Tsuji, T., White, T., Clausen, H., and Hakomori, S., Cloning and characterization of DNA complementary to human UDP-GalNAc: Fuc α1,2Gal α1,3GalNAc transferase (histo-blood group A transferase) mRNA, *J. Biol. Chem.*, 265, 1146–1151, 1990.

66. Yamamoto, F. and Hakomori, S., Sugar-nucleotide donor specificity of histo-blood group A and B transferases is based on amino acid substitutions, *J. Biol. Chem.*, 265, 19257–19262, 1990.

67. Okajima, T., Nakamura, Y., Uchikawa, M., Haslam, D.B., Numata, S.I., Furukawa, K., Urano, T., and Furukawa, K., Expression cloning of human globoside synthase cDNAs. Identification of β 3Gal-T3 as UDP-*N*-acetylgalactosamine, globotriaosylceramide β1,3-*N*-acetylgalactosaminyltransferase, *J. Biol. Chem.*, 275, 40498–40503, 2000.

68. Amado, M., Almeida, R., Schweintek, T., and Clausen, H., Identification and characterization of large galactosyltransferase gene families, galactosyltransferases for all functions, *Biochim. Biophys. Acta.*, 1473, 35–53, 1999.

69. Shiina, T., Kikkawa, E., Iwasaki, H., Kaneko, M., Narimatsu, H., Sasaki, K., Bahram, S., and Inoko, H., The β-1,3-galactosyltransferase-4 (B3GALT4) gene is located in the centromeric segment of the human MHC class II region, *Immunogenetics*, 51, 75–78, 2000.

70. Zhou, D., Berger, E.G., and Hennet, T., Molecular cloning of a human UDP- galactose, GlcNAcβ1,3GalNAc β1,3-galactosyltransferase gene encoding an *O*-linked core3-elongation enzyme, *Eur. J. Biochem.*, 263, 571–576, 1999.

71. Yeh, J.C., Hiraoka, N., Petryniak, B., Nakayama, J., Ellies, L.G., Rabuka, D., Hindsgaul, O., Marth, J.D., Lowe, J.B., and Fukuda, M., Novel sulfated lymphocyte homing receptors and their control by a Core 1 extension β1,3-*N*-acetylglucosaminyltransferase, *Cell*, 105, 957–969, 2001.

72. Shiraishi, N., Natsume, A., Togayachi, A., Endo, T., Akashima, T., Yamada, Y., Imai, N., Nakagawa, S., Koizumi, S., Sekine, S., Narimatsu, H., and Sasaki, K., Identification and characterization of three novel β1,3-*N*-acetylglucosaminyltransferases structurally related to the β1,3-galactosyltransferase family, *J. Biol. Chem.*, 276, 3498–3507, 2001.

73. Togayachi, A., Akashima, T., Ookubo, R., Kudo, T., Nishihara, S., Iwasaki, H., Natsume, A., Mio, H., Inokuchi, J., Irimura, T., Sasaki, K., and Narimatsu, H., Molecular cloning and characterization of UDP-GlcNAc, lactosylceramide β1,3-*N*-acetylglucosaminyltransferase (β3Gn-T5), an essential enzyme for the expression of HNK-1 and Lewis X epitopes on glycolipids, *J. Biol. Chem.*, 276, 22032–22040, 2001.

74. Ju, T., Brewer, K., D'Souza, A., Cummings, R.D., and Canfield, W.M., Cloning and expression of human core 1 β1,3-galactosyltransferase, *J. Biol. Chem.*, 277, 178–186, 2002.

75. Munster, A.-K., Eckhardt, M., Potvin, B., Muhlenhoff, M., Stanley, P., and Gerardy-Schahn, R., Mammalian cytidine 5-prime-monophosphate *N*-acetylneuraminic acid synthetase, a nuclear protein with evolutionarily conserved structural motifs, *Proc. Natl. Acad. Sci. USA*, 95, 9140–9145, 1998.

76. Kannagi, R., Regulatory roles of carbohydrate ligands for selectins in the homing of lymphocytes, *Curr. Opin. Struct. Biol.*, 12, 599–608, 2002.

77. Fisher, K.J. and Aronson, N.N.J., Cloning and expression of the cDNA sequence encoding the lysosomal glycosidase di-*N*-acetylchitobiase, *J. Biol. Chem.*, 267, 19607–19616, 1992.

78. Ahmad, W., Li, S., Chen, H., Tuck-Muller, C.M., Pittler, S.J., and Aronson, N.N.J., Lysosomal chitobiase (CTB) and the G-protein gamma(5) subunit (GNG5) genes co-localize to human chromosome 1p22, *Cytogenet. Cell Genet.*, 71, 44–46, 1995.

79. Yamagata, T., Tsuru, T., Momoi, M.Y., Suwa, K., Nozaki, Y., Mukasa, T., Ohashi, H., Fukushima, Y., and Momoi, T., Genome organization of human 48-kDa oligosaccharyltransferase (DDOST), *Genomics*, 45, 535–540, 1997.

80. Ahn, J., Ludecke, H.J., Lindow, S., Horton, W.A., Lee, B., Wagner, M.J., Horsthemke, B., and Wells, D.E., Cloning of the putative tumour suppressor gene for hereditary multiple exostoses (EXT1), *Nat. Genet.*, 11, 137–143, 1995.

81. McCormick, C., Duncan, G., Goutsos, K.T., and Tufaro, F., The putative tumor suppressors EXT1 and EXT2 form a stable complex that accumulates in the Golgi apparatus and catalyzes the synthesis of heparan sulfate, *Proc. Natl. Acad. Sci. USA*, 97, 668–673, 2000.

82. Wu, Y.Q., Heutink, P., de Vries, B.B., Sandkuijl, L.A., van den Ouweland, A.M., Niermeijer, M.F., Galjaard, H., Reyniers, E., Willems, P.J., and Halley, D.J., Assignment of a second locus for multiple exostoses to the pericentromeric region of chromosome 11, *Hum. Mol. Genet.*, 3, 167–171, 1994.

83. Stickens, D., Clines, G., Burbee, D., Ramos, P., Thomas, S., Hogue, D., Hecht, J.T., Lovett, M., and Evans, G.A., The EXT2 multiple exostoses gene defines a family of putative tumour suppressor genes. *Nat. Genet.*, 14, 25–32, 1996.

84. Wise, C.A., Clines, G.A., Massa, H., Trask, B.J., and Lovett, M., Identification and localization of the gene for EXTL, a third member of the multiple exostoses gene family, *Genet. Res.*, 7, 10–16, 1997.

85. Wuyts, W., Spieker, N., Van Roy, N., De Boulle, K., De Paepe, A.M., Willems, P.J., Van Hul, W., Versteeg, R., and Speleman, F., Refined physical mapping and genomic structure of the EXTL1 gene, *Cytogenet. Cell Genet.*, 86, 267–270, 1999.

86. Wuyts, W., Van Hul, W., Speleman, H. J. F., Wauters, J., De Boulle, K., Van Roy, N., Van Agtmael, T., Bossuyt, P., and Willems, P.J., Identification and characterization of a novel member of the EXT gene family, EXTL2, *Eur. J. Hum. Genet.*, 5, 382–389, 1997.

87. Saito, T., Seki, N., Yamauchi, M., Tsuji, S., Hayashi, A., Kozuma, S., and Hori, T., Structure, chromosomal location, and expression profile of EXTR1 and EXTR2, new members of the multiple exostoses gene family, *Biochem. Biophys. Res. Commun.*, 243, 61–66, 1998.

88. Van Hul, W., Wuyts, W., Hendrickx, J., Speleman, F., Wauters, J., De Boulle, K., Van Roy, N., Bossuyt, P., and Willems, P.J., Identification of a third EXT-like gene (EXTL3) belonging to the EXT gene family, *Genomics*, 47, 230–237, 1998.

89. Pastuszak, I., Ketchum, C., Hermanson, G., Sjoberg, E.J., Drake, R.D., and Elbein, A.D., GDP-L-fucose pyrophosphorylase. Purification, cDNA cloning, and properties of the enzyme, *J. Biol. Chem.*, 273, 30165–30174, 1998.

90. Haslam, D.B. and Baenziger, J.U., Expression cloning of Forssman glycolipid synthetase, a novel member of the histo-blood group ABO gene family, *Proc. Natl. Acad. Sci. USA*, 93, 10697–10702, 1996.

91. Xu, H., Storch, T., Yu, M., Elliott, S.P., and Haslam, D.B., Characterization of the human Forssman synthetase gene. An evolving association between glycolipid synthesis and host-microbial interactions, *J. Biol. Chem.*, 274, 29390–29398, 1999.

92. Hinderlich, S., Berger, M., Blume, A., Chen, H., Ghaderi, D., and Bauer, C., Identification of human L-fucose kinase amino acid sequence, *Biochem. Biophys. Res. Commun.*, 294, 650–654, 2002.

93. Larsen, R.D., Ernst, L.K., Nair, R.P., and Lowe, J.B., Molecular cloning, sequence, and expression of a human GDP-L-fucose, β-D-galactoside 2-α-L-fucosyltransferase cDNA that can form the H blood group antigen, *Proc. Natl. Acad. Sci. USA*, 87, 6674–6678, 1990.

94. Ball, S.P., Tongue, N., Gibaud, A., Le Pendu, J., Mollicone, R., Gerard, G., and Oriol, R., The human chromosome 19 linkage group FUT1 (H), FUT2 (SE), LE, LU, PEPD, C3, APOC2, D19S7 and D19S9, *Ann. Hum. Genet.*, 55, 225–233, 1991.

95.   Kelly, R.J., Rouquier, S., Giorgi, D., Lennon, G.G., and Lowe, J.B., Sequence and expression of a candidate for the human Secretor blood group $\alpha(1,2)$fucosyltransferase gene (FUT2). Homozygosity for and enzyme-inactivating nonsense mutation commonly correlates with the non-secretor phenotype, *J. Biol. Chem.*, 270, 4640–4649, 1995.

96.   Reguigne-Arnould, I., Couillin, P., Mollicone, R., Faure, S., Fletcher, A., Kelly, R.J., Lowe, J.B., and Oriol, R., Relative positions of two clusters of human α-L-fucosyltransferases in 19q (FUT1-FUT2) and 19p (FUT6-FUT3-FUT5) within the microsatellite genetic map of chromosome 19, *Cytogenet. Cell Genet.*, 71, 158–162, 1995.

97.   Weston, B.W., Nair, R.P., Larsen, R.D., and Lowe, J.B., Isolation of a novel human $\alpha(1,3)$ fucosyltransferase gene and molecular comparison to the human Lewis blood group $\alpha(1,3/1,4)$fucosyltransferase gene. Syntenic, homologous, nonallelic genes encoding enzymes with distinct acceptor substrate specificities, *J. Biol. Chem.*, 267, 4152–4160, 1992.

98.   Weston, B.W., Smith, P.L., Kelly, R.J., and Lowe, J.B., Molecular cloning of a fourth member of a human $\alpha(1,3)$fucosyltransferase gene family. Multiple homologous sequences that determine expression of the Lewis x, sialyl Lewis x, and difucosyl sialyl Lewis x epitopes, *J. Biol. Chem.*, 267, 24575–24584, 1992.

99.   Mollicone, R., Reguigne, I., Fletcher, A., Aziz, A., Rustam, M., Weston, B.W., Kelly, R.J., Lowe, J.B., and Oriol, R., Molecular basis for plasma $\alpha(1,3)$-fucosyltransferase gene deficiency (FUT6), *J. Biol. Chem.*, 269, 12662–12671, 1994.

100. Yanagidani, S., Uozumi, N., Ihara, Y., Miyoshi, E., Yamaguchi, N., and Taniguchi, N., Purification and cDNA cloning of GDP-L-Fuc, *N*-acetyl-beta-D-glucosaminide, α-1-6 fucosyltransferase (α-1-6 FucT) from human gastric cancer MKN45 cells, *J. Biochem.*, 121, 626–632, 1997.

101. Coullin, P., Crooijmans, R.P.M.A., Groenen, M.A.M., Heilig, Mollicone, R., Oriol, R., and Candelier, J.-J., Assignment of FUT8 to chicken chromosome band 5q1.4 and to human chromosome 14q23.2–q24.1 by *in situ* hybridization. Conserved and compared synteny between human and chicken, *Cytogenet. Genome Res.*, 97, 234–238, 2002.

102. Daude, N., Gallaher, T.K., Zeschnigk, M., Starzinski-Powitz, A., Petry, H.I.S, K.G., and Reichardt, J.K., Molecular cloning, characterization, and mapping of a full-length cDNA encoding human UDP-galactose 4'-epimerase, *Biochem. Mol. Med.*, 56, 1–7, 1995.

103. Ai, Y., Basu, M., Bergsma, D.J., and Stambolian, D., Comparison of the enzymatic activities of human galactokinase GALK1 and a related human galactokinase protein GK2, *Biochem. Biophys. Res. Commun.*, 212, 687–691, 1995.

104. Timson, D.J. and Reece, R.J. Identification and characterisation of human aldose 1-epimerase, *FEBS Lett.*, 543, 21–24, 2003.

105. Bennett, E.P., Hassan, H., Mandel, U., Hollingsworth, M.A., Akisawa, N., Ikematsu, Y., Merkx, G., van Kessel, A.G., Olofsson, S., and Clausen, H., Cloning and characterization of a close homologue of human UDP-*N*-acetyl-α-D-galactosamine, polypeptide *N*-acetylgalactosaminyltransferase-T3, designated GalNAc-T6. Evidence for genetic but not functional redundancy, *J. Biol. Chem.*, 274, 25362–25370, 1999.

106. Nagata, Y., Yamashiro, S., Yodoi, J., Lloyd, K.O., Shiku, H., and Furukawa, K., Expression cloning of β-1,4-*N*-acetylgalactosaminyltransferase cDNAs that determine the expression of G_{M2} and G_{D2} gangliosides, *J. Biol. Chem.*, 267, 12082–12089, 1992.

107. Reichardt, J.K. and Berg, P. Cloning and characterization of a cDNA encoding human galactose-1-phosphate uridyl transferase, *Mol. Biol. Med.*, 5, 107–122, 1988.

108. Tanizawa, Y., Koranyi, L.I., Welling, C.M., and Permutt, M.A., Human liver glucokinase gene, cloning and sequence determination of two alternatively spliced cDNAs, *Proc. Natl. Acad. Sci. USA*, 88, 7294–7297, 1991.

109. Yeh, J.C., Ong, E., and Fukuda, M., Molecular cloning and expression of a novel β-1, 6-N-acetylglucosaminyltransferase that forms core 2, core 4, and I branches, *J. Biol. Chem.*, 274, 3215–3221, 1999.
110. Kalz-Fuller, B., Bieberich, E., and Bause, E., Cloning and expression of glucosidase I from human hippocampus, *Eur. J. Biochem.*, 231, 344–351, 1995.
111. Kalz-Fuller, B., Heidrich-Kaul, C., Nothen, M., Bause, E., and Schwanitz, G., Localization of the human glucosidase I gene to chromosome 2p12–p13 by fluorescence *in situ* hybridization and PCR analysis of somatic cell hybrids, *Genomics*, 34, 442–443, 1996.
112. McKnight, G.L., Mudri, S.L., Mathewes, S.L., Traxinger, R.R., Marshall, S., Sheppard, P.O., and O'Hara, P.J., Molecular cloning, cDNA sequence, and bacterial expression of human glutamine:fructose-6-phosphate amidotransferase, *J. Biol. Chem.*, 267, 25208–25212, 1992.
113. Larsen, R.D., Rajan, V.P., Ruff, M.M., Kukowska-Latallo, J., Cummings, R.D., and Lowe, J.B., Isolation of a cDNA encoding a murine UDPgalactose, β-D-galactosyl-1,4-N-acetyl-D-glucosaminide α-1,3-galactosyltransferase, expression cloning by gene transfer, *Proc. Natl. Acad. Sci. USA*, 86, 8227–8231, 1989.
114. Joziasse, D.H., Shaper, N.L., Kim, D., Van den Eijnden, D.H., and Shaper, J.H., Murine α1,3-galactosyltransferase. A single gene locus specifies four isoforms of the enzyme by alternative splicing, *J. Biol. Chem.*, 267, 5534–5541, 1992.
115. Sullivan, F.X., Kumar, R., Kriz, R., Stahl, M., Xu, G.Y., Rouse, J., Chang, X.J., Boodhoo, A., Potvin, B., and Cumming, D.A., Molecular cloning of human GDP-mannose 4,6-dehydratase and reconstitution of GDP-fucose biosynthesis *in vitro*, *J. Biol. Chem.*, 273, 8193–8202, 1998.
116. Ohyama, C., Smith, P.L., Angata, K., Fukuda, M.N., Lowe, J.B., and Fukuda, M., Molecular cloning and expression of GDP-D-mannose-4,6-dehydratase, a key enzyme for fucose metabolism defective in Lec13 cells, *J. Biol. Chem.*, 273, 14582–14587, 1998.
117. Ning, B. and Elbein, A.D. Cloning, expression and characterization of the pig liver GDP-mannose pyrophosphorylase. Evidence that GDP-mannose and GDP-Glc pyrophosphorylases are different proteins, *Eur. J. Biochem.*, 267, 6866–6874, 2000.
118. Bistrup, A., Bhakta, S., Lee, J.K., Belov, Y.Y., Gunn, M.D., Zuo, F.-R., Chiao-Chain, H., Kannagi, R., Rosen, S.D., and Hemmerich, S., Sulfotransferases of two specificities function in the reconstitution of high endothelial cell ligands for L-selectin, *J. Cell Biol.*, 145, 899–910, 1999.
119. Uchimura, K., El-Fasakhany, F.M., Hori, M., Hemmerich, S., Blink, S.E., Kansas, G.S., Kanamori, A., Kumamoto, K., Kannagi, R., and Muramatsu, T., Specificities of N-acetylglucosamine-6-O-sulfotransferases in relation to L-selectin ligand synthesis and tumor-associated enzyme expression, *J. Biol. Chem.*, 277, 3979–3984, 2002.
120. Uchimura, K., Muramatsu, H., Kaname, T., Ogawa, H., Yamakawa, T., Fan, Q., Mitsuoka, C., Kannagi, R., Habuchi, O., Yokoyama, I., Yamamura, K., Ozaki, T., Nakagawara, A., Kadomatsu, K., and Muramatsu, T., Human N-acetylglucosamine-6-O-sulfotransferase involved in the biosynthesis of 6-sulfo sialyl Lewis X: molecular cloning, chromosomal mapping, and expression in various organs and tumor cells, *J. Biochem.*, 124, 670–678, 1998.
121. Bowman, K.G. and Bertozzi, C.R. Carbohydrate sulfotransferases, mediators of extracellular communication, *Chem. Biol.*, 6, R9–R22, 1999.
122. Hinderlich, S., Stache, R., Zeitler, R., and Reutter, W., A bifunctional enzyme catalyzes the first two steps in N-acetylneuraminic acid biosynthesis of rat liver: purification and characterization of UDP-N-acetylglucosamine 2-epimerase/N-acetylmannosamine kinase, *J. Biol. Chem.*, 272, 24313–24318, 1997.

123. Shevchenko, V., Hogben, M., Ekong, R., Parrington, J., and Lai, F.A., The human glucosamine-6-phosphate deaminase gene, cDNA cloning and expression, genomic organization and chromosomal localization, *Gene*, 216, 31–38, 1998.

124. Boehmelt, G., Fialka, I., Brothers, G., McGinley, M.D., Patterson, S.D., Mo, R., Hui, C.C., Chung, S., Huber, L.A., Mak, T.W., and Iscove, N.N., Cloning and characterization of the murine glucosamine-6-phosphate acetyltransferase EMeg32. Differential expression and intracellular membrane association, *J. Biol. Chem.*, 275, 12821–12832, 2000.

125. Xu, W., Lee, P., and Beutler, E., Human glucose phosphate isomerase: exon mapping and gene structure, *Genomics*, 29, 732–739, 1995.

126. Nishi, S., Seino, S., and Bell, G.I., Human hexokinase, sequences of amino- and carboxyl-terminal halves are homologous, *Biochem. Biophys. Res. Commun.*, 157, 937–943, 1988.

127. Murakami, K. and Piomelli, S., Identification of the cDNA for human red blood cell-specific hexokinase isozyme, *Blood*, 89, 762–766, 1997.

128. Lehto, M., Xiang, K., Stoffel, M., Espinosa, R.I., Groop, L.C., Le Beau, M.M., and Bell, G.I., Human hexokinase II, localization of the polymorphic gene to chromosome 2, *Diabeteologia*, 36, 1299–1302, 1993.

129. Colosimo, A., Calabrese, G., Gennarelli, M., Ruzzo, A.M., Sangiuolo, F., Magnani, M., Palka, G., Novelli, G., and Dallapiccola, B., Assignment of the hexokinase type 3 gene (HK3) to human chromosome band 5q35.3 by somatic cell hybrids and *in situ* hybridization, *Cytogenet. Cell Genet.*, 74, 187–188, 1996.

130. Furuta, H., Nishi, S., Le Beau, M.M., Fernald, A.A., Yano, H., and Bell, G.I., Sequence of human hexokinase III cDNA and assignment of the human hexokinase III gene (HK3) to chromosome band 5q35.2 by fluorescence *in situ* hybridization, *Genomics*, 36, 206–209, 1996.

131. Hong, K., Ma, D., Beverley, S.M., and Turco, S.J., The Leishmania GDP-mannose transporter is an autonomous, multi-specific, hexameric complex of LPG2 subunits, *Biochemistry*, 39, 2013–2022, 2000.

132. Tremblay, L.O., Campbell Dyke, N., and Herscovics, A., Molecular cloning, chromosomal mapping and tissue-specific expression of a novel human α-1,2-mannosidase gene involved in *N*-glycan maturation, *Glycobiology*, 8, 585–595, 1998.

133. Misago, M., Liao, Y.F., Kudo, S., Eto, S., Mattei, M.G., Moremen, K.W., and Fukuda, M.N., Molecular cloning and expression of cDNAs encoding human α-mannosidase II and a previously unrecognized α-mannosidase IIx isozyme, *Proc. Natl. Acad. Sci. USA*, 92, 11766–11770, 1995.

134. Kumar, R. and Stanley, P. Transfection of a human gene that corrects the Lec1 glycosylation defect, evidence for transfer of the structural gene for *N*-acetylglucosaminyltransferase I, *Mol. Cell Biol.*, 9, 5713–5717, 1989.

135. Kumar, R., Yang, J., Larsen, R.D., and Stanley, P., Cloning and expression of *N*-acetylglucosaminyltransferase I, the medial Golgi transferase that initiates complex *N*-linked carbohydrate formation, *Proc. Natl. Acad. Sci. USA*, 87, 9948–9952, 1990.

136. Tan, J., D'Agostaro, G.A.F., Bendiak, B., Reck, F., Sarkar, M., Squire, J.A., Leong, P., and Schachter, H., The human UDP-*N*-acetylglucosamine, α-6-D-mannoside-β-1,2-*N*-acetylglucosaminyltransferase II gene (MGAT2), cloning of genomic DNA, localization to chromosome 14q21, expression in insect cells and purification of the recombinant protein, *Eur. J. Biochem.*, 231, 317–328, 1995.

137. Ihara, Y., Nishikawa, A., Tohma, T., Soejima, H., Niikawa, N., and Taniguchi, N., cDNA cloning, expression, and chromosomal localization of human *N*-acetylglucosaminyltransferase III (GnT-III), *J. Biochem.*, 113, 692–698, 1993.

138.   Yoshida, A., Minowa, M.T., Takamatsu, S., Hara, T., Ikenaga, H., and Takeuchi, M., A novel second isoenzyme of the human UDP-*N*-acetylglucosamine, α1,3-D-mannoside β1,4-*N*-acetylglucosaminyltransferase family, cDNA cloning, expression, and chromosomal assignment, *Glycoconj. J.*, 15, 1115–1123, 1998.

139.   Appert, H.E., Rutherford, T.J., Tarr, G.E., Wiest, J.S., Thomford, N.R., and McCorquodale, D.J., Isolation of a cDNA coding for human galactosyltransferase, *Biochem. Biophys. Res. Commun.*, 139, 163–168, 1986.

140.   Masri, K.A., Appert, H.E., and Fukuda, M.N., Identification of the full-length coding sequence for human galactosyltransferase (β-*N*-acetylglucosaminide, β1,4-galactosyltransferase), *Biochem. Biophys. Res. Commun.*, 157, 657–663, 1988.

141.   Proudfoot, A.E., Turcatti, G., Wells, T.N., Payton, M.A., and Smith, D.J., Purification, cDNA cloning and heterologous expression of human phosphomannose isomerase, *Eur. J. Biochem.*, 219, 415–423, 1994.

142.   Hinderlich, S., Berger, M., Schwarzkopf, M., Effertz, K., and Reutter, W., Molecular cloning and characterization of murine and human *N*-acetylglucosamine kinase, *Eur. J. Biochem.*, 267, 3301–3308, 2000.

143.   Whitehouse, D.B., Putt, W., Lovegrove, J.U., Morrison, K., Hollyoake, M., Fox, M.F., Hopkinson, D.A., and Edwards, Y.H., Phosphoglucomutase 1, complete human and rabbit mRNA sequences and direct mapping of this highly polymorphic marker on human chromosome 1, *Proc. Natl. Acad. Sci. USA*, 89, 411–415, 1992.

144.   Pang, H., Koda, Y., Soejima, M., Kimura, H., and Iacomino, G., identification of human phosphoglucomutase 3 (PGM3) as *N*-acetylglucosamine-phosphate mutase (AGM1), *Ann. Hum. Genet.*, 66, 139–144, 2002.

145.   Li, C., Rodriguez, M., and Banerjee, D., Cloning and characterization of complementary DNA encoding human *N*-acetylglucosamine-phosphate mutase protein, *Gene*, 242, 97–103, 2000.

146.   Hansen, S.H., Frank, S.R., and Casanova, J.E., Cloning and characterization of human phosphomannomutase, a mammalian homologue of yeast SEC53, *Glycobiology*, 7, 829–834, 1997.

147.   Mitsuoka, C., Ohmori, K., Kimura, N., Kanamori, A., Komba, S., Ishida, H., Kiso, M., and Kannagi, R., Regulation of selectin binding activity by cyclization of sialic acid moiety of carbohydrate ligands on human leukocytes, *Proc. Natl. Acad. Sci. USA*, 96, 1597–1602, 1999.

148.   Manzi, A.E., Sjoberg, E.R., Diaz, S., and Varki, A., Biosynthesis and turnover of *O*-acetyl and *N*-acetyl groups in the gangliosides of human melanoma cells, *J. Biol. Chem.*, 265, 13091–13103, 1990.

149.   Hanai, N., Dohi, T., Nores, G.A., and Hakomori, S., A novel ganglioside, de-*N*-acetyl-GM3 (II3NeuNH2LacCer), acting as a strong promoter for epidermal growth factor receptor kinase and as a stimulator for cell growth, *J. Biol. Chem.*, 263, 6296–6301, 1988.

150.   Chammas, R., Sonnenburg, J.L., Watson, N.E., Tai, T., Farquhar, M.G., Varki, N.M., and Varki, A., De-*N*-acetyl-gangliosides in human, unusual subcellular distribution of a novel tumor antigen, *Cancer Res.*, 59, 1337–1346, 1999.

151.   Chang, M.-L., Eddy, R.L., Shows, T.B., and Lau, J.T.Y., Three genes that encode human β-galactoside α-2,3-sialyltransferases, structural analysis and chromosomal mapping studies, *Glycobiology*, 5, 319–325, 1995.

152.   Kitagawa, H. and Paulson, J.C. Differential expression of five sialyltransferase genes in human tissues, *J. Biol. Chem.*, 269, 17872–17878, 1994.

153.   Shang, J., Qiu, R., Wang, J., Liu, J., Zhou, R., Ding, H., Yang, S., Zhang, S., and Jin, C., Molecular cloning and expression of Galβ1,3GalNAc α2, 3-sialyltransferase from human fetal liver, *Eur. J. Biochem.*, 265, 580–588, 1999.

154. Kim, Y.J., Kim, K.S., Kim, S.H., Kim, C.H., Ko, J.H., Choe, I.S., Tsuji, S., and Lee, Y.C., Molecular cloning and expression of human Gal β1,3GalNAcα2,3-sialyltransferase (hST3Gal II), *Biochem. Biophys. Res. Commun.*, 228, 324–327, 1996.

155. Kitagawa, H., Mattei, M.G., and Paulson, J.C., Genomic organization and chromosomal mapping of the Gal β1,3GalNAc/Gal β1,4GlcNAc α2,3-sialyltransferase, *J. Biol. Chem.*, 271, 931–938, 1996.

156. Kitagawa, H. and Paulson, J.C. Cloning and expression of human Galβ1,3(4)GlcNAcα2,3-sialyltransferase, *Biochem. Biophys. Res. Commun.*, 194, 374–382, 1993.

157. Julien, S., Krzewinski-Recchi, M.A., Harduin-Lepers, A., Gouyer, V., Huet, G., Le Bourhis, X., and Delannoy, P., Expression of sialyl-Tn antigen in breast cancer cells transfected with the human CMP-Neu5Ac, GalNAc α2,6-sialyltransferase (ST6GalNac I) cDNA, *Glycoconj. J.*, 18, 883–893, 2001.

158. Ikehara, Y., Kojima, N., Kurosawa, N., Kudo, T., Kono, M., Nishihara, S., Issiki, S., Morozumi, K., Itzkowitz, S., Tsuda, T., Nishimura, S., Tsuji, S., and Narimatsu, H., Cloning and expression of a human gene encoding an *N*-acetylgalactosamine-α2,6-sialyltransferase (ST6GalNAc I), a candidate for synthesis of cancer-associated sialyl-Tn antigens, *Glycobiology*, 9, 1213–1224, 1999.

159. Sotiropoulou, G., Kono, M., Anisowicz, A., Gtenman, G., Tsuji, S., and Sager, R., Identification and functional characterization of human GalNAc α2,6 sialyltransferase with altered expression in breast cancer, *Mol. Med.*, 8, 42–55, 2002.

160. Samyn-Petit, B., Krzewinski-Recchi, M.A., Steelant, W.F., Delannoy, P., and Harduin-Lepers, A., Molecular cloning and functional expression of human ST6GalNAc II. Molecular expression in various human cultured cells, *Biochim. Biophys. Acta.*, 1474, 201–211, 2000.

161. Matsuda, Y., Nara, K., Watanabe, Y., Saito, T., and Sanai, Y., Chromosome mapping of the GD3 synthase gene (SIAT8) in human and mouse, *Genomics*, 32, 137–139, 1996.

162. Sasaki, K., Kurata, K., Kojima, N., Kurosawa, N., Ohta, S., Hanai, N., Tsuji, S., and Nishi, T., Expression cloning of a GM3-specific α-2,8-sialyltransferase (GD3 synthase), *J. Biol. Chem.*, 269, 15950–15956, 1994.

163. Ishii, A., Ohta, M., Watanabe, Y., Matsuda, K., Ishiyama, K., Sakoe, K., Nakamura, M., Inokuchi, J., Sanai, Y., and Saito, M., Expression cloning and functional characterization of human cDNA for ganglioside GM3 synthase, *J. Biol. Chem.*, 273, 31652–31655, 1998.

164. Mio, T., Yabe, T., Arisawa, M., and Yamada-Okabe, H., The eukaryotic UDP-*N*-acetylglucosamine pyrophosphorylases, gene cloning, protein expression, and catalytic mechanism, *J. Biol. Chem.*, 273, 14392–14397, 1998.

165. Moriarity, J.L., Hurt, K.J., Resnick, A.C., Storm, P.B., Laroy, W., Schnaar, R.L., and Snyder, S.H., UDP-glucuronate decarboxylase, a key enzyme in proteoglycan synthesis, cloning, characterization, and localization, *J. Biol. Chem.*, 277, 16968–16975, 2002.

166. Spicer, A.P., Kaback, L.A., Smith, T.J., and Seldin, M.F., Molecular cloning and characterization of the human and mouse UDP-glucose dehydrogenase genes, *J. Biol. Chem.*, 273, 25117–25124, 1998.

167. Holden, H.M., Rayment, I., and Thoden, J.B., Structure and function of enzymes of the Leloir pathway for galactose metabolism, *J. Biol. Chem.*, 278, 43885–43888, 2003.

168. Zachara, N.E. and Hart, G.W. The emerging significance of *O*-GlcNAc in cellular regulation, *Chem. Rev.*, 102, 431–438, 2002.

169. Hanover, J.A., Glycan-dependent signaling, *O*-linked *N*-acetylglucosamine, *FASEB J.*, 15, 1865–1876, 2001.

170.   Kearns, A.E., Vertel, B.M., and Schwartz, N.B.J., Topography of glycosylation and UDP-xylose production, *J. Biol. Chem.*, 268, 11097–11104, 1993.

171.   Angata, T. and Varki, A., Chemical diversity in the sialic acids and related α-keto acids, an evolutionary perspective, *Chem. Rev.*, 102, 439–469, 2002.

172.   Kayser, H., Geilen, C.C., Paul, C., Zeitler, R., and Reutter, W., New amino sugar analogues are incorporated at different rates into glycoproteins of mouse organs, *Experientia*, 49, 885–887, 1993.

173.   Yarema, K.J., Mahal, L.K., Bruehl, R.E., Rodriguez, E.C., and Bertozzi, C.R., Metabolic delivery of ketone groups to sialic acid residues. Application to cell surface glycoform engineering, *J. Biol. Chem.*, 273, 31168–31179, 1998.

174.   Yarema, K.J., New directions in carbohydrate engineering: a metabolic substrate-based approach to modify the cell surface display of sialic acids, *BioTechniq.*, 31, 384–393, 2001.

175.   Keppler, O.T., Horstkorte, R., Pawlita, M., Schmidt, C., and Reutter, W., Biochemical engineering of the *N*-acyl side chain of sialic acid, biological implications, *Glycobiology*, 11, 11R–18R, 2001.

176.   Hinderlich, S., Oetke, C., and Pawlita, M., Biochemical engineering of sialic acids, in *Handbook of Carbohydrate Engineering*, Yarema, K., Ed., Dekker/CRC Press, Boca Raton, 2005, chap. 13.

177.   Wilson, J.C., Kiefel, M.J., and von Itzstein, M., Sialic acids and sialylmimetics, useful chemical probes of sialic acid-recognising proteins, in *Handbook of Carbohydrate Engineering*, Yarema, K., Ed., Dekker/CRC Press, Boca Raton, 2005, chap. 24.

178.   Rapoport, E.M., Mochalova, L.V., Gabius, H.-J., Romanova, J., and Bovin, N.V., Patterning of lectins of Vero and MDCK cells and influenza viruses. The search for additional virus/cell interactions, in *Handbook of Carbohydrate Engineering*, Yarema, K., Ed., Dekker/CRC Press, Boca Raton, 2005, chap. 2.

179.   Chava, A.K., Mitali Chatterjee, M., and Mandal, C., *O*-Acetyl sialic acids in parasitic diseases, in *Handbook of Carbohydrate Engineering*, Yarema, K., Ed., Dekker/CRC Press, Boca Raton, 2005, chap. 3.

180.   Viswanathan, K. and Betenbaugh, M., Engineering sialic acid synthetic ability into insect cells: identifying metabolic bottlenecks and devising strategies to overcome them, in *Handbook of Carbohydrate Engineering*, Yarema, K., Ed., Dekker/CRC Press, Boca Raton, 2005, chap. 14.

181.   Abeijon, C., Mandon, E.C., and Hirschberg, C.B., Transporters of nucleotide sugars, nucleotide sulfate and ATP in the Golgi apparatus, *Trends Biochem. Sci.*, 22, 203–207, 1997.

182.   Presley, J.F., Smith, C., Hirschberg, K., Miller, C., Cole, N.B., Zaal, K.J.M., and Lippincott-Schwartz, J., Golgi membrane dynamics, *Mol. Biol. Cell.*, 9, 1617–1626, 1998.

183.   Puglielli, L., Mandon, E.C., and Hirschberg, C.B., Identification, purification, and characterization of the rat liver Golgi membrane ATP transporter, *J. Biol. Chem.*, 274, 12655–12669, 1999.

184.   Ishida, N., Ito, M., Yoshioka, S., Sun-Wada, G.-H., and Kawakita, M., Functional expression of human Golgi CMP-sialic acid transporter in the Golgi complex of a transporter-deficient Chinese hamster ovary cell mutant, *J. Biochem.*, 124, 171–178, 1998.

185.   Luhn, K., Wild, M.K., Eckhardt, M., Gerardy-Schahn, R., and Vestweber, D., The gene defective in leukocyte adhesion deficiency II encodes a putative GDP-fucose transporter, *Nat. Genet.*, 28, 69–72, 2001.

186.   Lubke, T., Marquardt, T., Etzioni, A., Hartmann, E., Von Figura, K., and Korner, C., Complementation cloning identifies CDG-IIc, a new type of congenital disorders of glycosylation, as a GDP-fucose transporter deficiency, *Nat. Genet.*, 28, 73–76, 2001.

187. Dean, N., Zhang, Y.B., and Poster, J.B., The VRG4 gene is required for GDP-mannose transport into the lumen of the Golgi in the yeast, *Saccharomyces cerevisiae, J. Biol. Chem.*, 272, 31908–31914, 1997.

188. Kamiyama, S., Suda, T., Ueda, R., Suzuki, M., Okubo, R., Kikuchi, N., Chiba, Y., Goto, S., Toyoda, H., Saigo, K., Watanabe, M., Narimatsu, H., Jigami, Y., and Nishihara, S., Molecular cloning and identification of 3′-phosphoadenosine 5′-phosphosulfate transporter, *J. Biol. Chem.*, 278, 25958–25963, 2003.

189. Mandon, E.C., Milla, M.E., Kempner, E., and Hirschberg, C.B., Purification of the Golgi adenosine 3′-phosphate 5′-phosphosulfate transporter, a homodimer within the membrane, *Proc. Natl. Acad. Sci. USA*, 91, 10707–10711, 1994.

190. Miura, N., Ishida, N., Hoshino, M., Yamauchi, M., Hara, T., Ayusawa, D., and Kawakita, M., Human UDP-galactose translocator, molecular cloning of a complementary DNA that complements the genetic defect of a mutant cell line deficient in UDP-galactose translocator, *J. Biochem.*, 120, 236–241, 1996.

191. Sprong, H., Degroote, S., Nilsson, T., Kawakita, M., Ishida, N., van der Sluijs, P., and van Meer, G., Association of the Golgi UDP-galactose transporter with UDP-galactose, ceramide galactosyltransferase allows UDP-galactose import in the endoplasmic reticulum, *Mol. Biol. Cell*, 14, 3482–3493, 2003.

192. Segawa, H., Kawakita, M., and Ishida, N., Human and Drosophila UDP-galactose transporters transport UDP-*N*-acetylgalactosamine in addition to UDP-galactose, *Eur. J. Biochem.*, 269, 128–138, 2002.

193. Norambuena, L., Marchant, L., Berninsone, P., Hirschberg, C.B., Silva, H., and Orellana, A., Transport of UDP-galactose in plants. Identification and functional characterization of AtUTr1, an *Arabidopsis thaliana* UDP-galactose/UDP-glucose transporter, *J. Biol. Chem.*, 277, 32923–32929, 2002.

194. Muraoka, M., Kawakita, M., and Ishida, N., Molecular characterization of human UDP-glucuronic acid/UDP-*N*-acetylgalactosamine transporter, a novel nucleotide sugar transporter with dual substrate specificity, *FEBS Lett.*, 495, 87–93, 2001.

195. Ishida, N., Yoshioka, S., Chiba, Y., Takeuchi, M., and Kawakita, M., Molecular cloning and functional expression of the human Golgi UDP-*N*-acetylglucosamine transporter, *J. Biochem.*, 126, 68–77, 1999.

196. Schweizer, F., Engineering carbohydrate scaffolds into the side chains of amino acids and use in combinatorial synthesis, in *Handbook of Carbohydrate Engineering*, Yarema, K., Ed., Dekker/CRC Press, Boca Raton, 2005, chap. 25.

197. Koscielak, J., Congenital disorders of glycosylation, in *Handbook of Carbohydrate Engineering*, Yarema, K., Ed., Dekker/CRC Press, Boca Raton, 2005, chap. 4.

198. Stanley, P. and Patnaik, S.K., Chinese hamster ovary (CHO) glycosylation mutants for glycan engineering, in *Handbook of Carbohydrate Engineering*, Yarema, K., Ed., Dekker/CRC Press, Boca Raton, 2005, chap. 12.

199. Helenius, A. and Aebi, M. Intracellular functions of *N*-linked glycans, *Science*, 291, 2364–2369, 2001.

200. Kornfeld, R. and Kornfeld, S. Assembly of asparagines-linked oligosaccharides, *Annu. Rev. Biochem.*, 54, 631, 1985.

201. Saito, H., Nishikawa, A., Gu, J., Ihara, Y., Soejima, H., Wada, Y., Sekiya, C., Niikawa, N., and Taniguchi, N., cDNA cloning and chromosomal mapping of human *N*-acetylglucosaminyltransferase V[+], *Biochem. Biophys. Res. Commun.*, 198, 318–327, 1994.

202. Brockhausen, I., Narasimhan, S., and Schachter, H., The biosynthesis of highly branched *N*-glycans, studies on the sequential pathway and functional role of *N*-acetylglucosaminyltransferases I, II, III, IV, V and VI, *Biochimie*, 70, 1521–1533, 1988.

203. Brockhausen, I., Carver, J.P., and Schachter, H., Control of glycoprotein synthesis. The use of oligosaccharide substrates and HPLC to study the sequential pathway for *N*-acetylglucosaminyltransferases I, II, III, IV, V, and VI in the biosynthesis of highly branched *N*-glycans by hen oviduct membranes, *Biochem. Cell. Biol.*, 66, 1134–1151, 1988.

204. Sandhoff, K. and Kolter, T., Biosynthesis and degradation of mammalian glycosphingolipids, *Philos. Trans. R. Soc. Lond. B Biol. Sci.*, 358, 847–861, 2003.

205. Watts, R.W.E., A historical perspective of the glycosphingolipids and sphingolipidoses, *Philos. Trans. R. Soc. Lond. B,* 358, 975–983, 2003.

206. Ichikawa, S. and Hirabayashi, Y., Glucosylceramide synthase and glycosphingolipid synthesis, *Trends Cell. Biol.*, 8, 198–202, 1998.

207. Ferguson, M.A.J. and Williams, A.F., Cell-surface anchoring of proteins via glycosylphosphatidyl-inositol structures, *Annu. Rev. Biochem.*, 57, 285, 1988.

208. Takeda, J. and Kinoshita, T., GPI-anchor biosynthesis, *Trends Biol. Sci.*, 20, 367–371, 1995.

209. Yeh, E.T.H., Kamitani, T., and Chang, H.M., Biosynthesis and processing of the glycosylphosphatidylinositol anchor in mammalian cells, *Semin. Immunol.*, 6, 73–80, 1994.

210. Ikezawa, H., Glycosylphosphatidylinositol (GPI)-anchored proteins, *Biol. Pharm. Bull.*, 25, 409–417, 2002.

211. Smith, T.K., Sharma, D., Crossman, A., Dix, A., Brimacombe, J.S., and Ferguson, M.A.J., Parasite and mammalian GPI biosynthetic pathways can be distinguished using synthetic substrate analogues, *EMBO J.*, 16, 6667–6675, 1997.

212. Bernfield, M., Gotte, M., Park, P.W., Reizes, O., Fitzgerald, M.L., Lincecum, J., and Zako, M., Functions of cell surface haparan sulfate proteoglycans, *Annu. Rev. Biochem.*, 68, 729–777, 1999.

213. Krusius, T., Finne, J., Margolis, R.K., and Margolis, R.U., Identification of an *O*-glycosidic mannose-linked sialylated tetrasaccharide and keratan sulfate oligosaccharides in the chondroitin sulfate proteoglycan of brain, *J. Biol. Chem.*, 261, 8237–8242, 1986.

214. Silbert, J.E. and Sugumaran, G., Biosynthesis of chondrotin/dermatan sulfate, *IUBMB Life*, 54, 177–186, 2002.

215. Funderbergh, J.L., Keratan sulfate biosynthesis, *IUBMB Life*, 54, 187–194, 2002.

216. Stone, B., Dispatches from the last frontier of molecular and cell biology, biosynthesis of polysaccharides and proteoglycans of the cell surface and extracellular matrix, *IUBMB Life*, 54, 161–162, 2002.

217. Sugahara, K., Mikami, T., Uyama, T., Mizuguchi, S., Nomura, K., and Kitagawa, H., Recent advances in the structural biology of chondroitin sulfate and dermatan sulfate, *Curr. Opin. Struct. Biol.*, 13, 612–620, 2003.

218. Pojasek, K., Raman, R., and Sasisekharan, R., Structural characterization of glycosaminoglycans, in *Handbook of Carbohydrate Engineering*, Yarema, K., Ed., Dekker/CRC Press, Boca Raton, 2005, chap. 6.

219. Kuhn, J., Muller, S., Schnolzer, M., Kempf, T., Schon, S., Brinkmann, T., Schottler, M., Gotting, C., and Kleesiek, K., High-level expression and purification of human xylosyltransferase I in High Five insect cells as biochemically active form, *Biochem. Biophys. Res. Commun.*, 312, 537–544, 2003.

220. Gotting, C., Kuhn, J., Zahn, R., Brinkmann, T., and Kleesiek, K., Molecular cloning and expression of human UDP-d-Xylose, proteoglycan core protein β-d-xylosyltransferase and its first isoform XT-II, *J. Mol. Biol.*, 304, 517–528, 2000.

221. Kusche-Gullberg, M. and Kjellen, L. Sulfotransferases in glycosaminoglycan biosynthesis, *Curr. Opin. Struct. Biol.*, 13, 605–611, 2003.

222. Itano, N. and Kimata, K., Mammalian hyaluronan synthases, *IUBMB Life*, 54, 195–199, 2002.
223. Watanabe, K. and Yamaguchi, Y., Molecular identification of a putative human hyaluronan synthase, *J. Biol. Chem.*, 271, 22945–22948, 1996.
224. Itano, N. and Kimata, K., Expression cloning and molecular characterization of HAS protein, a eukaryotic hyaluronan synthase, *J. Biol. Chem.*, 271, 9875–9878, 1996.
225. Spicer, A.P., Olson, J.S., and McDonald, J.A., Molecular cloning and characterization of a cDNA encoding the third putative mammalian hyaluronan synthase, *J. Biol. Chem.*, 272, 8957–8961, 1997.
226. Yoshida, M., Itano, N., Yamada, Y., and Kimata, K., *In vitro* synthesis of hyaluronan by a single protein derived from mouse HAS1 gene and characterization of amino acid residues essential for the activity, *J. Biol. Chem.*, 275, 497–506, 2000.
227. Itano, N., Sawai, T., Yoshida, M., Lenas, P., Yamada, Y., Imagawa, M., Shinomura, T., Hamaguchi, M., Yoshida, Y., Ohnuki, Y., Miyauchi, S., Spicer, A.P., McDonald, J.A., and Kimata, K., Three isoforms of mammalian hyaluronan synthases have distinct enzymatic properties, *J. Biol. Chem.*, 274, 25085–25092, 1999.
228. Camenisch, T.D., Spicer, A.P., Brehm-Gibson, T., Biesterfeldt, J., Augustine, M.L., Calabro, A.J., Kubalak, S., Klewer, S.E., and McDonald, J.A., Disruption of hyaluronan synthase-2 abrogates normal cardiac morphogenesis and hyaluronan-mediated transformation of epithelium to mesenchyme, *J. Clin. Invest.*, 106, 349–360, 2000.
229. Kim, B.T., Kitagawa, H., Tamura, J., Saito, T., Kusche-Gullberg, M., Lindahl, U., and Sugahara, K., Human tumor suppressor EXT gene family members EXTL1 and EXTL3 encode α1,4-*N*-acetylglucosaminyltransferases that likely are involved in heparan sulfate/ heparin biosynthesis, *Proc. Natl. Acad. Sci. USA*, 98, 7176–7181, 2001.
230. Bulik, D.A., Wei, G., Toyoda, H., Kinoshita-Toyoda, A., Waldrip, W.R., Esko, J.D., Robbins, P.W., and Selleck, S.B., sqv-3, -7, and -8, a set of genes affecting morphogenesis in *Caenorhabditis elegans*, encode enzymes required for glycosaminoglycan biosynthesis, *Proc. Natl. Acad. Sci. USA*, 97, 10838–10843, 2000.
231. Nilsson, T., Slusarewicz, P., Hoe, M.H., and Warren, G., Kin recognition. A model for the retention of Golgi enzymes, *FEBS Lett.*, 330, 1–4, 1993.
232. Nilsson, T., Hoe, M.H., Slusarewicz, P., Rabouille, C., Watson, R., Hunte, F., Watzele, G., Berger, E.G., and Warren, G., Kin recognition between medial Golgi enzymes in HeLa cells, *EMBO J.*, 13, 562–574, 1994.
233. Kazuyuki, S. and Hiroshi, K., Heparin and heparan sulfate biosynthesis, *IUBMB Life*, 54, 163–175, 2002.
234. Pinhal, M.A., Smith, B., Olson, S., Aikawa, J., Kimata, K., and Esko, J.D., Enzyme interactions in heparan sulfate biosynthesis, uronosyl 5-epimerase and 2-*O*-sulfotransferase interact *in vivo*, *Proc. Natl. Acad. Sci. USA*, 98, 12984–12989, 2001.
235. Kitagawa, H., Shimakawa, H., and Sugahara, K., The tumor suppressor EXT-like gene EXTL2 encodes an α1,4-*N*-acetylhexosaminyltransferase that transfers *N*-acetylgalactosamine and *N*-acetylglucosamine to the common glycosaminoglycan-protein linkage region. The key enzyme for the chain initiation of heparan sulfate, *J. Biol. Chem.*, 274, 13933–13937, 1999.
236. Rohrmann, K., Niemann, R., and Buddecke, E., Two *N*-acetylgalactosaminyl transferases are involved in the biosynthesis of chondroitin sulfate, *J. Biochem.*, 148, 463–469, 1985.
237. Uyama, T., Kitagawa, H., Tamura, J.J., and Sugahara, K., Molecular cloning and expression of human chondroitin *N*-acetylgalactosaminyltransferase, the key enzyme for chain initiation and elongation of chondroitin/dermatan sulfate on the protein linkage region tetrasaccharide shared by heparin/heparan sulfate, *J. Biol. Chem.*, 277, 8841–8846, 2002.

238.   Kitagawa, H., Uyama, T., and Sugahara, K., Molecular cloning and expression of a human chondroitin synthase, *J. Biol. Chem.*, 276, 38721–38726, 2001.

239.   Silbert, J.E. and Reppucci, A.C.J., Biosynthesis of chondroitin sulfate, independent addition of glucuronic acid and *N*-acetylgalactosamine to oligosaccharides, *J. Biol. Chem.*, 251, 3942–3947, 1976.

240.   Silbert, J.E., Biosynthesis of chondroitin sulfate: chain termination, *J. Biol. Chem.*, 253, 6888–6892, 1978.

241.   Otsu, K., Inoue, H., Tsuzuki, Y., Yonekura, H., Nakanishi, Y., and Suzuki, S., A distinct terminal structure in newly synthesized chondroitin sulphate chains, *Biochem. J.*, 227, 37–48, 1985.

242.   Funderbergh, J.L., Keratan sulfate: structure, biosynthesis, and function, *Glycobiology*, 10, 951–958, 2000.

243.   Christner, J.E., Distler, J.J., and Jourdian, G.W., Biosynthesis of keratan sulfate: purification and properties of a galactosyltransferase from bovine cornea, *Arch. Biochem. Biophys.*, 192, 548–558, 1979.

244.   Nakazawa, K., Takahashi, I., and Yamamoto, Y., Glycosyltransferase and sulfotransferase activities in chick corneal stromal cells before and after *in vitro* culture, *Arch. Biochem. Biophys.*, 359, 269–282, 1998.

245.   Morrison, S.L., The role of glycosylation in engineered antibodies, in *Handbook of Carbohydrate Engineering*, Yarema, K., Ed., Dekker/CRC Press, Boca Raton, 2005, chap. 20.

246.   Chen, H.Y. and Varki, A., *O*-acetylation of GD3: an enigmatic modification regulating apoptosis, *J. Exp. Med.*, 196, 1529–1533, 2002.

247.   Malisan, F., Franchi, L., Tomassini, B., Ventura, N., Condo, I., Rippo, M.R., Rufini, A., Liberati, L., Nachtigall, C., Kniep, B., and Testi, R., Acetylation suppresses the proapoptotic activity of GD3 ganglioside, *J. Exp. Med.*, 196, 1535–1541, 2002.

248.   Murrell, M.M., Yarema, K.J., and Levchenko, A., Sialic acid model, in *Handbook of Carbohydrate Engineering*, Yarema, K., Ed., Dekker/CRC Press, Boca Raton, 2005, chap. 8.

249.   Burdick, M.M., Bochner, B.S., Collins, B.E., Schnaar, R.L., and Konstantopoulos, K., Glycolipids support E-selectin-specific strong cell tethering under flow, *Biochem. Biophys. Res. Commun.*, 284, 42–49, 2001.

250.   No, H.K., Prinyawiwatkul, W., and Meyers, S.P., Treatment of wastewaters with the biopolymer chitosan, in *Handbook of Carbohydrate Engineering*, Yarema, K., Ed., Dekker/CRC Press, Boca Raton, 2005, chap. 19.

251.   Poulsen, M., Effects of short- and long-chained fructans on large intestinal physiology and development of pre-neoplastic lesions in rats, in *Handbook of Carbohydrate Engineering*, Yarema, K., Ed., Dekker/CRC Press, Boca Raton, 2005, chap. 26.

252.   Pandey, R., and Khuller, G.K., Alginate as drug delivery carrier, in *Handbook of Carbohydrate Engineering*, Yarema, K., Ed., Dekker/CRC Press, Boca Raton, 2005, chap. 27.

253.   Liu, W.G., Lu, W.W., and Yao, K.D., Chitosan-based nonviral vectors for gene delivery, in *Handbook of Carbohydrate Engineering*, Yarema, K., Ed., Dekker/CRC Press, Boca Raton, 2005, chap. 28.

254.   Hutmacher, D.W., Leong, D.T.W., and Chen, F., Polysaccharides in tissue engineering applications, in *Handbook of Carbohydrate Engineering*, Yarema, K., Ed., Dekker/CRC Press, Boca Raton, 2005, chap. 29.

2 Patterning of Lectins of Vero and MDCK Cells and Influenza Viruses: The Search for Additional Virus/Cell Interactions

E.M. Rapoport, L.V. Mochalova,
H.-J. Gabius, J. Romanova,
and N.V. Bovin

CONTENTS

I. INTRODUCTION

Influenza virus infection is initiated by specific interactions between the viral envelope glycoprotein hemagglutinin (HA) and cell surface receptors.[1-3] Terminal sialic
acid residue of glycoproteins or gangliosides is known to be the minimum binding
determinant of these receptors. Virus binding also depends on the type of the sialoside linkage to penultimate galactose and on the structure of more distant parts of
sialyl oligosaccharides.[1-7] Sialyl glycoconjugates have long been considered to be
the sole receptor for the influenza virus.

However, a mutant virus, NWS-Mvi, grows well in Madin–Darby canine kidney
(MDCK) cells continuously treated with exogenous neuraminidase to remove sialic
acid; binding of mutant and parent reassortant viruses to MDCK cells is indistinguishable and is only particularly reduced by sialidase treatment of the cells.[8] The
ability to infect desialylated cells was found to be shared by recent H3N2 clinical
isolates, suggesting that this may be a generic property of influenza A viruses. Thus,
it was proposed that influenza virus entry is a two-stage process, whereby initial
binding to surface sialic acid leads to an interaction with the second receptor
required for entry, and that the requirement for sialic acid can be bypassed[8] as was
demonstrated recently for reovirus attachment.[9] Evidence exists that other glycoconjugates such as sulfatide are also bound by the influenza virus,[10] and at the same
time, MDCK cells have been shown to express sulfatide.[11] According to Martin et
al.[12] infection of cells by influenza viruses is influenced by the abundance of the
receptors on the cell surface when the affinity of HA for sialic acid is reduced.
Finally, high-affinity binding of HA must be maintained for viral replication in cells
expressing low levels of sialic acid.[13] Taking this information into consideration we
made an attempt to detect a possible second receptor, hypothesizing the participation
of alternative (or additional/assistant) carbohydrate/lectin recognition in the
influenza virus attachment process, being first of all in situations when a lectin of the
host cell recognizes a carbohydrate ligand on the virus glycoconjugates. Using multivalent carbohydrate probes and antibodies to galectins and siglecs, we tested the
presence of galactose-binding lectins (galectins) and sialic acid-specific lectins
(siglecs) on MDCK and Vero cells as well as their mannose- and sulfo-N-acetyllac-
tosamine recognizing properties. We also took into consideration an alternative carbohydrate-mediated mechanism when lectin acquired by virion during its
assembling from the host cell can recognize carbohydrates of the target cell. The
selected two mammalian cell lines, i.e., MDCK and the African green monkey kidney (Vero) cell lines, have been approved recently for the production of inactivated
influenza vaccines.[14,15] Moreover, the MDCK cell line is widely used for primary
isolation and propagation of human influenza viruses. Vero cells were recently
shown to be useful for this purpose as well.[16] According to our data, human viruses

isolated on Vero cells had HA1 amino-acid sequences indistinguishable from original viruses present in clinical samples of influenza patients, while viruses isolated in MDCK cells sometimes harbored substitutions in HA1 subunit.[17,18]

II. MATERIALS AND METHODS

A. REAGENTS

Label-free glycoconjugates Sug-PAA, biotinylated glycoconjugates Sug-PAA-biot, and fluorescein-labeled probes Sug-PAA-fluo were obtained from Syntesome GmbH (Munich, Germany). 11-OS-PAA was obtained as described earlier,[19] and 7-OS was obtained by sequential neuraminidase and galactosidase digestion of 11-OS. Murine monoclonal antibodies to siglecs, i.e., Ser 4 (against siglec 1), aCD22 (against siglec 2), aCD33 (against siglec 3), anti-MAG (against siglec 4), aSiglec5 (against siglec 5), Ab7a (against siglec 7), Ab73 (against siglec 8), and Ab 2.2 (against siglec 10) were obtained from Dr. P. Crocker (Dundee, Scotland). Rabbit polyclonal antibodies against galectin-1 and galectin-3 lacking cross-reactivity against other family members such as galectin-2, -4, -5, -7, and -8 were obtained as described in Reference 20. Human anti-Galα1-3Gal antibodies were isolated from sera by affinity chromatography on Galα1-3Gal-Sepharose FF adsorbent as described in Reference 21. Digoxigenin-labeled lectins from *Maackia amurensis* (MAA) and *Sambucus nigra* (SNA), streptavidin-peroxidase conjugate, and anti-digoxigenin Fab fragment-FITC conjugate were obtained from Boehringer Mannheim (Germany). Phosphate-buffered saline (PBS) tablets, bovine serum albumin (BSA), fluorescein isothiocyanate (FITC)-labeled anti-mouse Ig and anti-rabbit Ig, and sialidase from *Vibrio cholerae* were purchased from Sigma (San Diego, USA). Dulbecco medium, fetal bovine serum (FBS), and penicillin/streptomycin were from Flow Laboratories (UK).

B. CELL CULTURE

Vero (WHO-certified) and MDCK cell lines were obtained from the American Type Culture Collection (ATCC). Vero cells were adapted and further cultivated in Dulbecco's modified Eagle's medium (DMEM)/Ham's F12 (Biochrom) protein-free medium.[22] MDCK cells were cultivated in DMEM/Ham's F12 (1:1) medium supplemented with 2% of heat-inactivated fetal calf serum (Gibco, UK). MDCK cells were cultured in a Dulbecco medium containing 10% FBS, 2 mM L-glutamine, and 100 U/mL penicillin/streptomycin. Vero cells were cultured in Dulbecco medium containing 2.0 mM L-glutamine and 100 U/mL penicillin/streptomycin.

C. SIALIDASE TREATMENT

Cells were harvested with a Versen solution, washed three times with Dulbecco's medium, and treated with neuraminidase (1:20 dilution) in Dulbecco's medium at 37°C for 3 h under agitation. The reaction was stopped with FBS (10%, by vol.). Cells were washed three times with PBS, containing 0.2% BSA (PBA) by centrifugation at 1200 r/min for 2 min. In the solid-phase cell assay, cells were incubated on plates in a Dulbecco's medium containing neuraminidase (dilution 1:20) for 3 h at 37°C in 5%

CO_2 atmosphere. The concentration of cells subject to flow cytometry analysis was $2 \times 10^5/100\ \mu L$.

D. BINDING OF SUG-PAA-FLUO PROBES TO THE CELLS

Cells were harvested with a cold Versene solution, washed three times in cold PBA, and incubated with fluorescein-labeled probes Sug-PAA-fluo (100 μg/mL) for 40 min at 4°C. All conjugates had the same molecular weight of 30 to 40 kDa and the same molar content of Sug ligand (20%) and fluorescein (1%). After washing with cold PBA, fluorescence analysis was performed using a flow cytometer (Dako Galaxy, Denmark) or a FACScan (Becton-Dickinson, USA). Carbohydrate-free PAA-fluo conjugate was used as a negative control.

E. ANTIBODY BINDING ASSAY

Cells were washed with PBA and incubated with anti-siglec mAbs or anti-galectin antibodies at 4°C for 1 h followed by incubation with FITC-labeled anti-mouse Ig (1:50 dilution) or anti-rabbit Ig (1:50 dilution) conjugates, respectively, for 30 min, then washed three times with PBA. FACS analysis was performed as described above. Anti-mouse or anti-rabbit IgG-FITC conjugates were used as a negative control.

F. BINDING OF CELLS WITH LECTINS SNA AND MAA

Cells were washed with PBA and incubated with digoxigenin-labeled MAA or SNA (1 μg/mL) at 4°C under agitation for 1 h, followed by incubation with anti-digoxigenin Fab-FITC conjugate (1:10 dilution) for 30 min, and then were washed three times with PBA. Anti-digoxigenin Fab-FITC conjugate without prior incubation with the first-step reagent was used as a negative control. Sialidase treatment of cells was performed as described above.

G. GALα1-3GAL-BINDING ASSAY

Virus purification: Virus suspension containing cultural fluid was incubated with 0.5 M NaCl for 30 min at room temperature followed by low-speed centrifugation. Supernatant was layered on top of 30% sucrose prepared in TN (0.02 M Tris HCl, pH 7.2, with 0.1 M NaCl) buffer and centrifuged at 24,000 r/min. for 1.5 h at 4°C (Beckman L-50, SW-27 rotor). The pellet of virus was resuspended in PBS or TN buffer and the virus suspension was cleared by low-speed centrifugation and stored at 4°C in buffer, containing 0.01% NaN_3.

Solid-phase direct binding assay:[21] Polystyrene 96-well plates (Costar, USA) were coated with purified influenza virus with a titer of 4 to 8 hemagglutination units (50 μL/well) at 4°C for 16 h followed by washing with PBS buffer. After that, 100 μL/well of blocking solution (PBS containing 2% of BSA) was added, then the plates were incubated at 37°C for 1 h and washed with PBS containing 0.05% of Tween-20 (washing solution). After the addition of affinity-purified human anti-Galα1-3Gal antibodies (45 μL/well)[21] in the working buffer (PBS with 0.3% of BSA), the plates were kept at 37°C for 1 h. The starting dilution of anti-Galα1-3Gal antibodies was 1:50 followed by two-fold serial dilutions. Plates were washed with the washing solution and incubated with anti-human antibodies conjugated with horseradish peroxidase (dilution in the working buffer was 1:2.000, 40 μL/well) at 37°C for 1 h. After washing, 50 μL/well of substrate solution (0.1 M sodium acetate, pH 5.0, containing 4 mM o-phenylenediamine and 0.004% H_2O_2) was added and the reaction was stopped with 2 M H_2SO_4. Optical density was read at 492 nm. To evaluate the affinity of anti-Galα1-3Gal antibodies for different isolates of human influenza viruses the dilutions of anti-Galα1-3Gal antibodies at the point $A_{max}/2$ of Scatchard plots were determined. The reported value of these dilutions represents an average of at least three independent measurements.

H. Probing the Influenza Virus with Sug-PAA-biot

Polystyrene 96-well plates were coated with purified viruses, 10 hemagglutinating units per well in PBS for 16 h at 4°C, washed with PBS, blocked with 0.4% gelatin in PBS for 2 h at room temperature, and finally washed with 0.05% Tween-20 in PBS. Double-diluted solution of Sug-PAA-biot (starting concentration 50 μM of Sug) in PBS containing 0.02% gelatin, 0.02% Tween, and neuraminidase (NA) inhibitor (buffer B) was added, plates were incubated for 2 h at 4°C, and washed. Streptavidin-peroxidase conjugate (1:3,000) in buffer B was added, plates were incubated for 30 min at 4°C, and washed. Development with o-phenylenediamine was performed as described above.

III. RESULTS

A. Revealing α2-3-and α2-6-Linked Sialic Acid on MDCK and Vero Cells

We investigated the expression of Neu5Acα2-6Gal and Neu5Acα2-3Gal carbohydrate motifs on the surface of MDCK and Vero cells (Figure 2.1) using SNA and MAA known to be specific lectins for these disaccharides. The presence of the

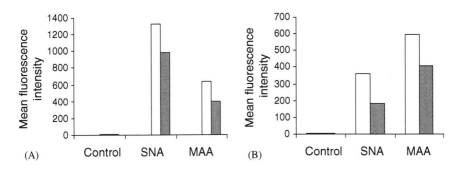

FIGURE 2.1 Assessment of presence of Neu5Acα2-3Gal and Neu5Acα2-6Gal on MDCK (A) and Vero (B) cells using SNA and MAA lectins. Cells were stained with digoxigenin-labeled lectins. Control cells were stained with anti-digoxigenin Fab-FITC. Intensity of fluorescence is provided in arbitrary units. Gray bars correspond to neuraminidase-treated cells. Typical data of three experiments with similar results are presented.

Neu5Acα2-3Gal epitope was observed on both cells, although its proportion compared with Neu5Acα2-6Gal was higher on Vero than on MDCK cells. On the contrary, the expression of Neu5Acα2-6Gal sequence, known to be a receptor for nonegg-adapted human viruses, was higher on MDCK cells. In absolute values, the intensity of MDCK cells staining with SNA was twofold higher than that of Vero cells. Staining decreased after neuraminidase treatment, but only partially (Figure 2.1, gray bars). Having characterized expression of these lectin-reactive sialylated epitopes in the first step, we next applied labeled neoglycoconjugates to map the profile of binding activity for distinct carbohydrate ligands.

B. CARBOHYDRATE-BINDING PATTERN OF MDCK AND VERO CELLS

We inspected the carbohydrate-binding profile of MDCK and Vero cells by using a panel of Sug-PAA-fluo probes, where Sug were galactoside-, sialic acid-, mannoside-, and sulfo-oligosaccharides. A number of OS, such as LacNAc, $T_{\beta\beta}$, GalNAcβ1-4GlcNAc, asialoGM1, and blood group A were known to be potent ligands for cellular lectins such as galectins.[23] Indeed, MDCK exhibited a significant level of binding to LacNAc, asialoGM1, GalNAcβ1-4GlcNAc, TF, blood group A, and Fucα1-2Galβ1-3GlcNAc (Le^d) oligosaccharides as shown in Figure 2.2. Binding of these probes to Vero cells was less pronounced. Fs disaccharide interacted intensely only with MDCK cells.

Galectin expression on MDCK and Vero cells was confirmed by an antibody binding assay. Purified rabbit polyclonal antibodies, which were raised against human

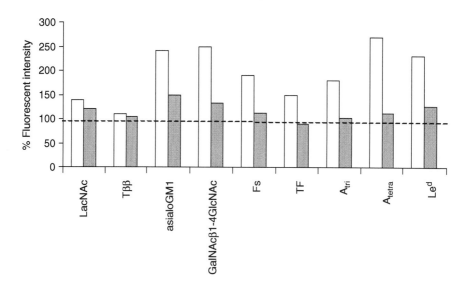

FIGURE 2.2.   Binding of galectin ligands as Sug-PAA-fluo probes to MDCK (white bars) and Vero cells (gray bars). Cells were stained with carbohydrate-free PAA-fluo as a negative control. Percentage of fluorescence intensity from cells incubated with probes relative to that of control cells. The dashed line indicates the level of 100% value. Typical data of three experiments with similar results are presented.

galectin-1 and murine galectin-3 and showed no mutual cross-reactivity[20] were used. As shown in Figure 2.3, the cells revealed dissimilar binding patterns with the two antibodies; Vero cells bound preferentially the antibody fraction to galectin-1, whereas MDCK bound only the antibody to galectin-3. Similar to Sug-PAA-fluo probing, antibody staining was more intense in the case of MDCK cells than for Vero cells.

Sialic acid-dependent binding of Vero and MDCK cells was studied by using Sug-PAA-fluo probes, where Sug was sialic acid monosaccharide, 3'SL, 6'SL, Neu5Acα2-6GalNAcα, or disialoside [Neu5Acα2-8]$_2$. We observed a lack of sialic acid binding even after the neuraminidase treatment of the cells (desialylation renders *cis*-connected siglecs accessible), only the Neu5Acα benzyl glycoside-based probe bound at a significant level (Figure 2.4). Similar results were obtained by solid-phase assays with the corresponding biotinylated sialoside probes (data not shown).

Notably, both Vero and MDCK cells bound to Neu5Acα in a benzyl glycoside spacered form, whereas binding of the cells with Neu5Acα as OCH$_2$CH$_2$CH$_2$-glycoside did not differ significantly from the background level. Interaction of Vero cells with Neu5AcαBn was three times stronger than with MDCK cells. Binding of Neu5AcαBn-PAA-fluo to both cell types was inhibited by a low-molecular-weight ligand Neu5Acα-OBn in a dose-dependent manner, indicating the specificity of the interaction (Figure 2.5). At the same time, neither Vero nor MDCK cells interacted with monoclonal antibodies against siglecs 1, 2, 3, 4, 5, 7, 8, and 10 (data not shown). These data indicate the presence of a sialic acid-specific binding site such as a lectin of nonsiglec nature on the cells.

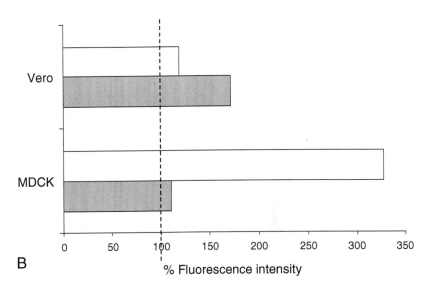

FIGURE 2.3.   Interaction of Vero and MDCK cells with rabbit polyclonal anti-galectin-1 (white bars) and anti-galectin-3 antibodies (gray bars). Cells were incubated with anti-rabbit IgG-FITC as a negative control. Mean fluorescence intensity from cells incubated with anti-galectin antibodies relative to that incubated with control cells. The dashed line indicates the level of 100% value.

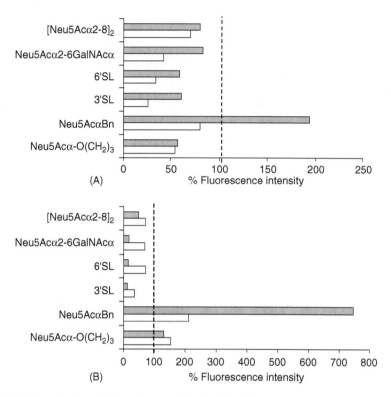

FIGURE 2.4. Binding of sialoside-containing Sug-PAA-fluo probes to MDCK (A) and Vero (B) cells. Gray bars correspond to neuraminidase-treated cells. Cells were exposed to carbohydrate-free PAA-fluo as a negative control (100% value, dashed line).

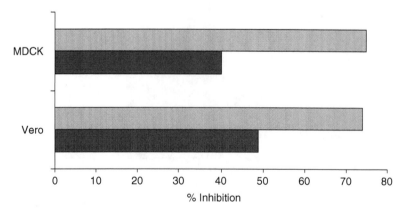

FIGURE 2.5. Inhibition of Neu5AcαBn-PAA-fluo binding to MDCK and Vero cells by free Neu5AcαBn. Cells were processed without or in the presence of 50 mM (gray bars) or a 12.5 mM (black bars) inhibitor. Inhibition degree was calculated as 100-(M_i/M) × 100 (%), where M_i are values of fluorescence in the presence of inhibitor, and M the value of fluorescence in the absence of the inhibitor.

Vero cells displayed significantly higher binding with all of the three sulfated oligosaccharides compared with MDCK cells (Figure 2.6). Apparently, the position of the sulfate group is not crucial; interaction with 6-HSO$_3$Galβ1-4GlcNAc (6'-HSO$_3$LacNAc), Galβ1-4(6-HSO$_3$)GlcNAc (6-HSO$_3$LacNAc), and 3-HSO$_3$Gal probes was practically equal, indicating a charge specific rather than a structure-dependent event.

Mannoside-binding potency of MDCK and Vero cells was studied with four fluorescein-labeled probes. Monomannoside-, trimannoside Manα1-6(Manβ1-3)Man- as well as the GlcNAc-terminated biantennary chain (GlcNAc-Man)$_2$-3, 6-Man-GlcNAc-GlcNAc (7-OS) demonstrated binding, especially with Vero cells, whereas the corresponding biantennary Neu5Ac-terminated chain (11-OS, negative control) was indifferent to both cell types (Figure 2.7). These data indicate the potency of Vero cells to recognize terminal Man/GlcNAc residues.

C. PROBING OF VIRUSES WITH SUG-PAA-BIOT

In order to reveal lectins possibly acquired by virions from host cells, human influenza virus B isolate and five influenza A viruses (passaged on Vero, MDCK, or Vero followed by MDCK cells) were probed with the following biotinylated glycoconjugates: 6'-HSO$_3$LacNAc-PAA-biot, 6-HSO$_3$LacNAc-PAA-biot, 3'HSO$_3$LacNAc-PAA-biot, 3-HSO$_3$Gal-PAA-biot, Manα-PAA-biot, 7-OS-PAA-biot, GalNAcα1-3GalNAcβ-PAA-biot, T$_{\beta\beta}$-PAA-biot, A$_{tri}$-PAA-biot, B$_{tri}$-PAA-biot, and Galα1-3Galβ1-4GlcNAc-PAA-biot. A solid-phase assay had recently been devised for studying the modern influenza virus strain with a low receptor affinity.[17,18] Manα and 7-OS probes demonstrated weak binding (data not shown), whereas 6'-HSO$_3$LacNAc (but not other sulfated probes) bound to both A and B viruses with affinities close to those of 6'-sialo probe (Table 2.1).

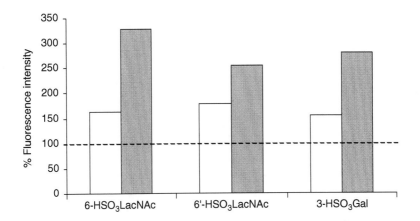

FIGURE 2.6. Binding of sulfated Sug-PAA-fluo probes to MDCK (white bars) and Vero (gray bars) cells. As a negative control (100% value, dashed line), cells were stained with carbohydrate-free PAA-flu.

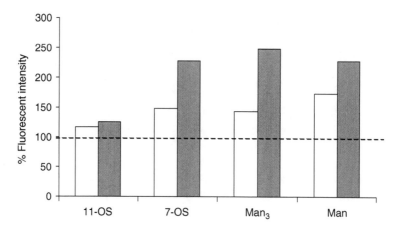

FIGURE 2.7. Binding of mannose-containing Sug-PAA-fluo probes to MDCK (white bars) and Vero (gray bars) cells. Cells were stained with carbohydrate-free PAA-fluo as a negative control (100% value, dashed line).

TABLE 2.1

Binding of Sulfated LacNAc-PAA-biot to Influenza Viruses as Compared with 6'SL-PAA-biot

Virus Strain, Passage History, Subtype	50% Binding of Sug-PAA-biot, as [Sug] (μM)		
	6'-HSO$_3$LacNAc	3'-HSO$_3$LacNAc	6'SL
B/Kirov/13/01, 1MDCK	2.0	40	0.3
A/Vienna/47/96, 4Vero (H3N2)	2.0	40	1.2
A/Vienna/47/96, 3MDCK (H3N2)	3.0	>50	0.8
A/HK/1180/99, 6Vero (H3N2)	3.0	>50	1.2
A/HK/1182/99 3Vero 5MDCK (H3N2)	0.8	>50	1.2
A/HK/1182/99 3MDCK 5Vero (H3N2)	0.6	>50	1.0

D.   REVEALING GALα1-3GAL EPITOPE ON MDCK- AND VERO-GROWN VIRUSES

Three tested influenza virus strains grown on MDCK cells displayed binding with human polyclonal antibodies against Galα1-3Gal epitope in a solid-phase assay, in contrast to the strains grown on Vero cells or chicken embryos. Synthetic antigen Galα1-3Gal-PAA coated as a positive control on the same plate showed binding, indicating the validity of the test. Vero-grown virus A/Hongkong/1182/99 after one passage on MDCK cells acquired the ability to bind the anti-Galα1-3Gal antibody.

IV.   DISCUSSION

The entry and penetration of influenza virus into the cell is very complex and only a partially understood process. Although the key event is the virus attachment to the target cell, it is necessary to take into consideration many accompanying factors, in particular virus interaction with extracellular inhibitors, such as sialoglycoproteins (mucins), anti-bodies to viral proteins,[24] and mannose-binding lectins such as conglutinin, mannoside-binding protein (MBP) and surfactant protein B (SP-D) ("β-inhibitors").[25,26] Mannose-binding alveolar surfactant protein A (SP-A) attaches to influenza viruses via its sialic acid residues, thereby conferring mannose-binding properties, for example, the ability to attach to mannose chains of the target cells to the virus.[27] Obviously, this list of extracellular factors affecting (not necessarily as inhibitors) the virus interaction with a cell is not comprehensive, and one can assume that this group can also include other lectins capable of binding carbohydrate chains of HA and NA. There are data available that demonstrate evidence that viral neuraminidase also plays a role in viral recep-tion;[28,29] thus, it is possible to suppose that endogenous extracellular glycosidases, such as human *trans*-sialidase,[30] can affect the infection process. In addition, the adhesion process is only one of many important events during virus penetration into the cell. Prior to the molecular rendezvous and specific recognition of the target cell, the virion has multiple opportunities to encounter various cell species having both similar and differ-ent glycosylation and lectin patterns as compared with the target cell. In order to reach its target, however, the virus must avoid an interaction with nontarget cells. For instance, due to the high density of terminal Gal and Man residues the influenza virus can be sub-ject of phagocytosis by macrophages, bearing the corresponding lectins. The virus should also avoid an interaction with NK cells bearing sialylated chains which act as a decoy for viral HA.[31] Indeed, the virus should not be attached firmly to nontarget cells due to its HA. At the same time it is possible to consider that some "third" cells assist virus movement into the desired direction, for example, as described for selectin-medi-ated leukocyte rolling.[30] It is known that in addition to basic cell surface receptors, HIV and Herpes simplex viruses require certain additional cellular factors that are not involved in primary binding in order to infect the target cells.[32,33]

In this investigation, we attempted to resolve the issue as to whether additional car-bohydrate-mediated events can be involved in the interaction of the influenza virus with host cells. In particular, the galactose-terminated chains of virion can be recognized by cell galectins; incompletely desialylated glycan chains via sialic acid by siglecs; high-mannose N-chains by MBP; and, finally, sulfated chains by the corresponding protein. In addition, the possibility of an insertion of host cell lectins into the membrane of *de novo* assembling virion cannot be excluded. If so, it is necessary to take into account this variant of carbohydrate-mediated virus–cell interaction, when besides its own HA, the virus can also bear proteins acquired from the host cell, as was documented for actin, for example.[34] The latter hypothesis can explain virus tropism to a very limited cell population: due to adsorption of specific adhesion molecules from the host cell, the virus acquires to some extent the specific adhesion properties of these cell species. Herein, we addressed the question whether MDCK and Vero cells can bind glycan lig-ands. Additionally, we characterized the carbohydrate-binding profile of viruses propa-gated in these cell lines owing to the "acquired cell lectins" hypothesis.

A. CARBOHYDRATE-BINDING PROPERTIES OF MDCK AND VERO CELLS

The human influenza viruses are capable of growing only in a limited number of mammalian cells, mainly originating from kidneys of a different species. MDCK is the most widely used cell line for laboratory cultivation of influenza viruses; recently,

Vero cells of monkey origin were shown to be a promising "substrate".[16–19] Importantly, influenza viruses grown on these cell cultures maintain 2–6 receptor-binding specificity of original human viruses.[5,16,35] According to Stray et al,[8] desialylation of MDCK cells by neuraminidase does not abolish the interaction of the influenza virus NWS-Mvi with these cells. Taking into account a high glycosylation rate of HA and NA it is possible to suppose that the target cell interacts with carbohydrate chains of the virion. In other words, cell lectin(s) can serve as a trigger of the influenza virus attachment followed by the penetration into a host cell. Galectins, siglecs, mannose-binding proteins, and sulfated saccharide-binding proteins are typical animal cell-membrane-associated lectins.[36] Consequently, it is necessary to cover their ligand spectrum in our lectinomic profiling. Thus, we used carbohydrate probes, namely Sug-PAA-fluo, which are capable of detecting *classes* of lectins (according to their specificity), including proteins that have not yet been defined biochemically. Additionally, antibodies directed to *particular* galectins and siglecs were used.

Flow cytometry analysis using galectin-specific probes demonstrated significant binding of asialoGM1 and other galectin-selective probes to MDCK cells, whereas only weak staining of Vero was observed under the same conditions, thus intimating only a low level of presentation of accessible galactose-binding lectins on Vero cells. Using polyclonal antibodies to galectin-1 and -3 we tested their interaction with the cells and observed the significant binding of anti-galectin-1 (but not anti-galectin-3) to MDCK, whereas Vero cells demonstrated weak binding only with anti-galectin-3 antibodies. It should be noted that galectin-3, formerly known as L-29, has been shown to be one of the major cytosolic proteins of MDCK cells which undergoes apical secretion.[37] Evidently, rebinding after secretion is not a major route for this protein but it establishes galactoside-binding potency of MDCK cells evidenced both from carbohydrate probes and antibodies, whereas Vero cells differ in this respect. Fittingly for an interaction, the presence of multiple galactose-terminated chains on the influenza virus HA and NA is a well-known fact: due to the action of its own sialidase influenza virion has an unusually high density of Gal residues when compared with other mammalian surface glycoproteins where most of the Gal residues are masked by sialic acids. Based on these results, it is of interest to characterize further MDCK cell galactose-specific lectins that are able to interact with virion glycoproteins to determine how relevant this event is for virus reception. It is even challenging to understand, taking into account the presence of galectins in bronchial and tracheal epithelium, whether galectin/galactoside interaction may play a role during the infection process.[38,39]

The next widespread group of animal lectins is the siglec family. Eleven characterized human siglecs are known to bind either 3'SL, 6'SL, or a disialic fragment.[40] We did not find evidence for the presence of siglecs on MDCK and Vero glycocytochemically using the trisaccharide probes 3'SL-PAA, 6'SL-PAA-flu, and [Neu5Acα2-8]$_2$-PAA-flu, as well as immunocytochemically with monoclonal antibodies. Siglecs are often presented on the cell membrane in a masked state, i.e., they are blocked by endogenous carbohydrate chains,[41] which become available for carbohydrate probes only after neuraminidase action. In our case, the treatment of cells with neuraminidase from *V. cholerae* did not increase the interaction with the mentioned probes, except for the Neu5AcBn probe. Notably, this probe is a potent binder for the most of siglecs, whereas the three conventional sialoside-containing probes mentioned are not.[42] Thus, questions arise concerning the expression of siglecs or other Sia-binding proteins on MDCK and Vero cells warrants further study. For instance, do siglecs in principle take part in an interaction with influenza viruses, normally lacking sialylated chains? Supposedly, in rare cases when the virion maintains a notable amount of sialylated chains due to low activity of viral NA, a situation is established where a concomitant decrease in HA affinity occurs[28,29] and one can then speculate that in order to retain cell-binding ability a virus requires the second, additional receptor.

The tandem-repeat-type mannose receptor of macrophages is involved in infection of these cells by influenza viruses, the infection being inhibited by yeast polysaccharide mannan.[43] It was shown[43] that this lectin is not expressed on MDCK cells, but MDCK cells have the 36 kDa vesicular-membrane protein (VIP36), an intracellular lectin recognizing high-mannose-type glycans.[44] Notably, this lectin localized also to the plasma membrane of MDCK cells.[45] In our flow cytometry experiments, three different mannoside probes demonstrated binding with both Vero (much more pronounced) and MDCK cells. Thus, high-mannose chains of modern highly glycosylated H3N2 influenza viruses may affect their binding to MDCK and Vero cells. This fact probably could explain the ability of modern viruses to infect desialylated cells. It should be noted that hemagglutinin of Vero cell-grown viruses is more heavily mannosylated as compared with MDCK viruses derived from the same isolate.[17] Effective participation of lectin/mannoside recognition system in viral adhesion remains to be fully investigated.

A number of proteins, such as CD44, CD54, I-type lectins, and scavenger receptor, recognize sulfated oligo- or polysaccharides and other anionic glycans; the macrophage mannose receptor is known to have a second binding site specific to sulfated saccharides.[46] At the same time, HA of influenza virus grown in MDCK cells carries sulfated N-acetyllactosamine-type carbohydrate N-glycan chains,[47] potential partners for sulfo-OS-binding cell lectins. As shown in Reference 48, the glycoproteins contain sulfate in composition of their glycans, additionally, sulfated mucins of host cells are strongly associated with Madin-Darby bovine kidney cells (MDBK) grown virus (no corresponding MDCK and Vero data). In our study, we found that sulfated probes bind to Vero cells, whereas the specific interaction with MDCK was weak. All probes tested, namely $6\text{-HSO}_3\text{Gal}$, $6\text{-HSO}_3\text{GlcNAc}$, and $3\text{-HSO}_3\text{Gal}$-terminated ones, yielded practically the same level of binding, intimating rather a charge-dependent unspecific binding than an interaction with a specific lectin.

Summarizing the cell-probing experiments, the two studied cell lines have different carbohydrate-binding patterns. MDCK cells bind galactoside probes stronger than Vero cells, whereas Vero cells bind preferentially sialoside, mannoside, and sulfo-OS probes.

B. PATTERN OF MDCK AND VERO CELLS SIALYLATION

Before inspection of an alternative interaction pathway was started, the pattern of cell sialylation was evaluated using the plant lectins SNA and MAA. It has been demonstrated by use of these probes[49] that both Neu5Acα2-3 and Neu5Acα2-6 sequences are present on MDCK cells. Here, using a slightly different experimental technique, we have confirmed that the expression of the α2-3 sequence is comparable with the α2-6 in our MDCK cells. Limited preference of Neu5Acα2-3Gal-exposing glycans has been demonstrated here for Vero cells. Predominance of the α2-3 motif was earlier observed also in the case of baby hamster kidney cells.[49] Reduced relative level of expression of α2-6 sialylation on Vero cells is noteworthy, because Vero cell line is one of the most suitable host systems for cultivation of human, Neu5Acα2-6-specific viruses [17] and, therefore, an adequate extent of α2-6 sialylation should be expected. Interestingly, we did not observe a change in HA specificity of influenza viruses from α2-6 to α2-3 after one or several passages of clinical viruses on Vero cells,[17] despite the prevalence of α2-3 receptors. Moreover, α2-6 binding specificity of Vero-derived viruses is generally more pronounced than that of MDCK ones. Briefly,[17] recent human H3N2 viruses isolated on Vero cells had HA amino acid sequences indistinguishable from original viruses present in clinical material, whereas some of the MDCK isolates had one or two amino acid substitutions. Despite these host-dependent mutations and differences in the structure of HA of individual strains, all of the Vero-isolated viruses studied bound preferentially to 6'SLN. The observed discrepancy between Neu5Acα2-6 receptor binding of the virus and the dominating Neu5Acα2-3 glycosylation of the Vero cells gives reason to study the effective binding partner in the cell in detail. Looking at the P-selectin a lesson emerges with potential relevance, i.e., a two-site interaction. P-selectin counter-receptor, PSGL-1, bears sulfo-tyrosine, essential for high-affinity binding, in addition to SiaLe^x oligosaccharide. Also, the OS must be presented on core type 2 O-chain but not type 1.[50] The proportion of "correctly" presented (interacting with P-selectin) SiaLe^x on cell surface is quite low: only 1% of the total amount of SiaLe^x. Analogously, it is not unlikely that HA of influenza virus is more discriminating than currently defined and that it has a glycoprotein receptor, presenting α2-6 sialylated carbohydrate chain to the virus in an optimal not yet characterized way. If so, the receptor could be expressed on the cell surface in a small number of copies, but as compensation, it could be highly reactive.

C. CARBOHYDRATE-BINDING PROPERTIES OF THE INFLUENZA VIRUS

As mentioned above, host cell lectins might also be acquired during viral particle formation raising the question of "What pattern of carbohydrate-binding activity can one anticipate for the influenza virus?" First, the well-known interaction with 6'SLN due to classical carbohydrate-binding site of HA. Secondly, HA is shown to have additional Sia-binding motif,[51] the role and fine specificity of this site remaining unclear. It may be speculated that the second Sia-binding motif is actually able to bind other negatively charged chains, for example, sulfated molecules such as sulfatide. Thirdly, viral NA might be considered as a weak galactoside-binding protein. Indeed, NA is able to discriminate oligosaccharide substrates and it can bind asialo part, as, for example, documented for *Tripanosoma* trans-sialidase. It cannot be excluded that HA or NA possess additional weakly expressed lectin properties. Finally, due to the acquisition of host cell lectins, the influenza virus might interact with a broader spectrum of carbohydrates. As shown in the present work, MDCK and especially Vero cells exhibit sialoside-, galactoside-, mannoside-, and also sulfo-OS-binding properties providing the reason why we tested MDCK- and Vero-grown viruses with respect to their carbohydrate-recognizing properties. Although binding of mannoside- and galactoside probes (including typical galectin ligands) was demonstrated, the interaction was shown to be basically weak and it is significant only in some strains. In contrast to other probes, the binding with $6'$-$HSO_3LacNAc$ is typical for all the strains tested, both A and B. The affinity was of the same order of magnitude as in the case of the sialylated probe 6′SL-PAA-biot under the same conditions (viruses originating from the same isolate were compared). The binding of $6'$-$HSO_3LacNAc$ probe to influenza viruses is independent of the nature of the host cells (MDCK or Vero), this conclusion emerging from an experiment where host cells were stained with this probe: only Vero cells (but not MDCK) have demonstrated significant affinity to the probe, thus it is unlikely that viruses acquire $6'$-$HSO_3LacNAc$-binding lectin from the host cell. Interaction of viruses with $6'$-$HSO_3LacNAc$ is specific; indeed, isomeric disaccharide probes 6-$HSO_3LacNAc$ and $3'$-$HSO_3LacNAc$ as well as monosaccharide probe 3-HSO_3Gal did not bind viruses. In the case of selectin/ligand interaction, the substitution of Sia in SiaLe^x to sulfate in position 3 of Gal moiety did not impair selectin-binding potency.[52] In our case, the structural similarity of $6'$-$HSO_3LacNAc$ and 6′SiaLacNAc (6′SLN) is obvious. However, we do not expect that a sulfated probe binds to the primary Sia-binding site of HA because this interaction has not been inhibited with 6′SLN (data not shown) and has no structurally (in terms of HA/OS complex) based rationale. It is also known that trachea tissues strongly express sulfated mucins,[53] which together with sulfatide-binding properties as mentioned above,[10, 11] and our data on $6'$-$HSO_3LacNAc$-binding properties of influenza virus provide reason to speculate that sulfated sugars (on cell) and corresponding lectins (on virus) might be involved in the infection process. The nature of $6'$-$HSO_3LacNAc$-binding molecule remains to be clarified in future research, as currently very little is known about cell proteins binding sulfated probes. However, the data obtained will stimulate further analysis of the sulfated chains' involvement in processes of influenza virus motility toward and eventually binding the host cells.

D.   POSSIBLE INVOLVEMENT OF XENOANTIGENS

Another aspect related to glycosylation of Vero and MDCK cells is the potential use of these cells for the production of anti-influenza vaccines. Vero cells originate from green macaque monkeys; these "Old World monkeys" are like humans, i.e., they fail to express active αGal-transferase and consequently do not have the Galα1-3Gal epitope.[54] This feature discriminates Vero from MDCK cells, which principally are capable of synthesizing Galα1-3Gal-xenoantigen. Two percent of total human IgG and IgM interact with this antigen, a part of them with high affinity.[54] Complement-mediated inactivation of Galα1-3Gal-coated envelope viruses in human serum triggered by anti-Galα1-3Gal represents a unique type of natural immunity. These antibodies recognize their target that is present on a variety of viruses, such as the retrovirus, equine encephalitis virus, LCMV, and Newcastle disease viruses grown in cell lines of nonprimate origin.[55] Influenza viruses grown on MDCK cells can theoretically bear this xeno-epitope due to the host glycosylation machinery. Unusually high level of Galα1-3Gal-epitopes was also shown for MDBK cells.[56] This is not the case in Vero cells and, thus, Vero-derived viruses. The presence or absence of the xeno-epitope on a virus can influence its *in vivo* destiny. Due to the mentioned natural antibodies, Galα1-3Gal-positive viruses will be subjected to rapid opsonization by antibodies followed by the elimination by the immune system, i.e., they should be less infective. However, from the point of view of inactivated vaccine's immunogeneity, the formation of an antigen–antibody complex with the corresponding antibody could result in an increase in the processing of these antigens by antigen-presenting cells increasing the immune response.[57] Thus, if MDCK cells are capable of realizing their potential to synthesize Galα1-3Gal-epitope on viral HA and NA, then MDCK-derived inactivated vaccines acquire additional advantages while viable vaccine could be neutralized by natural antibodies. Using affinity-isolated polyclonal antibodies to Galα1-3Gal, we have tested the presence of this epitope on viruses derived from MDCK and Vero cells as well as embryonated eggs (two H1N1 strains and two H3N2 strains: see Table 2.2). As expected, only MDCK-grown viruses presented the Galα1-3Gal epitope, and this should be taken into consideration when MDCK and Vero cells are used for vaccine production.

Another xenoantigen, the Forssman glycolipid GalNAcα1-3GalNAcβ1-3Galα1-4Galβ1-4Glc-Cer, is a major species detected in MDCK cells as reported for high passage cultures.[58] Interestingly, the Forssman antigen was identified in egg-grown H1N1, H3N2, and B virus strains.[59] Due to the presence of an acquired Fs glycolipid, the virion can be opsonized by anti-Fs antibodies that, similar to anti-Galα1-3Gal, are natural antibodies in humans. These factors mean that MDCK-grown viruses are able to acquire additional immunogeneity due to the Forssman xenoantigen.

TABLE 2.2

Interaction of Anti-Galα1-3Gal Antibodies with Human Influenza Viruses Isolated from MDCK and Vero Cells or Embryonated Chicken Eggs (CE)

Virus Strain, History	Dilution of Anti-Galα1-3Gal Antibodies at the Point $A_{max/2}^{459}$
H1N1	
A/HK/1134/98, MDCK	800
A/HK/1134/98, Vero	No binding
A/HK/1134/98, CE	No binding
A/HK/1035/98, CE	No binding
H3N2	
A/HK/1182/99, MDCK	200
A/HK/1182/99, Vero→MDCK	800
A/HK/1182/99, Vero	No binding
A/HK/1144/99, MDCK	200–400
A/HK/1144/99 Vero	No binding
Galα1–3Gal-PAA (positive control)	800–1200

V. CONCLUSIONS

In total, we have obtained and presented new data on the presence of galactose-binding molecules on the surface of Vero cells and have demonstrated preferential α2-3 vs. α2-6 sialylation of the Vero cells. We have also revealed the 6'-HSO$_3$LacNAc binding ability of human influenza viruses. In the current study, the functional investigation of detected molecules and their actual participation in virus/cell interaction was not addressed leaving important questions to be answered in future investigations. These questions include (i) What is the role of the Vero and MDCK cell *lectinome*[60] toward the influenza virus? (ii) What carbohydrate chains and what particular viral glycoprotein(s) interact with the host cells? (iii) Why do Vero-grown viruses retain α2-6 specificity in spite of a dominant α2-3 pattern of the cell sialylation? (iv) What is the role of the sulfooligosaccharide recognition by a virus? These questions represent only a selection of the many issues that remain to be addressed to gain a better understanding of the adhesion of the influenza to host cells, viral entry into the cell, and the ensuing events that result in cell intoxication and death with the corresponding deleterious effects to the host organism or human patient.

ACKNOWLEDGMENTS

We thank Dr. P.R. Crocker for the mAbs to siglecs. We also thank Dr. G.V. Pazynina, Dr. E.Y. Korchagina, and Dr. S.D. Shiyan for the synthesis of the oligosaccharides and glycoconjugates. This work was partially supported by a grant from the Russian Foundation for Basic Research (# 01-04-49253) and Russian Academy of Sciences program 'Molecular and Cell Biology.'

ABBREVIATIONS

biot, biotin residue; BSA, bovine serum albumin; DMEM, Dulbecco's modified Eagle's medium; FBS, fetal bovine serum; FITC, fluorescein isothiocyanate; fluo, fluorescein residue; HA, hemagglutinin; MBP, mannose-binding protein; MDBK, Madin–Darby bovine kidney cells; MDCK, Madin–Darby canine kidney cells; NA, neuraminidase; OS, oligosaccharide; PAA, polyacrylamide; PBS, phosphate-buffered saline; PBA, PBS containing 0.2% BSA; SP, surfactant protein; Sug, mono- or oligosaccharide residue; 3'SL, 3'-sialyllactose; LacNAc, N-acetyllactosamine; 6'SLN, 6'-sialyl-N-acetyllactosamine; 3'SLN, 3'-sialyl-N-acetyllactosamine; Neu5Ac, α-N-acetylneuraminic acid; asialoGM1, Galβ1-3GalNAcβ1-3Galβ1-4Glc; Fs, GalNAcα1-3GalNAcβ; Man$_3$, Manα1-6(Manα1-3)Man; A$_{tri}$, GalNAcα1-3(Fucα1-2)Gal; B$_{tri}$, Galα1-3(Fucα1-2)Gal; SiaLe^x, Neu5Acα2-3Galβ1-4(Fucα1-3)GlcNAc; 7-OS, (GlcNAcβ1-2Manα1-)$_2$-3,6-Manβ1-4GlcNAcβ1-4GlcNAc; 11-OS, (Neu5Acα2-6Galβ1-4GlcNAcβ1-2Manα1-)$_2$-3,6-Manβ1-4GlcNAcβ1-4GlcNAc; T$_{\beta\beta}$, Galβ1-3GalNAcβ.

REFERENCES

1. Paulson, J.C., Interaction of animal viruses with cell surface receptors, in *The Receptors*, vol. 2, Conn, M., Ed., Orlando: Academic Press, 1985, pp. 131–219.

2. Wiley, D.C. and Skehel, J.J., The structure and function of the hemagglutinin membrane glycoprotein of influenza virus, *Annu. Rev. Biochem.*, 56, 365–394, 1987.

3. Herrler, G., Hausmann, J., and Klenk, H.D., Sialic acid as receptor determinant of otho- and paramyxoviruses, in *Biology of the Sialic Acids*, Rosenberg, A., Ed., Plenum, New York, 1995, pp. 315–336.

4. Suzuki, Y., Gangliosides as influenza virus receptors. Variation of influenza viruses and their recognition of the receptor sialo-sugar chains, *Progr. Lipid Res.*, 33, 429–457, 1994.

5. Gambaryan, A.S., Tuzikov, A.B., Piskarev, V.E., Yamnikova, S.S., L'vov, D.K., Robertson, J.C., Bovin, N.V., and Matrosovich, M.N., Specification of receptor-binding phenotypes of influenza virus isolates from different hosts using synthetic sialyl-glycopolymers: non-egg-adapted human H1 and H3 influenza A, and influenza B viruses share a common high binding affinity for 6'-sialyl(*N*-acetyllactosamine), *Virology*, 233, 224–234, 1997.

6. Rogers, G.N. and D'Souza, B.L., Receptor-binding properties of human and animal H1 influenza virus isolates, *Virology*, 173, 317–322, 1989.

7. Gambaryan, A.S., Robertson, J.S., and Matrosovich, M.N., Effects of eggs-adaptation on the receptor-binding properties of human influenza A and B viruses, *Virology*, 258, 232–239, 1999.

8. Stray, S.J., Cummings, R.D., Air G.M., Influenza virus infection of desialylated cells, *Glycobiology*, 10, 649–658, 2000.

9. Barton, E.S., Connolly, J.L., Forrest, J.C., Chappell, J.D., and Dermody, T.S., Utilization of sialic acid as a coreceptor enhances reovirus attachment by multi-step adhesion strengthening, *J. Biol. Chem.*, 276, 2200–2211, 2001.

10. Suzuki, T., Sometani, A., Horiike, G., Mizutani, Y., Masuda, H., Yamada, M., Tahara, H., Xu, G., Myamoto, D., Oku, N., Okada, S., Kiso, M., Hasegawa, A., Ito, T., Kawaoka, Y., and Suzuki, Y., Sulphatide binds to human and animal influenza viruses and inhibits the viral infection, *Biochem. J.*, 318, 389–393, 1996.

11. Niimura, Y. and Ishizuka, I., Adaptive changes in sulfoglycolipids of kidney cell lines by culture in anisosmotic media, *Biochim. Biophys. Acta.*, 1052, 248–254, 1990.

12. Martin, J., Wharton, S.A., Lin, Y.P., Takemoto, D.K., Skehel, J.J., Wiley, D.C., and Steinhauer, D.A., Studies of the binding properties of influenza virus hemagglutinin receptor-site mutants, *Virology*, 241, 101–111, 1998.

13. Hughes, M.T., McGregor, M., Suzuki, T., Suzuki, Y., and Kawaoka, Y., Adaptation of influenza A viruses to cells expressing low levels of sialic acid leads to loss of neuraminidase activity, *J. Virol.*, 75, 3766–3770, 2001.

14. Kistner, O., Barrett, P.N., Mundt, W., Reiter, M., Schober-Bendixen, S., and Dorner, F., Development of a mammalian cell (Vero) derived candidate influenza virus vaccine, *Vaccine*, 16, 960–968, 1998.

15. Halperin, S.A., Smith, B., Mabrouk, T., Germain, M., Trepanier, P., Hassell, T., Treanor, J., Gauthier, R., and Mills, E.L., Safety and immunogenicity of a trivalent, inactivated, mammalian cell culture-derived influenza vaccine in healthy adults, seniors, and children, *Vaccine*, 20, 1240–1247, 2002.

16. Govorkova, E.A., Murti, G., Meignier, B., de Taisne, C., and Webster, R.G., African green monkey kidney (Vero) cells provide an alternative host cell system for influenza A and B viruses, *J. Virol.*, 70, 5519–5524, 1996.

17. Romanova, J., Katinger, D., Ferko, B., Voglauer, R., Mochalova, L., Bovin, N., Lim, W., Katinger, H., and Egorov, A., Distinct host range of influenza H3N2 virus isolates in Vero and MDCK cells is determined by cell specific glycosylation pattern, *Virology*, 307, 90–97, 2003.

18. Mochalova, L., Gambaryan, A., Romanova, J., Tuzikov, A., Chinarev, A., Katinger, D., Katinger, H., Egorov, A., and Bovin, N., Receptor-binding properties of modern

human influenza viruses primary isolated in Vero and MDCK cells and chicken embryonated eggs, *Virology*, 313, 473–480, 2003.

19. Tuzikov, A.B., Gambaryan, A.S., Juneja, L.R., and Bovin, N.V., Conversion of complex oligosaccharides into polymeric conjugates and their anti-influenza virus inhibitory potency, *J. Carbohydr. Chem.*, 19, 1191–1200, 2000.

20. Kopitz, J., von Reitzenstein, C., Burchert, M., Cantz, M., and Gabius, H.-J., Galectin-1 is a major receptor for ganglioside GM1, a product of the growth-controlling activity of a cell surface ganglioside sialidase, on human neuroblastoma cell in culture, *J. Biol. Chem.*, 273, 11205–11211, 1998.

21. Khraltsova, L.S., Sablina, M.A., Melikhova, T.D., Joziasse, D.H., Kaltner, H., Gabius, H.-J., and Bovin. N.V., An enzyme-linked lectin assay for $\alpha 1,3$-galactosyltransferase, *Anal. Biochem.*, 280, 250–257, 2000.

22. Kistner, O., Barrett, P.N., Mundt, W., Reiter, M., Schober-Bendixen, S., Eder, G., and Dorner, F., Development of a Vero cell-derived influenza whole virus vaccine, *Dev. Biol. Stand.*, 98, 101–110, 1999.

23. Kayser, K., Bovin, N.V., Zemlyanukhina, T.V., Donaldo-Jacinto, S., Koopmann, J., and Gabius, H.-J., Cell type-dependent alterations of binding of synthetic blood antigen-related oligosaccharides in lung cancer, *Glycoconj. J.*, 11, 339–344, 1994.

24. Anders, E.M., Hartley, C.A., and Jackson, D.C., Bovine and mouse serum beta inhibitors of influenza A viruses are mannose-binding lectins, *Proc. Natl. Acad. Sci. USA*, 87, 4485–4489, 1990.

25. Hartshorn, K.L., Crouch, E.C., White, M.R., Eggeleton, P., Tauber, A.I., Chang, D., and Sastry, K., Evidence for a protective role of pulmonary surfactant protein D (SP-D) against influenza A viruses, *J. Clin. Invest.*, 94, 311–319, 1994.

26. Anders, E.M., Hartley, C.A., Reading, P.C., and Ezekowitz, R.A.B., Complement-dependent neutralization of influenza virus by a serum mannose-binding lectin, *J. Gen. Virol.*, 74, 615–622, 1994.

27. Benne, C.A., Kraaijeveld, C.A., van Strijp, J.A.G., Brouwer, E., Harmsen, M., Verhoef, J., van Golde, L.M.G., and van Iwaarden, J.F., Interaction of surfactant protein A with influenza A viruses: binding and neutralization, *J. Infect. Dis.*, 171, 335–341, 1995.

28. Kaverin, N.V., Gambaryan, A.S., Bovin, N.V., Rudneva, I.A., Shilov, A.A., Khodova, O.M., Varich, N.L., Makarova, B.V., and Kropotkina, E.A., Postreassortment changes in influenza virus hemagglutinin restoring HA–NA functional match, *Virology*, 244, 315–321, 1998.

29. Kaverin, N.V., Matrosovich, M.N., Gambaryan, A.S., Rudneva, I.A., Shilov, A.A., Varich, N.L., Makarova, N.V., Kropotkina, E.A., and Sinitsin, B.V., Intergenic HA–NA interactions in influenza A virus: postreassortment substitutions of charged amino acid in the hemagglutinin of different subtype, *Virus Res.*, 66, 123–129, 2000.

30. Tertov, V.V., Kaplun, V.V., Sobenin, I.A., Boytsova, E.Y., Bovin, N.V., and Orekhov, A.N., Human plasma trans-sialidase causes atherogenic modification of low density lipoprotein, *Atherosclerosis*, 159, 103–115, 2001.

31. Mandelboim, O., Lieberman, N., Lev, M., Paul, L., Arnon, T.I., Bushkin, Y., Davis, D.M., Strominger, J.L., Yewdell, J.W., and Porgador, A., Recognition of haemagglutinins on virus-infected cells by NKp46 activates lysis by human NK cells, *Nature*, 409, 1055–1060, 2001.

32. Montgomery, R.I., Warner, M.S., Lum, B.J., and Spear, P.G., Herpes simplex virus-1 entry into cells mediated by a novel member of the TNF/NGF receptor family, *Cell*, 87, 427–436, 1996.

33. Margolis, L., HIV: from molecular recognition to tissue pathogenesis, *FEBS Lett.*, 433, 5–8, 1998.

34. Roberts, P.C. and Compans, R.W., Host cell dependence of viral morphology, *Proc. Natl. Acad. Sci. USA*, 95, 5746–5751, 1998.

35. Gambaryan, A.S., Marinina, V.P., Tuzikov, A.B., Bovin, N.V., Rudneva, I.A., Sinitsyn, B.V., Shilov, A.A., and Matrosovich, M.N., Effects of host-dependent glycosylation of hemagglutinin on receptor-binding properties of H1N1 human influenza A virus grown in MDCK cells and in embryonated eggs, *Virology*, 247, 170–177, 1998.

36. Gabius, H.-J., Animal lectins, *Eur. J. Biochem.*, 243, 543–576, 1997.

37. Lindstedt, R., Apodaca, G., Barondes, S.H., Mostov, K.E., and Leffler, H., Apical secretion of a cytosolic protein by Madin-Darby canine kidney cells, *J. Biol. Chem.*, 268, 11750–11757, 1993.

38. Sparrow, C.P., Leffler, H., and Barondes, S.H., Multiple soluble β-galactoside-binding lectins from human lung, *J. Biol. Chem.*, 262, 7383–7390, 1987.

39. Kaltner, H., Seyrek, K., Heck, A., Sinowatz, F., and Gabius, H.-J., Galectin-1 and galectin-3 in fetal development of bovine respiratory and digestive tracts, *Cell Tissue Res.*, 307, 35–46, 2002.

40. Crocker, P.R. and Varki, A., Siglecs in the immune system, *Immunology*, 103, 137–145, 2001.

41. Razi, N. and Varki, A., Cryptic sialic acid binding lectins on human blood leukocytes can be unmasked by sialidase treatment or cellular activation, *Glycobiology*, 9, 1225–1234, 1999.

42. Kelm, S., Brossmer, R., Isecke, R., Gross, H.-J., Strenge, K., and Schauer, R., Functional groups of sialic acids involved in binding to siglecs (sialoadhesins) deduced from interactions with synthetic analogues, *Eur. J. Biochem.*, 255, 663–672, 1998.

43. Reading, P.C., Miller, J.L., and Anders, E.M., Involvement of the mannose receptor in infection of macrophages by influenza virus, *J. Virol.*, 74, 5190–5197, 2000.

44. Hara-Kuge, S., Ohkura, T., Seko, A., and Yamashita, K., Vesicular-integral membrane protein, VIP36, recognizes high-mannose type glycans containing α1-2 mannosyl residues in MDCK cells, *Glycobiology*, 9, 833–839, 1999.

45. Hara-Kuge, S., Ohkura, T., Ideo, H., Shimoda, O., Atsumi, S., and Yamashita, K., Involvement of VIP36 in intracellular transport and secretion of glycoproteins in polarized Madin–Darby canine kidney (MDCK) cells, *J. Biol. Chem.*, 277, 16332–16339, 2002.

46. Fiete, D., Beranek. M.C., and Baenziger, J.U., The macrophage/endothelial cell mannose receptor cDNA encodes a protein that binds oligosaccharides termining with SO_4-4-GalNAcβ1,4GlcNAcβ or Man at independent sites, *Proc. Natl. Acad. Sci. USA*, 94, 11256–11261, 1997.

47. Karaivanova, V. and Spiro, R.G., Sulphation of *N*-linked oligosaccharides of vesicular stomatitis and influenza virus envelop glycoproteins: host cell specificity, subcellular localization and identification of substituted saccharides, *Biochem. J.*, 329, 511–518, 1998.

48. Compans, R.W. and Pinter, A., Incorporation of sulfate into influenza virus glycoproteins, *Virology*, 66, 151–160, 1975.

49. Govorkova, E.A., Matrosovich, M.N., Tuzikov, A.B., Bovin, N.V., Gerdil, C., Fanget, B., and Webster, R.G., Selection of receptor-binding variants of human influenza A and B viruses in Baby Hamster Kidney cells, *Virology*, 262, 31–38, 1999.

50. Aeed, P.A., Geng, J.-G., Asa, D., Raycroft, L., Ma, L., and Elhammer, A.P., Characterization of the *O*-linked oligosaccharide structures on P-selectin glycoprotein ligand-1 (PSGL-1), *Glycoconj. J.*, 15, 975–985, 1998.

51. Sauter, N.K., Glick, G.D., Crowther, R.L., Park, S.-J., Eisen, M.B., Skehel, J.J., Knowles, J.R., and Wiley, D.C., Crystallographic detection of a second ligand binding site in influenza virus hemagglutinin, *Proc. Natl. Acad. Sci. USA*, 89, 324–328, 1992.

52. Green, P.J., Tamatani, T., Watanabe, T., Myasaka, M., Hasegawa, A., Kiso, M., Stoll, M.S., and Feizi, T., High affinity binding of the leucocyte adhesion molecule L-selectin to 3′-sulphated-Le^a and –Le^x oligosaccharides and the predominance of sulphate in this interaction demonstrated by binding studies with a series of lipid-linked oligosaccharides, *Biochem. Biophys. Res. Commun.*, 188, 244–251, 1992.

53. Lo-Guidice, J.-M., Wieruszeski, J.-M., Lemoine, J., Verbert, A., Roussel, P., and Lamblin, G., Sialylation and sulfaton of the carbohydrate chains in respiratory mucins from a patient with cystic fibrosis, *J. Biol. Chem.*, 269, 18794–18813, 1994.

54. Galili, U., Rachmilewitz, E.A., Peleg, A., and Flechner, I., A unique natural human IgG antibody with anti-α-galactosyl specificity, *J. Exp. Med.*, 160, 1519–1531, 1984.

55. Rother, R.P. and Squinto, S.P., The α-galactosyl epitope: a sugar coating that makes viruses and cells unpalatable, *Cell*, 86, 185–188, 1996.

56. Mir-Shekari, S.Y., Ashford, D.A., Harvey, D.J., Dwek, R.A., and Schulze, I.T., The glycosylation of the influenza virus hemagglutinin by mammalian cells. A site-specific study, *J. Biol. Chem.*, 272, 4027–4036, 1997.

57. Galili, U., Repik, P.M., Anaraki, F., Mozdzanowska, K., Washko, G., and Gerhard, W., Enhancement of antigen presentation of influenza virus hemagglutinin by the natural human anti-Gal antibody, *Vaccine*, 14, 321–328, 1996.

58. Lala, P., Ito, S., and Lingwood, C.A., Retroviral transfection of Madin–Darby canine kidney cells with human MDR1 results in a major increase in globotriaosylceramide and 10^5–10^6-fold increased cell sensitivity to verocytotoxin. Role of P-glycoprotein in glycolipid synthesis, *J. Biol. Chem.*, 275, 6246–6251, 2000.

59. Nowinski, R., Berglund, C., Lane, J., Lostrom, M., Bernstein, I., Young, W., and Hakomori, S., Human monoclonal antibody against Forssman antigen, *Science*, 210, 537–539, 1980.

60. Gabius, H.-J., André, S., Kaltner, H., and Siebert, H.-C., The sugar code: functional lectinomics, *Biochim. Biophys. Acta.*, 1572, 165–177, 2002.

3 O-Acetyl Sialic Acids in Parasitic Diseases

Anil Kumar Chava, Mitali Chatterjee, and Chitra Mandal

CONTENTS

I. INTRODUCTION

Sialic acids are members of a structurally diverse family of nine-carbon polyhydroxy amino ketoacid of *N*- and *O*-substituted derivatives of neuraminic acid, a monosaccharide commonly referred to as *N*-acetyl neuraminic acid or Neu5Ac.[1] Although their existence has been known for over 50 years,[2] improvement in analytical techniques, in more recent years, have unraveled over 50 different modifications[3] (Figure 3.1). This diversity is generated following substitution of the amino groups by an acetyl, glycolyl, or hydroxyl groups by methylation or esterification with acetyl, lactyl, sulfate, or phosphate groups.[1,3] Among these, Neu5Ac along with its glycolyl (Neu5Gc) and *O*-acetylated derivative are the most frequently occurring derivatives. Neu5Ac is the most abundantly available monosaccharide present as terminal residues of cell surface sugar chains of most vertebrate oligosaccharides.[4] In the evolutionary ladder, echinoderms are considered as the first group of organisms to definitely contain sialic acids.[5,6] Unlike Neu5Ac that is ubiquitously present, all derivatives are not so universal. Possibly the best-investigated example is Neu5Gc, which occurs often in the animal kingdom and appears in certain tumors but not in bacteria or healthy human tissues.[4] The *O*-acetylation occurs at position C-7, C-8 and C-9 to form *N*-acetyl-7/-8/-9/-*O*-acetyl neuraminic acid, thus generating a family of *O*-acetylated sialoglycoconjugates (*O*-AcSGs).[1] *O*-acetylation at C-9 position of sialic acid appears to be developmentally regulated in a variety of systems. Altered expression of 9-*O*-AcSGs on cell surface glycoconjugates in diseases has been summarized in Table 3.1.[7–24]

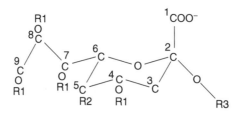

FIGURE 3.1 Structure of sialic acid. The chair confirmation structure of sialic acid. R1, R2, and R3 represent the usual substitutions. Achatinin-H recognizes this structure when R1 is replaced at C-9 position by *O*-acetyl moiety in an *α*2→6 linkage with GalNAc.

TABLE 3.1
Altered Expression of 9-*O*-AcSGs on Cell Surface Glycoconjugates of Patients in Disease Condition

Membrane	Diseases	Ref.
Erythrocytes (9-*O*-AcSA), peripheral blood mononuclear cells (9-*O*-AcSA)	Visceral Leishmaniasis (VL)	7–9,12
Lymphocytes of both T and B lineage (9-*O*-AcSA)	Acute Lymphoblastic Leukemia (ALL)	13–19
Melanoma cells (disialoganglioside GD3)	Melanoma	20,21
Basal and suprabasal keratinocytes (disialoganglioside GD3)	Psoriasis	22
Glomeruli (disialoganglioside GD3)	Congenital nephrotic syndrome of Finnish type	23
Breast carcinoma (disialoganglioside GD3 and GT3)	Breast cancer	24

This chapter highlights the occurrence of *O*-AcSGs in protozoal diseases and includes currently available methods for their detection and their potential clinical applications. In addition, the biological significance of *O*-AcSA has also been discussed.

II. OCCURRENCE OF *O*-ACETYLATED SIALIC ACIDS IN PARASITIC PROTOZOA

Protozoan parasites including *Plasmodia*, *Leishmania*, *Trypanosoma*, *Entamoeba*, *Trichomonas* show remarkable propensity to survive within hostile environments encountered during their life cycle. Identification of molecules that enable them to survive in such milieu is a subject of intense research. Currently available knowledge of the parasite cell surface architecture and biochemistry indicates that sialic acid and its principle derivatives are major components of the glycocalyx and assist the parasite to interact with its external environment through functions ranging from parasite survival, infectivity, and host-cell recognition.[25] The presence of sialic acid and its derivatives on parasitic protozoa (Table 3.2A) and their occurrence on host cells due to parasitic infection (Table 3.2B) has been summarized.[26–40]

III. PROBES FOR DETECTION OF *O*-ACSA

Although sialic acids are recognized by lectins (animal and plant origin), viruses, and also by certain naturally occurring or established antibodies, tools to detect its *O*-acetylated derivative are limited (see Table 3.3 and References 41–58). Accordingly, the analysis of *O*-AcSGs required both direct and indirect approaches for their assessment.

A. INFLUENZA C-VIRUS

The unique glycoprotein of influenza C virus, designated hemagglutinin, exhibits three functions, namely hemagglutination, esterase activity, and fusion factor (HEF).

TABLE 3.2
Occurrence of Sialoglycans/O-Acetyl Sialoglycans on Parasitic Protozoa and Host Cells Due to Parasitic Infection

Parasite	Sialoglycans	Ref.
(A) On Parasitic Protozoa		
Trypanosoma cruzi	Neu5Ac	26–28
Trypanosoma brucei	Neu5Ac	29
Entamoeba histolytica	Neu5Ac	30–32
Plasmodium falciparum	—	—
Trichomonas vaginalis and *Trichomonas foetus*	Neu5Ac	33,34
Toxoplasma gondii	Neu5Ac	35
L. donovani	Neu5Ac, Neu5Gc, Neu9Ac5Gc,	
Amastigotes	Neu5,9Ac$_2$, 9-OAcGD3	10
Promastigotes	Neu5Ac, Neu5Gc, Neu9Ac5Gc, Neu5,9Ac$_2$	11
Crithidia fasiculata	Neu5Ac, Neu5,9Ac$_2$	36–38
(B) On Host Cells Due to Parasitic Infection		
Trypanosoma cruzi	—	—
Trypanosoma brucei	—	—
Entamoeba histolytica	—	—
Plasmodium falciparum	EBA-175 on erythrocytes	39,40
Trichomonas vaginalis and *Trichomonas foetus*	—	—
Toxoplasma gondii	—	—
L. donovani	a) 9-O-AcSGs (112,107,103,57,51,48 kDa) on erythrocytes and PBMC	7–9,12

—, not reported.

TABLE 3.3
Probes to Detect O-Acetylated Sialic Acids

Probes	Source	Glycotope	Ref.
Cancer antennarius	Hemolymph from marine crab	4,9-O-AcSA	41,42
Achatinin-H	Hemolymph from *Achatina fulica* snail	9-O-AcSAα2→6 GalNAc	43–46
Liocarcinus depurator lectin	Hemolymph from *Liocarcinus depurator*	2,4,7,8,9-O-Ac-Neu5Ac	47
Paratelphusa jacquem- ontii (Rathbun)	Hemolymph	4,9-O-AcSA	48
Placenta lectin	Placenta from human and ovine	9(7,8)-O-acetyl and 4-O-acetylsialic acids	49,50
Anti-9O-AcSG	Serum from ALL patients	9-O-AcSAα2→6 GalNAc	19
Glycoprotein of Influenza C- virus	Virus	4,9-O-AcSA	4,51–56

Its hemagglutinating property allows it to bind to a wide spectrum of sialoglycoconjugates such as mucins, serum glycoproteins, or gangliosides containing naturally or synthetically O-acetylated sialic acid irrespective of the subterminal linkages.[51,52] The esterase activity of Influenza C virus cleaves the bond 9-AcSA to gain entry into the target cell at neutral pH at ambient temperatures.[51,52] While the *O*-acetyl ester at the C-9 position is essential for virus binding in all cases, the 4-*O*-acetylated derivatives are not recognized, thus making it a very specific tool for recognizing 9-*O*-AcSA.[53]

As the recombinant soluble form of Influenza C virus hemagglutinin esterase fusion protein (CHE-Fc) possesses hemagglutinin and esterase activities of Influenza C virus, detection of 9-*O*-AcSGs is achievable by incubation at 4°C. For binding studies at ambient temperature, prior treatment with (1.0 m*M*) di-isopropyl fluorophosphate (DFP), a serine protease inhibitor, can irreversibly block the esterase activity thus providing a probe (CHE-FcD), which can specifically detect sialoglycoconjugates with terminal 9-*O*-AcSA.[51,52,54–58] CHE-FcD has aided in the detection of sialoglycoconjugates having terminal *O*-AcSA in several cell lines and diseases.[4,51,54–58]

Another construct consisting of the influenza C virus HE1 domain has been reported to possess specific 9-*O*-acetyl hemagglutinin and esterase activity.[57,58] The construct consisting of the influenza C virus HE1 domain fused with the eGFP gene was originally cloned in a SV40 vector. The entire HE1 coding region was isolated as a *Sac*I/*Cla*I restriction fragment. The *Cla*I site was filled in to allow blunt-end ligation with the filled in *Bam*HI site immediately upstream of the eGFP gene derived from plasmid pEGFP-N3 (Clontech Laboratories, Austria). The resulting chimeric gene contains the entire HE1 domain and the first four codons of the HE2 domain linked via a five-codon spacer to the coding region of eGFP. This construct was ligated into the recombination vector pBakPAK8 (Clontech Laboratories, Austria). The resulting plasmid pBacPAK-CHE1-eGFP was co-transfected with baculovirus DNA (Pharmingen, San Diego, CA) into Sf9 cells. Recombinant baculovirus Bak-CHE1-eGFP was plaque-purified and used to express the recombinant HE1–eGFP fusion protein. The expression of HE-1 domain was sufficient to obtain a specific 9-*O*-acetylesterase activity that was determined with *p*-nitrophenyl acetate (pNPA).[11,58] Interestingly, this construct possesses increased esterase activity and minimal hemagglutinin activity. This construct has successfully been used for demonstration of 9-*O*-AcSGs on *Leishmania donovani* parasites (promastigotes and amastigotes) and hematopoietic cells of Visceral Leishmaniasis (VL) patients.[7–12]

Several strains of Corona viruses from various species including bovine and human, murine hepatitis virus, and porcine hemagglutinating encephalomyelitis virus (HEV), in common with influenza C virus, also employ 9-*O*-AcSA as their receptor determinants.[4,52,58,59] However, unlike the Influenza C virus wherein the hemagglutinin and esterase activities are encoded by the same viral coat protein,[51,52] Corona viruses express two proteins, namely, S and HE, wherein the former codes for its hemagglutinin activity, while the latter has the receptor destroying esterase activity with only a weak hemagglutinin activity.[59]

B. ACHATININ-H

Achatinin-H is a lectin purified from the African giant land snail, *Achatina fulica* and its sugar specificity has been established to be toward sialoglycoconjugates having terminal 9-*O*-AcSA in an α 2→6 linkage to a subterminal GalNAc of the underlying structure.[6, 43–46] The biological activity of Achatinin-H was evaluated by the hemagglutination assay (HA) using rabbit erythrocytes known to contain high amounts of 9-O-acetylated sialic acid. The carbohydrate specificity of Achatinin-H binding was confirmed using bovine submaxillary mucin (BSM) based on long-standing evidence of its high content of 9-O-AcSA.[60] In addition, other sugars were also used to demonstrate the lectin specificity (Table 3.4 and Reference 43). This restricted specificity has been successfully employed as a diagnostic probe to detect the selective presence of membrane 9-O-AcSGs on peripheral blood mononuclear cells of VL and acute lymphoblastic leukemia (ALL) patients[12,13–19] and erythrocytes of ALL[61] and VL patients[7–9].

Achatinin-H differs from Influenza C virus in that its binding glycotope is 9-O-AcSAα2→6GalNAc, whereas the CHE-FcD binds to terminal 9-O-AcSA, independent of the underlying linkage or subterminal sugar.

C. CANCER ANTENNARIUS LECTIN

A lectin isolated from the crab, *Cancer antennarius*, recognizes sialic acids that are O-acetylated both at C-4 and C-9 positions[41,42] and has been successfully used to identify an O-acetyl ganglioside, GD3 having O-AcSA in an α2→8 linkage on human melanoma cells.[62,63] The identification by lectin provides evidence that

TABLE 3.4
Binding Specificity of Achatinin-H by Hemagglutination Inhibition Assay

Saccharides/Sialoglycoproteins	Types/Nature of Terminal Linkages	I_{50}
Monosaccharides	Neu 5 Ac	30.48
	Neu 5,9 Ac$_2$[a]	1.30
	Neu 4,5 Ac$_2$[a]	NI[b]
Disaccharides	α-Neu 5 Ac-(2→6)-GalNAc-ol	NI[c]
Sialoglycoproteins		
BSM	α-Neu5,9Ac$_2$-(2→6)-β-DGalNAc-	0.0002
De-O-acetylated BSM	α-Neu5Ac-(2→6)-β-DGalNAc-	NI[c]
SSM	α-Neu5Ac-(2→6)-β-DGalNAc-	NI[c]
Asialo-BSM	D-GalNAc-	NI[c]
Fetuin	α-Neu5Ac-(2→6)-β-DGal-	NI[c]
Serotransferrin	α-Neu5Ac-(2→6)-β-DGal-	NI[c]
Lactotransferrin	α-Neu5Ac-(2→6)-β-DGal	NI[c]
Human choriogonadotropin	α-Neu5Ac-(2→6)-β-DGal	NI[c]

I_{50}:The minimal concentration of the monosaccharide required for 50% inhibition of 16 hemagglutination units of Achatinin-H; BSM=bovine submaxillary mucin; SSM=sheep submaxillary mucin.

[a]Purified Neu5,9AC$_2$ and Neu4,5Ac$_2$ were kindly provided by Prof. Dr. R. Schauer.

[b]Not inhibited up to a concentration of 100 mM.

[c]350-fold less inhibitory than BSM on the basis of 9-O-acetyl sialic acid.

Source: Data from Sinha, D., et al., *Br. J. Haematol.*, 110, 801–802, 2000; Sen, G. and Mandal, C., *Carbohydr. Res.*, 268, 115–125, 1995; Sharma, V., *Glycoconjugate* J., 7, 887–893, 2000. All with permission.

O-acetylated GD3 may represent an important tumor marker for detection and treatment of human melanoma.

D.   *Liocarcinus depurator* Lectin

A lectin from hemolymph of the marine crab *Liocarcinus depurator* was purified and characterized which recognizes *O*-AcSA. The lectin also had high affinity toward 2, 4, 7, 8, 9-penta-*O*-AcSA.[47]

E.   *Paratelphusa jacquemontii* (Rathbun) Lectin

A naturally occurring hemagglutinin has recently been detected in the serum of the freshwater crab, *Paratelphusa jacquemontii* (Rathbun) that showed specificity for both 9- and 4-*O*-acetylated sialic acid.[48]

F.   Human Placenta Lectin

A Ca^{2+}-independent sialic acid-specific lectin from two developmental stages of human placenta was affinity-purified from bovine submaxillary mucin and the specificity of the lectin was toward 9(7,8)-*O*-acetyl and 4-*O*-acetylsialic acids.[49,50]

IV.   METHODS TO IDENTIFY *O*-ACETYL SIALIC ACIDS

Analytical methods for quantification of *O*-AcSG's require their previous release from glycosidic linkages by either enzymatic or chemical hydrolysis. However, none of the currently available methods is totally satisfactory as potential pitfalls during analysis include the incomplete release of sialic acids, de-*O*-acetylation and spontaneous migration of *O*-acetyl groups. Therefore, an urgent need exists for development of methods to detect alkali-labile *O*-AcSGs, preferably bypassing their previous liberation. A major hurdle for detection of *O*-AcSA lies in the fact that they are easily saponifiable groups. Therefore, in practical terms, accurate detection of these saponifiable groups, irrespective of their linkage and subterminal sugar as also information especially with regard to their presence in the sterical context of the intact cell surface, is only feasible by using noninvasive approaches using the above-mentioned probes.

A.   Noninvasive Methods

1.   Hemagglutination Assay

Erythrocytes were collected from individual patients with VL by centrifugation (2000 r/min for 5 min). An HA was performed with a 2% erythrocyte suspension (v/v) in saline. Achatinin-H (25 μL of 50 mg/mL) was serially diluted 1:1 in 96-well U-bottom microtiter plates in Tris-buffered saline (TBS), pH 8.2, and was followed by the addition of 25 μL of 0.1 M $CaCl_2$ and 25 μL of 2% erythrocyte suspension. Hemagglutination titers were scored by visual assessment after 30 to 60 min of incubation at 25 to 30°C.[7] The reciprocal of the highest dilution of Achatinin-H that produced visible agglutination was taken as the hemagglutination titer expressed in hemagglutination units (HU). For the hemagglutination inhibition assay, 25 μL of sugar solution, serially diluted in saline was allowed to react for 30 min at 25 to 30°C with 25 μL of Achatinin-H and 25 μL of Ca^{2+}; this step was followed by the addition of 2% VL erythrocytes and an additional incubation for another 30 to 60 min and the hemagglutination titer recorded (Figure 3.2).

2. Parasite Agglutination Assay

To assess the degree of parasite binding,[11] *L. donovani* promastigotes were harvested and washed thrice with phosphate-buffered saline (PBS). Parasites (1.0×10^7/mL, 100 μL) were serially diluted with increasing concentrations of Achatinin-H

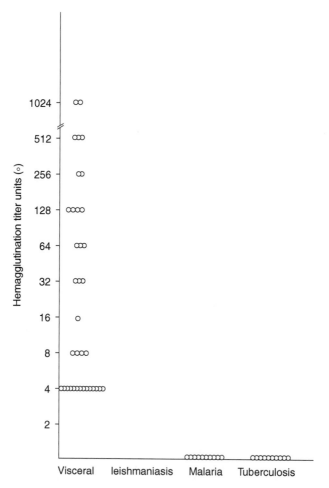

FIGURE 3.2 Status of *O*-acetylated sialoglycans on erythrocytes of patients suffering from V. Leishmaniasis by hemagglutination using Achatinin-H, a 9-*O*-acetylated sialic acid binding lectin. Hemagglutination titers (•) of patients with VL (pretreatment), malaria, and tuberculosis. (From Sharma, V. et al., *Am. J. Trop. Med. Hyg.*, 58, 551– 554,1998. With permission.)

(4–40 μg/well) and incubated at 20 to 25°C for 15 min. At higher concentrations of Achatinin-H, the degree of agglutination precluded counting of agglutinated cells; instead, we counted the number of nonagglutinated cells and accordingly the degree of agglutination was extrapolated (Figure 3.3A).

3. Quantitation of the Presence of O-AcSA by ELISA Using Achatinin-H as Coating Antigen

The exponential-phase cultures of *L. donovani* promastigotes were harvested, washed with PBS, and the cell pellet was resuspended in lysis buffer (20 mM TrisHCl, 40 mM NaCl, pH 7.4), containing 2.0 mM phenylmethylsulfonyl fluoride (PMSF), 1.0 mg/mL leupeptin, 5.0 mM ethylenediamine tetra acetic acid (EDTA) and 5.0 mM iodoacetamide. The crude parasite lysate was used as the coating antigen (5.0 μg/mL, 50 μL/well in 0.02 M phosphate buffer, pH 7.8), and binding of Achatinin-H[11] was determined by measuring binding of rabbit anti-Achatinin-H using horseradish peroxidase (HRP)-conjugated anti-rabbit IgG using azino-bis-thiosulfonic acid (ABTS) as the substrate (Figure 3.4B).

4. Erythrocyte Binding Assay

Optimization of the assay[8] required establishing conditions with regard to concentration of Achatinin-H as coating antigen. These conditions were determined by using normal rabbit and rat erythrocytes, based on available knowledge that they have high and low concentrations of cell surface 9-O-AcSA respectively.[64]

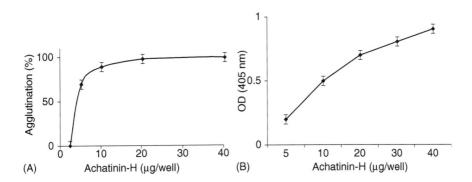

FIGURE 3.3 Identification of O-acetyl sialic acids on promastogote of *L. donovani* using Achatinin-H. (A) Parasites (1.0 × 10⁷/mL, 100 μL) were incubated with increasing concentrations of Achatinin-H at 20 to 25°C for 15 min. Cells were then examined microscopically and the number of nonagglutinated cells counted; accordingly, the degree of agglutination was extrapolated. Each point is the average of three independent experiments. (From Chatterjee, M. et al., *Glycobiology*, 13, 351–361, 2003. With permission.) (B) To demonstrate Achatinin-H binding to parasite membranes, membrane lysates were coated on 96-well plates and incubated with increasing concentrations of Achatinin-H. Binding of lectin was detected colorimetrically as described in section IV. A .3. Each point is the average of three independent experiments. (From Chatterjee, M. et al., *Glycobiology*, 13, 351–361, 2003. With permission.)

Increasing concentration of pre-and posttreated VL erythrocytes were used to select the optimal cell concentration for Achatinin-H binding (Figure 3.4A and B).

The 9-O-AcSA binding lectin, Achatinin-H, was immobilized on 96-well flat-bottomed polystyrene microtiter plates (0.5 μg/100 μL, diluted in TBS, pH 7.3; Nunc-Immunoplate, USA) After overnight incubation at 4°C, the unbound lectin was discarded and the wells were washed thrice with TBS. The nonspecific binding sites were blocked for 2.0 h at 4°C with TBS containing 2.0% fetal calf serum (FCS). Erythrocytes (4.0×10^8 cells/100 μL/well), washed in saline containing $CaCl_2$ (25 mM) were added and incubated overnight at 4°C. After removal of the non-specifically bound erythrocytes by a gentle wash with saline containing $CaCl_2$ (25 mM), the erythrocytes were fixed slowly with 0.25% glutaraldehyde (Sigma, St. Louis, MO, USA) for 10 min at 4°C. The wells were rigorously washed, and the extent of specifically bound erythrocytes was quantified using three approaches, namely (a) cell lysis with double-distilled water (100 μL/well) wherein the degree of hemolysis was quantified at 405 nm; (b) addition of a chromogenic substrate, 2,2'-azinobis(3-ethylbenzthiazoline-6-sulfonic acid) (ABTS), 100 μL/well (Roche diagnostics, Germany) and absorbances measured at 405 nm, and (c) using 2,7-diamino fluorene dihydrochloride (DAF) (Merck, Germany) and measuring absorbances at 620 nm. DAF (5.0 mg) was initially dissolved in 5.0 mL of glacial acetic acid (60%); 1.0 mL of the solution was then mixed with 1.0 mL of 30% of hydrogen peroxide (Qualigens, Glaxo India) and TrisHCl (0.2 M) buffer containing 6.0 M urea (Sigma, MO, USA). The assays using the substrates ABTS and DAF utilize the pseudoper-oxidase activity of hemoglobin to produce a colored product. In all three assays, absorbances greater than mean plus three times the standard deviation (SD) of nonendemic controls were considered as positive (Figure 3.4C).

Using this newly developed assay,[8] we measured the enhanced expression of this 9-O-acetylated glycotope on erythrocytes of amastigote-positive VL cases ($n=30$), their mean±SEM of OD_{405} value being 1.14±0.04 (Figure 3.4D). Erythrocytes from laboratory personnel were considered as nonendemic healthy controls and their mean±SEM of OD_{405} was 0.32±0.03. These levels were no different from the mean±SEM of the OD_{405} in controls from an endemic area (0.23±0.03). Interestingly, there was a marked decrease in the expression of this glycotope in VL patients following completion of chemotherapy. Because absorbance values decreased fourfold when compared with the levels measured during disease presentation (Figure 3.4D) this assay can be used both for initial diagnosis of VL and, subsequently, for monitoring the disease status.

5.   Quantitative Estimation of O-AcSG by Scatchard Plot

The purified Achatinin-H was iodinated following the method of Hunter (65), using chloramine T and Na [^{125}I]. Erythrocytes (1×10^6) from VL patients were incubated at 4°C for 60 min in a total reaction mixture (100 μL) that contained increasing amounts of [^{125}I]-Achatinin-H (specific activity=1.4×10^6 cpm/μg), $CaCl_2$ (0.3 M, 15 μL) and TBS containing BSA, pH 7.4. Nonspecific binding was removed by three washes with TBS–BSA, and bound reactivity in the cell pellet was quantified in a Gamma counter.[8] For evaluating the specific nature of binding, a 50-fold excess of

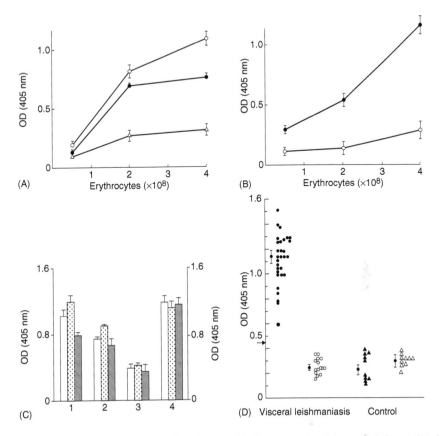

FIGURE 3.4 (A). Optimization of erythrocyte binding assay for detection of *O*-acetylated sialoglycans using mammalian erythrocytes. To optimize the cell density, erythrocytes at different dilutions from 0.5 to 4.0×10^8/well from rabbit (○), rat (●) and humans (△) were incubated with immobilized Achatinin-H and processed as described in Section IV.A.4. Each point is the average of duplicate determinations of three independent experiments. **(B).** Optimization of cell numbers. Increasing concentrations of erythrocytes from VL patients at presentation of disease (●—●) and after drug treatment (○—○) were incubated with immobilized Achatinin-H and processed as described in section IV.A.4. Each point is the average of duplicate determinations of three independent experiments. **(C).** Optimization of detection system of erythrocyte binding assay. In order to select the best probe for detecting the presence of *O*-acetylated sialoglycans, erythrocytes from normal rabbit (1), rat (2), and human (3), together with the erythrocytes of VL patients (4), were incubated in an Achatinin-H-coated plate. The extent of erythrocyte binding was detected by lysis with double-distilled water (□), absorbance of ABTS (▧) or DAF (▨). Each point is the average of duplicate determinations of three independent experiments. **(D).** Quantitation of linkage-specific *O*-acetylated sialoglycans on erythrocytes of VL patients. Scatter plot showing binding of individual patients with VL before (●, $n = 30$) and after completion of the treatment (○, $n = 15$). Also shown are data from normal humans living in areas endemic (▲, $n = 10$) and nonendemic (△, $n = 10$) for VL. Erythrocytes (4.0×10^8 cells/well) were added to immobilized Achatinin-H and assayed. Results are expressed as mean±SEM from duplicate determination. In both groups, mean $A_{405} \pm 3$ S.D. of normal humans was taken as the cutoff value as indicated (➜) on the *y*-axis. (From Chava, A.K. et al., *J.Immunol. Methods*, 270, 1–10, 2002. With permission.)

unlabeled Achatinin-H was added. The dissociation constant (K_d) and number of binding sites for Achatinin-H were calculated from the Scatchard plot (66) as shown in Figure 3.5.

6. Flow Cytometric Analysis

The presence of 9-O-AcSA on erythrocytes of VL patients and parasite surfaces was established by using FITC-labeled Achatinin-H.[8–11] Briefly, cells (1.0×10^6 cells/mL) were extensively washed and incubated with FITC-Achatinin-H (1.0 μg/tube) for 60 min at 4°C, fixed with 1.0% paraformaldehyde. The extent of binding was measured by flow cytometry (FACS Scan flow cytometer; Becton Dickinson, Mountain View, CA, USA), using normal erythrocytes and FITC-labeled desialylated BSA as the negative controls (Table 3.5).

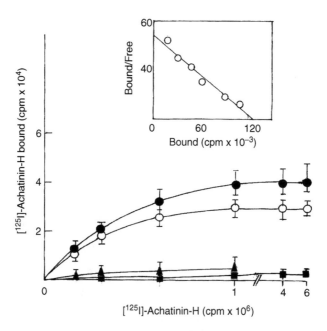

FIGURE 3.5 Analysis of receptor density of O-acetylated sialoglycans present on VL erythrocytes by using [^{125}I]-Achatinin-H by Scatchard plot. A fixed amount of erythrocytes (1×10^6) was incubated with increasing amount of [^{125}I]-Achatinin-H. For evaluating the specific nature of binding, a 50-fold excess of unlabeled Achatinin-H was added. The bound and unbound Achatinin-H were separated at 4°C and the bound radioactivity was determined as described in section IV.A.5. Specific binding (○) was calculated by the difference between total binding (●) and nonspecific binding (■). Posttreated erythrocytes (▲) were similarly processed for specific binding. Results are expressed as mean±SD from triplicate experiments. Significant differences in binding ligands of Achatinin-H with respect to pre- and posttreated erythrocytes were observed ($P<0.001$). The inset shows the Scatchard plot of the binding of [^{125}I]-Achatinin-H to erythrocytes from VL patients. (From Chava, A.K. et al., *J. Infect. Dis.*, 189, 1257–1264, 2004. With permission.)

TABLE 3.5

Status of O-acetylated Sialoglycans on Cell Surface of Mammalian Erythrocytes, Patients Suffering from VL and Parasite, _Leishmania donovani_ (Promastigotes and Amastigotes)

Cells	Achatinin-H Positive Cells[a] (%)	Ref.
Rabbit	97.53±1.4	8
VL (pretreatment)	76.07±6.31	9
VL (posttreatment)	5.07±2.86	9
Endemic controls	2.58±1.18	9
Nonendemic controls	2.38±1.37	9
Amastigotes (_L. donovani_)	49.3±4.5	10
Promastigotes (_L. donovani_)	44.3±3.4	11

[a]Mean±SD of Achatinin-H-positive cells (%) were determined by FACS analysis using FITC-Achatinin-H as an analytical probe.

The occurrence of sialic acids on the cell surface of _L. donovani_ parasite has been referred in Table 3.2A; this was established[10,11] by using two sialic acid binding lectins, _Sambucus nigra_ agglutinin (SNA) and _Maackia amurensis_ agglutinin (MAA) (Vector Labs, Burlingame, CA, USA). Parasites were extensively washed and resuspended (1.0×10^7 cells/mL) in prechilled RPMI medium supplemented with 2.0% BSA and 0.1% sodium azide (Medium A) for 1.0 h at 4°C in the dark. Parasites were then incubated with biotinylated SNA and MAA (5.0 µL, 5.0 µg/mL) for 30 min at 4°C. Cells were then washed and lectin binding detected by measuring the binding of FITC-conjugated streptavidin.

The presence of sialic acid was also detected by measuring the binding of Siglecs to _L. donovani_ promastigotes and amastigotes,[10,11] the Siglecs-Fc were initially complexed with biotinylated goat anti-human antibodies (Fc-specific) by incubating 25°C for 1 h. Subsequently, promastigotes (either untreated or pretreated with _Vibrio cholerae_ neuraminidase, [VCN]) were incubated with the complexed Siglecs for 30 min on ice. The cells were then washed, and the extent of binding detected by flow cytometry by using FITC-streptavidin.

For the detection of O-AcSA on the cell surface, they may also be pretreated with O-acetylesterase to remove O-acetyl group and followed by SNA and MAA binding. An increase in SNA and MAA binding indirectly supports the presence of O-AcSA. Alternatively, reduction in SNA binding after pretreatment of cells with O-acetylesterase esterase followed by sialidase treatment also indicated the presence of O-AcSA.[10]

7. Detection of Anti-O-AcSG in Serum as an Indirect Indicator of the Expression of Cell Surface O-Acetylated Sialic Acid

a. Probe to Detect Anti-O-AcSG: Bovine Submaxillary Mucin (BSM)

BSM was purified[7] from freshly collected bovine submaxillary glands. The glands were cleaned of connective tissue, blood clots, and debris and minced with one and half volume of chilled distilled water. The mixture was centrifuged at $12,000 \times g$ for 20 min at 4°C. The supernatant was retained and the pellet again centrifuged after reconstitution with 1 vol of chilled distilled water. The supernatants were pooled, its pH adjusted to 4.5 using 2.0 M acetic acid and centrifuged at $12,000 \times g$ for 20 min at 4°C. The pH of the clear supernatant was adjusted to 6.0 using 1.0 M NaOH, thoroughly dialyzed against chilled distilled water and centrifuged at $12,000 \times g$ for 20 min at 4°C. To the chilled supernatant was added barium acetate to a final concentration of 0.1 M. Chilled distilled methanol was added slowly to this mixture until it was 64% with respect to methanol. The mixture was kept at 4°C to allow complete precipitation. Following centrifugation at $12,000 \times g$ for 20 min at 4°C, the pellet was dissolved in minimum volume of 0.1 M EDTA and extensively dialyzed against distilled water. The percentage O-acetylation of total sialic acid was estimated by flourimetric analysis[7] and was found to be 22.5%. HPLC analysis of BSM (11) reveled well-resolved peaks of sialic acid and its main derivatives as shown in Figure 3.6 A (upper panel).

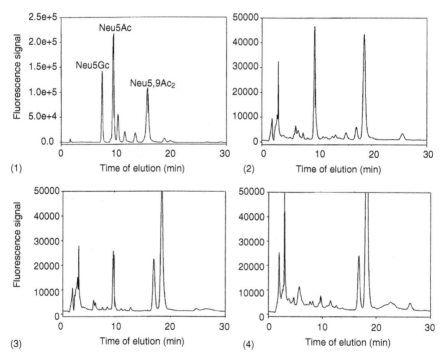

FIGURE 3.6 (A)   Determination of sialic acid (Neu5Ac) and its principal derivatives by fluorimetric HPLC. Representative profile of a HPLC chromatogram of fluorescent derivatives of free sialic acids derived from bovine submandibular gland mucin (1), *L. donovani* promastigotes before (2) and after preincubation with 0.1M ammonia vapor (3) and acyl neuraminate pyruvate lyase (4). Glycosidically bound sialic acids were subjected to acid hydrolysis, derivatized with 1,2 diamino-4,5-methylenedioxybenzole and analyzed (From Chatterjee, M. et al., *Glycobiology*, 13, 351–361, 2003. With permission.)

To confirm further the presence of 9-*O*-AcSA on BSM, Western-blot analysis was carried out using Achatinin-H, an established *O*-acetylated sialic acid binding lectin. BSM (10 μg) was resolved on SDS–PAGE (10%), transblotted onto nitrocellulose and probed with Achatinin-H. The reaction was chased with polyclonal anti-Achatinin-H. The presence of *O*-AcSA on BSM was confirmed by removal of the alkali-labile *O*-acetylated glycotopes by saponification using 0.1 *M* NaOH and incubating the blots for 45 min on ice and neutralizing the excess alkali with 0.1 *N* HCl (Figure 3.7, lane 1).

b. BSM as a Source of Capture Antigen for Detection of Anti-O-AcSG
BSM was used as a coating antigen (1.0 μg/uL, 50 μL/ well in 0.02 *M* phosphate buffer (PB, pH 7.8). Following an overnight incubation at 4°C, the wells were washed three times with PBS, pH 7.2, containing 0.1% Tween-20. The wells were then blocked with 2.0% FCS in PBS for 1 h at 25°C. Serum samples were diluted at 1:100, added to the well and incubated overnight at 4°C. Binding was assayed colorimetrically using HRP-conjugated goat anti-human IgG (diluted 1:5000) using ABTS as the substrate.[67] Optical density was estimated at 405 nm along with negative control sera obtained from asymptomatic individuals from both endemic and nonendemic areas. Absorbance greater than the mean plus three times the SD of endemic controls were considered positive (Figure 3.8).

B. INVASIVE METHODS

1. Quantitation of Sialic Acid and 9(8)-*O*-Acetyl Sialic Acid by the Acetyl Acetone Method

The assay was performed according to the classical method of Shukla and Schauer,[64] samples being divided into two sets. In the first set, sample dissolved in a buffer II (containing 3.0 g of NaCl, 0.2 g of KCl, 1.0 g of disodium hydrogen phosphate dihydrate, and 0.15 g of sodium dihydrogen phosphate monohydrate. 0.2 g of potassium dihydrogen phosphate per liter, pH 7.2) was hydrolyzed with 0.2 *M* NaOH (100 μL) and incubated at 4°C for 45 min to saponify the hydroxyl groups on sialic acids after which they were neutralized using 100 μL of 0.2 *N* HCl. In the other series, to the same amount of sample in buffer II, 200 μL of 0.1 *M* NaCl was added. All samples were simultaneously oxidized using sodium metaperiodate (200 μL of 2.5 m*M* in buffer II) at 4°C for 15 min in dark. Following oxidization, 400 μL of this reaction mixture was added to 100 μL of sodium arsenite (2.0% in 0.5 *M* HCl) and 300 μL of acetyl acetone solution (750 μL at glacial acetic acid, 500 μL of acetylacetone and 37.5 g of ammonium acetate in a final volume of 250 mL distilled water). The reaction mixtures were incubated at 60°C for 10 min and then diluted with distilled water to make a final volume of 2.5 mL. The relative fluorescence intensity of each reaction mixture at Ex_{410}–Em_{510} was recorded against a reagent blank. In parallel, free sialic acid (2 to 8 μg) was used to create a standard curve. On extrapolation of the readings, the values obtained for de-*O*-acetylation samples represented the total sialic acid content, while the other set of samples indicated the unmodified sialic acid; therefore, subtracting the modified sialic acid content from the total sialic acid gives the amount of 9-*O*-AcSA the samples contained[9,68] (Table 3.6).

TABLE 3.6
Estimation of Total 9(8)-O-Acetylated Sialic Acids on Erythrocytes

Cells[a]	9(8)-O-AcSA[b] (%)	HU of Achatinin-H[c]	Hemolysis[d] (%)
Rabbit	20±2.2	1024±4.7	70.0±5.7
Guinea-pig	22±1.8	32±2.2	23.0±2.8
Hamster	25±2.4	32±2.2	3.0±2.4
Rat	40±3.4	16±1.5	4.0±3.4
Mouse	60±1.5	0	30.0±2.2
VL (pretreatment)	30.6±5.1	0	36.0±2.26
VL (posttreatment)	7.5±3.4	0	7.5±1.48
Endemic controls	6.2±2.4	0	7.2±0.21
Nonendemic controls	5.2±2.2	0	7.1±0.11

[a]Blood was collected in Alsevers solution.

[b]Mean±SD of total percent of *9(8)-O*-acetylated sialic acid present on erythrocyte membranes was fluorimetrically estimated.[9,68]

[c]Mean±SD of Achatinin-H binding HU is defined as the reciprocal of the highest dilution of Achatinin-H that produced visible agglutination.[68]

[d]The percent hemolysis induced by the alternative complement pathway using 50 μL of guinea-pig serum as source of complement as described in section IV. A.4.[9,68]

2. Localization of Cell Surface O-Acetyl Sialoglycoconjugates

Localization of sialoglycoconjugates can be demonstrated by examining the localization of the fluorochrome-labeled lectins. Briefly, washed cells were fixed on polylysine-coated coverslips, and the nonspecific binding sites were blocked with appropriate blocker (BSA/goat serum/collagen for 30 min at 25°C). Following the addition of the fluorochrome-tagged lectins, incubation proceeds for 30 min on ice and then their localization is achieved by detection of the streptaividin-conjugated fluorochrome.

3. Fluorimetric HPLC for Estimation of Sialic Acid (Neu5Ac) and Its Principal Derivatives

L. donovani parasites[10,11] were extensively washed in PBS (0.02 M, pH 7.2), and the cell pellet was resuspended in 1.0 mL of double-distilled water. Cell lysis was completed by sonication (three pulses of 16 sec each and lysates maintained on ice in between). To release sialic acids, the glycoconjugates were then subjected to acid hydrolysis with an equal volume of 4.0 M propionic acid and samples were heated to 80°C for 4 h, they were then cooled on ice for 10 min, separated into three fractions and then lyophilized.[69] Controls included (i) specification of sialic acids by incubating the lyophilized sample with 100 μL of 0.1 M ammonia for 1.0 h at 37°C with subsequent lyophilization and (ii) lyase treatment by resolving the samples in 200 μL of 50 mM phosphate buffer (pH 7.2), including 25 mU acylsialate neuraminate lyase (EC 4.1.3.3) and incubated for 2.0 h at 37°C. Samples were then derivatized with 1,2 diamino-4,5-methylenedioxybenzole (DMB) for fluorimetric HPLC analysis.[70]

Analysis of the derivatized sialic acids was done on a RP-18 column (4×250 mm, Lichrospher RP-18, Merck, Darmstadt, Germany) using isocratic elution with water/acetonitrile/methanol (84:9:7, by vol) at a flow rate of 1.0 mL/min and compared with authentic standard sialic acids purified from BSM. Fluorescence detection was performed using an excitation wavelength of 373 nm and emission wavelength of 448 nm. In parallel, the three washes of the parasite cultures were analyzed similarly as described above (Figure 3.6A and B).

4. Gas Chromatography/Mass Spectrometric Analysis of Sialic Acids as TMS–Methylester Derivatives

Lyophilized samples were dissolved in dry methanol (0.5 mL) and 80 μL of Dowex H^+ in methanol was added. The samples were filtered over cotton wool to remove the Dowex and subsequently treated with diazomethane in ether for 5 min at room temperature. The solution was evaporated using a stream of nitrogen and dried over P_2O_5. The residue was dissolved in 6.0 μL of trimethysilane (TMS) reagent pyridine/hexamethyldisilisane (HMDS)/ trimethychlorosilane (TMCS) (5:1:1, by vol). After 2.0 h at room temperature, samples (3.0 μL) were analyzed by GC/MS, respectively.[59]

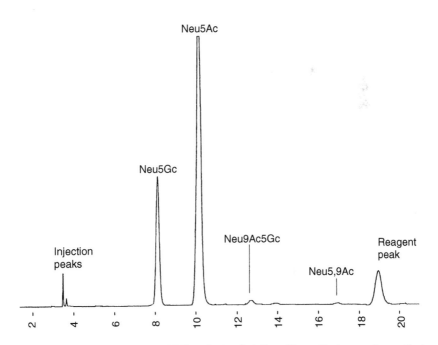

FIGURE 3.6 (B) Fluorimetric HPLC analysis of sialic acids on *L. donovani* amastigotes. Glycosidically bound sialic acids were released with 2.0 *M* acetic acid (3.0 h, 80°C), derivatized with 1,2,-diamino-4,5-methylenedioxybenzene. (From Chava, A.K. et al., *Bio. Chem. J.*, 385, 59–66, 2004. With permission.)

The following equipment and parameters were used for GC/MS analysis: a Fisons Instruments GC 8060/MD800 system (Interscience), an AT-1 column (30 m×0.25 mm; Alltech), the temperature program was 220°C for 25 min, 6°C min to 300°C, 6.0 min, the injector temperature was 230°C, the detection was done by EI mass spectrometry with a mass range of 150 to 800m/z. The mass fragment ions generated from *L. donovani* amastigotes are shown in Table 3.7.[10]

5. Molecular Identification of *O*ASGs on Erythrocytes of VL Patients and Parasites (*L. donovani*) by Western-Blot Analysis Using Achatinin-H

To identify the molecular determinants for Achatinin-H on erythrocytes of VL patients (Figure 3.7, lane 4) and *L. donovani* amastigotes (Figure 3.7, lane 2) and promastigotes (Figure 3.7, lane 3), Western-blot analysis was carried out using Achatinin-H as a probe.[9-11] Interestingly, an array of discreet *O*-AcSGs have been observed on erythrocytes of VL patients with M.w's of 112,107,103,57,51,48 kDa as compared with only two *O*-AcSGs on cell surface of both promastigotes and amastigotes being 109,123 and 150,158 respectively (Figure 3.7, lanes 2–4).

V. BIOLOGICAL ROLES OF *O*-ACETYL SIALIC ACID

This strategic terminal position provides *O*-AcSA accessibility, which is reflected in its regulation of a multitude of cellular and molecular interactions.[71, 72]

A. 9-*O*-ACETYL SIALIC ACID AND COMPLEMENT

Sialic acids have been identified as critical determinants for erythrocyte survival. They are known to prevent activation of the alternative pathway of complement by binding to factor H, the key regulatory molecule that normally restricts the amplification loop of the alternative complement pathway by preventing the formation of C3 convertase.[73] Studies have shown that the exocyclic chain of sialic acid is vital for its binding to factor H.[74-76] Substitution by the bulky acetyl group prevents the binding of sialic acid to factor H, which in turn triggers the cascade of the alternative pathway of complement. The role of 9-*O*-acetylated sialoglycans in triggering the alternative complement-mediated hemolysis has been extensively studied in

TABLE 3.7
Qualitative Analysis of Sialic Acids on Amastigotes by GC/EI-MS

Sialic Acid	Retention Time (min)	R_{Neu5Ac}	Mass Fragment Ions A–F (m/z)
Neu5Ac	13.6	1.00	668, 624, 478, 400, 317, 298
Neu5,9Ac$_2$	15.2	1.12	638, 594, 478, 400, 317, 298
Neu5Gc	23.6	1.74	756, 712, 566, 488, 386, 317
Neu9Ac5Gc	27.0	1.98	726, 682, 566, 488, 386, 317

Source: Chava, A.K. et al., *Bio. Chem. J.*, 385, 59–66, 2004. With permission.

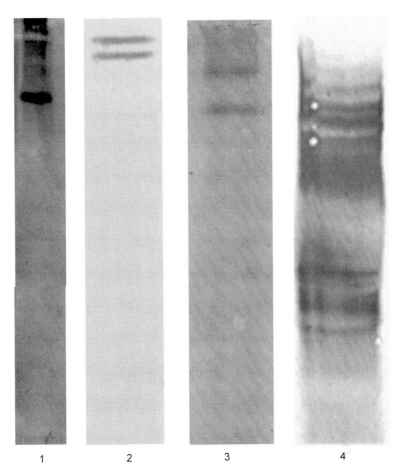

FIGURE 3.7 Molecular characterizations of *O*-acetyl sialoglycoproteins by Western blot analysis. BSM (10 μg, lane 1), membrane proteins (20 μg/lane) from *L. donovani* amastigotes (lane 2), promastigotes (lane 3), and crude membrane proteins (30 μg, lane 4) from erythrocytes of VL patients were electrophoresed on SDS-PAGE. These were transfered onto nitrocellulose membranes and incubated with Achatinin-H. Binding was detected with rabbit anti-Achatinin-H and peroxidase-conjugated anti-rabbit IgG. (From Chava, A.K. et al., *J. Infect. Dis.*, 189, 1257–1264, 2004.; Chava, A.K. et al., *Bio. Chem. J.*, 385, 59–66, 2004; Chatterjee, M. et al., *Glycobiology*, 13, 351–361, 2003. With permission.)

murine erythrocytes.[77,78] However, the case is not so simple, as we observed that the degree of complement-mediated hemolysis showed a relatively poor correlation ($r = 0.22$) with total 9-*O*-AcSA present on mammalian erythrocytes (Figure 3.9A and Table 3.6). Instead, a far better correlation ($r = 0.90$) was obtained between the presence of linkage-specific 9-*O*-AcSGs as defined by the binding of Achatinin-H and the degree of complement-mediated hemolysis.[68]

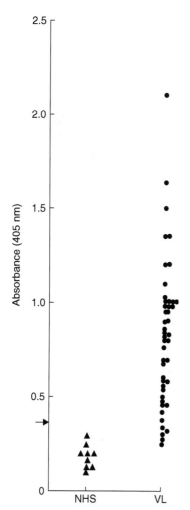

FIGURE 3.8 Detection of anti-OAcSG in VL serum using BSM as coating antigen. Binding of serum (1:100 dilution) from VL patients (●) and NHS (▲) to BSM-coated wells was assayed by BSM–ELISA as described in Reference 68. Each point is the average of duplicate determinations. Mean + 3 SD of NHS was taken as the cutoff value. (From Sharma, V. et al., *Glycoconjugate*, J., 7, 887–893, 2000. With permission.)

It may be postulated that in mammalian erythrocytes, those having 9-O-AcSA in an $\alpha2{\rightarrow}6$ linkage with GalNAc are more susceptible to complement-mediated hemolysis. In previous studies, we have reported an increased presence of the glycotope as recognized by Achatinin-H on erythrocytes from patients with VL as compared with erythrocytes from normal human donors (8,9,68). We were keen to address whether this increased expression of linkage-specific 9-O-AcSGs causes enhanced hemolysis. Indeed, the enhanced presence of this glycotope did cause enhanced hemolysis (Figure 3.9B, and Table 3.6) as evidenced by an excellent correlation ($r^2 = 0.90$) between the presence of this glycotope as recognized by binding of Achatinin-H and the degree of hemolysis mediated by the alternative pathway of complement. As anemia is a common manifestation in this disease and has been attributed to bone marrow infiltration, hypersplenism, and autoimmune hemolysis.[79] We, therefore, contend that these sialoglycans possibly contribute toward anemia associated with VL. As anemia contributes significantly to the

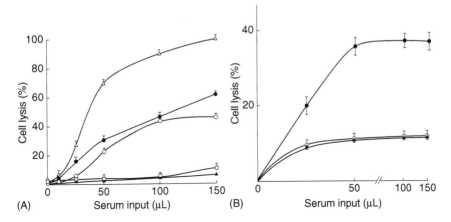

FIGURE 3.9 Alternative complement-mediated hemolysis of erythrocytes. (A) Cell lysis assay of different mammalian erythrocytes was carried out using increasing amounts of fresh human serum as the source of complement. Mammalian species included rabbit (△—△), guinea-pig (●—●), mouse (○—○), rat (☐—☐), and hamster (▲—▲). Each point is the average of triplicate determinations. (From Mawhinney, T.P. and Chance, D.L., *Anal. Biochem.*, 223, 164–167, 1994. With permission.) (B) The assay was carried out with erythrocytes obtained from different pools of VL patients using increasing amounts of fresh guinea-pig serum as the source of complement. Erythrocytes included were obtained from active VL patients (●—●), posttreated VL patients (○—○), and normal donors (▲—▲). Each point is the average of triplicate determinations. Results are expressed as mean ± SD from triplicate experiments. *P* < 0.001, induction of hemolysis in samples from active VL patients vs. normal/posttreated VL patients. (From Chava, A.K. et al., *J. Infect. Dis.*, 189, 1257–1264, 2004. With permission.)

morbidity associated with VL, it remains to be seen whether in the future, supportive treatment for VL could include methods where the selective expression of glycotopes can be reduced.

VI.   PERSPECTIVE

The constant variation of sialic acid and its *O*-acetylation has brought a big alert in the biomedical field. Detailed information on these biologically important molecules, their structural analysis, monitoring of their spatial and temporal patterns of expression and characterization of their complementary binding partners can make it possible to develope rationally strategies to combat diseases. *O*-acetylated sialoglycoconjugates, showing altered expression under various pathophysiological conditions, open new avenues of research and should be directed toward exploiting this expression as an index for evaluating the disease status. The biological significance of altered sialylation in disease and factors involve therein will also be an exciting field of future investigation. A better understanding regarding how these bioactive molecules have emerged will gain insights regarding its biological significance. The status of several enzymes both in the pathogen and the host cells along with their concentration in host serum will indirectly provide valuable information about the acquisition of various sialic acid derivatives. A balance

TABLE 3.8
Enzymes Involved in Cell Surface O-Acetylation

Enzymes	Assigned Function
Sialyl transferase	Transfer Neu5Ac from CMP-Neu5Ac
trans-Sialidase	Cleave Neu5Ac from sialoglycoconjugates and transfer to nascent glycoconjugates
O-acetyltransferase	Transfer O-acetyl moiety from O-acetyl sialoglycoconjugates
O-acetylesterase	Hydrolysis O-acetyl moiety from O-acetyl sialoglycoconjugates
Sialidase	Cleave the sialic acid from sialoglycoconjugates

between several enzymes (Table 3.8) ultimately decides the fate of the sialic acid derivatives.[80] As O-acetylation depends mainly on the balance between two most important enzymes, namely O-acetyltransferase and O-acetylesterase, quantitation of these are very critical to understand the status of these diverse molecules. Therefore, more knowledge about these enzymes and their regulation will be another way to assess the status of O-acetyl sialic acids. Elucidation of the three-dimensional structures of O-acetyl sialic acid binding glycotopes is a challenging area of the sialic acid research. In addition, lectins, viruses, and antibodies against O-acetylated sialoglycoconjugates will be helpful in selectively targeting therapeutic drugs. Future research should focus on searching for finer tools for the detection of altered expression and should be explored in other parasitic diseases. Therefore, sialic acids and their various modifications under different pathological conditions can no longer be considered as simple decorative molecules on the glycocalyx as being responsible for several biochemical changes are now stimulating glycobiologists to explore further their status.

ACKNOWLEDGMENTS

Financial support from the Department of Biotechnology and Indian Council of Medical Research, Government of India, is gratefully acknowledged. Anil Kumar Chava is a Senior Research Fellow of the Council of Scientific and Industrial Research, Government of India. Our sincere thanks to Asish Mullick for his excellent technical assistance and Prof. Shyam Sundar (Institute of Medical Sciences, Banaras Hindu University, Varanasi 221 005, India) for providing clinical samples. We acknowledge Prof. Dr. R. Schauer (Biochemisches Institut, Kiel, Germany) and Prof. R. Brossmer, (Biochemistry Centre, University of Heidelberg, Germany) for providing purified Neu5,9Ac$_2$ and Neu 4,5Ac$_2$, Dr. P. R. Crocker (Wellcome Trust Biocentre, Dundee, UK) for providing the Siglecs, Prof. A. Varki (Division of Cellular and Molecular Medicine, University of California, San Diego, CA) and Dr. R. Vlasak (Institute of Molecular Biology, Austrian Academy of Sciences, Salzburg, Austria) for their generous gift of influenza hemagglutinin esterase fusion protein (CHE-Fc) and recombinant HE 1-eGFP fusion protein. Our special thanks goes to Prof. R. Schauer, Prof. J. P. Kamerling, and Dr. G. Gerwig for their help in analyzing the samples by flourimetric HPLC and GC/MS. We also express our gratitude to the Director, National Institute of Immunology, New Delhi, for extending their flow cytometry facility. We also thank our ex and present research scholars Sujata Ghosh, Goutam Sen, Mala Chakrabarty, Vineeta Sharma, Diviya Sinha, Santanu Pal, Tanusree Das, Suman Bandyopadhyay, Shyamasree Ghosh, Kankana Mukherjee, Sumi Bandyopadhyay, Avijit Dutta, and Waliza Ansar for their continuous support in preparing this chapter.

REFERENCES

1. Schauer, R., Achievements and challenges of sialic acid research, *Glycoconjugate* J. 17, 485–99, 2000.
2. Roseman, S., The synthesis of complex carbohydrates by multiglycosyltransferase systems and their potential function in intercellular adhesion, *Chem. Phys. Lipids,* 5, 270–297, 1970.
3. Angata, T. and Varki, A., Chemical diversity in the sialic acids and related alpha-keto acids: an evolutionary perspective, *Chem. Rev.,* 102, 439–469, 2002.
4. Varki, A., Diversity in the sialic acids, *Glycobiology,* 2, 25–40, 1992.
5. Schauer, R., Kelm, S., and Reuter, G., In *Biology of Sialic Acids.* Rosenberg, A., Ed., Plenum Press, New York, 1995, pp. 7–67.
6. Mandal, C., Sialic acid binding lectins — a review, *Experientia,* 46, 433–441, 1990.
7. Sharma, V., Chatterjee, M., Mandal, C., Sen, S., and Basu, D., Rapid diagnosis of visceral leishmaniasis using Achatinin$_H$, a 9-*O*-acetyl sialic acid binding lectin, *Am. J. Trop. Med. Hyg.,* 58, 551–554, 1998.
8. Chava, A.K., Chatterjee, M., Sundar, S., and Mandal, C., Development of an assay for quantification of linkage-specific *O*-acetylated sialoglycans on erythrocytes; its application in Indian visceral leishmaniasis, *J. Immunol. Methods,* 270, 1–10, 2002.
9. Chava, A.K., Chatterjee, M., Sharma, V., Sundar, S., and Mandal, C., Differential expression of *O*-acetylated sialglycoconjugates induces variable degree of complement mediated hemolysis in Indian Visceral leishmaniasis. *J. Infect. Dis.,* 189, 1257–1264, 2004.
10. Chava, A.K., Chatterjee, M., Gerwig, G.J., Kamerling, J.P., and Mandal, C., Identification of sialic acids on *Leishmania donovani* amastigotes, *Biol. Chem. J.,* 385, 59–66, 2004.
11. Chatterjee, M., Chava, A.K., Kohla, G., Pal, S., Merling, A., Hinderlich, S., Unger, U., Strasser, P., Gerwig, G.J., Kamerling, J.P., Vlasak, R., Crocker, P.R., Schauer, R., Schwartz- Albiez, R., and Mandal, C., Identification and characterization of adsorbed serum sialoglycans on *Leishmania donovani* promastigotes, *Glycobiology,* 13, 351–361, 2003.
12. Bandyopadhyay, S., Chatterjee, M., Sundar, S., and Mandal, C., Identification of 9-*O*-acetylated sialoglycans on peripheral blood mononuclear cells in Indian visceral leishmaniasis, *Glycoconjugate J.,* 20, 531–536, 2004.
13. Mandal, C., Chatterjee, M., and Sinha, D., Investigation of 9-*O*-acetylated sialoglycoconjugates in childhood acute lymphoblastic leukemia, *Br. J. Haematol.,* 110, 801–812, 2000.
14. Sinha, D., Mandal, C., and Bhattacharya, D.K., Identification of 9- *O*-acetyl sialoglyco-conjugates (9-*O*-AcSGs) as biomarkers in childhood acute lymphoblastic leukemia using a lectin, AchatininH, as a probe, *Leukemia,* 13, 119–125, 1999.
15. Sinha, D., Mandal, C., and Bhattacharya, D.K., A novel method for prognostic evaluation of childhood acute lymphoblastic leukaemia, Leukaemia, 13, 309–312, 1999.
16. Sinha, D., Mandal, C., and Bhattacharya, D.K., Development of a simple, blood based lymphoproliferation assay to assess the clinical status of patients acute lymphoblastic leukaemia, *Leuk. Res.,* 23, 433–439, 1999.

17.   Sinha, D., Mandal, C., and Bhattacharya, D,K., A colorimetric assay to evaluate the chemotherapeutic response of children with acute lymphoblastic leukaemia (ALL) employing Achatinin-H: a 9-*O*-acetyl sialic acid binding lectin, *Leuk. Res.*, 23, 803–809, 1999.

18.   Sinha, D., Chatterjee, M., and Mandal, C., *O*-acetylation of sialic acids — their detection, biological significance and alteration in diseases, *Trends Glycosci. Glycotechnol.*, 12, 17–33, 2000.

19.   Pal, S., Chatterjee, M., Bhattacharya, D.K., Bandyopadhyay, S., and Mandal, C., Identification and purification of cytolytic antibodies directed against *O*-acetylated sialic acid in childhood acute lymphoblastic leukemia, *Glycobiology*, 10, 539–549, 2000.

20.   Ravindranaths, M.H., Paulson, J.C., and Irie, R.F., Human melanoma antigen *O*-acetylated ganglioside GD3 is recognized by *Cancer antennarius* lectin, *J. Biol. Chem.* 263,2079–2086, 1988.

21.   Ravindranath, M.H., Morton, D.L., and Irie, R.F., An epitope common to gangliosides *O*-acetyl-GD3 and GD3 recognized by antibodies in melanoma patients after active specific immunotherapy, *Cancer Res.*, 49,3891–3897, 1989.

22.   Skov, L., Chan, L.S., Fox, D.A., Larsen, J.K., Voorhees, J.J., Cooper, K.D., and Baadsgaard, O., Lesional psoriatic T cells contain the capacity to induce a T cell activation molecule CDw60 on normal keratinocytes, *Am. J. Pathol.*, 150, 675–683, 1997.

23,   Haltia, A., Solin, M.L., Jalanko, H., Holmberg, C., Miettinen, A., and Holthofer, H., Sphingolipid activator proteins in a human hereditary renal disease with deposition of disialogangliosides, *Histochem. J.*, 28, 681–687, 1996.

24.   Gocht, A., Rutter, G., and Kniep, B., Changed expression of 9-*O*-acetyl GD3 (CDw60) in benign and atypical proliferative lesions and carcinomas of the human breast, Histochem. Cell Biol., 110, 217–229,1998.

25.   Vimr, E., Lichtensteiger, C., To sialylate, or not to sialylate: that is the question, *Trends Microbiol.,*10, 254–257, 2002.

26.   Schenkman, S., Jiang, M.S., Hart, G.W., and Nussenzweig, V., A novel cell surface trans-sialidase of *Trypanosoma cruzi* generates a stage-specific epitope required for invasion of mammalian cells, *Cell*, 65, 117–125, 1991.

27.   Schenkman, S., Kurosaki, T., Ravetch, J.V., and Nussenzweig, V. Evidence for the participation of the Ssp-3 antigen in the invasion of nonphagocytic mammalian cells by *Trypanosoma cruzi*. *J. Exp. Med.,* 175, 1635–1641, 1992.

28.   Schenkman, S., Ferguson, M.A., Heise, N., de Almeida, M.L., Mortara, R.A., and Yoshida, N., Mucin-like glycoproteins linked to the membrane by glycosylphosphatidylinositol anchor are the major acceptors of sialic acid in a reaction catalyzed by *trans*-sialidase in metacyclic forms of *Trypanosoma cruzi, Mol. Biochem. Parasitol.,* 59, 293–303, 1993.

29.   Ferguson, M.A., The structure, biosynthesis and functions of glycosylphosphatidylinositol anchors, and the contributions of trypanosome research, *J. Cell Sci.,* 112, 2799–2809, 1999.

30.   Avron, B., Chayen, A., Stolarsky, T., Schauer, R., Reuter, G., and Mirelman, D., A stage-specific sialoglycoprotein in encysting cells of *Entamoeba invadens. Mol. Biochem. Parasitol.,* 25,257–266, 1987.

31.   Chayen, A., Avron, A., B., Nuchamowitz, Y., and Mirelman, D., Appearance of sialoglycoproteins in encysting cells of *Entamoeba histolytica. Infect. Immun.,* 56, 673–681, 1988.

32.   Ribiero, S., Alviano, C.S., Silva-Filho, F.C., da Silva, E.F., Angluster, J., and de Souza, W., Sialic acid is a cell surface component of *Entamoeba invadens* trophozoites, *Microbios,* 57,121–129,1989.

33. Dias Filho, B.P., Alviano, C.S., de Souza, W., and Angluster, J., Polysaccharide and glycolipid composition in *Tritrichomonas foetus. Int. J. Biochem.,* 20,329–335, 1988.

34. Dias Filho, B.P., Andrade, A.F., de Souza, W., Esteves, M.J., and Angluster, J., Cell surface saccharide differences in drug-susceptible and drug-resistant strains of *Trichomonas vaginalis, Microbios,* 71, 55–64, 1992.

35. Gross, U., Hambach, C., Windeck, T., and Heesemann, J., *Toxoplasma gondii*: uptake of fetuin and identification of a 15-kDa fetuin-binding protein. *Parasitol. Res.,* 79,191–194, 1993.

36. Motta, M.C., Soares, M.J., and de Souza, W., Intracellular lectin binding sites in symbiont-bearing *Crithidia* species, *Parasitol. Res.,* 79, 551–558, 1993.

37. Matta, M.A., Aleksitch, V., Angluster, J., Alviano, C.S., De Souza, W., Andrade, A.F., and Esteves, M.J., Occurrence of *N*-acetyl- and *N*-*O*-diacetyl-neuraminic acid derivatives in wild and mutant *Crithidia fasciculate, Parasitol. Res.,* 81, 426–433, 1995.

38. do Valle Matta, M.A., Sales Alviano, D., dos Santos Silva Couceiro, J.N., Nazareth, M., Meirelles, L., Sales Alviano, C., and Angluster, J., Cell-surface sialoglycoconjugate structures in wild-type and mutant *Crithidia fasciculate, Parasitol. Res.,* 85, 293–299, 1999.

39. Miller, L.H., Hudson, D., and Haynes, J.D., Identification of *Plasmodium knowlesi* erythrocyte binding proteins, *Mol. Biochem. Parasitol.,* 31, 217–222, 1988.

40. Orlandi, P.A., Klotz, F.W., and Haynes, J.D., A malaria invasion receptor, the 175-kilodalton erythrocyte-binding antigen of *Plasmodium falciparum* recognizes the terminal Neu5Ac (alpha 2-3) Gal-sequences of glycophorin A, *J. Cell Biol.* 116, 901–909, 1992.

41. Ravindranath, M.H., Higa, H.H., Cooper, E.L., and Paulson, J.C., Purification and characterization of an *O*-acetylsialic acid-specific lectin from a marine crab *Cancer antennarius, J. Biol. Chem.,* 260,8850–8856, 1985.

42. Ritter, G., Ritter-Boosfeld, E., Adluri, R., Calves, M., Ren, S., Yu, R.K., Oettgen, H.F., Old, L.J., and Livingston, P.O., Analysis of the antibody response to immunization with purified *O*-acetyl GD3 gangliosides in patients with malignant melanoma, *Int. J. Cancer.,* 62,668–672, 1995.

43. Sen, G. and Mandal, C., The specificity of the binding site of Achatinin-H a sialic-acid binding lectin from *Achatina fulica, Carbohydr. Res.,* 268,115–125, 1995.

44. Mandal, C., Basu, S., and Mandal, C., Physicochemical studies on Achatinin-H, a novel sialic acid binding lectin, *Biochem. J.,* 257, 65–71, 1989.

45. Basu, S., Mandal, C., and Allen, A.K., Chemical modification studies of a unique sialic acid -binding lectin from the snail *Achatina fulica. Biochem. J.,* 254, 195–202, 1988.

46. Mandal, C. and Basu, S., A unique specificity of a sialic acid binding lectin Achatinin-H, from the haemolymph of *Achatina fulica* snail, *Biochem. Biophys. Res. Commun.,* 148,795–801, 1987.

47. Fragkiadakis, G.A. and Stratakis, E.K., The lectin from the crustacean *Liocarcinus depurator* recognizes *O*-acetylsialic acids, *Comp. Biochem. Physiol.,* 117, 545–552,1997.

48. Maghil Denis, P.D., Mercy Palatty, N., Bai, R., Jeya Suriya, S., Purification and characterization of a sialic acid specific lectin from the hemolymph of the freshwater crab *Paratelphusa jacquemontii,Eur. J. Biochem.* 270,4348–4355, 2003.

49. Ahmed, H. and Gabius, H.J., Purification and properties of a Ca^{2+}-independent sialic acid-binding lectin from human placenta with preferential affinity to *O*-acetylsialic acids, *J. Biol. Chem.,* 264,18673–18678, 1989.

50. Iglesias, M.M., Cymes, G.D., and Wolfenstein-Todel, C., A sialic acid-binding lectin from ovine placenta: purification, specificity and interaction with actin, *Glycoconjugate J.,* 13,967–976, 1996.

51. Zimmer, G., Klenk, H.D., and Herrler, G., Identification of a 40-kDa cell surface sialoglycoprotein with the characteristics of a major influenza C virus receptor in a Madin–Darby canine kidney cell line, *J. Biol. Chem.,* 270, 17815–17822, 1995.

52. Herrler, G., Gross, H.J., and Brossmer, R., A synthetic sialic acid analog that is resistant to the receptor-destroying enzyme can be used by influenza C virus as a receptor determinant for infection of cells. *Biochem. Biophys. Res. Commun.,* 216, 821–827, 1995.

53. Zimmer, G., Reuter, G., and Shauer, R., Use of influenza C virus for detection of 9-*O*-acetylated sialic acids on immobilized glycoconjugates by esterase activity, *Eur. J. Biochem.,* 204, 209–215, 1992.

54. Muchmore, E.A. and Varki, A., Selective inactivation of influenza C esterase: a probe for detecting 9-*O*-acetylated sialic acids, *Science,* 236, 1293–1295, 1987.

55. Klein, A., Krishna, M., Varki, N.M., and Varki, A., 9-*O*-acetylated sialic acids have widespread but selective expression: analysis using a chimeric dual-function probe derived from influenza C hemagglutinin-esterase, *Proc. Natl. Acad. Sci. USA.,* 91,7782–7786, 1994.

56. Rogers, G.N., Herrler, G., Paulson, J.C., and Klenk, H.D., Influenza C virus uses 9-*O*-acetyl-*N*-acetylneuraminic acid as a high affinity receptor determinant for attachment to cells, *J. Biol. Chem.,* 26, 5947–5951, 1986.

57. Vlasak, R., Krystal, M., Nachat, M., and Palese, P., The influenza C virus glycoprotein (HE) exhibits receptor-binding (hemagglutinin) and receptor-destroying (esterase) activities, *Virology,* 160, 419–425, 1987.

58. Vlasak, A., Luytjes, W., Spaan, W., and Palese, P., Human and bovine coronaviruses recognize sialic acid-containing receptors similar to those of influenza C viruses, *Proc. Natl. Acad. Sci. USA.,* 85, 4526–4529, 1988.

59. Schauer, R. and Kamerling, J. P., Chemistry, biochemistry and biology of sialic acids, In: *Glycoproteins,* Montreuil, J., Vliegenthart, J. F. G., Schachter, H., Eds., Elsevier, Amsterdam, 1997, pp. 243–402.

60. Reuter, G., Pfeil, R., Stoll, S., Schauer, R., Kamerling, J.P., Versluis, C., and Vliegenthart, J.F.G., Identification of new sialic acids derived from glycoprotein of bovine submandibular gland, *Eur. J. Biochem.,* 134, 139–143, 1983.

61. Mandal, C., Sinha, D., Sharma, V., and Bhattacharya, D.K., *O*-acetyl sialic acid binding lectin as a probe for detection of subtle change on cell surface induced during acute lymphoblastic leukemia (ALL) and its clinical application, *Indian J. Biochem. Biophys.,* 34, 82–86,1997.

62. Ravindranath, M.H., Paulson, J.C., and Irie, R.F., Human melanoma antigen *O*-acetylated ganglioside GD3 is recognized by *Cancer antennarius lectin. J. Biol. Chem.,* 263,2079–2086, 1988.

63. Ravindranath, M.H., Morton, D.L., and Irie, R.F., An epitope common to gangliosides *O*-acetyl-GD3 and GD3 recognized by antibodies in melanoma patients after active specific immunotherapy, *Cancer Res.,* 49, 3891–3897, 1989.

64. Shukla, A.K. and Schauer, R., Fluorimetric determination of unsubstituted and 9(8)-*O*-acetylated sialic acids in erythrocyte membranes, *Hoppe-Seylers', Physiol. Chem.* 363, 255–262, 1982.

65. Hunter, W.M., in *Handbook of Experimental Medicine,* Weir, D.M., Ed., Blackwell Scientific Publication, Oxford: 1978, pp. 14.1–14.3.

66. Scatchard, G., The attractions of proteins for small molecules and ions, *Ann. NY. Acad. Sci.,* 51, 660–672, 1949.

67. Chatterjee, M., Sharma, V., Mandal, C. Sundar, C, S., and Sen, S., Identification of antibodies directed against *O*-Acetylated sialic acids in visceral leishmaniasis: its diagnostic and prognostic role, *Glycoconjugate J.,* 15,1141–1147, 1998.

68. Sharma, V., Chatterjee, M., Sen, G., Chava, A.K., and Mandal, C., Role of linkage specific 9-*O*-acetylated sialoglycoconjugates in activation of the alternative complement pathway in mammalian erythrocytes. *Glycoconjugate J.,* 7, 887–893, 2000.

69. Mawhinney, T.P., and Chance, D.L., Hydrolysis of sialic acids and *O*-acetylated sialic acids with propionic acid, *Anal. Biochem.,* 223, 164–167, 1994.

70. Hara, S., Yamaguchi, M., Takemori, Y., Furuhata, I.K., Ogura, H., and Nakamura, M., Determination of mono-*O*-acetylated *N*-acetylneuraminic acid in human and rat sera by fluorimetric high performance liquid chromatography, *Anal. Biochem.,*179, 162–166, 1989.

71. Kelm, S. and Schauer, R., Sialic acids in molecular and cellular interactions, *Inter. Rev. Cytol.,* 175, 135–240, 1997.

72. Klein, A. and Roussel, P., *O*-acetylation of sialic acids, *Biochimie.,* 80, 49–57, 1998.

73. Nydegger, U.W., Fearon, D.T., and Austen, K.F., Autosomal locus regulates inverse relationship between sialic acid content and capacity of mouse erythrocytes to activate human alternative complement pathway, *Proc. Natl. Acad. Sci. USA,* 75, 6078–6082, 1978.

74. Fearon, D.T., Activation of the alternative complement pathway, *CRC Crit. Rev. Immunol.,* 1, 1–32, 1979.

75. Kazatchkine, M.D., Fearon, D.T., and Austen, K.F., Human alternative complement pathway: membrane-associated sialic acid regulates the competition between B and beta1 H for cell-bound C3b, *J. Immunol.,* 122, 75–81, 1979.

76. Meri, S. and Pangburn, M.K., Discrimination between activators and nonactivators of the alternative pathway of complement: regulation via a sialic acid/polyanion binding site on factor H, *Proc. Natl. Acad. Sci. USA,* 87, 3982–3986, 1990.

77. Varki, A. and Kornfeld, S., An autosomal dominant gene regulates the extent of 9-*O*-acetylation of murine erythrocyte sialic acids. A probable explanation for the variation in capacity to activate the human alternate complement pathway, *J. Exp. Med.,* 152, 532–544, 1980.

78. Shi, W.X., Chammas, R., Varki, N.M., Powell, L., and Varki, A., Sialic acid 9-*O*-acetylation on murine erythroleukemia cells affects complement activation, binding to I-type lectins, and tissue homing, *J. Biol. Chem.,* 271, 31526–31532, 1996.

79. Herwaldt, B.L., Harrison's Principles of Internal Medicine, in *Infectious Diseases, Protozoal Infections.* Fauci S, Braunwald E, Isselbacher KJ, Wilson JD, Martin JB, Kasper DL, Hauser SL, Longo DL, Eds., 14th ed., McGraw-Hill Publishers, New York, 1998, pp. 1176–1205.

80. Shi, W.X., Chammas, R., Varki, A., Linkage-specific action of endogenous sialic acid *O*-acetyltransferase in Chinese hamster ovary cells. *J. Biol. Chem.,* 271, 15130–15138, 1996.

4 Congenital Disorders of Glycosylation

Jerzy Kościelak

CONTENTS

I.   INTRODUCTION

Congenital disorders of glycosylation (CDG) are a group of inherited, multisystemic diseases caused by deficiencies of key enzymes required for the biosynthesis of *N*- and *O*-linked glycans of glycoproteins and proteoglycans. Other glycoconjugates including glycosylphosphatidylinositol (GPI) anchors may also be affected. An aberrant biosynthesis of glycosphingolipids, another type of major glycoconjugates of mammalian cells, has not been found so far to cause human disease. From the standpoint of both medicine and science CDG are exceptionally interesting because of their novelty and intriguing pathogenesis, which allow an insight into previously unrecognized functions of carbohydrates. These diseases also signify the importance of glycosylation for human health and show pleiotropic function of genes in action. CDG may be also looked upon as an experiment in carbohydrate engineering performed by nature itself.

II.   BIOSYNTHETIC PATHWAYS IN CDG PATHOGENESIS

CDG history first began with a clinical description of children with severe developmental abnormalities including muscle weakness, mental retardation, dysmorphic features, and feeding difficulties.[1] It took several years to realize that in almost all patients many blood plasma glycoproteins were underglycosylated.[2,3] With the clinical picture often heterogenous, and the underlying molecular defect unresolved, the disease was initially known as carbohydrate-deficient glycoprotein syndromes (CDGS). Once the knowledge of the pathogenesis of CDGS had increased, the diseases were renamed as CDG, and divided into two groups[4,5] (see Table 4.1). Diseases of group I are defined as those resulting from a defective biosynthesis of lipid (dolichol) linked oligosaccharide (LLO) prior to the transfer of the oligosaccharide in *N*-glycosidic linkage onto asparagine residues of proteins. Disorders of group II include those caused by the malfunction of processing reactions. Thus, group II is more heterogenous, and comprises diseases of diverse

TABLE 4.1

Congenital Disorders of Glycosylation

Type	Gene Defect	OMIM	Enzyme or Function Affected
Ia	*PMM 2*	212065 601785	Phosphomannomutase
Ib	*PMI*	602579 154550	Phosphomannoisomerase
Ic	*hALG 6*	603147 604565 604566	α3-Glucosyltransferase I
Id	*hALG 3*	601110	α3-Mannosyltransferase
Ie	*DPM1*	603503	Dol-P-Man Synthase
If	*LEC 35*	604041	Dol-P-Man utilization
Ig	*hALG 12*	607143 607144	α6-Mannosyltransferase
Ih	*hALG 8*	608104	α3-Glucosyltransferase II
Ii	*hALG 2*	607905 607906	α3-Mannosyltransferase
Ij	*DPAGT1*		GPT
Ix	unknown	603585 212067	
IIa	*MGAT 2*	202066 602616	β2-*N*-acetylglucosaminyl- transferase II
IIb	*GCS1*	601336 606056	Glucosidase I
IIc	*GDP-Fuc* *transporter*	266265 605881	Import of GDP-Fuc into Golgi
IId	*β4GalT1*	607091 137060	β4-Galactosyltransferase I (β4-GalT1), (Gal→GlcNAc)
IIe			Sialic acid transporter
Progeroid variant, Ehlers–Danlos syndrome	*XGPT*	130070 604327	β4-Galactosyltransferase I (β4-GalT7), (Gal→xylose)
Hereditary multiple exostoses	*EXT1* *EXT2*	133700 133701	β4GlcAT & α-4-GlcNAcT
Galactosemia I	*GalT*	230400 606999	Gal-1-P uridyl transferase
Galactosemia I	*GalE*	230350 606953	Galactose epimerase
Galactosemia II	*GalK1*	604313 230200	Galactokinase

anabolic backgrounds, affecting not only *N*-linked but also *O*-linked glycosylation, import or export reactions, and so on. Letters a, b, c..... in both groups refer to the chronological order in which the diseases were recognized. Until the stage of

Man$_4$GlcNAc$_2$-PP-Dol the biosynthesis of LLOs occurs at the cytosolic side of the endoplasmic reticulum (ER) (see Figure 4.1). However, at the next biosynthetic step, LLO flip-flops to the ER lumen[6] wherein GDP-Man is unable to enter. Thus, in all the subsequent reactions, Dol-P-Man takes over as a mannose donor. The last step in the biosynthesis is the sequential addition of three glucose residues, each by a different glucosyltransferase.[7,8] The presence of those three glucose residues in LLO is regarded as essential for efficient transfer of the oligosaccharide to a nascent protein. The reaction, catalyzed by the heterooligomeric enzymatic complex known under the collective name of oligosaccharyl transferase,[9] is accompanied by the sequestration of the newly formed glycoprotein into the ER lumen. Still in the ER, the glycoprotein is freed of two glycosyl residues, and subsequently it binds to lectin-like molecules with chaperone functioning as a membrane-bound calnexin or soluble calreticulin. At this stage, the bound glycoprotein attains its proper conformation, and the last glucose residue is removed. Subsequently, the glycoprotein is released from the complex with the chaperones, packed into transport vesicles, and sent to the Golgi where the biosynthesis is completed. If the conformation is incorrect, the glycoprotein may be reglucosylated in the ER lumen by UDP-glucose:glycoprotein glucotransferase, thus allowing for its rebinding to chaperones, and another attempt at

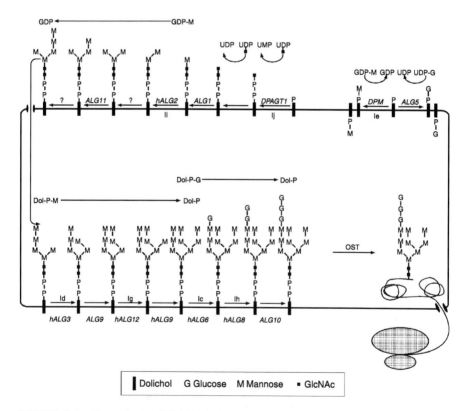

FIGURE 4.1 Biosynthesis of lipid-linked oligosaccharides. (Adapted from Aebi, M. and Hennet, T., Trends cell. Biol., 11, 136–141, 2001.)

the adoption of the proper conformation. Irreversibly misfolded glycoproteins are degraded. This process has been aptly named the quality control. En route to the Golgi, N-linked glycans may be trimmed by one or more mannose residues to form oligomannose (high mannose) or complex and hybrid types of glycans.[10] The oligo-mannosidic glycans may acquire lysosomal targeting signals.[11] Otherwise, more mannose residues are cleaved, and at the stage of five mannose residues, the N-linked glycan is acted upon by β2N-acetylglucosaminyltransferase I (GnT-I), which links GlcNAc residue to the 3′ mannose arm of the core[12] (see Figure 4.2). This modification gives a signal to mannosidase II to cleave two peripheral mannose residues from the 6′ mannose arm of the core, thereby providing space for β2N-acetylglucosaminyltransferase II (GnT-II) to incorporate the second residue of GlcNAc. Thus, GnT-I is a gateway for the action of all the subsequent glycosyl transfer reactions. Addition of galactose residues, and thereafter of sialic acid or fucose residues, with the latter occurring in the distal or *trans*-Golgi, completes a simple N-linked complex type of structure. In contrast to the N-linked glycosylation, which starts at the cytosolic side of the ER, the biosynthesis of O-linked chains is initiated in the Golgi by the addition of GalNAc to threonine or serine residues of proteins. The reaction occurs throughout the Golgi stack.[13] Glycosaminoglycans (GAGs) of the chondroitin or dermatan sulfate and heparan sulfate type are O-linked to core proteins by a common GlcAβ1-3Galβ1-3Galβ1-4Xylβ1-O-Ser tetrasaccharide.[14,15] Addition of xylose occurs at the inner surface of the ER, and is mediated by the UDP-xylose:polypeptide xylosyltransferase. The two galactosyl residues are subsequently attached stepwise by two distinct galactosyltransferases I and II, respectively, followed by the addition of a residue of glucuronic acid by glucuronyl-transferase I (GlcAT-I). In case of HS, a subsequent chain polymerization is initiated once a residue of αGlcNAc is added by the action of α4-GlcNAc transferase I encoded by gene *EXTL2* (*EXT2*-like)[16] (see below), while the addition of a residue of β-GalNAc triggers CS/DS biosynthesis.[17] GAG chains of some proteoglycans such as HS are membrane-bound through their protein cores, which are either trans-membrane proteins or are linked to glycosylphosphatidylinositol (GPI) anchors.[18,19] The minimum conserved GPI-core structure is a tetrasaccharide containing three mannose residues, and a single residue of glucosamine flanked on the nonreducing end by peptide-ethanolamine-PO$_4$, and on the reducing end by a myo-inositol residue of phosphatidylinositol: -6Manα1-2Manα1-6Manα1-4GclN-. The anchors

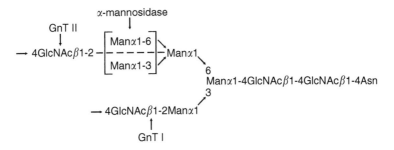

FIGURE 4.2   Initial steps of complex/hybrid glycans biosynthesis.

are molecules that have most likely evolved for attachment of glycoproteins and proteins to the apical cell surface and thereby provide them with a high degree of lateral mobility and a potential for rapid release through the action of phospholipases. *Trypanosoma* use this mechanism for defense against immune reactions directed at their GPI-anchored surface antigens. The GPI anchors are highly conserved in eukaryotic organisms. In mammals, they link to cell surface enzymes and molecules with adhesive, signaling, and regulatory functions. The examples include complement regulatory proteins and heparan sulfate proteoglycans, i.e., glypicans.[20]

Glycans attached to protein by the *O*-mannosidic linkage in the brain were described almost 25 years ago,[21] but their participation in the structure and function of dystroglycan of muscles and neuromuscular junctions has been described more recently.[22, 23] In rabbit brain, the *O*-linked mannosyl residues are monosubstituted at position 2 with GlcNAcβ1- or Galβ1-4GlcNAcβ1- or Galβ1-4 [Fuca1-3] GlcNcβ1- or NeuAcα2-3Galβ1-4GlcNacβ1- chains.[24] Otherwise, the mannose residues may be doubly substituted at positions 2 and 6 with NeuAcα2-3Galβ1-4GlcNAcβ1- chains or with a combination of NeuAcα2-3Galβ1-4 GlcNAcβ1-and Galβ1-4GlcNAcβ1- chains. The biosynthesis of GPI anchors, protein *O*–mannosyl, and *C*-mannosyl linkages requires activated mannose residues from Dol-P-Man.[25, 26]

Glycosylation, sulfation, and phosphorylation of glycoconjugates in the Golgi depend on a supply of appropriate nucleotide phosphate donors. The latter are, however, synthesized in the cytoplasm and must be imported from there to the Golgi lumen.[27] The import reactions are catalyzed by specific transporters, and are linked with the export of appropriate nucleoside monophosphates.

III. TYPES OF CDG

A. CDG-I

In CDG type I, underglycosylation of *N*-linked glycans may be due to an inefficient biosynthesis, or utilization of mannose donors, or a malfunction of transfer reactions. The disorders are genetically determined but, for the sake of brevity, the nature of mutations as well as their chromosomal locations will be discussed only briefly. For more comprehensive information the reader is referred to accession numbers for the Online Mendelian Inheritance in Man (OMIM), listed in Tables 4.1 and 4.2.

1. CDG-Ia and Ib

Deficiencies of phosphomannose mutase (PMM)[28, 29] and phosphomannoisomerase (PMI)[30–32] are responsible for CDG types Ia and Ib, respectively. Incidentally, the *PMM2* gene and its product, the PMM enzyme (see Table 4.1) were identified in mammals during the course of studies on the molecular basis of CDG. The PMM enzyme is obviously more important than the PMI and mannose kinase (both contribute to the formation of Man-6-P), because its diminished activity produces a bottleneck on the biosynthetic route to Man-1-P and GDP-Man (see Figure 4.3). The latter compound provides directly or indirectly activated mannose residues to LLO and other important reactions implicated in the pathogenesis of CDG. Thus,

TABLE 4.2
Congenital Muscular Dystrophies

Condition	Gene Defect	OMIM	Glycosy- lation of α- Dystroglycan	Function Affected	Clinical Manifestations
Walker–Warburg Syndrome (WWS)	POMTI	607423	Absent	O-mannosyl transferase	Hypotonia, mental retardation, cerebral and cerebellar hypoplasia, ocular abnormalities, and neural migration disorder
Fukuama type (FCMD)	Fukutin	253800 607440	Absent	Glycosyl- transferase?	Resemble WWS but less severe
Muscle–eye –brain disease (MEB)	POMGnTI (MGAT1.2)	253280 606822	Absent	Protein α- mannoside: β-N-acetylglu- cosaminyl- transferase	Resemble FCMD but with more pronounced ocular abnormalities
MDC1C[a] and LGMD2I[b]	FKRP	606612 607155	Absent or reduced	Glycosyl- transferase?	Hypotonia affecting proximal muscles, normal or reduced intelligence
Nonaka distal myopathy with rimmed vac- uoles (DMRV) and inclusion body myopathy (IBM2)	GNE	605620 600737	Presumably hyposia- lylated	Sialic acid biosynthesis[c]	Hypotonia of distal muscles in DMRV and of both distal and proximal mus cles in IBM2; rimmed vacuoles in muscle fibers in both condition, incl- usions only in IBM2
Myodystrophy mouse (myd)	LARGE	603590	Absent	Glycosyl- transferase?	Hypotonia, neuronal migration disorder

[a]Muscular dystrophy congenital 1c.

[b]Limb girdle muscular dystrophy 2I.

[c]UDP-N-acetylglucosamine-2-epimerase/N-acetylmannosamine kinase.

CDG-Ia is a more severe form of CDG, with the affected patients usually being unable to walk, and very often, to speak. Mental retardation, dysmorphism, muscular hypotonia, cerebellar hypoplasia, neuropathy, coagulopathy, liver dysfunction, inverted nipples, and esotropic strabismus are common clinical features in these patients. Nevertheless, the patients are cheerful and their disposition is pleasant. On the other hand, patients with CDG-Ib suffer primarily from a protein-losing enteropathy, diarrhea, and hypoglycemia, but do not show either neurological

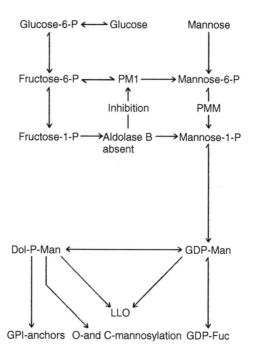

FIGURE 4.3 Simplified scheme of the biosynthesis of mannose donors.

abnormalities or mental retardation. It should be pointed out that the underglycosylation of proteins might involve either whole glycans or only their nonreducing ends. In the first instance, we speak of unglycosylation or nonglycosylation of potential glycosylation sites of glycoproteins and, in the second instance, of hypoglycosylation. A partial or total unglycosylation[33,34] is characteristic of many plasma glycoproteins in patients with CDG-Ia. The nonoccupancy of glycosylation sites in α1-antitrypsin is not random with a decreased glycosylation capacity and depends on structural features at each site.[35] It is, however, random in transferring.[33, 34]

2.   CDG-Ic

Clinically, patients with CDG type Ic resemble those with milder forms of CDG-Ia.[36–38] The patients are mentally retarded and have recurrent seizures, yet they do not display cerebellar hypoplasia, body dysmorphism, and strabismus. This particular syndrome is caused by a deficiency of α3-glucosyltransferase encoded by the *hAlg6* gene, which adds the first glucose to LLO. Mutations within the gene encoding this enzyme are leaky, and variable amounts of fully glucosylated LLO are synthesized.

3.   CDG-Id

Patients present severe neurological and psychomotor abnormalities, hypotonia, seizures, and optic nerve atrophy.[39] The disease, found so far only in a few patients, is due to a deficiency of α3-mannosyltransferases, which attaches the 6th Man residue from Dol-P-Man to $Man_5GlcNAc_2$-PP-Dol. The deficiency is never complete and a certain amount of fully glycosylated LLO is synthesized.

4. CDG-Ie

Dol-P-Man from GDP-Man is synthesized by dolichyl-mannose synthase (DPM), an enzyme consisting of three units, of which DPM1 is catalytic, and one (DPM2) is attached to the GPI anchor.[40] Mutations within the gene encoding DPM1 are responsible for a rare variant of CDG type Ie.[41, 42] Enzyme deficiency is only partial. In a few patients, it was reported that the disease was manifested by psychomotor retardation, muscle weakness, seizures, and dysmorphism. Interestingly, in two siblings, the dysmorphism was not observed directly at birth but it developed later, in early childhood.[42]

5. CDG-If

The disease is caused by a rare defect of a human gene known as *MPDUI*.[43,44] Its equivalent in hamsters is gene *Lec35*. Both genes encode the ER-located transmembrane proteins that are necessary for the utilization but not the synthesis of Dol-P-Man and Dol-P-Glc. Consequently, the efficiency of biosynthetic reactions requiring those sugar donors is reduced. The expression of the wild MPDUI gene in the Lec35 cells restores their ability to synthesize LLO, GPI-anchors as well as C- mannosyl linkages.[26] Surprisingly, a detergent treatment or a physical breakage of the Lec35 cells have a similar effect, suggesting that the Lec35 protein controls the organization of mannose-P-dolichol and glucose-P-dolichol in the ER. However, the precise mode of action of the MPDUI or Lec35 proteins is unknown. Clinically, the syndrome is characterized by delayed psychomotor development, muscular hypotonia, cerebral atrophy, blindness, dry skin, frequent vomiting, and seizures.[43,44]

6. CDG-Ig

This rare disease results from mutations in the *hAlg12* gene encoding dolichol-P-Man: $Man_7GlcNAc_2$-PP-Dol α 6-mannosyltransferase of the ER.[45–47] The mutated gene is leaky and some amount of normal $Glc_3Man_9GlcNac_2$ oligosaccharide is transferred to protein along with a $Glc_3Man_7GlcNAc_2$ variety. Surprisingly, the $Man_7GlcNAc_2$ oligosaccharide without any glucose residue is also transferred. At the red cell level, band 3 (anion transporter) has two types of glycans: one with only a slightly reduced number of carbohydrate residue, but with a full complement of three mannose residue and the other one with three residues of *N*-acetylglucosamine and four residues of mannose.[48] Clinically, the affected patients suffer from severe psychomotor retardation, facial dysmorphism, convulsions, and microcephaly.

7. CDG-Ih

This type of syndrome was identified in only one subject.[49] The patient, a 4-month-old infant, presented a protein-losing entereopathy, coagulopathy, and moderate hepatomegaly. There was no neurological involvement and her psychomotor development was normal. The symptoms were similar to those observed in CDG type Ib.

The underlying biochemical defect involved addition of the second glucosyl residue to LLO, a reaction catalyzed by the product of the *hAlg8* gene. In the presence of castanospermin, the patient's cells had accumulated monoglucosylated but otherwise fully mannosylated LLO. Northern-blot analysis showed that the cells contained only 10 to 20% of the normal amount of Dol-P-Glc:Glc$_1$Man$_9$GlcNAc$_2$-PP-Dol α3-glucosyltransferase. Mutations were identified in the two alleles of the patient's *hAlg8,* which both gave rise to premature stop codons predicted to generate severely truncated proteins. The biochemical cellular defect was successfully complemented with a wild-type *hAlg8* DNA.

8. CDG-Ii

A single patient with this type of CDG developed a multisystemic disease in the first year of life that resembled CDG-Ia with impaired mental and motor development, seizures, coloboma of the iris, and retarded myelinization.[50] Cells obtained from the patient accumulated two short LLOs with the composition Man$_1$GlcNAc$_2$-PP-Dol and Man$_2$GlcNac$_2$-PP-Dol. The mutated *hAlg2* was cloned and localized to chromosome 9p22. The normal gene encoded a protein of 416 amino acids. Two different mutations in each allele of the patient's *hAlg2* gene were identified but at the RNA level only one, D1040G, was found, derived from the mother. The mutation caused a frameshift, altering the sequence after amino acid 346, and a premature stop after amino acid 372. The missing paternal transcript was presumably unstable. Interestingly, only the maternal but not the paternal side of the family tree revealed cases of migraine, retarded psychomotor development, and seizures. It seems that the described *hAlg2* gene encodes GDP-Man:Man$_1$GlcNAc-PP-Dol α3-mannosyltransferase. The defect could be rescued by expression of the wild-type *hAlg2*. The other LLO accumulated in the patient's cells, Man$_2$GlcNAc$_2$-PP-Dol, is probably a product of as yet uncharacterized α6-mannosyltransferase. This defect is the first instance of CDG caused by the malfunction of an enzyme localized at the cytosolic side of the ER.

9. CDG-Ij

This recently identified variant of CDG is due to a deficiency of UDP-GlcNAc:Dol-P *N*-acetylglucosamine-1-P transferase (GPT) encoded by *DPAGT1*.[51] Thus far, it is the earliest reaction in the biosynthesis of LLO resulting in CDG type I. The patient's fibroblasts incorporated radioactive mannose into full-length LLO and glycoproteins at a reduced level. The cellular activity of GPT was reduced by 90%, and mutations were identified in both paternal and maternal alleles. The patient presented severe hypotonia, mental retardation, microcephaly, and exotropia.

B. CDG-II

1. CDG-IIa

This disease is the first CDG type fully explained on a molecular basis. The syndrome is due to a deficiency of *N*-acetylglucosaminyltransferase II (GnT-II) encoded by the *MGAT2* gene[53–54] (see also Figure 4.2). As a result, the second antenna on the

N-linked glycans cannot be initiated and the 6'-arm of the trimannosylcore of N-linked glycans remains free. Patients display psychomotor retardation, craniofacial dysmorphism, and stereotypic hand movements, but have no symptoms of peripheral neuropathy and cerebellar hypoplasia.

2. CDG-IIb

The disease, described so far in a single neonate, results from the markedly reduced activity of α-glucosidase (glucosidase I), which removes the distal α-linked glucose residue from the N-linked Glc_3Man_9 glycan.[55] The defect was compensated by endomannosidase, a Golgi resident enzyme, that cleaved from the N-linked glycan a Glc_3Man_1 tetrasaccharide. Consequently, approximately 80% of N-linked glycans in fibroblasts were correctly processed. The Glc_3Man_1 tetrasaccharide was excreted in urine. The condition was fatal and the child died 74 days after birth. The child's symptoms included dysmorphism, hypotonia, hepatomegaly, hypoventilation, feeding problems, and seizures.

3. CDG-IIc

CDG-IIc or the LAD II type of leukocyte adhesion deficiency is a rare syndrome that has been found in only four children of Arab,[56,57] one child of Turkish[58] and one child of Brazilian[59] descent. The disease is characterized by dwarfism, facial dysmorphism, mental retardation, neurological abnormalities, and immunological deficiency associated with an elevated leukocyte count. The affected individuals suffer from recurrent bacterial infections without pus formation. The deficiency in leukocyte adhesion concerns its first phase when neutrophils roll on the activated vascular endothelium during recruitment of these cells to the site of inflammation. Transmigration of neutrophils through the endothelium is not impaired.[60] The reverse is true for leukocyte adhesion type I caused by mutations in the gene encoding the integrin β_2 subunit.[61] The impaired rolling of neutrophils in LAD II results from hypofucosylation of glycoconjugates of neutrophils, which are ligands for the P and E selectins.[56,62] The hypofucosylation in LAD II affects glycoconjugates throughout the body including, among others, the H antigen of red cells with the resulting Bombay phenotype and Lewis antigens of the blood serum.[56, 60, 62] It has been demonstrated that hypofucosylation is due to a reduced import of GDP-Fuc from cytosol to the Golgi.[63,64] Genes encoding a defective GDP-Fuc transporter were cloned, and their normal allele was identified as the hydrophobic protein predicted to span the Golgi membrane tenfold (64, 65). Three types of homozygous mutations were identified in the GDP-Fuc transporter: the children of Arab origin showed a C–G transversion at base 923 leading to a threonine to arginine change at amino acid 308 (65, 66), the child of Turkish origin exhibited a C–T change at base 439 which produced an arginine to cysteine change at amino acid 147 (64, 65). The child from Brazil displayed a single base deletion at Gly588 producing a shift in the open reading frame after Ser195 with the introduction of 34 random amino acids followed by a stop codon (59). Addition of fucose to the culture medium restored fucosylation of glycoconjugates in fibroblasts from LAD II patients irrespective of their ethnic origin (59, 62, 67) but only the the Turkish and Brazilian patients responded to an oral fucose therapy (59, 67, 68) (see, Section V. Treatment).

4. CDG-IId

A deficiency of UDP-Gal:N-acetylglucosamine β4-galactosyltransferase I (GalT-I or β4-GalT1) has so far been identified only in a single child of Turkish origin (69). The disease, CDG-IId, manifested itself at birth with a macrocephaly due to a Dandy–Walker brain malformation and abdominal bleeding (70). During the child's further development mental retardation, myopathy and blood-clotting abnormalities became evident. The GalT-I is a member of a larger family of galactosyltransferases, all mediating transfer of Gal residues to the 4th position of GlcNAc, but differing in relative specificity to glycoproteins, glycolipids, and lactose. In the patient's fibroblasts and leukocytes, the activity of GalT-I was reduced to 5% and 9% of control, respectively. The mutated enzyme neither localized to the Golgi nor showed any activity in the lysates of the COS 7 cells overexpressing mutant GalT-I cDNA. Interestingly, the hypogalactosylation of glycoproteins clearly seen in the patient's serum was almost normal in fibroblasts, suggesting that other galactosyltransferases may compensate for the GalT-I deficiency.

5. CDG-IIe

A defective Golgi CMP-sialic acid transporter was found in two patients and has been listed as CDG-IIe.[71] One of the patients had suffered from severe thrombocytopenia, pulmonary, and gastrointestinal hemorrhages, diarrhea, and secondary infections. Both patients required frequent blood transfusions and died at the early age of 3 months and 3 years. Interestingly, glycoproteins of the blood serum from these patients were normally sialylated but lacked sialyl-Lex structures in leukocytes (R Oriol, personal communication, 2003)

C. Other Types of CDG

1. Progeroid Variant of Ehlers–Danlos Syndrome

Deficiency of galactosyltransferase I or β4-GalT7 (according to a more recent nomenclature),[72] which catalyzes the transfer of galactose to xylose residues of chondroitin/dermatan sulfate proteoglycans, leads to the progeroid form of Ehlers–Danlos syndrome.[73-77] The classical Ehlers–Danlos syndrome describes a heritable, heterogenous group of disorders of the connective tissue characterized by hypermobile joints and frequently hyperextensibility as well as thin skin.[78] The disease affects 1 in 5000 individuals. Basically, the syndrome is caused by mutations in the genes involved in fibrillar collagen metabolism, but in many patients the pathogenesis of the disease has expanded beyond the collagens and collagen-modified genes. The progeroid form of the syndrome has been reported only in a single subject. The affected individual showed an aged appearance, developmental delay, dwarfism, craniofacial disproportion, hypermobile joints, hypotonic muscles, and generalized osteopenia, but lacked other several characteristic features of progeria, including diminished subcutaneous fat, prominent scalp veins, alopecia, and joint contractures. Mental retardation was also less severe than in a classical progeria. Surprisingly, mutations in the *GALT-I* gene[76,77] affected predominantly decorin,[73-75] a small leucine-rich dermatan/chondroitin sulfate proteoglycan consisting of a single GAG chain and a core protein with M_r 36.319.[76] Decorin is a biologically active substance, essential, among others, in normal collagen fibrillogenesis[79] and myoblast migration during development.[80] A targeted disruption of decorin in mice induces an increase in collagen fibril diameters, and closely mimics the cutaneous defects found in patients with Ehlers–Danlos syndrome.[81] Interestingly, a reduced expression of the decorin gene in the patient with a progeroid variant of the Ehlers–Danlos syndrome has been associated with an increased expression of biglycan gene.[75] Biglycan is another member of the small leucine-rich dermatan/chondroitin sulfate proteoglycan family that has two GAG chains, is a positive regulator of bone growth, and binds both decorin and α-dystroglycan.[79] Deficiency of biglycan affects collagen type I fibril diameters in a way similar to that of decorin. Decorin/biglycan double-knockout mice phenotypically resemble the human progeroid variant of Ehlers–Danlos syndrome.[82] Biglycan expression is elevated also in muscles of mdx mice (a model for Duchenne muscular dystrophy),[83] suggesting that biglycan–α-dystroglycan interaction may play a role in the pathogenesis of hereditary myopathies. This interaction is mediated through GAG chains of biglycan that bind to the protein moiety of α-dystroglycan.

2. Hereditary Multiple Exostoses

Another disease of the connective tissue resulting from mutations in genes encoding putative glycosyltransferases is known as hereditary multiple exostoses.[84] The exostoses are cartilaginous caps at the ends of the long bones diaphyses. The formation of exostoses, considered to be benign tumors, starts early in infancy and continues until puberty. The disease is a hereditary autosomal dominant disorder, or sporadic, with an estimated incidence of 1 in 50,000 to 100,000. The candidate disease genes *EXT1* and *EXT2* have been mapped to chromosomes 8p24 and 11p11-12, respectively.[85, 86] In some cases, the disease has been linked to another locus *EXT3* on chromosome 19p.[87] The *EXT1* and *EXT2* genes may function as tumor suppressors, because exostosis formation at one or more *EXT* loci is often associated with a loss of heterozygosity.[85] The genes have been cloned and found to encode proteins of 746 and 718 amino acids with an overall homology of 30.9%.[88, 89] The proteins are ubiquitously expressed and have structural characteristics of glycosyltransferases. Both exhibit dual activity of β4-GlcAT and α4-GlcNAcT II transferases, and they localize to the Golgi as a highly active heterooligomeric complex of HS polymerase for chain elongation.[90,91] Mouse embryos that are homozygous for truncating mutation in *EXT1*, gene lack HS, and die during gastrulation.[92] There is another family of three *EXT*-like (*EXTL*) genes and EXTL proteins, which show homology to EXT1 and EXT2.[16,93–96] As already mentioned, the EXTL2 protein is α4-GlcNAc transferase I initiating the biosynthesis of HS/Hep and also showing the GalNAc transferase activity, whereas EXTL3 has the activity of both α4-GlcNAc I and II transferases, suggesting that it may both initiate and elongate the HS/Hep chain. These proteins have not been associated with hereditary multiple exostoses, but because of the involvement of HS in the growth and morphogenesis,[20,96] they may be potential candidate genes for yet unrecognized diseases of the connective tissue. The exact mechanism of exostosis formation is still unknown; however, it was postulated that under normal conditions, HS is a component of the negative feedback loop regulating chondrocyte proliferation and maturation. Thus, a decrease in HS level would promote the growth of exostoses.[96] A similar reasoning was applied to Simpson–Golabi–Behmel syndrome characterized by gigantism and a large number of dysmorphic features, including a distinct facial apperance, hypotonia, syndactyly, polydactyly, supernumerary nipples, and rib and vertebral abnormalities.[20,97,98] The disease is caused by a mutant X-linked gene encoding core 3 protein of membrane GPI anchored HS, i.e., glypican-3 or GPC3,[98] which may be a negative regulator of cell proliferation. The disease is not strictly a CDG variant, because the mutant gene affects the protein moiety of GPC3. Functionally, however, it may well be a variant because the deficiency of the core protein of GPC3 should decrease the HS content of cell membranes. Thus, it is tempting to speculate that the dysmorphic features in patients with CDG-I should be attributed, partly at least, to a diminished biosynthesis of the GPI anchors.

3. Galactosemia and I-Cell Disease

Three types of galactosemia are listed in Table 4.1.[99] Galactosemia is undoubtedly a CDG disease, yet it is described as a metabolic disease in most medical and biochemical textbooks. The reason is historical: galactosemia was discovered much earlier than CDG. Similarly, I-cell disease is usually and quite appropriately classified as a storage disease although it is in fact caused by the reduced activity of an anabolic enzyme (GlcNAc-phosphotransferase) responsible for the synthesis of the mannose-6-phosphate recognition marker on lysosomal enzymes.[11] Thus, the enzymes are not recognized by the specific receptor in transport vesicles, and are exported rather than directed to lysosomes. This phenomenon probably accounts for an increased level of the lysosomal enzymes in blood plasma of patients with CDG type I,[100] in whom mannosylation of the lysosomal enzymes should be reduced.

4. Congenital Muscular Dystrophies

a. Dystrophin Glycoprotein Complex

Hereditary myopathies or congenital muscular dystrophies (CMD) refer to a large group of diseases involving abnormalities of genes encoding proteins and glycoproteins of the dystrophin protein or glycoprotein complex (DPC or DGC) in myofibers (see References 101–104 for recent reviews on DGC). The complex links components of the extracellular matrix (the basal lamina) with a cytoskeletal actin (see Figure 4.4). Intermediaries are α- and β-dystroglycans and dystrophin. The latter also binds to α-dystrobrevin, and indirectly to syntrophins, nNOS (neural isoform of nitric oxide synthase), and aquaporin-4. Within the sarcolemma, the complex is associated with sarcoglycans and sarcospan. Each myofiber, i.e., a multinucleated cell formed through the fusion of myoblasts during embryonal development, is ensheathed by the basal lamina.[103] At the neuromuscular junction, dystrophin is replaced by utrophin,

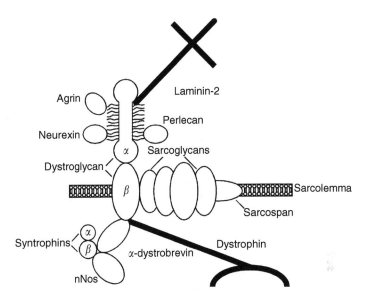

FIGURE 4.4   Dystrophin–glycoprotein complex.

a homologue of dystrophin,[105] and laminin 2 is replaced by laminins 4, 9 and 11.[106] Otherwise, DGC at both locations has a similar overall structure. The expression of utrophin at the neuromuscular junction is independent of that of dystrophin along the sarcolemma membrane. There are mutant mice that have utrophin at the junctions but lack dystrophin along the extrasynaptic sarcolemma[107] and, vice versa, mice lacking utrophin at the junction and a normal dystrophin and the associated glycoproteins at the extrasynaptic sarcolemma.[108,109] In addition, the neuromuscular junction contains glycans with terminal β-GalNAc residues.[110, 111] α- and β- Dystroglycans are periph- eral and transmembrane proteins respectively that arise by proteolytic cleavage of the dystroglycan precursor.[112] The α-dystroglycan is heavily glycosylated (in ~50%) pre- dominantly with O-mannose-linked chains with structures that are described in sec- tion II. Hence, the α-dystroglycan migrates on electrophoretograms as a broad band with an apparent molecular mass of 120 to 154 kDa, depending on the tissue of origin (the predicted molecular mass is ~74 kDa).[101–104] The oligosaccharide chains of α-dystroglycan are noncovalently linked to G domains on laminin-2, agrin, perlecan, and neurexin.[113] Apart from its function as an adhesive glycoprotein, laminin is known for its role in basement-membrane assembly, tissue organization, and cell survival.[114] Agrin and perlecan are basement-membrane-associated heparan sulfate proteoglycans, with the former being an organizer of postsynaptic differentiation at the neuromuscu- lar junction.[115,116] Neurexin is a neuron-specific protein receptor, which along with neuroligin forms an intercellular adhesion complex also involved in postsynaptic assembly and differentiation.[117–119] Dystroglycans are implicated in signal transduction involving tyrosine phosphorylation[120] and a number of signaling pathways.[121–126] Summarizing, the dystrophin-associated glycoprotein complex is an integral part of the myofiber and neuromuscular junction, which provides both mechanical stabilization to the sarcolemma and a scaffold for signaling proteins. Members of the complex are locked in position by dystroglycans. Given this pivotal role of dystroglycans in the development and function of the nervous and muscular systems, it is not surprising that null mutants of dystroglycan are lethal in mice,[127] and that any disturbance of the DGC has severe consequences on health. The most common disturbance concerns dystrophin, a member of the β-spectrin/α-actinin family.[128] A complete absence or a low level of this protein, primarily in the skeletal muscles, causes X-linked Duchenne/Becker-type muscular dystrophy manifested by proximal muscle weakness, hypotonia, and mental retardation. Proximal muscle weakness also occurs in less severe, autosomally transmitted, disorders known as limb girdle muscular dystrophies (LGMD). This type of dystrophy is due to mutations, among others, in genes encoding sarcoglycans[129, 130] and calpain 3.[131] On the other hand, a loss of merosin (α-laminin isoform in the skeletal muscles) results in a severe form of CMD.[132,133]

b. Muscular Dystrophies Caused by Presumed Underglycosylation of α-Dystroglycan

Recent discoveries strongly suggest that the hypoglycosylation or unglycosylation of α-dystroglycan in muscles and the brain result in a number of rare diseases (see reviews in References 134–136) listed in Table 4.2. Experimental evidence for the causative role of the α-dystroglycan underglycosylation is based mainly on a reduced immunostaining of α-dystroglycan in tissue biopsies or on Western blots with monoclonal antibodies against total α-dystroglycan,[137,138] its glycosylated epitope,[139] but not with those against the core protein.[137] The hypoglycosylation on Western blots is also manifested by a decreased apparent molecular mass of α-dystroglycan. These findings are supported by the evidence of mutations in genes encoding putative glycosyltransferases implicated in the biosynthesis of the carbohydrate portion of α-dystroglycan, but not in those affecting its protein moiety.

i. Walker–Warburg Syndrome, Fukuyama-Type Congenital Muscular Dystrophy and, Muscle–Eye–Brain Disease

These three muscular dystrophies all share important pathognomonic features including brain malformation with a neuronal migration defect and ocular abnormalities. These symptoms are emulated in mice with a selectively deleted α-dystroglycan in the brain.[140] A striking symptom of the defective neuronal migration is lissencephaly, i.e., a smooth brain appearance, occurring in patients with Walker-Warburg Syndrome (WWS) and muscle–eye–brain disease (MEB). All three diseases may be due to a loss of enzymatic activity involved in the early glycosylation steps of dystroglycan oligosaccharide chains. Thus, about 20% of WWS cases are caused by a deficiency of POMT1,[141] a putative O-mannosyltransferase, identified on the basis of its sequence similarity with the family of mannosyltransferases of *Saccharomyces cerevisiae*.[142] Both the disease and the putative mannosyltransferase map to the same chromosomal location 9q34.1, and a number of mutations in this locus have been characterized.[141] In other WWS cases, however, null mutations were found in the fukutin gene. The fukutin gene is responsible for hypoglycosylation of α-dystroglycan in Fukuyama-type congenital muscular dystrophy (FCMD).[143,144] The fukutin gene, localized to chromosome 9q31, encodes a protein (fukutin), which may be also a glycosyltransferase. The evidence is based upon the localization of fukutin protein to the Golgi[143] as well as upon its amino acid sequence similarity to bacterial and yeast proteins involved in modification of surface molecules.[145] There is also evidence that fukutin is a secreted protein,[143] but its function is still unknown. The third severe CMD with a neuronal migration disorder is MEB. It is caused by mutations in the gene encoding N-acetylglucosaminyltransferase (POMGnT1), which in terms of the primary sequence, is related to GnT-I but whose specificity is directed primarily to protein-linked O-mannoside.[146,147] Simple α-mannosides are also substrates, albeit at high concentrations.[147] The *POMGnT1* gene maps to the same chromosomal location (1p32-34) as the gene for MEB.[148]

ii. Congenital Muscular Dystrophy (MDC1C) and Limb Girdle Muscular Dystrophy 2I (LGMD2I)

A fukutin-related protein (FKRP) is encoded by the *FKRP* gene, whose mutations are responsible for allelic disorders MDC1C and LGMD2I.[149,150] The *FKRP* and *LGMD2I* map to the same chromosomal location, 19q13.3.[151] The FKRP protein shows structural similarity to fukutin, has putative transmembrane, stem, and catalytic domains, DxD motif, and has been localized to the medial Golgi.[152] Evidence was presented that in MDC1C the mutated FKRP (glycosyltransferase?) may not be targeted to the Golgi. Clinically, the patients with MDC1C and LGMD2I present milder symptoms than those with FCMD.[153,154]

iii. Hereditary Inclusion Body and Nonaka Myopathy

These are allelic diseases that most likely arise from hyposialylation of α-dystroglycan.[155–157] Responsible for the disease is a reduced activity of UDP-N-acetyl-glucosamine-2-epimerase or N-acetyl-mannosamine kinase (GNE), a bifunctional enzyme synthesizing mannose 1-P from N-acetylglucosamine. Interestingly, mutations of the same enzyme produce sialuria, a disease characterized by the overproduction of sialic acid.[158] The overproduction is caused by a dominant mutation of *GNE* resulting in a loss of its feedback inhibition. On the other hand, recessive mutations reduce GNE activity. A total inactivation of the *GNE* genes in mice is lethal.[159] The same is likely to apply to humans, because no patients with either HIBM or Nonaka myopathy have been described with homozygous truncating mutations in the *GNE* gene.[135]

iv. Mouse Myodystrophy (myd)

A mouse *LARGE* gene causes myodystrophy in the mouse that in many ways is similar to WWS, FCMD, and MEB.[160] The gene encodes a putative glycosyltransferase. A human ortholog of the mouse gene has been identified and mapped to chromosome 22q but is yet without a disease partner.[161] Interestingly, the catalytic domain of the human gene shows a similarity to two different glycosyltransferases: in its proximal portion to α- glycosyltransferases involved in the synthesis of lipopolysaccharides[162] and in the distal portion to i-β-3N-acetylglucosaminyltransferase (iGnT), which synthesizes linear segments of polylactosaminoglycans.[163]

IV. DIAGNOSIS OF CDG

A multisystemic presentation of CDG requires the cooperation of trained physicians, biochemists, and geneticists. Clinical symptoms are sometimes straightforward and allow establishing an instant diagnosis. An illustrative example is that of the abnormal distribution of fat on the buttocks of newborn babies, which is characteristic of CDG-Ia. Subsequent laboratory tests merely confirm the initial diagnosis. In other situations, especially in novel types of CDG, the diagnosis may be difficult to establish and may require many time-consuming laboratory tests. The cause of CDG remains undiagnosed in about 20% of new cases, which are listed as CDGx. Of the many different types of CDG, the Ia type is the archetype and at the same time the most common disease in about 300 cases reported so far.[164] The next most frequent types are CDG Ic and Ib. Thus, most currently used laboratory tests have been developed to diagnose CDG-I. For other, more recently recognized types of CDG, especially those affecting muscles and the brain, apart from a few exceptions, routine biochemically oriented diagnostics is not performed. Grünewald and co-workers[164] have developed a flowchart to help diagnose CDG. Isoelectric electrofocusing (IEF) of serum transferrin should be performed first, and while paying due attention to the limitations of this technique (see Section 6.1), the disease should be preliminarily diagnosed as type I or II. In a presumed type I, the activities of PMM and PMI should be determined; when either one is found to be deficient, the result should be confirmed by mutation analysis. With normal PMM and PMI activities LLO composition and size distribution should be assessed; in case of any irregularity, relevant enzymatic tests and mutation analysis are required. When LLO is normal, the diagnosis of CDG-II or CDG-x should be considered. To diagnose CDG-II, determination of glycan structure followed by enzymatic tests and mutation analysis should be performed. The diagnosis of yet unrecognized CDG variants requires identifying and cloning a disease gene as well as complementation analysis, which consists in rescuing the affected function in the patient's cultured cells by an overexpression of a normal gene or its yeast or animal ortholog. Obviously, the success much depends on the innovative mind of the researcher. Urinary oligosaccharides should be analysed in a presumed CDG-IIb.[55]

A. ANALYSIS OF ABNORMALLY GLYCOSYLATED REPORTER GLYCOPROTEINS BY SEPARATION TECHNIQUES

The most frequently examined reporter glycoprotein in CDG is blood serum transferrin. The glycoprotein is usually separated by IEF, which is followed by immunodetection of differently glycosylated isoforms. The method was first described by Stibler[165] for identifying chronic alcohol abuse. Its diagnostic value in CDG was realized later.[166] Transferrin has two N-linked glycans mostly of the complex biantennary type, which are terminated with sialic acid residues. Thus, in the serum of healthy subjects, tetrasialotransferrin will be the most abundant isoform. In CDG type I, however, a diminished supply of normal LLO will result in the synthesis of transferrin with one or two glycans missing.[33, 34] These isoforms will have a reduced charge and show a cathodical shift in the IEF. Therefore, in CDG-I, the most abundant isoforms are di- and asialotransferrin. The analysis requires the full saturation of transferrin with iron because the loss of one iron atom causes a cathodical shift of 0.2 pH units, which is almost equal to the loss of one sialic acid residue.[167] The pattern of transferrin isoforms in CDG-I will vary to some extent depending on the nature of mutation and the disease variant. A normal serum transferrin does not, however, exclude CDG. For example, in the subjects with CDG-Ia who survive to adulthood, the pattern of transferrin may normalize.[168] In CDG-IIa, the IEF pattern shows truncated, monoantennary forms of serum transferrin with disialotransferrin being the most conspicuous band.[52] Asialo- and monosialotransferrin are the major isoforms in blood serum from patients with CDG-IId [70] but in other CDG-II diseases, including those affecting connective tissues, bones, muscles, and brain, the isoforms transferrin show a normal pattern.[169] In contrast, an underglycosylated transferrin may be found in uncontrolled fructosemia, untreated galactosemia, liver diseases, and alcoholism. Underglycosylation of transferrin in fructosemia, due to the deficiency of aldolase B, is caused by the accumulation of fructose-1-P that inhibits PMI[169] (see Figure 4.3). In untreated galactosemia, serum transferrin is increased and hyposialylated due to a defective galactosylation of its glycans.[170] Serum transferrin is also increased and hyposialylated in chronic alcohol drinkers with IEF pattern resembling CDG-I.[165, 171] A major isoform of transferrin in alcoholics is disialotransferrin, principally resulting from the loss of an entire N-linked glycan.[172,173] The exact mechanisms of the unglycosylation of transferrin in alcoholics are not fully understood but both catabolism and biosynthesis of transferrin are probably affected. Interestingly, rats which were given ethanol for 8 weeks exhibited a decreased level of dolichyl phosphate in the ER and a reduced activity of terminal glycosyltransferases in the Golgi.[174] The former finding suggests that chronic and excessive ethanol consumption may affect the level of LLO and produce CDG-I-like symptoms. It should be pointed out that habitual heavy drinkers suffer from atrophy of the cerebellum, which is usually hypoplastic in many patients with CDG. In the reviewer's opinion, this observation suggests that the cerebellum may be unduly sensitive to underglycosylation. An abnormal IEF of transferrin may be also due to the polymorphism of its protein moiety.[175] Thus, exclusion of transferrin polymorphism by the predigestion of samples with neuraminidase should be performed done before evaluation of the IEF results.[176] Apart from the IEF, other techniques are employed to detect aberrant glycosylation of serum glycoproteins in CDG including immuno-turbidimetric method, anion-exchange chromatography, HPLC, and various types of gel electrophoresis. In diagnostically difficult cases, it is advisable to assess the glycosylation status of other serum glycoproteins such as α1-antitrypsin, antithrombin, thyroxine-binding globulin, α1-acid glycoprotein, and haptoglobins. The latter glycoproteins are often absent in the serum of CDG-I patients. Sometimes, however, serum glycoproteins in CDG-Ia patients may show a normal glycosylation pattern.[176] It seems that of all other serum glycoproteins analysis, determination of transferrin isoforms is the safest and most readily interpreted diagnostic marker of CDG.[178]

B. ENZYMATIC ASSAYS

The only enzymes whose activities are routinely performed in the extracts of leukocytes or fibroblasts from CDG patients are PMM and PMI. When the PMM activity is determined, leukocytes are preferred to fibroblasts because the latter frequently show a high residual PMM activity.[164] Both PMM and PMI activities are usually determined by a colorimetric method in which the reaction products, Man-6-P from

Man-1-P and fructose-6-P from Man-6-P, respectively, are converted into glucose-6-P with the aid of phosphoglucose isomerase.[28] In the case of PMM assay the reaction is carried out in the presence of exogenous PMI and glucose-1,6-biphosphate. Glucose-6-P is subsequently oxidized by yeast glucose-6-P dehydrogenase with a concomitant reduction in NADP to NADPH and the change in the absorbance A_{340}. Recently, another technique was developed in which the two enzymatic activities are measured together in a single sample by mass spectrometric determination of the reaction products.[179] However, phosphorylated monosaccharides are not readily distinguished by mass spectrometry and have to be converted into derivatives of different molecular mass. This step can be achieved by coupling the PMM- and PMI-catalyzed reactions with the synthesis of sedoheptulose-7-P from fructose-6-P and ribose-5-P, the reaction mediated by yeast transketolase. Fructose-6-P in PMM and PMI assays was generated in a similar manner as in the spectrophotometric method.[28] Ribose-5-P was attached through pyrophosphate linkage to a reporter molecule, biotin-N-methylglycine-amino alkyl phosphate, thus forming a conjugate substrate. Two kinds of conjugate substrates were synthesized, which differed by one methylene group in the amino alkyl phosphate moiety of the reporter molecule. Initial steps of the assays of PMM and PMI were carried out separately, each with a different conjugate substrate. Thereafter, the two reaction mixtures were pooled and quenched with an excess of ribose-5-P and fructose-6-P. Internal standards with structures identical with reaction products, i.e., sedoheptulose-7-P-conjugates containing either two or three methylene groups, but in addition, labeled with heavy carbon and hydrogen atoms were added; the reaction products together with internal standards were affinity-purified by passing through a column of immobilized streptavidin beads. The conjugates were then released from the column with an aqueous biotin methyl ester solution and the eluate was directly infused into the ESI-MS apparatus. The affinity purification procedure and the mass spectrometric analysis were fully automated allowing for high-throughput multiplex applications. Enzymatic assays are also performed to identify disorders of CDG type II. The GnT II assay on blood mononuclear cells with GlcNAc$_1$Man$_3$-octyl substrate will detect CDG type IIa.[53] Assay of β4-galactosyltransferase I (β4GalT-I) on fibroblasts with ovalbumin as substrate should be diagnostic for CGD-IId.[69] Determination of the activity of β4-galactosyltransferase I (β4-GalT7) on fibroblasts with xylose-serine as substrate should be helpful in the diagnosis of the progeroid variant of Ehlers–Danlos syndrome.[76] A relatively simple assay of the activity of POMGnT-I or GnT-I.2 has been recently developed using an inexpensive commercial substrate, Man(α1-)O-benzyl.[180] The test is performed on muscle biopsy samples and should allow for the diagnosis of MEB disease. Recently, however, POMGnT-I activity has been found in leukocytes, raising the possibility of elaborating a test on a more easily accessible material (H. Schachter, personal communication, 2003). An elevated serum creatine kinase confirms the diagnosis of a myopathy.

C. ANALYSIS OF LLOS AND PROTEIN-LINKED OLIGOSACCHARIDES

An important aspect in the diagnosis of CDGx is the determination of the size distribution of LLO in patients' fibroblasts. In yeasts and man, the most abundant LLO fraction is a fully mannosylated fraction with the formula $Glc_{1-3}Man_9GlcNAc_2$. In a defective glycosylation of LLO, its size is decreased with an accumulation of fractions just below the biosynthesis block. In CDG types Ia and Ib resulting from a low supply of GDP-Man, LLOs with 4 to 6 mannose residues[181] accumulate in fibroblasts, while a deficient synthesis of Dol-P-Man (as in CDG-Ie) results in a sharp increase in LLO with the composition of $Man_5GlcNAc2$.[41,42] Similarly, nonglucosylated $Man_9GlcNAc_2$ accumulates in patients' fibroblasts in CDG-Ic due to a deficiency of α3-glucosyltransferase.[36–38] Technically, the size distribution of LLO is determined after a short, 30 to 60 min labeling of fibroblasts at 70 to 90% confluency with [^{3}H] mannose.[181] Thereafter, the cells are extracted with 2:1 chloroform/methanol to remove Dol-P-Man and subsequently with water to remove free sugars and nucleotides. Finally, the pellet is first extracted with chloroform/methanol/water (10:10:3, by vol). Analysis of oligosaccharides released from LLOs by a mild acid hydrolysis is usually performed by HPLC. Intact LLOs may be separated by TLC in a propanol/acetic acid/water (3:3:2, by vol.) solvent mixture.[49] Sometimes, it is also important to determine the incorporation of [^{35}S] methionine into the protein because the proportion of [^{3}H]mannose/[^{35}S] methionine may be sharply decreased in CDG-I.[181] In some processing defects occurring in the ER, the size distribution of protein-linked oligosaccharides may be assessed. For example, in glucosidase I deficiency (CDG-IIb), an additional protein-linked oligosaccharide with the composition of $Glc_3Man_9GlcNAc$ is found among the endo-H digestion products of ^{3}H- mannose-labeled glycoproteins of fibroblasts.[182] The enzyme cleaves high-mannose-type oligosaccharides that are larger than $Man_5GlcNAc_2$. By contrast, PNGase F is less specific and cleaves almost all the *N*-linked oligosaccharides. To determine if smaller size oligosaccharides are also transferred onto proteins, the ratio of PNGase F to endo-H released oligosaccharides may be determined.[181] Usually, the results of LLO size distribution are straightforward, but there are a few exceptions. In CDG-Ih, fibroblasts accumulate unglucosylated fully mannosylated LLOs with only a small amount of the expected monoglucosylated form.[49] Accumulation of the latter compound together with the triglucosylated form occurs only in the presence of castanospermin, an ER glucosidase inhibitor. The authors explained these results by the operation of glucosyltransferase-glucosidase shuttle which had been proposed to regulate the pool size of mature triglucosylated LLO.[183] Another example of interpretation difficulties is provided by an accumulation in CDG-Ii fibroblasts of two LLOs instead of the usual one.[50]

D. IMMUNOCHEMICAL METHODS

These methods are used mainly for detection in skeletal muscles of underglycosylation of α-dystroglycan using antibodies against its glycosylated epitope.[139] Both immunochemical analysis and Western blotting are used.

E. STRUCTURAL METHODS

Structural methods were applied to purified blood serum and red cell membrane reporter glycoproteins as well as to total serum glycoproteins. The character of molecular abnormalities in type I and type II CDG demands different experimental approaches that are suitable to the nature of molecular lesions. In CDG type I the whole glycans are missing from serum glycoproteins with those remaining having a presumably normal structure. Thus, the analysis must be able to distinguish between unglycosylated, hypoglycosylated, and fully glycosylated isoforms. With most methods this requires the analysis of whole, intact glycoproteins or of defined glycopeptide fragments. In CDG-II, unglycosylation is absent but the structure of glycans is affected. Hence, in CDG type II, the analysis of glycans released from glycoproteins by enzymes or by chemical cleavage will suffice. Methods based on mass spectrometry such as electrospray ionization and MALDI-TOF allow the performance of an exact determination of the molecular masses of glycoprotein isoforms, and thus are well suited for the diagnosis of both CDG-I and CDG-II.[33, 34, 184] An automated ESI-MS with in-line purification of serum transferrin by immunoaffinity technique has been recently developed requiring less than 10 min for complete analysis.[185] Yet identifying nonoccupant sites of glycoproteins, the hallmark of CDG-I, may be difficult even employing mass spectrometry, especially when analyzing complex glycoprotein mixtures. Another factor that may affect the interpretation of results is glycans microheterogeneity. Also, as pointed out before, there are instances when reporter glycoproteins in CDG patients are normally glycosylated. Thus, researchers are developing techniques that would allow examination of the glycosylation status of glycoproteins, including nonoccupancy of glycosylation sites of a large number of glycoproteins in a single analysis. One such method consists in the selective isolation, identification, and quantitation of N-linked tryptic glycopeptide fragments of glycoproteins (186). First, the glycoproteins are oxidized with periodate; next, they are coupled to hydrazide beads through hydrazone bonds. Nonglycosylated proteins are washed away. The bound glycoproteins are digested with trypsin with non-N-linked fragments removed by washing. Subsequently, the remaining glycopeptides are released of glycans and the bondage with beads by digestion with peptide-N-glycosidase F, which cleaves the N-acetylglucosamine-asparagine linkage. During the cleavage, the asparagine residues of the former glycopeptides are converted into aspartate. Finally, the peptides are separated and fingerprinted by microcapillary HPLC with ESI or MALDI tandem MS. If quantitative analysis is required, the α-amino groups of the still bonded glycopeptides may be additionally tagged with succinic anhydride. The LC-MS-MS analysis procedure takes 2 h. The authors identified 145 glycopeptides from 57 proteins. A similar method was developed in which glycoproteins were immobilized on concanavalin A columns, eluted with α-methyl mannopyranoside, and subjected to digestion with trypsin.[187] Thereafter, the glycopeptides were again captured on a concanavalin A column, eluted, and digested with PGN-ase F in the presence of $H_2{}^{16}O$ or $H_2{}^{18}O$. The ^{18}O-tagged peptides were analyzed by multidimensional liquid chromatography, and identified using mass spectrometry. The method was applied to an extract of the *Caenorhabolitis elegans* glycoproteins; it helped in identifying 250 glycoproteins including 83 putative transmembrane proteins, and 400 unique N-glycosylation sites. To some extent, such methods are comparable with a high-throughput screening of mRNA expression by DNA microarray technology. Both techniques have not yet been applied for the analysis of glycoproteins from patients with CDG but certainly they will in the near future. Methods for the structural analysis of glycans from the whole serum were, however, developed and applied to material from patients with CDG-I and-II.

In one of them, requiring only a few micrograms of material, the glycoproteins immobilized on a polyvinylidene difluoride (PVDF) membrane in a 96-well multiscreen plate were subjected to reduction, alkylation, and deglycosylation with PNGase F.[188] This was followed by derivatization of glycans at the reducing end with a fluorescent 8-amino1,3,6-pyrenetrisulfonic acid. Lastly, the labeled glycans were subjected to polyacrylamide gel electrophoresis on a DNA sequencer. This integrated sample preparation is also compatible with capillary electrophoresis and MALDI-TOF-MS analysis. Derivatized glycans may be further characterized by sequential digestion with exoglycosidases and comparison of the product size, before and after digestion, with a reference oligosaccharide ladder. Employing this approach it was found that desialylated glycans from normal and CDG-I sera had a simple profile of nine oligosaccharides. Differences were only quantitative and concerned increased core fucosylation and reduced branching. The differences were most pronounced in CDG-Ia and decreased in magnitude through CDG-I types b–f. Another recent paper describes how glycans were released from whole serum by hydrazinolysis and from purified transferrin and IgG by digestion with PNGase F.[189] Fluorescent-labeled glycans were then subjected to treatment with multienzyme arrays of glycosidases followed by analysis by HPLC and structure confirmation by mass spectrometry. Glycans from a CDG-Ia patient were found to have a normal structure, while those from a patient with CDG-IIa lacked the complex type glycosylation. In yet another study, glycans released by PNGase F from the purified transferrin and isoforms of α1-antitrypsin were analysed by MALDI-TOF-MS.[190] It was found that glycans from all three patients with CDG-Ia, and from each patient with CDG types Ib and Ic exhibited increased fucosylation, both core and peripheral (α1-3). In contrast to the afore mentioned findings,[188] the glycans displayed increased branching. These results show that subtle glycosylation changes may vary in patients with CDG-I.

To evaluate both the hypoglycosylation and unglycosylation of glycoproteins, a simple screening method, based on the determination of their carbohydrate molar composition per mol of protein moiety, has been developed.[191] Glycoproteins (e.g., 100 μg of membrane proteins) are separated by sodium dodecyl sulfate–polyacrylamide gel electrophoresis, and then electroblotted onto PVDF membranes. Blots are stained; appropriate pieces of PVDF membranes are excised, destained, and subjected to sequential hydrolysis with weak, moderate, and strong acids to release sialic acids, neutral sugars with hexosamines, and amino acids, respectively. Carbohydrates, including sialic acids, are quantitated by high-pH anion-exchange chromatography with pulsed amperometric detection. Protein contents of the bands are determined as amino acids by the fluorescamine method. When a particular band of interest is contaminated with other proteins or glycoproteins, it should be subjected to a preliminary purification.[192] In certain reporter glycoproteins such as red cell band 3 and serum α-1-acid glycoprotein, the total number of N-linked glycans may be calculated from their mannose contents. In other glycoproteins, such as red cell glycophorin A, the total number of O-linked glycans may be calculated from their N-acetylgalactosamine contents. The contents of other carbohydrates, i.e. of sialic acid, galactose, and N-acetylglucosamine, will show whether the glycans are hypoglycosylated. This method showed that in the red cells of patients with CDG-Ia (193) and CDA-I and II (194, 195), a single N-linked glycan of band 3 was hypoglycosylated because it contained a decreased number of mol/mol of galactose and N-acetylglucosamine but a normal amount of 3 mol/mol of mannose. In some of these patients, glycophorin A was partly O-unglycosylated since it showed a reduced content of N-acetylgalactosamine, galactose, and sialic acid on the molar basis. On the other hand, an excessive mannose in band 3 as in CDG-Ig showed a pathological hybrid type of glycans.[48] The method, together with the analysis of red cell glycolipids, allowed establishing heterozygosity of the *CDAN-II* (HEMPAS) gene in healthy carriers in only one family so far (196).

F. GENETIC ANALYSIS

A detailed description of methods used to identify and characterize causative mutations in CDG is beyond the scope of this review. It should be pointed out, however, that genetic analysis is becoming an increasingly important component of the diagnosis and molecular dissection of various hereditary disorders. The pathogenesis of these diseases is often elucidated on the basis of genetic analysis alone even when neither the structure nor function of a causative gene is initially known. However, the diagnosis of CDG is at present so far advanced that biochemical tests frequently provide some clues as to the character of the molecular lesion. Nevertheless, the confirmation of the diagnosis relies firmly on genetic grounds, including complementation analysis. In CDG-I, the complementation has been often performed with the yeast *ALG* genes, since the biosynthesis of LLO is highly conserved, and yeast mutant strains can be directly used to characterize human molecular defects.[197] A microarray DNA biochip technology employing commercial high-density oligonucleotide arrays representing more than 22,000 genes was used to determine RNA expression in cells of patients with the CDG-I types e, g, and c.[198] It was found that about 300 genes involved in translational control, glycosylation, protein folding, vesical trafficking, lipid or inositol metabolism as well as peroxisome and membrane biogenesis, were overexpressed. About 200 genes were underexpressed. A similar technique, but employing individually prepared microchip arrays, was used to identify nonsense (stop) codons that are frequent products of causative mutations in CDG.[199] The arrays contained 240 450-bp cDNA probes, each belonging to one gene encoding a glycosyltransferase, sulfotransferase, glycosidase, and carbohydrate-binding protein. Some nonsense codons are known to initiate degradation of mutant transcripts through nonsense-mediated mRNA decay. The decay can be pharmacologically inhibited by stabilizing the nonsense transcripts, thus allowing for the comparison, for a given transcript, of control and disease cells in terms of "nonsense enrichment." The technique was validated by finding two premature stop codons in a patient with CDG-Ih in whom these mutations had been previously characterized.[49]

G. DETERMINATION OF GLYCOSPHINGOLIPIDS

In CDA-I – III, the accumulation of glycosphingolipids in red cells invariably accompanies the hypoglycosylation of glycoproteins.[194,195] An increased content of glycosphingolipids has also been found in the fibroblasts of patients with CDG-Ia.[200] The same paper provided evidence that inhibition of the glycoprotein synthesis in normal fibroblasts also resulted in the accumulation of GSLs suggesting its compensatory character. Yet in the red cells of a child with CDG-Ig, the contents of GSLs in red cells were normal though band 3 was hypoglycosylated.[48] Further studies on this point are required.

V. TREATMENT OF CDG

Attempts to treat patients with CDG started after the initial findings that culturing fibroblasts of type Ia patients in the presence of 1.0 mM mannose had almost normalized the GDP-Man pool size, corrected the LLO size distribution and the incorporation of a labeled mannose into glycoproteins.[181] The mechanism has not been completely resolved. The added mannose is probably transported across the cell membrane through a specific transporter,[201,202] and thereafter phosphorylated to Man-6-P, thus providing an additional building material for the biosynthesis of GDP-Man[202, 203] (see Figure 4.3). These observations prompted investigators to administer oral mannose treatment to patients.[204–206] Unfortunately, the therapy was not effective with respect to either the clinical symptoms or the glycosylation of serum glycoproteins. Some parents of the affected patients, however, expressed a different opinion and reported certain improvements[207] in their children. On the other hand, very favorable results were obtained in the CDG-Ib patients receiving an oral mannose therapy who recovered within a few months.[30] Those patients, however, did not exhibit mental retardation, and their main clinical symptom was protein-losing enteropathy. At present, the patients receive a high daily dose of 1g of mannose per kg of body weight divided into five doses.[168] With this regimen the patients recover rapidly, but unexpectedly, normalization of the transferrin patterns takes months to achieve the goal. On the basis of the assumption that, in CDG-Ia patients, mannose does not enter cells in sufficient amounts,[208] many investigators attempted to solve the problem by engineering derivatives of mannose phosphates. Man-1-P seems to be a reasonable choice as it is an immediate precursor of GDP-Man. However, the compound is highly hydrophilic and does not pass through the cell membrane; hence the attempts to synthesize membrane-permeable derivatives of Man-1-P in which polar groups would be protected with biologically reversible, lipophilic substituents. In the extension of the work on acetoxymethyl esters of organic phosphates,[209] a series of membrane-permeant Man-1-P derivatives was prepared and one of them, *bis*-acetoxymethyl-(2,3,4,6-tetra-*O*-butyryl-α-*D*-mannopyranosyl)-phosphate, was used as a model compound.[210] The assumption was that having entered the cell, the compound would be freed of substituents by endogenous esterases. MALDI-TOF analysis showed that the compound was indeed freed of all ester groups by incubation in a pure preparation of porcine liver esterase at 37°C for 3 h. A crude enzyme preparation, however, cleaved only acetoxymethyl groups and one butyryl group. Surprisingly, the phosphate-free mannnose appeared among by-products of the reaction. More recently, a series of membrane-permeable derivatives of mannose-1-P was developed, which normalized an impaired glycosylation in CDGI-a fibroblasts at 0.1 mM, i.e., a ten-fold lower concentration than that required with a free mannose.[211] In addition, the compound corrected the LLO size distribution in the CDG-Ie patients with the Dol-P-Man synthase deficiency.

Another simple sugar, fucose, is a drug of choice in the treatment of patients with the CDG-IIc or LAD II type of a leukocyte adhesion deficiency. The children of Arab descent, however, did not respond to the treatment.[212, 213] On the other hand, the response of the Turkish and the Brazilian patients was remarkable[59, 67] although there were some minor differences: in both patients the neutrophil count normalized rapidly, as early as 8 to 10 days from the initiation of the therapy with the appearance of sLe^X and P-selectin ligands on neutrophils. The restored ability of neutrophils of the Turkish patient to adhere to recombinant selectin-coated microtiter plates,[67] or of those from the Brazilian patient to roll *in vivo* in the cremasteric venules of NOD/SCID mice,[59] was found on days 16 and 40, respectively. The rolling of the LAD II neutrophils with the newly expressed P-selectin ligand was, however, weaker than that in the control neutrophils. The expression of the E-selectin ligand in the Turkish patient required a longer treatment with high doses of fucose up to more than 2.0 g per day than that of E-selectin. The latter did not appear in the Brazilian patient.[59] Unexpectedly, the Brazilian child not only responded well to the fucose therapy, but also expressed H antigen on erythrocytes, which fortunately did not elicit autoimmune anti-H antibodies. The auto-antibodies appeared against neutrophils. More significantly, psychomotor development improved in the Turkish patients but not in the Brazilian child. The difference should be attributed to a lower dose of fucose administered to the Brazilian patient in fear of possible complications due to the treatment-induced neutropenia. The molecular basis for the lack of response to fucose therapy in Arab patients is unknown at present. It is assumed that mutations in the GDP-Fuc transporter gene reduces the affinity of the transporter protein to GDP-Fuc. Under normal conditions most fucose in the GDP-fucose is synthesized by a *de novo* pathway from the GDP-mannose, and the minority (~10%) by a salvage pathway from Fuc1-P. The latter compound is formed from a free fucose by the fucose kinase.[214] Oral fucose therapy in LAD II patients increases blood fucose from an undetectable level up to 0.25 mM. It has been assumed that the success of oral fucose therapy was due to an elevated cytosolic GDP-Fuc, which compensated for a decreased affinity for the GDP-Fuc transporter. Kinetic studies confirmed this assumption because cell extracts from Arab and Turkish patients showed a decreased maximal velocity (V_{max}) of GDP-fucose import to preparations of Golgi vesicles.[63, 213] Interestingly, only the Golgi vesicles from the Turkish patient exhibited an unsaturable transport component, thus rationalizing the results of fucose therapy. The GDP-Fuc transporter gene in the Brazilian patient was, however, severely truncated and most likely a nonfunctional one. Thus, residual activity of the GDP-Fuc transporter, which can be compensated with an excess GDP-Fuc,[63, 213] may not account for the good response to the fucose therapy in the Brazilian patient. Therefore, Hidalgo and colleagues[59] assumed that another low-affinity GDP-Fuc transporter(s) may be present in humans. A faster restoration of the expression of P- over L-selectin ligands in the course of fucose therapy may be caused by differential modifying activities of fucosyltransferases VII and IV on the P-and L-selectin ligands in neutrophils.[215, 216] It should be pointed out that the discontinuation of the fucose therapy in the patients with LAD resulted in a rapid loss of the selectin ligands and recurrence of elevated neutrophil counts.[68]

Freeze proposed that oral sialic acid might ameliorate the condition of patients with hereditary inclusion body myopathy.[211] Otherwise, the patients could be treated with derivatized precursors of sialic acid such as ManNAc-6-P on the assumption that mutations in patients affect the function of either epimerase domain or ManNAc kinase domain of *GNE*. The reviewer would like to mention as possible beneficiaries also patients with a defective sialic acid transporter.[71]

VI. CONCLUSIONS

For almost a century since the discovery of carbohydrates and glycoconjugates, these substances had not been at the forefront of biological research, this distinction having been left to proteins and nucleic acids instead. Yet since the discovery of the carbohydrate nature of ABO (H) blood group antigens, interest and progress in the research into glycoconjugates has been steadily increasing. Starting from the mid-1970s it has been realized that, apart from their structural role, glycoconjugates are implicated in many other functions such as adhesion, recognition, migration, trafficking, signaling, immunity, development, and growth. Consequently, the whole area of research involving the biological functions of carbohydrates and glycoconjugates has been given the name of glycobiology. However, the new name is not enough: to gain the support of the general public, a scientific discipline has to find medical application, a reward that CDG have brought into glycobiology. It is one thing to discover a selectin, but another to find patients who lack the selectin. It may be of interest to biochemists to describe a new *O*-mannosyl protein linkage but it will certainly catch the public imagination to learn about patients who are severely ill because of the absence of this linkage in their muscles and brain. Thus, it is foreseeable that in the next decade or so, glycobiology and CDG will be in the limelight of medical and biological sciences. This will undoubtedly be advantageous to CDG patients, since the better the understanding of the function of carbohydrates and glycoconjugates, the more help they can receive from dedicated physicians and scientists. New types of CDG will be discovered and more will be learned about the pathogenesis of those already recognized. Less progress is expected in the field of treatment because of the involvement of the nervous system in most of the patients with CDG. In that case, the pathological changes may be irreversible. However there is hope even for the mentally retarded, as in some types of CDG, the disease develops after birth. We may also expect new therapeutic approaches such as novel procedures of gene therapy and new carbohydrate-based drugs. And, as it sometimes happens in science, unexpected discoveries may open new vistas for treatment. For the sake of the patients, let us hope that this possibility will come true.

REFERENCES

1. Jaeken, J., Vanderschueren-Lodeweyckx, M., Casaer, P., Snoeck, L., Corbeel, L., Eggermont, E., and Eeckels, R., Familial psychomotor retardation with markedly fluctuating serum prolactic, FSH and GH levels, partial TBG deficiency, increased serum arylsulphatase A and increased CSF protein: a new syndrome? *Pediatr. Res.*, 14, 179, 1980.
2. Jaeken, J., van Eijk, H.G., van der Heul, C., Corbeel, L., Eeckels, R., and Eggermont, E., Sialic acid-deficient serum and cerebrospinal fluid transferrin in a newly recognized genetic syndrome. *Clin. Chim. Acta*, 144, 245–247, 1984.

3. Jaeken, J., Stibler, H., and Hagberg, B., The carbohydrate-deficient glycoprotein syndrome. A new inherited multisystemic disease with severe nervous system involvement, *Acta Paediatr. Scand.* Suppl; 375, 1–71, 1991.
4. Aebi, M., Helenius, A., Schenk, B., Barone, R., Fiumara, A., Berger, E.G., Hennet, T., Imbach, T., Stutz, A., Bjursell, C., Uller, A., Wahlstrom, J.G., Briones, P., Cardo, E., Clayton, P., Winchester, B., Cormier-Dalre, V., de Lonlay, P., Cuer, M., Dupre, T., Seta, N., de Koning, T., Dorland, L., de Loos, F., Kupers, L., Fabritz, L., Hasilik, M., Marquardt, T., Niehues, R., Freeze, H., Grünewald, S., Heykants, L., Jaeken, J., Matthijs, G., Schollen, E., Keir, G., Kjaergaard, S., Schwartz, M., Skovby, F., Klein, A., Roussel, P., Körner, C., Lübke, T., Thiel, C., von Figura, K., Kościelak, J., Krasnewich, D., Lehle, L., Peters, V., Raab, M., Saether, O., Schachter, H., Van Schaftingen, E., Verbert, A., Vilaseca, A., Wevers, R., and Yamashita, K., Carbohydrate-deficient glycoprotein syndromes become congenital disorders of glycosylation: and updated nomenclature for CDG. First International Workshop on CDGS (letter), *Glycoconj. J.*, 16, 669–671, 1999.
5. Aebi, M., Helenius, A., Schenk, B., Barone, R., Fiumara, A., Berger, E.G., Hennet, T., Imbach, T., Stutz, A., Bjursell, C., Uller, A., Wahlstrom, J.G., Briones, P., Cardo, E., Clayton, P., Winchester, B., Cormier-Dalre, V., de Lonlay, P., Cuer, M., Dupre, T., Seta, N., de Koning, T., Dorland, L., de Loos, F., Kupers, L., Fabritz, L., Hasilik, M., Marquardt, T., Niehues, R., Freeze, H., Grünewald, S., Heykants, L., Jaeken, J., Matthijs, G., Schollen, E., Keir, G., Kjaergaard, S., Schwartz, M., Skovby, F., Klein, A., Roussel, P., Körner, C., Lübke, T., Thiel, C., von Figura, K., Kościelak, J., Krasnewich, D., Lehle, L., Peters, V., Raab, M., Saether, O., Schachter, H., Van Schaftingen, E., Verbert, A., Vilaseca, A., Wevers, R., and Yamashita, K., Carbohydrate-deficient glycoprotein syndromes become congenital disorders of glycosylation: and updated nomenclature for CDG. First International Workshop on CDGS (abstract), *Glycobiology*, 10, iii–v, 2000.
6. Helenius, J., Ng, D.T., Marolda, C.L., Walter, P., Valvano, M.A., and Aebi, M., Translocation of lipid-linked oligosaccharides across the ER membrane requires Rft1 protein, *Nature*, 415, 447–450, 2002.
7. Parodi, A.J., Protein glucosylation and its role in protein folding, *Ann. Rev. Biochem.*, 69, 69–93, 2000.
8. Spiro, R.G., Glucose residues as key determinants in the biosynthesis and quality control of glycoproteins with *N*-linked oligosaccharides, *J. Biol. Chem.*, 275, 35657–35660, 2000.
9. Dempski R.E., Jr. and Imperiali, A.B., Oligosaccharyl transferase: gatekeeper to the secretory pathway, *Curr. Opin. in Chem. Biol.*, 6, 844–850, 2002.
10. Herscovics, A., Importance of glycosidases in mammalian glycoprotein biosynthesis, *Biochem. Biophys. Acta*, 1473, 96–107, 1999.
11. Reitman, M.L., Varki, A., and Kornfeld, S., Fibroblasts from patients with I-cell disease and pseudo-Hurler polydystrophy are deficient in uridine 5'-diphosphate-*N*-acetylglucosamine:glycoprotein *N*-acetylglucosaminyl-phospho-transferase. *J. Clin. Invest.*, 67, 1574–1579, 1981.
12. Chen, S., Tan, J., Reinhold, V.N., Spence, A.M., and Schachter, H., UDP-*N*-acetylglucosamine:α-3-D-mannoside β-1,2-*N*-acetylglucosaminyltransferase I and UDP-*N*-acetylglucosamine: α-6-D-mannoside β-1,2-*N*-acetylglucosaminyltransferase II in *Caenorhabditis elegans*, *Biochim. Biophys. Acta*, 1573, 271–280, 2002.
13. Rottger, S., White, J., Wandall, H.H., Olivo, J.C., Stark, A., Bennett, E.P., Whitehouse, C., Berger, E.G., Clausen, H., and Nilsson, T., Localization of three human polypeptide GalNAc-transferases in HeLa cells suggests initiation of *O*-linked glycosylation throughout the Golgi apparatus, *J. Cell. Sci.*, 111, 45–60, 1998.

14. Silbert, J.E. and Sugumaran, G., Biosynthesis of chondroitin/dermatan sulfate, *IUBMB Life*, 54, 177–186, 2002.

15. Esko, J.D. and Selleck, S.B., Order out of chaos: assembly of ligand binding sites in heparan sulfate, *Ann. Rev. Biochem.*, 71, 435–471, 2002.

16. Kitagawa, H., Shimakawa, H., and Sugahara, K., The tumor suppressor EXT-like gene EXTL2 encodes an α1,4-N-acetylhexosaminyltransferase that transfers N-acetylgalactosamine and N-acetylglucosamine to the common glycosaminoglycan-protein linkage region. The key enzyme for the chain initiation of heparan sulfate, *J. Biol. Chem.*, 274, 13933–13937, 1999.

17. Nadanaka, S., Kitagawa, H., Goto, F., Tamura, J., Neumann, K.W., Ogawa, T., and Sugahara, K., Involvement of the core protein in the first beta-N-acetylgalactosamine transfer to the glycosaminoglycan–protein linkage-region tetrasaccharide and in the subsequent polymerization: the critical determining step for chondroitin sulphate biosynthesis, *Biochem. J.*, 340, 353–357, 1999.

18. Chatterjee, S. and Mayor, S., The GPI-anchor and protein sorting, *Cell. Mol. Life Sci.*, 58, 1969–1987, 2001.

19. Butikofer, P., Malherbe, T., Boschung, M., and Rodik, I., GPI-anchored proteins: now you see'em, now you don't, *FASEB. J.*, 13, 545–548, 2001.

20. Filmus, J. and Selleck, S.B., Glypicans: proteoglycans with a surprise, *J. Clin. Invest.*, 108, 497–501, 2001.

21. Finne, J., Krusius, T., Margolis, R.K., and Margolis, R.U., Novel mannitol-containing oligosaccharides obtained by mild alkaline borohydride treatment of chondroitin sulfate proteoglycan from brain, *J. Biol. Chem.*, 254, 10295–10300, 1979.

22. Sasaki, T., Yamada, H., Matsumura, K., Shimizu, T., Kobata, A., and Endo,T., Detection of O-mannosyl glycans in rabbit skeletal muscle alpha-dystroglycan, *Biochim. Biophys. Acta*, 1425, 599–606, 1998.

23. Smalheiser, N.R., Haslam, S.M., Sutton-Smith, M., Morris, H.R., and Dell, A., Structural analysis of sequences O-linked to mannose reveals a novel Lewis X structure in cranin (dystroglycan) purified from sheep brain, *J. Biol. Chem.*, 273, 23698–23703, 1998.

24. Chai, W., Yuen, C.-T., Kogelberg, H., Carruthers, R.A., Margolis, R.U., Feizi, T., and Lawson, A.M., High prevalence of 2-mono- and 2,6-di-substituted Manol-terminating sequences among O-glycans released from brain glycopeptides by reductive alkaline hydrolysis, *Eur. J. Biochem.*, 263, 879–888, 1999.

25. Doucey, M.A., Hess, D., Cacan, R., and Hofsteenge, J., Protein C-mannosylation is enzyme-catalysed and uses dolichyl-phosphate-mannose as a precursor, *Mol. Biol. Cell*, 9, 291–300, 1998.

26. Anand, M., Rush, J.S., Ray, S., Doucey, M., Weik, J., Ware, F.E., Hofsteenge, J., Waechter, C.J., and Lehrman, M.A., Requirement of the Lec35 gene for all known classes of monosaccharide-P-dolichol-dependent glycosyltransferase reactions in mammals, *Mol. Biol. Cell*, 12, 487–501, 2001.

27. Berninsone, P.M. and Hirschberg, C.B., Nucleotide sugar transporters of the Golgi apparatus, *Curr. Opin. Struct. Biol.*, 10, 542–547, 2000.

28. van Schaftingen, E. and Jaeken, J., Phosphomannomutase deficiency is a cause of carbohydrate-deficient glycoprotein syndrome type I, *FEBS, Lett.*, 377, 318–320, 1995.

29. Carchon, H., van Schaftingen, E., Matthijs, G., and Jaeken, J., Carbohydrate-deficient glycoprotein syndrome type IA (phosphomannomutase-deficiency), *Biochim. Biophys. Acta*, 1455, 155–165, 1999.

30. Niehues, R., Hasilik, M., Alton, G., Körner, C., Schiebe-Sukumar, M., Koch, H.G., Zimmer, K.P., Wu, R., Harms, E., Reiter, K., von Figura, K., Freeze, H.H., Harms, H.K., and Marquardt, T., Carbohydrate-deficient glycoprotein syndrome type Ib.

Phosphomannose isomerase deficiency and mannose therapy, *J. Clin. Invest.*, 101, 1414–1420, 1998.

31. de Köning, T.J., Dorland, L., van Diggelen, O.P., Boonman, A.M.C., de Jong, G.J., van Noort, W.L., de Schryver, J., Duran, M., *van den Berg*, I. E.T., Gerwig, G.J., Berger, R., and Poll-The, B.T., A novel disorder of *N*-glycosylation due to phosphomannomutase isomerase deficiency, *Biochem. Biophys. Res. Commun.*, 245, 38–42, 1998.

32. Jaeken, J., Mattthijs, G., Saudubray, J-M., Dionisi-Vici, C., Bertini, E., De Lonlay, P., Henri, H., Carchon, H., Schollen, E., and van Schaftingen, E., Phosphomannose isomerase deficiency: a carbohydrate-deficient glycoprotein syndrome with hepatic-intestinal presentation, *Am. J. Hum. Genet.*, 62, 1535–1539, 1998.

33. Wada, Y., Nishikawa, A., Okamoto, N., Inui, K., Tsukamoto, H., Okada, S., and Taniguchi, N., Structure of serum transferrin in carbohydrate-deficient glycoprotein syndrome, *Biochem. Biophys. Res. Commun.*, 189, 832–836, 1992.

34. Yamashita, K., Ideo, H., Ohkura, T., Fukushima, K., Yuasa, I., Ohno, K., and Takeshita, K., Sugar chains of serum transferrin from patients with carbohydrate deficient glycoprotein syndrome. Evidence of asparagine-*N*-linked oligosaccharide transfer deficiency, *J. Biol. Chem.*, 268, 5783–5789, 1993.

35. Mills, K., Mills, P.B., Clayton, P.T., Mian, N., Johnson, A.W., and Winchester, B.G., The underglycosylation of plasma α1-antitrypsin in congenital disorders of glycosylation type I is not random, *Glycobiology*, 13, 73–85, 2003.

36. Burda, P., Borsig, L., de Rijk-van Andel, J., Wevers, R., Jaeken, J., Carchon, H., Berger, E.G., and Aebi, M., A novel carbohydrate-deficient glycoprotein syndrome characterized by a deficiency in glycosylation of the dolichol-linked oligosaccharide, *J. Clin. Invest.*, 102, 647–652, 1998.

37. Körner, C., Knauer, R., Holzbach, U., Hanefeld, F., Lehle, L., von and Figura, K., Catbohydrate-deficient glycoprotein syndrome type V: deficiency of dolichol-P-Glc:Man9GlcNAc2-PP-dolichyl glucosyltransferase, *Proc. Natl. Acad. Sci. USA*, 95, 13200–13205, 1998.

38. Imbach, T., Burda, P., Kuhnert, P., Wevers, R.A., Aebi, M., Berger, E.G., and Hennet, T., A mutation in the human ortholog of the *Saccharomyces cerevisiae* ALG6 gene causes carbohydrate-deficient glycoprotein syndrome type-Ic, *Proc. Natl. Acad. Sci. USA*, 96, 6982–6987, 1999.

39. Körner, C., Knauer, R., Stephani, U., Marquardt, T., Lehle, L., and von Figura, K., Carbohydrate deficient glycoprotein syndrome type IV: deficiency of dolichyl-P-Man:Man(5)GlcNAc(2)-PP-dolichyl mannosyltransferase, *EMBO J.*, 18, 6816–6822, 1999.

40. Maeda, Y., Tanaka, S., Hino, J., Kangawa, K., and Kinoshita, T., Human dolichol-phosphate-mannose synthase consists of three subunits, DPM1, DPM2 and DPM3, *EMBO J.*, 19, 2475–2482, 2000.

41. Kim, S., Westphal, V., Srikrishna, G., Mehta, D.P., Peterson, S., Filiano, J., Karnes, P.S., Patterson, M.C., and Freeze, H.H., Dolichol phosphate mannose synthase (DPM1) mutations define congenital disorder of glycosylation Ie(CDG-Ie), *J. Clin. Invest.*, 105, 191–198, 2000.

42. Imbach, T., Schenk, B., Schollen, E., Burda, P, Stutz, A., Grunewald, S., Bailie, N.M., King, M.D., Jaeken, J., Matthijs, G., Berger, E.G., Aebi, M., and Hennet, T., Deficiency of dolichol-phosphate-mannose synthase-1 causes congenital disorder of glycosylation type Ie, *J. Clin. Invest.*, 105, 233–239, 2000.

43. Kranz, C., Denecke, J., Lehrman, M.A., Ray, S., Kienz, P., Kreissel, G., Sagi, D., Peter-Katalinic, J., Freeze, H.H., Schmid, T., Jackowski-Dohrmann, S., Harms, E., and Marquardt, T., A mutation in the human MPDU1 gene cause congenital disorder of glycosylation If (CDG-If), *J. Clin. Invest.*, 108, 1613–1619, 2001.

44.   Schenk, B., Imbach, T., Frank, C.G., Grubenmann, C.E., Raymond, G.V., Hurvitz, H., Kom-Lubetzki, I., Revel-Vik, S., Raas-Rotschild, A., Luder, A.S., Jaeken, J., Berger, E.G., Matthijs, G., Hennet, T., and Aebi, M., MPDU1 mutations underlie a novel human congenital disorder of glycosylation designated type If, *J. Clin. Invest.*, 108, 1687–1695, 2001.

45.   Chantret, I., Dupre, T., Delenda, C., Bucher, S., Dancourt, J., Barnier, A., Charollais, A., Heron, D., Bader-Meunier, B., Danos, O., Seta, N., Durand, G., Oriol, R., Codogno, P., and Moore, S.E., Congenital disorders of glycosylation type 1g is defined by a deficiency in dolichyl-P-mannose:Man7GlcNAc2-PP-dolichyl mannosyltransferase, *J. Biol. Chem.*, 277, 25815–25822, 2002, Epub April 30, 2002.

46.   Grubenmann, C.E., Frank, C.G., Kjaergaard, S., Berger, E.G., Aebi, M., and Hennet, T., ALG 12 mannosyltransferase defect in congenital disorder of glycosylation type 1g, *Hum. Mol. Genet.*, 11, 2331–2339, 2002.

47.   Thiel, C., Schwarz, M., Hasilik, M., Grieben, U., Hanefeld, F., Lehle, L., von Figura, K., and Körner, C., Deficiency of dolichyl-P-Man:Man7GlcNAc2-PP-dolichyl mannosyltransferase causes congenital disorder of glycosylation type 1g, *Biochem. J.*, 367, 195–201, 2002.

48.   Zdebska, E., Bader-Meunier, B., Schischmanoff, P.O., Dupre, T., Seta, N., Tschernia, G., Koscielak, J., and Delaunay, J., Abnormal glycosylation of red cell membrane band 3 in the congenital disorder of glycosylation type 1g, *Pediatr. Res.*, 54, 224–229, Epub, May 7, 2003.

49.   Chantret, I., Dancourt, J., Dupre, T., Delenda, C., Bucher, S., Vuillaumier-Barrot, S., Ogier de Baulny, H., Peletan, C., Danos, O., Seta, N., Durand, G., Oriol, R., Codogno, P., and Moore, S.E., A deficiency in dolichyl-P-glucose:Glc1Man9GlcNAc2-PP-dolichyl α-3glucosyltransferase defines a new subtype of congenital disorders of glycosylation, *J. Biol. Chem.*, 278, 9962–9971, 2003. Epub Dec. 11, 2002.

50.   Thiel, C., Schwarz, M., Peng, J., Grzmil, M., Hasilik, M., Braulke, T., Kohlschutter, A., von Figura, K., Lehle, L., and Körner, C., A new type of congenital disorders of glycosylation (CDG-Ii) provides new insights into early steps of dolichol-linked oligosaccharide biosynthesis, *J. Biol. Chem.*, 278, 22498–22505, Epub, April 8, 2003.

51.   Wu, X., Rush, J.S., Karaoglu, D., Krasnewich, D., Lubinsky, M.S., Waechter, C.J., Gilmore, R., and Freeze, H.H., Deficiency of UDP-GlcNAc: dolichol phosphate *N*-Acetylglucosamine-1 phosphate transferase (DPAGT1) causes a novel congenital disorder of glycosylation type Ij, *Hum. Mutat.*, 22, 144–150, 2003.

52.   Jaeken, J., Schachter, H., Carchon, H., De Cock, P., Coddeville, B., and Spik, G., Carbohydrate deficient glycoprotein syndrome type II: a deficiency in Golgi localised *N*-acetyl-glucosaminyltransferase II. *Arch. Dis. Child.*, 71, 123–127, 1994.

53.   Charuk, J.H., Tan, J., Bernardini, M., Haddad, S., Reithmeier, R.A., Jaeken, J., and Schachter, H., Carbohydrate-deficient glycoprotein syndrome type II. An autosomal recessive *N*-acetylglucosaminyltransferase II deficiency different from typical hereditary erythroblastic multinuclearity, with a positive acidified-serum lysis test (HEMPAS), *Eur. J. Biochem.*, 230, 797–805, 1995.

54.   Tan, J., Dunn, J., Jaeken, J., and Schachter, H., Mutations in the MGAT2 gene controlling complex *N*-glycan synthesis cause carbohydrate deficient glycoprotein syndrome type II, an autosomal recessive disease with defective brain development, *Am. J. Hum. Genet.*, 59, 810–817, 1996.

55.   De Praeter, C.M., Gerwig, G.J., Bause, E., Nuytinck, L.K., Vliegenthart, J.F., Breuer, W., Kamerling, J.P., Espeel, M.F., Martin, J.J., de Paepe, A.M., Chan, N.W.C., Dacremont, G.A., and van Coster, R.N., A novel disorder caused by defective biosynthesis of *N*-linked oligosaccharides due to glucosidase I deficiency, *Am. J. Hum. Genet.*, 66, 1744–1756, 2000.

56. Etzioni, A., Frydman, M., Pollack, S., Avidor, I., Phillips, M.L., Paulson, J.C., and
     Gerschoni-Baruch, R., Recurrent severe infections caused by a novel leukocyte adhe-
     sion deficiency, *N. Engl. J. Med.*, 327, 1789–1792, 1992.
57. Etzioni, A., Gerschoni-Baruch, R., Pollack, S., and Shehadeh, N., Leukocyte adhesion
     deficiency type II: long-term follow-up, *J. Allergy Clin. Immunol.*, 102, 323–324, 1998.
58. Marquardt, T., Brune, T., Lühn, K., Zimmer, K.P., Koerner, C., Fabritz, L., van der
     Werft, N., Vormoor, J., Freeze, H.H., Louwen, F., Biermann, B., Harms, E., von
     Figura, K., Vestveber, D., and Koch, H.G., A new patient with leukocute adhesion
     deficiency (LAD) II syndrome, a generalized defect in fucose metabolism, *J. Pediatr.*,
     134, 681–688, 1999.
59. Hidalgo, A., Ma, S., Peired, A.J., Weiss, L.A., Cunningham-Rundles, C., and
     Frenette, P.S., Insights into leukocyte adhesion deficiency type 2 from a novel muta-
     tion in the GDP-fucose transporter gene, *Blood*, 101, 1705–1712, 2003.
60. Becker, D.J. and Lowe, J.B., Leukocyte adhesion deficiency type II, *Bioch. Biophys.
     Acta*, 1455, 193–204, 1999.
61. Anderson, D.C. and Springer, T.A., Leukocyte adhesion deficiency: an inherited defect
     in the Mac-1, LFA-1, and p150,95 glycoproteins, *Ann. Rev. Med.*, 38, 175–194, 1987.
62. Karsan, A., Cornejo, C.J., Winn, R.K., Schwartz, B.R., Way, W., Lannir, N.,
     Gershoni-Baruch, R., Etzioni, A., Ochs, H.D., and Harlan, J.M., Leukocyte adhesion
     deficiency type II is a generalized defect of the novo GDP-fucose biosynthesis.
     Endothelial cell fucosylation is not required for neutrophil rolling an human nonlym-
     phoid endothelium, *J. Clin. Invest.*, 101, 2438–2445, 1998.
63. Lübke, T., Marquardt, T., von Figura, K., and Körner, C., A new type of carbohydrate-
     deficient glycoprotein syndrome due to a decreased import of GDP-fucose into the
     Golgi, *J. Biol. Chem.*, 274, 25986–25989, 1999.
64. Lühn, K., Wild, M.K., Eckhardt, M., Gerardy-Schahn, R., and Vestweber, D., The
     defective gene in leukocyte adhesion deficiency II encodes a putative GDP-fucose
     transporter, *Nat. Genet.*, 28, 69–72, 2001.
65. Lübke, T., Marquardt, T., Etzioni, A., Hartmann, E., von Figura, K., and Körner, C.,
     Complementation cloning identifies CDG-IIc (LAD II) a new type of congenital dis-
     order of glycosylation, as a GDP-fucose transporter deficiency, *Nat. Genet.*, 28,
     73–76, 2001.
66. Etzioni, A., Sturla, L., Antonellis, A., Green, E.D., Gershoni-Baruch, R., Berninsone,
     P.M., Hirschberg, C.B., and Tonetti, M., Leukocyte adhesion deficiency (LAD) type
     II/carbohydrate deficient glycoprotein (CDG) IIc founder effect and genotype/pheno-
     type correlation, *Am. J. Med. Genet.*, 110, 131–135, 2002.
67. Marquardt, T., Lühn, K., Srikrishna, G., Freeze, H.H., Harms, E., and Vestweber, D.,
     Correction of leukocyte adhesion deficiency type II with oral fucose, *Blood*, 94,
     3976–3985, 1999.
68. Lühn, K., Marquardt, T., Harms, E., and Vestveber, D., Discontinuation of fucose
     therapy in LAD II causes rapid loss of selectin ligands and rise of leukocyte counts,
     Blood, 97, 330–332, 2001.
69. Hanßke, B., Thiel, C., Lübke, T., Hasilik, M., Höning, S., Peters, V., Heidemann, P.H.,
     Hoffmann, G.F., Berger, E.C., von Figura, K., and Körner, C., Deficiency of UDP-
     galactose: *N*-acetylglucosamine β-1,4-galactosyltransferase I as cause of the congen-
     ital disorder of glycosylation type IId (CDG-IId), *J. Clin. Invest.*, 109, 725–733, 2002.
70. Peters, V., Penzien, J.M., Reiter, G., Körner, C., Hackler, R., Assmann, B., Fang, J.,
     Schaefer, J.R., Hoffmann, G.F., and Heidemann, P.H., Congenital disorder of glyco-
     sylation IId (CDG-IId) —a new entity: clinical presentation with Dandy-Walker mal-
     formation and myopathy, *Neuropediatrics*, 33, 27–32, 2002.

71.   Martinez-Duncker, I., Candelier, J.J., Dupre, T., de Lonlay, P., Elvim, C., Gougerot-Pocidalo, M.A., Youssef, N., Trichet, C., Tchernia, G., Oriol, R., and Mollicone, R., The Nucleotide Sugar Transporters in Relation to CDG-IIc and II-e, *Second International Meeting on Congenital Disorders of Glycosylation,* Acitrezza (Catania), Sicily, 2003, p. 31.

72.   Almeida, R., Levery, S.B., Mandel, U., Kresse, H., Schwientek, T., Bennett, E.P., and Clausen, H., Cloning and expression of a proteoglycan UDP-galactose β-xylose β1, 4-galactosyltransferase I. A seventh member of the human β4-galactosyltransferase gene family, *J. Biol. Chem.,* 274, 26165–26171, 1999.

73.   Kresse, H., Rosthoj, S., Quentin, E., Hollmann, J., Glossl, J., Okada, S., and Tonnesen, T., Glycosaminoglycan-free small proteoglycan core protein is secreted by fibroblasts from a patient with a syndrome resemblin progeroid. *Am. J. Hum. Genet.,* 41, 436–453, 1987.

74.   Beavan, L.A., Quentin-Hoffmann, E., Schonherr, E., Snigula, F., Leroy, J.G., and Kresse, H., Deficient expression of decorin in infantile progeroid patients, *J. Biol. Chem.,* 268, 9856–9862, 1993.

75.   Gu, J. and Wada, Y., Aberrant expression of decorin and biglycan genes in the carbohydrate-deficient glycoprotein syndrome, *J. Biochem.,* (Tokyo) 117, 1276–1279, 1995.

76.   Quentin, E., Gladen, A., Roden, L., and Kresse, H., A genetic defect in the biosynthesis of dermatan sulfate proteoglycan: galactosyltransferase I deficiency in fibroblasts from a patient with a progeroid syndrome, *Proc. Natl. Acad. Sci. USA,* 87, 1342–1346, 1990.

77.   Okajama, T., Fukumoto, S., Furukawa, K., and Urano, T., Molecular basis for the progeroid variant of Ehlers–Danlos syndrome. Identification and characterization of two mutations in galactosyltransferase I gene, *J. Biol. Chem.,* 274, 28841–28844, 1999.

78.   Mao, J.-R. and Bristow, J., The Ehlers–Danlos syndrome: on beyond collagens, *J. Clin. Invest.,* 107, 1063–1069, 2001.

79.   Ameye, L. and Young, M.F., Mice deficient in small leucine-rich proteoglycans: novel in vivo models for osteoporosis, osteoarthritis, Ehlers–Danlos syndrome, muscular dystrophy, and corneal diseases, *Glycobiol.,* 12, 107R–116R, 2002.

80.   Olguin, H.C., Santander, C., and Brandan, E., Inhibition of myoblast migration via decorin expression is critical for normal skeletal muscle differentiation, *Dev. Biol.,* 259, 209–224, 2003.

81.   Danielson, K.G., Baribault, H., Holmes, D.F., Graham, H., Kadler, K.E., and Iozzo, R.V., Targeted disruption of decorin leads to abnormal collagen fibril morphology and skin fragility, *J. Cell Biol.,* 136, 729–743, 1997.

82.   Corsi, A., Xu, T., Chen, X.D., Boyde, A., Liang, J., Mankani, M., Sommer, B., Iozzo, R.V., Eichstetter, I ., Robey, P.G., Bianco, P., and Young, M.F., Phenotypic effects of biglycan deficiency are linked to collagen fibril abnormalities, are synergized by decorin deficiency, and mimic Ehlers–Danlos-like changes in bone and other connective tissues, *J. Bone Miner. Res.,* 17, 1180–1189, 2002.

83.   Bowe, M.A., Mendis, D.B., and Fallon, J.R., The small leucine-rich repeat proteoglycan biglycan binds to α-dystroglycan and is upregulated in dystrophic muscle, *J. Cell Biol.,* 148, 801–810, 2000.

84.   Solomon, L., Hereditary multiple exostoses, *J. Bone Joint. Surg. Am.,* 45, 292–304, 1963.

85.   Lüdecke, H.J., Wagner, M.J., Nardmann, J., La Pillo, B., Parrish, J.E., Willems, P.J., Haan, E.A., Frydman, M., Hamers, G.J., Wells D.E., and Horsthemke, B., Molecular dissection of a contigous gene syndrome: localization of the genes involved in the Langer–Giedion syndrome, *Hum. Mol. Genet.,* 4, 31–36, 1995.

86. Wuyts, W., Ramlakhan, S., Van Hul, W., Hecht, J.T., van den Ouweland, A.M.W., Raskind, W.H., Hofstede, F.C., Reyniers, E., Wells, D.E., de Vries B., Conrad, E.U., Hill, A., Zalatayev, D., Weissenbach, J., Wagner, M.J., Bakker, E., Halley, D.J.J., and Willems, P.J., Refinement of the multiple exostoses locus (EXT2) to a 3-cM interval on chromosome 11, *Am. J. Hum. Genet.*, 57, 382–387, 1995.

87. Le Merrer, M., Legeai-Mallet, L., Jeannin, P.M., Horsthemke, B., Schinzel, A., Plauchu, H., Toutain, A., Achard, F., Munnich, A., and Maroteaux, P., A gene for hereditary multiple exostoses maps to chromosome 19p, *Hum. Mol. Genet.*, 3, 717–722, 1994.

88. Ahn, J., Ludecke, H.J., Lindow, S., Horton, W.A., Lee, B., Wagner, M.J., Horsthemke, B., and Wells, D.E., Cloning of the putative tumour suppressor gene for hereditary multiple exostoses (EXT1), *Nat. Genet.*, 11, 137–143, 1995.

89. Wuyts, W., Van Hul, W., Wauters, J., Nemtsova, M., Reyniers, E., Van Hul, E.V., De Boulle, K., de Vries, B.B., Hendrickx, J., Herrygers, I., Bossuyt, P., Balemans, W., Fransen, E., Vits, L., Coucke, P., Nowak, N.J., Shows, T.B., Mallet, L., van den Ouweland, A.M., McGaughran, J., Halley, D.J., and Willems, P.J., Positional cloning of a gene involved in hereditary multiple exostoses, *Hum. Mol. Genet.*, 5, 1547–1557, 1996.

90. Senay, C., Lind, T., Muguruma, K., Tone, Y., Kitagawa, H., Sugahara, K., Lindholt, K., Lindahl, U., and Kusche-Gullberg, M., The EXT1/EXT2 tumor suppressors: catalytic activities and role in heparan sulfate biosynthesis, *EMBO Rep.*, 1, 282–286, 2000.

91. McCormick, C., Duncan, G., Goutsos, K.T., and Tufaro, F., The putative tumor suppressors EXT1 and EXT2 form a stable complex that accumulates in the Golgi apparatus and catalyzes the synthesis of heparan sulfate, *Proc. Natl. Acad Sci. USA.*, 97, 668–673, 2000.

92. Lin, X., Wei, G., Shi, Z.Z., Dryer, L., Esko, J.D., Wells, D.E., and Matzuk, M.M., Disruption of gastrulation and heparan sulfate biosynthesis in EXT1-deficient mice, *Dev. Biol.*, 224, 299–311, 2000.

93. Wuyts, W., Van Hul, W., Hendrickx, J., Speleman, F., Wauters, J., De Boulle, K., Van Roy, N., Van Agtmael, T., Bossuyt, P.J., and Willems, P.J., Identification and characterization of a novel member of the EXT gene family, EXTL2, *Eur. J. Hum. Genet.*, 5, 382–389, 1997.

94. van Hul, W., Wuyts, W., Hendrickx, J., Speleman, F., Wauters, J., De Boulle, K., Van Roy, N., Bossuyt, P., and Willems, P.J., Identification of a third EXT-like gene (EXTL3) belonging to the EXT gene family, *Genomics*, 47, 230–237, 1998.

95. Kim, B.-T., Kitagawa, H., Tamura, J.-I., Saito, T., Kusche-Gullberg, M., Lindahl, U., and Sugahara, K., Human suppressor EXT gene family members EXTL1 and EXTL3 encode α1,4-*N*-acetylglucosaminyltransferases that likely are involved in heparan sulfate/heparin biosynthesis, *Proc. Natl. Acad. Sci. USA*, 98, 7176–7187, 2001.

96. Duncan, G, McCormick, C., and Tufaro, F., The link between heparan sulfate and hereditary bone disease: finding a function for the EXT family of putative tumor suppressor proteins, *J. Clin. Invest.*, 108, 511–516, 2001.

97. Neri, G., Gurrieri, F., Zanni, G., and Lin, A., Clinical and molecular aspects of the Simpson-Golabi-Behmel syndrome, *Am. J. Med. Genet.*, 79, 279–283, 1998.

98. Pilia, G., Hughes-Benzie, R.M., MacKenzie, A., Baybayan, P., Chen, E.Y., Huber, R., Neri, G., Cao, A., Forabosco, A., and Schlessinger, D., Mutations in GPC3 a glypican gene, cause the Simpson–Golabi–Behmel overgrowthsyndrome, *Nat. Genet.*, 12, 241–247, 1996.

99. Novelli, G. and Reichardt, J.K., Molecular basis of disorders of human galactose metabolism: past, present, and future, *Mol. Genet. Metab.*, 71, 62–65, 2000.

100. Jaeken, J., Kint, J., and Snaapen, L., Serum lysosomal abnormalities in galactosemia, *Lancet*, 340, 1472–1473, 1992.

101. Henry, M.D. and Campbell, K.P., Dystroglycan inside and out, *Curr. Opin. Cell Biol.*, 11, 602–607, 1999.

102. Blake, D.J., Weir, A., Newey, S.E., and Davis, K.E., Function and genetics of dystrophin and dystrophin-related proteins in muscle, *Physiol. Rev.*, 82, 291–329, 2002.

103. Michele, D.E. and Campbell, K.P., Dystrophin–glycoprotein complex: post-translational processing and dystroglycan function, *J. Biol. Chem.*, 278, 15457–15460, 2003.

104. Martin, P.T., Dystroglycan glycosylation and its role in matrix binding in skeletal muscles, *Glycobiology*, 13, 55R–66R, 2003.

105. Ohlendieck, K., Ervasti, J.M., Matsumura, K., Kahl, S.D., Leveille, C.J., and Campbell, K.P., Dystrophin-related protein is localized to neuromuscular junctions of adult skeletal muscle, *Neuron*, 7, 499–508, 1991.

106. Patton, B.L., Miner, J.H., Chiu, A.Y., and Sanes, J.R., Distribution and function of laminins in the neuromuscular system of developing, adult, and mutant mice, *J. Cell Biol.*, 139, 1507–1521, 1997.

107. Matsumura, K., Ervasti, J.M., Ohlendieck, K., Kahl, S.D., and Campbell, K.P., Association of dystrophin-related protein with dystrophin-associated proteins in mdx mouse muscle, *Nature*, 360, 588-591, 1992.

108. Grady, R.M., Merlie, J.P., and Sanes, J.R., Subtle neuromuscular defects in utrophin-deficient mice, *J. Cell. Biol.*, 136, 871–882, 1997.

109. Deconinck, A.E., Potter, A.C., Tinsley, J.M., Wood, S.J., Vater, R., Young, C., Metzinger, L., Vincent, A., Slater, C.R., and Davies, K.E., Postsynaptic abnormalities at the neuromuscular junctions of utrophin-deficient mice, *J. Cell Biol.*, 136, 883–894, 1997.

110. Sanes, J.R. and Cheney, J.M., Lectin binding reveals a synapse-specific carbohydrate in skeletal muscle, *Nature*, 300, 646–647, 1982.

111. Jayasinha, V., Hoyte, K., Xia, B., and Martin, P.T., Overexpression of the CT GalNAc transferase inhibits muscular dystrophy in a cleavage-resistant dystroglycan mutant mouse, *Biochim. Biophys. Res. Commun.*, 302, 831–836, 2003.

112. Ibraghimov-Beskrovnaya, O., Ervasti, J.M., Leveille, C.J., Slaughter, C.A., Sernett, S.W., and Campbell, K.P., Primary structure of dystrophin-aasociated glycoproteins linking dystrophin to the extracellular matrix, *Nature*, 355, 696–702, 1992.

113. Talts, J., Andac, Z., Göhring, W., Brancaccio, A., and Timpl, R., Binding of the G domains of laminin α1 and α2 chains and perlecan to heparin, sulfatides, α-dystroglycan and several extracellular matrix proteins, *EMBO. J.,* 18, 863–870, 1999.

114. Li, S., Edgar, D., Fassler, R., Wadsworth, W., and Yurchenco, P.D., The role of laminin in embryonic cell polarization and tissue organization, *Dev. Cell*, 4, 613–624, 2003.

115. Ruegg, M.A. and Bixby, J.L., Agrin orchestrates synaptic differentiation at the vertebrate neuromuscular junction, *Trends. Neurosci.*, 21, 22–27, 1998.

116. Bezakova, G. and Ruegg, M.A., New insights into the roles of agrin, *Nat. Rev. Mol. Cell Biol.,* 4, 295–308, 2003.

117. Cantallops, I. and Cline, H.T., Synapse formation: if it looks like a duck, and quacks like a duck, *Curr. Biol.*, 10, R620–R623, 2000.

118. Dean, C., Scholl, F.G., Choih, J., DeMaria, S., Berger, J., Isacoff, E., and Scheiffele, P., Neurexin mediates the assembly of presynaptic terminals, *Nat. Neurosci.*, 6, 708–716, 2003.

119. Sugita, S., Saito, F., Tang, J., Satz, J., Campbell, K., and Sudhoff, T.C., A stoichiometric complex of neurexins and dystroglycan in brain. *J. Cell Biol.,* 154, 435–445, 2001.

120. Ilsley, J.L., Sudol, M., and Winder, S.J., The interaction of dystrophin with â-dystroglycan is regulated by tyrosine phosphorylation, *Cell Signal,* 13, 625–632, 2001.

121. Oak, S.A., Russo, K., Petrucci, T.C., and Jarrett, H.W., Mouse α1-syntrophin binding to Grb2: further evidence of a role for syntrophin in cell signaling, *Biochemistry*, 40, 11270–11278, 2001.

122. Ilsley, J.L., Sudol, M., and Winder, S.J., The WW domain: linking cell signalling to the membrane cytoskeleton, *Cell Signal*, 14, 183–189, 2002.

123. Levi, S., Grady, R.M., Henry, M.D., Campbell, K.P., Sanes, J.R., and Craig, A.M., Dystroglycan is selectively associated with inhibitory GABAergic synapses but is dispensable for their differentiation, *J. Neurosci.*, 22, 4274–4285, 2002.

124. Chockalingam, P.S., Cholera, R., Oak, S.A., Zheng, Y., Jarrett, H.W., and Thomason. D.B., Dystrophin-glycoprotein complex and Ras and Rho GTPase signaling are altered in muscle atrophy, *Am. J. Physiol. Cell Physiol.*, 283, C500–C511, 2002.

125. Langenbach, K.J., and Rando, T.A., Inhibition of dystroglycan binding to laminin disrupts the PI3K/AKT pathway and survival signaling in muscle cells, *Muscle Nerve*, 26, 644–653, 2002.

126. Ferletta, M., Kikkawa, Y., Yu, H., Talts, J.F., Durbeej, M., Sonnenberg, A., Timpl, R., Campbell, K.P., Ekblom, P., and Genersch, E., Opposing roles of integrin α6Aβ1 and dystroglycan in laminin-mediated extracellular signal-regulated kinase activation, *Mol. Biol. Cell*, 14, 2088–2103, 2003.

127. Williamson, R.A., Henry, M.D., Daniels, K.J., Hrstka, R.F., Lee, J.C., Sunada, Y., Ibraghimov-Beskrovnaya, O., and Campbell, K.P., Dystroglycan is essential for early embryonic development: disruption of Reichart's membrane in Dag1-null mice, *Hum. Mol. Genet.*, 6, 831–841, 1997.

128. Matsumura, K., Burghes, A.H.M., Mora, M., Tome, F.M.S., Morandi, L., Cornello, F., Leturcq, F., Jeanpierre, M., Kaplan, J.C., Reinert, P., Fardeau, M., Mendell, J.R., and Campbell, K.P., Immunohistochemical analysis of dystrophin-associated proteins in Becker-Duchenne muscular dystrophy with huge in-frame deletions in the NH-2-terminal and rod domains of dystrophin. *J. Clin. Invest.* 93, 99–105, 1994.

129. Duggan, D.J., Gorospe, J.R., Fanin, M., Hoffman, E.P., and Angelini, C., Mutations in the sarcoglycan genes in patients with myopathy, *N. Engl. J. Med.*, 336, 618–624, 1997.

130. Coral-Vazquez, R., Cohn, R.D., Moore, S.A., Hill, J.A., Weiss, R.M., Davisson, R.L., Straub, V., Barresi, R., Bansal, D., Hrstka, R.F., Williamson, R., and Campbell, K.P., Disruption of the sarcoglycan-sarcospan complex in vascular smooth muscle: a novel mechanism for cardiomyopathy and muscular dystrophy, *Cell*, 98, 465–474, 1999.

131. Richard, I., Broux, O., Allamand, V., Fourgerousse, F., Chiannilkulchai, N., Bourg, N., Brenguier, L., Devaud, C., Pasturaud, P., Roudaut, C., Hillaire, D., Passos-Bueno, M.R., Zatz, M., Tischfield, J.A., Fardeau, M., Jackson, C.E., Cohen, D., and Beckmann, J.S., Mutations in the proteolytic enzyme calpain 3 cause limb-girdle muscular dystrophy type 2A, *Cell*, 81, 27–40, 1995.

132. Helbling-Leclerc, A., Zhang, X., Topaloglu, H., Cruaud, C., Tesson, F., Weissenbach, J., Tome, F.M.S., Schwartz, K., Fardeau, M., Tryggvason, K.S., and Guicheney, P., Mutations in the laminin α2-chain (LAMA2) cause merosin-deficient congenital muscular dystrophy, *Nat. Genet*, 11, 216–218, 1995.

133. Pegoraro, E., Marks, H., Garcia, C.A., Crawford, T., Mancias, P., Connolly, A.M., Fanin, M., Martinello, F., Trevisan, C.P., Angelini, C., Stella, A., Scavina, M., Munk, R.L., Servidi, S., Bonnemann, C.C., Bertorini, T., Acsadi, G., Thompson, C.E., Gagnon, D., Hoganson, G., Carver, V., Zimmerman, R.A., and Hoffman, E.P., Laminin α2 muscular dystrophy: genotype phenotype studies of 22 patients, *Neurology*, 51, 101–110, 1998.

134. Muntoni, F., Brockington, M., Blake, D.J., Torelli, S., and Brown, S.C., Defective glycosylation in muscular dystrophy, *Lancet*, 360, 1419–1421, 2002.

135. Hewitt, J.E. and Grewal, P.K., Glycosylation defects in inherited muscle disease. *Cell Mol. Life Sci.*, 60, 251–258, 2003.

136. Martin, P.T. and Freeze, H.H., Glycobiology of neuromuscular disorders, *Glycobiology*, 13, 67R–75R, 2003.

137. De Michele, D.E., Barresi, R., Kanagawa, M., Saito, F., Cohn, R.D., Satz, J.S., Dollar, J., Nishino, I., Kelley, R.I., Somer, H., Straub, V., Mathews, K.D., Moore, S.A., and Campbell, K.P., Post-translational disruption of dystroglycan-ligand interactions in congenital muscular dystrophies, *Nature*, 418, 417–422, 2002.

138. Hayashi, Y.K., Ogawa, M., Tagawa, K., Noguchi, S., Ishihara, T., Nonaka, I., and Arahata, K., Selective deficiency of α dystroglycan in Fukuyama-type congenital muscular dystrophy. *Neurology*, 57, 115–121, 2001.

139. Kano, H., Kobayashi, K., Herrmann, R., Tachikawa, M., Manya, H., Nishino, I., Nonaka, I., Straub, V., Talim, B., Voit, T., Topaloglu, H., Endo, T., Yoshikawa, H., and Toda, T., Deficiency of α-dystroglycan in muscle-eye-brain disease. *Biochem, Biophys. Res. Commun.*, 291, 1283–1286, 2002.

140. Moore, S.A., Saito, F., Chen, J., Michele, D.E., Henry, M.D., Messing, A., Cohn, R.D., Ross-Barta, S.E., Westra, S., Williamson, R.A., Hoshi, T., and Campbell, K.P., Deletion of brain dystroglycan recapitulates aspects of congenital muscular dystrophy. *Nature*, 418, 422–425, 2002.

141. Beltran-Valero de Bernabe, D., Currier, S., Steinbrecher, A., Celli, J., van Beusekom, E., van der Zwaag, B., Kayserili, H., Merlini, L., Chitayat, D., Dobyns, D.B., Cormand, B., Lehesjoki, A.E., Cruces, J., Voit, T., Walsch, C.A., van Bokhoven, H., and Brunner, H.G., Mutations in the *O*-mannosyltransferase gene POMT1 give rise to the severe neuronal migration disorder Walker-Warburg syndrome. *Am. J. Hum. Genet.* 71, 1033–1043, 2002.

142. Jurado, L.A., Coloma, A., and Cruces, J., Identification of a human homolog of the Drosophila rotated abdomen gene (POMT1) encoding a putative protein *O*-mannosyltransferase, and assignment to human chromosome 9q34.1. *Genomics*, 58, 171–180, 1999.

143. Kobayashi, K., Nakahori, Y., Miyake, M., Matsumura K., Kondo-Iida, E., Nomura, Y., Segawa, M., Yoshioka, M., Saito, K., Osawa, M., Hamano, K., Sakakihara, Y., Nonaka, I., Nokagome, Y., Kanazawa, I., Nakamura, Y., Tokunaga, K., and Toda, T., An ancient retrotransposal insertion causes Fukuyama-type congenital muscular dystrophy. *Nature*, 394, 388–392, 1998.

144. Toda, T., Kobayashi, K., Kondo-Iida, E., Sasaki, J., and Nakamura, Y., The Fukuyama congenital muscular dystrophy story, *Neuromusc. Disord.*, 10, 153–159, 2000.

145. Aravind, L. and Koonin, E.V., The fukutin protein family — predicted enzymes modifying cell-surface molecules, *Curr. Biol.*, 9, R836–R837, 1999.

146. Yoshida, A., Kobayashi, K., Manya, H., Taniguchi, K., Kano, H., Mizuno, M., Inazu, T., Mitsuhashi, H., Takahashi, S., Takeuchi, M., Herrmann, R., Straub, V., Talim, B., Voit, T., Topaloglu, H., Toda, T., and Endo, T., Muscular dystrophy and neuronal migration disorder caused by mutations in a glycosyltransferase, POMGnT1, *Dev. Cell*, 1, 717–724, 2001.

147. Schachter, H., The role of the GlcNAcβ1,2 Man α- moiety in mammalian development. Null mutations of the genes encoding UDP-*N*-acetylglucosamine: α-3-D-manosideβ-1,2-*N*-acetylglucosaminyltransferase I and UDP-*N*-acetylglucosamine: α-D-mannoside β-1,2-*N*-acetylglucosaminyl transferase 1,2 cause embryonic lethality and congenital muscular dystrophy in mice and men, respectively, *Biochim. Biophys. Acta*, 1573, 292–300, 2002.

148. Cormand, B., Avela, K., Pihko, H., Santavuori, P., Talim, B., Topaloglu, H., de la Chapelle, A., and Lehesjoki, A.E., Assignment of the muscle–eye–brain disease gene tp 1p32–p34 by linkage analysis and homozygosity mapping, *Am. J. Hum. Genet.,* 64, 126–135, 1999.

149. Brockington, M., Blake, D.J., Prandini, P., Brown, S.C., Torelli, S., Benson, M.A., Ponting, C.P., Estournet, B., Romero, N.B., Mercuri, E., Voit, T., Sewry, C.A., Guicheney, P., and Muntoni, F., Mutations in the fukutin-related protein gene (FKRP) cause a form of a congenital muscular dystrophy with secondary laminin $\alpha 2$ deficiency and abnormal glycosylation of α-dystroglycan, *Am. J. Hum. Genet.,* 69, 1198–1209, 2001.

150. Brockington, M., Yuva, Y., Prandini, P., Brown, S.C., Torelli, S., Benson, M.A., Herrmann, R., Anderson, L.V., Bashir, R., Burgunder, J.M., Fallet, S., Romero, N., Fardeau, M., Straub, V., Storey, G., Pollitt, C., Richard, I., Sewry, C.A., Bushby, K., Voit, T., Blake, D.J., and Muntoni, F., Mutations in the fukutin related protein gene (FKRP) identifies limb-girdle muscular dystrophy 2I as a milder allelic variant of congenital muscular dystrophy MDC1C, *Hum. Mol. Genet.,* 10, 2851–2859, 2001.

151. Driss, A., Amouri, R., Ben Hamida, C., Souilem, S., Gouider-Khouja, N., Ben Hamida, M., and Hentati, F., A new locus for autosomal recessive limbgirdle muscular dystrophy in a large consanguineous Tunisian family maps to chromosome 19q13.3, *Neuromusc. Disord.,* 10, 240–246, 2000.

152. Esapa, C.T., Benson, M.A., Schroeder, J.E., Martin-Rendon, E., Brockington, M., Brown, S.C., Muntoni, F., Kroger, S., and Blake, D.J., Functional requirements for fukutin-related protein in the Golgi apparatus, *Hum. Mol. Genet.,* 11, 3319–3331, 2002.

153. Topaloglu, H., Brockington, M., Yuva, Y., Talim, B., Haliloglu, G., Blake, D., Torelli, S., Brown, S.C., and Muntoni, F., FKRP gene mutations cause congenital muscular dystrophy, mental retardation, and cerebellar cysts, *Neurology,* 60, 988–992, 2003.

154. Mercuri, E., Brockington, M., Straub, V., Quijano-Roy, S., Yuva, Y., Herrmann, R., Brown, S.C., Torelli, S., Dubovitz, V., Blake, D.J., Romero, N.B., Estournet, B., Sewry, C.A., Guicheney, P., Voit, T., and Muntoni, F., Phenotypic spectrum associated with mutations in the fukutin-related protein gene, *Ann. Neurol.,* 53, 537–542, 2003.

155. Eisenberg, I., Avidan, N., Potikha, T., Hochner, H., Chen, M.,Olender, T., Barash, M., Shemesh, M., Sadeh, M., Grabov-Nardini, G., Shmilevich, I., Friedmann, A., Karpati, G., Bradley, W.G., Baumbach, L., Lancet, D., Asher, E.B., Beckmann, J.S., Argov, Z., and Mitrani-Rosenbaum, S., The UDP-*N*-acetylglucosamine 2-epimerase/*N*-acetylmannosamine kinase gene is mutated in recessive hereditary inclusion body myopathy, *Nat. Genet.,* 29, 83–87, 2001.

156. Nishino, I., Noguchi, S., Murayama, K., Driss, A., Sugie, K., Oya, Y., Nagata, T., Chida, K., Takahashi, T., Takusa, Y., Ohi, T., Nishimiya, J., Sunohara, N., Ciafaloni, E., Kawai, M., Aoki, M., and Nonaka, I., Distal myopathy with rimmed vacuoles is allelic to hereditary inclusion body myopathy, *Neurology,* 59, 1689–93, 2002.

157. Kayashima, T., Matsuo, H., Satoh, A., Ohta, T., Yoshiura, K., Matsumoto, N., Nakane, Y., Niikawa, N., and Kishino, T., Nonaka myopathy is caused by mutations in the UDP-*N*-acetylglucosamine-2-epimerase/*N*-acetylmannosamine kinase gene (GNE), *J. Hum. Genet.,* 47, 77–79, 2002.

158. Seppala, R., Lehto, V.P., and Gahl, W.A., Mutations in the human UDP-*N*-acetylglu-cosamine 2-epimerase gene define the disease sialuria and the allosteric site of the enzyme, *Am. J. Hum. Genet.,* 64, 1563–1569, 1999.

159. Schwarzkopf, M., Knobeloch, K.P., Rohde, E., Hinderlich, S., Wiechens, N., Lucka, L., Horak, I., Reutter, W., and Horstkorte, R., Sialylation is essential for early development in mice, *Proc. Natl. Acad. Sci. USA,* 99, 5267–5270, 2002.

160. Grewal, P. and Hewitt, J.E., Mutation of large, which encodes a putative glycosyl-transferase in an animal model of muscular dystrophy, *Biochim. Biophys. Acta*, 1573, 216–224, 2002.
161. Peyrard, M., Seroussi, E., Sandberg-Nordquist, A.C., Xie, Y.G., Han, F.Y., Fransson, I., Collins, J., Dunham, I., Kost-Alimova, M., Imreh, S., and Dumanski, J.P., The human LARGE gene from 22q12.3–q13.1 is a new distinct member of the glycosyl-transferase gene family, *Proc. Natl. Acad. Sci. USA*, 96, 598–603, 1999.
162. Heinrichs, D.E., Monteiro, M.A., Perry, M.B., Whitfield. C., The assembly system for the lipopolysaccharide R2 core-type of *Escherichia coli* is a hybrid of those found in *Escherichia coli* K-12 and *Salmonella enterica*, *J. Biol. Chem.*, 273, 8849–8859, 1998.
163. Sasaki, K., Kurata-Miura, K., Ujita, M., Angata, K., Nakagawa, S., Sekine, S., Nishi, T., Fukuda. M., Expression cloning of cDNA encoding a human β-1,3-N-acetylglu-cosaminyltransferase that is essential for poly-N-acetyllactosamine synthesis, *Proc. Natl. Acad. Sci. USA*, 94, 14294–14299, 1997.
164. Grünewald, S., Matthijs, G., and Jaeken, J., Congenital disorders of glycosylation: a review, *Ped. Res.*, 52, 618–624, 2002.
165. Stibler, H., Carbohydrate-deficient transferrin in serum: a new marker of potentially harmful alcohol consumption reviewed, *Clin. Chem.*, 37, 2029–2037, 1991.
166. Jaeken, J., Stibler, H., and Hagberg, B., The carbohydrate-deficient glycoprotein syndrome. A new inherited multisystemic disease with severe nervous system involvement, *Acta Paediatr. Scand. Suppl.*, 375, 1–71, 1991
167. Arndt, T., Carbohydrate-deficient transferrin as a marker of chronic alcohol abuse: a critical review of preanalysis, analysis, and interpretation, *Clin. Chem.*, 47, 13–27, 2001.
168. Marquardt, T. and Denecke, J., Congenital disorders of glycosylation: review of their molecular bases, clinical presentations and specific therapies, *Eur. J. Pediatr.*, 162, 359–379, 2003.
169. Jaeken, J., Pirard, M., Adamowicz, M., Pronicka, M., and van Schaftingen. E., Inhibition of phosphomannoisomerase by fructose 1-phosphate: an explanation for defective N-glycosylation in hereditary fructose intolerance, *Pediatr. Res.*, 40, 764–766, 1996.
170. Charlwood, J., Clayton, P., Keir, G., Mian, N., Young, E., and Winchester, B., Defective galactosylation of serum transferrin in galactosemia, *Glycobiology*, 8, 351–357, 1998.
171. Sillanaukee, P., Strid, N., Allen, J.P., Litten. R.Z., Possible reason why heavy drinking increases carbohydrate-deficient transferrin, *Alcohol. Clin. Exp. Res.*, 25, 34–40, 2001.
172. Inoue, T., Yamauchi, M., and Ohkawa, K., Structural studies on sugar chains of carbohydrate-deficient transferrin from patients with alcoholic liver disease using lectin affinity electrophoresis, *Electrophoresis*, 20, 452–457, 1999.
173. Flahaut, C., Michalski , J.C., Danel, T., Humbert, M.H., and Klein, A., The effects of ethanol on the glycosylation of human transferrin, *Glycobiology*, 13, 191–198, 2003.
174. Cottalasso, D., Gazzo, P., Dapino, D., Domenicotti, C., Pronzato, MA., Traverso, N., Bellochio, A., Nanni, G., and Marinari, U.M., Effect of chronic ethanol consumption on glycosylation processes in rat liver microsomes and Golgi apparatus, *Alcohol. Alcohol.*, 31, 51–59, 1996.
175. Kamboch, M.I. and Ferrell, R.E., Human transferrin polymorphism, *Hum. Hered.*, 37, 65–81, 1987.
176. Ohno, K., Yuasa, I., Akaboshi, S., Itoh, M., Yoshida, K., Ehara, H., Ochiai, Y., and Takeshita, K., The carbohydrate deficient glycoprotein syndrome in three Japanese children, *Brain. Dev.*, 14, 30–35, 1992.

177. Dupré, T., Cuer, M., Barrot, S., Barnier, A., Cormier-Daire, V., Munnich, A., Durand, G., and Seta, N., Congenital disorder of glycosylation Ia with deficient phosphomannomutase activity but with normal plasma glycoprotein pattern, *Clin. Chem.*, 47, 132–134, 2001.

178. Stibler, H., Holzbach, U., and Kristiansson, B., Isoforms and levels of transferrin, antithrombin, α1-antitripsin and thyroxine-binding globulin in 48 patients with carbohydrate-deficient glycoprotein syndrome type I, *Scand. J. Clin. Lab. Invest.*, 58, 55–61, 1998.

179. Li, Y., Ogata, Y., Freeze, H.H., Scott, C.R., Turecek, F., and Gelb, M.H., Affinity capture and elution/electrospray ionization mass spectrometry assay of phosphomannomutase and phosphomannose isomerase for the multiplex analysis of congenital disorders of glycosylation types Ia and Ib, *Anal. Chem.*, 75, 42–48, 2003.

180. Zhang, W., Betel, D., and Schachter, H., Cloning and expression of a novel UDP-GlcNAc:α-D-mannoside β1,2-*N*-acetylglucosaminyltransferase homologous to UDP-GlcNAc: α-3-D-mannoside β1,2-*N*-acetylglucosaminyltransferase I, *Biochem. J.*, 361, 153–162, 2002.

181. Panneerselvam, K., and Freeze, H.H., Mannose corrects altered *N*-glycosylation in carbohydrate-deficient glycoprotein syndrome fibroblasts. *J. Clin. Invest.*, 97, 1478–1487, 1996.

182. Volker, C., De Praeter, C.M., Hardt, B., Breuer, W., Kalz-Fuller, B., Van Coster, R.N., and Bause, E., Processing of *N*-linked carbohydrate chains in a patient with glucosidase I deficiency (CDG type IIb), *Glycobiology*, 12, 473–483, 2002.

183. Spiro, R.G., Spiro, M.J., and Bhoyroo, V.D., Studies on the regulation of the biosynthesis of glucose-containing oligosaccharide-lipids. Effect of energy deprivation, *J. Biol. Chem.*, 258, 9469–9476, 1983.

184. Coddeville, B., Carchon, H., Jaeken, J., Briand, G., and Spik, G., Determination of glycan structures and molecular masses of the glycovariants of serum transferrin from a patient with carbohydrate deficient syndrome type II, *Glycoconj. J.*, 15, 265–273, 1998.

185. Lacey, Y.M., Bergen, H.R., Magera, M.J., Naylor, S., and O'Brien, J.F., Rapid determination of transferrin isoforms by immunoaffinity liquid chromatography and electrospray mass spectrometry, *Clin. Chem.*, 47, 513–518, 2001.

186. Zhang, H., Li, X., Martin, D.B., and Aebersold, R., Identification and quantification of *N*-linked glycoproteins using hydrazide chemistry, stable isotope labeling and mass spectrometry, *Nat. Biotechnol.*, 21, 660–666, 2003.

187. Kaji, H., Saito, H., Yamauchi, Y., Shinkawa, T., Taoka, M., Hirabayashi, J., Kasai, K., Takahashi, N., and Isobe, T., Lectin affinity capture, isotope-coded tagging and mass spectrometry to identify *N*-linked glycoproteins, *Nat. Biotechnol.*, 21, 667–672, 2003.

188. Callewaert, N., Schollen, E., Vanhecke, A., Jaeken, J., Matthijs, G., and Contreras, R., Increased fucosylation and reduced branching of serum glycoprotein *N*-glycans in all known subtypes of congenital disorders of glycosylation I, *Glycobiology*, 13; 367–375, 2003.

189. Butler, M., Quelhas, D., Critchley, A.J., Carchon, H., Hebestreit, H.F., Hibbert, R.G., Vilarinho, L., Teles, E., Matthijs, G., Schollen, E., Argibay, P., Harvey, D.J., Dwek, R.A., Jaeken, J., and Rudd, P.M., Detailed glycan analysis of serum glycoproteins of patients with congenital disorders of glycosylation indicates the specific defective glycan processing step and provides an insight into pathogenesis, *Glycobiology*, 13, 601–622, 2003.

190. Mills, P., Mills, K., Clayton, P., Johnson, A., Whitehouse, D., and Winchester, B., Congenital disorders of glycosylation type I leads to altered processing of *N*-linked glycans, as well as underglycosylation, *Biochem. J.*, 359, 249–254, 2001.

191. Zdebska, E. and Kościelak J., A single-sample method for determination of carbohydrate and protein contents of glycoprotein bands separated by sodium dodecyl sulfate-polyacrylamide gel electrophoresis, *Anal. Biochem.*, 275, 171–179, 1999.

192. Zdebska, E., Adamczyk-Poplawska, M., and Kościelak, J., Glycophorin A in two patients with congenital dyserythropoietic anemia type I and type II is partly unglycosylated, *Acta Bioch. Polon.*, 47, 773–779, 2000.

193. Zdebska, E., Musielak, M., Jaeken, J., and Kościelak, J., Band 3 glycoprotein and glycophorin A from erythrocytes of children with congenital disorder of glycosylation type -Ia are underglycosylated, *Proteomics*, 1, 269–274, 2001.

194. Zdebska, E., Woźniewicz, B., Adamowicz-Salach, A., and Kościelak, J., Short report: erythrocyte membranes from a patient with congenital dyserythropoietic anaemia type I (CDA I) show identical, although less pronounced, glycoconjugate abnormalities to those from patients with CDA II (HEMPAS), *Br. J. Haematol.*, 110, 998–1001, 2000.

195. Zdebska, E., Golaszewska, E., Fabijańska-Mitek, J., Schachter, H., Shalev, H., Tamary, H., Sandström, H., Wahlin, A., and Kościelak, J., Glycoconjugate abnormalities in patients with congenital dyserythropoietic anemia type I, II and III, *Brit. J. Haematol.*, 114, 907–913, 2001.

196. Zdebska, E., Mendek-Czajkowska, E., Ploski, R., Woźniewicz, B., and Kościelak, J., Heterozygosity of CDAN II (HEMPAS) gene may be detected by the analysis of erythrocyte membrane glycoconjugates from healthy carriers, *Haematologica*, 87, 126–130, 2002.

197. Aebi, M. and Hennet, T., Congenital disorders of glycosylation: genetic model systems lead the way, *Trends. Cell. Biol.*, 11, 136–141, 2001.

198. Lecca, R., Berger, E.G., and Hennet, T., Transcriptome analysis of CDG I cells. Abstracts. *2nd International Meeting on Congenital Disorders of Glycosylation*, Acitrezza, Sicily, Apr. 2–6, 2003.

199. Forestier, L., Julien, R., and Lia-Baldini, A.S., Quantitation of transcripts and detection of mutations in CDG using a dedicated cDNA microarray. Abstracts. *2nd International Meeting on Congenital Disorders of Glycosylation*, Acitrezza, Sicily, Apr. 2–6, 2003.

200. Sala, G., Dupre, T., Seta, N., Codogno, P., and Ghidoni, R., Increased biosynthesis of glycosphingolipids in congenital disorder of glycosylation Ia (CDG-Ia) fibroblasts, *Pediatr. Res.*, 52, 645–651, 2002.

201. Panneerselvam, K. and Freeze, H.H., Mannose enters mammalian cells using a specific transporter that is insensitive to glucose, *J. Biol. Chem.*, 271, 9417–9421, 1996.

202. Panneerselvam, K., Etchison, J.R., and Freeze, H., Human fibroblasts prefer mannose over glucose as a source of mannose for *N*-glycosylation. Evidence for the functional importance of transported mannose, *J. Biol. Chem.*, 272, 23123–23129, 1997.

203. Rush, J.S., Panneerselvam, K., Waechter, C.J., and Freeze, H.H., Mannose supplementation corrects GDP-mannose deficiency in cultured fibroblasts from some patients with congenital disorders of glycosylation (CDG), *Glycobiology*, 10, 829–835, 2000.

204. Kjaergaard, S., Kristiansson, B., Stibler, H., Freeze, H.H., Schwartz, M., Martinsson, T., and Skovby, F., Failure of short-term mannose therapy of patients with carbohydrate-deficient glycoprotein syndrome type 1A, *Acta Paediatr.*, 87, 884–888, 1998.

205. Marquardt, T., Hasilik, M., Niehues, R., Herting, M., Muntau, A., Holzbach, U., Hanefeld, F., Freeze, H., and Harms, E., Mannose therapy in carbohydrate-deficient glycoprotein syndrome type I - first results, of the German multicenter study, *Amino. Acids*, 12, 389, 1997.

206. Mayatepek, E. and Kohlmüller, D., Mannose supplementation in carbohydrate-deficient glycoprotein syndrome type I and phosphomannomutase deficiency, *Eur. J. Pediatr.*, 157, 605–606, 1998.

207. Freeze, H.H., Update and perspectives on congenital disorders of glycosylation, *Glycobiology*, 11, 129R–143R, 2001

208. Duprė, T., Ogier-Denis, E., Moore, S.E., Cormier-Daire, V., Dehoux, M., Durand, G., Seta, N., and Codogno, P., Alteration of mannose transport in fibroblasts from type I carbohydrate deficient glycoprotein syndrome patients, *Biochim. Biophys. Acta.*, 1453, 369–377, 1999.

209. Schultz, C., Vajanaphanich, M., Harootunian, A.T., Sammak, P.J., Barrett, K.E., and Tsien, R.Y., Acetoxymethyl esters of phosphates, enhancement of the permeability and potency of cAMP, *J. Biol. Chem.*, 268, 6316–6322, 1993.

210. Rutschow, S., Thiem, J., Kranz, C., and Marquardt, T., Membrane-permeant derivatives of mannose-1-phosphate, *Bioorg. Med. Chem.*, 10, 4043–4049, 2002.

211. Freeze, H.H., Rush, J., Wu, X., deRossi, C., Norberg, T., Ichikawa, M., Karaoglu, D., Waechter, C., and Gilmore, R., CDG stories from th US: new defects, animal models, and potential therapy. Abstracts. *2nd International Meeting on Congenital Disorders of Glycosylation Acitrezza* Catania, Sicily, Apr. 2–6, 2003.

212. Etzioni, A. and Tonetti, M., Fucose supplementation in lekocyte adhesion deficiency type II, *Blood,* 95, 3641–3643, 2000.

213. Sturla, L., Puglielli, L., Tonetti, M., Berninsone, P., Hirschberg, C.B., de Flora, A., and Etzioni, A., Impairment of the Golgi GDP-L-fucose transport and unresponsiveness to fucose replacement therapy in LAD II patients, *Pediatr. Res.*, 49, 537–542, 2001.

214. Becker, D.J. and Lowe, J.B., Fucose: biosynthesis and biological functions in mammals, *Glycobiology,* 13, 41R–53R, 2003.

215. Niemela, R., Natunen, J., Majuri, M.L., Maaheimo, H., Helin, J., Lowe, J.B., Renkonen, O., and Renkonen, R., Complementary acceptor and site specificities of Fuc-TIV and T VII allow effective biosynthesis of sialyl-Tri Lex and related polylactosamines present on glycoprotein counterreceptors of selectins, *J. Biol. Chem*, 273, 4021–4026, 1998.

216. Huang, M.C., Zollner, O., Moll, T., Maly, P., Thall, A.D., Lowe, J.B., and Vestweber, D., P-selectin glycoprotein ligand-1 and E-selectin ligand-1 are differentially modified by fucosyltransferases Fuc-TIV and Fuc-TVII in mouse neutrophils, *J. Biol. Chem,* 275, 31353–31360, 2000.

NOTE ADDED IN PROOF

Three new types of CDG have been recently described: IK (Schwarz et al., *Am. J. Hum. Genet.*, 74, 472–481, 2004; Kranz et al., *Am. J. Hum. Genet.*, 74, 547–551, 2004) and 1L (Frank et al., *Am. J. Hum. Genet.*, 75, 146–150, 2004) which are due to deficiency of *hAlg1* and *hAlg9* gene, respectively. The third is caused by a mutation of the gene encoding a subunit of the conserved oligomeric golgi complex (COG7) (Wu et al., *Nat. Med.*, 10, 518–523, 2004). New therapeutic approaches were proposed (Shang and Lehrman, *J. Biol. Chem.*, 270, 9703–9712, 2004; Rano, N., *Engl. J. Med.*, 351, 1254–1255, 2004).

5 Role of Milk Carbohydrates in Preventing Bacterial Adhesion

Pablo Hueso, Samuel Martín-Sosa, and María-Jesús Martín

CONTENTS

I. MILK CARBOHYDRATES

Milk is the fundamental source of nutrients for the mammalian newborn. The milk from each species is specifically designed to fulfil the needs of its offspring: it is a complete food and is species specific.[1] Furthermore, the composition of milk changes markedly over the course of lactation in order to adapt to the new requirements of the neonate and is related to the growth rate of the newborn. For example, cow's milk has a high protein content (80% casein) since calves double their weight in only 35 days; by contrast, human milk has 3 to 4 times less protein than cow's milk (40 to 50% casein) and infants take 160 days to double their weight. The fat content of milk is mainly related to the energy needs of the species, which in turn are determined by factors such as climatic conditions. Although the carbohydrate content of milk has also been mainly considered from the energy point of view in the past, the possible role of milk carbohydrates as functional foods in the defense of the newborn is now under debate.

Milk contains different antimicrobial factors that protect the newborn from the most common infections and are crucial for survival while the immune system is still developing. Immunoglobulins are the main factor transferred in milk, although other compounds also cooperate in the newborn's defense, including proteins (lactoferrin and lysozyme), glycolipids, and carbohydrates.[2]

The starting point of milk carbohydrate research was the pioneer work of Polonovski and Lespagnol in the 1930s.[3] These authors isolated a levorotatory and reducing sugar from human milk that accompanied the common milk sugar lactose, which they called gynolactose. Several years ago it was reported that gynolactose was a mixture of complex oligosaccharides. This work was carried out in the laboratories of Jean Montreuil (Lille, France) and Richard Kuhn (Heidelberg, Germany) in the 1950s. Thereafter, several laboratories throughout the world have directed their efforts toward describing oligosaccharides in the milk of several mammals. Current advances in instrumentation and analytical techniques have allowed investigators to study a greater number of individual samples and lactational situations and accurately detect and quantify their major and minor components.

The carbohydrate content of milk varies among mammals. It is believed that diet or nutritional status does not influence the milk carbohydrate content.[1] The main carbohydrate in most types of milk from placental mammals is the disaccharide lactose. Human milk has a high lactose content, of 62 to 67g/L while, by contrast, cow's milk has a lower content of 45 to 50 g/L.[4] Other domestic ruminants such as goats and ewes have lactose contents similar to that of cows.[4] The lactose content seems to correlate with the development of the central nervous system. Marsupial milk contains no free lactose but has 3'-galactosyl-lactose-based oligosaccharides (from trisaccharides to heptasaccharides), and a gradual increase in monosaccharides occurs as lactation progresses.[5] Monotreme (egg-laying mammals) milk contains small amounts of lactose but high amounts of other oligosaccharides. The echidna secretes lactose (8%), sialyllactose (50%), fucosyllactose (26%), and difucosyllactose (13%): 90% of sialyllactose is 4-O-acetyl-N-acetylneuraminyl α2-3-linked to lactose;[6] and platypus milk has a high difucosyllactose content.[7] The presence of these simple small oligosaccharides in the milk of marsupials and monotremes has been explained in terms of osmotic pressure (osmolarity). The milk of these species contains high carbohydrate levels but remains isotonic with plasma.[7] The intestinal mucosa of these animals contains enzymes capable of splitting oligosaccharides into their component monosaccharides, which are used as an energy source.[8] Lactose and other oligosaccharides exert less osmotic pressure than individual monosaccharide constituents. Similar considerations could be made for other mammals, although several authors[8] do not consider human milk oligosaccharides to be an energy source since they are not digested by enzymes from milk or intestinal mucosa and are not nutritionally available. These results have been confirmed by others[9,10] who report that human milk oligosaccharides are either not digested or only minimally digested by human salivary amylase, human and porcine pancreas enzymes, and intestinal brush-border membrane preparations. Additionally, lactose has a β1–4 linkage between galactose and glucose. This unusual linkage is not broken by several potential infectious microorganisms, but by microflora living in the gastrointestinal tract. This bacterial antagonism protects the newborn from pathogenic bacteria.[4]

Milk oligosaccharides are mainly derived from lactose. Almost all carry the lactose unit at the reducing end; several glycosyltransferases add other monosaccharides to this core structure to synthesize complex oligosaccharides. This glycosyltransferase system, or closely related enzymes, seems to account for the biosynthesis of glycolipids and glycoproteins in epithelial cells, being responsible for the homology between cell surface glycoconjugates and milk oligosaccharides.[11] These enzymes appear to become activated with the onset of lactation since the urinary excretion of oligosaccharide by women is different, specifically, more complex, during pregnancy as compared with the lactation period.[12] D-glucose (Glc), D-galactose (Gal), L-fucose (Fuc), N-acetylglucosamine (GlcNAc), N-acetylgalactosamine (GalNAC) (occasionally), and sialic acid can be found in milk oligosaccharides.[13] N-acetylneuraminic acid (Neu5Ac) is the predominant sialic acid in milk oligosaccharides although occasionally, N-glycolylneuraminic acid (Neu5Gc)[14] and 4-, 7-, or 8-O-acetyl-N-acetylneuraminic acid [6] have also been found. Lactose is elongated enzymatically by the attachment of GlcNAc to a Gal residue in β1–3 or β1–6 linkages; after this residue is added, Gal can be attached by β1–3 or β1–4 linkages to the GlcNAc residue. Further addition of Fuc or sialic acid completes the process of oligosaccharide synthesis.[15] The degree of polymerization of these monomers ranges from 3 (monofucosyllactoses or monosialyllactoses) to 13 (pentafucosyllacto-N-octaoses), with a lower abundance of the more complex structures.[16]

Human milk is unique in its complex oligosaccharide content (12 to 14 g/L), including both neutral (90%) and sialylated (10%) forms, and is only comparable, both quantitatively and qualitatively, with that of the milk of the Asian elephant.[17] Moreover, oligosaccharides are the third solid component after lactose and fat, showing up in a higher amount than protein. Its content is higher in early milk (colostrum): 22 to 24 g/L. More than 100 different neutral and acidic oligosaccharides have been described so far.[13] Furthermore, recent studies[16] employing matrix-assisted laser desorption/ionization mass spectrometry (MALDI–MS) have revealed that the total number of naturally occurring human milk neutral (fucosylated) oligosaccharides could involve about 900 structures. Human milk mainly contains neutral fucosyloligosaccharides, which account for 65 to 70% of total oligosaccharides.[19] As reported,[15] lacto-*N*-tetraose is the most abundant oligosaccharide in human milk, followed by its monofucosylated derivatives lacto-*N*-fucopentaoses I and II and lacto-*N*-difucohexaose. In contrast, 2'fucosyllactose (2'FL) or 3'fucosyllactose (3'FL) (if 2'FL is not present) have been reported as the predominant oligosaccharides.[20,21] This discrepancy can be explained by recalling that the former work was carried out in pooled milk while the latter two involved the analysis of several samples from individual donors. The sialyloligosaccharide content is lower (about 1 g/L), 6'sialyllactose (6'SL), 3'sialyllactose (3'SL), and sialyl-lacto-*N*-tetraoses being the major components.[15] Other authors[22] have reported very high individual variations in the sialyloligosaccharide 3'sialyl 3 fucosyllactose (3'S3FL), the major component in some women's milk. It has long been known that the presence of certain oligosaccharides is related to blood groups.[23] In this sense, the oligosaccharide pattern and content of human milk is determined genetically and is related to the ABO, Lewis, and secretor status of the mother.[24] However, some oligosaccharides, such as 3'FL and lacto-*N*-tetraose, are present in human milk regardless of the blood group type of the mother.

Carbohydrates containing antigenic determinants of the Lewis blood group and secretor systems are synthesized by fucosyltransferases in a partially genetically determined process. They are present in glycoproteins and glycolipids from red blood cells and other tissues and biological fluids, such as saliva, plasma, and secretions. In human milk they are present in oligosaccharides. Fucose residues can be attached by at least three different fucosyltransferases, giving different oligosaccharide patterns. Fucosyltransferase II (FucT-II) is encoded by the secretor gene (Se) and produces high amounts of α1-2 fucosylated oligosaccharides (mainly 2'FL and lacto-*N*-fucopentaose I), indicative of a subject's secretor status (77% of the Caucasian population). Nonsecretor women present two inactive copies of the gene (Se) and do not produce oligosaccharides containing the α1-2 fucose residue. Fucosyltransferase III (FucT-III) is encoded by the Lewis gene (Le) and catalyses the transfer of fucose to the 3 and 4 positions of GlcNAc (20% of the population). Lewis-negative individuals have two inactive copies of the gene, (Le), while Lewis-positive individuals have at least one copy of the Le gene. A third fucosyltransferase related to neither the secretor nor the Lewis genes is responsible for the attachment of fucose residues to GlcNAc in α1-3 linkages (5%). Finally, several sialyltransferases can attach sialic acid residues to four different positions in core structures.[13] Regional variations in the oligosaccharide content of human milk have also been reported;[19] thus, 2'FL was found in 100% of the milk from Mexican and Swedish women, whereas only 46% of Philippine women have this oligosaccharide in their milk. This situation clearly reflects the above-mentioned variations in genetic status. Furthermore, the existence of four different oligosaccharide milk groups based on the Lewis blood group system has been proposed.[25]

Regarding the most common domestic mammals, bovine milk has an oligosaccharide content several times lower than that of human milk, sialylated species being predominant (80%, 1 g/L). Cow's milk has a very simple oligosaccharide pattern, lacking the complex oligosaccharides present in human milk. Thus, the former contains 3'SL, 6'sialyllactosamine (6'SLN), disialyllactose (DSL), 6'SL and 3'- or 6'NeuGc-containing SL as the most important constituents.[14,22,26,27] Eight neutral oligosaccharides (150 mg/L^{27}); including *N*-acetyllactosamine, *N*-acetylgalactosaminyl glucose, galactosyllactose derivatives, 3-fucosyllactosamine, and others, have also been reported.[26–28] *N*-acetyllactosamine is the main neutral oligosaccharide in colostrum (70%), although it disappears 7 days after calving.[28] Table 5.1 summarizes the main acidic and neutral oligosaccharides found in human and bovine milk. The milk of other domestic ruminants also contains low but appreciable amounts of oligosaccharides. In goat's milk, five different oligosaccharides have been described,[29] the major component being the trisaccharide GlcNAc *β*(1-6)Gal *β*(1-4)Glc, which contains a *β*1-6 linkage between GlcNAc and Gal that has not been found in human oligosaccharides. Moreover, a tetrasaccharide [plus Gal *β*(1-4)] and a pentasaccharide (plus Gal *β*(1-3) Fuc *α*(1-3)) derived from the trisaccharide have also been found. Fucose is present in three of the oligosaccharides. Ovine colostrum has been reported to contain three neutral (nonfucosylated) oligosaccharides (3'- *α*, 3'- *β*, and 6'- *β*-type isomers of galactosyllactose).[30] The *α*-glycosidic linkage is a newly discovered feature of mammalian milk. Trace amounts of tetra-, penta-, and hexasaccharides have also been described.[30]

TABLE 5.1
Main Oligosaccharide Species Found in Human and Bovine Milk

Trivial Name	Abbreviations	Structure	H/B
Neutral			
2'Fucosyllactose	2'FL	Fuc *α*(1-2)Gal *β*(1-4) Glc	H
3'Fucosyllactose	3'FL	Fuc *α*(1-3)Gal *β*(1-4) Glc	H
Lacto-*N*-tetraose	LNT	Gal *β*(1-3)GlcNAc *β*(1-3) Gal *β*(1-4) Glc	H
Lacto-*N*-neotetraose	LNnT	Gal *β*(1-4)GlcNAc *β*(1-3) Gal *β*(1-4) Glc	H
Lacto-*N*-fucopentaose I	LNFPI	Fuc *α*(1-2)Gal *β*(1-3) GlcNAc *β* (1-3)Gal *β*(1-4)	H
Lacto-*N*-fucopentaose II	LNFPII	Fuc *α*(1-4) [Gal *β*(1-3)] GlcNAc *β* (1-3)Gal *β*(1-4)	H
N-acetyllactosamine	—	Gal *β*(1-4)GlcNAc	B
N-acetylgalactosaminylglucose	—	GalNAc *β*(1-4)Glc	B
Acidic			
3'Sialyllactose	3'SL	NeuAc *α*(2-3)Gal *β*(1-4)Glc	H/B
6'Sialyllactose	6'SL	NeuAc *α*(2-6)Gal *β*(1-4)Glc	H/B
Sialyllacto-*N*-tetraose (a)	LST(a)	NeuAc *α*(2-3)Gal *β*(1-3)GlcNAc *β*(1-3) Gal *β*(1-4)Glc	H
Sialyllacto-*N*-tetraose (b)	LST(b)	Gal *β*(1-3) [NeuAc *α*(2-6)] GlcNAc *β*(1-3) Gal *β*(1-4)Glc	H
Sialyllacto-*N*-tetraose (c)	LST(c)	NeuAc *α*(2-6)Gal *β*(1-4) GlcNAc *β*(1-3) Gal *β*(1-4) Glc	H
Disialyllacto-*N*-tetraose	DSLNT	NeuAc *α*(2-3) Gal *β*(1-3) [NeuAc*α*2,6] GlcNAc *β*(1-3)Gal *β*(1-4) Glc	H
3'Sialyllactosamine	3'SLN	NeuAc *α*(2-3)Gal *β*(1-4) GlcNAc	B
6'Sialyllactosamine	6'SLN	NeuAc *α*(2-6)Gal *β*(1-4) GlcNAc	B
Disialyllactose	DSL	NeuAc *α*(2-8)NeuAc *β*(2-3) Gal *β*(1-4) Glc	B

H, human; B, bovine.

So far, no data about goat or sheep sialyloligosaccharides have emerged. Six neutral oligosaccharides have been isolated from horse colostrum:[31] the simplest are 3'- and 6'-galactosyllactose. Two pentasaccharides and two tetrasaccharides have also been described. Some of them have been previously reported in human (lacto-N-neotetraose and lacto-N-neohexaose) and goat milk.

The oligosaccharide content of milk changes over the course of lactation. An inverse relationship has been established between lactose and oligosaccharide contents in human milk.[32] The lactose content is low in early lactation, increasing during the first 3 months. Conversely, the oligosaccharide content is high in colostrum, decreasing as lactation progresses. It appears that at least for the first 3 to 4 months of lactation, the sum of lactose plus oligosaccharides is constant, giving very similar total carbohydrate values during this period.[24,32] The decrease in the oligosaccharide content is not accompanied by qualitative changes in individual compounds. Values of 20.9 g/L on day 4, decreasing to 12.9 g/L on day 120 have been reported;[33] these authors obtained less lactose (6%) and more oligosaccharides (4%) than others.[24] These changes are independent of the secretor status and Lewis group of the mother.[24] The same authors studied several milk oligosaccharide fractions: sialyloligosaccharides and lacto-N-tetraoses,-fucopentaoses and -difucohexaoses increase from day 2 to day 5 and then decrease. In contrast, fucosidolactoses and lactodifucotetraose decrease until day 5 and thereafter stabilized. The greatest changes seem to occur in the first 5 days postpartum. This date appears to mark the onset of stabilization.[24] It has also been reported[21] that the total oligosaccharide concentration declines along lactation; the mean concentration over 1 year was less than half of that observed in the first weeks postpartum. The authors also found that the ratio of α1-2-linked fucosyl oligosaccharides to α1-3/4-linked fucose changed along the first year of lactation from 5:1 to 1:1, with a significant decrease in 2'FL, lacto-N-fucopentaose I, and lacto-N-difucohexaose and an increase in 3'FL, lacto-N-fucopentaoses II and III, and lacto-N-difucohexaose II.

Most human acidic oligosaccharides identified decrease significantly during lactation, except for 3'SL, which did not vary.[22] The decrease coincides with that found for the total milk sialic acid content (71%) from week 1 to week 13 postnatally.[34]

Bovine sialyloligosaccharides decrease during lactation, colostrum being the stage with the highest values in most of the oligosaccharides identified.[22] No important qualitative changes in the sialyloligosaccharide content have been observed during lactation (from one stage to another). A small amount of 3'SLN in the four stages considered, transitional milk being the richest in this compound, has been found,[22] the presence of α2-3 isomers diminishing during lactation as compared with α2-6 isomers. A decrease in 3'SL, 6'SL, and 6'SLN contents takes place in the first 3 days postpartum.[35] These oligosaccharides were also determined in milk before parturition; their content increased from day 14 prepartum to the day of calving.[35]

The decrease in the sialyloligosaccharide content of bovine milk during lactation is consistent with that of the oligosaccharide-bound sialic acid (OBSA) previously reported,[36] from 230 mg/kg (colostrum) to 32.5 mg/kg (mature milk). A selective change in the relative content of OBSA was also observed. Colostrum has 35.3% of OBSA and 59% of glycoprotein-bound sialic acid (GBSA), calculated as percentages of total sialic acids. In mature milk, OBSA decreases to 20% and GBSA increases up to 80%. It seems that the decrease in OBSA is compensated by an increase in GBSA.

II. ANALYSIS AND CHARACTERIZATION OF MILK OLIGOSACCHARIDES

As previously mentioned, the main carbohydrate in most types of milk is lactose. The nonlactose carbohydrate fraction of milk was long ignored and, therefore, nonspecific methods used to quantify lactose wrongly included the oligosaccharide portion in the overall quantitation.[3] In recent decades, evidence for the biological relevance of milk oligosaccharides has been accompanied by the parallel development of several analysis and characterization techniques that have disclosed the great diversity of oligosaccharide contents, not just among mammalian species[4] but also among human populations.[21,25,37] The isolation of oligosaccharides from milk requires delipidation and deproteinization; after these steps, neutral and acidic oligosaccharides are separated by chromatographic techniques. Several types of high-performance liquid chromatography (HPLC), MS, and nuclear magnetic resonance (NMR) are the main tools employed for further identification and characterization.

A. ISOLATION OF THE CRUDE OLIGOSACCHARIDE FRACTION

The initial common step for obtaining oligosaccharides from milk is to eliminate the fat and proteins. Some authors[38] propose prior heat treatment (70°C, 30 min) of milk samples in order to inactivate possible contamination from hepatitis B and human immunodeficiency viruses. Nevertheless, in order to preserve the integrity of the oligosaccharides, measures should be taken to limit the extent of sample handling, and the use of high temperatures should be avoided.[19] Fat is easily eliminated by cold centrifugation and filtering the solidified lipids through loosely packed glass wool. The conditions of centrifugation vary considerably from some authors to others. Kobata's[23] preliminary work proposed 100g, 2°C, 15 min, but a whole set of different conditions can be found in the literature, most of which propose higher speeds (between 2000g and 15,000g) and longer centrifugation times (generally up to 30 min[22,39,40]). Apart from the fat layer, depending on the centrifugation conditions a casein pellet can also be formed and removed.[41] Proteins can be precipitated with ethanol (66%, stirred on ice, 3 h[9]), or with precooled acetone (50%, stirred 2 h and left overnight[17]). A 24% (w/v) solution of TCA can also be used.[39] The use of dialysis tubes to remove high-molecular-weight (MW) components is discouraged due to the possible formation of aggregates by acidic components.[13] However, some authors suggest the use of a dialysis MW cutoff size of 2000 to avoid proteins and glycoproteins.[32] One of the most common methods used to eliminate proteins is ultrafiltration, in which skimmed milk is filtered through a membrane (different cut-off sizes are used:[40–42]). Nevertheless, some authors choose a one-step method for delipidation and deproteinization: centrifuging through a 10,000 MW cut-off system for 2 h at 15 to 18°C, which would retain proteins and lipids and would yield a clear, colorless oligosaccharide-containing filtrate.[19] Other conditions for this combined method can be used.[43] Since some peptides may have a very low MW (close to that of oligosaccharides), paper chromatography has been proposed after filtration to eliminate not just peptides but also salts.[40] A fast-protein liquid chromatography (FPLC) gel filtration method (Superdex 200) can be used to separate glycoproteins from smaller peptides, and oligosaccharides can be extracted from the low MW (10 kDa) peptide-containing fractions by further gel filtration (Biogel P4).[41] Other molecular exclusion columns such as Sephadex G-25 are also widely used to eliminate residual proteins and peptides.[22,44] When HPLC techniques are to be implemented to identify the oligosaccharide components, some authors use a "micromethod," in which a very small amount (1 mL) of milk is mixed directly with the solvent (acetonitrile) prior to delipidation and deproteinization, and once these processes have been accomplished, samples are injected directly into the HPLC apparatus.[45]

The presence of lactose can pose an obstacle to purifying the oligosaccharide fraction, especially in species such as humans, in which lactose is so abundant. Although a large amount of lactose will precipitate with the proteins,[22] further steps are needed to prevent future interference, in chromatographic development or peak identification, on minor oligosaccharides that elute near lactose. Part of the lactose remaining in the supernatant can be removed by repeated crystallization in a rotary evaporator.[46] Several nanofiltration cellulose acetate membranes have been assayed with the aim of separating lactose from nearby-eluting bioactive trisaccharides, such as 3' and 6' sialyl derivatives of lactose, with partial success.[47] When gel permeation chromatography is used for further separation of the different oligosaccharide fractions (see below), lactose can be obtained separately in one of the fractions (Fractogel TSK HW 40 (S)[43]). The use of these exclusion chromatography columns for removing lactose is extensively reported in the literature: Bio-Gel P4,[48] Bio-Gel P2,[32,37] Toyopearl HW 40 (S),[25] Sephadex G-25.[9] Another possibility is passing over a charcoal column (adsorption chromatography).[49]

Treatment of delipidated, deproteinized whey with β-galactosidase (37°C, left overnight in a shaker) will lead to hydrolysis of lactose. The enzyme is later removed by ultrafiltration, giving a solution containing glucose, galactose, and the desired milk oligosaccharides.[47]

B. SEPARATION AND IDENTIFICATION OF DIFFERENT OLIGOSACCHARIDE COMPONENTS

Paper chromatography is a classic technique long used in the history of carbohydrate research, but unfortunately it requires too much time to accomplish the fractionation of large oligosaccharides.[50] A separation between acidic and neutral components can be successfully achieved by ion-exchange chromatography. Anion-exchange chromatography is widely used for this purpose: Dowex 1-×4 in acetate form,[43] Dowex 1-×2 in acetate form,[16] AG 1-×2 in acetate form,[51] DEAE-Sephadex A25,[13] or FPLC anion exchange (Resource Q,[9]). DEAE-cellulose has been used to separate mixtures of sialyloligosaccharides.[52] Medium-pressure anion-exchange chromatography has been used for the charge subfractioning of acidic oligosaccharides on MonoQ.[53]

As mentioned earlier, gel permeation chromatography can also be used to separate neutral and sialic acid-containing oligosaccharides (Toyopearl HW 40 (S)[25]), but this technique is more frequently used for the further separation of different-sized neutral oligosaccharides by eluting the different fractions with distilled water: by using Bio-Gel P2, the separation of tri- and tetrasaccharides, from penta- and hexa-saccharides, and from heavier neutral saccharides can be achieved.[24] The use of Fractogel TSK HW 40 (S)[16,38] yields five fucosyl-oligosaccharide fractions (monfu-cosyllactoses, difucosyllactose, lacto-*N*-tetraoses, monofucosyllacto-*N*-tetraoses, and difucosyllacto-*N*-tetraoses), one higher neutral oligosaccharide fraction, a frac-tion containing sialyloligosaccharides, salts and peptides, and finally the lactose fraction. Further use of Biogel P4 after this latter step, will provide separation for neutral oligosaccharides containing more than six sugar monomers.[44,48] Less fre-quently, the gel permeation technique can also be used to isolate different acidic fractions for which elution with buffer solutions is needed and, consequently, further desalting (Sephadex G-25) is required: Fractogel TSK HW 40 (S) yields monosia-lyllactose, monosialyllacto-*N*-tetraoses, monosialyllacto-*N*-hexaoses, monofucosyl-monosialyllacto-*N*-hexaoses, disialyllacto-*N*-tetraose, disialyllacto-*N*-hexaoses, and monofucosyl-disialyllacto-*N*-hexaoses.[16] Fractogel TSK HW 50 (S) has also been used for this purpose.[44]

Thin-layer chromatographic techniques can be useful for premonitoring the oligosaccharide content and composition of the different carbohydrate fractions.[41,51,54] Silica-HPTLC plates can be a useful tool for this purpose, using butanol–acetic acid–water (2.5:1:1, by vol) or butanol–ethanol–water–acetic acid–pyridine (5:50:15:1.5:5, by vol) as solvent systems. The plates can be sprayed with orcinol for carbohydrate detection, or with ninhydrin for monitoring the presence of residual pep-tides.[9,44] Specific silica gel TLC techniques for the separation of sialyloligosaccharides have been developed.[14] The purity of the fractions can also be checked by paper chro-matography (Whatman no. 3), using pyridine–ethyl acetate–water (1:2:2, by vol) as a solvent system, and developing with aniline oxalate reagent.[23]

Several colorimetric assays are available for monitoring the presence of oligosaccharides.[55] The orcinol–sulfuric acid and the phenol–sulfuric acid assays are good methods for detecting total oligosaccharides,[56] and the resorcinol assay is rec-ommended when working with acidic oligosaccharides.[57]

C. ANALYSIS OF INDIVIDUAL OLIGOSACCHARIDES

When GC analysis is carried out, prior methanolysis and trifluoroacetylation[58,59] of oligosaccharides is required, and knowing the molar ratio of the monosaccharide residues (fucose, galactose, glucose, etc) obtained in the GC analysis allows the identification of mono-, di-, tri-, and other substituted oligosaccharides.[14,23,43]

A variety of column and solvent systems has been used in HPLC systems. A combination of reverse- and normal-phase HPLC has been used[40] to identify milk oligosaccharides, comparing their glucose equivalent values with those of previously analyzed derivatized (via a reductive amination reaction with 2-aminoacridone) standards. Many oligosaccharides can be identified in this way, but some of the larger sialylated components require further MS studies for com-plete identification. In general, however, the use of derivatives not only has the advantage of improving HPLC separation and providing much greater detection sensitivity, it also allows direct analysis by fast-atom bombardment mass spectrometry (FAB-MS) after separation.[60]

In recent years, high-pH anion-exchange chromatography with pulsed amperometric detection (HPAEC-PAD), using the Dionex LC system, has been developed as a very powerful tool for carbohydrate research, and has also been widely used in milk oligosaccharide analysis. This method can be used for analytical or preparative purposes, and hence constitutes a final step in the purification process. An HPAEC-based method for the analysis of neutral and acidic human milk oligosaccharides has been described.[61] Several authors use a CarboPac PA 1 column (4 or 4.6×250 mm) equipped with a precolumn (CarboPac PA; 3×25 mm or 4×50 mm) and a PAD two-pulsed electrochemical detector with a gold electrode.[16,25,44,51,62] Others use a system equipped with two of the above CarboPac columns connected in series by no more than 15 cm of polyetheretherketone tubing.[19] Neutral and acidic oligosaccharides are analyzed using the following eluents: eluent A, NaOH; eluent B, NaOH and sodium acetate. The gradient conditions of the eluting program vary, depending on the authors,[19,25,44,62] and so do the flow rate (generally 1 mL/min) and detector conditions. Identification of the oligosaccharides from human milk separated by HPAEC is performed by comparing their absolute retention times or their retention times relative to an internal standard (stachyose) with that of oligosaccharide standards.[25] The respective amounts of the different oligosaccharides are automatically calculated on the calibration curves obtained using different concentrations of these single pure standards.

Prior to HPAEC-PAD analysis, some authors isolate certain structures by reverse-phase high performance liquid chromatography (rpHPLC) for better separation after the gel filtration process. The conditions for this step are as follows: 250×8 mm column, 30×8 mm precolumn, C_{18} reverse-phase material of 5 μm particle size, and UV detector.[51] Neutral structures are eluted isocratically with deionized water and acidic structures with 5 mM triethylamine hydrochloride. RpHPLC has also been used successfully for the separation of neutral milk oligosaccharides, using their perbenzoylated derivatives;[63] however, this technique does not always work so well with acidic oligosaccharides that have been separated by normal-phase HPLC.[64,65] Partition HPLC of neutral and acidic oligosaccharides on primary amine-bonded silica[60] has been used for the identification of milk carbohydrates. When using this method to identify milk sialyloligosaccharides, no previous separation of the carbohydrate fractions is needed, since under the conditions described for this purpose, only the acidic compounds are resolved and detected.[22] Figure 5.1 shows HPLC profiles of milk and formula sialylated oligosaccharides.

When analyzing oligosaccharides by HPLC, in order to confirm the correct assignation of the peaks to the different oligosaccharide species, parallel controls can be carried out: the use of several specific glycosidases and analysis by fluorophore-assisted carbohydrate electrophoresis (FACE) have been proposed.[37,47] For FACE analysis, oligosaccharides are labeled with a charged fluorophore and then subjected to electrophoresis along with labeled standards.

A very interesting method for analyzing milk oligosaccharides using capillary electrophoresis (specifically, micellar electrokinetic chromatography, MEKC) in a process that does not require derivatization with direct detection by UV absorbance, has been developed.[66] Since this technique separates molecules on the basis of ionic

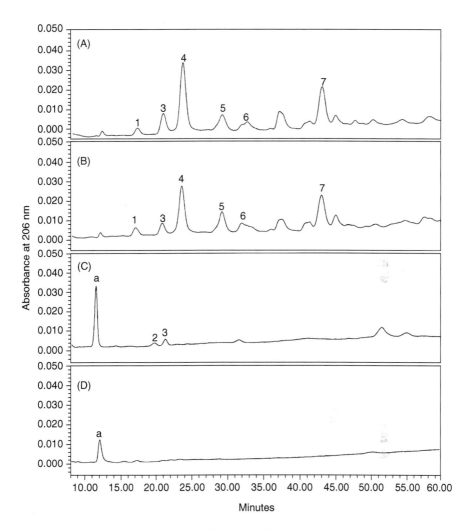

FIGURE 5.1   HPLC profiles of sialyloligosaccharides: (A) human colostrum; (B) human mature milk; (C) bovine mature milk; (D) infant formula. Peak 1, 3′sialyllactose; 2,6′sialyl-lactosamine; 3,6′sialyllactose; 4,3′sialyl-3 fucosyllactose; 5, sialyllacto-N-tetraose a; 6, sialyl-lacto-N-tetraose b + sialyllacto-N-tetraose c; 7, disialyllacto-N-tetraose. Peak "a", unidentified resorcinol-negative peak. (From Martin-Sosa et al., *J. Dairy Sci.*, 86, 52–59, 2003, American Dairy Science Association. With permission.)

charge, it is suitable for application to sialyloligosaccharides and has many advantages over the widely used HPAEC method: it does not require prior separation of the neutral and acidic fractions, and is a very sensitive (at fm level) and predictable method, with shorter running times. Structural isomers of sialylated oligosaccharides can be successfully separated by this method.[67]

Several individual oligosaccharides can be isolated from milk by immunoaffinity chromatography due to their bioactive epitopes, using the appropriate antibodies.[68,69]

In this sense, the use of specific column-bound lectins can also be used in the retention of certain types of oligosaccharides.[49,70] The incubation of milk oligosaccharides with monoclonal antibodies that recognize their ligand structures is a very
extended practice for the study of the possible biological role of these milk components.[71] We summarize the different processes used for the isolation and analysis of
milk oligosaccharides in Table 5.2.

D. FURTHER CHARACTERIZATION

To elucidate the structural configuration of milk oligosaccharides, information about
the type and number of sugar residues, the sequence, the site and type of anomeric
linkage, the conformation of sugar rings, and branching patterns is required.[13]

TABLE 5.2
Possible Steps for the Isolation and Analysis of Milk Oligosaccharides

Process	Method	Ref.
Delipidation	Centrifugation	22,39,40
Deproteinization	Precipitation	
	Ethanol	9
	Acetone	17
	TCA	39
	Ultrafiltration	40,41,42
Elimination of peptides	Paper chromatography	40
	Gel filtration	
	Biogel P4	41
	Sephadex G-25	44
Elimination of lactose	Crystallization	13
	Nanofiltration	47
	Gel filtration	
	Fractogel TSKHW(40)	16
	Biogel P4	18
	Biogel P2	32
	Toyopearl HW 40 (S)	25
	Adsorption	
	Charcoal column	49
	Enzymatic treatment	47
Separation between neutral	Ion-exchange	
and acidic components	Dowex 1X2	16
	AG1X2	51
	DEAE-Sephadex A 25	13
	Gel filtration	
	Toyopearl HW 40(S)	25
	Biogel P2	32
Individual analysis	HPAEC-PAD	
	Capillary electrophoresis	61
		66

Combinations of different types of MS and NMR spectroscopy have led to the characterization of many oligosaccharides and the elucidation of the core structures of these molecules in human milk;[72,73,74] further research in this field for a better characterization is in progress.[75] FAB-MS of native or derivatized oligosaccharides has been useful for identifying oligosaccharides or for confirming the data from HPLC analysis.[41,43,44,76] In recent years, development of the matrix-assisted laser desorption-time of the MALDI–TOF technique has also meant an improvement in milk oligosaccharide research owing to its great sensitivity and its rapid and easy use.[10,16,40,77] The use of this technique has helped in the analysis of high MW oligosaccharides from milk. Based upon information obtained from MALDI analysis, several authors have suggested that each molecular mass detected corresponds to a variety of isomeric structures, which would mean that the total number of oligosaccharides occurring in human milk could be much higher than previously supposed.[16,51]

III. BACTERIAL ADHESION AND MILK GLYCOCONJUGATES

The gastrointestinal tract provides a hospitable environment for hosting many types of bacteria and becomes colonized a few hours after birth. Although many of these bacteria are commensals, under immunodepressed conditions of the host they may cause several enteric diseases, such as diarrhea or gastritis. Additionally, there are several species or strains that are inherently pathogenic, such as *Salmonella* sp, *Vibrio cholerae,* or *Escherichia coli* O175:H7.

For infection to be achieved, the bacteria must first become firmly attached to the mucosa in order to avoid cleansing by intestinal movement and digestive fluids. The main way to bind to the host is through bacterial lectins, called adhesins, which recognize carbohydrate epitopes on the lumen of the gastrointestinal tract. This lectin–glycan interaction is specific to the extent that it allows bacteria to bind selectively to a particular host species or to a precise location along the tract. Protein–protein interactions can also contribute to the attachment.

In view of the rapid acquisition of antibiotic resistance by pathogenic strains, a new clinical approach involving adhesion blocking is currently being explored.[78] The new therapy would involve the design of a carbohydrate vaccine: a soluble fake receptor mimicking the actual one on the cell surface would block the bacterial adhesins and stop the infection at the very first stage. Milk glycoconjugates first provided the clue for this idea. For many decades, breast feeding has been recognized as the best way to prevent infections in newborns.[79] Besides its immunoglobulin content, breast milk also contains a variety of complex carbohydrates that promote the growth of beneficial bacteria (i.e., *Bifidobacterium*) and prevent the expansion of pathogenic microorganisms by blocking their anchoring to the mucosa. Most structures recognized by pathogenic bacteria have been found as soluble carbohydrates in human milk, one of the richest milks among mammalians in terms of the amount and diversity of oligosaccharides. Infant formulas are mainly prepared from bovine milk and, therefore, the oligosaccharide profile is fairly poor compared with human milk. Consequently, bottle-fed babies are more exposed to bacterial infections than breast-fed infants.

In order to achieve anti-microbial action, milk oligosaccharides must pass through the gastrointestinal tract without further modification. The appearance of milk oligosaccharide structures has been reported in both the feces and urine of newborns, indicating that these molecules can survive the acidic content of the stomach, reach the intestine, and be absorbed.[9,10] These properties clearly indicate that the potential protective effect of milk oligosaccharides can be exerted not only at a local level in the orogastrointestinal tract but also after absorption (in leukocyte endothelial interactions) and, later, in the urinary tract. The possible protection of part of the respiratory tract has also been suggested.[8]

A. PATHOGENIC BACTERIA CAUSING GASTROINTESTINAL INFECTIONS

Diarrhea is the main cause of death in infants below 12 months, especially in poverty-stricken areas of developing countries. Although some cases are due to rotaviruses, enteric bacteria are involved in most diarrheic diseases not only in infants but also in older children and adults. *E. coli*, in its many variants, invasive microbes such as *Salmonella, Shigella, Campylobacter,* or *Yersinia, Vibrio* species (*V. cholerae* and *V. parahaemolyticus*), *Aeromonas, Plesiomonas,* and *Clostridium difficile* as well as the protozoan *Giardia intestinalis* are often found in the stools of diarrhea patients.[80]

In order to understand how milk carbohydrates work in the prevention of bacterial adhesion, it is necessary to consider some general ideas about the mechanisms involved.

1. *E. coli*

E. coli is the most abundant Gram-negative bacillus that colonizes the mammalian large intestine. Although it is usually harmless, several strains of *E. coli* are associated with systemic diseases, meningitis, and mucosal infections when they occur in the intestinal or urinary tracts, and this microorganism is the main cause of diarrhea in children and young animals (Table 5.3). Its virulence is due to its ability to colonize or invade intestinal cells, to translocate into other organs, or to produce toxins. These pathogenic strains have been classified into five groups according to the characteristics of the infection they cause (Figure 5.2).

a. *Enterotoxigenic E. coli (ETEC)*
The different strains associated with this group are one of the main causes of diarrhea in children from developing countries, and are also responsible for traveler's diarrhea.[81] Animal strains are the cause of newborn diarrheas in young piglets, calves, and lambs.[82] ETEC strains are characterized by the release of toxins that enter mucosal cells and produce watery diarrhea. Other symptoms are abdominal cramps, vomiting, and malaise, although in the most severe form ETEC can cause a cholera-like dehydration.[83] Adherence to the mucosa is mediated by fimbriae (filamentous appendages).

b. *Enteropathogenic E. coli (EPEC)*
The *E. coli* strains first associated with summer diarrhea in children by Adam in 1923[83] corresponded with this group, being responsible for 95% of diarrhea cases in infants from developing countries. It is also related to septicemia in piglets and diarrhea in young dogs, cats, and rabbits (www.medvet.umontreal.ca/ecoli/english/pathogen.htm).

TABLE 5.3

Main *E. coli* Categories Associated with Human and Animal Diseases

Host	Disease	*E. coli* Categories	Localization	Toxins	Adhesins
Human	Watery diarrhea	ETEC	Small intestine	LT, ST	CFA/I, CFA/II, CFA/III, CFA/IV
	Nonspecific gastroenteritis	EPEC	Intestine	Protein F(?)	Bfp, intimin
	Bacillary dysenteria	EIEC	Large intestine	*Shigella* enterotoxin	Invasion-plasmid antigens (IpaC, . . .)
	Bloody diarrhea, hemolytic anemia, hemorrhagic uremic syndrome	EHEC	Large intestine, kidney	Stx-1, Stx-2	Intimin
	Persistent diarrhea	EAEC	Mainly large intestine	EAST, plasmid-encoded cytotoxin	AAF/I, . . .
Calf	Newborn diarrhea	ETEC	Small intestine	STa	F5 (K99), F41
	Hemorrhagic dysenteria	EPEC VPEC	Colon	VT-1 (SLT-I)	Intimin
	Septicemia	EPEC	Systemic		CS131A, F17
Piglet	Newborn diarrhea	ETEC	Small intestine	STa, STb, LT	F4 (K88), F5 (K99), F6 (987P), F41
	Postweaning diarrhea	ETEC EPEC	Small intestine	STa, STb, LT	F4 (K88), F18ab, F18ac
	Edema disease	VTEC	Small intestine	VTe (SLT-IIv)	F118ab
	Hemorrhagic gastroenteritis	ETEC	Small intestine	STa, STb, LT	
	Septicemia	EPEC ETEC	Systemic		F165 (P)
Cat, dog	Diarrhea	ETEC EPEC	Small intestine Small and large intestine		
Rabbit	Diarrhea	EPEC	Small and large intestine		AF/RI, intimin

Symptoms are similar to ETEC (watery diarrhea, no fever). EPEC does not produce toxins as does ETEC, although the production of a shiga-like toxin has been reported. The bacteria first attach to mucosal cells via a kind of fimbriae called "bundle forming pili" (BFP),[84] giving a characteristic adherence pattern termed "localized adherence."[85] The binding is followed by several signaling events that produce a huge rearrangement of the cytoskeleton, which leads to the disappearance of the villi and the accumulation of actin-F beneath the bacteria, forming the so-called pedestal. When this mechanism

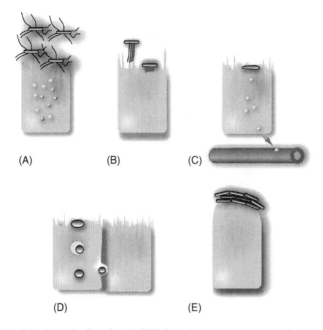

(A)   (B)   (C)

(D)   (E)

FIGURE 5.2 Diarrhegenic *E. coli*. (A) ETEC binds to the mucosa via fimbriae and secretes two types of toxins, ST and LT, into the cell. (B) EPEC binds first via BFP, and then produces the typical closer attachment and effacement of the villi. Bacteria modify the cytoskeleton and a pedestal is formed underneath the bound microbe. (C) EHEC produces toxins that are capable of entering the capillaries under the epithelium and cause severe hemorrhage. (D) EIEC are able to enter the enterocytes and pass to other cells. (E) EAEC forms tight aggregates on the cell surface and also induces a swollen morphology in the colonized cell.

was first described in pig and rabbit intestines, the authors coined the term "attachment/effacement" (A/E).[86]

c. Enterohemorrhagic E. coli (EHEC)

In 1993, an outbreak of EHEC from contaminated hamburgers served in a fast-food restaurant in the United States caused the deaths of four people and the infection of more than 600.[87] Previously, EHEC had been associated with hemolytic uremic syndrome (HUS).[88] EHEC infections vary from asymptomatic colonization or mild diarrhea to hemorrhagic colitis, neural disturbances, and renal damage. Although children and the elderly are more susceptible, EHEC can also infect healthy adults. The main cause of such a severe disease is the production of a toxin, identical to that of *Shigella dysenteriae*, that generates an intense inflammatory response. Similar to EPEC, EHEC also induces A/E lesions once it attaches to the mucosa.

d. Enteroinvasive E. coli (EIEC)

The clinical syndrome produced by EIEC resembles *Shigella* dysentery (watery diarrhea, sometimes blood and pus in the stools, fever, and inflammation), since both contain a closely related virulence plasmid encoding the genes required for cell invasiveness.[89] Similar to *Shigella*, EIEC are able to invade the colon via specialized lymphoepithelial cells[83] and then induce apoptosis in macrophages and the release of interleukins, which initiate an acute inflammation response. EIEC also release an enterotoxin (called Sen)[90] that is not genetically related to Shiga toxins.

e. Enteroaggregative E. coli (EAEC)

This category was first created to include some adherence patterns very different from the typical A/E, including bacteria that adhere in a diffuse way and other strains that form an arrangement that resembles a stacked brick wall (the enteroaggregative type). The association of this phenotype and diarrhea was difficult to pinpoint because EAEC is frequently isolated from apparently healthy people. However, more recent studies suggest that EAEC is an important cause of diarrhea in all ages and all countries. It has been related to persistent diarrhea in both children and adults, and is one of the main causes of diarrhea in AIDS patients.[91] Symptoms are abdominal pain and low-grade fever. The infected enterocytes exhibit a swelling of the villi and a characteristic cluster of bacteria within a thick layer of mucus.[92] Virulence has not been well characterized. EAEC strains (41%) express the enteroaggregative heat-stable toxin EAST-1, but this is also expressed by 100% of EHEC O157:H7. Regarding the fimbriae, two different fimbrial adhesins have been reported: AAFI and AAFII, both of which are likely to be responsible for the aggregative adherence phenotype.[93,94]

Recently, another type of diarrheagenic *E. coli* called cytolethal distending toxin-producing *E. coli* has been described, causes the distension and eventual disintegration of certain cell lines. This type shares some virulence factors with EPEC and EAEC, and it produces a toxin similar to that of *Campylobacter*. However, additional studies are required to further characterize this newly added group.[95]

In order to classify the increasing number of strains, Fritz Kauffmann adapted the serotyping system of *Salmonella enterica* to *E. coli*. According to this system, each *E. coli* isolate can be identified by three different antigens: O (somatic), which corresponds to the lipopolysaccharide structure of the cell wall, H (flagellar), which corresponds to the protein forming the flagella, and K (capsular). Each single O:H:K combination (generally, only the first two are used) defines a group or serotype. It turns out that most of the isolates from diarrhea patients belong to a few of these serotypes. A particular serotype is intrinsically associated with a specific pathotype; for example, O157:H55 *E. coli* are always EHEC, O55:H6 are human EPEC, and O101:K18 are bovine ETEC.

2. Bacterial Adhesins

When recognition between two molecules must be strictly specific, carbohydrates are usually involved because of their vast diversity of structures. Lectins, the proteins that bind carbohydrates, have evolved to be able to distinguish among anomers, stereoisomers, types of binding, branching, or closely related structures (such as Neu5Ac and Neu5Gc). The specificity of microbes allows them to attach to the susceptible host at the best place for infection or at the right time (i.e., young animals and not adults). Adhesins, the bacterial lectins, are mainly located in pili or fimbriae, although nonfimbrial adhesins have also been described.[96]

All fimbriae are assembled by following one of four general pathways.[97]

a. Chaperone–Usher Pathway

Fimbrial subunits leave the cytoplasmic membrane coupled to a chaperone protein, which prevents the already folded subunits from assembling in the periplasm. The chaperone releases the subunits at specific sites on the outer membrane, where assembly takes place using another protein, the usher, as an anchor. Many pili in Gram-negative bacteria are assembled by this mechanism, including UPEC and ETEC strains, *Haemophilus influenzae*, *Klebsiella pneumoniae*, *Proteus mirabilis*, *Bordetella pertussis*, *Salmonella*, or *Yersinia*.[98]

The crystal structure of PapD, the chaperone involved in the P pili assembly, has been determined,[99] showing two immunoglobulin-like domains oriented towards each other. All chaperones share this structure, although they have been organized into two groups, FGS and FGL, according to the amino acid differences in the cleft and the length of the loop between the F1 and G1 β-strands.[99] Each directs the assembly of a particular fimbrial architecture. FGS (short loop) mediates the formation of rod-like fimbriae, like P or type 1 pili from uropathogenic *E. coli*, and also thinner and flexible fimbriae such as K99 or K88 pili, from ETEC.[100] FGL (long loop) are involved in the assembly of very thin fimbriae, like CS3 pili[101] or AAFI fimbriae,[102] and nonfimbrial adhesins.

b. Alternate Chaperone Pathway

Another type of chaperone, with no sequence homology to the first one, has been described. CFA/I and CS1 pili from human ETEC follow this pathway. In CS1 pili formation, the periplasmic chaperone CooB forms complexes with the pilus components, CooD and CooA, transporting them to the outer membrane, where the protein CooC, a channel for CooD and A, is located.[103]

c. General Secretion Pathway

The type IV fimbriae from *Neisseria* spp., *V. cholerae*, *Mycobacterium bovis*, or *Pseudomonas aeruginosa* as well as BFP from EPEC follow the general secretion pathway. The prepilin is processed by a peptidase that cleaves the hydrophobic *N*-terminal segment, releasing the hydrophilic mature pilin subunit into the periplasm. There, a protein located in the inner membrane serves as a platform for fimbria assembly. The assembly also involves an ATPase and an outer component that forms a pore where the pilus can pass through the membrane.[104]

d. Extracellular Nucleation–Precipitation Pathway

Many pathogenic strains of *E. coli* and *S. enteritidis* produce thin, irregular, and clustered structures known as curli.[97] Curli formation occurs by the precipitation of subunits secreted into thin pili on the outer surface of the bacteria. This process is mediated by other proteins that act as inductors of the assembly of curli components.[105]

As shown in Figure 5.3, these fimbriae are formed by different types of subunits, only one being the actual adhesin: PapG in P pili, CooD in CS1 pili or PilC in type IV pili. This specificity means that one pilus can bind only one single epitope (or a group of related epitopes) on the host cell. However, K99 and K88 fimbriae from bovine and porcine ETEC are composed of only one type of protein, which is the actual adhesin: FanC (K99) and FaeA (K88).[106] In this case, each pilus

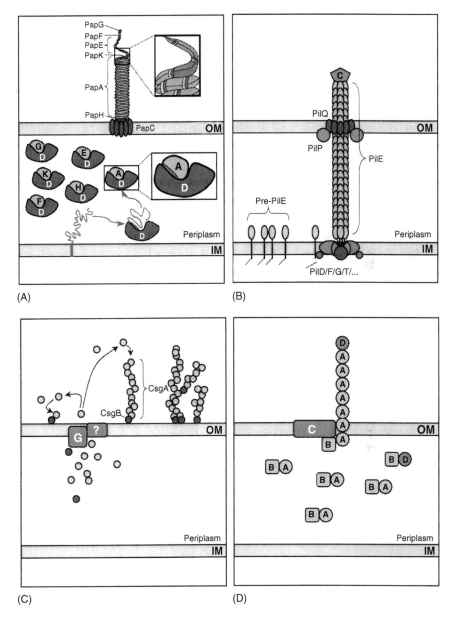

FIGURE 5.3   Different pathways of fimbria assembly. (A) Chaperone–usher pathway for P pili. Chaperone PapD drives the different components of the fimbria (PapA, PapF, PapH, PapK, PapG, and PapE) to the outer membrane, where the usher PapC acts as an anchor for the assembly of the fimbriae. (B) Alternate chaperone pathway for CS1 pili, showing chaperone CooB and usher CooC. (C) General secretion pathway for type IV pili. The prepilin subunits pre-PilE are cleaved by the peptidase PilD. PilC is the adhesion subunit on the tip of the pili. (D) Curli formation. Surface protein CsgB promotes the assembly of CsgA subunits and also participates in the branching of the curli. (From Soto and Hultgren S.J., *J. Bacteriol.*, 181, 1059–1071, 1999. American Society for Microbiology. With permission.)

can hypothetically bind to several molecules of the same epitope on the surface of the target cell, although it is likely that most of the binding sites on the fimbria will not be accessible to the ligand due to steric hindrance.

Many molecular studies on the P and type 1 pili have contributed to a better understanding of fimbrial adhesion. Thus, pilus subunits (PapA and FimA, respectively) have an incomplete immunoglobulin-like fold that lacks the C-terminal β-strand, resulting in a hydrophobic groove that interacts with the G1 strand of the N-terminal domain of the chaperone and, thereafter, with another subunit in the emerging fimbria. Thus, each subunit contributes to completing the fold of its neighbor and eventually to the stabilization of the whole fimbria.[98] Adhesins (PapG and FimH) share the same Ig-like structure of the pilins but they also have an *N*-terminal lectin domain.[107] Many efforts have been devoted to elucidating the crystal structure of these proteins.[108–110] Such analyses have shown that the lectin domain has a complete immunoglobulin-like folding, the G strand also fitting into the hydrophobic groove of the pilin subunit underneath. A comparative analysis with other proteins has also been performed,[110] revealing significant similarities with other fimbrial subunits as well as with animal proteins such as murine antibody Fab88 and murine sialoadhesin.[109]

Recent studies have pointed to the existence of a stronger adhesion under flow conditions, as occurs with selectin-dependent leukocyte rolling.[111] Others[112] have shown that a tenfold increase in shear force enhances the attachment of FimH-expressing *E. coli* to pig erythrocytes. Two models have been proposed to explain the phenomenon: the so-called catch bond, and another one that involves a conformational switch in the bacterial adhesin. In both, the shear force is required for the protein to strengthen the binding or for a conformational change into a high-affinity stage.[110] These biophysical properties allow bacteria to take advantage of intestinal movement to improve their binding to their targets.

3. Glycan Specificity of Bacterial Adhesion

Ligands for the best-known bacterial adhesins have been reported: both fimbrial and nonfimbrial. Table 5.4 summarizes the glycans recognized by the main bacterial species and the strains that cause human or animal diseases.

The first study addressed the type 1 fimbria, which is expressed by most of the members of the *Enterobacteriaceae* family.[113] Type 1 adhesin FimH binds specifically to D-mannose[114] and PapG, the adhesin in P pili, binds to galabiose.[115] S fimbriae, which are expressed in *E. coli* causing meningitis in the newborn, bind to terminal sialylα2-3galactoside on glycoproteins,[116] in particular to cellular fibronectin.[117] The gastric microorganism *Helicobacter pylori*, which is responsible for chronic active type B gastritis, also expresses sialyl-specific adhesins.[118] The most well known is BabA, which binds to Le[b], H antigen, and related blood group antigens.[119] Interestingly, during gastritis, *H. pylori* up-regulates the mucosal expression of inflammation-related sLe[x] antigens and binds to them via another adhesin called SabA, which mimics selectin binding to sialyl(di)Le[x(a)] for closer attachment to the mucosa.[120] The−SfaS adhesion (from S fimbriae) as well as the *H. pylori* adhesins BabA and SabA- recognize only

TABLE 5.4
Oligosaccharidic Epitopes Recognized by Different Microorganisms[1]

Organism	Fimbriae	Target Tissue	Oligosaccharide Structure
	Type 1	Urinary	Manα1-3Manα1-6Man
	P	Urinary	Galα1-4Gal
	S	Neural	Neu5Acα2-3Galβ1-3GalNAc
	CFA/I	Intestinal	NeuAcα2-8
	K88	Intestinal	GalNAcβ1-4Galβ1-4Glc[127]
			Galβ1-3GalNAcβ1-4Galβ1-4Glc
E. coli	F1C	Urinary	Lac-Cer, Gb$_3$Cer, iGb$_3$Cer[150]
	F17	Intestinal	GlcNAc[151]
	BFP	Intestinal	GalNAcβ1-4Gal[152]
			Galβ1-4GlcNAc
			Le^x, Le^y
	F41	Intestinal	GlcNAc, GalNac[153]
	K99	Intestinal	Neu5Gcα2-3Galβ1-4Glc
Haemophilus influenzae		Respiratory	(Neu5Acα2-3)Galβ1-4GlcNAc
			β1-3Galβ1-4GlcNAc
Helicobacter pylori		Stomach	Neu5Acα2-3Galβ1-3Glc(NAc)
			Fucα1-2Galβ1-4[Fucα1-4]Gal.
Klebsiella pneumoniae		Respiratory	Man
Mycoplasma pneumoniae		Respiratory	Neu5Acα2-3Galβ1-3Glc(NAc)
Neisseria gonorrhoea		Genital	Galβ1-3Glc(NAc)
N. meningitidis		Respiratory	(Neu5Acα2-3)Galβ1-4GlcNAc
			β1-3Galβ1-4GlcNAc
Pseudomonas aeruginosa		Respiratory	Galβ1-3/4Glc(NAc)β1-3Gal
			β1-4Glc[154]
Salmonella typhimurium		Intestinal	Man
S. enteritidis		Intestinal	(Neu5Acα2-6)Galβ1-4Glc
			NAcβ1-2Man[155]
Streptococcus pneumoniae		Respiratory	(Neu5Acα2-3/6)Galβ1[121]
			LNT, LNnT
S. suis		Neural	Galα1-4Galβ1-4Glc[156]
Campylobacter jejuni		Intestinal	Fucα1-2[157]
Bordetella pertussis		Respiratory	Gal on sulfated GSL[158]

[1] Mainly based on Reference 78 or Sharon and Ofek, 2000.

α2-3Neu5Ac and not α2-6Neu5Ac. In contrast, the adherence of *Streptococcus pneumoniae* to respiratory cells is inhibited by both α2-3 and α2-6 sialyllactose and sialyllactosamine.[121]

Due to this binding specificity, individual susceptibility to a particular infection is governed by the expression of the ligands for bacterial attachment. Thus, women belonging to the nonsecretor blood group are more likely to suffer from *E. coli* urinary-tract infections than secretor women, because only nonsecretor individuals express sialylgalactosylgloboside and disialylgalactosylgloboside, which specifically bind UPEC R45.[122]

Binding sites can also be expressed transiently at different stages of development leading to age-dependent susceptibility to infection by various pathogens. A well-known example is K99 fimbriae from animal ETEC. This strain infects newborn calves, lambs, and piglets and is a major cause of neonatal death in herds, but it does not affect adult animals or humans. K99 fimbriae mainly bind to Neu5Gc-GM3, but also recognize a plethora of Neu5Gc-containing gangliosides, including Neu5Gc-GD3,[123] Neu5Gc-GM2, Neu5Gc-GD1a, Neu5Gcα2-3nLcOse4Cer.[124] Adult pigs do not express Neu5Gc in their intestine and humans are not able to synthesize it at all;[125] therefore, ETEC K99 cannot attach and infect them.

Other porcine ETEC strains express fimbriae K88.[126] Three different variants of K88, called K88ab, K88ac, and K88ad, have been identified, showing slightly different binding specificities. Thus, the binding of K88ab to neolactosyl-tetraosylceramide is weaker than that of K88ac and K88ad to the same epitope, and the binding of K88ad to globotriosylceramide is much stronger than that of the other two variants,[127] although all three variants preferentially bind to β-linked N-acetylhexosamine, with or without a galactose in the terminal position. Because different breeds of pigs exhibit differences in the oligosaccharide chains expressed on their intestinal mucosa, this variety in bacterial adhesion preferences may be an evolutionary adaptation of the microbe to the polymorphism of the host.

Infection has been reported to be a likely cause of the huge diversity of animal glycoconjugate structures.[128] In the same way, microbes must rapidly evolve to adapt to these new structures on the host. A wider range of adhesins on a microbe provides it with a greater chance to succeed in its infection. Thus, many microorganisms exhibit more than one adhesin that would only enable binding to a single epitope. As mentioned above, *H. pylori* recognizes not only Neu5Acα2-3Gal but also other structures, including noncarbohydrates such as phosphatidylethanolamine.[129] Sialic acid recognition (both on gangliosides and on neutrophile-activating protein [NAP]) is apparently specific for interaction with neutrophils and produces an inflammatory response in the host that apparently benefits the bacteria.[130] *Plasmodium falciparum*, the protozoan responsible for the severe forms of malaria, also utilizes many different ways to infect human erythrocytes in both a sialic acid-dependent and -independent manner.[131]

In addition to fimbrial adhesions, many bacterial toxins also recognize carbohydrate epitopes. The best known example is the cholera toxin family, including the toxin produced by *V. cholerae* and *E. coli* heat-labile toxin (LT), which recognizes and binds to GM1.[132] Since EHEC cause such severe diseases in both developing and developed countries, many efforts are being made to characterize the binding specificity of the toxins produced by this type and also by *S. dysenteriae* type 1: shiga toxins 1 (Stx1) and 2 (Stx2, Stx2c, Stx2d) and shiga-like toxins (verotoxins) SLT-I and SLT-II. Like similar to the cholera toxin and LT, shiga toxins are constituted by an enzymatically active A subunit and a pentamer B_5 that recognizes the ligand on the cell surface and also creates a pore in the membrane for A to enter the cell.[133] The main ligand receptor for shiga toxins is the glycolipid Galα1-4Galβ1-4Glc-Cer (Gb3Cer).[134] Synthetic analogs have been shown to bind the toxin efficiently; Synsorb P^k is currently under clinical trials and many others are being investigated.[135] A soluble analog, called Starfish, protected mice when it was injected simultaneously with Stx1 but not Stx2, while a modified structure, Daisy, was able to bind both toxins.[136]

B.   THE ABILITY OF MILK TO PREVENT BACTERIAL INFECTIONS

As mentioned above, the bacterial colonization of the newborn occurs soon after birth, more so in babies delivered vaginally than in those born by Cesarean section.[137] That report also shows how an improvement in sanitary conditions decreases the levels of bacterial colonization in newborns, stressing that the number of *E. coli* strains found in children from developing countries is more than five-fold the number isolated from children from developed countries. Thus, although breast feeding is always recommended it becomes of far greater importance in less developed countries as an efficient way to prevent infections.

Milk immunoglobulins are the main antibacterial weapon. Secretory IgA (sIgA) in human milk neutralizes viruses, bacteria, and toxins.[138] It reflects the antigens the mother has been exposed to and helps to protect the baby from the most common microorganisms in the family environment. Bovine milk contains high levels of IgG, since the sIgA fraction is minor. In contrast to humans, the bovine placenta is not permeable to immunoglobulins and calves totally lack antibodies at the moment of birth, depending entirely on those transferred through the colostrum. Other proteins, such as lactoferrin and lysozyme, also exhibit bactericidal or bacteriostatic effects.[79]

The protective effects of milk glycoconjugates were first suggested by Kobata in 1978,[139] who argued that they could act as analogs of cell surface glycoproteins and glycolipids because they are synthesized by the same metabolic pathways. As shown in Figure 5.4, milk glycoconjugates resemble cellular ligands for both fimbriae and enterotoxins. In the lumen of the newborn, potentially infective bacteria are surrounded by a high density of milk glycocomponents that share the oligosaccharides structures with the cellular target they are seeking. It is likely that most of the time, as complexing with the ligand is a dynamic process, bacterial adhesins are bound to these soluble and "false" receptors, and are thus unable to attach to the cellular surface and initiate colonization. Bacteria blocked from adhesion in this manner are subsequently washed away by intestinal fluids and the baby remains healthy. The survival of milk oligosaccharides through the alimentary tract without further modification has been unambiguously proven by the detection of the same pattern of oligosaccharides in breast milk and baby's stool and urine.[11]

It is now well established that the oligosaccharide fraction from human milk does contribute to preventing bacterial infections, both intestinal and urinary, since many studies have been conducted both *in vitro* and *in vivo*.[11] It has been shown that sialylated oligosaccharides prevent the binding of certain septic *E. coli* strains,[140] while others[141] have reported the inhibition of *H. influenzae* binding to pharyngeal and buccal epithelium by human milk oligosaccharides. A human milk fucooligosaccharide was shown to inhibit ST in mice.[49] Studies on T84 cells showed that this fucooligosaccharide did not interact with the toxin by itself, but hindered the action of ST by binding to a carbohydrate-binding site on the extracellular domain of

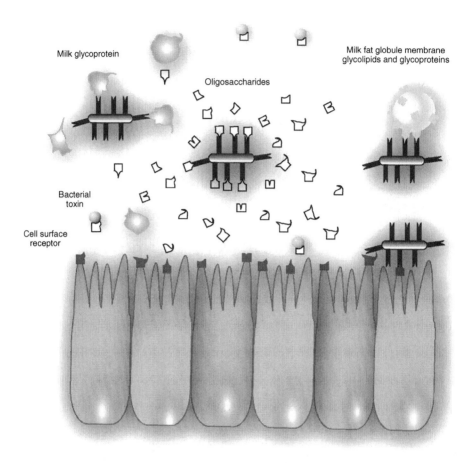

FIGURE 5.4 A model of the "glycanic antibacterial force" in milk. Fimbrial adhesins and toxins can be blocked by milk glycoconjugates that mimic targets on the epithelial cells. Milk glycolipids and glycoproteins act as decoys in most mammalian species. Owing to their abundance and variety, free oligosaccharides play an important role in the glycodefense of the human newborn.

guanilyl cyclase proximal to the ST binding site.[11] Another *in vivo* experiment, performed in colostrum-deprived calves, showed that ETEC K99 infection could be prevented by adding glycopeptides to the drinking water.[142]

Studies on the binding specificity of bacterial adhesins have usually been carried out using hemagglutination inhibition assays. With this technique, the blocking of *P. aeruginosa* lectins by fucosylated oligosaccharides from human milk has been reported,[143] and our group has recently reported the specificity of many human uropathogenic *E. coli* (UPEC) and human and bovine ETEC strains.[144,145] We selected a number of adhesin-expressing strains that are commonly found in newborn human and calf isolates, and checked the ability of the major oligosaccharides found in human and bovine milk to inhibit agglutination (Table 5.5). Whole oligosaccharide fractions from several stages of

TABLE 5.5

Hemagglutination by Several *E. coli* Adhesins from Human (Upper Cells) and Bovine Strains (Lower Cells)

Adhesin	PBS	3'SL	6'SL	3'SLN	6'SLN	LSTa	3'S3FL	DSL	DSLNT	LN	Lac
P	+	+	+	N	+	++	++	N	++	NT	+
P-like	+	−	±	N	+	−	±	N	−	NT	+
CFA/I	+	±	−	N	+	−	−	N	−	NT	+
CFA/II	+	−	±	N	+	−	−	N	−	NT	+
K99	+	±	±	+	±	N	N	+	N	±	+
F41+K99	+	−	−	±	−	N	N	+	N	+	+
F41	+	±	−	±	−	N	N	−	N	+	±
F17	+	+	±	±	±	N	N	−	N	+	+
B16	+	±	±	±	±	N	N	+	N	±	±
B23	+	±	+	±	+	N	N	−	N	+	±
B64	+	+	+	±	±	N	N	−	N	−	−

Note: Human erythrocytes were used for human strains, horse erythrocytes were used for bovine strains. 25 μg of oligosaccharides were added per well. Lac was tested up to 500 μg. −, no agglutination; ±, less agglutination than control (PBS); +, same agglutination as control; ++, more agglutination than control; N=not found in this species; NT=not tested

Source: Martin-Sosa, S., et al., *J. Nutr.*, 132, 3067–3072, 2002, ref. 144. American Society for Nutritional Science; With permission. Martin, M.J., et al., *Glycoconj. J.* 19, 5–11, 2002, ref. 145. Kluwer Academic Publishers. With permission.)

lactation were also assayed, with the observation that the human colostrum fraction was a better inhibitor than milk, while in bovine milk colostrum was a poorer inhibitor.

Human milk is unique among mammalian species for its variety and content of free oligosaccharides. In nonhuman mammalian newborns, milk glycans are part of glycoproteins and glycolipids. In fact, some of the protective effects of immunoglobulins are due to their glycan moiety. Thus, sIgA anti-adhesion effects on S-fimbriated *E. coli* have been reported to be mediated by the sialyloligosaccharides bound to the protein backbone of the sIgA.[138]

Many bacterial adhesins and toxins exhibit a preference for glycolipid ligands instead of glycoproteins. K99 mainly recognizes Neu5Gc-GM3; P and S fimbriae also recognize glycosphingolipids (GSLs) on the cell surface; LT and cholera toxin bind to GM1, and Stx to Gb3. Glycolipids in milk are a minor component found on the membrane that surrounds the fat globule (called milk fat globule membrane [MFGM]). However, their involvement with adhesion has been repeatedly demonstrated. Thus, it has been reported that human milk GM1 could bind to cholera toxin.[146] The blocking effect of bovine milk on the adhesion of *H. pylori* to cells is mainly due to the MFGM fraction, where the glycolipids are actually located.[147] When incubated with the shiga toxin, human milk lipids have been shown to bind ~85% of Stx, while the soluble fraction, containing the antibody pool, only binds 30%.[148] The ceramide moiety of GSLs, as well as the glycans chain, appears to play a role in binding, perhaps by modulating accessibility. According to our studies on bovine adhesins, an increase in ceramide flexibility results in stronger binding.[123]

It is plausible that the ceramide composition could control how the glycan is presented to the adhesion by determining the glycan density of a GSLs cluster on the cell surface.

IV. FINAL REMARKS

The benefits of breast feeding have been known for many years, but it has only been during the last decades that glycoconjugates have been implicated in the protection of newborns against infection. According to the World Health Organization (WHO), increasing breast feeding could reduce the number of deaths due to diarrhea by 66%. As shown in this review, human milk is very rich in oligosaccharides in comparison with other mammalian species and hence breast feeding should always be encouraged, although problems eventually arise when substitutes for human milk must be synthesized. Most infant formulas in the market have been developed from bovine milk, which is fairly poor in oligosaccharides. Additionally, bovine proteins are usually immunogenic in the newborn and are therefore usually removed. Taking into account the lack of potential targets, some manufacturers have included glycoconjugates in the formulas in order to mimic human milk as much as possible. Thus, it has been reported that infants fed with a ganglioside-supplemented formula have significantly lower contents of *E. coli* in their feces than infants fed with regular formula.[149] However, the unique variety of human milk oligosaccharides increases the difficulty of humanizing milk formulas. Milk from many other mammalian species is currently in progress in order to find a better source of the main structures involved in bacterial blocking. Chemical synthesis or the bioengineering of modified bacteria might also be an option, although this is not yet feasible. Nevertheless, all efforts aimed at developing better infant formulas are no doubt worthy.

ACKNOWLEDGMENTS

Financial support from the Programa de Apoyo a Proyectos de Investigación de la Junta de Castilla y León, España (SA 093/01) and Plan Nacional I+D+I (Ministerio de Ciencia y Tecnología, España; AGL 2001-2047) is gratefully acknowledged.

We are also grateful to Nicholas Skinner (from the Servicio de Idiomas, Universidad de Salamanca) for revising the English version of the manuscript.

REFERENCES

1. Emmett, P.M. and Rogers, I.S., Properties of human milk and their relationship with maternal nutrition, *Early Hum. Dev.,* 49, S7–S28, 1997.
2. Newburg, D.S., Oligosaccharides and glycoconjugates in human milk: their role in host defense, *J. Mam. Gland. Biol. Neoplasia,* 1, 271–283, 1996.
3. Montreuil, J., The saga of human milk gynolactose, in *New Perspectives in Infant Nutrition,* Sawaztki, G., Renner, B., Eds., Georg Thieme Verlag, Stuttgart, 1993, pp. 1–11.
4. Newburg, D.S. and Neubauer, S.H., Carbohydrates in milks: analysis, quantities and significance, in *Handbook of Milk Composition,* Jensen, R,G., Ed., Academic Press, San Diego, 1995, pp. 273–349.

5. Messer, M., Fitzgerald, P.A., Merchant, J.C., and Green, B., Changes in carbohydrates during lactation in the eastern quoll *Daysurus viverrinus* (Marsupialia), *Comp. Biochem. Physiol.*, 88B, 1083–1086, 1987.
6. Messer, M., Identification of *N*-acetyl-4-*O*-acetylneuraminyllactose in echidna milk, *Biochem. J.*, 139, 415–420, 1974.
7. Messer, M. and Kerry, K.R., Milk carbohydrates of the echidna and the platypus, *Science*, 180, 201–203, 1973.
8. McVeagh, P. and Brand Miller, J., Human milk oligosaccharides: only the breast, *J. Pediatr. Child Health*, 33, 281–286, 1997.
9. Gnoth, M.J., Kunz, C., Kinne-Saffran, E., and Rudloff, S., Human milk oligosaccharides are minimally digested *in vitro*, *J. Nutr.*, 130, 3014–3020, 2000.
10. Engfer, M.B., Stahl, B., Finke, B., Sawatzki, G., and Daniel, H., Human milk oligosaccharides are resistant to enzymatic hydrolysis in the upper gastrointestinal tract, *Am. J. Clin. Nutr.*, 71, 1589–1596, 2000.
11. Newburg, D.S., Do the binding properties of oligosaccharides in milk protect human infants from gastrointestinal bacteria? *J. Nutr.*, 127, 980S–984S, 1997.
12. Hallgren, P., Lindberg, B.S., and Lundblad, A., Quantitation of some urinary oligosaccharides during pregnancy and lactation, *J. Biol. Chem.*, 252, 1034–1040, 1977.
13. Egge, H., The diversity of oligosaccharides in human milk, in *New Perspectives in Infant Nutrition*, Renner, B. and Sawaztki, G., Eds., Georg Thieme Verlag, Stuttgart, 1993, pp. 12–26.
14. Veh, R.W., Michalski, J.C., Corfield, A.P., Sander-Weber, M., Gies, D., and Schauer, R., New chromatographic system for the rapid analysis and preparation of colostrum sialyloligosaccharides, *J. Chromatogr.*, 212, 313–322, 1981.
15. Kunz, C. and Rudloff, S., Biological functions of oligosaccharides in human milk, *Acta Paediatr.*, 82, 903–912, 1993.
16. Stahl, B., Thurl, S., Zeng, J., Karas, M., Hillencamp, F., Steup, M., and Sawatzki, G., Oligosaccharides from human milk as revealed by matrix-assisted laser desorption/ionization mass-spectrometry, *Anal. Biochem.*, 223, 218–226, 1994.
17. Kunz, C., Rudloff, S., Shaad, W., and Braun, D., Lactose-derived oligosaccharides in the milk of elephants: comparison with human milk, *Brit. J. Nutr.*, 82, 391–399, 1999.
18. Newburg, D.S., Human milk glycoconjugates that inhibit pathogens, *Curr. Med. Chem.*, 6, 117–127, 1999.
19. Erney, R.M., Malone, W.T., Skelding, M.B., Marcon, A.A., Kleman-Leyer, K.M., O'Ryan, M.L., Ruiz-Palacios, G., Hilty, M.D., Pickering, L.K., and Prieto, P.A., Variability of human milk neutral oligosaccharides in a diverse population, *J. Pediatr. Gastroenterol. Nutr.*, 30, 181–192, 2000.
20. Viverge, D., Grimmonprez, L., Cassanas, G., Bardet, L., and Solère, M., Discriminant carbohydrate components of human milk according to donor secretor types, *J. Pediatr. Gastroenterol. Nutr.*, 11, 365–370, 1990.
21. Chaturvedi, P., Warren, C.D., Altaye, M., Morrow, A.L., Ruiz-Palacios, G., Pickering, L.K., and Newburg, D.S., Fucosylated human milk oligosaccharides vary between individuals and over the course of lactation, *Glycobiology*, 11, 365–372, 2001.
22. Martín-Sosa, S., Martín, M.J., García-Pardo, L.A., and Hueso, P., Sialyloligosaccharides in human and bovine milk and in infant formulas: variations with the progression of lactation, *J. Dairy Sci.*, 86, 52–59, 2003.
23. Kobata, A., Isolation of oligosaccharides from human milk, *Methods Enzymol.*, 28, 262–271, 1972.
24. Viverge, D., Grimmonprez, L., Cassanas, G., Bardet, L., and Solère, M., Variations in oligosaccharides and lactose in human milk during the first week of lactation, *J. Pediatr. Gastroenterol Nutr.*, 11, 361–364, 1990.

25. Thurl, S., Henker, J., Siegel, M., Tovar, K., and Sawatzki, G., Detection of four human milk groups with respect to Lewis blood group dependent oligosaccharides, *Glycoconj. J.,* 14, 795–799, 1997.

26. Parkkinen, J., and Finne, J., Isolation of sialyloligosaccharides and sialyloligosaccharide phosphates from bovine colostrum and human urine, *Methods Enzymol.*, 138, 289–300, 1987.

27. Gopal, P.K. and Gill, H.S., Oligosaccharides and glycoconjugates in bovine milk and colostrum, *Brit. J. Nutr.* 84, S69–S74, 2000.

28. Saito, T., Itoh, T., and Adachi, S., Presence of two neutral disaccharides containing *N*-acetylhexosamine in bovine colostrum as free forms, *Biochim. Biophys. Acta* 801, 147–150, 1984.

29. Chaturvedi, P. and Sharma, C.B., Purification by high-performance liquid chromatography, and characterization, by high-field ^{1}H-NMR spectroscopy, of two fucose-containing pentasaccharides of goat's milk, *Carbohydr. Res.,* 203, 91–101, 1990.

30. Urashima, T., Saito, T., Nishimura, J., and Ariga, H., New galactosyllactose containing α-glycosidic linkage isolated from ovine (*Boorola dorset*) colostrum, *Biochim. Biophys. Acta,* 992, 375–378, 1989.

31. Urashima, T., Saito, T., and Kimura, T., Chemical structures of three neutral oligosaccharides obtained from horse (thoroughbred) colostrum, *Comp. Biochem. Physiol.*, 100B, 177–183, 1991.

32. Viverge, D., Grimmonprez, L., Cassanas, G., Bardet, L., Bonnet, H., and Solère, M., Variations of lactose and oligosaccharides in milk from women of blood types secretor A or H, secretor Lewis, and secretor H/nonsecretor Lewis during the course of lactation, *Ann. Nutr. Metab.*, 29, 1–11, 1985.

33. Coppa, G.V., Gabrielli, O., Pierani, P., and Giorgi, P.L., Oligosaccharides in human milk and their role in bacterial adhesion, in *New Perspectives in Infant Nutrition,* Renner, B. and Sawaztki, G., Eds., Georg Thieme Verlag, Stuttgart, 1993, pp. 43–48.

34. Brand Miller, J., Bull, S., Miller, J., and McVeagh, P., The oligosaccharide composition of human milk: temporal and individual variations in monosaccharide components, *J. Pediatr. Gastroenterol. Nutr.*, 19, 371–376, 1994.

35. Nakamura, T., Kawase, H., Kimura, K., Watanabe, Y., Ohtani, M., Arai, I., and Urashima, T., Concentrations of sialyloligosaccharides in bovine colostrum and milk during the prepartum and early lactation, *J. Dairy Sci.,* 86, 1315–1320, 2003.

36. Martín, M.J., Martín-Sosa, S., García-Pardo, L.A., and Hueso, P., Distribution of bovine milk sialoglycoconjugates during lactation, *J. Dairy Sci.,* 84, 995–1000, 2001.

37. Erney, R.M., Hilty, M., Pickering, L., Ruiz-Palacios, G., and Prieto, P., Human milk oligosaccharides. A novel method provides insight into human genetics, in *Bioactive Components in Human Milk*, Newburg, D., Ed., Kluwer Academic/Plenum Publishers, New York, 2001, pp. 281–297.

38. Thurl, S. and Sawatzki, G., Variations of neutral oligosaccharides in human milk during lactation: preliminary investigations, in *New Perspectives in Infant Nutrition,* Renner, B. and Sawaztki, G., Eds., Georg Thieme Verlag, Stuttgart, 1993, pp. 27–31.

39. Neeser, J.R., Golliard, M., and Del Vedovo, S., Quantitative determination of complex carbohydrates in bovine milk and in milk-based infant formulas, *J. Dairy Sci.,* 74, 2860–2871, 1991.

40. Charlwood, J., Tolson, D., Dwek, M., and Camilleri, P., A detailed analysis of neutral and acidic carbohydrates in human milk, *Anal. Biochem.*, 273, 261–277, 1999.

41. Schwertmann, A., Rudloff, S., and Kunz, C., Potential ligands for cell adhesion molecules in human milk, *Ann. Nutr. Metab.*, 40, 252–262, 1996.

42. Sabharwal, H., Sjöblad, S., and Lundblad, A., Sialylated oligosaccharides in human milk and feces of preterm, full-term, and weaning infants, *J. Pediatr. Gastroenterol. Nutr.*, 12, 480–484, 1991.

43. Thurl, S., Offermanns, J., Müller-Werner, B., and Sawatzki, G., Determination of neutral oligosaccharide fractions from human milk by gel permeation chromatography, *J. Chromatogr.* B, 568, 291–300, 1991.

44. Kunz, C., Rudloff, S., Hintelmann, A., Pohlentz, G., and Egge, H., High-pH anion exchange chromatography with pulsed amperometric detection and molar response factors of human milk oligosaccharides, *J. Chromatogr. B.*, 685, 211–221, 1996.

45. Coppa, G.V., Pierani, P., Zampini, L., Carloni, I., Carlucci, A., and Gabrielli, O., Oligosaccharides in human milk during different phases of lactation, *Acta. Paediatr. Suppl.*, 430, 89–94, 1999.

46. Egge, H., Dell, A., and von Nicolai, H., Fucose containing oligosaccharides from human milk, I. Separation and identification of new constituents, *Arch. Biochem. Biophys.*, 224, 235–253, 1983.

47. Sarney, D.B., Hale, C., Frankel, G., and Vulfson, E.N., A novel approach to the recovery of biologically active oligosaccharides from milk using a combination of enzymatic treatment and nanofiltration, *Biotechnol. Bioeng.*, 69, 461–467, 2000.

48. Kobata, A., Yamashita, K., and Tachibana, Y., Oligosaccharides from human milk, *Methods Enzymol.*, 50, 216–220, 1978.

49. Newburg, D.S., Pickering, L.K., McCluer, R.H., and Cleary, T.G., Fucosylated oligosaccharides of human milk protect suckling mice from heat-stable enterotoxin of *Escherichia coli*, *J. Infect. Dis.*, 162, 1075–1080, 1990.

50. Yamashita, K., Mizuochi, T., and Kobata, A., Analysis of oligosaccharides by gel filtration, *Methods Enzymol.*, 83, 105–126, 1982.

51. Finke, B., Stahl, B., Pfenninger, A., Karas, M., Daniel, H., and Sawatzki. G., Analysis of high-molecular-weight oligosaccharides from human milk by liquid chromatography and MALDI-MS, *Anal. Biochem.*, 71, 3755–3762, 1999.

52. Smith, D.F., Zopf, D., and Ginsburg, V., Sialyloligosaccharides from milk, *Methods Enzymol.*, 50, 221–226, 1978.

53. VanPelt, J., Damm, J.B.C., Kamerling, J.P., and Vliegenthart, J.F.G., Separation of sialyloligosaccharides by medium pressure anion exchange chromatography on Mono Q. *Carbohydr. Res.*, 169, 43–51, 1987.

54. Coppa, G.V., Orazio, G., Pierani, P., Catassi, C., Carlucci, A., and Giorgi, P.L., Changes in carbohydrate composition in human milk over 4 months of lactation, *Pediatrics*, 91, 637–641, 1993.

55. White, C.A. and Kennedy, J.F., Oligosaccharides, in *Carbohydrate Analysis. A Practical Approach*, Chaplin, M.F. and Kennedy, J.F., Eds., IRL Press, Oxford, 1986, pp. 37–54.

56. Dubois, M., Gilles, K., Hamilton, J.K., Rebers, P.A., and Smith, F., Colorimetric method for determination of sugars and related substances, *Anal. Chem.*, 28, 350–356, 1956.

57. Svennerholm, L., Quantitative estimation of sialic acids. II A colorimetric resorcinol-hydrochloric method, *Biochim. Biophys. Acta* 24, 604–611, 1957.

58. Zanetta, J.P., Breckenridge, W.C., and Vincendon, G.J., Analysis of monosaccharides by gas liquid chromatography of the *O*-methylglycosides as trifluoroacetate derivatives. Application to glycoproteins and glycolipids, *J. Chromatogr.* 69, 291–304, 1972.

59. Bryn, K. and Jantzen, E., Quantification of 2-keto-3-deoxyoctonate in (lipo)polysaccharides by methanolytic release, trifluoroacetylation and capillary gas chromatography, *J. Chromatogr.*, 370, 103–112, 1986.

60.   Michalski, J.C., Isolation of glycans by HPLC, in *Methods on Glycoconjugates*, Verbert, A., Ed., Harwood Academic Publisher, Chur, 1995, pp. 67–77.

61.   Thurl, S., Mueller-Werner, B., and Sawatzki, G., Quantification of individual oligosaccharide compounds from human milk using high-pH anion exchange chromatography, *Anal. Biochem.* 235, 202–206, 1996.

62.   Coppa, G.V., Pierani, P., Zampini, L., Bruni, S., Carloni, I., and Gabrielli, O., Characterization of oligosaccharides in milk and feces of breast-fed infants by high performance anion exchange chromatography, in *Bioactive Components in Human Milk*, Newburg, D., Ed., Kluwer Academic/Plenum Publishers, New York, 2001, pp. 307–314.

63.   Chaturvedi, P., Warren, C.D., Ruiz-Palacios, G., Pickering, L.K., and Newburg, D.S., Profiling of milk oligosaccharides using reversed phase HPLC of their perbenzoylated derivatives, *Anal. Biochem.*, 251, 89–97, 1997.

64.   Baenziger, J.U. and Natowicz, M., Rapid separation of anionic oligosaccharide species by high-performance liquid chromatography, *Anal. Biochem.*, 112, 357–361, 1981.

65.   Cardon, P., Parente, J.P., Leroy, Y., Montreuil, J., and Fournet, B., Separation of sialyloligosaccharides by high-performance liquid chromatography. Application to the analysis of mono-, di-, tri-, and tetrasialyloligosaccharides obtained by hydrazinolysis of α 1-acid glycoprotein, *J. Chromatogr.*, 356, 135–146, 1986.

66.   Shen, Z., Warren, C.D., and Newburg, D., High-performance capillary electrophoresis of sialylated oligosaccharides of human milk, *Anal. Biochem.*, 279, 37–45, 2000.

67.   Shen, Z., Warren, C.D., and Newburg, D., Resolution of structural isomers of sialylated oligosaccharides by capillary electrophoresis, *J. Chromatogr.* A, 921, 315–321, 2001.

68.   Kitagawa, H., Nakada, H., Kurosaka, A., Hiraiwa, N., Numata, Y., Fukui, S., Funakoshi, I., Kawasaki, T., and Yamashina, I., Three novel oligosaccharides with the sialyl-Le[a] structure in human milk: isolation by immunoaffinity chromatography, *Biochemistry* 28, 8891–8897, 1989.

69.   Kitagawa, H., Nakada, H., Fukui, S.H., Funakoshi, I., Kawasaki, T., and Yamashina, I., Novel oligosaccharides with the sialyl-Le[a] structure in human milk, *Biochemistry,* 30, 2869–2876, 1991.

70.   Kobata, A., Kochibe, N., and Endo, T., Affinity chromatography of oligosaccharides on *Psthyrella velutina* lectin column, *Methods Enzymol.*, 247, 228–237, 1994.

71.   Rudloff, S., Stefan, C., Pohlentz, G., and Kunz, C., Detection of ligands for selectins in the oligosaccharide fraction of human milk, *Eur. J. Nutr.*, 41, 85–92, 2002.

72.   Strecker, G., Wieruszeski, J.M., Michalski, J.C., and Montreuil, J., Primary structure of human milk nona- and decasaccharides determined by a combination of fast atom bombardment mass spectrometry and ^{1}H-/^{13}C-nuclear magnetic resonance spectroscopy. Evidence for a new core structure, iso-lacto-*N*-octaose, *Glycoconj. J.*, 6, 169–182, 1989.

73.   Haeuw-Fievre, S., Wieruszeski, J.M., Plancke, Y., Michalski, J.C., Montreuil, J., and Strecker, G., Primary structure of human milk octa-, dodeca- and tridecasaccharides determined by a combination of ^{1}H-NMR spectroscopy and fast-atom-bombardment mass spectrometry. Evidence for a new core structure, the para-lacto-*N*-octaose, *Eur. J. Biochem.*, 215, 361–371, 1993.

74.   Guérardel, Y., Morelle, W., Plancke, Y., Lemoine, J., and Strecker, G., Structural analysis of three sulfated oligosaccharides isolated from human milk, *Carbohydr. Res.*, 320, 230–238, 1999.

75.   Martin-Pastor, M., and Bush, A., Conformational studies of human milk oligosaccharides using ^{1}H-^{13}C one bond NMR residual dipolar couplings, *Biochemistry* 39, 4674–4683, 2000.

76.   Towbin, H., Schenenberger, C.A., Braun, D.G., and Rosenfelder, G., Chromogenic labeling of milk oligosaccharides: purification by affinity chromatography and structure determination, *Anal. Biochem.*, 173, 1–9, 1988.

77.   Suzuki, M. and Suzuki, A., Structural characterization of fucose-containing oligosaccharides by high-performance liquid chromatography and matrix-assisted laser desorption/ionization time-of-flight mass spectrometry, *J. Biol. Chem.*, 382, 251–257, 2001.

78.   Sharon, N. and Ofek, I., Safe as mother's milk: carbohydrates as future anti-adhesion drugs for bacterial diseases, *Glycoconj. J.*, 17, 659–664, 2000.

79.   Shah, N.P., Effects of milk-derived bioactives: an overview, *Brit. J. Nutr., 84* (suppl.1), S3–10, 2000.

80.   Gracey, M. and Walker-Smith, J.A., *Diarrheal Disease.* Nestlé Nutrition Workshop series, vol 38, Lippincott-Raven Publishers, Philadelphia, 1997.

81.   Nataro, J.P. and Kaper, J.B., Diarrheagenic *Escherichia coli, Clin. Microbiol. Rev.,* 11, 142–201, 1998.

82.   Gaastra, W. and de Graaf, G.F.K., Host-specific fimbrial adhesins of noninvasive enterotoxigenic *Escherichia coli* strains, *Microbiol. Rev.*, 46, 129–161, 1982.

83.   Bloom, P.D., Karaolis, D.K.R., and Boedeker, E.C., *Escherichia coli*-associated diarrhea, in *Gastrointestinal Infections: diagnosis and Management,* Thomas LaMont, J., Ed., Marcel Dekker Inc., New York NY, 1997, pp. 453–498.

84.   Giron, J.A., Ho, A.S., and Schoolnik, G.K., An inducible bundle-forming pilus of enteropathogenic *Escherichia coli, Science,* 254, 710–713, 1991.

85.   Scaletsky, I.C., Silva, M.L., and Trabulsi, L.R., Distinctive patterns of adherence of enteropathogenic *Escherichia coli* to HeLa cells, *Infect. Immun.*, 45, 534–536, 1984.

86.   Moon, H.W., Whipp, S.C., Argenzio, R.A., Levine, M.M., and Giannella, R.A., Attaching and effacing activities of rabbit and human enteropathogenic *Escherichia coli* in pig and rabbit intestines, *Infect. Immun.*, 41, 1340–1351, 1983.

87.   Bell, B.P., Goldoft, M., Griffin, P.M., Davis, M.A., Gordon, D.C., Tarr, P.I., Bartleson, C.A., Lewis, J.H., Barrett, T.J., and Wells, J.G., A multistate outbreak of *Escherichia coli* O157:H7-associated bloody diarrhea and hemolytic uremic syndrome from hamburgers. The Washington experience, *JAMA,* 272, 1349–1353, 1994.

88.   Riley, L.W., Remis, R.S., Helgerson, S.D., McGee, H.B., Wells, J.G., Davis, B.R., Hebert, R.J., Olcott, E.S., Johnson, L.M., Hargrett, N.T., Blake, P.A., and Cohen, M.L., Hemorrhagic colitis associated with a rare *Escherichia coli* serotype, *N. Engl. J. Med.* 308, 681–685, 1983.

89.   Hale, T.L., Sansonetti, P.J., Schad, P.A., Austin, S., and Formal, S.B., Characterization of virulence plasmids and plasmid-associated outer membrane proteins in *Shigella flexneri, Shigella sonnei,* and *Escherichia coli, Infect. Immun.,* 40, 340–350, 1983.

90.   Nataro, J.P., Seriwatana, J., Fasano, A., Maneval, D.R., Guers, L.D., Noriega, F., Dubovsky, F., Levine, M.M., and Morris, J.C.G., Identification and cloning of a novel plasmid-encoded enterotoxin of enteroinvasive *Escherichia coli* and *Shigella* strains, *Infect. Immun.* 63, 4721–4728, 1995.

91.   Okeke, I.N. and Nataro, J.P., Enteroaggregative *Escherichia coli, Lancet. Infect. Dis.,* 1, 304–313, 2001.

92.   Tzipori, S., Montanaro, J., Robins-Browne, R.M., Vial, P., Gibson, R., and Levine, M.M., Studies with enteroaggregative *Escherichia coli* in the gnotobiotic piglet gastroenteritis model, *Infect. Immun.*, 60, 5302–5306, 1992.

93.   Nataro, J.P., Deng, Y., Maneval, D.R., German, A.L., Martin, W.C., and Levine, M.M., Aggregative adherence fimbriae I of enteroaggregative *Escherichia coli* mediate adherence to HEp-2 cells and hemagglutination of human erythrocytes, *Infect. Immun.* 60, 2297–2304, 1992.

94.   Czeczulin, J.R., Balepur, S., Hicks, S., Phillips, A., Hall, R., Kothary, M.H., Navarro-Garcia, F., and Nataro, J.P., Aggregative adherence fimbria II, a second fimbrial antigen mediating aggregative adherence in enteroaggregative *Escherichia coli*, *Infect. Immun.*, 65, 4135–4145, 1997.

95.   Clarke, S.C., Diarrheagenic *Escherichia coli*-an emerging problem?, *Diagn. Microbiol. Infect. Dis.*, 41, 93–98, 2001.

96.   Monteiro-Neto, V., Bando, S.Y., Moreira-Filho, C.A., and Giron, J.A., Characterization of an outer membrane protein associated with haemagglutination and adhesive properties of enteroaggregative *Escherichia coli* O111:H12, *Cell Microbiol.*, 5, 533–547, 2003.

97.   Soto, G.E. and Hultgren, S.J., Bacterial adhesins: common themes and variations in architecture and assembly, *J. Bacteriol.*, 181, 1059–1071, 1999.

98.   Sauer, F.G., Pinkner, J.S., Waksman, G., and Hultgren, S.J., Chaperone priming of pilus subunits facilitates a topological transition that drives fiber formation, *Cell* 111, 543–551, 2002.

99.   Hung, D.L., Knight, S.D., Woods, R.M., Pinkner, J.S., and Hultgren, S.L., Molecular basis of two subfamilies of immunoglobulin-like chaperones, *EMBO.J.*, 15, 3792–3805, 1996.

100.   Bakker, D., Vader, C.E., Roosendaal, B., Mooi, F.R., Oudega, B., and de Graaf, G.F.K., Structure and function of periplasmic chaperone-like proteins involved in the biosynthesis of K88 and K99 fimbriae in enterotoxigenic *Escherichia coli*, *Mol. Microbiol.*, 5, 875–886, 1991.

101.   Jalajakumari, M.B., Thomas, C.J., Halter, R., and Manning, P.A., Genes for biosynthesis and assembly of CS3 pili of CFA/II enterotoxigenic *Escherichia coli*, novel regulation of pilus production by bypassing an amber codon, *Mol. Microbiol.*, 3, 1685–1695, 1989.

102.   Savarino, S.J., Fox, P., Deng, Y., and Nataro, J.P., Identification and characterization of a gene cluster mediating enteroaggregative *Escherichia coli* aggregative adherence fimbria I biogenesis, *J. Bacteriol.*, 176, 4949–4957, 1994.

103.   Voegele, K., Sakellaris, H., and Scott, J.R., CooB plays a chaperone-like role for the proteins involved in formation of CS1 pili of enterotoxigenic *Escherichia coli*. *Proc. Natl. Acad. Sci. USA.*, 94, 13257–13261, 1997.

104.   Alm, R.A. and Mattick, J.S., Genes involved in the biogenesis and function of type-4 fimbriae in *Pseudomonas aeruginosa*. *Gene* 192, 89–98, 1997.

105.   Bian, B. and Normark, S., Nucleator function of CsgB for the assembly of adhesive surface organelles in *Escherichia coli*, *EMBO. J.* 16, 5827–5836, 1997.

106.   Gaastra, W. and Svennerholm, A.M., Colonization factors of human enterotoxigenic *Escherichia coli* (ETEC), *Trends Microbiol.*, 4, 444–452, 1996.

107.   Sauer, F.G., Futterer, K., Pinkner, J.S., Dodson, K.W., Hultgren, S.J., and Waksman, G., Structural basis of chaperone function and pilus biogenesis, *Science* 285, 1058–1061, 1999.

108.   Dodson, K.W., Pinkner, J.S., Rose, T., Magnusson, G., Hultgren, S.J., and Waksman, G., Structural basis of the interaction of the pyelonephritic *Escherichia coli* adhesin to its human kidney receptor, *Cell* 105, 733–743, 2001.

109.   Choudhury, D., Thompson, A., Stojanoff, V., Langermann, S., Pinkner, J., Hultgren, S.J., and Knight, S.D., X-ray structure of the FimC-FimH chaperone–adhesin complex from uropathogenic *Escherichia coli*, *Science* 285, 1061–1066, 1999.

110.   Buts, L., Bouckaert, J., De, G.E., Loris, R., Oscarson, S., Lahmann, M., Messens, J., Brosens, E., Wyns, L., and De, G.H., The fimbrial adhesin F17-G of enterotoxigenic *Escherichia coli* has an immunoglobulin-like lectin domain that binds *N*-acetylglucosamine, *Mol. Microbiol.*, 49, 705–715, 2003.

111. Isberg, R.R. and Barnes, P., Dancing with the host; flow-dependent bacterial adhesion, *Cell* 110, 1–4, 2002.

112. Thomas, W.E., Trintchina, E., Forero, M., Vogel, V., and Sokurenko, E.V., Bacterial adhesion to target cells enhanced by shear force, *Cell* 109, 913–923, 2002.

113. Abraham, S.N., Sun, D., Dale, J.B., and Beachey, E.H., Conservation of the *D*-mannose-adhesion protein among type 1 fimbriated members of the family Enterobacteriaceae, *Nature* 336, 682–684, 1988.

114. Krogfelt, K.A., Bergmans, H., and Klemm, P., Direct evidence that the FimH protein is the mannose-specific adhesin of *Escherichia coli* type 1 fimbriae, *Infect. Immun.*, 58, 1995–1998, 1990.

115. Johnson, J.R., Virulence factors in *Escherichia coli* urinary tract infection, *Clin. Microbiol. Rev.* 4, 80–128, 1991.

116. Korhonen, T.K., Vaisanen-Rhen, V., Rhen, M., Pere, M., Parkkinen, A., and Finne, J., *Escherichia coli* fimbriae recognizing sialyl galactosides, *J. Bacteriol.* 159, 762–766, 1984.

117. Saren, A., Virkola, R., Hacker, J., and Korhonen, T.K., The cellular form of human fibronectin as an adhesion target for the S fimbriae of meningitis-associated *Escherichia coli*, *Infect. Immun.*, 67, 2671–2676, 1999.

118. Chaturvedi, G., Tewari, R., Mrigank, M., Agnihotri, N., Vishwakarma, R.A., and Ganguly, N.K., Inhibition of *Helicobacter pylori* adherence by a peptide derived from neuraminyl lactose binding adhesin, *Mol. Cell. Biochem.*, 228, 83–89, 2001.

119. Boren, T., Falk, P., Roth, K.A., Larson, G., and Normark, S., Attachment of *Helicobacter pylori* to human gastric epithelium mediated by blood group antigens, *Science*, 262, 1892–1895, 1993.

120. Mahdavi, J., Sonden, B., Hurtig, M., Olfat, F.O., Forsberg, L., Roche, N., Angstrom, J., Larsson, T., Teneberg, S., Karlsson, K.A., Altraja, S., Wadstrom, T., Kersulyte, D., Berg, D.E., Dubois, A., Petersson, C., Magnusson, K.E., Norberg, T., Lindh, F., Lundskog, B.B., Arnqvist, A., Hammarstrom, L., and Boren, T., *Helicobacter pylori* SabA adhesin in persistent infection and chronic inflammation, *Science*, 297, 573–578, 2002.

121. Barthelson, R., Mobasseri, A., Zopf, D., and Simon, P., Adherence of *Streptococcus pneumoniae* to respiratory epithelial cells is inhibited by sialylated oligosaccharides, *Infect. Immun.*, 66, 1439–1444, 1998.

122. Stapleton, A.E., Stroud, M.R., Hakomori, S.I., and Stamm, W.E., The globoseries glycosphingolipid sialosyl galactosyl globoside is found in urinary tract tissues and is a preferred binding receptor *in vitro* for uropathogenic *Escherichia coli* expressing pap-encoded adhesins, *Infect. Immun.*, 66, 3856–3861, 1998.

123. Martin, M.J., Martin-Sosa, S., Alonso, J.M., and Hueso, P., Enterotoxigenic *Escherichia coli* strains bind bovine milk gangliosides in a ceramide-dependent process, *Lipids*, 38, 761–768, 2003.

124. Teneberg, S., Willemsen, P.T., de Graaf, G.F.K., and Karlsson, K.A., Calf small intestine receptors for K99 fimbriated enterotoxigenic *Escherichia coli*, *FEMS Microbiol. Lett.*, 109, 107–112, 1993.

125. Varki, A., Loss of *N*-glycolylneuraminic acid in humans: mechanisms, consequences, and implications for hominid evolution, *Am. J. Phys. Anthropol.*, 33, 54–69, 2001.

126. Anderson, M.J., Whitehead, J.S., and Kim, Y.S., Interaction of *Escherichia coli* K88 antigen with porcine intestinal brush border membranes, *Infect. Immun.*, 29, 897–901, 1980.

127. Grange, P.A., Mouricout, M.A., Levery, S.B., Francis, D.H., and Erickson, A.K., Evaluation of receptor binding specificity of *Escherichia coli* K88 (F4) fimbrial adhesin variants using porcine serum transferrin and glycosphingolipids as model receptors, *Infect. Immun.*, 70, 2336–2343, 2002.

128. Varki, A., Selectin ligands: will the real ones please stand up? *J. Clin. Invest.*, 100, S31–S35, 1997.

129. Lingwood, C.A., Huesca, M., and Kuksis, A., The glycerolipid receptor for *Helicobacter pylori* (and exoenzyme S) is phosphatidylethanolamine, *Infect. Immun.*, 60, 2470–2474, 1992.

130. Karlsson, K.A., Meaning and therapeutic potential of microbial recognition of host glycoconjugates, *Mol. Microbiol.*, 29, 1–11, 1998.

131. Sherman, I.W., Eda, S., and Winograd, E., Cytoadherence and sequestration in *Plasmodium falciparum*: defining the ties that bind, *Microbes. Infect.*, 5, 897–909, 2003.

132. Donta, S.T. and Viner, J.P., Inhibition of the steroidogenic effects of cholera and heat-labile *Escherichia coli* enterotoxins by GM1 ganglioside: evidence for a similar receptor site for the two toxins, *Infect. Immun.*, 11, 982–985, 1975.

133. Merritt, E.A. and Hol, W.G., AB5 toxins *Curr. Opin. Struct. Biol.*, 5, 165–171, 1995.

134. Lindberg, A.A., Brown, J.E., Stromberg, N., Westling-Ryd, M., Schultz, J.E., and Karlsson, K.A., Identification of the carbohydrate receptor for Shiga toxin produced by *Shigella dysenteriae* type 1, *J. Biol. Chem.*, 262, 1779–1785, 1987.

135. Kitov, P.I., Sadowska, J.M., Mulvey, G.L., Armstrong, G.D., Ling, H., Pannu, N.S., Read, R.J., and Bundle, D.R., Shiga-like toxins are neutralized by tailored multivalent carbohydrate ligands, *Nature*, 403, 669–672, 2000.

136. Mulvey, G.L., Marcato, P., Kitov, P.I., Sadowska, J.M., Bundle, D.R.,and Armstrong G.D., Assessment in mice of the therapeutic potential of tailored, multivalent Shiga toxin carbohydrate ligands, *J. Infect. Dis.* 187, 640–649, 2003.

137. Nowrouzian, F., Hesselmar, B., Saalman, R., Strannegard, I.L., Aberg, N., Wold, A.E., and Adlerberth, I., *Escherichia coli* in infants' intestinal microflora: colonization rate, strain turnover, and virulence gene carriage, *Pediatr. Res.*, 54, 8–14, 2003.

138. Schroten, H., Stapper, C., Plogmann, R., Kohler, H., Hacker, J., and Hanisch, F.G., Fab-independent antiadhesion effects of secretory immunoglobulin A on S-fimbriated *Escherichia coli* are mediated by sialyloligosaccharides. *Infect. Immun.*, 66, 3971–3973, 1998.

139. Chaturvedi, P., Warren, C.D., Buescher, C.R., Pickering, L.K., and Newburg, D.S., Survival of human milk oligosaccharides in the intestine of infants, *Adv. Exp. Med. Biol.*, 501, 315–323, 2001.

140. Parkkinen, J. Finne, J., Achtman, M., Vaisanen, V., and Korhonen, T.K., *Escherichia coli* strains binding neuraminyl α 2-3 galactosides, *Biochem. Biophys. Res. Commun.*, 111, 456–461, 1983.

141. Andersson, B., Porras, O., Hanson, L.A., Lagergard, T., and Svanborg-Eden, C., Inhibition of attachment of *Streptococcus pneumoniae* and *Haemophilus influenzae* by human milk and receptor oligosaccharides, *J. Infect. Dis.*, 153, 232–237, 1986.

142. Mouricout, M., Petit, J.M., Carias, J.R., and Julien, R., Glycoprotein glycans that inhibit adhesion of *Escherichia coli* mediated by K99 fimbriae: treatment of experimental colibacillosis, *Infect. Immun.*, 58, 98–106, 1990.

143. Lesman-Movshovich, E., Lerrer, B., and Gilboa-Garber, N., Blocking of *Pseudomonas aeruginosa* lectins by human milk glycans, *Can. J. Microbiol.*, 49, 230–235, 2003.

144. Martin-Sosa, S., Martin, M.J., and Hueso, P., The sialylated fraction of milk oligosaccharides is partially responsible for binding to enterotoxigenic and uropathogenic *Escherichia coli* human strains, *J. Nutr.*, 132, 3067–3072, 2002.

145. Martin, M.J., Martin-Sosa, S., and Hueso, P., Binding of milk oligosaccharides by several enterotoxigenic *Escherichia coli* strains isolated from calves, *Glycoconj. J.*, 19, 5–11, 2002.

146. Laegreid, A., Otnaess, A.B., and Fuglesang, J.,. Human and bovine milk: comparison of ganglioside composition and enterotoxin-inhibitory activity, *Pediatr. Res.*, 20, 416–421, 1986.
147. Hata, Y., Kita, T., and Murakami, M., Bovine milk inhibits both adhesion of *Helicobacter pylori* to sulfatide and *Helicobacter pylori*-induced vacuolation of vero cells, *Dig. Dis. Sci.*, 44, 1696–1702, 1999.
148. Herrera-Insua, I., Gomez, H.F., Diaz-Gonzalez, V.A., Chaturvedi, P., Newburg, D.S., and Cleary, T.G., Human milk lipids bind Shiga toxin, *Adv. Exp. Med. Biol.*, 501, 333–339, 2001.
149. Rueda, R., Sabatel, J.L., Maldonado, J., Molina-Font, J.A., and Gil A., Addition of gangliosides to an adapted milk formula modifies levels of fecal *Escherichia coli* in preterm newborn infants, *J. Pediatr.*, 133, 90–94, 1998.
150. Bäckhed, F., Alsen, B., Roche, N., Angstrom, J., Von, E.A., Breimer, M.E., Westerlund-Wikstrom, B., Teneberg, S., and Richter-Dahlfors, A., Identification of target tissue glycosphingolipid receptor for uropathogenic F1C-fimbriated *Escherichia coli* and its role in mucosal inflammation, *J. Biol. Chem.*, 277, 18198–18205, 2002.
151. Bertin, Y., Girardeau, J.O., Darfeuille-Michaud, A., and Contrepois, M., Characterization of 20K fimbria, a new adhesin of septicemic and diarrhea-associated *Escherichia coli* strains, that belongs to a family of adhesins with *N*-acetyl-D-glucosamine recognition, *Infect. Immun.*, 64, 332–342, 1996.
152. Van Maele, R.P., Heerze, L.D., and Armstrong G.D., Role of lactosyl glycan sequences inhibiting enteropathogenic *Escherichia coli* attachment, *Infect. Immun.*, 67, 3302–3307, 1999.
153. Lindahl, M. and Wadström, T., Binding to erythrocyte membrane glycoproteins and carbohydrates specificity of F41 fimbriae of enterotoxigenic *Escherichia coli*, *FEMS Microbiol. Lett.*, 34, 297–300, 1986.
154. Kunz, C., Rudloff, S., Baier, W., Klein, N., and Strobel, S., Oligosaccharides in human milk:structural, functional and metabolic aspects, *Ann. Rev. Nutr.*, 20, 699–722, 2000.
155. Sugita-Konishi, Y., Sakanaka, S., Sasaki, K., Juneja, L.R., Noda,T., and Amano, F., Inhibition of bacterial adhesión and *Salmonella* infection in BALB/c mice by sialyloligosaccharides and their derivatives from chicken egg yolk, *J. Agric. Food. Chem.*, 50, 3607–3613, 2002.
156. Haataja, S., Tikkanen, K., Nilsson, U., Magnusson, G., Karlsson, K.A., and Finne, J., Oligosaccharide-receptor interaction of the Gal α1-4 Gal binding adhesin of *Streptococcus suis*, *J. Biol. Chem.*, 269, 27446–27472, 1994.
157. Ruiz-Palacios, G.M., Cervantes, L.E., Ramos, P., Chavez-Munguia, B., and Newburg, D.S., *Campylobacter jejuni* binds intestinal H(O) antigen (Fuc α1,2 Gal β 1,4 GlcNAc), and fucosyloligosaccharides of human milk inhibit its binding and infection, *J. Biol. Chem.*, 278, 14112–14120, 2003.
158. Sakarya, S. and Oncu, S., Bacterial adhesins and the role of sialic acid in bacterial adhesion, *Med. Sci. Monit.*, 9, RA 76–82, 2003.

6 Structural Characterization of Glycosaminoglycans

Kevin Pojasek, Rahul Raman, and Ram Sasisekharan

CONTENTS

Glycosaminoglycans (GAGs) are complex polysaccharides found on the surface of every cell in the body. These polysaccharides were long thought of as inert, structural supports occupying the space between cells and not as vital as other biopolymers, namely, DNA and proteins. The main reason for the neglect of GAGs in the past arose from their overwhelming structural and chemical heterogeneity, which complicated the study of these highly acidic carbohydrates. Recent advances in biochemical and analytical techniques have facilitated the isolation and characterization of GAGs in terms of their primary sequence and biological function. The biochemical methods include the cloning, recombinant expression, and characterization of a variety of GAG-degrading enzymes from both bacterial and mammalian sources with well-defined substrate specificities. The analytical techniques include the development of sensitive and accurate methods based on chromatography, electrophoresis, and mass spectrometry to detect very small amounts of GAG oligosaccharides. The use of GAG-degrading enzymes complement the analytical techniques by enabling an orchestrated degradation of these complex polysaccharides into smaller oligosaccharides that can be accurately detected and quantified. More recently, a novel bioinformatics framework was developed to incorporate the data obtained from the analytical techniques as constraints to determine the sequence of GAG oligosaccharides in a rapid and unbiased fashion. Together, these methods have generated enormous GAG sequence information and accelerated the progress in understanding their specific sequence–biological activity relationships. This chapter highlights the development of biochemical and analytical techniques for the characterization of GAGs with a particular emphasis on the impact of novel applications for understanding how specific carbohydrate sequences regulate biological function.

I. INTRODUCTION

The extracellular matrix (ECM), once thought of as an inert scaffold for cells within a tissue, has emerged as an important regulator of cell function through the active modulation of cell-to-cell signaling. Complex polysaccharides of the GAG family, found in the ECM as well as in the carbohydrate "coat" on the surface of cells, are central to this intercellular communication. GAGs at the cell surface and in the ECM bind to a multitude of proteins, such as growth factors, cytokines, chemokines, glycoproteins on microbial pathogens, and proteases and protease inhibitors, and thereby modulate their biological activity. The specificity of GAG protein interactions arises from the precise noncovalent interactions (ionic and surface contacts) between the GAG chains and the GAG-binding sites on different proteins. Therefore, the breadth of biological phenomena in which GAGs are involved is a direct consequence of the inherent heterogeneity in their chemical structure. GAGs are usually synthesized as proteoglycans with multiple polysaccharide chains attached to a single core protein; an exception is hyaluronic acid, which is synthesized as an independent carbohydrate polymer.[1] Since the GAG chains constitute the major component of the proteoglycans, it comes as no surprise that the majority of biological interactions mediated by proteoglycans involve noncovalent interactions with this carbohydrate component. In fact, individual cell types define the GAG component of their cell surface coat via tightly regulated biosynthesis machinery. By doing so, a cell regulates how it responds to various growth factors and cytokines present in its extracellular environment. An understanding of the molecular composition of the different GAG family members is critical to appreciating fully the biological events they mediate.

The GAG family of complex polysaccharides includes heparin and heparan sulfate (collectively abbreviated as HSGAGs), chondroitin sulfate (CS), dermatan sulfate (DS), keratan sulfate (KS), and hyaluronic acid (HA) (Figure 6.1). Each of

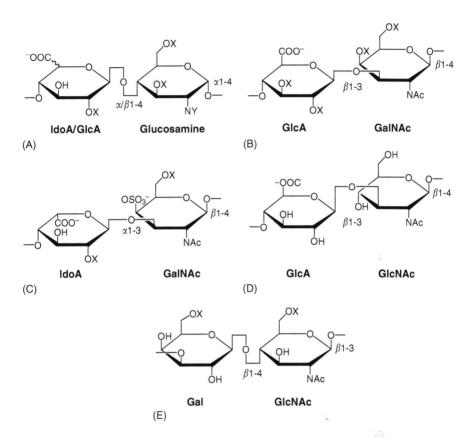

FIGURE 6.1 The basic GAG disaccharide units. (A) HSGAGs are composed of UA linked to a glucosamine. UA could be either an IdoA or GlcA epimer at the C5 position. The presence or absence of *O*-sulfation (X) along with the presence of an *N*-acetyl or *N*-sulfate (Y) on the glucosamine composes the different possible disaccharide modifications found in HSGAGs. (B) CS is composed of glucuronic acid linked to an GalNAc. Sulfation at the 4-*O* or 6-*O* position (X) defines CS A or CS C, respectively. Additional rare sulfation is found at the 3-*O* and 2-*O* positions (X). (C) DS is composed of primarily IdoA linked to *N*-acetyl galactosamine with primary sulfation at the 4-*O* position. Additional sulfation at the 2-*O* and 6-*O* position provides additional chemical heterogeneity. (D) Hyaluronic acid is a homopolymer of glucuronic acid linked to glucosamine with no additional modifications. (E) KS is composed of a galactose (Gal) linked to a *N*-acetyl glucosamine. Sulfation is found on the 6-*O* position of both the Gal and the GlcNAc monosaccharides.

these GAGs is synthesized in the Golgi as a co-polymer of a uronic acid linked to a hexosamine, thereby composing the common disaccharide repeat unit by which each chain is described. The sole exception is KS, which is composed of a disaccharide repeat of galactose linked to a hexosamine. Initiation of the synthesis of each of each of these GAG chains, with the sole exception of HA, begins with the attachment of a linkage sequence to a core protein at a serine residue, forming a proteoglycan. As synthesis of the GAG polymer proceeds, a variety of modifying enzymes refine the structure of the expanding backbone by selectively sulfating different positions of both the uronic acid and the hexosamine (H) as well as by epimerizing the C-5 position of the uronic acid (UA or U) from glucuronic acid (GlcA or G) to iduronic acid (IdoA or I). The wide range of biological events regulated by GAGs is a direct consequence of the structural and chemical diversity generated during their biosynthesis.

A. GLYCOSAMINOGLYCAN CHEMICAL STRUCTURE

HA, the simplest of all the GAGs, is composed of the basic disaccharide unit of a GlcA, $\beta(1\rightarrow3)$ linked to N-acetyl glucosamine.[1,2] This disaccharide unit is $\beta(1\rightarrow4)$ linked to form polymers of up to 25,000 disaccharides in length and does not contain any additional modifications (Figure 6.1). KS has the basic disaccharide repeat of galactose, $\beta(1\rightarrow4)$ linked to N-acetyl glucosamine.[3] These basic units are $\beta(1\rightarrow3)$ linked to adjacent disaccharides to form KS polymers of up to 100 disaccharides in length. KS can be modified at the 6-O position of both the Gal and the GlcNac moiety (Figure 6.1). The primary structures of HSGAGs and CS/DS are considerably more complex than those of HA and KS.

HSGAGs are synthesized as a linear chain of 20–100 disaccharide units of GlcA, $\beta(1\rightarrow4)$ linked to N-acetylated glucosamine[4] (Figure 6.1). The HSGAG polysaccharide consists of these disaccharide building blocks $\alpha(1$–$4)$ linked to each other. Modifications to the disaccharide building blocks include O-sulfation at the 6-O and 3-O positions of the glucosamine and at the 2-O position of the UA. Additionally, the glucosamine can be modified by N-deacetylation followed by N-sulfation. Finally, β-D-glucuronic acid or β-D-GlcA can be modified into α-L-IdoA through C5 epimerization (Figure 6.1). Taking into account all of the possible modifications yields 32 unique disaccharides that could be found in an HSGAG chain. While all of these disaccharide combinations have not been reported in nature, this observation underscores the fact that HSGAGs contain more informational density than either DNA (composed of four bases) or proteins (composed of 20 amino acids).[4] These modifications provide structural heterogeneity to an HSGAG chain that lead to diversity in the biological functions of HSGAGs that are mediated by their interactions with numerous proteins. The heterogeneity of HSGAGs also complicates the study of the underlying structure–activity relationships behind these biological functions.

Within a single HSGAG chain there exist regions of aggregate character that further assist in defining their biological role.[5] However, it is important to note that the HSGAG nomenclature traditionally used to describe these differentially modified regions is counterintuitive. The names would suggest that heparan sulfate is more sulfated than heparin when in fact the opposite is true. In a broad sense, "heparin-like" regions tend to be highly sulfated at the 2-O, 6-O, and N-positions as well as contain predominantly the IdoA epimer. On the other hand, regions of "heparan sulfate-like" character tend to have a lower degree of 2-O and N-sulfation and a higher GlcA content than heparin[5] (Figure 6.1). Therefore, a single HSGAG chain can be thought of as containing "heparin-like" and "heparan sulfate-like" regions each with different potential biological functionalities. An important additional distinction between heparin and heparan sulfate is drawn on their physiological location. Heparan sulfate is found on the surface of most cells and in the ECM where it plays a vital role in regulating intercellular communication, while heparin is synthesized solely in mast cells where it performs a relatively limited biological function.[4,5]

Unlike HSGAGs, CS/DS oligosaccharides are synthesized as a disaccharide repeat of GlcA $\beta(1{\to}3)$ linked to N-acetylated galactosamine, a defining chemical feature of CS/DS[6] (Figure 6.1). This basic repeat unit is in turn $\beta(1{\to}4)$ linked to adjacent disaccharides to form polymeric CS/DS. Similar to HSGAGs, the basic CS/DS chain is differentially modified through sulfation and epimerization. These modifications are used to divide CS/DS into three major subtypes. Chondroitin 4-SO_4 (CS A) is primarily sulfated at the 4-O position of the galactosamine residue and is composed of GlcA epimers at the C5 position of the UA moiety (Figure 6.1). Chondroitin 6-SO_4 (CS C) is sulfated at the 6-O position of the galactosamine residue and also contains primarily the GlcA epimer of the UA[1] (Figure 6.1B). Dermatan sulfate (CS B) is defined by a 4-O sulfated galactosamine and predominantly the IdoA epimer, although GlcA is present to differing degrees depending on the biological source of the DS.[1,7] In addition, minor fractions of 2-O, 4-O, and 4-O, 6-O disulfated disaccharides are found in DS and contribute to the biological activity of this GAG polymer[7] (Figure 6.1). Rare sulfation at the 3-O position of the UA moiety has also been reported in CS isolated from king crab cartilage, but has never been described for mammalian isolates[8] (Figure 6.1). In fact, quite a few aquatic species including hagfish,[9] squid, and shark,[10] as well as Ascidians[11] have been the source for a variety of highly sulfated CS/DS oligosaccharides. Finally, it is important to note that completely unsulfated CS, referred to simply as chondroitin, exists in lower organisms including *Caenorhabditis elegans.*[12,13] The presence of the IdoA epimer in DS is a factor distinguishing DS from CS and actually makes DS more chemically similar to IdoA-rich regions of HSGAGs.[7,14]

Given the diversity of the chemical structure of GAGs both in terms of the saccharide backbone and the sulfation pattern, the emerging view is that unique sequences of GAGs specifically bind to proteins and modulate their biological function. The specificity of GAG–protein interactions is generally attributed to the electrostatic contacts between the sulfate groups on the GAG chain and the basic amino acids on the protein. However, the three-dimensional (3D) structure and conformation of the GAG chains plays an important role in providing both charge and surface complementarity to the binding site on the protein. Conformational studies of IdoA in HSGAG oligosaccharides have revealed that this epimer can exist in a variety of energetically favorable conformations while the GlcA form is relatively rigid. Additionally, the conformation that a specific IdoA residue adopts appears to be dictated by its neighboring sulfation pattern as well as its overall location (internal vs. terminal) in the HSGAG oligosaccharide.[15] An initial insight into the importance of this was suggested by data showing that IdoA containing GAGs are more efficacious in inhibiting the proliferation of fibroblasts than GlcA containing GAGs.[16] It was recently shown that this structural property of the IdoA enhances the "flexibility" of GAG conformation, thus providing specific structural recognition motifs for the GAG-binding sites on a wide variety of proteins. In fact, this study confirmed that the flexibility of HSGAG chains at the IdoA residues plays a direct role in the binding of a specific oligosaccharide sequence to basic fibroblast growth factor (FGF), a model system often used for studying GAG–protein interactions.[17] Similarly, the presence of IdoA in DS probably plays a critical role in the specific binding of proteins to sequences in the oligosaccharide.

B. CHALLENGES OF GAG CHARACTERIZATION

The biosynthesis of GAGs follows a nontemplate-based process where enzymes act sequentially to polymerize and modify the growing GAG oligosaccharide. This is in contrast to DNA synthesis and protein translation in which enzymes rely on a template for generating the new polymer. Importantly, this lack of a template-driven synthesis for GAGs means that there is no molecular biological technique for their amplification from biological sources, such as polymerase chain reaction (PCR) or recombinant DNA technology that are used to amplify DNA and proteins, respectively. This inability to amplify tissue-derived GAG sequences results in limited sample amounts available for analysis. In addition, a limited understanding of the exact GAG biosynthetic mechanism has prevented the identification of the heterogeneity in the modifications along a specific GAG chain. Although it is known that the modifications are clustered, there is little understanding of the finer heterogeneities and how they correlate to the actual biosynthetic machinery.[4,6,18] This enormous heterogeneity of GAGs results in a lower abundance of functionally important sequences within the overall GAG chain. Therefore, the isolation and characterization of functionally important GAG sequences is complicated further by this "needle in a haystack" effect.

The extensive sulfation of GAG chains, particularly in HSGAGs and certain types of CS/DS oligosaccharides, has made these GAGs less amenable to conventional methods of analysis, namely chromatography and mass spectrometry. The chemical complexity of GAGs also leads to limitations in their isolation and fractionation. Thus, even fractionated and purified samples contain a heterogeneous mixture of sequences with different lengths and compositions with the heterogeneity becoming more pronounced at higher molecular weights. This limitation is particularly relevant to commercially available heparin preparations that are used as clinical anticoagulants.[19]

Owing to the heterogeneity in the sulfation pattern of the disaccharide repeat units of GAGs, the information density associated with GAG sequences is much higher than that of DNA and proteins. Thus, the use of alphabetic codes (similar to A, T, G, or C in DNA) for representing a GAG sequence is difficult and cumbersome. Representation of the information content in GAG sequences is important to correlate specific sequences to their associated biological activity.

Despite all of these inherent challenges in studying GAGs, there have been many successful attempts to isolate and sequence these important biopolymers. GAGs are usually isolated from different tissues by cleaving the polysaccharides from proteoglycans using either proteolytic cleavage or enzymes that specifically remove the polysaccharides from the protein core of the proteoglycan. The isolated GAGs are fractionated and subfractionated into oligosaccharides of different lengths and overall charge. The majority of the methods for characterizing GAGs involve degradation of the parent oligosaccharide using chemical and enzymatic techniques with subsequent detection of the fragments using sensitive analytical tools.

II. TECHNIQUES FOR THE DEPOLYMERIZATION OF GLYCOSAMINOGLYCANS

As mentioned above, to facilitate the characterization of GAGs, a variety of chemical and enzymatic methods have been developed to break down the parent oligomers into smaller fragments. The fragments derived from the depolymerization of the parent oligomers are separated based on mass or charge and detected using chromatography, electrophoresis, or mass spectrometry techniques. The sequence information of the parent oligomer is derived based on the knowledge of the orchestrated scission of the different glycosidic bonds. Some of the major depolymerization techniques for GAGs are discussed below.

A. CHEMICAL TECHNIQUES

Chemical treatment is the most traditional procedure for the depolymerization of GAGs into their mono-, di-, and higher-order oligosaccharide components. Complete acid hydrolysis can be used to degrade GAGs to their desulfated, deacetylated monosaccharides building blocks, thereby providing a useful technique for fundamental compositional analysis.[20] In addition, alkaline degradation of carbohydrates, known as the "peeling reaction," has been employed for the break down of GAGs. However, a prerequisite for this reaction is the replacement of the C5 carboxylate moiety with a benzyl group that serves as a more suitable leaving group. Subsequent incubation of the GAG chain with sodium hydroxide at elevated temperatures results in the sequential β-elimination of monosaccharides from the reducing end of the polymer. This technique is commonly employed for the synthesis of clinical anticoagulant preparations. Finally, activated oxygen species, generated through Fenton chemistry, also have been used to break down polymeric HSGAGs into lower molecular weight species.[21]

The most widely employed chemical technique for the break down of HSGAGs is known as nitrous acid degradation[20] (Figure 6.2). This reaction specifically targets an N-sulfated glucosamine residue and results in cleavage of the $\alpha/\beta(1\rightarrow4)$ bond to the adjacent UA with a concominant loss of the sulfate and the amine group, a process referred to as deaminative cleavage. While this reaction is useful for the degradation of HSGAGs; CS/DS, HA, and KS oligosaccharides are all resistant to nitrous acid degradation due to the ubiquitous presence of acetylation on the galactosamine (Figure 6.2). However, treatment of N-acetylated GAGs with hydrazine leads to the complete removal of the acetyl groups. Therefore, subsequent treatment of these free amine-containing GAGs with nitrous acid at a higher pH will lead to a complete depolymerization of GAGs via deaminative cleavage. Subsequent reduction of the products of deaminative cleavage with sodium borohydride results in the formation of an anhydromannitol on the terminal reducing end sugar (Figure 6.2).

The chemical techniques for the depolymerization of GAGs suffer from several drawbacks. First, the harsh reaction conditions for the different chemical treatments

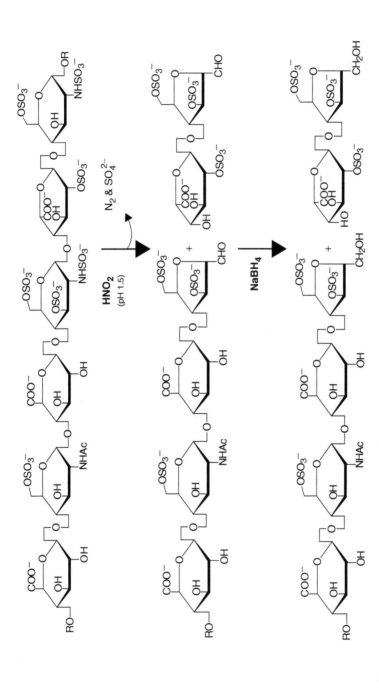

FIGURE 6.2 Nitrous acid cleavage of HSGAG oligosaccharides. Incubation of N-sulfated GAGs in the presence of nitrous acid (HNO$_2$) results in the cleavage of the $\alpha(1,4)$ bond between the N-sulfate containing glucosamine and the adjacent UA. The glucosamine is contracted to a 4-carbon ring and both nitrogen gas (N$_2$) and free sulfate (SO$_4^{2-}$) are released as by-products of the reaction. Subsequent addition of sodium borohydride (NaBH$_4$) results in the reduction of the ketone to a hydroxyl group on the contracted glucosamine. The reaction products from nitrous acid degradation also can be tritiated through addition of ^{3}H-labeled sodium borohydride for this reduction step.

can be difficult to control, often resulting in the formation of undesirable by-products. In addition, the majority of the chemical methods, perhaps with the exception of nitrous acid treatment, are nonspecific, resulting in the random cleavage of the GAG chain. Therefore, while they may be useful for generating GAG oligosaccharides of different sizes, these chemical techniques cannot be used for GAG sequence analysis. Finally, most of these chemical techniques, with the exception of the peeling reaction, do not result in the formation of an internal chromophore in the products. Therefore, an additional labeling step, such as reduction with [3H]sodium borohydride or introducing a fluorescent tag at the reducing end (see below), is required for the detection of the GAG products produced through chemical means. The development of GAG-depolymerizing enzymes (discussed in detail below) has enabled a more organized way of breaking down GAGs based on the specificities of these enzymes toward different modifications in the GAG chains. Therefore, the utilization of these enzymes, sometimes in conjunction with chemical methods, has improved the ability to obtain sequence information from a given GAG sample.

B. ENZYMATIC TECHNIQUES

GAG-depolymerizing enzymes, specifically those isolated from the soil bacterium *Flavobacterium heparinum*, have proven invaluable in the characterization of the structure and function of the different GAGs.[22] These bacterial enzymes degrade their respective substrates via a lytic mechanism that involves the abstraction of the C5 proton from the uronic acid of the basic disaccharide unit leading to the formation of a $\Delta4,5$ double bond in the product[22] (Figure 6.3). Importantly, this double bond serves as a chromophore with a UV absorbance at 232 nm, enabling the facile detection of depolymerized products using a variety of analytical techniques (see below). GAG-depolymerizing enzymes isolated from mammalian sources are predominantly hydrolyases, which cleave their respective substrates by hydrolysis (adding water to cleave the glycosidic bond). Cleavage of the glycosidic bond by hydrolyases retains the isomeric state of the UA and thus does not result in the formation of a chromophore (such as the $\Delta4,5$ double bond formed by lyases). Numerous bacterial and mammalian enzymes have been cloned, recombinantly expressed, and biochemically characterized, thereby significantly increasing their utility as enzymatic tools for analyzing GAG sequence information. In fact, employing *Escherichia coli*-based recombinant expression systems for several of the GAG-depolymerizing enzymes has circumvented many of the problems surrounding direct purification of the enzymes from native sources such as multiple purification steps, contaminating enzyme activities, and low yields.[23–25] As such, these enzymes can be employed in a fashion comparable to the way restriction endonucleases and proteases are used to degrade predictably DNA and proteins, respectively.

The best characterized and most widely applied of these bacterial GAG-depolymerizing enzymes are the three heparinases that specifically degrade HSGAGs with unique substrate specificities.[22] Heparinase I cleaves highly sulfated HSGAG sequences that contain 2-*O* sulfated uronic acids and are high in IdoA content (Table 6.1). Heparinase III is specific for regions with lower levels of 2-*O* sulfation that are

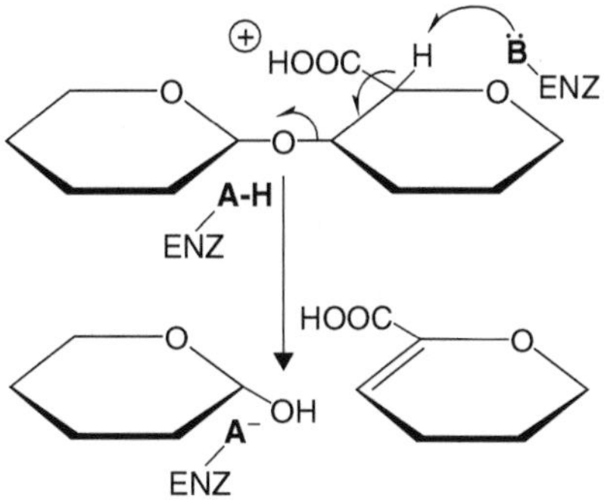

FIGURE 6.3   Proposed mechanism of eliminative cleavage by polysaccharide depolymerizing enzymes. A positive charge from the protein stabilizes the negative charge on the carboxylate group, thus lowering the pKa of the C5 proton. The proton is abstracted by an amino acid in the protein acting as a Brönsted base with subsequent protonation of the glycosidic oxygen by an amino acid acting as a Brönsted acid. The reaction product contains a Δ4,5 unsaturated bond on the UA. The additional hydroxyl and amine substituents of the sugar are not shown for simplicity. (See Figure 6.1 for the complete structure of the individual disaccharide units.)

TABLE 6.1
Summary of Heparin-Depolymerizing Enzyme Biochemistry

Enzyme	Size (kDa)	Specificity	Active Site Residues	Mode of Action
Heparinase I	41.7	GlcNS,6X~IdoA2S	C135, H203, K199, CB-1	Exolytic, processive
Heparinase II	84.1	GlcNY,6X~UA2X	H238, H451, H579, C348	Endolytic, nonprocessive
Heparinase III	70.8	GlcNY,6X~IdoA & GlcNac~IdoA	H295, H510	Endolytic
Heparanase	50.0	IdoA2S-GlcNS, 6S-GlcA~GlcNS,3S, 6S-IdoA2S-GlcNS,6S	E225, E348	Endolytic

X = sulfated or unsubstituted; Y = acetylated or sulfated; and
UA = GlcA or IdoA.

composed of a higher content of GlcA and *N*-acetylated glucosamine. Heparinase II has the broadest substrate specificity of the family of enzymes; and it is capable of cleaving HSGAG oligosaccharides with varying degrees of 2-*O* sulfation and with either of the uronic acid epimers[22] (Table 6.1).

In a similar fashion, the chondroitinase family of enzymes has proven useful in the analysis of biologically relevant CS/DS oligosaccharides. Chondroitinase AC is capable of depolymerizing both 4-*O* and 6-*O* sulfated CS while chondroitinase C degrades only 6-*O* sulfated regions (Table 6.2). Chondroitinase B from *F. heparinum* is the only known enzyme that degrades DS as its sole substrate. Chondroitinase ABC, isolated from a variety of bacterial sources, has the broadest substrate range

TABLE 6.2
Summary of CS/DS-Depolymerizing Enzymes

Enzyme	Size (kDa)	Specificity	Bacterial Source
Chondroitinase AC	77.7	GalNac,4S~GlcA and GalNac,6S~GlcA	*F. heparinum*
Chondroitinase B	54.0	GalNac,4S~IdoA2X	*F. heparinum*
Chondroitinase C	~ 70	GalNac,6S~GlcA	*F. heparinum*
Chondroitinase ABC	> 100	GalNac,4S~GlcA/IdoA and GalNac,6S~GlcA/IdoA	*F. heparinum, P. vulgaris*

~ = bond cleaved by enzyme; X = sulfated or unsubstituted.

of the chondroitinase family capable of depolymerizing CS/DS oligosaccharides regardless of sulfation pattern and UA epimerization[22] (Table 6.2). The availability of numerous crystal structures for chondroitinase AC,[26,27] B,[28] and ABC[29] has facilitated the biochemical characterization of these enzymes.

A variety of bacterial hyaluronate lyases, enzymes that degrade HA, have been isolated from different pathogenic bacteria including Group B *Streptococci*.[30,31] Hyaluronate lyase has a unique action pattern among the bacterial GAG-depolymerizing enzymes, first cleaving the substrate in a random, endolytic fashion and then proceeding to depolymerize the bound substrate in an exolytic, processive fashion.[32] This is in contrast to other GAG-depolymerizing enzymes, which tend to be either endolytic and nonprocessive (e.g., heparinase II, chondroitinases AC and B)[33,34] or exolytic and processive (e.g., heparinase I).[35] Similarly, several keratanases, bacterial hydrolyases that specifically degrade KS, have been used to explore the structure of KS from different biological sources.[36]

GAG-depolymerizing enzymes isolated from mammalian sources are characterized as hydrolyases. Heparanase is unique in comparison with the other mammalian GAG hydrolyases, i.e., it is primarily endolytic, found extracellularly, and is responsible for the degradation of *heparan sulfate* (Table 6.2). The pattern of expression of heparanase and its extracellular localization imply a role for the enzyme in the normal maintenance of tissue architecture as well as in the progression of the cancer.[37] However, heparanase has a limited substrate specificity that has proven difficult to define biochemically, thus decreasing its utility as a biochemical tool. Finally, hyaluronidases from mammalian sources have also been isolated and characterized, most notably testicular hyaluronidase.[30] On the whole, the bacterial-derived GAG-depolymerizing enzymes, owing to their relative amenability to biochemical characterization, have been the most useful enzymatic tools for the structural analysis of GAGs.

In addition to GAG-depolymerizing enzymes, other enzymes that specifically act at the ends of GAG chains have also been cloned and characterized. These enzymes, collectively referred to as "exo-enzymes" due to their exolytic action, include sulfatases that cleave specific sulfate groups and exo-glycosidases that cleave UA or hexosamine at the nonreducing end. In soil bacteria such as *F. heparinum*, the sulfatases and glycosidases act in conjunction with the other GAG lyases to metabolize GAGs. Some of the bacterial GAG sulfatases, including N-acetyl-galactosamine 4-sulfatase[38] and $\Delta 4,5$ uronic acid 2-O sulfatase,[39,40] have been characterized in terms of substrate specificity and catalytic mechanism. The $\Delta 4,5$ glycuronidase that cleaves the $\Delta 4,5$ uronic acid from the end of HSGAGs was also recently cloned and characterized from *F. heparinum*.[41] In the mammalian system, exo-enzymes are present in the lysosomes of cells where they are primarily involved in the catabolism of GAGs. Sequential application of these exo-enzymes coupled with high-resolution polyacrylamide gel electrophoresis (PAGE) has enabled the direct sequencing of an HSGAG oligosaccharide starting from the nonreducing end and proceeding to the reducing end (see below).[42] It is important to note that these enzymes are often membrane-associated and function at low pH, values making their application as biochemical tools more difficult than the bacterial enzymes.

III. ANALYTICAL TECHNIQUES FOR GLYCOSAMINOGLYCAN CHARACTERIZATION

There are four well-developed methods for analyzing the products of GAG degradation, namely: (1) chromatography, (2) electrophoresis, (3) nuclear magnetic resonance spectroscopy (NMR), and (4) mass spectroscopy (MS). Chromatography and electrophoresis techniques rely on the separation of the individual reaction products based on size and charge, thereby producing a profile of the degraded GAG polymer. Most of these separation techniques require the presence of a chromophore on the GAG oligosaccharide for detection. The $\Delta 4,5$ bond produced by the β-eliminative cleavage of GAGs by bacterial lyases is a common chromophore with UV absorbance at 232 nm. In addition, various fluorophores, such as 2-aminoacridone (AMAC), 1-aminopyrene-3,6,8-trisulfate (APTS), and 2-anthranilic acid, are often used to tag the reducing end of GAG oligosaccharides through reductive amination.[43] Metabolic labeling of a GAG chain using tritiated sodium borohydride is a more traditional approach for GAG detection and is still employed today.[20] MS techniques separate GAG oligosaccharides based on their mass-to-charge (m/z) ratio thereby providing direct mass information for a given sample. NMR can be used to determine directly sequence and structure information of GAG oligosaccharides. All four of these techniques, when coupled to either enzymatic or chemical degradation, have significantly contributed to understanding the structure of GAGs.

A. CHROMATOGRAPHY

Chromatographic methods for the separation of GAG oligosaccharides can be split into two general categories, namely, size-exclusion and ion-exchange chromatography.[20] GAG oligosaccharides resulting from β-elimination enzymatic cleavage can be detected using the absorbance of the $\Delta 4,5$ double bond at the nonreducing end. In the absence of this internal chromophore, GAG oligosaccharides are commonly labeled with a fluorophore or a radiolabel such as ^{3}H to facilitate their detection. Refractive index and pulsed amperometric detectors are also commonly used to monitor larger oligosaccharides that lack a chromophore.

Size-exclusion chromatography (also known as gel permeation chromatography) relies on the pore size of the column resin to separate GAG fragments based solely on their molecular weight. This technique has been used to purify different oligosaccharide fractions (ranging from di- to dodecasaccharides) derived from the partial enzymatic or chemical degradation of different GAGs.[20] In addition, size-exclusion columns coupled to high-performance liquid chromatography (HPLC) instruments have been employed to determine the molecular weight (MW) of intact GAG polymers as well as GAG oligosaccharides.[44] Oligosaccharide standards of known length or mass are used to create a chromatographic profile from which a standard curve (log MW vs. elution volume) can be generated. Using this standard curve, the mass of an unknown GAG oligosaccharide can be estimated based on its elution position.

Ion-exchange chromatography resolves GAG oligosaccharides based on their negative charge distribution resulting from differing sulfation patterns and the presence of carboxylate groups. HPLC-based strong anion exchange (SAX) and amine-based chemistries are commonly used for the analysis of GAG fractions. In fact, sequential use of size-exclusion and SAX chromatography is often employed to isolate individual GAG oligosaccharides of defined length and composition from a partial digestion of a GAG polymer.[45] However, SAX chromatography requires the use of high salt gradients (0.1–2.0 M) for elution of GAG oligosaccharides, thus mandating a desalting step prior to further characterization. In addition, weak anion-exchange resin such as DEAE can be used for the purification of intact GAG polymers or proteoglycans from cell extracts.[20]

B. ELECTROPHORESIS

Electrophoretic techniques separate GAG oligosaccharides based on their mass-to-charge ratio (m/z) and, similar to the chromatographic techniques discussed above, have been used for the analysis and sequencing of GAG oligosaccharides. PAGE has been used to separate GAG samples in a similar fashion for its use in the analysis of DNA and proteins. However, using PAGE for the analysis of GAGs has been complicated by the chemical heterogeneity of GAG samples isolated from biological sources, although recent advances such as discontinuous electrophoresis, enhanced buffering systems, and gradient gels have increased the utility of PAGE for the analysis of GAG mixtures. In addition, coupling PAGE with fluorescently labeled GAG oligosaccharides led to a technique known as fluorophore-assisted carbohydrate electrophoresis (FACE).[46] The FACE methodology has been used to characterize HSGAG, CS/DS, and KS oligosaccharides from various healthy and pathological tissue samples.[47-49]

Capillary electrophoresis (CE) has emerged as a powerful electrophoretic tool that offers high sensitivity at the expense of small amounts (picomoles) of sample.[50-55] Similar to the chromatography techniques, the absorbance of the Δ4,5 bond (resulting from cleavage with the bacterial lyases) is used to detect the different GAG di- and oligosaccharides. Labeling GAGs with sulfonate-containing fluorophores has enhanced the separation and detection of GAG oligosaccharides produced by other means or that contain a low level of sulfation (i.e., HA). The migration times of different disaccharide or oligosaccharide units through the capillary are compared with known disaccharide standards to provide a compositional analysis of an unknown sample. This technique was recently used to track a 3-*O* sulfate-containing tetrasaccharide that directly correlates to the anticoagulant activity of certain low-molecular-weight heparins (LMWHs).[19]

In addition to its need for significantly less sample than HPLC, CE offers several other advantages including a higher number of theoretical plates leading to an increase in the potential resolving power over HPLC-based techniques. CE-based separations can also be completed in either normal or reverse polarity yielding complementary information about the same sample. Under normal polarity, GAGs are injected at the anode and migrate to the cathode under electroosmotic flow usually in borate buffer containing sodium dodecyl sulfate at an elevated pH. Reverse polarity analysis relies on electrophoretic separation from the cathode to the anode in a lower pH buffer (usually Tris-HCl) containing dextran sulfate to suppress reverse, electroosmotic flow. Owing to its enhanced resolving power and ease of reproducibility, reverse polarity is the preferred mode for the analysis of GAGs from different sources and has been used extensively in characterizing the reaction products of the bacterial heparinases,[34,35] chondroitinases,[25,56,57] and hyaluronidases.[32,58,59] Finally, the use of functionalized capillaries in addition to standard fused silica provides an additional, approach for the separation and analysis of GAGs akin to the different types of chromatographic separations.

Chromatographic and electrophoretic analysis methods suffer from a number of limitations. To begin with, the analysis of GAG fractions is judged purely on electrophoretic mobility or HPLC elution volume, both of which can vary based on subtle changes in temperature, buffering conditions, and pH. Secondly, both methodologies may require the end labeling of the oligosaccharides resulting in the introduction of excess labeling reagent in the sample that could further complicate the analysis. Also, the laddering of peaks or gel bands produced by enzymatic treatment with mammalian exo-enzymes is not always fully reliable since these enzymes have not been well characterized in terms of their substrate specificities. Finally, chromatographic and electrophoretic techniques do not provide information about the C5 epimerization state of the UA moieties found in GAG oligosaccharides. One of the most reliable techniques for providing this detailed structural information for GAGs has been NMR spectroscopy.

C. NUCLEAR MAGNETIC RESONANCE

NMR spectroscopy is a valuable technique employed for the direct structural characterization of GAG oligosaccharides. NMR techniques can provide information on the three-dimensional orientation of GAG oligosaccharides and the conformation dynamics of individual monosaccharides within the GAG polymer.[8,60,61] This technique was recently used to explore alterations in the 3D structures of a GAG oligosaccharide upon its binding to different proteins thereby uncovering the structural basis for specificity of GAG–protein interactions.[62,63]

The characteristic anomeric signals (^{1}H and ^{13}C) for the commonly occurring monosaccharides with different sulfation patterns in GAGs have been identified and advances in 2D correlation spectroscopy have increased the amount of structural information obtained from NMR experiments.[64–66] Furthermore, distinct anomeric proton chemical shifts have been observed for glucosamine and galactosamine monosaccharides depending on their linkages to different UAs (IdoA or GlcA) with specific sulfation patterns.[67] Also, NMR provides the most accurate quantitative estimate of IdoA vs. GlcA content in a GAG sample.

Sequence determination of a purified GAG oligosaccharide by NMR usually involves the following steps. First, the identity of the individual monosaccharide components of the GAG is verified based on their characteristic anomeric proton chemical shifts from the 1D proton spectrum and coupling constants to the other protons from the two-dimensional correlation spectroscopy (COSY) or total correlation spectroscopy (TOCSY) spectra. The positions of the carbons in the monosaccharides are assigned by ^{13}C-NMR usually followed by the correlation of the carbon-proton signals using heteronuclear spin coupling. Linkage information between adjacent monosaccharides is accomplished using 1D and 2D nuclear Overhauser enhancement spectroscopy (NOESY) or rotating frame Overhauser enhancement spectroscopy (ROESY) experiments in which the relative position of the anomeric protons to the adjacent monosaccharides provides information on the linkage orientation. Chemical shift data for the UA and the hexosamine moieties are used to determine the presence of sulfation or acetylation on the GAG monosaccharides.

Despite these clear advantages of NMR, there are some important limitations in applying NMR to characterize tissue-derived GAGs. First, the amount of sample required (particularly for the ^{13}C and NOESY spectra) is in low milligram amounts, which is not well suited for the extremely low amounts of tissue-derived GAGs, thereby making the sensitivity of NMR lower than the detection limits for chromatographic, electrophoretic, and MS-based techniques. Although good sensitivity can be obtained for 1D proton and 2D COSY/TOCSY spectra using small sample amounts typical of HSGAGs, NOESY and ROESY spectra usually require more sample and high-field instruments for reasonable sensitivity and signal resolution. Also, NMR spectra of complex oligosaccharide sequences have overlapping proton signals and absent measurable coupling constants that complicate their interpretation.

D. MASS SPECTROMETRY

In recent years, advances in mass spectrometry have led to the facile characterization of GAG oligosaccharides.[68] The diverse number of evolving MS techniques all operate on the same biophysical principle, namely, the separation of ionized molecules based on their m/z ratio. Therefore, the mass of a compound can be directly calculated from the mass spectrum provided that the charge state of each species in the spectrum can be determined. Sample ionization is achieved using a variety of techniques, each with their pros and cons, which allow for the resolution and detection of the molecular ions in a sample.

Several mass spectrometric techniques have been developed for the analysis of GAG oligosaccharides including liquid secondary ion mass spectrometry (LSI-MS),[69] electrospray ionization mass spectrometry (ESI-MS),[70,71] and matrix-assisted laser desorption ionization mass spectrometry (MALDI-MS).[68] ESI-MS and LSI-MS are attractive due to the fact that they can be coupled to either HPLC[72,73] or to CE,[74,75] enabling the in-line separation and analysis of GAGs.

While these techniques have been successful for the analysis of unsulfated GAGs such as HA, analysis of sulfated GAGs has been complicated by the lability of the sulfate groups during ionization. Therefore, LSI-MS and ESI-MS techniques frequently rely on the coupling of GAGs with positively charged ion-pairing reagents such as sodium, ammonium, or calcium to help stabilize the sulfate groups and promote ionization.[70,76] Unfortunately, the presence of these ions can significantly complicate the mass spectra, often leading to difficulties when interpreting the results. However, ESI-based techniques allow for the use of collision-induced dissociation for the direct determination of sequence information based on predictable fragmentation patterns entirely within the mass spectrometer.

MALDI-MS involves the generation of molecular ions through laser-induced ionization of a sample that has been crystallized with a matrix. Upon ionization, the application of a voltage accelerates the ions through a time-of-flight tube where they are separated according to their m/z ratio. Similar to the ESI and LSI-MS techniques, ionization can lead to the random loss of sulfates in a GAG sample. This problem has been circumvented by noncovalently complexing GAG oligosaccharides with a basic peptide in the form $(Arg-Gly)_n$-Arg (where $n = 15, 17, 19$) in the crystalline spot.[77] The mass of the oligosaccharide is calculated by subtracting the mass of the peptide from the peptide–GAG complex for up to a tetradecasaccharide in length. Therefore, the peptide peak serves as an internal mass standard, enabling the facile calibration of mass spectra and the mass determination of unknown GAG samples. In addition, the mass spectra produced using this MALDI-based technique are relatively simple and devoid of adducts caused by the presence of salt or metals in the sample. An additional highlight of MALDI-MS GAG analysis is the requirement of only pico- to femtomolar concentration, making this technique highly suitable for small amounts of tissue-derived GAGs. However, an important drawback to this technique is its inability to perform collision-induced dissociation on the molecular ions due to the fact that they are not detected as free GAGs. Nevertheless, MALDI-MS has emerged as a leading technique for the analysis of oligosaccharides and has been incorporated into a variety of advanced analytical schemes that have shed new light on the structure–function of different families of GAGs.

IV.   SEQUENCING GLYCOSAMINOGLYCANS

As outlined above, the orchestrated decomposition of GAG oligosaccharides into smaller fragments, along with the application of separation and analytical techniques, has resulted in the development of diverse strategies for sequencing homogeneous preparations of GAG oligosaccharides. All of these strategies rely on the specificities of the GAG-degrading enzymes as well as the sensitivity and accuracy of the various analytical tools. While the sequencing methods have largely focused on HSGAGs due to the historical significance of heparin as an anticoagulant drug, some of these approaches have also been extended to study CS/DS GAGs. The recently developed sequencing methodologies for GAGs are discussed in the following section.

The majority of the sequencing techniques developed for GAGs rely on either chromatographic or electrophoretic separation of the oligosaccharide degradation products resulting from enzymatic or chemical scission. In one such technique,

HSGAG oligosaccharides are metabolically labeled at the reducing end with ^{3}H and then broken down using a combination of partial nitrous acid degradation and lysosomal exo-enzymes.[78] The products of each reaction are separated using SAX-HPLC and detected by the presence of the radioactive tag. Another technique known as internal glycan sequencing (IGS) also relies on nitrous acid degradation and exo-enzymes for the specific decomposition of the initial HSGAG oligosaccharide.[42] However, in the case of IGS, the reducing ends of the reaction products are labeled with a fluorophore and are separated using PAGE. Both of these techniques result in the relatively rapid and facile sequencing of nano- to picomoles of HSGAG oligosaccharides.

Similar techniques that rely on chromatographic separation of enzymatic degradation products have been developed for the sequence analysis of CS/DS oligosaccharides. The different chondroitinases along with sulfatases specific for each of the potential sites of sulfation in CS/DS are used to degrade the starting oligosaccharide in a predictable manner. The products are then analyzed using a combination of anion-exchange chromatography and NMR techniques to arrive at the ultimate sequence of the starting material.[66,79,80] These techniques have been used to characterize rare CS/DS modifications from GAGs isolated from a variety of sources, such as the 3-O sulfation of GlcA in CS isolated from king crab cartilage.[8] Additionally, similar HPLC-based techniques have been used to profile changes in cartilage CS as a function of the progression of osteoarthritis.[81,82]

FACE is a PAGE-based technique created for the analysis of HA and CS/DS oligosaccharides as well as their degradation products.[46,49] In a fashion similar to IGS, di-, and oligosaccharides are labeled with a reducing end fluorophore (usually AMAC) and resolved on a PAGE gel. By comparing the products of differential enzymatic treatment with known disaccharide standards, FACE has been used to examine changes in the sulfation pattern of the CS-containing proteoglycan, aggrecan, as a function of osteoarthritis progression.[83] The FACE methodology has also been recently adapted for characterizing sequences of KS found in corneal and skeletal tissues.[36]

Coupling enzymatic and chemical degradation to different MS techniques has also produced valuable sequence information for the various classes of GAGs. ESI and nano-ESI techniques have been used in tandem with HPLC-or CE-based separation techniques for the structural determination of different HSGAG oligosaccharides,[84] as well as CS/DS oligosaccharides.[85] In addition, collision-induced dissociation of parental molecular ions in ESI-MS allows for the MS/MS detection of the resulting products providing another technique for obtaining GAG sequence information.[70,86]

The different sequencing approaches described above rely on specific properties of GAGs such as mass, charge, and sulfation pattern for separation and sequence determination. While each of these analytical tools is valuable for characterizing GAGs, there are challenges involved in utilizing any single technique to develop a practical sequencing methodology for biologically important GAGs from tissues. Nevertheless, each of these methodologies provides data sets and constraints based on specific properties of a given GAG preparation. To address these challenges, informatics-based approaches have been developed, thus providing a practical and unbiased sequencing strategy for GAGs. These approaches utilize a framework that provides flexibility to incorporate information from a diverse set of analytical methods as constraints to arrive at the sequence of a purified GAG oligosaccharide.

V.   INFORMATICS-BASED APPROACHES FOR SEQUENCING GLYCOSAMINOGLYCANS

To enable the incorporation of experimental constraints provided by the data resulting from the different analytical tools, it is critical to have a framework to represent and manipulate sequence information content in GAGs. The use of single alphabet codes to represent sequence information in DNA and proteins has facilitated the creation of databases to store sequence information and led to the development of computational tools for pattern searching, sequence comparison, mutation mapping, and evolutionary evaluation. These databases and informatics-based computational tools have revolutionized molecular biology by capturing information at the whole genome and proteome level. While single-alphabet character representations are suitable for proteins and DNA, they are a bit cumbersome for complex glycans like GAGs, which have a much higher information density compared with other biopolymers. For example, in HSGAGs, the different sulfation patterns and epimeric states of UAs give rise to 32 theoretically possible disaccharide building blocks of which 24 have already been identified in nature.[87]

This variability in an HSGAG disaccharide unit arises from differential modification at four potential sites, namely, the 2-O position of the UA as well as the N-, 3-O, and 6-O position of the glucosamine. Importantly, each of these commonly occurring modifications can be described in a binary fashion. For example, the O positions of the disaccharide can be sulfated or unsulfated, whereas the N position can be sulfated or acetylated. The epimeric state of the UA is also binary, existing as either the glucuronic or iduronic epimer. Therefore, a binary framework was developed to capture the different possible modifications in an HSGAG disaccharide unit.[87] Importantly, this property-encoded nomenclature (PEN) accounts for both the mass as well as the charge distribution of the different disaccharides that compose an HSGAG polymer[87] (Figure 6.4). Thus, in the PEN scheme, the four binary digits that capture the sulfation pattern are represented using a single hexadecimal code (Figure 6.4). The 5th bit is represented as a sign to encode the epimeric state of the UA such that for the same sulfation pattern a positive hexadecimal code represents IdoA and the negative hexadecimal code represents GlcA (Figure 6.4).

The numerical basis of the PEN system enabled the use of mathematical and binary operations on the PEN representation to incorporate simultaneously rules for breaking down GAG oligosaccharides into smaller fragments as well as relating different physical properties of an oligosaccharide (i.e., mass and charge) to its composition and chain length. Thereby, the PEN framework facilitated the incorporation of data sets as unique constraints to sequence GAG oligosaccharides. Two such PEN-based sequencing approaches were developed recently and are discussed in detail below.

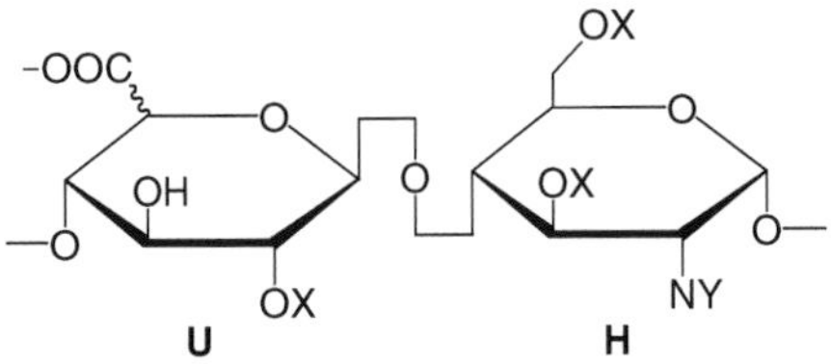

(A)

I/G	2X	6X	3X	NX	CODE	Unit	Mass(ΔU)
0	0	0	0	0	0	$I\text{-}H_{NAc}$	379.33
0	0	0	0	1	1	$I\text{-}H_{Ns}$	417.35
0	0	0	1	0	2	$I\text{-}H_{NAc,3S}$	459.39
0	0	0	1	1	3	$I\text{-}H_{NS,3S}$	497.41
0	0	1	0	0	4	$I\text{-}H_{NAc,6S}$	459.39
0	0	1	0	1	5	$I\text{-}H_{NS,6S}$	497.41
0	0	1	1	0	6	$I\text{-}H_{NAc,3S,6S}$	539.45
0	0	1	1	1	7	$I\text{-}H_{NS,3S,6S}$	577.47
0	1	0	0	0	8	$I_{2S}\text{-}H_{NAc}$	459.39
0	1	0	0	1	9	$I_{2S}\text{-}H_{NS}$	497.41
0	1	0	1	0	A	$I_{2S}\text{-}H_{NAc,3S}$	539.45
0	1	0	1	1	B	$I_{2S}\text{-}H_{NS,3S}$	577.47
0	1	1	0	0	C	$I_{2S}\text{-}H_{NAc,6S}$	539.45
0	1	1	0	1	D	$I_{2S}\text{-}H_{NS,6S}$	577.47
0	1	1	1	0	E	$I_{2S}\text{-}H_{NAc,3S,6S}$	619.51
0	1	1	1	1	F	$I_{2S}\text{-}H_{NS,3S,6S}$	657.53

(B)

FIGURE 6.4 Property-encoded nomenclature (PEN) for HSGAG disaccharides. (A) The basic HSGAG disaccharide repeat unit can be O-sulfated (X), epimerized at C5 position of the UA, as well as sulfated or acetylated (Y) at the N-position of the glucosamine. (B) The PEN hexadecimal code was developed to encode each potential disaccharide unit based on its sulfation and epimerization states. Binary digits for the 2X, 6X, 3X, and NY positions are employed such that 0 encodes for not sulfated and 1 for the presence of a sulfate; in the case of NY, 0 represents acetylation and 1 encodes for sulfation. The UA sign bit in the I/G column is 0 for the IdoA epimer and 1 for GlcA (not shown).

A. PEN-MALDI

MALDI-MS has been enormously useful for determining an accurate mass of intact GAG oligosaccharide as well as those generated by chemical and enzymatic depolymerization. Coupling the sensitivity of MALDI-MS with the PEN framework developed for HSGAGs produced a revolutionary sequencing methodology aptly named PEN-MALDI.[87] The theoretical limit of distinguishing the composition of unique GAG oligosaccharides based on their mass was determined by constructing a mass line using the PEN framework (Figure 6.5). This mass line comprised the masses of the possible di- through octadecasaccharides found in HSGAGs. Based on this theoretical mass line, it was determined that given the accuracy of the MALDI-MS (<1 Da), the length and the composition (number of sulfate and acetate groups) of any oligosaccharide, up to a tetradecasaccharide, could be determined solely from its

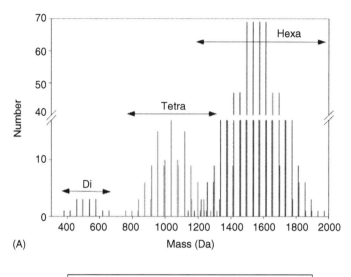

(A)

Fragment length	Minimum mass difference (Da)
1 (di)	101.13
2 (tetra)	13.03
3 (hexa)	13.03
4 (octa)	9.01
5 (deca)	9.01
6 (dodeca)	4.99
7 (tetradeca)	4.99
8 (hexadeca)	0.97

(B)

FIGURE 6.5 Theoretical mass line for HSGAG oligosaccharides. (A) A schematic representation of the possible masses of all the possible HSGAG di-, tetra-, and hexasaccharides. (B) Given the accuracy of MALDI-MS to resolve oligosaccharides down to a 1.0 Da difference in mass, this technique can be used to assign uniquely the length and number of sulfates and acetate groups up to a tetradecasaccharide solely from the mass.

mass (Figure 6.5). This mass-to-length/composition relationship enabled the application of MALDI-MS data as constraints to assist in the sequence assignment of HSGAGs using PEN-MALDI.

In this sequencing technique, the mass of an HSGAG oligosaccharide as determined by MALDI-MS is used to determine its overall length and total number of sulfate and acetate groups (Figure 6.6). In addition, exhaustive digestion of the HSGAG oligosaccharide with heparinases I, II, and III followed by CE separation is used to determine its overall disaccharide composition. At this point, a master list of

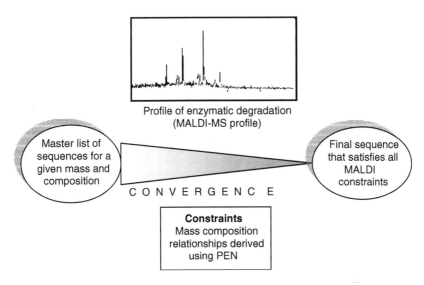

FIGURE 6.6 The framework for the PEN-MALDI sequencing methodology. The mass of the initial oligosaccharide (obtained using MALDI-MS) and its disaccharide composition (from CE analysis) are used to create a master list of all the possible sequences for the starting sample. From this master list, sequences are eliminated by iteratively analyzing the MALDI-MS profile of different enzymatic digests based on the unique specificities of each enzyme. The sequential application of these constraints results in the convergence of the master list to a single sequence describing the initial oligosaccharide.

all possible oligosaccharide sequences with the observed mass and disaccharide composition is assembled using the PEN framework (Figure 6.6). The sequence of the unknown oligosaccharide is then determined by applying different enzymatic treatments followed by MALDI-MS to eliminate sequences from the master list that do not meet the experimental constraints.[87] This iterative approach converges on the unique sequence of the unknown HSGAG oligosaccharide from the initial master list while using a minimal amount of material (Figure 6.6).

The PEN-MALDI methodology has been successfully applied to sequence several biologically important HSGAG oligosaccharides. This methodology was the first to point out that a heparin-derived decasaccharide (AT-10) with high affinity to antithrombin III (AT-III)[88] actually contained only a part of the well-established AT-III-binding pentasaccharide motif.

1. Sequencing AT-10

The mass of AT-10 was first determined by MALDI-MS data to be 2770.2 Da corresponding to a decasaccharide with 13 sulfate groups and 1 acetate group. CE-based compositional analysis of AT-10 indicated the presence of three building blocks, corresponding to $\pm D$ ($\Delta U_{2S}H_{NS,6S}$), ± 4 ($\Delta UH_{NAc,6S}$), and ± 7 ($\Delta UH_{NS,3S,6S}$) in the relative ratio of 3:1:1. The PEN scheme revealed a total of 320 possible decasaccharides with the observed disaccharide building blocks and mass[88] (Figure 6.7). The MALDI-MS profiles of the following chemical or enzymatic depolymerization were the constraints applied to reduce the master list. Partial cleavage of AT-10 with heparinase I revealed four unique products that reduced the master list to 52 sequences, which satisfied the constraint. Exhaustive digestion of the decasaccharide with heparinase I revealed two unique products identified based on their masses as a trisulfated disaccharide and an octasulfated hexasaccharide, thereby further reducing the master list to 28 unique sequences. The same exhaustive digest using a mass tag (semi-carbazide) at the reducing end further constrained the master list to two sequences[88] (Figure 6.7). Finally, nitrous acid degradation was combined with the application of specific exo-enzymes to converge on a unique sequence for AT-10. Interestingly, the final sequence of $\pm$DDD4-7 ($\Delta U_{2S}H_{NS,6S}I_{2S}$ $H_{NS,6S}I_{2S}H_{NS,6S}IH_{NAc,6S}GH_{NS,3S,6S}$) contains only a part of the intact AT-III-binding pentasaccharide ($H_{NAc,6S}GH_{NS,3S,6S}I_{2S}H_{NS,6S}$).[88] This novel sequence was independently confirmed using the PAGE-based IGS technique and NMR studies. Therefore, the application of the PEN-MALDI technique identified the actual sequence of the AT-10 decasaccharide and shed important light on the enzymatic generation of low-molecular-weight heparins with anticoagulant activity.[89]

B. PEN-NMR

Analogous to incorporating mass information from MALDI-MS and disaccharide composition from CE, the PEN framework was utilized to incorporate data from NMR spectroscopy to sequence HSGAG oligosaccharides. As mentioned earlier, NMR spectroscopy provides a quantitative monosaccharide composition based on the anomeric chemical shifts of the monosaccharides with specific sulfation pattern. The anomeric chemical shifts of hexosamines can further distinguish the type of UA linked to the reducing end of the hexosamine, thus providing H–U linkage abundance. The PEN framework used in the PEN-MALDI sequencing technique was originally developed to encode the U–H disaccharide building blocks.[87] To incorporate the monosaccharide abundance and H–U linkage information from NMR as constraints, some mathematical operations were performed to split the U–H PEN code into unique codes for the UA and hexosamine.[90]

In a fashion similar to the PEN-MALDI methodology, a series of NMR experiments are used to generate a master list of all possible sequences for an unknown oligosaccharide. The 1D proton NMR spectrum along with the 2D COSY, HSQC, and TOCSY spectra afford information on the chemical shifts and coupling constants of most of the ring protons of the constituent monosaccharides.[90] These data can be used to identify and quantify the UA and hexosamine constituents of a given oligosaccharide, as well as provide direct linkage information for adjacent monosaccharides, thereby generating a master list of all possible sequences that match the NMR data (Figure 6.8). CE-based disaccharide compositional analysis provides information on the U-H linkages and is used to reduce the NMR master list down to a single sequence. Incorporating the CE data into the PEN-NMR technique reduces the reliance on the 2D NOESY/ROESY experiments that are often noisy and require significantly more sample as well as advanced instrumentation.[90] (Figure 6.8).

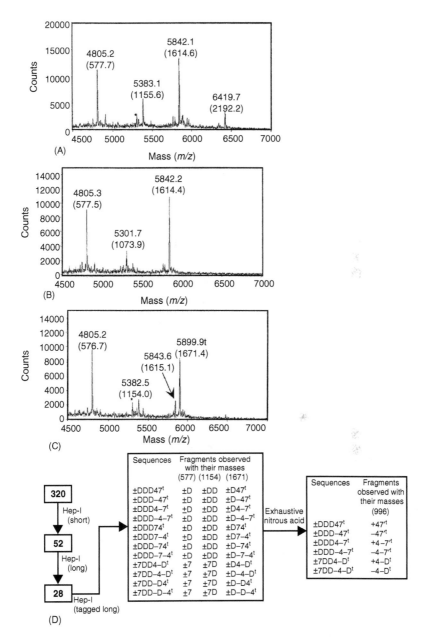

FIGURE 6.7   PEN-MALDI sequencing of AT-III-binding decasaccharide. The decasaccharide starting material was treated with a partial heparinase I digest (A); an exhaustive heparinase I digest (B); and an exhaustive heparinase I digest of the semicarbazide-labeled decasaccharide (C). The masses of the peptide peak and each oligosaccharide complexed with the peptide are shown for each mass spectrum with the actual mass of the oligosaccharide in parentheses. Sequential application of these experimental constraints (D) resulted in the progressive elimination of non-conforming PEN sequences (depicted as the number in each box) with the application of nitrous acid to narrow the list further. The final sequence from the list of six sequences was determined by applying the different exo-enzymes as further experimental constraints (not shown).

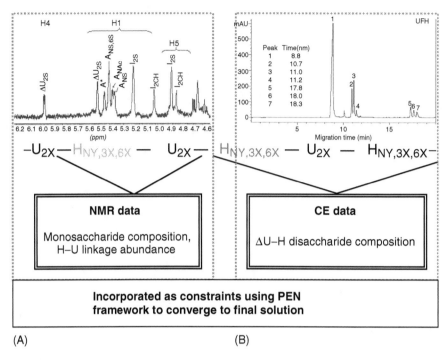

(A)   (B)

FIGURE 6.8 PEN-NMR sequencing strategy. (A) NMR spectroscopy provides quantitative information on the monosaccharide composition of an oligosaccharide as well as the relative abundance of specific linkages between glucosamine (H) and UA (U). (B) In parallel, data from CE analysis provide quantitative information on the sulfation pattern of different ΔU–H disaccharides that compose the oligosaccharide. The numerical framework of PEN was adapted to include this monosaccharide and linkage information from NMR as well as to incorporate NMR and CE data as constraints to arrive at the final sequence of an unknown oligosaccharide.

The PEN-NMR methodology was applied to sequence AT-10 successfully (Figure 6.9), thereby further validating the correct sequence of this decasaccharide. By incorporating the monosaccharide and H–U linkage abundance information provided by NMR spectroscopy with the U–H linkage information provided by CE as constraints using the PEN framework, the list of possible sequences is drastically reduced. Therefore, NMR data sets provide valuable constraints on the sequence parameters that significantly reduce the sample space even before the application of enzymatic constraints. This observation further highlights the utility of NMR as a powerful technique for the sequence analysis of GAGs.

C. ADDITIONAL INFORMATICS-BASED APPLICATIONS

The PEN-based approaches described above provide a practical solution for assigning the sequence of homogeneous GAG samples isolated from tissues. In many situations, the isolation process would potentially result in binary or multicomponent

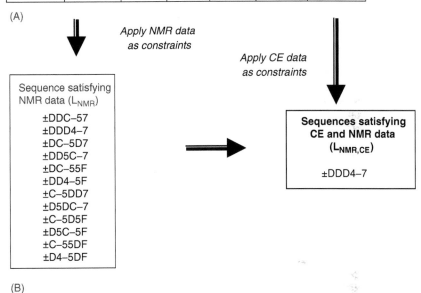

FIGURE 6.9 PEN-NMR sequencing of the AT-III-binding decasaccharide. (A) The information obtained from NMR and CE experiments is displayed in tabular format. Columns 1 to 3 indicate the number of linkages between the specific glucosamine residues and the specific UA residues in column 4 obtained from NMR data. Columns 5 to 7 represent the different linkages between UA's and glucosamines as obtained from exhaustive heparinase digest and CE analysis. X represents either the presence or absence of a sulfate at the 6-O position of glucosamine. (B) The sequences that satisfy the monosaccharide composition and H–U linkage information. Application of the U–H linkages from CE data reduces the master list (L_{NMR}) to the final unique sequence.

GAG mixtures. For example, the MALDI-MS analysis of AT-10 gave two oligosaccharide peaks where the major species (over 90% abundance) was assigned using PEN-MALDI and PEN-NMR techniques.[88] The minor species (<5% abundance) was also a decasaccharide with one less sulfate group. Despite the extremely low abundance of this species, most of its sequence was assigned using the PEN-MALDI approach as $\pm$D(+/−)5D4-7 ($\Delta U_{2S}H_{NS,6S}I/GH_{NS,6S}I_{2S}H_{NS,6S}IH_{NAc,6S}GH_{NS,3S,6S}$).[88] This further demonstrated the robustness of PEN-based approach for addressing minor contaminants in samples as well as the overall sensitivity of the technique for detecting extremely small quantities. It is easy to imagine the complexity in characterizing multicomponent GAG mixtures with comparable abundances between the different components.

Since the PEN-based approaches incorporate constraints provided by different analytical tools in an unbiased fashion, they can be readily extended to characterize oligosaccharide mixtures. CE and NMR accurately provide disaccharide composition, monosaccharide composition, and linkage abundances irrespective of whether the sample is homogeneous or a mixture. Thus, the PEN-NMR method that bridges these data sets can be applied to simulate the composition of the mixture in terms of complete disaccharide building blocks (with the epimeric state of the UA assigned based on NMR data). MALDI-MS analysis of a GAG mixture would provide a profile of different oligosaccharide masses in that mixture. Instead of treating the problem as looking for a single solution (using constraints) in a sample space defined by the disaccharide composition, the mixture characterization problem could be treated as looking for multiple solutions based on the sample space defined by the complete disaccharide composition. While convergence to a single solution is not guaranteed in the mixture analysis, the best characterization of the mixture in terms of the length, mass, composition, and perhaps even sequences of major species can be achieved. This would be possible due to integration of multiple datasets from CE, NMR, and MALDI-MS profiles of chemical and enzymatic degradation of the sample as constraints to probe the sample space. Developing PEN-based approaches to characterize mixtures is a part of ongoing research that aims to exploit the numerical nature and flexibility of the PEN framework to correlate properties of a GAG mixture to its biological or clinical output.

VI.   DIRECT ISOLATION AND SEQUENCING OF GLYCOSAMINOGLYCANS: SNA-MS

As mentioned earlier, one of the significant challenges faced in the analysis of GAGs is their relative scarcity from biological sources. This is further compounded by the lack of a technique for their amplification akin to PCR for DNA and recombinant expression for proteins. Previous efforts to evade the problem of limited starting materials have relied on the use of GAGs from commercial sources. While this has proven successful in certain cases, commercial sources of GAGs are often limited by the lack of biologically relevant sequences that would be found in GAGs directly isolated from different tissue or cell sources. Therefore, a technique that combined the direct isolation of bioactive GAGs with a sensitive analytical technique for GAG identification would be of enormous use in characterizing structure–function relationships for different oligosaccharides.

Surface noncovalent associated mass spectrometry (SNA-MS) combines several rigorous analytical methodologies, namely high-affinity fractionation, characterization by MS, and direct sequencing of bound GAG oligosaccharides into a single procedure[91] (Figure 6.10). This technique relies on adhering a GAG-binding protein to a MALDI-MS plate via a thin hydrophobic film or directly to the metal surface. A pool of GAG oligosaccharides with different masses and sulfation patterns are added directly to the MALDI spot containing the adhered protein of interest. After a salt

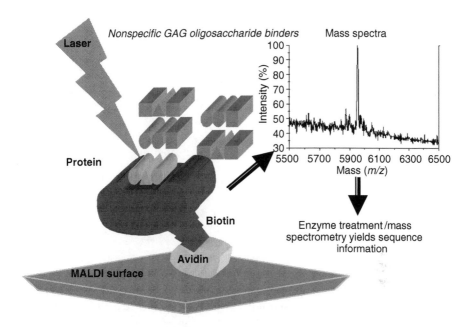

FIGURE 6.10 Surface noncovalent association mass spectrometry (SNA-MS). In this experimental technique, avidin is first adsorbed on a surface suitable for MALDI-MS analysis (gold plate or hydrophobic film). A GAG-binding, biotinylated protein is then immobilized on the avidin. Oligosaccharides (shown as disaccharide repeat units of three for hexasaccharides) are applied to the protein spot and washed with increasing salt concentrations to select for high-affinity binders (in grey). After application of $(RG)_{19}R$ peptide in a caffeic acid matrix, the sample is directly analyzed by MALDI-MS in a one-step procedure. Oligosaccharides are detected as noncovalent complexes with $(RG)_{19}R$; the mass of the oligosaccharide can be determined by subtraction of the mass of the peptide from the complex. Direct enzymatic treatment of the bound oligosaccharide with the heparinases followed by MALDI-MS analysis can yield sequence information on the high-affinity binders using the PEN-MALDI scheme.

wash (0.2 M NaCl) to remove oligosaccharides that bind the protein through nonspecific interactions, the peptide:matrix mixture is added directly to the spot and the oligosaccharide:peptide is detected using MALDI-MS (Figure 6.10). Additionally, the bound oligosaccharide can be enzymatically or chemically degraded prior to the addition of the peptide–matrix solution, thus enabling direct sequencing of the bound oligosaccharide on the MALDI plate.[91] As such, SNA-MS combines the isolation, enrichment, and sequencing of tissue or cell-derived GAGs with affinity to a known GAG-binding protein in a single technique (Figure 6.10).

SNA-MS has been used to probe several protein–oligosaccharide interactions, thus confirming the robustness of the technique and its ability to identify HSGAG sequences responsible for high-affinity protein binding. For example, comparing the prototypical interaction of AT-III with a high-affinity HSGAG pentasaccharide with other low-affinity oligosaccharides demonstrated that it was indeed possible to differentiate between specific and nonspecific ionic interactions using SNA-MS.[91] Additionally, HSGAGs with high-affinity for FGF-2 were enriched from a hexasaccharide pool. PEN-MALDI sequencing of these oligosaccharides revealed that the bound hexasaccharides have similar sequences to those seen in FGF–hexasaccharide co-crystal structure. SNA-MS analysis of a pool of HSGAGs isolated from the cell surface of smooth-muscle cells revealed the presence of the same hexasaccharide with high affinity to FGF, further extending the utility of the technique to analyze biologically derived GAGs.[91] Finally, SNA-MS was used to identify a hexasaccharide derived from porcine heparin with high affinity to endostatin, a collagen XVIII fragment with antiangiogenesis effects.[92] Therefore, SNA-MS has shown promise for the direct isolation and sequencing of bioactive HSGAGs that specifically interact with a variety of target proteins.

VII. FUTURE DIRECTIONS

The exciting progress in the field of glycobiology has been driven by innovations in the analysis of complex carbohydrates. The evolving methodologies for the sequence analysis of GAGs combining biochemical and analytical tools in novel ways have provided additional insights into the structure–activity relationship for these polysaccharides. Other techniques are also emerging as promising means to study the biology of GAGs. The development of oligosaccharide arrays in a fashion comparable with the "gene" chip has shown promise in studying the binding of antibodies to different carbohydrate antigens.[93] Such a carbohydrate array has the potential to probe thousands of different oligosaccharide spots in one experiment, making this technique an attractive tool for screening novel carbohydrate-binding proteins. Creating a carbohydrate chip with an array of defined GAG structures would be a useful tool for probing different cells and tissues for GAG-binding proteins.

Given the ongoing progress in the development of tools for GAG analysis discussed here, the emergence of methodologies to profile the temporal and spatial expression of GAGs on different cell types may not be too far away. The development of methods for identifying the GAG "glycotype" of specific cells will further enhance our understanding of GAG biology. In addition, such methods could prove useful in the detection and even prognostication of different diseases in which GAGs have been implicated in their etiology or pathology, such as cancer. In addition, our evolving knowledge of GAG biology also promises to fuel the development of novel carbohydrate-based therapeutics for treating these various diseases regulated by GAGs. Therefore, the ongoing growth of tools for the analysis of GAGs will continue to shed further light on the role of these complex carbohydrates in biology and provide additional promise for the diagnosis and treatment of GAG-related diseases.

REFERENCES

1. Knudson, C.B. and Knudson, W., Cartilage proteoglycans, *Semin. Cell. Dev. Biol.,* 12, 69–78, 2001.
2. Laurent, T.C. and Fraser, J.R., Hyaluronan, *FASEB J.,* 6, 2397–2404, 1992.
3. Funderburgh, J.L., Keratan sulfate: structure, biosynthesis, and function, *Glycobiology,* 10, 951–958, 2000.

4.   Sasisekharan, R. and Venkataraman, G., Heparin and heparan sulfate: biosynthesis, structure and function, *Curr. Opin. Chem. Biol.,* 4, 626–631, 2000.
5.   Esko, J.D. and Lindahl, U., Molecular diversity of heparan sulfate, *J. Clin. Invest.,* 108, 169–173, 2001.
6.   Silbert, J.E. and Sugumaran, G., Biosynthesis of chondroitin/dermatan sulfate, *IUBMB Life,* 54, 177–186, 2002.
7.   Trowbridge, J.M. and Gallo, R.L., Dermatan sulfate, new functions from an old glycosaminoglycan, *Glycobiology,* 12, 117–125, 2002.
8.   Sugahara, K., Tanaka, Y., Yamada, S., Seno, N., Kitagawa, H., Haslam, S.M., Morris, H.R., and Dell, A., Novel sulfated oligosaccharides containing 3-*O*-sulfated glucuronic acid from king crab cartilage chondroitin sulfate K. Unexpected degradation by chondroitinase ABC, *J. Biol. Chem.,* 271, 26745–26754, 1996.
9.   Ueoka, C., Nadanaka, S., Seno, N., Khoo, K.H., and Sugahara, K., Structural determination of novel tetra- and hexasaccharide sequences isolated from chondroitin sulfate H (oversulfated dermatan sulfate) of hagfish notochord, *Glycoconj. J.,* 16, 291–305, 1999.
10. Sugahara, K., Nadanaka, S., Takeda, K., and Kojima, T., Structural analysis of unsaturated hexasaccharides isolated from shark cartilage chondroitin sulfate D that are substrates for the exolytic action of chondroitin ABC lyase, *Eur. J. Biochem.,* 239, 871–880, 1996.
11. Gandra, M., Cavalcante, M., and Pavao, M., Anticoagulant sulfated glycosaminoglycans in the tissues of the primitive chordate *Styela plicata* (Tunicata), *Glycobiology,* 10, 1333–1340, 2000.
12. Mizuguchi, S., Uyama, T., Kitagawa, H., Nomura, K.H., Dejima, K., Gengyo-Ando, K., Mitani, S., Sugahara, K., and Nomura, K., Chondroitin proteoglycans are involved in cell division of *Caenorhabditis elegans, Nature,* 423, 443–448, 2003.
13. Hwang, H.Y., Olson, S.K., Esko J.D., and Horvitz, H.R., *Caenorhabditis elegans* early embryogenesis and vulval morphogenesis require chondroitin biosynthesis, *Nature,* 423, 439–443, 2003.
14. Venkataraman, G., Sasisekharan, V., Cooney, C.L., Langer, R., and Sasisekharan, R., A stereochemical approach to pyranose ring flexibility: its implications for the conformation of dermatan sulfate, *Proc. Natl. Acad. Sci. USA,* 91, 6171–6175, 1994.
15. Ernst, S., Venkataraman, G., Sasisekharan, V., Langer, R., Cooney, C.L., and Sasisekharan, R., Pyranose ring flexibility. Mapping of physical data for iduronate in continuous conformational space, *J. Am. Chem. Soc.,* 120, 2099–2107, 1998.
16. Westergren-Thorsson, G., Onnervik, P.O., Fransson, L.A., and Malmstrom, A., Proliferation of cultured fibroblasts is inhibited by L-iduronate- containing glycosaminoglycans, *J. Cell. Physiol.,* 147, 523–530, 1991.
17. Raman, R., Venkataraman, G., Ernst, S., Sasisekharan, V., and Sasisekharan, R., Structural specificity of heparin binding in the fibroblast growth factor family of proteins, *Proc. Natl. Acad. Sci. USA,* 100, 2357–2362, 2003.
18. Sugahara, K. and Kitagawa, H., Heparin and heparan sulfate biosynthesis, *IUBMB Life,* 54, 163–175, 2002.
19. Sundaram, M., Qi, Y., Shriver, Z., Liu, D., Zhao, G., Venkataraman, G., Langer, R., and Sasisekharan, R., Rational design of low-molecular weight heparins with improved *in vivo* activity, *Proc. Natl. Acad. Sci. USA,* 100, 651–656, 2003.
20. Conrad, H.E., Determination of heparinoid structures, in *Heparin-Binding Proteins.,* Academic Press, New York, 1998, pp. 61–114.
21. Nagasawa, K., Uchiyama, H., Sato, N., and Hatano, A., Chemical change involved in the oxidative–reductive depolymerization of heparin, *Carbohydr. Res.,* 236, 165–180, 1992.
22. Ernst, S., Langer, R., Cooney, C.L., and Sasisekharan, R., Enzymatic degradation of glycosaminoglycans, *Crit. Rev. Biochem. Mol. Biol.,* 30, 387–444, 1995.

23. Ernst, S., Venkataraman, G., Winkler, S., Godavarti, R., Langer, R., Cooney, C.L., and Sasisekharan, R., Expression in *Escherichia coli,* purification and characterization of heparinase I from *Flavobacterium heparinum, Biochem. J.,* 315, 589–597, 1996.

24. Godavarti, R., Davis, M., Venkataraman, G., Cooney, C., Langer, R., and Sasisekharan, R., Heparinase III from *Flavobacterium heparinum:* cloning and recombinant expression in *Escherichia coli, Biochem. Biophys. Res. Commun.,* 225, 751–758, 1996.

25. Pojasek, K., Shriver, Z., Kiley, P., Venkataraman, G., and Sasisekharan, R., Recombinant expression, purification, and kinetic characterization of chondroitinase AC and chondroitinase B from *Flavobacterium heparinum, Biochem. Biophys. Res. Commun.,* 286, 343–351, 2001.

26. Fethiere, J., Eggimann, B., and Cygler, M., Crystal structure of chondroitin AC lyase, a representative of a family of glycosaminoglycan degrading enzymes, *J. Mol. Biol.,* 288, 635–647, 1999.

27. Huang, W., Boju, L., Tkalec, L., Su, H., Yang, H.O., Gunay, N.S., Linhardt, R.J., Kim, Y.S., Matte, A., and Cygler, M., Active site of chondroitin AC lyase revealed by the structure of enzyme–oligosaccharide complexes and mutagenesis, *Biochemistry,* 40, 2359–2372, 2001.

28. Huang, W., Matte, A., Li, Y., Kim, Y.S., Linhardt, R.J., Su, H., and Cygler, M., Crystal structure of chondroitinase B from *Flavobacterium heparinum* and its complex with a disaccharide product at 1.7 A resolution, *J. Mol. Biol.,* 294, 1257–1269, 1999.

29. Huang, W., Lunin, V.V., Li, Y., Suzuki, S., Sugiura, N., Miyazono, H., and Cygler, M., Crystal structure of *Proteus vulgaris* chondroitin sulfate ABC lyase I at 1.9A resolution, *J. Mol. Biol.,* 328, 623–634, 2003.

30. Kreil, G., Hyaluronidases — a group of neglected enzymes, *Protein Sci.,* 4, 1666–1669, 1995.

31. Hynes, W.L. and Walton, S.L., Hyaluronidases of Gram-positive bacteria, *FEMS Microbiol. Lett.,* 183, 201–207, 2000.

32. Baker, J.R. and Pritchard, D.G., Action pattern and substrate specificity of the hyaluronan lyase from group B streptococci, *Biochem. J.,* 348, 465–471, 2000.

33. Jandik, K.A., Gu, K., and Linhardt, R.J., Action pattern of polysaccharide lyases on glycosaminoglycans, *Glycobiology,* 4, 289–296, 1994.

34. Rhomberg, A.J., Shriver, Z., Biemann, K., and Sasisekharan, R., Mass spectrometric evidence for the enzymatic mechanism of the depolymerization of heparin-like glycosaminoglycans by heparinase II, *Proc. Natl. Acad. Sci. USA,* 95, 12232–12237, 1998.

35. Ernst, S., Rhomberg, A.J., Biemann, K., and Sasisekharan, R., Direct evidence for a predominantly exolytic processive mechanism for depolymerization of heparin-like glycosaminoglycans by heparinase I, *Proc. Natl. Acad. Sci. USA,* 95, 4182–4187, 1998.

36. Plaas, A.H., West, L.A., and Midura, R.J., Keratan sulfate disaccharide composition determined by FACE analysis of keratanase II and endo-beta-galactosidase digestion products, *Glycobiology,* 11, 779–790, 2001.

37. Vlodavsky, I., Goldshmidt, O., Zcharia, E., Atzmon, R., Rangini-Guatta, Z., Elkin, M., Peretz, T., and Friedmann, Y., Mammalian heparanase: involvement in cancer metastasis, angiogenesis and normal development, *Semin. Cancer. Biol.* 12, 121–129, 2002.

38. Sugahara, K. and Kojima, T., Specificity studies of bacterial sulfatases by means of structurally defined sulfated oligosaccharides isolated from shark cartilage chondroitin sulfate D, *Eur. J. Biochem.,* 239, 865–870, 1996.

39. Myette, J., Shriver, Z., Claycamp, C., McLean, M., Venkataraman, G., and Sasisekharan, R., The heparin/heparan sulfate 2-*O*-sulfatase from *Flavobacterium heparinum:* molecular cloning, recombinant expression, and biochemical characterization, *J. Biol. Chem.,* 7, 7, 2003.

40. Raman, R., Myette, J., Shriver, Z., Pojasek, K., Venkataraman, G., and Sasisekharan, R., The heparin/heparan sulfate 2-*O*-sulfatase from *Flavobacterium heparinum:* a structural and biochemical study of the enzyme active site and saccharide substrate specificity, *J. Biol. Chem.,* 7, 7, 2003.

41. Myette, J.R., Shriver, Z., Kiziltepe, T., McLean, M.W., Venkataraman, G., and Sasisekharan, R., Molecular cloning of the heparin/heparan sulfate delta 4,5 unsaturated glycuronidase from *Flavobacterium heparinum,* its recombinant expression in *Escherichia coli,* and biochemical determination of its unique substrate specificity, *Biochemistry,* 41, 7424–7434, 2002.

42. Turnbull, J.E., Hopwood, J.J., and Gallagher, J.T., A strategy for rapid sequencing of heparan sulfate and heparin saccharides, *Proc. Natl. Acad. Sci. USA,* 96, 2698–2703, 1999.

43. Rudd, P.M. and Dwek, R.A., Rapid, sensitive sequencing of oligosaccharides from glycoproteins, *Curr. Opin. Biotechnol.* 8, 488–497, 1997.

44. Toida, T., Shima, M., Azumaya, S., Maruyama, T., Toyoda, H., Imanari, T., and Linhardt, R.J., Detection of glycosaminoglycans as a copper(II) complex in high-performance liquid chromatography, *J. Chromatogr. A.,* 787, 266–270, 1997.

45. Yang, H.O., Gunay, N.S., Toida, T., Kuberan, B., Yu, G., Kim, Y.S., and Linhardt, R.J., Preparation and structural determination of dermatan sulfate-derived oligosaccharides, *Glycobiology,* 10, 1033–1039, 2000.

46. Calabro, A., Midura, R., Wang, A., West, L., Plaas, A., and Hascall, VC., Fluorophore-assisted carbohydrate electrophoresis (FACE) of glycosaminoglycans, *Osteoarthritis Cartilage,* 9, 16–22, 2001.

47. Plaas, A.H., West, L., Midura, R.J., and Hascall, V.C., Disaccharide composition of hyaluronan and chondroitin/dermatan sulfate. Analysis with fluorophore-assisted carbohydrate electrophoresis, *Methods Mol. Biol.,* 171, 117–128, 2001.

48. Plaas, A.H., West, L.A., Wong-Palms, S., and Nelson, F.R., Glycosaminoglycan sulfation in human osteoarthritis. Disease-related alterations at the non-reducing termini of chondroitin and dermatan sulfate, *J. Biol. Chem.,* 273, 12642–12649, 1998.

49. Calabro, A., Benavides, M., Tammi, M., Hascall, V.C., and Midura, R.J., Microanalysis of enzyme digests of hyaluronan and chondroitin/dermatan sulfate by fluorophore-assisted carbohydrate electrophoresis (FACE), *Glycobiology,* 10, 273–281, 2000.

50. Ampofo, S.A., Wang, H.M., and Linhardt, R.J., Disaccharide compositional analysis of heparin and heparan sulfate using capillary zone electrophoresis, *Anal. Biochem.,* 199, 249–255, 1991.

51. Ruiz-Calero, V., Puignou, L., and Galceran, M.T., Analysis of glycosaminoglycan monosaccharides by capillary electrophoresis using indirect laser-induced fluorescence detection, *J. Chromatogr. A.,* 873, 269–282, 2000.

52. Kitagawa, H., Kinoshita, A., and Sugahara, K., Microanalysis of glycosaminoglycan-derived disaccharides labeled with the fluorophore 2-aminoacridone by capillary electrophoresis and high-performance liquid chromatography, *Anal. Biochem.,* 232, 114–121, 1995.

53. Mao, W., Thanawiroon, C., and Linhardt, R.J., Capillary electrophoresis for the analysis of glycosaminoglycans and glycosaminoglycan-derived oligosaccharides, *Biomed. Chromatogr.,* 16, 77–94, 2002.

54.   Lamari, F.N., Militsopoulou, M., Mitropoulou, T.N., Hjerpe, A., and Karamanos, N.K., Analysis of glycosaminoglycan-derived disaccharides in biologic samples by capillary electrophoresis and protocol for sequencing glycosaminoglycans, *Biomed. Chromatogr.,* 16, 95–102, 2002.

55.   Militsopoulou, M., Lamari, F.N., Hjerpe, A., and Karamanos, N.K., Determination of twelve heparin- and heparan sulfate-derived disaccharides as 2-aminoacridone derivatives by capillary zone electrophoresis using ultraviolet and laser-induced fluorescence detection, *Electrophoresis,* 23, 1104–1109, 2002.

56.   Mitropoulou, T.N., Lamari, F., Syrokou, A., Hjerpe, A., and Karamanos, N.K., Identification of oligomeric domains within dermatan sulfate chains using differential enzymic treatments, derivatization with 2-aminoacridone and capillary electrophoresis, *Electrophoresis,* 22, 2458–2463, 2001.

57.   Pojasek, K., Raman, R., Kiley, P., Venkataraman, G., and Sasisekharan, R., Biochemical characterization of the chondroitinase B active site, *J. Biol. Chem.,* 277, 31179–31186, 2002.

58.   Kinoshita, M., Okino, A., Oda, Y., and Kakehi, K., Anomalous migration of hyaluronic acid oligomers in capillary electrophoresis: correlation to susceptibility to hyaluronidase, *Electrophoresis,* 22, 3458–3465, 2001.

59.   Park, Y., Cho, S., and Linhardt, R.J., Exploration of the action pattern of *Streptomyces hyaluronate* lyase using high-resolution capillary electrophoresis, *Biochim. Biophys, Acta,* 1337, 217–226, 1997.

60.   Yamada, S., Yamane, Y., Sakamoto, K., Tsuda, H., and Sugahara, K., Structural determination of sulfated tetrasaccharides and hexasaccharides containing a rare disaccharide sequence, -3GalNAc(4,6-disulfate)beta1-4IdoAalpha1-, isolated from porcine intestinal dermatan sulfate, *Eur. J. Biochem.,* 258, 775–783, 1998.

61.   Chuang, W.L., Christ, M.D., and Rabenstein, D.L., Determination of the primary structures of heparin- and heparan sulfate-derived oligosaccharides using band-selective homonuclear-decoupled two-dimensional 1H NMR experiments, *Anal. Chem.,* 73, 2310–2316, 2001.

62.   Hricovini, M., Guerrini, M., Bisio, A., Torri, G., Naggi, A., and Casu, B., Active conformations of glycosaminoglycans. NMR determination of the conformation of heparin sequences complexed with antithrombin and fibroblast growth factors in solution, *Semin. Thromb. Hemost.,* 28, 325–334, 2002.

63.   Mulloy, B., Mourao, P.A., and Gray, E., Structure/function studies of anticoagulant sulphated polysaccharides using NMR, *J. Biotechnol.,* 77, 123–135, 2000.

64.   Mulloy, B. and Johnson, E.A., Assignment of the 1H-n.m.r. spectra of heparin and heparan sulphate, *Carbohydr. Res.,* 170, 151–165, 1987.

65.   Yates, E.A., Santini, F., Guerrini, M., Naggi, A., Torri, G., and Casu, B., 1H and 13C NMR spectral assignments of the major sequences of twelve systematically modified heparin derivatives, *Carbohydr. Res.,* 294, 15–27, 1996.

66.   Huckerby, T.N., Lauder, R.M., Brown, G.M., Nieduszynski, I.A., Anderson, K., Boocock, J., Sandall, P.L., and Weeks, S.D., Characterization of oligosaccharides from the chondroitin sulfates. (1)H-NMR and (13)C-NMR studies of reduced disaccharides and tetrasaccharides, *Eur. J. Biochem.,* 268, 1181–1189, 2001.

67.   Casu, B. and Torri, G., Structural characterization of low molecular weight heparins, *Semin. Thromb. Hemost.,* 25, 17–25, 1999.

68.   Harvey, D.J., Matrix-assisted laser desorption/ionization mass spectrometry of carbohydrates, *Mass Spectrom. Rev.,* 18, 349–450, 1999.

69.   Chai, W., Hounsell, E.F., Bauer, C.J., and Lawson, A.M., Characterisation by LSI-MS and 1H NMR spectroscopy of tetra-, hexa-, and octa-saccharides of porcine intestinal heparin, *Carbohydr. Res.,* 269, 139–156, 1995.

70. Chai, W., Luo, J., Lim, C.K., and Lawson, A.M., Characterization of heparin oligosaccharide mixtures as ammonium salts using electrospray mass spectrometry, *Anal. Chem.*, 70, 2060–2066, 1998.

71. Desaire, H. and Leary, J.A., Detection and quantification of the sulfated disaccharides in chondroitin sulfate by electrospray tandem mass spectrometry, *J. Am. Soc. Mass. Spectrom.*, 11, 916–920, 2000.

72. Oguma, T., Toyoda, H., Toida, T., and Imanari, T., Analytical method of chondroitin/dermatan sulfates using high performance liquid chromatography/turbo ionspray ionization mass spectrometry: application to analyses of the tumor tissue sections on glass slides, *Biomed. Chromatogr.*, 15, 356–362, 2001.

73. Oguma, T., Toyoda, H., Toida, T., and Imanari, T., Analytical method of heparan sulfates using high-performance liquid chromatography turbo-ionspray ionization tandem mass spectrometry, *J. Chromatogr. B.*, 754, 153–159, 2001.

74. Kuhn, A.V., Ruttinger, H.H., Neubert, R.H., and Raith, K., Identification of hyaluronic acid oligosaccharides by direct coupling of capillary electrophoresis with electrospray ion trap mass spectrometry, *Rapid Commun. Mass Spectrom.*, 17, 576–582, 2003.

75. Duteil, S., Gareil, P., Girault, S., Mallet, A., Feve, C., and Siret, L., Identification of heparin oligosaccharides by direct coupling of capillary electrophoresis/ionspray-mass spectrometry, *Rapid Commun. Mass Spectrom.*, 13, 1889–1898, 1999.

76. Siegel, M.M., Tabei, K., Kagan, M.Z., Vlahov, I.R., Hileman, R.E., and Linhardt, R.J., Polysulfated carbohydrates analyzed as ion-paired complexes with basic peptides and proteins using electrospray negative ionization mass spectrometry, *J. Mass Spectrom.*, 32, 760–772, 1997.

77. Juhasz, P. and Biemann, K., Utility of non-covalent complexes in the matrix-assisted laser desorption ionization mass spectrometry of heparin-derived oligosaccharides, *Carbohydr. Res.*, 270, 131–147, 1995.

78. Vives, R.R., Pye, D.A., Salmivirta, M., Hopwood, J.J., Lindahl, U., and Gallagher, J.T., Sequence analysis of heparan sulphate and heparin oligosaccharides, *Biochem. J.*, 339, 767–773, 1999.

79. Lauder, R.M., Huckerby, T.N., and Nieduszynski, I.A., A fingerprinting method for chondroitin/dermatan sulfate and hyaluronan oligosaccharides, *Glycobiology*, 10, 393–401, 2000.

80. Cheng, F., Yoshida, K., Heinegard, D., and Fransson, L.A., A new method for sequence analysis of glycosaminoglycans from heavily substituted proteoglycans reveals non-random positioning of 4- and 6-*O*- sulphated *N*-acetylgalactosamine in aggrecan-derived chondroitin sulphate, *Glycobiology*, 2, 553–561, 1992.

81. Lauder, R.M., Huckerby, T.N., and Nieduszynski, I.A., Increased incidence of unsulphated and 4-sulphated residues in the chondroitin sulphate linkage region observed by high-pH anion-exchange chromatography, *Biochem. J.*, 347, 339–348, 2000.

82. Lauder, R.M., Huckerby, T.N., Brown, G.M., Bayliss, M.T., and Nieduszynski, I.A., Age-related changes in the sulphation of the chondroitin sulphate linkage region from human articular cartilage aggrecan, *Biochem. J.*, 358, 523–528, 2001.

83. Calabro, A., Hascall, V.C., and Midura, R.J., Adaptation of FACE methodology for microanalysis of total hyaluronan and chondroitin sulfate composition from cartilage, *Glycobiology*, 10, 283–293, 2000.

84. Pope, R.M., Raska, C.S., Thorp, S.C., and Liu, J., Analysis of heparan sulfate oligosaccharides by nano-electrospray ionization mass spectrometry, *Glycobiology*, 11, 505–513, 2001.

85. Desaire, H., Sirich, T.L., and Leary, J.A., Evidence of block and randomly sequenced chondroitin polysaccharides: sequential enzymatic digestion and quantification using ion trap tandem mass spectrometry, *Anal. Chem.*, 73, 3513–3520, 2001.

86. Zaia, J. and Costello, C.E., Compositional analysis of glycosaminoglycans by electrospray mass spectrometry, *Anal. Chem.,* 73, 233–239, 2001.

87. Venkataraman, G., Shriver, Z., Raman, R., and Sasisekharan, R., Sequencing complex polysaccharides, *Science*, 286, 537–542, 1999.

88. Shriver, Z., Raman, R., Venkataraman, G., Drummond, K., Turnbull, J., Toida, T., Linhardt, R., Biemann, K., and Sasisekharan, R., Sequencing of 3-*O* sulfate containing heparin decasaccharides with a partial antithrombin III binding site, *Proc. Natl. Acad. Sci. USA,* 97, 10359–10364, 2000.

89. Shriver, Z., Sundaram, M., Venkataraman, G., Fareed, J., Linhardt, R., Biemann, K., and Sasisekharan, R., Cleavage of the antithrombin III binding site in heparin by heparinases and its implication in the generation of low molecular weight heparin, *Proc. Natl. Acad. Sci. USA,* 97, 10365–10370, 2000.

90. Guerrini, M., Raman, R., Venkataraman, G., Torri, G., Sasisekharan, R., and Casu, B., A novel computational approach to integrate NMR spectroscopy and capillary electrophoresis for structure assignment of heparin and heparan sulfate oligosaccharides, *Glycobiology,* 12, 713–719, 2002.

91. Keiser, N., Venkataraman, G., Shriver, Z., and Sasisekharan, R., Direct isolation and sequencing of specific protein-binding glycosaminoglycans, *Nat. Med.,* 7, 123–128, 2001.

92. Karumanchi, S.A., Jha, V., Ramchandran, R., Karihaloo, A., Tsiokas, L., Chan, B., Dhanabal, M., Hanai, J.I., Venkataraman, G., and Shriver, Z., et al., Cell surface glypicans are low-affinity endostatin receptors, *Mol. Cell.,* 7, 811–822, 2001.

93. Wang, D., Liu, S., Trummer, B.J., Deng, C., and Wang, A., Carbohydrate microarrays for the recognition of cross-reactive molecular markers of microbes and host cells, *Nat. Biotechnol.,* 20, 275–281, 2002.

7 Carrageenans: Structural and Conformational Studies

Carlos A. Stortz

CONTENTS

I. INTRODUCTION

Seaweed polysaccharides have been the subject of numerous studies regarding their structure, conformation, and physical and biological properties, with most of the attention devoted to alginates from brown seaweeds (Phaeophyceae) and galactans from red seaweeds (Rhodophyceae). The interest in these polysaccharides arises mainly from their capacity (depending on their structure and the medium) to form solutions of high viscosity or rigid gels, which can be stable in the presence of different chemical compounds. These properties make seaweed polysaccharides ideal for wide use in many industries including those related to food, paper, and textile production as well as petroleum extraction. Lately, fucoidans from brown seaweeds have also received close scrutiny due to their biological (antiviral, anticoagulant) properties.

Galactans are the main polysaccharides constituting the intercellular matrix and nonfibrillar cell walls of most Rhodophyceae. Their structure can be depicted by linear chains of alternating 3-linked β-galactopyranosyl residues and 4-linked α-galactopyranosyl residues. The galactose in the residues with the β-configuration always belongs to the D-series, whereas the α-galactose residues may include residues of the D- or the L-series, giving rise to a classification of red seaweed galactans as *carrageenans* (those with α-galactose units of the D-series) and *agarans* (those with 4-linked α-L-galactose residues). This simple picture is complicated by the fact that many galactans carry both D-and L- 4-linked galactose units interspersed in the same molecules, by the masking of this regularly repetitive structure by substitution with sulfate hemiester groups, pyruvic acid ketals, methoxylation of some hydroxyl groups, side chains, and mainly by the presence of a 3,6-anhydro ring as a cyclic ether of some or all of the α-galactose units. Actual polysaccharides isolated from natural sources seldom have a regularly repeating structure; instead they usually have different substitution patterns within the same molecule. Several reviews discussing different features of red seaweed galactans have appeared over the last few years.[1–6]

The first part of this chapter presents a short historical survey of the main studies carried out on carrageenans, followed by an overview of the latest structural studies on these polysaccharides. The second part presents, probably for the first time, conformational studies, carried out mainly by molecular modeling, on disaccharides representative of the carrageenan molecules, and their comparison with experimental data.

II. CARRAGEENANS: STRUCTURAL STUDIES

A. Steps and Landmarks in the History of Chemical Studies on Carrageenans: Discovery and First Attempts to Study Their Structure

The known history of carrageenans dates back to 1844 when Schmidt[7] reported the isolation of the mucilage of *Chondrus crispus* ("Carragheen"). It took about 40 years to discover the presence of galactose,[8] and about a century until the first structural studies were carried out.[9] In between, about 20 papers reported different features of this polysaccharide, then called "carragheen moss," "Irish moss," or "carrageenin."

During this period only a few of the structural features of carrageenans known today had been discovered. However, it was detected that the mucilages of other seaweeds, such as *Iridaea laminarioides*, *Gigartina stellata*, etc.,[10–12] had chemical properties similar to those of *C. crispus*. Thus, the name "carrageenin" was extended to all of them.

B. SECOND PERIOD: FRACTIONATION AND STRUCTURE

A landmark in carrageenan history was made when two components were separated by means of potassium chloride: the so-called κ-carrageenan formed gels in the presence of 0.125 M potassium ion, while the rest (λ-carrageenan), remained soluble.[13] Shortly thereafter, 3,6-anhydro-D-galactose was determined to be an important constituent of carrageenans,[14] leading to the correct structural determination of κ-carrageenan.[15] During the 1960s, it was determined that carrageenan is not a mixture of two or more compounds but a true family of structures. Under the leadership of David Rees, the structure of λ-carrageenan was determined.[16–19] This advance led to the structural elucidation of polysaccharide products from different seaweeds[19,20] and further established several new idealized structures (μ-carrageenan, ι-carrageenan, etc.). By using a different approach, the group led by Arne Haug tried to define carrageenans operationally (and not structurally) by means of potassium chloride precipitation. Thus, those insoluble at 0.125 M KCl were defined as "insoluble," those precipitating at higher concentrations of potassium chloride as "intermediate," and those soluble at all concentrations of this salt as "soluble."[21] Their results showed that no sharp separations can be achieved as molecules of carrageenans have continuously varying structures between idealized extremes.

C. THIRD PERIOD: LIFE STAGES AND FAMILIES

A new landmark for the study of carrageenans was set in the early 1970s when several groups discovered that different sexual stages of the seaweeds produced different carrageenans.[22–24] Those (and further) studies determined that in many species of the families Gigartinaceae and Phyllophoraceae, the carrageenans from the sporophytic cycle were viscous, devoid of 3,6-anhydrogalactose, and did not precipitate in the presence of potassium chloride, as occurs usually with λ-carrageenans, while the products that originated in gametophytic samples showed characteristics more alike those of κ-carrageenans (high amounts of 3,6-anhydrogalactose, high proportions of products precipitating at low concentrations of potassium chloride). These findings led to the necessity of sorting the seaweeds by their life cycle before studying their polysaccharides. In particular, a separation in families was proposed.[25] The κ-family includes most of the carrageenans actually found. It is characterized by having 4-sulfated β-D-galactose units (Table 7.1), and is produced by the gametophytic stage of the species of the families Gigartinaceae or the Phyllophoraceae, which yield different polysaccharides at each stage, or by taxonomic families that do not yield different products at each life stage.[1] This family comprises κ- and ι-carrageenans as well as μ- and ν-carrageenans. The former compounds can be obtained from the latter two by an alkaline or enzymatic treatment respectively, which converts α-galactose 6-sulfate into 3,6-anhydrogalactose (Figure 7.1). Most of the natural carrageenans of

TABLE 7.1
Idealized Repeating Units of Carrageenans, Grouped Out in Four Families

	3-Linked β-D-galactose	4-Linked α-D-galactose
κ-family		
κ	4-Sulfate	3,6-Anhydro
ι	4-Sulfate	3,6-Anhydro 2-sulfate
μ	4-Sulfate	6-Sulfate
ν	4-Sulfate	2,6-Disulfate
ο	4-Sulfate	2-Sulfate
λ-family		
λ	2-Sulfate	2,6-Disulfate
ξ	2-Sulfate	2-Sulfate
π	2-Sulfate, 4,6-(1-carboxyethyliden)	2-Sulfate
θ	2-Sulfate	3,6-Anhydro 2-sulfate
β-family		
β	—	3,6-Anhydro
α	—	3,6-Anhydro 2-sulfate
γ	—	6-Sulfate
δ	—	2,6-Disulfate
ω-family		
ω	6-Sulfate	3,6-Anhydro
ψ	6-Sulfate	6-Sulfate

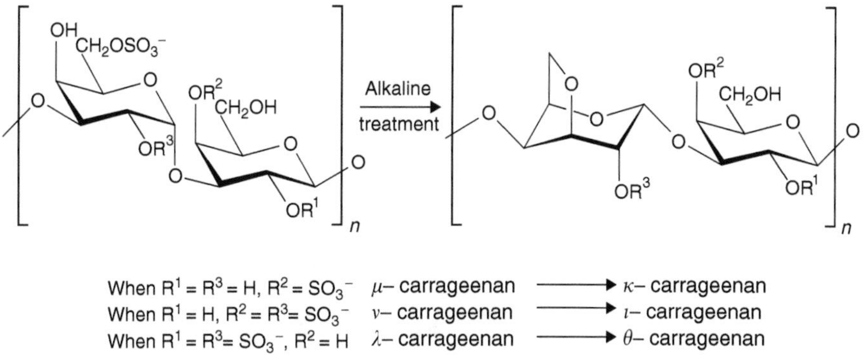

When $R^1 = R^3 = H$, $R^2 = SO_3^-$ μ– carrageenan ────▶ κ– carrageenan
When $R^1 = H$, $R^2 = R^3 = SO_3^-$ ν– carrageenan ────▶ ι– carrageenan
When $R^1 = R^3 = SO_3^-$, $R^2 = H$ λ– carrageenan ────▶ θ– carrageenan

FIGURE 7.1   Alkaline treatment of μ-, ν-, and λ-carrageenan.

the κ-family actually contain at least two of the four idealized disaccharide structures interspersed in the same molecules.[5,26] Precipitation with potassium chloride tends to yield fractions with higher κ-/ι-carrageenan content (and thus lower μ-/ν-proportion), while soluble fractions usually carry substantial amounts of κ-/ι-diads interspersed in a majority of μ-/ν-diads.[27–29] A fifth component of this family, *o*-carrageenan, was encountered in a few samples of the families Phyllophoraceae and Solieriaceae.[5,30] It corresponds to a ν-carrageenan without the 6-sulfate (Table 7.1) and is thus incapable of converting into a ι-carrageenan by alkaline treatment (Figure 7.1).

Another important family of carrageenans is the λ-family (Table 7.1), produced by the sporophytes of some species. λ-Carrageenans were once defined as being devoid of 3,6-anhydrogalactose and 4-sulfated units. Instead, the members of this family carry the β-galactose units sulfated at C-2 position. The alkali treatment of a λ-carrageenan generates a 3,6-anhydrogalactose ring (Figure 7.1). However, the product (θ-carrageenan) remains soluble in potassium chloride solutions, given the sulfation pattern of the β-galactose unit. By the operational definition,[13] λ-carrageenans were considered to be soluble in potassium chloride solutions. However, it was shown that actual λ-carrageenans do precipitate at high concentrations of this salt.[1,28,31,32] Two other structures (ξ- and π-carrageenans) also belong to this family (Table 7.1). Together with o-carrageenan, these diads are the only ones that contain neither 3,6-anhydrogalactose nor its precursor, galactose 6-sulfate.

Two other families of carrageenans have been added: the β-family comprises carrageenans with no sulfate on the β-D-galactose units. The β-carrageenan (Table 7.1), discovered first in *Eucheuma gelatinae* (now known as *Betaphycus gelatinum*),[33] is completely devoid of sulfate, and matches what Knutsen and co-workers[4,34] called carrageenose, by analogy with its diastereomer agarose. This diad has been found interspersed with those of its precursor (γ-carrageenan) and with κ-diads, in species of different families of the Gigartinales, being characteristic of the family Dicranemataceae.[5,35] Other products of the β-family have also been encountered.[5] The ω-family comprises carrageenans with a 6-sulfate on β-D-galactose units. Only two or three representatives of ω-carrageenan have been found in nature, as hybrids with κ-diads.[5,36,37]

D. APPEARANCE OF AGARAN/CARRAGEENAN HYBRIDS IN CARRAGEENOPHYTES: LATEST STUDIES ON CARRAGEENANS AND HYBRIDS

Seaweeds from some taxa, especially from the order previously recognized as Cryptonemiales (now most of its species belong to the Halymeniales), were known to produce carrageenan-predominant agaran–carrageenan hybrids.[1,3,38] These D/L-hybrids were regarded as a characteristic of certain orders, or of some species within a given order (as *Anatheca dentata* within the Solieriaceae).[1] However, by 1992, Craigie and Rivero-Carro reported the presence of agarans in known carrageenophytes, including *C. crispus*,[39] and shortly after, it was determined that gametophytes[40–42] and sporophytes[28,43] of the Argentinian carrageenophytes *G. skottsbergii* and *I. undulosa* (now *Sarcothalia crispata*) produce small amounts of agarans or agaran–carrageenan hybrids. These findings were (and will continue being) aided by the advent of new techniques available for the determination of small amounts of L-galactose and its derivatives.[44–46] Further studies revealed the presence of even larger proportions of 3,6-anhydro-L-galactose-containing polysaccharides in *Gymnogongrus torulosus*[47] and in one of the main sources of κ-carrageenan, *Kappaphycus alvarezii*,[48] among other species.[49] Although the coexistence of agaran and carrageenan diads or block copolymers in the same molecules was not proved and, thus, the presence of separate polysaccharides cannot be completely discarded,[6] evidence of the presence of copolymers exists.[47] 6-Linked α-D-glucans have also been found in small amounts with those products.[42,48]

Agarans or agaran–carrageenan hybrids have also been found with carrageenans, mannans, cellulose, and proteins as components of the skeletal cell walls of cysto-carpic *I. undulosa*.[50,51]

At the same time, several agarophytes were found to produce carrageenan or D/L-hybrids.[49] However, none of them are from the main agar-producing orders Gelidiales and Gracillariales.[49] Within the Gigartinales, the polysaccharides of most species from the genus *Gloiopeltis* (previously included in the Cryptonemiales) were regarded as agarans[3,5,30] with the particular name of funorans. A reinvestigation showed their coexistence with carrageenan diads.[2,52] Similar results were encountered for two species of Ceramiales,[49,53] two species of Rhodymeniales,[54,55] and a member of the Plocamiales.[56]

Other studies of the period tackled novel features of some species of carrageenans. For example, in tetrasporic samples of *G. pistillata*, carrageenans of the λ-family carried substantial amounts of 2,6-disulfated β-galactose units.[57] In several species of Australian samples of the genus *Callophycus* (Solieriaceae), a heavily pyruvylated α-carrageenan was encountered.[58] A similar diad was found interspersed with that of a ι-carrageenan in a sample of *Stenogramme interrupta* (Phyllophoraceae) from New Zealand, raising questions about its taxonomy.[59] In spite of the low proportions of naturally methoxylated galactose units usually found in carrageenans (in contrast with agarans), Chiovitti and colleagues have found a ι-carrageenan highly methoxylated on C-6 (and less on C-3) in samples from the genus *Rhabdonia*.[60] Later, they also found a similar, but more complex α/ι-hybrid with a similar methoxylation pattern on samples of the genus *Erythroclonium*.[61] The resemblance of both carrageenans to polysaccharides from samples of genera *Austroclonium* and *Areschougia*[62] led the authors to propose the gathering of all four genera into one family (Areschougiaceae or Rhabdoniaceae), separated from the family Solieriaceae, which does not yield methoxylated carrageenans.[5,62] This era was also rich in the consolidation of the use of spectroscopic techniques, both infrared[5] and different applications of nuclear magnetic resonance (NMR).[63–66]

III. CARRAGEENANS: CONFORMATIONAL STUDIES

A. CONFORMATIONAL ANALYSIS OF CARBOHYDRATES

Carbohydrates were first considered to have a biological role that was restricted to structural, reserve, and energy-producing functions, in contrast with proteins and nucleic acids which were regarded to be the key biological macromolecules. Despite the lesser role traditionally accorded to them, their economical importance made carbohydrates the subject of many studies since the 19th century. Today, it is recognized that their roles are not restricted to these coarse functions: carbohydrates, alone or linked to proteins or lipids, are known to be the key recognition signals in many biological processes.[67]

As these functions usually imply a receptor–host interaction, the spatial characteristics of carbohydrates are very important for fulfilling the functions as well as for the physical properties that lead to their industrial applications.[68] The growth of the interest in carbohydrates shapes is related to the advances in instrumental techniques for conformational studies (NMR spectroscopy, X-ray diffraction, molecular model-ing), which occurred especially over the last 30 years,[67,68] and are reflected in the knowledge that has been acquired of several conformational features of many car-bohydrates, including mono-, oligo,- and polysaccharides.

Studies by X-ray crystallography are probably the only ones which lead to a full spatial determination of molecules, including their bond lengths and angles, tor-sional angles, etc. However, the most successful results are restricted to molecules that give single crystals. Thus, only monosaccharides and lower oligomers can be studied by this method. X-ray diffraction analysis is possible for the much larger polysaccharides in cases where these molecules form oriented fibers, but their three-dimensional shape can only be assumed after building a model which includes the crystal data for the monomeric units, followed by further iterative refinements.[67] Besides the experimental difficulties encountered in X-ray diffraction analysis (the need of obtaining single crystals or good oriented fibers), it should be borne in mind that the final results (if the model is successfully arranged) reveal the three-dimen-sional structure only of the solid, and not of the one expected to appear in solution, which is where most of the biological and physical properties become patent.

NMR spectroscopy is one of the experimental tools that has helped most to deter-mine the spatial structure of many organic compounds in solution. For example, chemical shifts allow the differentiation of axially and equatorially linked protons; the measurements of different coupling constants ($^1J_{C,C}$, $^1J_{C,H}$, and $^3J_{H,H}$) permit the esti-mation of different torsional angles.[67,69] Even more important, is the measurement of the nuclear overhauser effect (NOE), especially by the way of a bidimensional exper-iment (NOESY, ROESY), which has a direct effect on conformational analysis, as it permits to observe proximity in space between two protons, regardless of their bond-ing pattern.[69–71] There are drawbacks to the use of NMR for the conformational analy-sis of carbohydrates associated with the clustering of resonances in a small region of the spectra. This problem may not be important for monosaccharides, but it grows with the size of the oligosaccharide, complicating the assignments needed for other measurements. Furthermore, carbohydrates only yield a few interglycosidic NOEs, thus complicating a full three-dimensional assessment.[67] In addition, many times more than one conformer is present in significant numbers for a given macromolecu-lar compound. In these cases, when the equilibrium is fast, spectroscopical data pro-vide a weighed average of those conformers. If these data are intended to be correlated with shape, a "virtual" conformation is obtained, with no true meaning.[72]

The measurement of optical rotation has also been correlated with the confor-mation around the glycosidic bond of di- and higher saccharides. Rees,[73] Rees and Thom[74] have established empirical equations, while Stevens and Sathyanaranyana[75] have followed up with a semiempirical treatment, which takes into account elec-tronic transitions and was applied to many disaccharides, in an effort to correlate molecular modeling with experimental data.[76]

Since the first application during the 1960s,[68] the use of computational methods to model carbohydrates has acquired a growing interest. Comparison with experi-ments has served to validate many computational methods in instances when experi-mental data are available. Furthermore, computational modeling allows properties of molecules that exist only scarcely in equilibrium to be calculated, thus disallowing their determination by experimental procedures,[68] The scope of modeling is limited only by the capacity of the computers and the reliability of the procedure employed.

B. INTRODUCTION TO COMPUTER MODELING OF CARBOHYDRATES: MONOSACCHARIDES

Broadly speaking, in its simplest approach, computer modeling is a way to calculate the energies of a given molecular system, and eventually to "move" the atoms from their current positions to other positions that yield a minimum energy. From this point on, several more complex tasks can be attempted, for example, plotting electrostatic potentials, observing the frontier orbitals, calculating the expected NMR chemical shifts, etc. However, the first key is always to find out the best functional manner to express the energy as a function of the coordinates of the atoms. This multidimensional function is known as a *potential energy surface.*

One can assume that the best results will be obtained from a rigorous solution of the Schrödinger equation with the least possible approximations.[67,68] Such studies are known as *ab initio* calculations, and have a large computational demand, as they require calculating a large number of integrals, even for small systems. Given this fact, only a few studies on monosaccharides have been carried out by *ab initio* procedures;[77-79] a few reports, however, of models based on a related quantum mechanical procedure, referred to as *density functional theory (DFT),* have started to appear.[79-82] The level of approximation (established by the basis sets used) significantly changes the results obtained by quantum mechanical methods. It was shown that the lower basis sets give unacceptable results.[77,82] However, it is not necessary to resort to the most sophisticated sets, as intermediate ones yield similar results.[77,82]

In order to save time and computer resources, the quantum mechanical procedures were simplified by the substitution of some terms of the integrals with empirical parameters, giving rise to the so-called *semiempirical methods.* Several studies applied the semiempirical Hamiltonians AM1[83] and PM3[84] to monosaccharides.[77,85-88] However, it was established that at least the PM3 method does not yield reliable results regarding the chair stability, giving rise to the $^1C_4$ form with a similar or even higher stability than the $^4C_1$ form.[77,88] This error was ascribed to a poor parameterization of the core-repulsion functions.[89] Both *ab initio* and semiempirical methods can add functions to calculate solvent effects,[77,85] which are very important in carbohydrate conformational studies. Furthermore, the major driving force in favor of the observed chairs of monosaccharides was ascribed to solvation effects,[90] thus giving a possible reason for the low-energy difference between both glucopyranose chairs found by the semiempirical methods.[77]

The methods most commonly used for carbohydrates are known as *molecular mechanics* (MM), empirical in nature. In this approach, relatively simple analytical energy functions are expressed in terms of the Cartesian coordinates of the atomic nuclei.[67,68] The empirical potential energy functions are expressed as sums of terms containing the contributions of bond stretching, angle bending, torsional strain, and nonbonded (van der Waals, electrostatic, and hydrogen-bonding) interactions. The parameters are obtained from model molecules, using experimental data or data obtained after *ab initio* calculations. The functions are considered to be transferable, so that similar functional groups will have the same interactions for all the molecules.[67] with careful selection of functions and parameters, good approximations of geometries and energies are obtained.[68] Initially developed for simple organic molecules, the method was rapidly extended to carbohydrates, even if carbohydrates presented particular problems originated in the presence of many polar groups, anomeric and exo-anomeric effects, and the possibility of hydrogen bonding. These problems led to the necessity of a special parameterization in order to comply with such interactions. The heart of MM is the force field, i.e., the functional form of each MM method. Several different general-purpose force fields have been used for carbohydrates, many of them with special revisions to cope with stereoelectronic effects and other issues, while other force fields were devised especially for carbohydrates.[67] It should be borne in mind that only quantum mechanical methods can be used to determine parameters such as transition states, energies, geometries, or reaction pathways. Using empirical methods, it is possible to study "stable" molecules or at most, conformational transition states. Furthermore, probably MM methods model the most stable conformers better, whereas modeling of conformers with higher energy yields poorer results, as stressed states are reached.

Most of the computational methods model isolated molecules, i.e., a gas state under high-vacuum conditions. As stated earlier, *ab initio* and semiempirical methods can add functions to emulate solvent effects.[77,85] The presence of solvent in MM is usually simulated indirectly by using a finite value of a dielectric constant, as a way to dampen electrostatic and hydrogen-bonding interactions. Distance-dependent dielectric constants are sometimes used, as it is arguable if a bulk dielectric constant is applicable to short distance interactions, and also because of the effect of a conformation-dependent screening produced by parts of the solute on parts of the solvent.[77]

Another approach, instead of searching for energy minima in a potential energy surface, is to use simulation methods, such as *Monte Carlo simulations* or *molecular dynamics* in order to explore the dynamic representation of the molecular system. These methods also require a functional expression of the energy vs. the position of the atoms. However, they give abundant thermodynamic information, time-related properties can be obtained, and the inclusion of solvent molecules is possible.[67] Molecular dynamics applications for monosaccharides[90,91] have received more attention than Monte Carlo simulations.[92]

Most effort so far devoted to the conformational analysis of monosaccharides has dealt with the chair equilibrium of pyranoses,[77,79,80,85,88,90,92] or the equivalent pathway in furanoses[68,93] and the rotameric equilibrium of the hydroxymethyl side chains in hexopyranoses.[77,81,85,87,88,94] Less work has been dedicated to the conformational preferences of the hydroxyl side chains,[85,87,88] the puckering of the chair conformations,[95] and the deprotonation energies of the different hydroxyl groups.[86]

C. Applications of Molecular Modeling to Disaccharides

Disaccharides (except those linked 1→6) contain two flexible bonds linking two monosaccharidic units (Figure 7.2A). The monomeric units usually have an established preferred ring conformation, so that the effect of its flexibility on the total energy of a disaccharide will be minimal when compared with the effect of the rotations around

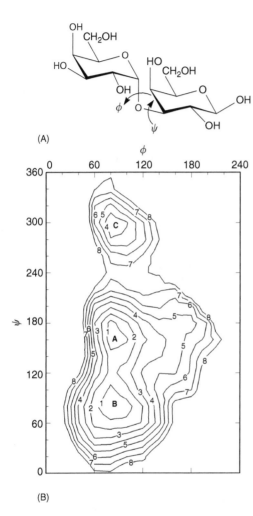

FIGURE 7.2 (A) A disaccharide (α-D-Gal$p$-(1$\rightarrow$3)-β-D-Gal$p$), showing its ϕ and ψ glycosidic angles. (B) MM3 potential energy surface (adiabatic conformational map) of that disaccharide. Isoenergy contours are graduated in 1 kcal/mol increments above the global minimum. Regions **A–C** are shown, whereas region **D** (around $\phi=-50°$, $\psi=95°$) higher energies, and thus is not shown.

these two bonds.[67] Thus, a typical modeling of disaccharides consists in determining the energy variation for all mutual orientations of the two monosaccharide residues,[68,96,97] expressed by the glycosidic angles ϕ and ψ (Figure 7.2A). The final output is usually a Ramachandran-like conformational map (Figure 7.2B), which shows isoenergy contour lines as a function of the glycosidic angles ϕ and ψ.[97]

The first studies were carried out using rigid residue analysis.[96] Ramachandran and co-workers developed the idea of "contact criteria"[67] by considering rigid pyranose rings composed of atoms treated like hard spheres. In this model, rotation around the glycosidic linkage gives rise to allowed or disallowed conformations, depending on whether nonbonded atoms approach beyond a certain distance of each other. Instead of a contour map with several isoenergy lines, these maps just showed one contour with the allowed region of minimum energy and the disallowed region of infinite energy.[67] Rees and co-workers[98,99] utilized this idea for several different linkages, and subdivided the conformational map into three regions: besides the disallowed conformations (highest energy), they showed a "fully allowed region" (less energy) and a "marginally allowed conformation" (high energy). At the same time, the first attempts to quantitate the potential energy (by the van der Waals contributions) appeared.[98] Shortly thereafter, Sathyanarayana and Rao[100,101] added new terms to the potential and compared the steric maps (by contact criteria) with contour maps where the potential energy was quantitated. Rigid residue analysis was improved as new or modified terms in the potential function were added in the force-fields[104,105] HSEA[102] and PFOS.[103] By allowing all variables to relax, Melberg and Rasmussen initiated flexible residue analysis by 1979 . Those studies were extended in the late 1980s[103,106–110] giving rise to the first fully relaxed energy maps of disaccharides. Carbohydrates have special features that complicate their study, such as the presence of hydrogen bonding, ring puckering, and anomeric and exo-anomeric effects. The parameterization of the force-field MM3[111] takes into consideration these facts.[96,112] Although the ability of this force field in modeling has been put in doubt,[113] the only disaccharide for which experimental data are not reproduced satisfactorily was sucrose,[97,112] probably due to the presence of overlapping anomeric effects.[114] In a series of landmark papers, Dowd and colleagues[96,115–118] applied this force field to several disaccharides with different linkages, and established a methodology to acquire the maps. The effects of the configurations of the linked carbons and those that were neighbors to the glycosidic linkage on the shape of the maps and on the calculated flexibility were also evaluated.[119,120]

Besides the difficulties in carbohydrate modeling, another issue has to be addressed regarding the Ramachandran maps: they depict the energy for each ϕ, ψ combination when actually, for such combinations, the authentic conformational hypersurface of the disaccharide carries the influence of the orientations of secondary hydroxyl groups, hydroxymethyl groups, and primary hydroxyl groups, all of which may affect the calculated final energy by several kilocalories. This problem is known as the "multiple minimum problem":[68,96,97] for a disaccharide formed by two hexopyranose residues and considering just three staggered rotamers for the ten exocyclic groups, 3^{10} (= 59,049) different starting points are possible for the construction of the map.[96,97] In other words, different starting points may yield different minima upon geometry optimization. Ideally, the Ramachandran map should be *adiabatic*,[97,121] i.e., it should plot the lower energy values found at each point. Thus, at least theoretically, the true adiabatic map can only be obtained after calculating 59,049 relaxed maps and determining, for each ϕ, ψ combination, the one with the lowest energy.[97] This problem has been recognized since the early flexible residue analysis,[104–106,109,110,121] and has been circumvented mainly by considering that the secondary hydroxyl groups are likely to form a crown of cooperative hydrogen bonds oriented either clockwise or anticlockwise, thus reducing the number of starting positions to a two-figure number of conformers.[115–118,122,123] However, it has been shown that these few starting points may not be adequate to produce an adiabatic energy surface, at least at high dielectric constants.[97] The multiple minimum problem was also tackled with less systematic procedures, such as iterative manual searches[97,109,121] or random sampling techniques.[124,125] Another way to elude the multiple minimum problem, and at the same time to consider the addition of solvent molecules, is to explore the conformational space of oligosaccharides by molecular dynamics (MD) simulations,[122,126–129] sometimes combined with Monte Carlo techniques in order to increase their efficiency at crossing energy barriers.[130] Other heuristic procedures have also been used.[112,131]

The uses of quantum mechanical methods have been mostly devoted to model compounds, sometimes with the purpose of validating a force field. Analogs of sucrose were studied by *ab initio* procedures in order to explain the failure of many force fields for modeling this sugar.[114,132] Similar studies were carried out on the backbone of a 4-linked disaccharide.[133] Momany and co-workers[134,135] studied by DFT several conformers of β-maltose and β-cellobiose. In the latter study, carried out at high levels of theory, an abnormally low energy for conformers with geometries in disagreement with experimental results was found. French and co-workers developed a hybrid QM/MM method:[136,137] adiabatic maps were produced by calculating the HF/6-31G* energies of analogs (containing the backbones) of the disaccharides at each ϕ,ψ grid point, and adding the calculated MM effects of the substituents (hydroxyl and hydroxymethyl groups).

D. X-RAY DIFFRACTION ANALYSIS OF CARRAGEENANS AND MODELS

As explained earlier, X-ray diffraction analysis is the only experimental technique that allows accounting fully for the three-dimensional shape of many carbohydrates. This assertion is even more important for highly hydrated polymer systems like carrageenans, where the gelling and thickening action may depend on their shapes and association possibilities.[138] The samples for fiber diffraction analysis are prepared by stretching hydrated films or fibers to induce molecular orientation. The diffraction pattern yields data directly in terms of the molecular pitch and helix symmetry. Other data have to be inferred constructing a model which uses some geometrical constraints: the main data deduced can be expressed in terms of the number of residues per pitch (n) and the distance advanced per residue (h).[67,138]

Each of the deduced models may lead to a different combination of the glycosidic angles ϕ and ψ. At this point, it should be mentioned that several different conventions were used to indicate those glycosidic angles (Figure 7.2). Hydrogen-related atoms were used more frequently,[67] but it has been suggested that torsion angles should be driven in terms of nonhydrogen atoms in molecular modeling studies, given the different motions of the three atoms during the driven rotation, and also due to the inaccuracy of hydrogen atom positions in diffraction studies.[136] In the current work, the dihedral angles ϕ and ψ for the α-(1→3) linkage are defined by atoms O5′–C1′–O3–C3 and C1′–O3–C3–C4, respectively, whereas those for the β-(1→4)-linkage are defined by atoms O5′–C1′–O4–C4 and C1′–O4–C4–C5, respectively. Sometimes, for the sake of clarity, a subscript indicating the type of linkage (α or β) is added to the symbol. When the literature data used another convention, an offset of ±120° was added in order to comply with the present conventions.

The first attempt to carry out X-ray fiber diffraction analysis on carrageenans was made by Bayley[139] with a λ- and a κ-carrageenan. Actually this work (albeit with a low-quality pattern) reports the only diffraction analysis published on λ-carrageenans. Most of the further work was done on κ- and, especially, on ι-carrageenan fibers. Two crystalline disaccharide models of carrageenans were also analyzed. Table 7.2 shows the results of the analysis.

TABLE 7.2
X-ray Diffraction Analysis Carried Out on Carrageenans and Models

Carrageenan	Ref.	Pitch and Other Data	ϕ_α,ψ_α (deg)	ϕ_β,ψ_β (deg)
κ- (K$^+$ salt)	140	24.6 Å, double helix	Not determined	
κ- (K$^+$ salt)	141,138	25.0 Å, double helix	61, 81	$-97,108$
ι- (K$^+$ salt)	140	26.0 Å, stag. double helix	Various possible	
ι- (Ca^{2+} salt)	142,138	26.6 Å, stag. double helix	77, 79	$-87,81$
	67,143	The earlier$^{(142)}$ refined to	75, 79	$-87,94$
ι- (Na$^+$ salt)	143	25.9 Å, stag. double helix	60, 77	$-89,109$
ι- (Ca^{2+} salt)	144	26.4 Å, stag. double helix	70, 76	$-91,104$
Neocarrabiose^a	145		94,142	
Carrabiose PA^b	146			$-92,72$

^{a}Neocarrabiose is the disaccharide 3,6-An-α-D-Gal-(1→3)-β-D-Gal.

^{b}The peracetylated dimethyl acetal of carrabiose (β-D-Gal-(1→4)-3,6-An-D-Gal).

The best-studied carrageenan molecule is the ι-carrageenan.[140,142] The presence of a molecular repeat distance of 13.3 Å and a periodicity of 26.6 Å can be explained in terms of a coaxial double-helical structure, where one of the chains is translated by exactly half the pitch.[138] This threefold, right-handed, parallel, half-staggered, double-helix model is shown to occur with both divalent and monovalent counterions. The double helix is stabilized by six interchain hydrogen bonds (involving O2 and O6 of the β-galactose units) per pitch length.[67,138] Recent studies[143,144] have confirmed this model, with slight changes on the numerical values of the pitch (Table 7.2) and other cell dimensions. It seems that calcium has no role in stabilizing the double helix, but it helps to stabilize higher structures by linking several double helices.[67,138,144]

The fibers of κ-carrageenan appear to be less crystalline than those of ι-carrageenan. The molecular repeat distance was found to be almost twice that of the ι-sample (Table 7.2), indicating that it does not correspond to a half-staggered double-helix model.[138] The best-fitted model assumes a coaxial threefold, right-handed parallel double helix, with only three interchain hydrogen bonds per pitch.[67] A rotation of 28° and a translation offset of 1Å from the half-staggered arrangement was observed.[67,138,141] Although other models, such as single strands, coaxial antiparallel double helices, or noncoaxial double helices, were also proposed, neither of them gives good agreement with the diffraction pattern.[138,141] There is no explanation of why κ-carrageenan does not adopt a half-staggered conformation (in view of its higher flexibility). However, it was found that a departure from the half-staggering for ι-carrageenan results in a strong destabilization of the double helix.[141]

As explained previously, the quality of the diffraction pattern for λ-carrageenan[139] was insufficient for structural determination.[138] Nevertheless a threefold helix with a pitch of 25.2Å was deduced. Molecular modeling (see below) suggested the presence of left-handed single helices stabilized by interresidual hydrogen bonding involving H(O)3 of the α-galactose unit and O2 of the β-galactose unit.[138,147] Further refinement of the earlier data indicated that left- and right-handed helices as well as other hydrogen-bonding sites are possible for λ-carrageenan.[138] However, a further study suggested that only right-handed helices are possible.[148] It was found that a λ-carrageenan is better described by a twisted ribbon structure than by an open helix, whereas a double-helical structure is sterically prohibited.[138]

Other data obtainable from the diffraction analysis models correspond to the glycosidic angles ϕ and ψ. The present studies (Table 7.2) as well as most of the crystal diffraction studies[67] led to ϕ angles restricted to the region predicted by the exoanomeric effect. With the present conventions, this corresponds to values of 60–80° for the α-(1→3)-linkage, and around −90° for the β(1→4)-linkage. The major variations of the glycosidic linkage are given by the ψ angle, which can usually fall into three different regions (Figure 7.2B), with two of them carrying low energies.[119,120] For the present structures, these minima correspond to ψ values around 80° (region **B**) and 160° (region **A**), for both the α- and the β-linkages (see Section E). This indicates that all the fiber diffraction studies (Table 7.2) showed both glycosidic conformations around the **B** regions. A similar result was encountered for the β-linkage of the acetylated carrabiose dimethyl acetal.[146] On the other hand, for neocarrabiose, it was found that its α-linkage has a conformation close to that of the **A** region.[145]

E. APPLICATION OF MOLECULAR MODELING TO CARRAGEENANS REPEATING UNITS

Computer-aided molecular modeling was applied to different carrageenan models. The main information provided by these methods arises from the shape of the potential energy surface, usually expressed as a contour map, where the energy is plotted against the two glycosidic angles ϕ and ψ. The geometries and energies of the different minima are usually the main numerical output of the calculations.

The first application of molecular modeling to carrageenans was made by Rees in 1969 by evaluating the allowed conformations of κ/ι- and λ-carrageenans[138] using rather rudimentary hard-sphere procedures, which were repeated in further work accompanying X-ray diffraction fiber analysis.[139,140] Tables 7.3 and 7.4 show the results of all the modeling studies carried out on the α-(1→3) and the β-(1→4)-linked units of carrageenans, respectively.

1. Analysis of the Conformation of α-Linked Disaccharides Containing 3,6-Anhydrogalactose (Neocarrabiose)

The early hard-sphere "calculations" for nonsulfated and 4′-sulfated neocarrabiose units indicated that the **A–B** region should be the only one sterically allowed.[141,142,147] For the ι-carrageenan repeating unit, an even smaller area in region **B** was determined to be allowed.[141,142] (Table 7.3). The crystal structure for neocarrabiose[145] shows angles located in the **A** region (Table 7.2). MM3 has also detected this region as the minimum-energy one,[151] although with very low-energy difference with the **B** region (which was the free energy-minimum region). A rigid-residue analysis of Lamba and colleagues[145] also identified **A** as the lowest-energy region, but it switched to **B** when a relaxed MM2CARB study was carried out (Table 7.3). A study using a CHARMM-type force field indicated **D** as the global minimum, and it failed to identify the **A–B** region as low energy region.[150] A different approach with a different version of the same force field found **B** region as

TABLE 7.3

Molecular Modeling Studies on the α-(1→3)-Linked Disaccharidic Repeating Units of Carrageenans

α-Gal	β-Gal	Carrageenans	Ref.	Method	Location of the Main Minima ϕ, ψ (deg), (Relative Energy, kcal/mol)
3,6-An	—	β	147	HS[a]	95,115[b] (in-between regions A and B)
3,6-An	—	β	142	HS	100,120[b] (regions A–B)
3,6-An	—	β	145	PFOS	100,170 (0.0) in A; 80,130 (0.5) in A–B
3,6-An	—	β	145	MM2CARB	40,70 (0.0) around B; -40,90 (0.8) in D
3,6-An	—	β	149	CHARMM22	55,62 (0.0) around B; 63,140 (1.1) in A
3,6-An	—	β	150	CHARMM	-39,99 in region D
3,6-An	—	β	151	MM3	86,160 (0.0) in A;.77,86 (0.1) in B
3,6-An	4S	κ	141	HS	100,120[b] (regions A–B)
3,6-An	4S	κ	152	MM[c]	80,90 in region B
3,6-An	4S	κ	148	Tripos	80,80 (0.0) in B; -40,90 (1.1) in D
3,6-An	4S	κ	151	MM3	86,159 (0.0) in A;.82,96 (0.9) in B
3,6-An	4S	κ	153	CHARMM25	68,72 (0.0) in B; 157,-56 (1.0)
3,6-An/2S	4S	ι	142	HS	70,80[b] in region B
3,6-An/2S	4S	ι	141	HS	70,95[b] in region B
3,6-An/2S	4S	ι	148	Tripos	60,75 (0.0) in B; -30,90 (0.5) in D
3,6-An/2S	4S	ι	151	MM3	72,87 (0.0) in B;.89,165 (0.4) in A
3,6-An/2S	4S	ι	154	CHARMM	42,54 (0.0) around B; 153,63 (2.1)
3,6-An/2S	4S	ι	154	CHARMM, ε=80	60,66 (0.0) in B; 173,140 (0.5)
3,6-An/2S	2S	θ	151	MM3	81,90 (0.0) in B;.95,163 (2.4) in A
–	—	Desulfated λ	147	HS	125,130[b] around regions A–B
–	—	Desulfated λ	155	MM2	89,169 (0.0) in A; 65,70 (2.4) in B
–	—	Desulfated λ	155	MM2, ε=80	72,80 (0.0) in B; 95,175 (1.6) in A
–	—	Desulfated λ	155	MM2 CARB, ε=80	74,80 (0.0) in B; 92,178 (1.4) in A
–	—	Desulfated λ	156	MM2	67,79 (0.0) in B;.87,178 (0.5) in A
–	—	Desulfated λ	97[d]	MM3	86,162 (0.0) in A;.82,82 (0.1) in B
–	—	Desulfated λ	97[d]	MM3, ε=80	79,83 (0.0) in B;.90,167 (1.1) in A
2,6-di-S	2S	λ	148	Tripos	60,90 (0.0) in B; 120,150 (1.6) close to A
2,6-di-S	2S	λ	156	MM2	78,93 (0.0) in B;.95,176 (1.3) in A
2,6-di-S	2S	λ	157[d]	MM3	89,81 (0.0) in B;.101,169 (3.3) in A
2,6-di-S	4S	ν	156	MM2	88,170 (0.0) in A; 62,79 (0.6) in B
2,6-di-S	4S	ν	157[d]	MM3	85,170 (0.0) in A; 68,84 (2.3) in B
6S	4S	μ	157[d]	MM3	83,164 (0.0) in A; 70,87 (2.6) in B

[a]Hard-sphere method.

[b]Estimated to be at the center of the allowed surface.

[c]Parameters for the force field were generated from MNDO calculations.

[d]Corrected after publication (Stortz, C.A., unpublished results).

the global minimum.[149] Using NMR studies, it is sometimes possible to show the conformational characteristics of the glycosidic linkages by the measurements of interresidual NOE interactions. For neocarrabiose, a conformation close to region **A** was suggested,[145] but further analysis indicates that the NOESY spectrum can be explained better by a minimum in region **B** or by an **A–B** equilibrium.

Fiber diffraction analyses of ι-carrageenan indicate an ordered conformation with angles into the **B** region. These data match with the global minimum found in all the calculations performed[141,142,148,151,154] (Table 7.3). However, in some calculations, particularly those effected by the Tripos and CHARMM force fields, a extending horizontally shape appears, giving rise to lower energies for conformers in the **D** and other unknown regions, and higher energies for the **A** rotamers.[148,154] The minimum obtained by MM3 calculations appears with a very close geometry to that obtained by the old hard-sphere calculations (Table 7.3). For κ-carrageenan, fiber-diffraction analysis suggests a conformation in the **B** region, matching most of the modeling results[148,152,153] (Table 7.3). MM3 gives **A** as the global minimum,[151] with an energy only slightly lower than that of the minimum **B**, a difference that is further reduced when free-energy calculations are carried out. The effect of calculating free energies was found to be similar as that considering high dielectric constants: it diminishes the strength of the hydrogen bonds.[119,151] NMR studies on neocarrabiose-4′-sulfate (repeating unit of κ-carrageenan) and on the polysaccharide itself agree with that result: the strong H1′–H3 and H1′–H4 NOE interactions indicate that κ-carrageenan and its disaccharidic repeating unit have at least a significant proportion of the **B** conformer in equilibrium.[153,160] Again, the Tripos[148] force field shows a minimum in the **D** region which is preferred to that in the **A** region, while CHARMM[153] fails to find the **A** minimum. In an experiment with reduced electrostatic interactions, the **D** region becomes the main one with an energy about 4 kcal lower than that of region **B**.[153] The presence of stable conformers in the **D** region has never been detected experimentally. Thus, it can be considered that the prediction of MM3 is closer to reality due to a better parameterization of the exoanomeric effect.[151] However, it can also be explained in terms of steric repulsions, considering that the rudimentary hard-sphere maps of 30 years ago, based solely on steric grounds, yielded results quite similar to those obtained by MM3 calculations (Tables 7.3 and 7.4). The small differences of energy between the **A** and **B** minima depend on subtle details in the parameterization. Figure 7.3 shows the MM3 adiabatic potential energy surfaces of the disaccharidic repeating units of β-, κ-, ι-, and θ-carrageenans. The effects of sulfation on the axial positions (O4 of the β-galactose unit and O2 of the 3,6-anhydrogalactose unit) are quite small (Table 7.5), thus giving a similar shape to the first three maps. On the other hand, sulfation of the equatorial O2 position of the β-galactose unit has the greatest effect by making **B** the global minimum (Table 7.3) and reducing the flexibility of the linkage.[151] Several hydrogen-bonding patterns can help to explain this shift in the stabilities of the minima.[151]

2. Analysis of the Conformation of α-Linked Disaccharides not Containing 3,6-Anhydrogalactose

For the base disaccharide (α-D-Galp-(1→3)-β-D-Galp), earlier hard-sphere calculations also indicated a minimum-energy region around the **A–B** zones.[147] MM3[97,157]

TABLE 7.4
Molecular Modeling Studies on the β-(1→4)-Linked Disaccharidic Repeating Units of Carrageenans

β-Gal	α-Gal	Carrageenans	Ref.	Method	Main Minima Deduced (ϕ,ψ)
–	3,6-An	β	147	HS[a]	−100,115[b] in region **B**, displaced to **A**
–	3,6-An	β	142	HS	−90,110[b] in region **B**
–	3,6-An	β	158	PFOS/MM2	−93,173 (0.00); −105,179 (0.13), both in **A**
–	3,6-An	β	149	CHARMM22	-60,175 (0.0) in **A**; 65,174 (0.2) in **D**
–	3,6-An	β	159	MM3	−81,168 (0.0) in **A**; −91,77 (0.8) in **B**
4S	3,6-An	κ	141	HS	−100,120[b](regions **A–B**)
4S	3,6-An	κ	152	MM[c]	−90,170 in **A**; 40,130 in **D**; -110,100 in **B**
4S	3,6-An	κ	148	Tripos	20,65 (0.0) and −35,−170 (0.3) in unknown regions; .−75,65 (0.3) in **B**
4S	3,6-An	κ	159	MM3	−81,168 (0.0) in **A**; -90,78 (0.9) in **B**
4S	3,6-An /2S	ι	142	HS	−90,110[b] in region **B**
4S	3,6-An /2S	ι	141	HS	−100,120[b] (regions **A–B**)
4S	3,6-An /2S	ι	148	Tripos	−35,-170 (0.0) in ?; 45,155 (0.3) in **D**
4S	3,6-An /2S	ι	159	MM3	−78,170 (0.0) in **A**; −93,76 (1.2) in **B**
4S	3,6-An /2S	ι	154	CHARMM	−64,74 (0.0) in **B**; −76,153 (1.2) in **A**
4S	3,6-An /2S	ι	154	CHARMM, $\varepsilon=80$	−61,176 (0.0) in **A**; 58,97 (0.8) in **D**
2S	3,6-An /2S	θ	159	MM3	−83,169 (0.0) in **A**; −94,77 (1.2) in **B**
–	–	Desulfated λ	155	HS	−80,140[b] in region **A**
–	–	Desulfated λ	159	MM3	−80,142 (0.0) in **A**; −134,112 (2.1) in **B**
2S	2,6-diS	λ	148	Tripos	−100,140 (0.0) in **A**; 30,140 (0.1) in **D**
2S	2,6-diS	λ	159	MM3	−85,147 (0.0) in **A**; −112,85 (3.8) in **B**
4S	2,6-diS	ν	159	MM3	−88,144 (0.0) in **A**; −152,117 (2.8) in **B**
4S	6S	μ	159	MM3	−85,145 (0.0) in **A**; −152,117 (2.9) in **B**

[a]Hard-sphere method.
[b]Estimated to be at the center of the allowed surface.
[c]Parameters for the force field were generated from MNDO calculations.

indicates very similar energies for the **A** and **B** minima (Table 7.3). When the electrostatic interactions are damped (by using a dielectric constant of 80), the **B** minimum becomes the global one. It was found that the use of a high dielectric constant[97] yields results more compatible with the NOE measurements determined for the disaccharide,[161] which is as expected considering that the NOE was measured in an aqueous solution. The gross features of the maps are not changed by sulfation.[157] Figure 7.4 depicts the MM3 conformational maps for the base disaccharide, and for the repeating units of λ, μ, and ν-carrageenans. The allowable surfaces appear more restricted than those for 3,6-anhydrogalactose-containing disaccharides.[151] The position of sulfation of the β-Gal unit exercises a major influence over the relative energies of the main conformers, while sulfation of the α-Gal unit has a less pronounced effect.[157] Table 7.5 shows the effects of sulfation on the relative energies of both

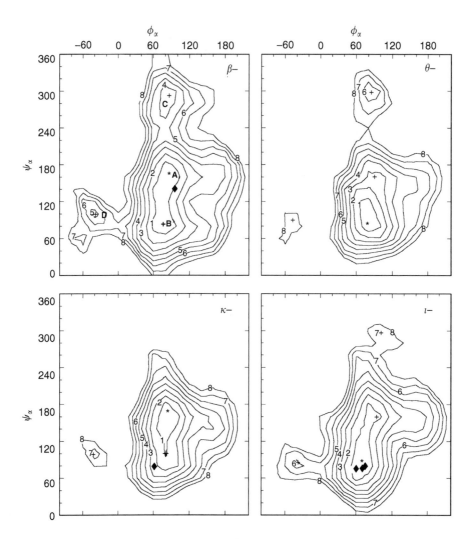

FIGURE 7.3 MM3 conformational maps of the α-linked repeating units of β-, θ-, κ-, and ι-carrageenan (adapted from Reference 151). Isoenergy contour lines are graduated at 1 kcal/mol increments above the global minimum, up to 8 kcal/mol. Regions **A–D** are shown. (From Lamba, D. et al., *Carbohydr. Res.,* 208, 215–230, 1990. With permission.)*, MM3 global minima; +, MM3 local minima; ◆ , published crystal (Reference 145) and fiber (Reference 138, 143 and 144) structures.

conformers: roughly speaking, sulfation of the β-galactose unit on O2 deepens the minimum **B** relative to **A** by about 2 kcal/mol. On the other hand, sulfation of the β-galactose unit on O4 stabilizes minimum **A** with respect to minimum **B** by about 2 kcal/mol. Sulfation of the α-galactose at O6 unit makes this effect even stronger (Table 7.5, Figure 7.4), leading to a difference higher than 5 kcal/mol in the relative

TABLE. 7.5

Effects of Sulfation on the MM3 Energy Differences (kcal/mol) between Minima of Four Base Disaccharides[a]

	3,6-An-α-Gal-(1→3)-β-Gal E_B-E_A	β-Gal-(1→4)-3,6-An-α-Gal E_B-E_A	α-Gal-(1→3)-β-Gal E_B-E_A	β-Gal-(1→4)-α-Gal E_B-E_A
At position 2 of β-Gal	−2.8	+0.4	−1.7	+1.0
At position 4 of β-Gal	+0.8	0.0	+2.1	0.0
β-Gal 2-sulfated				
At position 2 of α-Gal	+0.1	+0.0	−0.4	−1.8[c]
At position 6 of α-Gal			−1.3	−0.7[c]
β-Gal 4-sulfated				
At position 2 of α-Gal	−1.3	+0.2	−0.8[b]	−0.2
At position 6 of α-Gal			−0.5[b]	−0.8

[a]Adapted from Stortz, C.A., *Carbohydr. Res.,* 337, 2311–2323, 2002. With permission.

[b]an additional of 0.5 should be added if both sulfates are present.

[c]an additional of 0.5 should be subtracted if both sulfates are present.

energies of the minima **A** and **B** between the repeating units of λ- and ν-carrageenan, which only differ in the position of sulfation of the β-galactose unit (Table 7.1). This conformational difference can help to explain the marked displacement in the ^{13}C-NMR chemical shifts of the C1 of the α-Gal unit of a ν-carrageenan (98.8 ppm) and of a λ-carrageenan (92.0 ppm), even if both α-linked residues carry the same sulfation pattern.[29] It is known that the chemical shifts of the anomeric carbon have an inverse relationship with the distance between the anomeric proton and the proton of the carbon to which the anomeric carbon is linked.[157,162] The distance H1′–H3 is shorter in **A** conformers. Thus, it is expected that compounds with a larger proportion of **A** conformers (4-sulfated products) will exhibit a larger C1 chemical shift than those with predominant **B** conformation (2-sulfated products), whereas those with nonsulfated β-galactose units should have an intermediate value of δ. The experimental data[29,63] agree with these considerations, although the large displacement (7 ppm) is higher than that expected from these factors alone.[157,162]

3. Analysis of the Conformation of β-Linked Disaccharides Containing 3,6-Anhydrogalactose (Carrabiose)

The first hard-sphere studies indicated a conformational preference for the **A–B** region[142,147] in β-carrageenan models, somehow displaced toward the **B** region (Table 7.4). Further modeling studies[149,158,159] indicated the **A** minimum as the global one. While the PFOS/MM2 approach did not find an important **B** minimum,[158] CHARMM has found one, but with higher energy than other unexpected minima,[149] and MM3 yielded a secondary **B** minimum with a relative energy of 0.8 kcal/mol [159] (Table 7.4). By using MM3 a negligible effect of sulfation was found

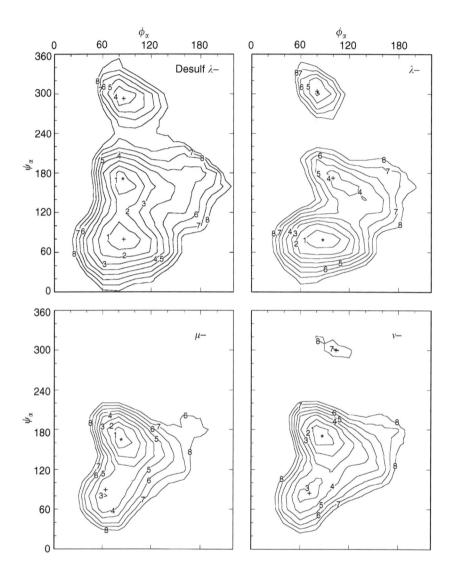

FIGURE 7.4 MM3 conformational maps of the α-linked repeating units of desulfated λ-, λ-, μ-, and ν-carrageenan (corrected from References 97 and 157). Iso-energy contour lines are graduated at 1 kcal/mol increments above the global minimum, up to 8 kcal/mol. *, MM3 global minima; +, MM3 local minima.

(Table 7.5). Figure 7.5 shows the similarity of the maps corresponding to four different carrabiose compounds. The Tripos force field fails to encounter the **A–B** region as the main one (Table 7.4): both in the repeating units of κ- and ι-carrageenan, unidentified regions appear as global minima. However, a **B** minimum appears in both cases with an energy slightly higher.[148] Urbani and co-workers[152]

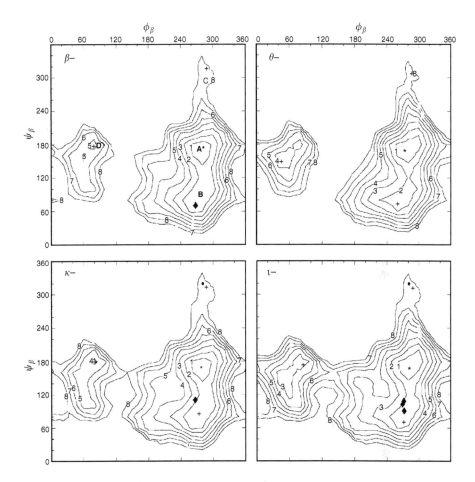

FIGURE 7.5   MM3 conformational maps of the β-linked repeating units of β-, θ-, κ-, and ι-carrageenan (adapted from Stortz, C.A., *Carbohydr. Res.*, 337, 2311–2323, 2002. With permission.) Isoenergy contour lines are graduated at 1 kcal/mol increments above the global minimum, up to 8 kcal/mol. Regions **A–D** are shown. *, MM3 global minima; +, MM3 local minima; ♦, published crystal (from Reference 146) and fiber (from References 138, 143 and 144) structures.

have encountered **A** as the global minimum on a 4′-sulfated carrabiose,[152] whereas Ueda and colleagues[154] found an important **A–B** region using the CHARMM force field for 4′-2-disulfated carrabiose: the **B** minimum is most important at a low dielectric constant (consistent with the X-ray fiber diffraction analysis, see above), while the **A** minimum becomes the global one at higher dielectric constants . MM3 calculations were carried out at an intermediate dielectric constant ($\varepsilon=3$), with results similar to those of CHARMM at a higher value. The further stabilization of minimum **A** under conditions of damped electrostatic/hydrogen-bonding interactions is shown in MM3 by its decreased free energy.[159] The CHARMM force field still seems to overestimate the stability of other minima, particularly those in the **D** region.[154]

4. Analysis of the Conformation of β-Linked Disaccharides not Containing 3,6-Anhydrogalactose

Figure 7.6 shows the MM3 conformational maps of four representatives. All the studies with different sulfation patterns indicate that region **A** is the most important (Table 7.4). For the nonsulfated disaccharide, earlier hard-sphere calculations [155] and

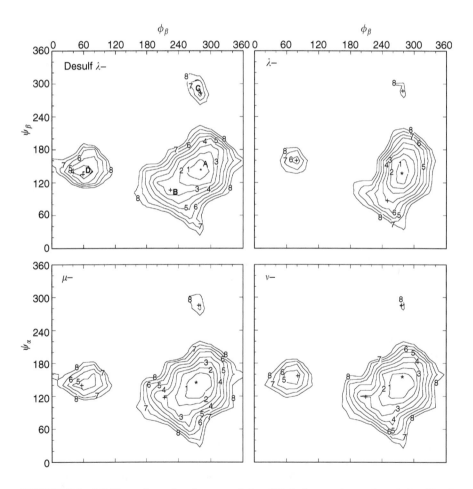

FIGURE 7.6 MM3 conformational maps of the β-linked repeating units of desulfated λ-, λ-, μ-, and ν-carrageenan. Isoenergy contour lines are graduated at 1kcal/mol increments above the global minimum, up to 8 kcal/mol. Regions **A–D** are shown. *, MM3 global minima; +, MM3 local minima. (Adapted from Stortz, C.A., *Carbohydr. Res.*, 337, 2311–2323, 2002. With permission.)

modern MM3 modeling gave rise to a striking coincidence in the global minimum region (Table 7.4). The Tripos force field[148] also gave the same global minimum as MM3 for a λ-carrageenan repeating unit. Sulfation at position 4 has a negligible effect on the potential energy surfaces, whereas sulfation at position 2 of the β-galactose unit stabilizes minimum **A**. This effect is even larger if the other position 2 is also sulfated, as in a λ-carrageenan (Table 7.5). Once more, the lack of 3,6-anhydrogalactose leads to smaller allowable surfaces (cf Figure 7.5 and Figure 7.6).

5. Application to Trisaccharides

Given the usually small variations of the ϕ angle in the low-energy regions of most disaccharides,[67,163] the adiabatic potential energy surfaces of trisaccharides can be depicted by single contour maps for which the energy is plotted against the two ψ glycosidic angles, while the ϕ angles are allowed to relax.[164] For cellotriose and maltotriose, the surfaces are those expected from the maps of the disaccharides containing the same linkages, i.e., each linkage is acting independently of the other one.[164] Six trisaccharides representative of carrageenan structure have also been studied.[165] In five of them, both linkages also behave in an almost independent manner. However, when a sulfate group is present on C2 of a β-galactose reducing end, a new low-energy minimum is produced, originating in a hydrogen bond between the first and third monosaccharidic moieties of the trisaccharide (Figure 7.7).

F. THE THREE-DIMENSIONAL STRUCTURES OF CARRAGEENANS: TERTIARY AND QUATERNARY STRUCTURES

Both κ- and ι-carrageenan form reversibly, temperature-dependent gels. However, ι-carrageenan gives clear and elastic gels that do not have syneresis or hysteresis effects, whereas κ-carrageenan gives hazy and brittle gels that exhibit syneresis and

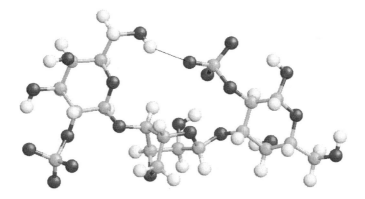

FIGURE 7.7   Molecular structure of a conformer of the trisaccharide β-D-Gal-(1→4)-3, 6-An-α-D-Gal-(1→3)- β-D-Gal 2,2″-bis(sulfate), showing a hydrogen bond between the first and the third monosaccharidic moieties. (From Stortz, C.A. and Cerezo, A.S., *Biopolymers*, 70, 227–239, 2003. With permission.)

hysteresis effects.[67] On the other hand, λ-carrageenan does not gel, but gives viscous solutions; it was postulated that it reduces the brittleness and syneresis effects of κ-carrageenan in their mixtures.[138] Gelling requires that the chains pass to an ordered system with associations of chain segments requiring junction zones, and solvent in the interstices.[166]

Fiber diffraction analysis of ι-carrageenan indicates the presence of a stabilized double-helix. The sulfate groups protrude from the helicoidal structure, while hydrogen bonds are buried inside the double-helix.[138] In order to form a three-dimensional network, cross-linking of different helices through water molecules or cations is needed. Divalent cations like calcium are ideal for this purpose:[138,144] they connect sulfate groups from different helices, and also bring the helices closer together.[144] For κ-carrageenan, a similar double-helix model has been proposed for the fiber.[138] Owing to the absence of the sulfate on the 3,6-anhydrogalactose, junction zones should be different from those in ι-carrageenan. In gels, it has been shown that calcium gives rise to better ι-carrageenan gels, while the best κ-carrageenan gels are promoted by potassium.[144,167]

A great deal of work has been carried out to elucidate the gelling mechanism of carrageenans. Gelation usually involves a temperature-dependent transition from a disordered state (random coil) in the sol to an ordered state as the first step. The second step comprises an association of ordered molecules to form a three-dimensional aggregate, thus requiring the presence of junction zones.[166] By analogy with the fiber diffraction results, it was first postulated that the order–disorder transition corresponds to the formation of double helices. Most of the early studies were performed on especially prepared ι-carrageenan segments, which do not gel but exhibit the order–disorder transitions. This transition was shown to occur by optical rotation measurement at different temperatures,[168] involving apparently a dimerization process,[168,169] which suggests strongly the formation of a double helix. In the second step of the gelation process, it is expected that double helices could cross-link if "kinks" are present in small amounts.[166] According to the "domain model",[170,171] Rees has slightly changed his original view: a limited number of chains are linked through double helices, leading to small clusters or domains that can then associate by cation-mediated aggregation. Some cations were found to promote association in ι- and κ-carrageenan (K^+ or Cs^+, but not Li^+ or Na^+) with a different mechanism: for ι-carrageenan, helices are observed even in the absence of potassium or equivalent cations, whereas for κ-carrageenan, to observe the helices, the specific cations are needed, i.e., the helices may only be stable if aggregated.[171] Some authors have questioned the requirement of double-helix formation as a prerequisite for gelation in the 1980s, and the problem still remains unsettled. Based on the work carried out at different ionic strengths, a conformational change was postulated as the first important step for gel formation.[172] A salt-mediated cross-linking of single helices has been proposed as the second step.[173,174] The role of iodide anion hindering the gelation process has also played a role in trying to explain its mechanism for κ-carrageenan.[174,175] Hydrogen bonding plays an important role in aggregation: it is practically suppressed in the presence of concentrated urea solutions.[167] Another mechanism, limiting the role of potassium in the cross-linking mechanism has also been proposed.[176] The questions of the helical structure involved in the aggregation as a double or single helix, and the role of the added salts are still unsettled.[144,154] Several of the last papers are supportive of the formation of double helices: whereas optical rotation agrees with the expectations of a double helix model.[177] A complete study has determined an agreement of the ordered conformations of κ- and ι-carrageenans with double helices, which tend to aggregate in a ι-carrageenan sample.[178] The results of this work are particularly important, as one of the authors is one of the creators of the single-helix model. Many conflicting studies on κ-carrageenan have arisen: Viebke and co-workers have clearly shown the presence of dimers in the ordered state,[179] while Slootmakers did not find any doubling of the molecular weight, thus concluding that single helices were involved.[180] Ciancia and co-workers encountered evidence of single and double helices, depending on the technique utilized for the measurement of the molecular weight.[181] They found that the dilution of the polymer and the concentration of the added salts might be a clue to explain many of these conflicting results.[179] Modeling of single and double helices of β-carrageenan[149] indicated that the double helix appeared stabilized, especially for van der Waals interactions, thus suggesting that electrostatic forces did not play an important role, and therefore the solvent and other polar molecules should not have a large influence on its stability. Furthermore, molecular dynamics studies preserved the double-helical structure.[149] Another modeling study based its calculations on stable single helices.[148]

Many mixed gels and fibers of carrageenans and other polysaccharides have been reported, and those carrageenans with galactomannans are especially important.[182,183] Fiber diffraction analysis of a mixed carob gum-κ-carrageenan fiber indicates a diffraction pattern similar to that of κ-carrageenan alone, indicating that the galactomannan chains are aggregated to the κ-carrageenan double-helical structures, probably through a region of an unsubstituted mannan backbone.[183] Mixed gels of κ-carrageenan and agarose[184,185] and of κ- and ι-carrageenan[186,187] were also reported. Attempts to find microscopical mixed gels were unsuccessful: all the evidence indicates independent conformational changes and gelling of each product.[184–187]

ACKNOWLEDGMENTS

The author is a Research Member of the National Research Council of Argentina (CONICET-CIHIDECAR). Dr. A.S. Cerezo is gratefully acknowledged for helpful discussions.

REFERENCES

1. Stortz, C.A., Cases, M.R., and Cerezo, A.S., Red seaweed galactans. Methodology for the structural determination of corallinan, a different agaroid, in: Townsend, R.R. and Hotchkiss, Jr, A.T., Eds, *Techniques in Glycobiology*, Marcel Dekker, New York, 1997, pp. 567–593.
2. Takano, R., Hayashi, K., and Hara, S., Variety in backbone modification of red algal galactan sulphates, *Recent Res. Dev. Phytochem.*, 1, 195–201, 1997.
3. Miller, I.J., The chemotaxonomic significance of the water-soluble red algal polysaccharides, *Recent Res. Dev. Phytochem.*, 1, 531–565, 1997.

4.   Usov, A.I., Structural analysis of red seaweed galactans of agar and carrageenan groups, *Food Hydrocolloids*, 12, 301–308, 1998.

5.   Chopin, T., Kerin, B.F., and Mazerolle, R., Gigartinales symposium. Phycocolloid chemistry as a taxonomic indicator of phylogeny in the Gigartinales, Rhodophyceae: a review and current developments using Fourier transform infrared diffuse reflectance spectroscopy, *Phycol. Res.*, 47, 167–188, 1999.

6.   Stortz, C.A. and Cerezo, A.S., Novel findings in carrageenans, agaroids and "hybrid" red seaweed galactans, *Curr. Top. Phytochem.*, 4, 121–134, 2000.

7.   Schmidt, C., Ueber Pflanzenschleim und Bassorin, *Ann*, 51, 29–62, 1844.

8.   Haedicke, J., Bauer, R.W., and Tollens, B., Ueber Galaktose aus Carragheen-Moos, *Ann*, 238, 302–318, 1887.

9.   Buchanan, J., Percival, E.E., and Percival, E.G.V., The polysaccharides of Carragheen moss (*Chondrus crispus*). Part I. The linkage of the D-galactose residues and the ethereal sulphate, *J. Chem. Soc.*, 51–54, 1943.

10. Hassid, W.Z., The isolation of a sodium sulfuric acid ester of galactan from *Irideae laminarioides* (Rhodophyceae), *J. Am. Chem. Soc.*, 55, 4163–4167, 1933.

11. Hassid, W.Z., The structure of sodium sulfuric acid ester of galactan from *Irideae laminarioides* (Rhodophyceae), *J. Am. Chem. Soc.*, 57, 2046–2050, 1935.

12. Dewar, T.E. and Percival, E.G.V., The polysaccharides of carragheen. Part II. The *Gigartina stellata* polysaccharides. *J. Chem. Soc.*, 1622–1626, 1947.

13. Smith, D.B. and Cook, W.H., Fractionation of carrageenin, *Arch. Biochem. Biophys.*, 45, 232–233, 1953.

14. O'Neill, A.N., 3,6-anhydro-D-galactose as a constituent of κ-carrageenin, *J. Am. Chem. Soc.*, 77, 2837–2839, 1955.

15. O'Neill, A.N., Derivatives of 4-*O*-β-D-galactopyranosyl-3,6-anhydro-D-galactose from κ-carrageenin, *J. Am. Chem. Soc.*, 77, 6324–6326, 1955.

16. Rees, D.A., The carrageenan system of polysaccharides. Part I. The relation between the κ-and λ-components, *J. Chem. Soc.*, 1821–1832, 1963.

17. Dolan, T.C.S. and Rees, D.A., The carrageenans. Part II. The positions of the glycosidic linkages and sulphate esters in λ-carrageenan, *J. Chem. Soc.*, 3534–3539, 1965.

18. Lawson, C.J. and Rees, D.A., Carrageenans. Part VI. Reinvestigation of the acetolysis products of λ-carrageenan. Revision of the structure of α-1,3-galactotriose and a further example of the reverse specificities of glycoside hydrolysis and acetolysis, *J. Chem. Soc.*, C, 1301–1304, 1968.

19. Anderson, N.S., Dolan, T.C.S., Lawson, C.J., Penman, A., and Rees, D.A., Carrageenans. Part V. The masked repeating structures of λ- and μ-carrageenans, *Carbohydr. Res.*, 7, 468–473, 1968.

20. Anderson, N.S., Dolan, T.C.S., and Rees, D.A., Carrageenans, Part VII. Polysaccharides from *Eucheuma spinosum* and *Eucheuma cottonii*. Covalent structure of iota-carrageenan, *J. Chem. Soc. Perkin Trans.* I, 2173–2176, 1973.

21. Pernas, A.J., Smidsrød, O., Larsen, B., and Haug, A., Chemical heterogeneity of carrageenans as shown by fractional precipitation with potassium chloride, *Acta. Chem. Scand.*, 21, 98–110, 1967.

22. Chen, L.C.M., McLachlan, J., Neish, A.C., and Shacklock, P.F., The ratio of kappa-to lambda-carrageenan in nuclear phases of the rhodophycean algae, *Chondrus crispus* and *Gigartina stellata, J. Mar. Biol. Assoc. UK*, 53, 11–16, 1973.

23. McCandless, E.L., Craigie, J.S., and Walter, J.A., Carrageenans in the gametophytic and sporophytic stages of *Chondrus crispus, Planta*, 112, 201–212, 1973.

24. Pickmere, S.E., Parsons, M.J., and Bailey, RW., Composition of *Gigartina* carrageenan in relation to sporophyte and gametophyte stages of the life cycle, *Phytochemistry*, 12, 2441–2444, 1973.

25. McCandless, E.L. and Craigie, J.S., Sulfated polysaccharides in red and brown algae, *Ann. Rev. Plant Physiol.*, 30, 41–53, 1979.

26. Bellion, C., Hamer, G.K., and Yaphe, W., Analysis of kappa–iota hybrid carrageenans with kappa-carrageenase, iota-carrageenase and ^{13}C-NMR, *Proc. Int. Seaweed Symp.*, 10, 379–384, 1981.

27. Ciancia, M., Matulewicz, M.C., Finch, P., and Cerezo, A.S., Determination of the structures of cystocarpic carrageenans from *Gigartina skottsbergii* by methylation analysis and NMR spectroscopy, *Carbohydr. Res.*, 238, 241–248, 1993.

28. Stortz, C.A. and Cerezo, A.S., The systems of carrageenans from cystocarpic and tetrasporic stages from *Iridaea undulosa*: fractionation with potassium chloride and methylation analysis of the fractions, *Carbohydr. Res.*, 242, 217–227, 1993.

29. Stortz, C.A., Bacon, B.E., Cherniak, R., and Cerezo, A.S., High-field NMR spectroscopy of cystocarpic and tetrasporic carrageenans from *Iridaea undulosa*, *Carbohydr. Res.*, 261, 317–326, 1994.

30. Craigie, J.S., Cell walls, *Biology of the Red Algae*, in: Cole, K.M. and Sheath, R.G., Eds., Cambridge University Press, Cambridge, 1990, pp. 221–257.

31. Stortz, C.A. and Cerezo, A.S., The λ-components of the "intermediate" fractions of the carrageenan from *Iridaea undulosa*, *Carbohydr. Res.*, 172, 139–146, 1988.

32. Matulewicz, M.C., Ciancia, M., Noseda, M.D., and Cerezo, A.S., The carrageenan system from tetrasporic and cystocarpic stages of *Gigartina skottsbergii*, *Phytochemistry*, 28, 2937–2941, 1989.

33. Greer, C.W. and Yaphe, W., Characterization of hybrid (beta–kappa–gamma) carrageenan from *Eucheuma gelatinae* J. Agardh. (Rhodophyta, Solieriaceae) using carrageenases, infrared and ^{13}C-nuclear magnetic resonance spectroscopy, *Bot. Mar.* 28, 473–478, 1984.

34. Knutsen, S.H., Myslabodski, D.E., Larsen, B., and Usov, A.I., A modified system of nomenclature for red algal galactans, *Bot. Mar.*, 37, 163–169, 1994.

35. Liao, M.L., Kraft, G.T., Munro, S.L.A., Craik, D.J., and Bacic, A., Beta/kappa-carrageenans as evidence for continued separation of the families Dicranemataceae and Sarcodiaceae (Gigartinales, Rhodophyta), *J. Phycol*, 29, 833–844, 1993.

36. Yarotskii, S.V., Shashkov, A.S., and Usov, A.I., Polysaccharides of algae. XXV. Use of ^{13}C NMR spectroscopy for analyzing the structures of polysaccharides of the κ-carrageenan type, *Bioorg. Khim*, 4, 745–751, 1978 (in Russian).

37. Mollion, J., Morvan, H., Bellanger, F., and Moreau, S., ^{13}C NMR study of heterogeneity in the carrageenan system from *Rissoella verrucosa*, *Phytochemistry*, 27, 2023–2026, 1988.

38. Chopin, T., Hanisak, M.D., and Craigie, J.S., Carrageenans from *Kallymenia westii* (Rhodophyceae) with a review of the phycocolloids produced by the Cryptonemiales, *Bot. Mar.*, 37, 433–444, 1994.

39. Craigie, J.S. and Rivero-Carro, H., Agarocolloids from carrageenophytes. *Abstract of the XIVth International Seaweed Symposium*, Brest, 1992, p. 71.

40. Ciancia, M., Matulewicz, M.C., and Cerezo, A.S., L-Galactose-containing galactans from the carrageenophyte *Gigartina skottsbergii*, *Phytochemistry*, 34, 1541–1543, 1993.

41. Ciancia, M., Matulewicz, M.C., and Cerezo, A.S., A L-galactose-containing carrageenan from cystocarpic *Gigartina skottsbergii*, *Phytochemistry*, 45, 1009–1013, 1997.

42. Flores, M.L., Cerezo, A.S., and Stortz, C.A., Alkaline treatment of the carrageenans from the cystocarpic stage of the red seaweed *Iridaea undulosa, J. Argent. Chem. Soc.*, 90, 65–76, 2002.

43. Stortz, C.A., Cases, M.R., and Cerezo, A.S., The system of agaroids and carrageenans from the soluble fraction of the tetrasporic stage of the red seaweed *Iridaea undulosa, Carbohydr. Polym.*, 34, 61–65, 1997.

44. Falshaw, R. and Furneaux, R.H., The structural analysis of disaccharides from red algal galactans by methylation and reductive partial-hydrolysis, *Carbohydr. Res.*, 269, 183–189, 1995.

45. Cases, M.R., Cerezo, A.S., and Stortz, C.A., Separation and quantitation of enantiomeric galactoses and their mono-*O*-methylethers as their diastereomeric acetylated 1-deoxy-1-(2-hydroxypropylamino)alditols, *Carbohydr. Res.*, 269, 333–341, 1995.

46. Navarro, D.A. and Stortz, C.A., Determination of the configuration of 3,6-anhydrogalactose and cyclizable α-galactose 6-sulfate units in red seaweed galactans, *Carbohydr. Res.*, 338, 2111–2118, 2003.

47. Estevez, J.M., Ciancia, M., and Cerezo, A.S., DL-Galactan hybrids and agarans from gametophytes of the red seaweed *Gymnogongrus torulosus, Carbohydr. Res.*, 331, 27–41, 2001.

48. Estevez, J.M., Ciancia, M., and Cerezo, A.S., The system of low-molecular-weight carrageenans and agaroids from the room-temperature-extracted fraction of *Kappaphycus alvarezii, Carbohydr. Res.*, 325, 287–299, 2000.

49. Takano, R., Shiomoto, K., Kamei, K., Hara, S., and Hirase, S., Occurrence of carrageenan structure in an agar from the red seaweed *Digenea simplex* (Wulfen) C. Agardh (Rhodomelaceae, Ceramiales) with a short review of carrageenan-agarocolloid hybrid in the Florideophycidae, *Bot. Mar.*, 46, 142–150, 2003.

50. Flores, M.L., Stortz, C.A., Rodríguez, M.C., and Cerezo, A.S., Studies on the skeletal cell wall and cuticle of the cystocarpic stage of the red seaweed *Iridaea undulosa* Bory, *Bot. Mar.*, 40, 411–419, 1997.

51. Flores, M.L., Stortz, C.A., and Cerezo,. A.S., Studies on the skeletal cell wall of the cystocarpic stage of the red seaweed *Iridaea undulosa* B. Part II. Fractionation of the cell wall and methylation analysis of the inner core-fibrillar polysaccharides, *Int. J. Biol. Macromol.*, 27, 21–27, 2000.

52. Takano, R., Iwane-Sakata, H., Hayashi, K., Hara, S., and Hirase, S., Concurrence of agaroid and carrageenan chains in funoran from the red seaweed *Gloiopeltis furcata* Post. et Ruprecht (Cryptonemiales, Rhodophyta), *Carbohydr. Polym.*, 35, 81–87, 1998.

53. Takano, R., Yokoi, T., Kamei, K., Hara, S., and Hirase, S., Coexistence of agaroid and carrageenan structures in a polysaccharide from the red seaweed *Rhodomela larix* (Turner) C.Ag, *Bot. Mar.*, 42, 183–188, 1999.

54. Takano, R., Nose, Y., Hayashi, K., Hara, S., and Hirase, S., Agarose–carrageenan hybrid polysaccharide from *Lomentaria catenata, Phytochemistry*, 37, 1615–1619, 1994.

55. Miller, I.J., Falshaw, R., and Furneaux, R.H., A polysaccharide fraction from the red seaweed *Champia novae-zelandiae* (Rhodymeniales, Rhodophyta), *Hydrobiologia*, 326/327, 505–509, 1996.

56. Falshaw, R., Furneaux, R.H., and Miller, I.J., The backbone structure of the sulfated galactan from *Plocamium costatum* (C. Agardh) Hook. f. et Harv. (Plocamiaceae, Rhodophyta), *Bot. Mar.*, 42, 431–435, 1999.

57. Amimi, A., Mouradi, A., Givernaud, T., Chiadmi, N., and Lahaye, M., Structural analysis of *Gigartina pistillata* carrageenans (Gigartinaceae, Rhodophyta), *Carbohydr. Res.*, 333, 271–279, 2001.

58. Chiovitti, A., Bacic, A., Craik, D.J., Munro, S.L.A., Kraft, G.T., and Liao, M.L., Cell-wall polysaccharides from Australian red algae of the family Solieriaceae (Gigartinales, Rhodophyta): novel, highly pyruvated carrageenans from the genus *Callophycus, Carbohydr. Res.*, 299, 229–243, 1997.

59. Miller, I.J., The structure of a pyruvylated carrageenan extracted from *Stenogramme interrupta* as determined by ^{13}C NMR spectroscopy, *Bot. Mar.*, 41, 305–315, 1998.

60. Chiovitti, A., Liao, M.L., Kraft, G.T., Munro, S.L.A., and Craik, D.J., Bacic, A Cell-wall polysaccharides from Australian red algae of the family Solieriaceae (Gigartinales, Rhodophyta): highly methylated carrageenans from the genus *Rhabdonia, Bot. Mar.*, 39, 47–59, 1996.

61. Chiovitti, A., Bacic, A., Craik, D.J., Munro, S.L.A., Kraft, G.T., and Liao, M.L., Carrageenans with complex substitution patterns from red algae of the genus *Erythroclonium, Carbohydr. Res.*, 305, 243–252, 1998.

62. Chiovitti, A., Kraft, G.T., Bacic, A., Craik, D.J., Munro, S.L.A., and Liao, M.L., Carrageenans from Australian representatives of the family Cystocloniaceae (Gigartinales, Rhodophyta), with description of *Calliblepharis celatospora* sp. nov., and transfer of *Austroclonium* to the family Areschougiaceae, *J. Phycol.*, 34, 515–535, 1998.

63. Stortz, C.A. and Cerezo, A.S., The ^{13}C NMR spectroscopy of carrageenans: calculations of chemical shifts and computer-aided structural determination, *Carbohydr. Polym.*, 18, 237–242, 1992.

64. Miller, I.J. and Blunt, J.W., New ^{13}C NMR methods for determining the structure of algal polysaccharides. Part 1. The effect of substitution on the chemical shifts of simple diad galactans, *Bot. Mar.*, 43, 239–250, 2000.

65. Miller, I.J. and Blunt, J.W., New ^{13}C NMR methods for determining the structure of algal polysaccharides. Part 2. Galactans consisting of mixed diads, *Bot. Mar.*, 43, 251–261, 2000.

66. van de Velde, F., Peppelman, H.A., Rollema, H.S., and Tromp, R.H., On the structure of κ/ι-hybrid carrageenans, *Carbohydr. Res.*, 331, 271–283, 2001.

67. Rao, V.S.R., Qasba, P.K., Balaji, P.V., and Chandrasekaran, R., *Conformation of Carbohydrates*, Harwood Academic Publishers, Amsterdam, 1998.

68. French, A.D. and Brady, J.W., Computer modeling of carbohydrates, *ACS Symp. Ser.*, 430, 1–19, 1990.

69. Poppe, L. and van Halbeek, H., Selective, inverse-detected measurements of long-range ^{13}C,^{1}H coupling constants. Application to a disaccharide, *J. Magn. Reson.*, 93, 214–217, 1991.

70. Poppe, L. Stuike-Prill, R., Meyer, B., and van Halbeek, H., The solution conformation of sialyl-$\alpha(2\rightarrow6)$-lactose studied by modern NMR techniques and Monte Carlo simulations, *J. Biomol. NMR*, 2, 109–136, 1992.

71. van Halbeek, H., NMR developments in structural studies of carbohydrates and their complexes, *Curr. Opin. Struct. Biol.*, 4, 697–709, 1994.

72. Jardetzky, O., On the nature of molecular conformations inferred from high resolution NMR, *Biochim. Biophys. Acta*, 621, 227–232, 1980.

73. Rees, D.A., Conformational analysis of polysaccharides. Part V. The characterization of linkage conformations (chain conformations) by optical rotation at a single wavelength. Evidence for distortion of cyclohexa-amylose in aqueous solution. Optical rotation and amylose conformation, *J. Chem. Soc.*, B, 877–884, 1970.

74. Rees, D.A. and Thom, D., Polysaccharide conformation. Part 10. Solvent and temperature effects on the optical rotation and conformation of model carbohydrates, *J. Chem. Soc. Perkin Trans. II,* 191–201, 1977.

75. Stevens, E.S. and Sathyanarayana, B.K., A semiempirical theory of the optical activity of saccharides, *Carbohydr. Res.*, 166, 181–193, 1987.
76. Stevens, E.S., Solution conformation of maltose from optical rotation: a procedure for evaluating carbohydrate force fields, *Biopolymers*, 32, 1571–1579, 1992.
77. Barrows, S.E., Dulles, F.J., Cramer, C.J., French, A.D., and Truhlar, D.G., Relative stability of alternative chair forms and hydroxymethyl conformations of β-D-glucopyranose, *Carbohydr. Res.*, 276, 219–251, 1995.
78. Polavarapu, P.L. and Ewig, C.S., *Ab initio* computed molecular structures and energies of the conformers of glucose, *J. Comput. Chem.*, 13, 1255–1261, 1992.
79. Csonka, G.I., Eliás, K., Kolossvary, I., Sosa, C.P., and Csizmadia, I.G., Theoretical study of alternative ring forms of α-L-fucopyranose, *J. Phys. Chem. A*, 102, 1219–1229, 1998.
80. Csonka, G.I., Eliás, K., and Csizmadia, I.G., Relative stability of $^1C_4$ and $^4C_1$ chair forms of β-D-glucose: a density functional study, *Chem. Phys. Lett.*, 257, 49–60, 1996.
81. Tvaroška, I., Taravel, F.R., Utille, J.P., and Carver, J.P., Quantum mechanical and NMR spectroscopy studies on the conformations of the hydroxymethyl and methoxymethyl groups in aldohexosides, *Carbohydr. Res.*, 337, 353–367, 2002.
82. Csonka, G.I., Proper basis sets for quantum mechanical studies of potential energy surfaces of carbohydrates, *J. Mol. Struct. (Theochem)*, 584, 1–4, 2002.
83. Dewar, M.J.S., Zoebisch, E.G., Healy, E.F., and Stewart, J.J.P., AM1: a new general purpose quantum mechanical molecular model, *J. Am. Chem. Soc.*, 107, 3902–3909, 1985.
84. Stewart, J.J.P., Optimization of parameters for semiempirical methods. I. Method. *J. Comput. Chem.*, 10, 209–220, 1989.
85. Cramer, C.J. and Truhlar, D.G., Quantum mechanical conformational analysis of glucose in aqueous solution, *J. Am. Chem. Soc.*, 115, 5745–5753, 1993.
86. Brewster, M.E., Huang, M., Pop, E., Pitha, J., Dewar, M.J.S., Kaminski, J.J., and Bodor, N., An AM1 molecular orbital study of α-D-glucopyranose and β-maltose: evaluation and implications, *Carbohydr. Res.*, 242, 53–67, 1993.
87. Zuccarello, F. and Buemi, G., A theoretical study of D-glucose, D-galactose, and parent molecules: solvent effect on conformational stabilities and rotational motions of exocyclic groups, *Carbohydr. Res.*, 273, 129–145, 1995.
88. Stortz, C.A., Full conformational search of monosaccharides using semiempirical and classical methods: application to α-D-galactopyranose, *An. Asoc. Quim. Argent.*, 86, 94–103, 1998.
89. Csonka, G.I. and Angyán, J.G., The origin of the problems with PM3 core repulsion function, *J. Mol. Struct. (Theochem)*, 393, 31–38, 1997.
90. Liu, Q. and Brady, J.W., Anisotropic solvent structuring in aqueous sugar solutions. *J. Am. Chem. Soc.*, 118, 12276–12286, 1996.
91. Madsen, L.J., Ha, S.N., Tran, V.H., and Brady, J.W., Molecular dynamics simulations of carbohydrates and their solvation, *ACS Symp. Ser.*, 430, 69–90, 1990.
92. Dunfield, L.G. and Whittington, S.G., A Monte Carlo investigation of the conformational free energies of the aldohexopyranoses, *J. Chem. Soc. Perkin Trans. II*, 654–658, 1976.
93. Serianni, A.S. and Barker, R., [^{13}C]-Enriched tetroses and tetrofuranosides: an evaluation of the relationship between NMR parameters and furanosyl ring conformations, *J. Org. Chem.*, 49, 3292–3300, 1984.
94. Bock, K. and Duus, J.Ø., A conformational study of hydroxymethyl groups in carbohydrates investigated by ^{1}H NMR spectroscopy, *J. Carbohydr. Chem.*, 13, 513–543, 1994.

95. Dowd, M.K., French, A.D., and Reilly P.J., Modeling of aldopyranosyl ring puckering with MM3(92), *Carbohydr. Res.*, 264, 1–19, 1994.

96. French, A.D. and Dowd, M.K., Exploration of disaccharide conformations by molecular mechanics, *J. Mol. Struct. (Theochem)*, 286, 183–201, 1993.

97. Stortz, C.A., Disaccharide conformational maps: how adiabatic is an adiabatic map? *Carbohydr. Res.*, 322, 77–86, 1999.

98. Rees, D.A. and Skerrett, R.J., Conformational analysis of cellobiose, cellulose, and xylan, *Carbohydr. Res.*, 7, 334–348, 1968.

99. Rees, D.A. and Scott, W.E., Polysaccharide conformation. Part VI. Computer model-building for linear and branched pyranoglycans. Correlations with biological function. Preliminary assessment of inter-residue forces in aqueous solution. Further interpretation of optical rotation in terms of chain conformation, *J. Chem. Soc.* B, 469–479, 1971.

100. Sathyanarayana, B.K. and Rao, V.S.R., Conformational studies of β-glucans, *Biopolymers*, 10, 1605–1615, 1971.

101. Sathyanarayana, B.K. and Rao, V.S.R., Conformational studies of α-glucans, *Biopolymers*, 11, 1379–1394, 1972.

102. Thøgersen, H., Lemieux, R.U., Bock, K., and Meyer, B., Further justification of the exo-anomeric effect. Conformational analysis based on nuclear magnetic resonance of oligosaccharides, *Can. J. Chem.*, 60, 44–57, 1982.

103. Tvaroška, I. and Pérez, S., Conformational-energy calculations for oligosaccharides. A comparison of methods and a strategy for calculation, *Carbohydr. Res.*, 149, 389–410, 1986.

104. Melberg, S. and Rasmussen, K., Conformations of disaccharides by empirical force-field calculations. Part I, β-maltose, *Carbohydr. Res.*, 69, 27–38, 1979.

105. Melberg, S. and Rasmussen, K., Conformations of disaccharides by empirical force-field calculations. Part II, β-cellobiose, *Carbohydr. Res.*, 71, 25–34, 1979.

106. French, A.D., Rigid- and relaxed-residue conformational analyses of cellobiose using the computer program MM2, *Biopolymers*, 27, 1519–1525, 1988.

107. Ha, S.N., Madsen, L.J., and Brady, J.W., Conformational analysis and molecular dynamics simulations of maltose, *Biopolymers*, 27, 1927–1952, 1988.

108. French, A.D., Comparison of rigid and relaxed conformational maps for cellobiose and maltose, *Carbohydr. Res.*, 188, 206–211, 1989.

109. Tran, V., Bulèon, A., Imberty, A., and Pérez, S., Relaxed potential energy surfaces of maltose, *Biopolymers*, 28, 679–690, 1989.

110. Imberty, A., Tran, V., and Pérez, S., Relaxed potential energy surfaces of *N*-linked oligosaccharides: the mannose-α-(1→3)-mannose case, *J. Comput. Chem.*, 11, 205–216, 1989.

111. Allinger, N.L., Yuh, Y.H., and Lii, J.H., Molecular mechanics. The MM3 force field for hydrocarbons, *J. Am. Chem. Soc.*, 111, 8551–8566, 1989.

112. Koča, J., Pérez, S. and Imberty, A., Conformational analysis and flexibility of carbohydrates using the CICADA approach with MM3, *J. Comput. Chem.*, 16, 296–310, 1995.

113. Momany, F.A. and Willett, J.L., Computational studies on carbohydrates: *in vacuo* studies using a revised AMBER force field, AMB99C, designed for α-(1→4) linkages, *Carbohydr. Res.*, 236, 194–209, 2000.

114. French, A.D., Schäfer, L., and Newton, S.Q., Overlapping anomeric effects in a sucrose analogue, *Carbohydr. Res.*, 239, 51–60, 1993.

115. Dowd, M.K., Zeng, J., French, A.D., and Reilly, P.J., Conformational analysis of the anomeric forms of kojibiose, nigerose, and maltose using MM3, *Carbohydr. Res.*, 230, 223–244, 1992.

116.   Dowd, M.K., French, A.D., and Reilly, P.J., Conformational analysis of the anomeric forms of sophorose, laminaribiose, and cellobiose using MM3, *Carbohydr. Res.*, 233, 15–34, 1992.

117.   Dowd, M.K., Reilly, P.J., and French, A.D., Conformational analysis of trehalose disaccharides and analogues using MM3, *J. Comput. Chem.*, 13, 102–114, 1992.

118.   Dowd, M.K., French, A.D., and Reilly, P.J., Molecular mechanics modeling of α-(1→2), α(1→3), and α-(1→6)-linked mannosyl disaccharides with MM3(92), *J. Carbohydr. Chem.*, 14, 589–600, 1995.

119.   Stortz, C.A., and Cerezo, A.S., Potential energy surfaces of α-(1→3)-linked disaccharides calculated with the MM3 force-field, *J. Carbohydr. Chem.*, 21, 355–371, 2002.

120.   Stortz, C.A., and Cerezo, A.S., 2D- and 3D-potential energy surfaces of β-(1→3)-linked disaccharides calculated with the MM3 force-field, *J. Carbohydr. Chem.*, 22, 217–239, 2003.

121.   Tran, V.H. and Brady, J.W., Disaccharide conformational flexibility. I. An adiabatic potential energy map for sucrose, *Biopolymers*, 29, 961–976, 1990.

122.   Engelsen, S.B., Pérez, S., Braccini, I., and Hervé du Penhoat, C., Internal motions of carbohydrates as probed by comparative molecular modeling and nuclear magnetic resonance of ethyl β-lactoside, *J. Comput. Chem.*, 16, 1066–1119, 1995.

123.   Engelsen, S.B. and Rasmussen, K., β-Lactose in the view of a CFF-optimized force field, *J. Carbohydr. Chem.*, 16, 773–788, 1997.

124.   Tvaroška, I., Kozár, T., and Hricovíni, M., Oligosaccharides in solution. Conformational analysis by NMR spectroscopy and calculation, *ACS Symp. Ser. A*, 430, 162–176, 1990.

125.   von der Lieth, C.W., Kozár, T., and Hull, W.E., A (critical) survey of modelling protocols used to explore the conformational space of oligosaccharides, *J. Mol. Struct. (Theochem)*, 395/396, 225–244, 1997.

126.   Homans, S.W. and Forster, M., Application of restrained minimization, simulated annealing and molecular dynamics simulations for the conformational analysis of oligosaccharides, *Glycobiology*, 2, 143–151, 1992.

127.   Ott, K.H. and Meyer, B., Molecular dynamics simulations of maltose in water, *Carbohydr. Res.*, 281, 11–34, 1996.

128.   Asensio, J.L., Martin-Pastor, M., and Jimenez-Barbero J., The use of CVFF and CFF91 force fields in conformational analysis of carbohydrate molecules. Comparison with AMBER molecular mechanics and dynamics calculations for methyl α-lactoside, *Int. J. Biol. Macromol.*, 17, 136–148, 1995.

129.   Cheetham, N.W.H., Dasgupta, P., and Ball, G.E., NMR and modelling studies of disaccharide conformation, *Carbohydr. Res.*, 338, 955–962, 2003.

130.   Bernardi, A., Raimondi, L., and Zanferrari, D., Conformational analysis of saccharides with Monte Carlo/stochastic dynamics simulations, *J. Mol. Struct. (Theochem)*, 395/396, 361–373, 1997.

131.   Engelsen, S.B., Koca, J., Braccini, I., Hervé du Penhoat, C., and Pérez, S., Travelling on the potential energy surfaces of carbohydrates: comparative application of an exhaustive systematic conformational search with an heuristic search, *Carbohydr. Res.*, 276, 1–29, 1995.

132.   van Alsenoy, C., French, A.D., Cao, M., Newton, S.Q., and Schäfer, L., *Ab initio*-MIA and molecular mechanics studies of the distorted sucrose linkage of raffinose, *J. Am. Chem. Soc.*, 116, 9590–9595, 1994.

133.   Bose, B., Zhao, S., Stenutz, R., Cloran, F., Bondo, P.B., Bondo, G., Hertz, B., Carmichael, I., and Serianni, A.S., Three-bond C-O-C-C spin-coupling constants in carbohydrates: development of a Karplus relationship, *J. Am. Chem. Soc.*, 120, 11158–11173, 1998.

134. Momany, F.A. and Willett, J.L., Computational studies on carbohydrates: I. Density functional *ab initio* geometry optimization on maltose conformations, *J. Comput. Chem.,* 21, 1204–1219, 2000.

135. Strati, G.L., Willett, J.L., and Momany, F.A., *Ab initio* computational study of β-cellobiose conformers using B3LYP/6-311+G**, *Carbohydr. Res.,* 337, 1833–1849, 2002.

136. French, A.D., Kelterer, A.M., Johnson, G.P., Dowd, M.K., and Cramer, C.J., Constructing and evaluating energy surfaces of crystalline disaccharides, *J. Mol. Graph. Model.,* 18, 95–107, 2000.

137. French, A.D., Kelterer, A.M., Johnson, G.P., Dowd, M.K., and Cramer, C.J., HF/6-31G* energy surfaces for disaccharide analogs, *J. Comput. Chem.,* 22, 65–78, 2001.

138. Millane, R.P., Nzewi, E.U., and Arnott, S., Molecular structures of carrageenans as determined by X-ray fiber diffraction, in, *Frontiers in Carbohydrate Research,* Millane, R.P., BeMiller, J.N., and Chandrasekaran, R., Eds., Elsevier, London, 1989, pp. 104–131.

139. Bayley, S.T., X-ray and infrared studies on carrageenin, *Biochim. Biophys. Acta,* 17, 194–205, 1955.

140. Anderson, N.S., Campbell, J.W., Harding, M.M., Rees, D.A., and Samuel. J.W.B., X-ray diffraction studies of polysaccharide sulphates: double helix models for κ- and ι-carrageenans, *J. Mol. Biol.,* 45, 85–99, 1969.

141. Millane, R.P., Chandrasekaran, R., Arnott, S., and Dea, I.C.M., The molecular structure of kappa-carrageenan and comparison with iota-carrageenan, *Carbohydr. Res.,* 182, 1–17, 1988.

142. Arnott, S., Scott, W.E., Rees, D.A., and McNab, C.G.A., ι-Carrageenan: molecular structure and packing of polysaccharide double helices in oriented fibres of divalent cation salts, *J. Mol. Biol.,* 90, 253–267, 1974.

143. Janaswamy, S. and Chandrasekaran, R., Three-dimensional structure of the sodium salt of iota-carrageenan, *Carbohydr. Res.,* 335, 181–194, 2001.

144. Janaswamy, S. and Chandrasekaran, R., Effect of calcium ions on the organization of iota-carrageenan helices: an X-ray investigation, *Carbohydr. Res.,* 337, 523–535, 2002.

145. Lamba, D., Segre, A.L., Glover, S., Mackie, W., Sheldrick, B., and Pérez, S., Molecular structure of 3-*O*-(3,6-anhydro-α-D-galactopyranosyl)-β-D-galactopyranose (neocarrabiose) in the solid state and in solution: an investigation by X-ray crystallography, n.m.r. spectroscopy, and molecular mechanics calculations, *Carbohydr. Res.,* 208, 215–230, 1990.

146. Lamba, D., Burden, C., Mackie, W., and Sheldrick, B., The crystal and molecular structure of hexa-*O*-acetylcarrabiose dimethyl acetal, *Carbohydr. Res.,* 155, 11–17, 1986.

147. Rees, D.A., Conformational analysis of polysaccharides. Part II. Alternating copolymers of the agar–carrageenan–chondroitin type by model building in the computer with calculation of helical parameters, *J. Chem. Soc. B,* 217–226, 1969.

148. Le Questel, J.Y., Cros, S., Mackie, W., and Pérez, S., Computer modelling of sulfated carbohydrates: applications to carrageenans, *Int. J. Biol. Macromol.,* 17, 161–175, 1995.

149. Ueda, K., Ochiai, H., Imamura, A., and Nakagawa, S., An investigation of the conformation of β-carrageenan by molecular mechanics and molecular dynamics simulations, *Bull. Chem. Soc. Jpn.,* 68, 95–106, 1995.

150. Ueda, K. and Brady, J.W., The effect of hydration upon the conformation and dynamics of neocarrabiose, a repeat unit of β-carrageenan, *Biopolymers,* 38, 461–469, 1996.

151. Stortz, C.A. and Cerezo, A.S., Conformational analysis of neocarrabiose and its sulfated and/or pyruvylated derivatives using the MM3 force-field, *J. Carbohydr. Chem.,* 19, 1115–1130, 2000.

152. Urbani, R., Di Blas, A., and Cesàro, A., Conformational features of carrabiose polymers: I. Configurational statistics of κ-carrageenan, *Int. J. Biol. Macromol.*, 15, 24–29, 1993.

153. Ueda, K., Saiki, M., and Brady, J.W., Molecular dynamics simulation and NMR study of aqueous neocarrabiose 4'-sulfate, a building block of κ-carrageenan, *J. Phys. Chem., B*, 105, 8629–8638, 2001.

154. Ueda, K., Iwama, K., and Nakayama H., Molecular dynamics simulation of neocarrabiose 2,4'-bis(sulfate) and carrabiose 4',2-bis(sulfate) as building blocks of ι-carrageenan in water, *Bull. Chem. Soc. Jpn.*, 74, 2269–2277, 2001.

155. Stortz, C.A. and Cerezo, A.S., Use of a general purpose force-field (MM2) for the conformational analysis of the disaccharide α-D-galactopyranosyl-(1→3)-β-D-galactopyranose, *J. Carbohydr. Chem.*, 13, 235–247,1994.

156. Stortz, C.A. and Cerezo, A.S., Conformational analysis of sulfated disaccharides using the MM2 force-field, *An. Asoc. Quim. Argent.*, 83, 171–181, 1995.

157. Stortz, C.A. and Cerezo, A.S., Conformational analysis of sulfated α-(1→3)-linked D-galactobioses using the MM3 force-field, *J. Carbohydr. Chem.*, 17,1405–1419, 1998.

158. Parra, E., Caro, H.N., Jiménez-Barbero, J., Martín-Lomas, M., and Bernabé, M., Synthesis and conformational studies of carrabiose and its 4'-sulphate and 2,4'-disulphate, *Carbohydr. Res.*, 208, 83–92, 1990.

159. Stortz C.A., Potential energy surfaces of carrageenan models: carrabiose, β-(1→4)-linked D-galactobiose, and their sulfated derivatives, *Carbohydr. Res.*, 337, 2311–2323, 2002.

160. Bosco, M., Segre, A.L., Miertus, S., and Paoletti, S., unpublished results (see ref. 154).

161. Lemieux, R.U., Bock, K., Delbaere, T.J., Koto, S., and Rao, V.S., The conformations of oligosaccharides related to the ABH and Lewis human blood group determinants, *Can. J. Chem.*, 58, 631–653, 1980.

162. Bock, K., Brignole, A., and Sigurskjold, B.W., Conformational dependence of ^{13}C nuclear magnetic resonance chemical shifts in oligosaccharides, *J. Chem. Soc. Perkin Trans.*, 2, 1711–1713, 1986.

163. Stortz, C.A. and Cerezo, A.S., Disaccharide conformational maps: 3D contours or 2D plots? *Carbohydr. Res.*, 337, 1861–1871, 2002.

164. Stortz, C.A. and Cerezo, A.S., Depicting the MM3 potential energy surfaces of trisaccharides by single contour maps: application to β-cellotriose and α-maltotriose, *Carbohydr. Res.*, 338, 95–107, 2003.

165. Stortz, C.A. and Cerezo, A.S., MM3 potential energy surfaces of trisaccharides. II. Carrageenan models containing 3,6-anhydro-D-galactose, *Biopolymers*, 70, 227–239, 2003.

166. Rees, D.A., Structure, conformation, and mechanism in the formation of polysaccharide gels and networks, *Adv. Carbohydr. Chem. Biochem.*, 24, 267–332, 1969.

167. Morris, V.J. and Belton, P.S., The influence of the cations sodium, potassium and calcium on the gelation of iota-carrageenan, *Prog. Food. Nutr. Sci.*, 6, 55–66, 1982.

168. McKinnon, A.A., Rees, D.A., and Williamson, F.B., Coil to double helix transition for a polysaccharide, *Chem. Commun.*, 701–702, 1969.

169. Jones, R.A., Staples, E.J., and Penman, A., A study of the helix-coil transition of ι-carrageenan segments by light scattering and membrane osmometry, *J. Chem. Soc. Perkin Trans. II*, 1608–1612, 1973.

170. Robinson, G., Morris, E.R., and Rees, D.A., Role of double helices in carrageenan gelation: the domain model, *Chem. Commun.*, 152–153, 1980.

171. Morris, E.R., Rees, D.A., and Robinson, G., Cation-specific aggregation of carrageenan helices: domain model of polymer gel structure, *J. Mol. Biol.,* 138, 349–362, 1980.

172. Smidsrød, O., Andresen, I.L., Grasdalen, H., Larsen, B., and Painter, T., Evidence for a salt-promoted "freeze-out" of linkage conformations in carrageenans as a pre-requisite for gel-formation, *Carbohydr. Res.,* 80, C11–C16, 1980.

173. Grasdalen, H. and Smidsrød, O., ^{133}Cs NMR in the sol–gel states of aqueous carrageenan. Selective site binding of caesium and potassium ions in kappa-carrageenan gels, *Macromolecules.* 14, 229–231, 1981.

174. Grasdalen, H. and Smidsrød, O., Iodide-specific formation of κ-carrageenan single helices. ^{127}I NMR spectroscopic evidence for selective site binding of iodide anions in the ordered conformations, *Macromolecules,* 14, 1842–1845, 1981.

175. Pelletier, E., Viebke, C., Meadows, J., and Williams, P.A., Solution rheology of κ-carrageenan in the ordered and disordered conformations, *Biomacromolecules,* 2, 946–951, 2001.

176. Rochas, C. and Rinaudo, M., Mechanism of gel formation in κ-carrageenan. *Biopolymers,* 23, 735–745, 1984.

177. Schafer, S.E. and Stevens, E.S., The optical rotation of ordered carrageenans, *Carbohydr. Polym.,* 21, 19–22, 1996.

178. Hjerde, T., Smidsrød, O., and Christensen, B.E., Analysis of the conformational properties of κ- and ι-carrageenan by size-exclusion chromatography combined with low-angle laser light scattering, *Biopolymers,* 49, 71–80, 1999.

179. Viebke, C., Borgström, J., and Piculell, L., Characterization of kappa- and iota-carrageenan coils and helices by MALLS/GPC, *Carbohydr. Polym.,* 27, 145–154, 1995.

180. Slootmakers, D., De Jonghe, C., Reynaeus, H., Varkevisser, F.A., and Bloys van Treslong, C.J., Static light scattering from kappa-carrageenan solutions, *Int. J. Biol. Macromol.,* 10, 160–168, 1988.

181. Ciancia, M., Milas, M., and Rinaudo, M., On the specific role of coions and counterions on κ-carrageenan conformation, *Int. J. Biol. Macromol.,* 20,35–41, 1997.

182. Miles, M.J., Morris, V.J., and Carroll, V., Carob gum-κ-carrageenan mixed gels: mechanical properties and X-ray fiber diffraction studies, *Macromolecules,* 17, 2443–2445, 1984.

183. Williams, P.A. and Langdon, M.J., The influence of locust bean gum and dextran on the gelation of κ-carrageenan, *Biopolymers,* 38, 655–664, 1996.

184. Zhang, J. and Rochas, C., Interactions between agarose and κ-carrageenans in aqueous solutions, *Carbohydr. Polym.,* 13, 257–271, 1990.

185. Amici, E., Clark, A.H., Normand, V., and Johnson, N.B., Interpenetrating network formation in agarose-κ-carrageenan gel composites, *Biomacromolecules,* 3, 466–474, 2002.

186. Parker, A., Brigand, G., Miniou, C., Trespoey, A., and Vallée, P., Rheology and fracture of mixed ι- and κ-carrageenan gels: two-step gelation, *Carbohydr. Polym.,* 20, 253–262, 1993.

187. Ridout, M.J., Garza, S., Brownsey, G.J., and Morris, V.J., Mixed iota–kappa carrageenan gels, *Int. J. Biol. Macromol.,* 18, 5–8, 1996.

8 Computational Modeling in Glycosylation

Michael P. Murrell, Kevin J. Yarema,
and Andre Levchenko

CONTENTS

Surface carbohydrates comprise the interface between a cell and its surroundings. They play key roles in governing cell–cell communication, transmitting external signals inside cells, and mediating interactions with the extracellular matrix; in short they are indispensable for the existence of multicellular organisms. Important challenges remain, however, to understand glycosylation at a fundamental level. In particular, a full explanation of how intracellular biosynthetic pathways produce diverse and multifunctional cell surface carbohydrates and regulate complex patterns of surface glycosylation remains unexplained. Additionally, it is almost completely unknown how a single cell can rapidly *modify* these surface patterns in response to internal or external stimuli. We propose that gaining a thorough understanding of the complexities of glycosylation lies beyond the scope of experiment alone. Instead, we suggest that common underlying properties developed as part of a larger concept of biological complexity,[1,2] including principles of regulation and control often referred to as "systems properties",[3] also apply to glycosylation. As a consequence, computational modeling can be pursued to generate experimentally driven hypotheses that elucidate how intracellular biosynthetic mechanisms relate to observable cell surface displays and vice versa.

I. THE COMPLEXITY OF CARBOHYDRATE STRUCTURE AND SYNTHESIS

In contrast to other cellular macromolecules assembled from smaller "building blocks" such as DNA, RNA, and proteins, carbohydrates form branching structures that geometrically increase their potential structural diversity. In addition, carbohydrates are extremely heterogeneous, for example, the prion protein (PrP) can display any one of over 52 different sugar structures at either of two attachment sites (Figure 8.1).[4,5] Interestingly, the normal cellular form of the prion protein (PrP^C) has a different complement of carbohydrates than the diseased form (PrP^Sc) and emerging evidence has implicated variation in glycosylation as a determining factor in the development of spongiform encephalopathies.[4] Furthermore, many glycosylated proteins have multiple (sometimes dozens of) sites for the attachment of carbohydrate chains; a sialomucin such as CD34, also shown in Figure 8.1, exemplifies this class of molecule.[6,7] As a result, the multiple potential sites of attachment for an oligosaccharide, along with the multiplicity of structures that can appear at each site, a single glycosylated protein can assume hundreds of different glycoforms.[8,9] This level of diversity makes characterizing even a single glycoprotein a formidable challenge and thorough elucidation of the entire complement of cellular glycans is virtually impossible at present (glycoproteomic techniques, however, are beginning to yield results in this area[10–14]).

In addition to simply characterizing the structural complexity of cell surface-displayed carbohydrates, one has to face the formidable challenge of understanding the internal regulatory processes that *produce* glycans. Once again, in comparison with nucleic acids and proteins, the study of glycan biosynthesis is complicated because oligosaccharide production is not a template-based process (and therefore PCR-type methods capable of analyzing minute quantities of material do not apply). Notwithstanding these difficulties, a basic framework of the "glycosylation

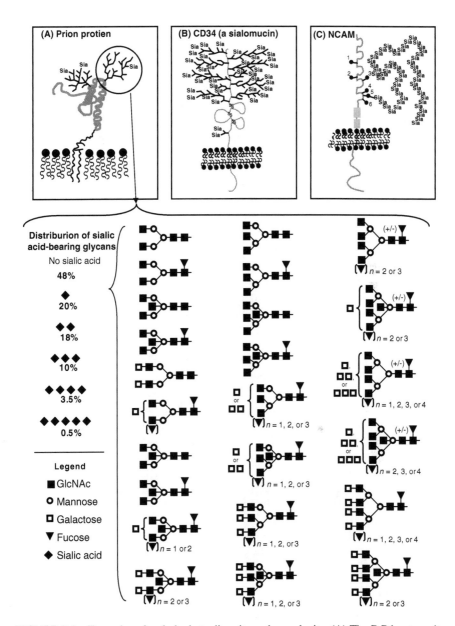

FIGURE 8.1 Examples of carbohydrate diversity and complexity. (A) The PrP has two sites of *N*-linked glycosylation; one of a set of over 50 different oligosaccharides can be attached at either site.[4,5] A representative sampling of these oligosaccharides is shown in the lower portion of the figure. (B) CD34 is a highly glycosylated protein (over 75% of the mass of this glycoprotein is carbohydrate) bearing dozens of individual oligosaccharide chains of both the *N*- and *O*-linked variety.[6,7] (C) The neural cell adhesion molecule (NCAM) exemplifies a third type of glycoprotein where polysaccharides, instead of shorter oligosaccharides, provide the glycosyl moiety. In this case, the polysaccharide is the *α*-2,8-linked polysialic acid (PSA) homopolymer, which can range in size from 55 to 200 residues, attached at one of six glycosylation sites.[84,85]

machinery" — thousands of intracellular components (2 to 3% of the human genome) that convert dietary sugars in to nucleotide sugar donors and assemble them in the endoplasmic reticulam (ER) or Golgi in a step-by-step process to form final products on the cell surface — is now in hand.[15] In its totality this framework, what we term the metabolic *system*, exhibits large-scale properties such as control and regulation that enable different cell types to display different complements of surface glycans. The precise mechanisms for cell-specific glycan display, however, remain largely unknown and we propose that computational methods will prove valuable in uncovering many of the hidden complexities of the glycosylation process. By examining the systems properties, the regulation, and control of fluxes in these pathways, we can understand better how various glycosylation-modulated behaviors occur in healthy organisms and how they contribute to the progression of diseased states when they go awry.

Computational methods in metabolism aim to discover underlying system elements of control and regulation by abstracting cellular processes into clearly defined mathematical models that simplify the analysis of their large-scale properties. Although very few efforts have been reported to date where computational methods have been applied to glycosylation, we propose that classical methods used in modeling metabolism will be powerful tools in unraveling the intricacies of oligosaccharide biosynthesis. More precisely, we predict they have the potential to reveal important mechanistic relationships within glycosylation pathways as well as imply larger scale interactions that govern carbohydrate synthesis and cell surface presentation. Among these tools are constraint-based approaches, such as metabolic flux analysis (MFA) and flux balance analysis (FBA) as well as more precise, quantitative methods such as metabolic control analysis (MCA) and biochemical systems theory (BST). Because of the relative infancy of the application of computational approaches to glycosylation, this chapter will not provide a mature perspective of the field; instead, the goal is to provide the glycobiologist with an overview of computational methods and offer illustrative examples of how these methods can be applied to specific aspects of glycosylation. In particular, emphasis will be placed on making connections between the intracellular metabolic processes and production of the surface sugars; this relationship is highly complex. It not only involves synthesis in the forward direction but also includes the impact of the recycling pathways by which the surface-displayed sugars are salvaged and reused as well as the signaling roles of these glycans, which can provide feedback on their biosynthetic processes.

II.   STEPS TOWARD MODELING GLYCOSYLATION

Up to now, 'systems' approaches to modeling in metabolism have focused on the output of secreted molecules or energy production rather than on persistent changes such as those seen in cell surface carbohydrate architecture. Moreover, modeling efforts undertaken to date have rarely, if ever, considered the complex interplay and crosstalk between intracellular biosynthetic processes and the biosynthetic product, in this case cell surface-displayed carbohydrates. Clearly, applying a computational approach to glycosylation involves unique and novel challenges and, consequently, in order to select an appropriate approach for the initial modeling of glycosylation, it is instructive to consider briefly the subsystems and basic elements of the glycosylation process.

For example, in the few reported attempts to apply computational approaches to glycosylation, it was assumed that consideration of the processes in the ER and Golgi, where the glycans are actually assembled and diversified, was sufficient. In one study, Monica and co-workers assumed that CMP-Neu5Ac, the building block for sialic acid biosynthesis, was maintained at saturating concentrations in the Golgi, and competition between the various sialyltransferases that produce α-2,3-linked, α-2,6-linked, or α-2,8-linked sialosides (Figure 8.2) was the primary factor responsible for the product distribution of sialylated oligosaccharides.[16] It is now known that this assumption was oversimplified and additional factors, which include the cytoplasmic metabolic networks that produce nucleotide sugar donors and transport them into the Golgi as well as the regulation of glycosyltransferases expression, must be considered. These points are discussed in the following two paragraphs, respectively.

In-depth analysis of the mechanisms responsible for glycosylation indicates that the flux of nucleotide sugar donors into the Golgi or ER can play a defining role in determining the finer details of cell surface carbohydrate architecture. For example, one study reported that when UDP-galactose levels were limited, certain galactose-containing products were made normally (heparan sulfate and chondroitin 4-sulfate) while others were severely reduced (glycoproteins, glycolipids, and keratan sulfate; Figure 8.3A).[17] Similarly, the congenital disease leukocyte adhesion deficiency II (LADII) is characterized by defective GDP-L-fucose transport into the Golgi. The ensuing decrease in availability of this nucleotide sugar donor to the glycosyltransferases that collectively assemble cell surface-displayed oligosaccharides [18–22] results in decreased N-linked glycosylation.[23] Thus, although bulk levels of fucosylation remain relatively unchanged, the altered *pattern* of glycosylation, where certain crucial epitopes such as blood group antigens or selectin ligands are defective, leads to clinical manifestation of disease. Interestingly, dietary supplementation with fucose can alleviate symptoms of LAD in some cases, thereby suggesting that factors outside of the ER or Golgi play an important role in glycosylation (Figure 8.3C). Two possibilities exist to account for fucose supplementation results; in some cases exogenous fucose may complement a defect in the endogenous pathway where GMDS or TSTA1 activity is reduced.[24,25] In other clinical cases of LAD II, the defect occurs in the Golgi transporter; if the defect is a "K_m" problem, then dietary supplementation can alleviate disease symptoms whereas other transporter defects do not respond.[26,27] In either case, LAD illustrates that assumptions, such as those made by Monica and co-workers[16] that nucleotide sugar donor levels within the Golgi are the primary factor that need consideration during modeling, are inadequate for accurately determining the systems properties of glycosylation.

The galactose and fucose examples illustrated in Figure 8.3 and cited above serve as evidence that efforts to understand glycosylation must encompass factors beyond glycosyltransferases reactions that occur in the ER and Golgi compartments. The sialic acid biosynthetic pathway (Figure 8.3D), a set of enzymes located in the cytosol and nucleus that produces CMP-Neu5Ac for transport into the Golgi, provides an intriguing system to begin a detailed computational modeling study of the "pre-Golgi" aspects of glycosylation. After it is produced, CMP-Neu5Ac is transported into the Golgi where it is used as a nucleotide sugar donor "building block" for the sialylation of glycoproteins and glycolipids that appear on the cell surface. A subtle

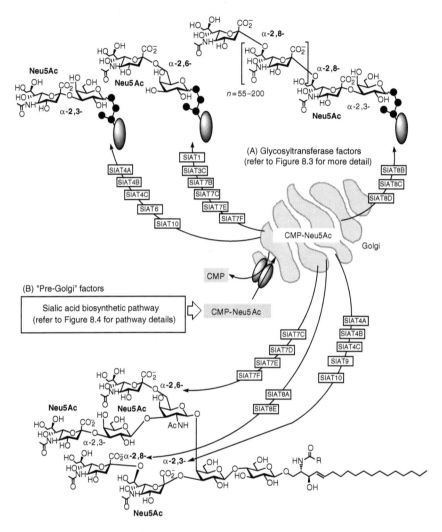

FIGURE 8.2 Oligosaccharide biosynthesis and diversity is determined in the Golgi. The basic biosynthetic mechanisms responsible for the structural diversity of cell surface oligosaccharides are illustrated, for simplicity, by only focusing on sialic acid. (A) Diversity is believed to result largely from the action of glycosyltransferases in the ER and Golgi; in the case of sialic acid, this final residue is added to growing oligosaccharide chains during their biosynthesis and processing in the medial and trans-Golgi compartments by a family of sialyltransferases. As shown, distinct, but overlapping, complements of these enzymes install α-2,3-, α-2,6-, and α-2,8-linked sialic acids onto glycoproteins (top) and glycolipids (bottom; note that more information on each of these enzymes is provided in Table 8.1). (B) There is increasing evidence that factors outside the ER or Golgi, such as "Pre-Golgi" cytoplasmic monosaccharide processing reactions (illustrated by the sialic acid biosynthetic pathway that produces CMP-Neu5Ac as well as the transport of sugar nucleotide donor "building blocks" into the Golgi), also substantially contributes to the exact patterns of cellular carbohydrate display.

TABLE 8.1
Enzymes Referred to throughout This Chapter

Designation	Full Name	Ref.
2.3.1.45[a]	N-acetylneuraminate 7-O-(or 9-O)-acetyltransferase	103,104
3.1.3.29[a]	N-acylneuraminic acid 9-phosphate phosphatase	105
5.1.3.9[a]	N-acylglucosamine-6-phosphate 2-epimerase	106
AGS	O-acetyl ganglioside synthetase	107,108
CMAH	CMP-Neu5Ac hydroxylase	109
CMPNS	CMP-Neu5Ac synthetase	110
CMPST	CMP-sialic acid Golgi transporter	111
CSE-C	Cytosolic sialic acid 9-O-acetylesterase homolog	112,113
CST	Cerebroside sulfotransferase	92,114
FPGT	Fucose-1-phosphate guanylyltransferase	115
FUCT1	GDP-fucose transporter 1 (aka SLC35C1)	18,116
FUK	L-fucose kinase	117
GALE	UDP-galactose-4-epimerase	118–120
GMDS	GDP-mannose-4,6-dehydratase	121,122
GNE	UDP-GlcNAc 2-epimerase/ManNAc 6-kinase	123
HE	Influenza C virus glycoprotein	100,124,125
NAGK	N-acetylglucosamine kinase	126
NEU1	Sialidase 1; lysosomal sialidase	127
NEU2	Sialidase 2; cytoplasmic sialidase	128
NEU3	Sialidase 3; membrane sialidase	129,130
NEU4	Sialidase 4	131
NPL	N-acetylneuraminate pyruvate lyase	132,133
OGT	O-GlcNAc transferase	134,135
PGM3	Phosphoglucomutase 3	136,137
RENBP	Renin-binding protein; a GlcNAc 2-epimerase	56,138
SANAE	Sialic acid N-acetylesterase	96,139–141
SAS	N-acetylneuraminic acid phosphate synthase	142
SIAT1	Sialyltransferase 1 (β-galactoside- α-2,6-sialyltransferase)	143
SIAT3C	α-2,6-Sialyltransferase (aka ST6GalNAc IV or SIAT7D)	144,145
SIAT4A	α-2,3-Sialyltransferase (aka ST3GAL IA)	146–148
SIAT4B	α-2,3-Sialyltransferase (aka ST3GAL II)	146,149
SIAT4C	α-2,3-Sialyltransferase (aka ST3Gal IV)	146,150
SIAT6	α-2,3-Sialyltransferase (aka ST3Gal III; or SIAT3 in mouse)	151
SIAT7B	α-2,6-Sialyltransferase (aka ST6GalNAc II)	152,153
SIAT7C	α-2,6-Sialyltransferase (aka ST6GalNAc III)	131
SIAT7D	α-2,6-Sialyltransferase (aka ST6GalNAc IV or SIAT3C)	144,145
SIAT7E	α-2,6-Sialyltransferase (aka ST6GalNAc V)	154,155
SIAT7F	α-2,6-Sialyltransferase (in humans, is STGalNAc VI)	156,157
SIAT8A	α-2,8 Ganglioside GD3 synthase	158,159
SIAT8B	α-2,8-Polysialyltransferase (aka STX)	160

(Continued)

TABLE 8.1   (Continued)

Designation	Full Name	Ref.
SIAT8C	α-2,8-Polysialyltransferase	161,162
SIAT8D	α-2,8-Polysialyltransferase 1 (aka PST)	163,164
SIAT8E		
SIAT9	Sialyltransferase 9; forms α-2,3 ganglioside linkages	165
SIAT10	α-2,3-sialyltransferase VI	166–167
TSTA3	Tissue-specific transplantation antigen P35B	121
UAP1	UDP-N-GlcNAc Pyrophosphorylase 1	168
UGT	UDP-galactose Golgi transporter	169–171

aka = also known as.

[a]EC (Enzyme Commission) numbers are given for situations where an enzyme activity is known but no gene has yet been cloned.

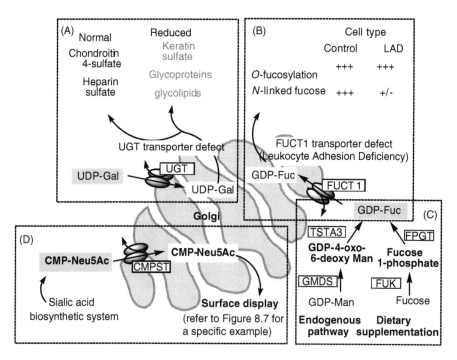

FIGURE 8.3 Nucleotide sugar donor availability in the Golgi determines cell surface oligosaccharide composition. (A) Defective transport of UDP-Gal into the Golgi alter reduces the production of keratan sulfate, glycolipids, and glycoproteins but does not measurably affect chondroitin 4-sulfate or heparan sulfate production. (B) Similarly, when GDP-Fuc transport into the Golgi is diminished, certain forms of surface fucose, such as those occurring on N-linked glycans, are reduced while others, such as O-linked fucose, remain relatively unchanged. (C) The fucose-related transporter defect can be complemented by dietary supplementation of fucose, indicating the factors outside of the Golgi play important roles in determining final product composition; in this chapter, we focus on the sialic acid biosynthetic system as a model for exploring this concept in more detail (D).

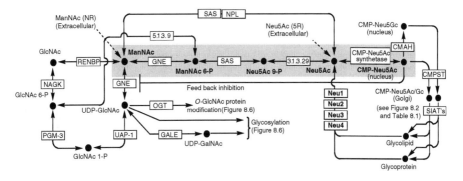

FIGURE 8.4   The sialic acid biosynthetic system. The "core" sialic acid biosynthetic pathway, usually regarded as the enzymes that sequentially convert ManNAc into CMP-Neu5Ac (shaded), is used in this chapter as a subset of the entire glycosylation system to illustrate the application of computational modeling methods to oligosaccharide biosynthesis. Thorough understanding of sialic acid metabolism, however, requires the consideration of additional enzymes that directly connect with the core pathway (such as those shown on the left of the diagram) as well as those that produce and recycle surface forms of sialic acid (bottom right). The full name of each enzyme and relevant references are given in Table 8.1. In addition, the pathway has many secondary connections with larger cellular metabolism; of few of these links, including O-GlcNAc protein modification and "underlying" glycosylation processes, are also outlined in Figure 8.6.

but important effect of flux through the "core" sialic acid pathway, which occurs outside of the Golgi, is secondary effects on the expression of the Golgi-localized sialyltransferases. Accordingly, a second factor to consider when producing a computational model of glycosylation is the effects of metabolic flux on the constituent biosynthetic enzymes. This consideration is important because in the past, attempts to model metabolic flux without considering the regulation of the enzymes involved has led to the construction of models with poor predictive abilities.[28–30] Together, these intracellular and cell surface aspects of sialylation (Figure 8.4) exemplify why a systems approach that incorporates a wide range of cellular elements is required to construct accurately a realistic model of glycosylation.

III.   COMPUTATIONAL MODELING IN METABOLISM

The emergence of the field of "systems biology"[31–33] has been driven by a realization that the full exploration of the capabilities of cellular processes requires more than the isolated characterization of the individual components. In this context, properties that are exhibited on a larger scale are referred to as *systems properties* and include robustness,* control, and mechanisms of regulation. In metabolic systems, these properties are frequently applied to study metabolic flux distributions, the most significant variables of the pathway. Moreover, systems properties can be used to elucidate pathway ("network") topology,[34,35] which is defined as the precise

*Here, by robustness, we understand the property of the metabolic system to withstand various perturbations, including mutation of enzymes or variable metabolic inputs. Controllability is defined as the flexibility of the system to more specific perturbations, usually used to control metabolic output. Thus, the notions of robustness and controllability are often complementary.

sequence of reactions that connect a metabolic input with metabolic output. Glycosylation is governed by the same principles of regulation and control as any other metabolic system and therefore, although glycosylation has not yet been the subject of intense systems level study, the same conceptual framework used to model other systems applies. In glycosylation, however, there is an additional layer of complexity in that the intracellular processes that synthesize carbohydrates are not solely responsible for carbohydrate diversity and heterogeneity at the cell surface. Instead, numerous dynamic processes modulate the surface presentation of carbohydrates and these molecules remain continually subjected to postsynthetic modification and recycling, as well as other regulated, systemic processes. The discovery of these systems properties in glycosylation contributes to contemporary goals of metabolic modeling such as the engineering and design of metabolic pathways. Systems-level approaches can help identify strategies for constructing or modifying metabolic networks with desired properties, including optimizing the yield of a metabolic product (i.e., cell surface glycosylation), or the enhancement of system properties, such as adaptation and modulation of enzyme activities. For these purposes, computational infrastructures that can process large-scale algorithms complement systems analysis by making such studies feasible. As mentioned, there are few precedents for the modeling of glycosylation processes in human cells. We will therefore next briefly outline the extended metabolic analysis toolbox and evaluate which of the available methods appear well suited for modeling the control mechanisms for glycan biosynthesis.

Metabolism has been investigated extensively as the first primary target for systems biology due in part to the availability of high-throughput technologies and annotated genome sequences for reconstructing whole cell metabolic networks. Mathematical frameworks to deal with systems-level metabolism vary in their level of sophistication, but are frequently based on either constraint-based approaches, such as MFA and FBA, or more precise methods, such as MCA or BST. Both classes are used to derive information on metabolite flux distributions. MFA classifies the contribution of various pathways to an overall metabolic process, and can be used to determine maximum theoretical yields. It can also be used to identify potential branch points and alternative pathways.[36,37] MCA is a more quantitative method for studying regulation in metabolic systems. With *a priori* knowledge of molecular effects and enzyme kinetics, it can be used to identify the sensitive elements in a pathway for altering flux.[38,39] These two differ fundamentally in how they determine flux information, contrasted with the level of knowledge they each assume. MFA (and likewise FBA) rely on constraints, in the absence of kinetics, while methods like MCA hinge on the perturbation of specific enzyme activities. To elaborate, MFA begins with the stoichiometry of the pathway reactions, and imposes constraints on the system, to define where the system resides in terms of its flux. When the system residence is defined, MCA can determine how it can be controlled to yield a desired or optimal flux. BST is an approach at modeling nonlinear systems, where both steady-state and transient responses are taken into account.[40,41] The primary focus of this review, however, is in the comparison of constraint and control-based methods, and how they can theoretically be used to connect the intracellular mechanisms of glycosylation to the cell surface aspects.

IV. MATHEMATICAL REPRESENTATION

Metabolic flux is one of the critical determinants of cellular function and is therefore the primary focus of modeling in metabolism. Fluxes are the rates of metabolite conversion expressed at steady state, which connect metabolites throughout entire pathways. Each pathway therefore has a *distribution* of fluxes, which characterize it uniquely, as the sequence of biochemical reactions in each pathway is distinct. Accordingly, flux distributions from pathways that are highly branched will differ from those that are purely linear. Specifically, the complexity of pathways with a large number of intermediates with multiple potential fates is better appreciated by identifying each of the pathway fluxes. It is therefore clear that the identification of flux distributions is central to metabolic engineering, where knowledge of *specific* pathway elements and routes facilitate manipulating metabolic output. Similarly, metabolic flux identification has been established as a paradigm for systems-level approaches, allowing us to take a look at *groups* of interacting biochemical reactions, and how they contribute to overall metabolic, and even cellular processes. Moreover, the elucidation of the mechanisms responsible for the control of flux has proven equally valuable. Therefore, considering the significance of metabolic flux distributions, the next step is to motivate a precise mathematical framework to represent them. While this is not a minor task, considering that pathways vary in size and complexity, there exist basic, underlying commonalities from which we can develop any framework. Regardless of the complexity, the basic conversion of substrates into metabolic products and the conversion of energy are subject to basic physical laws and internal constraints common to all metabolic systems, and can serve as the fundamental building blocks from which one can develop any formalism. More technically, we first rely on basic reaction stoichiometry to lump together large numbers of biochemical reactions to build a comprehensive model of reaction networks.[37] We can then pursue varying levels of sophistication, to incorporate explicit, pathway-specific enzyme kinetics, or retain this level of generality, for a constraint-based, scalable mathematical model.

A. STOICHIOMETRY: REPRESENTING REACTION EQUATIONS

As previously stated, all metabolic reactions share the same principle: the conversion of substrates into metabolic products and free energy. These conversions often include a large number of metabolites, where many metabolites are shared between different reactions, and potentially different pathways. Conversely, the sensitivity of an entire metabolic pathway or cellular process can hinge on the availability of sparsely "connected" metabolites. Consequently, to analyze fully the apparent complexity of product synthesis and modification, we must first define their stoichiometry, the molar quantities of every substrate and product. More generally, we can account for the net change in metabolites within certain system boundaries. These boundaries become mass-balance constraints on the rates of conversion, the basis for constraint-based modeling in metabolism (see Section V). To enumerate, each metabolite (substrate X, co-factor C, or product P) is present in the model along with other metabolites in a linear combination, weighed by its coefficient α, β, and γ (Equation 8.1). The coefficients relate the level of *contribution* to each reaction, whose sign will depend on the type of contribution. Furthermore, the indices i and j correspond to the reaction, and the compound respectively, of which there are N total substrates, M co-factors, and K products:

$$\sum_{j=1}^{N} \alpha_{ij} X_j + \sum_{j=1}^{M} \beta_{ij} C_j + \sum_{j=1}^{K} \gamma_{ij} P_j = 0 \tag{8.1}$$

For each of the reactions in the pathway, there is an equation, and thus the system can be expressed in matrix form, where each matrix of stoichiometric coefficients is represented as a single constant, A, B, and G, and is related to its respective metabolite variable vector, X, C, and P:

$$AX + BC + GP = 0 \tag{8.2}$$

We can express the formation of CMP-sialic acid, for example, from fructose and N-acetylglucosamine in the sialic acid biosynthesis with this formalism, totaling eight reactions and sets of substrates, co-factors, and products (Example 1).

Example 1:   *Stoichiometry of the sialic acid biosynthesis*

$$
\begin{vmatrix}
-1 & 1 & 0 & 0 & 0 & 0 & 0 & 0 & 0 \\
0 & -1 & 1 & 0 & 0 & 0 & 0 & 0 & 0 \\
0 & 0 & -1 & 1 & 0 & 0 & 0 & 0 & 0 \\
0 & 0 & 0 & -1 & 1 & 0 & 0 & 0 & 0 \\
0 & 0 & 0 & 0 & -1 & 1 & 0 & 0 & 0 \\
0 & 0 & 0 & 0 & 0 & -1 & 1 & 0 & 0 \\
0 & 0 & 0 & 0 & 0 & 0 & -1 & 1 & 0 \\
0 & 0 & 0 & 0 & 0 & 0 & 0 & -1 & 1 \\
0 & 0 & 0 & 0 & 0 & 0 & 0 & 0 & -1
\end{vmatrix}
\begin{vmatrix}
X_{\text{Fru-6p}} \\
X_{\text{GlcN-6P}} \\
X_{\text{GlcNAc-6P}} \\
X_{\text{GlcNAc-1P}} \\
X_{\text{UDP-GlcNAc}} \\
X_{\text{ManNAc}} \\
X_{\text{ManNAc-6P}} \\
X_{\text{Neu5Ac-9P}} \\
X_{\text{Neu5Ac}}
\end{vmatrix}
$$

$$
+\begin{vmatrix}
0 & 0 & 0 & 0 & 0 & 0 & 0 & 0 & 0 \\
0 & 0 & 0 & 0 & 0 & 0 & 0 & 0 & 0 \\
0 & 0 & 0 & 0 & 0 & 0 & 0 & 0 & 0 \\
-1 & 0 & 0 & 0 & 0 & 1 & 0 & 0 & 0 \\
0 & 1 & 0 & 0 & 0 & 0 & 0 & -1 & 0 \\
0 & 0 & -1 & 1 & 0 & 0 & 0 & 0 & 0 \\
0 & 0 & 0 & 0 & 0 & 0 & 1 & 0 & -1 \\
0 & 0 & 0 & 0 & 0 & 0 & 1 & -1 & 0 \\
0 & 0 & 0 & -1 & 1 & 0 & 0 & 0 & 0
\end{vmatrix}
\begin{vmatrix}
C_{\text{UTP}} \\
C_{\text{UDP}} \\
C_{\text{ATP}} \\
C_{\text{ADP}} \\
C_{\text{CTP}} \\
C_{\text{PPi}} \\
C_{\text{Pi}} \\
C_{\text{H2O}} \\
C_{\text{PEP}}
\end{vmatrix}
+\begin{vmatrix}
0 \\
0 \\
0 \\
0 \\
0 \\
0 \\
0 \\
0 \\
1
\end{vmatrix}
\left| P_{\text{CMP-Neu5Ac}} \right| =
\begin{vmatrix}
0 \\
0 \\
0 \\
0 \\
0 \\
0 \\
0 \\
0 \\
0
\end{vmatrix}
$$

In the above equation, (Fru-6P = D-fructose-6-phosphate; GlcN-6P-D-glucosamine-6-phosphate; ; GlcNAc-6P = N-acetyl-D-glucosamine-6-phosphate; GlcNAc-1P = N-acetyl-D-glucosamine-1-phosphate; UDP-GlcNAc = UDP-N-acetyl-D-glucosamine; ManNAc = N-acetyl-D-mannosamine; ManNAc-6P = N-acetyl-D-mannosamine-6-phosphate;

Neu5Ac-9P = N-acetylneuraminic acid 9-phosphate; Neu5Ac (or Sia) = N-acetylneu-raminic acid [also known as N-acetylneuraminate]; and CMP-Neu5Ac = CMP-N-acetylneuraminic acid.)

With this representation, we can get a conceptual idea of how various compounds are involved in an overall process, as conferred by the location and frequency of occurrence of each compound's individual coefficients in the overall stoichiometric matrices. The organizations of these matrices reveal important "structural features" of metabolic pathways, or networks.[3] For example, by the apparent sparsity and con-centration of the coefficients along the diagonal of the substrate stoichiometric matrix in Example 1, we can see that the biosynthesis is a linear chain. The stoi-chiometric matrices therefore represent static characteristics, or constraints on meta-bolic models, and do not depend on the knowledge of reaction rates. As will be described in the following section however, reaction rates are also bound by the same constraints, and stoichiometric analysis is the first step in constructing constraint-based metabolic models.

B. Assumptions and Basic Laws

Systems of stoichiometric equations describing the relative production and con-sumption of metabolites do not provide information on the rates of these reactions. To preface, reaction rates are expressed as changes in concentrations per unit time. While concentrations are defined explicitly as the number of moles of a given sub-stance per unit volume, we tend to simplify the system by presuming all volumes are small and uniform.[42] Furthermore, the concentrations can be represented as occupy-ing "metabolic pools" connected by the reaction rates or fluxes of metabolite trans-port (mass flow). Certain assumptions are made to validate the macroscopic approach, in comparison with microscopic approaches, which quantify interactions at the molecular level. For instance, the number of particles is considered to be so large that stochastic effects can be ignored. If this assumption fails, concentration can still be defined as above keeping in mind that changes in concentration might become discontinuous with sudden step-like increases and decreases. Other assump-tions include stating that the system has a unique steady state, meaning that for every system, the concentrations will reach some constant value, if left to evolve long enough. By convention, if these chemical equilibria are reached, we represent reac-tion rates as reaction fluxes.

1. Balancing Equations

If we assume that mass flow is only due to the biochemical reactions and factors such as diffusion are ignored, the behavior of the concentrations over time can be expressed in terms of mass balance equations. This assumption follows the con-servation of mass, which reflects the physical law that input into a metabolic pool must be balanced with efflux from the pool. This principle is seen in Equation 8.3, where we choose to represent the rate of metabolite transfer through a particular metabolite pool. Specifically, we can express the rate of conversion of any substrate (X_i) by its corresponding fluxes (v_j) and stoichiometric constants (α_{ij}). The fluxes themselves are functions of the time-dependent substrate concentrations:

$$\frac{\mathrm{d}X_j}{\mathrm{d}t} = \sum_{i=1}^{k} \alpha_{ji} v_i \tag{8.3}$$

Balance equations are one of the most general ways of describing chemical reactions due to their dependance on the conservation laws. From here, simplifying assumptions can be made to reduce the size or complexity of the system equations. Often metabolic pools are lumped together, and represented as individual variables, if two or more metabolites are sufficiently correlated in their behavior. Also, certain reactions can be eliminated from consideration is some cases to reduce the total number of reactions in a pathway; an example is the analysis of the time scales, as described in Section VI. Additionally, Equation 8.3 allows us to make assumptions that render the system equations more explicit, and represent system detail by expanding the above reaction rates. If we maintain this level of generality and express Equation 8.3 in matrix notation, we form the basis for constraint-based modeling, as discussed in Section V.

2. Rate Laws

Rate laws are detailed forms of the balance equations (Equation 8.3), whereby the rates or fluxes become dependent on other factors, such as metabolite concentrations or thermodynamics.[43] Concentration dependencies are usually referred to as kinetics, in that specific enzymatic phenomena, such as cooperativity, saturation, or the influence of effector proteins are incorporated. Kinetics can be presented in various forms, although the most common include Michaelis–Menten kinetics and the generalized mass-action rate law (similar to BST; see Section V).[42] Moreover, the choice of rate laws may also depend on the extent an experimental method can characterize enzyme kinetics. Alternatively, it is often useful to consider the influence of thermodynamics in metabolic flux determination. Typically, thermodynamic rate laws depend on reaction affinities as the primary thermodynamic forces acting upon metabolite conversion. Such an approach is favored when the internal mechanisms are completely unknown.[44] Mathematical formulations of these rate laws will be omitted in this review, but can be found in standard textbooks on modeling in metabolism.[37]

V. MODELING AND CONTROL IN METABOLIC SYSTEMS

Balance equations and stoichiometric constraints can be used as a platform to motivate further the methods of flux identification, and if reaction rates are specified — how metabolic flux is controlled. Methods of flux identification fall primarily into two classes: constraint-based approaches that determine precise (or unique) flux distributions, and approaches that quantify the influence of reactions on given distributions (control of flux). Constraint-based methods, such as MFA and FBA rely purely on readily available stoichiometries and simple conservation laws to identify metabolic flux. Moreover, they can be used to identify flux boundaries and "extreme modes" when unique solutions cannot be found. In contrast, methods such as MCA and BST can elucidate how to achieve an optimal flux. Furthermore, they reflect pathway-specific knowledge of enzyme kinetics, and thus yield stronger implications regarding metabolic regulation.

A. METABOLIC FLUX ANALYSIS

MFA is a constraint-based methodology for identifying metabolic flux distributions using the stoichiometry-based approach introduced in Section IV, and extends it to the determination of metabolic fluxes.[36, 45] To do so, entire biochemical systems are abstracted into linear systems, whereby the intracellular fluxes can be calculated by applying mass balancing around the intracellular metabolites. More technically, the intracellular metabolic fluxes are the system variables, which can be uniquely defined by the system parameters, the reaction stoichiometries (S matrix, Equation 8.4). The system takes as its inputs measured rates of uptake or secretion of extracellular metabolites and the final outcome is a set of uniquely determined intracellular fluxes for every reaction. Traditionally, quantification of metabolic fluxes was critical for metabolic engineering, in the context of manipulating metabolic output. From an engineering perspective, knowledge of rigid or flexible points of flux in a pathway is important for altering product yields. Yields are determined by the stoichiometry of the reactions and, thus, MFA may uncover alternative pathways, more optimized for desired output. Conversely, flux determination is important from a systems perspective, as it communicates the level of metabolite contribution on a local and a global scale, as well as relating structural features of pathways, such as metabolite connectivity and sparsity. Because MFA can calculate the flux about different points in the pathway, it can identify split fluxes or branch points. Analysis of these branch points allows one to calculate the rigidity of the system along different points in the pathway. In this chapter, we merely present the basic theory behind MFA, and later FBA for modeling in metabolism, and provide rudimentary examples demonstrating how to connect it to modeling in glycosylation.

1. Assumptions of MFA

MFA assumes that metabolic networks will reach a steady state if constrained by their reaction stoichiometries. As the primary constraints, the stoichiometries represent static, structural confinements to the metabolic flux, and illustrate the more topological structure of metabolic pathways, ignoring features that vary temporally. Therefore, constraint-based modeling typically assumes complete ignorance of precise enzymatic or kinetic information. The implications for such omission are that the kinetic properties of metabolic pathways, which often evolve quickly, and occur almost ubiquitously in metabolism have no direct bearing on the identification of flux distributions (with this formality). Flux distributions are identified purely by the dependence on conservation and mass balances. By maintaining this generalization, we retain linearity, and solutions to even large systems are computationally tractable. However, with fewer assumptions and less knowledge of the system, the system may be underdetermined because multiple steady-state conditions may result. In order to solve the system, the number of unknown fluxes needs to equal the number of metabolites in the pathway (the S matrix in Equation 8.4 must be square). Typically, experimental determination of some of the unknown fluxes can aid in the identification of a unique set of flux distributions.[46] In the absence of experimental determination however, the system is left with too many degrees of freedom. Consequently, alternative methods of flux identification must be used, but they come at the cost of computational difficulty.[30,44,47]

2. Connecting Flux to Reaction Stoichiometry

As previously stated, we can represent biochemical pathways as linear systems, and the reaction stoichiometries are mass balance constraints on the system variables, the metabolic fluxes (v). We calculate these by representing the stoichiometric rate constants as a square matrix $(S_{m,n})$ of m metabolites and n reactions, which forms a linear transformation of the flux vector (Equation 8.4). This formulation is derived from the stoichiometric approach as described in Section IV, now focusing on the rates of metabolite conversion (of select metabolites — the conversion of GlcNAc 6-P to CMP-Neu5Ac). In this respect, we preserve the same conservation and balances as the metabolite representation, while attempting to quantify flux distributions. We additionally presume that the mass balance constraints (the parameters of the system) will uniquely define a steady-state solution, known as the *nullspace*:[48]

$$Sv=0 \qquad\qquad (8.4)$$

Example 2:   *Stoichiometry as mass-balance constraints on fluxes*

$$
\begin{vmatrix}
1 & -1 & 0 & 0 & 0 & 0 & 0 \\
0 & 1 & -1 & 0 & 0 & 0 & 0 \\
0 & 0 & 1 & -1 & 0 & 0 & 0 \\
0 & 0 & 0 & 1 & -1 & 0 & 0 \\
0 & 0 & 0 & 0 & 1 & -1 & 0 \\
0 & 0 & 0 & 0 & 0 & 1 & -1
\end{vmatrix}
\begin{vmatrix}
v_1 \\ v_2 \\ v_3 \\ v_4 \\ v_5 \\ v_6 \\ v_7
\end{vmatrix}
=
\begin{vmatrix}
0 \\ 0 \\ 0 \\ 0 \\ 0 \\ 0 \\ 0
\end{vmatrix}
$$

This simplified representation allows us to analyze the rates of metabolite transfer without the need for precise kinetic or enzymatic information. Instead, we rely on the simple notion of metabolic fluxes, and how to reduce the number of known fluxes to a "solvable" number. The solution to this very basic, linear system at the steady state is the set of fluxes (v vector), whose distribution will characterize the metabolic pathway under study. In the case of a square matrix, there is a unique solution. However, the solution is seldom unique. In the case of a nonunique solution, computational methods can be used to reduce the total solution set, the most common of these methods is known as linear programming, characteristic to methods such as FBA.[3,47,49]

3. Flux Balancing

The number of intracellular fluxes is always greater than the number of metabolites between them, and therefore there is always a certain degree of freedom, given by their difference. If the number of measurable fluxes is equal to the degrees of freedom, then MFA has been satisfied, and matrix algebra can be used to compute the solution or set of flux distributions. Usually, however, the number of measurable fluxes and the degrees of freedom are unequal. In the case where there are a greater number of measurable fluxes than degrees of freedom, the system is redundant or "overdetermined." Redundant systems can be used to optimize estimates of both calculated and measured fluxes. More commonly, however, the degrees of freedom are greater than the number of measurable fluxes, and the system is underdetermined. In MFA, the goal is to measure unknown fluxes in order to eliminate the degrees of freedom; these measurements, however, may or may not be feasible. In the event that it is not possible to reduce experimentally the number of unknowns sufficiently to compute a unique solution and, as a result, there are infinitely many solutions, the remaining alternative is to find the extreme behaviors. This alternative approach can be pursued by drawing upon additional assumptions concerning the flux distributions, namely that they exist for a cellular purpose or objective, and are restricted by both internal and external constraints.

All physiological processes are subject to physicochemical, environmental, and regulatory constraints.[30,43,44,50] For metabolic pathways, we have thus far imposed mass-balance stoichiometric constraints, due to their ubiquity and applicability to all metabolic fluxes. We can further constrain the fluxes however, to reflect additional constraints, such as thermodynamics, environmental restrictions, and reaction capacities. In FBA, there has been great success in relying on the latter to constrain the solution space further.[51] Specifically, we can place bounds on fluxes where we have prior knowledge of the enzyme turnover rates. The implication is that in the determination of a set of flux distributions, each flux is bound by experimentally derived estimates of enzyme activity (maximum uptake rates), the reaction capacities:

$$lb_i < v_i < ub_i \qquad (8.5)$$

The values of lb and ub can first be estimated broadly, as they will depend on the reversibility, or thermodynamics of each reaction. Irreversible reactions, for example, will always have a positive flux, and therefore are bounded below by zero. Consequently, additional thermodynamic constraints can further reduce the total set of solutions, and facilitate computational analysis (i.e., linear programming) considerably.

Given sufficient constraints, and a suitable objective function, we can use linear programming to identify the intracellular flux distributions for cases when the number of unknown fluxes exceeds the number of pathway metabolites. Accordingly, we assume that the metabolic system has an objective that it serves a purpose which can be exploited to determine a precise set of flux distributions. One example, commonly used in FBA of bacterial metabolism, is optimization for the increase in cell mass. In a noteworthy study performed by Palsson and co-workers,[52, 53] the metabolic flux distribution of the reconstructed *Escherichia coli* MG1655 metabolic network was determined by FBA. Using the hypothesis that the fluxes are optimized to increase biomass, the study was able to generate *in silico* predictions of bacterial growth as functions of oxygen and nutrient uptake rates. The model predictions correlated well with the experimentally measured growth rates and nutrient and oxygen uptake rates, illustrating that constraint-based computational modeling can be valid for the representation of large-scale biological processes in the absence of precise kinetic or enzymatic information. Consistent with the concept of flux balancing, the Palsson study relied on the application of known constraints, namely the systemic mass balances and reaction capacities to bound the fluxes at steady state (Equations 8.4 and 8.5). Subject to these constraints, linear optimization routines were applied to find the precise fluxes that correspond to maximum cellular growth, the objective function used in the study. Placing bounds on the system, as opposed to defining the precise kinetic information, not only renders the system scalable to metabolic pathways of different sizes and complexities, but makes it generally applicable to any pathways, including glycosylation pathways. Thus, while there are few precedents for applying flux balancing specifically to glycosylation, there are numerous examples in other metabolic systems. Although the same formalism for declaring flux boundaries apply, the objective function will depend on a particular goal, such as the maximization of cell surface presentation of a particular glycoform.

B. METABOLIC CONTROL ANALYSIS (MCA)

While MFA is useful for quantifying metabolic flux distributions, it is often important to analyze how flux is controlled or how to obtain an optimal flux. The control of flux is also important from a systems perspective; it can be used to describe robustness quantitatively or how homeostasis is maintained across varying external conditions.[38,39] MCA is a mathematical framework for describing the *control* exerted by enzymatic activities on the metabolic flux of a given pathway. This method applies only to the steady-state conditions (or pseudo steady state — see Section VI), under the assumption that the enzyme activities can uniquely define the equilibrium. Again, the system variables in the equation set are the fluxes through the pathway, which differ from the reaction rate when branched or cyclic pathways are considered. Enzyme activities are therefore considered to be system parameters, and each enzyme confers a specific *sensitivity* to the variables in the system. Specifically, the influence of each enzyme on a flux or metabolite concentration (at steady state) is quantified by the calculation of the corresponding flux control coefficient (FCC):

$$C_v = \frac{\partial v}{\partial E} \frac{E}{v} \tag{8.6}$$

Each FCC reflects the relative change in flux (v) with infinitesimal changes in enzyme activity (E). The differences are normalized by each respective flux and activity so that all control coefficients (n) sum to unity, as reflected in the *summation theorem* of MCA

$$\sum_{i=1}^{n} C_i = 1 \tag{8.7}$$

It is evident from the summation theorem that large pathways will have small control coefficients, and thus the magnitudes of each of the control coefficients are negligible. Each coefficient, however, does confer sensitivity *within* the pathway based on the convention that larger control coefficients exert greater control over flux under steady-state conditions than do smaller coefficients. Moreover, the value of the control coefficient depends significantly on the location of the enzyme in the overall pathway. Elasticity coefficients (ECs) can be used to reflect local behavior with increased precision and greater molecular detail (Equation 8.8). ECs relate the influence of substrate concentrations on the control of flux. In this respect, they represent properties of the enzymes themselves, adding more precise kinetic influence to the control of flux:

$$\varepsilon_X = \frac{\delta v}{\delta X} \frac{X}{v} \tag{8.8}$$

ECs complement FCCs, further completing a systems perspective, as global control retains molecular detail and precision. A quantitative relationship between the two can be seen in the *connectivity theorem* (Equation 8.9). Often considered to be the most important of the control theorems, the connectivity theorem illustrates how enzyme kinetics influence FCCs (it can be shown that the magnitudes of FCCs and ECs are inversely related):

$$\sum_{i=1}^{n} C_i^v \varepsilon_X^i = 0 \tag{8.9}$$

Consequently, these coefficients are the core of MCA, as they allow us to analyze the fractional change of fluxes at the system level, at different levels of specificity and detail. Moreover, from a systems perspective, MCA ties together both local and global properties; to this effect, systemic behaviors can be related to the influence of individual network components. In this respect, MCA is similar to *Sensitivity Analysis*, a method used in many disciplines for determining the sensitivity of a solution to perturbations. For more explicit descriptions of the mathematical formulation and experimental considerations, refer to more detailed reviews.[37, 38]

1. Assumptions of MCA

Sensitivity of metabolic flux to changes in enzymatic activity requires knowledge of model equations and parameters, and thus a great deal of assumptions must be made to identify the model structure. Certain assumptions are flexible but increase the complexity of the solution if violated. First, MCA assumes that the metabolic system can be represented as an interconnected set of metabolic pools, with at least one metabolite input source, and one metabolite output. Also, MCA applies only to systems that are in steady state or "pseudo-steady state," meaning that there must be an observable constant metabolic state. This stipulation implies that the rate of synthesis of a metabolic intermediate must be balanced by its rate of conversion or degradation. The steady state must also be uniquely defined by the enzymatic activities along each step in the pathway. Additional information is usually required to identify further the metabolic system. Furthermore, the rates of transition between stages are usually modified to represent various forms of enzyme kinetics, including the effects of inhibitors and activators. In the case that the system is unidentifiable however, MCA is inadequate.

2. Objections to MCA

Although FCCs are fundamental to the theory of metabolic control, they have little predictive value. FCCs do not remain constant; instead they describe a given steady state and are expected to change given sufficient perturbation. Moreover, as the perturbation significantly drives the magnitude of the enzyme activity further away from the original activity, the coefficients become less accurate, as enzyme kinetics can be largely nonlinear. Therefore, FCCs are only valid under experimental conditions similar to those under which the original measurements were made to obtain these values; this limitation can be overcome to a degree, however, by measuring

FCCs at multiple steady states that result from metabolic perturbations of various magnitudes. Another drawback of this method is that FCCs cannot differentiate regulatory enzymes from control enzymes. Yet, convention has allowed us to make conclusions carefully in this respect. Enzymes that may be subject to inhibition, for example, can be considered regulatory, as they can modulate flux in response to downstream metabolite production (such as the UDP-GlcNAc 2-epimerase/ManNAc 6-kinase gene in the sialic acid biosynthesis).[54] Paradoxically, the magnitude of "regulatory" FCCs was often found to be small, implying they exert little control over metabolic flux. Therefore, classifying the value of control coefficients as the relative level of control exerted over the flux at steady state can be misleading. To address this issue, alternative definitions of the concepts of a "regulatory enzyme" arose, including the idea that such enzymes can be defined outside of the equilibrium state.[55] If there were one rate-limiting step, it would have a very large control coefficient, and by the summation theorem, all other control coefficients must be close to zero, which by experimental analysis, has been proven false. Researchers in MCA now agree with the concept of distributed control; specifically that all of the enzymes control the output of the pathway to a degree, depending primarily on their location in the pathway. Consequently, the distribution of control among several pathway elements allows regulatory enzymes to be distinguished from "rate-limiting enzymes."

3. MCA Applied to the Analysis of Sialic Acid Biosynthesis

In sialic acid biosynthesis, the primary known control point (or point of regulation) is the UDP-GlcNAc 2-epimerase/ManNAc 6-kinase enzyme, referred to as "GNE" for short. This bifunctional enzyme converts UDP-GlcNAc into ManNAc, in turn ManNAc is a compound considered to be "committed" to entry into the sialic acid pathway as it has no other known metabolic fate within a cell[56] (Figure 8.4). GNE is subject to allosteric feedback inhibition upon binding of the downstream metabolite CMP-Neu5Ac, and thus represents the most obvious regulatory element in the pathway.[54] Evidence has emerged, however, that additional control points govern biosynthesis of cell surface-displayed sialic acid. For example, mutant forms of GNE found in human disease do not measurably alter overall levels of cell surface sialic acid,[57] and changes in subsets of overall populations of surface sialosides occur during apoptosis thereby suggesting that regulatory mechanisms beyond GNE are at work to fine-tune the levels of surface presentation of this sugar.[58] Currently, control points in sialic acid biosynthesis remain unresolved and present an attractive opportunity to apply an MCA modeling approach to solve an important aspect of the glycosylation process. Complicating an MCA analysis of sialic acid metabolism, however, is the possibility that confining the analysis to the "core" components shown in Figure 8.4, may not provide an adequate explanation of, or ability to predict, experimentally observed responses. If so, such a result would imply that other metabolic connections that sialic acid metabolism forms with other metabolic networks, such as those outlined in Figure 8.6, must play additional important roles in the control and regulation of this system. As can been seen from Figure 8.6, sialic acid biosynthesis has the potential to influence, and in turn be influenced by, hundreds (or thousands) of different metabolic components of a cell and may there-fore exceed current abilities, both from a modeling and experimental standpoint of rigorous MCA. In this scenario, a constraint-based approach, discussed in Section V.A, may provide a way of unraveling the larger scale interactions.

4. Comparing MFA to MCA

By applying mass balances around intracellular metabolites and applying con-straints to metabolic fluxes, we can gain an understanding of the natural boundaries that limit flux distributions. However, to understand the maintenance of metabolic homostasis (the insensitivity of flux distributions to internal perturbations), a con-trol approach is more appropriate. Specifically, to comprehend fully how the rates of synthesis and conversion of metabolites are maintained over a wide range of internal (or external) conditions, it is critical to understand how metabolic systems respond to disturbances. This task is more suited for MCA than for constraint-based methods. Moreover, from the perspective of carbohydrate engineering, knowledge of flux control can increase our ability to manipulate glycosylation pathways to optimize carbohydrates at the cell surface. In this respect, certain enzymes will have greater control over the synthesis or presentation of particular glycoforms, and a control approach can potentially identify these enzymes. Unfortunately, MCA suf-fers from the drawback of requiring almost complete knowledge of the pathway (enzyme kinetics), in which case a quantitative description of control is difficult, and analysis is limited to flux identification. Generally speaking, this is true for large systems, or even genome-scale systems, where complete kinetic information is certainly unavailable.

C. Biochemical Systems Theory (BST)

One of the more successful frameworks for systems analysis of biochemical processes, known as "BST", was developed by Michael Savageau in 1969.[40,41] This theory operates under the assumption that metabolic processes are inherently non-linear, and can be represented by the power-law relationships. Within this framework (considering the variables to be metabolic concentrations), the sets of fluxes (v_i) pro-ducing and consuming metabolites ($X_1, ..., X_k$) are related as

$$v_i = \alpha_i \prod_{j=1}^{k} X_j^{g_{ij}}$$
(8.10)

where the terms that determine the rate of change for each metabolite are the prod-uct of all influences leading to the production, and the product of all influences leading to the removal of X_j. The influences are independent variables, related by their kinetic orders g_{ij}, which reflect the effect of each parameter on the flux (e.g., a negative kinetic order would represent an inhibitory effect). The magnitude of each given process toward influencing the flux is represented by the nonnegative rate parameters, α_i. Collectively, this is known as the *power-law formalism*, a use-ful framework for the modeling and analysis of nonlinear systems. BST is similar but not identical to the generalized mass action rate law, as the kinetic orders are purely phenomenological influences, as opposed to stoichiometric rate constants. BST is also similar to MCA, in that it can be used to derive information on flux distributions, although it quantifies them at both transient and steady-state conditions.[37]

VI. METABOLIC TIME SCALES

Glycosylation pathways, similar to many metabolic systems, have the potential to exhibit wide temporal variation across their various biochemical reactions. More precisely, different steps in the monosaccharide processing reactions and in the subsequent oligosaccharide biosynthesis can occur over large gradients, often differing by orders of magnitude. For purposes of initial model building, it is useful to group rate constants into three categories based on comparative time scales given by (or estimated from) experimental measurements: (1) rate constants that are much greater (rapid chemical reactions); (2) much smaller (slow chemical reactions); and (3) close to the time constant of interest. Consequently, knowledge of the gradients between these rate parameters can aid in simplifying the model equations, and enhance the computational feasibility of the solution. This information is useful from a systems perspective, as analysis of large metabolic or signaling systems often requires simplifying assumptions to increase the tractability of a precise solution. Moreover, as systems become increasingly complex, for example if they are comprised of many short or transient responses or if they involve a large number of metabolites and intermediates, then knowledge of the critical or limiting reactions becomes increasingly relevant. To illustrate, we next cover a few of the more general simplifications that can be made to reduce the complexity of a biochemical system.

Depending on the relative magnitudes of rate constants in a given metabolic pathway, simplifying assumptions can be made in treatment of the corresponding mathematical model. An important simplification often made in the cases when differences in kinetic rates are significant is the so-called "quasi-steady-state approximation".[59–61] This approximation states that certain metabolites instantaneously achieve their steady states and the same value applies to all time points analyzed; in this case, the critical steady-state values for model construction are determined by slower reactions involving other metabolites of the pathway. In practical terms, the quasi-steady-state approximation results in conversion of ordinary differential equations into algebraic equations establishing dependencies between concentrations of constituent metabolites. Again considering sialic acid biosynthesis as an illustrative example, CMP-sialic acid is produced via a number of sequential metabolic conversions in a linear pathway (Figure 8.4). Focusing on the intermediate conversions of ManNAc-6-P through CMP-sialic acid production, it is established that ManNAc-6-P is converted into CMP-sialic acid via two intermediate stages (X_1 = ManNAc-6-P; X_2 = sialic acid-9-P; X_3 = sialic acid; X_4 = CMP-sialic acid in Example 3). The second step, the conversion of sialic acid-9-P to sialic acid, is assumed to be a very rapid reaction in comparison with the other reactions in the pathway because of the lack of build-up of this intermediate when the pathway is challenged with increased flux (Figure 8.5A).[62]

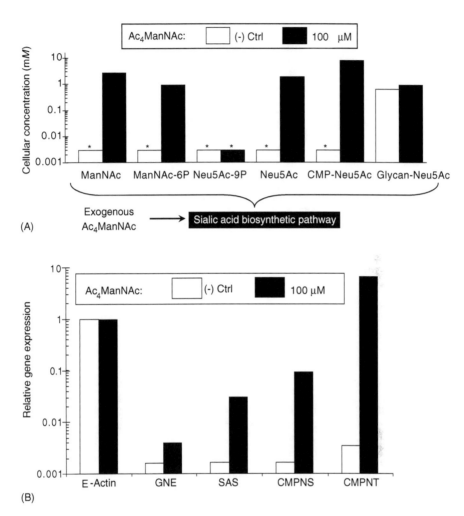

FIGURE 8.5 Metabolic flux through the sialic acid pathway and the resulting patterns of gene expression can be regulated by exogenous agents. (A) Metabolic flux through the sialic acid pathway can be increased by addition of exogenous metabolites, such as Ac$_4$ManNAc.[82,86] Without exogenous Ac$_4$ManNAc, pathway intermediates are maintained at low levels of less than ~ 5.0 μM in human Jurkat cells[62] (note that the "*" indicates the intermediate was not experimentally detectable, and was therefore present at less than ~1 – 5 μM) but glycoconjugate-bound sialic acid is easily measurable. When exogenous Ac$_4$ManNAc is added, or flux is increased by genetic methods,[57] the concentrations of several intracellular intermediates increase significantly but overall levels of cell surface sialic acid increase by less than twofold. (B) Increased metabolic flux through the sialic acid pathway alters the expression of the enzymes responsible for the production of cell surface-displayed sialic acids. Data from a representative experiment is shown where human embryoid-body-derived stem cells[87] are incubated with 100 μM Ac$_4$ManNAc for 3 days and the expression of pathway enzymes (Figure 8.4) responsible for the production of the metabolites shown in (A) are measured by real-time PCR as described in more detail in a separate publication.[88]

Example 3:   Time gradients in the sialic acid biosynthesis:

$$X_1 \xrightarrow{k_1} X_2 \xrightarrow{k_2} X_3 \xrightarrow{k_3} X_4$$

$$\frac{dX_3}{dt} = k_2 X_2 - k_3 X_3 \qquad\qquad (8.11a)$$

$$\frac{dX_2}{dt} = k_1 X_1 - k_2 X_2 \qquad\qquad (8.11b)$$

In Example 3, the rates of conversion for sialic acid-9-P and sialic acid can be described by the following equations, where each of the fluxes contributes to an overall rate equation. Using the knowledge that k_2 is much greater than k_1, we can construct a quasi-steady-state approximation for sialic acid-9-P (Equation 8.13):

$$k_2 >> k_1 \qquad\qquad (8.12)$$

$$X_2 = \frac{k_1 X_1}{k_2} \qquad\qquad (8.13)$$

If for a sufficiently measurable time period, it appears that the concentrations of these two metabolites are significantly correlated, they can be presumed proportional, related by a constant, here the ratio of the constants of the influx and efflux of sialic acid-9-P. The system can now be described by the rate equation for sialic acid, and a simple algebraic equation for sialic acid-9-P, as it exhibits little diversity in overall behavior due to the predominance of its rate constant of the conversion. In effect, sialic acid-9-P is considered to be at equilibrium, and the number of equations to solve has been reduced. In large networks, where they are many equations, such simplifications can greatly reduce computationally demanding solutions.

Similar assumptions can be made for the subsystems of metabolic pathways that reach steady state over a wider differential time scale. In the case that forward and reverse rate constants for a pair of sequentially altered metabolites are very high, the corresponding interconversion reaction will reach equilibrium quickly leading to a constant metabolite ratio or proportion, which can be assumed for the entire time course of their behavior. This scenario, referred to as the "rapid-equilibrium" assumption, reduces the dimension of the system, which in turn reduces the complexity of the mathematical model.[63,64] Such an assumption significantly simplifies a system, given that the rates of production and consumption of metabolites operate on significantly disparate time scales. Additionally, the validity of the rapid-equilibrium assumption increases when the initial transient response (before sequential metabolites are sufficiently balanced) is short. For more detailed information on time constants in metabolic systems, refer to standard textbooks on modeling in biology.[42,65]

VII.   EXPERIMENTAL METHODS

A.   DETERMINING METABOLIC FLUX

As mentioned, the concept of metabolic flux plays an important role in elucidating how pathways are controlled, and thus accurate methods of intracellular flux determination become critical to the modeling of metabolic systems. Extracellular flux measurement is the most common experimental method for determining intracellular flux distributions, and can be accomplished in simple pathways by isotope labeling. Isotopic labeling is attractive for flux determination, due to the diversity of radiolabeled substrates and the sensitivity of the radioactivity measurements. It is also preferred when metabolic flux balancing alone is insufficient.[29] Typical methods involve ^{13}C or ^{14}C isotopic labeling at a defined metabolic stationary state, and observing the resulting metabolic distributions via MS and NMR instruments.[66–68] The type of information retrieved will depend on the labeled compound as well as the size and complexity of the metabolic pathway. When pathways are cyclic and more complex, flux determination by labeling can be insufficient. For these cases, more complicated applications may be necessary, involving the complete enumeration of metabolite isotopomers or various numerical methods.[66] Additionally, it should be noted that extracellular flux measurements are limited in defining pathway connectivity. It can provide information about the distribution of metabolites and control at a small number of branch points, but is poorly suited for cases of metabolic cycling and pathways with 'split points' that reconverge at a later point.[37]

B.   METHODS OF CONTROL COEFFICIENT DETERMINATION

There exist a number of different methods for determining FCCs.[39,69] Direct enzymatic manipulation via controlling gene expression is the most basic way to determine metabolic product sensitivity to enzymatic activity.[45] Classical molecular biological techniques aim at increasing the copy number of an enzyme by various methods of transfection or by genetic engineering. These techniques are limited in that the former method does not allow for decreasing the gene activity level easily, and often results in large perturbations, obfuscating subtle relationships.[38] Recently, these shortcomings have been increasingly alleviated by the development of efficient techniques for altering enzymatic gene expression include inserting regulatory promoters sensitive to small molecule (such as tetracycline) regulation in front of the genes of interest. Variations on this method can modulate enzyme expression in either direction, as well as produce more refined perturbations.[70] In addition, small interfering RNA (siRNA or RNAi) technology now allows any defined gene sequence to be downregulated, and is proving to be a valuable tool for cell wide and systematic determination of gene function.[71–73]

VIII. CHALLENGES AND FUTURE DIRECTIONS IN MODELING GLYCOSYLATION

A. EXPERIMENTAL AND COMPUTATIONAL CHALLENGES

As previously illustrated, metabolic systems often exhibit complex, transient responses evolving over short time scales. These factors can make quantitative experimental observation difficult when one wants to test systematically a metabolic system with various inputs into the pathway or observe the metabolic flux distribution before the asymptotic or steady-state response. The experimental data required for building and validating a metabolic model need to be precise enough to *resolve* the unique behavior of various metabolic elements involved. High-order modeling and sophisticated techniques by themselves cannot extract complex behavior from simple dynamics. Ideally, a frequently sampled time course of enzymatic activities and metabolic concentrations would provide a reasonable framework for computational analysis. An obstacle to acquiring these data is the incomplete characterization of enzyme kinetics; for example, at present, kinetic parameters for the complete set of enzymes in even a subset of overall glycosylation, such as the sialic acid system (Figure 8.4), have not been determined. If the rate laws in the model depend on enzymatic kinetic and equilibrium constants that have not been determined experimentally, the constants become 'free parameters' in the model. An abundance of free parameters can make identification of unique system behavior difficult. To identify the true control architecture of the pathway, a reduction in the parametric freedom becomes necessary.

To cope with the indeterminacy of models with free parameters, we can switch to constraint-based modeling (such as FBA) that is less reliant upon enzymatic information, but is often computationally demanding. The calculation of precise solutions to large reaction networks can be impractically large. In this case, various methods to reduce the computational complexity need to be employed. There are many methods to tackle this problem, including the reduction of the pathway into smaller subsystems, and exploration of the *elementary modes* and *extreme pathways* of the system.[3, 49] As discussed further in the next section, a primary value of models is their ability to rule out false interpretations of data or dispute false metabolic connections.

B. MODEL VALIDATION

We once again consider the sialic acid metabolic system to provide an example of model validation. The possibility exists that abnormal diversion of UDP-GlcNAc into or out of the sialic acid pathway could perturb large-scale cellular physiological processes including the production of underlying oligosaccharide structures that are ultimately capped by terminal sialic acid residues (see Figure 8.6). Another possibility is that the O-GlcNAc protein modification signaling system, which also uses UDP-GlcNAc as a metabolic intermediate, could be disrupted (Figure 8.6). Based on our current understanding, the magnitude of these effects remains uncertain because UDP-GlcNAc levels are maintained in the 5 to 10 mM range in the cell while steady-state levels of sialic acid pathway intermediates are much lower (< 20 μM). On the

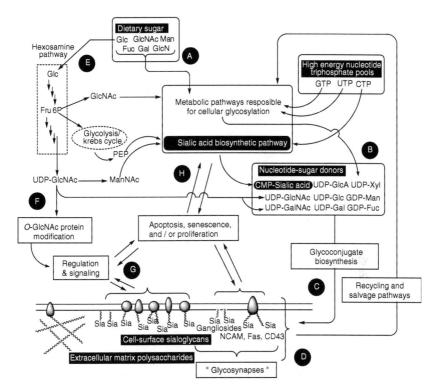

FIGURE 8.6   Glycosylation and sialic acid metabolism form many metabolic connections with other cellular systems. The first step in glycosylation is the import of dietary sugars such as glucose into a cell (A) followed by a series of phosphorylation, epimerization, and acetylation reactions that diversify these sugars and convert them into high-energy nucleotide sugar donors (B). These compounds serve as the "building blocks" for the assembly of complex carbohydrates in the ER and Golgi apparatus (C). The newly synthesized carbohydrates, which are generally attached to proteins or lipids, are then transported to the cell surface and contribute to the interface between a cell and its environment by playing important structural, adhesion, and signaling roles after they assemble in cell surface domains that have been dubbed "glycosynapses"[89] (D). In addition to their contributions to "core" glycan biosynthesis, cellular glycosylation processes and the sialic acid biosynthetic pathway form numerous additional connections with other cellular systems. To illustrate, glucose can also enter the hexosamine pathway (E). Fructose-6-phosphate, an intermediate in this pathway, can be diverted for the production of the high-energy intermediate phosphoenol pyruvate (PEP) used in sialic acid biosynthesis via the glycolysis/citric acid pathway. Otherwise, the hexosamine pathway produces UDP-GlcNAc, a compound needed for ManNAc production to supply metabolic flux into the sialic acid pathway (as shown in more detail in Figure 8.4). UDP-GlcNAc is also used in the biosynthesis of the underlying oligosaccharide chains, both directly and after epimerization to UDP-GalNAc. Diversion of UDP-GlcNAc for use in biosynthetic processes has the potential to perturb its role as a glucose sensor for the cell (90), as mediated through O-GlcNAc protein modification, a cellular signaling mechanism that affects hundreds of different molecules (F).[91] Interestingly, one of the regulatory and signaling roles of O-GlcNAc protein modification is to modulate apoptosis (G) and growing evidence is also establishing direct links between apoptosis and sialic acid biosynthesis (H). Deciphering these complex relationships, as well as many not shown here, provides a challenging task for both the glycobiologist and computational modeler.

surface, these numbers suggest that the input of flux into the sialic acid pathway probably has a negligible effect on UDP-GlcNAc levels; a static picture of the data, however, may obscure dynamic features that prove otherwise. A modeling approach, incorporating kinetic factors, could support this premise, or rule it out, thereby directing future experimental directions. For example, if an effect is seen, then genetic methods such as the transfection of the sialuria form of GNE to increase flux[74] or the GlcNAc 2-epimerase to divert flux away [56] from the pathway can be used for the experimental modulation cell-wide effects of sialic acid metabolism.

Mathematical models are always simplifications of observed metabolic processes and their outcomes, and thus make various assumptions either to interpret the results or to facilitate computational analysis. Incorporating these assumptions into a model requires the knowledge of the *ranges* of their validity, and often represents a compromise between adequacy of biological representation and mathematical simplicity. In this sense, the most critical part of model development is model validation. Model validation, however, is a misnomer, as models can generally only be invalidated; we therefore restrict the notion of validity to the ability of a model to resist falsification within its intended scope. Moreover, given the dependence of validity on assumptions, the validity can also hinge critically on factors such as the resolution of the observed behavior (precision of experimental data), and its suitability for supporting the hypothesis.

The most relevant data for model development are those which can resolve unique behaviors of the system, which only certain mathematical formulations can represent. In this sense, even comprehensive, high-throughput data now readily available from the emergence and persistence of the "omics" era may not suit the model hypothesis, or assert validity within its intended scope.[75] Consequently, it is important that experiments are carefully designed to support desired conclusions. Experimental perturbation of enzyme activities designed to elucidate mechanisms of control, for example, may not reveal the absolute boundaries of metabolic flux. Similarly, constraint-based approaches may not directly support conclusions related to metabolic regulation, and therefore pathway alteration may not be useful in this respect.[28] Thus, tight integration of modeling with experiment increases the reliability of the theoretical framework and the parameters used in the model toward the reproduction of the acquired data, increasing the confidence in model predictions, if only to reject previous hypotheses as false.

Further model development requires precise definition of the allowable "states" of the metabolic system, the ranges of metabolite concentrations and parameter values, as well as the definition of the initial conditions or external inputs. These are the system parameters and variables, which must exist within a physiological range, as determined by the observed data set, in order to make valid predictions. Moreover, they contribute to define uniquely a solution that reproduces experimental data with certain accuracy and a corresponding level of confidence. The confidence (or uncertainty) may or may not be determined rigorously. Again depending on the scope, it may be sufficient that model solutions merely reproduce dynamic phenomena. Often it is the case that certain models reproduce certain system behaviors, such as certain stability criteria (as in bifurcation analysis), or nonlinear phenomena (such as sigmoidal responses). Quantitative methods include statistical methods which deal with the precision of the model predictions and thresholds of regression procedures. Employing a statistical method of this sort for validation requires small confidence intervals for fitted parameters and the high experimental data reproducibility. In general, both phenomenological observations as well as precise statistical verification contribute to model validity. The definition of validity, however, is subject to change during model development because it is an iterative process, continually withdrawing incorrect assumptions and making new ones in order to increase the predictive power.

C. GLYCOSYLATION LIES AT THE INTERSECTION OF SIGNALING AND METABOLISM

Glycosylation in general and sialic acid in particular lies at the intersection of metabolism and cellular signaling. As mentioned above and outlined in Figure 8.6, the possibility arises that altered flux through the sialic acid pathway could indirectly perturb cellular regulatory systems such as O-GlcNAc protein modification. In turn, O-GlcNAc protein modification impacts many metabolic events, for example, overall levels of O-GlcNAc increase during times of cellular stress to provide a prosurvival mechanism to thwart apoptotic impulses of a cell.[76] In addition to intracellular metabolic effects, specific cell surface-displayed products of the sialic acid biosynthetic pathway are now known to modulate directly apoptosis as well. During the early stages of apoptosis, *reduced* sialylation of Fas increases sensitivity to inducting agents [77,78] and *increased* sialylation of CD43 allows this glycoprotein to cap, a response which leads to phagocytosis of the host cell.[79] Changes in flux through the pathway, in response to exogenously supplied metabolic intermediates,[80] can also influence the composition of a set of gangliosides that has been described as a switch between senescence, proliferation, and apoptosis (see Figure 8.7). A challenging problem is to determine how individual sialic acid epitopes achieve selective up- or downregulation, even in the face of the overall downregulation of sialic acid display that occurs during apoptosis.[58,81]

Adding to the complexity of attempts to understand the role of sialic acid in apoptosis are observations that not only do the products of the pathway mediate important steps in the apoptotic process, but also that the reverse appears to be true. More specifically, the apoptotic process and growth state of a cell determines the metabolic capacity of the sialic acid pathway as seen by a connection that has been established between robust cell growth and maximized flux through the pathway.[82] Conversely, cells undergoing apoptosis are unable to support high levels of, and may actually shut down, metabolic flux through the sialic acid pathway.[58] As of now, many questions remain whether these changes to sialic acid metabolism are the cause or result of apoptosis; for example is the inhibition of metabolic flux through the sialic acid pathway in Ac_4ManNLev-treated cells[58] the cause or result of apoptosis in the treated cells?[†] Answering this type of question through the assistance of computation methods will provide valuable cues to direct future research; for instance in the example listed above establishing that Ac_4ManNLev is the *cause*

[†]Ac_4 ManNLev is a nonnatural metabolic intermediate used in "biochemical engineering" of sialic acid synthesis as described further in Chapter 13 of this book.

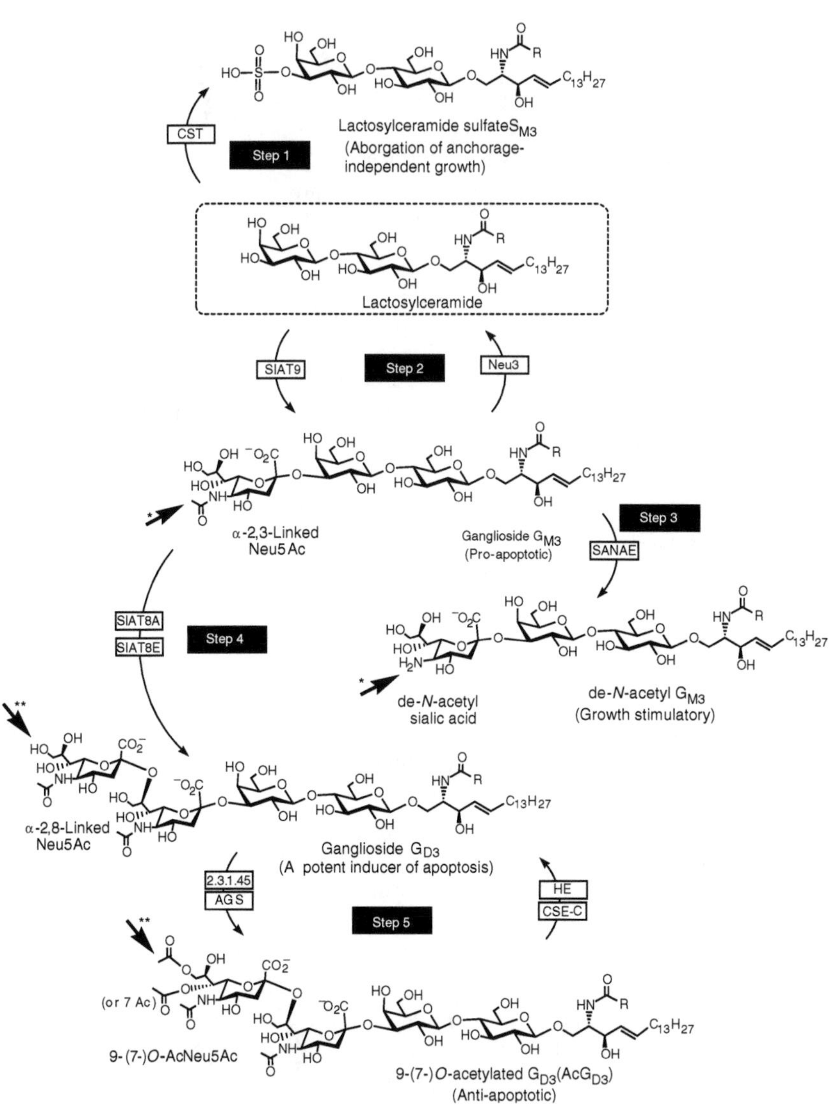

FIGURE 8.7. The cellular signaling properties of the lactosylceramide glycolipid (LacCer) are governed by sulfation and sialic acid. *Step 1*. Sulfation of LacCer on the 3-position of the terminal galactose suppresses the ability of cancer cells to undergo anchorage-independent growth.[92] *Step 2*. Addition of a single α-2,3-linked sialic acid to the same position of LacCer produces the G_{M3} ganglioside, which has been described as having proapoptotic properties,[93–95] or in other cases, G_{M3} has been reported to increase the proliferation of cancer cells.[92] *Step 3*. The seemingly disparate responses of G_{M3} may be a result of the further modification of this molecule as de-*N*-acetylation of the α-2,3-linked sialic acid residue stimulates cell growth,[96] while elaboration of G_{M3} with a second, α-2,8-linked sialic acid (*Step 4*) creates the G_{D3} ganglioside, which is a potent inducer of apoptosis.[97–99] Finally, addition of a 9-(or 7-)*O*-acetyl group to the terminal sialic acid produces AcG_{D3} (*Step 5*); this ganglioside, in contrast to G_{M3} and G_{D3}, has anti-apoptotic properties.[100–102]

(as opposed to the result) of apoptosis opens the door to the development of a new class of cancer treatments based on the growing importance of apoptosis-inducers as therapeutic agents.

IX. CONCLUDING REMARKS

Efforts to use computational modeling approaches to decipher human glycosylation processes are in their infancy. Consequently, the purpose of this chapter was not to provide an extensive set of results and conclusions, which, in any event, are not available at this time. Instead, it is intended to provide an introduction to metabolic modeling for the nonspecialist in computational methods, and hopefully stimulate efforts that coalesce at the intersection of glycobiology and mathematics. Rapid progress in amassing experimental data on the many components of the glycosylation machinery, combined with ever-increasing modeling capabilities, suggest that the combination of these two approaches now have the capability of greatly facilitating both a basic understanding of the biological underpinnings of glycosylation as well as providing powerful new tools for the carbohydrate engineer.

REFERENCES

1. Bailey, J.E., Complex biology with no parameters, *Nat. Biotechnol.*, 19, 503–504, 2001.
2. Adami, C., Ofria, C., and Collier, T.C., Evolution of biological complexity, *Proc. Natl. Acad. Sci. USA*, 97, 4463–4468, 2000.
3. Papin, J.A., Price, N.D., Wiback, S.J., Fell, D.A., and Palsson, B.O., Metabolic pathways in the post-genome era, *Trends Biochem. Sci.*, 28, 250–258, 2003.
4. Rudd, P.M., Endo, T., Colominas, C., Groth, D., Wheeler, S.F., Harvey, D.J., Wormald, M.R., Serban, H., Prusiner, S.B., Kobata, A., and Dwek, R.A., Glycosylation differences between the normal and pathogenic prion protein isoforms, *Proc. Natl. Acad. Sci. USA*, 96, 13044–13049, 1999.
5. Endo, T., Groth, D., Prusiner, S.B., and Kobata, A., Diversity of oligosaccharide structures linked to asparagines of the scrapie prion protein, *Biochemistry*, 28, 8380–8388, 1989.
6. Lanza, F., Healy, L., and Sutherland, D.R., Structural and functional features of the CD34 antigen: an update, *J. Biol. Regul. Homeost. Agents*, 15, 1–13, 2001.
7. Puri, K.D., Finger, E.B., Gaudernack, G., and Springer, T.A., Sialomucin CD34 is the major L-selectin ligand in human tonsil high endothelial venules, *J. Cell. Biol.*, 131, 261–270, 1995.
8. Gervais, A., Hammel, Y.A., Pelloux, S., Lepage, P., Baer, G., Carte, N., Sorokine, O., Strub, J.M., Koerner, R., Leize, E., and Van Dorsselaer, A., Glycosylation of human recombinant gonadotrophins: characterization and batch-to-batch consistency, *Glycobiology*, 13, 179–189, 2003.
9. Golhke, M., Mach, U., Nuck, R., Zimmermann-Kordmann, M., Grunow, D., Fieger, C., Volz, B., Tauber, R., Thomas, P., Debus, N., and Reutter, W., Carbohydrate structures of soluble human L-selectin recombinantly expressed in baby-hamster kidney cells, *Biotechnol. Appl. Biochem.*, 32, 41–51, 2000.
10. Cooper, C.A., Gasteiger, E., and Packer, N.H., GlycoMod — A software tool for determining glycosylation compositions from mass spectrometric data, *Proteomics*, 1, 340–349, 2001.

11. Hirabayashi, J. and Kasai, K.-I., Separation technologies for glycomics, *J. Chromatogr. B*, 771, 67–87, 2002.
12. Kaufmann, H., Bailey, J.E., and Fussenegger, M., Use of antibodies for detection of phosphorylated proteins separated by two-dimensional gel electrophoresis, *Proteomics*, 1, 194–199, 2001.
13. Kuster, B., Krogh, T.N., Mortz, E., and Harvey, D.J., Glycosylation analysis of gel-separated proteins, *Proteomics*, 1, 350–361, 2001.
14. Taniguchi, N., Ekuni, A., Ko, J.H., Miyoshi, E., Ikeda, Y., Ihara, Y., Nishikawa, A., Honke, K., and Takahashi, M., A glycomic approach to the identification and characterization of glycoprotein function in cells transfected with glycosyltransferase genes, *Proteomics*, 1, 239–247, 2001.
15. Chen, H., Wang, Z., Sun, Z., Kim, E.J., and Yarema, K.J. Mammalian glycosylation: an overview of carbohydrate biosynthesis, in *Hand book of Carbohydrate Engineering*, Yarema K., Ed., Dekkar CRC Press, Boca Raton, 2005, chap. 1.
16. Monica, T.J., Andersen, D.C., and Goochee, C.F., A mathematical model of sialylation of *N*-linked oligosaccharides in the *trans*-Golgi network, *Glycobiology*, 7, 515–521, 1997.
17. Toma, L., Pinhal, M.A.S., Dietrich, C.P., Nader, H.B., and Hirschberg, C.B., Transport of UDP-galactose into the Golgi lumen regulates the biosynthesis of proteoglycans, *J. Biol. Chem.*, 271, 3897–3901, 1996.
18. Luhn, K., Wild, M.K., Eckhardt, M., Gerardy-Schahn, R., and Vestweber, D., The gene defective in leukocyte adhesion deficiency II encodes a putative GDP-fucose transporter, *Nat. Genet.*, 28, 69–72, 2001.
19. Marquardt, T., Brune, T., Luhn, K., Zimmer, K.-P., Körner, C., Fabritz, L., Van der Werft, N., Vormoor, J., Freeze, H.H., and Louwen, F., Leukocyte adhesion deficiency II syndrome, a generalized defect in fucose metabolism, *J. Pediatr.*, 134, 681–688, 1999.
20. Hirschberg, C.B., Golgi nucleotide sugar transport and leukocyte adhesion deficiency II, *J. Clin. Invest.*, 108, 3–6, 2001.
21. Lübke, T., Marquardt, T., Von Figura, K., and Körner, C., A new type of carbohydrate-deficient glycoprotein syndrome due to a decreased import of GDP-fucose in the Golgi, *J. Biol. Chem.*, 274, 25986–26989, 1999.
22. Wild, M.K., Lühn, K., Marquardt, T., and Vestweber, D., Leukocyte adhesion deficiency II: therapy and genetic defect, *Cells Tissues Organs*, 172, 161–173, 2002.
23. Sturla, L., Rampal, R., Haltiwanger, R.S., Fruscione, F., Etzioni, A., and Tonetti, M., Differential terminal fucosylation of *N*-linked glycans versus protein *O*-fucosylation in leukocyte adhesion deficiency type II (CDG IIc), *J. Biol. Chem.*, 278, 26727–26733, 2003.
24. Korner, C., Linnebank, M., Koch, H.G., Harms, E., von Figura, K., and Marquardt, T., Decreased availability of GDP-L-fucose in a patient with LAD II with normal GDP-D-mannose dehydratase and FX protein activities, *J. Leukoc. Biol.*, 66, 95–98, 1999.
25. Sturla, L., Etzioni, A., Bisso, A., Zanardi, D., De Flora, G., Silengo, L., De Flora, A., and Tonetti, M., Defective intracellular activity of GDP-D-mannose-4,6-dehydratase in leukocyte adhesion deficiency type II syndrome, *FEBS Lett.*, 429, 274–278, 1998.
26. Etzioni, A., Sturla, L., Antonellis, A., Green, E.D., Gershoni-Baruch, R., Berninsone, P.M., Hirschberg, C.B., and Tonetti, M., Leukocyte adhesion deficiency (LAD) type II/carbohydrate deficient glycoprotein (CDG) IIc founder effect and genotype/phenotype correlation, *Am. J. Med. Genet.*, 110, 131–135, 2002.

27. Sturla, L., Puglielli, L., Tonetti, M., Berninsone, P., Hirschberg, C.B., De Flora, A., and Etzioni, A., Impairment of the Golgi GDP-L-fucose transport and unresponsiveness to fucose replacement therapy in LAD II patients, *Pediatr. Res.*, 49, 537–542, 2001.

28. Wiechert, W., Modeling and simulation: tools for metabolic engineering, *J. Biotechnol.*, 94, 37–63, 2002.

29. Morgan, J.A. and Rhodes, D., Mathematical modeling of plant metabolic pathways, *Metab. Eng.*, 4, 80–89, 2002.

30. Covert, M.W., Schilling, C.H., and Palsson, B., Regulation of gene expression in flux balance models of metabolism, *J. Theoret. Biol.*, 213, 73–88, 2001.

31. Kitano, H., Systems biology: a brief overview, *Science*, 295, 1662–1664, 2002.

32. Ideker, T., Systems biology 101 — what you need to know, *Nat. Biotechnol.*, 22, 473–475, 2004.

33. Ideker, T., Galitski, T., and Hood, L., A new approach to decoding life: systems biology, *Annu. Rev. Genomics Hum. Genet.*, 2, 343–372, 2001.

34. Ravasz, E., Somera, A.L., Mongru, D.A., Oltvai, Z.N., and Barabasi, A.L., Hierarchical organization of modularity in metabolic networks, *Science*, 297, 1551–1555, 2002.

35. Jeong, H., Tombor, B., Albert, R., Oltvai, Z.N., and Barabasi, A.L., The large-scale organization of metabolic networks, *Nature*, 407, 651–654, 2000.

36. Nielsen, J., Metabolic engineering: techniques for analysis of targets for genetic manipulations, *Biotechnol. Bioeng.*, 58, 125–132, 1998.

37. Stephanopoulos, G., Aristidou, A.A., and Nielsen, J.H.i., *Metabolic engineering : principles and Methodologies*, Vol xxi, Academic Press, San Diego, 1998, p. 725.

38. Fell, D.A., Metabolic control analysis: a survey of its theoretical and experimental development, *Biochem. J.*, 286 (Part 2), 313–330, 1992.

39. Fell, D., Understanding the Control of Metabolism, in *Frontiers in metabolism*, Vol 2. Miami Brookfield, VT, Portland Press London (Distributed by Ashgate Pub. Co. in North America. Vol xii, 301.) 1997.

40. Savageau, M.A., Biochemical systems analysis. I. Some mathematical properties of the rate law for the component enzymatic reactions, *J. Theor. Biol.*, 25, 365–369, 1969.

41. Savageau, M.A., Biochemical systems analysis. II. The steady-state solutions for an n-pool system using a power-law approximation, *J. Theor. Biol.*, 25, 370–379, 1969.

42. Heinrich, R. and Schuster, S., *The Regulation of Cellular Systems*, Vol xix, Chapman & Hall, New York, 1996, p. 372.

43. Covert, M.W., Famili, I., and Palsson, B.O., Identifying constraints that govern cell behavior: a key to converting conceptual to computational models in biology? *Biotechnol. Bioeng.*, 84, 763–772, 2003.

44. Beard, D.A., Liang, S.D., and Qian, H., Energy balance for analysis of complex metabolic networks. *Biophys. J.*, 83, 79–86, 2002.

45. Morgan, J.A. and Rhodes, D., Mathematical modeling of plant metabolic pathways, *Metab. Eng.*, 4, 80–89, 2002.

46. Wahl, S.A., Dauner, M., and Wiechert, W., New tools for mass isotopomer data evaluation in ^{13}C flux analysis: mass isotope correction, data consistency checking, and precursor relationships, *Biotechnol. Bioeng.*, 85, 259–268, 2004.

47. Kauffman, K.J., Prakash, P., and Edwards, J.S., Advances in flux balance analysis, *Curr. Opin. Biotechnol.*, 14, 491–496, 2003.

48. Anton, H. and Rorres, C., *Elementary Linear Algebra : Applications Version*. 8th ed., Wiley, New York, Vol xvi, 2000, p. 822.

49. Schilling, C.H., Letscher, D., and Palsson, B.O., Theory for the systemic definition of metabolic pathways and their use in interpreting metabolic function from a pathway-oriented perspective, *J. Theor. Biol.*, 203, 229–248, 2000.

50. Palsson, B., The challenges of *in silico* biology, *Nat. Biotechnol.*, 18, 1147–1150, 2000.

51. Edwards, J.S. and Palsson, B.O., The *Escherichia coli* MG1655 *in silico* metabolic genotype: its definition, characteristics, and capabilities, *Proc. Natl. Acad. Sci. USA.*, 97, 5528–5533, 2000.

52. Ibarra, R.U., Edwards, J.S., and Palsson, B.O., *Escherichia coli* K-12 undergoes adaptive evolution to achieve *in silico* predicted optimal growth, *Nature*, 420, 186–189, 2002.

53. Edwards, J.S., Ibarra, R.U., and Palsson, B.O., *In silico* predictions of *Escherichia coli* metabolic capabilities are consistent with experimental data, *Nat. Biotechnol.*, 19, 125–130, 2001.

54. Keppler, O.T., Hinderlich, S., Langner, J., Schawartz-Albiez, R., Reutter, W., and Pawlita, M., UDP-GlcNAc 2-epimerase: a regulator of cell surface sialylation, *Science*, 284, 1372–1376, 1999.

55. Newsholme, E.A. and Start, C., *Regulation in Metabolism.* Vol xiii, Wiley, London, New York, 1973, p. 349.

56. Luchansky, S.J., Yarema, K.J., Takahashi, S., and Bertozzi, C.R., GlcNAc 2-epimerase can serve a catabolic role in sialic acid metabolism, *J. Biol. Chem.*, 278, 8036–8042, 2003.

57. Yarema, K.J., Goon, S., and Bertozzi, C.R., Metabolic selection of glycosylation defects in human cells, *Nat. Biotechnol.*, 19, 553–558, 2001.

58. Kim, E.J., Sampathkumar, S.-G., Jones, M.B., Rhee, J.K., Baskaran, G., and Yarema, K.J., Characterization of the metabolic flux and apoptotic effects of *O*-hydroxyl- and *N*-acetylmannosamine (ManNAc) analogs in Jurkat (human T-lymphoma-derived) cells, *J. Biol. Chem.*, 279, 18342–18352, 2004.

59. Schauer, M. and Heinrich, R., Analysis of the quasi-steady-state approximation for an enzymatic one-substrate reaction, *J. Theor. Biol.*, 79, 425–442, 1979.

60. Segel, L.A., On the validity of the steady state assumption of enzyme kinetics, *Bull. Math. Biol.*, 50, 579–593, 1988.

61. Briggs, G.E. and Haldane, J.B.S., A note on the kinetics of enzyme action, *Biochem. J.*, 19, 338–339, 1925.

62. Jacobs, C.L., Goon, S., Yarema, K.J., Hinderlich, S., Hang, H.C., Chai, D.H., and Bertozzi, C.R., Substrate specificity of the sialic acid biosynthetic pathway, *Biochemistry*, 40, 12864–12874, 2001.

63. Moss, M.L., Kuzmic, P., Stuart, J.D., Tian, G., Peranteau, A.G., Frye, S.V., Kadwell, S.H., Kost, T.A., Overton, L.K., and Patel, I.R., Inhibition of human steroid 5alpha reductases type I and II by 6-aza-steroids: structural determinants of one-step vs two-step mechanism, *Biochemistry*, 35, 3457–3464, 1996.

64. Waas, W.F. and Dalby, K.N., Transient protein–protein interactions and a random-ordered kinetic mechanism for the phosphorylation of a transcription factor by extra-cellular-regulated protein kinase 2, *J. Biol. Chem.*, 277, 12532–12540, 2002.

65. Murray, J.D., Interdisciplinary applied mathematics, in *Mathematical Biology.* 3rd ed., Springer, New York, 2002.

66. Wiechert, W. and de Graaf, A.A., *In vivo* stationary flux analysis by 13C labeling experiments, *Adv. Biochem. Eng. Biotechnol.*, 54, 109–154, 1996.

67. Wiechert, W. and Wurzel, M., Metabolic isotopomer labeling systems. Part I: global dynamic behavior, *Math. Biosci.*, 169, 173–205, 2001.

68. Wittmann, C., Metabolic flux analysis using mass spectrometry, *Adv. Biochem. Eng. Biotechnol.*, 74, 39–64, 2002.

69. Kacser, H. and Acerenza, L., A universal method for achieving increases in metabolite production, *Eur. J. Biochem.*, 216, 361–367, 1993.

70. Ruyter, G.J., Postma, P.W., and van Dam, K., Control of glucose metabolism by enzyme IIGlc of the phosphoenolpyruvate-dependent phosphotransferase system in *Escherichia coli. J. Bacteriol.*, 173, 6184–6191, 1991.

71. Chi, J.T., Chang, H.Y., Wang, N.N., Chang, D.S., Dunphy, N., and Brown, P.O., Genome wide view of gene silencing by small interfering RNAs, *Proc. Natl. Acad. Sci. USA*, 100, 6343–6346, 2003.

72. Harborth, J., Elbashir, S., Bechert, M, K., Tuschl, T., and Weber, K., Identification of essential genes in cultured mammalian cells using small interfering RNAs, *J. Cell. Sci.*, 114, 4557–4565, 2001.

73. Kuwabara, P.E. and Coulson, A., RNAi — prospects for a general technique for determining gene function, *Parasitol. Today*, 16, 347–349, 2000.

74. Seppala, R., Lehto, V.P., and Gahl, W.A., Mutations in the human UDP-*N*-acetylglucosamine 2-epimerase gene define the disease sialuria and the allosteric site of the enzyme, *Am. J. Hum. Genet.*, 64, 1563–1569, 1999.

75. Nicholson, J.K. and Wilson, I.D., Opinion: understanding 'global' systems biology: metabonomics and the continuum of metabolism, *Nat. Rev. Drug Discov.*, 2, 668–676, 2003.

76. Zachara, N.E., Butkinaree, C., and Hart, G.W., O-GlcNAc: a new paradigm for modulating cellular responses to stress, *Glycobiology*, 13, 833, 2003.

77. Keppler, O.T., Peter, M.E., Hinderlich, S., Moldenhauer, G., Stehling, P., Schmitz, I., Schwartz-Albiez, R., Reutter, W., and Pawlita, M., Differential sialylation of cell surface glycoconjugates in a human B lymphoma cell line regulates susceptibility for CD95 (APO-1/Fas)-mediated apoptosis and for infection by a lymphotropic virus, *Glycobiology*, 9, 557–569, 1999.

78. Suzuki, O., Nozawa, Y., and Abe, M., Sialic acids linked to glycoconjugates of Fas regulate the caspase-9-dependent and mitochondria-mediated pathway of Fas-induced apoptosis in Jurkat T cell lymphoma, *Int. J. Oncol.*, 23, 769–774, 2003.

79. Eda, S., Yamanaka, M., and Beppu, M., Carbohydrate-mediated phagocytic recognition of early apoptotic cells undergoing transient capping of CD43 glycoprotein, *J. Biol. Chem.*, 279, 5967–5974, 2004.

80. Ghosh, P., Ender, I., and Hale, E.A., Long-term ethanol consumption selectively impairs ganglioside pathway in rat brain, *Alcohol. Clin. Exp. Res.*, 22, 1220–1226, 1998.

81. Azuma, Y., Taniguchi, A., and Matsumoto, K., Decrease in cell surface sialic acid in etoposide-treated Jurkat cells and the role of cell surface sialidase, *Glycoconj. J.*, 17, 301–306, 2000.

82. Jones, M.B., Teng, H., Rhee, J.K., Baskaran, G., Lahar, N., and Yarema, K.J., Characterization of the cellular uptake and metabolic conversion of acetylated *N*-acetylmannosamine (ManNAc) analogs to sialic acids, *Biotechnol. Bioeng.*, 85, 394–405, 2004.

83. Hinderlich, S., Oetke, C., and Pawlita, M., Biochemical engineering of sialic acids, in *Handbook of Carbohydrate Engineering*, Yarema, K., Ed., Dekkar CRC Press, Boca Raton, 2005, chap. 13.

84. Nakayama, J., Angata, K., Ong, E., Katsuyama, T., and Fukuda, M., Polysialic acid, a unique glycan that is developmentally regulated by two polysialyltransferases, PST and STX, in the central nervous system: from biosynthesis to function, *Pathol. Int.*, 48, 665–677, 1998.

85. Muhlenhoff, M., Eckhardt, M., and Gerardy-Schahn, R., Polysialic acid: three-dimensional structure, biosynthesis, and function, *Curr. Opin. Struct. Biol.*, 8, 558–564, 1998.

86. Yarema, K.J., Mahal, L.K., Bruehl, R.E., Rodriguez, E.C., and Bertozzi, C.R., Metabolic delivery of ketone groups to sialic acid residues. Application to cell surface glycoform engineering, *J. Biol. Chem.*, 273, 31168–31179, 1998.

87. Shamblott, M.J., Axelman, J., Littlefield, J.W., Blumenthal, P.D., Huggins, G.R., Cui, Y., Cheng, L., and Gearhart, J.D., Human embryonic germ cell derivatives express a broad range of developmentally distinct markers and proliferate extensively *in vitro*, *Proc. Natl. Acad. Sci. USA*, 98, 113–118, 2001.

88. Murrell, M.P., Li, A., Baskaran, G., and Yarema, K.J., Modulation of metabolic flux regulates the expression of sialic acid processing enzymes in human cells, Manuscript in preparation.

89. Hakomori, S.-I., The glycosynapse, *Proc. Natl. Acad. Sci. USA*, 99, 225–232, 2002.

90. Hanover, J.A., Glycan-dependent signaling: *O*-linked *N*-acetylglucosamine, *FASEB J.*, 15,1865–1876, 2001.

91. Zachara, N.E. and Hart, G.W., The emerging significance of *O*-GlcNAc in cellular regulation, *Chem. Rev.*, 102, 431–438, 2002.

92. Uemura, S., Kabayama, K., Noguchi, M., Igarashi, Y., and Inokuchi, J.-i., Sialylation and sulfation of lactosylceramide distinctly regulate anchorage-independent growth, apoptosis, and gene expression in 3LL Lewis lung carcinoma cells, *Glycobiology*, 13, 207–216, 2003.

93. Noll, E.N., Lin, J., Nakatsuji, Y., Miller, R.H., and Black, P.M., GM3 as a novel growth regulator for human gliomas, *Exp. Neurol.*, 168, 300–309, 2001.

94. Zhou, J., Shao, H., Cox, N.R., Baker, H.J., and Ewald, S.J., Gangliosides enhance apoptosis of thymocytes, *Cell. Immunol.*, 183, 90–98, 1998.

95. Molotkovskaya, I.M., Kholodenko, R.V., and Molotkovsky, J.G., Influence of gangliosides on the IL-2- and IL-4-dependent cell proliferation, *Neurochem. Res.*, 27, 761–770, 2002.

96. Hanai, N., Dohi, T., Nores, G.A., and Hakomori, S., A novel ganglioside, de-*N*-acetyl-GM3 (II3NeuNH2LacCer), acting as a strong promoter for epidermal growth factor receptor kinase and as a stimulator for cell growth, *J. Biol. Chem.*, 263, 6296–6301, 1988.

97. Castro-Palomino, J.C., Simon, B., Speer, O., Leist, M., and Schmidt, R.R., Synthesis of ganglioside GD3 and its comparison with bovine GD3 with regard to oligodendrocyte apoptosis mitochondrial damage, *Chemistry*, 7, 2178–2184, 2001.

98. Scorrano, L., Petronilli, V., Di Lisa, F., and Bernard, P., Commitment to apoptosis by GD3 ganglioside depends on opening of the mitochondrial permeability transition pore, *J. Biol. Chem.*, 274, 22581–22585, 1999.

99. Melchiorri, D., Martini, F., Lococo, E., Gradini, R., Barletta, E., De Maria, R., Caricasole, A., Nicoletti, F., and Lenti, L., An early increase in the disialoganglioside GD3 contributes to the development of neuronal apoptosis in culture, *Cell. Death. Differ.*, 9, 609–615, 2002.

100. Malisan, F., Franchi, L., Tomassini, B., Ventura, N., Condo, I., Rippo, M.R., Rufini, A., Liberati, L., Nachtigall, C., Kniep, B., and Testi, R., Acetylation suppresses the proapoptotic activity of GD3 ganglioside, *J. Exp. Med.*, 196, 1535–1541, 2002.

101. Cheresh, D.A., Reisfeld, R.A., and Varki, A., *O*-acetylation of disialoganglioside GD3 by human melanoma cells creates a unique antigenic determinant, *Science.*, 225, 844–846, 1984.

102. Chen, H.Y. and Varki, A., *O*-acetylation of GD3: an enigmatic modification regulating apoptosis, *J. Exp. Med.*, 196, 1529–1533, 2002.

103. Schauer, R., Biosynthese von *N*-Acetyl-*O*-acetylneuraminsauren, I. Inkorporation von [14C]Acetate in Schnitte der Unterkieferspeicheldruse von Rind und Pferd, *Hoppe-Seyler's Z. Physiol. Chem.*, 351, 595–602, 1970.

104. Schauer, R., Biosynthese von *N*-Acetyl-*O*-acetylneuraminsauren, II. Untersuchungen uber Substrat und intracellulare Lokaslisation der Acetyl-Coenzym A: *N*-Acetylneuraminat-7- und 8-*O*-Acetyltransferase vom Rind, *Hoppe-Seyler's Z. Physiol. Chem.*, 351, 749–758, 1970.

105. Jourdian, G.W., Swanson, A., Watson, D., and Roseman, S., *N*-Acetylneuraminic (sialic) acid 9-phosphatase, *Methods Enzymol.*, 8, 205–208, 1966.

106. Ghosh, S. and Roseman, S., The sialic acids. IV. *N*-Acyl-D-glucosamine 6-phosphate 2-epimerase, *J. Biol. Chem.*, 240, 1525–1530, 1965.

107. Shi, W.X., Chammas, R., and Varki, A., Induction of sialic acid 9-*O*-acetylation by diverse gene products: implications for the expression cloning of sialic acid *O*-acetyltransferases, *Glycobiology*, 8, 199–205, 1998.

108. Ogura, K., Nara, K., Watanabe, Y., Kohno, K., Tai, T., and Sanai, Y., Cloning and expression of cDNA for *O*-acetylation of GD3 ganglioside, *Biochem. Biophys. Res. Commun.*, 225, 932–938, 1996.

109. Chou, H.H., Takematsu, H., Diaz, S., Iber, J., Nickerson, E., Wright, K.L., Muchmore, E.A., Nelson, D.L., Warren, S.T., and Varki, A., A mutation in human CMP-sialic acid hydroxylase occurred after the Homo-Pan divergence, *Proc. Natl. Acad. Sci. USA*, 95, 11751–11756, 1998.

110. Munster, A.-K., Eckhardt, M., Potvin, B., Muhlenhoff, M., Stanley, P., and Gerardy-Schahn, R., Mammalian cytidine 5-prime-monophosphate *N*-acetylneuraminic acid synthetase: a nuclear protein with evolutionarily conserved structural motifs, *Proc. Natl. Acad. Sci. USA*, 95, 9140–9145, 1998.

111. Ishida, N., Ito, M., Yoshioka, S., Sun-Wada, G.-H., and Kawakita, M., Functional expression of human Golgi CMP-sialic acid transporter in the Golgi complex of a transporter-deficient Chinese hamster ovary cell mutant, *J. Biochem.*, 124, 171–178, 1998.

112. Takematsu, H., Diaz, S., Stoddart, A., Zhang, Y., and Varki, A., Lysosomal and cytosolic sialic acid 9-*O*-acetylesterase activities can be encoded by one gene via differential usage of a signal peptide-encoding exon at the *N* terminus, *J. Biol. Chem.*, 274, 25623–25631, 1999.

113. Shen, Y., Kohla, G., Lrhorfi, A.L., Sipos, B., Kalthoff, H., Gerwig, G.J., Kamerling, J.P., Schauer, R., and Tiralongo, J., *O*-acetylation and de-*O*-acetylation of sialic acids in human colorectal carcinoma, *Eur. J. Biochem.*, 271, 281–290, 2004.

114. Honke, K., Tsuda, M., Hirahara, Y., Ishii, A., Makita, A., and Wada, Y., Molecular cloning and expression of cDNA encoding human 3'-phosphoadenylyl-sulfate:galactosylceramide 3'-sulfotransferase, *J. Biol. Chem.*, 272, 4864–4868, 1997.

115. Pastuszak, I., Ketchum, C., Hermanson, G., Sjoberg, E.J., Drake, R.D., and Elbein, A.D., GDP-L-fucose pyrophosphorylase. Purification, cDNA cloning, and properties of the enzyme, *J. Biol. Chem.*, 273, 30165–30174, 1998.

116. Lübke, T., Marquardt, T., Etzioni, A., Hartmann, E., Von Figura, K., and Körner, C., Complementation cloning identifies CDG-IIc, a new type of congenital disorders of glycosylation, as a GDP-fucose transporter deficiency, *Nat. Genet.*, 28, 73–76, 2001.

117. Hinderlich, S., Berger, M., Blume, A., Chen, H., Ghaderi, D., and Bauer, C., Identification of human L-fucose kinase amino acid sequence, *Biochem. Biophys. Res. Commun.*, 294, 650–654, 2002.

118. Glaser, L., The biosynthesis of *N*-acetylgalactosamine, *J. Biol. Chem.*, 234, 2801–2805, 1959.
119. Kornfeld, S. and Glaser, L.T., The synthesis of thymidine-linked sugars. V. Thymidine diphosphate-amino sugars, *J. Biol. Chem.*, 237, 3052–3059, 1962.
120. Daude, N., Gallaher, T.K., Zeschnigk, M., Starzinski-Powitz, A., Petry, K.G., Haworth, I. S., and Reichardt, J.K., Molecular cloning, characterization, and mapping of a full-length cDNA encoding human UDP-galactose 4′-epimerase, *Biochem. Mol. Med.*, 56, 1–7, 1995.
121. Sullivan, F.X., Kumar, R., Kriz, R., Stahl, M., Xu, G.Y., Rouse, J., Chang, X.J., Boodhoo, A., Potvin, B., and Cumming, D.A., Molecular cloning of human GDP-mannose 4,6-dehydratase and reconstitution of GDP-fucose biosynthesis *in vitro*, *J. Biol. Chem.*, 273, 8193–8202, 1998.
122. Ohyama, C., Smith, P.L., Angata, K., Fukuda, M.N., Lowe, J.B., and Fukuda, M., Molecular cloning and expression of GDP-D-mannose-4,6-dehydratase, a key enzyme for fucose metabolism defective in Lec13 cells, *J. Biol. Chem.*, 273, 14582–14587, 1998.
123. Hinderlich, S., Stache, R., Zeitler, R., and Reutter, W., A bifunctional enzyme catalyzes the first two steps in *N*-acetylneuraminic acid biosynthesis of rat liver: Purification and characterization of UDP-*N*-acetylglucosamine 2-epimerase/*N*-acetylmannosamine kinase, *J. Biol. Chem.*, 272, 24313–24318, 1997.
124. Vlasak, R., Krystal, M., Nacht, M., and Palese, P., The influenza C virus glycoprotein (HE) exhibits receptor-binding (hemagglutinin) and receptor-destroying (esterase) activities, *Virology*, 160, 419–425, 1987.
125. Ariga, T., Blaine, G.M., Yoshino, H., Dawson, G., Kanda, T., Zeng, G.C., Kasama, T., Kushi, Y., and Yu, R.K., Glycosphingolipid composition of murine neuroblastoma cells: *O*-acetylesterase gene downregulates the expression of *O*-acetylated GD3, *Biochemistry*, 34, 11500–11507, 1995.
126. Hinderlich, S., Berger, M., Schwarzkopf, M., Effertz, K., and Reutter, W., Molecular cloning and characterization of murine and human *N*-acetylglucosamine kinase, *Eur. J. Biochem.*, 267, 3301–3308, 2000.
127. Pshezhetsky, A.V., Richard, C., Michaud, L., Igdoura, S., Wang, S., Elsliger, M.A., Qu, J., Gravel, L. D. R., Dallaire, L., and Potier, M., Cloning, expression and chromosomal mapping of human lysosomal sialidase and characterization of mutations in sialidosis, *Nat. Genet.*, 15, 316–320, 1997.
128. Monti, E., Preti, A., Rossi, E., Ballabio, A., and Borsani, G., Cloning and characterization of NEU2, a human gene homologous to rodent soluble sialidases, *Genomics.*, 57, 137–143, 1999.
129. Wada, T., Yoshikawa, Y., Tokuyama, S., Kuwabara, M., Akita, H., and Miyagi, T., Cloning, expression, and chromosomal mapping of a human ganglioside sialidase, *Biochem. Biophys. Res. Commun.*, 261, 21–27, 1999.
130. Monti, E., Bassi, M.T., Papini, N., Riboni, M., Manzoni, M., Venerando, B., Croci, G., Preti, A., Ballabio, A., Tettamanti, G., and Borsani, G., Identification and expression of NEU3, a novel human sialidase associated to the plasma membrane, *Biochem. J.*, 349, 343–351, 2000.
131. Strausberg, R.L., Feingold, E.A., Grouse, L.H., and et al, Generation and initial analysis of more than 15,000 full-length human and mouse cDNA sequences, *Proc. Natl. Acad. Sci .USA.*, 99, 16899–16908, 2002.
132. Traving, C., Bruse, P., Wachter, A., and Schauer, R., The sialate-pyruvate lyase from pig kidney. Elucidation of the primary structure and expression of recombinant enzyme activity, *Eur. J. Biochem.*, 268, 6473–6486, 2001.

133. Sood, R., Bonner, T.I., Makalowska, I., Stephan, D.A., Robbins, C.M., Connors, T.D., Morgenbesser, S.D., Su, K., Faruque, M.U., Pinkett, H., Graham, C., Baxevanis, A.D., Klinger, K.W., Landes, G.M., Trent, J.M., and Carpten, J.D., Cloning and characterization of 13 novel transcripts and the human RGS8 gene from the 1q25 region encompassing the hereditary prostate cancer (HPC1) locus, *Genomics*, 73, 211–222, 2001.

134. Haltiwanger, R.S., Blomberg, M.A., and Hart, G.W., Glycosylation of nuclear and cytoplasmic proteins: purification and characterization of a uridine diphospho-*N*-acetylglucosamine:polypeptide beta-*N*-acetylglucosaminyltransferase, *J. Biol. Chem.*, 267, 9005–9013, 1992.

135. Lubas, W.A., Frank, D.W., Krause, M., and Hanover, J.A., *O*-linked GlcNAc transferase is a conserved nucleocytoplasmic protein containing tetratricopeptide repeats, *J. Biol. Chem.*, 272, 9316–9324, 1997.

136. Pang, H., Koda, Y., Soejima, M., Kimura, H., and Iacomino, G., Identification of human phosphoglucomutase 3 (PGM3) as *N*-acetylglucosamine-phosphate mutase (AGM1), *Ann. Hum. Genet.*, 66, 139–144, 2002.

137. Li, C., Rodriguez, M., and Banerjee, D., Cloning and characterization of complementary DNA encoding human *N*-acetylglucosamine-phosphate mutase protein, *Gene*, 242, 97–103, 2000.

138. Takahashi, S., Inoue, H., and Miyake, Y., The human gene for renin-binding protein, *J. Biol. Chem.*, 267, 13007–13013, 1992.

139. Manzi, A.E., Sjoberg, E.R., Diaz, S., and Varki, A., Biosynthesis and turnover of *O*-acetyl and *N*-acetyl groups in the gangliosides of human melanoma cells, *J. Biol. Chem.*, 265, 13091–13103, 1990.

140. Chammas, R., Sonnenburg, J.L., Watson, N.E., Tai, T., Farquhar, M.G., Varki, N.M., and Varki, A., De-*N*-acetyl-gangliosides in human: unusual subcellular distribution of a novel tumor antigen, *Cancer Res.*, 59, 1337–1346, 1999.

141. Kannagi, R., Regulatory roles of carbohydrate ligands for selectins in the homing of lymphocytes, *Curr. Opin. Struct. Biol.*, 12, 599–608, 2002.

142. Lawrence, S.M., Huddleston, K.A., Pitts, L.R., Nguyen, N., Lee, Y.C., Vann, W.F., Coleman, T.A., and Betenbaugh, M.J., Cloning and expression of the human *N*-acetylneuraminic acid phosphate synthase gene with 2-keto-3-deoxy-D-glycero-D-galacto-nononic acid biosynthetic ability, *J. Biol. Chem.*, 275, 17869–17877, 2000.

143. Grundmann, U., Nerlich, C., Rein, T., and Zettlmeissl, G., Complete cDNA sequence encoding human β-galactoside α-2,6-sialyltransferase, *Nucleic Acids Res.*, 18, 667, 1990.

144. Harduin-Lepers, A., Stokes, D.C., Steelant, W.F., Samyn-Petit, B., Krzewinski-Recchi, M.A., Vallejo-Ruiz, V., Zanetta, J.P., Auge, C., and Delannoy, P., Cloning, expression and gene organization of a human Neu5Acα2-3Gal β1-3GalNAc α2,6-sialyltransferase: hST6GalNAc IV, *Biochem. J.*, 352, 37–48, 2000.

145. Kim, S.W., Kang, N.Y., Lee, S.H., Kim, W., Kim, K.S., Lee, J.H., Kim, C.H., and Lee, Y.C., Genomic structure and promoter analysis of human NeuAc α2,3Gal β1,3GalNAc α2,6-sialyltransferase (hST6GalNAc IV) gene, *Gene*, 305, 112–120, 2003.

146. Chang, M.-L., Eddy, R.L., Shows, T.B., and Lau, J.T.Y., Three genes that encode human β-galactoside α-2,3-sialyltransferases: structural analysis and chromosomal mapping studies, *Glycobiology*, 5, 319–325, 1995.

147. Kitagawa, H. and Paulson, J.C., Differential expression of five sialyltransferase genes in human tissues, *J. Biol. Chem.*, 269, 17872–17878, 1994.

148. Shang, J., Qiu, R., Wang, J., Liu, J., Zhou, R., Ding, H., Yang, S., Zhang, S., and Jin, C., Molecular cloning and expression of Galβ1,3GalNAc α2, 3-sialyltransferase from human fetal liver, *Eur. J. Biochem.*, 265, 580–588, 1999.

149. Kim, Y.J., Kim, K.S., Kim, S.H., Kim, C.H., Ko, J.H., Choe, I.S., Tsuji, S., and Lee, Y.C., Molecular cloning and expression of human Gal β1,3GalNAc α2,3-sialytransferase (hST3Gal II), *Biochem. Biophys. Res. Commun.*, 228, 324–327, 1996.

150. Kitagawa, H., Mattei, M.G., and Paulson, J.C., Genomic organization and chromosomal mapping of the Gal β1,3GalNAc/Gal β1,4GlcNAc α2,3-sialyltransferase, *J. Biol. Chem.*, 271, 931–938, 1996.

151. Kitagawa, H. and Paulson, J.C., Cloning and expression of human Galβ1,3(4)GlcNAcα2,3-sialyltransferase, *Biochem. Biophys. Res. Commun.*, 194, 374–382, 1993.

152. Sotiropoulou, G., Kono, M., Anisowicz, A., Gtenman, G., Tsuji, S., and Sager, R., Identification and functional characterization of human GalNAc α2,6 sialyltransferase with altered expression in breast cancer, *Mol. Med.*, 8, 42–55, 2002.

153. Samyn-Petit, B., Krzewinski-Recchi, M.A., Steelant, W.F., Delannoy, P., and Harduin-Lepers, A., Molecular cloning and functional expression of human ST6GalNAc II. Molecular expression in various human cultured cells, *Biochim. Biophys. Acta.*, 1474, 201–211, 2000.

154. Ikehara, Y., Shimizu, N., Kono, M., Nishihara, S., Nakanishi, H., Kitamura, T., Narimatsu, H., Tsuji, S., and Tatematsu, M., A novel glycosyltransferase with a polyglutamine repeat; a new candidate for GD1αsynthase (ST6GalNAc V)(1), *FEBS Lett.*, 463, 92–96, 1999.

155. Okajima, T., Fukumoto, S., Ito, H., Kiso, M., Hirabayashi, Y., Urano, T., and Furukawa, K., Molecular cloning of brain-specific GD1α synthase (ST6GalNAc V) containing CAG/Glutamine repeats, *J. Biol. Chem.*, 274, 30557–30562, 1999.

156. Tsuchida, A., Okajima, T., Furukawa, K., Ando, T., Ishida, H., Yoshida, A., Nakamura, Y., Kannagi, R., Kiso, M., and Furukawa, K., Synthesis of disialyl Lewis a (Le(a)) structure in colon cancer cell lines by a sialyltransferase, ST6GalNAc VI, responsible for the synthesis of alpha-series gangliosides, *J. Biol. Chem.*, 278, 22787–22794, 2003.

157. Okajima, T., Chen, H.-H., Ito, H., Kiso, M., Tai, T., Furukawa, K., Urano, T., and Furukawa, K., Molecular cloning and expression of mouse GD1α/GT1aα/GQ1bα synthase (ST6GalNAc VI) gene, *J. Biol. Chem.*, 275, 6717–6723, 2000.

158. Matsuda, Y., Nara, K., Watanabe, Y., Saito, T., and Sanai, Y., Chromosome mapping of the GD3 synthase gene (SIAT8) in human and mouse, *Genomics*, 32, 137–139, 1996.

159. Sasaki, K., Kurata, K., Kojima, N., Kurosawa, N., Ohta, S., Hanai, N., Tsuji, S., and Nishi, T., Expression cloning of a GM3-specific α-2,8-sialyltransferase (GD3 synthase), *J. Biol. Chem.*, 269, 15950–15956, 1994.

160. Scheidegger, E.P., Sternberg, L.R., Roth, J., and Lowe, J.B., A human STX cDNA confers polysialic acid expression in human cells, *J. Biol. Chem.*, 270, 22685–22688, 1995.

161. Lee, Y.C., Kim, Y.J., Lee, K.Y., Kim, K., U, K., Kim, H.N., Kim, C.H., and Do, S.I., Cloning and expression of cDNA for a human Sia α2,3Gal β1, 4GlcNA:α2,8-sialyltransferase (hST8Sia III), *Arch. Biochem. Biophys.*, 360, 41–46, 1998.

162. Angata, K., Suzuki, M., McAuliffe, J., Ding, Y., Hindsgaul, O., and Fukuda, M., Differential biosynthesis of polysialic acid on neural cell adhesion molecule (NCAM) and oligosaccharide acceptors by three distinct α2,8-sialyltransferases, ST8Sia IV (PST), ST8Sia II (STX), and ST8Sia III, *J. Biol. Chem.*, 275, 18594–18601, 2000.

163. Eckhardt, M., Muhlenhoff, M., Bethe, A., Koopman, J., Frosch, M., and Gerardy-Schahn, R., Molecular characterization of eukaryotic polysialyltransferase-1, *Nature*, 373, 715–718, 1995.

164. Nakayama, J., Fukuda, M.N., Fredette, B., Ranscht, B., and Fukuda, M., Expression cloning of a human polysialyltransferase that forms the polysialylated neural cell adhesion molecule present in embryonic brain, *Proc. Natl. Acad. Sci. USA*, 92, 7031–7035, 1995.

165. Ishii, A., Ohta, M., Watanabe, Y., Matsuda, K., Ishiyama, K., Sakoe, K., Nakamura, M., Inokuchi, J., Sanai, Y., and Saito, M., Expression cloning and functional characterization of human cDNA for ganglioside GM3 synthase, *J. Biol. Chem.*, 273, 31652–31655, 1998.

166. Taniguchi, A., Kaneta, R., Morishita, K., and Matsumoto, K., Gene structure and transcriptional regulation of human Gal β-1,4(3) GlcNac α-2,3-sialyltransferase VI (hST3Gal VI) gene in prostate cancer cell line, *Biochem. Biophys. Res. Commun.*, 287, 1148–1156, 2001.

167. Okajima, T., Fukumoto, S., Miyazaki, H., Ishida, H., Kiso, M., Furukawa, K., Urano, T., and Furukawa, K., Molecular cloning of a novel α-2,3-sialyltransferase (ST3Gal VI) that sialylates type II lactosamine structures on glycoproteins and glycolipids, *J. Biol. Chem.*, 274, 11479–11486, 1999.

168. Mio, T., Yabe, T., Arisawa, M., and Yamada-Okabe, H., The eukaryotic UDP-*N*-acetylglucosamine pyrophosphorylases: gene cloning, protein expression, and catalytic mechanism, *J. Biol. Chem.*, 273, 14392–14397, 1998.

169. Ishida, N., Miura, N., Yoshioka, S., and Kawakita, M., Molecular cloning and characterization of a novel isoform of the human UDP-galactose transporter, and of related complementary DNAs belonging to the nucleotide-sugar transporter gene family, *J. Biochem.*, 120, 1074–1078, 1996.

170. Miura, N., Ishida, N., Hoshino, M., Yamauchi, M., Hara, T., Ayusawa, D., and Kawakita, M., Human UDP-galactose translocator: molecular cloning of a complementary DNA that complements the genetic defect of a mutant cell line deficient in UDP-galactose translocator, *J. Biochem.*, 120, 236–241, 1996.

171. Sprong, H., Degroote, S., Nilsson, T., Kawakita, M., Ishida, N., van der Sluijs, P., and van Meer, G., Association of the Golgi UDP-galactose transporter with UDP-galactose:ceramide galactosyltransferase allows UDP-galactose import in the endoplasmic reticulum, *Mol. Biol. Cell.*, 14, 3482–3493, 2003.

9 Biosynthetic Pathways for Differential Expression of Functional Chondroitin Sulfate and Heparan Sulfate

Shuji Mizumoto*, Toru Uyama*, Tadahisa Mikami, Hiroshi Kitagawa, and Kazuyuki Sugahara

CONTENTS

I. INTRODUCTION

Sulfated glycosaminoglycan (GAG) chains are covalently attached to various core proteins forming proteoglycans (PGs), which are ubiquitously distributed at the cell surfaces and to extracellular matrices (Figure 9.1). PGs play important roles via the GAG side chains in a variety of biological processes such as cell adhesion, proliferation, tissue morphogenesis, neurite outgrowth, infections with viruses and bacteria, and the

* These authors contributed equally to this chapter.

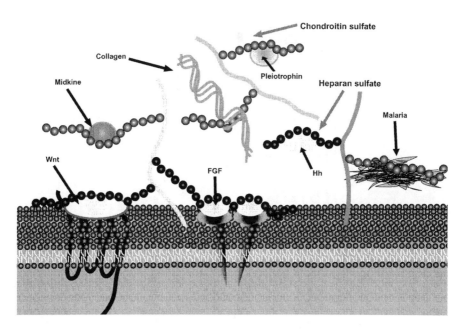

FIGURE 9.1   Various functions of GAG chains. GAGs including HS and CS expressed on the cell surface and in the extracellular matrix interact with various proteins such as growth factors, morphogens, and adhesive molecules.

regulation of various growth factors and cytokine effectors. Linear sulfated GAGs are classified based on structural units in chondroitin sulfate/dermatan sulfate (CS/DS) and heparan sulfate/heparin (HS/Hep), which are composed of sulfated disaccharide units $[4\text{GlcA}\beta1(\text{IdoA}\alpha1)\text{-}3\text{GalNAc}\beta1\text{-}]_n$ and $[\text{-}4\text{GlcA}\beta1(\text{IdoA}\alpha1)\text{-}4\text{GlcNAc}\alpha1\text{-}]_n$, respectively, where GlcA, IdoA, GalNAc, and GlcNAc represent D-glucuronic acid, L-iduronic acid, N-acetyl-D-galactosamine, and N-acetyl-D-glucosamine, respectively (Figure 9.2).

Both types of GAG chains are built on the so-called common GAG–protein linkage region, $\text{GlcA}\beta1\text{-}3\text{Gal}\beta1\text{-}3\text{Gal}\beta1\text{-}4\text{Xyl}\beta1\text{-}O\text{-Ser}$, of their respective core proteins via the stepwise addition of each monosaccharide residue by the corresponding specific glycosyltransferase (see Figure 9.3 and Tables 9.1–9.3). The transfer of β-GalNAc to the GlcA residue of the tetrasaccharide linkage region initiates the synthesis of the repeating disaccharide region of CS/DS. In contrast, the transfer of α-GlcNAc to the tetrasaccharide linkage region initiates the synthesis of the repeating disaccharide region of HS/Hep through the alternate addition of GlcA and GlcNAc (Figure 9.3). The first hexosamine transfer, therefore, is critical in the differential assembly of CS/DS and HS/Hep.

Following the synthesis of the backbone, various modifications such as sulfation and epimerization occur in both CS/DS and HS/Hep through the actions of sulfotransferases and epimerases. The cDNA cloning of glycosyltransferases, sulfotransferases, and GlcA-specific C5 epimerase involved in the biosynthesis of GAGs has recently been achieved.[23,183] Functional analyses of these genes using knockout mice in addition to mutants of the fruit fly and nematode have revealed important roles for CS/DS and

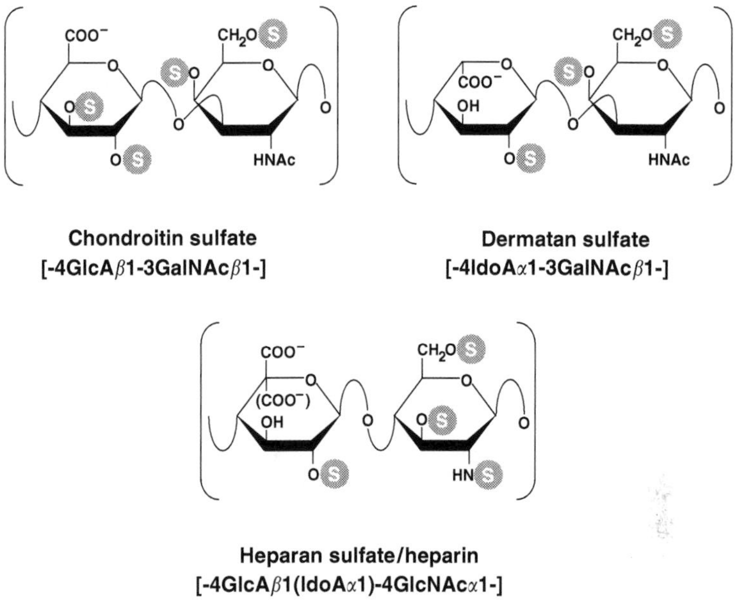

FIGURE 9.2 Typical repeating disaccharide units in CS/DS and HS/Hep and their potential sulfation sites. CS and DS are constituted of uronic acid and GalNAc. DS is a stereoisomer of CS including IdoA instead of or in addition to GlcA. HS and Hep consist of uronic acid and GlcNAc residues with varying proportions of IdoA. These sugar residues can be esterified by sulfate at various positions as indicated by "S."

HS/Hep. In this chapter, we focus on recent advances in the study of the biosynthetic pathways for the differential expression of functional CS/DS and HS/Hep chains.

II. BIOSYNTHESIS OF THE GAG–PROTEIN LINKAGE REGION

All the four glycosyltransferases responsible for the biosynthesis of the linkage region have been cloned (see Figure 9.3 and Table 9.1).[1] The transfer of a Xyl residue to specific Ser residues in the core proteins by XylT[2] takes place in the endoplasmic reticulum (ER) and the *cis*-Golgi compartments (Figure 9.4).[3] XylT is important in determining the position of the GAG attachment site in the core proteins. Next, GalT-I and GalT-II successively transfer two Gal residues to the xylosylated Ser residue.[4–6] These reactions are catalyzed in the *cis*-and *medial*-Golgi compartments.[3] GlcAT-I completes the biosynthesis of the linkage region by transferring a GlcA residue to the Gal-Gal-Xyl trisaccharide sequence[7] in the *medial*- and *trans*-Golgi compartments.[3] The three glycosyltransferases, except for XylT, which are responsible for synthesizing the linkage region are unique in that they exist as a single gene product, indicating that a deficiency of each glycosyltransferase would lead to the complete elimination of GAG chains. Recent experiments have shown that two XylTs, XylT-I and XylT-II, exist in rats and humans.[2] Although amino acid sequences of these XylTs show similarity to each other, XylT activity is only detected for XylT-I and not

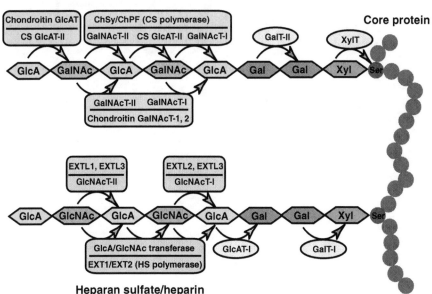

FIGURE 9.3 Schema of biosynthetic assembly of the GAG backbones by various glycosyl-transferases. A number of glycosyltransferases are required for the synthesis of the backbones of GAGs. XylT, Xyl transferase; GalT-I, Gal transferase-I; GalT-II, Gal transferase-II, GlcAT-I; GlcA transferase-I; GalNAcT-I, GalNAc transferase-I; GlcAT-II, GlcA transferase-II; GalNAcT-II, GalNAc transferase-II; CS polymerase, GlcA/GalNAc transferase; GlcNAcT-I, GlcNAc transferase-I; GlcNAcT-II, GlcNAc transferase-II; HS polymerase, GlcA/GlcNAc transferase; ChSy, chondroitin synthase; and ChPF, chondroitin polymerizing factor.

TABLE 9.1
Human Linkage Region Glycosyltransferases

Name	Abbreviation	Chromosome Location	Amino Acid	mRNA Expression	mRNA Accession
Xyl transferase	XylT	16p13.1	959	Ubiquitous	AJ539163
Gal transferase-I	GalT-I	5q35.1–q35.3	327	Ubiquitous	AB028600
Gal transferase-II	GalT-II	1p36.3	329	Ubiquitous	AY050570
GlcA transferase-I	GlcAT-I	11q12–q13	335	Ubiquitous	AB009598

for XylT-II. Thus, the function of XylT-II remains unclear.[2] Notably, overexpression of GlcAT-P, whose amino acid sequence is similar to that of GlcAT-I, rescued the biosyn-thesis of GAGs in a Chinese hamster ovary (CHO) cell line that lacked GlcAT-I.[8] GlcAT-P is a $\beta1,3$-GlcAT that utilizes glycoproteins with the terminal Gal$\beta1$-4GlcNAc sequence and is required for the formation of human natural killer cell car-bohydrate antigen-1 (HNK-1) (GlcA(SO$_4$) $\beta1$-3Gal$\beta1$-4GlcNAc$\beta1$-R) abundant on the surface of neural cells.[9] Conversely, overexpression of GlcAT-I in COS-1 cells,

TABLE 9.2
Human CS Glycosyltransferases and CS/DS Sulfotransferases

Name	Abbreviation	Chromosome Location	Amino Acid	mRNA Expression	mRNA Accession
Chondroitin synthase	ChSy (GalNAcT-II/GlcAT-II)	15q26.3	802	Ubiquitous	AB071402
Chondroitin GalNAcT-1	ChGn-1 (GalNAcT-I/II)	8p21.3	532	Ubiquitous	AB071403
Chondroitin GalNAcT-2	ChGn-2 (GalNAcT-I/II)	10q11.22	542	Ubiquitous	AB090811
Chondroitin GlcAT	GlcAT-II	7q35	772	Ubiquitous	AB037823
Chondroitin polymerizing factor	ChPF	1p11-p12	775	Ubiquitous	AB095813
Chondroitin 4-O-sulfotransferase-1	C4ST-1	12q23	352	Ubiquitous	AF239820
Chondroitin 4-O-sulfotransferase-2	C4ST-2	7p22	414	Ubiquitous	AF239822
Chondroitin 4-O-sulfotransferase-3	C4ST-3	3q21.3	341	Liver	AY120869
Dermatan 4-O-sulfotransferase-1	D4ST-1	15q14	376	Ubiquitous	AB066595
Chondroitin 6-O-sulfotransferase	C6ST	10q21.3	479	Ubiquitous	AB017915
Uronyl 2-O-sulfotransferase	CS/DS2ST	—	406	Ubiquitous	AB020316
GalNAc 4-sulfate 6-O-sulfotransferase	GalNAc4S-6ST	10q26	561	—	AB062423

—, not reported.

which do not endogenously express the HNK-1 epitope due to a deficiency of GlcAT-P expression, leads to the formation of the HNK-1 epitope to some extent.[10] Thus, the functions of GlcAT-P and GlcAT-I, both of which transfer a GlcA residue to the terminal β-linked Gal, may be partially redundant.

Modifications in the linkage region of GAGs have been reported. One such modification is 2-O-phosphorylation of the Xyl residue in the linkage region of both CS and HS chains.[3,11] Although this phosphorylation is not always found, it seems to affect the transfer of GlcA by GlcAT-I in the biosynthesis of the linkage region.[12] The phosphorylation of Xyl is most prominent after the addition of two Gal residues. Then, the addition of GlcA is followed by rapid dephosphorylation. Although the biological role of the phosphorylation of the Xyl residue and the enzymes responsible for this modification have not been investigated, the phosphorylated trisaccharide is one of the best acceptors for GlcAT-I *in vitro* (Tone, Y., Pedersen, L., Yamamoto, T., Kitagawa, H., Nishihara, J.,Tamura, J., Darden, T.A., Negishi, M., and Sugahara, K., unpublished data), suggesting that phosphorylation of the Xyl residue accelerates the transfer of a GlcA residue, thereby upregulating the biosynthesis of GAGs. The sulfation of Gal residues in the linkage region has also been reported. The potential sites for sulfation are C6 of the first Gal and C4 or C6 of the second Gal residue.[184–186] The sulfation

TABLE 9.3
Human HS Glycosyltransferases, Sulfotransferases, and Epimerase

Name	Abbreviation	Chromosome Location	Amino Acid	mRNA Expression	mRNA Accession
GlcA/GlcNAc transferase	EXT1	8q24.11–q24.13	746	Ubiquitous	S79639
	EXT2	11p12–p11	718	Ubiquitous	U62740
GlcNAc transferase-I	GlcNAcT-I (EXTL2)	1p21	330	Ubiquitous	AF000416
GlcNAc transferase-II	GlcNAcT-II (EXTL1)	1p36.1	676	Skeletal muscle, brain	U67191
GlcNAc transferase-I/II	GlcNAcT-I/II (EXTL3)	8p21	919	Ubiquitous	AF001690
GlcNAc N-deacetylase/	NDST1	5q33.1	882	Ubiquitous	U18918
N-sulfotransferase	NDST-2	10q22	883	Ubiquitous	U36601
	NDST-3	4q27	873	Brain, kidney, liver	AF074924
	NDST-4	4q25–q26	872	—	AB036429
Uronyl C5 epimerase	None	15q22.31	618	Ubiquitous	XM_290631
HS 2-O-sulfotransferase	HS2ST	1p31.1–p22.2	356	Ubiquitous	AB024568
HS 6-O-sulfotransferase	HS6ST-1	2q21	401	Ubiquitous	AB006179
	HS6ST-2	Xq26.2	499	Brain	AB067776
	HS6ST-2S	Xq26.2	459	Ovary, placenta, fetal kidney	AB067777
	HS6ST-3	13q32.2	471		AF539426
HS 3-O-sulfotransferase	HS3ST-1	4p16	307	Kidney, brain, heart, lung	AF019386
	HS3ST-2	16p12	367	Brain	AF105375
	HS3ST-3A	17p12–p11.2	406	Ubiquitous	AF105376
	HS3ST-3B	17p12–p11.2	390	Ubiquitous	AF105377
	HS3ST-4	16p11.2	>250	Brain	AF105378
	HS3ST-5	6q22.31	346	Skeletal muscle	AF503292

—, not reported.

occurs in the linkage region of CS, but not HS. Although it has not been proven that the linkage region of all CS chains is modified with sulfates, it has been suggested that the sulfation of the linkage region is involved in the selective chain assembly of CS and HS in the linkage region, playing a significant role in CS biosynthesis.[11]

Dysfunction of the glycosyltransferases responsible for the biosynthesis of the linkage region causes disease. The deficiency of GalT-I is responsible for the progeroid variant of Ehlers–Danlos syndrome (Table 9.4).[14] Fibroblasts, derived from a patient bearing a point mutation in GalT-I, synthesize proteoglycans devoid of GAGs, the core proteins of which are substituted with only a Xyl residue. This mutant GalT-I is abnormally thermolabile; incubation of fibroblasts at 41°C leads to a reduction in the production of mature proteoglycans.[14] In addition, the mutant GalT-I shows a diffuse cytoplasmic expression pattern, whereas the normal GalT-I shows localization to the Golgi.[15] It is interesting to investigate possible mutations in the three other enzyme genes, involved in the biosynthesis of the linkage region,

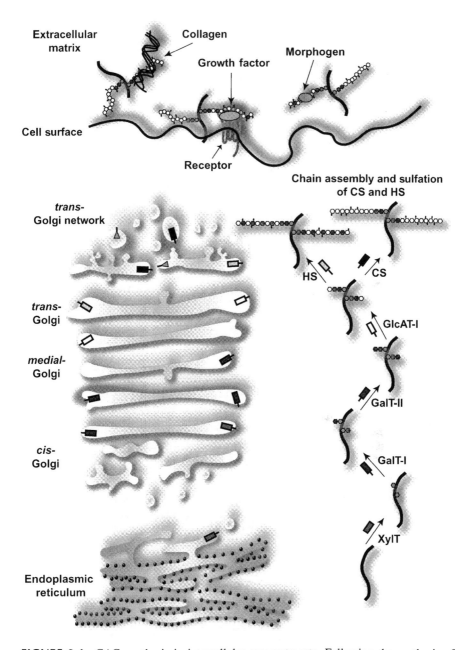

FIGURE 9.4 GAG synthesis in intracellular compartments. Following the synthesis of core proteins, GAG chain synthesis is initiated by XylT in ER. During the translocation of core proteins with immature GAG chains, GAG chains are gradually assembled by various glycosyltransferases and sulfotransferases. Mature proteoglycans composed of a core protein and multiple GAG chains are expressed on the cell surface and in the extracellular matrix, and involved in various events such as cell adhesion, cell differentiation, and cell division.

TABLE 9.4

The Model Organisms with Defects in the Biosynthesis of GAGs and Core Proteins

Mutants and Knockout Mouse	Name	Phenotypes	Ref.
C. elegans			
sqv-1	UDP-GlcA decarboxylase	Defects in cytokinesis and vulval morphogenesis	145
sqv-2	GalT-II	Defects in cytokinesis and vulval morphogenesis	146
sqv-3	GalT-I	Defects in cytokinesis and vulval morphogenesis	4,20
sqv-4	UDP-Glc dehydrogenase	Defects in cytokinesis and vulval morphogenesis	147
sqv-5	ChSy	Defects in cytokinesis and vulval morphogenesis	21,22
sqv-6	XylT	Defects in cytokinesis and vulval morphogenesis	146
sqv-7	UDP-GlcA, UDP-GalNAc, UDP-Gal transporter	Defects in cytokinesis and vulval morphogenesis	37
sqv-8	GlcAT-I	Defects in cytokinesis and vulval morphogenesis	20
unc-52	Perlecan	Defects in the formation or maintenance of the muscle myofilament lattice, the affects of the regulation of distal tip cell migration	148,149
rib-2	EXTL3	Developmental abnormalities in embryonic stage	94,95
hst-6	HS6ST	Suppression of kal-1-dependent axon-branching phenotype	150
D. melanogaster			
sugarless	UDP-Glc dehydrogenase	Defects in Wg , FGF signalings	34,35, 36,60
frc	UDP-sugar transporter	Defects in Wg, Hh, FGF, Notch signalings	38,39
slalom	PAPS transporter	Defects in Wg, Hh signalings	56, 57
oxt	XylT	–	151
beta4GalT7	GalT-I	Abnormal wing and leg morphology similar to flies with defective Hh and Dpp signalings	16,152
ttv	EXT1	Defects in Hh, Wg, Dpp signalings	89,90
sotv	EXT2	Defects in Hh, Wg, Dpp signalings	91
botv	EXTL3	Defects in Hh, Wg, Dpp signalings	91, 92
sulfateless	NDST	Defects in Wg, FGF signalings	59, 60
pipe	HS2ST	Defects in the formation of embryonic dorsal–ventral polarity	127
dHS2ST	HS2ST	—	128
dHS6ST	HS6ST	Defects in FGF signaling	134
dally	Glypican	Defects in Wg, Dpp, Hh signalings	59,153–157
dally-like	Glypican	Defects in Wg, Hh signalings	157–159
dSyndecan	Syndecan	—	160
trol	Perlecan	Defects in neuroblast proliferation in the CNS	161

(Continued)

TABLE 9.4   (Continued)

Mutants and Knockout Mouse	Name	Phenotypes	Ref.
Zebrafish (Danio rerio)			
jekyll	UDP-Glc dehydrogenase	Defects in cardiac valve formation	33
b3gat3	GlcAT-I	Defective branchial arches and jaw	162
uxs1	UDP-GlcA decarboxylase	Defective cartilage unstained with Alcian Blue	162
zHS6ST	HS6ST	Defects in muscle differentiation	163
knypek	Glypican	Defects in Wnt signaling	164
syndecan-2	Syndecan-2	Defects in angiogenesis	165
Mouse			
lzme (lazy mesoderm	UDP-Glc dehydrogenase	Defects in FGF signaling	32
brachymorphic mouse	PAPS synthase 2	Dome-shaped skull, shortened but not widened limbs, short tail	54
cmd (cartilage matrix deficiency	Aggrecan	Perinatal lethal dwarfism, craniofacial abnormalities	166
hdf (Versican$^{-/-}$)	Versican	Embryonic lethality with heart defect	167
Syndecan-1$^{-/-}$	Syndecan-1	Defects in the repair of skin and corneal wounds, low susceptibility to Wnt-1 signaling	168,169
Syndecan-3$^{-/-}$	Syndecan-3	Reduction of reflex hyperphagia following food deprivation	170
Syndecan-4$^{-/-}$	Syndecan-4	Impairment of focal adhesion under restricted conditions	171
Glypican-2$^{-/-}$	Glypican-2	No phenotypes	172
Glypican-3$^{-/-}$	Glypican-3	Developmental overgrowth typical of human Simpson–Golabi–Behmel–syndrome	173
Agrin$^{-/-}$	Agrin	Perinatal lethality owing to breathing failure, defects of neuromuscular synaptogenesis	174
Perlecan$^{-/-}$	Perlecan	Defective cephalic development	175,176
Decorin$^{-/-}$	Decorin	Abnormal collagen morphology in skin and tendon	177
Biglycan$^{-/-}$	Biglycan	Reduced growth rate and decreased bone mass	178
Neurocan$^{-/-}$	Neurocan	No phenotypes	179
PTP$^{-/-}$	PTP	Resistance to gastric ulcer induction by VacA of Helicobacter pylori	180
Thrombo-modulin$^{-/-}$	Thrombo-modulin	Embryonic lethality with dysfunctional maternal– embryonic interaction	181
EXT1$^{-/-}$	EXT1	Disruption of gastrulation	75
EXT1$^{-/-}$ (specific for brain)	EXT1	Defects in the midbrain–hindbrain region, disturbed Wnt-1 distribution	76
EXT2$^{-/-}$	EXT2	Disruption of gastrulation	77

(Continued)

TABLE 9.4   (Continued)

Mutants and Knockout Mouse	Name	Phenotypes	Ref.
NDST-1$^{-/-}$	NDST-1	Neonatal lethality due to respiration defects	102,103
NDST-2$^{-/-}$	NDST-2	Loss of heparin, abnormal mast cell	110,111
HS2ST$^{-/-}$	HS2ST	Renal agenesis, defects in the eye and the skeleton	120
HS3ST-1$^{-/-}$	HS3ST-1	Genetic background-specific lethality, intrauterine growth retardation	144
Uronyl C5 epimerase$^{-/-}$ (Hsepi)	Uronyl C5 epimerase	Neonatal lethality with renal agenesis, lung defects skeletal malformation	119
C6ST$^{-/-}$	C6ST	Decrease in naive T lymphocytes	43
Human			
Ehlers–Danlos syndrome	GalT-I	Aged appearance, developmental delay, dwarfism, craniofacial disproportion, and generalized osteopenia	14,15
Hereditary multiple exostoses	EXT1 EXT2	An autosomal dominant disorder characterized by the formation of cartilage-capped tumors (exostoses) that develop from the growth plate of endochondral bones, especially of long bones	67,182
Spondyloep-imetaphyseal dysplasia	PAPS synthase 2	Short, bowed lower limbs, enlarged knee joints, kyphoscoliosis, a mild generalized brachydactyly	55
Achondro-genesis type 1B	DTDST (diastrophic dysplasia sulfate transporter)	Autosomal recessive, lethal chondrodysplasia with severe underdevelopment of skeleton, extreme micromelia, and death before or immediately after birth	51

——, not reported.

in patients with progeroid variants of Ehlers–Danlos syndrome. So far, no progeroid variant of Ehlers–Danlos syndrome has been reported, where other glycosyltransferases responsible for the synthesis of the linkage region are involved. Functions of GalT-I have been investigated in *Drosophila melanogaster* (*D. melanogaster*) using the RNA interference (RNAi) technique, where double-stranded RNA (dsRNA) of the target gene sequence is used to inhibit the functions of the target gene. GalT-I dsRNA-treated *D. melanogaster* shows that the biosynthesis of GAGs in these flies is drastically downregulated, and consequently Hedgehog (Hh) and Decapentaplegic (Dpp) signalings are impaired.[16] These signaling molecules play crucial roles in morphogenesis and organogenesis in early developmental stages and disruption of these signaling pathways leads to the abnormal formation of organs such as the wing. Thus, GAGs are apparently required for morphogenesis and organogenesis in the regulation of signaling molecules.

Caenorhabditis elegans (*C. elegans*) expresses HS and chondroitin, a nonsulfated form of CS, and has been used for the functional analysis of GAGs. Recently, mutants defective in the biosynthesis of GAGs including chondroitin and HS, have

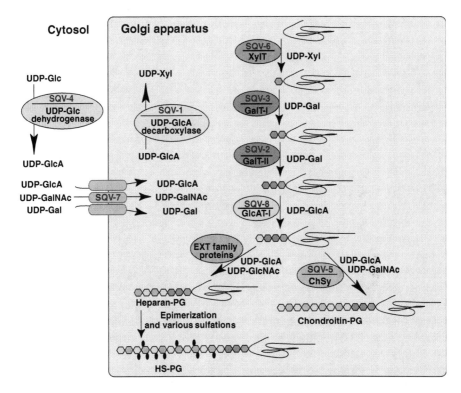

FIGURE 9.5 Schematic image of SQV proteins. All *sqv* genes encode the proteins involved in GAG biosynthesis, especially chondroitin. SQV-1, UDP-GlcA decarboxylase; SQV-2, Gal transferase-II; SQV-3, Gal transferase-I; SQV-4, UDP-Glc dehydrogenase; SQV-5, Chondroitin synthase; SQV-6, Xyl transferase; SQV-7, UDP-sugar transporter, SQV-8, GlcA transferase-I; EXT family proteins (rib-1 and rib-2), and HS polymerase.

been identified and named *sqv* (squashed vulva) mutants based on a perturbation of the vulval invagination (see Figure 9.5 and Table 9.4).[17] These worms have mutations in the genes *sqv-1* to *sqv-8* coding for eight different functional proteins responsible for GAG biosynthesis.[18] Among them, *sqv-6, sqv-3, sqv-2,* and *sqv-8* encode the glycosyltransferases XylT, GalT-I, GalT-II, and GlcAT-I, respectively, which are involved in the biosynthesis of the linkage region (Table 9.4).[19] Since these enzymes are required for the biosynthesis of GAGs, the mutants deficient in the corresponding genes show a reduced synthesis of both chondroitin and HS, and exhibit phenotypes of abnormal cell division, particularly cytokinesis and aberration of vulval morphogenesis.[20] Thus, GAGs are required for the early developmental stage in *C. elegans*, although further investigation is needed to clarify the specific roles of GAGs.

III. BIOSYNTHESIS OF CHONDROITIN BACKBONE

CS proteoglycan (CS–PG) molecules consist of a core protein and at least one covalently attached GAG chain. Most of the physiological functions of CS–PGs are largely expressed through the CS side chains, with the core protein seeming to act as a scaffold. There is increasing evidence that CS plays crucial role in various biological phenomena including signaling, cell differentiation, cell–cell or cell–matrix interactions, and morphogenesis. Recently, gene manipulations of agents responsible for the biosynthesis of CS have been used for model organisms of both vertebrates and invertebrates. The *sqv* mutants of *C. elegans, sqv-1* to *sqv-8*, show not only a perturbation of vulval formation but also abnormal cytokinesis in early embryonic stages due to a defect of the agents responsible for GAG biosynthesis (Table 9.4).[19] Recently performed elegant experiments have revealed that the perturbation of vulval morphogenesis is caused by a defect in the synthesis of chondroitin and not HS (Figure 9.5).[21,22]

CS is a linear, sulfated polysaccharide composed of repeating disaccharide units (GlcAβ1-3GalNAc) and is synthesized as a proteoglycan, being linked to a specific Ser residue in the core protein. Although the mechanism by which CS is produced has long been a mystery, recent experiments using molecular biological techniques have revealed various glycosyltransferases responsible for the biosynthesis (Table 9.2).[23] Following completion of the synthesis of the linkage region, the first GalNAc residue is transferred to the linkage region built on specific Ser residues in the core protein by GalNAcT-I, which initiates the biosynthesis of CS (Figure 9.3).[3,11,23] Then, polymerization of the CS chain occurs to build up repeating disaccharide units composed of GalNAcβ1-4GlcA. This polymerization reaction is performed by successive additions of GlcA and GalNAc, leading to chain elongation. The enzyme activities for the addition of GlcA and GalNAc are called CS GlcAT-II and GalNAcT-II activities, respectively (Figure 9.3). Although the glycosyltransferases involved in the biosynthesis of CS had for a long time not been characterized, it was recently shown that both the transfer of the first GalNAc residue and the formation of disaccharide units are catalyzed by the enzyme complex, CS polymerase, consisting of chondroitin synthase (ChSy) and chondroitin polymerizing factor (ChPF).[24,25] ChSy is a bifunctional glycosyltransferase with GalNAcT-II and CS GlcAT-II activities required for the formation of the disaccharide unit, whereas ChPF possesses only weak GalNAcT-II activity. In spite of the dual enzymatic activities of ChSy, ChSy itself cannot achieve polymerization reactions to build up the repeating disaccharide units of CS. However, the association of ChSy with ChPF results in a dramatic augmentation of the glycosyltransferase activities of ChSy. Furthermore, this enzyme complex can polymerize a CS chain onto the linkage region tetrasaccharide, attached to the core protein. Thus, ChPF may function as a chaperone, which confers the stronger glycosyltransferase activities on ChSy or stabilizes ChSy by forming an enzyme complex, CS polymerase.[25] Interestingly, CS polymerase cannot efficiently catalyze the polymerization reactions when a linkage region, glycopeptide with a serine or small peptide but without a core protein, is used as an acceptor substrate, suggesting that the core protein facilitates the efficient polymerization by CS polymerase. In addition to ChSy and ChPF, chondroitin GalNAcT-1 and GalNAcT-2, both of which possess GalNAcT-I and GalNAcT-II activities, exist and chondroitin GlcAT, which harbors CS GlcAT-II activity probably involved in the transfer of a GlcA residue to the disaccharide repeating region, has also been reported (Figure 9.3).[26–30] Thus, various glycosyltransferases appear to be involved in the biosynthesis of CS. After the formation of the disaccharide units, sulfation at various positions in the CS backbone is catalyzed by a number of sulfotransferases (see Section IV).

Another modification of CS chains, the epimerization of GlcA to IdoA, results in the conversion of CS into DS (Figure 9.2 and Figure 9.6).[3] This reaction is catalyzed by GlcA C5 epimerase, which converts GlcA into IdoA by epimerizing the C5 carboxy group of GlcA accompanied by an anomeric change of the glycosidic linkage of GlcA from β to α. This anomeric configuration confers not only structural flexibility but also the ability to bind various growth factors to DS chains. For instance, Hep cofactor II, which is homologous to antithrombin (AT) III and regulates blood coagulation, tightly binds to DS chain, but not to CS.[31] Thus, although CS and DS show structural similarity except for the IdoA content, the biological functions of these GAGs are quite different. In addition, CS is synthesized in virtually all cells, whereas DS is mainly expressed in endothelial and epidermal cells. Since DS synthesis requires GlcA C5 epimerase, endothelial and epidermal cells may highly express GlcA C5 epimerase, compared with other cells. Although GlcA C5 epimerase has yet to be cloned, the cloning and characterization of CS/DS-specific GlcA C5 epimerase should provide insights into the regulation of DS biosynthesis.

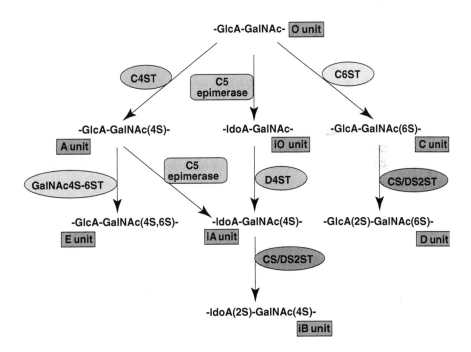

FIGURE 9.6 Pathways of biosynthetic modifications of CS and DS chains. C4ST, chondroitin 4-O-sulfotransferase; C6ST, chondroitin 6-O-sulfotransferase; D4ST, dermatan 4-O-sulfotransferase; CS/DS2ST, uronyl 2-O-sulfotransferase; GalNAc4S-6ST, GalNAc 4-sulfate 6-O-sulfotransferase; O unit, GlcAβ1-3GalNAc; iO unit, IdoAα1-3GalNAc; A unit, GlcAβ1-3GalNAc(4S); iA unit, IdoAα1-3GalNAc(4S); iB unit, IdoA(2S)α1-3GalNAc(4S); C unit, GlcAβ1-3GalNAc(6S); D unit, GlcA(2S)β1-3GalNAc(6S); and E unit, GlcAβ1-3GalNAc(4S,6S).

All glycosyltransferases responsible for the biosynthesis of GAGs use UDP-sugar as a donor substrate, and have at least one DXD motif for UDP-sugar binding, which is characteristic of Golgi-residing glycosyltransferases. UDP-sugars are synthesized in the cytosol by a panel of enzymes, except for UDP-Xyl, which is produced from UDP-GlcA via UDP-GlcA decarboxylase in the Golgi lumen (Figure 9.7). The UDP-sugars are transported into the Golgi and ER lumen by the corresponding UDP-sugar transporters, where they are utilized by a number of glycosyltransferases to build up various glycans including GAGs. Recently, mutants of the enzymes required for the biosynthesis of UDP-sugars have been isolated in various model organisms including mouse, zebrafish, *D. melanogaster,* and *C. elegans*, and have shown not only reduced GAG synthesis but also abnormal development (Table 9.4).[32–36] These results suggest that GAGs are essential for morphogenesis and organogenesis in embryonic development. For instance, embryonic development of the mouse mutant of UDP-Glc dehydrogenase, which catalyzes the formation of UDP-GlcA from UDP-Glc, is arrested during gastrulation with defects in mesoderm and endoderm migration, probably due to a disruption of fibroblast growth factor (FGF) signaling.[32] In addition to the mutants of enzymes involved in producing UDP-sugars, mutants of UDP-sugar transporters, such as *sqv-7* in *C. elegans* and *fringe connection* in *D. melanogaster*, have also been isolated and characterized (Table 9.4).[37–39] These transporters take various

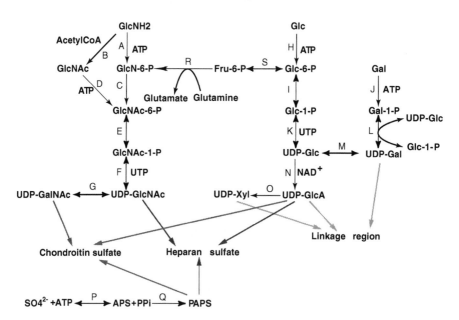

FIGURE 9.7 Synthetic pathways of UDP-sugars and PAPS. A, hexokinase; B, GlcNH$_2$ acetyltransferase; C, GlcNH$_2$ 6-phosphate N-acetyltransferase; D, GlcNAc kinase; E, GlcNAc phosphomutase; F, UDP-GlcNAc pyrophosphorylase; G, UDP-GalNAc 4-epimerase; H, hexokinase; I, phosphoglucomutase; J, galactokinase; K, UDP-Glc pyrophosphorylase; L, Gal-1-phosphate uridylyltransferase; M, UDP-Glc 4-epimerase; N, UDP-Glc dehydrogenase; O, UDP-GlcA decarboxylase; P, ATP sulfurylase; Q, APS kinase; R, glutamine-Fru-6-phosphate aminotransferase; S, Glc-6-phosphate isomerase.

UDP-sugars from the cytosol into the Golgi and ER lumen. The mutants of these transporters display drastic decrease in GAG contents due to a deficiency of the donor substrates required for the biosynthesis of GAGs in the ER and Golgi compartments, and show defects in morphogenesis early in embryonic development. Thus, enzymes responsible for the biosynthesis of UDP-sugars and UDP-sugar transporters are required not only for the biosynthesis of GAGs but also for the development of organisms, particularly morphogenesis in early developmental stages.

IV. SULFATION OF CHONDROITIN BACKBONE

Following the formation of the repeating disaccharide region of CS, the CS backbone is modified with a number of sulfates. Various positions of GalNAc and GlcA in CS potentially become sulfated and the resultant sulfated CS displays diverse structures. Sulfation occurs mainly at positions 4 and 6 of GalNAc and position 2 of GlcA. Sulfated disaccharides in CS are divided into four groups according to the sulfation pattern: A [GlcAβ1-3GalNAc(4S)], C [GlcAβ1-3GalNAc(6S)], D [GlcA(2S)β1-3GalNAc(6S)], and E [GlcAβ1-3GalNAc(4S,6S)] units, where 2S, 4S, and 6S represent 2-O-, 4-O-, and 6-O-sulfates, respectively (Figure 9.2 and Figure 9.6).[13] Specific sulfation of CS chains may be involved in the high-affinity binding of various proteins including Hep-binding growth factors and adhesion molecules, although the physiological significance of the binding has not been firmly established.[40] In the brain, highly sulfated CS structures appear to be involved in the regulation of axonal growth and pathfinding of various neurons. For example, neurons cultured on a substrate coated with CS-D, composed of D units, exhibit a flattened cell soma with multiple neurites, whereas neurons cultured on a substrate coated with CS-E composed of E units shows a round-shaped cell soma with a single prominent long neurite.[19] These differences are caused by the distinct sulfation patterns of the CS substrate. As exemplified by these observations, specifically sulfated CS chains in the extracellular matrix appear to have significant effects on cellular processes such as cell proliferation and differentiation, directly or indirectly mediated by growth factors. Since the sulfation is regulated by a number of sulfotransferases, the cloning and characterization of these sulfotransferases are essential for investigating the biological roles in CS.

Seven sulfotransferases responsible for the sulfation of CS have been cloned to date (Table 9.2).[41] Each sulfotransferase transfers sulfate from 3$'$-phosphoadenosine 5$'$-phosphosulfate (PAPS) to a specific position of GalNAc, GlcA, or IdoA in CS/DS. Chondroitin 6-O-sulfotransferase (C6ST) transfers sulfate to position 6 of the GalNAc residue and is involved in the formation of C and D units.[42,43] For sulfation of position 4 of the GalNAc residue, four sulfotransferases have been cloned. Chondroitin 4-O-sulfotransferases-1, -2, and –3 (C4ST-1, -2, and –3) are responsible for 4-O-sulfation of a GalNAc residue next to GlcA in CS, whereas dermatan 4-O-sulfotransferase-1 (D4ST-1) is involved in 4-O-sulfation of a GalNAc residue next to IdoA in DS.[44–47] Thus, four sulfotransferases are responsible for the formation of the A unit. 2-O-sulfation of GlcA and IdoA is catalyzed by uronyl 2-O-sulfotransferase.[48] This sulfotransferase is essential for the formation of the D unit, which is formed in two steps. C6ST catalyzes 6-O-sulfation of the GalNAc residue first and then 2-O-sulfation of the GlcA or IdoA residue is catalyzed by uronyl 2-O-sulfotransferase (Figure 9.6). GalNAc 4-sulfate 6-O-sulfotransferase (GalNAc4S-6ST) transfers sulfate to position 6 of GalNAc(4S) formed by C4ST, and is responsible for the formation of the E unit (Figure 9.6).[49]

PAPS serves as the universal sulfate donor for all sulfotransferase reactions and is synthesized from two modes of ATP and an inorganic sulfate in the cytosol (Figure 9.7).[50] Inorganic sulfate is taken up from extracellular milieu into the cytosol by diastrophic dysplasia sulfate transporter, which is expressed on the cell surface.[51,52] The PAPS-synthesizing reactions are catalyzed by PAPS synthases 1 and 2, which are bifunctional enzymes composed of an N-terminal adenosine 5′-phosphosulfate (APS) kinase domain and a C-terminal ATP sulfurylase domain.[53–55] In the first step, the ATP sulfurylase domain of PAPS synthase combines an inorganic sulfate with ATP to form APS (Figure 9.7). In the second step, the APS kinase domain combines APS with another ATP to form PAPS (Figure 9.7). Then, PAPS synthesized in the cytosol is transported to the Golgi apparatus by PAPS transporter, where it is used for the sulfation of various molecules including GAGs.[56,57]

Dysfunctional PAPS synthesis is related to diseases such as osteochondrodysplasias (Table 9.4).[58] The brachymorphic mouse has been identified to have a deficiency of APS kinase and the gene affected is for PAPS synthase 2.[54] This mouse has a dome-shaped skull, a short thick tail, and shortened but not widened limbs, and produces undersulfated CS–PG. A deficiency of PAPS synthase 2 in humans leads to spondyloepimetaphyseal dysplasia, the clinical features of which are short and bowed lower limbs, enlarged knee joints, and early onset of degenerative joint disease.[55] These results together suggest that PAPS synthase 2 plays a crucial role in bone and cartilage development. Interestingly, a disease caused by a defect of PAPS synthase 1 has not been reported to date. Hence, a compensatory mechanism may exist for the PAPS synthase 1 deficiency, but not for the deficiency of PAPS synthase 2. Alternatively, the tissue-specific expression of PAPS synthases 1 and 2 may explain the dysfunctional formation of bone and cartilage, where PAPS synthase 2 is abundantly expressed.[54] It has recently been revealed that a defect in the *D. melanogaster* gene *slalom*, which encodes a PAPS transporter with ten transmembrane regions, leads to a perturbation of the signaling pathway of morphogens including Wingless (Wg) and Hh.[56] The phenotypes are reminiscent of the fly mutant of *sulfateless*, which encodes NDST required for the sulfation of HS chains.[59,60] Thus, sulfation of GAGs has a decisive role in modulating the signaling pathways of morphogens such as Wg and Hh, and PAPS synthase and PAPS transporter, in addition to glycosyltransferases and sulfotransferases are essential for GAGs to mature and carry out their biological functions.

V. BIOSYNTHESIS OF HEPARAN SULFATE AND HEPARIN BACKBONES

Chain polymerization of HS and Hep is initiated by the transfer of an α-GlcNAc residue from UDP-GlcNAc to the linkage region tetrasaccharide, GlcAβ1-3Galβ1-3Galβ1-4Xylβ1-O-Ser, by α-GlcNAc transferase-I (GlcNAcT-I). The resultant nascent pentasaccharide is elongated further by alternate additions of GlcA and GlcNAc from UDP-GlcA and UDP-GlcNAc, respectively, to form a HS or Hep backbone by HS polymerase(s), which exhibit the dual catalytic activities of HS GlcA transferase-II (HS GlcAT-II) and GlcNAc transferase-II (GlcNAcT-II) (Figure 9.3). Such HS/Hep precursor chains are subsequently modified through a series of reactions involving the NSDT of GlcNAc, epimerization of GlcA, and *O*-sulfation at various positions of both residues, which confer a structural diversity to HS/Hep (Figure 9.8).[11]

Two independent studies revealed a relation between HS-synthesizing glycosyl-transferases and the *hereditary multiple exostoses*(HMEs) (*EXT*) gene family of tumor suppressors. First, using a functional assay based on the ability of herpes simplex virus to infect cells by attaching to cell surface HS, the putative tumor suppressor *EXT1* was isolated from a HeLa-cell cDNA library.[61] Human EXT1, the gene product, has a type II transmembrane topology and consists of 746 amino acids. Transfection of mouse sog9 cells, which are mutant L cells and are unable to synthesize GAGs, with the human *EXT1* cDNA restored the susceptibility of sog9 to the viral infection, suggesting an association of the expression of cell surface HS with

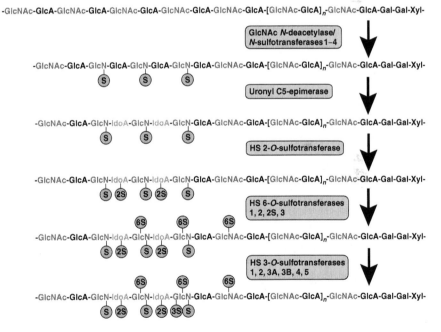

FIGURE 9.8 Pathways of biosynthetic modifications of HS and Hep chains. Following synthesis of the backbone of HS or Hep by HS polymerases belonging to the EXT gene family, modifications of the precursor HS/Hep chains are conducted by various sulfotransferases and a single epimerase. The first modifications, *N*-deacetylation and *N*-sulfation, are essential for all subsequent reactions. Next, GlcA residues adjacent to GlcNS residues are converted into IdoA residues by uronyl C5 epimerase. Thereafter, sulfation at C2 of IdoA residues, and at C6 and C3 of GlcNS and GlcNAc residues takes place through the actions of specific sulfotransferases. S, 2-*N*-sulfate; 2S, 2-*O*-sulfate; 3S, 3-*O*-sulfate; 6S, 6-*O*-sulfate; and n, the number of repeating disaccharide units.

EXT1.[61] Overexpression of *EXT1* in sog9 cells caused a slight yet significant increase in both HS GlcAT-II and GlcNAcT-II activities, indicating that EXT1 harbors both enzyme activities.[62]

Secondly, EXT2 was identified as a HS polymerase by direct peptide sequencing of the enzyme protein purified from bovine serum.[62] The amino acid sequence of a tryptic peptide of the purified enzyme matched a human expressed sequence tag (EST) cDNA fragment. The bovine cDNA encoded a type II transmembrane protein composed of 718 amino acids. The cDNA sequence was 94% identical to the sequence of human *EXT2*. The recombinant bovine EXT2 exhibited both GlcA and GlcNAc transferase activities towards GlcNAc-$[GlcA-GlcNAc]_n$ and $[GlcA-GlcNAc]_n$ oligosaccharide acceptors, respectively. Human *EXT2* has strong sequence homology (31% identity and 69.5% similarity) to human *EXT1*, whose gene product also showed weak yet appreciable GlcA and GlcNAc transferase activities.[62]

EXT1 and EXT2 proteins form a heterooligomeric complex *in vivo*, which leads to the accumulation of both proteins in the Golgi apparatus, although separately overexpressed, EXT1 and EXT2 proteins are located predominantly in the ER.[63] Co-expression of the two proteins results in stronger GlcA and GlcNAc transferase activities compared with the individually expressed proteins.[63,64] These findings suggest that the cooperation of EXT1 and EXT2 in the Golgi is required for the polymerization of HS chains. Significant *in vitro* HS polymerization has recently been demonstrated using soluble forms of recombinant enzymes expressed by co-transfection of *EXT1* and *EXT2*.[65] On the other hand, it has also been reported that recombinant EXT1 alone, as well as the EXT1/EXT2 heterocomplex, catalyzes the *in vitro* polymerization to a limited extent of the HS backbone structure on an oligosaccharide primer.[66]

HME is an autosomal dominant disorder characterized by the formation of cartilage-capped tumors (exostoses) that develop from the growth plate of endochondral bones, especially of long bones.[67] This pathologic condition leads to skeletal abnormalities and short stature. Malignant transformation from exostoses to chondrosarcomas[68,69] or osteosarcomas[70,71] occurs in approximately 2% of HME patients as a result of the loss of heterozygosity in *EXT1* or *EXT2*. Genetic linkage analysis of HME has identified three different loci, 8q24.1, 11p11–13, and 19p, which include *EXT1*, *EXT2*, and *EXT3*, respectively.[72–74] In most HME cases, missense or frameshift mutations in *EXT1* and *EXT2* have been identified, although an association of *EXT3* with HME has not been shown. Targeted disruption of *EXT1* in mice results in embryonic lethality caused by a failure to form a mesoderm and causes defects in egg cylinder elongation.[75] ES cells derived from the homozygous null mice show marked decreases in GlcNAc and GlcA transferase activities and a complete lack of HS. Furthermore, the embryonic mouse brain, where *EXT1* alleles are conditionally disrupted, shows serious fallacy in midline axon guidance through Slit, Sonic hedgehog (Shh), and FGF8 signalings.[76] These findings indicate that HS is essential for mammalian brain development in addition to early embryogenesis. *EXT2* homozygous null embryos also fail to develop at the gastrulation stage.[77] Heterozygous embryos of both *EXT1* and *EXT2* appear normal but form exostoses at low frequency. On the other hand, compound heterozygotes (*EXT1*$^{+/-}$*EXT2*$^{+/-}$) show a high frequency of exostoses.[77]

The *EXT* family contains three additional members, designated *EXTL1-3* (*EXT*-like genes 1–3), based on the amino acid sequence similarity of their gene products to EXT1 and EXT2.[78–81] The *EXTL* genes have not been co-related to HME, although the chromosomal loci of the genes imply that they might also be tumor suppressors. Among the products of the three *EXTL* genes, EXTL2 was first demonstrated to have the activity of α-GalNAc transferase (α-GalNAcT) and GlcNAcT-I, which transfer GalNAc and GlcNAc to the linkage region tetrasaccharide, respectively.[82] GlcNAcT-I is essential for the initiation of HS/Hep biosynthesis. On the other hand, no structure containing α1-4GalNAc has been found in any naturally occurring GAG chains. Products of the other *EXTL* family genes, *EXTL1* and *EXTL3*, were subsequently demonstrated to exhibit α-GlcNAc transferase activity likely involved in the biosynthesis of HS and probably Hep.[83] EXTL1 shows only GlcNAcT-II activity, which may also be involved in HS chain elongation. On the other hand, EXTL3 shows both GlcNAcT-I and GlcNAcT-II activities, which appear to be involved in the biosynthetic initiation and elongation, respectively, of HS/Hep precursor chains. HS GlcAT-II activity, however, is not detected for either EXTL1 or EXTL3. Thus, as summarized in Figure 9.3, the gene products of all five EXT family members harbor glycosyltransferase activity, which appears to contribute to HS/Hep biosynthesis. EXTL3 also has other properties such as a receptor function for a β-cell regeneration factor (Reg)[84] and a modulator function for nuclear factor kappaB (NF-κB) activity in response to TNF-α.[85]

HS and CS are present in genetically tractable model animals as well.[86–88] *EXT* genes are well conserved in humans, *D. melanogaster* (*ttv*, *sotv/DEXT2*, *botv/DEXT3*),[89–92] and *C. elegans* (*rib-1*, *rib-2*)[93–95] (Table 9.4). However, *in vitro* glycosyltransferase activities have been demonstrated only for Botv/DEXT3 and Rib-2, both of which have GlcNAcT-I and -II activities.[92,94] *Drosophila* homologs of *EXT* genes, *tout-velu* (*ttv*), *sister of tout-velu* (*sotv*), and *brother of tout-velu* (*botv*), appear to be involved in the signaling pathways mediated by Hh, Wg, and Dpp, all of which function as morphogens and organize the wing patterning in *Drosophila*, since mutants of these genes show severe morphological phenotypes.[89–91] *C. elegans* *rib-2* shows the highest homology to *EXTL3* among the human *EXT* gene family.[94] Its mutants exhibit abnormal characteristics such as a developmental delay and egg-laying defects, which are most likely caused by defects in HS biosynthesis.[95]

VI. MODIFICATIONS OF HEPARAN SULFATE/HEPARIN BACKBONE

A. *N*-DEACETYLATION AND *N*-SULFATION

N-Deacetylation and *N*-sulfation of GlcNAc residues are initial modifications of nonsulfated precursor HS/Hep chains and are prerequisites for all the subsequent modification reactions (Figure 9.8). These reactions are catalyzed by bifunctional enzymes, NDSTs. The acetylgroup of GlcNAc is hydrolytically released by the *N*-deacetylase activities of NDSTs, yielding unsubstituted glucosamine (GlcN), to which a sulfate group is then transferred from PAPS by the *N*-sulfotransferase activities of NDSTs. The vertebrate NDST family consists of four isozymes (Table 9.3), but only one is found in *D. melanogaster* and *C. elegans* (Table 9.4). NDST-1 and

NDST-2 were originally purified from rat liver and Hep-producing mouse mastocytoma, respectively,[96,97] and the cDNAs of both NDSTs were cloned based on the amino acid sequences of the corresponding purified enzymes,[98,99] *NDST-3* and *NDST-4* have been found in the EST database using the nucleotide sequences of NDST-1 and NDST-2.[100,101] *NDST-1* and *NDST-2* are widely expressed in mouse tissues, whereas *NDST-3* is strongly expressed in the brain and more so at embryonic day 11 than at other developmental stages, and *NDST-4* is expressed exclusively in the brain among adult tissues and throughout the embryonic development.[101] Comparison of the enzymatic properties of the four isozymes has revealed marked differences in the ratios between *N*-deacetylase and *N*-sulfotransferase activities. Among the four NDSTs, NDST-1 and NDST-2 have high levels of both activities. NDST-3 shows a strong *N*-deacetylase activity and a very weak *N*-sulfotransferase activity. NDST-4 is peculiar in that its *N*-deacetylase activity is extremely weak when compared with that of NDST-1–3, whereas its *N*-sulfotransferase activity is comparable with that of NDST-1 and NDST-2.[101]

Approximately two thirds of *NDST-1*-deficient mouse embryos survive until birth. However, they turn cyanotic and die during the neonatal period with symptoms resembling infant respiratory distress syndrome in humans.[102,103] In addition, one third of *NDST-1*-deficient mice die as embryos, showing skull defects and a disturbed eye development.[104] Interestingly, these skull and eye abnormalities resemble the phenotypes observed in mouse embryos that are deficient in HS-binding morphogens including Wnt-1,[105,106] Shh,[107] or Chordin and Noggin.[108] An analysis of *NDST-1* knockout mice has also elucidated that HS is involved in Ca^{2+} kinetics of skeletal muscles.[109] Structural analysis of HS from cultured *NDST-1* homozygous (*NDST-1$^{-/-}$*) fibroblasts has revealed that the degree of *N*-sulfation is considerably less than that in HS obtained from heterozygous and wild-type fibroblasts.[103] On the other hand, mice carrying a targeted disruption of *NDST-2* are unable to synthesize Hep.[110,111] Although they are viable and fertile, connective tissue type mast cells are greatly reduced in the knockout mice. The mast cells have an altered morphology and contain severely reduced amounts of histamine and specific granule proteases. These observations indicate that the storage of specific granule proteases in mast cells is regulated by Hep. Moreover, a lack of both *NDST-1* and *NDST-2* results in early embryonic lethality in mice.[104] This outcome is similar to mice homozygous for mutations in the *EXT1* gene, which die during gastrulation due to an unorganized mesoderm and a lack of extraembryonic tissue.[75] These results suggest that sulfated HS chains are needed for development and that one isozyme may partly compensate for the loss of the other in the *NDST-1* and *NDST-2* knockout strains. The *Drosophila* gene *sulfateless*, which encodes a homolog of vertebrate NDSTs, is essential for the Wg and FGF receptor signalings.[59,60]

B. URONYL C5 EPIMERIZATION

Uronyl C5 epimerase, which catalyzes the reversible interconversion of GlcA into IdoA in HS/Hep but not in CS/DS in a soluble *in vitro* system, was originally purified to homogeneity from bovine liver.[112] It should be noted, however, that a recent study has revealed that C5 epimerization is an efficiently irreversible reaction in cultured cells.[113] Full-length cDNA for epimerase has been cloned from bovine, mouse, and human, and its *C. elegans* and *D. melanogaster* counterparts have also been cloned.[114–116] Database searches do not show the additional homologs, suggesting that there may be only one HS/Hep C5 epimerase. The full-length epimerase is predicted to have a type II transmembrane topology and reside in the Golgi, where the synthesis of HS takes place, while an *N*-terminal truncated form of epimerase, which shows 75% less activity than the full-length protein, accumulates in the cytosol.[116] Possible functions of the cleaved *N*-terminal peptide are unknown, but apparently remain associated with the catalytic portion of the enzyme. C5 epimerase of a microsomal preparation acts on uronic acids when the target residues are located on the nonreducing side of *N*-sulfated glucosamine (GlcNS) residues, while it does not attack uronic acids that are *O*-sulfated or flanked by *O*-sulfated GlcN residues.[117] These observations suggest that the epimerization begins after NDST but before *O*-sulfation of uronic acid and GlcN residues (Figure 9.8). It has been proposed that the presence of IdoA residues confer conformational flexibility on the GAG chain to facilitate interactions with proteins.[118] Targeted disruption of the murine HS/Hep C5-epimerase gene (*Hsepi*) leads to the production of a structurally altered HS lacking IdoA.[119] The observed phenotype is lethal, with renal agenesis, lung defects, and skeletal malformations. The C5-epimerase homozygous (*Hsepi*$^{-/-}$) littermates show an immature lung phenotype and die immediately after birth as in the case of *NDST-1*$^{-/-}$ pups. Unexpectedly, major organs including the brain, liver, gastrointestinal tract, skin, and heart, appear normal. In contrast to the absence of the renal structure in *Hsepi*$^{-/-}$, *NDST-1*$^{-/-}$ mice have apparently normal kidneys. Renal agenesis is also found in homozygous mice (*Hs2st*$^{-/-}$) deficient in HS 2-*O*-sulfotransferase (HS2ST) that catalyzes 2-*O*-sulfation of both IdoA and GlcA in HS chains.[120] (described below). The *Hsepi*$^{-/-}$ mice further show skeletal abnormalities indistinguishable from those observed in *Hs2st*$^{-/-}$ animals. On the other hand, contrary to *Hsepi*$^{-/-}$ and *NDST-1*$^{-/-}$ mice, *Hs2st*$^{-/-}$ mutants have apparently normal lungs, and can survive for several hours after birth. Although the biosynthesis of HS/Hep is presumed to occur as shown in Figure 9.8, based on the substrate specificity of each enzyme involved in modifications of its precursor's chains, it remains enigmatic that the renal phenotypes found in *Hs2st*$^{-/-}$ and *Hsepi*$^{-/-}$ mice are not observed in *NDST-1*$^{-/-}$ mice.

C. *O*-SULFATION

HS2ST has been purified from cultured CHO cells to apparent homogeneity.[121] The purified enzyme catalyzes the transfer of sulfate from PAPS to the C-2 position of IdoA residues in HS/Hep, whereas the enzyme exhibits significantly less activity toward GlcA residues in HS/Hep. Based on the amino acid sequence obtained from the purified enzyme, the full-length cDNA, which encodes a predicted type II transmembrane protein, has been cloned.[122] The recombinant HS2ST catalyzes 2-*O*-sulfation of IdoA residues in [-IdoA-GlcNS-], but not in [-IdoA-GlcNS(6S)-] sequences. A CHO mutant cell, pgsF-17, is defective in *HS2ST*.[123] The HS chains in pgsF-17 lack [-IdoA(2S)-GlcNS-] and [-IdoA(2S)-GlcNS(6S)-], but contain a higher proportion of [-GlcA-GlcNS(6S)-] and [-IdoA-GlcNS(6S)-]. Moreover, HS2ST and C5 epimerase interact with each other *in vivo*,[124] suggesting that their physical interaction may be required for stability, enzymatic activity, and residing in the Golgi apparatus. These observations may support that HS2ST prefers IdoA to GlcA residues in HS/Hep. Mice homozygous for the *HS2ST* gene trap allele die during the neonatal period, exhibiting bilateral renal agenesis and defects in the eye and skeleton.[120] In addition, loss of 2-*O*-sulfation in HS leads to a significant reduction in cell proliferation but not in cell migration in the developing cerebral cortex.[125] Analysis of HS isolated from fibroblasts of *Hs2st*[−/−] mutants has revealed compensatory increases in *N*- and *O*-sulfation to maintain the overall charge density.[126] The apparent affinities of the mutant HS for hepatocyte growth factor and fibronectin are unchanged but are reduced for FGF-1 and -2. Surprisingly, the *Hs2st*[−/−] cells are capable of mounting an apparently normal signaling in response to FGF-1 and -2 as well as hepatocyte growth factor. *Drosophila* mutants of the *pipe* gene, which encodes a putative HS2ST, show defects in the establishment of dorsal–ventral polarity in the embryo.[127] Another putative *Drosophila HS2ST* gene, *dHS2ST*, homologous to mammalian *HS2ST* has also been identified at the *Segregation distorter* locus associated with the meiotic drive system.[128] These observations suggest essential functions of 2-*O*-sulfation coupled with epimerization of uronic acid in HS in embryonic development.

Heparan sulfate 6-*O*-sulfotransferase (HS6ST), which catalyzes the transfer of sulfate to the C-6 position of GlcNS residues in HS, was originally purified to apparent homogeneity from the serum-free medium of CHO cells.[129] On the basis of its amino acid sequence, cDNA encoding HS6ST-1 has been cloned,[130] and two additional HS6ST isoforms, designated as HS6ST-2 and HS6ST-3, have also been identified [131] (Table 9.3). Mouse HS6ST-1, HS6ST-2, and HS6ST-3 are type II transmembrane proteins composed of 401, 506, and 470 amino acid residues, respectively. The amino acid sequence of mouse HS6ST-1 is 51 and 57% identical to that of HS6ST-2 and HS6ST-3, respectively. In humans, there are two splicing forms of HS6ST-2: the original form (HS6ST-2) and a short form (HS6ST-2S), lacking 40 amino acids encoded by exons 2 and 3[132] (Table 9.3). These mouse isoforms show different preferences toward the isomeric hexuronic acid adjacent to the targeted GlcNS residues. Recombinant mouse HS6ST-1 appears to prefer IdoA residues adjacent to the targeted GlcNS residues, whereas HS6ST-2 prefers GlcA to IdoA, and HS6ST-3 can act on either of the structures.[131] Furthermore, three recombinant mouse HS6STs are capable of catalyzing the 6-*O*-sulfation of GlcNAc, which is embedded in an AT-binding sequence of HS/Hep, as well as GlcNS residues.[133] *Drosophila* contains only one HS6ST, which is 54, 46, and 53% identical to mouse HS6ST-1, -2 and -3, respectively.[134] RNAi experiments have demonstrated that the reduced HS6ST activity causes embryonic lethality and disruption of the primary branching of the tracheal system. These phenotypes are similar to the defects observed in mutants of FGF signaling components.

Heparan sulfate 3-*O*-sulfotransferase (HS3ST), which catalyzes the transfer of sulfate to the C-3 position of GlcN residues in HS/Hep, is an essential enzyme for the production of the AT-binding domain in HS/Hep.[135] HS3ST was initially purified to homogeneity from serum-free medium of the mouse L cell line LTA,[136] and a cDNA encoding HS3ST-1 was cloned based on the partial amino acid sequence of the purified enzyme.[137] The five isoforms, designated as HS3ST-2, -3A, -3B, -4, and -5, have also been isolated based on sequence homology to human HS3ST-1 (Table 9.3).[138–140] These HS3STs, except for HS3ST-1, are putative type II transmembrane proteins. Human *HS3STs* show differential expression patterns in various tissues and cell types, testifying to their distinct functional roles[138,140] (see below). Sulfotransferase activity has not been examined for HS3ST-4 among these isoforms because of an incomplete knowledge of its cDNA. The other isoforms show divergent substrate specificities toward the internal sequences of HS/Hep. Human recombinant HS3ST-1 transfers sulfate to the C-3 position of GlcNS or GlcNS(6S) adjacent to the reducing side of GlcA. On the other hand, HS3ST-2 catalyzes 3-*O*-sulfation of GlcN residues in [-GlcA(2S)-GlcNS-] and [-IdoA(2S)-GlcNS-] sequences in HS/Hep. Additionally, HS3ST-3A prefers [-IdoA(2S)-GlcN-] and HS3ST-5 selectively acts on the [-GlcA-GlcNS(6S)-], [-IdoA-GlcNS-], and [-IdoA-GlcNS(6S)-] sequences.[139,140] Although the 3-*O*-sulfated GlcN is a rare constituent in the HS from natural sources, it plays critical roles in the binding of various proteins such as AT,[135] envelope glycoprotein D of the herpes simplex virus-1,[141] FGF receptor,[142] and FGF-7.[143] The HS chains modified by HS3ST-1 and -5 can bind AT,[137,140] while those modified by HS3ST-3A, -3B, and -5 can bind glycoprotein D to mediate viral entry.[140,141] The *in vivo* roles of anticoagulant HS have been elucidated by the generation of *HS3ST-1* knockout mice.[144] *HS3ST-1⁻/⁻* mice show negligible HS3ST activity in plasma and tissue extracts, and dramatic reductions in tissue levels of anticoagulant HS while maintaining wild-type levels of tissue fibrin accumulation. *HS3ST-1⁻/⁻* mice, however, do not show an obvious procoagulant phenotype, suggesting that the other *HS3ST* family members compensate for the loss of function of *HS3ST-1*.

VII.   PERSPECTIVES

cDNA cloning of enzymes involved in GAG biosynthesis has led to a knowledge of not only the biosynthetic mechanisms but also *in vivo* functions of CS/DS and HS/Hep chains. However, the molecular cloning of the C5 epimerase and GlcA 3-*O*-sulfotransferase specific for CS/DS chains remains to be achieved. In addition, investigation of the regulatory mechanisms for the expression of these genes and the reciprocal relationship between these glycosyltransferases and modifying enzymes including sulfotransferases and epimerases is required to gain a better understanding of the mechanisms of differential assembly of CS/DS and HS/Hep chains, their effective polymerization followed by various modifications, and chain termination to regulate the chain lengths. The gene knockout of the ChSy family members and multiple sulfotransferases will also provide direct evidence for the biological functions of CS/DS. Thus, further study of the biosynthesis of CS/DS and HS/Hep will help clarify the pathological mechanisms of diverse human diseases and provide insights into possible therapeutic applications of the knowledge developed with GAG glycobiology.

ACKNOWLEDGMENTS

This work was supported in part by the Science Research Promotion Fund of the Japan Private School Promotion Foundation and by grants from the Ministry of Education, Science, Sports, and Culture of Japan.

REFERENCES

1. Esko, J.D. and Selleck, S.B., Order out of chaos: assembly of ligand binding sites in heparan sulfate, *Annu. Rev. Biochem.*, 71, 435, 2002.
2. Götting, C., Kuhn, J., Zahn, R., Brinkmann, T., and Kleesiek, K., Molecular cloning and expression of human UDP-D-xylose:proteoglycan core protein β-D-xylosyltransferase and its first isoform XT-II, *J. Mol. Biol.*, 304, 517, 2000.
3. Prydz, K. and Dalen, K.T., Synthesis and sorting of proteoglycans, *J. Cell Sci.*, 113, 193, 2000.
4. Okajima, T., Yoshida, K., Kondo, T., and Furukawa, K., Human homolog of *Caenorhabditis elegans sqv-3* gene is galactosyltransferase I involved in the biosynthesis of the glycosaminoglycan-protein linkage region of proteoglycans, *J. Biol. Chem.*, 274, 22915, 1999.
5. Almeida, R., Levery, S.B., Mandel, U., Kresse, H., Schwientek, T., Bennett, E. P., and Clausen, H., Cloning and expression of a proteoglycan UDP-galactose: β-xylose β1,4-galactosyltransferase I: A seventh member of the human β4-galactosyltransferase gene family, *J. Biol. Chem.*, 274, 26165, 1999.
6. Bai, X., Zhou, D., Brown, J.R., Crawford, B.E., Hennet, T., and Esko, J.D., Biosynthesis of the linkage region of glycosaminoglycans: cloning and activity of galactosyltransferase II, the sixth member of the β1,3-galactosyltransferase family (β3GalT6), *J. Biol. Chem.*, 276, 48189, 2001.
7. Kitagawa, H., Tone, Y., Tamura, J., Newmann, K.W., Ogawa, T., Oka, S., Kawasaki, T., and Sugahara, K., Molecular cloning and expression of glucuronyltransferase I involved in the biosynthesis of the glycosaminoglycan-protein linkage region of proteoglycans, *J. Biol. Chem.*, 273, 6615, 1998.
8. Wei, G., Bai, X., Sarkar, A.K., and Esko, J.D., Formation of HNK-1 determinants and the glycosaminoglycan tetrasaccharide linkage region by UDP-GlcUA: galactose β1,3-glucuronosyltransferases, *J. Biol. Chem.*, 274, 7857, 1999.
9. Terayama, K., Oka, S., Seiki, T., Miki, Y., Nakamura, A., Kozutsumi, Y., Takio, K., and Kawasaki, T., Cloning and functional expression of a novel glucuronyltransferase involved in the biosynthesis of the carbohydrate epitope HNK-1, *Proc. Natl. Acad. Sci. USA*, 94, 6093, 1997.
10. Tone, Y., Kitagawa, H., Imiya, K., Oka, S., Kawasaki, T., and Sugahara, K., Characterization of recombinant human glucuronyltransferase I involved in the biosynthesis of the glycosaminoglycan-protein linkage region of proteoglycans, *FEBS Lett.*, 459, 415, 1999.
11. Sugahara, K. and Kitagawa, H., Recent advances in the study of the biosynthesis and functions of sulfated glycosaminoglycans, *Curr. Opin. Struct. Biol.*, 10, 518, 2000.
12. Moses, J., Oldberg, Å., and Fransson, L.-Å., Initiation of galactosaminoglycan biosynthesis: separate galactosylation and dephosphorylation pathways for phosphoxylosylated decorin protein and exogenous xyloside, *Eur. J. Biochem.*, 260, 879, 1999.
13. Sugahara, K. and Yamada, S., Structure and function of oversulfated chondroitin sulfate variants: unique sulfation patterns and neuroregulatory activities, *Trends Glycosci. Glycotechnol.*, 12, 321, 2000.
14. Quentin, E., Gladen, A., Rodén, L., and Kresse, H., A genetic defect in the biosynthesis of dermatan sulfate proteoglycan: galactosyltransferase I deficiency in fibroblasts from a patient with a progeroid syndrome, *Proc. Natl. Acad. Sci. USA*, 87, 1342, 1990.

15. Okajima, T., Fukumoto, S., Furukawa, K., Urano, T., and Furukawa, K., Molecular basis for the progeroid variant of Ehlers–Danlos syndrome: identification and characterization of two mutations in galactosyltransferase I gene, *J. Biol. Chem.*, 274, 28841, 1999.

16. Nakamura, Y., Haines, N., Chen, J., Okajima, T., Furukawa, K., Urano, T., Stanley, P., Irvine, K.D., and Furukawa, K., Identification of a *Drosophila* gene encoding xylosylprotein β4-galactosyltransferase that is essential for the synthesis of glycosaminoglycans and for morphogenesis, *J. Biol. Chem.*, 277, 46280, 2002.

17. Herman, T., Hartwieg, E., and Horvitz, H.R., *sqv* mutants of *Caenorhabditis elegans* are defective in vulval epithelial invagination, *Proc. Natl. Acad. Sci. USA*, 96, 968, 1999.

18. Herman, T. and Horvitz, H.R., Three proteins involved in *Caenorhabditis elegans* vulval invagination are similar to components of a glycosylation pathway, *Proc. Natl. Acad. Sci. USA*, 96, 974, 1999.

19. Sugahara, K., Mikami, T., Uyama, T., Mizuguchi, S., Nomura, K., and Kitagawa, H., Recent advances in the structural biology of chondroitin sulfate and dermatan sulfate, *Curr. Opin. Struct. Biol.*, 13, 612, 2003.

20. Bulik, D.A., Wei, G., Toyoda, H., Kinoshita-Toyoda, A., Waldrip, W.R., Esko, J.D., Robbins, P.W., and Selleck, S.B., *sqv-3*, *-7*, and *-8*, a set of genes affecting morphogenesis in *Caenorhabditis elegans*, encode enzymes required for glycosaminoglycan biosynthesis, *Proc. Natl. Acad. Sci. USA*, 97, 10838, 2000.

21. Mizuguchi, S., Uyama, T., Kitagawa, H., Nomura, K.H., Dejima, K., Gengyo-Ando, K., Mitani, S., Sugahara, K., and Nomura, K., Chondroitin proteoglycans are involved in cell division of *Caenorhabditis elegans*, *Nature*, 423, 443, 2003.

22. Hwang, H.-Y., Olsen, S.K., Esko, J.D., and Horvitz, H.R., *Caenorhabditis elegans* early embryogenesis and vulval morphogenesis require chondroitin biosynthesis, *Nature*, 423, 439, 2003.

23. Silbert, J.E. and Sugumaran, G., Biosynthesis of chondroitin/dermatan sulfate, *IUBMB Life*, 54, 177, 2002.

24. Kitagawa, H., Uyama, T., and Sugahara, K., Molecular cloning and expression of a human chondroitin synthase, *J. Biol. Chem.*, 276, 38721, 2001.

25. Kitagawa, H., Izumikawa, T., Uyama, T., and Sugahara, K., Molecular cloning of a chondroitin polymerizing factor that cooperates with chondroitin synthase for chondroitin polymerization, *J. Biol. Chem.*, 278, 23666, 2003.

26. Uyama, T., Kitagawa, H., Tamura, J., and Sugahara, K., Molecular cloning and expression of human chondroitin *N*-acetylgalactosaminyltransferase: the key enzyme for chain initiation and elongation of chondroitin/dermatan sulfate on the protein linkage region tetrasaccharide shared by heparin/heparan sulfate, *J. Biol. Chem.*, 277, 8841, 2002.

27. Gotoh, M., Sato, T., Akashima, T., Iwasaki, H., Kameyama, A., Mochizuki, H., Yada, T., Inaba, N., Zhang, Y., Kikuchi, N., Kwon, Y.-D., Togayachi, A., Kudo, T., Nishihara, S., Watanabe, H., Kimata, K., and Narimatsu, H., Enzymatic synthesis of chondroitin with a novel chondroitin sulfate *N*-acetylgalactosaminyltransferase that transfers *N*-acetylgalactosamine to glucuronic acid in initiation and elongation of chondroitin sulfate synthesis, *J. Biol. Chem.*, 277, 38189, 2002.

28. Uyama, T., Kitagawa, H., Tanaka, J., Tamura, J., Ogawa, T., and Sugahara, K., Molecular cloning and expression of a second chondroitin *N*-acetylgalactosaminyltransferase involved in the initiation and elongation of chondroitin/dermatan sulfate, *J. Biol. Chem.*, 278, 3072, 2003.

29. Sato, T., Gotoh, M., Kiyohara, K., Akashima, T., Iwasaki, H., Kameyama, A., Mochizuki, H., Yada, T., Inaba, N., Togayachi, A., Kudo, T., Asada, M., Watanabe, H., Imamura, T., Kimata, K., and Narimatsu, H., Differential roles of two N-acetylgalactosaminyltransferases, CSGalNAcT-1, and a novel enzyme, CSGalNAcT-2: initiation and elongation in synthesis of chondroitin sulfate, *J. Biol. Chem.*, 278, 3063, 2003.

30. Gotoh, M., Yada, T., Sato, T., Akashima, T., Iwasaki, H., Mochizuki, H., Inaba, N., Togayachi, A., Kudo, T., Watanabe, H., Kimata, K., and Narimatsu, H., Molecular cloning and characterization of a novel chondroitin sulfate glucuronyltransferase that transfers glucuronic acid to N-acetylgalactosamine, *J. Biol. Chem.*, 277, 38179, 2002.

31. Liaw, P.C., Becker, D.L., Stafford, A.R., Fredenburgh, J.C., and Weitz, J.I., Molecular basis for the susceptibility of fibrin-bound thrombin to inactivation by heparin cofactor II in the presence of dermatan sulfate but not heparin, *J. Biol. Chem.*, 276, 20959, 2001.

32. Garcia-Garcia, M.J. and Anderson, K.V., Essential role of glycosaminoglycans in fgf signaling during mouse gastrulation, *Cell*, 114, 727, 2003.

33. Walsh, E.C. and Stainier, D.Y.R., UDP-glucose dehydrogenase required for cardiac valve formation in zebrafish, *Science*, 293, 1670, 2001.

34. Binari, R.C., Staveley, B.E., Johnson, W.A., Godavarti, R., Sasisekharan, R., and Manoukian, A.S., Genetic evidence that heparin-like glycosaminoglycans are involved in *wingless* signaling, *Development,* 124, 2623, 1997.

35. Haerry, T.E., Heslip, T.R., Marsh, J.L., and O'Connor, M.B., Defects in glucuronate biosynthesis disrupt wingless signaling in *Drosophila*, *Development,* 124, 3055, 1997.

36. Häcker, U., Lin, X., and Perrimon, N., The *Drosophila sugarless* gene modulates wingless signaling and encodes an enzyme involved in polysaccharide biosynthesis, *Development,* 124, 3565, 1997.

37. Berninsone, P., Hwang, H.-Y., Zemtseve, I., Horvitz, H.R., and Hirschberg, C.B., *Proc.* SQV-7, a protein involved in *Caenorhabditis elegans* epithelial invagination and early embryogenesis, transports UDP-glucuronic acid, UDP-N-acetylgalactosamine, and UDP-galactose, *Natl. Acad. Sci. USA,* 98, 3738, 2001.

38. Selva, E.M., Hong, K., Baeg, G.-H., Beverley, S.M., Turco, S.J., Perrimon, N., and Häcker, U., Dual role of the *fringe connection* gene in both heparan sulphate and *fringe*-dependent signalling events, *Nat. Cell Biol.*, 3, 809, 2001.

39. Goto, S., Taniguchi, M., Muraoka, M., Toyoda, H., Sado, Y., Kawakita, M., and Hayashi, S., UDP–sugar transporter implicated in glycosylation and processing of Notch, *Nat. Cell Biol.*, 3, 816, 2001.

40. Deepa, S.S., Umehara, Y., Higashiyama, S., Itoh, N., and Sugahara, K., Specific molecular interactions of oversulfated chondroitin sulfate E with various heparin-binding growth factors: implications as a physiological binding partner in the brain and other tissues, *J. Biol. Chem.*, 277, 43707, 2002.

41. Kusche-Gullberg, M., and Kjellén, L., Sulfotransferases in glycosaminoglycan biosynthesis, *Curr. Opin. Struct. Biol.*, 13, 605, 2003.

42. Fukuta, M., Kobayashi, Y., Uchimura, K., Kimata, K., and Habuchi, O., Molecular cloning and expression of human chondroitin 6-sulfotransferase, *Biochim. Biophys. Acta,* 1399, 57, 1998.

43. Uchimura, K., Kadomatsu, K., Nishimura, H., Muramatsu, H., Nakamura, E., Kurosawa, N., Habuchi, O., El-Fasakhany, F.M., Yoshikai, Y., and Muramatsu, T., Functional analysis of the chondroitin 6-sulfotransferase gene in relation to lymphocyte subpopulations, brain development, and oversulfated chondroitin sulfates, *J. Biol. Chem.*, 277, 1443, 2002.

44. Hiraoka, N., Nakagawa, H., Ong, E., Akama, T.O., Fukuda, M.N., and Fukuda, M., Molecular cloning and expression of two distinct human chondroitin 4-O-sulfotransferases that belong to the HNK-1 sulfotransferase gene family, *J. Biol. Chem.*, 275, 20188, 2000.

45. Kang, H.-G., Evers, M. R., Xia, G., Baenziger, J.U., and Schachner, M., Molecular cloning and characterization of chondroitin-4-O-sulfotransferase-3: a novel member of the HNK-1 family of sulfotransferases, *J. Biol. Chem.*, 277, 34766, 2002.

46. Evers, M.R., Xia, G., Kang, H.-G., Schachner, M., and Baenziger, J.U., Molecular cloning and characterization of a dermatan-specific N-acetylgalactosamine 4-O-sulfotransferase, *J. Biol. Chem.*, 276, 36344, 2001.

47. Mikami, T., Mizumoto, S., Kago, N., Kitagawa, H., and Sugahara, K., Specificities of three distinct human chondroitin/dermatan N-acetylgalactosamine 4-O-sulfotransferases demonstrated using partially desulfated dermatan sulfate as an acceptor: implication of differential roles in dermatan sulfate biosynthesis, *J. Biol. Chem.*, 278, 36115, 2003.

48. Kobayashi, M., Sugumaran, G., Liu, J., Shworak, N.W., Silbert, J.E., and Rosenberg, R.D., Molecular cloning and characterization of a Human Uronyl 2-sulfotransferase that sulfates iduronyl and glucuronyl residues in dermatan/chondroitin sulfate, *J. Biol. Chem.*, 274, 10474, 1999.

49. Ohtake, S., Ito, Y., Fukuta, M., and Habuchi, O., Human N-acetylgalactosamine 4-sulfate 6-O-sulfotransferase cDNA is related to Human B Cell recombination activating gene-associated gene, *J. Biol. Chem.*, 276, 43894, 2001.

50. Venkatachalam, K. V., Human 3'-phosphoadenosine 5'-phosphosulfate (PAPS) synthase: biochemistry, molecular biology and genetic deficiency, *IUBMB Life*, 55, 1, 2003.

51. Superti-Figra, A., Hästbacka, J., Wilcox, W.R., Cohn, D.H., van der Harten, J.J., Rossi, A., Blau, N., Rimoin, D.L., Steinmann, B., Lander, E.S., and Gitzelmann, R., Achondrogenesis type IB is caused by mutations in the diastrophic dysplasia sulphate transporter gene, *Nat. Genet.*, 12, 100, 1996.

52. Rossi, A., Bonaventure, J., Delezoide, A.-L., Cetta, G., and Superti-Furga, A., Undersulfation of proteoglycans synthesized by chondrocytes from a patient with achondrogenesis type 1B homozygous for an L483P substitution in the diastrophic dysplasia sulfate transporter, *J. Biol. Chem.*, 271, 18456, 1996.

53. Li, H., Deyrup, A., Mensch, J.R., Domowicz, M., Konstantinidis, A.K., and Schwartz, N.B., The isolation and characterization of cDNA encoding the mouse bifunctional ATP sulfurylase-adenosine 5'-phosphosulfate kinase. *J. Biol. Chem.*, 270, 29453, 1995.

54. Kurima, K., Warman, M.L., Krishnan, S., Domowicz, M., Krueger, R.C., Deyrup. A., and Schwartz, N.B., A member of a family of sulfate-activating enzymes causes murine brachymorphism, *Proc. Natl. Acad. Sci. USA*, 95, 8681, 1998.

55. ul Haque, M.F., King, L.M., Krakow, D., Cantor, R.M., Rusiniak, M.E., Swank, R.T., Superti-Furga, A., Haque, S., Abbas, H., Ahmad, W., Ahmad, M., and Cohn, D.H., Mutations in orthologous genes in human spondyloepimetaphyseal dysplasia and the brachymorphic mouse, *Nat. Genet.*, 20, 157, 1998.

56. Lüders, F., Segawa, H., Stein, D., Selva, E.M., Perrimon, N., Turco, S.J., and Häcker, U., *slalom* encodes an adenosine 3'-phosphate 5'-phosphosulfate transporter essential for development in *Drosophila*, *EMBO J.*, 22, 3635, 2003.

57. Kamiyama, S., Suda, T., Ueda, R., Suzuki, M., Okubo, R., Kikuchi, N., Chiba, Y., Goto, S., Toyoda, H., Saigo, K., Watanabe, H., Narimatsu, H., Jigami, Y., and Nishihara, S., Molecular cloning and identification of 3'-phosphoadenosine 5'-phosphosulfate transporter, *J. Biol. Chem.*, 278, 25958, 2003.

58. Schwartz, N.B. and Domowicz, M., Chondrodysplasias due to proteoglycan defects, *Glycobiology,* 12, 57R, 2002.

59. Lin, X. and Perrimon, N., Dally cooperates with *Drosophila* frizzled 2 to transduce wingless signalling, *Nature,* 400, 281, 1999.

60. Lin, X., Buff, E.M., Perrimon, N., and Michelson, A.M., Heparan sulfate proteoglycans are essential for FGF receptor signaling during *Drosophila* embryonic development, *Development,* 126, 3715, 1999.

61. McCormick, C., Leduc, Y., Martindale, D., Mattison, K., Esford, L.E., Dyer, A.P., and Tufaro, F., The putative tumour suppressor EXT1 alters the expression of cell-surface heparan sulfate, *Nat. Genet.,* 19, 158, 1998.

62. Lind, T., Tufaro, F., McCormick, C., Lindahl, U., and Lidholt, K., The putative tumor suppressors EXT1 and EXT2 Are glycosyltransferases required for the biosynthesis of heparan sulfate, *J. Biol. Chem.,* 273, 26265, 1998.

63. McCormick, C., Duncan, G., Goutsos, T., and Tufaro, F., The putative tumor suppressors EXT1 and EXT2 form a stable complex that accumulates in the Golgi apparatus and catalyzes the synthesis of heparan sulfate, *Proc. Natl. Acad. Sci. USA,* 97, 668, 2000.

64. Senay, C., Lind, T., Muguruma, K., Tone, Y., Kitagawa, H., Sugahara, K., Lidholt, K., Lindahl, U., and Kusche-Gullberg, M., The EXT1/EXT2 tumor suppressors: catalytic activities and role in heparan sulfate biosynthesis, *EMBO Rep.,* 1, 282, 2000.

65. Kim, B. -T., Kitagawa, H., Tanaka, J., Tamura, J., and Sugahara, K., *In vitro* heparan sulfate polymerization: crucial roles of core protein moieties of primer substrates in addition to the EXT1–EXT2 interaction, *J. Biol. Chem.,* 278, 41618, 2003.

66. Busse, M. and Kusche-Gullberg, M., *In vitro* polymerization of heparan sulfate backbone by the EXT proteins, *J. Biol. Chem.,* 278, 41333, 2003.

67. Solomon, L., Hereditary multiple exostosis, *J. Bone Joint Surg.,* 45, 292, 1963.

68. Leone, N.C., Shupe, J.L., Gardner, E.J., Millar, E.A., Olson, A.E., and Phillips, E.C., Hereditary multiple exostosis. A comparative human-equine-epidemiologic study, *J. Hered.,* 78, 171, 1987.

69. Hennekam, R.C., Hereditary multiple exostosis, *J. Med. Genet.,* 28, 262, 1991.

70. Schmale, G.A., Conrad, E.U., and Raskind, W.H., The natural history of hereditary multiple exostoses, *J. Bone Joint Surg.,* 76, 986, 1994.

71. Luckert-Wicklund, C., Pauri, R., Johnston, D., and Hecht, J., Natural history of hereditary multiple exostoses, *Am. J. Med. Genet.,* 55, 43, 1995.

72. Cook, A., Raskind, W., Blanton, S.H., Pauli, R.M., Gregg, R.G., Francomano, C.A., Puffenberger, E., Conrad, E.U., Schmale, G., Schellenberg, G., Wijsman, E., Hecht, J.T., Wells, D., and Wagner, M., Genetic heterogeneity in families with hereditary multiple exostoses, *J. Am. J. Hum. Genet.,* 53, 71, 1993.

73. LeMerrer, M., Legeai-Mallet, L., Jeannin, P.M., Horsthemke, B., Schinzel, A., Plauchu, H., Toutain, A., Achard, F., Munnich, A., and Maroteaux, P., A gene for hereditary multiple exostoses maps to chromosome 19p, *Hum. Mol. Genet.,* 3, 717, 1994.

74. Wu, Y.Q., Heutink, P., De Vries, B.B., Sandkuijl, L.A., Van den Ouweland, A.M., Niermeijer, M.F., Galjaard, H., Reyniers, E., Willems, P.J., and Halley, D., Assignment of a second locus for multiple exostoses to the pericentromeric region of chromosome 11, *J. Hum. Mol. Genet.,* 3, 167, 1994.

75. Lin, X., Wei, G., Shi, Z., Dryer, L., Esko, J.D., Wells, D.E., and Matzuk, M.M., Disruption of gastrulation and heparan sulfate biosynthesis in EXT1-Deficient Mice, *Dev. Biol.,* 224, 299, 2000.

76. Inatani, M., Irie, F., Plump, A.S., Tessier-Lavigne, M., and Yamaguchi, Y., Mammalian brain morphogenesis and midline axon guidance require heparan sulfate, *Science,* 302, 1044, 2003.

77. Zak, B.M, Stickens, D., Wells, D., Evans, G., and Esko, J.D., A murine model for hereditary multiple exostoses (HME), *Glycobiology,* 12, 642, 2002.

78. Wise, C.A., Clines, G.A., Massa, H., Trask, B.J., and Lovett, M., Identification and localization of the gene for EXTL, a third member of the multiple exostoses gene family, *Genome Res.,* 7, 10, 1997.

79. Wuyts, W., Van Hul, W., Hendrickx, J., Speleman, F., Wauters, J., De Boulle, K., Van Roy, N., Van Agtmael, T., Bossuyt, P., and Willems, P.J., Identification and characterization of a novel member of the EXT gene family, EXTL2, *Eur. J. Hum. Genet.,* 5, 382, 1997.

80. Van Hul, W., Wuyts, W., Hendrickx, J., Speleman, F., Wauters, J., De Boulle, K., Van Roy, N., Bossuyt, P., and Willems, P.J., Identification of a third EXT-like gene (EXTL3) belonging to the EXT gene family, *Genomics,* 47, 230, 1998.

81. Saito, T., Seki, N., Yamauchi, M., Tsuji, S., Hayashi, A., Kozuma, S., and Hori, T., Structure, chromosomal location, and expression profile of EXTR1 and EXTR2, new members of the multiple exostoses gene family, *Biochem. Biophys. Res. Commun.,* 243, 61, 1998.

82. Kitagawa, H., Shimakawa, H., and Sugahara, K., The tumor suppressor EXT-like gene EXTL2 encodes an α1,4-*N*-acetylhexosaminyltransferase that transfers *N*-acetylgalactosamine and *N*-acetylglucosamine to the common glycosaminoglycan-protein linkage region, *J. Biol. Chem.,* 274, 13933, 1999.

83. Kim, B.-T., Kitagawa, H., Tamura, J., Saito, T., Kusche-Gullberg, M., Lindahl, U., and Sugahara, K., Human tumor suppressor EXT gene family members EXTL1 and EXTL3 encode α1,4-*N*-acetylglucosaminyltransferases that likely are involved in heparan sulfate/ heparin biosynthesis, *Proc. Natl. Acad. Sci. USA,* 98, 7176, 2001.

84. Kobayashi, S., Akiyama, T., Nata, K., Abe, M., Tajima, M., Shervani, N.J., Unno, M., Matsuno, S., Sasaki, H., Takasawa, S., and Okamoto, H., Identification of a receptor for reg (regenerating gene) protein, a pancreatic β-cell regeneration factor, *J. Biol. Chem.,* 275, 10723, 2000.

85. Mizuno, K., Irie, S., and Sato, T., Overexpression of EXTL3/EXTR1 enhances NF-*k*B activity induced by TNF-α, *Cell. Signal.,* 13, 125, 2001.

86. Yamada, S., Van Die, I., Van den Eijnden, D.H., Yokota, A., Kitagawa, H., and Sugahara, K., Demonstration of glycosaminoglycans in *Caenorhabditis elegans,* *FEBS Lett.,* 459, 327, 1999.

87. Toyoda, H., Kinoshita-Toyoda, A., and Selleck, S.B., Structural analysis of glycosaminoglycans in *Drosophila* and *Caenorhabditis elegans* and demonstration that *tout-velu,* a *Drosophila* gene related to EXT tumor suppressors, affects heparan sulfate *in vivo, J. Biol. Chem.,* 275, 2269, 2000.

88. Toyoda, H., Kinoshita-Toyoda, A., Fox, B., and Selleck, S.B., Structural analysis of glycosaminoglycans in animals bearing mutations in *sugarless, sulfateless,* and *tout-velu*: *Drosophila* homologues of vertebrate genes encoding glycosaminoglycan biosynthetic enzymes, *J. Biol. Chem.,* 275, 21856, 2000.

89. Bellaiche, Y., The, I., and Perrimon, N., *Tout-velu* is a *Drosophila* homologue of the putative tumour suppressor EXT-1 and is needed for Hh diffusion, *Nature,* 394, 85, 1998.

90. The, I., Bellaiche, Y., and Perrimon, N., Hedgehog movement is regulated through *tout velu*–dependent synthesis of a heparan sulfate proteoglycan, *Mol. Cell,* 4, 633, 1999.

91. Takei, Y., Ozawa, Y., Sato, M., Watanabe, A., and Tabata, T., Three *Drosophila* EXT genes shape morphogen gradients through synthesis of heparan sulfate proteoglycans, *Development,* 131, 73, 2003.

92. Kim, B.-T., Kitagawa, H., Tamura, J., Kusche-Gullberg, M., Lindahl, U., and Sugahara, K., Demonstration of a novel gene DEXT3 of *Drosophila melanogaster* as the essential *N*-acetylglucosamine transferase in the heparan sulfate biosynthesis: chain initiation and elongation, *J. Biol. Chem.*, 277, 13659, 2002.

93. Clines, G.A., Ashley, J.A., Shah, S., and Lovett, M., The structure of the human multiple *Exostoses* 2 gene and characterization of homologs in mouse and *Caenorhabditis elegans*, *Genome Res.*, 7, 359, 1997.

94. Kitagawa, H., Egusa, N., Tamura, J., Kusche-Gullberg, M., Lindahl, U., and Sugahara, K., *rib-2*, a *Caenorhabditis elegans* homolog of the human tumor suppressor EXT genes encodes a novel α1,4-*N*-acetylglucosaminyltransferase involved in the biosynthetic initiation and elongation of heparan sulfate, *J. Biol. Chem.*, 276, 4834, 2001.

95. Morio, H., Honda, Y., Toyoda, H., Nakajima, M., Kurosawa, H., and Shirasawa, T., EXT gene family member *rib-2* is essential for embryonic development and heparan sulfate biosynthesis in *Caenorhabditis elegans*, *Biochem. Biophys. Res. Commun.*, 301, 317, 2003.

96. Brandan, E. and Hirschberg, C.B., Purification of rat liver *N*-heparan-sulfate sulfotransferase, *J. Biol. Chem.*, 263, 2417, 1988.

97. Pettersson, I., Kusche, M., Unger, E., Wlad, H., Nylund, L., Lindahl, U., and Kjellén, L., Biosynthesis of heparin: purification of a 110-kDa mouse mastocytoma protein required for both glucosaminyl *N*-deacetylation and *N*-sulfation. *J. Biol. Chem.*, 266, 8044, 1991.

98. Hashimoto, Y., Orellana, A., Gil, G., and Hirschberg, C.B., Molecular cloning and expression of rat liver *N*-heparan sulfate sulfotransferase, *J. Biol. Chem.*, 267, 15744, 1992.

99. Eriksson, I., Sandbäck, D., Ek, B., Lindahl, U., and Kjellén, L., cDNA cloning and sequencing of mouse mastocytoma glucosaminyl *N*-deacetylase/*N*-sulfotransferase, an enzyme involved in the biosynthesis of heparin, *J. Biol. Chem.*, 269, 10438, 1994.

100. Aikawa, J. and Esko, J.D., Molecular cloning and expression of a third member of the heparan sulfate/heparin GlcNAc *N*-deacetylase/*N*-sulfotransferase family, *J. Biol. Chem.*, 274, 2690, 1999.

101. Aikawa, J., Grobe, K., Tsujimoto, M., and Esko, J.D., Multiple isozymes of heparan sulfate/heparin GlcNAc *N*-Deacetylase/GlcN *N*-sulfotransferase: structure and activity of the fourth member, NDST4, *J. Biol. Chem.*, 276, 5876, 2001.

102. Fan, G., Xiao, L., Cheng, L., Wang, X., Sun, B., and Hu, G., Targeted disruption of NDST-1 gene leads to pulmonary hypoplasia and neonatal respiratory distress in mice, *FEBS Lett.*, 467, 7, 2000.

103. Ringvall, M., Ledin, J., Holmborn, K., Van Kuppevelt, T., Ellin, F., Eriksson, I., Olofsson, A.-M., Kjellén, L., Forsberg, E., Olofsson, A.-M., Kjellén, L., and Forsberg, E. Defective heparan sulfate biosynthesis and neonatal lethality in mice lacking *N*-deacetylase/*N*-sulfotransferase-1, *J. Biol. Chem.*, 275, 25926, 2000.

104. Grobe, K., Ledin, J., Ringvall, M., Holmborn, K., Forsberg, E., Esko, J.D., and Kjellén, L., Heparan sulfate and development: differential roles of the *N*-acetylglucosamine *N*-deacetylase/*N*-sulfotransferase isozymes, *Biochim. Biophys. Acta*, 1573, 209, 2002.

105. Thomas, K.R. and Capecchi, M.R., Targeted disruption of the murine *int-1* proto-oncogene resulting in severe abnormalities in midbrain and cerebellar development, *Nature*, 346, 847, 1990.

106. McMahon, A.P. and Bradley, A., The *Wnt-1* (*int-1*) proto-oncogene is required for development of a large region of the mouse brain, *Cell*, 62, 1073, 1990.

107. Chiang, C., Litingtung, Y., Lee, E., Young, K.E., Corden, J.L., Westphal, H., and Beachy, P.A., Cyclopia and defective axial patterning in mice lacking *Sonic hedgehog* gene function, *Nature*, 383, 407, 1996.

108. Bachiller, D., Klingensmith, J., Kemp, C., Belo, J.A., Anderson, R.M., May, S.R., McMahon, J.A., McMahon, A.P., Harland, R.M., Rossant, J., and De Robertis, E.M., The organizer factors Chordin and Noggin are required for mouse forebrain development, *Nature*, 403, 658, 2000.

109. Jenniskens, G.J., Ringvall, M., Koopman, W.J.H., Ledin, J., Kjellén, L. Willems, P.H. G.M., Forsberg, E., Veerkamp, J.H., and Van Kuppevelt, T.H., Disturbed Ca^{2+} kinetics in *N*-deacetylase/*N*-sulfotransferase-1 defective myotubes, *J. Cell Sci.*, 116, 2187, 2003.

110. Humphries, D.E., Wong, G.W., Friend, D.S., Gurish, M.F., Qiu, W.-T., Huang, C., Sharpe, A.H., and Stevens, R.L., Heparin is essential for the storage of specific granule proteases in mast cells, *Nature*, 400, 769, 1999.

111. Forsberg, E., Pejler, G., Ringvall, M., Lunderius, C., Tomasini-Johansson, B., Kusche-Gullberg, M., Eriksson, I., Ledin, J., Hellman, L., and Kjellén, L., Abnormal mast cells in mice deficient in a heparin-synthesizing enzyme, *Nature*, 400, 773, 1999.

112. Campbell, P., Hannesson, H.H., Sandbäck, D., Rodén, L., Lindahl, U., and Li, J.-P., Biosynthesis of heparin/heparan sulfate: Purification of the D-glucuronyl C-5 epimerase from bovine liver, *J. Biol. Chem.*, 269, 26953, 1994.

113. Hagner-McWhirter, Å., Li, J.-P., Oscarson, S., and Lindahl, U., Irreversible glucuronyl C5-epimerization in the biosynthesis of heparan sulfate, *J. Biol. Chem.*, 279, 1463, 2004.

114. Li, J.-P., Hagner-McWhirter, Å., Kjellén, L., Palgi, J., Jalkanen, M., and Lindahl, U., Biosynthesis of heparin/heparan sulfate: cDNA cloning and expression of D-Glucuronyl C5-epimerase from bovine lung, *J. Biol. Chem.*, 272, 28158, 1997.

115. Li, J.-P., Gong, F., Darwish, K.E., Jalkanen, M., and Lindahl, U., Characterization of the D-glucuronyl C5-epimerase involved in the biosynthesis of heparin and heparan sulfate, *J. Biol. Chem.*, 276, 20069, 2001.

116. Crawford, B.E., Olson, S.K., Esko, J.D., and Pinhal, M.A.S., Cloning, golgi localization, and enzyme activity of the full-length heparin/heparan sulfate-glucuronic acid C5-epimerase, *J. Biol. Chem.*, 276, 21538, 2001.

117. Jacobsson, I., Bäckström, G., Höök, M., Lindahl, U., Feingold, D.S., Malmström, A., and Rodén, L., Biosynthesis of heparin: assay and properties of the microsomal uronosyl C-5 epimerase, *J. Biol. Chem.*, 254, 2975, 1979.

118. Casu, B., Petitou, M., Provasoli, M., and Sinaÿ, P., Conformational flexibility: a new concept for explaining binding and biological properties of iduronic acid-containing glycosaminoglycans, *Trends Biochem. Sci.*, 13, 221, 1988.

119. Li, J.-P. Gong, F., Hagner-McWhirter, Å., Forsberg, E., Åbrink, M., Kisilevsky, R., Zhang, X., and Lindahl, U., Targeted disruption of a murine glucuronyl C5-epimerase gene results in heparan sulfate lacking L-iduronic acid and in neonatal lethality, *J. Biol. Chem.*, 278, 28363, 2003.

120. Bullock, S.L., Fletcher, J.M., Beddington, R.S.P., and Wilson, V.A., Renal agenesis in mice homozygous for a gene trap mutation in the gene encoding heparan sulfate 2-sulfotransferase, *Genes Dev.*, 12, 1894, 1998.

121. Kobayashi, M., Habuchi, H., Habuchi, O., Saito, M., and Kimata, K., Purification and characterization of heparan sulfate 2-sulfotransferase from cultured chinese hamster ovary cells, *J. Biol. Chem.*, 271, 7645, 1996.

122. Kobayashi, M., Habuchi, H., Yoneda, M., Habuchi, O., and Kimata, K., Molecular cloning and expression of chinese hamster ovary cell heparan-sulfate 2-sulfotransferase, *J. Biol. Chem.*, 272, 13980, 1997.

123. Bai, X. and Esko, J.D., An animal cell mutant defective in heparan sulfate hexuronic acid 2-*O*-sulfation, *J. Biol. Chem.*, 271, 17711, 1996.

124. Pinhal, M.A.S., Smith, B., Olson, S., Aikawa, J., Kimata, K., and Esko, J.D., Enzyme interactions in heparan sulfate biosynthesis: Uronosyl 5-epimerase and 2-*O*-sulfotransferase interact *in vivo*, *Proc. Natl. Acad. Sci. USA*, 98, 12984, 2001.

125. McLaughlin, D., Karlsson, F., Tian, N., Pratt, T., Bullock, S.L., Wilson, V.A., Price, D.J., and Mason, J.O., Specific modification of heparan sulphate is required for normal cerebral cortical development, *Mech. Dev.*, 120, 1481, 2003.

126. Merry, C.L.R., Bullock, S.L., Swan, D.C., Backen, A.C., Lyon, M., Beddington, R.S.P., Wilson, V.A., and Gallagher, J.T., The molecular phenotype of heparan sulfate in the *Hs2st*$^{-/-}$ mutant mouse, *J. Biol. Chem.*, 276, 35429, 2001.

127. Sen, J., Golts, J.S., Stevens, L., and Stein, D., Spatially restricted expression of *pipe* in the *Drosophila* egg chamber defines embryonic dorsal–ventral polarity, *Cell*, 95, 471, 1998.

128. Merrill, C., Bayraktaroglu, L., Kusano, A., and Ganetzky, B., Truncated RanGAP encoded by the *Segregation Distorter* locus of *Drosophila*, *Science*, 283, 1742, 1999.

129. Habuchi, H., Habuchi, O., and Kimata, K. J., Purification and characterization of heparan sulfate 6-sulfotransferase from the culture medium of chinese Hamster ovary cells, *Biol. Chem.*, 270, 4172, 1995.

130. Habuchi, H., Kobayashi, M., and Kimata, K.J., Molecular characterization and expression of heparan-sulfate 6-Sulfotransferase: complete cDNA cloning in human and partial cloning in chinese hamster ovary cells, *Biol. Chem.*, 273, 9208, 1998.

131. Habuchi, H., Tanaka, M., Habuchi, O., Yoshida, K., Suzuki, H., Ban, K., and Kimata, K.J., The occurrence of three isoforms of heparan sulfate 6-*O*-sulfotransferase having different specificities for hexuronic acid adjacent to the targeted *N*-sulfoglucosamine, *Biol. Chem.*, 275, 2859, 2000.

132. Habuchi, H., Miyake, G., Nogami, K., Kuroiwa, A., Matsuda, Y., Kusche-Gullberg, M., Habuchi, O., Tanaka, M., and Kimata, K., Biosynthesis of heparan sulphate with diverse structures and functions: two alternatively spliced forms of human heparan sulphate 6-*O*-sulphotransferase-2 having different expression patterns and properties, *Biochem. J.*, 371, 131, 2003.

133. Smeds, E., Habuchi, H., Do, A.-T., Hjertson, E., Grundberg, H., Kimata, K., Lindahl, U., and Kuche-Gullberg, M., Substrate specificities of mouse heparan sulphate glucosaminyl 6-*O*-sulphotransferases, *Biochem. J.*, 372, 371, 2003.

134. Kamimura, K., Fujise, M., Villa, F., Izumi, S., Habuchi, H., Kimata, K., and Nakato, K.J., *Drosophila* heparan sulfate 6-*O*-sulfotransferase (*dHS6ST*) gene: structure, expression, and function in the formation of the tracheal system, *Biol. Chem.*, 276, 17014, 2001.

135. Lindahl, U., Bäckström, G., Thunberg, L., and Leder, I. G., Evidence for a 3-*O*-sulfated D-glucosamine residue in the antithrombin-binding sequence of heparin, *Proc. Natl. Acad. Sci. USA*, 77, 6551, 1980.

136. Liu, J., Shworak, N.W., Fritze, L.M.S., Edelberg, J.M., and Rosenberg, R.D., Purification of heparan sulfate D-glucosaminyl 3-*O*-sulfotransferase, *J. Biol. Chem.*, 271, 27072, 1996.

137. Shworak, N.W., Liu, J., Fritze, L.M.S., Schwartz, J.J., Zhang, L., Logeart, D., and Rosenberg, R.D., Molecular cloning and expression of mouse and human cDNAs encoding heparan sulfate D-glucosaminyl 3-*O*-sulfotransferase, *J. Biol. Chem.*, 272, 28008, 1997.

138. Shworak, N.W., Liu, J., Petros, L.M., Zhang, L., Kobayashi, M., Copeland, N.G., Jenkins, N.A., and Rosenberg, R.D., Multiple isoforms of heparan sulfate D-glucosaminyl 3-O-sulfotransferase: isolation, characterization, and expression of human cDNAs and identification of distinct genomic loci, *J. Biol. Chem.*, 274, 5170, 1999.

139. Liu, J., Shworak, N.W., Sinaÿ, P., Schwartz, J.J., Zhang, L., Fritze, L.M.S., and Rosenberg, R.D., Expression of heparan sulfate D-glucosaminyl 3-O-sulfotransferase isoforms reveals novel substrate specificities, *J. Biol. Chem.*, 274, 5185, 1999.

140. Xia, G., Chen, J., Tiwari, V., Ju, W., Li, J. -P., Malmström, A., Shukla, D., and Liu, J., Heparan sulfate 3-O-sulfotransferase isoform 5 generates both an antithrombin-binding site and an entry receptor for herpes simplex virus, type 1, *J. Biol. Chem.*, 277, 37912, 2002.

141. Shukla, D., Liu, J., Blaiklock, P., Shworak, N.W., Bai, X., Esko, J.D., Cohen, G.H., Eisenberg, R.J., Rosenberg, R.D., and Spear, P.G., A novel role for 3-O-sulfated heparan sulfate in herpes simplex virus 1 entry, *Cell*, 99, 13, 1999.

142. McKeehan, W.L., Wu, X., and Kan, M., Requirement for anticoagulant heparan sulfate in the fibroblast growth factor receptor complex, *J. Biol. Chem.*, 274, 21511, 1999.

143. Ye, S., Luo, Y., Lu, W., Jones, R.B., Linhardt, R.J., Capila, I., Toida, T., Kan, M., Pelletier, H., and McKeehan, W.L., Structural basis for interaction of FGF-1, FGF-2, and FGF-7 with different heparan sulfate motifs, *Biochemistry*, 40, 14429, 2001.

144. HajMohammadi, S., Enjoji, K., Princivalle, M., Christi, P., Lech, M., Beeler, D., Rayburn, H., Schwartz, J.J., Barzeger, S., De Agostini, A.I., Post, M.J., Rosenberg, R.D., and Shworak, N.W., Normal levels of anticoagulant heparan sulfate are not essential for normal hemostasis, *J. Clin. Invest.*, 111, 989, 2003.

145. Hwang, H.-Y. and Horvitz, H.R., The SQV-1 UDP-glucuronic acid decarboxylase and the SQV-7 nucleotide-sugar transporter may act in the Golgi apparatus to affect *Caenorhabditis elegans* vulval morphogenesis and embryonic development, *Proc. Natl. Acad. Sci. USA*, 99, 14218, 2002.

146. Hwang, H,-Y., Olson, S.K., Brown, J.R., Esko, L.D., and Horvitz, H.R., The *Caenorhabditis elegans* genes *sqv-2* and *sqv-6*, which are required for vulval morphogenesis, encode glycosaminoglycan galactosyltransferase II and xylosyltransferase, *J. Biol. Chem.*, 278, 11735, 2003.

147. Hwang, H.-Y. and Horvitz, H.R., The *Caenorhabditis elegans* vulval morphogenesis gene *sqv-4* encodes a UDP-glucose dehydrogenase that is temporally and spatially regulated, *Proc. Natl. Acad. Sci. USA*, 99, 14224, 2002.

148. Merz, D.C., Alves, G., Kawano, T., Zheng, H., and Culotti, J.G., UNC-52/Perlecan affects gonadal leader cell migrations in *C. elegans* hermaphrodites through alterations in growth factor signaling, *Dev. Biol.*, 256, 174, 2003.

149. Rogalski, T.M., Mullen, G.P., Bush, J.A., Gilchrist, E.J., and Moerman, D.G., UNC-52/perlecan isoform diversity and function in *Caenorhabditis elegans*, *Biochem. Soc. Trans.*, 29, 171, 2001.

150. Bülow, H.E., Berry, K.L., Topper, L.H., Peles, E., and Hobert, O., Heparan sulfate proteoglycan-dependent induction of axon branching and axon misrouting by the Kallmann syndrome gene *kal-1*, *Proc. Natl. Acad. Sci., USA*, 99, 6346, 2002.

151. Wilson, I.B.H., Functional characterization of *Drosophila melanogaster* peptide O-Xylosyltransferase, the key enzyme for proteoglycan chain initiation and member of the core 2/I N-acetylglucosaminyltransferase family, *J. Biol. Chem.*, 277, 21207, 2002.

152. Takemae, H., Ueda, R., Okubo, R., Nakato, H., Izumi, S., Saigo, K., and Nishihara, S., Proteoglycan UDP-galactose: β-xylose β1,4-galactosyltransferase I is essential for viability in *Drosophila melanogaster*, *J. Biol. Chem.*, 278, 15571, 2003.

153. Nakato, H., Futch, T.A., and Selleck, S.B., The *division abnormally delayed (dally)* gene: a putative integral membrane proteoglycan required for cell division patterning during postembryonic development of the nervous system in *Drosophila, Development,* 121, 3687, 1995.

154. Jackson, S.M., Nakato, H., Sugiura, M., Jannuzi, A., Oakes, R., Kaluza, V., Golden, C., and Selleck, S.B., *dally, a Drosophila* glypican, controls cellular responses to the TGF-β-related morphogen, Dpp, *Development,* 124, 4113, 1997.

155. Tsuda, M., Kamimura, K., Nakato, H., Archer, M., Staatz, W., Fox, B., Humphrey, M., Olson, S., Futch, T., Kaluza, V., Siegfried, E., Stam, L., and Selleck, S.B., The cell-surface proteoglycan Dally regulates Wingless signalling in *Drosophila, Nature,* 400, 276, 1999.

156. Fujise, M., Takeo, S., Kamimura, K., Matsuo, T., Aigaki, T., Izumi, S., and Nakato, H., Dally regulates Dpp morphogen gradient formation in the *Drosophila wing, Development,* 130, 1515, 2003.

157. Han, C., Belenkaya, T.Y., Wang, B., and Lin, X., *Drosophila* glypicans control the cell-to-cell movement of Hedgehog by a dynamin-independent process, *Development,* 131, 601, 2004.

158. Baeg, G.-H., Lin, X., Khare, N., Baumgartner, S., and Perrimon, N., Heparan sulfate proteoglycans are critical for the organization of the extracellular distribution of Wingless, *Development,* 128, 87, 2001.

159. Desbordes, S.C. and Sanson, B., The glypican Dally-like is required for Hedgehog signalling in the embryonic epidermis of *Drosophila, Development,* 130, 6245, 2003.

160. Spring, J., Paine-Saunders, S.E., Hynes, R.O., and Bernfield, M., *Drosophila* syndecan: conservation of a cell-surface heparan sulfate proteoglycan, *Proc. Natl. Acad. Sci. USA.,* 91, 3334, 1994.

161. Voigt, A., Pflanz, R., Schäfer, U., and Jäckle, H., Perlecan participates in proliferation activation of quiescent *Drosophila* neuroblasts, *Dev. Dyn.,* 224, 403, 2002.

162. Golling, G., Amsterdam, A., Sun, Z., Antonelli, M., Maldonado, E., Chen, W., Burgess, S., Haldi, M., Artzt, K., Farrington, S., Lin, S.-Y., Nissen, R.M., and Hopkins, N., Insertional mutagenesis in zebrafish rapidly identifies genes essential for early vertebrate development, *Nat. Genet.,* 31, 135, 2002.

163. Bink, R.J., Habuchi, H., Lele, Z., Dolk, E., Joore, J., Rauch, G.-J., Geisler, R., Wilson, S.W., Den Hertog, J., Kimata, K., and Zivkovic, D., Heparan sulfate 6-*O*-sulfotransferase is Essential for muscle development in zebrafish, *J. Biol. Chem.,* 278, 31118, 2003.

164. Topczewski, J., Sepich, D.S., Myers, D.C., Walker, C., Amores, A., Lele, Z., Hammerschmidt, M., Postlethwait, J., and Solnica-Krezel, L., The zebrafish glypican knypek controls cell polarity during gastrulation movements of convergent extension, *Dev. Cell,* 1, 251, 2001.

165. Chen, E., Hermanson, S., and Ekker, S.C., Syndecan-2 is essential for angiogenic sprouting during zebrafish development, *Blood,* 103, 1710, 2004.

166. Watanabe, H., Kimata, K., Line, S., Strong, D., Gao, L.Y., Kozak, C.A., and Yamada, Y., Mouse cartilage matrix deficiency (*cmd*) caused by a 7 bp deletion in the aggrecan gene, *Nat. Genet.,* 7, 154, 1994.

167. Mjaatvedt, C.H., Yamamura, H., Capehart, A.A., Turner, D., and Markwald, R.R., The *Cspg2* gene, disrupted in the *hdf* mutant, is required for right cardiac chamber and endocardial cushion formation, *Dev. Biol.,* 202, 56, 1998.

168. Alexander, C.M., Reichsman, F., Hinkes, M.T., Lincecum, J., Becker, K.A., Cumberledge, S., and Bernfield, M., Syndecan-1 is required for Wnt-1-induced mammary tumorigenesis in mice, *Nat. Genet.,* 25, 329, 2000.

169. Bernfield, M., Götte, M., Park, P.W., Reizes, O., Fitzgerald, M.L., Lincecum, J., and Zako, M., Functions of cell surface heparan sulfate proteoglycans, *Annu. Rev. Biochem.*, 68, 729, 1999.

170. Reizes, O., Lincecum, J., Wang, Z., Goldberger, O., Huang, L., Kaksonen, M., Ahima, R., Hinkes, M.T., Barsh, G.S., Rauvala, H., and Bernfield, M., Transgenic expression of syndecan-1 uncovers a physiological control of feeding behavior by syndecan-3, *Cell*, 106, 105, 2001.

171. Ishiguro, K., Kadomatsu, K., Kojima, T., Muramatsu, H., Tsuzuki, S., Nakamura, E., Kusugami, K., Saito, H., and Muramatsu, T., Syndecan-4 deficiency impairs focal adhesion formation only under restricted conditions, *J. Biol. Chem.*, 275, 5249, 2000.

172. Perrimon, N. and Bernfield, M., Specificities of heparan sulphate proteoglycans in developmental processes, Nature, 404, 725, 2000.

173. Cano-Gauci, D.F., Song, H.H., Yang, H., McKerlie, C., Choo, B., Shi, W., Pullano, R., Piscione, T.D., Grisaru, S., Soon, S., Sedlackova, L., Tanswell, A.K., Mak, T.W., Yeger, H., Lockwood, G.A., Rosenblum, N.D., and Filmus, J., Glypican-3–deficient mice exhibit developmental overgrowth and some of the abnormalities typical of simpson-golabi-behmel syndrome, *J. Cell Biol.*, 146, 255, 1999.

174. Gautam, M., Noakes, P.G., Moscoso, L., Rupp, F., Scheller, R.H., Merlie, J.P., and Sanes, J.R., Defective neuromuscular synaptogenesis in agrin-deficient mutant mice, *Cell*, 85, 525, 1996.

175. Aikawa-Hirasawa, E., Watanabe, H., Takami, H., Hassell, J.R., and Yamada, Y., Perlecan is essential for cartilage and cephalic development, *Nat. Genet.*, 23, 354, 1999.

176. Costell, M., Gustafsson, E., Aszódi, A., Mörgelin, M., Bloch, W., Hunziker, E., Addicks, K., Timpl, R., and Fässler, R., Perlecan maintains the integrity of cartilage and some basement membranes, *J. Cell Biol.*, 147, 1109, 1999.

177. Danielson, K.G., Baribault, H., Holmes, D.F., Graham, H., Kadler, K.E., and Iozzo, R.V., Targeted disruption of decorin leads to abnormal collagen fibril Morphology and skin fragility, *J. Cell Biol.*, 136, 729, 1997.

178. Xu, T., Bianco, P., Fisher, L.W., Longenecker, G., Smith, E., Goldstein, S., Bonadio, J., Boskey, A., Heegaard, A. -M., Sommer, B., Satomura, K., Dominguez, P., Zhao, C., Kulkarni, A.B., Robey, P.G., and Young, M.F., Targeted disruption of the biglycan gene leads to an osteoporosis-like phenotype in mice, *Nat. Genet.*, 20, 78, 1998.

179. Zhou, X.-H., Brakebusch, C., Matthies, H., Oohashi, T., Hirsch, E., Moser, M., Krug, M., Seidenbecher, C.I., Boeckers, T.M., Rauch, U., Buettner, R., Gundelfinger, E.D., and Fässler, R., Neurocan is dispensable for brain development, *Mol. Cell Biol.*, 21, 5970, 2001.

180. Fujikawa, A., Shirasaka, D., Yamamoto, S., Ota, H., Yahiro, K., Fukuda, M., Shintani, T., Wada, A., Aoyama, N., Hirayama, T., Fukumachi, H., and Noda, M., Mice deficient in protein tyrosine phosphatase receptor type Z are resistant to gastric ulcer induction by VacA of *Helicobacter pylori*, *Nat. Genet.*, 33, 375, 2003.

181. Healy, A.M., Rayburn, H.B., Rosenberg, R.D., and Weiler, H., Absence of the blood-clotting regulator thrombomodulin causes embryonic lethality in mice before development of a functional cardiovascular system, *Proc. Natl. Acad. Sci. USA,* 92, 850, 1995,

182. Zak, B.M., Crawford, B.E., and Esko, J.D., Hereditary multiple exostoses and heparan sulfate polymerization, *Biochim. Biophys. Acta*, 1573, 346, 2002.

183. Sugahara, K. and Kitagawa, H., Heparin and heparan sulfate biosynthesis, *IUBMB Life*, 54, 163, 2002.

184.   Sugahara, K., Yamashina, I., de Waard, P., Van Halbeek., and Vliegenthart, J.F.G., Structural studies on sulfated glycopeptides from the carbohydrate- protein linkage region of chondroitin 4-sulfate proteoglycans of swarm rat chondrosarcoma: demonstration of the structure GAL(4-O-sulfate) β1-3Galβ1-4XYL β1-O-ser, *J. Biol. Chem.*, 263, 10168, 1988.

185.   Sugahara, K., Ohi, Y., Harada, T., de Waard, P., and Vliegenthart, J.F.G., Structural studies on sulfated oligosaccharides derived from the carbohydrate-protein linkage region of chondroitin 6-sulfate proteoglycans of shark cartilage: I. Six compounds containing 0 or 1 sulfate and/or phosphate residues, *J. Biol. Chem.*, 267, 6027, 1992.

186.   de Waard, P., Vliegenthart, J.F.G., Harada, T., and Sugahara, K., Structural studies on sulfated oligosaccharides derived from the carbohydrate-protein linkage region of chondroitin 6-sulfate proteoglycans of shark cartilage: II. Seven compounds containing 2 or 3 sulfate residues, *J. Biol. Chem.*, 267, 6036 1992.

10 Large-Scale Production of Oligosaccharides Using Engineered Bacteria

Satoshi Koizumi

CONTENTS

I. INTRODUCTION

It is well known that oligosaccharides on the cell surface are very important in biochemical recognition processes such as cell adhesion.[1,2] Therefore, these oligosaccharides are potential pharmaceuticals for preventing infections by pathogens, neutralizing toxins, regulating inflammation, and cancer immunotherapy.[2] The development of oligosaccharides of clinical utility, however, has been restricted because it is very difficult to synthesize oligosaccharides and only very limited quantities of oligosaccharides can be obtained by chemical or enzymatic methods or by extraction from natural sources.

A variety of chemical and enzymatic methods for the synthesis of oligosaccharides have been developed.[3–5] Chemical synthesis of oligosaccharides requires multiple protection and deprotection steps, and this complexity does not render the chemical synthesis as realistic as an industrial method. On the other hand, enzymatic synthesis using glycosyltransferases of the Leloir pathway could circumvent the drawbacks of the chemical methods, because glycosyltransferases involved in the biosynthesis of oligosaccharides show rigid stereo- and regiospecific bond formation.

Moreover, almost no side products are formed during enzymatic reactions even without the use of protecting groups. Disadvantages of oligosaccharides synthesis using glycosyltransferases, however, are the limited availability and the expense of the enzymes and their substrates.

Until recently the preparation of glycosyltransferases had been a significant problem, but recent progress in genetic engineering and recombinant protein production methods has now made several glycosyltransferases available in large quantities. The other problem involves the preparation of sugar nucleotides, the substrates for glycosyltransferases; these compounds are very expensive and are not readily available. This second limitation could be overcome, however, if the large-scale synthesis of sugar nucleotides was made possible by the use of bacterial activities.

This chapter focuses on the recent progress in the production of oligosaccharides using genetically engineered bacteria.

II. BACTERIAL GLYCOSYLTRANSFERASES

A large number of mammalian glycosyltransferases have been isolated and shown to be useful for oligosaccharide synthesis; however, they are often rare and expensive largely because they are usually membrane proteins, which makes it difficult for them to express their genes in *Escherichia coli*. Consequently, the application of *mammalian* glycosyltransferases for the large-scale synthesis of oligosaccharides has been restricted. As microbial genomes analysis has rapidly progressed in recent years, more and more studies suggest that *E. coli* itself, or other bacterial species, may be attractive sources of glycosyltransferase genes. Continuous research using sophisticated screening techniques and structure analysis of bacterial lipopolysaccharides has identified a growing number of bacterial genes of glycosyltransferases, mainly pathogenic bacteria. Upon cloning, most of the genes can be expressed as soluble and active proteins in *E. coli*.

Examples of bacterial glycosyltransferases whose activities were detected in *E. coli* are shown in Table 10.1. The galactosyltransferase genes were cloned from *Neisseria gonorrhoeae*,[6] *N. meningitidis*,[7,8] *Helicobacter pylori*,[9,10] *Streptococcus pneumoniae*,[11] *S. agalactiae*,[12,13] and *Campylobacter jejuni*.[14] Of these genes, the galactosyltransferases of *N. gonorrhoeae* and *N. meningitidis* have been highly expressed in *E. coli* and used for the synthesis of galactosylated oligosaccharides. Genes encoding *N*-acetylglucosaminyltransferases were also cloned from *N. gonorrhoeae*,[6] *N. meningitidis*[7] and *C. jejuni*[14] and sialyltransferase genes were cloned from *C. jejuni*,[14] *N. gonorrhoeae*,[15] *N. meningitides*,[15] *Haemophilus influenzae*,[16,17] *H. ducreyi*,[18] *S. agalactiae*,[19] *Photobacterium damsela*,[20] and *E. coli*.[21] The sialyltransferase from *C. jejuni* had the unique characteristic that it could transfer *N*-acetylneuraminic acid (NeuAc) to both the O-3 position of galactose (Gal) and to the O-8 of NeuAc that linked Gal with an α2,3-linkage.[14] Three additional sialyltransferases were identified in *H. influenzae*[16,17] and fucosyltransferases have been cloned from *H. pylori*.[22-25] The wide variety of glycosyltransferases already cloned, and the many more anticipated to be identified in bacterial genomes, promise to be valuable for oligosaccharide synthesis.

TABLE 10.1
Bacterial Glycosyltransferase

Enzyme	Source	Ref.
GalT		
α1,4-Galactosyltransferase	N. gonorrhoeae, N. meningitidis	6,7
β1,4-Galactosyltransferase	H. pylori	9,10
β1,4-Galactosyltransferase	N. gonorrhoeae, N. meningitidis	6,7,8
β1,4-Galactosyltransferase	S. pneumoniae	11
β1,4-Galactosyltransferase	S. agalactiae	12
β1,3-Galactosyltransferase	C. jejuni	14
β1,3-Galactosyltransferase	S. agalactiae	13
GlcNAcT		
β1,4-GlcNAc transferase	C. jejuni	14
β1,3-GlcNAc transferase	N. gonorrhoeae, N. meningitidis	6,7
GalNAcT		
β1,3-GalNAc transferase	C. jejuni	13
β1,3-GalNAc transferase	N. gonorrhoeae, N. meningitidis	6,7
SiaT		
α2,3-Sialyltransferase	C. jejuni	14
α2,3-Sialyltransferase	H. influenzae	16,17
α2,3-Sialyltransferase	H. ducreyi	18
α2,3-Sialyltransferase	N. gonorrhoeae, N. meningitidis	15
α2,3-Sialyltransferase	S. agalactiae	19
α2,3/8-Sialyltransferase	C. jejuni	13
α2,6-Sialyltransferase	P. damsela	20
α2,8/9-Sialyltransferase	E. coli	21
FucT		
α1,3-Fucosyltransferase	H. pylori	22,23
α1,2-Fucosyltransferase	H. pylori	24
α1,3/4-Fucosyltransferase	H. pylori	25

III. OLIGOSACCHARIDE SYNTHESIS WITH COFACTOR REGENERATION

Prospects for oligosaccharide synthesis using bacterial glycosyltransferases obtained through expression in *E. coli* have been examined intensively,[26–28] but the supply of the substrates, the sugar nucleotides, remains a problem. One solution that allows sugar nucleotides to be used efficiently and escape the product inhibition on the glycosyl-transferases due to the resulting nucleoside diphosphates or monophosphates is the development of production systems that enable the *in situ* regeneration of sugar nucleotides using enzymes such as pyruvate kinase (Figure 10.1 and Figure 10.2).[29–31] In another approach, an innovative UDP-Gal-recycling bead composed of four enzymes involved in the biosynthesis of UDP-Gal with His-tag has been constructed.[32] Finally, because UDP-glucose (UDP-Glc) is a cheaper and more easily available sugar

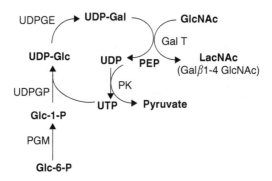

FIGURE 10.1   Synthesis of *N*-acetyllactosamine with UDP-Gal recycling. GalT, β1,4-galactosyltransferase; PGM, phosphoglucomutase; PK, pyruvate kinase; UDPGE, UDP-Gal 4′-epimerase, UDPGP, UDP-Glc pyrophosphorylase; PEP, phosphoenolpyruvate. (From Wong, C.-H. et al., *J. Org. Chem.*, 47, 5416–5918, 1982.With permission.)

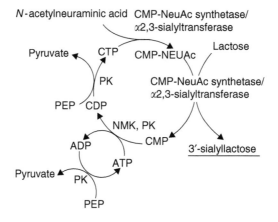

FIGURE 10.2   Synthesis of 3′-sialyllactose with CMP-NeuAc recycling. PK, pyruvate kinase; NMK, nucleoside monophosphate kinase; PEP, phosphoenolpyruvate. (From Gilbert, M. et al., *Nat. Biotechnol.*, 16, 769–772, 1998. With permission.)

nucleotide, oligosaccharide syntheses using a fusion enzyme composed of UDP-Gal 4′-epimerase and a galactosyltransferase from UDP-Glc have been examined.[33–35]

A fusion enzyme consisting of CMP-NeuAc synthetase and α2,3-sialyltransferase was created for the synthesis of sialylated oligosaccharides.[36] The fusion enzyme was more stable than the original enzyme and could be easily purified. 3′-Sialyllactose, which exists in human milk, was synthesized at a scale of 100 g using the fusion enzyme.[36]

In addition to the recycling system using pyruvate kinase and phosphoenolpyruvate (PEP), sugar nucleotide recycling systems using sucrose synthetase[37–39] or an inexpensive kinase system with polyphosphate kinase and polyphosphate[40] were also developed. P1 trisaccharide (Galα1-4Galβ1-4GlcNAc) was produced and accumulated up to

50 m*M* using *E. coli* cells which coexpressed the four genes for *E. coli* UDP-Gal 4'-epimerase: *H. pylori* β1,4-galactosyltransferase, *N. meningitidis* α1,4-galactosyltransferase, and *Anabena* sp. sucrose synthase (Figure 10.3).[39] These examples demonstrate that bacterial glycosyltransferases can be readily expressed in *E. coli,* enabling the large-scale synthesis of oligosaccharides with co-factor recycling to be accomplished. However, this strategy has its drawbacks, as these methods require purified enzymes such as pyruvate kinase and expensive substrates such as PEP.

IV. PRODUCTION OF OLIGOSACCHARIDES BY BACTERIAL COUPLING

A. SUGAR NUCLEOTIDE PRODUCTION BY BACTERIAL COUPLING

The establishment of a method for the large-scale synthesis of oligosaccharides that does not require enzyme purification and the addition of high-energy phosphates would be highly attractive. The microbial synthesis of ATP from adenine using *Corynebacterium ammoniagenes* cells as an enzyme source has been reported to be part of a method that meets such requirements.[41] In this system, cells were permeabilized by treatment with surfactants or organic solvents and used as enzyme bags. High-energy phosphates such as phosphoribiosylpyrophosphate and ATP were supplied by the metabolism of glucose in the cells. In the same manner, CDP-choline was produced by the combination of *C. ammoniagenes* and recombinant *E. coli* cells expressing the genes involved in the biosynthesis of CDP-choline.[42]

These microbial methods can be applied to the synthesis of sugar nucleotides. For instance, UDP-Gal has been efficiently produced by the combination of recombinant *E. coli* and *C. ammoniagenes.*[43] Recombinant *E. coli* overexpressed the UDP-Gal biosynthetic genes, and *C. ammoniagenes* contributed to the formation of UTP from orotic acid, an inexpensive precursor of UTP. UDP-Gal accumulated to 44 g/L

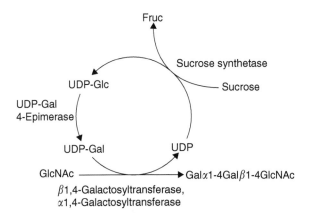

FIGURE 10.3 Synthesis of Galα1-4Galβ1-4GlcNAc with UDP-Gal recycling by sucrose synthetase. (From Liu, Z. et al., *Appl. Environ. Microbiol.,* 69, 2110–2115, 2003. With permission.)

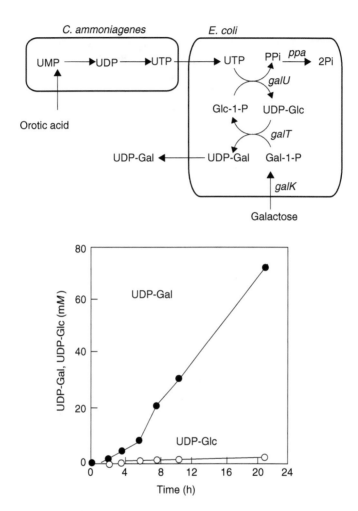

FIGURE 10.4 Production of UDP-galactose by bacterial coupling. galK, Galactokinase; galT, Gal-1-phosphate uridyltransferase; galU, UDP-Glc pyrophosphorylase; ppa, pyrophosphatase. (From Koizumai, S. et al., *Nat. Biotechnol.*, 16, 847–850, 1998. With permission.)

after a 21 h reaction starting with orotic acid and galactose (Figure 10.4).[43] In a similar manner, by the coupling of *E. coli* expressing the genes involved in the biosynthesis of GDP-Fuc and *C. ammoniagenes*, 18.4 g/L of GDP-Fuc was formed in 22 h from GMP and mannose.[44] Similarly, 17 g/L of CMP-NeuAc was produced after a 27 h reaction starting with orotic acid and NeuAc through the coupling of recombinant *E. coli* cells overexpressing the genes of CMP-NeuAc synthetase and CTP synthetase, with *C. ammoniagenes*.[45] These examples have indicated that by means of the combination of a nucleoside 5′-triphosphate producing microorganism with recombinant *E. coli* cells expressing the genes involved in sugar nucleotide biosynthesis, a large-scale production system of sugar nucleotides can be established.

B.   OLIGOSACCHARIDE PRODUCTION BY BACTERIAL COUPLING

A system of oligosaccharide production by the combination of a bacterial glycosyl-transferase and sugar nucleotide production was proposed on the basis that sugar nucleotides could be synthesized by bacterial coupling. For instance, by coupling recombinant *E. coli* overexpressing the α1,4-galactosyltransferase gene from *N. gonorrhoeae* and the UDP-Gal production system, globotriose (Galα1-4Galβ1-4Glc), an oligosaccharide portion of the receptor of vero toxin produced by *E. coli* O157 accumulated to a level of 188 g/L for 36 h from orotic acid, galactose, and lactose (Figure 10.5).[43] When *E. coli* cells overexpressing the β1,4-galactosyltransferase gene of *H. pylori* were introduced to the above system, *N*-acetyllactosamine (Galβ1-4GlcNAc) accumulated to 60 g/L after an incubation period of 20 h.[46] In another example, when *E. coli* cells overexpressing the α1,3-fucosyltransferase gene

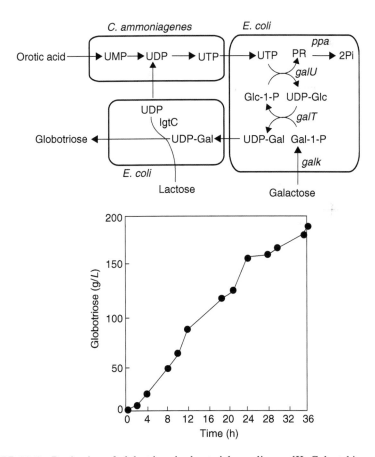

FIGURE 10.5   Production of globotriose by bacterial coupling. galK, Galactokinase; galT, Gal-1-phosphate uridyltransferase; galU, UDP-Glc pyrophosphorylase; ppa, pyrophosphatase; lgtC, α1,4-galactosyltransferase. (From Koizumai, S. et al., *Nat. Biotechnol.*, 16, 847–850, 1998. With permission.)

of *H. pylori* were introduced to the GDP-Fuc production system, Lewis X [Galβ1-4(Fucα1-3)GlcNAc] accumulated to 21 g/L after 30 h.[44]

The demonstration that sialic acid-containing oligosaccharides can be produced by the combination of a CMP-NeuAc production system and *E. coli* overexpressing the α2,3-sialyltransferase of *N. gonorrhoeae* is a valuable advance. In one experiment, 33 g/L of 3'-sialyllactose was produced after 11 h starting with orotic acid, NeuAc, and lactose.[45] Similarly, a level of 45 g/L of sialylTn oligosaccharide (NeuAcα2-6GalNAc), one of cancer-associated antigens, accumulated in 25 h of incubation by using the α2,6-sialyltransferase of *P. damsela* (Figure 10.6).[47] NeuAc is important as a starting material for sialylated oligosaccharides; however, it remains expensive, and cheaper methods for the synthesis have been sought, although enzymatic synthesis procedures using NeuAc aldolase along with chemical or enzymatic epimerization of GlcNAc were previously established.[48,49] Cells expressing GlcNAc 2-epimerase from *Synechcystis* sp. and NeuAc synthetase from *E. coli* were combined to develop as a microbial method of supplementing cell-free systems (Figure 10.7).[50] In this case, NeuAc was produced at a level of 12 g/L from GlcNAc and glucose in 22 h.

The coupling of the sugar nucleotide production system with bacterial glycosyltransferases successfully constructs an oligosaccharide production system that can be applied to the industrial manufacture of carbohydrates.[51]

V. OLIGOSACCHARIDE PRODUCTION BY FERMENTATION

An alternative way for the production of oligosaccharides is *in vivo* production in recombinant *E. coli* cells expressing glycosyltransferases. In this fermentation procedure, oligosaccharides can be produced extracellularly and intracellularly during cell growth using the machinery for sugar nucleotide synthesis in the cells.

When *E. coli* cells lacking the ability to degrade lactose and expressing the β1, 3-*N*-acetylglucosaminyltransferase and the β1,4-galactosyltransferase of *N. meningitidis* were cultivated at a high density using glycerol as a carbon source, lacto-*N*-neotetraose (Galβ1-4GlcNAcβ1-3Galβ1-4Glc) and lacto-*N*-neohexaose (Galβ1-4GlcNAcβ1-3Galβ1-4GlcNAcβ1-3Galβ1-4Glc) accumulated to a level of more than 5 g/L after 35 h.[52] In addition, oligosaccharides containing fucose were formed by the improvement in the biosynthesis of GDP-Fuc and introduction of fucosyltransferase of *H. pylori*.[53]

Similarly, an *E. coli* strain that was devoid of NeuAc aldolase and β-galactosidase activities and overexpressed the α2,3-sialyltransferase and CMP-NeuAc synthetase of *N. meningitidis* efficiently produced 3'-sialyllactose at an intracellular level of 1.1 g/L, and an extracellular level of 1.5 g/L.[52] In addition, an *E.coli* strain that additionally overexpressed the genes for UDP-GlcNAc C4 epimerase, β1, 4-*N*-acetylgalactosaminyltransferase, and β1,3-galactosyltransferase was used for the *in vivo* formation of the carbohydrate moieties of gangliosides GM2 and GM1 (Figure 10.8).[54] The xenotransplantation antigen, Galα1-3Galβ1-4GlcNAc epitope, has also been successfully produced by *E. coli*.[55]

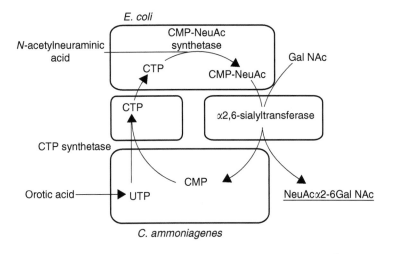

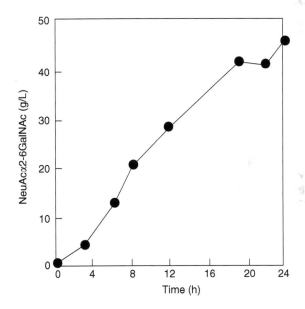

FIGURE 10.6   Production of NeuAcα2-6GalNAc by bacterial coupling. (From Endo, T. et al., *Carbohydr. Res.*, 316, 179–183, 1999. With permission.)

VI.   CONCLUSIONS

The large-scale production of oligosaccharides is becoming feasible by the use of bacterial glycosyltransferases and the improvement in the supply of sugar nucleotides. The supply of oligosaccharides and sugar nucleotides at a lower cost promises to strongly facilitate the functional analysis of these oligosaccharides and expand their use in both research and practical application. Furthermore, because

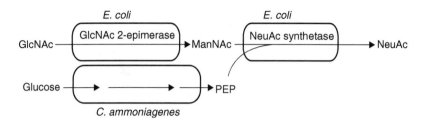

FIGURE 10.7 Production of *N*-acetylneuraminic acid by bacterial coupling. ManNAc, *N*-acetylmannosamine; PEP, phosphoenolpyruvate. (From Tabata, K. et al., *Enzyme. Microbiol. Technol.*, 30, 327–333, 2002. With permission.)

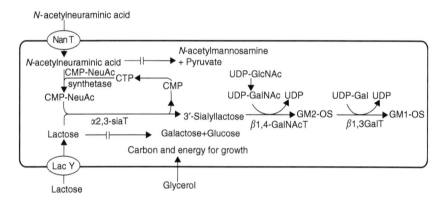

FIGURE 10.8 Production of ganglioside GM1 by fermentation. LacY, lactose permease; NanT, *N*-acetylneuraminic acid permease; GM2-OS, carbohydrate moiety of ganglioside GM2; GM1-OS, carbohydrate moiety of ganglioside GM1. (From Antoine, T. et al., *Chem. Biol. Chem.*, 4, 406–412, 2003. With permission.)

microbial methods using the whole cell are very simple, this strategy is attractive for industrial applications.

REFERENCES

1. Gagneux, P. and Valki, A., Evolutionary considerations in relating oligosaccharide diversity to biological function, *Glycobiology*, 9, 747–755, 1999.
2. McAuliffe, J.C. and Hindsgaul, O., Carbohydrates in medicine, in Fukuda, M. and Hindsgaul, O., Eds., *Molecular and Cellular Glycobiology*, Oxford University Press, New York, 2000, pp. 249–285.
3. Flowers, H.M., Chemical synthesis of oligosaccharides, *Methods Enzymol.*, 50, 93–121, 1978.
4. Palcic, M.M., Biocatalytic synthesis of oligosaccharides, *Curr. Opin. Biotechnol.*, 10, 616–624, 1999.
5. Vic, G. and Crout, D.H.G., Glycosidases and glycosyltransferase in glycoside and oligosaccharide synthesis, *Curr. Opin. Chem. Biol.*, 2, 98–111, 1998.

6. Gotschlich, E.C., Genetic locus for the biosynthesis of the variable portion of *Neisseria gonorrhoeae* lipooligosaccharide, *J. Exp. Med.*, 180, 2181–2190, 1994.

7. Jennings, M.P., Hood, D.W., Peak, I.R.A., Virji, M., and Moxon, E.R., Molecular analysis of a locus for the biosynthesis and phase-variable expression of the lacto-*N*-neotetraose terminal lipopolysaccharide structure in *Neisseria meningitides*, *Mol. Microbiol.*, 18, 729–740, 1995.

8. Wakarchuk, W.W., Martin, A., Jennings, M.P., Moxon, E.R., and Richards, J.C., Functional relationships of the genetic locus encoding the glycosyltransferase enzymes involved in expression of the lacto-*N*-neotetraose terminal lipopolysaccharide structure in *Neisseria meningitides*, *J. Biol. Chem.*, 271, 19166–19173, 1996.

9. Logan, S.M., Conlan, J.W., Monteiro, M.A., Wakarchuk, W.W., and Altman, E., Functional genomics of *Helicobacter pylori*: identification of a beta-1,4 galactosyltransferase and generation of mutants with altered lipopolysaccharide, *Mol. Microbiol.*, 35, 1156–1167, 2000.

10. Endo, T., Koizumi, S., Tabata, K., and Ozaki, A., Cloning and expression of beta 1,4-galactosyltransferase gene from *Helicobacter pylori*, *Glycobiology*, 10, 809–813, 2000.

11. Kolkman, M.B.A., Wakarchuk, W.W., Nuijten, P.J.M., and van der Zeijst, B.A.M., Capsular polysaccharide synthesis in *Streptococcus pneumoniae* serotype 14: molecular analysis of the complete cps locus and identification of genes encoding glycosyltransferases required for the biosynthesis of the tetrasaccharide subunit, *Mol. Microbiol.*, 26, 197–208, 1997.

12. Yamamoto, S., Miyake, K., Koike, Y., Watanabe, M., Machida, Y., Ohta, M., and Iijima, S., Molecular characterization of type-specific capsular polysaccharide biosynthesis genes of *Streptococcus agalactiae* type Ia, *J. Bacteriol.*, 181, 5176–5184, 2000.

13. Watanabe, M., Miyake, K., Yanae, K., Kataoka, Y., Koizumi, S., Endo, T., Ozaki, A., and Iijima, S., Molecular characterization of a novel beta 1,3-galactosyltransferase for capsular polysaccharide synthesis by *Streptococcus agalactiae* type Ib, *J. Biochem.*, 131, 183–191, 2002.

14. Gilbert, M., Brisson, J.R., Karwaski, M.F., Michniewicz, J., Cunningham, A.M., Wu, Y., Young, M., and Wakarchuk, W.W., Biosynthesis of ganglioside mimics in *Campylobacter jejuni* OH4384. Identification of the glycosyltransferase genes, enzymatic synthesis of model compounds, and characterization of nanomole amount by 600 mHz ^{1}H and ^{13}C NMR analysis, *J. Biol. Chem.*, 275, 3896–3906, 2000.

15. Gilbert, M., Watson, D.C., Cunningham, A.M., Jennings, M.P., Young, N.M., and Wakarchuk, W.W., Cloning of the lipooligosaccharide α-2,3-sialyltransferase from the bacterial pathogens *Neisseria meningitides* and *Neisseria gonorrhoeae*, *J. Biol. Chem.*, 271, 28271–28276, 1996.

16. Hood, D.W., Cox, A.D., Gilbert, M., Makepeace, K., Walsh, S., Deadman, M.E., Cody, A., Martin, A., Mansson, M., Schweda, E.K.H., Brisson, J.-R., Richards, J.C., Moxon, E.R., and Wakarchuk, W.W., Identification of a lipopolysaccharide α-2,3-sialyltransferase from *Haemophilus influenzae*, *Mol. Microbiol.*, 39, 341–350, 2001.

17. Jones, P.A., Samuels, N.M., Phillips, N.J., Munson Jr, R.S., Bozue, J.A., Arseneau, J.A., Nichols, W.A., Zaleski, A., Gibson, B.W., and Apicella. M.A., *Haemophilus influenzae* type b strain A2 has multiple sialyltransferases involved in lipooligosaccharide sialylation, *J. Biol. Chem.*, 277, 14598–14611, 2002.

18. Bozue, J.A., Tullius, M.V., Wang, J., Gibson, B.W., and Munson R.S., Jr. *Haemophilus ducreyi* produces a novel sialyltransferase. Identification of the sialyltransferase gene and construction of mutants deficient in the production of the sialic acid-containing gycoform of the lipooligosaccharide, *J. Biol. Chem.*, 274, 4106–4114, 1999.

19. Chaffin, D.O., McKinnon, K., and Rubens, C.E., CpsK of *Streptococcus agalactiae* exhibits alpha2,3-sialyltransferase activity in *Haemophilus ducreyi*, *Mol. Microbiol.*, 45, 109–122, 1996.

20. Yamamoto, T., Nakashizuka, M., and Terada, I., Cloning and expression of a marine bacterial β-galactoside α2,6-sialyltransferase from *Photobacterium damsela* JT0160, *J. Biochem.*, 123, 94–100, 1998.

21. Shen, G.J., Datta, A.K., Izumi, M., Koeller, K.M., and Wong, C.-H., Expression of α2,8/2,9-polysialyltransferase from *Escherichia coli* K92, *J. Biol. Chem.*, 274, 35139–35146, 1999.

22. Martin, S.L., Edbrook, M.R., Hodgman, T.C., van den Eijnden, D.H., and Bird, M.I., Lewis, X., biosynthesis in *Helicobacter pylori*, *J. Biol. Chem.*, 272, 21349–21356, 1997.

23. Ge, Z., Chan, N.W.C., Palcic, M.M., and Taylor, D.E., Cloning and heterologous expression of an α1,3-fucosyltransferase gene from the gastric pathogen *Helicobacter pylori*, *J. Biol. Chem.*, 272, 21357–21363, 1997.

24. Wang, G., Boulton, P.G., Chan, N.W.C., Palcic, M.M., and Taylor, D.E., Novel *Helicobacter pylori* α1,2-fucosyltransferase, a key enzyme in the synthesis of Lewis antigens, *Microbiology*, 145, 3245–3253, 1999.

25. Rasko, D.A., Wang, G., Palcic, M.M., and Taylor, D.E., Cloning and characterization of the α(1,3/4) fucosyltransferase of *Helicobacter pylori*, *J. Biol. Chem.*, 275, 4988–4994, 2000.

26. Blixt, O., van Die, I., Norberg, T., and van den Eijnden, D.H., High-level expression of the *Neisseria meningitidis* lgtA gene in *Escherichia coli* and characterization of the encoded N-acetylglucosaminyltransferase as a useful catalysis in the synthesis of GlcNAcβ1→3Gal and GalNAcβ1→3Gal linkages, *Glycobiology*, 9, 1061–1071, 1999.

27. Johnson, K.F., Synthesis of oligosaccharides by bacterial enzymes, *Glycoconj. J.*, 16, 141–146, 1999.

28. Izumi, M., Shen, G.-J., Wacowich-Sgarbi, S., Nakatani, T., Plettenburg, O., and Wong, C.-H., Microbial glycosyltransferases for carbohydrate synthesis: α2,3-sialyltransferase from *Neisseria gonorrhoeae*, *J. Am. Chem. Soc.*, 123, 10909–10918, 2001.

29. Wong, C.-H., Haynie, S.L., and Whitesides, G.M., Enzyme-catalyzed synthesis of N-acetyllactosamine with *in situ* regeneration of uridine 5′-diphosphate glucose and uridine 5′-diphosphate galactose, *J. Org. Chem.*, 47, 5416-5418, 1982.

30. Ichikawa, Y., Shen, G.-J., and Wong, C.-H., Enzyme-catalyzed synthesis of sialyl oligosaccharide with *in situ* regeneration of CMP-sialic acid, *J. Am. Chem. Soc.*, 113, 4698–4700, 1991.

31. Ichikawa, Y., Look, G.C., and Wong, C.-H., Enzyme-catalyzed oligosaccharide synthesis, *Anal. Biochem.*, 202, 215–238, 1992.

32. Chen, X., Fang, J., Zhang, J., Liu, Z., Shao, J., Kowal, P., Andreana, P., and Wang. P.G., Sugar nucleotide regeneration beads (superbeads): a versatile tool for the practical synthesis of oligosaccharides, *J. Am. Chem. Soc.*, 123, 2081–2082, 2001.

33. Fang, J., Chen, X., Zhang, W., Janczuk, A., and Wang, P.G., Synthesis of α-Gal epitope derivatives with a galactosyltransferase–epimerase fusion enzyme, *Carbohydr. Res.*, 329, 873–878, 2000.

34. Yan, F., Wakarchuk, W.W., Gilbert, M., Richards, J.C., and Whitfield, D.M., Polymer-supported and chemoenzymatic synthesis of the *Neisseria meningitidis* pentasaccharide: a methodological comparison, *Carbohydr. Res.*, 328, 3–16, 2000.

35. Blixt, O., Brown, J., Schur, M.J., Wakarchuk, W.W., and Paulson, J.C., Efficient preparation of natural and synthetic galactoside with a recombinant β1,4-galactosyl-transferase-/UDP-4'-Gal epimerase fusion protein, *J. Org. Chem.*, 66, 2442–2448, 2001.

36. Gilbert, M., Cunningham, A.M., DeFrees, S., Gao, Y., Watson, D.C., Young, N.M., and Wakarchuk, W.W., The synthesis of sialylated oligosaccharides using a CMP-Neu5Ac synthetase/sialyltransferase fusion, *Nat. Biotechnol.*, 16, 769–772, 1998.

37. Hokke, CH., Zervosen, A., Elling, L., Joziasse, D.H., and van den Eijnden, D.H., One-pot enzymatic synthesis of the Galα1→3Galβ1→4GlcNAc sequence with *in situ* UDP-Gal regeneration, *Glycoconj. J.*, 13, 687–692, 1996.

38. Chen, X., Zhang, J., Kowal, P., Liu, Z., Andeana, P.R., Lu, Y., and Wang, P.G., Transferring a biosynthetic cycle into a productive *Escherichia coli* strain: large-scale synthesis of galactosides, *J. Am. Chem. Soc.*, 123, 8866–8867, 2001.

39. Liu, Z., Lu, Y., Zhang, J., Pardee, K., and Wang, P.G., P1 trisaccharide (Galα1,4Galβ1,4GlcNAc) synthesis by enzyme glycosylation reactions using recombinant *Escherichia coli*. *Appl., Environ. Microbiol.*, 69, 2110–2115, 2003.

40. Noguchi, T. and Shiba, T., Use of *Escherichia coli* polyphosphate kinase for oligosaccharide synthesis, *Biosci. Biotechnol. Biochem*, 62, 1594–1596, 1998.

41. Fujio, T. and Furuya, A., Production of ATP from adenine by *Brevibacterium ammoniagenes*, *J. Ferment. Technol.*, 61, 261–267, 1983.

42. Fujio, T. and Maruyama, A., Enzymatic production of pyrimidine nucleotides using *Corynebacterium ammoniagenes* cells and recombinant *Escherichia coli* cells: enzymatic production of CDP-choline from orotic acid and choline chloride (part I), *Biosci. Biotechnol. Biochem.*, 61, 956–959, 1997.

43. Koizumi, S., Endo, T., Tabata, K., and Ozaki, A., Large-scale production of UDP-galactose and globotriose by coupling metabolically engineered bacteria, *Nat. Biotechnol.*, 16, 847–850, 1998.

44. Koizumi, S., Endo, T., Tabata, K., Nagano, H., Ohnishi, J., and Ozaki, A., Large-scale production of GDP-fucose and Lewis X by bacterial coupling, *J. Ind. Micriobiol. Biotechnol.*, 25, 213–217, 2000.

45. Endo, T., Koizumi, S., Tabata, K., and Ozaki, A., Large-scale production of CMP-NeuAc and sialylated oligosaccharides through bacterial coupling, *Appl. Microbiol. Biotechnol.*, 53, 257–261, 2000.

46. Endo, T., Koizumi, S., Tabata, K., and Ozaki, A., Large-scale production of *N*-acetyl-lactosamine through bacterial coupling, *Carbohydr. Res.*, 316, 179–183, 1999.

47. Endo, T., Koizumi, S., Tabata, K., and Ozaki, A., Large-scale production of the carbohydrate portion of the sialyl-Tn epitope, α-Neu$p$5Ac-(2→6)-D-Gal$p$NAc, through bacterial coupling, *Carbohydr. Res.*, 330, 439–443, 2001.

48. Kragl, U., Gygax, D., Ghisalba, O., and Wandrey, C., Enzymatic two step synthesis of *N*-acetylneuraminic acid in the enzyme membrane reactor, *Angew. Chem. Int. Ed. Eng.*, 30, 827–828, 1991.

49. Maru, I., Ohnishi, J., Ohta, T., Tsukada,Y., Simple and large-scale production of *N*-acetylneuraminic acid from *N*-acetyl-D-glucosamine and pyruvate using *N*-acetyl-D-glucosamine 2-epimerase and *N*-acetylneuraminate lyase, *Carbohydr. Res.*, 306, 575–578, 1998.

50. Tabata, K., Koizumi, S., Endo, T., and Ozaki, A., Production of *N*-acetyl-D-neuraminic acid by coupling bacteria expressing *N*-acetyl-D-glucosamine 2-epimerase and *N*-acetyl-D-neuraminic acid synthetase, *Enzyme Microbiol Technol.*, 30, 327–333, 2002.

51. Endo, T. and Koizumi, S., Production of oligosaccharides using engineered bacteria, *Curr. Opin. Struct. Biol.*, 10, 536–541, 2000.

52. Priem, B., Gilbert, M., Wakarchuk, W.W., Heyraud, A., and Samain, E., A new fermentation process allows large-scale production of human milk oligosaccharides by metabolically engineered bacteria, *Glycobiology*, 12, 235–240, 2002.
53. Dumon, C., Priem, B., Martin, S.L., Heyraud, A., Bosso, C., and Samain, E., *In vivo* fucosylation of lacto-*N*-neotetraose and lacto-*N*-neohexaose by heterologous expression of *Helicobacter pylori* α-1,3 fucosyltransferase in engineered *Escherichia coli*, *Glycoconj. J.*, 18, 465–474, 2001.
54. Antoine, T., Priem, B., Heyraud, A., Greffe, L., Gilbert, M., Wakarchuk, W.W., Lam, J.S., and Samain, E., Large-scale *in vivo* synthesis of the carbohydrate moieties of gangliosides GM1 and GM2 by metabolically engineered *Escherichia coli*, *Chem. Biol. chem.*, 4, 406–412, 2003.
55. Bettler, E., Imberty, A., Priem, B.,Chazalet, V., Heyraud, A., Joziasse, D.H., and Geremia, R.A., Production of recombinant xenoplantation antigen in *Escherichia coli*,. *Biochem. Biophys. Res. Commun.*, 302, 620–624, 2003.

11 Fructan Biosynthesis in Genetically Modified Plants

Andrew J. Cairns and Sophie J. Perret

CONTENTS

339

I.   INTRODUCTION

A.   AN OVERVIEW OF FRUCTAN

Fructans (oligo- and polyfructosyl sucrose), rather than starch, occur naturally as
the primary reserve carbohydrate in the leaves, stems, and roots of 12–15% of
higher plants, which together represent more than 40,000 species.[1–5] In some
species, fructan is the sole reserve polymer that accumulates; in others, starch and
fructan occur together.[5] Carbon partitioning in natural fructan-accumulators (NFA)
and its regulation are distinct from those in natural starch-accumulators (SA) in a
number of important respects. Fructan is synthesized directly from sucrose, apparently without the involvement of phosphorylated sugars or nucleotide cofactors. Its
synthesis is extrachloroplastic, it is water soluble, and it accumulates in the vacuole
of both photosynthetic and heterotrophic storage cells. In addition to its reserve
role, it is supposed by some authorities to confer stress resistance on plant tissues.[6,7]
Fructan occurs naturally in many economically important species, for example, in
the vegetative tissues of the temperate grasses and cereals, such as barley, wheat,
and oat, but it does not occur in the cereals of warmer regions, such as maize, rice,
sorghum, and millet. Fructan is also the natural reserve carbohydrate in the peren-
niating organs of chicory, artichoke, asparagus, dahlia, and the onion family. The
extracellular polysaccharide of some classes of bacteria is elaborated as fructan.
With molecular masses of $1–5 \times 10^6$ Da, the bacterial polymer, termed levan, is
larger than plant fructans by 2 to 3 orders of magnitude. Apparently, fructans are
not synthesized or directly degraded by animals although they are used as a carbon
and energy source by the intestinal microflora. The nonspecialist reader is referred
to recent accounts of the structure, physiology, biosynthesis, and enzymology of
fructan (see references 4–8 and references therein), a knowledge of which is
assumed here.

B. Aims of This Review

An earlier critique[9] dealt with some difficulties of interpretation that arose when conclusions from enzymological studies performed in isolation were placed into a physiological context. When parameters such as substrate concentration, enzyme affinities, carbon flux, accumulation rates, enzyme activity, enzyme concentrations, and the structure of fructan products were considered in parallel *in vitro* and *in vivo,* a number of ambiguities were identified that were associated with two central factors: firstly, the unusually low affinity (high apparent K_m) of plant fructosyltransferases for their substrates and, secondly, the difficulty in distinguishing physiologically relevant *in vitro* fructan synthesis from side reactions of invertase. It has since become apparent that this lack of discrimination between transferases and hydrolases is also reflected at the sequence level and, hence, these issues remain pertinent to the assessment of more recent studies involving genetic modification. Cairns[9] also identified some confusion in the definition of the term "synthesis" and how it is distinguished from the related concepts of "primer glycosylation" and "disproportionation." Despite an extensive literature describing the enzymology of plant fructan synthesis, to date there remain only a few reports of its *de novo* net polymerization *in vitro.*[8] In spite of these difficulties, considerable progress has been made in our understanding of fructan synthesis in the last decade. Recent work has amply confirmed the peculiarity of plant fructosyltransferases with regard to their kinetic properties (low affinity), and high enzyme concentration is emerging as an important requirement for the *in vitro* polymerization of plant fructans.

The rather odd properties of enzymes determined *in vitro* have interesting physiological consequences for transgenics. Clearly, the subcellular environment in which a transgene is expressed will have consequences for its function since, for example, host plants may not contain the high intracellular concentrations of sucrose and express sufficient enzyme necessary for effective fructan synthesis. Such physiological considerations may have been overlooked in the haste to produce transgenic plants. As with the earlier studies of enzymology, results from molecular biology need to be carefully assessed with regard to their physiological consequences in order to determine their significance adequately. In our view this has generally not been the case in primary reports of fructan transgenesis. Hence, this review updates and extends the earlier critique[9] by examining results from transgenic fructan-accumulating plants from the physiological viewpoint.

C. A Physiological Context for Transgenic Fructan Synthesis

The basis of expression for quantities of fructan product varies widely, the choice of units clearly being influenced by the type of information to be communicated. For transgenics, there has been a tendency to present "plus or minus" and "single endpoint" values. In such studies, the smallest trace of oligosaccharide is "positive," regardless of its significance in terms of physiological fluxes. Physiologists and biochemists, in contrast, have an interest in the details of concentrations, affinities, and rates. In many cases, it is difficult or impossible to estimate such parameters from data reported in studies of transgenesis. Additionally, transgenic fructan concentrations in tissues are rarely compared with those of natural endogenous reserve polysaccharide accumulation. Thus, although reports of "high-level" fructan accumulation in transgenics are common, there is a lack of definition of the context and meaning of the term "high level". Further, physiological rates of transgenic product formation *in planta,* expressed as a function of time and some measure of tissue quantity, have never been reported. To address these omissions and to allow for comparisons between transgenics themselves and between transgenics and NFA, values have been calculated from the primary literature into a common format as described previously[10] and are summarized in Tables 11.1 and 11.2.

There has also been a tendency to not report the full details of the conditions of culture of transgenic plants. Field data presented in Reference 10 provide a natural physiological context that can be summarized as follows: a clear sunny day in mid-May (early summer) in the U.K. (52° 30′N 04° 00′W) had a photosynthetic integral (PI) of ca. 29 moL/m^2 quanta, photosynthetically active radiation (PAR) with maximum and mean irradiances of 1200 and 510 μmoL/m^2/sec. Leaves of *Lolium perenne,* an NFA ryegrass grown under these conditions, exported ca. 260 mg sugar/g during the photoperiod, equivalent to a minimum flux of carbohydrate of 37 nkat/g as hexose. The maximal rate of fructan accumulation was 22.5 mg/g/h (39 nkat/g) at a substrate concentration of 22 mg/g (~90 mM sucrose in the vacuole). Leaf fructan content transiently reached 60 mg/g. These parameters are summarized in Tables 11.1 and 11.2 for comparison with the transgenics.

II. PLANTS TRANSFORMED WITH BACTERIAL FRUCTOSYLTRANSFERASE GENES

A. OVERVIEW

Historically, fructan biosynthesis was first elucidated in bacteria rather than in higher plants. By the early 1990s, the bacterial fructan polymerase, levansucrase, had been purified and characterized, antibody probes had become available, and the genes had been isolated, sequenced, and cloned. In contrast, the first plant fructosyltransferase genes did not become available until the mid-1990s. It was an early logical step to incorporate the available bacterial gene into the available, transformable, fructan-nonaccumulating plants. The stated rationales for the production of the bacterial transformants were:

(a) *Experimental.* To produce plants with altered source–sink balance for the experimental investigation of carbon partitioning.[12]
(b) *Experimental.* To examine the effects of differential subcellular location of exogenous levan synthesis.[12,13]
(c) *Industrial.* To provide sources of fructan qualitatively and quantitatively improved by comparison with that from natural sources,[20] particularly the generation of high-fructose feedstock for chemical syntheses.
(d) *Agricultural.* Aimed at crop improvement via enhanced carbohydrate retention and hence yield.[13,19,21]
(e) *Agricultural.* Aimed crop improvement via enhanced stress tolerance.[16,17]

The extent to which these aims have been achieved may be assessed on the basis of the following discussion.

B. QUANTITATIVE ASPECTS OF NONFRUCTAN PLANTS TRANSFORMED WITH BACTERIAL FRUCTOSYLTRANSFERASES

Levansucrases, mainly the *Sac B* gene from *Bacillus* spp., generally under the control of a constitutive promoter (usually CaMV 35S) and a putative subcellular targeting sequence, were used to produce stable transformations of SA such as potato, maize, tobacco, and clover. Table 11.1 summarizes a number of parameters pertinent to reserve carbohydrate accumulation for these transformants. An initial striking feature of Table 11.1 is how little attention has been given to the reporting of the cultural conditions that determine primary productivity, even in these studies, where reserve accumulation and biomass productivity are of primary concern. Irradiance and photoperiod, hence PI, were ignored as important experimental variables in 13 out of 17 studies. Reported values were in the range of 8 to 29% of the natural PI as presented in Table 11.1. In three of the four instances where irradiance was specified (for potato and tobacco), photosynthetic flux density (PFD) was low at 42 μmoL/m^2/s. This level approximates to the light compensation point for C3 photosynthesis,[28] where net carbon fixation should have been zero. Hence, the origin of the carbohydrate for growth and fructan synthesis in these studies is not clear.

In terms of the endpoint concentrations of levan, reports varied between 0.04 and 160 mg/g (ca. 0.02 to 80% of dry biomass) with the majority lying between 5 and 20 mg/g (ca. 2.5 to 10% of dry biomass). The maximum value was obtained for leaves grown *in vitro* under unspecified conditions, but presumably involved an exogenous carbon source. This value is remarkable because it is roughly twice the maximal reserve carbohydrate concentration observed in leaves in nature[5,10] and is more similar to the generally reported *total* dry matter content of ~200 mg/g in leaves.[10] The highest endpoint concentrations for autotrophic transformants were 66 and 20 mg/g (~33 and 10% of dry biomass) reported by the same group. Overall, these values for leaf tissue were very impressive, comparing well with the maximum natural fructan concentration of 60 mg/g in grass leaves (Table 11.1), with high concentrations of endogenous starch accumulated in leaves of SA, for example, in *Trifolium pratense* (62 mg/g),[29] and with natural fructan accumulation in tubers of *Helianthus tuberosus* (150 mg/g).[30] The data superficially suggest a marked shift in carbon resource allocation in the transformants.

C. ENDPOINT CONCENTRATIONS AND FLUXES

The plants were grown for weeks or months (Table 11.1) with the transgene under the control of a *constitutive* promoter. Hence, we may infer that the endpoint fructan concentrations represented the sum of levansucrase activity occurring over the entire growth period and were representative of the total flux of carbon into levan. This interpretation is confirmed by one data set[18] where levan concentration in the leaves increased steadily from young, to middle aged, to old leaves (absolute age unspecified), and is consistent with results from clover leaves[21] in which levansucrase activity increased with developmental age, implying both (i) a stable enzyme increasing in total activity by accumulation with time and (ii) continued enzyme function throughout the life of the leaf. Interestingly, many authors[15,18,20] pointed out that the leaves were "*harvested at the end of the light period.*" This statement could be

TABLE 11.1

Summary of Parameters Associated with Fructan Accumulation for Plants Transformed with Bacterial Fructosyltransferases

Host	Tissue/Putative Compartment	PFD (μmoL/m²/sec)	Photoperiod (h)	Photo-Synthetic Integral (mol/m²/day)	Accumulation Period (days)	Endpoint Fructan (mg/g)	Calculated Rate (nkat/g)	Estimated Sucrose Substrate mM	Ref.
Maize	Kernel/vacuole[h]	NR	NR	ID	30	18	0.03	ID	3
Potato	Tuber / plastid	NR[g]	NR	ID	135	66	0.03	ID	11
Potato[f]	Tuber /vacuole	NR	16	ID	56	17.5	0.02	25	12
Maize	Kernel/cytoplasm	NR	NR	ID	55	7.5	0.01	ID	3
Potato[a]	Tuber/cytoplasm	NR	NR	ID	70	9.8	0.01	ID	13
Potato	Leaf/vacuole	NR[c]	10	ID	49	5.0	0.007	28	14
Tobacco	Leaf	NR	NR	ID	77	6	0.006	54	15
Tobacco	Droughted leaf	42	24	3.6	28	0.35	0.0009	3	16
Sugar beet	Droughted leaf	100	14	5.0	101	1	0.0007	ID	17
Potato	Leaf	42[e]	16	2.4	NR	160	ID	ID	18
Tobacco[b]	Leaf/cytoplasm	NR[d]	NR	ID	1.5	NR	ID	ID	13
Potato	Tuber	NR[d]	16	ID	NR	11	ID	48	14
Ryegrass	Leaf	145	16	8.4	NR	0.04	ID	24	19
Tobacco	Leaf	NR	8	ID	NR	2.8	ID	5	20
Potato	Tuber	42[e]	16	2.4	NR	10	ID	ID	18

Tobacco	Leaf / plastid	NR	NR	ID	NR	20	ID	32	11
Clover	Leaf / vacuole[i]	NR[j]	12	ID	NR	3.9	ID	15	21
L.perenne	Endogenous metabolism	510[k]	12	29	0.3	60	39	90	10

Except where indicated otherwise, the structural gene was Sac B from *B. subtilis* under the control of a constitutive promoter. The Fructan concentrations are expressed on a fresh mass basis. Where several values were reported, the maximum is presented. Data are ranked in decreasing order of levan accumulation rate. Equivalent data for the (nontransgenic) NFA *L. perenne* are included for comparison. All calculations were performed as described in Reference (10).

[a] Levansucrase gene from *B. amyloliquefaciens*: Patatin promoter: Cultured tubers: Assumes 80% water.

[b] Levansucrase gene from *B. amyloliquefaciens*: SSU / 2-2 promoters

[c] Glasshouse: winter.

[d] Glasshouse: spring.

[e] 3000 lux: Converted. From Mc Cree, K.J., *Agric. Meteorol.*, 10, 443–453, 1972.

[f] Levan sucrase gene from *E. amylovora*: B33 tuber-specific promoter.

[g] Glasshouse: time of year unspecified.

[h] Zein promoter.

[i] ftf Structural gene from *S. salivarius*, tobacco chitinase A vacuolar targeting sequence.

[j] Glasshouse: time of year unspecified.

[k] Field data. Early summer. From Cairns, A.J., *J. Exp. Bot.*, 54, 549–567, 2003.

NR: appropriate data not reported.

ID: insufficient data for calculation.

Source: Modified from Cairns, A.J., *J. Exp. Bot.*, 54, 549–567, 2003. (With permission.)

TABLE 11.2
Summary of Parameters Associated with Fructan Accumulation in Plants Transformed with Plant-Derived Fructosyltransferases

Host/Tissue	Gene/Source	PFD (μmoL/m²/sec)	Photoperiod (h)	Photo-Synthetic Integral (moL/m²/day)	Accum. Period (days)	Endpoint Transgenic Fructan (mg/g)	Calculated Rate (nkat/g)	Estimated Tissue Substrate (mM)	Km of Gene Product (mM)	Ref.
Sugar beet/Root	SST/H.tuberosus	NR[a]	14	ID	180	62.3	0.025	32	280	23
Potato/Tuber	1-SST/1-FFT/C. scolymus	NR[b]	NR[b]	ID	84	6.1	0.005	15	NR	24
Potato/Tuber	1-SST/C. scolymus	NR[b]	NR[b]	ID	84	4.9	0.004	17	NR	24
Tobacco/Root	6-SFT/H. vulgare	220	16	12.6	60	0.6	0.0007	38	>300	25
Petunia/Leaf	SST/H.tuberosus	25–40	16	2.3	90	0.47	0.0004	2	280	26
Sugar beet/Leaf	SST/H.tuberosus	NR	14	ID	180	0.6	0.0002	1	280	23
Tobacco/Leaf	6-SFT/H. vulgare	220	16	12.6	60	0.06	0.00007	8	>300	25
Petunia/Leaf	SST + FFT/H. tuberosus	25–40	16	2.3	90	0.08	0.00006	2	280	26
Potato/Leaf	1-SST/C. scolymus	NR[b]	NR[b]	ID	45	0.1	0.00002	6	NR	24
Chicory/Leaf	6-SFT/H. vulgare	550	24	47.5	2	Trace	ID	NR	>300	25
Potato/Tuber	SST/C. scolymus[c]	NR	NR	ID	NR	9.7	ID	22	NR	27
L.perenne/Leaf	Endogenous Metabolism	510[d]	12	29	0.3	60	39	90	NR	10

Fructan concentrations are expressed on a fresh mass basis. Where several values were reported, the maximum is presented. Equivalent data for the (nontransgenic) *L. perenne* are included for comparison. All calculations were performed as described in Reference 10. Data are ranked in decreasing order of fructan accumulation rate.

[a] Glasshouse, winter.

[b] Glasshouse.

[c] Patatin promoter.

[d] Field data. Early summer,U.K. From Cairns, A.J., *J. Exp. Bot.*, 54, 549–567, 2003.

NR, appropriate data not reported; ID, Insufficient data for calculation.

Source: Modified from Cairns, A.J., *J. Exp. Bot.*, 54, 549–567, 2003. (With permission.)

misleading, because it could be taken to imply that the recorded levan contents were the product of photosynthesis in that photoperiod alone, rather than the sum of constitutive accumulation over the entire growth period. The probability of the reader adopting this interpretation is increased because, in general, levan accumulation was presented along with instantaneous concentrations of other endogenous carbohydrates. Such presentation does not compare equivalent parameters because the endogenous sugars are subject to accumulation, turnover, and transport on hourly and daily time scales,[10,21] whilst the levan accumulated over weeks and months. To assess the magnitude of levan accumulation as a proportion of primary metabolism, we need to compare it with the *flux* of sugar through photoassimilation, rather than with instantaneous pools of endogenous sugars.

When endpoint concentrations of levan were expressed as a function of accumulation period, the rates of accumulation were low, between 0.0007 and 0.03 nkat/g (Table 11.1). These values are 1200 to 53,000-fold lower than the daily flux of carbon from the leaves of NFA (37 nkat/g, Table 11.1). When expressed proportionally, flux into levan was equivalent to 0.0002 to 0.08% of this value. This contrasts with the emphasis in statements such as "we observed that bacterial fructosyltransferase can compete with this process [the sink demand of starch synthesis] and divert normal carbohydrate flow".[18] Clearly, in some cases and over long accumulation periods, such quantitatively minor leakage can cumulatively result in substantial instantaneous concentrations of levan, but this does not imply strong competition with endogenous sinks, major modification of flux or significant changes in patterns of partitioning. In contrast, there is one report, for tobacco leaves, where levan was detected within 36 h of the chemical induction of levansucrase expression.[13] This finding was suggestive of rapid accumulation, but levan concentrations were not reported, precluding the calculation of a rate. This apparent rapid accumulation was accompanied by equally rapid tissue necrosis (discussed below).

D. EXPLAINING THE LOW RATES OF PRODUCT ACCUMULATION IN THE LEVAN TRANSFORMANTS

It is of interest to speculate on the cause of these low rates of levan accumulation. One possibility is product turnover. However, a generally stated reason for the choice of host is that they were not expected to contain enzymes for levan hydrolysis[3,15–18] or were explicitly shown not to contain extractable levan hydrolase.[20] In instances where leaves of tobacco[20] and clover[21] were subjected to incubation in the dark, accumulated levan was not degraded. Therefore, turnover does not appear to explain the low accumulation rates.

Sucrose substrate concentration appeared to influence levan accumulation rate, since transformation of high-sucrose mutants of maize resulted in increased levan accumulation in kernels, although the sucrose concentrations were not quantified and related to levansucrase activity.[3] Hence, a possible explanation for low rates of levan accumulation is limiting substrate. For the levansucrase from *B. subtilis*, the sucrose concentration for half-maximal activity (apparent K_m) is 20 mM.[31] Estimates of sucrose concentrations at the putative site of levansucrase expression, the vacuole, are listed in Table 11.1. Seven of the nine values were at ~15 mM or above, i.e., sufficient to cause approximately half-maximal rate. The activity would have been substantial and is thus unlikely to provide a general explanation for the observed low rates of accumulation. Table 11.1 also shows that in two reports, leaf tissue sucrose concentrations were lower than 5 mM, at which the enzyme activity would have been $< \sim7\%$ of V_{max}. These values correlated with lower endpoint levan concentrations and could have contributed to the low total accumulation in these transformants.

Apparently high rates of levan accumulation occurred when expression was under the control of an inducible promoter.[13] This expression was associated with tissue necrosis that occurred within 36 h. When otherwise identical plants were transformed with a normally successful constitutive promoter, no plants were regenerated. Both results could indicate levan toxicity. Hence, the occurrence of only low-levan accumulators may be explained as a function of lethal negative selection of high-rate transformants.

Although not explicitly emphasizing low absolute rates of accumulation, many studies of levan transformants tacitly alluded to the issue by reference to other, related expression parameters. It was generally noted that (a) mRNA for the levan-sucrase could not be detected by Northern blots, although low-level expression of the mRNA was detected in transformed ryegrass by the more sensitive method of RT-PCR[21] (also, Perret, S.J., Morris, P., Cairns, A.J., unpublished observations). (b) Levansucrase protein could not be detected in the transformants by immunological probes. (c) Levansucrase activity could not be detected by conventional fructosyl-transferase assays. Instead, sensitive detection using ^{14}C-labeled sucrose was necessary. In general when this strategy was successful, qualitative data were presented (as Thin layer chromatography (TLC) autoradiographs) and trace amounts were synthesized. Given that milligram quantities of polymeric fructan (sufficient for chemical, chromatographic, and structural analyses) can be polymerized in several hours by endogenous enzymes from NFA,[8] in comparison, the levansucrase activity in the transformants was extremely low. Taken together, the above evidence suggests that the primary reasons for the low rates of levan accumulation were low expression rates and low enzyme quantity.

In contrast, a recent report[21] found an exceptionally high rate of 120 nmoL/min/g for radiochemically measured transgenic fructosyltransferase (equivalent to 2 nkat/g or 1.2 mg/h/g). This rate is comparable with enzyme rates reported for some NFA grown in controlled environments[5,8,32] but is ~70 to 2900-fold greater than the *in vivo* rates of fructan accumulation for levan transgenics (Table 11.1). The same report[21] gives an endpoint levan concentration of 3.9 mg/g in clover leaves, whereas leaves from "young plants" contained only 0.2 mg/g. Hence, we may infer that, as with the other transformants, levan accumulated slowly in the tissue over the time scale of leaf development. The comparison of *in vitro* and *in vivo* results is interesting because at the reported rate (1.2 mg/h/g), the enzyme would have synthesized the total levan accumulated in the leaves (3.9 mg/g) in 3.3 h. This result conflicts with their stated view that fructan accumulation was limited in the tissue by low enzyme activity. There may be a problem of overestimation of the enzyme activity or underestimation of tissue levan content for these clover transformants; otherwise, we must hypothesize a marked downregulation of the activity *in vivo* by some previously unrecognized and unknown mechanism.

E. Attempts to Increase Levansucrase Expression in Transformants

In tacit acknowledgement of the general low rates of levan expression, a number of approaches to increase expression were attempted, such as the use of tandem promoters and a translational enhancer.[20] In addition, a range of promoters[3] and a variety of putative subcellular targeting sequences were also used with levansucrase genes. Compared with the generally observed levansucrase expression rates, conspicuous improvements have not been demonstrated using combinations of these sequences. This has been interpreted to indicate a more fundamental problem resulting from the characteristics of the structural gene. In this connection, codon usage is different in eukaryotes compared with the Gram-positive bacteria such as *B. subtilis* and *Streptococcus salivarius*.[12,19] It is possible, in an eukaryotic host, that the activity of the gene product is impaired due to inappropriate amino acid substitutions. Further, localized regions of AT richness and the presence of mRNA-destabilizing ATTTA sequences may cause the rapid turnover of the *Bacillus* message.[19] All of these factors could contribute to the observed low expression rates.

The levansucrase gene from the Gram-negative bacterium *Erwinia amylovora*, which does not exhibit the rare codon usage, has been used for plant transformation.[12] Table 11.1 compares the performance of the *E. amylovora* transformant with those using levansucrase from *Bacillus* spp. At 17.5 mg/g, the endpoint concentration of levan was in the general range of 5 to 20 mg/g with an accumulation rate of 0.02 nkat/g. These values were amongst the highest for all the transformants, but not higher than the highest results obtained using the *B. subtilis* SacB gene. Hence, differential codon usage does not seem to provide a general explanation for low rates of levan expression. Attempts to explain low expression rates have focussed on the bacterial origin of the structural gene, but as we shall see, the problem is not exclusive to the bacterial genes. Plant-derived fructosyltransferases are also expressed at low rates in transformants (Table 11.2, discussed below). Expression rates of all fructan transgenes are low in all surviving transformants. This limitation is indicative of a more fundamental and general problem.

F. Structural Identification of Levan in Transformants

Bacterial levan is a very large polymer. Its large size presents some difficulties for routine identification. In the absence of detailed derivatization and structural analyses,[8,33] or the use of direct physical measurement techniques such as low-angle laser light scattering,[34] assessment of molecular size is not straightforward. Techniques reported for the identification of transgenic levan include TLC, size-exclusion chromatography (SEC), and nuclear magnetic resonance spectroscopy (NMR). Levan does not migrate from the origin on TLC and this has been used widely to show its presence in transformants or in enzyme digests. The size threshold for nonmigration is DP = ca.10 for 2,6-linked fructans.[8] Hence, strictly interpreted, nonmigration on TLC is evidence only for the presence of levan larger than 2×10^3 Da and when presented alone[11,13,17] it is not indicative of large polymers equivalent to those formed naturally by bacteria ($1-5 \times 10^6$ Da). SEC on materials such as Sepharose 4B have been used to estimate molecular size, in some cases by reference to protein size standards and an excluded volume marked by Blue Dextran. However, it is noteworthy that Sepharose 4B did not separate transgenic levan from the endogenous fructan of *Lolium multiflorum*: both were apparently present in the excluded volume.[19] The large fructan of *Lolium* spp. has a maximum size of up to 6.5 × 10³ Da as shown by HPAEC.[35] Conclusion of levan size by exclusion on SEC appears ambiguous.

NMR data have been presented in the form of comparisons of spectra obtained for standard levan and levan isolated from transformants.[12,20] NMR spectra give information about the electronic configurations of individual atoms within a compound. Although the spectra are indicative of bond structure, they give no indication of molecular size for polymers unless, for example, signals from atoms in terminal sugars are determined in a ratio with signals from equivalent atoms in the linkage sugars. It is concluded that for the majority of studies of transgenic levan synthesis in plants, coidentity with the bacterial polymer, especially with respect to size, has not been unambiguously demonstrated.

G. HAS SUBCELLULAR TARGETING BEEN ESTABLISHED UNEQUIVOCALLY IN BACTERIAL LEVAN TRANSFORMANTS?

The native *SacB* gene contains a bacterial secretion sequence.[13,21] Plant vectors have been designed to contain targeting sequences to replace or override this internal signal and send the levansucrase to different subcellular locations: the vacuole, cytoplasm, extracellular matrix, endomembrane system, and most recently, to plastids. In general, however, the actual targeting of the transgene product has been assumed rather than demonstrated.

The most convincing evidence for actual targeting to predetermined locations was reported by Gerrits and co-workers.[11] Using subcellular fractionation, levan was unambiguously shown to be associated with chloroplasts isolated from protoplasts of tobacco plants transformed with SacB fused with a ferredoxin chloroplast targeting sequence. In control transformants targeted with the sporamin vacuolar targeting sequence (shown to target the GUS reporter gene to the vacuole, but also to target levansucrase only to the endomembrane system[15]) no levan was associated with the chloroplasts. Unfortunately, having elegantly demonstrated levan association with chloroplasts in leaves of tobacco, the paper[11] subsequently contains a detailed analysis of the physiology of potato tubers. Although localization of transgenic invertase was determined for potato chloroplasts, there were no data demonstrating levan association with amyloplasts in potato tubers. The validity of the conclusions concerning the effects of plastidic expression of levansucrase in potato awaits the demonstration of actual targeting in this species.

Although immunological detection of levansucrase failed because of the low expression of the enzyme protein (discussed above), sufficient polysaccharide product did accumulate in some transformants for visualization by immunofluorescence microscopy. Putative apoplastic and putative vacuolar targeting sequences were fused with levansucrase and used to transform starch-deficient potato.[12] Micrographs convincingly showed enhanced fluorescence in tuber cell walls of apoplastically targeted levan transformants relative to nonlevan transformed controls. However, in similarly transformed leaves, there were considerable additional signals from cell interiors, implying the additional presence of levan in vacuoles. Further, in the vacuole-targeted transformants there was a strong signal in the cell walls similar to that in the apoplastically targeted tissue, and the fluorescence signal was clearly present in the vacuoles of only three of the 12 cells in the image. Clear segregation of independently targeted levan synthesis was not demonstrated in these experiments.

Similar fluorescence experiments[14] clearly showed that transgenic levan from a yeast carboxypeptidase-Y (CPY) vacuolar-targeted enzyme did not localize in the vacuole, but rather, at the cell perimeter in potato leaves. Hence, work from two independent laboratories indicates mislocalization of the transgene product, regardless of the putative specificity of the targeting sequence. The CPY construct has been used in subsequent studies and described as containing "the CPY yeast vacuolar targeting sequence."[17] Although strictly speaking, this statement is accurate, to an uncritical reader this could be taken to imply that "the gene product went to the vacuole." Subcellular localization was not demonstrated in this study. The vacuolar targeting sequence for tobacco chitinase A has recently been used with the fructosyltransferase for transformation of clover. Localization of the enzyme was attempted, but the authors described the results as "inconclusive".[21]

In studies of maize transformation, a seed-specific promoter was fused separately with vacuolar-targeting sequences from barley lectin and sweet potato sporamin.[3] The latter has been shown to mislocalize the levansucrase in the endomembrane system in tobacco (Reference 15, discussed above). Neither was shown to actually target the gene product to the vacuole although the results were discussed under this assumption. A third construct[3] contained no targeting sequence and was assumed to target the gene product to the cytoplasm, but again in the absence of any demonstration. The description of the construction of the vector made no reference to the internal bacterial secretion sequence. The possibility that the putative cytosolic target may have placed the SacB gene product at the cell surface was not discussed, although in a subsequent study it was made explicit that the internal sequence had been removed from the gene.[13] The actual subcellular localization of the transgene products was not established in these studies.

The modified SacB gene, without the internal secretory sequence, was used under the control of the chemically controlled maize 2-2 gene in the absence of a targeting sequence and had a presumed cytosolic localization.[13] When levan accumulation was induced in tobacco leaves containing this construct, instead of their usual position at the cell perimeter, the chloroplasts aggregated at the cell center. While no explanation was offered, this could be explained if levan synthesized in the cytoplasm interacts with the chloroplast outer membrane, crosslinking and thereby agglutinating the chloroplasts. This explanation is consistent with recent reports that indicate that levan interacts directly with membranes.[36,37] If we hypothesize that cytoplasmically synthesized levan can bind to and agglutinate chloroplasts, then levan found in the chloroplastic fraction[15] (discussed above) need not be synthesized within the chloroplast, but may have occurred by surface interaction of the outer membrane with levan synthesized elsewhere, such as in the cytoplasm. This observation could invalidate the conclusions that (i) the levansucrase was actually targeted to, and active in, the chloroplast and (ii) that a sucrose pool exists in this organelle. The control was the targeted accumulation of levan within the endomembrane system and the resulting absence of levan in the chloroplast fraction. This result is consistent with the hypothesis, because accumulation within, or binding to, the endomembrane system could have sequestered the levan away from the chloroplasts, resulting in the absence of a levan signal in the control chloroplast fraction.

In conclusion, there are no unambiguous demonstrations of successful predetermined targeting, but some good indicators that the levansucrase has been inadvertently sent to the cell surface and to the endomembrane system. For the other studies it is not clear where the enzyme was localized. The observation that levan may be synthesized in the endomembrane system sets an interesting precedent for the organization of endogenous fructan polymerization, since there are indications of the microsomal localization of this process.[8,38]

H. ABERRANT DEVELOPMENT OF PLANTS TRANSFORMED WITH BACTERIAL LEVANSUCRASE

As already noted, in some studies it proved impossible to recover levan transformants using an otherwise successful transformation system, indicating that the expression of levansucrase can be lethal.[3] In other reports of successfully recovered transformants, levan accumulated but phenotypic problems were not noted. Some of these studies used low irradiance[16] and it is possible that the resulting low assimilate concentrations and in turn, low levan accumulation rate, precluded the attainment of concentrations sufficient to cause developmental problems. In subsequent reports, a number of developmental aberrations, some quite dramatic, were associated with the expression of levansucrase activity; as a general observation, higher concentrations of levan correlated with more severe phenotypic problems.

At the whole plant level, above-ground stature was often reported to be diminished and root size reduced relative to controls. Leaves exhibited bleaching or necrosis and in some instances specific subcellular aberrations such as chloroplast agglutination. Effects on leaf morphology occurred in older tissues, which contained relatively higher levan concentrations, and a gradient of severity of the effect was clearly visible from lower (older) to higher (younger) leaves.[11] In some instances, the occurrence of the aberrations correlated with specific developmental phases, for example, the onset of tuberization in potato plants[12] and the switch to reproductive development in ryegrass.[19] Leaf tissues accumulating levan generally exhibited lowered starch concentrations relative to controls, and in one instance diurnal turnover of starch and sucrose was reported to be abolished.[14] Elsewhere, decreased translocation of photoassimilates from leaves and an associated increase in both starch and hexose were reported.[15] Clover plants exhibited a 60% reduction in photosynthesis, lower leaf carbohydrate contents overall, and reduced growth rates.[21]

Because of its potential as a conveniently harvested source of industrial products, the potato tuber has been of particular interest as a vehicle for levan transformation. Transgenic tubers also exhibit aberrant phenotypes. In general, transformants produced a total yield lower by up to 80 to 90%,[11] fewer and smaller tubers with reduced starch concentration, and in some cases, modified starch granule morphology. A brown cortex was a common phenotype in levan-accumulating potato tubers.[12,14] Under microtuber-inducing conditions, a major developmental shift caused the formation of florets instead of microtubers.[13]

The use of a seed-specific promoter[3] clearly showed that in maize transformed with levansucrase fused to a putative vacuolar-targeting sequence, seeds developed normally and accumulated fructan, whereas a putatively untargeted levansucrase caused devastating retardation of seed development to <10% of the dry mass of controls. The results were interpreted to show that levan accumulation in the cytosol is disruptive, while vacuolar localization sequesters levan into a space where its disruptive effects are minimized. However, the study is open to criticism because (a) actual localization was not demonstrated for either subcellular compartment; (b) in another system, the sporamin vacuolar targeting sequence is known to mislocalize levansucrase into the endomembrane system rather than the vacuole[15]; (c) the bacterial secretion sequence was not taken into account and it is possible that the untargeted (putatively cytosolic) gene product was actually localized elsewhere, such as on the cell perimeter, as shown in References 12 and 14. Regardless of the actual location, the study does provide an indication that different targeting sequences modulate the detrimental effects of levansucrase transformation.

I. MECHANISM OF LEVANSUCRASE-MEDIATED DEVELOPMENTAL EFFECTS

The mechanism by which tissues are damaged by transgenic levan accumulation is by no means clear. However, a useful initial observation is that levansucrase has been used as a (negative-) selectable marker in bacterial transformation. Levansucrase contains an internal restriction site. When other genes of interest are inserted within the levansucrase structural gene, its function is abolished. In instances where insertion does not disrupt the structural gene, the activity is expressed, killing the cell. The expression of a large levan polymer in the cytoplasm presumably disrupts cellular spatial organization so drastically that the cell dies, thereby offering a potential explanation for the effects observed in plants. Additionally, expression of levan appears to result in the crosslinking of membranes (discussed above), which could also be disruptive of cellular function and explain aberrant or lethal phenotypes in the transformants. Vacuolar localization may negate these effects and explain the survival of transformants so targeted. Inadvertent targeting to the cell perimeter or sequestration within the endomembrane could also prevent direct effects on cytoplasmic organization and reduce toxicity.

Based on the water solubility of commercially available levan of ~1% (w/v), *in situ* levan precipitation has been proposed as a mechanism, for example, to explain the browning of potato tuber cortex.[14] This model has the virtue that it explains the threshold effects, where phenotypic aberration occurs only after continued accumulation. However, ryegrass plants reported to contain a maximum of 0.04 mg/g (ca. 0.005%, w/v) levan exhibited adverse phenotypic effects,[19] whereas potato plants reported to contain 160 mg/g (ca.20%, w/v) levan had no reported aberrant phenotype.[18] It would not appear that levan concentration or precipitation *per se* explains all of the observations.

The necrotic lesions formed in some levan transgenics are reminiscent of the hypersensitive response of plants to pathogen infection. Some levan-producing bacteria belong to taxa such as *Erwinia*, which include plant pathogens. It is possible that levan, levansucrase, or the message, are recognized by plants as a warning of pathogenesis. Subsequent triggering of the hypersensitive response in transformants could explain the necrosis. When levan was supplied to cut petioles in potato, necrosis was not induced. Injecting the levansucrase caused some necrosis but the results were inconclusive since controls from bacteria without levansucrase also produced necrosis.[12] It would appear that exogenous levan or levansucrase alone do not trigger necrosis and that occurrence within the cell is necessary, at least in potato.

Because levan is a soluble polymer, its detrimental effects have been attributed to its osmotic contribution,[3] although no analyses of tissue osmotic adjustment were reported. However, assuming the transgenic levan molecule to be as large as is generally thought, the molar concentration of the polymer would be negligible compared with an equivalent mass of sucrose or hexose, especially in view of the low mass generally accumulated. The osmotic contribution of levan and its physiological effect should, in theory, be minimal.

A further suggestion to explain developmental effects is metabolic disruption of sucrose metabolism caused by the diversion of carbon flow from primary metabolism.[3] Given the estimates of flux into levan of between 0.0002 and 0.08% of primary carbon metabolism (above), which are minute compared with natural hourly and daily variations in carbon flux,[10] it is unlikely that this minor leakage into levan *per se* had any significant detrimental impact on the steady-state sucrose pool or fluxes of photosynthate.

A number of reports found a negative correlation between levan and starch concentrations. The latter were generally depressed when compared with wild-type or empty-vector controls. For example, a greater than ten-fold reduction in starch accumulation was observed in potato.[18] It was hypothesized that the mechanism for the reduced starch concentration was efficient competition for carbon by levansucrase compared with starch and a resulting diversion of the flow of photoassimilate. As discussed above, the low rates of flux of carbon into levan synthesis make metabolic competition an unlikely primary mechanism for reduced starch accumulation and imply that secondary factors are more influential.

J. BENEFICIAL PHENOTYPES RESULTING FROM LEVAN ACCUMULATION

One of the rationales for levan transgenesis is rooted in the suggestion that fructan accumulation confers stress resistance on plants, particularly to drought. For example, tobacco plants expressing levan accumulation are reported to show improved performance (determined as growth rate, fresh mass, root size, and dry mass), correlating with increased levan content under polyethylene glycol-mediated osmotic stress.[16] Physiological aspects of these experiments are discussed elsewhere.[4] Briefly, the results are ambiguous because plants were grown at the light compensation point, and substrate concentration and rate of accumulation were low. Endpoint concentrations of levan were also low and were sensitive to substrate concentration. Both sucrose and levan varied between treatments and correlated with the deduced stress tolerance. It was not possible to distinguish the levan accumulation as a direct or indirect effect of the stress treatment. Similar experiments were reported for sugar beet[17] with similar results. Although parameters such as biomass and leaf area were quantified for the plants in the beet drought trial, levan concentration was not.[17]

Similarly, conventional measurements of osmotic adjustment[39] have not been reported for any of these transformants. It is difficult to assess mechanistically, the involvement of levan in the claimed drought resistance in the absence of such measurements.[40]

III. PLANTS TRANSFORMED WITH PLANT FRUCTOSYLTRANSFERASE GENES

A. OVERVIEW

The enzymology of plant fructan synthesis is more complex than bacterial levan synthesis. Plant fructans exhibit a wide range of species-specific linkage structures and size distributions.[32] Multiple enzymes appear to be involved in polymerization; the enzymes generally catalyze more than one fructosyltransferase or hydrolase reaction and, hence, the interpretation of enzymological data can be rendered ambiguous.[9] Partly for these reasons, plant enzymes were isolated, sequenced, and used for transformation later than their bacterial counterparts. The reasons for the production of plant–plant fructan transformants have mainly been to (i) confirm an established two-enzyme model for fructan biosynthesis (see e.g. Reference 24, 27 and 40), and (ii) create crops of biotechnological utility (see e.g. Reference 23). This section examines the results of these transformations and begins with an assessment of the primary technique used for their analysis.

B. HIGH-PERFORMANCE ANION-EXCHANGE CHROMATOGRAPHY-PULSED AMPEROMETRIC DETECTION

Fructans in higher plants are smaller than bacterial levan, are more amenable to chromatographic analyses, and unlike levan, can be fractionated by TLC and SEC. In addition, in the last 15 years, high-performance anion-exchange chromatography coupled with pulsed amperometric detection (HPAEC-PAD; Dionex Corporation) has revolutionized the study of plant fructan polymers and has become a standard technique. Before considering the results obtained for transgenic plants in which HPAEC-PAD has been used, it is instructive to first consider its strengths and weaknesses.

The Carbopac columns have exquisitely high resolution, giving separations of individual oligo- and polysaccharides differing in size by only one hexose moiety (162 Da). The technique can reproducibly resolve members of a homologous series and distinguish structural isomers differing in chain linkage type and branching. Manipulation of elution conditions permits the resolution of glycans in the range of DP between 1 and ~100. The PAD data capture system has a wide sensitivity and attenuation range, and quantities of polymer in the low-nanogram range can be clearly detected with a high signal-to-noise ratio (illustrated in Reference 10). The quantitative response of the PAD is different for each individual carbohydrate species, and the detector response factor decreases with increase in DP. Hence, in the absence of calibrated responses for each individually detected compound, the method is at best semiquantitative. The units of the output from the PAD are the coulomb or the ampere, although PAD data are sometimes reported in arbitrary units, often without any indication of the total mass of carbohydrate or tissue fresh mass equivalent injected, or the inclusion of standards. Consequently, reports of HPAEC-PAD data can be difficult to interpret quantitatively.

High sensitivity *per se* can be problematic. Traces of carbohydrate can produce apparently large signals, which could appear to an uncritical reader to be quantitatively very impressive. A further problem is that masses of carbohydrate in the low-nanogram range, easily detected and isolated by HPAEC-PAD, can be insufficient for subsequent analysis by other confirmatory methods. For example, TLC and structural determination by glycosyl linkage analysis or GC-MS require ~1 to 10 μg minimum, whereas NMR spectroscopy requires masses in the low-mg range. In the absence of confirmatory data, we are often totally dependent on HPAEC-PAD as the basis for our conclusions. Because of the unarguable power of HPAEC-PAD, we risk being seduced into regarding it as an absolute determination of fructan, which it is not. Although powerful, it is a relative technique, and the data should be interpreted with appropriate caution. The demerits of the technique are especially pertinent when considering reports of plant–plant transgenesis, where rates of fructan product accumulation and instantaneous concentrations, both *in planta* and *in vitro,* are generally low.

C. QUANTITATIVE ASPECTS OF NONFRUCTAN PLANTS TRANSFORMED WITH PLANT FRUCTOSYLTRANSFERASES

Table 11.2 summarizes quantitative parameters of relevance to reserve accumulation in plant–plant transformants. As for the bacterial transformants, more than half of the studies did not report conditions of irradiance. Two of the transformants were apparently grown close to the light compensation point for C3 plants, so the observations set out above may apply. The instantaneous concentrations of transgenic fructan were generally below 9 mg/g, lower than the maxima found for endogenous reserve accumulation (60–160 mg/g). The exception was roots of sugar beet transformed with sucrose:sucrose fructosyltransferase (SST) from *H.tuberosus*, where 62 mg/g of oligofructan of DP=3–5 accumulated. This concentration parallels the maximal natural concentrations of starch and fructan in photosynthetic tissue,[5] although in storage roots, for example, of *H. tuberosus,* natural fructan content can reach 160 mg/g.[30] Similar to levan accumulation, when rates of accumulation were estimated, values were low, in the range 0.00002 to 0.025 nkat/g. This range is equivalent to 0.00005 to 0.07% of the primary flux calculated for the NFA. The maximal value was exceptional; all others were below 0.005 nkat/g (0.014% of flux).

D. EXPLAINING THE LOW RATES OF PRODUCT ACCUMULATION IN THE PLANT–PLANT TRANSFORMANTS

With the exception of chicory and onion, the host plants were chosen because they lacked endogenous fructan metabolism. As reasoned for levan transformants, it was assumed that fructan hydrolase was absent in these plants and that product hydrolysis did not provide a general explanation for the observed low rates of accumulation.

However, invertase is ubiquitous in higher plants. Small oligofructans (DP = ca.3–7) are degraded by invertases[9] and can hydrolyze transgenic fructan products. It is pertinent that acid invertases are thought to be naturally present in the vacuole, the putative site of natural and much of the transgenic fructan accumulation. Fructan hydrolysis certainly occurs in some instances where enzyme preparations from transformants and controls have been tested against oligofructan.[26] From the available data it is not clear if this is a problem *in planta*.

Unlike the levan transformants, accumulation rates *in planta* are likely to be adversely influenced by the kinetics of the plant-derived synthetic enzymes because of the universally low substrate affinity of plant fructosyltransferases.[10] Table 11.2 lists the approximate substrate affinities of the plant fructosyltransferases that were used for plant transformation. All values were 280 mM or higher. These affinities are compared with estimates of substrate concentration in the transformants, all of which were 38 mM or lower. Of the ten values, eight concentrations were $\leq$22 mM, i.e.,$<$8% of K_m. The enzymes would have been severely substrate-limited in the transformants, the rate of fructan synthesis would have been sensitive to changes in substrate concentration, and the limitation would have significantly contributed to the observed low rates of product accumulation.

Similar to the levan transformants, other features of expression such as detection of enzyme activity, immunological detection of enzyme protein, and detection of expressed message were generally low or nonexistent in the plant–plant transformants. Hence, absolute concentrations of gene products and total enzymatic activity were also low. In combination with substrate limitation of synthesis and the possibility for invertase-mediated hydrolysis of product, this strategy provided circumstances favoring extremely low product accumulation rates.

E. TRANSFORMATION OF PLANTS WITH FRUCTOSYLTRANSFERASES FROM *HELIANTHUS TUBEROSUS*

The most pervasive model for fructan synthesis in plants is based on studies of *H. tuberosus,* and involves two monofunctional enzymes: (a) SST, proposed to generate trisaccharide (only), from sucrose and (b) fructan:fructan fructosyltransferase (FFT), which by disproportionation, is proposed to elongate and shorten polymers, using trisaccharide as the entry point for fructose. The corresponding enzymes were purified, the genes were obtained by RT-PCR, and subsequently transformed into petunia.[26,41] Sequence analysis indicated the presence of signal sequences in both genes with the implication of vacuolar targeting and localization for the gene products. SST and FFT were separately introduced into petunia and later combined into the same plant by crossing. Significantly, when analyzed with the sensitive HPAEC–PAD as the sole analytical method, an equivalent signal corresponding to trisaccharide was detected in both the SST transformant and the untransformed control (see Figures 11.5a,b and 11.7 in Reference 26). Traces of material coeluting with inulin of DP = 4 and 5 in the transformant were the primary differences between the material and the control. By definition, these are not products of SST. Strictly interpreted, the petunia SST transformant did not provide evidence for transformation with SST, but rather indicated an FFT-or sucrase-type activity acting upon endogenous trisaccharide and sucrose.

Transformation with the FFT was clearly demonstrated *in vitro* since enzyme extracts formed traces of product of up to DP = 8 by disproportionation from tetrasaccharide. Only enzyme extracts and not tissue sugars from transformants were assessed for FFT expression because of the authors' expectation that the enzyme could not function in the absence of SST. Given the apparent presence of trisaccharide in the untransformed plants, it would have been interesting to see what fructan products accumulated *in planta* in the FFT-only transformant.

Petunia plants containing both the SST and FFT constructs did not contain fructan in any tissues, except for trace quantities in yellow senescent leaves. No explanation for this pattern of expression (of constitutively controlled genes) was offered. The fructan detected by sensitive HPAEC-PAD in plant extracts was concentrated ten-fold relative to the samples from the SST transformants; hence, instantaneous product concentration in the tissue must have been extremely low. The discontinuous size distribution pattern of the transgenic fructan products was interesting. There was a general reduction in abundance of products with size in the range DP = 3 to 8, followed by an abrupt increase in abundance at DP = 9 to 10, which was in turn followed by a smooth pattern of size decrease to ca. DP = 40. This overall pattern was quite distinct from the chicory–inulin standard, which exhibited a smooth decay over the whole size range of DP = 3 to 40. The authors offer a number of plausible scenarios to explain the pattern based on invertase hydrolysis of the smaller products and differences in the balance of expression of SST and FFT activities. An explanation not considered is as follows: the pattern from DP = 9 to 40 in the transformant is strongly reminiscent of the water-soluble products of starch hydrolysis during debranching.[33,42] It is conceivable that the pattern of sugars reported for the transformant could be explained as a combination of the trisaccharide apparently occurring naturally in petunia, oligosaccharides resulting from "SST" transformation (DP = 4 and 5) and by starch mobilization in senescing leaves. Since HPAEC data were the sole evidence presented, we are unable to distinguish these possibilities and decide whether the evidence actually represents polyfructan synthesis and hence a role for FFT *in planta*, in the petunia transformants. Since it is also difficult to decide if SST (trisaccharide synthesis) was actually expressed in either the single- or double transformant and whether the evidence for fructan polymerization is open to alternative interpretation, their view that they had unambiguously cloned the pathway of fructan biosynthesis in Jerusalem artichoke may have been premature.

The *H.tuberosus* SST structural gene was subsequently used to transform sugar beet.[23] From the perspective of enzyme kinetics, the sugar beet root vacuole is a highly appropriate environment for a low-affinity plant fructosyltransferase, because sucrose concentrations may reach $500\,mM$.[43] The transformant was convincingly shown to accumulate small fructans to high instantaneous concentration, although at low rate (Table 11.2). Unlike the petunia transformation, this demonstrated SST activity (i.e., trisaccharide synthesis), but since roughly half of the mass of the product was present as oligofructan of DP = 4 and 5, the enzyme was not monofunctional, exhibiting additional transferase activity. Although no data were shown, independent enzyme incubations with substrates of DP = 3, 4, and 5 produced higher homologues by disproportionation, confirming that the SST had FFT activity. Enzyme extracts from wild-type controls were stated to contain invertase activity and to hydrolyze small fructans (data not shown). Unfortunately, rates of enzymatic hydrolysis were not presented, so their potential quantitative contribution to net fructan accumulation *in planta* could not be assessed. Invertase was implicated as the primary cause of low fructan concentrations in leaves, even though the SST activity could not be detected in protein extracts of this tissue. Invertase may certainly have had an effect, but low expression and low sucrose concentration were equally likely to have caused low fructan accumulation in leaves. No phenotypic aberrations were reported for beet, but the tubers shown were described as "young" and were grown for 2 months in the glasshouse, in winter (R. Sevenier, personal communication). Since the primary aim of the work was ostensibly the industrial production of oligofructan as a sweetener, results from mature beet grown under agricultural conditions (presumably in the summer) would have been more informative.

F. TRANSFORMATION OF PLANTS WITH FRUCTOSYLTRANSFERASES FROM *CYNARA SCOLYMUS*

One consequence of modern molecular biology is that genes can be isolated from a given species and transformed into another without any necessity for the *in vitro* understanding of the enzymatic properties of the gene product. However, this methodology can lead to interpretive difficulties because the functional and sequence distinctions between invertases and plant fructosyltransferases are blurred. An example is the work of Hellwege and co-workers,[24,27,40] where a *C. scolymus* cDNA library from blossom discs was screened with an RT-PCR product for 6-SFT obtained from barley leaves. Positive clones were sequenced, and a gene (Cy21) with homology to both β-fructosyl hydrolases and fructosyltransferases was subsequently transformed into nonfructan plants. Cy21 was designated as encoding an SST and was subsequently used to isolate an FFT. There are no published kinetic data for the gene products or direct indication of their substrate and product specificities, responses to inhibitors, and other functions. The working assumption is that the model for fructan synthesis proposed for *H. tuberosus* holds for *C. scolymus* and that the enzymes have similar properties. The difficulty with this gene isolation procedure is that 6-SFT (used as the primary probe) is predominantly an invertase (80% of the total activity against sucrose discussed below), and catalyzes, in addition, the 6-transfer of fructose to sucrose, and to a range of acceptor inulins, and also 6-polymerization. Conversely, SST supposedly catalyzes 1-transfer to sucrose only and has no hydrolytic activity, yet the gene for the one apparently identifies the other. Further, the gene for the supposedly specific SST was subsequently used to isolate the (differentially) specific FFT.

In their discussion, Hellwege and co-workers[40] interpreted their results to "*exclude the possibility*" that Cy21 encoded invertase activity. However, when the Cy21 gene product was transiently expressed in tobacco protoplasts, lysates incubated for 20 h with sucrose in the presence and absence of an invertase inhibitor (pyridoxal-HCl) produced trace amounts of fructan trisaccharide in addition to free fructose. It was not possible from the data to assign the source of the fructose to the endogenous tobacco invertase, the Cy21 gene product, or both. When expressed in potato tubers, some Cy21 lines accumulated free fructose in the tissue while other transformants and the wild-type control did not. These results could be interpreted as evidence for the introduction of invertase or fructan hydrolase activity encoded by Cy21, at least in some lines. Hence, on closer inspection, neither the studies *in vitro* nor *in planta* exclude the possibility that the Cy21 gene product has hydrolytic activity, as claimed.

When stably transformed into potato, Cy21 also resulted in the accumulation of trisaccharide and oligomers of DP up to 7. In common with SST from *H. tuberosus* expressed in sugar beet, the gene product had substantial additional FFT or fructan–sucrase activity. Hence, by definition, the Cy21 gene product is not an SST. Given the mixed hydrolase and transferase activities of 6-SFT used as the probe, it is perhaps not surprising that the Cy21 gene product should also exhibit these functions.

C. scolymus FFT was transiently expressed in tobacco protoplasts. Lysates were incubated with substrates of DP = 3 to 7 for 4 days at 22°C. As with the SST, the activity was extremely low and only traces of the products of disproportionation up to DP = 20 could be detected by sensitive HPAEC-PAD.[27] The experiment was repeated using an FFT gene isolated from *H. tuberosus*. The results were similar except that the products reached only DP = 12. This result was used as evidence that the product specificity of the two enzymes was different and that it explained the natural product size distribution *in vivo*: larger in *C. scolymus* and smaller in *H. tuberosus*. However, this conclusion will require further substantiation. Since the total FFT activity in both lysates was extremely low and since no quantification of product was reported, rates of fructosyltransfer could not be calculated or normalized. The latter is important, as the authors point out, since it has been shown that maximal fructan size is related to enzyme concentration.[8,44,45]

Uncalibrated SEC was used in an attempt to show that fructan of up to DP = 200 was present in the double-transformant potato tubers; however, the evidence for these large polymers was lacking: SEC was standardized using a soluble extract of roots of *C. scolymus* and Reference 46 was cited as the authority for a maximum native polymer size of DP = 200. The similar SEC profiles obtained for the roots of *C. scolymus* and for tubers of the transformant was taken as proof that the two contained the same fructan size distribution, and by implication, fructan of DP up to 200. This interpretation is misleading for two reasons: first, Praznik and Beck[46] characterized extracts of *blossom discs*, unlike Hellwege and co-workers,[24] who used *roots* of *C. scolymus*. Because the fructans of the roots have not (apparently) been characterized, it has not been shown that fructans of DP up to 200 were present in the material used as the standard. Secondly, the blossom disc fructan exhibited a steep and continuous increase in the abundance of chains in the range DP = ca. 15–90 (Figure 11.5 in Reference 46). If the root sample used as a standard by Hellwege and co-workers[24] was representative of the blossom fructan, this marked increase should have been apparent when the root sample was analyzed by HPAEC. However this was not the case even though fructans in the range DP = 15 to 90 can be resolved by HPAEC.[8,10,24] It is concluded that the root fructan used as the standard was not representative of blossom fructan characterized by Praznik and Beck[46], and the large fructan content of transformed roots was not established. In the absence of adequate standardization or size calibration for SEC there is no compelling evidence that the large transgenic polymers were synthesized as claimed.

G. TRANSFORMATION OF NONFRUCTAN PLANTS WITH 6-SUCROSE FRUCTOSYLTRANSFERASE FROM BARLEY

Fructan biosynthesis in grass leaves is inducible in response to sucrose accumulation.[5] The parallel induction of a fructosyltransferase led to the identification and purification of 6-SFT enzyme from barley leaves.[48,49] With sucrose as sole substrate, roughly 80% of its activity is hydrolytic (i.e., invertase) and ca. 20% is fructosyltransferase, catalyzing the formation of the trisaccharide, 6-kestose. The apparent K_m for of the invertase activity of sucrose is ca. <10 mM, and for trisaccharide synthesis is >300 mM. 6-SFT has a third activity *in vitro*: it transfers fructose from sucrose to the 6-carbon of fructosyl residues of inulin oligosaccharides, i.e., it catalyzes 6-branching of small 2,1-linked primers. As a consequence of these properties *in vitro*, transforming 6-SFT into a nonfructan plant where sucrose would be the sole substrate, we would expect 6-kestose accumulation. Transformation into an inulin accumulator such as chicory should result in primer glycosylation to form branched oligofructans. Both transformations would also be expected to co-introduce invertase activity to the site of transgenic fructosyl transfer.

When 6-SFT was transiently expressed in tobacco,[50] incubations of intact protoplasts allowed detection of trisaccharide synthesis from sucrose and 6-branch glycosylation of inulin trisaccharide. High background invertase in the protoplasts precluded determination of transgenic sucrose hydrolysis (N. Sprenger, personal communication). The enzyme was subsequently stably expressed at low rates in tobacco. This species contained sucrose as sole substrate and, as predicted, contained traces of 6-kestose.[25] Endpoint concentrations were low, but were higher in roots than in leaves and positively correlated with sucrose concentration. Yellow senescing leaves contained the highest product concentrations and highest sucrose concentrations of the leaves sampled. Analyses of the products of roots by HPAEC revealed, as predicted, 6-kestose, but also traces of two branched tetrasaccharides and a series of higher oligomers up to DP = 12. Hence, *in planta*, 6-SFT clearly has a fourth activity that could be either 6-FFT or a 6-fructan sucrase (6-polymerase). In addition, one of the branched tetrasaccharide products was identified as bifurcose. This contains a 2,1 glycosidic linkage that 6-SFT does not form *in vitro*. The enzyme appears to have fifth activity *in planta*, that of 1-SST.

H. TRANSFORMATION OF INULIN-ACCUMULATING PLANTS WITH BARLEY 6-SUCROSE FRUCTOSYLTRANSFERASE

Chicory naturally accumulates 2,1-linked inulin of DP $\leq$ 40 in roots. In 6-SFT transformants of chicory, no transgenic fructan products occurred in the roots.[25] Sprenger and co-workers[25] demonstrated that the formation of traces of inulins up to DP $\leq$ 16 can be induced in excised, illuminated leaves of wild-type chicory. When chicory leaves expressing the 6-SFT gene were similarly induced, additional trace shoulders occurred on the HPAEC peaks of the natural inulins of DP = 4 to 7. This result was indicative of limited 6-branch fructosylation (primer glycosylation) of the endogenous oligosaccharides. The gene was under the control of a constitutive promoter, but no explanation was offered for the differential expression in roots and leaves.

Based on sequence homologies between fructosyltransferases, 6-SFT was used to isolate a 6^G-fructosyltransferase gene from onion.[51] Oligofructans of the onion family naturally contain 6-fructosylated glucose residues, collectively termed the "neoseries" after the parent trisaccharide "neokestose." When transformed into chicory, the onion gene resulted in the formation of trace quantities of neoseries fructans of up to DP = 7 by primer glycosylation of the natural linear inulins. This is another example of a probe with one transferase specificity identifying a gene for another. It is not clear if the onion enzyme exhibited invertase activity.

I. HAS SUBCELLULAR LOCALIZATION OF PLANT TRANSGENE PRODUCTS BEEN ESTABLISHED?

Fructosyltransferase genes from plants are reported to contain naturally putative signal and targeting sequences. Hence, heterologous targeting sequences have not been included in vectors constructed for plant–plant transgenesis. It has been generally assumed that vacuolar targeting of transgenes took place because fructan synthesis occurred in transformants. However, in only one instance has vacuolar localization been demonstrated, i.e., for onion 6^G-FFT in tobacco, where 70 to 100% of the activity in protoplasts was reported to be found in isolated vacuoles.[51] Indirect evidence for vacuolar localization *in planta* was adduced from the studies of transformed chicory because of the observed primer glycosylation of endogenous inulins. The logic of the argument is as follows: because natural inulin synthesis is thought to be vacuolar and branched inulins were synthesized, natural synthesis and transgenic primer glycosylation must have been colocalized in the vacuole. The possibilities that inulin synthesis or inulin synthesis plus transgenic primer glycosylation could have taken place in the endomembrane system prior to entering the vacuole[8,15,38] were not considered. Endomembrane-associated synthesis in transit to the vacuole would not affect results showing vacuolar localization of fructan or fructosyltransferases and conditions permitting levan synthesis seem to exist in the endomembrane system of tobacco.[15] It is worth noting that (i) the direct evidence for vacuolar localization of natural fructan synthesis rests on only a few studies,[47,52–54] (ii) with the exception of the study of Darwen and John,[54] the protoplasting enzymes used contained fructosyltransferases, which may confound the interpretation of data,[55] and (iii) the evidence for vacuolar localization is for the synthesis of only the smallest fructans, and not the full size range of natural polymers. Recent evidence based on the requirements for high enzyme and substrate concentrations to enable polymerization of fructan indicates that the vacuole may not be an ideal compartment for fructan polymerization.[8,45]

J. PHENOTYPIC ABERRATIONS ARE NOT REPORTED FOR PLANT–PLANT TRANSGENICS

Unlike the bacterial transformants, no aberrant developmental effects have been reported in transgenics containing plant-derived fructosyltransferases. In the case of beet tap root, very high endpoint concentrations of oligofructan were obtained, and the lack of phenotypic effects was attributed to vacuolar localization (although this was not demonstrated).[23] Equally, the lack of phenotypic effects could be a function of the small size and chemical and structural- similarity of the transgenic oligosaccharides to sucrose. Further, the plants were apparently grown at low irradiance and were harvested young, as were the potato tubers in the study of Hellwege and co-workers.[27] Developmental problems in the bacterial transformants correlated with older tissues and longer accumulation periods. All other plant–plant transformants contained only low concentrations of transgenic fructan and the lack of phenotypic effects may simply reflect this fact. An additional possibility is that bacterial levan exhibits a specific phytotoxicity that is not a property of plant fructan.

K.   DO RESULTS FROM PLANT–PLANT TRANSGENESIS ADVANCE OUR UNDERSTANDING OF NATURAL FRUCTAN SYNTHESIS?

Overall, the understanding acquired from plant–plant transgenesis is largely confirmatory of that obtained previously from enzymology. A general criticism of work in plant–plant transgenesis is that the experiments have been performed with the emphasis on confirming, rather than testing, the SST/FFT hypothesis. Thus, some authors appear surprised that by synthesizing higher oligofructans *in planta*, sucrose-fructosyltransferases do not behave as expected, observations for which they cannot offer an explanation (see e.g. References 40 and 23). The answer is straightforward: the data do not fit the theory and the theory needs to be changed. Further, theory dictates that FFT cannot function in the absence of the product of SST. Hence, no attempt has yet been made to observe the products, *in planta*, of FFT-only transformants. Given the apparently anomalous behavior of sucrose-fructosyltransferases *in planta*, why not also FFT? Besides, the experiment is of intrinsic value since it tests the hypothesis: a negative result would strengthen evidence for a monofunctional FFT.

The results of transgenesis with 6-SFT confirmed two of the activities observed *in vitro*, namely, 6-kestose synthesis and 6-primer fructosylation. However, transgenesis provided no information about the expression and effects of its major activity, invertase. This subject is of interest because under the conditions where sucrose is the sole or predominant substrate (as during the initiation of natural fructan synthesis *in vivo*), hydrolysis predominates. Under these conditions, total endogenous invertase activity can be similar to the rate of sucrose synthesis, although it is not clear how sucrose is able to accumulate in the vacuole if, as is generally accepted, these hydrolytic enzymes are also vacuolar. The invertase activity of 6-SFT could be an additional factor explaining the low rates of expression in transgenics.

Two additional activities of 6-SFT, namely, 6-polymerase (sucrase or 6-FFT) and 1-SST, were also manifested *in planta*. The transgenic results confirm *in vitro* data and extend the view of 6-SFT as a multifunctional enzyme with a wide substrate and product specificity. Although the physiological studies of the transformants do not generally provide information about the hydrolytic functions of the fructosyl-transferases, sequence data certainly show universally agreed similarities with invertases. We speculate on how the sequences would have been identified and named, had the genes been isolated without the preconception that they were involved in fructan synthesis. In view of the continued inability to distinguish 6-SFT from invertase and consideration of the quantitatively minor proportion of its activity which is synthetic *in vitro*, it still seems pertinent to question if some plant fructosyltransferase polypeptides function primarily as invertases. There is no doubt that the 6-SFT transgene product catalyzes trisaccharide synthesis and primer fructosylation, but equally, so do other invertases *in vitro*. Neither observation rules out the possibility that the enzyme involved is an invertase or provides proof of its role in natural fructan synthesis. In this connection, no studies of antisense repression or sense co-suppression have yet been reported for any plant fructosyltransferase.

L. CELLULAR EFFECTS ON TRANSGENE EXPRESSION

The interaction between essentially hydrolytic enzymes and the environment in which they function may be a significant factor in determining the extent of fructan polymerization. From the authors' perspective, the most exciting observations from fructan transgenics concern the differential patterns of product formation observed *in vitro* and *in planta*. As alluded to above, SST enzymes from *H. tuberosus* and *C. scolymus* synthesize larger fructans *in planta* than *in vitro*. The 6-SFT, which generates only trisaccharide from sucrose *in vitro*, catalyzes the formation of 2,6-linked fructans of up to DP = 12 from sucrose, i.e., it acts as a polymerase in transgenic potato tubers. Finally, a mutant bacterial levansucrase that synthesizes only trisaccharide *in vitro*, forms a polymer of apparently high molecular mass when expressed *in planta*.[3,56] Taken together, these observations indicate that the fructosyltransferases have the capacity for polymerization and that expression in intracellular conditions favor the expression of this activity. The observations *in planta* parallel with the effects of enzyme concentration *in vitro*, and it is well documented that conditions of reaction affect the qualitative function of fructosyltransferases.[44,45,57] In the authors' view, the most convincing demonstrations of enzymatic fructan polymerization require high enzyme and substrate concentration and the mechanism *in vitro* may be a function of (i) increased rate of intermediate synthesis, (ii) closer spatial proximity of enzyme, substrate and intermediates, and (iii) reduced water availability, leading to reduced invertase activity of fructosyltransferases (i.e., reduced competition caused by fructosyl transfer to water).[8] In addition, it is known that the reduction in water content in enzyme reactions by increased solvent content also favors polymerization.[56] It is possible that similar physicochemical effects modify the activity of transgene products *in planta* resulting in reduced substrate hydrolysis and enhanced polymer synthesis. These effects may also explain the historical difficulty in establishing *in vitro* polymerization with (generally dilute) plant enzymes.

IV. CONCLUSIONS

Scrutiny from a physiological perspective reveals that the transgenics have resulted in a quantitatively minor modification of carbon flux. Transformants exhibited unpredictable additional secondary phenotypic effects, including yield reductions, nonstoichiometric alterations in reserve metabolism, and other adverse developmental effects. Studies of targeting and subcellular localization were, on the whole, inconclusive. The results were not simple to interpret, mainly because of the influence of uncontrolled secondary phenotypic effects. Hence, the general utility of these transgenics as robust models for the investigation of the physiology of carbon partitioning appears limited.

In terms of commercial possibilities, the general rates of synthesis and total accumulation of levan under field conditions will have to be improved in order to realize the goal of high-yield, high-fructose feedstocks from, for example, potato. In turn, this will require a more definitive understanding of levan-mediated phenotypic aberrations and total yield reductions as well as a more effective predetermined targeting of levan to cellular compartments which may contain these adverse effects. The larger plant fructans, such as those from *C. scolymus,* also have potential as a source of high-fructose feedstock that may avoid levan-associated aberrations, although a claim for the synthesis of such large plant fructans in transgenics remains to be substantiated. At the moment, the transgenics do not appear to provide any unqualified advantages over the NFA crops as sources of fructose.

In terms of advancing our understanding of natural fructan synthesis, data for plant–plant transformants have been largely confirmatory of earlier evidence from traditional physiology and enzymology. The results have not resolved the issue of invertase and fructosyltransferase co-identity since the two classes of enzymes are now known to exhibit sequence homology in addition to functional homology. The results have reemphasized that the sucrose-fructosyltransferases are multifunctional and do not conform to the simple SST-FFT model for fructan synthesis. A significant positive observation is that expression *in planta* enhances the polymerase function of sucrose-fructosyltransferases. This may have important consequences for the understanding of natural fructan biosynthesis.

Because of the ambiguity resulting from (i) the apparent incompatibility of the properties of plant fructosyltransferases with the physiology of endogenous fructan accumulators, (ii) the problem of invertase coidentity, and (iii) the consequences of these factors for polymerization, our understanding of endogenous, natural fructan polymerization remains incomplete. As in many other biological disciplines in the last decade, the emphasis in studies of fructan synthesis has shifted towards genes and transgenics and away from conventional physiology and enzymology. It is arguable that the study of fructan metabolism has, to a considerable degree, become the study of fructan genes. For the reasons outlined, the information gained from the transgenic plants has not greatly enhanced our understanding of natural fructan polymerization. It is our view that an integrated and appropriate investigation at all levels of bioloigical organization will be neccesary to elucidate the process fully. This will in turn facilitate effective genetic modification. Recent emphasis on genes and on artificial fructan accumulators produced by transgenesis may have served as a distraction from the consideration of other, equally important issues in natural fructan synthesis.

ACKNOWLEDGMENTS

We thank Dr. Zahid Latif (Molecular Nature Ltd., Plas Gogerddan, Aberystwyth, U.K.) for helpful discussions concerning the interpretation of data from nuclear magnetic resonance spectroscopy and to Dr. Arnt Heyer (Max Planck Institute for Molecular Plant Physiology, Germany) for his critical assessment of some parts of the text. This work was supported by a Post-Doctoral Research Grant (to S.J.P.) from the Agri-Food Committee of the BBSRC (U.K.).

ABBREVIATIONS

CaMV 35S, cauliflower mosaic virus 35S promoter; CPY, yeast carboxypeptidase Y; DP, degree of polymerisation; FFT, fructan:fructan fructosyltransferase (EC 2.4.1.100); GUS, β-glucuronidase reporter gene; HPAEC-PAD, high-performance anion-exchange chromatography-pulsed amperometric detection; NFA, natural fructan accumulator; PAR, photosynthetically active radiation; PFD, photosynthetic flux density; PI, photosynthetic integral; RT-PCR, reverse transcriptase polymerase chain reaction; SA, starch accumulator; SEC, size-exclusion chromatography; 6-SFT, sucrose:fructan 6-fructosyltransferase; SST, sucrose:sucrose fructosyltransferase (EC 2.4.1.99); NMR, nuclear magnetic resonance spectroscopy; TLC, thin-layer chromatography.

REFERENCES

1. Pollock, C.J. and Cairns, A.J., Fructan metabolism in grasses and cereals, *Ann. Rev. Plant Physiol. Plant. Mol. Biol.,* 42, 77–101, 1991.
2. Hendry, G.A.F. and Wallace, R.K., The origin, distribution and evolutionary significance of fructans, in Science and Technology of Fructans, Suzuki, M. and Chatterton, N.J., Eds., CRC Press, Boca Raton, 1993, pp. 119–139.
3. Caimi, P.G., McCole, L.M., Klein, T.M., and Kerr, P.S., Fructan accumulation and sucrose metabolism in transgenic maize endosperm expressing a *Bacillus amyloliquefaciens* sacB gene, *Plant Physiol.,* 110, 355–363, 1996.
4. Cairns, A.J., Pollock, C.J., Gallagher, J.A., and Harrison, J., Fructans: synthesis and regulation, in Advances Photosynthesis, Leegood, R.C., Sharkey, T.D., and von Caemmerer, S., Eds., Kluwer, The Netherlands vol. 9, 2000, pp. 301–318.
5. Cairns, A.J., Cookson, A., Thomas, B.J., and Turner, L.B., Starch metabolism in the fructan-grasses: patterns of starch accumulation in excised leaves of *Lolium temulentum* L., *J. Plant Physiol.,* 159, 293–305, 2002.
6. Vijn, I. and Smeekens, S.C.M., Fructan: more than a reserve carbohydrate? *Plant Physiol,* 120, 351–359, 1999.
7. Ritsema, T. and Smeekens, S.C.M., Fructans: beneficial for plants and humans, *Curr. Opin. Plant Biol.,* 6, 223–230, 2003.
8. Cairns, A.J., Nash, R., Machado De Carvalho, M-A., and Sims, I.M., Characterisation of the enzymatic polymerisation of 2,6-linked fructan by leaf extracts from Timothy grass *(Phleum pratense), New Phytol.,* 142, 79–91, 1999.
9. Cairns, A.J., Evidence for the *de novo* synthesis of fructan by enzymes from higher plants; a reappraisal of the SST/FFT model, *New Phytol.,* 123, 15–24, 1993.
10. Cairns, A.J., Fructan biosynthesis in transgenic plants, *J. Exp. Bot.,* 54, 549–567, 2003.
11. Gerrits, N., Turk, S.C.H.J., van Dun, K.P.M., Hulleman, S.H.D., Visser, R.G.F., Weisbeek, P.J., and Smeekens, S.C.M., Sucrose metabolism in plastids, *Plant Physiol.,* 125, 926–934, 2001.
12. Röber, M., Geider, K., Muller-Röber, B., and Willmitzer, L., Synthesis of fructans in tubers of transgenic starch-deficient potato plants does not result in an increased allocation of carbohydrates, *Planta,* 199, 528–536, 1996.

13. Caimi, P.G., McCole, L.M., Klein, T.M., and Hershey, H.P., Cytosolic expression of the *Bacillus amyloliquefaciens* SacB protein inhibits tissue development in transgenic tobacco and potato, *New Phytol.,* 136, 19–28, 1997.

14. Pilon-Smits, E.A.H., Ebskamp, M.J.M., Jeuken, M.J.W., van der Meer, I.M., Visser, R.G.F.V., Weisbeek, P.J., and Smeekens, S.C.M., Microbial fructan production in transgenic potato plants and tubers, *Industr. Crops. Prod.,* 5, 35–46, 1996.

15. Turk, S.C.H.J., de Roos, K., Scott, P.A., van Dun, K., Weisbeek, P., and Smeekens, S.C.M., The vacuolar sorting domain of sporamin transports GUS, but not levansucrase, to the plant vacuole, *New Phytol.* 136, 29–38,1997.

16. Pilon-Smits, E.A.H., Ebskamp, M.J.M., Paul, M.J., Jeuken, M.J.W., Weisbeek, P.J., and Smeekens, S.C.M., Improved performance of transgenic fructan-accumulating tobacco under drought stress, *Plant Physiol.,* 107, 125–130, 1995.

17. Pilon-Smits, E.A.H., Terry, N., Sears, T., and van Dun, K., Enhanced drought resistance in fructan-producing sugar beet, *Plant Physiol. Biochem.,* 37, 313–317, 1999.

18. van der Meer, I.M., Ebskamp, M.J.M., Visser, R.G.F., Weisbeek, P.J., and Smeekens, S.C.M., Fructan as a new carbohydrate sink in transgenic potato plants, *Plant Cell,* 6, 561–570, 1994.

19. Ye, X.D., Wu, X.L., Zhao, H., Frehner, M., Nösberger, J., Potrykus, I., and Spangenberg, G., Altered fructan accumulation in transgenic *Lolium multiflorum* plants expressing a *Bacillus subtilis* sacB gene, *Plant Cell.* Rep., 20, 205–212, 2001.

20. Ebskamp, M.J.M., van der Meer, I.M., Spronk, B.A., Weisbeek, P.J., and Smeekens, S.C.M., Accumulation of fructose polymers in transgenic tobacco, *Bio/Technol.,* 12, 272–275, 1994.

21. Jenkins, C.L.D., Snow, A.J., Simpson, R.J., Higgins, T.J., Jacques, N.A., Pritchard, J., Gibson, J., and Larkin, P.J., Fructan formation in transgenic white clover expressing a fructosyltransferase from *Streptococcus salivarius, Funct. Plant. Biol.,* 29 1287–1298, 2002.

22. McCree, K.J., Test of current definitions of photosynthetically active radiation against leaf photosynthetic data, *Agric. Meteorol.,* 10, 443–453, 1972.

23. Sévenier, R., Hall, R.D., van der Meer, I.M., Hakkert, H.J.C., van Tunen, A.J., and Koops, A.J., High level fructan accumulation in a transgenic sugar beet, *Nat. Biotechnol.,* 16, 843–846, 1998

24. Hellwege, E.M., Czapla, S., Jahnke, A., Willmitzer, L., and Heyer, A.G., Transgenic potato (*Solanum tuberosum*) tubers synthesize the full spectrum of inulin molecules naturally occurring in globe artichoke *(Cynara scolymus)* roots, *Proc. Natl. Acad. Sci. USA,* 97, 8699–8704, 2000.

25. Sprenger, N., Schellenbaum, L., van Dun, K., Boller, T., and Wiemken, A., Fructan synthesis in transgenic tobacco and chicory plants expressing barley sucrose:fructan 6-fructosyltransferase, *FEBS Lett.,* 400, 355–358, 1997.

26. van der Meer, I.M., Koops, A.J., Hakkert, J.C., and van Tunen, A.J., Cloning of the fructan biosynthesis pathway of Jerusalem artichoke, *Plant J.,* 15, 489–500, 1998.

27. Hellwege, E.M., Raap, M., Gritscher, D., Willmitzer, L., and Heyer, A.G., Differences in chain length distribution of inulin from *Cynara scolymus and Helianthus tuberosus* are reflected in a transient plant expression system using the respective 1-FFT cDNAs, *FEBS Lett.,* 427, 25–28, 1998.

28. Milthorpe, F.L. and Moorby, J., *An Introduction to Crop Physiology,* Cambridge University Press, Cambridge, 1974.

29. Jones, T.W.A., Use of a flowering mutant to investigate changes in carbohydrates during floral transition in red clover, *J. Exp. Bot.,* 41, 1013–1019, 1990.

30. Schubert, S. and Fuerle, R., Fructan storage in tubers of Jerusalem artichoke: characterization of sink strength, *New Phytol.,* 136, 115–122, 1997.

31. Dedonder, R., Levansucrase from *Bacillus subtilis, Meths. Enzymol.,* 8, 500–505, 1966.

32. Cairns, A.J. and Ashton, J.E., Species-dependent patterns of fructan synthesis by enzymes from excised leaves of oat, wheat, barley and timothy, *New Phytol.,* 124, 381–388, 1993.

33. Cairns, A.J., Begley, P., and Sims, I.M., The structure of starch from seeds and leaves of the fructan-accumulating ryegrass, *Lolium temulentum* L., *J. Plant Physiol.,* 159, 221–230, 2002.

34. Fishman, M.L., Rodriguez, L., and Chau, H.K., Molecular masses and sizes of starches by high performance size-exclusion chromatography with on line multi-angle laser light scattering detection, *J. Agric. Food Chem.,* 44, 3182–3188, 1996.

35. Turner, L.B., Humphreys, M.O., Cairns, A.J., and Pollock, C.J., Carbon assimilation and partitioning into non-structural carbohydrate in contrasting varieties of *Lolium perenne, J. Plant Physiol.,* 159, 257–263, 2002.

36. Demel, R.A., Dorrepaal, E., Ebskamp, M.J.M., Smeekens, J.C.M., and de Kruijff, B., Fructans interact strongly with model membranes, *Biochim. Biophys. Acta,* 1375, 36–42, 1998.

37. Vereyken, I.J., Chupin, V., Demel, R.A., Smeekens, S.C.M., and de Kruijff, B., Fructans insert between the headgroups of phospholipids, *Biochim. Biophys. Acta,* 1510, 307–320, 2001.

38. Kaeser, W., Ultrastructure of storage cells in Jerusalem artichoke tubers (*Helianthus tuberosus* L.). Vesicle formation during inulin synthesis, *Z. Pflanzenphysiol.,* 111, 253–260, 1983.

39. Blum, A., Munns, R., Passioura, J.B., and Turner, N.C., Genetically engineered plants resistant to soil drying and salt stress: how to interpret osmotic relations? *Plant Physiol.,* 110, 1051–1053, 1996.

40. Hellwege, E.M., Gritscher, D., Willmitzer, L., and Heyer, A.G., Transgenic potato tubers accumulate high levels of 1-kestose and nystose: functional identification of a sucrose–sucrose 1-fructosyltransferase of artichoke (*Cynara scolymus*) blossom discs, *Plant J.,* 12, 1057–1065, 1997.

41. Koops, A.J. and Jonker, H.H., Purification and characterisation of the enzymes of fructan biosynthesis in tubers of *Helianthus tuberosus* (Colombia). 2. Purification of sucrose–sucrose 1-fructosyltransferase and reconstitution of fructan synthesis *in vitro* with purified sucrose–sucrose 1-fructosyltransferase and fructan-fructan 1-fructosyltransferase, *Plant Physiol.,* 110, 1167–1175, 1996.

42. Tomlinson, K.L., Lloyd, J.R., and Smith, A.M., The importance of isoforms of starch-branching enzyme in determining the structure of starch in pea leaves, *Plant J.,*11, 31–43, 1997.

43. Saftner, R.A., Dale, J., and Wyse, R.E., Sucrose uptake and compartmentation in sugar beet tap root tissue, *Plant Physiol.,* 72, 1–6, 1983.

44. Cairns, A.J., Effects of enzyme concentration on oligofructan synthesis from sucrose, *Phytochemistry,* 40, 705–708, 1995

45. van den Ende, W. and van Laere, A., Variation in the *in vitro* generated fructan pattern from sucrose as a function of the purified chicory root 1-SST and 1-FFT concentrations, *J. Exp. Bot.,* 47, 1797–1803, 1996.

46. Praznik, W. and Beck, R.H.F., Application of gel-permeation chromatographic systems to the determination of the molecular-weight of inulin, *J. Chromatogr.,* 348, 187–197, 1985.

47. Cairns, A.J., Winters, A., and Pollock, C.J., Fructan biosynthesis in excised leaves of *Lolium temulentum* L. III. A comparison of the *in vitro* properties of fructosyltransferase activities with the characteristics of *in vivo* fructan accumulation, *New Phytol.*, 112, 343–352, 1989.

48. Simmen, U., Obenland, D., Boller, T., and Wiemken, A., Fructan synthesis in excised barley leaves. Identification of two sucrose–sucrose fructosyltransferases induced by light and their separation from constitutive invertases, *Plant Physiol.*, 101, 459–468, 1993.

49. Duchateau, N., Bortlik, K., Simmen, U., Wiemken, A., and Bancal, P., Sucrose–fructan 6-fructosyltransferase: a key enzyme for diverting carbon from sucrose to fructan in barley leaves, *Plant Physiol.*, 107, 1249–1255, 1995.

50. Sprenger, N., Bortlik, K., Brandt, A., Boller, T., and Wiemken, A., Purification, cloning and functional expression of sucrose–fructan 6-transferase, a key enzyme of fructan synthesis in barley, *Proc. Natl. Acad. Sci. USA*, 92, 11652–11656, 1995.

51. Vijn, I., van Dijken, A., Sprenger, N., van Dun, K., Weisbeek, P., Wiemken, A., and Smeekens, S., Fructan of the inulin neoseries is synthesized in transgenic chicory plants (*Chichorium intybus* L.) harbouring onion (*Allium cepa* L.) fructan:fructan 6G-fructosyltransferase, *Plant J.*, 11, 387–398, 1997.

52. Wagner, W., Keller, F., and Wiemken, A., Fructan metabolism in cereals: induction in leaves and compartmentation in protoplasts and vacuoles, *Z. Pflanzenphysiol.*, 112, 359–372, 1983.

53. Frehner, M., Keller, F., and Wiemken, A., Localisation of fructan metabolism in vacuoles isolated from protoplasts of Jerusalem artichoke tubers, *J. Plant Physiol.*, 116, 197–208, 1984.

54. Darwen, C.E. and John, P., Localization of the enzymes of fructan metabolism in vacuoles isolated by a mechanical method from tubers of Jerusalem artichoke (*Helianthus tuberosus* L.), *Plant Physiol.*, 89, 658–663, 1989.

55. Winters, A.L., Smeekens, S.C.M., and Cairns, A.J., fructosyltransferase activity in the tissue-macerating preparation, pectolyase Y-23; physiological role of fructosyl transfer in *Aspergillus* and significance for studies of fructan synthesis in grasses, *New Phytol.*, 121, 525–533, 1992.

56. Chambert, R. and Petit-Glatron, M., Modification of the transfructosylation activity of *Bacillus subtilis* levansucrase by solvent effect and site-directed mutagenesis, in Inulin and Inulin Containing Crops, Fuchs, Eds., Elsevier Science Publishers, Amsterdam, pp. 259–266, 1993.

57. Cairns, A.J. and Ashton, J.E., The interpretation of *in vitro* measurements of fructosyltransferase activity: an analysis of patterns of fructosyl transfer by fungal invertase, *New Phytol.*, 118, 23–34, 1991.

12 Chinese Hamster Ovary (CHO) Glycosylation Mutants for Glycan Engineering

Pamela Stanley and Santosh K. Patnaik

CONTENTS

I. INTRODUCTION

It is now well accepted that the nature of the *N*- and *O*-glycans on glycoproteins or proteoglycans may profoundly affect their biophysical, pharmacological, and biological properties. A key challenge is to identify optimal glycan structures for each glycoprotein and to engineer their synthesis. A convenient strategy in mammalian cells is to produce recombinant glycoproteins from cells that carry a mutation in glycosylation. Such mutations may be loss-of-function (LOF) and lead to immature, truncated glycans on a glycoprotein. Alternatively, they may be gain-of-function (GOF) and lead to a glycoprotein with more complex glycans than would normally be synthesized in that cell. Chinese hamster ovary (CHO) glycosylation mutants exist in both classes[1-7] and have been widely used to produce glycoproteins with a desired complement of *N*- and *O*-glycans.

Most CHO mutants altered in *N*- and *O*-glycan synthesis have been isolated by selection for resistance to plant lectins.[8] Toxic lectins kill cells that bind them at the cell surface, whereas rare mutant cells with altered glycans bind less well and consequently survive toxic lectin concentrations. An advantage of this is that mutants resistant to low, intermediate, or high levels of toxin may be obtained, and these may all be different. Another advantage is that when a cell becomes resistant to one plant lectin due to the production of altered glycans, it usually becomes hypersensitive to different lectins that bind the altered glycans that are present at low levels or not at all on parent CHO cells. This allows for the selection of revertants and also for expression cloning of genes that rescue the mutant phenotype. Specific examples of the unique lectin resistance phenotypes of the most commonly used CHO glycosylation mutants are given in Table 12.1. The mutants Lec1, Lec2, and Lec8 are available from the American Type Culture Collection (ATCC). Finally, complementation studies in cell hybrids based on the ability of a fusion partner to rescue lectin resistance can be easily used to identify genetic complementation groups.[9]

The best illustration of the points outlined above is the most extensively characterized CHO mutant termed Lec1.[8] It was first isolated by selecting for resistance to the leukoagglutinin from *Phaseolus vulgaris* (L-PHA).[10] Mutations in the same gene were subsequently isolated in selections for resistance to ricin, wheat germ agglutinin (WGA), or the agglutinin from *Lens culinaris* (LCA)[9] and a similar mutant, Clone 15B, was independently isolated from a ricin selection.[11] Each independent isolate selected with a different lectin was shown by complementation analysis to have a mutation in the same gene,[12] and thus the generic name of Lec1 was coined.[8] All Lec1 mutants are hypersensitive to the cytotoxicity of concanavalin A (Con A), providing a strategy for selecting revertants. Biochemical analyses showed that Lec1 mutants lack N-acetylglucosaminyltransferase I (GlcNAc-TI) activity, the enzyme that initiates the synthesis of complex and hybrid N-glycans.[13,14] Analyses of the N-glycans synthesized in Lec1 cells identified the substrate of GlcNAc-TI as $Man_5GlcNAc_2Asn$,[15,16] thereby explaining the increased sensitivity of Lec1 cells to killing by Con A (Table 12.1), a lectin that binds Man residues. Further biochemical studies revealed the existence of α-mannosidase II which acts after GlcNAc-TI.[17]

The reduced lectin binding of Lec1 cells allowed expression cloning of the complementing gene that encodes GlcNAc-TI by screening for revertants.[18,19] Analyses of independent Lec1 isolates have now shown that all have inactivating mutations in the *Mgat1* gene that encodes GlcNAc-TI.[20] Unfortunately, none of these mutations are missense and thus they shed little light on structure/function relationships governing

TABLE 12.1
Lectin Resistance Phenotype of CHO Glycosylation Mutants

Cell Line	L-PHA (μg/mL)	WGA (μg/mL)	ConA (μg/mL)	Ricin (ng/mL)	LCA (μg/mL)	PSA (μg/mL)	E-PHA (μg/mL)	MOD (pg/mL)	Abrin (ng/mL)
CHO[a]	5	2	18	5	18	40	35	2.5	2.5
Lec1	>1000(R)	30(R)	6(S)	100(R)	>200(R)	9(R)	>10(R)	4(R)	300(R)
Lec1A	>300(R)	9(R)	5(S)	10(R)	35(R)	5(R)	>10(R)	(R)	3(R)
Lec2	(S)	11(R)	(−)	100(S)	2(S)	2(S)	(−)	5(S)	>10(S)
Lec8	10(R)	100(R)	(S)	(R)	10(S)	2(S)	>10(R)	2(R)	(R)

Note: Fold resistance (R) or sensitivity (S) compared with wild-type parental CHO D_{10} value; (S) or (R) are <2-fold different from wild-type CHO; (−) are same as wild-type CHO. L-PHA=leukophytohemagglutinin from *P. vulgaris;* WGA=wheat germ agglutinin; ConA=concanavalin A, Ricin=*Ricinus communis* lectin II; LCA=*L. culinaris* lectin; PSA=*Pisum sativum* lectin E-PHA=erythrophytohemagglutinin from *P. vulgaris* ; MOD=modeccin.

[a]D_{10} is the lectin concentration at 10% relative plating efficiency.

GlcNAc-TI activity. In contrast, Lec1A mutants that belong to the same genetic complementation group but have a milder phenotype have missense mutations in the *Mgat1* gene that give rise to K_m mutants of GlcNAc-TI.[21] These mutations are informative in relation to the known crystal structure of GlcNAc-TI.[22] It would now be possible by using different mutagens and a specific selection for Lec1 mutants to isolate a vast panel of functional *Mgat1* gene mutations that result in the Lec1A or Lec1 lectin-resistant phenotypes in CHO cells.[23,24]

The cloning of the *Mgat1* gene allowed construction of targeting vectors and the generation of mice lacking GlcNAc-TI.[25,26] Homozygous mutant embryos die at ~E9.5 but mutant blastocysts are unaffected and are rescued by maternal *Mgat1* mRNA from the oocyte.[27,28] A cell type-specific defect implicating complex or hybrid *N*-glycans was identified in chimeras generated with *Mgat1*$^{-/-}$ embryonic stem (ES) cells. The mutant ES cells contributed to essentially all tissues in chimeras with the dramatic exception of the organized layer of bronchial epithelium.[29] Studies currently in progress are examining the effects of deleting the *Mgat1* gene in a tissue or cell type-specific manner using a Cre/*lox*P strategy.

II.   BIOCHEMICAL BASES OF CHO GLYCOSYLATION MUTANTS WITH AN ALTERED ARRAY OF CELL SURFACE GLYCANS

Most of the CHO mutants selected for resistance to plant lectins have proven to be glycosylation mutants that synthesize altered glycans on cell surface glycoconjugates. Mutants defective in the biosynthesis of dolichol-P-P-oligosaccharide[30,31] or the formation of glycolipid anchors[32] have also been obtained. The first group gives rise to underglycosylated glycoproteins rather than to glycoproteins that carry altered structures. The latter give rise to cells with reduced expression of cell surface GPI-anchored proteins. Thus, for glycosylation engineering purposes, the mutants in which the processing or maturation of *N*-glycans is altered are most useful. These mutants fall into two major categories: LOF and GOF mutants. LOF mutants have a mutation that weakens or inactivates a gene required for optimal glycosylation. Most GOF mutants have a mutation that activates a silent gene such as a glycosyltransferase gene, thereby generating more complex glycan structures. However, GOF mutants may also arise from the LOF of a negative regulator that is required for silencing of a glycosyltransferase gene.[33] LOF and GOF mutants are defined based on phenotypic complementation assays in somatic cell hybrids formed between the mutant and wild-type CHO. If the hybrid cells behave like CHO, the mutation is recessive and the mutant is in the LOF class. If the hybrid cells behave like the mutant, the mutation is dominant and the mutant is in the GOF class.

The biochemical bases of mutation in CHO glycosylation mutants have been determined from structural analyses of glycans on glycoproteins and glycolipids that identify the block in a pathway, and biochemical assays of candidate glycosyltransferases, glycosidases, nucleotide-sugar synthases, and nucleotide-sugar transporters. Molecules that are required for optimal glycoprotein trafficking have also been shown to cause altered glycosylation in CHO cells.[34] The biochemical defects in CHO mutants selected to have a single glycosylation mutation are summarized in Table 12.2. The established (or predicted in some cases) alterations in *N*- and *O*-glycan structures

TABLE 12.2

Glycosylation Defects in Lectin-Resistant CHO Mutants

CHO Line	Biochemical Alteration	Genetic Alteration	Pedicted N-Glycans[a]	Predicted O-Glycans	Most Recent Ref.
Gat⁻2 (parent)	—	—	Complex	◄●■-S/T	50
Pro⁻5 (parent)	↓ Gal on N-glycans	No expression of β4GalT-6 gene		◄●□-S/T	35
Lec1	↓ GlcNAc-TI	Insertion/deletion in Mgat1 gene ORF	Oligomannosyl	◄●□-S/T	20
Lec1A	Kₘ mutant of GlcNAc-TI	Point mutation in Mgat1 gene ORF	↓ Complex ↑ Oligomannosyl and hybrid	◄●□-S/T	21
Lec2	↓ CMP-sialic acid Golgi translocase	Mutation in CMST gene ORF		●□-S/T	41
Lec3	↓ UDP-GlcNAc 2-epimerase	Mutation in Gne gene ORF (epimerase domain)		(◄)●□-S/T	42
Lec4	↓ GlcNAc-TV	Deletion in Mgat5 gene ORF		◄●□-S/T	43
Lec4A	Mislocalized GlcNAc-TV	Point mutation in Mgat5 gene ORF		◄●□-S/T	43,44
Lec8	↓ UDP-Gal Golgi translocase	Mutation in UGT gene ORF		□-S/T	45
Lec9	? ↓ polyprenol reductase synthesis	?	Decreased occupancy	◄●□-S/T	59
LEC10	↑ GlcNAc-TIII	Activation of Mgat3 gene		◄●□-S/T	60,75

TABLE 12.2   (Continued)

CHO Line	Biochemical Alteration	Genetic Alteration	Pedicted N-Glycans[a]	Predicted O-Glycans	Most Recent Ref.
LEC11 LEC11B LEC11A	↑ Fuc-TVIB ↑ Fuc-TVIB ↑ Fuc-TVIA	Activation of *Fut6* gene (*Fut6A* or *Fut6B*)	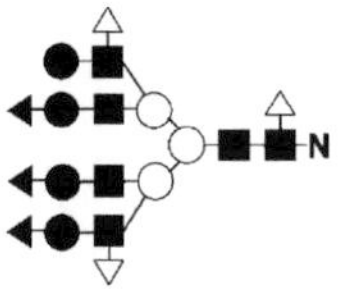	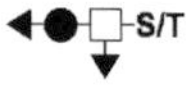	33
LEC12	↑ Fuc-TIX	Activation of *Fut9* gene	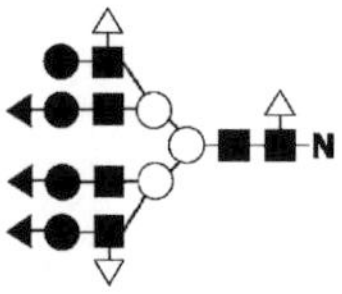	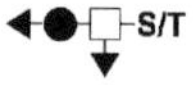	51
Lec13	↓ GDP-Man-4,6-dehydratase	↓ Expression of *Gmd* gene	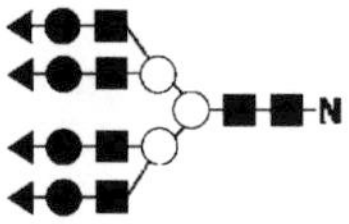	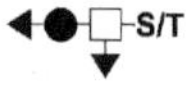	46,47
LEC14	↑ GlcNAc-TVII	?	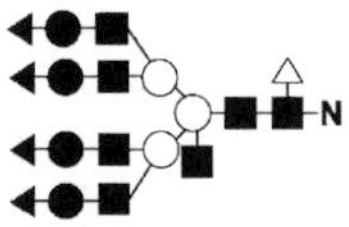	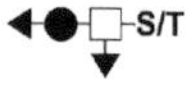	61
Lec15	↓ Dol-P-Man synthase	Allelic loss and *DPM2* gene inactivation	Decreased occupancy	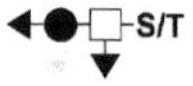	62
LEC18	↑ GlcNAc-TVIII	?	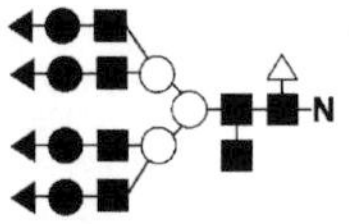	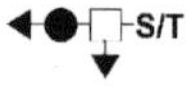	61
Lec19	↓ β4GalT's	↓ Expression of six *β4GalT* genes	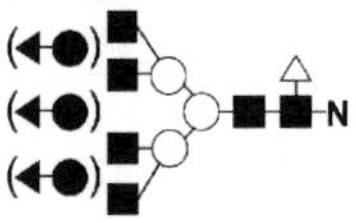		50
Lec20	↓ β4GalT-1	Inactivation or ↓ expression of *β4GalT-1* gene	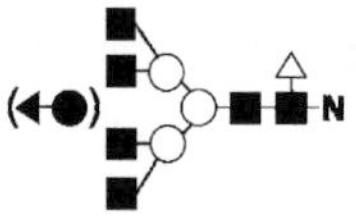		35
Lec23	↓ α-gluco-sidase I	Point mutation in *Gcs* gene ORF	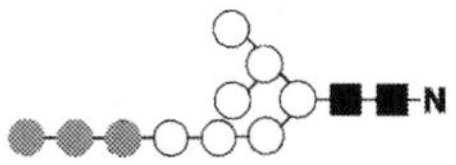	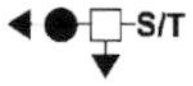	63,76

(*Continued*)

TABLE 12.2   (Continued)

CHO Line	Biochemical Alteration	Genetic Alteration	Predicted N-Glycans[a]	Predicted O-Glycans	Most Recent Ref.
LEC29	↑ Fuc-TIX	Activation of *Fut9* gene	(glycan structure) —N	(O-glycan structure) –S/T	51
LEC30	↑ Fuc-TIV and Fuc-TIX	Activation of *Fut4* and *Fut9* genes	(glycan structure) —N	(O-glycan structure) –S/T	51
LEC31	↑ α-1, 3-fuco-syltransferase	Activation of *Fut9* gene	(glycan structure) —N	(O-glycan structure) –S/T	b
Lec32	↓ CMP-sialic acid synthetase	↓ expression of gene encoding CMP-sialic acid synthetase	(glycan structure) —N	(O-glycan structure) –S/T	48,77
Lec35	Man$_5$ GlcNAc$_2$-P-P-dolichol accumulates in cell	Inactivation of *SL15 (MPDU1)* gene	Decreased occupancy	(O-glycan structure) –S/T	64
ldlD	↓ UDP-Glc-4-epimerase	?	(glycan structure) —N	–S/T	40

○ = mannose; ● = galactose; ◕ = glucose; □ = N-acetylgalactosamine; Δ = fucose; ▲ = sialic acid; ■ = N-acetylglucosamine.

[a] Complex N-glycans of CHO cells have polylactosamines.[35]

[b] Unpublished observations (From P. Stanley's laboratory).

expressed by the mutant cell are also shown. Table 12.3 describes a number of mutants that were selected to have more than one glycosylation mutation and the corresponding effects of those mutations on glycan structure. In both tables the general name of the mutant line is given, followed by the enzymic or transport activity that is altered, followed by the major structural alteration known or predicted to occur on N- and O-glycans.

Our parental CHO cells make a wide range of branched N-glycans best evidenced from matrix-assisted laser desorption time-of-flight (MALDI-TOF) mass

TABLE 12.3
CHO Mutants with Multiple Glycosylation Defects

CHO Line	Biochemical Alteration	Genetic Alteration	Predicted N-Glycans[a]	Predicted O-Glycans	Most Recent Ref.
Lec3.2	↓ UDP-GlcNAc 2-epimerase, ↓ CMP-sialic acid Golgi translocase	Mutations in *Gne* gene (and *CMST* gene)[b]	(structure) -N	(◄)●-□-S/T ▼	42,52
Lec3.2.1	↓ UDP-GlcNAc 2-epimerase, ↓ CMP-sialic acid Golgi translocase, ↓ GlcNAc-TI	Mutations in *Gne*, and *Mgat1* genes (and *CMST* gene)[b]	(structure) -N	(◄)●-□-S/T ▼	20,42,52
Lec3.2.8	↓ CMP-sialic acid Golgi translocase, ↓ UDP-Gal Golgi translocase	Mutations in *Gne* and *UGT* genes (and ?*CMST* gene)[b]	(structure) -N	□-S/T (▼)	42,45,52
Lec3.2.8.1	↓ UDP-GlcNAc 2-epimerase, ↓ CMP-sialic acid Golgi translocase, ↓ UDP-Gal Golgi translocase, ↓ GlcNAc-TI	Mutation in *Gne, Mgat1* and *UGT* genes (and ? *CMST* genes)[b]	(structure) -N	□-S/T (▼)	20,42,45,52
Lec4.8	↓ GlcNAc-TV, ↓ UDP-Gal Golgi translocase	Mutation in *Mgat5* and *UGT* genes	(structure) -N	□-S/T ▼	43,45
Lec4A.8	Mislocalized GlcNAc-TV, ↓ UDP-Gal Golgi translocase	Mutation in *Mgat5* and *UGT* genes	(structure) -N	□-S/T ▼	43,45
LEC10.8	↑ GlcNAc-TIII, ↓ UDP-Gal Golgi translocase	Mutation in *Mgat3* and *UGT* genes	(structure) -N	□-S/T ▼	45,65
ldlD.Lec1	↓ UDP-Glc-4 epimerase, ↓ GlcNAc-TI	Mutations in *Mgat1* gene and ?	(structure) -N	−S/T	20,40
Lec35.Lec1	↓ GlcNAc-TI and no synthesis of Man₆GlcNAc₂-P-P-dolichol	Inactivation of *SL15 (MPDU1)* gene and mutation in *Mgat1* gene?	(structure) -N	◄-●-□-S/T ▼	20,64

○ = mannose; ● = galactose; ◉ = glucose; □ = N-acetylgalactosamine; Δ = fucose;
▲ = sialic acid; ■ = N-acetylglucosamine.

[a] Complex N-glycans of CHO cells have polylactosamines.[35]

[b] Based on genetic complementation analysis.[52]

spectrometry.[35] The *O*-glycans however are quite simple, the largest being a tetrasaccharide as shown from characterization of erythropoietin[36] and interleukin 2[37] produced in CHO cells. The glycolipids of CHO cells are also quite simple with the major species being GM$_3$, a minor amount of lactosylceramide and glucosylceramide.[38] Glycolipids are affected in the lectin-resistant CHO mutants that have altered synthesis or transport of CMP-sialic acid or UDP-Gal (Table 12.2). Proteoglycans of CHO cells are primarily heparan sulfate and chondroitin sulfate (reviewed in References 5 and 39). Synthesis of glycosaminoglycans (GAG) will also be affected in the lectin-resistant CHO mutants that have altered synthesis or transport of CMP-sialic acid or UDP-Gal (Table 12.2). There are a large number of glycosaminoglycan (GAG) CHO mutants that are defective in a specific glycosyltransferase or other enzyme required for GAG synthesis.[5,39] The ldlD mutant defective in UDP-Glc-4-epimerase[40] reduces the addition of Gal and GalNAc to all *N*- and *O*-glycans, glycolipids, and proteoglycans.

III. MOLECULAR GENETIC BASES OF LECTIN-RESISTANT CHO GLYCOSYLATION MUTANTS

Many CHO mutants have now been characterized at the molecular genetic level to determine the precise mutational basis of their phenotype. Most LOF mutants have a mutation in the coding region of the gene that encodes the defective biochemical activity. Thus, all mutants in the Lec1 group have a mutation in the *Mgat1* gene that encodes GlcNAc-TI[20] similar to mutants in the Lec1A group.[21] The latter are K_m mutants of GlcNAc-TI.[23] Mutants in the Lec2 group have a mutation in the CMP-sialic acid transporter gene.[41] Mutants in the Lec3 group have an inactivating mutation in the *Gne* gene.[42] Mutants in the Lec4 and Lec4A groups have a mutation in the *Mgat5* gene.[43] Interestingly, the *lec4A* mutation is a point mutation that prevents GlcNAc-TV from localizing to the Golgi compartment and thus from access to substrate.[44] Lec8 mutants have a mutation in the UDP-Gal transporter gene.[45]. Lec13 mutants have reduced transcripts of the gene encoding GDP-Man-4,6-dehydratase[46,47] and Lec32 mutants lack transcripts of CMP-sialic acid synthetase.[48] Interestingly, our parental CHO line Pro$^-$5 is also a glycosylation mutant; it has no detectable transcripts from the gene encoding *β*4GalT-6.[35] However, the glycosylation defect is extremely subtle affecting only a small subset of *N*-glycans. Lec20 mutants have an inactivating mutation in the gene encoding *β*4GalT-1 and have severely affected *N*-glycans.[35] Lec20 mutants also synthesize *O*-fucose glycans lacking Gal[49] and potentially mucin-type *O*-glycans and glycolipids may be affected. One LOF mutant, Lec19, appears to have a regulatory mutation that is intriguing because it causes a reduction in the transcript levels of six *β*4GalT genes.[50]

   All GOF CHO glycosylation mutants characterized to date express a glycosyltransferase activity that is not expressed in parent CHO cells (Table 12.2). In some cases this appears to be due to transcriptional upregulation of the relevant glycosyltransferase gene. Thus, LEC10 CHO mutants express the CHO gene encoding GlcNAc-TIII, whereas parent CHO cells do not (X. Yang and P. Stanley, unpublished observations). LEC11A mutants express transcripts from the Chinese hamster *Fut6A* gene, while LEC11 and LEC11B mutants express transcripts from the Chinese hamster *Fut6B* gene.[33] Gene rearrangements correlate with upregulation of transcription in LEC11 and LEC11A. However, LEC11B mutants have lost a negative regulator of gene expression.[33] LEC12 and LEC29 mutants express the hamster *Fut9* gene that is silent in CHO cells.[51] However, these mutants differ markedly and inversely in the level of *Fut9* gene expression compared with their Fuc-TIX transferase activity. LEC30 mutants express both the *Fut9* and the *Fut4* hamster genes due to transcriptional upregulation of these genes.[51] LEC30 cells have extremely high FucT activity and would be a good cell line to use in order to obtain a high level of fucosylation of a recombinant glycoprotein.

While the mutants in Table 12.2 were isolated following a single selection strategy and should in theory have arisen from a single mutational event, a number of CHO lines with multiple glycosylation mutations have been isolated using sequential selections.[52] These are described in Table 12.3. In most lines, the mutations are now characterized at both the biochemical and genetic levels but some, such as Lec6, are characterized only at the structural level.[15] A major advantage of these lines is that a very simplified complement of both *N*- and *O*-glycans is synthesized in some mutants, allowing membrane glycoproteins to have the advantages of normal ER glycosylation but to finally mature with a very low glycan content that facilitates crystallization.

IV. A MULTITUDE OF USES FOR CHO GLYCOSYLATION MUTANTS

From the moment the first CHO glycosylation mutants were biochemically characterized they were used as tools to address a variety of questions. The first assays used the mutants to investigate the mechanisms of intracellular vesicular trafficking.[53] The Lec23 mutant was used to investigate roles for *N*-glycans in glycoprotein folding and secretion[54] and many CHO mutants were used to identify the Gal-binding lectin activity of *Entamoeba histolytica*.[55] Examples of other uses of CHO glycosylation mutants are summarized in Table 12.4. These applications represent a subset chosen to give an idea of the range of questions that may be investigated. They include strategies to determine roles for sugars on individual glycoprotein receptors such as Notch receptors or Notch ligands, on recombinant glycoproteins such as soluble complement receptor or immunoglobulins, functions of a particular sugar residue on a range of glycoproteins expressed at the cell surface that may be recognized as a receptor for viruses or bacteria, and intracellular roles for sugars in folding and trafficking. In addition, the mutants can be used to improve glycoprotein therapeutics. For example, glucocerebrosidase is used in enzyme replacement therapy for Gaucher's disease and is most effective when targeted to mannose receptors (reviewed in Reference 56). Exposing mannose is greatly facilitated by producing the recombinant glycoprotein in Lec1 cells so that all complex *N*-glycans are oligomannosyl (Table 12.2) and is the basis for the latest version of glucocerebrosidase, Cerezyme™.[57] Another example of improved targeting of a therapeutic is from a mouse model of stroke. In this case, soluble complement receptor was produced in LEC11 CHO cells so it would carry the sialylated Lewis X determinant recognized by selectins.[58] This modification improved the effectiveness of soluble complement receptor threefold, presumably because it was targeted to sites of inflammation induced by stroke.

TABLE 12.4

Some Applications of CHO Cell Lines with Glycosylation Mutations

Mutant	Biochemical Alteration	Application	Ref.
Lec1, Lec2 and Lec8	No GlcNAcT-I ↓ CMP-sialic acid Golgi translocase ↓ UDP-Gal Golgi translocase	An adhesin of *Candida glabrata* specifically recognizes asialo- lactosyl *N*-glycans on human epithelial cells	66
		In vivo trafficking and catabolism of IgG1 antibodies with Fc associated carbohydrates of differing structure	74
Lec2	↓ CMP-sialic acid Golgi translocase	Adenovirus type 37 uses sialic acid as a cellular receptor	67
Lec8	↓ UDP-Gal Golgi translocase	Expression cloning of *Caenorhabditis elegans* β4GalNAc-transferase	68
LEC10	Activation of GlcNAcT-III	Bisecting GlcNAc of complex *N*-glycans of human IgG1 is not important for antibody-dependent cellular cytotoxicity	69
LEC11	Activation of FucT-VI	Neuronal protection in stroke by an sLex-glycosylated complement inhibitory protein	58
Lec13	↓ GDP-Man-4,6-dehydratase	*N*-glycan core fucose on human IgG1 inhibits binding to FcγR-III and ADCC	70
Lec15	↓ Dol-P-Man synthase	Protein C-mannosylation uses dolichyl-phosphate mannose as a precursor	71
Lec20	↓ β4GalT-I	Fringe modulation of Jagged1-induced Notch signaling requires the action of β4GalT-I	49
Lec23	↓ α-glucosidase I	ERp57 interacts with both calnexin and calreticulin in the absence of their glycoprotein substrates	72
Lec3.2.8.1	↓ UDP-GlcNAc 2-epimerase, ↓ CMP-sialic acid Golgi translocase, ↓ UDP-Gal Golgi translocase, no GlcNAc-TI	Effect of sialylation on structures of *O*-glycosylated stalk-like peptides	73

In conclusion, the mutants are very useful for producing glycoproteins with a defined sugar complement, for expression cloning of sugar-modifying enzymes, and for the production of glycoproteins lacking sugars that inhibit a reaction or carrying sugars that facilitate a reaction. The myriad applications of CHO and related glycosylation mutants for research into the roles of sugars, for understanding fundamental mechanisms in biology, and for practical reasons are limited only by the imagination.

REFERENCES

1. Briles, E.B., Lectin-resistant cell surface variants of eukaryotic cells, *Int. Rev. Cytol.,* 75, 101–165, 1982.
2. Stanley, P., Glycosylation mutants of animal cells, *Annu. Rev. Genet.,* 18, 525–552, 1984.

3. Stanley, P., Glycosylation mutants and the functions of mammalian carbohydrates, *Trends Genet.,* 3, 77–81, 1987.

4. Stanley, P., Glycosylation engineering, *Glycobiology,* 2, 99–107, 1992.

5. Esko, J.D., Stewart, T.E., and Taylor, W.H., Animal cell mutants defective in glycosaminoglycan biosynthesis, *Proc. Natl. Acad. Sci. USA,* 82, 3197–3201, 1985.

6. Stanley, P. and Ioffe, E., Glycosyltransferase mutants: key to new insights in glycobiology, *FASEB J.,* 9, 1436–1444, 1995.

7. Stanley, P., Raju, T.S., and Bhaumik, M., CHO cells provide access to novel *N*-glycans and developmentally regulated glycosyltransferases, *Glycobiology,* 6, 695–699, 1996.

8. Stanley, P., Selection of lectin-resistant mutants of animal cells, *Methods Enzymol.,* 96, 157–184, 1983.

9. Stanley, P., Caillibot, V., and Siminovitch, L., Selection and characterization of eight phenotypically distinct lines of lectin-resistant Chinese hamster ovary cell, *Cell,* 6, 121–128, 1975.

10. Stanley, P., Caillibot, V., and Siminovitch, L., Stable alterations at the cell membrane of Chinese hamster ovary cells resistant to the cytotoxicity of phytohemagglutinin, *Somatic Cell Genet.,* 1, 3–26, 1975.

11. Gottlieb, C., Skinner, A.M., and Kornfeld, S., Isolation of a clone of Chinese hamster ovary cells deficient in plant lectin-binding sites, *Proc. Natl. Acad. Sci. USA,* 71, 1078–1082, 1974.

12. Stanley, P., Membrane mutants of animal cells: rapid identification of those with a primary defect in glycosylation, *Mol. Cell Biol.,* 5, 923–929, 1985.

13. Stanley, P., Narasimhan, S., Siminovitch, L., and Schachter, H., Chinese hamster ovary cells selected for resistance to the cytotoxicity of phytohemagglutinin are deficient in a UDP-*N*-acetylglucosamine-glycoprotein *N*-acetylglucosaminyltransferase activity, *Proc. Natl. Acad. Sci. USA,* 72, 3323–3327, 1975.

14. Gottlieb, C., Baenziger, J., and Kornfeld, S., Deficient uridine diphosphate-*N*-acetylglucosamine:glycoprotein *N*-acetylglucosaminyltransferase activity in a clone of Chinese hamster ovary cells with altered surface glycoproteins, *J. Biol. Chem.,* 250, 3303–3309, 1975.

15. Robertson, M.A., Etchison, J.R., Robertson, J.S., Summers, D.F., and Stanley, P., Specific changes in the oligosaccharide moieties of VSV grown in different lectin-resistant CHO cells, *Cell,* 13, 515–526, 1978.

16. Kornfeld, S., Li, E., and Tabas, I., The synthesis of complex-type oligosaccharides. II. Characterization of the processing intermediates in the synthesis of the complex oligosaccharide units of the vesicular stomatitis virus G protein, *J. Biol. Chem.,* 253, 7771–7778, 1978.

17. Tabas, I. and Kornfeld, S., The synthesis of complex-type oligosaccharides. III. Identification of an alpha-D-mannosidase activity involved in a late stage of processing of complex-type oligosaccharides, *J. Biol. Chem.,* 253, 7779–7786, 1978.

18. Kumar, R. and Stanley, P., Transfection of a human gene that corrects the Lec1 glycosylation defect: evidence for transfer of the structural gene for *N*-acetylglucosaminyltransferase I, *Mol. Cell Biol.,* 9, 5713–5717, 1989. [Erratum: *Mol. Cell. Biol.,* 10, 3857, 1990.]

19. Kumar, R., Yang, J., Larsen, R.D., and Stanley, P., Cloning and expression of *N*-acetylglucosaminyltransferase I, the medial Golgi transferase that initiates complex *N*-linked carbohydrate formation, *Proc. Natl. Acad. Sci. USA,* 87, 9948–9952, 1990.

20. Chen, W. and Stanley, P., Five Lec1 CHO cell mutants have distinct Mgat1 gene mutations that encode truncated *N*-acetylglucosaminyltransferase I, *Glycobiology,* 13, 43–50, 2003.

21. Chen, W., Unligil, U.M., Rini, J.M., and Stanley, P., Independent Lec1A CHO glycosylation mutants arise from point mutations in *N*-acetylglucosaminyltransferase I that reduce affinity for both substrates. Molecular consequences based on the crystal structure of GlcNAc-TI, *Biochemistry*, 40, 8765–8772, 2001.

22. Unligil, U.M., Zhou, S., Yuwaraj, S., Sarkar, M., Schachter, H., and Rini, J.M., X-ray crystal structure of rabbit *N*-acetylglucosaminyltransferase I: catalytic mechanism and a new protein superfamily, *EMBO J.*, 19, 5269–5280, 2000.

23. Chaney, W. and Stanley, P., Lec1A Chinese hamster ovary cell mutants appear to arise from a structural alteration in *N*-acetylglucosaminyltransferase I, *J. Biol. Chem.*, 261, 10551–10557, 1986.

24. Stanley, P. and Chaney, W., Control of carbohydrate processing: the Lec1A CHO mutation results in partial loss of *N*-acetylglucosaminyltransferase I activity, *Mol. Cell Biol.*, 5, 1204–1211, 1985.

25. Ioffe, E. and Stanley, P., Mice lacking *N*-acetylglucosaminyltransferase I activity die at mid-gestation, revealing an essential role for complex or hybrid *N*-linked carbohydrates, *Proc. Natl. Acad. Sci. USA*, 91, 728–732, 1994.

26. Metzler, M., Gertz, A., Sarkar, M., Schachter, H., Schrader, J.W., and Marth, J.D., Complex asparagine-linked oligosaccharides are required for morphogenic events during post-implantation development, *EMBO J.*, 13, 2056–2065, 1994.

27. Campbell, R.M., Metzler, M., Granovsky, M., Dennis, J.W., and Marth, J.D., Complex asparagine-linked oligosaccharides in Mgat1-null embryos, *Glycobiology*, 5, 535–543, 1995.

28. Ioffe, E., Liu, Y., and Stanley, P., Complex *N*-glycans in Mgat1 null preimplantation embryos arise from maternal Mgat1 RNA, *Glycobiology*, 7, 913–919, 1997.

29. Ioffe, E., Liu, Y., and Stanley, P., Essential role for complex *N*-glycans in forming an organized layer of bronchial epithelium, *Proc. Natl. Acad. Sci. USA*, 93, 11041–11046, 1996.

30. Rosenwald, A.G., Stanley, P., McLachlan, K.R., and Krag, S.S., Mutants in dolichol synthesis: conversion of polyprenol to dolichol appears to be a rate-limiting step in dolichol synthesis, *Glycobiology*, 3, 481–488, 1993.

31. Camp, L.A., Chauhan, P., Farrar, J.D., and Lehrman, M.A., Defective mannosylation of glycosylphosphatidylinositol in Lec35 Chinese hamster ovary cells, *J. Biol. Chem.*, 268, 6721–6728, 1993.

32. Kinoshita, T., Ohishi, K., and Takeda, J., GPI-anchor synthesis in mammalian cells: genes, their products, and a deficiency, *J. Biochem. (Tokyo)*, 122, 251–257, 1997.

33. Zhang, A., Potvin, B., Zaiman, A., Chen, W., Kumar, R., Phillips, L., and Stanley, P., The gain-of-function Chinese hamster ovary mutant LEC11B expresses one of two Chinese hamster FUT6 genes due to the loss of a negative regulatory factor, *J. Biol. Chem.*, 274, 10439–10450, 1999.

34. Podos, S.D., Reddy, P., Ashkenas, J., and Krieger, M., LDLC encodes a brefeldin A-sensitive, peripheral Golgi protein required for normal Golgi function, *J. Cell. Biol.*, 127, 679–691, 1994.

35. Lee, J., Sundaram, S., Shaper, N.L., Raju, T.S., and Stanley, P., Chinese hamster ovary (CHO) cells may express six β4- galactosyltransferases (β4GalTs). Consequences of the loss of functional β4GalT-1, β4GalT-6, or both in CHO glycosylation mutants, *J. Biol. Chem.*, 276, 13924–13934, 2001.

36. Sasaki, H., Bothner, B., Dell, A., and Fukuda, M., Carbohydrate structure of erythropoietin expressed in Chinese hamster ovary cells by a human erythropoietin cDNA, *J. Biol. Chem.*, 262, 12059–12076, 1987.

37. Conradt, H.S., Geyer, R., Hoppe, J., Grotjahn, L., Plessing, A., and Mohr, H., Structures of the major carbohydrates of natural human interleukin-2, *Eur. J. Biochem.*, 153, 255–261, 1985.

38. Stanley, P., Sudo, T., and Carver, J.P., Differential involvement of cell surface sialic acid residues in wheat germ agglutinin binding to parental and wheat germ agglutinin-resistant Chinese hamster ovary cells, *J. Cell. Biol.*, 85, 60–69, 1980.

39. Esko, J.D., Animal cell mutants defective in heparan sulfate polymerization, *Adv. Exp. Med. Biol.*, 313, 97–106, 1992.

40. Kingsley, D.M., Kozarsky, K.F., Hobbie, L., and Krieger, M., Reversible defects in O-linked glycosylation and LDL receptor expression in a UDP-Gal/UDP-GalNAc 4-epimerase deficient mutant, *Cell*, 44, 749–759, 1986.

41. Eckhardt, M., Gotza, B., and Gerardy-Schahn, R., Mutants of the CMP-sialic acid transporter causing the Lec2 phenotype, *J. Biol. Chem.*, 273, 20189–20195, 1998.

42. Hong, Y. and Stanley, P., Lec3 CHO mutants lack UDP-GlcNAc 2-epimerase activity due to mutations in the epimerase domain of the Gne gene, *J. Biol. Chem.*, 2004, 278, 53045–53054.

43. Weinstein, J., Sundaram, S., Wang, X., Delgado, D., Basu, R., and Stanley, P., A point mutation causes mistargeting of Golgi GlcNAc-TV in the Lec4A Chinese hamster ovary glycosylation mutant, *J. Biol. Chem.*, 271, 27462–27469, 1996.

44. Chaney, W., Sundaram, S., Friedman, N., and Stanley, P., The Lec4A CHO glycosylation mutant arises from miscompartmentalization of a Golgi glycosyltransferase, *J. Cell. Biol.*, 109, 2089–2096, 1989.

45. Oelmann, S., Stanley, P., and Gerardy-Schahn, R., Point mutations identified in Lec8 Chinese hamster ovary glycosylation mutants that inactivate both the UDP-galactose and CMP-sialic acid transporters, *J. Biol. Chem.*, 276, 26291–26300, 2001.

46. Ohyama, C., Smith, P.L., Angata, K., Fukuda, M.N., Lowe, J.B., and Fukuda, M., Molecular cloning and expression of GDP-D-mannose-4,6-dehydratase, a key enzyme for fucose metabolism defective in Lec13 cells, *J. Biol. Chem.*, 273, 14582–14587, 1998.

47. Sullivan, F.X., Kumar, R., Kriz, R., Stahl, M., Xu, G.Y., Rouse, J., Chang, X.J., Boodhoo, A., Potvin, B., and Cumming, D.A., Molecular cloning of human GDP-mannose 4,6-dehydratase and reconstitution of GDP-fucose biosynthesis *in vitro*, *J. Biol. Chem.*, 273, 8193–8202, 1998.

48. Potvin, B., Raju, T.S., and Stanley, P., Lec32 is a new mutation in Chinese hamster ovary cells that essentially abrogates CMP-N-acetylneuraminic acid synthetase activity, *J. Biol. Chem.*, 270, 30415–30421, 1995.

49. Chen, J., Moloney, D.J., and Stanley, P., Fringe modulation of Jagged1-induced Notch signaling requires the action of beta-4-galactosyltransferase-1, *Proc. Natl. Acad. Sci. USA*, 98, 13716–13721, 2001.

50. Lee, J., Park, S.H., Sundaram, S., Raju, T.S., Shaper, N.L., and Stanley, P., A mutation causing a reduced level of expression of six beta-4-galactosyltransferase genes is the basis of the Lec19 CHO glycosylation mutant, *Biochemistry*, 42, 12349–12357, 2003.

51. Patnaik, S.K., Zhang, A., Shi, S., and Stanley, P., Alpha(1,3)fucosyltransferases expressed by the gain-of-function Chinese hamster ovary glycosylation mutants LEC12, LEC29, and LEC30, *Arch. Biochem. Biophys.*, 375, 322–332, 2000.

52. Stanley, P., Chinese hamster ovary cell mutants with multiple glycosylation defects for production of glycoproteins with minimal carbohydrate heterogeneity, *Mol. Cell Biol.*, 9, 377–383, 1989.

53. Brandli, A.W., Mammalian glycosylation mutants as tools for the analysis and reconstitution of protein transport, *Biochem. J.*, 276, 1–12, 1991.

54. Hammond, C. and Helenius, A., Folding of VSV G protein: sequential interaction with BiP and calnexin, *Science*, 266, 456–458, 1994.

55. Ravdin, J.I., Stanley, P., Murphy, C.F., and Petri, Jr., W.A., Characterization of cell surface carbohydrate receptors for *Entamoeba histolytica* adherence lectin, *Infect. Immun.*, 57, 2179–2186, 1989.

56. Brady, R.O. and Barton, N.W., Enzyme replacement and gene therapy for Gaucher's disease, *Lipids*, 31 (Suppl. S), 137–139, 1996.

57. Grabowski, G.A., Barton, N.W., Pastores, G., Dambrosia, J.M., Banerjee, T.K., McKee, M.A., Parker, C., Schiffmann, R., Hill, S.C., and Brady, R.O., Enzyme therapy in type 1 Gaucher disease: comparative efficacy of mannose-terminated glucocerebrosidase from natural and recombinant sources, *Ann. Intern. Med.*, 122, 33–39, 1995.

58. Huang, J., Kim, L.J., Mealey, R., Marsh, Jr., H.C., Zhang, Y., Tenner, A.J., Connolly, Jr., E.S., and Pinsky, D.J., Neuronal protection in stroke by an sLex-glycosylated complement inhibitory protein, *Science*, 285, 595–599, 1999.

59. Rosenwald, A.G. and Krag, S.S., Lec9 CHO glycosylation mutants are defective in the synthesis of dolichol, *J. Lipid. Res.*, 31, 523–533, 1990.

60. Campbell, C. and Stanley, P., A dominant mutation to ricin resistance in Chinese hamster ovary cells induces UDP-GlcNAc:glycopeptide beta-4-*N*-acetylglucosaminyltransferase III activity, *J. Biol. Chem.*, 259, 13370–13378, 1984.

61. Raju, T.S. and Stanley, P., Gain-of-function Chinese hamster ovary mutants LEC18 and LEC14 each express a novel *N*-acetylglucosaminyltransferase activity, *J. Biol. Chem.*, 273, 14090–14098, 1998.

62. Maeda, Y., Tomita, S., Watanabe, R., Ohishi, K., and Kinoshita, T., DPM2 regulates biosynthesis of dolichol phosphate-mannose in mammalian cells: correct subcellular localization and stabilization of DPM1, and binding of dolichol phosphate, *EMBO J.*, 17, 4920–4929, 1998.

63. Ray, M.K., Yang, J., Sundaram, S., and Stanley, P., A novel glycosylation phenotype expressed by Lec23, a Chinese hamster ovary mutant deficient in alpha-glucosidase I, *J. Biol. Chem.*, 266, 22818–22825, 1991.

64. Anand, M., Rush, J.S., Ray, S., Doucey, M.A., Weik, J., Ware, F.E., Hofsteenge, J., Waechter, C.J., and Lehrman, M.A., Requirement of the Lec35 gene for all known classes of monosaccharide-P-Dolichol-dependent glycosyltransferase reactions in mammals, *Mol. Biol. Cell.*, 12, 487–501, 2001.

65. Stanley, P., Sundaram, S., and Sallustio, S., A subclass of cell surface carbohydrates revealed by a CHO mutant with two glycosylation mutations, *Glycobiology*, 1, 307–314, 1991.

66. Cormack, B.P., Ghori, N., and Falkow, S., An adhesin of the yeast pathogen *Candida glabrata* mediating adherence to human epithelial cells, *Science*, 285, 578–582, 1999.

67. Arnberg, N., Edlund, K., Kidd, A.H., and Wadell, G., Adenovirus type 37 uses sialic acid as a cellular receptor, *J. Virol.*, 74, 42–48, 2000.

68. Kawar, Z.S., Van Die, I., and Cummings, R. D., Molecular cloning and enzymatic characterization of a UDP-GalNAc:GlcNAc(β)-R β1,4-*N*-acetylgalactosaminyltransferase from *Caenorhabditis elegans*, *J. Biol. Chem.*, 277, 34924–34932, 2002.

69. Shinkawa, T., Nakamura, K., Yamane, N., Shoji-Hosaka, E., Kanda, Y., Sakurada, M., Uchida, K., Anazawa, H., Satoh, M., Yamasaki, M., Hanai, N., and Shitara, K., The absence of fucose but not the presence of galactose or bisecting *N*-acetylglucosamine of human IgG1 complex-type oligosaccharides shows the critical role of enhancing antibody-dependent cellular cytotoxicity, *J. Biol. Chem.*, 278, 3466–3473, 2003.

70. Shields, R.L., Lai, J., Keck, R., O'Connell, R.L.Y., Hong, K., Meng, Y.G., Weikert, S.H., and Presta, L.G., Lack of fucose on human IgG1 *N*-linked oligosaccharide improves binding to human Fcgamma RIII and antibody-dependent cellular toxicity, *J. Biol. Chem.,* 277, 26733–26740, 2002.

71. Doucey, M.A., Hess, D., Cacan, R., and Hofsteenge, J., Protein C-mannosylation is enzyme-catalysed and uses dolichyl-phosphate-mannose as a precursor, *Mol. Biol. Cell.,* 9, 291–300, 1998.

72. Oliver, J.D., Roderick, H.L., Llewellyn, D.H., and High, S., ERp57 functions as a subunit of specific complexes formed with the ER lectins calreticulin and calnexin, *Mol. Biol. Cell.,* 10, 2573–2582, 1999.

73. Merry, A.H., Gilbert, R.J., Shore, D.A., Royle, L., Miroshnychenko, O., Vuong, M., Wormald, M.R., Harvey, D.J., Dwek, R.A., Classon, B.J., Rudd, P.M., and Davis, S.J., *O*-glycan sialylation and the structure of the stalk-like region of the T cell co-receptor CD8, *J. Biol. Chem.,* 278, 27119–27128, 2003.

74. Wright, A., Sato, Y., Okada, T., Chang, K., Endo, T., and Morrison, S., *In vivo* trafficking and catabolism of IgG1 antibodies with Fc associated carbohydrates of differing structure, *Glycobiology,* 10, 1347–1355, 2000.

75. Stanley, P., Sundaram, S., Tang, J., and Shi, S., Molecular analysis of three gain-of-function CHO mutants that add the bisecting GlcNAc to *N*-glycans, *Glycobiology,* 15, 43–53, 2005.

76. Hong, Y., Sundaram, S., Shin, D.J., and Stanley, P., The Lec23 Chinese hamster ovary mutant is a sensitive host for detecting mutations in alpha-glucosidase I that give rise to congenital disorder of glycosylation IIb (CDG IIb), *J. Biol. Chem.,* 279, 49894–49901, 2004.

77. Munster, A.K., Eckhardt, M., Potvin, B., Múhlenhoff, M., Stanley, P., and Gerardy-Schah, R., Mammalian CMP-*N*-Acetylneuraminic Acid Synthetase: A Nuclear Protein with Evolutionary Conserved Structural Motifs, *Proc. Natl. Acad. Sci. USA,* 95, 9140–9145, 1998.

13 Biochemical Engineering of Sialic Acids

Stephan Hinderlich, Cornelia Oetke, and Michael Pawlita

CONTENTS

I. INTRODUCTION

Sialic acids are essential components of glycoconjugates. Their high structural versatility strongly contributes to the diversity of oligosaccharides *in vivo*. To date more than 50 naturally occurring members of the sialic acid family are known.[1] Sialic acids have numerous biological functions in mammalian organisms including mediation of cell–cell interactions,[2] formation or masking of recognition determinants,[3] and stabilization of glycoproteins.[4] To investigate the different functional roles of sialic acids, artificial modification of their structures is an essential tool. One possibility is the desialylation of glycoconjugates and subsequent resialylation by sialyltransferases using CMP-activated modified sialic acids as substrates.[5,6] A cheaper and more elegant method is the treatment of cells with modified metabolic precursors of sialic acids. These sugars are taken up by cells, metabolized intracellularly to sialic acids and finally bound to glycoconjugates. A first hint on the functionality of this method was found in a study by Grünholz and co-workers,[7] where it was shown that mannosamines with extended *N*-acyl side chains are capable of competitively inhibiting sialic acid biosynthesis in rat liver cytosol. Kayser and colleagues[8] then demonstrated that radiolabeled variants of these mannosamine derivatives can be

387

incorporated into membrane and serum proteins of rats. Treatment of cultivated cells with millimolar concentrations of the derivatives resulted in a high percentage of modified sialic acids on their surfaces.[9] Finally, it was shown that analogs of sialic acids are incorporated and metabolized by cells and can therefore also be used to modulate the structure of sialic acids.[10] The aim of this review is to give a detailed insight into metabolization of unnatural sialic acid precursors and an overview of the recent developments made in this technique for biomedical applications.

II.   BIOSYNTHESIS AND METABOLISM OF SIALIC ACIDS

A.   BIOSYNTHESIS OF NATURAL SIALIC ACIDS

To understand how unnatural sialic acid precursors are metabolized by cells it is important to describe first the biosynthetic pathways of natural sialic acids (Figure 13.1). Nearly all sialic acids are derived from a common precursor, N-acetylneuraminic acid (Neu5Ac). Neu5Ac is synthesized in the cytosol of mammalian cells, activated to CMP-Neu5Ac in the nucleus, and then transported to the Golgi apparatus where the sugar is transferred to the nascent oligosaccharide chains by sialyltransferases. All modifications of sialic acids are synthesized after binding of the sugar to glyco-conjugates, except for N-glycolylneuraminic acid, which derives from the activated nucleotide sugar in the cytosol.[11]

UDP-GlcNAc is the starting compound for the biosynthesis of sialic acids. Additionally it is the key metabolite of amino sugar metabolism. Synthesis of UDP-GlcNAc branches from glycolysis at fructose-6-phosphate, where glutamine-fructose-6-phosphate amidotransferase, a highly regulated enzyme, builds the first amino sugar glucosamine-6-phosphate.[12] Three further enzymatic steps form UDP-GlcNAc, which serves as substrate for the synthesis of protein- or lipid-bound oligosaccharides,[13,14] proteoglycans,[15] glycosylphosphatidylinositol membrane anchors,[16] O-GlcNAc modification of cytosolic and nuclear proteins,[17] and finally the synthesis of sialic acids.

CMP-Neu5Ac is synthesized in five enzymatic steps from UDP-GlcNAc in the cytosol and the nucleus of mammalian cells (Figure 13.1). The first step in this pathway is the formation of N-acetylmannosamine (ManNAc) from UDP-GlcNAc. It is catalyzed by UDP-GlcNAc 2-epimerase.[18] ManNAc is then phosphorylated by a specific ManNAc kinase.[19,20] In mammals, both enzymes, UDP-GlcNAc 2-epimerase and ManNAc kinase, are combined to one bifunctional enzyme.[21,22] Cloning of the enzyme cDNA from rat,[22] mouse,[23] and man[24] revealed that it consists of two functional domains, an N-terminal epimerase domain and a C-terminal kinase domain. This was underlined by introducing point mutations of conserved amino acids, which resulted in a selective loss of enzymatic activity of the respective domain.[25]

UDP-GlcNAc 2-epimerase is the key enzyme of sialic acid biosynthesis. The main control mechanism for its activity is a strong allosteric feedback inhibition by CMP-Neu5Ac.[26] Inhibitory concentrations of about 60 μM result in total loss of enzyme activity. This is very close to the concentration of CMP-Neu5Ac in the cytosol of cells,[27] resulting in a nearly complete inactivity of UDP-GlcNAc 2-epimerase in cells. Only the consumption of CMP-Neu5Ac in oligosaccharide synthesis leads to activation of the enzyme until the CMP-Neu5Ac pool is refilled.

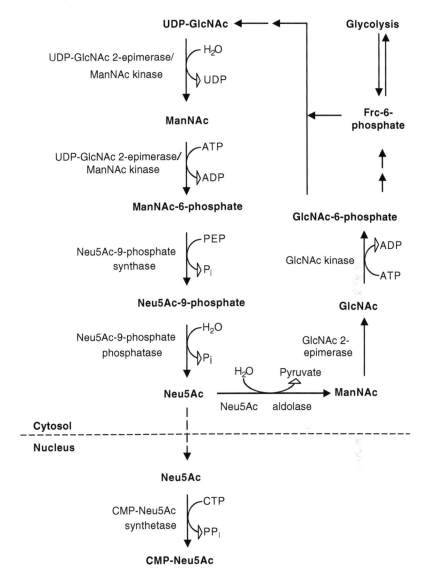

FIGURE 13.1   Sialic acid biosynthesis in mammals.

Thus, the availability of UDP-GlcNAc for biosynthesis of sialic acid depends on the consumption of CMP-Neu5Ac in oligosaccharide biosynthesis. Two more mechanisms are considered for additional regulation of the enzyme activity. Phosphorylation by protein kinase C leads to an increase in epimerase activity[28] and the protein is capable of assembling as a hexamer and also as a dimer.[21] The hexamer shows both enzyme activities, whereas the dimer possesses only ManNAc kinase activity. The ManNAc kinase activity is not affected by all these mechanisms, underlining the independent functions of the two domains of the bifunctional enzyme and the important regulatory role of UDP-GlcNAc 2-epimerase.

ManNAc-6-phosphate, which emerges by the consecutive reactions of UDP-GlcNAc 2-epimerase and ManNAc kinase, is used for synthesis of the first sialic acid in the pathway, Neu5Ac-9-phosphate. The sugar phosphate is condensed with phospho*enol*pyruvate (PEP) by Neu5Ac-9-phosphate synthase.[29] Before formation of the final product, CMP-Neu5Ac, the phosphate group is released from Neu5Ac-9-phosphate.[30] It has not been clear until now whether this reaction is performed by a specific or a nonspecific phosphatase. CMP-Neu5Ac synthetase then condenses Neu5Ac and CTP with release of pyrophosphate. Surprisingly, this enzyme is localized in the nucleus.[31,32] The amino acid sequence of the protein contains three nuclear localization signals consisting of clusters of basic amino acids.[33] Although the nuclear localization of CMP-Neu5Ac has been known for many years, we can still only speculate about its functional implication. Ferwerda and co-workers[34] postulated a mechanism where Neu5Ac is dephosphorylated immediately before transport into the nucleus, in order to protect Neu5Ac from the *de novo* synthesis against degradation by the catabolic pathway. Alternatively, CMP-Neu5Ac could be a substrate for, a so far unidentified, nuclear sialyltransferase, which could modify and regulate proteins of the nucleus or nuclear membrane in a manner similar to the known *O*-GlcNAc modification.[17]

Besides the *de novo* synthesis of Neu5Ac described above, a cytosolic pathway for the degradation of the sugar also exists (Figure 13.1). Presumably, Neu5Ac from the anabolic pathway is not degraded in the cytosol because it is protected by 9-phosphorylation. Another source for free Neu5Ac could be hydrolyzed CMP-Neu5Ac, but the presence of a CMP-Neu5Ac hydrolase has never be shown.[35] However, Neu5Ac from the degradation of plasma membrane oligosaccharides is transported to the cytosol by the lysosomal anion transporter sialin.[36] This Neu5Ac, when not directly recycled in sialic acid biosynthesis, is cleaved to ManNAc and pyruvate by Neu5Ac aldolase,[37] whereas pyruvate is directly fed to glycolysis. ManNAc can be further metabolized to GlcNAc by the action of GlcNAc 2-epimerase.[38] GlcNAc is finally yielded in amino sugar metabolism, where it is either used for the synthesis of UDP-GlcNAc or further degraded in glycolysis.

B.   PATHWAYS FOR UNNATURAL SIALIC ACIDS

It is widely accepted that unnatural sialic acid precursors can go through the sialic acid biosynthetic pathway (Figure 13.2). The analogs used most commonly for modulation of the structure of sialic acids are derivatives of mannosamine with extended or altered *N*-acyl side chains (Figure 13.3). These modifications are obviously tolerated by enzymes involved in sialic acid metabolism. The molecular basis for this promiscuity is most likely the natural *N*-glycolyl modification of sialic acids. It is the only modification enzymatically formed before the sugar is bound to the glycoconjugate. A specific hydroxylase forms CMP-Neu5Gc from CMP-Neu5Ac.[11] CMP-Neu5Gc therefore has to be accepted by the CMP-sialic acid transporter and the sialyltransferases. Furthermore, as already described for Neu5Ac, Neu5Gc is also recycled from degradation of oligosaccharides. The sugar is transported from the lysosome to the cytosol and is either directly activated to form CMP-Neu5Gc or further degraded to *N*-glycolylmannosamine (ManNGc) and pyruvate, whereby ManNGc can be recycled during sialic acid synthesis. Finally, Neu5Gc or ManNGc

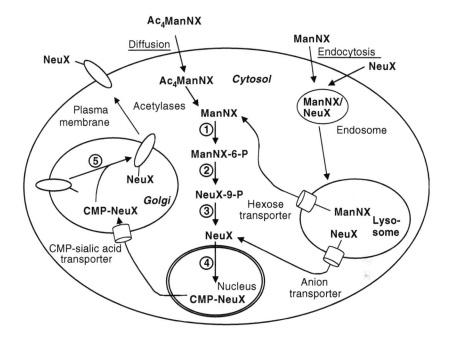

FIGURE 13.2 Metabolism of unnatural sialic acids and sialic acid precursors. "ManNX" represents mannosamine derivatives with altered N-acyl side chains. "NeuX" represents sialic acid derivatives with modified C-5 or C-9 positions.

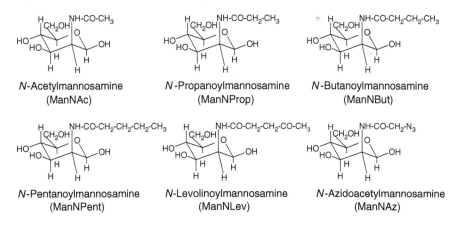

FIGURE 13.3 Structures of unnatural mannosamine derivatives found to be incorporated into cellular sialic acids.

are likely to be derived from nutritional sources. All enzymes involved in the pathway for metabolization of the sugars, ManNAc kinase,[39] Neu5Ac-9-phosphate synthase,[29] CMP-Neu5Ac synthetase,[40] and Neu5Ac aldolase[37] have been found to accept the N-glycolyl as well as the N-acetyl derivative as a substrate. It can be concluded that the N-acyl side chain is not important for substrate recognition of the different enzymes and therefore tolerates artificial structure modifications.

Although several mannosamine derivatives have been found to be metabolized by sialic acid biosynthesis and incorporated into oligosaccharides (Figure 13.3), it is necessary to discuss the limitations of this method. Extension of N-acyl side chains to six or more carbon atoms or the addition of bulky, branched, or ring substituents to the amino group avoids metabolization.[41] In contrast to the N-acyl side chain the amino group seems to be essential for substrate recognition. Substitution of the amino group with a methylene group inhibits metabolization of this mannose derivative.[41] For some enzymes of sialic acid biosynthesis, the promiscuity for different substrates was investigated in more detail. A critical role within the metabolization of unnatural sialic acid precursors seems to be their phosphorylation by ManNAc kinase. As mentioned above, the enzyme is capable of accepting ManNGc as well as ManNAc and also ManNProp as substrates. However, sugars with longer side chains are not phosphorylated by the kinase (S. Hinderlich, unpublished observation). Nevertheless, metabolization of ManNBut and ManNPent was unequivocally shown, therefore these sugars have to be phosphorylated for further metabolization. Presumably, phosphorylation of these sugars is taken over by another kinase. For ManNBut, ManNPent, and ManNLev it was shown that they can be phosphorylated by GlcNAc kinase *in vitro* (S. Hinderlich, unpublished observation).

Metabolic sialic acid precursors can also be incorporated into polysialic acid. Whereas the incorporation of ManNProp, as well as ManNLev,[42] into polysialic acid appears to occur readily, surprising results were found for cells treated with ManNBut and ManNPent. These cells lost their ability to express polysialic acid,[43,44] supposedly due to inhibition of the polysialyltransferases by the respective CMP-sialic acid derivatives. A similar observation was made for the CMP-Neu5Ac hydroxylase, which also does not accept modifications of the N-acyl side chain of its substrate,[45] most likely due to the fact that this functional group is modified by the enzyme.

N-acyl derivatives of mannosamine are nearly exclusively metabolized to sialic acids. A conversion into the respective glucosamine derivatives, followed by incorporation into glycoconjugates, cannot be observed.[46] *Vice versa*, the incorporation of glucosamine derivatives into oligosaccharides after cell treatment is much less effective than that of the respective mannosamine derivatives.[9,47,48] Interestingly, it was recently found that glucosamines with modified N-acyl side chains can be metabolized intracellularly. Vocadlo and co-workers[49] showed that N-azidoacetylglucosamine (GlcNAz) is converted into UDP-GlcNAz by the enzymes of the amino sugar pathway. This nucleotide sugar then serves as a substrate for the O-GlcNAc transferase, and azido-modified O-GlcNAc can be found on proteins. In contrast, the modified glucosamine was hardly incorporated into oligosaccharides,[48] indicating that either the UDP-GlcNAc transporter or the GlcNAc transferases do not tolerate the modification. This result is presumably due to the lack of naturally occurring modifications of GlcNAc and demonstrates the advantages of the sialic acid biosynthetic pathway for modulating the structure of glycoconjugates.

As previously mentioned, biochemical engineering of sialic acids by modified metabolic precursors is based on the extracellular application of the compounds and its intracellular metabolization. Thus, the first important step is uptake of the sugars by cells. There are no plasma membrane transporters known for amino sugars or sialic acids. Furthermore, the hexose transporters have a very high substrate specificity.[50] Therefore it is unlikely that sugar analogs reach the cytosol by direct transport. It should be noted that millimolar concentrations are needed for substantial incorporation of analogs into glycoconjugates.[9] It can be speculated that the compounds are taken up by endocytosis. Indirect evidence for this assumption can be taken from the use of peracetylated sugar analogs for cell treatment. All free hydroxyl groups of the sugars are modified by acetyl groups in order to increase hydrophobicity of the compounds, promoting their diffusion through the plasma membrane. Using ManNGc as an example, Collins and co-workers[51] demonstrated that peracetylation drastically increases cellular uptake. Once reaching the cytosol it is thought that the acetyl groups of peracetylated compounds are released by nonspecific acetylases,[52,53] and the sugar analogs can then enter the sialic acid biosynthetic pathway. Peracetylated compounds can be used in concentrations 10- to 100-fold lower than the nonmodified counterparts giving similar incorporation rates.[51,54] It can therefore be concluded that peracetylated sugars diffuse through the plasma membrane, whereas the only rational possibility for nonacetylated sugars is the endocytotic pathway.

The efficiency of incorporation of the sialic acid precursors into oligosaccharides strongly depends on the type of cells used. The duration of cell treatment with the sugars for maximal incorporation differs from 2 days for the B cell line BJA-B[9] to 6 days for the Jurkat T-cell line.[47] These differences are most likely due to different turnover rates of oligosaccharide-bound sialic acids because the time period for complete degradation of modified sialic acids is the same as for maximal incorporation.[47] The rate of unnatural sialic acids incorporated into glycoconjugates varies from 10 to 90% between cell lines (see Reference 55 for overview). The most crucial factor for incorporation seems to be the endogenous cellular production of ManNAc. ManNAc and its products are in competition with the analogs during metabolization, and the natural compounds are preferred by the different enzymes. The enzyme responsible for the intracellular production of ManNAc is UDP-GlcNAc 2-epimerase. Cells lacking this enzyme showed an incorporation rate of unnatural sialic acid precursors near to 100%.[56] On the other hand, UDP-GlcNAc 2-epimerase is inhibited by CMP-Neu5Ac which limits the production of ManNAc. CMP-Neu5Prop and CMP-Neu5But are also capable of inhibiting UDP-GlcNAc 2-epimerase (S. Hinderlich, unpublished observation). By employing this mechanism, metabolized analogs can suppress the synthesis of endogenous ManNAc and promote their own metabolization. This explanation may account for observation that substantial analog incorporation can be achieved even in cells expressing high amounts of UDP-GlcNAc 2-epimerase.

Modulation of the structure of sialic acids by mannosamine analogs is an effective method but is limited to modifications of the *N*-acyl side chain. For example, modifications at C-6 prevent phosphorylation at this position and were therefore not accepted for metabolization.[10] The C-6 position of ManNAc becomes C-9 in sialic acids after addition of PEP (Figure 13.4). The C-9 position of sialic acid is important for several biological functions but is not accessible for modulation by mannosamine analogs. Therefore an alternative method was developed based on the observation that free sialic acids can be taken up by cells.[57] Different derivatives of sialic acids were synthesized (Figure 13.5) and applied to cells, an incorporation comparable

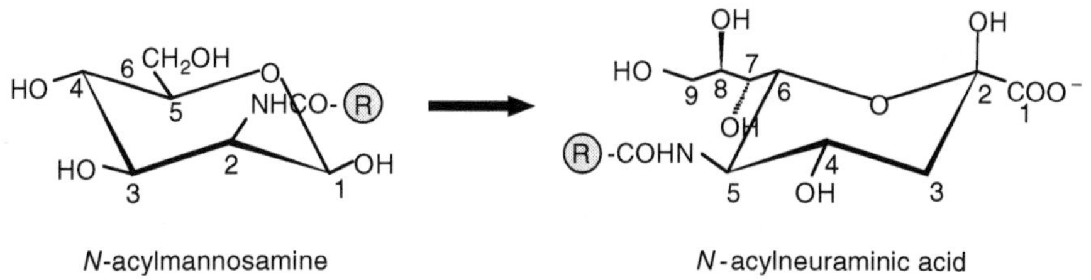

N-acylmannosamine   N-acylneuraminic acid

FIGURE 13.4 Structure of N-acyl-modified neuraminic acid and the corresponding sialic acid precursor N-acylmannosamine. Substituents R used for the N-acyl group are propanoyl-, butanoyl-, pentanoyl-, levolinoyl-, glycolyl- and azidoacetyl residues. The black area in the N-acylneuraminic acid structure originates from N-acylmannosamine and the gray area from PEP.

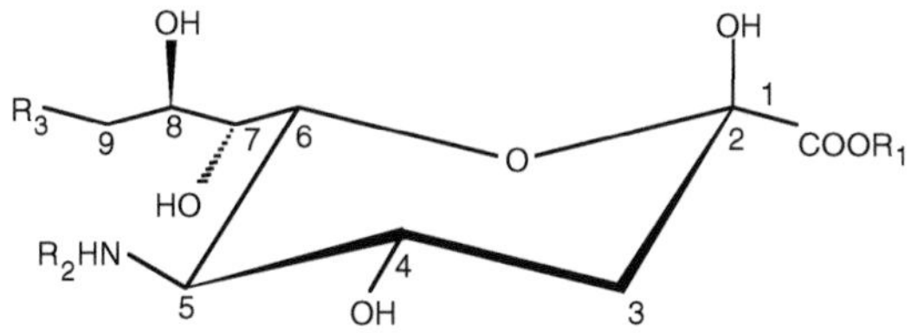

(A)

Sialic Acids	Substituents			Effect on Lectin Binding	Incorporation shown by HPLC Analysis
	R_1	R_2	R_3		
NeuAc	H		HO		
9-Deoxy-NeuAc	H	CH_3CO	H	+	
9-Amino-NeuAc	H	CH_3CO	H_2N	±	
9-Acetamido-NeuAc	H	CH_3CO	CH_3CO-NH	±	
9-N-Gly-NeuAc	H	CH_3CO	H_2NCH_2CO-NH	±	
9-N-Succ-NeuAc	H	CH_3CO	$HOOC(CH_2)_2CO$-NH	±	
9-Iodo-NeuAc	H	CH_3CO	I	+	+
9-Thio-NeuAc	H	CH_3CO	HS	+	
9-SCH_3-NeuAc	H	CH_3CO	CH_3S	+	
9-SO_2CH_3-NeuAc	H	CH_3CO	CH_3SO_2	+	
5-N-fluoroac-Neu	H	FCH_2CO	HO	+	+
5-N-trifluoroac-Neu	H	CF_3CO	HO	+	
5-N-Gly-Neu	H	H_2NCH_2CO	HO	±	
5-N-Succ-Neu	H	$HOOC(CH_2)_2CO$	HO	±	
5-N-thioac-Neu	H	CH_3CS	HO	+	
5-N-Gc-Neu	H	OH	HO	+	
NeuAc-Me-ester	H_3C	CH_3CO	HO	+	
NeuAc-Et-ester	H_5C_2	CH_3CO	HO	+	

(B)

FIGURE 13.5 Synthetic sialic acid analogs investigated for incorporation into cellular oligosaccharides. (A) All derivatives based on Neu5Ac. (B) Sialic acids used are substituted either in position C-1 (R_1), C-5 (R_2), or C-9 (R_3). Lectin-binding studies were performed using *Limax flavus* agglutinin, *Tritrichomonas mobilensis* lectin and *Vicia villosa* agglutinin.[10] (From Oetke, C. et al., *J. Biol. Chem.*, 277, 6688–6695, 2002. With permission.)

with mannosamine analogs was observed.[10] By using this method, new positions of sialic acids can be modified and investigated based on their biological effects. To date, only C-9- and C-5-modified sialic acids have been tested, although it would be interesting to try modifications at other sites too. The striking advantages of sialic acid derivatives over mannosamines are that fewer enzymes are involved in metabolization and that the acyl group is already present at the C-5 position. However, extending the modifications and modification sites is not only dependent on the acceptance by the cellular enzymes but also requires a patient and skilled chemist. Despite the fact that C-9- and C-5-modified sialic acids are incorporated, some modification at these positions prevented metabolization. Although some of them were bulky or charged it is not predictable whether the modification will interfere with incorporation. As already discussed, enzymes are expected to be more restricted at substrate positions other than the *N*-acyl group in C-5. Finally, it needs to be mentioned that naturally occurring acetylation of the sialic acid, found at the C-7, C-8, and C-9 position, cannot be metabolically incorporated into glycoconjugates since these would be cleaved intracellularly by acetylases.

III. APPLICATIONS OF BIOCHEMICAL ENGINEERING OF SIALIC ACIDS

A. BIOLOGICAL EFFECTS OF UNNATURAL SIALIC ACIDS

There are numerous biological processes in which sialic acids are directly or indirectly involved (Table 13.1). Nevertheless, in most cases it is unclear which part of the sugar is important for the respective function. The technique of biochemical engineering of parts of the sialic acid molecule by the use of modified metabolic precursors offers a good opportunity to answer this question. The first system investigated by this method was the interaction of viruses with cells. The cellular receptors recognized by some viruses often contain sialic acids. Treatment of cells with mannosamines containing extended *N*-acyl chains have enabled us to investigate the role of this functional group in virus—receptor interaction. For the human polyoma virus BK, it was found that expression of Neu5Prop and Neu5But on the cell surface results in an increased virus permissivity, whereas modifications with longer alkyl chains reduce this process.[9] These results indicate the presence of a hydrophobic binding pocket for the *N*-acyl chain on the virus. This virus—receptor interaction is therefore increased by more hydrophobic sialic acids. However, *N*-acyl side chains longer than four carbon atoms seemed to be too bulky for binding. Binding of B-lymphotropic polyoma virus,[9] murine polyoma virus,[58] and influenza A virus[59] was always inhibited when the susceptible cell lines were treated with sialic acid precursors with elongated *N*-acyl side chains. Molecular modeling studies of the sialic acid binding site of the murine polyoma virus[58] and influenza A virus[59] showed steric hindrance of the virus–sialic acid interaction by addition of only one methylene group to the side chain of the sugar, indicating a sugar binding pocket too small for binding these modified sugars. In addition to elongation of the *N*-acyl side chain, virus–cell interaction was also investigated by treatment of the cells with sialic acid derivatives. Interestingly and in contrast to the results using mannosamine analogs,

TABLE 13.1
Biological Effects of Unnatural Sialic Acids

Precursor	Cells Analyzed	Biological Effect	Ref.
ManNProp ManNBut ManNPent	Monkey kidney epithelium (Vero)	Increased (ManNProp and ManBut) or decreased (ManNPent) permissivity for human polyoma virus BK	9
ManNProp ManNBut ManNPent	Human B-lymphoma (BJA-B)	Decreased infection by B-lymphotropic polyoma virus	9
9-Iodo-Neu5Ac 5-N-Fluoro-Neu5Ac	Human B-lymphoma (BJA-B)	Increased infection by B-lymphotropic polyoma virus	10
ManNProp ManNBut ManNPent	Mouse fibroblasts (NIH-3T6)	Decreased binding of murine polyoma virus	58
ManNProp ManNBut ManNPent	Dog kidney epithelium (MDCK-II)	Decreased binding of Influenza A virus	59
ManNProp	Rat pheochromocytoma (PC12)	Increased half-life time of glycoproteins	62
ManNGc	Rodent neuroblastomaglioma hybrid cells (NG108-15)	Abolition of binding of myelin-associated glycoprotein	51
ManNProp ManNBut ManNPent	Human lung fibroblasts	Abolition of contact inhibin-mediated inhibition of cell growth	63
ManNProp	Human T-lymphocytes	Stimulation of proliferation, increased IL-2 expression	64
ManNProp	Primary rat brain cells	Stimulation of proliferation of astrocytes and microglia, induced expression of A2B5 epitope on oligodendrocytes	65
ManNProp	Rat oligodendrocytes	Initiation of calcium oscillation after stimulation with γ-aminobutyric acid	66
ManNLev	Terminally differentiated human neurons (NT2)	Incorporation into polysialic acid, protection against sialidases	42
ManNProp	Primary rat neurons and PC12	Increased axonal outgrowth	67

in these experiments, infection of cells by the B-lymphotropic polyoma virus was increased by presentation of 9-iodo-Neu5Ac and 5-N-fluoro-Neu5Ac on the cell surface.[10] This result shows that it is possible in general to increase viral—receptor binding by distinct modifications of the structure of sialic acids. Together these data display the important role of the N-acyl side chain and the acyl side chain in the

C-6 position of sialic acid for the binding of viruses and may be an important structural basis for antiviral strategies based on sialic acid analogs, as has been shown for potent inhibitors of influenza A virus neuraminidase.[60,61]

A further example of the biological effects of sialic acids with altered structures is the increase in glycoprotein stability.[62] The biological half-life of many glycoproteins is regulated via terminal sialic acids. Engineering of the highly sialylated transmembrane glycoprotein CEACAM 1 by ManNProp in PC12 cells resulted in an estimated 50% increase in half-life. The α1-integrin subunit was not affected by engineering, indicating that the stability of some, but not all, glycoproteins is influenced by the structure of sialic acids.

Conversion of sialic acids of neuronal cells into Neu5Gc by treatment with peracetylated ManNGc resulted in about 80% modified gangliosides. Gangliosides are the presumed ligands of the myelin-associated glycoprotein (MAG) and, consequently, interaction of the protein with its ligand was abolished.[51] This interaction is implicated in the inhibition of axon regeneration following injury and treatment with ManNGc, which may enhance the possibility of posttraumatic nerve regeneration. In a way similar to the described MAG — ligand interaction, the contactinhibin-mediated inhibition of cell growth of human lung fibroblasts is also suppressed by treatment of the cells with ManNAc analogs of different N-acyl chain length.[63]

The effects of treatment of the cells with unnatural sialic acid precursors described so far can be attributed to known receptor-ligand interactions; in addition, less-defined effects were observed. Human T-lymphocytes are stimulated to proliferate by ManNProp in a concentration-dependent manner. At concentrations of 10 mM, proliferation was in the same range as observed for the commonly used strong mitogen concanavalin A.[64] In addition to proliferation, IL-2 secretion and expression of the IL-2-receptor by T-cells were also induced. Treatment of neonatal rat brain cultures with ManNProp also increased proliferation of astrocytes and microglia.[65] On the other hand, oligodendrocytes, although not induced to proliferate, expressed a specific marker for the nonmature cell stage after sugar treatment. Furthermore, ManNProp treatment of oligodendrocytes initiated calcium oscillation after stimulation with γ-aminobutyric acid.[66] Finally, axonal outgrowth of primary neurons and the neuronal PC12 cell line was significantly increased by ManNProp.[67] It is unclear however, whether all of these observations are caused by a direct effect of the unnatural sugars in the cytosol or by secondary effects after expression of novel sialic acids on the cell surface. It was found that several cytosolic proteins with regulatory functions are differentially expressed in neuronal cells after treatment with mannosamine analogs, and a direct influence of the sugars or of the respective CMP-sialic acid derivatives on gene expression was suggested.[67]

B.   BIOMEDICAL APPLICATIONS

Besides the investigation of the functional roles of sialic acids, biochemical engineering of the sugar can also be used for technological applications (Table 13.2). Modulating the structure of sialic acids by metabolic precursors allows the expression of functional groups on the cell surface not present under normal conditions. These functional groups can undergo specific chemical reactions and allow binding

TABLE 13.2
Applications of Biochemical Engineering of Sialic Acids

Precursor	Cell System	Biological Effect	Ref.
ManNLev	Human T-lymphoma (Jurkat), human myeloid leukemia (HL-60), human cervix carcinoma (HeLa)	Expression of keto groups on cell surface for selective chemical addition of ligands	68, 47
ManNAz	Human T-lymphoma (Jurkat), human cervix carcinoma (HeLa)	Expression of azido groups on cell surface for selective chemical addition of ligands	54
ManNLev	Human T-lymphoma (Jurkat)	Binding of contrast reagents on cell surface keto groups	69
ManNLev	Mouse fibroblasts (NIH-3T3), human umbilical vein endothelial cells (HUVEC)	Binding of adenovirus antibody and increased virus permissivity	72
ManNLev	Human T-lymphoma (Jurkat)	Identification of glycosylation defects	73
ManNProp	Rat leukemia (RBL-3H3), mouse leukemia (RMA)	Engineering of polysialic acid, immunotargeting for complement-mediated cell lysis	75, 76
ManNBut ManNPent	Human cervix carcinoma (HeLa), human neuroblastoma (SH-SY5Y), human small cell lung carcinoma (H345), human neurons (NT2), chick dorsal root ganglion neurons	Abolition of polysialic acid expression, reduced neurite outgrowth	43, 44

of distinct molecules on the cell surface. Two different tools for this technique have been developed by Bertozzi and colleagues.[54,68] The first method is based on the addition of a levolinoyl group to the amino group of mannosamine. Cells treated with this sugar, N-levolinoylmannosamine (ManNLev), express a keto group within the N-acyl side chain of cell surface sialic acids. Although keto groups do exist in the cytosol, they are not present on the cell surface. Therefore, they can undergo highly specific chemical reactions, as for example with hydrazides to form stable hydrazones. Different molecules can be covalently bound to cell surface sialic acids by this reaction.[47,68] In the case of biotin, it can then be detected by fluorescently labeled avidin. This method has been used for quantitative detection of keto groups on the cell surface by FACS analysis. Similar to the keto group, the N-acyl side chain of mannosamine can also be modified by an azido group. Cells treated with N-azidoacetylmannosamine (ManNAz) express the reactive azido group linked to sialic acids on the cell surface. It can then react with a phosphine to yield a stable amide. Again, the azido group can be detected by coupling of biotin to the modified sialic acid followed by FACS analysis.

Lemieux and co-workers[69] used the expression of unique functional groups for the binding of contrast reagents to the cell surface. Cells were treated with

ManNLev, and the contrast reagent, modified by an aminooxy group, was bound to Neu5Lev presented on the cell surface. Expression levels of modified sialic acids depend on the total amount of sialic acid present on the cells. Cells with a high-sialylation phenotype are capable of binding significantly more contrast reagents than cells of a low-sialylation phenotype. Since several cancer cells belong to the high-sialylation phenotype,[70,71] an application of this method for detection of cancer tissues *in vivo* was suggested. Reactive groups engineered on the cell surface can also be used to facilitate entry and uptake of therapeutic macromolecules. Biotin was bound to the surface of cells treated with ManNLev. An avidin-conjugated antibody directed against adenoviruses then allowed infection of naturally nonpermissive NIH-3T3 cells. By this method adenoviral gene transfer was made possible.[72]

Yarema and co-workers[73] used the advantages of cell treatment with ManNLev followed by easy detection of Neu5Lev by FACS analysis for the identification of metabolic glycosylation defects in cells. This technique is based on metabolization of the unnatural sugar by the sialic acid pathway. Defects of enzymes involved in this pathway or altered levels of acceptor sites for sialic acids, glycoproteins or glycolipids, lead to changes in the expression of Neu5Lev on the cell surface. For instance, cells expressing sialuria mutations of the UDP-GlcNAc 2-epimerase showed drastically decreased incorporation of unnatural sialic acid. Sialuria is a human disease characterized by the synthesis of high amounts of sialic acids. It is caused by point mutations of the CMP-sialic acid binding site of UDP-GlcNAc 2-epimerase, resulting in abolition of the feedback inhibition mechanism.[74] Excess of sialic acids in the cytosol of these cells competes with the unnatural sialic acid precursors and reduces their metabolization. Altered expression of sialylated cell surface proteins can also be detected by this method. Overexpression of neural cell adhesion molecule (N-CAM), which is the main carrier of polysialic acid, leads to an enhanced presentation of sialic acids, especially polysialic acid, and consequently leads to an increased exposure of ketone groups when the cells are treated with ManNLev.

Polysialic acid is also the target of other biomedical applications using biochemical engineering of sialic acids. Antibodies can be raised against sialic acid with artificially altered structures, such as when polysialic acid is modified by *N*-propanoyl groups, especially when several of these epitopes are expressed back to back in the polymer. Pon and co-workers[75] used poly-Neu5Prop for the generation of a specific antibody against this unusual epitope. Cells treated with ManNProp present the novel polysialic acid on the cell surface and can therefore be detected by the antibody.[76] In adults, polysialic acid is primarily found in the brain but also on a number of cancer tissues, such as small cell lung carcinoma or Wilms' tumor.[77,78] Expression of poly-Neu5Prop on these tumors might become part of a new anticancer strategy. Either cytotoxic agents could be coupled to the antibody or the bound antibody could initiate cell lysis directly by activating the complement cascade.[76] A different approach of tumor therapy is based on the observation that cells treated with ManNBut or ManNPent lose their capability to synthesize polysialic acid.[43] Polysialic acid is thought to promote tumor metastasis.[79,80] Treatment of tumor cells with the unnatural sialic precursors might therefore be helpful for therapy in combination with other anticancer agents.

IV. CONCLUSIONS

The method of biochemical engineering of sialic acids is a powerful tool for the modulation of the structures of cellular glycoconjugates. Several applications for this method are presented in this chapter. On the one hand, there is an academic interest to learn more about the functions of sialic acids; on the other lies the foundation for technological and biomedical applications. One major advantage of biochemical engineering is the utilization of the intracellular metabolism of sialic acids for incorporation of modified sugars into glycoconjugates. Therefore, only small molecules have to be synthesized and the more complex compounds, such as nucleotide sugars, are made by the cell. The synthesis of mannosamine analogs is highly achievable for scientists not equipped with a complete chemical lab and some of the compounds are even commercially available. For the more sophisticated chemistry of sialic acids assistance from a skillful chemist is advisable.

Advances have been made in the efficiency of biochemical glycoconjugate engineering. Several years ago, millimolar concentrations of analogs had to be applied to get substantial incorporation by cells. Therefore, often gram amounts of the sugars had to be synthesized for functional studies. Generation of peracetylated variants of the sugars drastically improved cellular incorporation and reduced the necessary amounts of substance by several orders of magnitude. Although the sialic acid biosynthetic pathway seems to be well suited for modulation of the structure of glycoconjugates, the method appears not to be restricted to this monosaccharide. Analogs of *N*-acetylgalactosamine, having a methylene group instead of the amino group, are incorporated into glycoconjugates by metabolization of the respective pathways.[81] Very recently, it was also shown that the amino sugar pathway is also permissive for GlcNAc analogs and at least intracellular glycoproteins can be modified.[49] Thus, biochemical engineering might be extended to other kinds of sugars and perhaps to other biomolecules in the future.

For biochemical engineering of sugars and sialic acids to meet its final goal of biomedical application, several major questions must be addressed. So far, very few *in vivo* experiments have employed artificial sialic acids or precursors in complex organisms like the mouse or rat. There is also a need to investigate whether some tissues or cell types incorporate the derivatives at a higher rate than others, and whether selective uptake is particularly true for normal vs. cancer cells. Even more important will be the effects of incorporated artificial sialic acids on sialic acid-dependent interactions within a complex organism. The consequences of the above intersection need to be carefully investigated, especially with respect to the immune system. However, some interesting ideas are in progress and biochemical engineering may become as useful and established *in vivo* as it already is *ex vivo*.

ACKNOWLEDGMENTS

The authors gratefully acknowledge financial support of the Fonds der Chemischen Industrie (Frankfurt, Germany), the Sonnenfeld-Stiftung (Berlin, Germany), and the Wilhelm Sander-Stiftung (München, Germany).

REFERENCES

1. Angata, T. and Varki, A., Chemical diversity in the sialic acids and related alpha-keto acids: an evolutionary perspective, *Chem. Rev.,* 102, 439–469, 2002.
2. Edelman, G.M. and Crossin, K.L., Cell adhesion molecules: implications for a molecular histology, *Annu. Rev. Biochem.,* 60, 155–190, 1991.
3. Varki, A., Diversity in the sialic acids, *Glycobiology,* 2, 25–40, 1992.
4. Rens-Domiano, S. and Reisine, T., Structural analysis and functional role of the carbohydrate component of somatostatin receptors, *J. Biol. Chem.,* 266, 20094–20102, 1991.
5. Gross, H.J., Merling, A., Moldenhauer, G., and Schwartz-Albiez, R., Ecto-sialyltransferase of human B lymphocytes reconstitutes differentiation markers in the presence of exogenous CMP-*N*-acetyl neuraminic acid, *Blood,* 87, 5113–5126, 1996.
6. Bergler, W., Riedel, F., Schwartz-Albiez, R., Gross, H.J., and Hormann, K., A new histobiochemical method to analyze sialylation on cell-surface glycoproteins of head and neck squamous-cell carcinomas, *Eur. Arch. Otorhinolaryngol,* 254, 437–441, 1997.
7. Grünholz, H.J., Harms, E., Opetz, M., Reutter, W., and Cerny, M., Inhibition of *in vitro* biosynthesis of *N*-acetylneuraminic acid by *N*-acyl- and *N*-alkyl-2-amino-2-deoxyhexoses, *Carbohydr. Res.,* 96, 259–270, 1981.
8. Kayser, H., Zeitler, R., Kannicht, C., Grunow, D., Nuck, R., and Reutter, W., Biosynthesis of a nonphysiological sialic acid in different rat organs, using *N*-propanoyl-D-hexosamines as precursors, *J. Biol. Chem.,* 267, 16934–16938, 1992.
9. Keppler, O.T., Stehling, P., Herrmann, M., Kayser, H., Grunow, D., Reutter, W., and Pawlita, M., Biosynthetic modulation of sialic acid-dependent virus–receptor interactions of two primate polyoma viruses, *J. Biol. Chem.,* 270, 1308–1314, 1995.
10. Oetke, C., Brossmer, R., Mantey, L.R., Hinderlich, S., Isecke, R., Reutter, W., Keppler, O.T., and Pawlita, M., Versatile biosynthetic engineering of sialic acid in living cells using synthetic sialic acid analogues, *J. Biol. Chem.,* 277, 6688–6695, 2002.
11. Schauer, R., Achievements and challenges of sialic acid research, *Glycoconj. J.,* 17, 485–499, 2000.
12. Milewski, S., Glucosamine-6-phosphate synthase — the multi-facets enzyme, *Biochim. Biophys. Acta.,* 1597, 173–192, 2002.
13. Hakomori, S., Traveling for the glycosphingolipid path, *Glycoconj, J.,* 17, 627–647, 2000.
14. Schachter, H., The joys of HexNAc. The synthesis and function of *N*- and *O*-glycan branches, *Glycoconj. J.,* 17, 465–483, 2000.
15. Esko, J.D. and Selleck, S.B., Order out of chaos: assembly of ligand binding sites in heparan sulfate, *Annu. Rev. Biochem.,* 71, 435–471, 2002.
16. Eisenhaber, B., Maurer-Stroh, S., Novatchkova, M., Schneider, G., and Eisenhaber, F., Enzymes and auxiliary factors for GPI lipid anchor biosynthesis and post-translational transfer to proteins, *Bioessays,* 25, 367–385, 2003.
17. Wells, L., Vosseller, K., and Hart, G.W., Glycosylation of nucleocytoplasmic proteins: signal transduction and *O*-GlcNAc, *Science,* 291, 2376–2378, 2001.
18. Comb, D.G. and Roseman, S., Enzymatic synthesis of *N*-acetyl-D-mannosamine, *Biochim. Biophys. Acta.,* 29, 653–654, 1958.
19. Gosh, S. and Roseman, S., Enzymatic phosphorylation of *N*-acetyl-D-mannosamine, *Proc. Natl. Acad. Sci. USA,* 47, 955–958, 1961.
20. Warren, L. and Felsenfeld, H., *N*-acetylmannosamine-6-phosphate and *N*-acetylneuraminic acid-9-phosphate as intermediates in sialic acid biosynthesis, *Biochem. Biophys. Res. Commun.,* 5, 185–190, 1961.

21. Hinderlich, S., Stäsche, R., Zeitler, R., and Reutter, W., A bifunctional enzyme cat-
     alyzes the first two steps in *N*-acetylneuraminic acid biosynthesis of rat liver.
     Purification and characterization of UDP-*N*-acetylglucosamine 2-epimerase/*N*-
     acetylmannosamine kinase, *J. Biol. Chem.,* 272, 24313–24318, 1997.

22. Stäsche, R., Hinderlich, S., Weise, C., Effertz, K., Lucka, L., Moormann, P., and
     Reutter, W., A bifunctional enzyme catalyzes the first two steps in *N*-acetylneu-
     raminic acid biosynthesis of rat liver. Molecular cloning and functional expression of
     UDP-*N*-acetyl-glucosamine 2-epimerase/*N*-acetylmannosamine kinase, *J. Biol.
     Chem.,* 272, 24319–24324, 1997.

23. Horstkorte, R., Nöhring, S., Wiechens, N., Schwarzkopf, M., Danker, K., Reutter, W.,
     and Lucka, L., Tissue expression and amino acid sequence of murine UDP-*N*-acetyl-
     glucosamine-2-epimerase/*N*-acetylmannosamine kinase, *Eur. J. Biochem.,* 260,
     923–927, 1999.

24. Lucka, L., Krause, M., Danker, K., Reutter, W., and Horstkorte, R., Primary structure
     and expression analysis of human UDP-*N*-acetyl-glucosamine-2-epimerase/*N*-acetyl-
     mannosamine kinase, the bifunctional enzyme in neuraminic acid biosynthesis, *FEBS
     Lett.,* 454, 341–344, 1999.

25. Effertz, K., Hinderlich, S., and Reutter, W., Selective loss of either the epimerase or
     kinase activity of UDP-*N*-acetylglucosamine 2-epimerase/*N*-acetylmannosamine
     kinase due to site-directed mutagenesis based on sequence alignments, *J. Biol. Chem.,*
     274, 28771–28778, 1999.

26. Kornfeld, S., Kornfeld, R., Neufeld, E., and O'Brien, P.J., The feedback control of
     sugar nucleotide biosynthesis in liver, *Proc. Natl. Acad. Sci. USA,* 52, 371–379, 1964.

27. Harms, E., Kreisel, W., Morris, H.P., and Reutter, W., Biosynthesis of *N*-acetylneu-
     raminic acid in Morris hepatomas, *Eur. J. Biochem.,* 32, 254–262, 1973.

28. Horstkorte, R., Nöhring, S., Danker, K., Effertz, K., Reutter, W., and Lucka, L.,
     Protein kinase C phosphorylates and regulates UDP-*N*-acetylglucosamine-2-
     epimerase/*N*-acetylmannosamine kinase, *FEBS Lett.,* 470, 315–318, 2000.

29. Roseman, S., Jourdian, G.W., Watson, D., and Rood, R., Enzymatic synthesis of sialic
     acid 9-phosphates, *Proc. Natl. Acad. Sci. USA,* 47, 958–961, 1961.

30. Warren, L. and Felsenfeld, H., The biosynthesis of sialic acids, *J. Biol. Chem.,* 237,
     1421–1431, 1962.

31. Kean, E.L., Sialic acid activating enzyme in ocular tissue, *Exp. Eye. Res.,* 8, 44–54, 1969.

32. Kean, E.L., Nuclear cytidine 5′ -monophosphosialic acid synthetase, *J. Biol. Chem.,*
     245, 2301–2308, 1970.

33. Münster, A.K., Eckhardt, M., Potvin, B., Mühlenhoff, M., Stanley, P., and Gerardy-
     Schahn, R., Mammalian cytidine 5′-monophosphate *N*-acetylneuraminic acid syn-
     thetase: a nuclear protein with evolutionarily conserved structural motifs, *Proc. Natl.
     Acad. Sci. USA,* 95, 9140–9145, 1998.

34. Ferwerda, W., Blok, C.M., and van Rinsum, J., Synthesis of *N*-acetylneuraminic acid
     and of CMP-*N*-acetylneuraminic acid in the rat liver cell, *Biochem. J.,* 216, 87–92,
     1983.

35. Kean, E.L., Sialic acid activation, *Glycobiology,* 1, 441–447, 1991.

36. Mancini, G.M., de Jonge, H.R., Galjaard, H., and Verheijen, F.W., Characterization of
     a proton-driven carrier for sialic acid in the lysosomal membrane. Evidence for a
     group-specific transport system for acidic monosaccharides, *J. Biol. Chem.,* 264,
     15247–15254, 1989.

37. Brunetti, P., Jourdian, G.W., and Roseman, S., The sialic acids III. Distribution and
     properties of animal *N*-acetylneuraminic acid aldolase, *J. Biol. Chem.,* 237,
     2447–2453, 1962.

38. Gosh, S. and Roseman, S., The sialic acids. V. *N*-Acyl-D-glucosamine 2-epimerase, *J. Biol. Chem.*, 240, 1531–1536, 1965.
39. Kundig, W., Gosh, S., and Roseman, S., The sialic acids. VII. *N*-Acyl-D-mannosamine kinase from rat liver, *J. Biol. Chem.*, 241, 5619–5626, 1966.
40. Kean, E.L. and Roseman, S., The sialic acids. X. Purification and properties of cytidine 5′-monophosphosialic acid synthetase, *J. Biol. Chem.*, 241, 5643–5650, 1966.
41. Jacobs, C.L., Goon, S., Yarema, K.J., Hinderlich, S., Hang, H.C., Chai, D.H., and Bertozzi, C.R., Substrate specificity of the sialic acid biosynthetic pathway, *Biochemistry*, 40, 12864–12874, 2001.
42. Charter, N.W., Mahal, L.K., Koshland, Jr., D.E., and Bertozzi, C.R., Biosynthetic incorporation of unnatural sialic acids into polysialic acid on neural cells, *Glycobiology*, 10, 1049–1056, 2000.
43. Mahal, L.K., Charter, N.W., Angata, K., Fukuda, M., Koshland, Jr., D.E., and Bertozzi, C.R., A small-molecule modulator of poly-alpha 2,8-sialic acid expression on cultured neurons and tumor cells, *Science*, 294, 380–381, 2001.
44. Charter, N.W., Mahal, L.K., Koshland, Jr., D.E., and Bertozzi, C.R., Differential effects of unnatural sialic acids on the polysialylation of the neural cell adhesion molecule and neuronal behavior, *J. Biol. Chem.*, 277, 9255–9261, 2002.
45. Humphrey, A.J., Fremann, C., Critchley, P., Malykh, Y., Schauer, R., and Bugg, T.D., Biological properties of *N*-acyl and *N*-haloacetyl neuraminic acids: processing by enzymes of sialic acid metabolism, and interaction with influenza virus, *Bioorg. Med. Chem.*, 10, 3175–3185, 2002.
46. Stehling, P., Grams, S., Nuck, R., Grunow, D., Reutter, W., and Gohlke, M., *In vivo* modulation of the acidic *N*-glycans from rat liver dipeptidyl peptidase IV by *N*-propanoyl-D-mannosamine, *Biochem. Biophys. Res. Commun.*, 263, 76–80, 1999.
47. Yarema, K.J., Mahal, L.K., Bruehl, R.E., Rodriguez, E.C., and Bertozzi, C.R., Metabolic delivery of ketone groups to sialic acid residues. Application to cell surface glycoform engineering, *J. Biol. Chem.*, 273, 31168–31179, 1998.
48. Saxon, E., Luchansky, S.J., Hang, H.C., Yu, C., Lee, S.C., and Bertozzi, C.R., Investigating cellular metabolism of synthetic azidosugars with the Staudinger ligation, *J. Am. Chem. Soc.*, 124, 14893–14902, 2002.
49. Vocadlo, D.J., Hang, H.C., Kim, E.J., Hanover, J.A., and Bertozzi, C.R., A chemical approach for identifying *O*-GlcNAc-modified proteins in cells, *Proc. Natl. Acad. Sci. USA*, 21, 21, 2003.
50. Silverman, M., Structure and function of hexose transporters, *Annu. Rev. Biochem.*, 60, 757–794, 1991.
51. Collins, B.E., Fralich, T.J., Itonori, S., Ichikawa, Y., and Schnaar, R.L., Conversion of cellular sialic acid expression from *N*-acetyl- to *N*-glycolylneuraminic acid using a synthetic precursor, *N*-glycolylmannosamine pentaacetate: inhibition of myelin-associated glycoprotein binding to neural cells, *Glycobiology*, 10, 11–20, 2000.
52. Sarkar, A.K., Fritz, T.A., Taylor, W.H., and Esko, J.D., Disaccharide uptake and priming in animal cells: inhibition of sialyl Lewis X by acetylated Gal beta 1→4GlcNAc beta-*O*-naphthalenemethanol, *Proc. Natl. Acad. Sci. USA*, 92, 3323–3327, 1995.
53. Sarkar, A.K., Rostand, K.S., Jain, R.K., Matta, K.L., and Esko, J.D., Fucosylation of disaccharide precursors of sialyl LewisX inhibit selectin-mediated cell adhesion, *J. Biol. Chem.*, 272, 25608–25616, 1997.
54. Saxon, E. and Bertozzi, C.R., Cell surface engineering by a modified Staudinger reaction, *Science*, 287, 2007–2010, 2000.

55. Keppler, O.T., Horstkorte, R., Pawlita, M., Schmidt, C., and Reutter, W., Biochemical engineering of the *N*-acyl side chain of sialic acid: biological implications, *Glycobiology*, 11, 11R–18R, 2001.

56. Mantey, L.R., Keppler, O.T., Pawlita, M., Reutter, W., and Hinderlich, S., Efficient biochemical engineering of cellular sialic acids using an unphysiological sialic acid precursor in cells lacking UDP-*N*-acetylglucosamine 2-epimerase, *FEBS Lett.*, 503, 80–84, 2001.

57. Oetke, C., Hinderlich, S., Brossmer, R., Reutter, W., Pawlita, M., and Keppler, O.T., Evidence for efficient uptake and incorporation of sialic acid by eukaryotic cells, *Eur. J. Biochem.*, 268, 4553–4561, 2001.

58. Herrmann, M., von der Lieth, C.W., Stehling, P., Reutter, W., and Pawlita, M., Consequences of a subtle sialic acid modification on the murine polyomavirus receptor, *J. Virol.*, 71, 5922–5931, 1997.

59. Keppler, O.T., Herrmann, M., von der Lieth, C.W., Stehling, P., Reutter, W., and Pawlita, M., Elongation of the *N*-acyl side chain of sialic acids in MDCK II cells inhibits influenza A virus infection, *Biochem. Biophys. Res. Commun.*, 253, 437–442, 1998.

60. von Itzstein, M., Wu, W.Y., Kok, G.B., Pegg, M.S., Dyason, J.C., Jin, B., Van Phan, T., Smythe, M.L., White, H.F., and Oliver, S.W., et al., Rational design of potent sialidase-based inhibitors of influenza virus replication, *Nature*, 363, 418–423, 1993.

61. Cheer, S.M. and Wagstaff, A.J., Zanamivir: an update of its use in influenza, *Drugs* 62, 71–106, 2002.

62. Horstkorte, R., Lee, H.Y., Lucka, L., Danker, K., Mantey, L., and Reutter, W., Biochemical engineering of the side chain of sialic acids increases the biological stability of the highly sialylated cell adhesion molecule CEACAM1, *Biochem. Biophys. Res. Commun.*, 283, 31–35, 2001.

63. Wieser, J.R., Heisner, A., Stehling, P., Oesch, F., and Reutter, W., *In vivo* modulated *N*-acyl side chain of *N*-acetylneuraminic acid modulates the cell contact-dependent inhibition of growth, *FEBS Lett.*, 395, 170–173, 1996.

64. Reutter, W., Keppler, O.T., Pawlita, M., Schüler, C., Horstkorte, R., Schmidt, C., Hoppe, B., Lucka, L., Herrmann, M., Stehling, P., and Mickeleit, M., Biochemical engineering by new *N*-acylmannosamines of sialic acids creates new biological characteristics and technical tools for its *N*-acyl side chain. in: Inoue, Y., Lee, Y.C., Troy II, F.A., *Sialobiology and Other Forms of Glycosylation*, Eds., Gakushin Publishing Company, Osaka, 1999, pp. 281–288.

65. Schmidt, C., Stehling, P., Schnitzer, J., Reutter, W., and Horstkorte, R., Biochemical engineering of neural cell surfaces by the synthetic *N*-propanoyl-substituted neuraminic acid precursor, *J. Biol. Chem.*, 273, 19146–19152, 1998.

66. Schmidt, C., Ohlemeyer, C., Kettenmann, H., Reutter, W., and Horstkorte, R., Incorporation of *N*-propanoylneuraminic acid leads to calcium oscillations in oligodendrocytes upon the application of GABA, *FEBS Lett.*, 478, 276–280, 2000.

67. Büttner, B., Kannicht, C., Schmidt, C., Löster, K., Reutter, W., Lee, H.Y., Nöhring, S., and Horstkorte, R., Biochemical engineering of cell surface sialic acids stimulates axonal growth, *J. Neurosci.*, 22, 8869–8875, 2002.

68. Mahal, L.K., Yarema, K.J., and Bertozzi, C.R., Engineering chemical reactivity on cell surfaces through oligosaccharide biosynthesis, *Science*, 276, 1125–1128, 1997.

69. Lemieux, G.A., Yarema, K.J., Jacobs, C.L., and Bertozzi, C.R., Exploiting differences in sialoside expression for selective targeting of MRI contrast reagents, *J. Am. Chem. Soc.*, 12, 663–672, 1999.

70. Hakomori, S., Aberrant glycosylation in tumors and tumor-associated carbohydrate antigens, *Adv. Cancer Res.,* 52, 257–331, 1989.
71. Bhavanandan, V.P., Cancer-associated mucins and mucin-type glycoproteins, *Glycobiology,* 1, 493–503, 1991.
72. Lee, J.H., Baker, T.J., Mahal, L.K., Zabner, J., Bertozzi, C.R., Wiemer, D.F., and Welsh, M.J., Engineering novel cell surface receptors for virus-mediated gene transfer, *J. Biol. Chem.,* 274, 21878–21884, 1999.
73. Yarema, K.J., Goon, S., and Bertozzi, C.R., Metabolic selection of glycosylation defects in human cells, *Nat. Biotechnol.,* 19, 553–558, 2001.
74. Seppala, R., Lehto, V.P., and Gahl, W.A., Mutations in the human UDP-*N*-acetylglucosamine 2-epimerase gene define the disease sialuria and the allosteric site of the enzyme, *Am. J. Hum. Genet.,* 64, 1563–1569, 1999.
75. Pon, R.A., Lussier, M., Yang, Q.L., and Jennings, H.J., *N*-Propionylated group B meningococcal polysaccharide mimics a unique bactericidal capsular epitope in group B *Neisseria meningitidis, J. Exp. Med.,* 185, 1929–1938, 1997.
76. Liu, T., Guo, Z., Yang, Q., Sad, S., and Jennings, H.J., Biochemical engineering of surface alpha 2-8 polysialic acid for immunotargeting tumor cells, *J. Biol. Chem.,* 275, 32832–32836, 2000.
77. Troy, F.A., 2nd, Polysialylation: from bacteria to brains, *Glycobiology,* 2, 5–23, 1992.
78. Fukuda, M., Possible roles of tumor-associated carbohydrate antigens, *Cancer Res.,* 56, 2237–2244, 1996.
79. Michalides, R., Kwa, B., Springall, D., van Zandwijk, N., Koopman, J., Hilkens, J., and Mooi, W., NCAM and lung cancer, *Int. J. Cancer. Suppl.,* 8, 34–37, 1994.
80. Scheidegger, E.P., Lackie, P.M., Papay, J., and Roth, J., *In vitro* and *in vivo* growth of clonal sublines of human small cell lung carcinoma is modulated by polysialic acid of the neural cell adhesion molecule, *Lab Invest,* 70, 95–106, 1994.
81. Hang, H.C. and Bertozzi, C.R., Ketone isosteres of 2-*N*-acetamidosugars as substrates for metabolic cell surface engineering, *J. Am. Chem. Soc.,* 123, 1242–1243, 2001.

14 Glycosylation in Native and Engineered Insect Cells

Karthik Viswanathan
and Michael J. Betenbaugh

CONTENTS

I. INTRODUCTION

Many recombinant proteins are produced today in eukaryotic hosts. These eukaryotes range from relatively simple yeasts to insect cells to mammalian cell lines. Members of the insecta class appear to be intermediate between mammalian and yeast in their posttranslational processing capabilities. Insect cells used in conjunction with a recombinant baculovirus form the basis of the baculovirus expression vector system (BEVS). BEVS can be used to produce recombinant proteins with yields which can be much higher than those obtained in mammalian cells.[1,2] Also, the recombinant proteins generated in insect cells undergo various posttranslational processing modifications including chemical covalent modifications such as phosphorylation, prenylation, glycosylation, acylation, methylation, sulfation, palmitoylation, and myristoylation as well as other processing events such as proteolytic cleavage and disulfide bond formation. In spite of these advantages, none of the therapeutic proteins produced today are synthesized in insect cells. One of the primary reasons insect cells are not used for producing therapeutics is because the posttranslational modifications, principally glycosylation, obtained from BEVS are different from those present in humans.

Glycosylation is the attachment of an oligosaccharide to a protein or lipid molecule followed by its modification during its transport through the secretory apparatus. Although insect cell derived proteins undergo glycosylation, their final structures often differ significantly from those obtained from mammalian cells. During N-linked glycosylation, insect cells typically generate oligosaccharides terminating in a few mannose (Man) residues (paucimannosidic), or occasionally high Man (oligo Man) or hybrid glycans terminating in N-acetylglucosamine (GlcNAc), while mammalian cells produce glycans typically terminating in galactose (Gal) and sialic acid (SA) (Figure 14.1). These differences are significant since the nature of attached glycan can have a pronounced impact on the biological properties of the glycoprotein including folding, stability, and solubility. The terminal residue is especially important for the *in vivo* circulatory lifetime of glycoproteins.[3,4] Glycoproteins produced in insect cells are removed from circulation of mammals rapidly[5] due to the presence of receptors in hepatocytes and macrophages that bind and remove from the bloodstream structures terminating in Man, GlcNAc, and Gal.[4,6–8] In contrast, glycoproteins produced in mammalian cells remain in circulation longer since many of these have terminal sialic acid residues that are not recognized by *in vivo* clearance mechanisms. These glycosylational differences lead to much lower *in vivo* activities for insect-derived glycoproteins as compared with their mammalian counterparts.

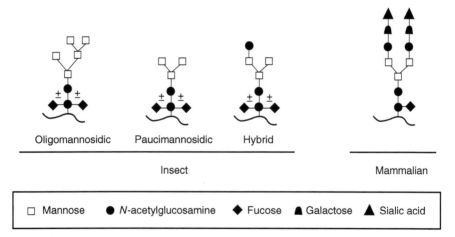

FIGURE 14.1 Typical N-glycans present on glycoproteins from insect and human cell lines. Insects produce paucimannosidic, high mannose or occasionally hybrid N-glycans. Human and other mammalian cell lines produce complex N-glycans with terminal sialic acid.

To make BEVS suitable for production of therapeutic glycoproteins, a key requirement will be its capacity to glycosylate recombinant proteins similar to those obtained in humans. In order to address this issue, the differences in the glycosylation processing between insects and human must be understood. Then a suitable metabolic engineering strategy involving substrate feeding and genetic engineering can be implemented in order to allow insect cells to mimic mammalian glycosylation. The differences between insect and mammalian glycosylation can be identified by measuring the activities of the different processing enzymes and by observing differences in the glycoforms obtained from insect and mammalian hosts. In this chapter we describe the glycosylation potential of native insect cells compared with mammalian cells, the bottlenecks that exist in their glycosylation pathway, and various genetic engineering efforts to "humanize" insect glycosylation.

II. GLYCOSYLATION IN NATIVE INSECT CELLS

The first few studies of glycosylation in insect cells were carried out in *Drosophila*[9] and mosquito.[10] While the initial steps of glycosylation in insect cells were found to be similar to those in mammalian cells, the subsequent modifications in insects were different, yielding glycans that were not as complex as those obtained in their mammalian counterparts.[9–11] With the advent of baculovirus technology, many more insect cell lines were developed and the glycosylation potential of these baculovirus-compatible cell lines has been studied. This knowledge has improved our understanding of the different glycosylational activities present in insect cells greatly and provided insights into the similarities and differences in the glycosylational potential among the insect cell lines.

N-glycosylation in both mammalian and insect cells starts with the transfer of a $Glc_3Man_9GlcNAc_2$ oligosaccharide from a donor dolichol molecule, onto an asparagine amino acid on the acceptor sequon (-Asn-Xaa-Ser/Thr-) of the nascent glycoprotein. The synthesis of this dolichol-linked oligosaccharide (DLO) in insect cells is similar to its synthesis in mammalian cells.[12] DLO is synthesized by the mevalonate-dependent pathway that involves the synthesis of polyprenol by sequential addition of isoprene units. The terminal double bond is then reduced to synthesize dolichol phosphate (Dol-P) on the membrane of the endoplasmic reticulum (ER). Monosaccharides are then added to this Dol-P. First, two GlcNAc residues are added to yield $GlcNAc_2$-P-P-Dol followed by sequential addition of five Man residues to yield $Man_5GlcNAc_2$-P-P-Dol. The donor molecules for this addition are UDP-GlcNAc and GDP-Man, respectively, and this addition takes place on the cytosolic side of the ER membrane. Next, the dolichol-linked $Man_5GlcNAc_2$ structure flips into the lumen of the ER and the subsequent monosaccharide attachments takes place in the ER. Four Man residues and three glucose residues are added to the DLO with dolichol-phosphate-glucose (Dol-P-Glc) and dolichol-phosphate-mannose (Dol-P-Man) as the donor molecules for these transfers. The $Glc_3Man_9GlcNAc_2$ oligosaccharide attached to the Dol-P is the molecule that is transferred to the nascent polypeptide in the ER. The transfer of the oligosaccharide from the DLO is catalyzed by the oligosaccharyl transferase

(OST) enzyme complex. Once transferred, the $Glc_3Man_9GlcNAc_2$ oligosaccharide is trimmed by the sequential action of a number of enzymes in the ER and Golgi compartments.

The differences in the final glycan structures produced in insect and mammalian cells are due to the differences in the trimming and subsequent modifications of the $Glc_3Man_9GlcNAc_2$ oligosaccharide (Figure 14.2). We will first describe mammalian N-glycan processing. We will then examine the insect N-glycan processing pathway and identify specific differences in the processing events between the two hosts. Finally, we will describe methods in progress to modify insect cell processing so that these cells can generate mammalian-type glycoproteins.

A. MAMMALIAN *N*-GLYCAN PROCESSING

In mammalian cells, N-glycan processing begins with the trimming of Glc and Man residues on $Glc_3Man_9GlcNAc_2$ by the action of various glycosidases. Several intermediates are generated in this process. The enzyme $\alpha(1,2)$-glucosidase I, found in all eukaryotic cells[13] except trypanosomes,[14] initiates the modification of the glycan in the ER by the removal of the terminal $\alpha(1,2)$-linked Glc residue to yield $Glc_2Man_9GlcNAc_2$. Human $\alpha(1,2)$-glucosidase I has been cloned and expressed and has been found to localize on the nuclear envelope and the ER.[15] Next, the two $\alpha(1,3)$-linked Glc residues are removed by the action of $\alpha(1,3)$-glucosidase II in the ER. These initial processing steps are common for the generation of high Man, hybrid, and complex-type N-glycans. Following the removal of the three Glc residues, $\alpha(1,2)$-mannosidases cleave four $\alpha(1,2)$-linked Man residues ultimately yielding a $Man_5GlcNAc_2$ glycan. Several $\alpha(1,2)$-mannosidases have been purified and

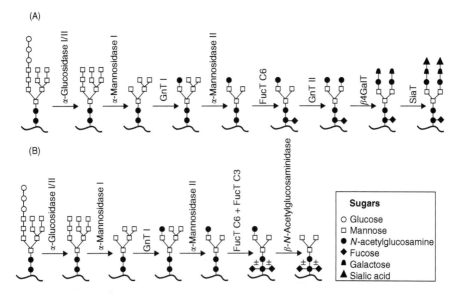

FIGURE 14.2 *N*-Glycan processing pathway in (a) mammalian cells and (b) insect cells.

cloned,[16–26] which differ in the Man residues they cleave. Different $\alpha(1,2)$-mannosidases are present in both ER and Golgi. The ER resident $\alpha(1,2)$-mannosidases I removes the $\alpha(1,2)$-Man from the middle branch yielding $Man_8GlcNAc_2$ glycan.[27,28] The ER also has another $\alpha(1,2)$-mannosidase, $\alpha(1,2)$-mannosidase II, that removes a single $\alpha(1,2)$-Man residue from the $Man\alpha(1,6)$-branch. The Golgi resident $\alpha(1,2)$-mannosidases termed $\alpha(1,2)$-mannosidases IA, IB, and IC cleave additional Man residues to yield $Man_5GlcNAc_2$.[25,29]

The $Man_5GlcNAc_2$ glycan structure is then acted upon by various glycosyltransferases present in the Golgi to yield complex N-glycans. The first steps toward the synthesis of complex N-glycans is the attachment of a single GlcNAc residue to the $Man\alpha(1,3)$-branch of $Man_5GlcNAc_2$. The GlcNAc is transferred from the donor sugar-nucleotide UDP-GlcNAc to the glycan by the action of N-acetylglucosaminyltransferase I (GnT I) (reviewed in Reference 30), which has been cloned from several mammalian species.[31–39]

The $GlcNAcMan_5GlcNAc_2$ glycan structure can then be modified in two different ways in mammalian cells. The core GlcNAc residue attached to the Asn residue of the protein can be modified by attachment of an $\alpha(1,6)$-linked fucose. This modification is catalyzed by the action of $\alpha(1,6)$-fucosyltransferase.[40,41] The presence of $\beta(1,2)$-linked GlcNAc on the $Man\alpha(1,3)$-branch is essential for the action of this enzyme. Biantennary glycans, with GlcNAc on both $Man\alpha(1,3)$- and $Man\alpha(1,2)$-branches, also serve as substrates for this core $\alpha(1,6)$ fucosyltransferase.[41–43] Alternatively, $GlcNAcMan_5GlcNAc_2$ or $GlcNAcMan_5[Fuc\alpha(1,6)]GlcNAc_2$ can be modified by the action of Golgi-resident α-mannosidase II (reviewed in Reference 44). This enzyme is responsible for the cleavage of the terminal $\alpha(1,6)$- and terminal $\alpha(1,3)$-linked Man residues on the $Man\alpha(1,6)$-branch of $GlcNAcMan_5[\pm Fuc\alpha(1,6)]GlcNAc_2$. The resulting structure is $GlcNAcMan_3[\pm Fuc\alpha(1,6)]GlcNAc_2$. While the Golgi localization of the α-mannosidase II has been confirmed in rodent cells by immunoelectron microscopy, the sub-Golgi localization is found to be different in different cell types.[45] The α-mannosidase II enzyme has been purified and characterized from rat liver[46,47] and mung bean seedlings[48] and the genes encoding α-mannosidase II have been cloned from human[49] and mouse[50,51] among others.

Other than the commonly observed processing involving the mannosidases described above, various mannosidases have been found which remove $\alpha(1,3)$- and $\alpha(1,6)$-linked terminal Man residues from $Man_{5-6}GlcNAc_2$ (but not from $GlcNAcMan_5GlcNAc_2$).[52] A Golgi resident α-mannosidase has been identified in rat liver that can convert $(Man)_{4-9}GlcNAc$ into $Man_3GlcNAc$.[52] A similar mannosidase that removes Man residues from $Man_5GlcNAc$-pyridylamino substrate to yield $Man_{2-4}GlcNAc$-pyridylamino has been found in various mouse tissues.[53] This enzyme, referred to as α-mannosidase III, is found to be involved in the alternate processing of $Man_5GlcNAc_2$ to yield $Man_3GlcNAc_2$.[53] Another human mannosidase, named α-mannosidase IIx, has been cloned,[49] which, along with endogenous α-mannosidase I, is responsible for the cleavage of $\alpha(1,3)$- and $\alpha(1,6)$-linked terminal Man residues from $Man_6GlcNAc_2$ to yield $Man_3GlcNAc_2$.[54] This $Man_3GlcNAc_2$ can also serve as a substrate for the mammalian GnT I.[55,56]

Following the action of α-mannosidase II, a second GlcNAc is added to the $Man\alpha(1,6)$-branch of $GlcNAcMan_3[\pm Fuc \alpha(1,6)]GlcNAc_2$ to yield a biantennary complex glycan. The attachment of this GlcNAc by $\beta(1,2)$-linkage is catalyzed by N-acetylglucosaminyltransferase II (GnT II). GnT II gene has been cloned from human[57] and rat.[58] There are other known N-acetylglucosaminyltransferases such as GnT IV, GnT V, and GnT VI that add GlcNAc to the $\alpha(1,3)$- or $\alpha(1,6)$-linked Man to generate multiantennary glycans.

The biantennary/multiantennary glycans are then modified by the addition of Gal on the GlcNAc termini. The enzyme involved, β-1,4-galactosyltransferase (β4GalT), catalyzes the transfer of Gal to the terminal GlcNAc residues. Several different β-1,4-galactosyltransferases have been found in human and mouse (reviewed in References 59,60) and these differ in their acceptor branch.[61] The N-glycans on mammalian glycoproteins are often capped with sialic acid. N-acetyl-neuraminic (Neu5Ac) is the sialic acid most commonly found in humans. This sialylation step involves the transfer of Neu5Ac from the donor sugar-nucleotide CMP-Neu5Ac onto the Gal residues on the acceptor glycan. The Neu5Ac is mostly $\alpha(2,6)$-linked and the transfer is catalyzed by β-galactoside $\alpha(2,6)$-sialyltrans-ferase (ST6Gal). Two different $\alpha(2,6)$-sialyltransferases, ST6Gal I[62] and ST6Gal II,[63,64] have been cloned from humans, but ST6Gal I is the principal enzyme involved in the sialylation of the $\beta(1,4)$-linked Gal on the Gal$\beta(1,4)$GlcNAc sequence.[63,64] The other known families of sialyltransferases are β-galactoside $\alpha(2,3)$-sialyltransferase (ST3Gal), GalNAc $\alpha(2,6)$-sialyltransferase (ST6GalNAc), and $\alpha(2,8)$-sialyltransferase (ST8Sia),[65,66] and they differ based on their preference for acceptor, substrates, or linkage types.

B. Insect Cells N-Glycan Processing

Analyses of N-glycans from various lepidopteran insect cells have been performed and these studies indicate the formation of primarily high Man and paucimannosidic glycan structures. The presence of these structures indicates that the initial process-ing enzymes such as $\alpha(1,2)$-glucosidase I, $\alpha(1,3)$-glucosidase II, and $\alpha(1,2)$-man-nosidase are present. Indeed, experiments involving glycosidase inhibitors have confirmed the presence of α-glucosidase I, II, and α-mannosidases in insect cells.[67] As yet, no $\alpha(1,2)$-glucosidase I or $\alpha(1,3)$-glucosidase II has been cloned from insect cells. On the other hand, α-mannosidase I, which can utilize Man$_6$GlcNAc$_2$, Man$_7$GlcNAc$_2$, Man$_8$GlcNAc$_2$, and Man$_9$GlcNAc$_2$ as substrates, with Man$_6$GlcNAc$_2$ being the preferred substrate, has been purified from the Golgi fraction of Sf21 cells.[68] Also, a Golgi-resident α-mannosidase I has been cloned from Sf9 cells (SfMan I).[69,70] The amino acid sequence of this enzyme has homology with all the three known human Golgi α-mannosidases IA, IB, and IC and its substrate speci-ficity is similar to human $\alpha(1,2)$-mannosidase IC.[25]

The next step in mammalian glycan processing involves the addition of a GlcNAc residue to the Man$\alpha(1,3)$-branch of Man$_5$GlcNAc$_2$ by N-acetylglu-cosaminyltransferase I (GnT I) activity. GnT I has been cloned from *Drosophila melanogaster*[71] and its activity has been observed in various insect cell lines including Sf9, Sf21, Mb0503, and Bm-N.[72,73] Following the addition of GlcNAc to the Man$\alpha(1,3)$-branch, the glycan is acted upon by α-mannosidase II, which removes $\alpha(1,3)$Man and $\alpha(1,6)$Man from the GlcNAcMan$_5$GlcNAc$_2$ structure.

α-Mannosidase II activity has also been detected in Sf21, Mb0503, and Bm-N cells.[74] Similar to the mammalian enzyme, the presence of GlcNAc on the Manα(1,3)-branch was found to be essential for this mannosidase activity.[74]

N-Glycans with terminal GlcNAc are often modified in mammalian cells by core α(1,6)-fucosyltransferase (FucT C6), which adds a α(1,6)-linked Fuc to the innermost GlcNAc attached to the Asn residue on the polypeptide. In addition to this FucT C6 activity, some insect cells have a core α(1,3)-fucosyltransferase (FucT C3)[75] not present in mammalian cells. N-glycans with Fucα(1,6) and a Fucα(1,3) attachment to the Asn-linked GlcNAc have been observed on membrane glycoproteins from Mb0503, Sf21, and Bm-N cells. Similarly, α(1,3)-fucosylation has been observed on recombinant proteins such as human serum transferrin[76] and human IgG[77] from Tn-5B1-4 cells. However, human serum transferrin produced in Ld652Y cells,[78] and human interferon ω1 expressed in Sf9 cells showed no α(1,3)-fucosylated N-glycan.[79] FucT C3 and FucT C6 activities have been studied in different cell lines. In Mb0503 cells, FucT C3 activity was found to be much greater than FucT C6,[75] whereas in Sf9 cells, no FucT C3 activity was observed.[75] In another study, FucT C6 activity was detected in Sf9, Mb0503, and Bm-N cells but FucT C3 activity was found only in Mb0503 cells.[80] Recently, a gene encoding a functional FucT C3 has been cloned from *D. melanogaster*,[81] although no FucT C6 gene has as yet been cloned. The presence of an α(1,3)-linked fucose attachment on the innermost GlcNAc has been suggested to represent a potential epitope that may cause allergenic reactions in human.[82–86] Thus, the presence of FucT C3 activity is another limitation in the potential use of insect cells to express human glycoproteins.

The GlcNAcMan$_5$GlcNAc$_2$[$\pm$Fucα(1,6)] structure generated in mammalian cells following α-mannosidase II processing and FucT C6 activity acts as the substrate for GnT II to yield biantennary structures. But unlike mammalian cells, insect cells studied to date were found to have insufficient GnT II activity. In Sf9, Sf21, Mb0503, and Bm-N cells, the GnT II activity was found to be less than 1% of that of mammalian cells.[72] The lack of sufficient GnT II activity in insect cells precludes them from generating complex biantennary structures. A homologue of GnT II has been detected in the *Drosophila* genome,[81,87] but the expression of a functional gene is apparently limited in many insect cell lines.

In mammalian cells, glycans terminating in GlcNAc typically serve as the substrate for the addition of β(1,4)-Gal by β(1,4)-galactosyltransferase (β4GalT). However, measurements in Sf9, Tn-5B1-4, and Mb0503 cells have indicated the presence of neglible β4GalT activity.[88–90] In one study, a high-sensitivity Eu-fluorescence-based assay found the β4GalT activity in Tn-5B1-4 cells to be 10% of the activity found in Chinese hamster ovary (CHO) cells.[91] The same study also revealed the absence of detectable GalT activity in Sf9 cells. In addition, structural analysis of N-glycans from insect cells has only rarely indicated the presence of Gal-containing structures. The lack of sufficient GalT activity presents another bottleneck to the generation of complex glycans in insect cells.

The next step in mammalian cell N-glycosylation is the attachment of Neu$_5$Ac to the terminal β(1,4)-linked galactose residues by sialyltransferases ST6Gal I and ST6Gal II. Studies performed in Sf9,[90,92,93] Sf21,[93] Tn-5B1-4,[92] Mb0503,[92] and Ea4 cells[93] using radiolabeled or fluorescent-labeled CMP-Neu5Ac as the donor substrate revealed negligible sialyltransferase activity. However, a recent study reported the presence of a gene encoding $\alpha(2,6)$-sialyltransferase in *D. melanogaster* and its expression was found during embryonic development in a tissue- and stage-specific fashion.[94] Indeed, tissue and stage specificity of this enzyme may explain the lack of sufficient SiaT activity in many insect cell lines, and this deficiency represents another bottleneck to the synthesis of complex glycans with terminal sialic acid in insect cells.

In addition to the absence of certain transferases, another hindrance to the synthesis of mammalian-like glycans in insect cells is the presence of other enzymatic activities not typically observed in mammalian cells such as core FucT C3 described previously and β-*N*-acetylglucosaminidase. β-*N*-acetylglucosaminidase, involved in the degradation of intermediate oligosaccharides, is present in insect cells but not mammalian cells. This β-*N*-acetylglucosaminidase activity found in Sf21, Bm-N, and Mb0503 cells, specifically cleaves the terminal $\beta(1,2)$-linked GlcNAc on the Man$\alpha(1,3)$-branch.[95] β-*N*-acetylglucosaminidase-like activity has been detected in both the lysate and culture supernatant of insect cells derived from *Spodoptera frugiperda*, *Trichoplusia ni*, *Bombyx mori*, and *Malacosoma disstria*.[96] The analyses of *N*-glycans from human IgG[77] and hTf[76] expressed in Tn-5B1-4 cells and hTf expressed in Ld652Y cell[78] also suggest the presence of significant β-*N*-acetylglucosaminidase activity in Tn-5B1-4 and Ld652Y cells. The removal of the $\beta(1,2)$-linked GlcNAc from the glycan precludes further mammalian-like complex modifications such as addition of another GlcNAc by GnT II,[97] attachment of fucose by core α-fucosyltransferase(s),[41–43] and attachment of Gal by β4GalT. Often, the removal of this GlcNAc is followed by further removal of additional Man residues by α-mannosidase(s). This process generates paucimannosidic structures with fewer than the three Man residues commonly observed on glycoproteins from insect cells.

C. AVAILABILITY OF SUGAR NUCLEOTIDES INVOLVED IN *N*-LINKED GLYCOSYLATION

In order to generate complex *N*-glycans, a cell line requires a sufficient pool of the donor sugar-nucleotides in addition to the necessary glycosyltransferases. The most common sugars associated with *N*-linked glycosylation are Glc, Gal, Man, Fuc, GlcNAc, and Neu5Ac, and the corresponding sugar-nucleotides donors are UDP-Glc, UDP-Gal, GDP-Man, GDP-Fuc, UDP-GlcNAc, and CMP-Neu5Ac. The presence of these donor sugar nucleotides is essential for the synthesis of the complex glycans. Tomiya et al.[94] developed an HPAEC analytical method to separate and quantify all these sugar nucleotides involved in *N*-linked glycosylation. Using this method they measured the levels of sugar-nucleotides in Sf9 and Tn-5B1-4 cells grown in serum-free media and compared these levels to those in mammalian cells. All cells contained significant levels of UDP-glucose, UDP-galactose, UDP-GlcNAc, UDP-GalNAc, GDP-Fuc, and GDP-Man with the exact amount varying between cell lines. However, unlike mammalian CHO cells, Sf9 and Tn-5B1-4 lacked significant levels of CMP-Neu5Ac (Figure 14.3). Hooker et al.[93] found a similar deficiency in CMP-Neu5Ac levels in Sf9 cells.

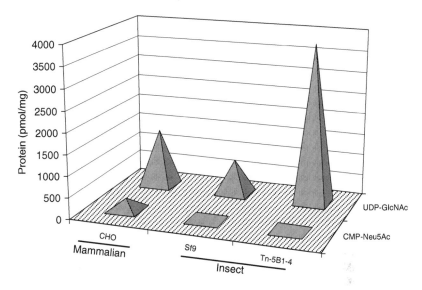

FIGURE 14.3   CMP-Neu5Ac and its precursor UDP-GlcNAc sugar nucleotide levels in
CHO, Sf9 and Tn-5B1-4 cell lines.

D.   CMP-NEU5AC SYNTHESIS PATHWAY IN MAMMALS AND INSECTS

CMP-Neu5Ac is the donor sugar-nucleotide for the sialylation step and is synthesized from UDP-GlcNAc in mammalian cells by a series of enzymatic reactions
(Figure 14.4A). UDP-GlcNAc is first epimerized to ManNAc, a dedicated precursor
for the synthesis of Neu5Ac, which is then phosphorylated to yield ManNAc-6-
phosphate (ManAc-6-P). This two-step reaction is catalyzed by a bifunctional
enzyme UDP-GlcNAc-2-epimerase/ManNAc kinase, which has been isolated and
cloned from rat liver.[98] Next, the ManNAc-6-P combines with phosphoenol pyruvate
(PEP) to yield Neu5Ac-9-phosphate (Neu5Ac-9-P) by the enzyme N-acetylneuram-
inate-9-phosphate synthase or sialic acid 9-phosphate synthase (SAS). Synthesized
Neu5Ac-9-P is then dephosphorylated by N-acetylneuraminate-9-phosphate phosphatase followed by activation to CMP-Neu5Ac by CMP-Neu5Ac synthetase, also
called CMP-sialic acid synthetase (CMP-SAS).

   In the studies performed with insect cells grown in serum-free conditions, little or
no Neu5Ac and CMP-Neu5Ac were detected.[99,100] Effertz et al.[101] reported that the
specific activity of UDP-N-acetylglucosamine-2-epimerase in Sf9 cells was about 30
times lower than that in the rat liver cytosol fraction and Sf9 cells were found to have
50 times higher ManNAc kinase activity than the 2-epimerase activity.[101] In studies
conducted in our laboratories, Sf9 cells were found to have negligible N-acetylneu-
raminate-9-phosphate synthase and CMP-SAS activities.[99] Thus, metabolic pathways
responsible for generating the donor sugar-nucleotide, CMP-Neu5Ac, required for
sialylation appear to be absent in insect cells. Recently, a functional sialic acid phosphate synthase gene was identified by our group in the genome of *Drosophila*,[102] but
the enzyme activity levels are extremely low in *D. schneider* S2 cell line.

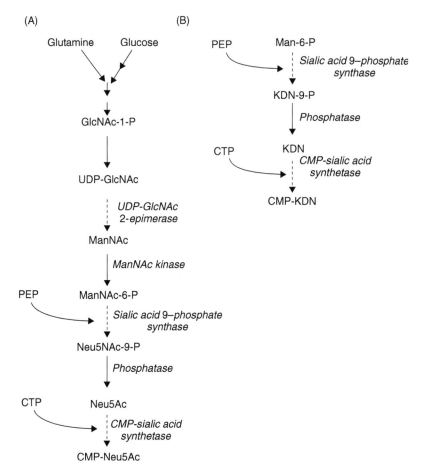

FIGURE 14.4 (A) CMP-Neu5Ac and (B) CMP-KDN synthesis pathway in mammalian cells. Dotted arrows indicate metabolic activities that are insufficient in insect cells.

To summarize, the bottlenecks and limitations in the synthesis of complex mammalian-like glycans in insect cells can thus be attributed to (i) lack of GnT II, GalT, and SiaT glycosyltransferase activities, (ii) the presence of high FucT C3 and β-N-acetylglucosaminidase activities in certain insect cell lines, and (iii) the absence of CMP-Neu5Ac synthesis pathway enzymes.

III. METABOLIC ENGINEERING OF GLYCOSYLATION PATHWAYS IN INSECT CELLS

N-linked glycans obtained from insect cells such as *S. frugiperda*,[79,93,103–116] *T. ni*,[77,103,106,114] *M. brassica*,[105,117] *B. mori*,[105,118] *Estigmene acrea*,[108,119] *Lymantria dispar*,[78,118] and *Heliothis virescens*[118] are typically paucimannosidic with one to three Man and one or two GlcNAc residues with or without Fuc attached to the Asn-linked

GlcNAc. In contrast, mammalian cells usually produce terminally sialylated complex-type *N*-glycans. In order for insect cells to produce mammalian-like complex glycans, the limitations described in the previous section must be overcome. A general strategy for humanizing glycoproteins produced by the insect cell–baculovirus expression system is summarized in Figure 14.5

A. Engineering Glycosyltransferase and Inhibiting Glucosaminidase Activities in Insect Cells

N-glycans containing GlcNAc on the Manα(1,3)-branch typically represent a very small fraction of all the glycans found on insect cell proteins.[76] While insect cells have GnT I activity responsible for the addition of GlcNAc to the Manα(1,3)-branch, the presence of *β-N*-acetylglucosaminidase activity removes this terminal GlcNAc from the Manα(1,3)-branch.[95] Cell lines such as *E. acrea,* which lack this glucosaminidase activity, produced *N*-glycans with terminal GlcNAc residues but other cell lines, such as the Sf9 line that has high glucosaminidase activity, produced primarily paucimannosidic *N*-glycans.[119] Watanabe et al.[120] reported that the use of a *β-N*-acetylglucosaminidase inhibitor, 2-acetamide-1,2-dideoxynojirimycin (2-ADN), resulted in sialylated glycans on bovine interferon-γ in Tn-5B1-4 cells. The authors suggested that the inhibition of *β-N*-acetylglucosaminidase in turn allowed for the formation of a glycan possessing a *β*(1,2)-linked GlcNAc,

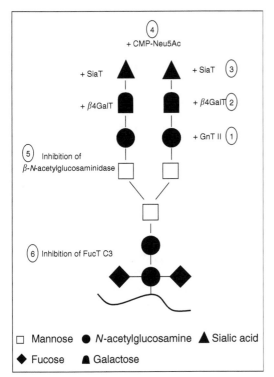

FIGURE 14.5 Engineering efforts required for the synthesis of 'humanized' *N*-glycan on glycoproteins produced in insect cells. Expression of (1) GnT II, (2) *β*4GalT, (3) SiaT glycosyltransferases, engineering (4) CMP-Neu5Ac synthesis pathway, and inhibition of (5) *β-N*-acetylglucosaminidase, and (6) FucT C3.

which was then modifie further by the GalT and SiaT activities to generate sialy-lated *N*-glycans. These results are indeed surprising because insufficient activities of GalT and SiaT and CMP-SA synthesis pathway genes have been found in the cell lines of this study. As an alternative to the inhibition of β-*N*-acetylglu-cosaminidase, the overexpression of GnT I has been reported to improve the expression of GlcNAc on the Manα(1,3)-branch of fowl plague virus hemagglu-tinin from Sf9 cells.[121]

The GlcNAc-terminating glycan serves as an acceptor for the addition of Gal in mammalian cells. In order to address the limitation in GalT activity,[88–90,122] a mam-malian β4GalT was expressed in insect cells using a baculovirus vector under the control of an immediate early (ie1) promoter.[123] Glycans terminating in Gal on gp64 glycoprotein were detected by *Ricinus communis* agglutinin (RCA) lectin stain-ing.[123] A subsequent study demonstrated that the Gal residue added with recombi-nant β4GalT expression was exclusively on the GlcNAc on the Manα(1,3)-branch.[76] Interestingly, in this study, the percentage of *N*-glycans with terminal GlcNAc on the Manα(1,3)-branch increased by more than twofold in the presence of β4GalT to demonstrate that the addition of a Gal residue protected the GlcNAc on Manα(1,3)-branch from cleavage by β-*N*-acetylglucosaminidase. Sf9 and Tn-5B1-4 cells have subsequently been stably transformed with β4GalT to create Sfβ4GalT[88] and Tn5β4GalT,[124] respectively. These cell lines were capable of generating galactosy-lated gp64[88,124] and t-PA.[88]

The absence of Gal on the Manα(1,6)-branch with β4GalT expression[76] indi-cated a limitation in the available GlcNAc on the Manα(1,6)-branch of the acceptor glycan. Indeed, endogenous GnT II activity has been found to be 1% or less in the different insect cell lines studied.[72] In order to increase the levels of biantennary gly-cans terminating in GlcNAc, GnT II activity was increased by generating stable transgenic insect cell lines (SfSWT-1)[125] or transiently expressed by using a bac-ulovirus expression vector carrying GnT II.[126] Tn5β4GalT cells coinfected with GnT II and hTf baculoviruses generated hTf *N*-glycans, with more than 50% of the struc-tures containing Gal residues on both termini according to 3D HPLC and mass spec-troscopy analyses.[126] Thus, the coexpression of GnT II and β4GalT in insect cells, yield biantennary fully galactosylated *N*-glycans.

For insect cells to produce sialylated mammalian-like glycoproteins, the follow-ing components are required: (i) an acceptor glycan terminating in Gal in one or more branches, (ii) the donor sugar-nucleotide CMP-sialic acid (CMP-Neu5Ac), and (iii) sialyltransferase enzymatic activity. The overexpression of GalT and GnT II over-came the limitation in the Gal acceptor substrate.[126] However, previous studies con-firmed the absence of both CMP-sialic acid substrate and sialyltransferase.[90,92,93,100] In order to address the limitation in sialyltransferase activity, mammalian sialyltrans-ferase was expressed in insect cell lines. In one study, sialylation was observed on the viral glycoprotein gp64 when Sf9 cells, grown in serum-bearing media, were infected with a baculovirus containing both mammalian β4GalT and α(2,6)-sialyltransferase (ST6Gal).[127] Sialylation was also observed in the Sfβ4GalT cell line infected with a baculovirus containing ST6Gal.[128] Stable cell lines from Sf9 and Tn-5B1-4 constitu-tively expressing both ST6Gal and GalT have also been developed (Sfβ4GalT/ST6[90] and Tn5β4GalT/ST6[124]) and found to generate sialylated glycoproteins.

However, sialylation in these cells lines required growth in serum-bearing culture media, known to be rich in sialic acid.[129] More recently, it has been shown that Sfβ4GalT/ST6 cells could sialylate their proteins in the absence of serum in the media, if the serum-free media contains sialylated glycoproteins such as fetuin instead.[130] In contrast, when fetuin in the media was replaced by asialofetuin, the insect cells were unable to sialylate glycoproteins. These results suggested the presence of a sialic acid salvaging pathway in insect cells, but the direct addition of sialic acid to the media was not an effective strategy either possibly due to a limitation in the uptake of sialic acid by the cells.

B.   GENERATION OF CMP-NEU5AC SIALYLATION SUBSTRATES IN INSECT CELLS

The inability of engineered insect cell lines to generate sialylated glycoproteins without external supplementation of sialic acid through serum or fetuin was not surprising since previous analyses revealed insignificant levels of Neu5Ac[93,100] and CMP-Neu5Ac.[93,100] Indeed, the Neu5Ac content of insect cells was found to be nearly 50 times lower than that in mammalian CHO cells.[99] In order to address this limitation, genes encoding the CMP-Neu5Ac synthesis pathway enzymes have been engineered into insect cells.

The biosynthetic pathway for generating Neu5Ac and CMP-Neu5Ac in mammalian cells is shown in Figure 14.4A. Insect cells have been observed to lack sufficient UDP-GlcNAc-2-epimerase, SAS, and CMP-SAS activites. ManNAc is a dedicated precursor for the synthesis of Neu5Ac, which is synthesized by the 2-epimerization of UDP-GlcNAc in mammalian cells. UDP-GlcNAc, a sugar-nucleotide, has been found in significant quantities in insect cells.[100] However, insect cell lines have been found to be incapable of synthesizing ManNAc from UDP-GlcNAc.[101] This deficiency in UDP-*N*-acetylglucosamine 2-epimerase activity limits the synthesis of ManNAc in insect cells. To overcome this limitation, insect cell culture media have been supplemented with ManNAc. Alternatively, insect cells have been infected with a baculovirus containing the mammalian UDP-*N*-acetylglucosamine 2-epimerase/ManNAc kinase bifunctional gene (AcEpimKin) in order to synthesize ManNAc from intracellular UDP-GlcNAc.[99,131] Indeed, the cells co-infected with AcEpimKin and a baculovirus containing the human sialic acid 9-phosphate synthase gene (AcSAS) were able to generate Neu5Ac from endogenous substrates at levels that were comparable to those obtained with AcSAS-infected cells cultured in media supplemented with 10 m*M* ManNAc. The ability of Sf9 cells to synthesize Neu5Ac with AcSAS infection and ManNAc supplementation, indicated that Sf9 cells had endogenous nonspecific kinase activity present for the conversion of intracellular ManNAc into ManNAc-6-P, the substrate of SAS. The ManNAc kinase activity was later attributed to a secondary activity derived from endogenous insect GlcNAc kinase enzyme.[131]

Addition of 10 m*M* GlcNAc, a precursor of UDP-GlcNAc, to the media of cells co-infected with AcSAS and AcEpimKin, resulted in a sixfold increase in Neu5Ac levels. Subsequent experiments revealed metabolic bottlenecks in both the uptake of ManNAc from the medium and phosphorylation of ManNAc.[131] Expression of the

ManNAc kinase domain of the EpimKin bifunctional gene overcame the limitation in the phosphorylation of ManNAc, and the tetra-*O*-acetylation of ManNAc overcame the transport bottleneck, although this chemical modification was lethal to the cells at high concentrations.[131]

Even when insect cells were engineered to express the human SAS gene and supplemented with ManNAc, the levels of CMP-Neu5Ac remained ten times below those of CHO cells.[132] To overcome the deficiency in CMP-Neu5Ac synthesis, human CMP-SAS was identified, cloned into a baculovirus (AcCMP-SAS), and expressed in Sf9 cells coinfected with AcSAS in concert with ManNAc supplementation to the media. CMP-Neu5Ac levels obtained from the engineered insect cells were six times higher than endogenous levels from CHO cells.[132]

Interestingly, in the absence of ManNAc supplementation, Sf9 cells infected with the AcSAS produced an alternative sialic acid, KDN (2-keto-3-deoxy-D-glycero-D-galactononic acid), formed from endogenous Man with Man-6-P and KDN-9-P as pathway intermediates (Figure 14.4B). In addition, coinfection of Sf9 cells with AcSAS and AcCMP-SAS in the absence of ManNAc supplementation resulted in synthesis of CMP-KDN, albeit at low levels.[132] Thus, by suitable baculovirus infections, the potential of generating both CMP-Neu5Ac and CMP-KDN intracellularly from endogenous metabolites through the expression of SAS, CMP-SAS, and EpimKin (for CMP-Neu5Ac) was demonstrated. Aumiller et al.[133] followed up on these results by developing a cell line, SfSWT-3, which included SAS and CMP-SAS genes along with the GnT I, GnT II, GalT, and sialyltransferases genes present in SfSWT-1. This cell line was capable of generating the complex biantennary *N*-glycans with sialic acid on one branch, when grown in serum-free media supplemented with ManNAc. *N*-glycans containing sialic acid on both branches were also observed in smaller quantities.

C. NONMETABOLIC-ENGINEERING APPROACHES TO SYNTHESIZE COMPLEX-TYPE GLYCANS IN INSECT CELLS

Researchers have also explored different cell lines as an alternative strategy for producing complex sialylated glycoproteins. Indeed, such a strategy may be possible given the recent genomic evidence of a complex sialylation pathway in *D. melanogaster*.[94,102] A clonal isolate of Tn-5B1-4, Tn-4h, was capable of expressing sialylated glycoproteins in certain culture conditions. For example, sialylated glycoproteins were found when these cell lines were grown in high aspect ratio vessel (HARV) bioreactor in serum-bearing medium.[134] This cell line was also capable of generating sialylated glycoproteins when grown in T-flasks when the serum-bearing media were supplemented with ManNAc, the precursor for CMP-Neu5Ac. Another clonal isolate, Tn-4s, was also identified that could generate sialylated glycoproteins in spinner flasks,[135] while its parent cell line Tn-5B1-4 was unable to generate similar glycans when cultured in normal serum-bearing medium. However, in another study, it was found that Tn-5B1-4 cells were capable of generating fully sialylated glycoproteins when their cell culture media was supplemented with silkworm hemolymph (SH). An insect cell line isolated from

Monarch butterfly, *Danaus plexippus* (DpN1), was also observed to generate sialylated *N*-glycans on recombinant alkaline phosphatase (SEAP) expressed using baculovirus vectors.[136]

The reasons why these few insect cells are capable of generating complex glycans under certain culture conditions are unknown. Such activities may be related to the induction of repressed genes under specific cell culture conditions. However, even in cell lines capable of generating complex glycoproteins, the fraction of all *N*-glycans that are complex is often low. It should also be noted that the vast majority of cell lines studied to date lack sufficient endogenous enzymatic activities required for generating complex *N*-glycans similar to those in humans.

IV. CONCLUDING REMARKS

Structural analyses of glycans and studies on the glycosylational pathway enzymes have provided a much improved understanding of the *N*-glycan processing in insect cells. Unlike mammalian cells, most insect cell lines are unable to synthesize complex sialylated glycans. Insufficient intracellular activities for critical processing enzymes, along with the presence of a unique degradation enzyme, β-*N*-acetylglucosaminidase, and core FucT C3 in most insect cell lines lead to expression of non-mammalian-type *N*-glycans containing primarily paucimannosidic structures of the form $Man_{1-3}GlcNAc\beta(1,4)GlcNAc[\pm Fuc\alpha(1,6) \pm Fuc\alpha(1,3)]$-Asn. Currently, little is known about the biological function of such paucimannosidic *N*-glycans in insect cells and insects. However, the inability of insect cell lines to generate humanized *N*-glycans on their proteins greatly restricts the use of many insect cell lines for the production of therapeutic glycoproteins.

In order to adapt insect cell lines to synthesize mammalian-type complex glycoproteins, metabolic engineering approaches have been combined with suitable substrate-feeding strategies. The expression of multiple recombinant mammalian glycosyltransferases along with the genes required for CMP-sialic acid metabolism has been successfully used to modify the glycosylation pattern in order to generate fully humanized *N*-glycans.[133] The study of oligosaccharide structures and the endogenous enzyme activities in native cell lines has been essential to identifying which enzymes are needed in order to humanize insect cell glycosylation patterns.

Cell culture conditions may also stimulate certain insect cell lines to synthesize more complex glycans. Recent studies in *Drosophila* have revealed the existence of certain complex glycosylational processing and metabolic enzymes during specific stages of insect development. Particular cell culture stimuli may also activate the genes in cell lines to generate more complex glycoforms. The presence of widely differing glycosylation patterns from different tissues, developmental stages, and cell lines also stimulates further questions concerning the function of variable glycosylation processing in insects. Evaluating the biological roles that these different *N*-glycans play during development and tissue function represents an interesting challenge in the future study of the glycobiology of insects.

REFERENCES

1. Farrell, P.J., Behie, L.A., and Iatrou, K., Transformed Lepidopteran insect cells: new sources of recombinant human tissue plasminogen activator, *Biotechnol. Bioeng.*, 64, 426–433, 1999.
2. O'Reilly, D.R., Miller, L.K., and Luckow, V.A., *Baculovirus Expression Vectors: A Laboratory Manual*, W. H. Freeman and Company, New York, 1992.
3. Cumming, D.A., Glycosylation of recombinant protein therapeutics: control and functional implications, *Glycobiology*, 1, 115–130, 1991.
4. Schauer, R., Kelm, S., Reuter, G., Roggentin, P., and Shaw, L., Biochemistry and role of sialic acids, in *Biology of the Sialic Acids*, Abraham Rosenberg, Ed., Plenum Press, New York, 1995, pp. 7–67.
5. Grossmann, M., Wong, R., Teh, N.G., Tropea, J.E., East-Palmer, J., Weintraub, B.D., and Szkudlinski, M.W., Expression of biologically active human thyrotropin (hTSH) in a baculovirus system: effect of insect cell glycosylation on hTSH activity *in vitro* and *in vivo*, *Endocrinol.*, 138, 92–100, 1997.
6. Ashwell, G. and Morell, A., The dual role of sialic acid in the hepatic recognition and catabolism of serum glycoproteins, *Biochem. Soc. Symp.*, 117–124, 1974.
7. Goochee, C.F., Gramer, M.J., Andersen, D.C., Bahr, J.B., and Rasmussen, J.R., The oligosaccharides of glycoproteins: bioprocess factors affecting oligosaccharide structure and their effect on glycoprotein properties, *Biotechnology, (N Y)* 9, 1347–1355, 1991.
8. Oganah, O.W., Freedman, R.B., Jenkins, N., Patel, K., and Rooney, B.C., Isolation and characterization of an insect cell line able to perform complex *N*-Linked glyco-sylation on recombinant proteins, *Bio/technology*, 14, 197–202, 1996.
9. Parker, G.F., Williams, P.J., Butters, T.D., and Roberts, D.B., Detection of the lipid-linked precursor oligosaccharide of *N*-linked protein glycosylation in *Drosophila melanogaster*, *FEBS Lett.*, 290, 58–60, 1991.
10. Hsieh, P. and Robbins, P.W., Regulation of asparagine-linked oligosaccharide pro-cessing. Oligosaccharide processing in *Aedes albopictus* mosquito cells, *J. Biol. Chem.*, 259, 2375–2382, 1984.
11. Butters, T.D. and Hughes, R.C., Isolation and characterization of mosquito cell mem-brane glycoproteins, *Biochim. Biophys. Acta.*, 640, 655–671, 1981.
12. Sagami, H. and Lennarz, W.J., Glycoprotein synthesis in Drosophila Kc cells. Biosynthesis of dolichol-linked saccharides, *J. Biol. Chem.*, 262, 15610-15617, 1987.
13. Dairaku, K. and Spiro, R.G., Phylogenetic survey of endomannosidase indicates late evolutionary appearance of this *N*-linked oligosaccharide processing enzyme, *Glycobiology*, 7, 579–586, 1997.
14. Parodi, A.J., *N*-glycosylation in trypanosomatid protozoa, *Glycobiology*, 3, 193–199, 1993.
15. Kalz-Fuller, B., Bieberich, E., and Bause, E., Cloning and expression of glucosidase I from human hippocampus, *Eur. J. Biochem.*, 231, 344–351, 1995.
16. Tabas, I. and Kornfeld, S., Purification and characterization of a rat liver Golgi alpha-mannosidase capable of processing asparagine-linked oligosaccharides, *J. Biol. Chem.*, 254, 11655–11663, 1979.
17. Tulsiani, D.R., Hubbard, S.C., Robbins, P.W., and Touster, O., α-D-Mannosidases of rat liver Golgi membranes. Mannosidase II is the GlcNAcMAN5-cleaving enzyme in glycoprotein biosynthesis and mannosidases Ia and IB are the enzymes converting Man9 precursors to Man5 intermediates, *J. Biol. Chem.*, 257, 3660–3668, 1982.
18. Schweden, J., Legler, G., and Bause, E., Purification and characterization of a neutral processing mannosidase from calf liver acting on (Man)9(GlcNAc)2 oligosaccha-rides, *Eur. J. Biochem.*, 157, 563–570, 1986.

19. Tulsiani, D.R. and Touster, O., The purification and characterization of mannosidase IA from rat liver Golgi membranes, *J. Biol. Chem.,* 263, 5408–5417, 1988.

20. Forsee, W.T., Palmer, C.F., and Schutzbach, J.S., Purification and characterization of an α-1,2-mannosidase involved in processing asparagine-linked oligosaccharides, *J. Biol. Chem.,* 264, 3869–3876, 1989.

21. Schweden, J. and Bause, E., Characterization of trimming Man9-mannosidase from pig liver. Purification of a catalytically active fragment and evidence for the transmembrane nature of the intact 65 kDa enzyme, *Biochem. J.,* 264, 347–355, 1989.

22. Bause, E., Bieberich, E., Rolfs, A., Volker, C., and Schmidt, B., Molecular cloning and primary structure of Man9-mannosidase from human kidney, *Eur. J. Biochem.,* 217, 535–540, 1993.

23. Herscovics, A., Schneikert, J., Athanassiadis, A., and Moremen, K.W., Isolation of a mouse Golgi mannosidase cDNA, a member of a gene family conserved from yeast to mammals, *J. Biol. Chem.,* 269, 9864–9871, 1994.

24. Lal, A., Schutzbach, J.S., Forsee, W.T., Neame, P.J., and Moremen, K.W., Isolation and expression of murine and rabbit cDNAs encoding an α-1,2-mannosidase involved in the processing of asparagine-linked oligosaccharides, *J. Biol. Chem.,* 269, 9872–9881 1994.

25. Tremblay, L.O. and Herscovics, A., Characterization of a cDNA encoding a novel human Golgi α1,2-mannosidase (IC) involved in *N*-glycan biosynthesis, *J. Biol. Chem.,* 275, 31655–31660, 2000.

26. Tremblay, L.O., Campbell Dyke, N., and Herscovics, A., Molecular cloning, chromosomal mapping and tissue-specific expression of a novel human α1,2-mannosidase gene involved in *N*-glycan maturation, *Glycobiology,* 8, 585–595, 1998.

27. Tremblay, L.O. and Herscovics, A., Cloning and expression of a specific human α-1,2-mannosidase that trims Man9GlcNAc2 to Man8GlcNAc2 isomer B during *N*-glycan biosynthesis, *Glycobiology,* 9, 1073–1078, 1999.

28. Gonzalez, D.S., Karaveg, K., Vandersall-Nairn, A.S., Lal, A., and Moremen, K.W., Identification, expression, and characterization of a cDNA encoding human endoplasmic reticulum mannosidase I, the enzyme that catalyzes the first mannose trimming step in mammalian Asn-linked oligosaccharide biosynthesis, *J. Biol. Chem.,* 274, 21375–21386, 1999.

29. Lal, A., Pang, P., Kalelkar, S., Romero, P.A., Herscovics, A., and Moremen, K.W., Substrate specificities of recombinant murine Golgi α1,2-mannosidases IA and IB and comparison with endoplasmic reticulum and Golgi processing α-1,2-mannosidases, *Glycobiology,* 8, 981–995, 1998.

30. Schachter, H., The joys of HexNAc. The synthesis and function of *N*- and *O*-glycan branches, *Glycoconj. J.,* 17, 465–483, 2000.

31. Kumar, R., Yang, J., Larsen, R.D., and Stanley, P., Cloning and expression of *N*-acetylglucosaminyltransferase I, the medial Golgi transferase that initiates complex *N*-linked carbohydrate formation, *Proc. Natl. Acad. Sci. USA,* 87, 9948–9952, 1990.

32. Kumar, R., Yang, J., Eddy, R.L., Byers, M.G., Shows, T.B., and Stanley, P., Cloning and expression of the murine gene and chromosomal location of the human gene encoding *N*-acetylglucosaminyltransferase I, *Glycobiology,* 2, 383–393, 1992.

33. Sarkar, M., Hull, E., Nishikawa, Y., Simpson, R.J., Moritz, R.L., Dunn, R., and Schachter, H., Molecular cloning and expression of cDNA encoding the enzyme that controls conversion of high-mannose to hybrid and complex *N*-glycans: UDP-*N*-acetylglucosamine: α-3-D-mannoside β-1,2-*N*-acetylglucosaminyltransferase I, *Proc. Natl. Acad. Sci. USA,* 88, 234–238, 1991.

34. Pownall, S., Kozak, C.A., Schappert, K., Sarkar, M., Hull, E., Schachter, H., and Marth, J.D., Molecular cloning and characterization of the mouse UDP-*N*-acetylglu-cosamine:α-3-D-mannoside β-1,2-*N*-acetylglucosaminyltransferase I gene, *Genomics,* 12, 699–704, 1992.

35. Schachter, H., Hull, E., Sarkar, M., Simpson, R.J., Moritz, R.L., Hoppener, J.W., and Dunn, R., Molecular cloning of human and rabbit UDP-*N*-acetylglucosamine: α-3-D-mannoside β-1,2-*N*-acetylglucosaminyltransferase I, *Biochem. Soc. Trans.,* 19, 645–648, 1991.

36. Hull, E., Sarkar, M., Spruijt, M.P., Hoppener, J.W., Dunn, R., and Schachter, H., Organization and localization to chromosome 5 of the human UDP-*N*-acetylglu-cosamine:α-3-D-mannoside β-1,2-*N*-acetylglucosaminyltransferase I gene, *Biochem. Biophys. Res. Commun.,* 176, 608–615, 1991.

37. Fukada, T., Iida, K., Kioka, N., Sakai, H., and Komano, T., Cloning of a cDNA encod-ing *N*-acetylglucosaminyltransferase I from rat liver and analysis of its expression in rat tissues, *Biosci. Biotechnol. Biochem.,* 58, 200–201, 1994.

38. Puthalakath, H., Burke, J., and Gleeson, P.A., Glycosylation defect in Lec1 Chinese hamster ovary mutant is due to a point mutation in *N*-acetylglucosaminyltransferase I gene, *J. Biol. Chem.,* 271, 27818–27822, 1996.

39. Opat, A.S., Puthalakath, H., Burke, J., and Gleeson, P.A., Genetic defect in *N*-acetyl-glucosaminyltransferase I gene of a ricin-resistant baby hamster kidney mutant, *Biochem. J.* 336 , 593–598, 1998.

40. Wilson, J.R., Williams, D., and Schachter, H., The control of glycoprotein synthe-sis: *N*-acetylglucosamine linkage to a mannose residue as a signal for the attach-ment of L-fucose to the asparagine-linked *N*-acetylglucosamine residue of glycopeptide from α1-acid glycoprotein, *Biochem. Biophys. Res. Commun.,* 72, 909–916, 1976.

41. Longmore, G.D. and Schachter, H., Product-identification and substrate-specificity studies of the GDP-L-fucose:2-acetamido-2-deoxy-β-D-glucoside (FUC goes to Asn-linked GlcNAc) 6-α-L-fucosyltransferase in a Golgi-rich fraction from porcine liver, *Carbohydr. Res.,* 100, 365–392, 1982.

42. Voynow, J.A., Kaiser, R.S., Scanlin, T.F., and Glick, M.C., Purification and charac-terization of GDP-L-fucose-*N*-acetyl β-D-glucosaminide α 1→6fucosyltransferase from cultured human skin fibroblasts. Requirement of a specific biantennary oligosaccharide as substrate, *J. Biol. Chem.,* 266, 21572–21577, 1991.

43. Shao, M.C., Sokolik, C.W., and Wold, F., Specificity studies of the GDP-[L]-fucose: 2-acetamido-2-deoxy-β-[D]-glucoside (Fuc—>Asn-linked GlcNAc) 6-α-[L]-fuco-syltransferase from rat-liver Golgi membranes, *Carbohydr. Res.,* 251, 163–173, 1994.

44. Moremen, K.W., Golgi α-mannosidase II deficiency in vertebrate systems: implica-tions for asparagine-linked oligosaccharide processing in mammals, *Biochim. Biophys. Acta.,* 1573, 225–235, 2002.

45. Velasco, A., Hendricks, L., Moremen, K.W., Tulsiani, D.R., Touster, O., and Farquhar, M.G., Cell type-dependent variations in the subcellular distribution of α-mannosidase I and II, *J. Cell Biol.,* 122, 39–51, 1993.

46. Moremen, K.W., Touster, O., and Robbins, P.W., Novel purification of the catalytic domain of Golgi α-mannosidase II. Characterization and comparison with the intact enzyme, *J. Biol. Chem.,* 266, 16876–16885, 1991.

47. Tulsiani, D.R., Opheim, D.J., and Touster, O., Purification and characterization of α-D-mannosidase from rat liver golgi membranes, *J. Biol. Chem.,* 252, 3227–3233, 1977.

48. Kaushal, G.P., Szumilo, T., Pastuszak, I., and Elbein, A.D., Purification to homogeneity and properties of mannosidase II from mung bean seedlings, *Biochemistry,* 29, 2168–2176, 1990.

49. Misago, M., Liao, Y.F., Kudo, S., Eto, S., Mattei, M.G., Moremen, K.W., and Fukuda, M.N., Molecular cloning and expression of cDNAs encoding human α-mannosidase II and a previously unrecognized α-mannosidase IIx isozyme, *Proc. Natl. Acad. Sci. USA,* 92, 11766–11770, 1995.

50. Moremen, K.W., Isolation of a rat liver Golgi mannosidase II clone by mixed oligonucleotide-primed amplification of cDNA, *Proc. Natl. Acad. Sci. USA,* 86, 5276–5280, 1989.

51. Moremen, K.W. and Robbins, P.W., Isolation, characterization, and expression of cDNAs encoding murine α-mannosidase II, a Golgi enzyme that controls conversion of high mannose to complex N-glycans, *J. Cell. Biol.,* 115, 1521–1534, 1991.

52. Bonay, P. and Hughes, R.C., Purification and characterization of a novel broad-specificity ($\alpha 1 \rightarrow 2$, $\alpha 1 \rightarrow 3$ and $\alpha 1 \rightarrow 6$) mannosidase from rat liver, *Eur. J. Biochem.,* 197, 229–238, 1991.

53. Chui, D., Oh-Eda, M., Liao, Y.F., Panneerselvam, K., Lal, A., Marek, K.W., Freeze, H.H., Moremen, K.W., Fukuda, M.N., and Marth, J.D., α-mannosidase-II deficiency results in dyserythropoiesis and unveils an alternate pathway in oligosaccharide biosynthesis, *Cell,* 90, 157–167, 1997.

54. Oh-Eda, M., Nakagawa, H., Akama, T.O., Lowitz, K., Misago, M., Moremen, K.W., and Fukuda, M.N., Overexpression of the Golgi-localized enzyme α-mannosidase IIx in Chinese hamster ovary cells results in the conversion of hexamannosyl-N-acetyl-chitobiose to tetramannosyl-N-acetylchitobiose in the N-glycan-processing pathway, *Eur. J. Biochem.,* 268, 1280–1288, 2001.

55. Schachter, H., Biosynthetic controls that determine the branching and microheterogeneity of protein-bound oligosaccharides, *Biochem. Cell. Biol.,* 64, 163–181, 1986.

56. Schachter, H., The 'yellow brick road' to branched complex N-glycans, *Glycobiology,* 1, 453–461, 1991.

57. Tan, J., D'Agostaro, A.F., Bendiak, B., Reck, F., Sarkar, M., Squire, J.A., Leong, P., and Schachter, H., The human UDP-N-acetylglucosamine: α-6-D-mannoside-β-1,2-N-acetylglucosaminyltransferase II gene (MGAT2). Cloning of genomic DNA, localization to chromosome 14q21, expression in insect cells and purification of the recombinant protein, *Eur. J. Biochem.,* 231, 317–328, 1995.

58. D'Agostaro, G.A., Zingoni, A., Moritz, R.L., Simpson, R.J., Schachter, H., and Bendiak, B., Molecular cloning and expression of cDNA encoding the rat UDP-N-acetylglucosamine: α-6-D-mannoside β-1,2-N-acetylglucosaminyltransferase II, *J. Biol. Chem.,* 270, 15211–15221, 1995.

59. Furukawa, K. and Sato, T., β-1,4-Galactosylation of N-glycans is a complex process, *Biochim. Biophys. Acta.,* 1473, 54–66, 1999.

60. Hennet, T., The galactosyltransferase family, *Cell Mol. Life Sci.,* 59, 1081–1095, 2002.

61. Guo, S., Sato, T., Shirane, K., and Furukawa, K., Galactosylation of N-linked oligosaccharides by human β-1,4-galactosyltransferases I, II, III, IV, V, and VI expressed in Sf-9 cells, *Glycobiology,* 11, 813–820, 2001.

62. Grundmann, U., Nerlich, C., Rein, T., and Zettlmeissl, G., Complete cDNA sequence encoding human β-galactoside α-2,6-sialyltransferase, *Nucleic Acids Res.,* 18, 667, 1990.

63. Takashima, S., Tsuji, S., and Tsujimoto, M., Characterization of the second type of human β-galactoside α2,6-sialyltransferase (ST6Gal II), which sialylates

Gal β1,4GlcNAc structures on oligosaccharides preferentially. Genomic analysis of human sialyltransferase genes, *J. Biol. Chem.*, 277, 45719–45728, 2002.

64. Krzewinski-Recchi, M.A., Julien, S., Juliant, S., Teintenier-Lelievre, M., Samyn-Petit, B., Montiel, M.D., Mir, A.M., Cerutti, M., and Harduin-Lepers, A., and Delannoy, P., Identification and functional expression of a second human β-galactoside α2,6-sialyltransferase, ST6Gal II, *Eur. J. Biochem.*, 270, 950–961, 2003.

65. Harduin-Lepers, A., Vallejo-Ruiz, V., Krzewinski-Recchi, M.A., Samyn-Petit, B., Julien, S., and Delannoy, P., The human sialyltransferase family, *Biochimie.*, 83, 727–737, 2001.

66. Tsuji, S., Datta, A.K., and Paulson, J.C., Systematic nomenclature for sialyltransferases, *Glycobiology*, 6, v-vii, 1996.

67. Davis, T.R., Schuler, M.L., Granados, R.R., and Wood, H.A., Comparison of oligosaccharide processing among various insect cell lines expressing a secreted glycoprotein, *In Vitro Cell. Dev. Biol. Anim.*, 29A, 842–846, 1993.

68. Ren, J., Bretthauer, R.K., and Castellino, F.J., Purification and properties of a Golgi-derived (α1,2)-mannosidase-I from baculovirus-infected lepidopteran insect cells (IPLB-SF21AE) with preferential activity toward mannose$_6$-N-acetylglucosamine$_2$, *Biochemistry*, 34, 2489–2495, 1995.

69. Kawar, Z., Herscovics, A., and Jarvis, D.L., Isolation and characterization of an α1,2-mannosidase cDNA from the lepidopteran insect cell line Sf9, *Glycobiology*, 7, 433–443, 1997.

70. Kawar, Z. and Jarvis, D.L., Biosynthesis and subcellular localization of a lepidopteran insect α1,2-mannosidase, *Insect Biochem. Mol. Biol.*, 31, 289–297, 2001.

71. Sarkar, M., and Schachter, H., Cloning and expression of *Drosophila melanogaster* UDP-GlcNAc:α-3-D-mannoside β1,2-N-acetylglucosaminyltransferase I, *Biol. Chem.*, 382, 209–217, 2001.

72. Altmann, F., Kornfeld, G., Dalik, T., Staudacher, E., and Glossl, J., Processing of asparagine-linked oligosaccharides in insect cells. N-acetylglucosaminyltransferase I and II activities in cultured lepidopteran cells, *Glycobiology*, 3, 619–625, 1993.

73. Velardo, M.A., Bretthauer, R.K., Boutaud, A., Reinhold, B., Reinhold, V.N., and Castellino, F.J., The presence of UDP-N-acetylglucosamine:α-3-D-mannoside β1,2-N-acetylglucosaminyltransferase I activity in *Spodoptera frugiperda* cells (IPLB-SF-21AE) and its enhancement as a result of baculovirus infection, *J. Biol. Chem.*, 268, 17902–17907, 1993.

74. Altmann, F. and Marz, L., Processing of asparagine-linked oligosaccharides in insect cells: evidence for α-mannosidase II, *Glycoconj. J.*, 12, 150–155, 1995.

75. Staudacher, E. and Marz, L., Strict order of (Fuc to Asn-linked GlcNAc) fucosyltransferases forming core-difucosylated structures, *Glycoconj. J.*, 15, 355–360, 1998.

76. Ailor, E., Takahashi, N., Tsukamoto, Y., Masuda, K., Rahman, B.A., Jarvis, D.L., Lee, Y.C., and Betenbaugh, M.J., N-glycan patterns of human transferrin produced in *Trichoplusia ni* insect cells: effects of mammalian galactosyltransferase, *Glycobiology*, 10, 837–847, 2000.

77. Hsu, T.A., Takahashi, N., Tsukamoto, Y., Kato, K., Shimada, I., Masuda, K., Whiteley, E.M., Fan, J.Q., Lee, Y.C., and Betenbaugh, M.J., Differential N-glycan patterns of secreted and intracellular IgG produced in *Trichoplusia ni* cells, *J. Biol. Chem.*, 272, 9062–9070, 1997.

78. Choi, O., Tomiya, N., Kim, J.H., Slavicek, J.M., Betenbaugh, M.J., and Lee, Y.C., N-glycan structures of human transferrin produced by *Lymantria dispar* (gypsy moth) cells using the LdMNPV expression system, *Glycobiology*, 13, 539–548, 2003.

79. Voss, T., Ergulen, E., Ahorn, H., Kubelka, V., Sugiyama, K., Maurer-Fogy, I., and Glossl, J., Expression of human interferon omega 1 in Sf9 cells. No evidence for complex-type *N*-linked glycosylation or sialylation, *Eur. J. Biochem.,* 217, 913–919, 1993.

80. Staudacher, E., Kubelka, V., and Marz, L., Distinct *N*-glycan fucosylation potentials of three lepidopteran cell lines, *Eur. J. Biochem.,* 207, 987–993, 1992.

81. Fabini, G., Freilinger, A., Altmann, F., and Wilson, I.B., Identification of core α1,3-fucosylated glycans and cloning of the requisite fucosyltransferase cDNA from *Drosophila melanogaster.* Potential basis of the neural anti-horseradish peroxidase epitope, *J. Biol. Chem.,* 276, 28058–28067, 2001.

82. Hemmer, W., Focke, M., Kolarich, D., Wilson, I.B., Altmann, F., Wohrl, S., Gotz, M., and Jarisch, R., Antibody binding to venom carbohydrates is a frequent cause for double positivity to honeybee and yellow jacket venom in patients with stinging-insect allergy, *J. Allergy Clin. Immunol.,* 108, 1045–1052, 2001.

83. Wilson, I.B., Harthill, J.E., Mullin, N.P., Ashford, D.A., and Altmann, F., Core α1,3-fucose is a key part of the epitope recognized by antibodies reacting against plant *N*-linked oligosaccharides and is present in a wide variety of plant extracts, *Glycobiology,* 8, 651–661 1998.

84. Wilson, I.B., Zeleny, R., Kolarich, D., Staudacher, E., Stroop, C.J., Kamerling, J.P., and Altmann, F., Analysis of Asn-linked glycans from vegetable foodstuffs: widespread occurrence of Lewis a, core α1,3-linked fucose and xylose substitutions, *Glycobiology,* 11, 261–274, 2001.

85. Prenner, C., Mach, L., Glossl, J., and Marz, L., The antigenicity of the carbohydrate moiety of an insect glycoprotein, honey-bee (*Apis mellifera*) venom phospholipase A2. The role of α1,3-fucosylation of the asparagine-bound *N*-acetylglucosamine, *Biochem., J.,* 284, 377–380, 1992.

86. Bencurova, M., Hemmer, W., Focke-Tejkl, M., Wilson, I.B., and Altmann, F., Specificity of IgG and IgE antibodies against plant and insect glycoprotein glycans determined with artificial glycoforms of human transferrin, *Glycobiology,* 14, 457–466, 2004.

87. Schachter, H., *N*-Acetylglucosaminyltransferase-II, in *Handbook of Glycosyltransferases and Related Genes*, Vol., Taniguchi KH, N., and Fukuda, M., Eds., Springer, Tokyo, Japan, 2002, pp. 70–79.

88. Hollister, J.R., Shaper, J.H., and Jarvis, D.L., Stable expression of mammalian β1,4-galactosyltransferase extends the *N*-glycosylation pathway in insect cells, *Glycobiology,* 8, 473–480, 1998.

89. van Die, I., van Tetering, A., Bakker, H., van den Eijnden, D.H., and Joziasse, D.H., Glycosylation in lepidopteran insect cells: identification of a β1→4-*N*-acetylgalactosaminyltransferase involved in the synthesis of complex-type oligosaccharide chains, *Glycobiology,* 6, 157–164, 1996.

90. Hollister, J.R. and Jarvis, D.L., Engineering lepidopteran insect cells for sialoglycoprotein production by genetic transformation with mammalian β1,4-galactosyltransferase and α-2,6-sialyltransferase genes, *Glycobiology,* 11, 1–9 2001.

91. Abdul-Rahman, B., Ailor, E., Jarvis, D.L., Betenbaugh, M.J., and Lee, Y.C., β-1,4-Galactosyltransferase activity in native and engineered insect cells measured with time-resolved Eu-fluorescence, *Carbohydr. Res.,* 337, 2181–2186, 2002.

92. Lopez, M., Tetaert, D., Juliant, S., Gazon, M., Cerutti, M., Verbert, A., and Delannoy, P., *O*-glycosylation potential of lepidopteran insect cell lines, *Biochim. Biophys. Acta,* 1427, 49–61, 1999.

93. Hooker, A.D., Green, N.H., Baines, A.J., Bull, A.T., Jenkins, N., Strange, P.G., and James, D.C., Constraints on the transport and glycosylation of recombinant IFN-gamma in Chinese hamster ovary and insect cells, *Biotechnol. Bioeng.,* 63, 559–572, 1999.

94. Koles, K., Irvine, K.D., and Panin, V.M., Functional characterization of Drosophila sialyltransferase, *J. Biol. Chem.,* 279, 4346–4357, 2004.

95. Altmann, F., Schwihla, H., Staudacher, E., Glossl, J., and Marz, L., Insect cells contain an unusual, membrane-bound β-*N*-acetylglucosaminidase probably involved in the processing of protein *N*-glycans, *J. Biol. Chem.,* 270, 17344–17349, 1995.

96. Licari, P.J., Jarvis, D.L., and Bailey, J.E., Insect cell hosts for baculovirus expression vectors contain endogenous exoglycosidase activity, *Biotechnol. Prog.,* 9, 146–152, 1993.

97. Bendiak, B. and Schachter, H., Control of glycoprotein synthesis. Kinetic mechanism, substrate specificity, and inhibition characteristics of UDP-*N*-acetylglucosamine:α-D-mannoside β1-2 *N*-acetylglucosaminyltransferase II from rat liver, *J. Biol. Chem.,* 262, 5784–5790, 1987.

98. Hinderlich, S., Stasche, R., Zeitler, R., and Reutter, W., A bifunctional enzyme catalyzes the first two steps in *N*-acetylneuraminic acid biosynthesis of rat liver. Purification and characterization of UDP-*N*-acetylglucosamine 2-epimerase/*N*-acetylmannosamine kinase, *J. Biol. Chem.,* 272, 24313–24318, 1997.

99. Lawrence, S.M., Huddleston, K.A., Pitts, L.R., Nguyen, N., Lee, Y.C., Vann, W.F., Coleman, T.A., and Betenbaugh, M.J., Cloning and expression of the human *N*-acetylneuraminic acid phosphate synthase gene with 2-keto-3-deoxy-D-glycero- D-galacto-nononic acid biosynthetic ability, *J. Biol. Chem.,* 275, 17869–17877, 2000.

100. Tomiya, N., Ailor, E., Lawrence, S.M., Betenbaugh, M.J., and Lee, Y.C., Determination of nucleotides and sugar nucleotides involved in protein glycosylation by high-performance anion-exchange chromatography: sugar nucleotide contents in cultured insect cells and mammalian cells, *Anal. Biochem.,* 293, 129–137, 2001.

101. Effertz, K., Hinderlich, S., and Reutter, W., Selective loss of either the epimerase or kinase activity of UDP-*N*-acetylglucosamine 2-epimerase/*N*-acetylmannosamine kinase due to site-directed mutagenesis based on sequence alignments, *J. Biol. Chem.,* 274, 28771–28778, 1999.

102. Kim, K., Lawrence, S.M., Park, J., Pitts, L., Vann, W.F., Betenbaugh, M.J., and Palter, K.B., Expression of a functional *Drosophila melanogaster N*-acetylneuraminic acid (Neu5Ac) phosphate synthase gene: evidence for endogenous sialic acid biosynthetic ability in insects, *Glycobiology,* 12, 73–83, 2002.

103. Kulakosky, P.C., Shuler, M.L., and Wood, H.A., *N*-glycosylation of a baculovirus-expressed recombinant glycoprotein in three insect cell lines, *In Vitro Cell Dev. Biol. Anim.,* 34, 101–108, 1998.

104. Grabenhorst, E., Hofer, B., Nimtz, M., Jager, V., and Conradt, H.S., Biosynthesis and secretion of human interleukin 2 glycoprotein variants from baculovirus-infected Sf21 cells. Characterization of polypeptides and posttranslational modifications, *Eur. J. Biochem.,* 215, 189–197, 1993.

105. Kubelka, V., Altmann, F., Kornfeld, G., and Marz, L., Structures of the *N*-linked oligosaccharides of the membrane glycoproteins from three lepidopteran cell lines (Sf-21, IZD-Mb-0503, Bm-N), *Arch. Biochem. Biophys.,* 308, 148–157, 1994.

106. Rudd, P.M., Downing, A.K., Cadene, M., Harvey, D.J., Wormald, M.R., Weir, I., Dwek, R.A., Rifkin, D.B., and Gleizes, P.E., Hybrid and complex glycans are linked to the conserved *N*-glycosylation site of the third eight-cysteine domain of LTBP-1 in insect cells, *Biochemistry* 39, 1596–1603, 2000.

107. Wagner, R., Liedtke, S., Kretzschmar, E., Geyer, H., Geyer, R., Klenk, H.D., Elongation of the *N*-glycans of fowl plague virus hemagglutinin expressed in *Spodoptera frugiperda* (Sf9) cells by coexpression of human β1,2-*N*-acetylglucosaminyltransferase I, *Glycobiology,* 6, 165–175, 1996.

108. Ogonah, O.W., Freedman, R.B., Jenkins, N., Patel, K., and Rooney, B., Isolation and characterization of an insect cell line able to perform complex N-linked glycosylation on recombinant proteins, *Bio/Technology,* 14, 197–202, 1996.

109. Kretzschmar, E., Geyer, R., and Klenk, H.D., Baculovirus infection does not alter N-glycosylation in *Spodoptera frugiperda* cells, *Biol. Chem. Hoppe-Seyler,* 375, 23–27, 1994.

110. Aeed, P.A. and Elhammer, A.P., Glycosylation of recombinant prorenin in insect cells: the insect cell line Sf9 does not express the mannose 6-phosphate recognition signal, *Biochemistry,* 33, 8793–8797, 1994.

111. Manneberg, M., Friedlein, A., Kurth, H., Lahm, H.W., and Fountoulakis, M., Structural analysis and localization of the carbohydrate moieties of a soluble human interferon gamma receptor produced in baculovirus-infected insect cells, *Protein Sci.,* 3, 30–38, 1994.

112. Hogeland, K.E., Jr., and Deinzer, M.L., Mass spectrometric studies on the N-linked oligosaccharides of baculovirus-expressed mouse interleukin-3, *Biol. Mass Spectrom.,* 23, 218–224, 1994.

113. James, D.C., Freedman, R.B., Hoare, M., Ogonah, O.W., Rooney, B.C., Larionov, O.A., Dobrovolsky, V.N., Lagutin, O.V., and Jenkins, N., N-glycosylation of recombinant human interferon-gamma produced in different animal expression systems, *Biotechnology, (N Y)* 13, 592–596, 1995.

114. Joshi, L., Davis, T.R., Mattu, T.S., Rudd, P.M., Dwek, R.A., Shuler, M.L., and Wood, H.A., Influence of baculovirus-host cell interactions on complex N-linked glycosylation of a recombinant human protein, *Biotechnol. Prog.,* 16, 650–656, 2000.

115. Ogonah, O.W., Freedman, R.B., Jenkins, N., and Rooney, B.C., Analysis of human interferon-gamma glycoforms produced in baculovirus infected insect cells by matrix assisted laser desorption spectrometry, *Biochem. Soc. Trans.,* 23, 100S, 1995.

116. Wathen, M.W., Aeed, P.A., and Elhammer, A.P., Characterization of oligosaccharide structures on a chimeric respiratory syncytial virus protein expressed in insect cell line Sf9, *Biochemistry,* 30, 2863–2868, 1991.

117. Lopez, M., Coddeville, B., Langridge, J., Plancke, Y., Sautiere, P., Chaabihi, H., Chirat, F., Harduin-Lepers, A., Cerutti, M., Verbert, A., and Delannoy, P., Microheterogeneity of the oligosaccharides carried by the recombinant bovine lactoferrin expressed in *Mamestra brassicae* cells, *Glycobiology,* 7, 635–651, 1997.

118. Kulakosky, P.C., Hughes, P.R., and Wood, H.A., N-Linked glycosylation of a baculovirus-expressed recombinant glycoprotein in insect larvae and tissue culture cells, *Glycobiology,* 8, 741–745, 1998.

119. Wagner, R., Geyer, H., Geyer, R., and Klenk, H.-D., N-Acetyl-β-glucosaminidase accounts for differences in glycosylation of influenza virus hemagglutinin expressed in insect cells from a baculovirus vector, *J. Virol.,* 70, 4103–4109, 1996.

120. Watanabe, S., Kokuho, T., Takahashi, H., Takahashi, M., Kubota, T., and Inumaru, S., Sialylation of N-glycans on the recombinant proteins expressed by a baculovirus-insect cell system under β-N-acetylglucosaminidase inhibition, *J. Biol. Chem.,* 277, 5090–5093, 2002.

121. Wagner, R., Liedtke, S., Kretzschmar, E., Geyer, H., Geyer, R., and Klenk, H.D., Elongation of the N-glycans of fowl plague virus hemagglutinin expressed in *Spodoptera frugiperda* (Sf9) cells by coexpression of human β1,2-N-acetylglucosaminyltransferase I, *Glycobiology,* 6, 165–175, 1996.

122. Abdul-Rahman, B., Ailor, E., Jarvis, D., Betenbaugh, M., and Lee, Y.C., β-(1→4)-Galactosyltransferase activity in native and engineered insect cells measured with time-resolved europium fluorescence, *Carbohydr. Res.,* 337, 2181–2186, 2002.

123. Jarvis, D.L. and Finn, E.E., Modifying the insect cell *N*-glycosylation pathway with immediate early baculovirus expression vectors, *Nat. Biotechnol.*, 14, 1288–1292, 1996.

124. Breitbach, K. and Jarvis, D.L., Improved glycosylation of a foreign protein by Tn-5B1-4 cells engineered to express mammalian glycosyltransferases, *Biotechnol. Bioeng.*, 74, 230–239, 2001.

125. Hollister, J., Grabenhorst, E., Nimtz, M., Conradt, H., and Jarvis, D.L., Engineering the protein *N*-glycosylation pathway in insect cells for production of biantennary, complex *N*-glycans, *Biochemistry*, 41, 15093–15104, 2002.

126. Tomiya, N., Howe, D., Aumiller, J.J., Pathak, M., Park, J., Palter, K., Jarvis, D.L., Betenbaugh, M.J., and Lee, Y.C., Complex-type biantennary *N*-glycans of recombinant human transferrin from *Trichoplusia ni* insect cells expressing mammalian β-1,4-galactosyltransferase and β-1,2-*N*-acetylglucosaminyltransferase II, *Glycobiology*, 13, 23–34 2003.

127. Jarvis, D.L., Howe, D., and Aumiller, J.J., Novel baculovirus expression vectors that provide sialylation of recombinant glycoproteins in lepidopteran insect cells, *J. Virol.*, 75, 6223–6227, 2001.

128. Seo, N-S., Hollister, J.R., and Jarvis, D.L., Mammalian glycosyltransferase expression allows sialoglycoprotein production by baculovirus-infected insect cells, *Protein Expr. and Purif.*, 22, 234–241, 2001.

129. Hara, S., Yamaguchi, M., Takemori, Y., Furuhata, K., Ogura, H., and Nakamura, M., Determination of mono-*O*-acetylated *N*-acetylneuraminic acids in human and rat sera by fluorometric high-performance liquid chromatography, *Anal. Biochem.*, 179, 162–166, 1989.

130. Hollister, J., Conradt, H., and Jarvis, D.L., Evidence for a sialic acid salvaging pathway in lepidopteran insect cells, *Glycobiology*, 13, 487–495, 2003.

131. Viswanathan, K., Lawrence, S., Hinderlich, S., Yarema, K.J., Lee, Y.C., and Betenbaugh, M.J., Engineering sialic acid synthetic ability into insect cells: identifying metabolic bottlenecks and devising strategies to overcome them, *Biochemistry*, 42, 15215–15225, 2003.

132. Lawrence, S.M., Huddleston, K.A., Tomiya, N., Nguyen, N., Lee, Y.C., Vann, W.F., Coleman, T.A., and Betenbaugh, M.J., Cloning and expression of human sialic acid pathway genes to generate CMP-sialic acids in insect cells, *Glycoconj. J.*, 18, 205–213, 2001.

133. Aumiller, J.J., Hollister, J.R., and Jarvis, D.L., A transgenic insect cell line engineered to produce CMP-sialic acid and sialylated glycoproteins, *Glycobiology*, 13, 497–507, 2003.

134. Joshi, L., Shuler, M.L., and Wood, H.A., Production of a sialylated *N*-linked glycoprotein in insect cells, *Biotechnol. Prog.*, 17, 822–827, 2001.

135. Joosten, C.E. and Shuler, M.L., Effect of culture conditions on the degree of sialylation of a recombinant glycoprotein expressed in insect cells, *Biotechnol. Prog.*, 19, 739–749, 2003.

136. Palomares, L.A., Joosten, C.E., Hughes, P.R., Granados, R.R., and Shuler, M.L., Novel insect cell line capable of complex *N*-glycosylation and sialylation of recombinant proteins, *Biotechnol. Prog.*, 19, 185–192, 2003.

15 N-Glycan Engineering in Yeasts and Fungi: Progress toward Human-Like Glycosylation

Nico Callewaert, Wouter Vervecken, Steven Geysens, and Roland Contreras

CONTENTS

I. INTRODUCTION

Asparagine-linked glycosylation, a structurally highly complex co- and posttranslational modification of proteins, is of critical importance in the recombinant production of glycoproteins. The N-linked glycans modulate the folding and sorting of the proteins to which they are attached in the secretory pathway.[1] The structure of these glycans influences a therapeutic glycoprotein's properties (by shielding of sometimes substantial parts of the protein surface,[2] and by influencing the protein's tissue distribution, peak levels, and residence time in the blood stream.[3,4] Moreover, the type of N-glycan structure that is generated can make the difference between obtaining an extremely heterogenous mixture of glycoforms that is both hard to purify and to characterize in detail, or obtaining a very homogenous preparation that is easily manufactured in a reproducible way. In many cases, where intravenous administration of a therapeutic glycoprotein is needed, success critically depends on assuring that the N-glycan structures do not contain immunogenic determinants and on providing a high degree of sialylation to avoid the very rapid clearance that is mediated by the hepatic receptors for Gal-terminated and Man/GlcNAc-terminated glycoproteins.[5] The importance of proper glycosylation is one of the main reasons that expensive mammalian cell culture techniques are still the mainstay for the production of glycoprotein therapeutics, although in these systems too, homogeneity and level of sialylation are often suboptimal.[6]

Alternatives are being sought in insect cell culture, where glycan engineering efforts have recently provided proof of principle that the generation of sialylated glycans is possible.[7] However, insect cell culture does not solve many of the problems and costs associated with mammalian cell culture. Another area of intensive research and development is the use of transgenic plants and animals for biopharmaceutical production. Both of these approaches also face problems with glycan modifications. Plants can modify their glycoproteins with substituents that are not present in humans and therefore have immunogenic/allergenic potential.[8] From the few published data, it seems that glycoproteins produced in the milk of transgenic mammals substantially more high-mannose glycan structures than natural counterparts.[9,10] Moreover, the sialylation on at least some of the proteins produced in milk is a mixture of N-acetyl- and N-glycolyl-neuraminic acid,[9] the latter being immunogenic to humans. It is clear that neither plants nor transgenic animals provide recombinant proteins produced in them with "human serum-type" glycosylation (see below), and only little or no data are so far available to evaluate the impact of these differences on the long-term efficacy and safety of these products.

For all of the above reasons, it is sensible to explore the possibility of reengineering the *N*-glycosylation pathway of eukaryotic microorganisms, especially of the commonly used yeasts (*Saccharomyces cerevisiae* and *Pichia pastoris*) and also of filamentous fungi (mainly of the genera *Aspergillus*, *Penicillium* and *Trichoderma*). There is long-standing experience with protein production in all of these fungal organisms, and they possess the basic biosynthetic machinery to construct *N*-glycans and efficiently transfer them to recombinant glycoproteins.[11,12] Moreover, fungal cells do not substitute the protein-proximal "Man$_3$GlcNAc$_2$ core" of their *N*-glycans with potentially immunogenic groups (as plants do). The main problem is that fungal cells lack the capability to construct fully modified complex-type *N*-glycans of the human type, and instead generate a very complicated hyper-mannosylated glycoform mixture. These hypermannosylated glycans are efficiently recognized by receptors in the liver and on reticuloendothelial cells, and in some cases contain immunogenic determinants. However, the knowledge that has been gained concerning the yeast *N*-glycosylation pathway has allowed the rational reengineering of this pathway in a number of fungal organisms, especially in the yeast *P. pastoris*. This reengineering is the subject of this chapter.

This chapter first gives an overview of the *N*-glycosylation pathway as it has been unravelled over the last three decades by studies on *S. cerevisiae*. This endeavor is still ongoing, and some of the reaction steps in the pathway still remain uncharacterized at the genetic level. However, this large body of fundamental knowledge has been instrumental in devising engineering approaches to the pathway. In particular, the finding that the pathway in other yeasts resembles the pathway in *S. cerevisiae* was important in this regard. We also provide fairly extensive information on the relevant parts of the human *N*-glycosylation pathway, both to clarify the "engineering goal" of fungal *N*-glycan engineering experiments (i.e., human serum-type *N*-glycans), and to provide a "parts list" of mammalian genes that are essential tools in these experiments.

In the second part of the chapter, we provide an account of *N*-glycan pathway engineering in *S. cerevisiae*, followed by a discussion of *N*-glycan engineering studies on the methylotrophic yeast *P. pastoris*, which is currently the preferred yeast for recombinant protein production.

II.　*S. CEREVISIAE* AND HUMAN *N*-GLYCOSYLATION PATHWAYS

A.　Biosynthesis of the Sugar-Nucleotide Donor Substrates of Glycosylation Reactions: Cytoplasmic Events

A first prerequisite for the biosynthesis of eukaryotic *N*-glycans is the generation of activated monosaccharides. These activated monosaccharides take the form of sugar nucleoside mono- or diphosphates, with the exception of some reactions in the endoplasmic reticulum (ER) that require the dolichol derivatives Dol-P-Man and Dol-P-Glc (see Section II.C). The sugar nucleotides are generated starting from the non-activated monosaccharides that are either directly taken up from the extracellular environment or generated from other monosaccharides during the cell's metabolism. In the following sections, we discuss the biogenesis of the six most common sugar nucleotide donors in *N*-glycan synthesis. The metabolic diagrams were generated with the metabolic pathway tools provided by the Kegg project (http://www.genome.ad.jp/kegg/metabolism.html). The diagrams were curated and cross-checked with the biological databases so that they represent only the pathways for which there is factual evidence for their existence in *S. cerevisiae* and *Homo sapiens* (Table 15.1).

TABLE 15.1
Yeast (*S. cerevisiae*) and Human (*Homo sapiens*) Enzymes Responsible for the Production of Sugar-Nucleotide Donor Substrates Used in Glycosylation

	Enzyme Name	EC Number	OMIM	Human Gene(s)	Yeast Gene(s)
UDP-GlcNAc	Glutamine-fructose-6-phosphate transminase	EC 2.6.1.16	138292	BC000012	GFA1
	Glucosamine-phosphate *N*-acetyltransferase	EC 2.3.1.4	—	XM_085119	GNA1
	Phosphoacetylglucosamine mutase	EC 5.4.2.3	605135	AB032081	PCM1
	UDP-*N*-Acetylhexosamine diphosphorylase	EC 2.7.7.23	602862	NM_003115	UAP1
	N-Acetylglucosamine kinase	EC 2.7.1.59	606828	AJ242910	—
	N-Acetylglucosamine-6-phosphate deacetylase	EC 3.5.1.25	—	AF132948	—
	Glucosamine 6-phosphate deaminase	EC 3.5.99.6	601798	AF048826	—
UDP-GalNAc	UDP-Glucose 4-epimerase	EC 5.1.3.7	230350	L41668	GAL10
	GalNac kinase	EC 2.7.1-	137028	M84443	—
	UDP-*N*-Acetylhexosamine diphosphorylase	EC 2.7.7.23	602862	AB011004	a
CMP-NANA	UDP-*N*-Acetylglucosamine 2-epimerasel/ ManNac kinase	EC.1.3.14/ EC 2.7.1.60	603824/ 269921 (Sialuria)/ 600737 (HIBM)	NM_005476	—
	N-Acylneuraminate-9-phosphate synthase	EC 4.1.3.20	605202	NM_018946	—
	N-Acylneuraminate-9-phosphatase	EC 3.1.3.29	—		—
	N-Acylneuraminate cytidylytransferase	EC 2.7.7.43	603316	NM_018686	—
	N-Acetyl-glucosamine 2-epimerase	EC 5.1.3.8	—	NM_002910	—
	N-Acetyl-Imannosamine kinase	EC 2.7.1.60	606828	AJ242910	—

(Continued)

TABLE 15.1   (Continued)

	Enzyme Name	EC Number	OMIM	Human Gene(s)	Yeast Gene(s)
GDP-Man	Mannose-6-phosphate isomerase	EC 5.3.1.8	154550/ 602579 (CDG-lb)	X76057	PMI40
	Phosphomannomutase	EC 5.4.2.8	601785/ 212065 (CDG-la)	U85773 and NM_002676	ALG4
	Mannose-1-phosphate guanylyltransferase	EC 2.7.7.13	—	NP_037466	MPG1
GDP-Fuc	GDPmannose 4, 6-dehydratase	EC 4.2.1.47	602884	AF042377	—
	GDP-keto-6-deoxymannose 3,5-epimerase, 4-reductase	N.A.	137020	U58766	—
	Fucokinase	EC 2.7.1.52	—	Uncharacterized	—
	Fucose-1-phosphate guanylyltransferase	EC 2.7.7.30	603609	NM_003838	—
UDP-Gal	UTP-Hexose-1-phosphate uridylyltransferase	EC 2.7.7.10	230400 (galactosemia I)	M60091	GAL7
	Galactokinase	EC 2.7.1.6	604313 and 230200 (galactosemia II)	U26401 and M84443	GAL1
	UDP-Glucose-hexose-1-phosphate uridylyltransferase	EC 2.7.7.12	—	Uncharacterized	—
	UDPglucose 4-epimerase	EC 5.1.3.2	230350 (galactosemia III)	AF022382	GAL10

[a] Unknown whether *S. cerevisiae* UAPI can use GalNAc-1-Pi as a substrate.

1. UDP-GlcNAc and UDP-GalNAc

There are two routes to the synthesis of UDP-GlcNAc, either starting from GlcNAc or starting from the glycolytic intermediate fructose-6-phosphate (Pathways: Figure 15.1). The latter pathway is the more general one and involves transamidation of fructose-6-phosphate to glucosamine-6-phosphate with glutamine as the amino donor.[13,14] The next step in the pathway acetylates glucosamine-6-phosphate with acetyl-CoA as the donor. The yeast gene (GNA1) encoding this activity has been cloned[15] and the enzyme has been crystallized.[16] The human gene(s) have not been characterized thus far. Subsequently, GlcNAc-6-phosphate is isomerized to GlcNAc-1-phosphate[17,18] and this intermediate then reacts with UTP to form UDP-GlcNAc and pyrophosphate.[19] An alternative pathway in which glucosamine-6-phosphate is first isomerized to glucosamine-1-phosphate by phosphomannomutase (EC 5.4.2.8) also exists.

The pathway that starts from GlcNAc is best studied in mammalian cells and can be considered as a salvage pathway, as most free GlcNAc in the cytosol comes from lysosomal degradation of glycoconjugates. GlcNAc is activated by phosphorylation

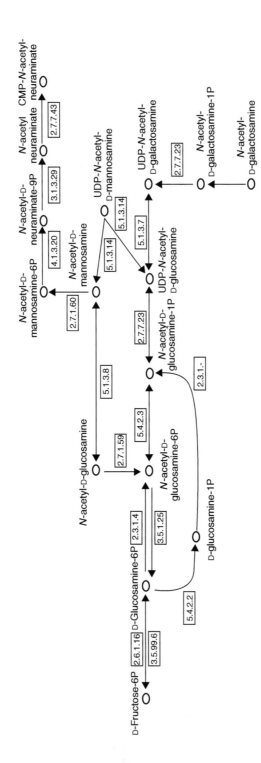

FIGURE 15.1 UDP-HexNAc and CMP-NANA biosynthetic pathways. See Table 15.1 for current state of knowledge on the genes underlying these enzymatic activities.

to GlcNAc-6-phosphate[20] and then isomerized to GlcNAc-1-phosphate, followed by reaction with UTP as mentioned above. There is no homolog of N-acetylglucosamine kinase in yeast and thus it is not clear whether yeast can use GlcNAc from the medium directly for incorporation in its glycoconjugates. UDP-GalNAc can be biosynthesized by 4-epimerization of UDP-GlcNAc. Little is known about the enzyme, but it has been purified from porcine submaxillary gland,[21] and judging from the enzymatic characterization of the purified heterodimeric enzyme, it is almost certainly the same enzyme as UDP-Glc 4-epimerase (EC 5.1.3.2). Another route to UDP-GalNAc involves the phosphorylation of GalNAc (derived mostly from lysosomal degradation) to GalNAc-1-phosphate by a specific kinase.[22] Subsequently, this compound reacts with UTP to form UDP-GalNAc, catalyzed by the same UDP-N-acetyl-hexosamine diphosphorylase[23] as for GlcNAc-1-phosphate.

2. CMP-NANA

None of the steps involved in CMP-N-acetylneuraminic acid synthesis has been reported for yeast. However, we provide an overview of the pathway here to illustrate the complexity that would be involved in building this biosynthetic capability in fungal organisms (Pathways: Figure 15.1). For this reason, N-glycan engineering research in fungi has focused on generating glycan structures with terminal Gal. The UDP-GlcNAc and UDP-Gal sugar donors that are necessary to reach this stage have now been demonstrated to be present in the yeast's secretory system, at least to a certain extent (see below).

Efficient and scaleable *in vitro* sialylation technology is available to provide the sialic acid residues during downstream processing (e.g., GlycoAdvance of Neose Inc., Horsham, PA). The biosynthesis of CMP-N-acetylneuraminic acid derives from the UDP-GlcNAc synthetic pathway. The best characterized pathway takes UDP-GlcNAc as the starting compound, and converts it into N-acetylmannosamine-6-phosphate by a bifunctional enzyme,[24–26] through the UDP-N-acetylmannosamine and N-acetylmannosamine intermediates. As the first committed reaction in CMP-NANA biosynthesis, this homohexameric enzyme is potently feedback-inhibited by the final product.[27] UDP-N-acetylglucosamine 2-epimerase/ManNAc kinase is also the rate-limiting step in the sialylation reactions, at least in hematopoietic cell lines.[28]

Upon formation of N-acetylmannosamine-6-phosphate, it is converted into N-acylneuraminate-9-phosphate by the corresponding phosphoenolpyruvate-utilizing synthase. The human and murine genes have been cloned [29,30] and the rat liver enzyme was purified and enzymatically characterized.[31] The human enzyme can also use mannose-6-phosphate as a suboptimal substrate, which is converted into 2-keto-3-deoxy-D-glycero-D-galacto-nononic acid (KDN). This sialic acid species is mainly present as poly-KDN on a very restricted number of proteins.[32,33]

The next step is the dephosphorylation of N-acylneuraminate-9-phosphate by N-acylneuraminate-9-phosphatase, an enzyme on which virtually no information is available, except that it is cytosolically localized and present in rat liver lysates.[34] Finally, N-acetyl neuraminic acid is converted into CMP-NANA by N-acylneuraminate cytidylyltransferase.[35] The human enzyme can also use N-glycolylneuraminic acid as a substrate, but as no active CMP-N-acetylneuraminic acid hydroxylase enzymes are present in human cells, very little or no CMP-*N*-glycolylneuraminic acid is available for sialylation redundant. Remarkably, *N*-acylneuraminate cytidylyltransferase is localized in the nucleus. Moreover, its sequence is similar to prokaryotic equivalents and the mammalian enzyme is capable of rescuing a defect in its equivalent in *Escherichia coli*, providing evidence for an ancestral relation between the bacterial and mammalian CMP-NANA synthesis systems.

Alternative pathways to CMP-NANA probably exist. The bifunctional key enzyme in the normal pathway (UDP-GlcNAc-2-epimerase/ManNAc kinase) is expressed to different extents in different tissues,[26] and in cell lines that lack the bifunctional enzyme, a ManNAc kinase activity can still be detected, copurifying with GlcNAc kinase.[20,36] Together with *N*-acetyl-D-glucosamine-2-epimerase,[37] the Glc/ManNAc kinase may provide a pathway from GlcNAc to CMP-NANA in some tissues. More research is needed to fully establish the relevance of this pathway.

3. GDP-Man

In human cells, almost all GDP-Man is generated from the glycolytic intermediary fructose-6-phosphate (Pathways: Figure 15.2). The first step is conversion into mannose-6-phosphate by the key enzyme phosphomannose isomerase (PMI).[38] In patients with a partial PMI deficiency (congenital disorders of glycosylation (CDG) type Ib), mannose supplementation of the diet rescues the defect. This exogenous mannose enters the GDP-Man biosynthetic pathway via hexokinase-mediated phosphorylation. In yeast, deletion of the gene coding for PMI is lethal, if the growth medium doen not contain mannose (either from the culture broth or from

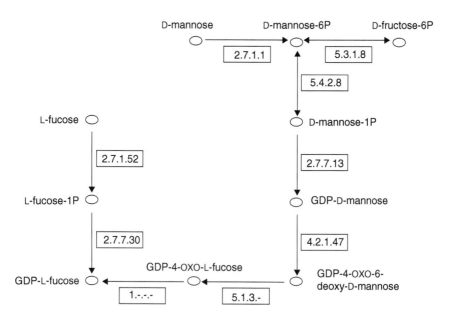

FIGURE 15.2 GDP-Fuc and GDP-Man biosynthetic pathways. See Table 15.1 for current state of knowledge on the genes underlying these enzymatic activities.

lysosomal degradation of mannoproteins). Mannose-6-phosphate is further converted into mannose-1-phosphate by phosphomannomutase (PMM). Two genes code for this activity in humans, with PMM2[39] having a more widespread expression[40] than PMM1.[41]

The next step in the pathway is the reaction of mannose-1-phosphate with GTP to form GDP-Man. The gene coding for the enzyme in this step, mannose-1-phosphate guanylyltransferase, was first cloned from *S. cerevisiae*,[42] but homologs have been described for a range of organisms, including *Homo sapiens* (see Table 15.1).

4. GDP-Fuc

GDP-Fucose (GDP-Fuc) is physiologically most robustly generated from GDP-Man in a three-step pathway that involves two enzymes. First, C4 of Man is oxidized by GDP-Man-4,6-dehydratase, resulting in the ketone 4-dehydro-6-deoxy-mannose.[43] Subsequently, the bifunctional enzyme GDP-keto-6-deoxymannose 3,5-epimerase, 4-reductase converts this ketone to GDP-Fuc.[44] Thus, this highly complex reaction first involves epimerization at C3 and C5 to form GDP-4-keto-6-deoxyglucose, followed by reduction at C4 to the GDP-Fuc end product.

As shown in Figure 15.2, there is also a pathway leading directly from fucose to GDP-Fuc,[45] but as fucose is present in only low concentrations in the blood, it is unlikely that this pathway contributes very significantly to the generation of GDP-Fuc under normal circumstances. As the precursor GDP-Man is relatively abundant in fungi, and just two enzymes are needed to convert this into GDP-Fuc, the engineering of *S. cerevisiae* for the production of GDP-Fuc was recently reported. [46]

5. UDP-Gal

A first pathway to UDP-Gal involves the conversion of galactose into Gal-1-phosphate by galactokinase (Pathways: Figure 15.3).[47,48] This product can then be converted in either of two ways to UDP-Gal: (1) by reaction with UTP, catalyzed by UTP-hexose-1-phosphate uridylyltransferase,[49,50] or (2) via a uridyl transferase reaction in which UDP-Glc is the uridinemonophosphate donor.[51,52] In an alternative pathway to UDP-Gal, the C4 of UDP-Glc is epimerized.[53,54] Both pathways from

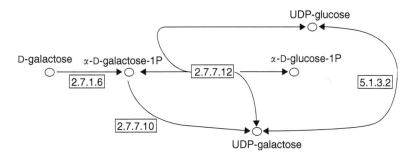

FIGURE 15.3 UDP-Gal biosynthetic pathways. See Table 15.1 for current state of knowledge on the genes underlying these enzymatic activities.

galactose and from UDP-Glc exist in yeast until recently. The main question concerning the availability of UDP-Gal for the biosynthesis of *N*-glycans whether it was at all transported into the *S. cerevisiae* and *P. pastoris* Golgi apparatus. In *Schizosaccharomyces pombe*, α-linked galactose residues are part of the mannan structure, and accordingly, both the galactosyltransferase[55] and the UDP-Gal transporter[56] have been identified. A UDP-Gal transporter (Hut1p) has also been characterized in *S. cerevisiae*.[57] Our recent demonstration that *in vivo* galactosylation of *N*-glycans is possible in *P. pastoris* without expressing a UDP-Gal transporter has taken away any doubt that an endogenous transport activity for UDP-Gal is also present in this yeast.[149]

B. Transport of Sugar Nucleotides to the ER/Golgi Lumen and "Flipping" of the Dol-P-Man and Dol-P-Glc Substrates to the Luminal Membrane Side

The only glycosylation reactions in the entire *N*-glycan biosynthesis pathway that take place on the cytosolic side of the secretory system membranes are the first reactions in the synthesis of the dolichol-oligosaccharide precursor (Figure 15.4). All the other glycosylation reactions take place within the lumen of the ER and the Golgi apparatus. Consequently, a crucial part of the biochemistry of the *N*-glycan synthesis pathway is the transport of the sugar nucleotides from the cytosol into the ER/Golgi. Over the last 5 years, the genes coding for the yeast and human transporters have been cloned and characterized (summarized in Table 15.2). The cloning

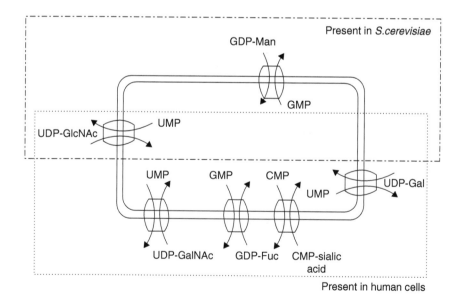

FIGURE 15.4 Sugar nucleotide transport activities. Only the activities relevant for this chapter and present in *S. cerevisiae* and *Homo sapiens* are represented. See Table 15.2 for the encoding genes and OMIM accession numbers, where applicable.

TABLE 15.2

Genes Encoding Sugar Nucleotide Transport Activities and OMIM Accession Numbers, where Applicable, in Yeast and Humans

Gene Name	Human Gene	S. cerevisiae Gene	OMIM	Human Disease Caused by Deficiency
GDP-Man transporter	—	VRG4	—	—
UDP-GlcNAc transporter	NM_012243	YEA4	605632	—
UDP-GalNAc transporter	NM_015139 and D88146	—	314375	—
UDP-Gal transporter	D88146	—	314375	—
GDP-Fuc transporter	NM_018389	—	605881 and 266265	CDG-IIc (or LADII)
CMP-sialic acid trasnporter	D87969	—	605634	CDG-IIe (unpublished)

was generally accomplished using complementation of yeast and CHO lectin-resistant mutants with defects in glycosylation that could not be ascribed to sugar nucleotide synthesis or glycosyltransferase deficiency. Their transport mechanism is essentially an antiport with the cognate nucleotide by-products of the glycosylation reaction (Figure 15.4). Most of these transporters reside in the Golgi apparatus, but detailed information on their targeting awaits further experimentation. The genes encode highly hydrophobic proteins with a large number of predicted transmembrane domains.[58–60]

The mannosylation and glucosylation reactions that involve Dol-P-Man and Dol-P-Glc in the ER occur on the luminal side of the membrane, whereas the synthesis of these molecules occurs on the cytoplasmic side. Therefore, there must be a mechanism by which these glycoconjugates change orientation in the membrane. This process is referred to as "flipping." To date, there are no studies that have conclusively attributed the "flippase" reaction to a particular protein. However, there is evidence that the human MPDU1 gene could be involved in the flipping reaction of the monosaccharide derivatives of Dol-P, as its product is required for the utilization of both Dol-P-Man and Dol-P-Glc by the luminal ER transferases, and as MPDU1p is not involved in the glycosyltransferase activity itself.[61]

C. ASSEMBLY OF THE DOLICHOL-OLIGOSACCHARIDE PRECURSOR: ER MEMBRANE-BOUND EVENTS

Remarkably, eukaryotes generate a tetradecasaccharide precursor, carried by the polyisoprenoid dolichol, in a pathway that is highly conserved from yeast to humans (Pathways: Figure 15.5; Table 15.3). Subsequently, this oligosaccharide is transferred from dolichol to Asn-X-Ser/Thr sequons of nascent glycoproteins as they enter the ER.

First, dolichol is phosphorylated (yeast SEC59 gene) and substituted with the first GlcNAc residue in a reaction that requires UDP-GlcNAc and generates a pyrophosphate linkage between the oligosaccharide part and the dolichol part of the glycoconjugate. In yeast, this reaction is catalyzed by ALG7.

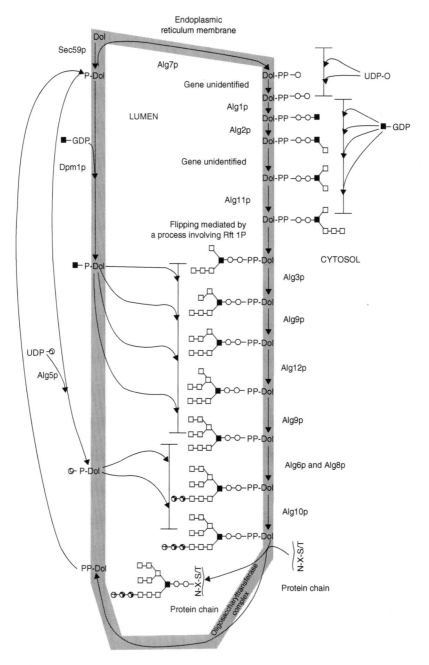

FIGURE 15.5   Biosynthesis of the Glc$_3$Man$_9$GlcNAc$_2$-PP-Dolichol precursor oligosaccharide and transfer to nascent glycoproteins. See Table 15.3 for the encoding genes and OMIM accession numbers, where available. In the figure, the names of the corresponding yeast proteins are given for clarity. Table 15.4 gives an overview of the characterized oligosaccharyltransferase subunits in *S. cerevisiae* and *Homo sapiens*.

TABLE 15.3

Yeast and Human Genes Encoding Biosynthetic Acitivites Required for the Production of the $Glc_3Man_9GlcNAc_2$-PP-Dolichol Precursor Oligosaccharide and Transfer to Nascent Glycoproteins

Enzyme Name	EC Number	OMIM	Human Gene(s)	Human Disease	Yeast Gene(s)
Dolichol kinase	EC 2.7.1.108	—	Several SEC53-homologous ESTs such as BE672680	—	SEC59
Dolicholphosphate β-D-mannosyltransferase	EC 2.4.1.83	603503 (DPM1) and 603564 (DPM2) and 605951 (DPM3)	D86198 (DPM1) and NM_003863 (DPM2) and (AB028128 (DPM3)	CDG-Ie	DPM1
Dolichol-phosphate β-D-glucosyltransferase	EC 2.4.1.117	604565	NM_013338	—	ALG5
UDP-N-acetyglucosamine-dolichol-phosphate N-acetylglucosamine phosphotransferase	EC 2.7.8.15	—	Z82022	—	ALG7
N-acetylglucosaminyl-PP-dolichol N-acetylglucosaminyl transferase	EC 2.4.1.141	—	Unidentified	—	Unidentified
Chitobiosyl-PP-dolichol β-mannosyltransferase	EC 2.4.1.142	605907	AB019038	—	ALG1
GDP-Man:Man GlcNAc(2)-PP-dolichol mannosyltransferase	EC 2.4.1.13	607906	NM_033087	CDG-Ii	ALG2
GDP-Man:Man(2) GlcNAc(2)-PP-dolichol mannosyltransferase	EC 2.4.1.x	—	Unidentified	—	Unidentified
GDP-Man:Man(3/4) GlcNAc(2)-PP-dolichol mannosyltransferase	EC 2.4.1.131	—	AK025456 (by homology)	—	ALG11
Dolichol-P-Man:Man(5) GlcNAc(2)-PP-dolichol mannosyltransferase	EC 2.4.1.130	601110	Y09022	CDG-Id	ALG3
Dolichol-P-Man: oligosac-charide-PP-colichol α-1,2-mannosyltransferase	EC 2.4.1.130	606941	AF395532	—	ALG9
Dolichol-P-Man:Man (7)GlcNAc(2)-PP-dolicholm annosyltransferase	EC 2.4.1.130	607143	AJ303120	CDG-Ig	ALG12

(Continued)

TABLE 15.3   (Continued)

Enzyme Name	EC Number	OMIM	Human Gene(s)	Human Disease	Yeast Gene(s)
Dolichol-P-Glc:Man (9)GlcNAc(2)-PP-dolicholglucosyltransferase	EC 2.4.1.x	603147	NP_037471	CDG-Ic	ALG6
Dolichol-P-Glc:Glc(1) Man(9)GlcNAc(2)-PP-dolicholglucosyltransferase	EC 2.4.1.x	—	BC001133	CDG-Ih	ALG8
Dolichol-P-Glc:Glc(2) Man(9)GlcNAc(2)-PP-dolicholglucosyltransferase	EC 2.4.1.x	—	AJ312278	—	ALG10

Note: *S. cerevisiae* asparagine linked glycosylation mutants

The *S. cerevisiae* asparagine linked glycosylation (ALG) mutants have been instrumental in shaping our understanding of the early steps in the eukaryotic *N*-glycosylation pathways, and several of them will be encountered in this section. Most of these mutants were obtained by mannose-suicide selection, a technique that involves feeding *S. cerevisiae* with radioactive mannose. Only cells that incorporate lower amounts of this "hot" mannose into their mannoproteins can survive.[62] To date, twelve complementation groups have been obtained using this and several other selection strategies (tunicamycin resistance yielded ALG7,[63] additive negative effect on yeast viability with an oligosaccharyltransferase mutant yielded ALG9[64] and ALG10,[65] sodium vanadate resistance yielded ALG11[66] and searching the *S. cerevisiae* sequence for homologs of ALG9 yielded ALG12.[67]

The second GlcNAc residue is incorporated by an as yet unidentified protein, followed by the addition of five mannose residues, all with GDP-Man as donor and occurring on the cytoplasmic face of the ER membrane (see References 68 and references therein). Alg2p is the α-1,3-mannosyltransferase that takes the monomannosylated structure as substrate.[69–71] At the $Man_5GlcNAc_2$ stage of the dolichol-linked precursor synthesis, the precursor changes its orientation, with the oligosaccharide part switching from a cytoplasmic orientation to an ER-luminal orientation. This process is referred to as "flipping" and has remained enigmatic for a very long time. However, recent progress has been made with the discovery that yeast Rft1p is required in the process of $Man_5GlcNAc_2$-PP-Dol flipping.[68] The human homolog has the accession number AJ318099. However, whether Rft1p is sufficient for this flipping reaction remains formally unproven, so it is still possible that additional gene products are necessary for this flipping reaction.

After the $Man_5GlcNAc_2$-PP-Dol reaches the luminal side of the ER membrane, the mannosyltransferases that add further four mannose residues take Dol-P-Man as the donor substrate. Dol-P-Man is synthesized from dolicholphosphate and GDP-Man in a reaction catalyzed by DPM1.

The four mannose residues are added in a specific sequence to the α-1,6 arm of the Man$_5$GlcNAc$_2$ carbohydrate: first, Alg3p adds the residue in α-1,3-linkage,[72,73] and the branch formed by this residue is then elongated with an α-1,2-linked mannose residue by Alg9p.[64] Then, the α-1,6-linked residue is built in by Alg12p[67] and this branch is also elongated by Alg9p with an α-1,2-linked mannose residue.

The last steps of the dolichol-oligosaccharide biosynthesis are glucosylation reactions with Dol-P-Glc as the donor substrate. Dol-P-Glc is formed from Dol-P and UDP-Glc in an Alg5p catalyzed reaction.[74] Each of the three glucose units is added by a different enzyme, in the sequence Alg6p,[75,76] Alg8p,[77,78] and Alg10p.[65] Alg6p and Alg8p are α-1,3-glucosyltransferases, whereas Alg10p is an α-1,2-glucosyltransferase.

Upon completion of the Glc$_3$Man$_9$GlcNAc$_2$-PP-Dol precursor, the next step is the transfer of the oligosaccharide to Asn-X-Ser/Thr sequons as they enter the ER. This reaction is catalyzed by oligosaccharyltransferase, a heterooligomeric complex.[79] The available information on this subunit's composition in yeast and man is summarized in Table 15.4. Using an *in vitro* system for reconstitution of the yeast oligosaccharyltransferase, it was found that Glc$_3$Man$_9$GlcNAc$_2$-PP-Dol allosterically activates the enzyme.[80]

D. DEGLUCOSYLATION AND REGLUCOSYLATION: ROLE IN GLYCOPROTEIN FOLDING

Upon transfer of the precursor oligosaccharide to the glycoprotein, the three glucose residues are removed by glucosidase I (α-1,2-specific) and glucosidase II (α-1, 3-specific see Table 15.5 and Figure 15.6). Remarkably, the first two glucose residues are quickly removed, while the third one is removed only minutes later. Two specific binding proteins for this monoglucosylated glycan have been identified: calnexin and calreticulin.[81,82] Moreover, in the ER, a glucosyltransferase that can reglucosylate unfolded glycoproteins is present.[83] Remarkably, however, this glycoprotein glucosylation activity is absent in *S. cerevisiae*.[84] Thus, at least in higher eukaryotes, a mechanism can be envisaged in which the retention of unfolded glycoproteins by

TABLE 15.4
The Characterized Oligosaccharyltransferase Subunits in *S. cerevisiae* and Man, *Homo sapiens*

Yeast gene	Human Homolog	Accession Number for Human Gene	OMIM
OST1	Ribophorin I	Y00281	180470
OST2	DAD1	D15057	600243
OST3	N33 and IAP	AH003689 and NM_032121	—
OST4	—	—	—
OST5	—	—	—
OST6	—	—	—
WBP1	OST48	D89060	602202
SWP1	Ribophorin II	Y00282	180490
STT3	STT3-A and STT3-B	L47337	601134

TABLE 15.5

Genes in the N-linked Glycoprotein Biosynthesis Pathway after Precursor Oligosaccharide Transfer to the Protein, but before Export from the ER

Enzyme Name	Yeast Gene	Human Gene Accession Number	Human Disease	OMIM
Glucosidase I	CWH41	X87237	CDG-IIb	601336 and 606056 (CDG-IIb)
Glucosidase II	Z36098	AF-144074 (α-subunit) and AF-144075 (β-subunit)	—	601862 (α-subunit) and 601864 (β-subunit)
UDP-glucose:glycoprotein glucosyltransferase	—	AF227905 and AF227906	—	605897 and 605898
ER mannosidase I	MNS 1	AF148509	—	604346

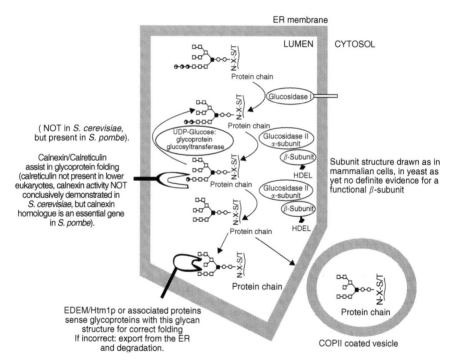

FIGURE 15.6   Simplified representation of the events after precursor oligosaccharide transfer to the protein, but before export from the ER. The folding-related reglucosylation cycle of the glycan is not present in *S. cerevisiae*. Certain aspects of the exact glycan structures involved in these steps are still unclear, especially the structures recognized by the export machinery for misfolded proteins. Therefore, this aspect of the diagram must not be considered as definitive. The genes in these stages of the glycoprotein biosynthesis pathway can be found in Table 15.5.

calnexin and calreticulin gives the ER folding machinery sufficient time to fold the glycoprotein correctly, [85] after which the glucosyltransferase does not modify the *N*-glycans anymore, and the protein is released for further processing. However, if the glycoprotein is not folded correctly at the time of removal of the last glucose unit, it can be recognized by the glucosyltransferase, be reglucosylated, and given a "second chance" to fold correctly. Proteins that fail to fold correctly and fail to be reglucosylated are exported from the ER back into the cytoplasm and are degraded by ER-associated proteasome complexes.[86]

Upon removal of the three glucose units, one mannose residue is removed from the $Man_9GlcNAc_2$ N-glycans by the ER mannosidase I.[87] This mannosidase seems to play an important role in targeting unfolded proteins for proteasomal degradation, and most probably the presence of $Man_8GlcNAc_2$ on incorrectly folded proteins is in some way a recognition tag for the degradation machinery.[88] The recognizing protein may be the catalytically inactive mannosidase I-homologous protein EDEM,[89] as overexpression of this protein accelerates unfolded glycoprotein degradation. A similar protein (Htm1p) has been detected in yeast,[90] and inactivation of its gene leads to a slower degradation of unfolded glycoproteins. However, the presence of $Man_8GlcNAc_2$ on folded proteins leads to their export from the ER.

E. *S. CEREVISIAE* GOLGI *N*-GLYCAN PROCESSING PATHWAY

Morphologically, the structure of *S. cerevisiae* Golgi apparatus is a less well-defined than in higher eukaryotes, as no "cisternal stacking' is generally observed (Figure 15.7). Functionally, however, there are at least four biochemically distinct subcompartments, each with its own complement of glycosyltransferases.[91] In analogy with the Golgi apparatus in higher eukaryotes, these four subcompartments could be called *cis-*, medial- and *trans*-Golgi and *trans*-Golgi network.

Upon arrival in the *S. cerevisiae cis*-Golgi, the N-glycans are very efficiently modified by the Och1p α-1,6-mannosyltransferase,[92] initiating the hallmark of yeast N-glycosylation: an α-1,6-linked mannose oligomer or polymer, substituted with oligomannose branches.[12] This α-1,6-linked mannose polymerization occurs by far on the largest fraction of yeast-produced N-glycans. However, on a minor fraction it is blocked because an as yet unidentified α-1,2-mannosyltransferase acts on the Och1p-modified branch faster than the polymerizing α-1,6-mannosyl-transferase complexes.[93] In these cases, the resulting glycan is said to be of the "core type." Glycans that are effectively modified by the polymerizing complexes are said to be "hyperglycosylated." Two of the polymerizing complexes have been characterized, one consisting of Van1p and Mnn9p, the other consisting of Mnn9p/Anp1p/ Mnn10p/Mnn11p/Hoc1p.[94] The first complex appears to polymer-ize the first ten or so residues, after which the latter complex finishes the job by extending the α-1,6-linked mannose oligomer to a maximal size in the range of 50 residues.

On this α-1,6-linked polymer, side branches are added by the α-1,2-mannosyl-transferase Mnn2p,[95] and extended to two or three (rarely more) α-1,2 residues by Mnn5p.[96] Phosphorylation on these side branches can be quite extensive and is responsible for a large part of the anionic character of the yeast cell wall (mutants in this phosphorylation can be selected by their reduced binding of cationic dyes such as Alcian Blue). Two proteins have been shown to be involved in this phosphoryla-tion process: Mnn4p[97] and Mnn6p.[98] Most data point to a transferase role for Mnn6p

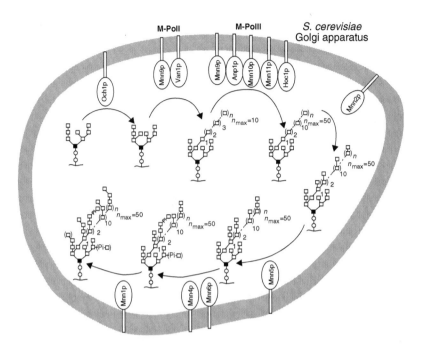

FIGURE 15.7 Yeast Golgi-localized N-glycan biosynthetic pathway. Och1p initiates the yeast "outer chain" of the glycan with an α-1,6-mannose residue. Glycans that are modified with this residue are then used as a substrate for two distinct polymerization complexes that generate a poly-α-1,6-mannose backbone, upon which side branches are synthesized that consist of α-1,2-linked mannoses. Terminating modifications are either a phosphodiester-linked mannose residue or an α-1,3-linked mannose cap.

and a supporting function for Mnn4p, although Mnn4p has homology to a known phospholigand transferase.[97]

Finally, a portion of the α-1,2-linked side branches are capped with an α-1, 3-linked mannose residue by Mnn1p.[99] These residues are immunogenic and hamper the use of glycoproteins produced in wild-type *S. cerevisiae* for therapeutic purposes.

The consequence of all of these (phospho)mannosyltransferase reactions is that the total size of the N-glycan can reach 150 to 200 residues, and as they concern polymerizing reactions, this results in a very broad range of structures that can show a broad distribution in size. Obviously, this extensive heterogeneity forms a substantial hindrance for the production of glycosylated biopharmaceuticals in yeast, and the removal of this heterogeneity has been primary engineering goal, as we discuss below.

F. HUMAN GOLGI N-GLYCAN PROCESSING I: PATHWAY TOWARD BIANTENNARY SIALYLATED GLYCANS

In contrast to the immediate extension of the glycan structures in the yeast Golgi apparatus, the Golgi apparatus of higher eukaryotic cells contains a number of α-1,

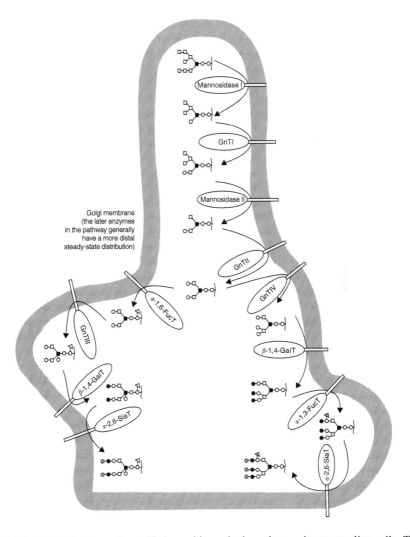

FIGURE 15.8 Most prominent N-glycan biosynthetic pathways in mammalian cells. The pathways shown in this figure are by no means exhaustive, but show the steps that are involved in the biosynthesis of the N-glycan that constitute > 95% of the N-glycan structures detected on the total mixture of human serum glycoproteins. Therefore, these glycan structures are the most relevant ones in the context of biopharmaceutical protein production for intravenous administration. The end products are α-2,6-sialylated bi or triantennary glycans, with or without a bisecting GlcNAc residue and with or without fucose residues, either attached to the glycan branches or to the protein-proximal GlcNAc residue.

2-mannosidases that first trim back the $Man_8GlcNAc_2$ glycan to the $Man_5GlcNAc_2$ structure (Figure 15.8). In man, three such Golgi class I mannosidase genes have been characterized, and each has a subtly different fine specificity in the preferred order in which the three α-1,2-mannose residues are removed. Moreover, they have a different tissue distribution.[100] The physiological relevance of these differences, if any, is unknown.

Note: Modifications that target a glycoprotein to the lysosomes

Glycoproteins that function in the lysosomal compartment of the cell (mainly in degradation pathways) are modified with a special carbohydrate recognition tag. This involves the substitution of the 6-position of subterminal α-1,2-linked mannose residues of the $Man_8GlcNAc_2$ ER-exit structure with $GlcNAc(\beta$-1)-phosphate, leading to the formation of a phosphodiester linkage. This reaction is catalyzed by N-acetylglucosamine-1-phosphotransferase, after which the GlcNAc residue is removed by N-acetylglucosamine-1-phosphodiester α-N-acetylglucosaminidase,[101] leaving a phosphate group on the high-mannose glycan. These phosphorylated high-mannose glycans are recognized by the mannose-6-phosphate receptor[102] in the medial-Golgi apparatus, upon which the receptor-glycoprotein complexes are selectively transported to the lysosomes. Yeast does not use this lysosomal targeting mechanism, which makes it an attractive host organism for the production of recombinant forms of this type of enzymes, as they are efficiently secreted.[103] Such enzyme preparations are in high demand for the treatment of a category of glycosylation-related hereditary disorders, the lysosomal storage diseases (most of them are defects in the catabolism of complex carbohydrates).[104] The reason for the success in the treatment of some of these disorders with intravenously injected recombinant forms of the required enzyme[105–107] is that the mannose-6-phosphate receptor is not only present in the Golgi apparatus, but also on the surface of most cells. Thus, if the recombinant enzyme is at least partially glycosylated with the described phosphorylated high-mannose glycans, it can be efficiently internalized and transported to the lysosomes of a substantial number of cells and cell types in the patient. This approach to therapy has the general name of enzyme replacement therapy (ERT).

The most common pathway (see References 11)[108,109], upon the generation of the $Man_5GlcNAc_2$ structure, is the transfer of a GlcNAc residue in β-1,2-linkage to the α-1,3-arm of the glycan by N-acetylglucosaminyltransferase I (GnTI). This activity is coded for by a single, essential gene in the mammalian genome. GnTI activity on an N-glycan is truly a turning point in this pathway, as the generated $GlcNAcMan_5$-$GlcNAc_2$ structure is the substrate for several further reactions that convert the glycan into a "hybrid-type" or "complex-type" structure.

The next of these reactions is the removal of the α-1,3- and α-1,6-linked mannose residues from the α-1,6 branch of the glycan by the class II Golgi mannosidases (two isozymes are expressed in most human cells, designated as ManII and ManIIx). These are very large homodimeric proteins (>1000 amino acids per subunit) and there is some evidence[110] that they form a direct complex with GnTI. A note of importance is that mannosidase IIx (also sometimes called mannosidase III) does not have the requirement of prior modification of its substrate glycans by GnTI and can thus work directly on $Man_5GlcNAc_2$ to generate $Man_3GlcNAc_2$. This glycan is also a good substrate for GnTI, so that both pathways ultimately result in the same GlcNAcMan$_3$GlcNAc$_2$ structure.

N-acetylglucosaminyltransferase II in the next step substitutes the α-1,6-branch mannose residue with a β-1,2-linked GlcNAc. Glycans that are not modified by GntII are said to be of the "hybrid type."

Prior to elaborating on the subsequent buildup of the glycan branches, it should be mentioned that fucosylation of the glycan on the protein-proximal fucose residue is very common in mammalian N-glycans. The enzyme that catalyzes the α-1,6-fucosylation requires prior modification of the glycan by GnTI. Core fucosylation is blocked by prior addition of a "bisecting GlcNAc residue" to the β-linked mannose residue of the N-glycan core. The addition of this bisecting GlcNAc residue is catalyzed by N-acetylglucosaminyltransferase III, and also blocks further branching of N-glycans beyond the two antennae that have been described already.

After the addition of GlcNAc residues, these are further substituted with galactose, and for serum protein N-glycans the β-1, 4-linkage is by for the most common. In this way, the LacNAc substructure that is extremely common in N-glycans is formed (the so-called type 2 chains, in contrast to Galβ1,3GlcNAc, the type 1 chain). In most human tissues, β-1,4-galactosyltransferase I is responsible for most of the N-glycan galactosylation capacity.

Sialylation is the main chain-terminating modification for serum protein N-glycans, and in humans, sialic acid is mostly of the N-acetylneuraminic acid form, in contrast to most mammalian species, where a mixture of N-glycolylneuraminic acid and N-acetylneuraminic acid is generally found. Both α-2,6 and α-2,3-sialyltransferases are present in most human cell types, and it is common to find a glycan with sialic acid in a different linkage on the different branches of the same N-glycan. However, human serum protein N-glycans have almost exclusively α-2,6-linked sialic acid residues.

G.   HUMAN GOLGI N-GLYCAN PROCESSING II: FURTHER COMPLEXITY BY HIGHER BRANCHING, BRANCH ELONGATION AND BRANCH SUBSTITUTION

Further branching of N-glycans occurs early in the pathway as described in the previous section, immediately after the synthesis of the biantennary GlcNAc-terminated structure and before galactosylation is complete. Triantennary glycans are formed by GnTIV activity. Tetra-antennary glycans can subsequently be formed by GnTV; this activity is upregulated in cancer cells. GnTVI activity is uncommon and leads to penta-antennary glycans.

Prior to terminal sialylation, the branches can also be elongated with LacNAc repeats. β-1,3-N-acetylglucosaminyltransferases add GlcNAc to β-1,4-Gal-terminated branches, after which the β-1,4-galactosyltransferases can add Gal again and so on, until a "polylactosamine" structure is formed.

To add to the complexity of N-glycan structures, N-glycan branches can be substituted with a number of groups. The most common on serum protein N-glycans is the substitution of the 3- or 4-position of GlcNAc residues with fucose. These fucosylation reactions are again catalyzed by a whole family of transferases.[110] Other substituents include sulphate and GalNAc groups, but these are not the main components of the abundant human serum glycoproteins.

H.   PECULIARITIES OF THE *N*-GLYCOSYLATION PATHWAY IN DIFFERENT YEASTS AND FUNGI, AS COMPARED WITH THE *S. CEREVISIAE* PATHWAY

1.   Yeasts other than *S. cerevisiae*

The major yeast species currently in use for the manufacturing of recombinant proteins is the methylotroph *P. pastoris*. It is a very attractive expression system for several reasons. First, the molecular biological techniques involved in manipulating this yeast are very similar to those used for *S. cerevisiae* and, thus, many investigators are familiar with these procedures.[111] Second, extremely powerful gene regulatory elements are available to drive the expression of the gene of interest. In particular, the methanol-inducible AOX-I promoter is used for batch fermentations.[112] The endogenous AOX-I protein can constitute up to 30% of total cellular protein under induced conditions and, consequently, extremely high levels of several cytoplasmically expressed heterologous proteins (such as TNFα)[113] have been obtained. Naturally, the expression levels are attainable for proteins secreted into the growth medium are more limited due to secretory system capacity constraints. However, also for secreted proteins, robust gram per liter yields can be obtained in some cases such as human serum albumin.[114,115] Convenient kits containing vectors, strains, and primers are available from Invitrogen Inc. (Carlsbad, CA) under licensing agreement with Research Corporation Technologies (Tucson, AZ), and the availability of these kits has greatly enhanced the spread of *Pichia* expression technology throughout academia and industry. A further important point is that the upscaling of a protein production process in *Pichia* to very large volumes (over 1000 L) is feasible at lower cost than with mammalian cell culture. Moreover, *Pichia* can be grown to immensely dense cultures (up to 400 g wet weight/L culture) in fermentors, which typically leads to a 5- to 10-fold increased expression level when a process is transferred from shake flasks to optimized fermentor cultures.[116]

When we turn to glycoproteins, the picture for *Pichia* looks brighter than for *S. cerevisiae*, for several reasons. Importantly, there are no highly immunogenic α-1, 3-linked terminal mannose residues on the side branches of the *Pichia*-produced *N*-glycans.[117–119] Furthermore, in a significant number of cases, hyperglycosylation does not occur or occurs only to a limited extent.[120,121] In general, one can expect the majority of the *N*-glycans to be in the size range of $Man_8GlcNAc_2$-$Man_{14}GlcNAc_2$ on proteins secreted by *Pichia*. However, in a growing number of reported cases, this does not hold true and hyperglycosylation is indeed observed, be it to a lower extent than on the same proteins expressed in baker's yeast.[122–124] Sometimes, a relatively homogenous protein preparation can still be obtained during downstream processing by specifically purifying the lower glycosylated forms, but this is detrimental for the final protein yield. Phosphate and mannose phosphate substituents are also frequent on *Pichia N*-glycans, and their abundance can be influenced by the constitution of the growth medium.[125]

Hansenula polymorpha is another methylotrophic yeast that is closely related to *P. pastoris*,[126] and the same considerations as above seem to hold true for this yeast, with the exception that *Hansenula* glycosylation remains relatively poorly studied as compared with *Pichia.*

S. *pombe* is sometimes used for protein expression, but for glycosylation, it has the significant drawback of adding α-1,2-galactose residues to its glycans[127] as well as of strong hypermannosylation.[128]

2.   Filamentous Fungi

The structural knowledge of fungal *N*-glycans has been reviewed in depth.[129] There are several general trends. First, hypermannosylation as observed in *S. cerevisiae* is rare in filamentous fungi and by far most of the *N*-glycans are in the range of $Man_5GlcNAc_2$ to $Man_9GlcNAc_2$, with substituents. The observation that glycans smaller than $Man_8GlcNAc_2$ occur on glycoproteins produced in these fungi indicates the presence in these organisms of class IB α-1,2-mannosidase activity, be it in the secretory system or secreted in the growth medium (in which case the trimming of α-1,2-linked mannose residues would occur postsecretion). Genes coding for enzymes with this activity have indeed been cloned from *P. citrinum, Aspergillus saitoi* and *Trichoderma reesei,* and were enzymatically characterized to have an acidic pH optimum (around pH 5) and to be relatively thermophilic (optimum around 50°C, although much more stable at lower temperatures).[130-132] The availability of these fungal class IB mannosidase genes has been exploited for the bioengineering of the *N*-glycosylation pathways of yeasts.[133,134]

A second general trend is the frequent presence of nonmammalian-type monosaccharidic substituents on these high-mannose glycans. In *Aspergilli*, a substituent of major concern is the immunogenic α-linked galactofuranose,[135-136] which is added to apparently varying extent, depending on the specific *Aspergillus* strain and its growth conditions. Other substituents are mannose residues linked to the rest of the glycan moiety via a phosphodiester linkage or terminal phosphate groups with mannose as a carrier.[137] α-1,3-Linked terminal glucose units that apparently result from incomplete glucosidase II activity in the fungal ER have also been detected in proteins secreted by *T. reesei.*[138]

Finally, in some cases, a large percentage of the *N*-glycosylation sites are occupied by a single GlcNAc residue.[139] Most probably, these sites are effectively glycosylated with the normal precursor oligosaccharide, but are subsequently processed by an endoglycosidase that cleaves the chitobiose core of the oligosaccharide.

III.   *N*-GLYCAN ENGINEERING IN YEASTS AND FILAMENTOUS FUNGI

A.   Engineering Goals: *N*-Glycan Homogenization and Generation of Human-Complex-Type Structures

The first goal for all *N*-glycan engineering in fungi, and especially in the yeast species, is the homogenization of the *N*-glycans they produce from an almost uncharacterizable pool of hyperglycosylated structures to one or a few small high-mannose structures. Efforts to restrict product diversity are beneficial for various reasons. In the field of therapeutic glycoprotein manufacturing, the benefits are two fold. First, the presence of a bewildering variety of large, often charged (phosphate esters) glycans makes postfermentation downstream processing of the glycoprotein difficult and severely affects the yield of the production process. One will typically lose a very significant percentage of the product on any chromatography-based separation and focus on obtaining the sometimes small amount of core-glycosylated product.[140] Engineering the producing yeast for the production of a more homogenously glycosylated product will obviously increase the yield of the final purified product. A second important point with regard to biopharmaceutical production is that the regulatory authorities require that the glycosylation of a therapeutic glycoprotein be demonstrably reproducible. This requirement would be hard to meet if one were to characterize fully the N-glycan structures of a yeast-type hypermannosylated glycoprotein. Therefore, homogenization of the glycosylation in itself can significantly contribute to the ease of product characterization and thus help in the regulatory approval process.

Apart from these considerations, just engineering the glycosylation for small, uniform high-mannose glycans could already be useful when glycosylation is mainly needed for the purpose of efficient folding of the protein, and where modest high-mannose glycosylation is not too detrimental to the therapeutic effect of the molecule. For example, subunit vaccines against a number of viruses could benefit from this approach. It is clear that hyperglycosylated proteins can easily have most of the protein part covered by the glycans.[141,142] Trimming down these glycans should yield more immunogenic and therefore more efficacious vaccines.

A third field that can benefit from homogenizing the fungal N-glycosylation pathway is structural biology, because glycosylation sometimes introduces a lot of heterogeneity and glycosylated proteins are notoriously difficult to crystallize.[143,144] The simplest approach to solving this problem is to mutagenize the N-glycosylation sites. However, many eukaryotic proteins need N-glycosylation for their efficient folding and expression, and thus mutagenesis is not a solution in these cases. Expression hosts with a uniform, low-molecular-weight N-glycosylation should be very useful in producing protein preparations with strongly increased chances of crystallization. For this purpose, we have recently collaborated with the group of Dr. Khorana at MIT to generate an inducible mammalian expression system with uniform $Man_5GlcNAc_2$ glycosylation. This expression system was capable of expressing the prototype G-protein-coupled receptor rhodopsin at >5 mg/L cell culture.[144] The recent advances in *Pichia* glycoengineering (see below) now also make it possible to obtain the same glycan structure in this easily scaleable expression host (demonstrated for a series of heterologous proteins).

Once homogenization to $Man_5GlcNAc_2$ is obtained in a fungal species, one can start to engineer the enzymes required to synthesize more human-type glycans. It must be clarified here that "human-type" in this context is understood by all workers in the field to mean "terminally β-1,4-galactosylated" glycans. A monoantennary structure should, in principle, already be beneficial as it can be sialylated, and it is expected that this single sialylation per N-glycan would significantly reduce, if not abolish, recognition by the hepatic asialoglycoprotein receptors. Further engineering toward biantennary N-glycans could further improve this clearing behavior, and that remains to be studied. So far, no groups have reported data on the engineering of sialylation yeast, an endeavor that is complicated because it requires the expression of a large number of genes involved in the synthesis of CMP-NANA (see earlier in this chapter), a transporter to get this substrate into the secretory pathway, and a sialyltransferase.

Note: Engineering of sialylation into yeast

A problem that arises in such extensive engineering, and even with the more modest steps described above, is the availability of a sufficient number of orthogonal marker genes to assist in the transformation of the host. In *S. cerevisiae*, quite a few of these markers are useful, but this number is much smaller in other yeasts such as *P. pastoris* (but see Reference 145) and the choice is even more restricted for most filamentous fungi. To engineer the pathway with any number of genes above five or so, one would have to resort to cotransformation of a number of marker-free expression plasmids in molar excess to a selection marker-carrying DNA fragment, with high-throughput screening of the ensuing clones. Such an approach depends on high transformation efficiency, which could be problematic for filamentous fungi. Alternatively, a recoverable marker could solve this problem.

B.   ENGINEERING IN THE MODEL: *S. CEREVISIAE*

Most of the knowledge of *N*-glycan biosynthesis pathways is available for *S. cerevisiae*. Along with the availability of all the relevant gene sequences (see Section II) and the ease of manipulating multiple genes in this organism, it was obvious that this would be the first fungal organism to become a target for *N*-glycan engineering for the purpose of biopharmaceutical production. The group of Dr. Jigami in Tsukuba (Japan), in collaboration with Kirin Brewery's Central Laboratories for Key Technology, have published most extensively in this field.

A comparison of the yeast and human *N*-glycosylation pathways (see Section II) reveals that the first real point of divergence occurs upon transport of a glycoprotein to the *cis*-Golgi apparatus. Whereas yeast glycans are quickly modified by the Och1p α-1,6-mannosyltransferase and then elongated, human glycans are trimmed back to $Man_5GlcNAc_2$ by class Ib α-1,2-mannosidases. This step is a logical starting point for the reengineering of the yeast pathway. Fortunately, the Och1p activity is encoded by a single gene. The knockout of this gene in *S. cerevisiae* is viable, but its vitality is severely affected. Upon Och1p inactivation in *S. cerevisiae*, the *N*-glycan patterns produced are not homogenous,[146] as the Mnn1p α-1,3-mannosyltransferase can use the core $Man_8GlcNAc_2$ glycan as a substrate, and the same holds true for the addition of phosphomannose residues by the enzyme system, which involves Mnn6p and Mnn4p.[97,98] Consequently, Dr. Jigami's group inactivated both MNN1 and MNN4 in an Och1 background, to yield a homogenous $Man_8GlcNAc_2$ *N*-glycan profile on yeast carboxypeptidase Y.[133] Once this point was reached, it was possible to envisage the incorporation of a heterologous class Ib α-1,2-mannosidase activity into the *S. cerevisiae* secretory pathway. Jigami's group chose the enzyme from

A. saitoi, and tagged it with the HDEL C-terminal sequence to keep the protein in the yeast's ER. The resulting strain produced about 20% of carboxypeptidase Y glycans as $Man_5GlcNAc_2$.[133] This result provided an important proof of concept that the switch from hypermannosylation to the substrate for *N*-acetylglucosaminyltransferase I was possible in a yeast cell. Dr. Jigami's group has since built further on their work, which yielded the phosphorylated $Man_8GlcNAc_2$ structure to produce human lysosomal α-galactosidase.[147] This enzyme (produced in mammalian cells) is currently approved for the treatment of the lysosomal storage disease called Fabry's disease.[148] Importantly, a bacterial α-mannosidase has been described that can remove the mannose residues that are linked to the yeast glycans via phosphodiester bonds.[147] This enzyme leaves the phosphate groups on the glycan, yielding essentially the same structure as is generated in the mammalian Golgi apparatus on lysosomally targeted proteins. A similar strategy could potentially be applied to glycosylation-homogenous *P. pastoris* strains. α-Galactosidase has indeed been produced at high levels in *Pichia*,[122] but as the product is hyperglycosylated, it would definitely benefit from *N*-glycan homogenization.

Most recently, Dr. Jigami's group has also shown that it is possible to synthesize GDP-fuc in the yeast cytoplasm at levels that are 3.5 times higher than the level of GDP-Man in these cells (GDP-Mannose is the precursor for GDP-Fuc synthesis).[43] Now that it has been shown that *N*-acetylglucosaminyltransferase I can be functionally expressed in fungal cells (see below), the fucose-related work breaks the ground for further engineering of core-α-1,6-fucosylation these strains (the core fucosyltransferase is dependent on prior modification of the glycan by *N*-acetylglucosaminyltransferase I). This modification occurs on virtually all IgG Fc *N*-glycans, and the fucose residue on these glycans can influence the interaction of the IgG with Fc receptors. In at least one case it has been shown that an IgG with a low degree of core fucosylation had a much higher ADCC activity.[149]

C. *N*-GLYCAN ENGINEERING IN *P. PASTORIS*

Today, the methylotrophic yeast *P. pastoris* is the predominant yeast system used for the production of protein-based biopharmaceuticals,[116] with some products currently close to regulatory approval (such as Albrec™, recombinant serum albumin) and many others in different stages of clinical development. Several of them are glycoproteins. Therefore, we[134,150] and others[151,152] are adapting and modifying the glycoengineering principles that were validated in *S. cerevisiae* for use in *P. pastoris*, and are going farther than was so far possible with baker's yeast.

Fundamental research data that were available before the engineering studies started indicated that the Golgi form of $Man_5GlcNAc_2$ is not a substrate for any of the *P. pastoris* Golgi mannosyltransferases *in vitro*.[153] Based on this information, we hoped that if one could introduce sufficient α-1,2-mannosidase activity at the *P. pastoris* ER-to Golgi transition to generate $Man_5GlcNAc_2$ before the secretory cargo protein reached the Golgi mannosyltransferases, then one could potentially block hyperglycosylation. To test this hypothesis, we expressed both a murine and a *T. reesei* α-1,2-mannosidase, fused to the *S. cerevisiae* α-mating factor signal sequence at the *N*-terminus and to the HDEL sequence at the *C*-terminus.[134] The

HDEL sequence is recognized in yeast by a specific receptor in the *cis*-Golgi apparatus, inducing retrieval to the ER.[154] The murine protein was evaluated because its optimum temperature and pH are expected to be relatively close to the conditions in the *P. pastoris* ER (30°C and assumed pH = 7). However, these beneficial factors were apparently offset by a low expression level or a low folding efficiency, and the *T. reesei* enzyme (with an optimum temperature at about 50°C and an optimum pH of 5.0) yielded better efficiency in removing α-1,2-linked mannose residues from the *N*-glycans of two coexpressed heterologous glycoproteins: influenza virus hemagglutinin and *Trypanosoma cruzi trans*-sialidase (see Figure 15.9; glycan analysis is shown for the *trans*-sialidase). We concluded from the results obtained that the engineered *T. reesei* mannosidase was indeed very active *in vivo* (approaching full removal of α-1,2-linked mannose residues), but that this did not preclude a substantial amount of the glycans from modification by *P. pastoris* Och1p before the mannosidase had a chance to cut back the glycan to the $Man_5GlcNAc_2$ "nonsubstrate." The *T. reesei* mannosidase-Myc-HDEL protein was localized by immunofluorescence microscopy to a circonuclear staining pattern, with some staining in the cell periphery, compatible with the ER. In a collaboration with the group of Marc Claeyssens at Ghent University, we developed an easy screening tool for *P. pastoris* clones that express this relatively thermophilic mannosidase: we found that heat denaturation of *P. pastoris* cell lysates at 50°C allowed us to measure the *T. reesei* mannosidase activity in this very crude background (also containing swainsonine to inhibit class II mannosidase activity) using 2′,4′-dinitrophenyl-α-mannoside (DNPM) as a chromogenic substrate. This easy assay allows the selection of those clones with the highest level of mannosidase catalytic activity.[155]

This engineered mannosidase proved to be a very efficient tool in our further engineering studies, which involved inactivation of the *P. pastoris OCH1* (Figure 15.10 provides a schematic presentation of the *Pichia N*-glycan engineering strategies reported to date by our group and by Glycofi Inc.). In the Japanese patent literature, a document (JP07145005) is available that describes the generation of *Pichia* strains in which a gene homologous to *S. cerevisiae OCH1* had been inactivated by homologous recombination. By Western-blot analysis, it was shown how the heterogeneity of a coexpressed heterologous glycoprotein (IgE-R) was very significantly reduced. During our own experiments, we found that the *Pichia OCH1* gene is very recalcitrant to double homologous recombination-mediated inactivation, even when large flanking homologous regions are used in the knockout construct. As a consequence, we figured that an *N*-glycan engineering strategy involving such an inefficient step would make it very tedious to apply on well-developed expression strains that many users already have. Consequently, we turned to a knock-in strategy: in *P. pastoris*, expression constructs are most often integrated in the host genome via such a knock-in in the Alcohol Oxidase I (AOXI) or HIS4 locus, with concomitant duplication of the homologous sequence that is present on the plasmid. The specificity of integration at the exact locus is improved manyfold by linearizing the construct in this plasmid-borne genome-homologous sequence. In contrast to the situation in *S. cerevisiae*, such integrations are very stable in *P. pastoris*. Consequently, we devised the strategy shown in Figure 15.11: we cloned bp 73–467 of the *P. pastoris OCH1*-homologous ORF, preceded by two in-frame nonsense codons to avoid read-through from potential

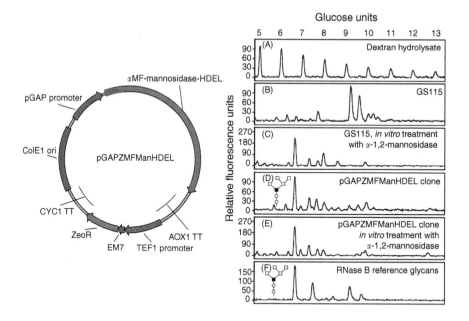

FIGURE 15.9 *N*-glycan analysis of *Trypanosoma cruzi trans*-sialidase coexpressed with *T. reesei* mannosidase-HDEL. Top: the vector used for the constitutive strong expression of the mannosidase. The gene was fused to the prepro signal sequence of the *S. cerevisiae* α-mating factor to direct translocation in the yeast's ER, and C-terminus fused to the coding sequence for HDEL, to keep the enzyme in the ER. Expression was driven by the *Pichia* promoter of the glyceraldehyde-3-phosphate dehydrogenase gene (glycolytic). Bottom: *N*-glycan analysis of *Trypanosoma cruzi trans*-sialidase. (A) Malto-oligosaccharide size reference ladder. Sizes of the glycans are expressed in Glucose Units (GU) by comparison of their electrophoretic mobility to the mobility of these malto-oligosaccharides. (B) *N*-glycans derived from recombinant *T. cruzi trans*-sialidase expressed in *P. pastoris*. The peak at GU=9,2 corresponds to Man$_8$GlcNAc$_2$. (C) Same analytes as in (B), but after overnight treatment with 3U/mL purified recombinant *T. reesei* α-1,2-mannosidase. (D) *N*-glycans derived from recombinant *trans*-sialidase coexpressed in *P. pastoris* with *T. reesei* mannosidase-HDEL (under control of the GAP promoter). The peak at GU=7,6 corresponds to the Man$_5$GlcNAc$_2$ peak in the profile of RNase B (F). (E) Same analytes as in (D) but after overnight treatment with 3 mU/mL purified recombinant *T. reesei* α-1,2-mannosidase. (F) *N*-glycans derived from bovine RNase B. These glycans consist of Man$_5$GlcNAc$_2$ to Man$_9$GlcNAc$_2$. Different isomers are resolved, accounting for the number of small peaks for Man$_7$GlcNAc$_2$. (Reprinted from Callewaert, N. et al., *FEBS Lett.*, 503, 173–178, 2001. With permission.)

earlier translation start sites in the vector. This fragment contains a centrally located *Bst*BI site that is conveniently used for vector linearization before transformation, and was 3′ linked to the HIS4 transcription terminator sequence. Upon integration in the genomic *OCH1* locus of *Pichia*, this vector would duplicate the *OCH1* sequence present in the vector. The resulting first *OCH1* copy can give a translation product that is maximally 161 amino acids long (of which six amino acids result from vector sequences), not including the catalytic domain of this type-II transmembrane protein.

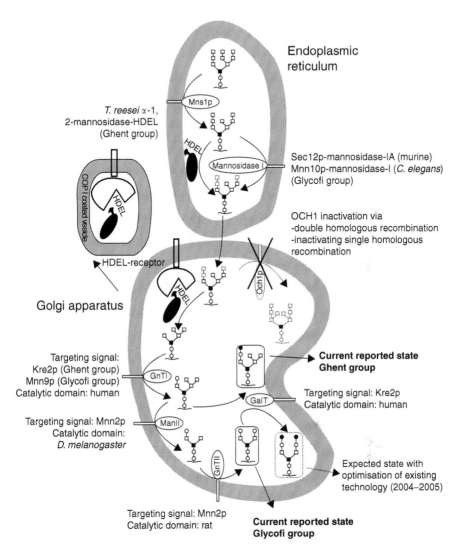

Endoplasmic reticulum

T. reesei α-1, 2-mannosidase-HDEL (Ghent group)

Mns1p

HDEL

Mannosidase I

Sec12p-mannosidase-IA (murine)
Mnn10p-mannosidase-I (*C. elegans*)
(Glycofi group)

OCH1 inactivation via
-double homologous recombination
-inactivating single homologous recombination

COPI coated vesicle

HDEL

HDEL-receptor

Och1p

Golgi apparatus

HDEL

Targeting signal:
Kre2p (Ghent group)
Mnn9p (Glycofi group)
Catalytic domain: human

GnTI

**Current reported state
Ghent group**

Targeting signal: Kre2p
Catalytic domain: human

GalT

Targeting signal: Mnn2p
Catalytic domain:
D. melanogaster

ManII

GnTII

**Expected state with
optimisation of existing
technology (2004–2005)**

Targeting signal: Mnn2p
Catalytic domain: rat

**Current reported state
Glycofi group**

FIGURE 15.10 Strategy for conversion of the *P. pastoris* N-glycan biosynthesis pathway toward biosynthesis of a mammalian-type hybrid and complex N-glycans. The pathway representation is based on knowledge derived from the *S. cerevisiae* N-glycan biosynthetic pathway. The figure represents the strategies used by the two groups that have been published in this field (see text). The approaches are similar but distinct in several aspects. The boxed glycan structures represent the current "state of the art" in *P. pastoris*, and the structure boxed with a dotted line is the probable result of optimization of the existing technology. This biantennary, bigalactosylated glycan is the main glycan on human serum glycoproteins.

The second copy lacks the coding sequence of the first 25 amino acids, and two in-frame stop codons prevent any read-through from potential upstream translation initiation sites. Thus, this strategy should in effect abolish the activity of this *OCH1*

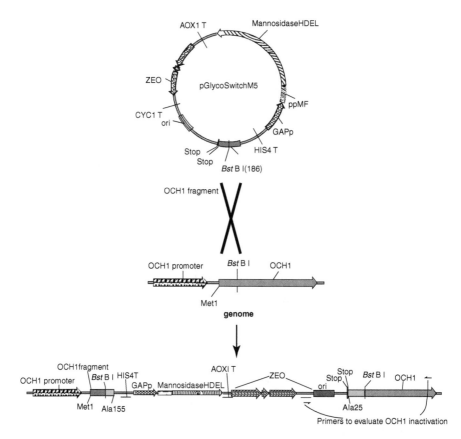

FIGURE 15.11 Strategy used for concomittant efficient OCH1 inactivation and ER-retained α-1,2-mannosidase overexpression engineering of *P. pastoris*. Upon linearization of pGlyco-SwitchM5 with *Bst*BI, correct integration of this construct in the *P. pastoris* OCH1 locus is achieved in > 50% of drug-resistant clones.

gene. To make the engineering even more efficient, we integrated the expression cassette for the *T. reesei* mannosidase (see above) on the same vector. In several experiments, we found that this construct integrates correctly in the *OCH1* locus of *P. pastoris* in 50% of drug-resistant clones. An easy PCR screen on a very small number of clones will always identify correct integrants, in stark contrast to strategies that incorporate *OCH1* knockout via double homologous recombination. The analysis of the *N*-glycans present on the *P. pastoris* cell wall mannoproteins of strains engineered with the construct revealed only one major peak (almost 100%): $Man_5GlcNAc_2$. The same conclusion was obtained with heterologous proteins expressed in this strain. Because our construct enables the one-step conversion of hyperglycosylation into a very homogenous $Man_5GlcNAc_2$ structure, we named this plasmid pGlycoSwitchM5. Importantly, the doubling time of the resulting strain is virtually identical to the parent strain (only tested in shake flasks so far), and reaches stationary phase at an OD_{600} that is only 20% lower than for the parent.[149] These results are very different from those seen in the *S. cerevisiae OCH1* strains, which have a very severely impaired viability.

Subsequently, the pGlycoSwitchM5-engineered strain was further provided with expression plasmids for human *N*-acetylglucosaminyltransferase I and human *β*-1, 4-galactosyltransferase chimeras with the *N*-terminal Golgi-targeting signals of the yeast mannosyltransferase Kre2p.[156] About 90% conversion of Man$_5$GlcNAc$_2$ into GlcNAcMan$_5$GlcNAc$_2$ was obtained upon expression of the GnTI chimera, and about 20% of this was further converted into GalGlcNAcMan$_5$GlcNAc$_2$ upon coexpression of the galactosyltransferase chimera (see Figure 15.12). The latter result is the first demonstration that *in vivo* biosynthesis of a mammalian-type glycan branch is possible in a fungal organism, thereby also establishing that the galactose donor-substrate UDP-Gal is biosynthesized in *P. pastoris*, and transported to the Golgi apparatus, where it is available for the mammalian galactosyltransferase. Further experiments are ongoing to improve the still rather modest yield of galactosylation in *P. pastoris* (Figure 15.12).

In parallel to these experiments, researchers at Glycofi Inc. (Lebanon, NH) have taken a combinatorial approach to the problem of selecting those fungal targeting signal or higher eukaryotic catalytic domain chimeras with the highest *in vivo* efficiency. These researchers inactivated *OCH1* via the classical, double homologous recombination approach which had rather low efficiency in our hands, and then focused on obtaining the biantennary backbone of complex *N*-glycans, by screening combinatorial libraries of fungal targeting signals and Mannosidase I, GnTI, Mannosidase II, and GnTII catalytic domains. Moreover, they incorporated extra UDP-GlcNAc transport capacity in their strain by overexpressing the UDP-GlcNAc transporter from *Kluyveromyces lactis*.[151] This approach has very recently resulted in a strain that produces homogenous GlcNAc$_2$Man$_3$GlcNAc$_2$, a truly remarkable feat,[152] (see Figure 15.10). Combined with our demonstration of the possibility of functionally expressing the galactosyltransferase in the *P. pastoris* Golgi apparatus, all of this bodes well for the near future. Consequently, we expect that *P. pastoris* strains that express high levels of glycoproteins modified with homogenous mammalian-type hybrid or biantennary bi-*β*-1,4-galactosylated *N*-glycans will become available for biopharmaceutical protein production within the next 2 years.

Up to this time, there have been no reports on developing *P. pastoris* for the production of humanized or fully human monoclonal antibodies. The availability of the described glycan-engineered strains should stimulate such explorations. IgG in serum is virtually not sialylated and only partially galactosylated. Consequently, this almost perfectly matches the obtained glycosylation potential of the engineered strains.

D. *N*-Glycan Engineering in Filamentous Fungi

Filamentous fungi, such as several *Trichoderma* and *Aspergillus* strains, have the capacity to secrete large amounts of proteins, metabolites, and organic acids into the growth medium. This property has been widely used by the food, beverage, paper, and textile industries, where compounds secreted by these fungi have been used for several years. During the past decades, several new strains with higher protein production and secretion capacities have been obtained through conventional mutagenesis strategies or

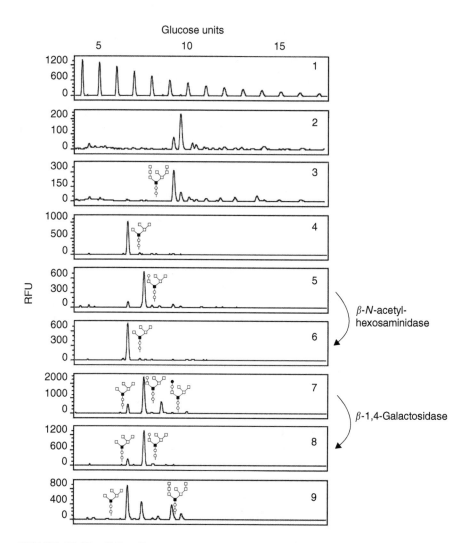

FIGURE 15.12 Cell wall mannoprotein *N*-glycan analysis of glycoengineered *P. pastoris* strains (Ghent group). Panel 1: malto-oligosaccharide glycan size standard. Panel 2: WT *P. pastoris* glycans. Panel 3: *P. pastoris* och1 strain. Panel 4: pGlycoSwitch M5 strain (note: virtually 100% homogeneity). Panel 5: pGlycoSwitch M5 strain expressing Kre2p-GnTI fusion. Panel 6: same as in Panel 5, but after β-*N*-acetylhexosaminidase digestion, which removes the GlcNAc residue transferred by GnTI. Panel 7: pGlycoSwitch M5 strain expressing both the Kre2p-GnTI and Kre2p-GalT fusions. Panel 8: same as in Panel 7, but after digestion with a specific β-1,4-galactosidase. Panel 9: *N*-glycans of bovine RNase B, as a reference for the Man₅GlcNAc₂ glycan.

genetic engineering.[157,158] As a result, the yield of secreted cellulases in the growth medium of some *Trichoderma* strains exceeds 40 g/L.[159]

Until now, there have not been many reports concerning the successful production of heterologous proteins in filamentous fungi. Moreover, the production yield

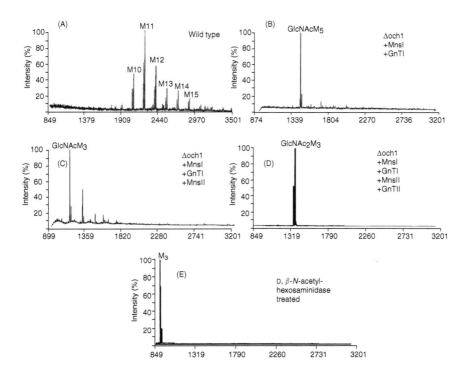

FIGURE 15.13 Glycan analysis of recombinant kringle-3 domain of tissue plasminogen activator, expressed in various glycoengineered *Pichia* strains (Glycofi, Inc.). The glycan profiles here were obtained by MALDI-TOF-MS in the positive-ion mode from kringle-3 domain, a recombinant protein expressed at levels of several 100 mg/L in *P. pastoris*. The overexpressed mannosidase and glycosyltransferases are the optimal combination obtained by screening libraries of chimeras consisting of one of several fungal secretory pathway localization signals and one of several catalytic domains of different origin. (Reprinted with minor adaptation from Hamilton, S.R. et al., *Science*, 301, 1244–1246, 2003. With permission. Copyright 2003 American Association for the Advancement of Science.)

for proteins of mammalian origin was low when compared with that of endogenous glycosylhydrolases.[160–162] However, some small successes were obtained in the production of for example, calf chymosin[163] and murine antibody Fab fragments[164] in *T. reesei*. Also, earlier work describing the secretion of more than 2 g/L of recombinant human lactoferrin by *Aspergillus awamori*[165] and the recent press release by Genencor International concerning the g/L production of antibodies in *Aspergillus niger* clearly indicate that filamentous fungi are able to secrete large amounts of foreign proteins.

Because of their high productivity, it would be beneficial to create fungal strains that are capable of synthesizing *N*-glycans of a more mammalian-like type. In a first attempt to evaluate whether *N*-glycans of *T. reesei* could be converted into more mammalian-like oligosaccharides, *in vitro* glycosylation assays were performed using recombinant *N*-acetylglucosaminyltransferase I, human β-1,4-galactosyltransferase, and rat α-2,6-sialyltransferase.[166] Only a modest amount of *N*-glycans could be modified, ranging from 0.25 to 1.8% (depending on the strain used). This initial result was not surprising, since earlier analysis of the oligosaccharides on *T. reesei* RutC30-secreted cellobiohydrolase I (CBH I) had indicated that only a very small amount of the GlcNAc-transferase I acceptor substrate Man$_5$GlcNAc$_2$ was synthesized.[167] Trimming of the more elaborate glycosyl structures to Man$_5$GlcNAc$_2$ by α-1,2-mannosidases seemed to be hampered by the presence of end-capping residues such as α-1,3-linked glucose[168] and phosphodiesters.[167–170]

To evaluate whether *Trichoderma* N-glycans could be converted *in vivo* into more mammalian-like structures, the complete coding sequence of the human GlcNAc-transferase I was introduced into the genome of the fungus under the control of the inducible CBH I promoter. When grown on cellulase-inducing conditions, successful *in vivo* transfer of GlcNAc was demonstrated after the analysis of the neutral N-glycans synthesized on CBH I secreted by the transformed strain.[171] Final proof of the formation of GlcNAcMan$_5$GlcNAc$_2$ was obtained through NMR analysis. Although GlcNAcMan$_5$GlcNAc$_2$ represents only a very small fraction of the total N-glycan pool of the transformant, this was the first report showing *in vivo* transfer of a GlcNAc residue to fungal glycans through the action of a recombinant β-1, 2-GlcNAc-transferase I. This result also indicated the presence of a functional Golgi UDP-GlcNAc transporter in at least some fungal organisms. Earlier attempts in *A. nidulans* to demonstrate GlcNAc transfer after the introduction of a functional rabbit GlcNAc-transferase I had been unsuccessful.[172] This failure was probably due to mislocalization of the recombinant enzyme or to the absence of a suitable acceptor–donor substrate.

As discussed earlier, most of the high-mannose N-glycans synthesized by the majority of *T. reesei* strains contain phosphodiester linkages similar to those observed on the N-glycans of *S. cerevisiae*. Abolishing these capping residues would increase the amount of Man$_5$GlcNAc$_2$, and therefore also the potential to create larger amounts of mammalian-like glycan structures on heterologous proteins secreted by *T. reesei*. Moreover, these phosphodiester structures are known to be immunogenic,[173] increasing the need to remove them from the fungal N- and O-glycans. In this regard, we are trying to clone the fungal genes involved in the transfer of those phosphodiester structures. Since the genetics of phosphomannosyl transfer have only been studied in *S. cerevisiae*,[97,98,174] early efforts focused on the cloning of *Trichoderma* genes with homology to either yeast MNN4 or MNN6. Recently, a partial ORF with homology to the yeast MNN4 gene was cloned from a *Trichoderma* cDNA library (Steven Geysens and Roland Contreas, unpublished results).

Several lines of evidence suggest that phosphomannosyl transfer occurs in the Golgi apparatus of *S. cerevisiae*.[175,176] The same holds true for *T. reesei*, since glycoproteins leaving the fungal ER display homogenous behavior in isoelectric focusing, while a complex pattern of more acidic species is observed for the same proteins upon secretion.[177] Furthermore, studies in yeast reveal that the phosphomannosyltransferase needs an α-1,2-mannobiose structure as a minimal prerequisite for the transfer to occur .[176, 178] Therefore, we pursued the idea of introducing sufficient α-1, 2-mannosidase activity in the ER or early Golgi vesicles to generate Man$_5$GlcNAc$_2$ before the secretory proteins reached the phosphomannosyl transferase(s) (similar approach as attempted before in *Pichia* to obviate Och1p activity). We used the

QM9414 strain and again expressed the *T. reesei* α-1,2-mannosidase, fused *N*-terminally to the *S. cerevisiae* α-mating factor signal sequence and C-terminally to the HDEL ER-retrieval signal[134] (S. Geysens et al., unpublished). Transformants were evaluated by analyzing the *N*-glycans on secreted glycoproteins. While the untransformed strain exhibited a wide variety of charged (phosphorylated) and neutral high-mannose *N*-glycans, transformants expressing a functional HDEL-tagged α-1,2-mannosidase produced almost exclusively $Man_5GlcNAc_2$ (see Figure 15.14). Under conditions where the secretion of glycosylhydrolases was induced (much higher glycoprotein flux through the secretory system), the $Man_5GlcNAc_2$ structure was the most dominant structure, with only trace amounts of charged and neutral high-mannose *N*-glycans (not shown). The obtained strain creates an ideal platform for the generation of more mammalian-like *N*-glycan structures, since it almost exclusively synthesizes the acceptor substrate for GlcNAc-transferase I.

For high-level protein secretion, the *T. reesei* RutC30 strain looks very appealing, as it is a hypersecretor of the glycosylhydrolases that it produces and thus could be expected to secrete heterologous proteins at increased levels as well. As for glycosylation, however, this strain has a significant drawback as it not only decorates its *N*-glycans with phosphodiesters (which could be solved as discussed above for QM9414), but also with α-1,3-linked glucose residues.[167] The gene encoding the RutC30 glucosidase II catalytic α-subunit was cloned by our group and proved to contain a frameshift, resulting in a premature translation stop (Geysens et al., unpublished). A repaired version of the coding sequence of the RutC30 glucosidase II α- subunit was placed under the control of a constitutive gpdA promoter and transformed to the RutC30 strain. Transformants were screened based on changes in their

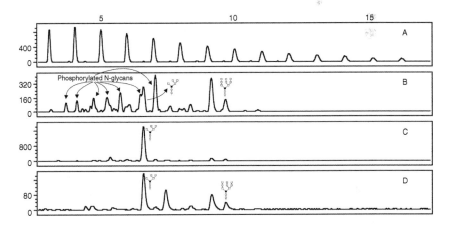

FIGURE 15.14 *N*-glycan analysis of *T. reesei* QM9414-secreted total glycoproteins, effect of expressing the ER-targeted *T. reesei* α-1,2-mannosidase. (A) Malto-oligosaccharide reference ladder. (B) parent QM9414 strain, *N*-glycans of total secreted proteins, noninducing conditions for cellulase secretion. (C) ER-targeted mannosidase expressing strain, noninducing conditions. (D) ER-targeted mannosidase expressing strain, inducing conditions. The *N*-glycan profile of the QM9414-secreted glycoproteins is significantly homogenized to $Man_5GlcNAc_2$ upon expression of the ER-targeted mannosidase.

glycosylation profile. One of them had the expression cassette of the repaired glucosidase II integrated in its genome. Thus, in this strain, the repaired and the mutant version of the glucosidase II α-subunit were expressed under the transcriptional control of the gpdA promoter and the endogenous promoter, respectively. Growth on glucose as the main carbon source resulted in a partial conversion of monoglucosylated N-glycans into high-mannose N-glycans (especially $Man_5GlcNAc_2$). However, when grown on cellulase-inducing conditions (thus enormously inducing cellulase protein flow through the fungal secretory system), far less monoglucosylated structures were trimmed to $Man_5GlcNAc_2$. From these results, it is clear that the repaired glucosidase II α- subunit is not expressed at sufficient levels to complement the defect fully (especially not in cellulase-inducing conditions) and that the variants of the glucosidase II α-subunit are somehow competing with each other. A plausible explanation could be a competition for interaction with the glucosidase II β-subunit, which provides for the ER retention of the glucosidase II activity. Further optimization of this strategy to convert most of the monoglucosylated structures into $Man_5GlcNAc_2$ is in progress.

IV. CONCLUSIONS

With some maturation of the recent fungal glycan engineering breakthroughs, yeasts (especially *P. pastoris*) and eventually also filamentous fungi should finally be able to realize fully their next great benefit to humankind, after assisting in the production of such commonplace things as bread, wine, beer, sake, cheese, and washing powder enzymes: the production of safe, cheaper, therapeutic glycoproteins.

ACKNOWLEDGMENTS

N.C. is a postdoctoral fellow of the Fund for Scientific Research-Flanders (FWO), W.V. is supported by a predoctoral grant of the IWT. We thank Research Corporation Technologies for supporting our *Pichia*-related work. Our research is also financed by FWO (Grant no. G005201N).

REFERENCES

1. Hebert, D.N., Zhang, J.X., Chen, W., Foellmer, B., and Helenius, A., The number and location of glycans on influenza hemagglutinin determine folding and association with calnexin and calreticulin, *J. Cell Biol.*, 139, 613–623, 1997.
2. Pantophlet, R., Wilson, I.A., and Burton, D.R., Hyperglycosylated mutants of human immunodeficiency virus (HIV) type 1 monomeric gp120 as novel antigens for HIV vaccine design, *J. Virol.*, 77, 5889–5901, 2003.
3. Sheffield, W.P., Marques, J.A., Bhakta, V., and Smith, I.J., Modulation of clearance of recombinant serum albumin by either glycosylation or truncation, *Thromb. Res.*, 99, 613–621, 2000.
4. Elliott, S., Lorenzini, T., Asher, S., Aoki, K,. Brankow, D., Buck, L., Busse, L., Chang, D., Fuller, J., Grant, J., Hernday, N., Hokum, M., Hu, S., Knudten, A., Levin, N., Komorowski, R., Martin, F., Navarro, R., Osslund, T., Rogers, G., Rogers, N., Trail, G., and Egrie, J., Enhancement of therapeutic protein *in vivo* activities through glyco-engineering, *Nat. Biotechnol.*, 21, 414–421, 2003.

5. Ashwell, G. and Harford, J., Carbohydrate-specific receptors of the liver, *Annu. Rev. Biochem.*, 51, 531–554, 1982.
6. Weikert, S., Papac, D., Briggs, J., Cowfer, D., Tom, S., Gawlitzek, M., Lofgren, J., Mehta, S., Chisholm, V., Modi, N., Eppler, S., Carroll, K., Chamow, S., Peers, D., Berman, P., and Krummen, L., Engineering Chinese hamster ovary cells to maximize sialic acid content of recombinant glycoproteins, *Nat. Biotechnol.*, 17, 1116–1121, 1999.
7. Aumiller, J.J., Hollister, J.R., and Jarvis, D.L., A transgenic insect cell line engineered to produce CMP-sialic acid and sialylated glycoproteins, *Glycobiology*, 13, 497–507, 2003.
8. Cabanes-Macheteau, M., Fitchette-Laine, A.C., Loutelier-Bourhis, C., Lange, C., Vine, N.D., Ma, J.K., Lerouge, P., and Faye, L., *N*-Glycosylation of a mouse IgG expressed in transgenic tobacco plants, *Glycobiology*, 9, 365–372, 1999.
9. Edmunds, T., Van Patten, S.M., Pollock, J., Hanson, E., Bernasconi, R., Higgins, E., Manavalan, P., Ziomek, C., Meade, H., McPherson, J.M., and Cole, E.S., Transgenically produced human antithrombin: structural and functional comparison to human plasma-derived antithrombin, *Blood*, 91, 4561–4571, 1998.
10. van Berkel, P.H., Welling, M.M., Geerts, M., van Veen, H.A., Ravensbergen, B., Salaheddine, M., Pauwels, E.K., Pieper, F., Nuijens, J.H., and Nibbering, P.H., Large scale production of recombinant human lactoferrin in the milk of transgenic cows, *Nat. Biotechnol.*, 20, 484–487, 2002.
11. Kornfeld, R. and Kornfeld, S., Assembly of asparagine-linked oligosaccharides, *Annu. Rev. Biochem.*, 54, 631–664, 1985.
12. Kukuruzinska, M.A., Bergh, M.L., and Jackson, B.J., Protein glycosylation in yeast, *Annu. Rev. Biochem.*, 56, 915–944, 1987.
13. Broschat, K.O., Gorka, C, Page, J.D., Martin-Berger, C.L., Davies, M.S., Huang Hc, H., Gulve, E.A., Salsgiver, W.J., and Kasten, T.P., Kinetic characterization of human glutamine: fructose-6-phosphate amidotransferase I: potent feedback inhibition by glucosamine-6-phosphate, *J. Biol. Chem.*, 277, 14764–14770, 2002.
14. McKnight, G.L., Mudri, S.L., Mathewes, S.L., Traxinger, R.R., Marshall, S., Sheppard, P.O., and O'Hara, P.J., Molecular cloning, cDNA sequence, and bacterial expression of human glutamine:fructose-6-phosphate amidotransferase, *J. Biol. Chem.*, 267, 25208–25212, 1992.
15. Mio, T., Kokado, M., Arisawa, M., and Yamada-Okabe, H., Reduced virulence of *Candida albicans* mutants lacking the GNA1 gene encoding glucosamine-6-phosphate acetyltransferase, *Microbiology*, 146, 1753–1758, 2000.
16. Peneff, C., Mengin-Lecreulx, D., and Bourne, Y., The crystal structures of Apo and complexed *Saccharomyces cerevisiae* GNA1 shed light on the catalytic mechanism of an amino-sugar *N*-acetyltransferase, *J. Biol. Chem.*, 276, 16328–16334, 2001.
17. Hofmann, M., Boles, E., and Zimmermann, F.K., Characterization of the essential yeast gene encoding *N*-acetylglucosamine-phosphate mutase, *Eur. J. Biochem.*, 221, 741–747, 1994.
18. Li, C., Rodriguez, M., and Banerjee, D., Cloning and characterization of complementary DNA encoding human *N*-acetylglucosamine-phosphate mutase protein, *Gene*, 242, 97–103, 2000.
19. Mio, T., Yabe, T., Arisawa, M., and Yamada-Okabe, H., The eukaryotic UDP-*N*-acetylglucosamine pyrophosphorylases. Gene cloning, protein expression, and catalytic mechanism, *J. Biol. Chem.*, 273, 14392–14397, 1998.
20. Hinderlich, S., Berger, M., Schwarzkopf, M., Effertz, K., and Reutter, W., Molecular cloning and characterization of murine and human *N*-acetylglucosamine kinase, *Eur. J. Biochem.*, 267, 3301–3308, 2000.

21. Piller, F., Hanlon, M.H., and Hill, R.L., Co-purification and characterization of UDP-glucose 4-epimerase and UDP-*N*-acetylglucosamine 4-epimerase from porcine submaxillary glands, *J. Biol. Chem.*, 258, 10774–10778, 1989.

22. Pastuszak, I., O'Donnell, J., and Elbein, A.D., Identification of the GalNAc kinase amino acid sequence, *J. Biol. Chem.*, 271, 23653–23656, 1996.

23. Wang-Gillam, A., Pastuszak, I., and Elbein, A.D., A 17-amino acid insert changes UDP-*N*-acetylhexosamine pyrophosphorylase specificity from UDP-GalNAc to UDP-GlcNAc, *J. Biol. Chem.*, 273, 27055–27057, 1998.

24. Hinderlich, S., Stasche, R., Zeitler, R., and Reutter, W., A bifunctional enzyme catalyzes the first two steps in *N*-acetylneuraminic acid biosynthesis of rat liver. Purification and characterization of UDP-*N*-acetylglucosamine 2-epimerase/*N*-acetylmannosamine kinase, *J. Biol. Chem.*, 272, 24313–24318, 1997.

25. Horstkorte, R., Nohring, S., Wiechens, N., Schwarzkopf, M., Danker, K., Reutter, W., and Lucka, L., Tissue expression and amino acid sequence of murine UDP-*N*-acetylglucosamine-2-epimerase/N-acetylmannosamine kinase, *Eur. J. Biochem.*, 260, 923–927, 1999.

26. Lucka, L., Krause, M., Danker, K., Reutter, W., and Horstkorte, R., Primary structure and expression analysis of human UDP-*N*-acetyl-glucosamine-2-epimerase/*N*-acetyl-mannosamine kinase, the bifunctional enzyme in neuraminic acid biosynthesis, *FEBS Lett.*, 454, 341–344, 1999.

27. Seppala, R., Lehto, V.P., and Gahl, W.A., Mutations in the human UDP-*N*-acetylglucosamine 2-epimerase gene define the disease sialuria and the allosteric site of the enzyme, *Am. J. Hum. Genet.*, 64, 1563–1569, 1999.

28. Keppler, O.T., Hinderlich, S., Langner, J., Schwartz-Albiez, R., Reutter, W., and Pawlita, M., UDP-GlcNAc 2-epimerase: a regulator of cell surface sialylation, *Science*, 284, 1372–1376, 1999.

29. Lawrence, S.M., Huddleston, K.A., Pitts, L.R., Nguyen, N., Lee, Y.C., Vann, W.F., Coleman, T.A., and Betenbaugh, M.J., Cloning and expression of the human *N*-acetylneuraminic acid phosphate synthase gene with 2-keto-3-deoxy-D-glycero-D-galacto-nononic acid biosynthetic ability, *J. Biol. Chem.*, 275, 17869–17877, 2000.

30. Nakata, D., Close, B.E., Colley, K.J., Matsuda, T., and Kitajima, K., Molecular cloning and expression of the mouse *N*-acetylneuraminic acid 9-phosphate synthase which does not have deaminoneuraminic acid (KDN) 9-phosphate synthase activity, *Biochem. Biophys. Res. Commun.*, 273, 642–648, 2000.

31. Chen, H., Blume, A., Zimmermann-Kordmann, M., Reutter, W., and Hinderlich, S., Purification and characterization of *N*-acetylneuraminic acid-9-phosphate synthase from rat liver, *Glycobiology*, 12, 65–71, 2002.

32. Ziak, M., Meier, M., and Roth, J., Megalin in normal tissues and carcinoma cells carries oligo/poly alpha2,8 deaminoneuraminic acid as a unique posttranslational modification, *Glycoconj. J.*, 16, 185–188, 1999.

33. Ziak, M., Qu, B., Zuo, X., Zuber, C., Kanamori, A., Kitajima, K., Inoue, S., Inoue, Y., and Roth, J., Occurrence of poly(alpha2,8-deaminoneuraminic acid) in mammalian tissues: widespread and developmentally regulated but highly selective expression on glycoproteins, *Proc. Natl. Acad. Sci. USA*, 93, 2759–2763, 1996.

34. Van Rinsum, J., Van Dijk, W., Hooghwinkel, G.J., and Ferwerda, W., Subcellular localization and tissue distribution of sialic acid-forming enzymes. *N*-acetylneuraminate-9-phosphate synthase and *N*-acetylneuraminate 9-phosphatase, *Biochem. J.*, 223, 323–328, 1984.

35. Munster, A.K., Eckhardt, M., Potvin, B., Muhlenhoff, M., Stanley, P., and Gerardy-Schahn, R., Mammalian cytidine 5′-monophosphate *N*-acetylneuraminic acid synthetase: a nuclear protein with evolutionarily conserved structural motifs, *Proc. Natl. Acad. Sci. USA*, 95, 9140–9145, 1998.

36. Hinderlich, S., Berger, M., Keppler, O.T., Pawlita, M., and Reutter, W., Biosynthesis of *N*-acetylneuraminic acid in cells lacking UDP-*N*-acetylglu-cosamine 2-epimerase/*N*-acetylmannosamine kinase, *Biol. Chem.*, 382, 291–297, 2001.

37. Takahashi, S., Takahashi, K., Kaneko, T., Ogasawara, H., Shindo, S., and Kobayashi, M., Human renin-binding protein is the enzyme *N*-acetyl-D-glucosamine 2-epimerase, *J. Biochem., Tokyo*, 125, 348–353, 1999.

38. Proudfoot, A.E., Turcatti, G., Wells, T.N., Payton, M.A., and Smith, D.J., Purification, cDNA cloning and heterologous expression of human phosphomannose isomerase, *Eur. J. Biochem.*, 219, 415–423, 1994.

39. Matthijs, G., Schollen, E., Pardon, E., Veiga-Da-Cunha, M., Jaeken, J., Cassiman, J.J., and Van Schaftingen, E., Mutations in PMM2, a phosphomannomutase gene on chromosome 16p13, in carbohydrate-deficient glycoprotein type I syndrome (Jaeken syndrome), *Nat. Genet.*, 16, 88–92, 1997.

40. Pirard, M., Achouri, Y., Collet, J.F., Schollen, E., Matthijs, G., and Van Schaftingen, E., Kinetic properties and tissular distribution of mammalian phosphomannomutase isozymes. *Biochem. J.*, 339, 201–207, 1999.

41. Matthijs, G., Schollen, E., Pirard, M., Budarf, M.L., Van Schaftingen, E., and Cassiman, J.J., PMM (PMM1), the human homologue of SEC53 or yeast phosphomannomutase, is localized on chromosome 22q13, *Genomics*, 40, 41–47, 1997.

42. Hashimoto, H., Sakakibara, A., Yamasaki, M., and Yoda, K., *Saccharomyces cerevisiae* VIG9 encodes GDP-mannose pyrophosphorylase, which is essential for protein glycosylation, *J. Biol. Chem.*, 272, 16308–16314, 1997.

43. Sullivan, F.X., Kumar, R., Kriz, R., Stahl, M., Xu, G.Y., Rouse, J., Chang, X.J., Boodhoo, A., Potvin, B., and Cumming, D.A., Molecular cloning of human GDP-mannose 4,6-dehydratase and reconstitution of GDP-fucose biosynthesis *in vitro*, *J. Biol. Chem.*, 273, 8193–8202, 1998.

44. Tonetti, M., Sturla, L., Bisso, A., Benatti, U., and De Flora, A., Synthesis of GDP-L-fucose by the human FX protein, *J. Biol. Chem.*, 271, 27274–27279, 1996.

45. Park, S.H., Pastuszak, I., Drake, R., and Elbein, A.D., Purification to apparent homogeneity and properties of pig kidney L-fucose kinase, *J. Biol. Chem.*, 273, 5685–5691, 1998.

46. Nakayama, K., Maeda, Y., and Jigami, Y., Interaction of GDP-4-keto-6-deoxymannose-3,5-epimerase-4-reductase with GDP-mannose-4,6-dehydratase stabilizes the enzyme activity for formation of GDP-fucose from GDP-mannose, *Glycobiology*, 13, 673–680, 2003.

47. Lee, R.T., Peterson, C.L., Calman, A.F., Herskowitz, I., and O'Donnell, J.J., Cloning of a human galactokinase gene (GK2) on chromosome 15 by complementation in yeast, *Proc. Natl. Acad. Sci. USA*, 89, 10887–10891, 1992.

48. Stambolian, D., Ai, Y., Sidjanin, D., Nesburn, K., Sathe, G., Rosenberg, M., and Bergsma, D.J., Cloning of the galactokinase cDNA and identification of mutations in two families with cataracts, *Nat. Genet.*, 10, 307–312, 1995.

49. Christacos, N.C., Marson, M.J., Wells, L., Riehman, K., and Fridovich-Keil, J.L., Subcellular localization of galactose-1-phosphate uridylyltransferase in the yeast Saccharomyces cerevisiae, *Mol. Genet. Metab.*, 70, 272–280, 2000.

50. Reichardt, J.K. and Berg, P., Cloning and characterization of a cDNA encoding human galactose-1-phosphate uridyl transferase, *Mol. Biol. Med.*, 5, 107–122, 1988.

51. Banroques, J., Gregori, C., and Dreyfus, J.C., Purification of human liver uridylyl transferase and comparison with the erythrocyte enzyme, *Biochimie*, 65, 7–13, 1983.

52. Banroques, J., Gregori, C., and Schapira, F., Purification and characterization of human erythrocyte uridylyl transferase, *Biochim. Biophys. Acta*, 657, 374–382, 1981.

53. Daude, N., Gallaher, T.K., Zeschnigk, M., Starzinski-Powitz, A., Petry, K.G., Haworth, I.S., and Reichardt, J.K., Molecular cloning, characterization, and mapping of a full-length cDNA encoding human UDP-galactose 4'-epimerase, *Biochem. Mol. Med.*, 56, 1–7, 1995.

54. Maceratesi, P., Daude, N., Dallapiccola, B., Novelli, G., Allen, R., Okano, Y., and Reichardt, J., Human UDP-galactose 4' epimerase (GALE) gene and identification of five missense mutations in patients with epimerase-deficiency galactosemia, *Mol. Genet. Metab.*, 63, 26–30, 1998.

55. Kainuma, M., Ishida, N., Yoko-o, T., Yoshioka, S., Takeuchi, M., Kawakita, M., and Jigami, Y., Coexpression of alpha-1,2-galactosyltransferase and UDP-galactose transporter efficiently galactosylates *N*- and *O*-glycans in *Saccharomyces cerevisiae*, *Glycobiology*, 9, 133–141, 1999.

56. Tanaka, N., Konomi, M., Osumi, M., and Takegawa, K., Characterization of a *Schizosaccharomyces pombe* mutant deficient in UDP-galactose transport activity, *Yeast*, 18, 903–914, 2001.

57. Kainuma, M., Chiba, Y., Takeuchi, M., and Jigami, Y., Overexpression of HUT1 gene stimulates *in vivo* galactosylation by enhancing UDP-galactose transport activity in *Saccharomyces cerevisiae*, *Yeast*, 18, 533–541, 2001.

58. Hirschberg, C.B., Robbins, P.W., and Abeijon, C., Transporters of nucleotide sugars, ATP, and nucleotide sulfate in the endoplasmic reticulum and Golgi apparatus, *Annu. Rev. Biochem.*, 67, 49–69, 1998.

59. Berninsone, P.M. and Hirschberg, C.B., Nucleotide sugar transporters of the Golgi apparatus, *Curr. Opin. Struct. Biol.*, 10, 542–547, 2000.

60. Gerardy-Schahn, R., Oelmann, S., and Bakker, H., Nucleotide sugar transporters: biological and functional aspects, *Biochimie*, 83, 775–782, 2001.

61. Anand, M., Rush, J.S., Ray, S., Doucey, M.A., Weik, J., Ware, F.E., Hofsteenge, J., Waechter, C.J., and Lehrman, M.A., Requirement of the Lec35 gene for all known classes of monosaccharide-P-dolichol-dependent glycosyltransferase reactions in mammals, *Mol. Biol. Cell*, 12, 487–501, 2001.

62. Huffaker, T.C. and Robbins, P.W., Temperature-sensitive yeast mutants deficient in asparagine-linked glycosylation, *J. Biol. Chem.*, 257, 3203–3210, 1982.

63. Barnes, G., Hansen, W.J., Holcomb, C.L., and Rine, J., Asparagine-linked glycosylation in *Saccharomyces cerevisiae:* genetic analysis of an early step, *Mol. Cell Biol.*, 4, 2381–2388, 1984.

64. Burda, P., te Heesen, S., Brachat, A., Wach, A., Dusterhoft, A., and Aebi, M., Stepwise assembly of the lipid-linked oligosaccharide in the endoplasmic reticulum of *Saccharomyces cerevisiae*: identification of the ALG9 gene encoding a putative mannosyl transferase, *Proc. Natl. Acad. Sci. USA*, 93, 7160–7165, 1996.

65. Burda, P. and Aebi, M., The ALG10 locus of *Saccharomyces cerevisiae* encodes the alpha-1,2 glucosyltransferase of the endoplasmic reticulum: the terminal glucose of the lipid-linked oligosaccharide is required for efficient *N*-linked glycosylation, *Glycobiology*, 8, 455–462, 1998.

66. Cipollo, J.F., Trimble, R.B., Chi, J.H., Yan, Q., and Dean, N., The yeast ALG11 gene specifies addition of the terminal alpha-1,2-Man to the Man$_5$GlcNAc$_2$-PP-dolichol N-glycosylation intermediate formed on the cytosolic side of the endoplasmic reticulum, *J. Biol. Chem.*, 276, 21828–21840, 2001.

67. Burda, P., Jakob, C.A., Beinhauer, J., Hegemann, J.H., and Aebi, M., Ordered assembly of the asymmetrically branched lipid-linked oligosaccharide in the endoplasmic reticulum is ensured by the substrate specificity of the individual glycosyltransferases, *Glycobiology*, 9, 617–625, 1999.

68. Helenius, J., Ng, D.T., Marolda, C.L., Walter, P., Valvano, M.A., and Aebi, M., Translocation of lipid-linked oligosaccharides across the ER membrane requires Rft1 protein, *Nature*, 415, 447–450, 2002.

69. Jackson, B.J., Kukuruzinska, M.A., and Robbins, P., Biosynthesis of asparagine-linked oligosaccharides in *Saccharomyces cerevisiae*: the alg2 mutation, *Glycobiology*, 3, 357–364, 1993.

70. Takeuchi, K., Yamazaki, H., Shiraishi, N., Ohnishi, Y., Nishikawa, Y., and Horinouchi, S., Characterization of an alg2 mutant of the zygomycete fungus *Rhizomucor pusillus*, *Glycobiology*, 9, 1287–1293, 1999.

71. Thiel, C., Schwarz, M., Peng, J., Grzmil, M., Hasilik, M., Braulke, T., Kohlschutter, A., von Figura, K., Lehle, L., and Korner, C., A new type of congenital disorders of glycosylation (CDG-Ii) provides new insights into the early steps of dolichol-linked oligosaccharide biosynthesis, *J. Biol. Chem.*, 278, 22498–22505, 2003.

72. Aebi, M., Gassenhuber, J., Domdey, H., and te Heesen, S., Cloning and characterization of the ALG3 gene of *Saccharomyces cerevisiae*, *Glycobiology*, 6, 439–444, 1996.

73. Sharma, C.B., Knauer, R., and Lehle, L., Biosynthesis of lipid-linked oligosaccharides in yeast: the ALG3 gene encodes the Dol-P-Man:Man5GlcNAc2-PP-Dol-mannosyltransferase, *Biol. Chem.*, 382, 321–328, 2001.

74. Heesen, S., Lehle, L., Weissmann, A., and Aebi, M., Isolation of the ALG5 locus encoding the UDP-glucose:dolichyl-phosphate glucosyltransferase from *Saccharomyces cerevisiae*, *Eur. J. Biochem.*, 224, 71–79, 1994.

75. Reiss, G., te Heesen, S., Zimmerman, J., Robbins, P.W., and Aebi, M., Isolation of the ALG6 locus of *Saccharomyces cerevisiae* required for glucosylation in the N-linked glycosylation pathway, *Glycobiology*, 6, 493–498, 1996.

76. Runge, K.W., Huffaker , T.C., and Robbins, P.W., Two yeast mutations in glucosylation steps of the asparagine glycosylation pathway, *J. Biol. Chem.*, 259, 412–417, 1984.

77. Runge, K.W. and Robbins, P.W., A new yeast mutation in the glucosylation steps of the asparagine-linked glycosylation pathway. Formation of a novel asparagine-linked oligosaccharide containing two glucose residues, *J. Biol. Chem.*, 261, 15582–15590, 1986.

78. Stagljar, I., te Heesen, S., and Aebi, M., New phenotype of mutations deficient in glucosylation of the lipid-linked oligosaccharide: cloning of the ALG8 locus, *Proc. Natl. Acad. Sci. USA*, 91, 5977–5981, 1994.

79. Karaoglu, D., Silberstein, S., Kelleher, D.J., and Gilmore, R., The *Saccharomyces cerevisiae* oligosaccharyltransferase: a large hetero-oligomeric complex in the endoplasmic reticulum, *Cold Spring Harb. Symp. Quant. Biol.*, 60, 83–92, 1995.

80. Karaoglu, D., Kelleher, D.J., and Gilmore, R., Allosteric regulation provides a molecular mechanism for preferential utilization of the fully assembled dolichol-linked oligosaccharide by the yeast oligosaccharyltransferase, *Biochemistry*, 40, 12193–12206, 2001.

81. Hebert, D.N., Foellmer, B., and Helenius, A., Calnexin and calreticulin promote folding, delay oligomerization and suppress degradation of influenza hemagglutinin in microsomes, EMBO J., 15, 2961–2968, 1996.

82. Wada, I., Calnexin is involved in the quality-control mechanism of the ER, Seikagaku, 67, 1133–1137, 1995.

83. Arnold, S.M., Fessler, L.I., Fessler, J.H., and Kaufman, R.J., Two homologues encoding human UDP-glucose:glycoprotein glucosyltransferase differ in mRNA expression and enzymatic activity, Biochemistry, 39, 2149–2163, 2000.

84. Jakob, C.A., Burda, P., te Heesen, S., Aebi M., and Roth, J., Genetic tailoring of N-linked oligosaccharides: the role of glucose residues in glycoprotein processing of Saccharomyces cerevisiae in vivo, Glycobiology, 8, 155–164, 1998.

85. Hebert, D.N., Foellmer, B., and Helenius, A., Glucose trimming and reglucosylation determine glycoprotein association with calnexin in the endoplasmic reticulum, Cell, 81, 425–433, 1995.

86. Parodi, A.J., Role of N-oligosaccharide endoplasmic reticulum processing reactions in glycoprotein folding and degradation, Biochem. J., 348, 1–13, 2000.

87. Tremblay, L.O. and Herscovics, A., Cloning and expression of a specific human alpha 1,2-mannosidase that trims $Man_9GlcNAc_2$ to $Man_8GlcNAc_2$ isomer B during N-glycan biosynthesis, Glycobiology, 9, 1073–1078, 1999.

88. Fagioli, C. and Sitia, R., Glycoprotein quality control in the endoplasmic reticulum. Mannose trimming by endoplasmic reticulum mannosidase I times the proteasomal degradation of unassembled immunoglobulin subunits, J. Biol. Chem., 276, 12885–12892, 2001.

89. Hosokawa, N., Wada, I., Hasegawa, K., Yorihuzi, T., Tremblay, L.O., Herscovics, A., and Nagata, K., A novel ER alpha-mannosidase-like protein accelerates ER-associated degradation, EMBO Rep., 2, 415–422, 2001.

90. Jakob, C.A., Bodmer, D., Spirig, U., Battig, P., Marcil, A., Dignard, D., Bergeron, J.J., Thomas, D.Y., and Aebi, M., Htm1p, a mannosidase-like protein, is involved in glycoprotein degradation in yeast, EMBO Rep., 2, 423–430, 2001.

91. Brigance, W.T., Barlowe, C., and Graham, T.R., Organization of the yeast Golgi complex into at least four functionally distinct compartments, Mol. Biol. Cell, 11, 171–182, 2000.

92. Nagasu, T., Shimma, Y., Nakanishi, Y., Kuromitsu, J., Iwama, K., Nakayama, K., Suzuki, K., and Jigami, Y., Isolation of new temperature-sensitive mutants of Saccharomyces cerevisiae deficient in mannose outer chain elongation, Yeast, 8, 535–547, 1992.

93. Munro, S., What can yeast tell us about N-linked glycosylation in the Golgi apparatus? FEBS Lett., 498, 223–227, 2001.

94. Jungmann, J., Rayner, J.C., and Munro, S., The Saccharomyces cerevisiae protein Mnn10p/Bed1p is a subunit of a Golgi mannosyltransferase complex, J. Biol. Chem., 274, 6579–6585, 1999.

95. Devlin, C. and Ballou, C.E., Identification and characterization of a gene and protein required for glycosylation in the yeast Golgi, Mol. Microbiol., 4, 1993–2001, 1990.

96. Rayner, J.C. and Munro, S., Identification of the MNN2 and MNN5 mannosyltransferases required for forming and extending the mannose branches of the outer chain mannans of Saccharomyces cerevisiae, J. Biol. Chem., 273, 26836–26843, 1998.

97. Odani, T., Shimma, Y., Wang, X.H., and Jigami, Y., Mannosylphosphate transfer to cell wall mannan is regulated by the transcriptional level of the MNN4 gene in Saccharomyces cerevisiae, FEBS Lett., 420, 186–190, 1997.

98. Wang, X.H., Nakayama, K., Shimma, Y., Tanaka, A., and Jigami, Y., MNN6, a member of the KRE2/MNT1 family, is the gene for mannosylphosphate transfer in *Saccharomyces cerevisiae, J. Biol. Chem.*, 272, 18117–18124, 1997.

99. Yip, C.L., Welch, S.K., Klebl, F., Gilbert, T., Seidel, P., Grant, F.J., O'Hara, P.J., and MacKay, V.L., Cloning and analysis of the *Saccharomyces cerevisiae* MNN9 and MNN1 genes required for complex glycosylation of secreted proteins, *Proc. Nat. Acad. Sci. USA*, 91, 2723–2727, 1994.

100. Herscovics, A., Importance of glycosidases in mammalian glycoprotein biosynthesis, *Biochim. Biophys. Acta*, 1473, 96–107, 1999.

101. Kornfeld, R., Bao, M., Brewer, K., Noll, C., and Canfield, W., Molecular cloning and functional expression of two splice forms of human *N*-acetylglucosamine-1-phosphodiester alpha-*N*-acetylglucosaminidase, *J. Biol. Chem.*, 274, 32778–32785, 1999.

102. Natowicz, M., Hallett, D.W., Frier, C., Chi, M., Schlesinger, P.H., and Baenziger, J.U., Recognition and receptor-mediated uptake of phosphorylated high mannose-type oligosaccharides by cultured human fibroblasts, *J. Cell Biol.*, 96, 915–919, 1983.

103. Chen, Y., Jin, M., Egborge, T., Coppola, G., Andre, J., and Calhoun, D.H., Expression and characterization of glycosylated and catalytically active recombinant human alpha-galactosidase A produced in *Pichia pastoris*, *Protein Expr. Purif.*, 20, 472–484, 2000.

104. Winchester, B., Vellodi, A., and Young, E., The molecular basis of lysosomal storage diseases and their treatment, *Biochem. Soc.Trans.*, 28, 150–154, 2000.

105. Chen, Y.T. and Amalfitano, A., Towards a molecular therapy for glycogen storage disease type II (Pompe disease), *Mol. Med. Today*, 6, 245–251, 2000.

106. Schiffmann, R., Murray, G.J., Treco, D., Daniel, P., Sellos-Moura, M., Myers M., Quirk, J.M., Zirzow, G.C., Borowski, M., Loveday, K., Anderson, T., Gillespie, F., Oliver, K.L., Jeffries, N.O., Doo, E., Liang, T.J., Kreps, C., Gunter, K., Frei, K., Crutchfield, K., Selden, R.F., and Brady, R.O., Infusion of alpha-galactosidase A reduces tissue globotriaosylceramide storage in patients with Fabry disease, *Proc. Natl. Acad. Sci. USA,* 97, 365–370, 2000.

107. Vogler, C., Levy, B., Galvin, N.J., Thorpe, C., Sands, M.S., Barker, J.E., Baty, J., Birkenmeier, E.H., and Sly, W.S., Enzyme replacement in murine mucopolysaccharidosis type VII: neuronal and glial response to beta-glucuronidase requires early initiation of enzyme replacement therapy, *Pediatr. Res.*, 45, 838–844, 1999.

108. Varki, A., Cummings, R., Esko, J., Freeze, H., Hart, G., and Marth, J., Essentials of Glycobiology, *Cold Spring Harbor Laboratory Press*, New York, 1999.

109. Field, M.C. and Wainwright, L.J., Molecular cloning of eukaryotic glycoprotein and glycolipid glycosyltransferases: a survey, *Glycobiology*, 5, 463–472, 1995.

110. Nilsson, T., Slusarewicz, P., Hoe, M.H., and Warren, G., Kin recognition: a model for the retention of Golgi enzymes, *FEBS Lett.*, 330, 1–4, 1993.

111. Higgins, D.R. and Cregg, J.M., Introduction to *Pichia pastoris*, *Methods Mol. Biol.*, 103, 1–15, 1998.

112. Cregg, J.M., Madden, K.R., Barringer, K.J., Thill, G.P., and Stillman, C.A., Functional characterization of the two alcohol oxidase genes from the yeast *Pichia pastoris*, *Mol. Cell. Biol.*, 9, 1316–1323, 1989.

113. Sreekrishna, K., Nelles, L., Potenz, R., Cruze, J., Mazzaferro, P., Fish, W., Fuke, M., Holden, K., Phelps, D., and Wood, P., et al., High-level expression, purification, and characterization of recombinant human tumor necrosis factor synthesized in the methylotrophic yeast *Pichia pastoris, Biochemistry*, 28, 4117–4125, 1989.

114. Kobayashi, K., Nakamura, N., Sumi, A., Ohmura, T., and Yokoyama, K., The development of recombinant human serum albumin, *Ther. Apher.*, 2, 257–262, 1998.

115. Watanabe, H., Yamasaki, K., Kragh-Hansen, U., Tanase, S., Harada, K., Suenaga, A., and Otagiri, M., *in vitro* and *in vivo* properties of recombinant human serum albumin from *Pichia pastoris* purified by a method of short processing time, *Pharm. Res.*, 18, 1775–1781, 2001.

116. Cereghino, J.L. and Cregg, J.M., Heterologous protein expression in the methylotrophic yeast *Pichia pastoris*, *FEMS Microbiol. Rev.*, 24, 45–66, 2000.

117. Kalidas, C., Joshi, L., and Batt, C., Characterization of glycosylated variants of beta-lactoglobulin expressed in *Pichia pastoris*, *Protein Eng.*, 14, 201–207, 2001.

118. Miele, R.G., Nilsen, S.L., Brito, T., Bretthauer, R.K., and Castellino, F.J., Glycosylation properties of the Pichia pastoris-expressed recombinant kringle 2 domain of tissue-type plasminogen activator, *Biotechnol. Appl. Biochem.*, 25, 151–157, 1997.

119. Trimble, R.B., Atkinson, P.H., Tschopp, J.F., Townsend, R.R., and Maley, F., Structure of oligosaccharides on *Saccharomyces* SUC2 invertase secreted by the methylotrophic yeast *Pichia pastoris*, *J. Biol. Chem.*, 266, 22807–22817, 1991.

120. Bretthauer, R.K. and Castellino, F.J., Glycosylation of *Pichia pastoris*-derived proteins, *Biotechnol, Appl. Biochem.*, 30, 193–200, 1999.

121. Montesino, R., Garcia, R., Quintero, O., and Cremata, J.A., Variation in *N*-linked oligosaccharide structures on heterologous proteins secreted by the methylotrophic yeast Pichia pastoris, *Protein. Expr. Purif.*, 14, 197–207, 1998.

122. Chen, Y., Jin, M., Egborge, T., Coppola, G., Andre, J., and Calhoun, D.H., Expression and characterization of glycosylated and catalytically active recombinant human alpha-galactosidase A produced in *Pichia pastoris*, *Protein Expr. Purif.*, 20, 472–484, 2000.

123. Kang, H.A., Sohn, J.H., Choi, E.S., Chung, B.H., Yu, M.H., and Rhee, S.K., Glycosylation of human alpha 1-antitrypsin in *Saccharomyces cerevisiae* and methylotrophic yeasts,. *Yeast*, 14, 371–381, 1998.

124. Martinet, W., Saelens, X., Deroo, T., Neirynck, S., Contreras, R., Min Jou, W., and Fiers, W., Protection of mice against a lethal influenza challenge by immunization with yeast-derived recombinant influenza neuraminidase, *Eur. J. Biochem.*, 247, 332–338, 1997.

125. Montesino, R., Nimtz, M., Quintero, O., Garcia, R., Falcon, V., and Cremata, J.A., Characterization of the oligosaccharides assembled on the *Pichia pastoris*-expressed recombinant aspartic protease, *Glycobiology*, 9, 1037–1043, 1999.

126. van Dijk, R., Faber, K.N., Kiel, J.A., Veenhuis, M., and van der Klei, I., The methylotrophic yeast *Hansenula polymorpha*: a versatile cell factory, *Enzyme Microb. Technol.*, 26, 793–800, 2000.

127. Ziegler, F.D., Cavanagh, J., Lubowski, C., and Trimble, R.B., Novel *Schizosaccharomyces pombe* *N*-linked GalMan$_9$GlcNAc isomers: role of the Golgi GMA12 galactosyltransferase in core glycan galactosylation, *Glycobiology*, 9, 497–505, 1999.

128. Zarate, V. and Belda, F., Characterization of the heterologous invertase produced by *Schizosaccharomyces pombe* from the SUC2 gene of *Saccharomyces cerevisiae*, *J. Appl. Bacteriol.*, 80, 45–52, 1996.

129. Maras, M., van Die, I., Contreras, R., and van den Hondel, C.A., Filamentous fungi as production organisms for glycoproteins of bio-medical interest, *Glycoconj. J.*, 16, 99–107, 1999.

130. Ichishima, E., Taya, N., Ikeguchi, M., Chiba, Y., Nakamura, M., Kawabata, C., Inoue, T., Takahashi, K., Minetoki, T., Ozeki, K., Kumagai, C., Gomi, K., Yoshida, T., and Nakajima, T., Molecular and enzymic properties of recombinant 1, 2-alpha-mannosidase from *Aspergillus saitoi* overexpressed in *Aspergillus oryzae* cells, *Biochem. J.*, 339, 589–597, 1999.

131. Inoue, T., Yoshida, T., and Ichishima, E., Molecular cloning and nucleotide sequence of the 1,2-alpha-D- mannosidase gene, msdS, from *Aspergillus saitoi* and expression of the gene in yeast cells, *Biochim. Biophys. Acta*, 1253, 141–145, 1995.

132. Maras, M., Callewaert, N., Piens, K., Claeyssens, M., Martinet, W., Dewaele, S., Contreras, H., Dewerte, I., Penttila, M., and Contreras, R., Molecular cloning and enzymatic characterization of a *Trichoderma reesei* 1,2-alpha-D-mannosidase, *J. Biotechnol.*, 77, 255–263, 2000.

133. Chiba, Y., Suzuki, M., Yoshida, S., Yoshida, A., Ikenaga, H., Takeuchi, M., Jigami, Y., and Ichishima, E., Production of human compatible high mannose-type (Man$_5$GlcNAc$_2$) sugar chains in *Saccharomyces cerevisiae, J. Biol. Chem.*, 273, 26298–26304, 1998.

134. Callewaert, N., Laroy, W., Cadirgi, H., Geysens, S., Saelens, X., Min Jou, W., and Contreras, R., Use of HDEL-tagged *Trichoderma reesei* mannosyl oligosaccharide 1,2-alpha-D-mannosidase for *N*-glycan engineering in *Pichia pastoris*, *FEBS Lett.*, 503, 173–178, 2001.

135. Reiss, E. and Lehmann, P.F., Galactomannan antigenemia in invasive aspergillosis, *Infect. Immun.*, 25, 357–365, 1979.

136. Stynen, D., Sarfati, J., Goris, A., Prevost, M.C., Lesourd, M., Kamphuis, H., Darras, V., and Latge, J.P., Rat monoclonal antibodies against *Aspergillus galactomannan*, *Infect. Immun.,* 60, 2237–2245, 1992.

137. Takayanagi, T., Kimura, A., Chiba, S., and Ajisaka, K., Novel structures of *N*-linked high-mannose type oligosaccharides containing alpha-D-galactofuranosyl linkages in *Aspergillus niger* alpha- D-glucosidase, *Carbohydr. Res.*, 256, 149–158, 1994.

138. Maras, M., De Bruyn, A., Schraml, J., Herdewijn, P., Claeyssens, M., Fiers, W., and Contreras, R., Structural characterization of *N*-linked oligosaccharides from cellobiohydrolase I secreted by the filamentous fungus *Trichoderma reesei* RUTC 30, *Eur. J. Biochem.*, 245, 617–625, 1997.

139. Klarskov, K., Piens, K., Stahlberg, J., Hoj, P.B., Van Beeumen, J., and Claeyssens, M., Cellobiohydrolase I from *Trichoderma reesei*: identification of an active-site nucleophile and additional information on sequence including the glycosylation pattern of the core protein, *Carbohydr. Res.*, 304, 143–154, 1997.

140. Boer, H., Teeri, T.T., and Koivula, A., Characterization of *Trichoderma reesei* cellobiohydrolase CeI7A secreted from *Pichia pastoris* using two different promoters, *Biotechnol. Bioeng.*, 69, 486–494, 2000.

141. Johnson, W.E., Sauvron, J.M., and Desrosiers, R.C., Conserved, *N*-linked carbohydrates of human immunodeficiency virus type 1 gp41 are largely dispensable for viral replication, *J. Virol.*, 75, 11426–11436, 2001.

142. Reitter, J.N., Means, R.E., and Desrosiers, R.C., A role for carbohydrates in immune evasion in AIDS, *Nat. Med.*, 4, 679–684, 1998.

143. Van Petegem, F., Contreras, H., Contreras, R., and Van Beeumen, J., *Trichoderma reesei* alpha-1,2-mannosidase: structural basis for the cleavage of four consecutive mannose residues, *J. Mol. Biol.*, 312, 157–165, 2001.

144. Reeves, P.J., Callewaert, N., Contreras, R., and Khorana, H.G., Structure and function in rhodopsin: High-level expression of rhodopsin with restricted and homogeneous *N*-glycosylation by a tetracycline-inducible *N*-acetylglucosaminyltransferase I-negative HEK293S stable mammalian cell line, *Proc. Natl. Acad. Sci. USA*, 99, 13419–13424, 2002.

145. Lin, Cereghino, G.P., Lin, Cereghino, J., Sunga, A.J., Johnson M.A., Lim, M., Gleeson, M.A., and Cregg, J.M., New selectable marker/auxotrophic host strain combinations for molecular genetic manipulation of *Pichia pastoris, Gene*, 263, 159–169, 2001.

146. Nakanishi-Shindo, Y., Nakayama, K.I., Tanaka, A., Toda, Y., and Jigami, Y., Structure of the *N*-linked oligosaccharides that show the complete loss of alpha-1,6-polymannose outer chain from och1, och1 mnn1, and och1 mnn1 alg3 mutants of *Saccharomyces cerevisiae, J. Biol. Chem.*, 268, 26338–45, 1993.

147. Chiba, Y., Sakuraba, H., Kotani, M., Kase, R., Kobayashi, K., Takeuchi, M., Ogasawara, S., Maruyama, Y., Nakajima, T., Takaoka, Y., and Jigami, Y., Production in yeast of alpha-galactosidase A, a lysosomal enzyme applicable to enzyme replacement therapy for Fabry disease, *Glycobiology*, 12, 821–828, 2002.

148. Lee, K., Jin, X., Zhang, K., Copertino, L., Andrews, L., Baker-Malcolm, J., Geagan, L., Qiu, H., Seiger, K., Barngrover, D., McPherson, J.M., and Edmunds, T., A biochemical and pharmacological comparison of enzyme replacement therapies for the glycolipid storage disorder Fabry disease, *Glycobiology*, 13, 305–313, 2003.

149. Shinkawa, T., Nakamura, K., Yamane, N., Shoji-Hosaka, E., Kanda, Y., Sakurada, M., Uchida, K., Anazawa, H., Satoh, M., Yamasaki, M., Hanai, N., and Shitara, K., The absence of fucose but not the presence of galactose or bisecting *N*-acetylglucosamine of human IgG1 complex-type oligosaccharides shows the critical role of enhancing antibody-dependent cellular cytotoxicity, *J. Biol. Chem.*, 278, 3466–3473, 2003.

150. Vervecken, W., Kaigorodov, V., Callewaert, N., De Vusser, K., and Contreras, R., *In vivo* synthesis of mammalian like hybrid type *N*-glycans in *Pichia pastoris*, *Appl. Environ. Microbiol.*, 70, 2639–2646, 2004.

151. Choi, B.K., Bobrowicz, P., Davidson, R.C., Hamilton, S.R., Kung, D.H., Li, H., Miele, R.G., Nett J.H., Wildt, S., and Gerngross, T.U., Use of combinatorial genetic libraries to humanize *N*-linked glycosylation in the yeast *Pichia pastoris, Proc. Natl. Acad. Sci. USA*, 100, 5022–5027, 2003.

152. Hamilton, S.R., Bobrowicz, P., Bobrowicz, B., Davidson, R.C., Li, H., Mitchell, T., Nett, J.H., Rausch, S., Stadheim, T.A., Wischnewski, H., Wildt, S., and Gerngross, T.U., Production of complex human glycoproteins in yeast, *Science*, 301, 1244–1246, 2003.

153. Verostek, M.F. and Trimble, R.B., Mannosyltransferase activities in membranes from various yeast strains, *Glycobiology*, 5, 671–681, 1995.

154. Lewis, M.J, Sweet, D.J, and Pelham, H.R., The ERD2 gene determines the specificity of the luminal ER protein retention system, *Cell*, 61, 1359–1363, 1990.

155. Desmet, T., Nerinckx, W., Stals, I., Callewaert, N., Contreras, R., and Claeyssens, M., Novel tools for the study of class I alpha-mannosidases: a chromogenic substrate and a substrate-analog inhibitor, *Anal. Biochem.*, 307, 361–367, 2002.

156. Lussier, M., Sdicu, A.M., Ketela, T., and Bussey, H., Localization and targeting of the *Saccharomyces cerevisiae* Kre2p/Mnt1p alpha-1,2-mannosyltransferase to a medial-Golgi compartment, *J. Cell. Biol.*, 131, 913–927, 1995.

157. Nevalainen, H., Penttila, M., Harkki, A., Teeri, T., and Knowles, J.K.C., The molecular biology of *Trichoderma* and its application to the expression of both homologous and heterologous genes, in *Molecular Industrial Mycology*, Leong, S.A. and Berka, R., Eds., Marcel Dekker, New York, 1991, pp. 129–148.

158. Nevalainen, H. and Penttila, M., Molecular biology of cellulolytic fungi, The Mycotica II Genetics and Biotechnology, in Kuck, U. Ed., Springer, Berlin, 1995, pp. 303–319.

159. Durand, H., Clanet, M., and Tiraby, G., Genetic improvement of *Trichoderma reesei* for large-scale cellulase production, *Enzyme Microb. Technol.*, 10, 341–346, 1998.

160. Gouka, R.J., Punt, P.J., and van den Hondel, C., Efficient production of secreted proteins by *Aspergillus*: progress, limitations and prospects, *Appl. Microbiol. Biotechnol.*, 47, 1–11, 1997.

161. Radzio, R. and Kuck, U., Synthesis of biotechnologically relevant heterologous proteins in filamentous fungi, *Process Biochem.*, 32, 529–539, 1997.

162. Punt, P.J., van Biezen, N., Conesa, A., Albers, A., Mangnus, J., and van den Hondel, C., Filamentous fungi as cell factories for heterologous protein production, *Trends Biotechnol.*, 20, 200–206, 2002.

163. Uusitalo, J.M., Nevalainen, K.M.H., Harkki, A.M., Knowles, J.K.C., and Penttila, M.E., Enzyme-production by recombinant *Trichoderma reesei* strains, *J. Biotechnol.*, 17, 35–49, 1991.

164. Nyyssonen, E., Penttila, M., Harkki, A., Saloheimo, A., Knowles, J.K.C., and Keranen, S., Efficient production of antibody fragments by the filamentous fungus *Trichoderma reesei*, *Bio-Technology*, 11, 591–595, 1993.

165. Ward, P.P., Piddington, C.S., Cunningham, G.A., Zhou, X.D., Wyatt, R.D., and Conneely, O.M., A system for production of commercial quantities of human lactoferrin - a broad-spectrum natural antibiotic, *Bio-Technology*, 13, 498–503, 1995.

166. Maras, M., Saelens, X., Laroy, W., Piens, K., Claeyssens, M., Fiers, W., and Contreras, R., *In vitro* conversion of the carbohydrate moiety of fungal glycoproteins to mammalian-type oligosaccharides — Evidence for *N*-acetylglucosaminyltransferase-I-accepting glycans from *Trichoderma reesei*, *Eur. J. Biochem.*, 249, 701–707, 1997.

167. Maras, M., DeBruyn, A., Schraml, J., Herdewijn, P., Claeyssens, M., Fiers, W., and Contreras, R., Structural characterization of *N*-linked oligosaccharides from cellobiohydrolase I secreted by the filamentous fungus *Trichoderma reesei* RUTC 30, *Eur. J. Biochem.*, 245, 617–625, 1997.

168. DeBruyn, A., Maras, M., Schraml, J., Herdewijn, P., and Contreras, R., NMR evidence for a novel asparagine-linked oligosaccharide on cellobiohydrolase I from *Trichoderma reesei* RUTC 30, *FEBS Lett.*, 405, 111–113, 1997.

169. Garcia, R., Cremata, J.A., Quintero, O., Montesino, R., Benkestock, K., and Stahlberg, J., Characterization of protein glycoforms with *N*-linked neutral and phosphorylated oligosaccharides: studies on the glycosylation of endoglucanase 1 (Cel7B) from *Trichoderma reesei*, *Biotechnol. Appl. Biochem.*, 33, 141–152, 2001.

170. Harrison, M.J., Nouwens, A.S., Jardine, D.R., Zachara, N.E., Gooley, A.A., Nevalainen, H., and Packer, N.H., Glycosylation of acetyhylan esterase from *Trichoderma reesei*, *Glycobiology*, 12, 291–298, 2002.

171. Maras, M., De Bruyn, A., Vervecken, W., Uusitalo, J., Penttila, M., Busson, R., Herdewijn, P., and Contreras, R., *In vivo* synthesis of complex *N*-glycans by expression of human *N*-acetylglucosaminyltransferase I in the filamentous fungus *Trichoderma reesei*, *FEBS Lett.*, 452, 365–370, 1999.

172. Kalsner, I., Hintz, W., Reid, L.S., and Schachter, H., Insertion into Aspergillus nidulans of Functional UDP-GlcNAc - Alpha-3-D-Mannoside Beta-1,2-*N*-Acetylgluco-saminyltransferase-I, the enzyme catalyzing the first committed step from oligomannose to hybrid and complex *N*-glycans, *Glycoconj. J.*, 12, 360–370, 1995.

173. Raschke, W. and Ballou, C., Immunochemistry of the phosphomannan of the yeast *Kloeckera brevis*, *Biochemistry*, 10, 4130–4135, 1971.

174. Nakayama, K., Feng, Y., Tanaka, A., and Jigami, Y., The involvement of mnn4 and mnn6 mutations in mannosylphosphorylation of *O*-linked oligosaccharide in yeast *Saccharomyces cerevisiae*, *Biochim. Biophys. Acta-Gen. Subj.*, 1425, 255–262, 1998.

175.   Jigami, Y. and Odani, T., Mannosylphosphate transfer to yeast mannan, *Biochim. Biophys. Acta-Gen. Subj.*, 1426, 335–345, 1999.

176.   Pakula, T.M., Uusitalo, J., Saloheimo, M., Salonen, K., Aarts, R.J., and Penttila, M., Monitoring the kinetics of glycoprotein synthesis and secretion in the filamentous fungus *Trichoderma reesei*: cellobiohydrolase I (CBHI) as a model protein, *Microbiology-UK*, 146, 223–232, 2000.

177.   Karson, E.M. and Ballou, C.E., Biosynthesis of yeast mannan — properties of a mannosylphosphate transferase in *Saccharomyces cerevisiae*, *J. Biol. Chem.*, 253, 6484–6492, 1978.

16 Bacterial Capsules: A Route for Polysaccharide Engineering

Clare M. Taylor and Ian S. Roberts

CONTENTS

I. INTRODUCTION

A number of bacteria produce an extracellular polysaccharide (EPS) that is linked to the cell surface.[1] Such polysaccharides may be in the form of a discrete structure termed a capsule or they may be in the form of an amorphous layer termed slime, which is loosely associated and easily sloughed off from the cell surface. Bacterial capsules have long been recognized as important virulence determinants in isolates capable of causing infection in humans and animals[2] and it is the physicochemical properties of capsules that confer advantageous properties on the bacteria that possess them (Table 16.1). EPS are highly hydrated and constitute more than 95% water.[3] This property is thought to aid transmission of encapsulated bacteria from host to host by providing resistance against desiccation.[4,5] As the polysaccharide capsule represents the outermost layer of the cell, it is not surprising that the capsule mediates interactions between the bacterium and its immediate environment. Adhesion to abiotic surfaces may result in the establishment of biofilms and EPS-mediated interspecies coaggregation within biofilms enhances the colonization of

TABLE 16.1
Functions of Bacterial Capsules

Function	Relevance
Prevention of desiccation	Transmission from environment/host to host
	Survival outside host
Adherence	Colonization of surfaces
	Colonization of medical devices, e.g. in-dwelling catheters
	Bacteria–plant interactions
Resistance to nonspecific host immunity	Complement-mediated killing
	Complement-mediated phagocytosis
Resistance to specific host immunity	Poor antibody response to capsular polysaccharide

Source: From Roberts, I.S., *Annu. Rev. Microbiol.,* 50, 285–315, 1996. With permission.

various ecological niches. Examples include the colonization of food preparation surfaces and machinery, industrial pipelines, waterpipes, in-dwelling catheters, and prostheses.[6] Growth as a biofilm may offer protection from phagocytic protozoa and present nutritional advantages, while it is thought that the presence of EPS acts as a permeability barrier against antimicrobial agents.[7]

During invasive infections of humans and animals, interactions between the bacterial capsule and the host's immune system play a critical role in determining the fate of the infection.[8] During a nonspecific host response, i.e., in the absence of specific antibody, the bacterial capsule may confer resistance to complement-mediated killing by masking the bacterial surface and underlying structures that would normally be potent activators of the alternative pathway.[9] In addition, certain capsular polysaccharides may modulate the ability of the host to mount an immune response by effecting the release of certain cytokines, resulting in the disruption of the host's cell-mediated immune response.[10] Furthermore, a certain small set of capsular polysaccharides are capable of confering some resistance to the specific immune response of the host. Capsules such as those containing NeuNAc, for example, *Escherichia coli* K1 and *Nesisseria meningitidis* serogroup B[11] in addition to the *E. coli* K5 polysaccharide, which is identical to *N*-acetyl heparosan (precursor in heparin/heparan sulfate biosynthesis),[12] are poorly immunogenic. Infected hosts only mount a poor immune response to these antigens as a consequence of the structural similarities of these capsules to host polysaccharides encountered in the extracellular matrices.[13,14] As a result, the expression of these capsules provides protection against the specific arm of the host's immunity.

II. INDUSTRIAL APPLICATIONS OF MICROBIAL POLYSACCHARIDES

A number of bacterial polysaccharides are already established as industrial polysaccharides. Hyaluronan derived from Group *C. streptococci*, is used for surgical, ophthalmic, and viscoelastic applications,[15] while EPS of lactic acid bacteria are used widely in the food industry to improve the rheology, texture, and "mouthfeel" of fermented milk products such as yoghurt.[16] While these EPS have no taste of their own, their addition to milk products increases the time the product spends in the mouth and therefore impart an enhanced perception of taste.[17] In addition, it is thought that they may persist longer in the gastrointestinal tract and thus promote colonization by probiotic bacteria.[18]

Alginate polysaccharides (polymers of β-ManA and α-GulA) are routinely used in applications such as textile printing, paper treatments, ceramic production as well as in pharmaceutical applications such as wound healing, dermatology, and treatment of esophageal reflux.[19,20] As a gelling agent, alginate may also be used to alter the texture of foods. Alginate is mostly derived from brown seaweeds; however, alginate is also produced by microbes belonging to the genera *Pseudomonas* and *Azotobacter.* Recombinant alginate biosynthetic enzymes such as the hexuronyl C5 -epimerases (capable of conversion of β-D-ManA into α-D-GulA) of these genera may be used commercially to modify low-grade alginates derived from seaweeds or for the direct production of high-quality alginates directly from bacteria.[21] The widest nonfood application of bacterial polysaccharides is the use of microbial xanthan in the oil industry.[22,23] Xanthan has unique rheological properties and is capable of maintaining its physical properties over a broad temperature range.[24] As such, this polymer has been used in drilling fluids to lubricate the drill head, to remove rock cuttings, and in enhanced oil recovery. Another microbial polysaccharide, succinoglycan, has also been used in these processes.[25]

III. BIOMEDICAL APPLICATIONS OF MICROBIAL POLYSACCHARIDES

In the biomedical industry, the most significant exploitation of bacterial polysaccharides to date has been their use as vaccines. In general, capsular polysaccharides, with a few exceptions, are immunogenic in noninfants, nontoxic, and have none of the deleterious side effects associated with whole-organism vaccines.[26] Furthermore, the use of purified polysaccharide allows the vaccine agent to be precisely defined both physically and chemically. A number of capsular polysaccharides form the basis of vaccines that are currently used to treat bacterial infections in man. In the case of *Streptococcus pneumoniae*, which is the major cause of bacterial pneumonia in all age groups and pneumococcal meningitis in all age groups apart from infants of age 3 months to 2 years, the polysaccharide-based vaccine consists of 23 capsular serotypes, designed to provide maximum protection.[27] In the case of meningococcal disease, 90% of all infections are caused by only five serogroups of *N. meningitidis*; A, B, C, W135, and Y and the current tetravalent vaccine comprises polysaccharides of serogroups A, C, W135, and Y.[28] However, despite the successful use of these polysaccharide-based vaccines in noninfants, improvements have been made to increase their immunogenicity in infants. One approach has been to conjugate the polysaccharides to protein carriers. This strategy allows an immune response to be raised in infants and importantly, these conjugates act as T-dependent antigens thereby allowing boosting to occur upon reexposure.[29] This approach has been used successfully for vaccines against *Haemophilus influenzae* type B[30] and will soon be available for *S. pneumoniae*.[27]

In economic terms, the most biomedically important polysaccharide is heparin. The anticoagulant and antithrombic properties of heparin are already exploited to treat thromboembolic disease;[31] however, heparin is principally derived from animal tissue with obvious ethical considerations in a consumer-conscious climate. Given the structural similarity between heparin and certain bacterial capsular polysaccharides, particularly the capsule of *E. coli* K5,[12] in future it may be feasible to generate a "biotechnological" heparin from microbial sources.

In this chapter, we will describe Group 2 capsules of *E. coli* and discuss ways in which they may be exploited to unlock their potential as candidates for polysaccharide engineering and the synthesis of biomedically important compounds.

IV. *E. COLI* CAPSULES

E. coli produces more than 80 different serologically and chemically distinct types of polysaccharide capsules.[32] Termed K antigens, these have been classified into four functional groups (Table 16.2), based on a number of biochemical and genetic criteria.[33] The ability to express certain K antigens has been associated with certain specific infections; however, most pathogenic extraintestinal *E. coli* express group 2 K antigens.[1] This group of K antigens probably represents the most intensively studied capsular polysaccharides in terms of biochemistry, expression, and their role in disease. Of these, the K1 and K5 antigens are the best-documented examples. Other *E. coli* capsule groups are described in detail elsewhere.[34,35]

Group 2 capsular polysaccharides represent a heterogenous group in terms of composition, while in terms of structure and cell surface assembly they resemble the capsules of other Gram-negative pathogens, *N. meningitidis* and *H. influenzae*. Group 2 polysaccharides are thought to be anchored to the outer membrane via hydrophobic interactions mediated by α-glycero-phosphatidic acid, which is linked to the reducing terminus and plays a role in formation and stabilization of the capsule structure.[36] The K4 antigen has a chondroitin backbone consisting of -(4)-βGlcA-(1,3)-βGalNAc-(1)-, while the K1 antigen is a polymer of α-2,8-linked-N-acetyl neuraminic acid that is very similar to sialic acid structures present on glycoconjugates of the host.[37] In addition, the K5 antigen has the structure -(4)-βGlcA-(1,4)-αGlcNAc-(1)- that is identical to N-acetyl-heparosan, the precursor of heparin and heparan sulfate.[12] The similarity between these bacterial polysaccharides and host structures results in a poor specific immune response in individuals.

V. GENETIC ORGANIZATION AND REGULATION OF *E. COLI* GROUP 2 CAPSULE GENE CLUSTERS

A large number of group 2 capsule gene clusters have been cloned, and analysis has revealed that they have a conserved modular genetic organization, consisting of three functional regions (Figure 16.1). Furthermore, it appears that this modular organization

TABLE 16.2
Classification of *E. coli* Capsules

Characteristic	Group 1	2	3	4
Former K antigen group	IA	II	I/II or III	IB (O-antigen capsules)
Co-expressed with O serogroups	Limited range (O8, O9, O20, O101)	Many	Many	Often O8, O9 but sometimes none
Co-expressed with colanic acid	No	Yes	Yes	Yes
Thermostability	Yes	No	No	Yes
Terminal lipid moiety	Lipid A core in K_{LPS}; unknown for K antigen	α-Glycerophosphate	α-Glycerophosphate	†Lipid A core in K_{LPS}; unknown for K antigen
Direction of chain growth	Reducing terminus	Nonreducing terminus	Nonreducing terminus?	Reducing terminus
Polymerization system	Wzy-dependent	Processive	Processive?	Wzy-dependent
Transplasma membrane export	Wzx (PST)	ABC-2 exporter	ABC-2 exporter?	Wzx (PST)
Elevated levels of CMP-Kdo synthetase	No	Yes	No	No
Genetic locus	*cps* near *his* and *rfb*	*kps* near *serA*	*kps* near *serA*	*rfb* near *his*
Thermoregulated (not expressed below 20°C)	No	Yes	No	No
Positively regulated by Rcs system	Yes	No	No	No
Model system	K30	K1, K5	K10, K54	K40, O111
Similar to	*Klebsiella, Erwinia*	*Neisseria, Haemophilus*	*Neisseria, Haemophilus*	Many genera

Source: From Whitfield, C. and Roberts, I.S., *Mol. Microbiol.*, 31, 1307–1319, 1999. With Permission.

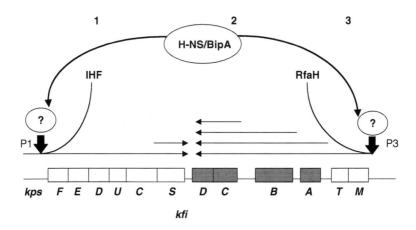

FIGURE 16.1 Genetic organization and regulation of *E. coli* group 2 capsule gene clusters. In this example, the gene cluster of *E. coli* K5 is shown. The numbers at the top refer to the three functional regions, the serotype-specific region, region 2, is shaded. P1 and P3 represent region 1 and 3 promoters, respectively, and the straight arrows denote the major transcripts.

is applicable to capsule gene clusters of other bacteria.[1] Gene expression is achieved following transcription from two convergent promoters P1 and P3 which flank regions 1 and 3, respectively. Regions 1 and 2 of the gene cluster are conserved among group 2 isolates and encode proteins necessary for the transport of the polysaccharide from its site of synthesis to the cell surface. The central region, region 2, is serotype-specific and encodes the enzymes responsible for biosynthesis (where necessary) and polymerization of the individual monosaccharides that comprise the particular polysaccharide. The size of this region is variable and is thought to reflect the complexity of the polysaccharide.[38] For instance, region 2 of the K4 gene cluster is 14 kb in size consisting of seven genes *kfoA-G*, in addition to one insertion sequence (IS2), which are necessary for the synthesis of the chondroitin backbone that is additionally substituted with β-fructose at C-3 of GlcA.[39] In contrast, region 2 of K1 is 5.8 kb.[38]

Region 1 comprises six *kps* genes, organized in a single transcriptional unit (Figure 16.1) that encode proteins involved in transport of the polysaccharide. The individual functions of these genes are shown in Table 16.3. A single *E. coli* σ^{70} promoter has been mapped 225 bp upstream of *kpsF*, and transcription from this promoter generates an 8.0 kb polycistronic transcript that is subsequently processed to generate a stable 1.3 kb *kpsS*-specific transcript.[41] This genomic organization may facilitate the differential expression of *kpsS*, the result of which may influence the attachment of phosphatidyl-Kdo to nascent polysaccharide and regulate its entry into the export machinery. An intragenic Rho-dependent transcriptional terminator has also been identified within *kpsF*. This terminator may play a role in regulating transcription by preventing synthesis of untranslated region 1 transcripts under conditions of physiological stress.[42] Mutations in region 1 genes that abolish export of

TABLE 16.3
Functions of the Conserved Kps Proteins *of E. coli* Group 2 Capsules

Protein	Function
KpsF	Arabinose-5-phosphate isomerase
KpsE	Inner-membrane protein involved in the transport of group 2 polysaccharides across the periplasmic space
KpsD	Periplasmic protein involved in the export of group 2 polysaccharides across the periplasmic space
KpsU	CMP-Kdo synthetase
KpsC, KpsS	Synthesis and attachment of phosphatidyl Kdo to the reducing end of nascent poly saccharide prior to export across the inner membrane
KpsT	ATPase component of the inner-membrane ABC transporter for the export of group 2 polysaccharides across the inner membrane
KpsM	Integral membrane component of the ABC transporter for the export of group 2 polysaccharides across the inner membrane

polysaccharide to the cell surface have been shown to result in reduced membrane transferase activity,[43] Therefore, the net effect of terminating transcription within *kpsF* would be a decrease in synthesis of polysaccharide.

Region 3 of the gene cluster contains two genes, *kpsM* and *kpsT,* organized in a single transcriptional unit.[44,1] The promoter, which has a typical *E. coli* $\sigma^{70}-10$ consensus sequence but no -35 motif, has been mapped 741 bp upstream of the initiation codon of *kpsM*. No consensus binding sites for other alternative σ factors or other DNA-binding proteins have been identified.[45] However, region 3 is subject to control by an antitermination process, conferred by RfaH and *ops* elements. A *cis*-acting regulatory sequence termed *ops*, which is essential for the function of RfaH has been identified 33 bp upstream of the initiation codon of *kpsM*.[45] This *ops* element, with the sequence GGCGGTAC, is contained within a larger regulatory element of 39 bp termed just upstream from many polysaccharide-associated gene starts (JUMPstart).[46] RfaH is known to regulate a number of gene clusters in *E. coli* including the hemolysin operon and the gene clusters for LPS core and O-antigen biosynthesis.[47,48] In addition, RfaH is a homolog of NusG, an essential transcription elongation factor that is necessary for Rho-dependent transcription termination and bacteriophage λN-mediated antitermination. RfaH is thought to act as a transcriptional elongation factor that allows transcription to proceed over long distances.[47] As such, mutations in *rfaH* give rise to increased transcription polarity throughout RfaH-regulated operons without disrupting initiation from operon promoters.[47] To act, *ops* elements must be located on the nascent mRNA transcript, where they recruit RfaH, and perhaps other proteins, to the transcription complex, promoting transcription elongation. It is thought that the JUMPstart sequence on the mRNA molecule may permit the formation of stem-loop structures at the 5′-end that mediate the interaction with RfaH and any other proteins.[46] A mutation in *rfaH* or deletion of the JUMPstart sequence has been shown to abolish capsule production in *E. coli* K1 and K5[45] and serves to confirm the role of RfaH in the regulation of group 2 capsule gene clusters.

Transcription from the region 1 and 3 promoters is temperature-regulated, enabling capsule expression at 37°C but not at 18°C.[41,49] Temperature regulation is in part achieved via the action of the global regulatory protein H-NS (histone-like nucleoid-associated protein), since *hns* mutants show comparable levels of transcription from the region 1 promoter at both 18 and 37°C, albeit lower than those seen in a wild-type strain at 37°C, indicating that H-NS is required for maximal transcription at 37°C as well as repression at 18°C.[50] This pattern of gene expression is analogous to the H-NS-mediated thermoregulation of the *virB* promoter in *Shigella flexneri*.[51] In this system, however, activation of the *virB* promoter has an absolute requirement for the AraC-like protein VirF.[51] It is not yet clear whether an AraC-like transcriptional activator is involved in activating transcription from the region 1 and 3 promoters.

Mutations in *bipA* also result in increased transcription at 18°C and reduced transcription at 37°C.[50] Although this phenomenon mirrors the effect of a mutation in *hns*, the phenotype of a *bipA* mutant cannot be explained by a loss of H-NS function as this is unaltered in a *bipA* mutant. BipA was first described as a tyrosine-phosphorylating protein in enteropathogenic *E. coli* (EPEC).[52] EPEC *bipA* mutants are unable to trigger cytoskeletal rearrangements in host cells, are hypersensitive to bactericidal permeability increasing (BPI) protein, and demonstrate increased flagella expression and motility.[52] Furthermore, BipA is a GTPase with similarity to the TetO resistance protein and elongation factor G (EF–G), both of which interact with the ribosome. These observations have led to the suggestion that BipA may represent a novel class of regulators that interact directly with the ribosome by regulating translation elongation.[52] It is therefore likely that BipA does not regulate the region 1 and 3 promoters directly, but that regulation is achieved via interaction with other proteins that do modulate transcription from regions 1 and 3. This hypothesis is currently under investigation.

At 37°C, the mechanism of temperature regulation is further complicated by the interaction of integration host factor (IHF) with the region 1 promoter. IHF is required for optimal capsule gene expression and IHF binding sites that flank the region 1 promoter have been identified.[45] IHF usually acts to facilitate the activity of other regulatory proteins[53] and as such it is likely that IHF also acts in concert with an as-yet unidentified regulatory protein or proteins that act to control transcription from regions 1 and 3 at 37°C. However, the lack of IHF binding sites in the region 3 promoter[45] demonstrates that there is no absolute requirement for IHF.

The genetic organization of region 2 is serotype-specific and differs among group 2 K antigens. In the case of the K5, region 2 comprises four genes *kfiABCD*,[54] while K4 consists of seven genes *kfoABCDEFG* in addition to an IS2,[39] and K1 comprises six genes *neuDBACES*.[44] In each case, transcription of region 2 proceeds in the same direction as that of region 3, which is important in the RfaH-mediated regulation of region 2 expression.[45] In *E. coli* K5, promoters have been mapped 5′ to the *kfiA*, *kfiB* and *kfiC* genes. Transcription from the *kfiA* promoter generates an 8.0 kb polycistronic transcript, while transcription from the *kfiB* and *kfiC* promoters results in transcripts of 6.5 and 3.0 kb, respectively.[54] The genetic organization of region 2 of the K1 gene cluster suggests that it may be transcribed as a single polycistronic unit,[55] and it has been demonstrated that the capsule biosynthetic genes of *N. meningitidis* are also transcribed as a single operon.[56] Thus, the transcriptional organization in K5 is much more complex, with overlapping transcripts and large intergenic regions that appear to be untranslated.

VI. *E. COLI* CAPSULES AND GLYCOSAMINOGLYCANS

Glycosaminoglycan (GAG) polysaccharides are linear polysaccharides composed of repeating disaccharide units containing a derivative of an amino sugar (either glucosamine or galactosamine). For example, hyaluronan, chondroitin, and heparin/heparan sulfate contain an uronic acid as the other constituent of the disaccharide backbone while keratan contains a galactose[15] (Table 16.4). In vertebrates, the polysaccharide chain is often further modified following polymerization of the chain. One or more modifications, such as *O*-sulfation of certain hydroxyls, deacetylation, *N*-sulfation, and epimerization of GlcA to IdoA, are found in most GAGs except hyaluronan.[57] Furthermore, an astonishing diversity of distinct structures has been reported for chondroitin sulfate and heparan sulfate/heparin, even within a single polymer chain.[57] We described earlier that certain *E. coli* capsules resemble heparin/heparan sulfate (K5) and chondroitin (K4); however, none of these microbial GAGs have been shown to undergo sulfation or epimerization, although, the known chondroitin and heparin/heparan sulfate glycosyltransferases responsible for synthesis of GAG backbones in both bacteria and vertebrates utilize uridine diphospho-(UDP)-sugar precursors. Depending on the specific GAG and the particular organism or tissue examined, the degree of polymerization may vary from ~25 to ~10,000.[15] The rest of this chapter will focus on the microbial GAG glycosyltransferases and discuss their nature and potential biotechnological applications.

VII. CHONDROITIN SYNTHASES

E. coli K4 synthesizes a polysaccharide capsule that comprises a chondroitin backbone substituted with a β-linked fructose attached to C3 of GlcA.[58] The capsule locus from *E. coli* K4 was cloned some time ago,[59] however the nucleotide sequence was only recently reported.[39] The chondroitin polymerase, KfoC (686 amino acids, calculated

TABLE 16.4
Composition of Some Glycosaminoglycans

Glycosaminoglycan	Disaccharide Repeat	Modification Following Polymerization	
		Vertebrates	Bacteria
Hyaluronan	β3GlcNAcβ4GlcA	None	None
Chondroitin	β3GalNAcβ4GlcA	*O*-sulfation, epimerization	None or fructose (β1,3)GlcA
Heparin/heparan sulfate	α4GlcNAcβ4GlcA	*O*-,*N*-sulfation, epimerization	None
Keratan	β4GlcNAcβ3Gal	*O*-sulfation	Not reported

Source: From Angelis, P.L., *Glycobiology,* 12, 9R–16R, 2002. With, permission.

$M_r = 79,256$) has been shown to be a bifunctional glycosyltransferase responsible for the alternate addition of GalNAc and GlcA to the nonreducing terminus of the polysaccharide chain, as recombinant His-tagged KfoC protein was shown to be capable of synthesising polysaccharide that was completely digested by chondroitinase ABC.[39] In an earlier study, before the identity of the chondroitin polymerase was reported, it was shown that defructosylated chondroitin oligosaccharides could act as substrates for both the GalNAc and GlcA glycosyltransferase activities, however, fructosylated oligosaccharide could not serve as an acceptor for GlcA activity. This observation suggests that fructose is either added postpolymerization or as a second step after the addition of GlcA and of the GalNAc units further downstream in the polymer.[59] These data indicate that chondroitin chain elongation probably proceeds in the same direction as that of chondroitin synthesized in vertebrates and makes biosynthesis of K4 polysaccharide a useful parallel system for the study of chondroitin sulfate biosynthesis. Interestingly, the addition of the fructose branch in K4 polysaccharide makes the polymer more antigenic,[15] with obvious implications for the microbe, however, presently, an *E. coli* K4 capsular type without the fructose addition has not been reported.

The chondroitin polymerase KfoC also shows significant similarity to GAG glycosyltransferases of *P. multocida*. The pmHAS protein and pmCS protein of *P. multocida* serotypes A and F, respectively, are dual-action glycosyltransferases responsible for polymerization of hylauronan (β3GlcNAcβ4GlcA) and chondroitin (β4GalNAcβ4GlcA), respectively.[60,61] At the amino acid level, KfoC has 59% identity to pmHAS and 61% identity to pmCS and all of these proteins contain two glycosyltransferase domains per molecule. In pmCS, an upstream domain termed A1 is probably responsible for the *N*-acetyl hexosamine β1-4 glycosyl bond and the downstream domain A2 for the uronic acid β1-3 glycosyl bond.[62] The A1 domain (residues 153–258) of KfoC shows close resemblance to that of pmCS, while the A2 domain (residues 435–539) resembles that (70% identity) of both pmHAS and pmCS. In contrast, however, KfoC lacks a C-terminal membrane association domain (present in both pmHAS and pmCS) required for interaction with the polysaccharide transport machinery or a membrane bound partner.[39] In *E. coli* K4, the KfoC polymerase may participate in a membrane-bound complex in a way analogous to that seen in *E. coli* K5.[63]

VIII.   HEPAROSAN SYNTHASES

Heparin/heparan sulfate and related GAGs contain alternating β- and α-glycosidic linkages and are therefore distinct from the very different β-linked hyaluronan and chondroitin polymers. The UDP-sugar precursor molecules are α-linked, therefore heparosan (and subsequent heparin biosynthesis) requires two reaction pathways: a retaining mechanism to produce the α-linkage and an inverting mechanism that results in the β-linkage. It might be expected that heparosan biosynthesis requires either two distinct α- and β-glycosyltransferases, or that one protein with two functional domains might be present, in a manner similar to that of the KfoC chondroitin polymerase. In reality, both exist in different bacterial systems.

E. coli K5 produces the capsular polysaccharide with the structure α4GlcNAc β4GlcA, i.e., identical to that of heparosan (unmodified heparin/heparan sulfate). The K5 biosynthetic locus, within region 2 of the capsule gene cluster, consists of four genes *kfiABCD*.[54] The α- and β-glycosyltransferases are the KfiA (α-UDP-GlcNAc) and KfiC (β-UDP-GlcA) proteins, respectively.[64,65] Based on its amino acid sequence, KfiA can be assigned to the C family of glycosyltransferases, all of which have a conserved acidic motif, often DDD (or EDD or DSD).[64] These acidic motifs have been implicated in catalysis of glycosidic bond formation and the importance of the DDD motif (residues 79–81) of KfiA has been demonstrated by site-directed mutagenesis, as mutations, both conservative and nonconservative, abolished α-UDP-GlcNAc transferase activity.[64] Similarly, mutation of two acidic aspartates of KfiC (residues 301 and 352) abolished β-UDP-GlcA transferase activity. It is likely that these residues, which are highly conserved among β-glycosyltransferases, are the catalytically active amino acids for β-transferase activity.[65] Furthermore, these two enzymes, which act in concert to permit polymer chain elongation, have been shown to require each other in order to form a stable complex at the site of K5 polysaccharide synthesis.[64] This association also requires the KfiB protein, whose function has not yet been assigned, but which the data suggest may be the initiator for K5 synthesis at the cytoplasmic membrane (M. Pourhossein and I.S. Roberts, unpublished results).

The capsular polysaccharide of *P. multocida* type D is also heparosan and comprises the same structure as the capsule of *E. coli* K5.[15] In this organism, however, polysaccharide chain elongation is carried out by a single bifunctional enzyme termed pmHS.[66] This enzyme is capable of the alternate α- and β-addition of the corresponding UDP-sugars, and recombinant pmHS is capable of synthesizing a polymer that is sensitive to heparin lyase III, confirming its identity as heparosan.[66]

The pmHS protein consists of two regions, HS1 (residues 91–240) and HS2 (residues 491–540) that have similarity to KfiC and KfiA of *E. coli* K5, respectively. This organization suggests that pmHS has two domains; however, the pmHS sequence is quite distinct from that of pmHAS and pmCS.[66] Interestingly, none of the microbial heparosan synthases resemble the EXT1 and EXT2 proteins that synthesize the heparin/heparan sulfate backbone in vertebrates.[67]

IX. APPROACHES TO SYNTHESIS OF BIOTECHNOLOGICAL HEPARIN

Following assembly of the *N*-acetyl heparosan precursor, polymerization of heparin/heparan sulfate proceeds as the backbone is modified. These modifications include GlcNAc *N*-deacetylation and *N*-sulfation, C5 epimerization of GlcA to IdoA, and variable *O*-sulfation at C2 of IdoA and GlcA, at C6 of GlcNAc and GlcNS units, and occasionally at C3 of GlcN residues. Currently, all of these modifications can be carried out chemically with the exception of the epimerization of GlcA to IdoA, which can only be carried out enzymatically by the glucuronyl C5 epimerase.[68] Chemically modified derivatives of K5 polysaccharide, such as highly *O*-sulfated and highly *N,O*-sulfated K5 derivatives, already show promise as therapeutic compounds, including use as potential topical microbicides capable of preventing attachment of HIV-1 virions to human cell lines.[69] Furthermore, highly *N,O*-sulfated may also provide the basis for design of novel angiostatic compounds.[70] Through the combined chemical and enzymatic modification of K5 polysaccharide, it may be possible to produce novel heparin-like structures with important therapeutic properties. In the laboratory, we have used a rational approach to design a strategy for the *in vivo* synthesis of both K5 oligosaccharides and heparin biosynthetic enzymes in heterologous hosts. Most of the enzymes involved in heparin/heparan sulfate have been purified and molecularly cloned and in our laboratory, we have been able to achieve functional small-scale expression of isoforms of the NDST (*N*-deacetylase, sulfotransferase) enzymes and, crucially, the C5 epimerase enzyme in yeast hosts (B.R. Clarke and I.S. Roberts, unpublished results, C.M. Taylor and I.S. Roberts, unpublished results). This accomplishment suggests that potential exists for the large-scale production of recombinant heparin biosynthetic enzymes and is an important step toward production of biotechnological heparin.

The yeast *Saccharomyces cerevisiae* has Generally Regarded As Safe (GRAS) status and as such is a desirable host for the production of biomolecules designed for human use. Furthermore, *S. cerevisiae* has been intensively studied as a model eukaryote and this has enabled us to make predictions about the behavior of heterologous proteins expressed in yeast. *S. cerevisiae* synthesizes oligosaccharides as the first step in the production of *N*-linked glycoproteins and thus potential exists to exploit this pathway for the synthesis of nonnative oligosaccharides in yeast. As we have a great deal of understanding about the genetic regulation and synthesis of K5 capsular polysaccharide, we have been able to devise strategies for the exploitation of this knowledge and, in our laboratory, we have been able to achieve expression of the *E. coli* K5 glycosyltransferases responsible for synthesis of K5 polysaccharide in *S. cerevisiae*. The KfiA (UDP-GlcNAc transferase) and KfiC (UDP-GlcA transferase) proteins have been expressed along with the KfiD protein (UDP-Glc dehydrogenase), which may result in the ability to build K5 oligosaccharides *in vivo* (C.M. Taylor and I.S. Roberts, unpublished results). Furthermore, we have also achieved expression of the *P. multocida* bifunctional transferase, pmHS, as a fusion to a native yeast protein in order to bring the recombinant protein in close proximity to the *N*-glycosylation machinery of the host (C.M. Taylor and I.S. Roberts, unpublished results). The success of these strategies for the synthesis of K5 oligosaccharides in a eukaryote is currently under investigation. However, in theory it should be possible to transfect the K5-synthesizing host with other constructs for the expression of recombinant heparin biosynthetic enzymes, permitting *in vivo* modification of the oligosaccharides. The construction of such a recombinant strain would certainly facilitate large-scale production of heparin-like molecules and represent a major achievement in the field of polysaccharide engineering.

Finally, apart from providing systems for parallel studies of vertebrate GAG synthesis, understanding and exploiting microbial GAG biosynthesis may also aid in the design of antimicrobial compounds for the treatment of invasive infections caused by encapsulated bacteria, without disruption to the host, particularly when infection is caused by clever microbes whose capsules mimic host molecules.

REFERENCES

1.   Roberts, I.S., The biochemistry and genetics of capsular polysaccharide production in bacteria, *Annu. Rev. Microbiol.,* 50, 285–315, 1996.

2. Moxon, E.R. and Kroll, J.S., The role of bacterial polysaccharide capsules as virulence factors, *Curr. Microbiol. Immunol.,* 21, 221–231, 1990.

3. Costerton, J.W., Irvin, R.T., and Cheng, K.-J., The bacterial glycocalyx in nature and disease, *Annu. Rev. Microbiol.,* 35, 299–324, 1981.

4. Roberson, E. and Firestone, M., Relationship between desiccation and exopolysaccharide production in soil *Pseudomona*s sp., *Appl. Environ. Microbiol.,* 58, 1284–1291, 1992.

5. Ophir, T. and Gutnick, D., A role for exopolysaccharides in the protection of microorganisms from desiccation, *Appl. Environ. Microbiol.,* 60, 740–745, 1994.

6. Costerton, J.W., Cheng, K-J., Geesey, G.G., Ladd, T.I., Nickel, J.C., Dasgupta, M., and Marrie, T., Bacterial biofilms in nature and disease, *Annu. Rev. Microbiol.,* 41, 435–464, 1987.

7. Costerton, J.W., Stewart, P.S., and Greenberg, E.P., Bacterial biofilms: a common cause of persistent infections, *Science,* 284, 1318–1322, 1999.

8. Roberts, I.S., Saunders, F.K., and Boulnois, G.J., Bacterial capsules and interactions with complement and phagocytes, *Biochem. Soc. Trans.,* 17, 462–464, 1989.

9. Howard, C.J. and Glynn, A.A., The virulence for mice of strains of *Escherichia coli* related to the effects of K antigens on their resistance to phagocytosis and killing by complement, *Immunology,* 20, 767–777, 1971.

10. Cross, A., The biological significance of bacterial encapsulation, *Curr. Top. Microbiol. Immunol.,* 150, 87–95, 1990.

11. Bhattacharjee, A., Jennings, H., Kenny, C., Martin, A., and Smith, I., Structural determination of the sialic acid polysaccharide antigens of *Neisseria meningitidis* serogroups B and C with carbon 13 nuclear magnetic resonance, *J. Biol. Chem.,* 250, 1926–1932, 1975.

12. Vann, W.F., Schmidt, M., Jann, B., and Jann, K., The structure of the capsular polysaccharide (K5 antigen) of urinary tract infective *Escherichia coli* O10:K5:H4. A polymer similar to desulfo-heparin, *Eur. J. Biochem.,* 116, 359–364, 1981.

13. Roberts, I.S., Saunders, F.K., and Boulnois, G.J., Bacterial capsules and interactions with complement and phagocytes, *Biochem. Soc. Trans.,* 17, 462–464, 1989.

14. Wyle, F., Artenstein, M., Brandt, B.L., Tramont, E.C., and Kasper, D.L., Immunological response of man to group B meningococcal polysaccharide vaccines, *J. Infect. Dis.,* 126, 514–521, 1972.

15. DeAngelis, P.L., Microbial glycosaminoglycan glycosyltransferases, *Glycobiology,* 12, 9R–16R, 2002.

16. Welman, A.D. and Maddox, I.S., Exoploysaccharides from lactic acid bacteria: perspectives and challenges, *Trends Biotechnol.,* 21, 269–274, 2003.

17. Duboc, P. and Mollet, B., Applications of exopolysaccharides in the dairy industry, *Int. Dairy. J.,* 11, 759–768, 2001.

18. German, B., The development of functional foods: lessons from the gut, *Trends Biotechnol.,*17, 492–499, 1999.

19. Onsøien, E., Commercial applications of alginates, *Carbohydr. Eur.,*14, 26–31, 1996.

20. Skjåk-Braek, G. and Espevik, T., Application of alginate gels in biotechnology and biomedicine, *Carbohydr. Eur.,* 14, 19–24, 1996.

21. Valla, S., Li, J.P., Ertesvag, H., Barbeyron, T., and Lindahl, U., Hexuronyl C5-epimerases in alginate and glycosaminoglycan biosynthesis, *Biochimie,* 83, 819–830, 2001.

22. Moradi-Araghi, A., Beadmore, D.H., and Stahl, G.A., The application of gels in enhanced oil recovery: theory, polymers and crosslinker systems, in Stahl, G.A., Schulz, D.N., *Water-Soluble Polymers for Petroleum Recovery,* Eds., Plenum Publishers, New York, 1988, pp. 299–312.

23. Linton, J.D., Ash, S.G., and Huybrechts, L., Microbial polysaccharides, in *Novel Materials from Biological Sources,* Byrom, D., Ed., Macmillan Publishers, Basingstoke, UK, 1991, pp. 215-262.
24. Morris, V.J., Franklin, D., and I'Anson, K., Rheology and microstructure of dispersions and solutions of the microbial polysaccharide from *Xanthamonas campestris* (xanthan gum), *Carbohydr. Res.,* 121, 12–30, 1983.
25. Clarke-Sturman, A.J., den Ottelander, D., and Sturla, P.L., Succinoglycan. A new biopolymer for the oilfield, in, *Oil-field Chemistry: Enhanced Recovery and Production Stimulation,* Borchardt, J.K., Yen, T.F., Eds., American Chemical Society, Washington, DC, 1989, pp. 157–168.
26. Jennings, H.J., Capsular polysaccharides as vaccine candidates, *Curr. Top. Microbiol. Immunol.,* 150, 97–128, 1990.
27. Mulholland, K., Strategies for the control of pneumococcal diseases, *Vaccine,* 17, S79–S84, 1999.
28. Cadoz, M., Armand, J., Arminjon, F., Gire, R., and Lafaix, C., Tetravalent (A, C, Y, W135) meningococcal vaccine in children: immunogenicity and safety, *Vaccine,* 3, 340–342, 1985.
29. Robbins, J.B. and Scneerson, R., Polysaccharide-protein conjugates: a new generation of vaccines, *J. Infect. Dis.,* 161, 821–832, 1990.
30. Moxon, E.R. and Kroll, J.S., The role of bacterial polysaccharide capsules as virulence factors, *Curr. Top. Microbiol. Immunol.,* 150, 65–86, 1990.
31. Hemker, H.C., Ed., Heparin — Present and Future, *Proceedings of the Symposium on Heparin,* Vol 20, S. Karger Medical and Scientific Publishers, Florence, 1990.
32. Jann, K. and Jann, B., Capsules of *Escherichia coli*, expression and biological significance, *Can. J. Microbiol.,* 38, 705–710, 1992.
33. Whitfield, C. and Roberts, I.S., Structure, assembly and regulation of expression of capsules in *Escherichia coli*, *Mol. Microbiol.,* 31, 1307–1319, 1999.
34. Roberts, I.S., The expression of polysaccharide capsules in *Escherichia coli*: a molecular genetic perspective, in: *Glycomicrobiology,* Doyle, R., Ed., Kluwer Academic/Plenum Publishers, New York, 2000, pp. 441–464.
35. Whitfield, C., Drummelsmith, J., Rahn, A., and Wugeditsch, T., Biosynthesis of group 1 capsules in *Escherichia coli* and related extracellular polysaccharides in other bacteria, in *Glycomicrobiology,* Doyle, R., Ed., Kluwer Academic/Plenum Publishers, New York, 2000, pp. 275–297.
36. Jann, B. and Jann, K., Structure and biosynthesis of capsular antigens of *Escherichia coli,* *Curr. Top. Microbiol. Immunol.,* 150, 19–42, 1990.
37. Finne, J., Occurrence of unique polysialosyl carbohydrate units in glycoproteins of developing brain, *J. Biol. Chem.,* 257, 11966–11970, 1982.
38. Boulnois, G., Drake, R., Pearce, R., and Roberts, I., Genome diversity at the serA-linked capsule locus in *Escherichia coli, FEMS Microbiol. Lett.,* 100, 121–124, 1992.
39. Ninomiya, T., Sugiura, N., Tawada, A., Sugimoto, K., Watanabe, H., and Kimata, K., Molecular cloning and characterization of chondroitin polymerase from *Escherichia coli strain* K4, *J. Biol. Chem.,* 277, 21567–21575, 2003.
40. Taylor, C.M. and Roberts, I.S., The regulation of capsule expression, in, *Bacterial Adhesion to Host Tissues, Mechanisms and consequences,* Wilson, M., Ed., Cambridge University Press, Cambridge, UK, 2002, pp. 115–138.
41. Simpson, D.A., Hammarton, T.C., and Roberts, I.S., Transcriptional organization and regulation of expression of region 1 of the *Escherichia coli* K5 capsule gene cluster, *J. Bacteriol.,* 178, 6466–6474, 1996.

42. Richardson, J.P., Preventing the synthesis of unused transcripts by Rho factor, *Cell,* 64, 1047–1049, 1991.

43. Bronner, D., Sieberth, V., Pazzani, C., Roberts, I.S., Boulnois, G.J., Jann, B., and Jann, K., Expression of the capsular K5 polysaccharide of *Escherichia coli*: biochemical and electron microscopic analyses of mutants with defects in region 1 of the K5 gene cluster, *J. Bacteriol.,*175, 5984–5992, 1993.

44. Bliss, J.M. and Silver, R.P., Coating the surface: a model for expression of capsular polysialic acid in *Escherichia coli* K1, *Mol. Microbiol.,* 21, 221–231, 1996.

45. Stevens, M.P., Clarke, B.R., and Roberts, I.S., Regulation of the *Escherichia coli* K5 capsule gene cluster by transcription antitermination, *Mol. Microbiol.,* 24, 1001–1012, 1997.

46. Hobbs, M. and Reeves, P.R., The JUMPstart sequence: a 39 bp element common to several polysaccharide gene clusters, *Mol. Microbiol.,* 12, 855–856, 1994.

47. Bailey, M.J., Hughes, C., and Koronakis, V., RfaH and the *ops* element, components of a novel system controlling bacterial transcription elongation, *Mol. Microbiol.,* 26, 845–851, 1997.

48. Marolda, C.L. and Valvano, M.A., Promoter region of the *Escherichia coli* O7-specific lipopolysaccharide gene cluster: structural and functional characterization of an upstream untranslated mRNA sequence, *J. Bacteriol.,* 180, 3070–3079, 1998.

49. Cieslewicz, M. and Vimr, E., Thermoregulation of *kpsF*, the first region 1 gene in the *kps* locus for polysialic acid biosynthesis in *Escherichia coli* K1, *J. Bacteriol.,* 178, 3212–3220, 1996.

50. Rowe, S., Hodson, N., Griffiths, G., and Roberts, I.S., Regulation of the *Escherichia coli* K5 capsule gene cluster; evidence for the role of H-NS, BipA and IHF in the regulation of group II capsule gene clusters in pathogenic *E. coli, J. Bacteriol.,* 182, 2741–2745, 2000.

51. Dorman, C.J. and Porter, M.E., The *Shigella* virulence gene regulatory cascade: a paradigm of bacterial gene control mechanisms, *Mol. Microbiol.,* 29, 677–684, 1998.

52. Farris, M., Grant, A., Richardson, T.B., and O'Connor, C.D., BipA: a tyrosine-phosphorylated GTPase that mediates interactions between enteropathogenic *Escherichia coli* (EPEC) and epithelial cells, *Mol. Microbiol.,* 28, 265–279, 1998.

53. Freundlich, M., Ramani, N., Mathew, E., Sikiro, A., and Tsui, P., The role of integration host factor in gene expression in *Escherichia coli, Mol. Microbiol.,* 6, 2557–2563, 1992.

54. Petit, C., Rigg, G.P., Pazzani, C., Smith, A., Sieberth, V., Boulnois, G., Jann, K., and Roberts, I.S., Region 2 of the *Escherichia coli* K5 capsule gene cluster encoding proteins for the biosynthesis of the K5 polysaccharide, *Mol. Microbiol.,* 17, 611–620, 1995.

55. Silver, R.P., Annunziato, P., Pavelka, M., Pigeon, R.P., Wright, L.F., and Wunder, D.E., Genetic and molecular analyses of the polysialic acid gene cluster of *Escherichia coli* K1, in, *Polysialic Acid: from Microbes to Man,* Roth, J., Rutishauser, U., Troy, F.A., Eds., Birkhauser-Verlag, Basel, 1993, pp. 59–73.

56. Edwards, U., Muller, A., Hammerschmidt, S., Gerady-Schan, R., and Frosch, M., Molecular analysis of the biosynthesis of the α-2,8 polysialic acid capsule by *Neisseria meningitidis* serogroup B, *Mol. Microbiol.,* 14, 141–150, 1994.

57. Esko, J.D. and Lindahl, U., Molecular diversity of heparan sulphate, *J. Clin. Invest.,* 108, 169–173, 2001.

58. Rodriguez, M.L., Jann, B., and Jann, K., Structure and serological characteristics of the capsular K4 antigen of *Escherichia coli* O5:K4:H4, a fructose-containing polysaccharide with a chondroitin backbone, *Eur. J. Biochem.,* 177, 117–124, 1988.

59. Lidholt, K. and Fjelstad, M., Biosynthesis of the *Escherichia coli* K4 capsule polysaccharide. A parallel system for studies of glycosyltransferases in chondroitin formation, *J. Biol. Chem.*, 272, 2682–2687, 1997.

60. DeAngelis, P.L., Jing, W., Drake, R.R., and Achyuthan, A.M., Identification and molecular cloning of a unique hyaluronan synthase from *Pasteurella multocida*, *J. Biol. Chem.*, 273, 8454–8458, 1988.

61. DeAngelis, P.L. and Padgett-McCue, A.J., Identification and molecular cloning of a chondroitin synthase from *Pasteurella multocida* Type F, *J. Biol. Chem.*, 275, 24124–24129, 2000.

62. Jing, W. and DeAngelis, P.L., Dissection of the two transferase activities of the *Pasteurella multocida* hyaluronan synthase: two active sites exist in one polypeptide, *Glycobiology*, 10, 883–889, 2000.

63. Rigg, G.P., Barrett, B., and Roberts, I.S., The localization of KpsC, S and T, and KfiA, C and D proteins involved in the biosynthesis of the *Escherichia coli* K5 capsular polysaccharide: evidence for a membrane-bound complex, *Microbiology*, 144, 2905–2914, 1998.

64. Hodson, N., Griffiths, G., Cook, N., Pourhossein, M., Gottfridson, E., Lind, T., Lidholt, K., and Roberts, I.S., Identification that KfiA, a protein essential for the biosynthesis of the *Escherichia coli* K5 capsular polysaccharide, is an alpha-UDP-GlcNAc glycosyltransferase. The formation of a membrane-associated K5 biosynthetic complex requires KfiA, KfiB and KfiC, *J. Biol. Chem.*, 275, 27311–27315, 2000.

65. Griffiths, G., Cook, N.J., Gottfridson, E., Lind, T., Lidholt, K., and Roberts, I.S., Characterization of the glycosyltransferase enzyme from *Escherichia coli* K5 capsule gene cluster and identification and characterization of the glucuronyl active site, *J. Biol. Chem.*, 273, 11752–11757, 1998.

66. DeAngelis, P.L. and White, C.L., Identification and molecular cloning of a heparosan synthase from *Pasteurella multocida* Type D, *J. Biol. Chem.*, 277, 7209–7213, 2002.

67. Duncan, G., McCormick, C., and Tufaro, F., The link between heparan sulphate and hereditary bone disease: finding a function for the EXT family of the putative tumor suppressor proteins, *J. Clin. Invest.*, 108, 511–516, 2001.

68. Naggi, A., Torri, G., Casu, B., Oreste, P., Zoppetti, G., Li, J.P., and Lindahl, U., Toward a biotechnological heparin through combined chemical and enzymatic modification of the *Escherichia coli* K5 polysaccharide, *Semin. Thromb. Hemost.*, 27, 437–443, 2001.

69. Vicenzi, E., Gatti, A., Ghezzi, S., Oreste, P., Zoppetti, G., and Poli, G., Broad spectrum inhibition of HIV-1 infection by sulfated K5 *Escherichia coli* polysaccharide derivatives, *AIDS*, 17, 177–181, 2003.

70. Leali, D., Belleri, M., Urbinati, C., Coltrini, D., Oreste, P., Zoppetti, G., Ribatti, D., Rusnati, M., and Presta, M., Fibroblast growth factor-2 antagonist activity and angiostatic capacity of sulphated *Escherichia coli* K5 polysaccharide derivatives, *J. Biol. Chem.*, 276, 37900–37908, 2001.

17 Chemo-Biological Approach to Modification of the Bacterial Cell Wall

Reiko Sadamoto
and Shin-Ichiro Nishimura

CONTENTS

I. INTRODUCTION

G. P. Smith's report on the display of target proteins on phage surfaces,[1] a technique now generally referred to as phage display, was the precursor of many later studies that sought to apply the same technique to the display of large proteins on microbial cells. Subsequently, target proteins fused with a cellular anchor protein have been successfully displayed on the cell surface of *Escherichia coli* and yeast.[2,3] This strategy is based on a genetic method using recombinant DNA corresponding to the target and anchor proteins.[4,5]

On the other hand, chemical rather than genetic methods are more useful for displaying biofunctional compounds, such as synthetic compounds or nonnatural oligosaccharides, that are not specified by the genetic code. For mammalian cells, Bertozzi and colleagues[6–8] have developed an effective chemical display method that utilizes the metabolic incorporation of artificial sugars having anchor groups such as ketones.[6–8]

Since this strategy is based on the biosynthetic route from *N*-acetyl mannose to sialic acid, the application is limited, however, to only those cells that display sialic acid.

We have reported bacterial cell-surface engineering using chemically synthesized cell-wall precursor derivatives.[9–11] As almost all bacteria have a cell wall consisting of similar conserved structures independent of strain, the bacterial cell wall provides a good platform for surface display using chemical techniques (Figure 17.1). UDP-MurNAc pentapeptide derivatives can be incorporated into the cell wall to attach and display functional compounds, such as ketone groups, on the bacterial surface. Here, we extend this method to determine the control of adhesion properties of living lactic acid bacteria through the display of target oligosaccharides. Lactic acid bacteria are Gram-positive bacteria, having a thick cell-wall layer with no sig-

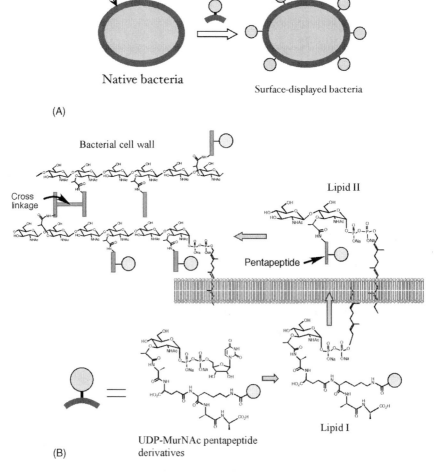

FIGURE 17.1   (a) Bacterial cell-wall engineering. (b) Surface display through the cell-wall biosynthesis.

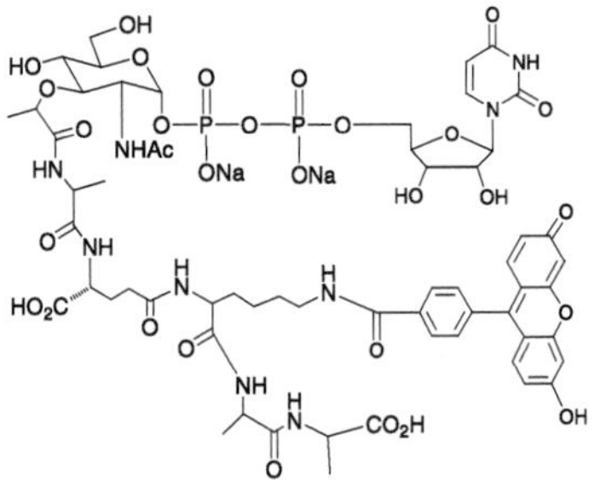

SCHEME 1

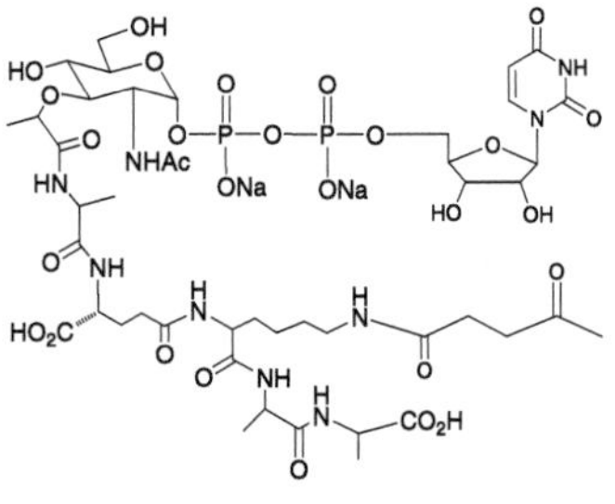

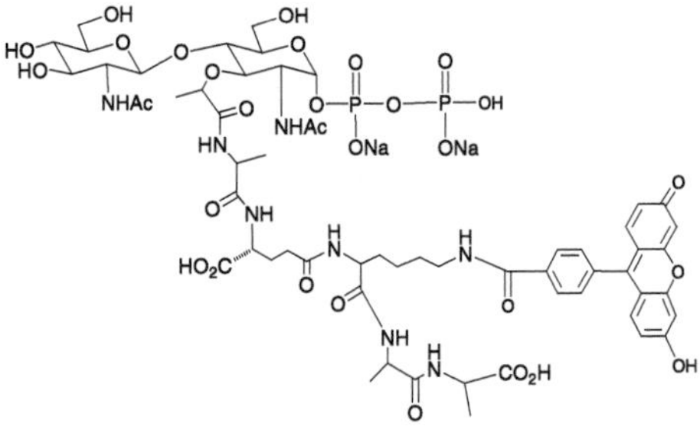

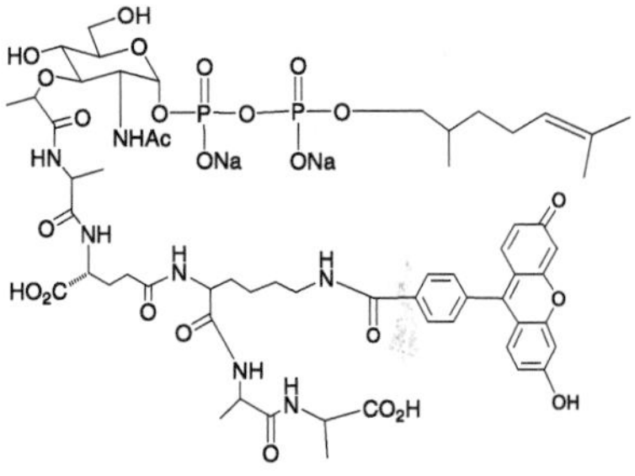

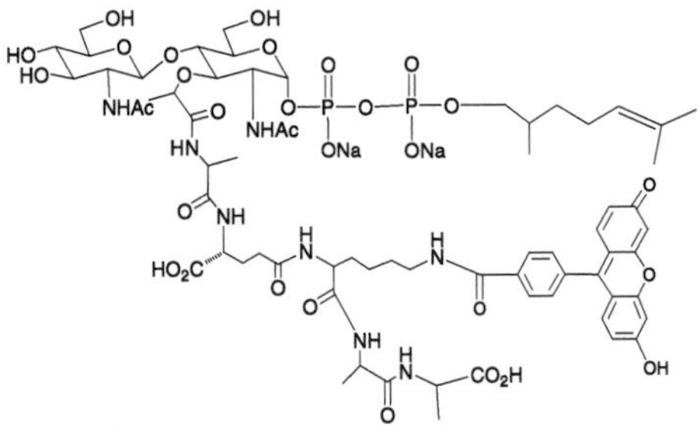

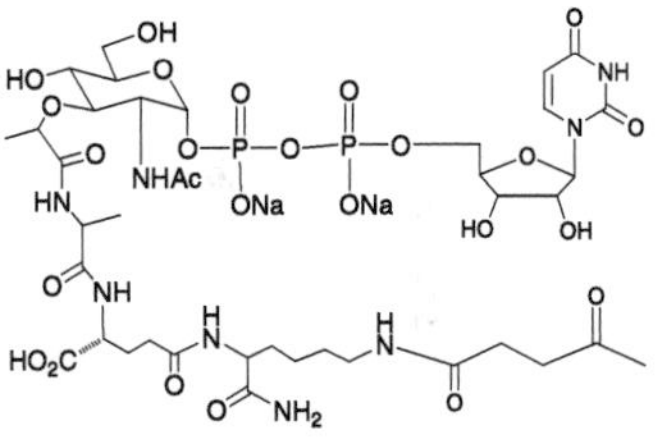

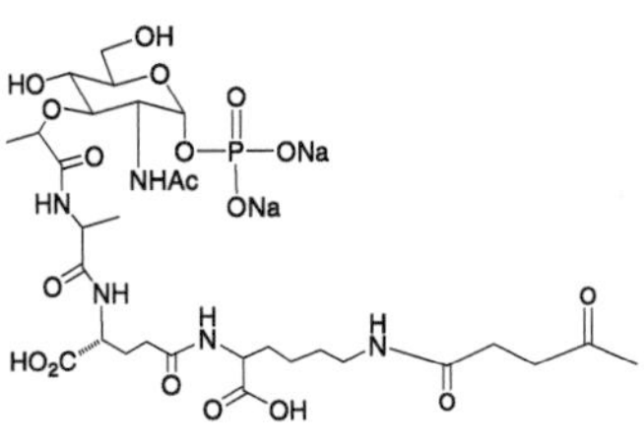

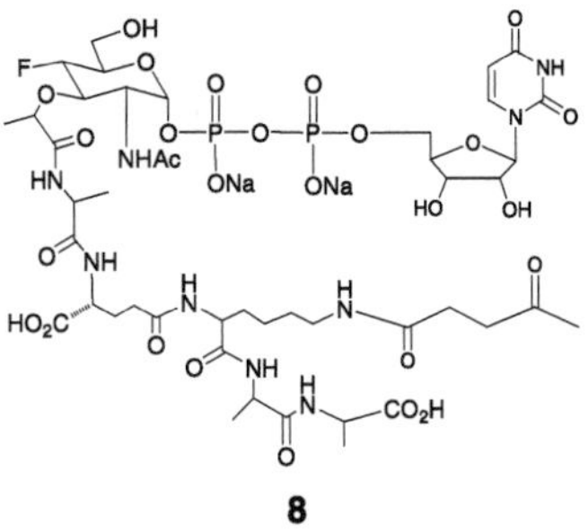

Gram negative

Gram positive

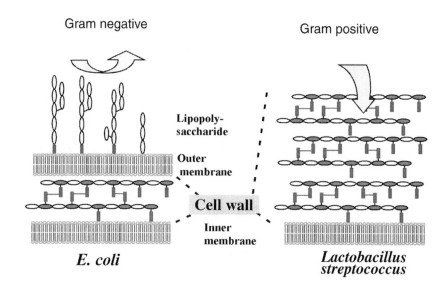

Lipopoly-
saccharide

Outer
membrane

Cell wall

Inner
membrane

E. coli

*Lactobacillus
streptococcus*

FIGURE 17.2   Representation of Gram-negative and -positive bacterial surfaces.

nificant membrane outside the cell wall (Figure 17.2). Therefore, lactic acid bacteria also have certain advantages over Gram-negative bacteria with regard to surface display in that they are generally recognized as safe (GRAS). Unlike other bacteria, lactic acid bacteria they can be easily used in food or for medical purposes. Isolauri and co-workers[12] have reported the potential benefits of the use of lactobacilli in the development of vaccines against infectious diseases, and in the treatment of autoimmune or other immune disorders.[12] Our method is not only useful for the immobilization of lactic acid bacteria onto a solid substrate, but also promises to be widely applicable to the development of vaccine carriers against cancers and allergies.

In this chapter, we describe our technique for the incorporation of bacterial cell-wall precursors and, UDP-MurNAc pentapeptide derivatives (**1** and **3**, Scheme 1) into the cell wall via the biosynthetic route. We also present an extension of this technique that allows the control of the adhesion of living lactic acid bacteria through the display of target oligosaccharides.

A.   INCORPORATION OF SYNTHETIC PRECURSORS INTO THE BACTERIAL CELL WALL

As Gram-positive bacteria such as lactic acid bacteria have a thick cell wall, but unlike Gram-negative bacteria, no outer membrane, we expected the incorporation of the target compounds into the cell wall to be more efficient due to the higher permeability of the cell wall. Lactic acid bacteria were incubated in Lactobacilli MRS broth (Difco Laboratories, Detroit, MI) medium containing **1** or **3** (5.0 m*M*) under anaerobic conditions at 37°C overnight. The bacteria were collected by centrifugation and washed with Phospate-buffered Saline (PBS). After lysation by ultrasonication, the residue was applied to a centrifuge column (Microcon YM-10,

Millipore) and washed with PBS. The washed residue was then treated with 4.0 mL of lysozyme solution for 5 h at 37°C. After the addition of a PBS buffer containing 10% dimethyl sulfoxide to the residue, the mixture was filtered through a Microcon YM-10. The filtrate, containing the digested cell wall fragment, was diluted to the same volume for each sample with the buffer and used for fluorescence measurement. Background fluorescence intensity was minimized using lysozyme digestion, which allows the incorporated fluorophore to be separated from the nonincorporated fluorophore (Figure 17.3). The bacterial cell wall, after incubation with **1**, generated significantly higher fluorescence than a control experiment, showing that the fluorescence-attaching precursor had been incorporated into the cell wall.

On the other hand, for *E. coli*, incorporation into the cell wall was achieved only after the bacteria were treated with an EDTA-containing buffer for about 1 h at 37°C. As is common in Gram-negative bacteria, *E. coli* cells have a tough outer membrane that has to be made permeable by the Ethylenediamine tetra acetate (EDTA) treatment for effective incorporation of the precursors into the cell wall. We then tested the other analogs of the cell-wall precursors, lipid I and lipid II derivatives **4**, and **5**: these compounds were incorporated into the bacterial cellwall of Gram-positive and negative bacteria.

We were able to display the hydrazine-attached fluorescent dye on the surface of engineered lactic acid bacteria via ketone-hydrazine coupling. Briefly, lactic acid bacteria were cultured in the presence of the desired concentration of **2** for 15 h and then collected. The hydrazine-attached fluorescent dye was then added to the bacterial suspension. The fluorescent images of lactic acid bacteria are shown in Figure 17.4. As a control experiment, the fluorescent dye was added to natural bac-

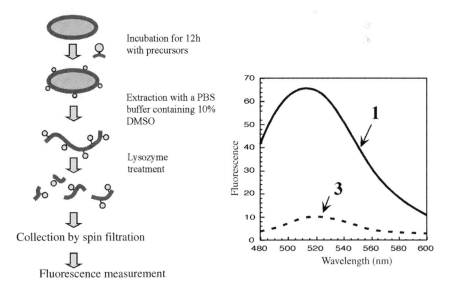

FIGURE 17.3 Experimental procedure for incorporation of fluorescent precursors into lactic acid bacteria in Lactobacilli MRS broth and collection of the cell-wall fraction.

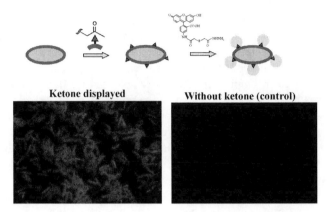

FIGURE 17.4   Fluorescent images of lactic acid bacteria. Left panel: Lactic acid bacteria after incubation with 2 and coupling with the fluorescent dye. Right panel: Lactic acid bacteria after treatment with the dye (negative control).

teria. The ketone-displaying bacteria showed a higher fluorescence intensity than control samples, indicating that the ketone group is useful for further modifications of the bacterial surface.

B.   ADHESION OF SURFACE-ENGINEERED BACTERIA TO THE CON A-IMMOBILIZED FILM BY DISPLAYING CARBOHYDRATES

The further application of this method allows the display of a sugar moiety, rather than a fluorescent dye, on the bacterial surface. We chose a mannose derivative **9** as the target oligosaccharide. The aminooxyl group located on the opposite end of the linker compound could react selectively with the ketone group on the bacterial surface. Changes in bacterial adhesion could be observed via the strong, specific interaction between this oligosaccharide and concanavalin A (Con A). The ketone-displaying bacteria were prepared by incubating lactic acid bacteria (*Lactobacillus plantarum* JCM1149) in a Lactobacilli MRS broth (Difco-Laboratories, MI) containing precursor **2** (0.5 mM) under anaerobic conditions for 15 h at 37°C. The addition of the artificial precursor into the broth was found, on the basis of colony counting, to have no effect on bacterial growth.

A surface plasmon resonance (SPR) spectrometer (BIAcore 2000, Biacore, Sweden) with an HPA gold sensor chip, covered by a 1-octadecanethiol self-assembled monolayer designed to facilitate liposome-mediated hydrophobic adsorption, was used in the adhesion experiments.[13] Modification of bacterial adhesion by cell-wall engineering was tested using a Con A (mannose-binding protein)-immobilized surface through glycolipid monolayers (Figure 17.5). As Con A has four binding sites, at least two are still available after binding to the sensor chip. At time zero, the mannose-displaying bacteria in 15 mM PBS buffer at 25°C were injected onto the sensor surface, and binding (shown by the black line) was recorded. Results show an initial small sigmoidal response followed by a large response with good reproducibility. The injection of natural bacteria, however, gave no response. These results indicate that mannose-displaying bacteria readily adhered to the Con A film.

Binding was confirmed directly by microscopic observation (Figure 17.6) in which a lectin plate was prepared on a slide glass with slight modifications as in the SPR experiment. The mannose-displaying bacteria was found to bind more readily to the Con A-immobilized substrate than did the natural bacteria, indicating that manipulation of bacterial adhesion could be achieved through surface display.

C. Modification of Cell-Wall Precursors

We speculated that the use of more penetrative precursors would allow efficient incorporation. Therefore, we synthesized an artificial cell-wall precursor with a shorter peptide moiety, UDP-MurNAc tripeptide derivative **6**. Due to its smaller molecular weight, we expected this precursor to be accessible to the cytoplasm

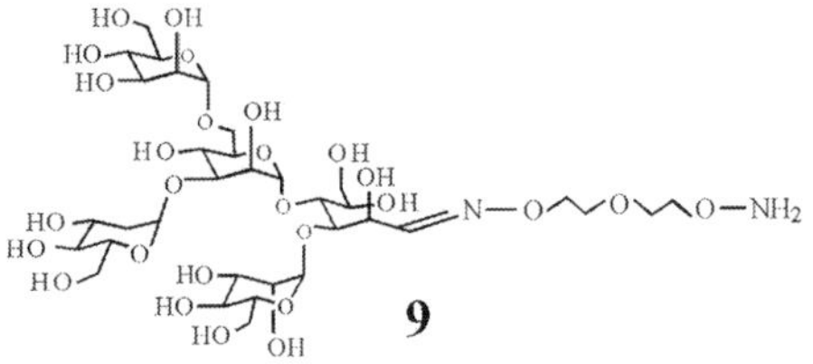

9

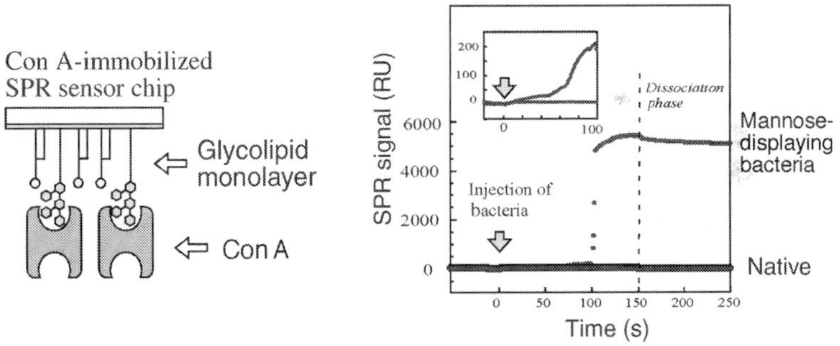

FIGURE 17.5 SPR detection of the adhesion of oligomannose-displaying bacteria and native bacteria onto a ConA-immobilized film.

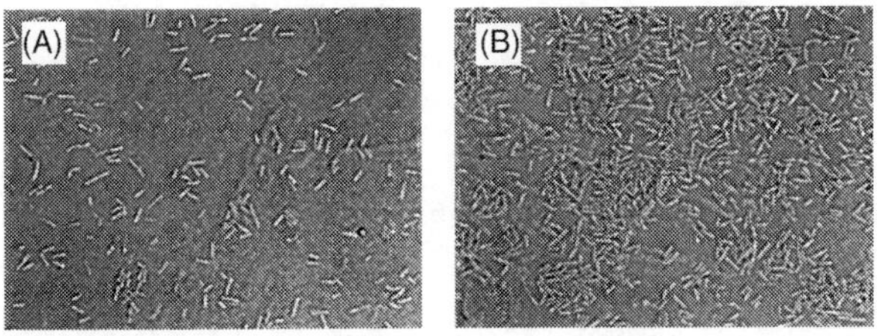

FIGURE 17.6 Binding of (a) native bacteria and (b) mannose-displaying bacteria onto a ConA-immobilized substrate.

through the cell wall and inner membrane. Lactic acid bacteria (*L. plantarum* JCM1149) were incubated for 15 h in a Lactobacilli MRS broth in the presence of the synthesized cell-wall precursor **6** (0.5 m*M*). As a control, the bacteria were also incubated with **7**, which, as shown by prior studies, could not be incorporated into the cell wall. After the fluorophore labeling of the ketone group, the cells were collected and the cell-wall component isolated prior to the measurement of the fluorescence intensity of the degraded cell-wall components. Figure 17.7 shows the incorporation of the pentapeptide-type precursor **2** and the tripeptide-type precursor **6**. The fluorescence intensity of the cell-wall fraction of the tripeptide-type precursor was not significantly stronger than the negative control, indicating that truncation of the peptide moiety did not improve incorporation. UDP-MurNAc tripeptide has been reported to be a poor substrate of translocase,[14] which transfers naturally occurring precursors to the outside of the inner membrane of bacteria. Therefore, precursor **6** needs to attach to a D-Ala- D-Ala dipeptide for incorporation into the cell wall. This reaction might not occur in the bacteria. It is considered that the structure of the precursor was merely not suitable for surface display.

In order to confirm that the cell-wall precursors are taken up inside the bacteria, we synthesized 4-fluorinated UDP-MurNAc-pentapeptide 8, which was unable to form glycosyl bonds with GlcNAc via the biosynthetic route.[15] If compound 8 is incorporated inside the cell as a natural occurring incorporation of UDP-MurNAc pentapeptide, the bacteria cannot survive due to inability to produce the mature cell wall. Fluorine-substituted analogs of biologically active organic compounds have been widely studied. Fluorinated carbohydrate derivatives have widespread medical applications as substrate mimics for the inhibition of enzymatic processes because of the inability of enzymes to differentiate between the fluorinated and original compounds. Recently, Walker and co-workers[16] reported that fluorinated Lipid I analogs can be a potent inhibitor of MurG, which catalyzes the transfer of *N*-acetyl glucosamine from

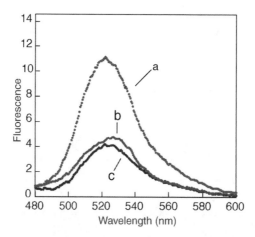

FIGURE 17.7   Fluorescence intensity of the cell-wall fraction of lactic acid bacteria. Lactic acid bacteria were incubated with the synthesized precursors and labeled with hydrazine-attached fluorescent dye. (a) Precursor 2, (b) Precursor 6, and (c) Precursor 7 (negative control).

UDP to the C4 hydroxyl residue of lipid I. Lactic acid bacteria (*L. plantarum* JCM1149) were incubated at 37°C in lactobacilli MRS broth (Difco-Laboratories) in the presence of 8 at concentrations of 0.01 and 0.1 mg/mL. Compound 8 was dissolved in buffered solution and sterilized before it was added to the bacteria. After 3, 5, 7 and 10 h, the number of bacteria in the culture was counted by the conventional plate counting method. Figure 17.8 shows the time curve of growth of the bacteria in the presence of 8. Compound 8 is not sufficiently active as an antibiotic to kill the bacteria completely at this concentration. However, bacterial growth was significantly inhibited as compared with the negative control that was cultured without adding any antibiotics. This result can be interpreted to mean that UDP-MurNAc mimetic was incorporated into the biosynthetic route of the bacterial cell wall.

D. ENHANCEMENT OF INCORPORATION EFFICIENCY BY ANTIBIOTICS

In order to enhance the incorporation of the artificial precursor, we next limited the availability of native cell-wall precursors. We speculated that the incubation of lactic acid bacteria in the presence of antibiotics known to inhibit the biosynthesis of cell-wall precursors would result in a significant reduction in the availability of the native precursors. Under these conditions, bacteria must incorporate the artificial precursor added to the medium in order to survive. Fosfomycin, a well-known inhibitor of the transformation of UDP-GlcNAc to UDP-MurNAc,[17] was used as the antibiotic. Lactic acid bacteria (*L. plantarum* JCM1149, *L. salivarious* JCM1044, and *L. fermentum* #20) were incubated for 8 h in a Lactobacilli MRS broth containing the ketone-attached UDP-MurNAc pentapeptide (0.2 mM) and fosfomycin (1.8 mg/mL), at concentrations at which bacterial growth was not totally inhibited.

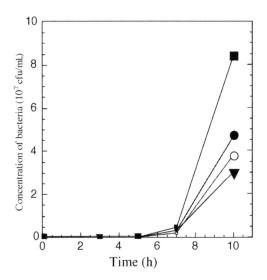

FIGURE 17.8 Changes in bacterial proliferation in the presence of compound 8, fosfomycin, and compound 8 with fosfomycin: ■, control; ●, with compound 8; ○, with fosfomycin; ▲, with compound 8 and fosfomycin.

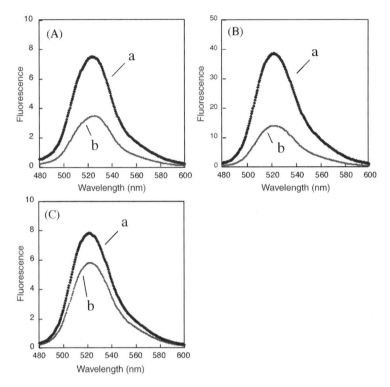

FIGURE 17.9 Fluorescence intensity of the cell-wall fraction from three strains of bacteria incubated with and without fosfomycin. (a) Fluorescence intensity after incubation with fosfomycin (b) Fluorescence intensity after incubation without fosfomycin. Panel (A) *L. plantarum* JCM1149 (B) *L. salivarious* JCM1044 and (C) *L. fermentum* #20

Figure 17.9 shows the fluorescent intensity of the cell-wall fractions of the three strains of bacteria. The fluorescence intensity of the bacteria incubated with the antibiotic was greater than that from the bacteria incubated without the antibiotic for each strain, indicating that incubation with fosfomycin enhanced the incorporation of the precursor into the cell wall.

II. CONCLUSIONS

By utilizing the modified cell-wall precursors as carriers, we successfully displayed the target oligosaccharide sugars on the surface of living bacteria. Further, a fosfomycin-mediated increase in the incorporation of the modified cell-wall precursor was also observed. Our results show that the display of oligosaccharides on the bacterial surface allows the manipulation of bacterial adhesion to the target substrate through protein–carbohydrate interactions. This method has widespread applications to a variety of ligand–receptor interactions. The ability to control bacterial surface adhesion also offers potential benefits in the development of novel bacterial drugs, particularly oral vaccines. The ability of the engineered bacteria to interact effectively with the mucosal wall of the intestine also strongly points to possible uses in antitumor and antiallergy pharmaceuticals that utilize the mucosal immune system.

REFERENCES

1. Smith, G.P., Filamentous fusion phage: novel expression vectors that display cloned antigens on the virion surface, *Science*, 228, 1315–1317, 1985.
2. Lee, S.Y., Choi, J.H., and Xu, Z., Microbial cell-surface display, *Trends Biotechnol.*, 21, 45–52, 2003.
3. Samuelson, P., Gunneriusson, E., Nygren, P.A., and Stahl, S., Display of proteins on bacteria, *J. Biotechnol.*, 96, 129–154, 2002.
4. Kondo, A. and Ueda, M., Yeast cell-surface display-applications of molecular display, *App.l Microbiol. Biotechnol.*, 64, 28–40, 2004.
5. Wernerus, H., Lehtio, J., Samuelson, P., and Stahl, S., Engineering of Staphylococcal surfaces for biotechnological applications, *J. Biotechnol.*, 96, 67–78, 2002.
6. Yarema, K.J., Mahal, L.K., Bruehl, R.E., Rodriguez, E.C., and Bertozzi, C.R., Metabolic delivery of ketone groups to sialic acid residues. Application to cell surface glycoform engineering, *J. Biol. Chem.*, 273, 31168–31179, 1998.
7. Saxon, E. and Bertozzi, C.R., Cell surface engineering by a modified Staudinger reaction, *Science* 287, 2007–2010, 2000.
8. Mahal, L.K., Yarema, K.J., and Bertozzi, C.R., Engineering chemical reactivity on cell surfaces through oligosaccharide biosynthesis, *Science*, 276, 1125–1128, 1997.
9. Sadamoto, R., Niikura, K., Ueda, T., Monde, K., Fukuhara, N., and Nishimura, S.-I., Control of bacteria adhesion by cell-wall engineering, *J. Am. Chem. Soc.*, 126, 3755–3761, 2004.
10. Sadamoto, R., Niikura, K., Monde, K., and Nishimura, S.-I., Cell wall engineering of living bacteria through biosynthesis, *Methods Enzymol.*, 362, 273–286, 2003.
11. Sadamoto, R., Niikura, K., Sears, P.S., Liu, H., Wong, C.-H., Suksomcheep, A., Tomita, F., Monde, K., and Nishimura, S.-I., Cell-wall engineering of living bacteria, *J. Am. Chem. Soc.*, 124, 9018–9019, 2002.
12. Isolauri, E., Joensuu, J., Suomalainen, H., Luomala, M., and Vesikari, T., Improved immunogenicity of oral D x RRV reassortant rotavirus vaccine by *Lactobacillus casei* GG, *Vaccine*, 13, 310–312, 1995.
13. Mann, D.A., Kanai, M., Maly, D.J., and Kiessling, L.L., Probing low affinity and multivalent interactions with surface plasmon resonance: ligands for concanavalin A, *J. Am. Chem. Soc.*, 120, 10575–10582, 1998.
14. Rogers, H.J., Perkins, H.R., and Ward, J.B., *Microbial Cell Walls and Membranes*, Chapman Hall, London, 1980, pp. 253–254.
15. Ueda, T., Feng, F., Sadamoto, R., Niikura, K., Monde, K., and Nishimura, S.-I., Synthesis of 4-fluorinated UDP-MurNAc pentapeptide as an inhibitor of bacterial growth, *Org. Lett.*, 6, 1753–1756, 2004.
16. Chen, L., Men, H., Ha, S., Ye, X.Y., Brunner, L., Hu, Y., and Walker, S., Intrinsic lipid preferences and kinetic mechanism of *Escherichia Coli* MurG, *Biochemistry*, 41, 6824–6833, 2002.
17. Ritter, T.K. and Wong, C.-H., Carbohydrate-based antibiotics: a new approach to tackling the problem of resistance, *Angew. Chem. Int. Ed.*, 40, 3508–3533, 2002.

18 Engineering Bacterial Biopolymers for the Biosorption of Heavy Metals

Horacio Bach and David L. Gutnick

CONTENTS

I. INTRODUCTION

One of the major impacts of microbial metabolic processes on environmental biotechnology has been the manipulation and exploitation of pathways for biodegradation and the consequent bioremediation of organic pollutants in the environment.[1-4] Both aerobic and anaerobic processes have been shown to lead to the efficient combustion and biomineralization of the pollutant to a nontoxic form. In contrast, heavy metal or radionuclide contamination, which is one of the major sources of pollution both in terrestrial and aquatic environments, presents quite a different challenge. In the case of metals and radionuclides, the metal ion is not completely eliminated, but may be converted to the base metal,[5] methylated,[5,6] precipitated,[7–14] volatilized,[5,6,14] or complexed with an organic ligand. The development of technologies involving many of the processes listed above has been the subject of a host of basic and more applied projects.[8,15–24] Bioremediation technologies in general should be relatively inexpensive and simple because of the low added value associated with their commercial application.[25] One such approach for heavy metal remediation involves the formation of stable complexes between heavy metals and radionuclides with microbial biomass.[26–29] These complexes are generally the result of electrostatic interactions between the metal ligands and negatively charged cellular biopolymers. For this application, microbial biomass may simply constitute dried cellular material, which can be used in a variety of physical configurations (i.e., columns, biofilters, packed resins, slurries, etc.). Advantages of this technology include the ready availability of inexpensive biomass and the development of technology for its immobilization. In contrast, the efficient exploitation of electrostatic interactions for binding heavy metals depends to a large extent on the specificity of interaction, a rather unpredictable and far from reproducible feature of ill-defined biomass. Moreover, the cation-binding capacity of a biomass preparation may depend on the nature and relative abundance of specific biopolymers. The characteristics of such materials are likely to vary over a wide range, depending on the physiology and history of the biomass preparation. Nevertheless, the ease of application and the demonstrated efficacy in certain field trials[30] have led to a large number of reports favoring this approach. Limited space does not permit a thorough treatment of this system within the context of this review. The reader is referred to a number of excellent recent publications dealing with the subject of cation binding to microbial biomass and its application in bioremediation.[27,31–37]

Another approach to metal ion binding involves exploiting the high affinity of cations for specific protein molecules, such as metallothionein. The gene encoding metallothionein has been cloned into several microorganisms with a view toward engineering organisms that exhibit an enhanced binding of specific cations.[38,39] The approach has been used to generate overexpressed fusion proteins in which the metal-binding protein is fused to an extracellular domain of an outer membrane protein to enhance the binding capacity of the microbial biomass.[39,40]

In recent years, great advances have been made in characterizing the biosynthetic genes associated with polysaccharide production in both eukaryotic and prokaryotic systems. Moreover, new developments have led to a deeper understanding of the regulatory processes controlling polysaccharide synthesis and gene expression.[4–51] In addition, since biosynthetic regulons encoding polysaccharide biosynthetic functions frequently contain orthologs of the various genes for monomer activation, precursor synthesis, polymerization, and biopolymer export, all clustered within a contiguous region on the chromosome, genomic analysis enables the rapid identification of putative polysaccharide-producing organisms from a wide variety of environmental samples.[52,53] Finally, the vast array of genetic, physiological, and biochemical techniques for modifying microbial polysaccharides suggests new possibilities for generating new or modified biopolymers with improved characteristics. These modifications might affect stability, specificity, compatibility with other materials used in the application, and so on.

This chapter will focus on the following features of metal biosorption to bacterial biopolymers:

1. Cation binding to specific bacterial biopolymers; presenting data related to range of cations bound, specificity, and extent of binding.
2. Binding of cations by amphipathic biopolymers either free in solution or oriented at an oilwater interface in a stable emulsion. This feature allows for concentration and recovery of cations such as Cd^{2+} or UO_2^{2+} from a relatively large volume of aqueous phase into an emulsified cream, by at least two orders of magnitude.[54]
3. Molecular approaches to biopolymer modification and formulation enhancing the range of preparations, which can be used to bind metal ligands at the oil water interface.
4. Potential applications for metal remediation and recovery.

II. CATION BINDING

A. CATION BINDING TO BACTERIAL BIOPOLYMERS

A wide variety of microorganisms have been shown to produce various polysaccharides and other biopolymers, which exhibit metal-binding properties.[55-58] A representative sample of such biomolecules is shown in Table 18.1, from which it can be seen that biopolymers are produced in both Gram-negative and positive microorganisms. Prominent among the various polysaccharides and other organic biopolymers are peptidoglycan, water-soluble and amphipathic exopolysaccharides (EPS), capsular polysaccharides, capsular polyglutamic acid, teichoic and teichuronic acids, and lipopolysaccharides (LPS). For electrostatic interactions, the binding of cations to bacterial biopolymers generally occurs through interaction with negatively charged functional groups such as (1) uronic acids (EPS from *Bradyrhizobium japonicum*, alginate, teichuronic acid, emulsan, or LPS from various sources), (2) phosphoryl groups associated with membrane components, or (3) carboxylic groups of amino acids. In addition to electrostatic interactions, there may also be cation-binding by positively charged polymers[66] or coordination with hydroxyl groups.[54,67] Cation binding has been observed for eukaryotic polymers such as chitin or chitosan, presumably by chelation and coordination with hydroxyl groups.[66] These forms of nonelectrostatic interaction may account for the greater-than-stoichiometric binding of cations at an oil/water interface, discussed below.[54]

Despite the relatively few functional groups potentially involved in cation binding, microbial polymers differ widely both in specificity and in their metal-binding capacity. Table 18.2 presents some quantitative data on the binding of different cations (grouped according to their position in the periodic table) to several bacterial biopolymer preparations. It should be noted that the various biopolymeric preparations presented in Table 18.2 may differ in their degree of purity. For example, in one case, cation binding to a purified preparation of peptidoglycan from *Escherichia coli* was investigated,[68] while in a second case, the peptidoglycan was from a less purified preparation containing a mixture of peptidoglycan, protein, and LPS from

TABLE 18.1

Representative Examples of Microbial Polysaccharides

Polymer	Strain	Subunit Composition	Ref.
Apoemulsan	*Acinetobacter venetianus* RAG-1	D-galactosamine, L-galactosamine uronic acid, bacillosamine (2,4) diamino, 6-deoxy glucose	47
EPS[a]	*Arthrobacter viscosus*	D-glucose, D-galactose, D-mannuronic acid	59
	Enterobacter sp. 11870	Glucose, fucose, glucuronic acid	60
	Klebsiella aerogenes type 54 strain A3(sl)	Glucose, fucose, glucuronic acid, acetate	60
	Rhizobium meliloti YE-2	Glucose, galactose, acetate, pyruvate, succinate	60
Gellan	*Sphingomonas elodea*	L-rhamnose-(α-1-3)-D-glucose-(β-1-4)- D-glucuronic acid-(β-1-4)-D-glucose-(β-1 with O(2) L-glyceryl and O(6) acetyl substituents	61
γ-Glutamyl capsular polymer	*Bacillus liqueniformis*	γ-Glutamic acid	62
Peptidoglycan	*Escherichia coli* K-12	β(1,4)-*N*-acetyl glucosamine-*N*-acetylmuramic acid crosslinked (L)-Ala-D-glucose (L)-meso-diamino pimelic acid-D-alanine	63
Pullulan	*Aureobasidium pullulans*	(1-6)-{α–D-glucose-(1-4)-α–D-glucose-(1-4)-α–D-glucose}-(1-6)-{α–D-glucose-(1-4)-α-D-glucose-(1-4)-α–D-glucose -(1-4)-α–D-glucose}	64
Xanthan	*Xanthomonas campestris*	Glucose, mannose, glucuronate, acetate, pyruvate	65
Zooglan	*Zooglea ramigera*	Glucose, galactose, acetate, pyruvate, succinate	60

[a] Exopolysaccharide.

the same organism.[63] Surprisingly, the two preparations differed both in their cation-binding specificity and in their cation-binding capacity (discussed below). In addition, in some cases, binding to specific groups within the polymer was determined directly on isolated material, while in other instances the binding to specific functional groups was inferred from the results of measurements of binding before and after specific blocking or extraction of such groups from cell walls.[67] For example, a wall fraction from *Bacillus subtilis* was found to bind magnesium (Mg) at about 8 mmol/g of cell wall. After treatment with ethylene diamine to block carboxyl groups, Mg binding was reduced to only 160 μmol/g of cell wall. In sharp contrast, prior extraction with NaOH to remove teichoic acids resulted in a doubling of Mg binding to the walls, suggesting that not all of the available binding sites were exposed on native walls, but were normally shielded by teichoic acid residues.

Unlike the case with Mg binding, the blocking of carboxylic acid groups had little effect on calcium (Ca) binding in the same system. However, Ca binding to

TABLE 18.2
Binding of Cations by Microbial Biopolymers

Metal	Strain	Biopolymer	Bound Metal		Ref.
			mg/g polymer	μmol/g polymer	
		Group II Alkaline Earth Metals			
Mg^{2+}	*Bacillus liqueniformis*	γ-Glutamyl capsular polymer[g]	1.8	73	62
	Escherichia coli K-12	Peptidoglycan[a]	0.9	35	68
		Peptidoglycan,[a] LPS,[b] proteins	6	256	63
	Enterobacter sp. 11870	EPS[c]	8	329	60
	Klebsiella aerogenes type 54 strain A3 (sl)	EPS[d]	7.4	304	60
	Marine pseudomonad B-16	Peptidoglycan[a]	5.6	230	69
	Rhizobium meliloti YE-2	EPS[e]	5.4	222	60
	Z. ramigera	Zooglan[f]	5.5	226	60
Ca^{2+}	*B. liqueniformis*	γ-Glutamyl capsular polymer[g]	41.8	1044	62
	B. subtilis 168	Peptidoglycan[a]	30	750	70
	E. coli K-12	Peptidoglycan[a]	1.5	38	68
		Peptidoglycan,[a] LPS,[b] proteins	1.4	35	63
	Enterobacter sp. 11870	EPS[c]	11.8	295	60
	K. aerogenes type 54 strain A3 (sl)	EPS[d]	9.1	228	60
	R. meliloti YE-2	EPS[e]	7.2	180	60
	Zooglea ramigera	Zooglan[f]	4.5	113	60
Sr^{2+}	*E. coli* K-12	Peptidoglycan[a]	2.2	25	68
		Peptidoglycan,[a] LPS,[b] proteins	0.1	1	63
Ba^{2+}	*E. coli* K-12	Peptidoglycan[a]	9.7	71	68
		Transition Elements I and II			
Sc^{3+}	*E. coli* K-12	Peptidoglycan[a]	110.8	2464	68
		Peptidoglycan,[a] LPS,[b] proteins	4.3	96	63
La^{3+}	*E. coli* K-12	Peptidoglycan[a]	299.5	2156	68
		Peptidoglycan,[a] LPS,[b] proteins	10.8	78	63
	P. aeruginosa type A^+B^-	LPS[b]	31.8	229	71
Ce^{3+}	*E. coli* K-12	Peptidoglycan[a]	308	2198	68
		Peptidoglycan,[a] LPS,[b] proteins	14	100	63
Pr^{3+}	*E. coli* K-12	Peptidoglycan,[a] LPS,[b] proteins	8.2	58	63
Sm^{3+}	*E. coli* K-12	Peptidoglycan,[a] LPS,[b] proteins	1.7	11	63
UO^{2+}	*Acinetobacter venetianus* RAG-1	Apoemulsan[h]	243	1021	54
		Apoemulsan[h] in emulsanosol	958	4025	54
	E. coli K-12	Peptidoglycan[a]	2.4	10	68
		Peptidoglycan,[a] LPS[b], proteins	15.7	66	63
	Pseudomonas sp. EPS-5028	EPS[i]	96	403	72
		EPS[j]	46	193	72
	Thiobacillus ferrooxidans TF1-35	LPS[b]	0.2	1	73
	Z. ramigera	Zooglan[f]	370	1554	74

(Continued)

TABLE 18.2 (Continued)

Metal	Strain	Biopolymer	Bound Metal		Ref.
			mg/g polymer	μmol/g polymer	
ZrO^{2+}	E. coli K-12	Peptidoglycan[a]	134	1469	68
		Peptidoglycan,[a] LPS,[b] proteins	19.3	212	63
HfO^{2+}	E. coli K-12	Peptidoglycan[a]	1416	7932	68
		Peptidoglycan,[a] LPS,[b] proteins	167.8	940	63
Cr^{3+}	B. liqueniformis	γ-Glutamyl capsular polymer[g]	48.9	940	62
	E. coli K12 LE392	Cell surface	16	307	75
	Micrococcus luteus ATCC 381	Cell surface	15.5	297	75
	P. aeruginosa ATCC 14886	Cell surface	16.6	320	75
Cr^{6+}	Bacillus sp.	Cells	2.84	54.6	76
MoO^{2+}	E. coli K-12	Peptidoglycan[a]	532	5545	68
		Peptidoglycan,[a] LPS,[b] proteins	21.6	225	63
Mn^{2+}	B. liqueniformis	γ-Glutamyl capsular polymer[g]	3.9	71	62
	B. subtilis 168	Peptidoglycan[c]	40.7	741	70
	E. coli K-12	Peptidoglycan[a]	2.9	52	68
		Peptidoglycan,[a] LPS,[b] proteins	7.7	140	63
	Gloeothece magna	EPS[i]	115–425	2061–7616	77

Transition elements III

Metal	Strain	Biopolymer	mg/g	μmol/g	Ref.
Fe^{3+}	B. liqueniformis	γ-Glutamyl capsular polymer[g]	74.8	1340	62
	Bradyrhizobium japonicum USDA 110	EPS[i]	59.2	1060	78
	Bradyrhizobium (chamae-cytisus) strain BGA-1	EPS[i]	24	430	78
	E. coli K-12	Peptidoglycan[a]	5.6	100	68
		Peptidoglycan,[a] LPS,[b] proteins	11.2	200	63
	P. aeruginosa type A^-B^+	LPS[b]	90	1611	71
Fe^{2+}	E. coli K-12	Peptidoglycan,[a] LPS,[b] proteins	3.2	57	63
		Alginic acid beads	7.1	120	79
		Gellan gum	41.8	710	80
Co^{2+}	B. liqueniformis	γ-Glutamyl capsular polymer[g]	5.9	100	62
	E.coli K-12	Peptidoglycan[a]	2.5	42	68
		Peptidoglycan,[a] LPS,[b] proteins	10.5	178	63
	E.coli K-12 LE392	Cell surface	7.8	131	75
	Micrococcus luteus ATCC 381	Cell surface	9.9	168	75
	P. aeruginosa ATCC 14886	Cell surface	4	68	75
Ni^{2+}		Agar	2.3	40	80
		Alginic acid beads	4.1	70	79

(Continued)

TABLE 18.2 (Continued)

Metal	Strain	Biopolymer	Bound Metal		Ref.
			mg/g polymer	μmol/g polymer	
		Ca alginate	7.6	130	80
		Carrageenan	1.7	30	80
		Gellan gum	149	720	80
	Activated sludge		4.7	80.5	81
		EPS[i]	4	69.6	82
	B. liqueniformis	γ-Glutamyl capsular polymer[g]	4.7	80	62
	B. subtilis 168	Peptidoglycan[a]	37.5	639	70
	E. coli K-12	Peptidoglycan[a]	1.1	19	68
		Peptidoglycan,[a] LPS,[b] proteins	0.1	2	63
	E.coli K-12 LE392	Cell surface	4.9	83	75
	Micrococcus luteus ATCC 381	Cell surface	5.6	96	75
	Nocardia amarae	Biomass	6.4	109	83
	P. aeruginosa ATCC 14886	Cell surface	6.4	109	75
Ru^{3+}	*E. coli* K-12	Peptidoglycan,[a] LPS,[b] proteins	9.1	90	63
OsO_4	*E. coli* K-12	Peptidoglycan[a]	3.8	20	68
		Peptidoglycan,[a] LPS,[b] proteins	198	1040	63
Pt^{4+}	*E. coli* K-12	Peptidoglycan,[a] LPS,[b] proteins	0.4	2	63
		Transition elements IV			
Cu^{2+}		Alginic acid beads	17	270	79
		Gellan gum	155	750	80
	Activated sludge	EPS[i]	127–255	2020–4050	84
		EPS[i]	14	66.6	82
	Bacillus sp.	Cells	3.5	16.6	76
	B. liqueniformis	γ-Glutamyl capsular polymer[g]	56.5	890	62
	Brevibacterium sp. HZM-1	Cell mass	32.4	510	85
	Brevibacterium sp. PBZ	Biomass	20.2	318	86
	E. coli K-12	Peptidoglycan,[a] LPS,[b] proteins	5.7	90	63
	E. coli K-12 LE 392	Cell surface	7.6–12	119–188	75
	Fresh-water sediments	EPS[i]	16	253	65
	K. aerogenes	EPS[i]	13.2	207	87
	Micrococcus luteus ATCC 381	Cell surface	12.6	198	75
	Nocardia amarae	Biomass	56.8	902	83
	P. aeruginosa ATCC 14886	Cell surface	5–10.5	80–166	75
	P. aeruginosa type A^-B^+	LPS[b]	14	220	71
	Xanthomonas campestris	Xanthan[k]	7.81	123	65
	Z. ramigera	Zooglan[f]	323	5083	74

(*Continued*)

TABLE 18.2   (Continued)

Metal	Strain	Biopolymer	Bound Metal		Ref.
			mg/g polymer	μmol/g polymer	
Au[3+]	E.coli K-12	Peptidoglycan,[a] LPS,[b] proteins	11	56	63
	P. aeruginosa type A⁻B⁺	LPS[b]	108	548	71
		Gellan gum	46.4	710	80
Zn[2+]	Alteromonas macleodii subs. fijiensis	EPS[i]	75	1150	88
	B. liqueniformis	γ-Glutamyl capsular polymer[g]	9.7	149	62
	Brevibacterium sp. HZM-1	Cell mass	41.6	640	85
	Citrobacter sp. MCM B-181	Biomass	22.7	349	89
	E. coli K-12	Peptidoglycan,[a] LPS,[b] proteins	25.5	390	63
	S. rimosus	Biomass	2.9	44.5	81
Cd[2+]	A. venetianus RAG-1	Alginic acid beads	12.4	110	79
		Gellan gum	69.2	620	80
		Apoemulsan[h]	141	1250	90
		Apoemulsan in apoemulsanosol[h]	282	2250	90
	Actinomycete R27	Biomass	1120	8960	91
	Activated sludge	EPS[i]	11.3–28	90–220	84
	A. macleodii subs. fijiensis	EPS[i]	125	1120	88
	Arthrobacter viscosus	EPS[l]	0.9	8	59
	Brevibacterium sp. PBZ	Biomass	12.5	112	86
	Citrobacter sp. MCM B-181	Biomass	25.4	227	89
	Fomitopsis pinicola CCBAS 535	Biomass	130	1040	92
	Gloeothece magna	EPS[i]	473–906	4216–8075	77
	K. aerogenes	EPS[i]	11	98	87
	Nocardia amarae	Biomass	34.7	310	83
	Z. ramigera 115	Zooglan[f]	1.9	19	93
	P. aeruginosa PU21	Biomass	58	464	94
	P. fluorescens 4F39	Biomass	28	224	95
Hg[2+]	E. coli K-12	Peptidoglycan,[a] LPS,[b] proteins	12.8	64	63
In[3+]	E. coli K-12	Peptidoglycan,[a] LPS,[b] proteins	0.114	1	63
Pb[2+]	Sphingomonas elodea	Alginic acid beads	294	1420	79
		Gellan gum	176	850	80
	Activated sludge		74	357	96
		EPS[i]	8	38.6	82
		EPS[i]	288–646	1390–3120	84
	A. macleodii subs. fijiensis	EPS	316	1520	88
	Aureobasidium pullulans KFCC110245	Pullulan	300	1449	96
	Bacillus sp.	Cells	1.75	8.45	76
	Blast furnace sludge		64	309	88
	Brevibacterium sp. PBZ	Biomass	34.6	167	86

TABLE 18.2 (Continued)

Metal	Strain	Biopolymer	Bound Metal (mg/g polymer)	Bound Metal (μmol/g polymer)	Ref.
	Citrobacter sp. MCM B-181	Biomass	67	324	89
	E.coli K-12	Peptidoglycan[a]	10.3	49.7	68
		Peptidoglycan,[a] LPS,[b] proteins	31.5	152	63

[a] $\beta(1,4)$-Linked *N*-acetyl glucosamine-*N*-acetylmuramic acid crosslinked (L)-Ala-D-glucose (L)-meso-diamino pimelic acid-D-alanine residues.

[b] Lipopolysaccharide.

[c] Glucose, fucose, and glucuronic acid; in the ratio 1.5:1:0.8.

[d] Glucose, fucose, glucuronic acid, and acetate; in the ratio 2.1:1:0.83:0.43.

[e] Glucose, galactose, acetate, pyruvate, and succinate; in the ratio 7:1:0.63:0.76:1.

[f] Glucose, galactose, acetate, pyruvate, and succinate; in the ratio 2:1:0.64:0.44.

[g] γ-Glutamic acid.

[h] (D-Galactosamine, D-galactosamine uronic acid, bacillosamine (2,4) diamino, 6-deoxy glucose) apo-emulsanosol concentrated emulsion (70% [w/v] hexadecane in water); see text.

[i] Exopolysaccharide.

[j] Deacylated exopolysaccharide.

[k] Glucose, mannose, glucuronate, acetate, and pyruvate.

[l] D-Glucose, D-galactose and D-mannuronic acid.

B. subtilis walls lacking teichoic acid was reduced by a factor of 10 once the teichoic acid was extracted. Similar approaches were used to examine cation binding to different walls in other organisms,[60,70,71,97–100] although binding differed both in terms of specificity and binding capacity. For example, both Ca and Mg were found to bind to native walls of *Enterobacter* sp., *Klebsiella aerogenes*, *Rhizobium meliloti*, and *Zoogloea ramigera*.[60] However, the deacylation of these walls only slightly impaired the binding of either Ca or Mg when compared with results with *B. subtilis*. Although comparative studies of cation binding to functional groups present in wall-associated polymers yielded some information regarding specificity, there have been only a few reports dealing with binding to walls prepared from specific mutants with altered surface properties. In one such report,[100] ^{13}C- and ^{31}P-NMR nuclear magnetic resonance spectroscopy was used to characterize metal binding to LPS from a heptoseless mutant of *E. coli* K12. Low concentrations of Ca^{2+}, Cd^{2+}, Gd^{3+}, La^{3+}, and Yb^{3+}, all affected the ^{31}P-NMR spectrum at low concentrations. The authors concluded that the LPS from this mutant contained a high-affinity metal-binding site, which involves the participation of a glycosidic diphosphate moiety. Langley and Beveridge[71] exploited the fact that *Pseudomonas aeruginosa* PAO1 normally produces two chemically distinct types of LPS, the A- and B-band LPS. A series of isogenic strains were used including A^+B^-, A^-B^+, and A^-B^- mutants. All strains bound small amounts of copper (Cu) onto the cell surface, suggesting that the binding may be to a common surface-binding site, such as the phosphoryl groups on the core lipid A region. Mutants lacking the A-band LPS caused precipitation of iron (Fe) onto the cell surface, while mutants lacking B-band LPS gave rise to La crystals. The authors proposed that the binding of metal ions to LPS did not involve the direct involvement of O-antigen side chains, but that the B-band LPS might affect cell surface properties, which enhance the precipitation of metals in specific regions on the cell surface.

B. CATION BINDING TO EMULSAN AND OTHER AMPHIPATHIC BIOPOLYMERS

As illustrated in Table 18.2, negatively charged biopolymers bind cations with different specificities and with different overall metal-binding capacities. Although the basis for such specificity is not clear, in one class of biopolymers there is a clear case of metal-binding enhancement under a condition, which modifies the conformation of the biopolymer. The best studied of this type of amphipathic, polyanionic polysaccharide is emulsan (molecular mass, 10^6 Da), a galactosamine-containing capsular bioemulsifier produced by the hydrocarbon-degrading organism, *Acinetobacter venetianus* RAG-1.[101–104] As shown in Figure 18.1, the emulsan backbone is assembled from a trisaccharide subunit consisting of L-galactosamine, D-galactosamine uronic acid, and 2,3 diamino, 2-deoxy glucosamine.[105] The amphipathicity of emulsan (pK_a 3.05) is mediated by the presence of fatty acids (about 25%, by wt) present in both ester and amide linkages[106] and, in crude form, by the noncovalent association with several exocellular proteins.[101] Protein-free emulsan, termed apoemulsan, retains partial emulsifying activity toward more polar hydrocarbons such as mixtures of aliphatics and aromatics, or crude or machine oil.[107] Nevertheless, the apoemulsan does not emulsify very hydrophobic substances such as hexadecane or long-chain waxes. Emulsification and emulsion stabilization are due to the tight affinity of emulsan for the oil/water interface, a property that

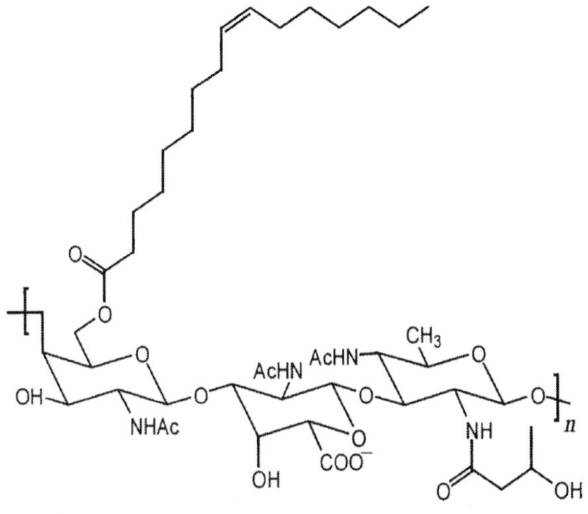

FIGURE 18.1   The trisaccharide structure of the emulsan subunit.

allows the negatively charged water-soluble polymer to partition into a cream layer either after standing or following centrifugation. Such concentrated cream layers are oil-in-water emulsions in which the oil content can be as high as 70% (by wt), yet water remains the bulk solvent. Such concentrated emulsan-stabilized creams are termed emulsanosols.[108] Emulsanosols are formed from the binding of emulsan to the oil/water interface, resulting in a polymer orientation such that the hydrophilic sugar residues (including the negative charges on the galactosamine uronic acid residues) face outward toward the aqueous solvent, while the hydrophobic groups are oriented toward the oil. The stabilization of the emulsion in the emulsanosol is thought to be due to the electrostatic charge repulsion between these uronic acid residues, thereby preventing droplet coalescence. Both chemical and biological evidence support the notion that the orientation of the polymer brings about a conformational change in the polysaccharide backbone. For example, emulsan has been shown to serve as a cell-surface receptor for a specific bacteriophage, ap3, which binds to the cells of *A. venetianus* RAG-1. However, no phage binding to emulsan is observed with the cell-free, water-soluble emulsan polymer. Phage binding is restored, however, when the polymer is on the surface of oil droplets in an oil/water emulsion. Interestingly, the data not only provide evidence for specific orientation at an oil/water interface, but also strongly suggest that this conformation resembles the conformation of emulsan on the surface of the bacterial cell.[109,110] One prediction of this hypothesis is that removal of the emulsan from the cell surface should yield cells, that are actually more hydrophobic than those of their parent organism. In fact, it has been shown that using mutants defective in the emulsan capsule are more hydrophobic than the wild-type strains.[110] Similar results were obtained when the biopolymer was physically removed from the cell surface using either mechanical shearing[111] or enzymatic breakdown.[112]

Chemical evidence for a change in polymer conformation at the oil/water interface comes from studies on cation binding to emulsan and apoemulsan preparations. Interestingly, when purified apoemulsan was mixed with positively charged organic cations such as rhodamine, almost none of the cation remained bound to the biopolymer after dialysis. However, when the same experiment was performed, but this time substituting the water-soluble apoemulsan with an apoemulsanosol of hexadecane-in-water containing the same amounts of the bioemulsifier, about 3 μmol rhodamine/mg apoemulsan was bound at the oil/water interface.[108] It should be noted that if all the negative charges were saturated with cation, one would have expected only about 1.5 μmol rhodamine to have been bound. The results suggest that the orientation of the polymer not only stabilizes the cation–biopolymer interaction, but also results in the coordination of additional cations, perhaps through the interaction with hydroxyl groups on the amino sugars in the polymer backbone. Enhanced binding of metal ions such as Cd^{2+} or UO_2^{2+} to apoemulsan was also observed at the oil/water interface. In a subsequent set of experiments, it was shown that cations bound to the emulsanosol could be completely removed to the aqueous phase when the pH was lowered to below the pK of the uronic acid residues (<3.05). Under these conditions, the emulsanosol remained stable as a concentrated oil-in-water emulsion and could be resuspended in an aqueous solution containing cations at neutral pH to generate an oil-in-water emulsion.

This suggests the possibility of using emulsanosols as recyclable, water-soluble, cation-exchange complexes as illustrated in Figure 18.2. The scheme represents a

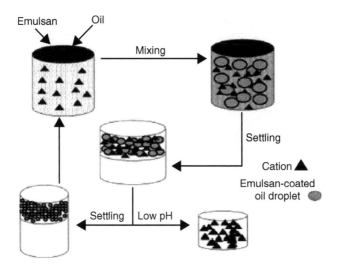

FIGURE 18.2 Scheme for metal recovery and recycling of emulsanosols. An oil/water emulsion is formed by adding a preparation of emulsan and oil to a tank or other source of water containing heavy metals. After emulsification a cream layer, the emulsanosol, is formed during the settling process. The metal-containing emulsanosol is subjected to low pH that causes the cations to dissociate into a separate container, while the emulsanosols are allowed to reform by settling. They are then returned directly to the polluted source to continue the cycle. Note that emulsan needs to be added only once to this system. Once formed, the emulsanosols are recycled through the metal-containing site for additional recovery. Metal ions are represented by black triangles and oil droplets by gray ellipses.

potential application for metal removal from large dilute aqueous volumes by concentration into a small, concentrated, water-in-oil emulsion layer. In this system, a mixture of emulsan or apoemulsan and oil (composing about $0.1 \pm 2\%$ of the total volume) can be used to generate an oil-in-water emulsion in a body of water polluted with heavy metal ions. At these concentrations of emulsan, more than 90% of the polymer adheres to the oil/water interface. Initial emulsion formation requires some input energy to form the emulsion. In our experience, continuous pumping of the emulsan-treated material can be used to generate the emulsion. Upon settling, the emulsion and adsorbed cations separate into a small concentrated emulsanosol. This cation-associated phase can be skimmed off the surface and transferred to a new container where the cations are removed by lowering the pH. The emulsanosol can then be recycled for another round of metal binding. The emulsanosols offer a number of potential advantages including stability, water compatibility, the ability to be formed from waste oils and crude sludges with concomitant viscosity reduction of the oil, and biodegradability.

C.   EMULSIFICATION ENHANCEMENT, NOVEL ESTERASE FORMULATIONS

Emulsan release from the cell surface is mediated by a cell-surface esterase enzyme,[113] which has been purified, cloned and sequenced.[105,114–118] The overexpression of the esterase in *E. coli* led to its appearance in inclusion bodies, which could be isolated. The enzyme could be refolded using chaotropic agents, which were then removed by slow dialysis to yield an active enzyme.[118] Modeling studies have shown that the enzyme most closely resembles proteins exhibiting an $\alpha-\beta$ hydrolase fold in their tertiary conformation.[116] This esterase is also one of several proteins released from the cell surface together with the emulsan biopolymer and in fact can partially deesterify emulsan. Removal of the emulsan-associated proteins using any of a number of techniques, including hot phenol, proteases, or selective precipitation of the deproteinated polymer with cationic surfactants such as cetyl-trimethyl-ammonium bromide, yields a product termed apoemulsan. Apoemulsan exhibits a lower emulsifying activity, particularly toward very hydrophobic substrates such as very waxy oils, or straight-chain aliphatics such as hexadecane. The addition of denatured protein to the apoemulsan restores the emulsifying activity toward hydrophobic substrates. Surprisingly, the addition of purified recombinant esterase to apoemulsan resulted in a formulation that showed emulsifying activity toward a host of hydrophobic substrates.[118] Moreover, esterase-containing emulsanosols could be generated by centrifugation and exhibited the same stability as the concentrated emulsan-generated oil/water emulsions. This protein-mediated enhancement of emulsifying activity is accompanied by the enhanced binding of cations such as Cd^{2+} or UO_2^{2+} at the oil/water interface. It was also found that site-directed mutant esterase molecules, which are catalytically inactive, are still capable of enhancing the emulsifying activity and cation-binding properties of esterase-containing emulsanosols.[114,118] In an attempt to map the portion of the esterase responsible for enhancing the emulsification, a series of peptides were generated by subjecting the emulsan to a series of proteolytic enzymes (Bach, unpublished observations). Both chymotrypsin and papain yielded peptides of about 10 and 14 kDa, respectively. These peptides were also capable of interacting with apoemulsan to reconstitute emulsification and to enhance cation binding at the oil/water interface. In addition, a series of fusion proteins were generated in which the maltose-binding protein (MBP) was fused at its *C*-terminus to various esterase fragments.[114] These studies demonstrated that after purification, fused fragments from the terminal third of the esterase were active in reconstituting apoemulsan-mediated emulsification toward hydrophobic substrates. This emulsification enhancement was absolutely dependent on the presence of the terminal 15 amino acids. Removal of this peptide rendered the protein inactive. Apoemulsanosols formed by emulsification-enhancing peptides (EEPs) were also capable of binding the cations as efficiently as any other formulation. Remarkably, no other esterase, lipase, or any other protein preparation were able to enhance the apoemulsan-mediated emulsification of very hydrophobic materials. Perhaps the most significant feature of the EEP activity involves the finding that in addition to enhancing the activity of apoemulsan, both the recombinant esterase and an MBP fusion protein bound to the terminal third of the esterase molecule were capable of interacting with a series of polysaccharides such as cellulose, starch, pectin, alginic acid, xanthan as well as the additional compounds listed in Table 18.3. These unique combinations were actually good emulsifiers toward a variety of pure and crude hydrophobic substrates including waxy oils and sludges, which were removed from old storage tanks.[114] We refer to the combination of polysaccharides and EEPs generated either by proteolysis of the RAG-1 esterase or by subcloning of the esterase gene as EEPosans. The resulting

TABLE 18.3
Converting Polysaccharides into Bioemulsifiers Using Recombinant Esterase[a]

Polysaccharide	Maximum Emulsifying Activity (U/mg polysaccharide/ mg esterase)
Emulsan	6752
Apoemulsan	5430
Agarose	963
Alginic acid	496
BD-4 exopolysaccharide	3396
Carrageenan	3345
Cellobiose	626
Cellulose	766
Chitin	540
Colanic acid	2050
Dextran	583
Gum arabic	1895
Pectin	1830
Polyvinyl pyrrolydone	1950
Pullulan	3400
Starch	544
Stewartan	1196
Xanthan	2720
Xylan	1854

[a] With the exception of emulsan, none of the polysaccharides listed exhibited any emulsification in the absence of the esterase.

water-in-oil emulsions are EEPosanosols. As expected, divalent cations such as Cd^{2+} or UO_2^{2+} bound to the polysaccharide in the EEPosanosols and could be concentrated in a process identical to that which occurred in the apoemulsanosols. The ability to generate EEPosanosols offers the possibility of utilizing a host of low-cost biopolymers, which in the presence of an EEP can bind cations and subsequently concentrate them in a concentrated oil-in-water emulsion as described in Figure 18.2.

D. POLYSACCHARIDE ENGINEERING AND CATION BINDING

In this section, we suggest that recombinant DNA technology can be used to modify polysaccharide-encoding biosynthetic operons with a view toward producing biopolymers with enhanced cation-binding capacity. Genes encoding enzymes for the biosynthesis of polysaccharides usually occur in clusters.[43,119–122] Generally, the biosynthetic regulons encode housekeeping enzymes for monomer synthesis, subunit synthesis (transglycosylases), polymerization, decoration with acetyl, succinyl, pyruvylyl, and phosphoryl groups, and translocation to the outer surface of the cell.[121] It was noted that some of the xanthan-defective mutants of *Xanthomonas campestris* still produced small amounts of biopolymer with a modified structure. By introducing specific combinations of mutations into the chromosome, the authors were able to generate a family of viscous polymeric xanthan derivatives containing fewer numbers of sugars in the backbone with or without decoration, i.e., deacety-lated, depyruvylated, etc. Similar approaches were used to produce mutant and recombinant derivatives of the viscous polyanionic biopolymer acetan produced by *Acetobacter xylinum*.[123] In another case,[124] substitutions were made between biosynthetic genes for the production of EPS from *Erwinia amylovora* (amylovoran) and *Pantoea stewartii* (stewartan). Mutants of *P. stewartii*, defective in stewartan synthesis, could be complemented by cosmid libraries encoding the biosynthetic genes of amylovoran in order to produce recombinant polysaccharides. Since many of the water-soluble polymers and EPS listed above are capable of interacting with recombinant esterase to generate amphipathic emulsifying formulations as described above, cation binding was examined at the oil/water interface and was found to be enhanced. It would be of interest in this regard to test the properties of recombinant polysaccharide derivatives that exhibit such enhanced cation-binding properties.

A similar approach was used to generate a new polysaccharide in place of the biopolymer acetan normally produced by *A. xylinum* strain KE5.[125] The new polymer P2 was obtained as a result of knocking out the *ace*P gene to yield a pentasaccharide repeating unit in place of the normal heptasaccharide present in acetan itself. In principle, such modifications to generate new biopolymers might yield polysaccharides with enhanced metal-binding characteristics.

The genes encoding the biosynthetic pathway for apoemulsan have recently been localized to a 27 kbp cluster termed the *wee* regulon.[47] The entire cluster was sequenced and shown to encode 23 putative open reading frames arranged in two divergent operons and separated by a nontranslated region (Figure 18.3). Mutations in any of the genes resulted in a defect in emulsan production. All of these defects could be complemented by a wild-type allele.[51] Several approaches have been used to prepare engineered derivatives of emulsan. In the first approach, mutants were prepared by modifying the fatty acid composition of the bioemulsifier, using a fatty acid auxotroph and feeding fatty acids directly into the medium.[4] Four derivatives were generated, which contained 2-OH C_{14}, palmitoleic, linoleic, or linolenic acid, respectively, in place of the 2-OH C_{12}, which is the main fatty acid of wild-type emulsan.[106,126] The derivatives were tested for emulsification, ability to remove microbes from hydrophobic surfaces,[111] and binding of UO_2^{2+} at the oil/water interface. The incorporation of linolenic acid into the polymer gave rise to a derivative that exhibited only about 10% of the emulsifying activity, compared with the wild-type polymer. The palmitoleic acid derivative was about half as active, while the linoleic acid mutant emulsan was about a third as active. As regards UO_2^{2+} binding at the oil/water interface, all of the derivatives showed enhanced binding of the cations. These derivatives were recently tested using Cd binding and were found to be as effective as the parental bioemulsifier (Avigad, unpublished observations). These experiments were conducted with the standard mixture of hexadecane-2-methyl naphthalene. However, when hexadecane or octadecane were used as hydrocarbon substrates for emulsification, the mutant biopolymers containing elevated levels of palmitoleic acid were about twice as active as those containing the hydroxy dodecanoic acids. It was recently found that significant fatty acid changes in the acyl side chains of emulsan could be achieved by growing the cells of *A. venetianus* RAG-1 in the presence of fatty acids in the growth medium,[126] although none of the activities of the biopolymer product was reported. These results may account for earlier observations that the properties of the emulsan produced on oil substrates (the so-called β-emulsans) differ significantly from those produced on ethanol.

The second approach is similar to those described above for other EPS and is currently in progress. Recently, the cluster encoding the biosynthetic pathway for emulsan synthesis was cloned and sequenced.[47] On the basis of homology to other known sequences, emulsan most closely resembles the O-antigen of LPS, lacking ketodesoxyoctulosonic acid residues as well as the lipid A moiety of LPS. A series of emulsan-defective mutants were generated in the various biosynthetic genes and complemented with a genomic library from the organism *A. calcoaceticus* BD413. This strain produces a rhamnose-containing, extracellular, water-soluble capsular polysaccharide,[127] which normally does not exhibit any emulsifying activity.[128] This polymer, however, does exhibit emulsifying activity when it interacts with crude extracellular protein associated with it, or with recombinant esterase from *A. venetianus* RAG-1 (Bach and Gutnick, in preparation). The biosynthetic pathway of the BD413 polymer includes a series of transglycosylation reactions, one of which transfers a UDP-glucose to the membrane during subunit assembly.[129] One of the complemented mutants was found to contain a glucose moiety in place of galactosamine. It is tempting to consider that by removing the galactosamine, potential sites for additional acylation could be lost, thereby altering the polarity of the molecule. In fact, this mutant derivative of emulsan exhibits about half as much emulsifying activity as the native bioemulsifier, but is still active in cation binding at the oil/water interface and in the removal of adherent organisms from hydrophobic surfaces (preliminary observations). It should be noted, however, that the overall yield of glucose-containing emulsan derivative is only about 10% of the wild-type yield, which is in agreement with previous findings for xanthan derivatives.[121] It is yet to be determined whether this low productivity can be increased by optimizing fermentation conditions for production. The ability to modify a polysaccharide backbone by genetic techniques may pave the way to enhance the cation binding of the biopolymer, using derivatives, which have a higher charge density.

As indicated in Figure 18.3, the genes encoding the biosynthetic enzymes are encoded in the rightward operon from a single promoter. The leftward operon encodes three proteins apparently involved in the control of production.[51] As shown in Figure 18.4, genes *wza, wzb,* and *wzc* are thought to encode a porin (or channel through which the emulsan polysaccharide is exported), a protein tyrosine phosphatase, and a protein tyrosine kinase, respectively. Wzc is an autophosphorylase, which is thought to inhibit the export of the polysaccharide. Wzb has been shown to be a phosphatase, which dephosphorylates the Wzc allowing the polymer to be exported through a membrane-bound complex. Mutants lacking either the kinase or the phosphatase were found to be defective in emulsan. As shown for other orthologs of Wzc, the site of phosphorylation is located in a C-terminal region of Wzc, which contains five tyrosine residues, putative sites for phosphorylation. Deletion of these residues led to the production of a product that exhibited five times the reduced viscosity of emulsan itself as well as a molecular mass of about 25×10^6 Da. The results support the notion that Wzc controls the size of the extracellular emulsan polymer. Although this viscous product was inactive as an emulsifier, it did form an EEPosanosol in the presence of esterase derivatives. Interestingly, a micromole of viscoemulsan bound up to twice as much

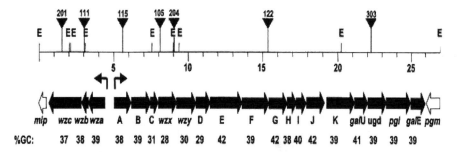

FIGURE 18.3   The *wee* cluster encoding the biosynthetic genes of emulsan.The *wee* gene cluster for the biosynthesis of emulsan. The scale of the cluster size is in kilobases. The black arrows represent putative orf sequences. White arrows represent partially sequenced orf's. Putative promoter sites are indicated with thin black arrows. The names of the genes are shown below the corresponding orf's. Orf's labeled solely with capital letters are putative pathway-specific genes encoding Wee A-K.

UO_2^{2+} as did 1 μmol of apoemulsan. Since the polysaccharide composition of vis-coemulsan is the same as that of apoemulsan, the difference in binding capacity may be due to polymer conformation at the oil/water interface.

Finally, the possibility of fusing active peptide fragments from the esterase with other proteins, such as metallothionein,[130] may lead to the generation of novel EEPs. These new fusions might lead to preparations and new formulations, which both stabilize EEPosanosols and exhibit enhanced binding of cations. This approach is currently being considered in our laboratory.

III.   POTENTIAL APPLICATIONS

Thus far, isolated biopolymers for heavy-metal remediation have not been applied on a large scale, although synthetic polymers have been used for various precipitation treatments. It seems likely that incorporating polysaccharides into biofilter technology may provide useful applications for remediation, although much depends on the economics of such treatments. A biofilter for heavy metal removal from polluted water may involve the use of crude biomass, or it could involve a more selective cation-exchange process employing immobilized polysaccharides. Of course, the economics will depend on the added value associated with enhanced specificity and cation-binding capacity. In addition, the use of defined materials in such a system enables the process to be modeled and simulated with better predictability.

An alternative approach could involve the use of emulsanosol technology as illustrated in Figure 18.2. Emulsanosols are highly stable and can be recycled once the initial emulsion is formed. Previous results have demonstrated that once formed, emulsanosols can withstand the enormous shear forces of over 500 pump transits at a low rate of 6000 L/h at a pressure of 500 kPa through a 9 cm pipe. In terms of scaling-up, this stability could accommodate about 250,000 barrels of oil per day in a 50 ± 75 cm pipe without any inversion of the oil-in-water emulsion. Moreover, the use of emulsanosols for viscosity reduction in heavy oils and in the generation of

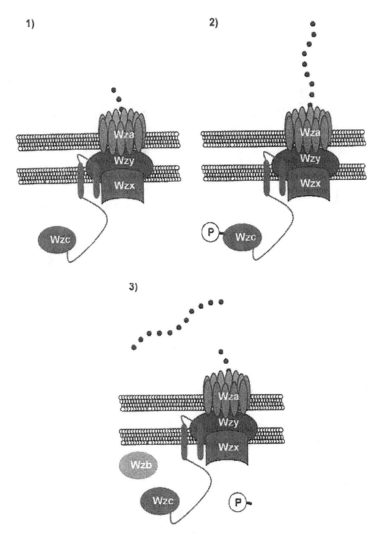

FIGURE 18.4   Model for role of protein complex in emulsan export.Hypothetical model for the role of protein tyrosine kinase (Wzc) and protein tyrosine phosphatase (Wzb) in emulsan export: (1) Dephosphorylated Wzc allows for polymerization and translocation of emulsan. (2) Phosphorylation of Wzc halts the process, thereby determining the size of the exported polymer. (3) Emulsan release and beginning of a new round of polymerization, translocation, and release. Wza-translocation channel; Wzb-protein tyrosine phosphatase; Wzc-Protein tyrsoine kinase; Wzx-polymerase; Wzy-translocase.

homogenous emulsions for the efficient combustion of crude sludges and heavier materials has been scaled-up and demonstrated on a semiindustrial scale.[131]

In line with this approach, novel formulation technology using highly active recombinant protein (or peptide)–polysaccharide complexes may provide an application for inexpensive excess polysaccharides, which need not be produced by fermentation and for which there is little industrial use. For example, the generic formulation might be applicable in designing specific emulsanosols in the remediation of particular polluted sites. The approach also offers the possibility of recovering heavy metals during the cleanup of sludges, tank bottoms, and other oily wastes, using the same materials for both purposes. In addition, as emulsions with relatively low viscosity, emulsanosols may be more stable for applications on sites such as ships, barges, and tank farms where the handling of pumpable liquids is generally easier than the handling of solids.[59,62,65,68–70,72–74,78,90,93]

A. POLYSACCHARIDE-COATED MATRIXES

The effectiveness of heavy metal removal using polysaccharides has been well demonstrated, although new technologies are constantly being introduced.[79,132–138] In general, these are focused primarily on applications where inert matrixes assist the performance of polysaccharides in sorption processes.[79,136,138] In these systems, two characteristics are generally affected. First, the distribution of the polysaccharide at the matrix surface is more uniform, thereby significantly increasing the biosorption capacity of the system by expansion of the adsorption surface. Second, there is a decrease in resistance to internal diffusion facilitating the treatment of high volumes of fluids. Matries are becoming increasingly popular owing to their availability, coupled with the fact that they are often inexpensive, which may enable their application in industrial waste treatments.[136,138] For example, an adsorption of 124 mg Cd^{2+}/g by alginate-coated sponge discs has been reported recently.[138] Another suitable matrix with a high adsorption surface is granular activated carbon (GAC), used commonly in drinking water treatment. By coating GAC with a bacterial biofilm of *Arthrobacter viscosus,* Quintelas and Tavares[139] obtained a removal of 8.5 mg Cr and 4.2 mg Cd/g GAC with a residence time of only 1.2 min.

B. DRIED PACKED BIOMASS

Extracellular polymeric substances are rich in polysaccharides. Microorganisms coexisting in a microenvironment generate these substances and form a three-dimensional biofilm matrix. The polysaccharide content in such biofilms may be as high as 95%.[140] Therefore, the use of such dried packed biomass may serve as a useful inexpensive sorbent with a high capacity for heavy metal adsorption in biofilters. For example, dried polysaccharide extracts from the cyanobacteria *Gloeothece magna* bound Cd^{2+} and Mn^{2+} by complexing with capsular polysaccharides.[77] However, the biosorption capacity of dried biomass is limited by the natural composition of the microorganisms, which constitute the biomass. In this regard, the use of organisms engineered to contain higher quantities of suitable biosorbent is an interesting approach.

The capacity of many microorganisms to grow in high concentrations of heavy metals is often due to enhanced metal-binding capacity in such strains. These heavy metal-resistant microorganisms may provide an additional advantage as constituents in packed biomass. Such organisms are frequently found in environments rich in heavy metals. An interesting example[85] involves the preparation of lyophilized biomass from a *Brevibacterium* species isolated from an abandoned zinc mine. This strain was capable of growing at concentrations of >15 mM of zinc. The biomass exhibited characteristics suitable for its use in the bioremediation of heavy metals.

C. CHITIN AND CHITOSAN SORBENTS

The binding of heavy metals and radionuclides to fungal biomass such as that of *Rhizopus arrhizus* has generally been attributed to the ease with which the polymer chitin binds to such metals and radionuclides.[141–145] In addition to chitin, chitosan, the deacylated derivative of chitin, also avidly binds cations. In order to enhance the binding capacity of chitin and chitosan preparations from fungi, finely ground material was mixed with copper ferrocyanide, which was crystallized within the chitin mixture in the presence of ammonia. Columns containing 10 g of adsorbent (100 mL bed volume) were prepared and tested for their ability to remove ^{137}Cs from waste effluents collected from a nuclear reactor site.[146] Filtration of the concentrated radioactive Cs solution through 100 column volumes removed 99.99% of the radioactivity. Interestingly, in the absence of the microcrystalline ferrocyanide, only about 6% of the material was removed. Such columns were also found to be very efficient in removing various cations from aqueous solutions as well as radionuclides such as uranium, plutonium, americium, and curium.

D. CHEMICAL MODIFICATIONS

The diversity of functional groups, particularly hydroxyl groups, opens a new way to increase the heavy metal biosorption capacity of polysaccharides. Oxidants modify hydroxyl groups generating new carboxylic groups, which have been shown to be one of the crucial factors involved in metal complexation.[147,148] Potassium permanganate, a potent oxidant, generated four new carboxylic groups in the subunit of alginic acid. As a result, the modified polysaccharide increased its metal-binding capacity by 0.8 mmol Pb^{2+}/g dry mass as compared with unmodified alginic acid.[149] This treatment enhanced the sensitivity of the derivative to harsh conditions. By using a PVA–boric and glutaraldehyde method, Hashimoto and Furukawa,[150] developed new alginic acid beads, which were shown to be stable under strong acidic conditions (pH$<$1.0) and high temperatures (170°C). Therefore, successive chemical modifications, such as oxidation and cross-linking, might endow polysaccharides with new properties, expanding their utilization under different conditions.

E. ACTIVATED SLUDGE

Owing to the relative abundance of municipal sewage treatment facilities, activated sludge is an inexpensive source of biosorption matrix. Although activated sludge does not show a uniform composition, it contains a mixture of potential metal biosorbents, for example living and dead cells, carbohydrates, proteins, polysaccharides, biofilms, and extracellular polymeric substances. Recently, the use of pilot plants for heavy metal removal has been reported, showing biosorptions of 90 mg Cu^{2+} and 4.7 mg Ni^{2+}/g biomass.[135,151]

The introduction of suitable strains in the activated sludge might improve its performance. For example, the considerable increase in the growth of a *Nocardia amarae* isolate in activated sludge was accompanied by heavy metal biosorption due to the formation of a very intricate network of cells that led to a large increase in adsorption surface.[83] It is tempting to consider employing this approach using recombinant organisms with enhanced metal-combining capacity.

REFERENCES

1. Alexander, M., *Biodegradation and Bioremediation*, Academic Press, San Diego, 1999.
2. Colwell, R.R., Scientific foundation of bioremediation and gaps remaining to be filled, *Res. Microbiol.*, 145, 40–41, 1994.
3. Gutnick, D.L., Microbiological treatment of contaminated storage containers, *Res. Microbiol.*, 145, 56–60, 1994.
4. Gutnick, D.L., Engineering polysaccharides for biosorption of heavy metals at oil/water interfaces, *Res. Microbiol.*, 148, 519–521, 1997.
5. Lovley, D.R. and Coates, J.D., Bioremediation of metal contamination, *Curr. Opin. Biotechnol.*, 8, 285–289, 1997.
6. Silver, S., Exploiting heavy metal resistance systems in bioremediation, *Res. Microbiol.*, 145, 61–67, 1994.
7. Beveridge, T.J., Role of cellular design in bacterial metal accumulation and mineralization, *Ann. Rev. Microbiol.*, 43, 147–171, 1989.
8. Diels, L., De Smet, M., Hooyberghs, L., and Corbisier, P., Heavy metals bioremediation of soil, *Mol. Biotechnol.*, 12, 149–158, 1999.
9. Macaskie, L.E., Dean, A.C.R., Cheetham, A.K., Jakeman, R.J.B., and Skarnulis, A.J., Cadmium accumulation by a *Citrobacter* sp.: the chemical nature of the accumulated metal precipitate and its location on the bacterial cells, *J. Gen. Microbiol.*, 133, 539–544, 1987.
10. Macaskie, L.E., Empson, R.M., Cheetham, A.K., Grey, C.P., and Skarnulis, A.J., Uranium bioaccumulation by a *Citrobacter* sp. as a result of enzymically mediated growth of polycrystalline HUO2PO4, *Science*, 257, 782–784, 1992.
11. McLean, R.J., Fortin, D., and Brown, D.A., Microbial metal-binding mechanisms and their relation to nuclear waste disposal, Can. J. Microbiol., 42, 392–400, 1996.
12. Taghavi, S., Mergeay, M., Nies, D., and van der Lelie, D., *Alcaligenes eutrophus* as a model system for bacterial interactions with heavy metals in the environment, *Res. Microbiol.*, 148, 536–551, 1997.
13. Tolley, M.R., Strachan, L.F., and Macaskie, L.E., Lanthanum accumulation from acidic solutions using a *Citrobacter* sp. immobilized in a flow-through bioreactor, *J. Ind. Microbiol.*, 14, 271–280, 1995.
14. White, C., Sayer, J.A., and Gadd, G.M., Microbial solubilization and immobilization of toxic metals: key biogeochemical processes for treatment of contamination, *FEMS Microbiol. Rev.*, 20, 503–516, 1997.
15. Brown, M.J. and Lester, J.N., Metal removal in activated sludge: the role of bacterial extracellular polymers, *Water Res.*, 13, 817–837, 1979.
16. Cheng, M.H., Patterson J.W., and Minear, R.A., Heavy metals uptake by acitvated sludge, *J. Water Pollut. Control Fed.*, 47, 362–376, 1975.
17. Diels, L., Vanroy, S., Somers, K., Willems, I., Mergeay, M., Springael, D., and Leysen, R., The use of bacteria immobilized in tubular membrane reactors for heavy-metal recovery and degradation of chlorinated aromatics, *J. Membr. Sci.*, 100, 249–258, 1995.

18. Fristoe, B.R. and Nelson, P.O., Equilibrium chemical modelling of heavy metals in activated sludge, *Water Res.*, 17, 771–778, 1983.

19. Kasan, H.C. and Baecker, A.A.W., An assessment of toxic metal biosorption by activated sludge from the treatment of coal-gasification effluent of a petrochemical plant, *Water Res.*, 23, 795–800, 1989.

20. Matis, K.A., Zouboulis, A.I., Grigoriadou, A.A., Lazaridis, N.K., and Ekateriniadou, L.V., Metal biosorption-flotation. Application to cadmium removal, *Appl. Microbiol. Biotechnol.*, 45, 569–573, 1996.

21. Matis, K.A. and Zouboulis, A.I., Biosorptive flotation for separation of zinc and cadmium cations from dilute solutions, *Resour. Environ. Biotechnol.*, 2, 117–136, 1998.

22. Roane, T.M., Pepper, I.L., and Miller, R.M., Microbial remediation of metals, in *Bioremediation Principles and Application*, Crawford, R. and Crawford, D., Eds., Cambridge University Press, New York, 1996, pp. 312–340.

23. Sterritt, R.M. and Lester, J.N., The influence of sludge age on heavy metal removal in the activated sludge process, *Water Res.*, 15, 59–65, 1981.

24. Zouboulis, A.I., Solari, P., Matis, K.A., and Stalidis, G.A., Toxic metals removal from dilute solutions by biosorptive flotation, *Water Sci. Technol.*, 32, 211–220, 1995.

25. Nies, D.H., Microbial heavy-metal resistance, *Appl. Microbiol. Biotechnol.*, 51, 730–750, 1999.

26. Volesky, B., *Biosorption of Heavy Metals*, CRC Press, Boca Raton, 1990.

27. Volesky, P. and Holan, Z.R., Biosorption of heavy metals, *Biotechnol. Prog.*, 11, 235–250, 1995.

28. Krachtovil, D. and Volesky, B., Advances in the biosorption of heavy metals, *Trends Biotechnol.*, 16, 291–300, 1998.

29. Ledin, M. and Pedersen, K., The environmental impact of mine wastes — roles of microorganisms and their significance in treatment of mine wastes, *Earth-Sci. Rev.*, 41, 67–108, 1996.

30. Glombitza, F., Eckhart, L., and Hummel, A., Fundamentals of the application of biosorption to the separation of uranium from mining drainage waters, *Res. Microbiol.*, 148, 517–518, 1997.

31. Asthana, R.K., Chatterjee, S., and Singh, S.P., Investigations on nickel biosorption and its remobilization, *Process. Biochem.*, 30, 729–734, 1995.

32. Avery, S.V. and Tobin. J.M., Mechanism of adsorption of hard and soft metal ions to *Saccharomyces cerevisiae* and influence of hard and soft anions, *Appl. Environ. Microbiol.*, 59, 2851–2856, 1993.

33. Foureste, E. and Roux, C., Heavy-metal biosorption by fungal mycelial by-products — mechanisms and influence of pH, *Appl. Microbiol. Biotechnol.*, 37, 399–403, 1992.

34. Gutnick, D.L. and Bach, H., Engineering bacterial biopolymers for the biosorption of heavy metals; new products and novel formulations, *Appl. Microbiol. Biotechnol.*, 54, 451–460, 2000.

35. Gadd, G.M., Bioremedial potential of microbial mechanisms of metal mobilization and immobilization, *Curr. Opin. Biotechnol.*, 11, 271–279, 2000.

36. Malik, A., Metal bioremediation through growing cells, *Environ. Int.*, 30, 261–278, 2004.

37. Sandrin, T.R. and Maier, R.M., Impact of metals on the biodegradation of organic pollutants, *Environ. Health Perspect.*, 111, 1093–1101, 2003.

38. Sousa, C., Kotrba, P., Ruml, T., Cebolla, A., and De Lorenzo, V., Metalloadsorption by *Escherichia coli* cells displaying yeast and mammalian metallothioneins anchored to the outer membrane protein LamB, *J. Bacteriol.*, 180, 2280–2284, 1998.

39. Valls, M., Gonzalez Duarte, R., Atrian, S., and De Lorenzo, V., Bioaccumulation of heavy metals with protein fusions of metallothionein to bacterial OMPs, *Biochimie,* 80, 855–861, 1998.

40. Sousa, C., Cebolla, A., and De Lorenzo, V., Enhanced metalloadsorption of bacterial cells displaying poly-His peptides, *Nat. Biotechnol.,* 14, 1017–1020, 1996.

41. Glucksmann, M.A., Reuber, T.L., and Walker, G.C., Family of glycosyl transferases needed for the synthesis of succinoglycan by *Rhizobium meliloti, J. Bacteriol.,* 175, 7033–7044, 1993.

42. Bernhard, F., Coplin, D.L., and Geider, K., A gene cluster for amylovoran synthesis in Erwinia amylovora: characterization and relationship to cps genes in *Erwinia stewartii, Mol. Gen. Genet.,* 239, 158–168, 1993.

43. Stevenson, G., Andrianopoulos, K., Hobbs, M., and Reeves, P.R., Organization of the *Escherichia coli* K-12 gene cluster responsible for production of the extracellular polysaccharide colanic acid, *J. Bacteriol.,* 178, 4885–4893, 1996.

44. Aldridge, P., Bernhard, F., Bugert, P., Coplin, D.L., and Geider, K., Characterization of a gene locus from *Erwinia amylovora* with regulatory functions in exopolysaccharide synthesis of *Erwinia* sp, *Can. J. Microbiol.,* 44, 657–666, 1998.

45. Wehland, M., Kiecker, C., Coplin, D.L., Kelm, O., Saenger, W., and Bernhard, F., Identification of an RcsA/RcsB recognition motif in the promoters of exopolysaccharide biosynthetic operons from *Erwinia amylovora* and *Pantoea stewartii* subspecies stewartii, *J. Biol. Chem.,* 274, 3300–3307, 1999.

46. Wang, L., Briggs, C.E., Rothemund, D., Fratamico, P., Luchansky, J.B., and Reeves, P.R., Sequence of the *E. coli* O104 antigen gene cluster and identification of O104 specific genes, *Gene,* 270, 231–236, 2001.

47. Nakar, D. and Gutnick, D.L., Analysis of the wee gene cluster responsible for the biosynthesis of the polymeric bioemulsifier from the oil-degrading strain *Acinetobacter lwoffii* RAG-1, *Microbiology,* 147, 1937–1946, 2001.

48. Jiang, S.M., Wang, L., and Reeves, P.R., Molecular characterization of *Streptococcus pneumoniae* type 4, 6B, 8, and 18C capsular polysaccharide gene clusters, *Infect. Immunol,* 69, 1244–1255, 2001.

49. Wang, L., Huskic, S., Cisterne, A., Rothemund, D., and Reeves, P.R., The O-antigen gene cluster of *Escherichia coli* O55:H7 and identification of a new UDP-GlcNAc C4 epimerase gene, *J. Bacteriol.,* 184, 2620–2625, 2002.

50. Rahn, A. and Whitfield, C., Transcriptional organization and regulation of the Escherichia coli K30 group 1 capsule biosynthesis (cps) gene cluster, *Mol. Microbiol.,* 47, 1045–1060, 2003.

51. Nakar, D. and Gutnick, D.L., Involvement of a protein tyrosine kinase in production of the polymeric bioemulsifier emulsan from the oil-degrading strain *Acinetobacter lwoffii RAG-1, J. Bacteriol.,* 185, 1001–1009, 2003.

52. Kuhnert, P., Nicolet, J., and Frey, J., Rapid and accurate identification of *Escherichia coli K-12* strains, *Appl. Environ. Microbiol.,* 61, 4135–4139, 1995.

53. Kang, S.K., Lee, S.O., Lim, Y.S., Jang, K.L., and Lee, T.H., Purification and characterization of a novel levanoctaose-producing levanase from *Pseudomonas* strain K-52, *Biotechnol. Appl. Biochem.,* 27, 159–166, 1998.

54. Zosim, Z., Gutnick, D.L., and Rosenberg, E., Uranium binding by emulsan and emulsanosols, *Biotechnol. Bioeng.,* 25, 1725–1735, 1983.

55. Chen, J.H., Czajka, D.R., Lion, L.W., Shuler, M.L., and Ghiorse, W.C., Trace metal mobilization in soil by bacterial polymers, *Environ. Health. Perspect.,* 103 (Suppl. 1), 53–58, 1995.

56.   Chen, J.H., Lion, L.W., Ghiorse, W.C.,and Shuler, M.L., Mobilization of adsorbed cadmium and lead in aquifer material by bacterial extracellular polymers, *Water Res.*, 29, 421–430, 1995.

57.   Cozzi, D., Desideri, P.G., Lepri, L., and Coas, V., Ion-exchange thin-layer chromatographic separation of amino acids on alginic acid, *J. Chromatogr.*, 40, 138–144, 1969.

58.   Kaplan, D., Christiaen, D., and Arad S., Chelating properties of extracellular polysaccharide from *Chlorella spp.*, *Appl. Environ. Microbiol.*, 53, 2953–2956, 1987.

59.   Scott, J.A. and Palmer S.J., Cadmium bio-sorption by bacterial exopolysaccharide, *Biotechnol. Lett.*, 10, 21–24, 1988.

60.   Geddie, J.L. and Sutherland, I.W., Uptake of metals by bacterial polysaccharides. *J. Appl. Bacteriol.*, 74, 467–472, 1993.

61.   http://www.lsbu.ac.uk/water/hygellan.html.

62.   McLean, R.J., Beauchemin, D., Clapham, L., and Beveridge, T.J., Metal-binding characteristics of the gamma-glutamyl capsular polymer of *Bacillus licheniformis* ATCC 9945, *Appl. Environ. Microbiol.*, 56, 3671–3677, 1990.

63.   Beveridge, T.J., and Koval S.F., Binding of metals to cell envelopes of *Escherichia coli* K-12, *Appl. Environ. Microbiol.*, 42, 325–335, 1981.

64.   Yalpani, M., Ed., Structures, in *Polysaccharides — Syntheses, Modifications and Structures/Property Relations*, Elsevier Science Publisher, The Netherlands, 1988, pp. 35.

65.   Mittelman, M.W. and Geesey, G.G., Copper-binding characteristics of exopolymers from a freshwater-sediment bacterium, *Appl. Environ. Microbiol.*, 49, 846–851, 1985.

66.   Muzzarelli, R.A. and Tubertini, O., Chitin and chitosan as chromatographic supports and adsorbents for collection of metal ions from organic and aqueous solutions and sea-water, *Talanta*, 16, 1571–1577, 1969.

67.   Beveridge, T.J. and Murray, R.G., Sites of metal deposition in the cell wall of *Bacillus subtilis*, *J. Bacteriol.*, 141, 876–887, 1980.

68.   Hoyle, B.D. and Beveridge, T.J., Metal binding by the peptidoglycan sacculus of *Escherichia coli* K-12, *Can. J. Microbiol.*, 30, 204–211, 1984.

69.   Rayman, M.K. and MacLeod, R.A., Interaction of Mg^{2+} with peptidoglycan and its relation to the prevention of lysis of a marine pseudomonad, *J. Bacteriol.*, 122, 650–659, 1975.

70.   Doyle, R.J., Matthews, T.H., and Streips, U.N., Chemical basis for selectivity of metal ions by the *Bacillus subtilis* cell wall. *J. Bacteriol.*, 143, 471–480, 1980.

71.   Langley, S. and Beveridge, T.J., Effect of O-side-chain-lipopolysaccharide chemistry on metal binding, *Appl. Environ. Microbiol.*, 65, 489–498, 1999.

72.   Marques, A.M., Bonet, R., Simon-Pujol, M.D., Fuste, M.C., and Congregado, F., Removal of uranium by an exopolysaccharide from *Pseudomonas* sp., *Appl. Microbiol. Biotechnol.*, 34, 429–431, 1990.

73.   Dispirito, A.A., Talgani Jr, J.W., and Tuovinen, O.H., Accumulation and cellular distribution of uranium in *Thiobacillus ferrooxidans*, *Arch. Microbiol.*, 135, 250–253, 1983.

74.   Norberg, A.B. and Persson, H., Accumulation of heavy-metal ions by *Zooglea ramigera*, *Biotechnol. Bioeng.*, 26, 239–246, 1984.

75.   Churchill, S.A., Walters, J.V., and Churchill, P.F., Sorption of heavy metals by prepared bacterial cell surfaces, *J. Environ. Eng.*, 121, 706–711, 1995.

76.   Nourbakhsh, M.N., Kilicarslan, S., Ilhan, S., and Ozdag, H., Biosorption of Cr^{6+}, Pb^{2+} and Cu^{2+} ions in industrial waste water on *Bacillus sp.*, *Chem. Eng. J.*, 85, 351–355, 2002.

77.   Mohamed, Z., Removal of cadmium and manganese by a non-toxic strain of the freshwater cyanobacterium *Gloeothece magna*. *Water Res.*, 18, 4405–4409, 2001.

78.   Corzo, J., Leoon-Barrios, M., Hernando-Rico, V., and Gutierrez-Navarro, A.M., Precipitation of metallic cations by the acidic exopolysaccharides from *Bradyrhizobium japonicum* and *Bradyrhizobium (Chamaecytisus)* strain BGA-1, *Appl. Environ. Microbio.*, 60, 4531–4536, 1994.

79.   Jeon, C., Park, J.Y., and Yoo, Y.J., Novel immobilization of alginic acid for heavy metal removal, *Biochem. Eng. J.*, 11, 159–166, 2002.

80.   Lazaro, N., Sevilla, A.L., Morales, S., and Marques, A.M., Heavy metal biosorption by gellan gum gel beads. *Water Res.*, 37, 2118–2126, 2003.

81.   Addour, L., Belhocine, D., Boudries, N., Comeau, Y., Pauss, A., and Mameri, N., Zinc uptake by *Streptomyces rimosus* biomass using a packed-bed column, *J. Chem. Technol. Biotech.*, 74, 1089–1095, 1999.

82.   Fukushi, K., Chang, D., and Ghosh, S., Enhanced heavy metal uptake by activated sludge cultures growth in the presence of biopolymer stimulators, *Water Sci. Technol.*, 34, 267–272, 1996.

83.   Kim, D.W., Cha, D.K., Wang, J., and Huang, C.P., Heavy metal removal by activated sludge: influence of *Nocardia amarae*, *Chemosphere*, 46, 137–142, 2002.

84.   Guibaud, G., Tixier, N., Bouju, A., and Baudu, M., Relation between extracellular polymers' composition and its ability to complex Cd, Cu and Pb, *Chemosphere*, 52, 1701–1710, 2003.

85.   Taniguchi, J., Hemmi, H., Tanahashi, K., Amano, N., Nakayama, T., and Nishino, T., Zinc biosorption by a zinc-resistant bacterium, *Brevibacterium* sp. strain HZM-1, *Appl. Microbiol. Biotechnol.*, 54, 581–588, 2000.

86.   Vecchio, A., Finoli, C., Di Simine, D., and Andreoni, V., Heavy metal biosorption by bacterial cells, *Fresenius J. Anal. Chem.*, 361, 338–342, 1998.

87.   Bitton, G. and Freihofer, V., Influence of extracellular polysaccharides on the toxicity of copper and cadmium toward *Klebsiella aerogenes*, *Microb. Ecol.*, 4, 119–125, 1978.

88.   Loaec, M., Olier, R., and Guezennec, J., Uptake of lead, cadmium and zinc by a novel bacterial exopolysaccharide, *Water Res.*, 31, 1171–1179, 1997.

89.   Puranik, P.R. and Paknikar, K.M., Biosorption of lead, cadmium, and zinc by *Citrobacter* strain MCM B-181: characterization studies, *Biotechnol. Prog.*, 15, 228–237, 1999.

90.   Solomon, O., Cadmium binding to Apoemulsan and Apoemulsanosol Produced by the Bacteria *Acinetobacter lwoffii* RAG-1, M.S. thesis, Tel-Aviv University, 1997.

91.   Amoroso, M.J., Castro, G.R., Carlino, F.J., Romero, N.C., Hill, R.T., and Oliver, G., Screening of heavy metal-tolerant actinomycetes isolated from the Sali River, *J. Gen. Appl. Microbiol.*, 44, 129–132, 1998.

92.   Gabriel, J., Vosahlo, J., and Baldrian, P., Biosorption of cadmium to mycelial pellets of wood-rotting fungi, *Biotechnol. Lett.*, 10, 345–348, 1996.

93.   Park, J.K., Jin, Y.B., and Chen, J.H., Reusable biosorbents in capsules from Zooglea ramigera cells for cadmium removal, *Biotechnol. Bioeng.*, 63, 116–121, 1999.

94.   Chang, J.S., Law, R., and Chang, C.C., Biosorption of lead, copper and cadmium by biomass of *Pseudomonas aeruginosa PU21*, *Water Res.*, 31, 1651–1658, 1997.

95.   Lopez, A., Lazaro, N., Priego, J.M., and Marques, A.M., Effect of pH on the biosorption of nickel and other heavy metals by *Pseudomonas fluorescens* 4F39, *J. Ind. Microbiol. Biotechnol.*, 24, 146–151, 2000.

96.   Suh, J.H. and Kim, D.S., Comparison of different sorbents (inorganic and biological) for the removal of $Pb2+$ from aqueous solutions, *J. Chem. Technol. Biotechnol.*, 75, 279–284, 2000.

97.   Beveridge, T.J., Forsberg, C.W., and Doyle, R.J., Major sites of metal binding in *Bacillus licheniformis* walls, *J. Bacteriol.*, 150, 1438–1448, 1982.

98. Ferris, F.G. and Beveridge, T.J., Site specificity of metallic ion binding in *Escherichia coli* K-12 lipopolysaccharide, *Can. J. Microbiol.*, 32, 52–55, 1986.

99. Mullen, M.D., Wolf, D.C., Ferris, F.G., Beveridge, T.J., Flemming, C.A., and Bailey, G.W., Bacterial sorption of heavy metals, *Appl. Environ. Microbiol.*, 55, 3143–3149, 1989.

100. Strain, S.M., Fesik, S.W., and Armitage, I.M., Structure and metal-binding properties of lipopolysaccharides from heptoseless mutants of *Escherichia coli* studied by ^{13}C and ^{31}P nuclear magnetic resonance, *J. Biol. Chem.*, 258, 13466–13477, 1983.

101. Rosenberg, E., Zuckerberg, A., Rubinovitz, C., and Gutnick, D.L., Emulsifier of *Arthrobacter* RAG-1: isolation and emulsifying properties, *Appl. Environ. Microbiol.*, 37, 402–408, 1979.

102. Gutnick, D.L., The emulsan polymer: perspectives on a microbial capsule as an industrial product, *Biopolymers*, 26, S223–S240, 1987.

103. Shabtai, Y., Pines, O., and Gutnick, D.L., Emulsan: a case study of microbial capsules as industrial products, *Dev. Indust. Microbiol.*, 26, 291–217, 1985.

104. Zuckerberg, A., Diver, A., Peeri, Z., Gutnick, D.L., and Rosenberg, E., Emulsifier of *Arthrobacter* RAG-1: chemical and physical properties, *Appl. Environ. Microbiol.*, 37, 414–420, 1979.

105. Bach, H. and Gutnick, D.L., Potential applications of bioemulsifiers in the oil industry, in *Petroleum Biotechnology*, Duhalt, R.V. and Ramirez, R.Q., Eds., Studies in Surface Science and Catalysis, 151, 2005, chap. 8.

106. Belsky, I., Gutnick, D.L., and Rosenberg, E., Emulsifier of Arthrobacter RAG-1: determination of emulsifier-bound fatty acids. *FEBS Lett.*, 101, 175–178, 1979.

107. Rosenberg, E., Perry, A., Gibson, D.T., and Gutnick, D.L., Emulsifier of *Arthrobacter* RAG-1: specificity of hydrocarbon substrate, *Appl. Environ. Microbiol.*, 37, 409–413, 1979.

108. Zosim, Z., Gutnick, D.L., and Rosenberg, E., Properties of hydrocarbon-in-water emulsions stabilized by *Acinetobacter* RAG-1 emulsan, *Biotechnol. Bioeng.*, 24, 281–292, 1982.

109. Pines, O. and Gutnick, D.L., Specific binding of a bacteriophage at a hydrocarbon–water interface, *J. Bacteriol.*, 157, 179–183, 1984.

110. Pines, O. and Gutnick, D.L., A role for emulsan in growth of *Acinetobacter calcoaceticus* on crude oil, *Appl. Environ. Microbiol.*, 51, 661–663, 1986.

111. Rosenberg, M., Gutnick, D., and Rosenberg, E., Adherence of bacteria to hydrocarbons: a simple method for measuring cell-surface hydrophobicity, *FEMS Microbiol. Lett.*, 9, 29–33, 1980.

112. Shoham, Y., Rosenberg, M., and Rosenberg, E., Bacterial degradation of emulsan, *Appl. Environ. Microbiol.*, 46, 573–579, 1983.

113. Shabtai, Y. and Gutnick, D.L., Exocellular esterase and emulsan release from the cell surface of *Acinetobacter calcoaceticus, J. Bacteriol.*, 161, 1176–1181, 1985.

114. Gutnick, D.L. and Bach, H., U.S. Patent 6,512,014, 2003.

115. Alon, R., Esterase from the Oil-Degrading *Acinetobacter lwoffii* RAG-1: Expression of the *est* Gene and Protein Characterization, Ph. D. thesis, Tel-Aviv University, 1993.

116. Alon, R.N., Mirny, L., Sussman, J.L., and Gutnick, D.L., Detection of α/β-hydrolase fold in the cell surface esterases of *Acinetobacter* species using an analysis of 3D profiles, *FEBS Lett.*, 371, 231–235, 1995.

117. Reddy, P.G., Allon, R., Mevarech, M., Mendelovitz, S., Sato, Y., and Gutnick, D.L., Cloning and expression in *Escherichia coli* of an esterase-coding gene from the oil-degrading bacterium *Acinetobacter calcoaceticus* RAG-1, *Gene*, 76, 145–152, 1989.

118. Bach, H., Berdichevsky, Y., and Gutnick, D.L., An exocellular protein from the oil-degrading microbe *Acinetobacter venetianus* RAG-1 enhances the emulsifying activity of the polymeric bioemulsifier emulsan, *Appl. Environ. Microbiol.,* 69, 2608–2615, 2003.

119. Bugert, P. and Geider, K., Molecular analysis of the ams operon required for exopolysaccharide synthesis of *Erwinia amylovora, Mol. Microbiol.,* 15, 917–933, 1995.

120. Reuber, T.L. and Walker, G.C., Biosynthesis of succinoglycan, a symbiotically important exopolysaccharide of *Rhizobium meliloti, Cell,* 74, 269–280, 1993.

121. Vanderslice, R.W., Doherty, D.H., Capage, M.A., Betlach, M.R., Hassler, R.A., Henderson, N.M., Graniero, R., and Tecklenburg, J.M., Genetic engineering of polysaccharide structure in *Xantomonas campestris,* Crescenzi, V., Dea, I., Paoletti, S., Stivala, S., and Sutherland, I., Eds., in *Biomedical and Biotechnological Advances in Industrial Polysaccharides,* Gordon and Breach, New York, 1989, pp. 145–156.

122. Whitfield, C. and Roberts, I.S., Structure, assembly and regulation of expression of capsules in *Escherichia coli, Mol. Microbiol.,* 31, 1307–1319, 1999.

123. Edwards, K.J., Jay, A.J., Colquhoun, I.J., Morris, V.J., Gasson, M.J., and Griffin, A.M., Generation of a novel polysaccharide by inactivation of the aceP gene from the acetan biosynthetic pathway in *Acetobacter xylinum, Microbiology,* 145, 1499–1506, 1999.

124. Bernhard, F., Schullerus, D., Bellemann, P., Nimtz, M., Coplin, D.L., and Geider, K., Genetic transfer of amylovoran and stewartan synthesis between *Erwinia amylovora and Erwinia stewartii, Microbiology,* 142, 1087–1096, 1996.

125. Colquhoun, I.J., Jay, A.J., Eagles, J., Morris, V.J., Edwards, K.J., Griffin, A.M., and Gasson M.J., Structure and conformation of a novel genetically engineered polysaccharide P2, *Carbohydr. Res.,* 330, 325–333, 2001.

126. Gorkovenko, A., Zhang, J., Gross, R.A., Allen, A.L., and Kaplan, D.L., Bioengineering of emulsifier structure: emulsan analogs, *Can. J. Microbiol.,* 43, 384–390, 1997.

127. Juni, E. and Heym, G.A., Pathways for biosynthesis of a bacterial polysaccharide, *J. Bacteriol.,* 87, 461–467, 1964.

128. Kaplan, N., Zosim, Z., and Rosenberg, E., Reconstitution of emulsifying activity of *Acinetobacter calcoaceticus* BD4 emulsan by using pure polysaccharide and protein, *Appl. Environ. Microbiol.,* 53, 440–446, 1987.

129. Kaplan, N. and Rosenberg, E., Exopolysaccharide distribution and bioemulsifier production in *Acinetobacter calcoaceticus* BD-4 and BD413, *Appl. Environ. Microbiol.,* 44, 1335–1341, 1982.

130. Pazirandeh, M. and Campbell, J.R., Bacteria Expressing Metallothionein Gene into the Periplasmic Space, and Method of Using Such Bacteria in Environment Cleanup, U.S. Patent 5, 824, 512, 1998.

131. Gutnick, D.L., Allon, R., Levy, C., Petter, R., and Minas, W., Applications of *Acinetobacter* as an industrial microorganism, in *The Biology of Acinetobacter,* Towner, K., Bergogne-Berezin, E., and Fewson, C., Eds., Plenum Press, New York, 1991, pp. 411–441.

132. Chen, D., Lewandowski, A., Roe, F., and Surapaneni, P., Diffusivity of Cu^{2+} in calcium alginate gel beads, *Biotechnol. Bioeng.,* 41, 755–760, 1993.

133. Chang, J.S. and Huang, J.C., Selective adsorption/recovery of Pb, Cu, and Cd with multiple fixed beds containing immobilized bacterial biomass, *Biotechnol. Prog.,* 14, 735–741, 1998.

134. Xiang, L., Chan, L.C., and Wong, J.W., Removal of heavy metals from anaerobically digested sewage sludge by isolated indigenous iron-oxidizing bacteria, *Chemosphere,* 41, 283–287, 2000.

135.   Artola, A., Martin, M.J., Balaguer, M. D., and Rigola, M., Pilot plant biosorption in an integrated contact-settling system: application to Cu(II) removal by anaerobically digested sludge, *J. Chem. Technol. Biotechnol.*, 76, 1141–1146, 2001.

136.   Xu, Z., Bae, W., Mulchandani, A., Mehra, R.K., and Chen, W., Heavy metal removal by novel CBD-EC20 sorbents immobilized on cellulose, *Biomacromolecules*, 3, 462–465, 2002.

137.   Zhuang, J.M., Walsh, T., and Lam, T., A new technology for the treatment of mercury contaminated water and soils, *Environ. Technol.*, 24, 897–902, 2003.

138.   Iqbal, M. and Edyvean, R.G., Alginate coated loofa sponge discs for the removal of cadmium from aqueous solutions, *Biotechnol. Lett.*, 26, 165–169, 2004.

139.   Quintelas, C. and Tavares, T., Removal of chromium(VI) and cadmium(II) from aqueous solutions by a bacterial biofilm supported on granular activated carbon, *Biotechnol. Lett.*, 23, 1349–1353, 2001.

140.   Flemming, H.C. and Wingender, J., Relevance of microbial extracellular polymeric substances (EPSs) — Part II: Technical aspects, *Water Sci. Technol.*, 43, 9–16, 2001.

141.   Volesky, B., Ed., Biosorption by fungal biomass, in *Biosorption of Heavy Metals*, CRC Press, Boca Raton, FL,1990, pp. 139–172.

142.   Zhou, J.L., Zn biosorption by *Rhizopus arrhizus* and other fungi, *Appl. Microbiol. Biotechnol.*, 51, 686–693, 1999.

143.   Vieira, R.H. and Volesky, B., Biosorption: a solution to pollution?, *Int. Microbiol.*, 3, 17–24, 2000.

144.   Boddu, V.M., Abburi, K., Talbott, J.L., and Smith, E.D., Removal of hexavalent chromium from wastewater using a new composite chitosan biosorbent, *Environ. Sci. Technol.*, 37, 4449–4456, 2003.

145.   Rae, I.B. and Gibb, S.W., Removal of metals from aqueous solutions using natural chitinous materials, *Water Sci. Technol.*, 47, 189–196, 2003.

146.   Gorovoj, L.F. and Kosyakov, V.N., Adsorption Means for Radionuclides, U.S. Patent 6,402,953, 2002.

147.   Kuyucak, N. and Volesky, B., Desorption of cobalt-laden algal biosorbent, *Biotechnol. Bioeng.*, 33, 815–822, 1989.

148.   Crist, R.H., Oberholster, K., McGarrity, J., Crist, D.R., Johnson, J.K., and Brittsan, J.M., Interaction of metals and protons with algae-3. Marine algae, with emphasis on lead and aluminium, *Environ. Sci. Technol.*, 26, 496–502, 1992.

149.   Jeon, C., Park, J.Y., and Yoo, Y.J., Characteristics of metal removal using carboxy-lated alginic acid, *Water Res.*, 36, 1814–1824, 2002.

150.   Hashimoto, S. and Furukawa, K., Immobilization of activated sludge by PVA-boric acid method, *Biotechnol. Bioeng.*, 30, 52–59, 1986.

151.   Arican, B., Gokcay, C.F., and Yetis, U., Mechanistics of nickel sorption by activated sludge, *Process. Biochem.*, 37, 1307–1315, 2002.

19 Treatment of Wastewaters with the Biopolymer Chitosan

Hong Kyoon No, Witoon Prinyawiwatkul, and Samuel P. Meyers

CONTENTS

I. INTRODUCTION

Significant volumes of wastewaters, containing organic and inorganic contaminants such as suspended solids, dyes, pesticides, toxicants, and heavy metals, are discharged from various industries. These wastewaters create a serious environmental problem and pose a threat to water quality when discharged into rivers and lakes. Thus, the contaminants must be effectively removed to meet increasingly stringent environmental quality standards. It is being recognized more and more that the nontoxic and biodegradable biopolymer chitosan can be used for purposes of wastewater treatment.[1]

Chitosan is a modified, natural carbohydrate, biodegradable polymer derived by deacetylation of chitin, a major component of the shells of crustacea such as crab, shrimp, and crawfish, and the second most abundant natural biopolymer after cellulose[2] (Figure 19.1). During the past several years, chitosan has received increased attention for use in commercial applications in biomedical, food, and chemical industries.[3–5] Currently, its major applications are in industrial wastewater treatment.

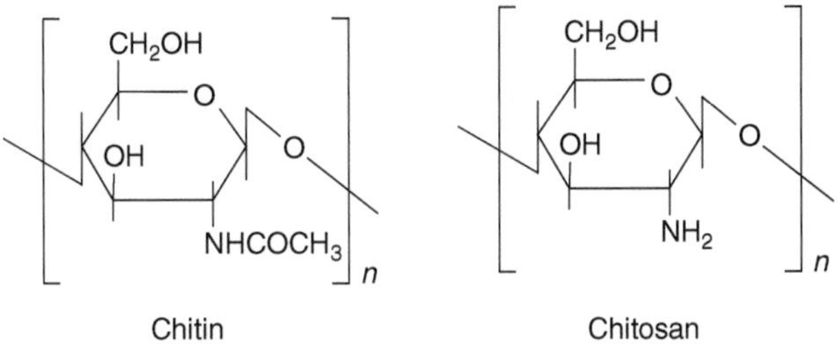

FIGURE 19.1   Chemical structures of chitin and chitosan.

Chitosan with high free amino groups can effectively function as a coagulant and as an adsorbent in wastewater treatment.[1]

Earlier investigations have demonstrated the effectiveness of chitosan for the coagulation and recovery of suspended solids in food-processing wastes bringing about a reduction in suspended solids of 65 to 99%.[6–8] Chitosan has also been applied as a chelating polymer for binding harmful metal ions, such as copper, lead, mercury, and uranium, from wastewater.[4,9] The effectiveness of chitosan for its ability to chelate transition metal ions has been reported by numerous authors.[4,10,11] In other areas, chitosan has been employed as an excellent adsorbent for the sorption of dyes,[12,13] phenols,[14,15] and polychlorined biphenyls (PCBs)[16,17] from wastewater.

This chapter focuses on the application of chitosan for the treatment of wastewaters containing suspended solids, metals, dyes, phenols, and PCBs, and covers work reported mostly from 1970 to the present.

II.   WASTEWATER TREATMENT

A.   FOOD-PROCESSING WASTEWATER

The major commercial applications for chitosan are currently found in industrial wastewater treatment and in the recovery of feedgrade material from food-processing plants. Chitosan carries a partial positive charge; thus, it functions effectively as a polycationic coagulating agent in wastewater treatment applications.[1]

A mechanism of action of polymeric flocculating agents was described by LaMer and Healy,[18] in which the polymer destabilizes a colloidal suspension by adsorption of particles with subsequent formation of particle–polymer–particle bridges. This action generally applies to both anionic and nonionic polyelectrolytes that are used to coagulate negative colloid. Positively charged cationic polymers can destabilize a negative colloidal suspension by charge neutralization as well as by bridge formation.[19]

Various studies have demonstrated the effectiveness of chitosan for coagulation and recovery of suspended solids in processing wastes from a variety of industries including poultry,[20] eggs,[21] cheese,[22,23] meat and fruit cakes,[24] seafood,[2,8,24,25] vegetable operations,[6,26] and tofu.[7] These extensive studies indicate that chitosan can reduce the suspended solids in such food-processing wastes by as much as 65 to 99%. Table 19.1 summarizes data from various studies on the reduction of turbidity (TB), suspended

TABLE 19.1

Reduction of Turbidity (TB), Suspended Solids (SS), and Chemical Oxygen Demand (COD) in Food-Processing Waste Effluents by Coagulation with Chitosan and Gravity Settling (GS)

Effluent	Chitosan Concentration (mg/L)	GS Time (h)	% Reduction			Ref.
			TB	SS	COD	
Egg-Breaking Composite	150[a]	1.5	78	72	76	21
	200[b]	1.5	77	74	57	21
Egg Washer Waste	100[c]	1.5	95	94	60	21
Greens Washer	10[d]	3	90	99	62	6
Filler	5[e]	1.5	83	93	10	6
Composite	10	1	85	90	—	6
Spinach Composite	20	1	94	90	—	6
Pimiento Peeling	40	1	82	96	—	6
Coring	10	1	64	84	—	6
Composite	20	1	77	89	—	6
Green Bean Blancher	5[f]	1	91	95	—	6
Meat Packing	30	1.5	96	89	55	24
Meat Processing and Curing	10[g]	1.5	84	95	72	24
	5	—[i]	—	92	79	24
Shrimp Processing	10[h]	1	97	94	76	24
	60–360	1	85	65[j]	—	8
	30	0.5	98	98	—	25
Crab Processing	30	0.5	96	98	—	25
Salmon Processing	30	0.5	97	98	—	25
Fruit Cake Processing	2	2	89	94	47	24
Cheese Whey	53	3	—	92	4	22
Poultry Composite	5	1	93	74–94	13	20
Chiller	6	1	75	75	62	20
Scalder	30	1.5	87	88	49	20
Crawfish Processing	150	1	83	97	45	2
Tofu	300	1	97	—	—	7

[a] Plus 10 mg/L Betz 1130.

[b] Plus 15 mg/L Betz 1130.

[c] Plus 20 mg/L Betz 1130.

[d] Plus 15 mg/L NJAL-240.

[e] Plus 10 mg/L NJAL-240 and 40 mg/L alum.

[f] Plus 80 mg/L $CaCl_2$.

[g] Plus 40 mg/L $FeCl_3$.

[h] Plus 5 mg/L WT-3000.

[i] Not reported.

[j] Percentage of protein removal.

solids (SS), and chemical oxygen demand (COD) in food-processing waste effluents by coagulation with chitosan followed by gravity settling. In addition to the reduction of waste load, the coagulated by-products recovered from food-processing wastes with

TABLE 19.2

Proximate Composition of Coagulated Solids Recovered from Food-Processing Waste Effluents by Coagulation with Chitosan and Gravity Settling

Effluent	Solids Composition (%)			Ref.
	Protein	Fat	Ash	
Egg Breaking Composite	42.2–42.9[a]	40.8–45.3	2.9–8.8	21
Egg Washer Waste	49.4[a]	36.1	4.2	21
Meat Packing	41[a]	17	11	24
Fruit Cake	13[a]	—[d]	—	24
Cheese Whey	72.3[b]	0.2	9.5	22
Poultry Composite	54.0[a]	29.4	4.3	20
Chiller	34.4[a]	57.7	1.3[e]	20
Scalder	70.6[a]	1.2	15.7	20
Crawfish Processing	27.1[a]	51.7	3.3	2
Tofu	41.9[c]	1.5	0.9	7

[a] Based on Kjeldahl N $\times$ 6.25.

[b] Based on Kjeldahl N $\times$ 6.4.

[c] Based on total amino acid content.

[d] Not determined.

[e] Ash value of solids recovered by dissolved air flotation.

chitosan contain significant amounts (13–72%) of protein (Table 19.2), and offer potential sources of protein in animal feeds.[27]

Bough and co-workers[20] observed that the treatment of poultry processing wastes with 5 mg/L chitosan as a coagulating agent reduced suspended solids in the composite effluents by 74 to 94%. The dry, coagulated solids contained 54.0% crude protein, 29.4% fat, and 4.3% ash. Proteinaceous solids recovered from cheese whey, with and without chitosan as a coagulating agent, had protein efficiency ratios (PER) equivalent to that of the casein control.[22]

In some instances, chitosan has been used in conjunction with a synthetic poly-electrolyte or an inorganic salt to increase treatment effectiveness. With egg-breaking wastes,[21] 100 to 200 mg/L chitosan and 10 to 20 mg/L of the synthetic coagulant Betz 1130 effectively coagulated suspended solids in wastewater. These solids were reduced by 72 to 94%, and COD by 57 to 76%. The amino acid composition of coagulated egg by-products was comparable with that of whole eggs.

In general, suspended solids in food-processing wastes are often reduced by 90% or more with coagulation. On the other hand, COD reductions are usually much less because of the high content of dissolved organic substances that are not removed by coagulation.[20] However, in certain instances, i.e., poultry,[20] eggs,[21] and meat and seafood wastes,[24] COD reductions of 60 to 80% have been obtained.

No and Meyers[28] investigated the recovery of amino acids from crawfish-processing wastewater using amino copper-chitosan. The amino acids recovered by this treatment had potential application as seafood flavors, based on their sensory attributes.

Chitosan is also effective for conditioning activated sludge in conjuction with centrifugal dewatering.[24] Treatment of activated brewery and vegetable sludge with

75 and 40 mg/L chitosan resulted in the removal of 95% and 99% of suspended solids, respectively. Asano and co-workers[29] found that chitosan is effective for dewatering municipal and industrial sludges. Among various polyelectrolytes, chitin–chitosan derived polymers, known as Flonac in Japan, have been widely used for sludge-dewatering applications, mainly because of their effectiveness in sludge conditioning, rapid biodegradability in soil environments, and economic advantages in centrifugal sludge dewatering.[29]

Manipulation in the chitosan-manufacturing process produces chitosans with varying chemical characteristics and molecular weight distributions.[30] These products differ in their effectiveness as waste treatment agents for conditioning of activated sludge and coagulation of food-processing wastewater. For example, Wu and Bough[30] found that high-viscosity chitosan products that were more effective for vegetable sludge conditioning were less effective for coagulation of cheese whey. These workers postulated that this was primarily due to the difference in particle size and charge characteristics of the two wastewater systems, thereby suggesting that different chitosan samples can be produced for particular wastewater systems. Bough and co-authors[31] compared various chitosan samples prepared under different conditions for effectiveness of coagulation of an activated vegetable sludge suspension. Treatment effectiveness did not correlate linearly with viscosity and molecular weight distribution of products; however, higher values for molecular weight were predictive of greater effectiveness for coagulation of activated sludge suspensions.

B. METAL-CONTAMINATED WASTEWATER

Wastewater containing a variety of heavy metals is discharged from many sources, such as mining operations, metal-plating facilities, fertilizer manufactures, or electronic-device manufacturing operations.[11,32] Since these metals are often toxic at low concentrations and are not biodegradable, they must be physically removed from the contaminated water to meet increasingly stringent environmental quality standards.[32] Many methods including chemical precipitation, electrodeposition, ion exchange, membrane separation, and adsorption have been used to treat such streams.[33] Of these, adsorption techniques, specifically those using chelating resin, have been widely promoted, and have been demonstrated to be feasible technologies.[11,34]

Chitosan is very attractive for heavy metal ion separations from wastewater because it selectively binds to virtually all group III transition metal ions at low concentrations but does not bind to groups I and II alkali and alkaline earth metal ions.[4,35] The amino groups on the chitosan chain serve as a chelation site for transition metal ions.[32] Studies on chemical interactions between three metal ions [Cu(II), Mo(VI), Cr(VI)] and chitosan by X-ray photoelectron spectroscopy confirmed that sorption occurred on amino functional groups for all three metals.[36]

Muzzarelli[4] documented the effectiveness of chitin, chitosan, and other polymers for chelating of transition metal ions. He indicated that chitosan was a powerful chelating agent and exhibited the best collection ability of all the tested polymers because of its high amino group content. Masri and co-workers[9] compared the chelating ability of chitosan with several other materials, such as bark, activated sewage sludge, and poly(p-aminostyrene), and confirmed that chitosan had a higher chelation ability. Yang and Zall[11] found that chitin, chitosan, and scales from three species of fish (porgy, flounder, and cod) were potentially useful materials to remove metals [Cu, Zn, Cr(III), Cd, and Pb] from contaminated water; however, the diffusion of solutes through internal particles of chitosan was relatively faster than through the other adsorbents tested.

Jha and colleagues[37] studied the uptake of cadmium with time at pH 6.5 at initial concentrations of 1.5, 5.0, and 10.0 mg/L, and found that the rate of sorption was very rapid initially, decreasing markedly after 4 h. According to them, amino groups of chitosan act initially as cadmium coordination sites, and the slower removal rate of metal after an initial rapid uptake can be due to the binding of cadmium by complexed metal ions. A relatively rapid initial rate of adsorption has also been observed for mercuric ions.[38]

There are essentially three consecutive stages associated with the adsorption of materials from solution by porous adsorbents: transport of the adsorbate to the exterior surface of the adsorbent, diffusion of the adsorbate into the pores of the adsorbent, and adsorption of the solute on the interior surfaces of adsorbent. In general, the adsorption of solute onto the surface of an adsorbent is relatively fast compared with the other two processes. In addition, external resistance is small if sufficient mixing is provided.[11] The study of adsorption kinetics of copper by chitosan showed that the intraparticle diffusion was the rate-determining step.[11] A similar mechanism was observed subsequently with Cu, Hg, and Cd ions.[38–40]

The free amino group in chitosan is much more effective for binding metal ions than the acetyl groups in chitin,[11,41–44] leading us to postulate that the higher free amino group content of chitosan should give higher metal ion adsorption rates. Yang and Zall[11] noted that chitosan has many free amino groups, chelating as much as five- to sixfold greater amounts of metals than does chitin. Similarly, Eiden and colleagues[41] reported that the uptake of Pb(II) on chitin was approximately 21% compared with chitosan. However, Deans and Dixon[34] studied comparative efficiencies of a series of different functionalized biopolymers for removing Pb(II) and Cu(II) ions from water at 1, 10, and 100 ppm and observed that there is no optimal biopolymer for metal ion uptake. The uptake results varied with the metal ion identity, its concentration, and adsorbent structure. For example, for Cu and Pb at 1.0 ppm, the best adsorbent was found to be chitosan and chitin, respectively. For Pb at 10 and 100 ppm, the carboxymethylcellulose hydroxamic acid was the best performer. These investigators concluded that the choice of an ideal adsorbent should be determined for each application.

The adsorption ability of chitosan is dependent on many other factors, such as crystallinity, deacetylation, hydrophilicity,[45] particle size, and agitation rate.[40] Earlier, Muzzarelli and Tubertini[43] also mentioned that adsorption depends on the time of contact, temperature, pH, concentration of the ion under examination, and concentration of other ions present.

It is interesting to note the relationship between physicochemical properties or sources of chitosan and metal-binding property. Madhavan and Nair[46] studied metal-binding (Zn, Ni, Cr, Cu, Fe, and Mn) properties of three chitosans with viscosities of 155, 430, and 902 cP in a 1% solution in 1% acetic acid. It was found that the metal-binding property of chitosan was not affected by viscosity characteristics. Nair and

Madhavan[47] investigated the metal binding (Fe, Co, Ni, Hg, and Cu) property of chitosans prepared from four different sources, i.e., crab (*Scylla serrata*), prawn (*Penaeus indicus*), squid (*Loligo* sp.), and squilla (*Oratosquilla nepa*). These four chitosans studied did not show notable variations in the adsorption rate (Table 19.3). Similarly, the adsorption capacity of Cu(II) on flake and bead type of chitosans prepared from three different sources (shrimp, crab, and lobster) was reported to be comparable.[48] However, Tseng and colleagues[49] reported that the metal binding (Cu, Ni, Cd) capacity of chitosans prepared from shrimp, lobster, crab, and cuttlebone was different due to their dissimilar pore structures and differences in pore size distribution and pore diameter. Kurita and co-workers[45] found that a homogenous hydrolysis process could give a chitosan product with a higher adsorption rate for metal ions than one prepared by a heterogeneous process with the same degree of deacetylation.

Particle size may also affect the metal adsorption capacity of chitosan. A smaller particle size increased the adsorption capacity for Cd(II),[32,37,40] Cr(III),[42] Cr(VI),[50] and U(VI).[51,52] The increase in adsorption capacity with decreasing particle size suggests that the metal became preferentially adsorbed on the outer surface and did not fully penetrate the particle. Because adsorption initially takes place at the outside surface of

TABLE 19.3
Rate of Adsorption of Metal Ions by Chitosan from Different Sources

Metal Ions	Chitosan Source	mg of Metal Adsorbed/g of Chitosan			
		10 min[a]	30 min	60 min	120 min
Fe^{3+}	Crab	11.7	17.6	23.4	23.4
	Prawn	5.9	11.7	15.7	23.4
	Squid	17.6	17.6	20.5	23.4
	Squilla	11.7	14.6	17.6	29.3
Co^{2+}	Crab	0.0	4.1	4.7	5.9
	Prawn	4.7	5.3	7.1	7.1
	Squid	4.7	4.7	4.7	7.4
	Squilla	4.7	4.7	4.7	4.7
Ni^{2+}	Crab	11.7	35.2	55.2	64.6
	Prawn	29.3	47.0	64.6	82.1
	Squid	11.7	29.3	64.6	82.1
	Squilla	5.8	29.3	52.8	76.3
Hg^{2+}	Crab	161	241	281	321
	Prawn	251	311	331	341
	Squid	200	321	346	366
	Squilla	261	351	381	411
Cu^{2+}	Crab	15.1	21.1	36.2	39.3
	Prawn	27.2	30.2	42.3	66.4
	Squid	12.1	27.2	45.3	51.3
	Squilla	9.0	42.3	60.4	60.4

[a] Shaking time.

Source: From Nair, K.R. and Madhavan, P., *Proceedings of the 2nd International Conference on Chitin and Chitosan*, Sapporo, Japan, 1982, pp. 187–190. With permission.

a sorbent, the dependence of the uptake rate on size is most likely because the smaller particles have greater outside surface area per weight, resulting in faster initial uptake.[40] On the other hand, several workers found only a small variation in Cu,[39] Pt,[53] and V[51] adsorption capacity per unit mass of chitosan with particle size. Cervera and co-workers[54] studied the effect of molecular weight (150, 400 and 2000 kDa) and particle size (flake and powder) of chitosan on the retention of Cd(II) and Cr(III). These authors claimed that medium-molecular-weight (400 kDa) chitosan in flakes provided the best adsorption capacity for both Cd (66%) and Cr (83%). These values were comparable with those found by using 2000 kDa chitosan in fine powders.

Evans and colleagues[40] compared four different agitation rates of 50, 75, 100, and 200 r/min for Cd adsorption at an initial concentration of 2.0 ppm and found that from 50 to 100 r/min, cadmium uptake rate increased with an increase in agitation rate. The uptake rates at 100 and 200 r/min were similar, indicating that the agitation rate at this range and above is not a limiting factor for uptake. Stirring speed (200, 400 and 600 r/min) had no influence on the adsorption of Cr[55] or vanadium (IV)[56] but the adsorption rate of Cu(II)[55] increased slightly at a higher stirring speed. Findon and colleagues[57] suggested that an increase in turbulence reduces the boundary layer thickness of the particle, thus increasing the uptake for certain metal ions.

Regarding the effect of contact time on metal adsorption, Cervera and co-workers[54] reported that an increase in contact time between metal solution and chitosan up to 30 min increased the retention percentages of Cd(II) and Cr(III); a further increase in contact time up to 90 min did not improve the metal retention. McKay and co-workers[39,58] measured adsorption isotherms for Cu^{2+}, Hg^{2+}, Ni^{2+}, and Zn^{2+} on chitosan powder as a function of temperature (25–60°C) and found that adsorption capacity decreased with increasing temperature.

Differences in the pH of the solution have been reported to influence the metal ion adsorption capacity of chitosan. The uptake of Cd(II),[37,40,44,54] Cr(III),[42,54] Cu(II),[59,60] Ni(II),[59,61] Zn(II),[59] and U(VI)[51] ions by chitosan decreased with decreasing pH. These results suggest that hydronium ions compete with metal ions for available amino sites on chitosan.[32] However, a drastic decrease in Cr(VI) removal has been observed with an increase in pH to neutral and above; in one study, sorption was almost 90% at pH 3 at an initial Cr concentration of 5 mg/L, and was reduced to 10% at pH 7 and above.[50] In contrast, the pH (2.9–8.4 for Co^{2+}, 2.9–6.5 for Ni^{2+}, and 3.0–6.0 for Hg^{2+}) of the solution did not significantly influence either the rate of adsorption or the quantity of these metals adsorbed by chitosan.[47] Schmuhl and co-workers[55] studied adsorption over the pH range of 2 to 11 using 1.0 g chitosan and 100 mg/L Cr(VI) solution and found that the maxium adsorption of Cr(VI) occurred at pH 5, and decreased at lower and higher pH values. In a further study, the effect of pH (2, 3, 4, and 5) on the removal of Cu(II)[55] was studied; it was found that pH did not play a significant role in the removal of Cu, except at pH 2, where excellent removal was achieved.

Schmuhl and co-workers[55] studied the removal of Cr(VI) and Cu(II) by chitosan as a function of time at pH 5, at various initial concentrations. The results showed that the time taken to reach equilibrium was longer as the concentration of Cr(VI) and Cu(II) increased. It also appeared that at lower concentrations, less Cr(VI) and Cu(II) was adsorbed onto the chitosan than at higher concentrations. At an initial concentration of 1000, 500, 200, 100, 50, and 10 mg/L, the Cr(VI) removal was 200, 150, 78, 45, 24, and 5 mg/L, while the amount of Cu(II) adsorbed was 285, 176, 60, 32, 24, and 7.6 mg/L, respectively. An increase in the adsorption of Cu(II) with increasing initial concentration was also observed by Wu and colleagues.[62]

The presence of counterions or other substances in solution may, or may not, affect the adsorption of metals by chitosan. The presence of the sulfate ion enhanced adsorption of Co, Cu, and Ni ions by swollen chitosan beads,[63,64] and uptake of Cr(III) by chitosan was enhanced in the presence of 0.05 M phosphate at pH 5.0.[42] Conversely, the presence of appreciable quantities of chloride did not have any effect on cadmium removal.[37] Furthermore, removal of cadmium by chitosan was not significantly affected by the presence of calcium ions at 100 ppm, but was significantly decreased by zinc or complexing agents such as ethylenediaminetetraacetic acid (EDTA). On the other hand, competitor anions, such as chloride and nitrate, induced a large decrease in the platinum sorption efficiency of glutaraldehyde cross-linked chitosan.[53]

Bassi and co-workers[65] studied the effects of citric and oxalic acids (pH adjusted to 5.5), both commonly used to leach heavy metals from contaminated soils, on the adsorption of Zn^{2+}, Cu^{2+}, Cd^{2+}, and Pb^{2+} ions by chitosan flakes. The adsorption capacity of chitosan for these ions was not affected in the range of 10^{-5} to $10^{-2} M$ of both organic acids, but was significantly reduced at concentrations higher than $10^{-2} M$. Decreased uptake of metal ions by chitosan in the presence of high concentrations of organic acids is attributed to the fact that organic acids compete with metal ions for free amino groups.

The use of chitosan for the removal of Cu(II) from simulated rinse solutions (0.3–5.0 mol/m^3) containing chelating agents (EDTA, citric acid, tartaric acid, and sodium gluconate) has been studied.[66] In the absence of chelating agents, chitosan showed an excellent ability for Cu(II) adsorption with a capacity of 2.75 mol/kg. However, the adsorption capacity in the presence of chelating agents sharply decreased, except in the case of gluconate.

The affinity of chitosan for different metals in a single-solute system[46,47,67,68] as well as in a multi-solute system[11] can vary greatly. In a single-solute system, for example, the order of chitosan removal (%) was Cu (98.3)>Ni (78.5)>Co (21.0)>Mn (7.0) when each metal was tested at a concentration of 100 ppm.[67] In a similar study, Nair and Madhavan found significant variation in the affinity of the different metal ions (Fe, Co, Ni, Hg, and Cu) for chitosan, with minimum adsorption for Co^{2+} (7.4 mg/g) and maximum for Hg^{2+} (411 mg/g).[47] Comparable results were observed[68] with swollen chitosan beads at metal ion concentrations of 100 ppm as follows: Cu (98.4)>Ni (82.2)>Zn (70.5)>Co (36.8)>Mn (13.8). In a multi-solute system, Yang and Zall[11] found that the selectivity sequence in the uptake of metals by chitosan was Cu>Cr=Cd>Pb>Zn. The percentage removal of metals by chitosan was shown to be approximately 10 to 15% less in a multi-solute system than in a single-solute system, because competitive adsorption occurs in a multi-solute system. However, metal ions can be removed almost completely under competitive conditions using amino acid ester-substituted chitosans.[67]

Jha and co-workers[37] demonstrated that 88% of the cadmium adsorbed on chitosan powder could be released in 0.01 N HCl. Under low-pH conditions, protons effectively compete for the active sites, thus releasing the metal ion into the suspending medium. Ni, Cu, and Cr(III) adsorbed on chitosan flake can also be desorbed at low pH, even under repeated adsorption/desorption cycles.[69] Thus, the metal ion adsorption process on chitosan is reversible, making adsorbent regeneration and metal recovery at low pH feasible.[32] On the other hand, Masri and Randall[10] indicated that the copper bound to chitosan was readily desorbed with a solution of ammonium hydroxide–ammonium chloride at pH 10. Minimal desorption of chromium was observed on washing previously chromium-equilibrated chitosan with distilled water.[42]

Masri and Randall[10] demonstrated the effectiveness of chitosan for the removal of toxic metal ions in the actual waste streams. Wastes treated were (a) from electroplating and metal-finishing operations (with disposal problems mainly of cyanide and salts of Cr, Cd, Zn, Pb, Cu, Fe, and Ni); (b) from a nickel-salt manufacturing plant (disposal of nickel and alkali); (c) from a lead-battery manufacturing plant (disposal of sulfuric acid and lead salts); and (d) from exhausted dyebath for wool fabrics in which dichromate is included in the bath (disposal of chromium). Chitosan was effective in reducing the content of copper, cadmium, iron, zinc, lead and nickel salts, and sulfuric acid. For example, treatment of nickel waste with chitosan (500 mL/2.0 g for 1.0 h) reduced the nickel ion content from 7.2 to 1.2 ppm. Treatment of lead-battery wastes with chitosan (50 mL/2.0 g for 1.0 h) reduced the lead concentration from 1.8 to less than 0.5 ppm and the iron concentration from 30 to 8.7 ppm.

Chitosan has potential for purifying electroplating wastewater.[69] Partially deacetylated chitin (PDC, deacetylated at the outer periphery of the particles without substantially deacetylating the interior regions of the particles) has been used as a metal ion-complexing material. PDC was found to be a useful sorbent for transition metal ions (Cu, Ni, and Cr), being nearly as effective as relatively pure chitosan. Coughlin and co-workers[69] concluded that this approach was economically more favorable than the conventional precipitation process under reasonable assumptions. Another study found that water-soluble chitosan (WsCs) was effective in removing heavy metal ions (Cr, Fe, and Cu) from plating wastewater.[70] When WsCs was blended with either sodium N,N-diethyldithiocarbamate trihydrate (SDDC$_T$) or sodium salicylate (SS$_C$), the removal efficiency was further increased primarily due to the excess amount of hydrophilic sulfonic and carboxylic groups.

Other workers[10,11] have suggested that although chitosan as such is useful for treating wastewaters, it may be advantageous to use the cross-linked chitosan with di- or polyfunctional reagents to impart increased resistance to solubilization in acidic pH effluents. Chitosan is insoluble in aqueous alkali solution of pH up to 13 and in inorganic solvents, but is soluble in weak acid solutions, except sulfuric acid.[71] These factors undoubtedly limit its use as an adsorbent in low-pH regions.

More recent studies have shown that the chelating ability of chitosan could be further improved by chemical modifications such as cross-linking with glutaraldehyde[72,73] (Figure 19.2A), acylation[73–75] (Figure 19.2B), and amino acid conjugates[67] (Figure 19.3). For example, Koyama and Taniguchi[72] reported that homogeneously cross-linking chitosan with glutaraldehyde at the aldehyde–amino group ratio of 0.7 increased adsorption of Cu ion to as high as 96% in an aqueous solution of pH 5.2, although the original chitosan collected only 74% of Cu ion. This increase in adsorption was interpreted in terms of the increase in hydrophilicity and accessibility of chelating groups as a result of partial destruction of the crystalline structure of chitosan by cross-linking under homogeneous conditions. The X-ray diffraction

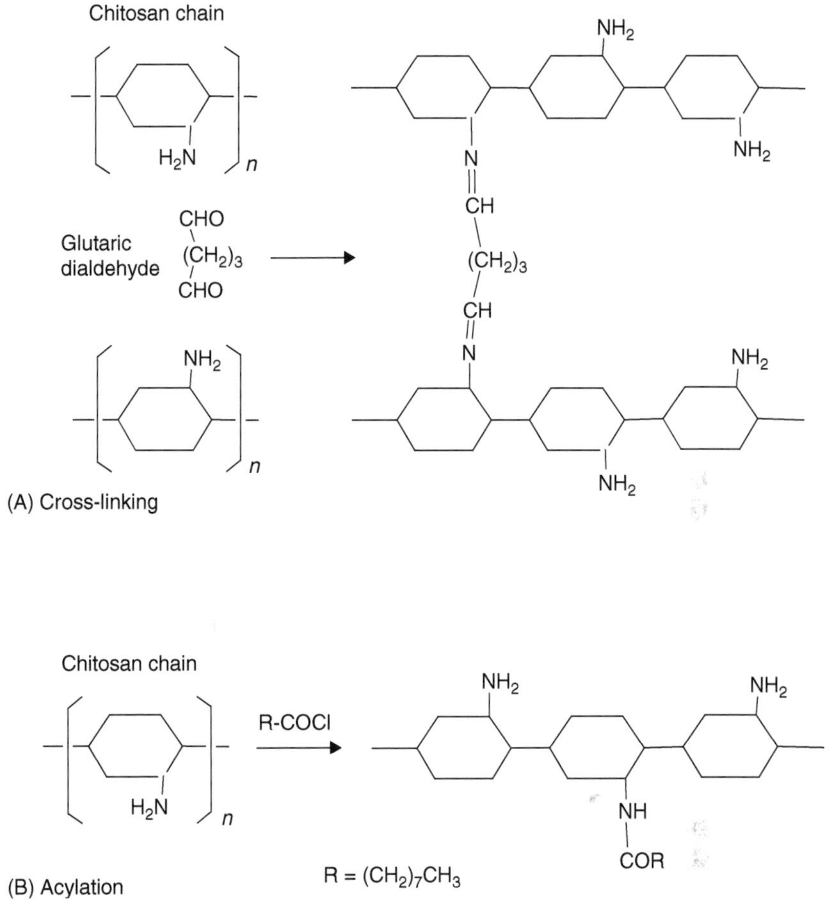

FIGURE 19.2   Chemical structures. (A) Crosslinking of chitosan with glutaraldehyde. (B) N-acylation of chitosan (From Hsien, T.Y. and Rorrer, G.L., *Sep. Sci. Technol.*, 30, 2455–2475, 1995. With permission.)

diagrams revealed a marked reduction in crystallinity by cross-linking, and the cross-linked chitosan was found to be almost amorphous.

Masri and Randall,[10] on the other hand, attempted the cross-linking of chitosan with glutaraldehyde under heterogeneous conditions and found that the adsorption was decreased by cross-linking, although cross-linking enabled the use of chitosan in acidic media as a result of insolubilization. Consequently, controlled cross-linking with an appropriate amount of glutaraldehyde under homogeneous conditions effectively enhanced the adsorption capacity of chitosan, producing high-potential adsorbents, practically applicable in all pH regions without a dissolution problem.[72]

Chitosan–amino acid conjugates have been prepared by coupling amino acid esters to the carboxyl group of glyoxylic acid-substituted chitosan[67] and have been shown to increase the removal of heavy metals (Cu, Ni, Co, and Mn). These metals

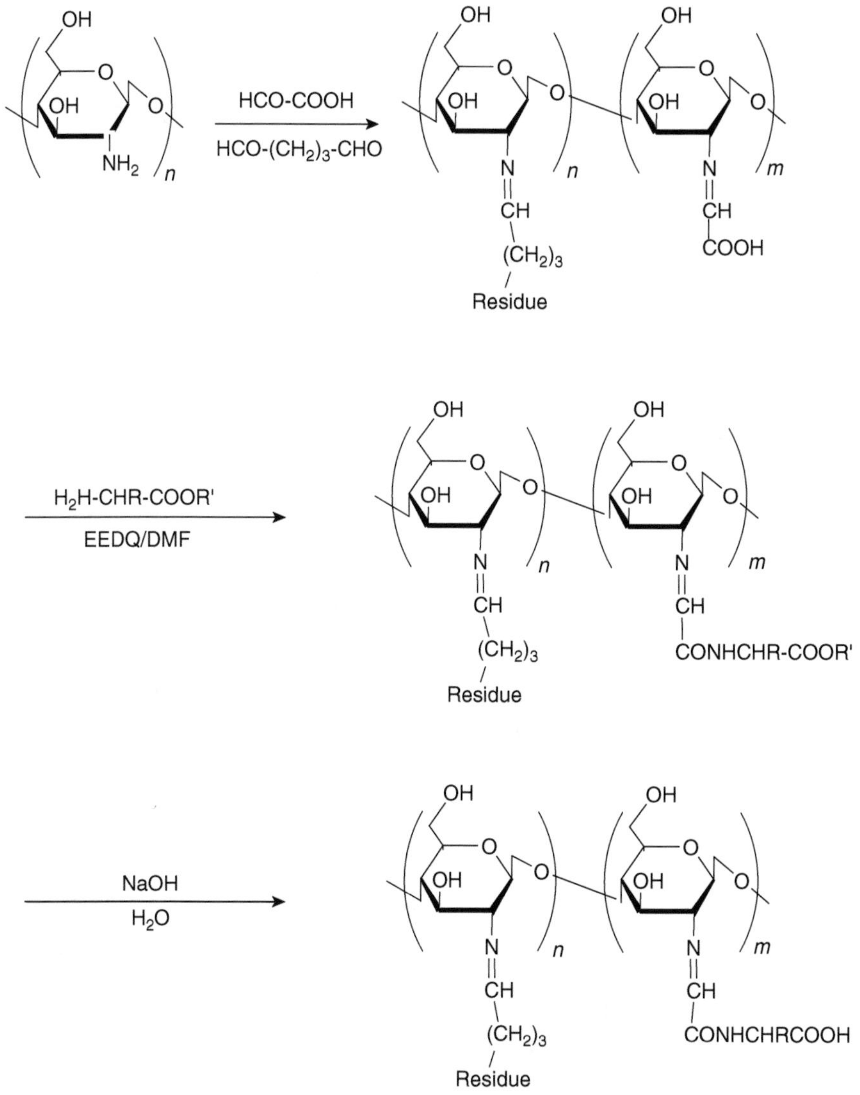

FIGURE 19.3 Synthetic scheme for production of chitosan–amino acid conjugates. R, side chain of amino acid; R′, alcohol component of ester. (From Ishii, H. et al., *Int. J. Biol. Macromol.*, 17, 21–23, 1995. With permission.)

were almost completely removed from solution by chitosan–amino acid conjugates at 100 ppm. The extent of removal, however, was not affected by the type of amino acid introduced.

Many researchers have clearly demonstrated that chitosan has an intrinsically high affinity and selectivity for transition metal ions. However, the adsorbent raw material is not suitable for processing aqueous waste streams. Chitosan is usually obtained in a flaked or powdered form that is both nonporous and soluble in acidic media. The low internal surface area of the nonporous material limits access to interior adsorption sites and hence lowers metal ion adsorption capacities and adsorption rates. The solubility of chitosan in acidic media prevents its use in the recovery of metal ions from wastewater at low pH. Furthermore, the flaked or powdered form of chitosan swells and crumbles easily and does not function ideally in packed-column configurations common to pump-and-treat adsorption processes.[32]

Recently, porous beads of chitosan were synthesized for the adsorption of ppm-level transition metal ions from aqueous solution.[32,74,76,77] For example, highly porous, magnetic beads of chemically cross-linked chitosan have been used for the removal of cadmium ions from dilute aqueous solution.[32] These highly porous chitosan beads were prepared by the dropwise addition of an acidic chitosan solution into a sodium hydroxide solution precipitation bath. The gelled chitosan beads were cross-linked with glutaraldehyde and then freeze-dried. Beads of 1 and 3 mm diameter possessed internal surface areas of 156 and 94 m^2/g and mean pore sizes of 560 and 540 Å, respectively, and were insoluble in acid media at pH 2. By comparison, the surface area of the nonporous chitosan flake was only 0.35 m^2/g.

Kawamura and co-workers[76] synthesized polyaminated highly porous chitosan chelating resin and found that polyamination of the amine group significantly improved adsorption capacity for mercury ions. The adsorption capacity of Hg(II) on polyaminated highly porous chitosan beads was about 1.4 times greater than that on Unicellex UR-120H, which was the best commercial chelate resin for removal of Hg(II) in Japan.[78] Hsien and Rorrer[74] synthesized N-acylated, cross-linked highly porous chitosan beads insoluble in 1 M acetic acid solution (pH 2.4) with an internal surface area of 224 m^2/g.

The behavior of the adsorption isotherms suggests that both the size and the porous structure of the bead have a profound effect on adsorption capacity.[32] The internal surface area and porosity of the 1 mm beads were greater than the 3 mm beads, and maximum adsorption capacities for the 1 and 3 mm beads were 518 and 188 mg of Cd/g of bead, respectively. An increase in adsorption capacity with a decrease in particle size also suggests that the metal ions do not completely penetrate the particle and that the metal preferentially adsorbs near the outer surface of the bead by the pore-blockage mechanism.

Table 19.4 provides a list of references for further reading on adsorption of metals by chitosan and its derivatives.

C. DYE WASTEWATER

Textile effluents usually contain very small amounts of dyes; however, these are highly detectable and have become a pollution concern. Several difficulties are encountered in the removal of dyes from wastewaters because they are highly stable molecules, made to resist degradation by light, chemical, biological, and other exposures. Commercial dyes are usually mixtures of a large complex with often uncertain molecular structure and properties.[13]

The adsorption of dyestuffs by solid polymers has been widely investigated, mainly for industrial applications. At least 70 dyes have been studied so far in terms of

TABLE 19.4

References for Further Reading on Adsorption of Metals by Chitosan and Its Derivatives

Metal	Ref.
Ag	9,10,43,88
Au	9,10,43
Cd	9–11,32,37,40,44,49,54,65,71,74,76,84–86
Co	9,10,43,47,64,67,68,82,84,85
Cr	9–11,41,42,46,50,54,55,69–71,85
Cu	9–11,33,34,38,39,43,46–49,55,58,59,61–63,65–73,75,76,79,81,82,84,85
Fe	9,10,46,47,70,86
Hg	9,10,43,47,58,76,78,81,85,89
Ir	43,80
Mn	9,10,46,67,68,84,86
Mo	9,43,77
Ni	9,10,33,46,47,49,58,59,61,63,64,67–69,76,82–85
Pb	9–11,34,41,65,71,85,89
Pd	9,43,80
Pt	9,10,53,80
Sb	43
U	51,52,76,84,85,87
V	51,56
Zn	9–11,33,43,44,46,58,59,65,68,71,76,84–86

their adsorption on chitin and chitosan.[90] Extensive studies on the adsorption of dyes on chitin[91–97] have revealed that chitin has the ability to adsorb substantial quantities of dyestuffs from aqueous solutions. However, chitosan with higher amino group content was reported to be more effective for binding dyes than chitin.[13,98–101] For example, the dye adsorption capacity of chitosan toward Basic Blue 69, Acid Blue 25, Reactive Red 123, and Reactive Yellow 145 was about 3 to 6 times higher than that of chitin.[98] The binding behaviors of chrome violet (C.I. Mordant Violet 5, a monoazo dye) to chitin and partially deacetylated chitin (degree of deacetylation=65%) were investigated in another study that found that the dye-binding capacity of partially deacetylated chitin was much greater than that of chitin.[101] The content of amino groups in partially deacetylated chitin increased by a factor of 8.6 compared with that in chitin.

The interaction of chitosan with dyes has been studied by several workers. Yamamoto[102] studied the chiral interaction of chitosan with azo dyes (Orange I, Alizarin Yellow GG, and Congo Red) and suggested that intermolecular interactions of the dye molecules are most probable in the chitosan–dye systems. Shimizu and co-workers,[101] and Juang and colleagues[99] reported that electrostatic interactions are involved in the binding of chrome violet and reactive dyes (Reactive Red 222, Reactive Yellow 145, and Reactive Blue 222) to chitosan, respectively. The interaction of dyes with chitosan derivatives was investigated by Seo and co-workers,[103] who explored the predominant contribution of the hydrophobic interaction in the systems of butyl orange/N-octanoyl chitosan gels.

Recently, the interaction of several cationic and anionic dyes with water-soluble chitosan derivatives (*N*-carboxymethyl chitosan, *N*-carboxybutyl chitosan, and the reduction product of the aldimine obtained from chitosan and 5-hydroxymethyl-2-furaldehyde) and the parent chitosan was studied in water.[90,104] The occurrence of an interaction in water among the hydrophilic chitosan derivatives, the parent chitosan, and the anionic dyes Orange II, Alizarin S, Alizarin GG, and Congo Red was proven by optical techniques[90] and by optical and thermodynamic approaches.[104] Binding depended on pH, with the pH range 3.5 to 5 being the most effective in this regard. However, the cationic dyes Neocuproin, 1-(2-pyridilazo)-2-naphthol, Ethidium bromide, and Ruthenium Red did not interact with the polysaccharides studied in the pH range 5 to 6.[90]

Representative classes of anionic dyes include the acid, direct, mordant, and reactive dyes. Acidic dyes are water-soluble molecules containing one or more anionic groups (most often sulfonic acid).[105] Maghami and Roberts[106] investigated adsorption characteristics of three anionic dyes (C.I. Acid Orange 7, C.I. Acid Red 13, and C.I. Acid Red 27) on chitosan under acid conditions and demonstrated a 1:1 stoichiometry for the interaction of sulfonic acid groups on the dyes with protonated amino groups of the chitosan for mono-, di-, and trisulfonated dyes. Treatment of chitosan with acid produces protonated amine groups along the chain, and these can act as dye sites for anionic dyes. Chitosan, although soluble in dilute acetic acid, will not dissolve if the acetic acid solution contains an excess of an anionic dye. Venkatrao and co-workers[107] studied the adsorptive behavior of chitosan with respect to two dyes (C.I. Reactive Red 73 and C.I. Direct Red 31) and found that intraparticle diffusion was a major rate-controlling step in the adsorption of direct and reactive dyes from aqueous solutions.

The dye adsorption ability of chitosan is influenced by many factors, such as contact time, dye concentration, particle size, temperature, types of chitosan, and metal ions. Bhavani and Dutta[108] reported that the adsorption of dye on chitosan increased with an increase in contact time of up to 4 h for Direct Jacophix T-Blue and 5 h for Reactive T-Blue and Direct Inchromine Brown-2G, respectively. Studies on the variation of adsorption with different dye concentrations (0.25, 0.50, 0.75 and 1.0 mg/100 mL) at their equilibrium times (4 or 5 h) showed that dye adsorption was more effective at higher concentrations.

Annadurai and co-workers[109] investigated the effects of adsorbent particle size (1.651, 0.384, and 0.177 mm) and dye solution temperature (30, 45, and 60°C) on basic dye (Methylene Blue) adsorption by chitosan. As the particle size decreased and the temperature increased, adsorption was found to increase. Increase in temperature may produce a swelling effect within the internal structure of chitosan, enabling the dyes to penetrate further. Wu and co-workers[62] also reported that adsorption of Reactive Red 222 dye decreased with the increased size of chitosan particles (average diameter of 0.335, 0.505, and 0.715 mm). An increase in dye capacity with decreasing particle size suggests that the dye does not completely penetrate the particles or that the dye preferentially adsorbs near the outer surface of the particle.[99]

Wu and co-workers[48] studied the adsorption capacity of reactive dye (Reactive Red 222) on flaked and bead type of chitosans prepared from three different sources (shrimp, crab, and lobster). It was found that the bead type of chitosan had a higher adsorption capacity than the flake type by a factor of 2.0 to 3.8, depending on the chitosan sources. These investigators explained that this is probably due to the looser pore structure of the bead type of chitosan compared with the flake type of chitosan.

Shimizu and co-workers[101] examined the effect of metal ions on the binding of chrome violet by chitin and partially deacetylated chitin. The results indicated that Zn^{2+} and Cu^{2+} ions did not perceptively influence the binding affinity of chrome violet to chitin. In contrast, Co^{2+} ion enhanced binding and Ni^{2+} ion suppressed it. At lower free dye concentrations, dye uptake by partially deacetylated chitin was greatly enhanced by the addition of Co^{2+} ion in the buffer solution of pH 5. Dye uptake was considerably increased by the addition of Cu^{2+} ion at pH 5, becoming much larger at pH 6, the amount corresponding to that in the presence of Co^{2+} ion.

Smith and co-workers[13] extensively studied the decolorization of dye wastewater using chitosan. In isotherm studies performed on sorption of Acid Red 1 dye onto chitin, cold batch chitin, and chitosan, the latter revealed an extremely high sorption capacity for this dye. A study on the effect of flow velocity through a bed of chitosan using a sorption unit demonstrated a complex relationship between flow rate and rate of dye removal from an aqueous dye solution; this probably resulted from complex factors related to internal pore structure, penetration, and diffusion. Nine different dyes were also tested on the prototype decolorization unit using chitosan as a sorbent (Table 19.5). The molecular size of the dye was a major factor in sorption characteristics, with small, low-molecular-weight dyes sorbing best on chitosan, while dyes with metal in the structure did not appear to sorb any better than their nonmetallized counterparts.

In equilibrium adsorption measurements with three anionic dyes, Maghami and Roberts[106] found that equilibrium was reached more rapidly with the smallest dye, being attained in less than 2 h with C.I. Acid Orange 7, while approximately 9 h were required with the other two dyes (C.I. Acid Red 13 and C.I. Acid Red 27). However, the ionic charge on the dye ion appeared to have negligible effect on the time to equilibrium.

Removal of color from Navy 106 slack wash water discharge was tested by using 22 different adsorbents at pH 10.4 (the pH of the process water), 7.5, and 4.6 by

TABLE 19.5
Sorption Characteristics for Several Dyes on Chitosan

Dye	Molecular Weight	Chemical Type	Original Concentration (g/L)	Sorption Capacity (mg/g)
Acid Red 1	509	Monoazo	0.0908	7.13
Acid Blue 25	401	Anthraquinone	0.1510	13.39
Acid Blue 193	479	Monoazo	0.1410	8.43
Mordant Black 17	375	Monoazo	0.1120	9.21
Direct Blue 86	781	Phthalocyanine	0.0955	3.94
Direct Red 81	699	Diazo	0.1299	—
Direct Green 26	1344	Triazo	0.0510	—
Reactive Red 120	1466	Diazo	0.0972	2.61
Reactive Violet 5	733	Monoazo	0.1770	—

Source: From Smith, B. et al., *Am. Dyest. Rep.*, 82, 18–36, 1993. With permission.

Michelsen and co-workers.[12] In general, the color of the Navy 106 was removed more effectively by sorption at decreased pH. With chitosan, 36, 72, and 86% of the color was removed at pH 10.4, 7.5, and 4.6, respectively. In a study on color removal from slack wash water as a function of sorbent concentration for TM-399 (bentonite clay modified with a quaternary ammonium surfactant), activated carbon, chitosan, and pure chitin, chitosan was found to be less effective than activated carbon. Similarly, color removal by the adsorption of dyes from Navy 106 jet dye cycle effluent diluted 1 to 20 and adjusted to pH 7.0 was performed with varying amounts of alumina, activated carbon, TM-399, and chitosan. In this test, activated carbon performed best with color changes of 4000 to 5 ADMI color units at a dosage of 2500 ppm, and chitosan was the poorest performer.

Park and co-workers[110] applied chitosan as an adsorbent for the dye Toluidine Blue O. The adsorption of Toluidine Blue O by chitosan was found to be affected by the particle size and mass of chitosan, initial dye concentration, reaction time, and pH of the solution. Dye was adsorbed more efficiently with chitosan of a smaller particle size and with an increase in the pH of the solution. When the initial ratio of dye to chitosan was more than 1:500, the adsorption of dye rapidly declined.

In a further study, Park and co-workers[100] applied chitin and chitosan prepared from red crab and squid pen as adsorbents for trapping dyes in wastewater from dyeworks. They found that chitin and chitosan were effective adsorbents for such dyes, with chitosan being more effective in dye adsorption than chitin. In a continuous-elution column experiment, the researchers claimed that 1.0 kg of chitosan could be used for the treatment of up to 120 L of wastewater containing 0.05% dye wasted from dyeworks at 75% efficacy, i.e., 45 g of dye absorbed/kg of chitosan.

D. PHENOL- AND PCB-CONTAMINATED WASTEWATER

1. Phenol-Contaminated Wastewater

Phenols represent one of the most important classes of synthetic industrial chemicals and are often found in effluents from various manufacturing operations.[111] Furthermore, phenols are common components of pulp and paper wastes and have been observed to be groundwater contaminants. Despite the importance of treating phenol-containing wastewaters, current methods, in the form of physical, chemical, and microbiological treatments, are not satisfactory.[14]

A two-step approach for removing phenols from wastewater has been investigated by Sun and co-workers.[14] In the first step, weakly adsorbable phenols are converted into quinones by the mushroom enzyme tyrosinase. The tyrosinase-generated quinones are then chemisorbed onto chitosan. This proposed approach is illustrated by the following:

Step 1. Tyrosinase reaction:

    phenol → *o*-quinone + other intermediates

Step 2. Chemisorption:

    o-quinone + other intermediates + sorbent → chemisorbed compounds

When mushroom tyrosinase and chitosan were simultaneously added to dilute, phenol-containing solutions, a nearly complete removal of phenols was observed.

Sun and co-workers[14] mentioned three potential benefits of the two-step tyrosinase reaction–chitosan adsorption approach. As the adsorption of quinones onto chitosan is very strong, the two-step approach is more effective in removing traces of phenols from wastewater. The second benefit is that tyrosinase can react with a wide range of phenols and is less sensitive to changes in waste stream composition and strength. The final benefit is that chitosan is obtained from chitin, a waste product of the shellfish industry. However, as the mushroom enzyme is quite expensive, a less costly tyrosinase should be available for practical applications.

Payne and co-workers[112] also found that a two-step tyrosinase reaction–chitosan adsorption approach could be used to remove selectively phenols from aqueous mixtures containing nonphenols (anisole and benzyl alcohol). The mushroom tyrosinase was specific for the phenol and did not react with either the anisole or the benzyl alcohol. Chitosan effectively adsorbed the tyrosinase-generated products without adsorbing nonphenols.

The removal of phenols from wastewater by soluble and immobilized tyrosinase was investigated by Wada and co-workers.[15] These workers found that phenols in an aqueous solution were removed after treatment with mushroom tyrosinase, with the reduction rate of phenols accelerated in the presence of chitosan. They further found that by treatment with tyrosinase immobilized on cation-exchange resins, 100% of phenol was removed after 2 h, and enzymatic activity was reduced only slightly even after ten repeated treatments.

The use of peroxidase enzymes to catalyze the oxidation of phenols by hydrogen peroxide also has been studied extensively.[113-115] Wagner and Nicell[115] attempted to reduce the phenol content of petroleum refinery wastewater by treatment with horseradish peroxidase and H_2O_2. As a result of the treatment, phenols were transformed into less biodegradable compounds, which could then be removed by subsequent coagulation and precipitation. The use of chitosan as a protective additive resulted in a 25-fold reduction in enzyme requirements.

While peroxidases have the potential to treat a variety of compounds over wide ranges of pH and temperature, one of the major concerns regarding the use of peroxidase is the prohibitive cost of enzyme and hydrogen peroxide. Tyrosinase catalyzes the oxidation of phenols using oxygen as an oxidant. Thus, tyrosinase may eventually represent a less expensive alternative to peroxidases for the treatment of phenolic wastes.[116]

2. PCB-Contaminated Wastewater

PCBs are synthetic organic molecules widely used in various industrial sectors (e.g., plastics, electricity, lubricants, and hydraulic systems), notably to develop, combined with other organic substances, insulating fluids used in electric transformers and capacitors. PCBs have always been used in complex mixtures of congeners characterized by their different chlorine contents. PCB mixtures are sold under trade names such as Aroclor® (Monsanto, U.S.A), Clophen® (Bayer, Germany), and Phenoclor® (France). The Aroclor mixtures, especially Aroclor 1254 and Aroclor 1260 (respective chlorine contents: 54% and 60%, by wt), have been widely used in Europe and North America.[117]

In aquatic ecosystems, PCBs are considered as priority pollutants due to their high lipophility, low water solubility, and low biodegradability. Standard methods for water purification used in water-softening plants remain largely ineffective in eliminating these highly persistent toxic compounds. Until now, expensive treatments leading only to an incomplete elimination of PCBs, such as adsorption onto activated charcoal or synthetic resins, have been used. Several investigators have observed that chitosan could perhaps fulfill the requirements of being an efficient treatment for PCB-contaminated wastewater.[17,118]

Thomé and Van Daele[16] studied PCB adsorption capabilities of chitosan and other adsorbing agents by filtration of distilled water supplemented with 0.5 ppb PCB (Aroclor 1260) through cartridges filled with the adsorbents. As seen in Table 19.6, sorption abilities of chitosan were more efficient than those of other organic substances tested. Up to 84% of the PCB present in water was adsorbed by 100 mg of chitosan. C_{18} could be used as an efficient PCB-adsorbing medium, but its high cost makes it unsuitable for application on an industrial scale.

Thomé and Van Daele[16] also studied the elimination of PCB from contaminated water (0.5 ppb) in a closed-loop system using a chitosan filter (75 g) provided with an activated charcoal filter (1.5 kg). Water circulation was achieved by a pump at a rate of 180 L/h. It was found that complete removal of PCB from water was rapidly obtained after less than 120 h. With an activated charcoal filter, without chitosan, concentration decreased only to 0.2 ppb after the same time. These results suggest that the highly efficient purification of PCB-contaminated water using a chitosan filter could find favorable application in environmental situations. Van Daele and Thomé[17] reported that filtration of the PCB-contaminated water (0.5 ppb) through chitosan was incomplete but quite sufficient to decrease the PCB contamination to less than 1.0 ppm level. This change was sufficient to protect effectively fish (*Barbus barbus*) against serious metabolic diseases, such as growth inhibition, liver and kidney volume increase, and decrease in the hemoglobin ratio in blood.

To improve the PCB sorption property of chitosan and to elucidate the role of amino groups in the PCB adsorption process, Thomé and co-workers[118] chemically

TABLE 19.6
Efficiency of Various Materials to Remove PCB from Distilled Water (100 mL) Spiked with 0.5 ppb PCB

Adsorbing Material	Mean Percentage of Adsorption ± S.D.	Concentration of PCB on Adsorbent ± SD (ng/g)
Chitosan (100 mg)	83.3 ± 7.5	416 ± 35
Chitin (100 mg)	66.1 ± 33	330 ± 153
Activated charcoal (200 mg)	66.6 ± 1.2	166.4 ± 3
Sand (200 mg)	59.2 ± 3.6	148 ± 9
C_{18}(150 mg)	97 ± 4.1	323 ± 14

Source: From Thomé, J.P. and Van Daele, Y., *Proceedings of the 3rd International Conference on Chitin and Chitosan*, Senigallia, Italy, 1986, pp. 551–554. With permission.

TABLE 19.7

Efficiency of PCB (Aroclor 1260)-Binding Ability of Various Chitosan Derivatives in a Batch System (24 h PCB/Chitosan Contact)

Initial PCB concentration in Water(ppb)	% Recovery			
	CHT[a]	CHT-Glu-Red[b]	CHT-Glu[c]	CHT-BQ[d]
1	93.0	94.4	70.7	48.3
10	96.9	96.5	87.6	89.1
100	96.2	94.0	59.3	50.8
1000	39.9	82.3	73.3	58.8

[a] Unmodified chitosan.

[b] Glutaraldehyde cross-linked NaBH$_3$CN reduced chitosan.

[c] Glutaraldehyde cross-linked chitosan.

[d] Benzoquinone cross-linked chitosan.

Source: From Thomé, J.P. et al., *Proceedings of the 5th International Conference on Chitin and Chitosan*, Princeton, NJ, 1992, pp. 639–647. With permission.

modified chitosan (CHT) by means of a cross-linking procedure with benzoquinone (CHT-BQ), glutaraldehyde (CHT-Glu), and glutaraldehyde–sodium cyanoborohydride (reductive amination: CHT-Glu-Red). In a batch system, unmodified chitosan, the glutaraldehyde cross-linked derivative, and NaBH$_3$CN-reduced chitosan (CHT-Glu-Red) were the most effective PCB adsorbents (Table 19.7). In a flow-through cartridge system, CHT-Glu-Red appeared to bind PCBs most efficiently, followed by unmodified chitosan (60% of PCB bonded on chitosan). CHT-Glu and CHT-BQ generally remained less effective. Concerning the role of amino groups in the PCB adsorption process, it was found that the presence of the amino group (primary or secondary) in the chemically modified chitosan was prerequisite for chitosan to maintain its adsorptive properties for PCBs.

III. CONCLUSIONS

Researchers have focused considerable attention on the development of methods for the treatment and disposal of industrial discharge waters, especially those containing toxic substances. In many instances, expensive treatments have led to the incomplete elimination of such toxic discharges. Thus, a need exists for more efficient and less expensive treatments of targeted wastewaters.

Chitosan is a renewable, natural, and environmentally safe biopolymer that can be obtained abundantly, and at low cost, through the processing of chitinous wastes from a variety of crustacean-processing industries. This biopolymer is often more effective, or more economically attractive, than other polymers such as synthetic resins, activated charcoal, cellulosic derivatives, and synthetic polyelectrolytes.

The physicochemical characteristics of chitosan can be variously affected by preparation methods and crustacean species. Therefore, the relationship between process conditions and the particular characteristics of chitosan products must be constantly monitored to achieve uniformity and proper quality control. This approach can considerably strengthen the biotechnology process supplying products of assorted grades of chitosan, selected for their particular intended use in various wastewater discharges containing real and potential hazardous compounds.

Because of its relatively low cost combined with its high degree of efficiency, chitosan has been shown to be a competitive and powerful decontaminating agent. However, some present limitations exist for the general use of chitosan in the treatment of specific wastewaters. Chitosan is usually produced in a flaked or powdered form that is both nonporous and soluble in acidic media. The low internal surface area of the nonporous material limits access to interior adsorption sites, thus lowering both adsorption capacities and uptake rates of metal ions and other pollutants. The solubility of chitosan in acidic media presently limits its use as an adsorbent in low-pH wastewaters. However, chemical modifications of the biopolymer, such as cross-linking with glutaraldehyde, and the synthesis of highly porous beads of chemically cross-linked chitosan will, in all likelihood, overcome usage limitations by imparting increased resistance to solubilization in acidic pH effluents, thus improving the overall adsorption capacity. A broad variety of modified chitosans are available commercially.

To date, most research involved in wastewater treatments using chitosan has been conducted in batch systems. For high removal efficiency of toxic wastes such as heavy metals, dyes, and PCBs, continuous column operation using an immobilized chitosan is preferred to the batch system. Research should be directed toward further studies of effective column-immobilized microbe reactors for continuous removal of such toxicants. Research at Louisiana State University[119,120] has effectively combined principles of microbial bioremediation with the development of chitinous immobilization for use in effective hazardous waste detoxification and biodegradation. Successful trials with chlorinated phenol mixtures have been extended to a variety of other hazardous organic compounds. In essence, chitinous products may be utilized as an inexpensive support surface for immobilization of whole cell-adapted microbes for single and multiple toxic transformations. Regeneration of column-packing materials needs to be considered to reduce wastewater treatment costs.

Additional pilot-scale studies are needed to apply adequately appropriate chitosan systems to actual waste streams from the particular "target" plant because levels of organic or inorganic compounds present in the discharge stream will vary notably from plant to plant.

ACKNOWLEDGMENTS

The authors appreciate the cooperation of Springer-Verlag in allowing the use of previously published information on this general subject (*Reviews of Environmental Contamination and Toxicology*, 163,1–28, 2000).

REFERENCES

1.   Peniston, Q.P. and Johnson, E.L., Method for treating an aqueous medium with chitosan and derivatives of chitin to remove an impurity, *U.S. patent* 3,533,940, 1970.

2. No, H.K. and Meyers, S.P., Crawfish chitosan as a coagulant in recovery of organic compounds from seafood processing streams, *J. Agric. Food. Chem.*, 37, 580–583, 1989.

3. Knorr, D., Use of chitinous polymers in food—a challenge for food research and development, *Food Technol.*, 38, 85–97, 1984.

4. Muzzarelli, R.A.A., Natural Chelating Polymers, Pergamon Press, 1973.

5. Sandford, P.A. and Hutchings, G.P., Chitosan—a natural, cationic biopolymer: commercial applications, in *Industrial Polysaccharides: Genetic Engineering, Structure/Property Relations and Applications*, Yalpani, M., Ed., Elsevier, Amsterdam, 1987, pp. 363–376.

6. Bough, W.A., Reduction of suspended solids in vegetable canning waste effluents by coagulation with chitosan, *J. Food Sci.*, 40, 297–301, 1975.

7. Jun, H.K., Kim, J.S., No, H.K., and Meyers, S.P., Chitosan as a coagulant for recovery of proteinaceous solids from tofu wastewater, *J. Agric. Food Chem.*, 42, 1834–1838, 1994.

8. Senstad, C. and Almas, K.A., Use of chitosan in the recovery of protein from shrimp processing wastewater, in *Proceedings of the Third International Conference on Chitin and Chitosan*, Muzzarelli, R., Jeuniaux, C. and Gooday, G.W., Eds. Senigallia, Italy, 1986, pp. 568–570.

9. Masri, M.S., Reuter, F.W., and Friedman, M., Binding of metal cations by natural substances, *J. Appl. Polym. Sci.*, 18, 675–681, 1974.

10. Masri, M.S. and Randall, V.G., Chitosan and chitosan derivatives for removal of toxic metallic ions from manufacturing-plant waste streams, in *Proceedings of the First International Conference on Chitin/Chitosan*, Muzzarelli, R.A.A. and Pariser, E.R., Eds., MIT Sea Grant Program, Cambridge, MA, 1978, pp. 277–287.

11. Yang, T.C. and Zall, R.R., Absorption of metals by natural polymers generated from seafood processing wastes, *Ind. Eng. Chem. Prod. Res. Dev.*, 23, 168–172, 1984.

12. Michelsen, D.L., Fulk, L.L., Woodby, R.M., and Boardman, G.D., Adsorptive and chemical pretreatment of reactive dye discharge, in Tedd, D.W. and Pohland, F.G., Eds., *Emerging Technologies in Hazardous Waste Management* III: ACS Symposium Series 518, American Chemical Society, Washington, DC, 1993, pp. 119–136.

13. Smith, B., Koonce, T., and Hudson, S., Decolorizing dye wastewater using chitosan, *Am. Dyest. Rep.*, 82, 18–36, 1993.

14. Sun, W.Q., Payne, G.F., Moas, M.S.G.L., Chu, J.H., and Wallace, K.K., Tyrosinase reaction/chitosan adsorption for removing phenols from wastewater, *Biotechnol. Prog.*, 8, 179–186, 1992.

15. Wada, S., Ichikawa, H., and Tatsumi, K., Removal of phenols from wastewater by soluble and immobilized tyrosinase, *Biotechnol. Bioeng.*, 42, 854–858, 1993.

16. Thomé, J.P. and Van Daele, Y., Adsorption of polychlorinated biphenyls (PCB) on chitosan and application to decontamination of polluted stream waters, in *Proceedings of the Third International Conference on Chitin and Chitosan*, Muzzarelli, R., Jeuniaux, C., and Gooday, G.W., Eds., Senigallia, Italy, 1986, pp. 551–554.

17. Van Daele, Y. and Thomé, J.P., Purification of PCB contaminated water by chitosan: A biological test of efficiency using the common barbel, *Barbus barbus, Bull. Environ. Contam. Toxicol.*, 37, 858–865, 1986.

18. LaMer, V.K. and Healy, T.W., Adsorption-flocculation reactions of macromolecules at the solid-liquid interface, *Rev. Pure Appl. Chem.*, 13, 112, 1963.

19. O'Melia, C.A., Coagulation and flocculation, in *Physicochemical Processes for Water Quality Control*, Weber, W.J., Ed., Wiley, New York, 1972, pp. 61–109.

20. Bough, W.A., Shewfelt, A.L., and Salter, W.L., Use of chitosan for the reduction and recovery of solids in poultry processing waste effluents, *Poultry Sci.*, 54, 992–1000, 1975.

21. Bough, W.A., Coagulation with chitosan — an aid to recovery of by-products from egg breaking wastes, *Poultry Sci.*, 54, 1904–1912, 1975.

22. Bough, W.A. and Landes, D.R., Recovery and nutritional evaluation of proteinaceous solids separated from whey by coagulation with chitosan, *J. Dairy Sci.*, 59, 1874–1880, 1976.

23. Wu, A.C.M., Bough, W.A., Holmes, M.R., and Perkins, B.E., Influence of manufacturing variables on the characteristics and effectiveness of chitosan products. III. Coagulation of cheese whey solids, *Biotechnol. Bioeng.*, 20, 1957–1966, 1978.

24. Bough, W.A., Chitosan — a polymer from seafood wastes, for use in treatment of food processing wastes and activated sludge, *Process Biochem.*, 11, 13–16, 1976.

25. Johnson, R.A. and Gallanger, S.M., Use of coagulants to treat seafood processing wastewaters, *J. Water Pollut. Control Fed.*, 56, 970–976, 1984.

26. Moore, K.J., Johnson, M.G., and Sistrunk, W.A., Effect of polyelectrolyte treatments on waste strength of snap and dry bean wastewater, *J. Food Sci.*, 52, 491–492, 1987.

27. Bough, W.A. and Landes, D.R., Treatment of food-processing wastes with chitosan and nutritional evaluation of coagulated by-products, in *Proceedings of the First International Conference on Chitin/Chitosan*, Muzzarelli, R.A.A. and Pariser, E.R., Eds., MIT Sea Grant Program, Cambridge, MA, 1978, pp. 218–230.

28. No, H.K. and Meyers, S.P., Recovery of amino acids from seafood processing wastewater with a dual chitosan-based ligand-exchange system, *J. Food Sci.*, 54, 60–62 and 70, 1989.

29. Asano, T., Havakawa, N., and Suzuki, T., Chitosan applications in wastewater sludge treatment, in *Proceedings of the First International Conference on Chitin/Chitosan*, Muzzarelli, R.A.A. and Pariser, E.R., Eds., MIT Sea Grant Program, Cambridge, MA, 1978, pp. 231–252.

30. Wu, A.C.M. and Bough, W.A., A study of variables in the chitosan manufacturing process in relation to molecular-weight distribution, chemical characteristics and waste-treatment effectiveness, in *Proceedings of the First International Conference on Chitin/Chitosan*, Muzzarelli, R.A.A. and Pariser, E.R., Eds., MIT Sea Grant Program, Cambridge, MA, 1978, pp. 88–102.

31. Bough, W.A., Wu, A.C.M., Campbell, T.E., Holmes, M.R., and Perkins, B.E., Influence of manufacturing variables on the characteristics and effectiveness of chitosan products. II. Coagulation of activated sludge suspensions, *Biotechnol. Bioeng.*, 20, 1945–1955, 1978.

32. Rorrer, G.L., Hsien, T.Y., and Way, J.D., Synthesis of porous-magnetic chitosan beads for removal of cadmium ions from waste water, *Ind. Eng. Chem. Res.*, 32, 2170–2178, 1993.

33. Juang, R.S. and Shao, H.J., A simplified equilibrium model for sorption of heavy metal ions from aqueous solutions on chitosan, *Water Res.*, 36, 2999–3008, 2002.

34. Deans, J.R. and Dixon, B.G., Uptake of Pb^{2+} and Cu^{2+} by novel biopolymers, *Water Res.*, 26, 469–472, 1992.

35. Muzzarelli, R.A.A., *Chitin*, Pergamon Press, Oxford, UK, 1977.

36. Dambies, L., Guimon, C., Yiacoumi, S., and Guibal, E., Characterization of metal ion interactions with chitosan by X-ray photoelectron spectroscopy, *Coll. Surf. A*, 177, 203–214, 2001.

37. Jha, I.N., Iyengar, L., and Prabhakara Rao, A.V.S., Removal of cadmium using chitosan, *J. Environ. Eng.*, 114, 962–974, 1988.

38. Peniche-Covas, C., Alvarez, L.W., and Argüelles-Monal, W., The adsorption of mercuric ions by chitosan, *J. Appl. Polym. Sci.*, 46, 1147–1150, 1992.
39. McKay, G., Blair, H., and Findon, A., Kinetics of copper uptake on chitosan, in *Proceedings of the Third International Conference on Chitin and Chitosan*, Muzzarelli, R., Jeuniaux, C. and Gooday, G.W., Eds., Senigallia, Italy, 1986, pp. 559–565.
40. Evans, J.R., Davids, W.G., MacRae, J.D., and Amirbahman, A., Kinetics of cadmium uptake by chitosan-based crab shells, *Water Res.*, 36, 3219–3226, 2002.
41. Eiden, C.A., Jewell, C.A., and Wightman, J.P., Interaction of lead and chromium with chitin and chitosan, *J. Appl. Polym. Sci.*, 25, 1587–1599, 1980.
42. Maruca, R., Suder, B.J., and Wightmen, J.P., Interaction of heavy metals with chitin and chitosan. III. Chromium, *J. Appl. Polym. Sci.*, 27, 4827–4837, 1982.
43. Muzzarelli, R.A.A. and Tubertini, O., Chitin and chitosan as chromatographic supports and adsorbents for collection of metal ions from organic and aqueous solutions and sea-water, *Talanta*, 16, 1571–1577, 1969.
44. Suder, B.J. and Wightman, J.P., Interaction of heavy metals with chitin and chitosan. II. Cadmium and zinc, in *Adsorption from Solution*, Ottewill, R.H., Rochester, C.H., and Smith, A.L., Eds., Academic Press, London, 1983, pp. 235–244.
45. Kurita, K., Sannan, T., and Iwakura, Y., Studies on chitin. VI. binding of metal cations, *J. Appl. Polym. Sci.*, 23, 511–515, 1979.
46. Madhavan, P. and Nair, K.G.R., Metal-binding property of chitosan from prawn waste, in *Proceedings of the First International Conference on Chitin/Chitosan*, Muzzarelli, R.A.A. and Pariser, E.R., Eds., MIT Sea Grant Program, Cambridge, MA, 1978, pp. 444–448.
47. Nair, K.R. and Madhavan, P., Metal binding property of chitosan from different sources, in *Proceedings of the Second International Conference on Chitin and Chitosan*, Hirano, S. and Tokura, S., Eds., Sapporo, Japan, 1982, pp. 187–190.
48. Wu, F.C., Tseng, R.L., and Juang, R.S., Comparative adsorption of metal and dye on flake- and bead-types of chitosans prepared from fishery wastes, *J. Hazard. Mater.*, B73, 63–75, 2000.
49. Tseng, R.L., Wu, F.C., and Juang, R.S., Pore structure and metal adsorption ability of chitosans prepared from fishery wastes, *J. Environ. Sci. Health*, A34, 1815–1828, 1999.
50. Udaybhaskar, P., Iyengar, L., and Prabhakara Rao, A.V.S., Hexavalent chromium interaction with chitosan, *J. Appl. Polym. Sci.*, 39, 739–747, 1990.
51. Guibal, E., Saucedo, I., Jansson-Charrier, M., Delanghe, B., and Le Cloirec, P., Uranium and vanadium sorption by chitosan and derivatives, *Wat. Sci. Tech.*, 30, 183–190, 1994.
52. Jansson-Charrier, M., Guibal, E., Roussy, J., Surjous, R., and Le Cloirec, P., Dynamic removal of uranium by chitosan: influence of operaing parameters, *Wat. Sci. Tech.*, 34, 169–177, 1996.
53. Guibal, E., Larkin, A., Vincent, T., and Tobin, J.M., Chitosan sorbents for platinum sorption from dilute solutions, *Ind. Eng. Chem. Res.*, 38, 4011–4022, 1999.
54. Cervera, M.L., Arnal, M.C., and de la Guardia, M., Removal of heavy metals by using adsorption on alumina or chitosan, *Anal. Bioanal. Chem.*, 375, 820–825, 2003.
55. Schmuhl, R., Krieg, H.M., and Keizer, K., Adsorption of Cu(II) and Cr(VI) ions by chitosan: kinetics and equilibrium studies, *Water SA* 27, 1–7, 2001.
56. Jansson-Charrier, M., Guibal, E., Roussy, J., Delanghe, B., and Le Cloirec, P., Vanadium (IV) sorption by chitosan: kinetics and equilibrium, *Water Res.*, 30, 465–475, 1996.
57. Findon, A., McKay, G., and Blair, H.S., Transport studies for the sorption of copper ions by chitosan, *J. Environ. Sci. Health*, A28, 173–185, 1993.

58. McKay, G., Blair, H.S., and Hindon, A., Equilibrium studies for the sorption of metal ions onto chitosan, *Indian J. Chem.*, 28A, 356–360, 1989.

59. Juang, R.S. and Shao, H.J., Effect of pH on competitive adsorption of Cu(II), Ni(II), and Zn(II) from water onto chitosan beads, *Adsorption*, 8, 71–78, 2002.

60. El-Sawy, S.M., Abu-Ayana, Y.M., and Abdel-Mohdy, F.A., Some chitin/chitosan derivatives for corrosion protection and waste water treatments, *Anti-Corrosion Methods Materi.*, 48, 227–234, 2001.

61. Huang, C., Chung, Y.C., and Liou, M.R., Adsorption of Cu(II) and Ni(II) by pelletized biopolymer, *J. Hazard. Mater.*, 45, 265–277, 1996.

62. Wu, F.C., Tseng, R.L., and Juang, R.S., Kinetic modeling of liquid-phase adsorption of reactive dyes and metal ions on chitosan, *Water Res.*, 35, 613–618, 2001.

63. Mitani, T., Fukumuro, N., Yoshimoto, C., and Ishii, H., Effects of counter ions (SO_4^{2-} and Cl^-) on the adsorption of copper and nickel ions by swollen chitosan beads, *Agric. Biol. Chem.*, 55, 2419, 1991.

64. Mitani, T., Nakajima, C., Sungkono, I.E., and Ishii, H., Effects of ionic strength on the adsorption of heavy metals by swollen chitosan beads, *J. Environ. Sci. Health*, A30, 669–674, 1995.

65. Bassi, R., Prasher, S.O., and Simpson, B.K., Effects of organic acids on the adsorption of heavy metal ions by chitosan flakes, *J. Environ. Sci. Health*, A34, 289–294, 1999.

66. Juang, R.S., Wu, F.C., and Tseng, R.L., Adsorption removal of copper(II) using chitosan from simulated rinse solutions containing chelating agents, *Water Res.*, 33, 2403–2409, 1999.

67. Ishii, H., Minegishi, M., Lavitpichayawong, B., and Mitani, T., Synthesis of chitosan-amino acid conjugates and their use in heavy metal uptake, *Int. J. Biol. Macromol.*, 17, 21–23, 1995.

68. Mitani, T., Moriyama, A., and Ishii, H., Heavy metal uptake by swollen chitosan beads, *Biosci. Biotech. Biochem.*, 56, 985, 1992.

69. Coughlin, R.W., Deshaies, M.R., and Davis, E.M., Chitosan in crab shell wastes purifies electroplating wastewater, *Environ. Prog.*, 9, 35–39, 1990.

70. Seo, S.B., Kajiuchi, T., Kim, D.I., Lee, S.H., and Kim, H.K., Preparation of water soluble chitosan blendmers and their application to removal of heavy metal ions from wastewater, *Macromol. Res.*, 10, 103–107, 2002.

71. Hauer, A., The chelating properties of Kytex H chitosan, in *Proceedings of the First International Conference on Chitin/Chitosan*, Muzzarelli, R.A.A. and Pariser, E.R., Eds., MIT Sea Grant Program, Cambridge, MA, 1978, pp. 263–276.

72. Koyama, Y. and Taniguchi, A., Studies on chitin. X. Homogeneous cross-linking of chitosan for enhanced cupric ion adsorption, *J. Appl. Polym. Sci.*, 31, 1951–1954, 1986.

73. Kurita, K., Binding of metal cations by chitin derivatives: improvement of adsorption ability through chemical modifications, in *Industrial Polysaccharides: Genetic Engineering, Structure/Property Relations and Applications*, Yalpani, M., Ed., Elsevier, Amsterdam, 1987, pp. 337–346.

74. Hsien, T.Y. and Rorrer, G.L., Effects of acylation and crosslinking on the material properties and cadmium ion adsorption capacity of porous chitosan beads, *Sep. Sci. Technol.*, 30, 2455–2475, 1995.

75. Kurita, K., Chikaoka, S., and Koyama, Y., Improvement of adsorption capacity for copper(II) ion by *N*-nonanoylation of chitosan, *Chem. Lett.*, pp. 9–12, 1988.

76. Kawamura, Y., Mitsuhashi, M., Tanibe, H., and Yoshida, H., Adsorption of metal ions on polyaminated highly porous chitosan chelating resin, *Ind. Eng. Chem. Res.*, 32, 386–391, 1993.

77. Milot, C., Guibal, E., Roussy, J., and Cloirec, P., Chitosan gel beads as a new biosorbent for molybdate removal, *Min. Pro. Ext. Met. Rev.*, 19, 293–308, 1998.

78. Kawamura, Y., Yoshida, H., Asai, S., and Tanibe, H., Breakthrough curve for adsorption of mercury(II) on polyaminated highly porous chitosan beads, *Wat. Sci. Tech.*, 35, 97–105, 1997.

79. Holme, K.R. and Hall, L.D., Novel metal chelating chitosan derivative: attachment of iminodiacetate moieties via a hydrophilic spacer group, *Can. J. Chem.*, 69, 585–589, 1991.

80. Inoue, K., Yamaguchi, T., Iwasaki, M., Ohto, K., and Yoshizuka, K., Adsorption of some platinum group metals on some complexane types of chemically modified chitosan, *Sep. Sci. Technol.*, 30, 2477–2489, 1995.

81. Ohga, K., Kurauchi, Y., and Yanase, H., Adsorption of Cu^{2+} or Hg^{2+} ion on resins prepared by crosslinking metal-complexed chitosans, *Bull. Chem. Soc. Jpn.*, 60, 444–446, 1987.

82. Muzzarelli, R.A.A., Rocchetti, R., and Muzzarelli, M.G., The isolation of cobalt, nickel, and copper from manganese nodules by chelation chromatography on chitosan, *Sep. Sci. Technol.*, 13, 153–163, 1978.

83. Randal, J.M., Randal, V.G., McDonald, G.M., Young, R.N., and Masri, M.S., Removal of trace quantities of nickel from solution, *J. Appl. Polym. Sci.*, 23, 727–732, 1979.

84. Sakaguchi, T. and Nakajima, A., Recovery of uranium by chitin phosphate and chitosan phosphate, in *Proceedings of the Second International Conference on Chitin and Chitosan*, Hirano, S. and Tokura, S., Eds., Sapporo, Japan, 1982, pp. 177–182.

85. Muzzarelli, R.A.A. and Tanfani, F., N-(carboxymethyl) chitosans and N-(o-carboxybenzyl) chitosans: novel chelating polyampholytes, in *Proceedings of the Second International Conference on Chitin and Chitosan*, Hirano, S. and Tokura, S., Eds., Sapporo, Japan, 1982, pp. 45–53.

86. Seo, H. and Kinemura, Y., Preparation and some properties of chitosan beads, in *Proceedings from the Fourth International Conference on Chitin and Chitosan*, Skjåk-Braek, G., Anthonsen, T., and Sandford, P., Eds., Trondheim, Norway, 1989, pp. 585–588.

87. Gerente, C., Andres, Y., and Le Cloirec, P., Uranium removal onto chitosan: competition with organic substances, *Environ. Technol.*, 20, 515–521, 1999.

88. Lasko, C.L. and Hurst, M.P., An investigation into the use of chitosan for the removal of soluble silver from industrial wastewater, *Environ. Sci. Technol.*, 33, 3622–3626, 1999.

89. Debbaudt, A., Zalba, M., Ferreira, M.L., and Gschaider, M.E., Theoretical and experimental study of Pb^{2+} and Hg^{2+} adsorption on biopolymers, 2, *Macromol. Biosci.*, 1, 249–257, 2001.

90. Stefancich, S., Delben, F., and Muzzarelli, R.A.A., Interaction of soluble chitosans with dyes in water. I. Optical evidence, *Carbohydr. Polym.*, 24, 17–23, 1994.

91. Giles, C.H., Hassan, A.S.A., and Subramanian, R.V.R., Adsorption at organic surfaces IV—adsorption of sulphonated azo dyes by chitin from aqueous solution, *J. Soc. Dyers Colour.*, 74, 682–688, 1958.

92. McKay, G., Mass transport processes for the adsorption of dyestuffs onto chitin, *Chem. Eng. Process.*, 21, 41–51, 1987.

93. McKay, G., Blair, H.S., and Gardner, J.R., Adsorption of dyes on chitin. I Equilibrium studies, *J. Appl. Polym. Sci.*, 27, 3043–3057, 1982.

94. McKay, G., Blair, H.S., and Gardner, J., The adsorption of dyes on chitin. III Intraparticle diffusion processes, *J. Appl. Polym. Sci.*, 28, 1767–1778, 1983.

95. McKay, G., Blair, H.S., and Gardner, J.R., The adsorption of dyes onto chitin in fixed bed columns and batch adsorbers, *J. Appl. Polym. Sci.*, 29, 1499–1514, 1984.

96. McKay, G., Blair, H.S., Gardner, J.G., and McConvey, I.F., Two-resistance mass transfer model for the adsorption of various dyestuffs onto chitin, *J. Appl. Polym. Sci.*, 30, 4325–4335, 1985.

97. McKay, G., Blair, H.S., and Gardner, J.R., Two resistance mass transport model for the adsorption of acid dye onto chitin in fixed beds, *J. Appl. Polym. Sci.*, 33, 1249–1257, 1987.

98. Juang, R.S., Tseng, R.L., Wu, F.C., and Lin, S.J., Use of chitin and chitosan in lobster shell wastes for color removal from aqueous solutions, *J. Environ. Sci. Health*, A31, 325–338, 1996.

99. Juang, R.S., Tseng, R.L., Wu, F.C., and Lee, S.H., Adsorption behavior of reactive dyes from aqueous solutions on chitosan, *J. Chem. Tech. Biotechnol.*, 70, 391–399, 1997.

100. Park, R.D., Cho, Y.Y., La, Y.G., and Kim, C.S., Application of chitosan as an adsorbent of dyes in wastewater from dyeworks, *Agric. Chem. Biotechnol.*, 38, 452–454, 1995.

101. Shimizu, Y., Kono, K., Kim, I.S., and Takagishi, T., Effects of added metal ions on the interaction of chitin and partially deacetylated chitin with an azo dye carrying hydroxyl groups, *J. Appl. Polym. Sci.*, 55, 255–261, 1995.

102. Yamamoto, H., Chiral interaction of chitosan with azo dyes, *Makromol. Chem.*, 185, 1613–1621, 1984.

103. Seo, T., Hagura, S., Kanbara, T., and Iijima, T., Interaction of dyes with chitosan derivatives, *J. Appl. Polym. Sci.,* 37, 3011–3027, 1989.

104. Delben, F., Gabrielli, P., Muzzarelli, R.A.A., and Stefancich, S., Interaction of soluble chitosans with dyes in water. II Thermodynamic data, *Carbohydr. Polym.*, 24, 25–30, 1994.

105. Laszlo, J.A., Removing acid dyes from textile wastewater using biomass for decolorization, *Am. Dyest. Rep.*, 83, 17–18,20–21,48, 1994.

106. Maghami, G.G. and Roberts, G.A., Studies on the adsorption of anionic dyes on chitosan, *Makromol. Chem.*, 189, 2239–2243, 1988.

107. Venkatrao, B., Baradarajan, A., and Sastry, C.A., Adsorption of dyestuffs on chitosan, in *Proceedings of the Third International Conference on Chitin and Chitosan*, Muzzarelli, R., Jeuniaux, C., and Gooday, G.W., Eds., Senigallia, Italy, 1986, pp. 554–559.

108. Bhavani, K.D. and Dutta, P.K., Physico-chemical adsorption properties on chitosan for dyehouse effluent, *Am. Dyest. Rep.*, 88, 53–58, 1999.

109. Annadurai, G., Chellapandian, M., and Krishnan, M.R.V., Adsorption of basic dye from aqueous solution by chitosan : equilibrium studies, *Indian J. Environ. Prot.*, 17, 95–98, 1997.

110. Park, R.D., Cho, Y.Y., Kim, K.Y., Bom, H.S., Oh, C.S., and Lee, H.C., Adsorption of toluidine blue O onto chitosan, *Agric. Chem. Biotechnol.*, 38, 447–451, 1995.

111. Keith, L.H. and Telliard, W.A., Priority pollutants I: a perspective view, *Environ. Sci. Technol.,* 13, 416–423, 1979.

112. Payne, G.F., Sun, W.Q., and Sohrabi, A., Tyrosinase reaction/chitosan adsorption for selectively removing phenols from aqueous mixtures, *Biotechnol. Bioeng.*, 40, 1011–1018, 1992.

113. Klibanov, A.M., Alberti, B.N., Morris, E.D., and Felshin, L.M., Enzymatic removal of toxic phenols and anilines from wastewaters, *J. Appl. Biochem.*, 2, 414–421, 1980.

114. Nicell, J.A., Bewtra, J.K., Biswas, N., and St. Pierre, C.C., Enzyme catalyzed polymerization and precipitation of aromatic compounds from aqueous solution, *Can. J. Civ. Eng.*, 20, 725–735, 1993.

115.   Wagner, M. and Nicell, J.A., Peroxidase-catalyzed removal of phenols from a petro-
        leum refinery wastewater, *Wat. Sci. Technol.*, 43, 253–260, 2001.
116.   Ikehata, K. and Nicell, J.A., Color and toxicity removal following tyrosinase-
        catalyzed oxidation of phenols, *Biotechnol. Prog.*, 16, 533–540, 2000.
117.   Thomé, J.P., Jeuniaux, C., and Weltrowski, M., Applications of chitosan for the elim-
        ination of organochlorine xenobiotics from wastewater, in *Applications of Chitin and
        Chitosan*, Goosen, M.F.A., Ed., Technomic, Lancaster, PA, 1997, pp. 309–331.
118.   Thomé, J.P., Hugla, J.L., and Weltrowski, M., Affinity of chitosan and related deriv-
        atives for PCBs, in *Proceedings from the Fifth International Conference on Chitin
        and Chitosan*, Brine, C.J., Sandford, P.A., and Zikakis, J.P., Eds., Princeton, NJ, 1992,
        pp. 639–647.
119.   Portier, R.J., Chitin immobilization systems for hazardous waste detoxification and
        biodegradation, in *Immobilization of Ions by Naturally Occurring Materials*, Eecles, I.I.
        Ed., Norwood, London, 1986, pp. 230–243.
120.   Portier, R.J., Nelson, J.A., and Christianson, J.C., Biotreatment of dilute contami-
        nated ground water using an immobilized microbe packed bed reactor, *Environ.
        Prog.*, 8, 120–125, 1989.

20 The Role of Glycosylation in Engineered Antibodies

Sherie L. Morrison

CONTENTS

I. INTRODUCTION

While it is widely appreciated that the amino acid sequence of a protein is critically important for determining its functional properties, it is less well appreciated that posttranslational modifications, such as glycosylation, can also have an impact on protein function. All immunoglobulins are glycoproteins with at least one N-linked carbohydrate moiety on the constant region of their heavy (H) chain. Numerous studies have now shown that the presence as well as the structure of the H chain N-linked glycans can have a significant impact on immunoglobulin function. A significant percentage of immunoglobulins also have carbohydrate present in the variable regions of their H or light (L) chains. The presence of variable (V)-region carbohydrate can also have significant functional consequences.

Both N- and O-linked carbohydrates are present on immunoglobulins. For N-linked glycosylation, after the formation of dolichol-P-P-GlcNAc$_2$, five mannoses are added yielding dolichol-P-P-GlcNAc$_2$Man$_5$. Subsequently, four mannoses and three glucoses are added to form dolichol-P-P-GlcNAc$_2$Man$_9$Glc$_3$, which is attached to Asn residues on nascent polypeptide chains in the lumen of the rough endoplasmic reticulum (ER). The Asn occurs in the sequon Asn-Xaa-Ser/Thr; when Pro is present at the Xaa or immediately after a sequon, carbohydrate addition does not occur.[1] In addition, the sequons Asn-Trp-Ser, Asn-Asp-Ser, Asn-Glu-Ser, and Asn-Leu-Ser are poor oligosaccharide acceptors, whereas the comparable Asn X-Thr sequons are efficiently glycosylated.[1] Removal of the terminal glucose residues must take place before the immunoglobulin can exit the ER. The carbohydrate structures can either remain high mannose or be further processed to complex. The degree of processing can be quite variable depending on the protein, the position of the carbohydrate, and the expression system. A biantennary, complex structure is shown schematically in Figure 20.1A. O-glycosylation occurs in the *cis*-Golgi compartment and involves the posttranslational transfer of an oligosaccharide to a Ser or Thr residue (Figure 20.1B). There is no well-defined motif for the acceptor site other than the proximity of Pro residues (reviewed in Reference 2).

II. *N*-GLYCOSYLATION OF CONSTANT REGIONS: POSITION AND STRUCTURE

A. IgG

All IgGs share a conserved glycosylation site at position 297 within the second constant region (C$_H$2) domain. The carbohydrate attached to C$_H$2 is a biantennary complex (Figure 20.1A), but can be of many different structures with approximately 30 different biantennary oligosaccharides associated with total human serum IgG.[3] However, there are no disialylated structures and only a low incidence of monosialylated ones (about 10%). There is also a low incidence of core structures that carry a bisecting GlcNAc. An important issue is what effect this variation in carbohydrate structure has on the functional properties of antibodies, including the recognition by receptors for the Fc portion of γ chain (FcγR), complement activation, and *in vivo* persistence. In multiple myeloma, the altered pattern of carbohydrate present on the IgG paraprotein is also present on the polyclonal IgG, suggesting that the abnormal physiological environment of the bone marrow in this disease may also affect normal plasma cells producing polyclonal IgG as well as the malignant plasma cells.[4]

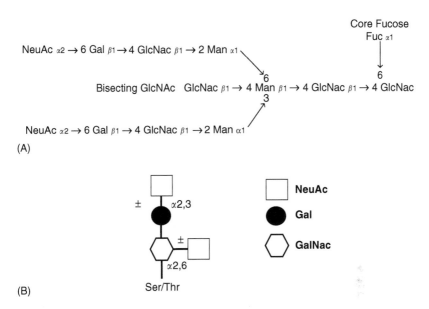

(A)

(B)

FIGURE 20.1 A. Structure of biantennary complex carbohydrates. The degree of substitution of the $Man_3GlcNac_2$ core is quite variable. Fucose and a bisecting GlcNac may or may not be present. The structure of the sialic acid added depends on the cells in which the immunoglobulin is produced. Human immunoglobulins contain oligosaccharides with N-acetylneuraminic acid (NANA), whereas mouse IgGs contain oligosaccharides with N-glycolylneuraminic acid (NGNA). The bisecting N-acetylglucosamine is absent when proteins are produced in murine myelomas. Wild-type CHO cells appear to lack the glycosyltransferases necessary for generating both the bisecting GlcNac and $\alpha2 \rightarrow 6$-linked sialic acid and produce sugars ending in NeuAc $\alpha2 \rightarrow 3$ Gal. B. Structure of the O-linked carbohydrate added to the hinge of IgA1 and IgD.

B. IgA

Human IgA exists as two isotypes: IgA1 and IgA2. Three allotypes of IgA2 have been described: IgA2m(1), IgA2m(2), and IgA2(n).[5] Only IgA1 possesses a 13-amino-acid hinge region containing 3- to 5-O-linked carbohydrate moieties. IgA1m(1) has two N-linked carbohydrate addition sites, one in C_H2 (Asn[263]) and the other in the tail-piece extension of C_H3 (Asn[459]).[6] IgA2 has two additional N-linked sites in C_H1 (Asn[166]) and in C_H2 (Asn[337]). The IgA2m(2) and the IgA2(n) allotypes have a fifth N-linked site in C_H1 (Asn[211]).[5,6]

Human serum IgA comprises 90% IgA1 and 10% IgA2, while in external secretions the proportion of IgA2 can be as high as 50%. This characteristic distribution may reflect the fact that plasma cells in the bone marrow are the source of serum IgA1, and plasma cells in the lamina propria are the source of secreted IgA. Alternatively, differences in the serum half-life of IgA1 and IgA2 could contribute to this distribution.[7] Over 80% of the N-glycans present on pooled monomeric serum IgA1 were digalactosylated biantennary complex oligosaccharides with less than 10% of the structures triantennary and none tetraantennary.[8] Surprisingly, when compared with IgG, over 90% of the N-glycans were sialylated, with sialic acid predominantly in an $\alpha2{\to}6$ linkage.

C.   IgM

The human IgM H chain contains five N-linked glycosylation sites (Asn[171], Asn[332], Asn[395], Asn[402], and Asn[563]). Early studies of human myeloma proteins showed that the carbohydrates attached at Asn[402] and Asn[563] were high mannose,[9,10] while the carbohydrates attached at the other three positions were complex.[11] Similarly, in a human hybridoma cell line producing IgM, the sugar chains at Asn[402] and Asn[563] were high mannose, whereas the predominant structures at Asn[171], Asn[332], and Asn[395] were fully galactosylated biantennary complex types with more sialic acid, bisecting GlcNAc and core fucose the closer they are to the C-terminus.[12]

D.   IgD

Analysis of an IgD myeloma protein showed that N-linked carbohydrate was present at Asn[354], Asn[445], and Asn[496]. The oligosaccharide at Asn[354] was high mannose, while the remainder were complex.[13] In addition, three or four Thr and one Ser residue in the hinge region bear O-glycosidically linked oligosaccharides. Approximately half of these molecules have the structure Gal$\beta1{\to}3$GalNAc, while the remainder have one or two residues of N-acetylneuraminic acid.[14]

E.   IgE

IgE has seven potential N-linked glycosylation sites. Analysis of a human IgE myeloma protein showed that four oligosaccharide units were present.[15,16] The carbohydrate at Asn[304], the position homologous to Asn[297] in IgG, remains high mannose while the remainder are complex. Asn[383] appears not to be glycosylated.[17]

III.   INFLUENCE OF GLYCOSYLATION ON THE PROPERTIES OF IMMUNOGLOBULINS

A.   GLYCOSYLATION OF IgG AND INTERACTIONS WITH FcγRs

Fc receptors bind immunoglobulins in their Fc region. Fc receptors for IgG play an important role in triggering effector responses such as phagocytosis, antigen-dependent cellular cytoxicity by NK cells, activation of neutrophils, macrophages, and dendritic cells, and inhibition of B cell activation by IgG immune complexes. There are three functionally distinct classes of FcγRs: FcγRI, FcγRII, and FcγRIII (CD64, CD32, and CD16, respectively). FcγRI is unique among FcγRs in that it binds monomeric IgG, whereas both FcγRII and FcγRIII recognize IgG with low affinity and essentially only bind IgG complexes. Human FcγRIII exists as both an integral membrane protein (FcγRIIIA) and a GPI-linked form (FcγRIIIB).

FcγRs function either as activation receptors characterized by the presence of a cytoplasmic immunoreceptor tyrosine-based activation motif (ITAM) sequence or inhibitory receptors containing a cytoplasmic immunoreceptor tyrosine-based inhibitory motif (ITIM) sequence. Both FcγRI and the non-GPI-linked FcγRIII associate with homodimers of the γ(or ζ) –chain, which has an ITAM that mediates signal transduction, and have similar biological functions: phagocytosing IgG-coated particles, mediating a respiratory burst, and triggering the release of inflammatory mediators. For FcγRII, the signaling sequence is part of the IgG binding chain of the receptor. FcγRIIB, which contains an ITIM, modulates B cell activation by down-regulating proliferation and inhibiting antibody production. On macrophages, neutrophils, and mast cells, FcγRIIB is often coligated to FcγRIII, and can prevent FcγRIII activation of these cells. Under some circumstances, engagement of FcγRs can have negative consequences as FcγRIIIA plays a critical role in the first-dose cytokine-release syndrome seen with some therapeutic monoclonal antibodies.[18]

Crystal structures of human FcγRIIB[19] and FcγRIIIB,[20] complexed with IgG, show that the receptor binds asymmetrically to the lower hinge region of both Fcs creating a 1:1 receptor–ligand stoichiometry. A similar interaction is seen between murine IgG2b and murine FcγRII.[21] The interaction breaks the dyad symmetry of the Fc creating an asymmetric interface where identical residues from Hinge-a and Hinge-b interact with different, unrelated surfaces of the receptor.[22] The interactions with the receptor are dominated by residues Leu^{234}-Pro^{238} of the lower hinge with the carbohydrate making only a small contribution to the contact surface.[22]

Given that the carbohydrate makes only a minor contribution to the contact area, it plays a remarkably important role in determining the interaction between IgG and the FcγRs. IgGs lacking the carbohydrate at position 297 in C_H2 fail to interact with the FcγRs[23,24] and are deficient in antibody-dependent cellular cytotoxicity (ADCC).[25,26] The absence of carbohydrate at Asn^{297} results in a localized structural change in the vicinity of His^{268} [27] and decreased thermal stability, suggesting that the absence of the oligosaccharides causes disorder and a closed disposition of the two C_H2 domains, thereby impairing FcγR binding.[28,29] Indeed, efforts to crystallize deglycosylated IgG1-Fc have failed to date, probably because of the increased internal disorder of the C_H2 domains.[30]

Deglycosylated Fc showed a more compact shape and reduced affinity for protein G compared with native Fc.[31] The reduced protein G binding suggests that there is a structural change in the C_H2–C_H3 hinge angles. The carbohydrate also plays a role in stabilizing the IgG lower hinge in an active receptor binding conformation with the removal of the carbohydrate, causing a conformational change in the relative orientation of the two C_H2 domains that affects the interface between the Fc and FcγRs.[30] The removal of sugar residues permits the mutual approach of C_H2 domains, resulting in the generation of a "closed" conformation.[30] This alteration in conformation appears to make murine IgG1 and IgG2a as well as chimeric human IgG1 and IgG3 more sensitive to proteolytic enzymes following the removal of the N-linked carbohydrate.[24,32]

It is apparent that the C_H2 carbohydrate serves to maintain proper spacing and orientation of the two C_H2 domains. In the crystal of the FcγRIII–Fc complex, the horseshoe-shaped Fc opens up on complex formation, with the distance between the Pro^{329} residues increasing by 7 Å. With a decasaccharide structure on each C_H2, the distance between the C_H2 domains was maximal (26.6 Å), whereas with only a tetrasaccharide structure on each C_H2, it was minimal.[30] The closer approach of the

C_H2 domains interferes with effective FcγR binding. Residue 265 is a contact for N-acetylglucosamine of the core oligosaccharide and an Asp265Ala mutation in human IgG3 results in a loss of human FcγRI and FcγRII recognition.[33] The presence of carbohydrate in C_H2 influences the thermal transition temperature of the C_H2 but not the C_H3 domain.[34] The presence of just the ManGlcNAc$_2$ trisaccharide was sufficient to provide some degree of stabilization that was lost in the aglycosylated protein. The influence of the ManGlcNAc$_2$ trisaccharide is consistent with the prediction that these three sugars have the potential to make a total of 31 contacts with seven amino acid residues on the FcγR.

Mannose on the $\alpha1 \rightarrow 3$ arm is the contact sugar residue between the two oligosaccharide moieties, and its presence defines a minimal C_H2–C_H2 distance. Although the mannose on the $\alpha1 \rightarrow 6$ arm does not contribute to the oligosaccharide/oligosaccharide interactions, it may stabilize the position of the carbohydrate via its interaction with Phe^{241} and Phe^{243} of the C_H2 domain.[30] Truncation of $\alpha1 \rightarrow 3$ and $\alpha1 \rightarrow 6$ arm mannoses resulted in a significantly decreased affinity for FcγRIIB.[28] Interruption of GlcNAc addition, leading to a pentamannose structure, resulted in Abs with significant but reduced binding to FcγRI.[35,36]

Addition of GlcNAc to each arm of the oligosaccharide further influences the open conformation with GlcNAc of the $\alpha1 \rightarrow 6$ arm found in contact with Phe^{243}, Thr^{260}, and Lys^{246}.[30] Outer-arm GlcNAc residues contributed significantly to thermal stability of the C_H2 domains but only slightly to FcγRIIB-binding affinity,[28] although the data suggest that some increase in the FcγRIIB affinity of glycoforms containing GlcNAc on the arms results from a rigid-loop conformation that does not change upon FcγR binding.[30] It is noteworthy that the observed movement of the C_H2 domains in the different structures is asymmetric upon oligosaccharide truncation.[30] Mouse–human chimeric IgG1, lacking terminal sialic acid or galactose produced in mutant Chinese hamster ovary (CHO) cell lines showed an affinity comparable with wild-type for FcγRI[36] and enzymatic removal of terminal galactose residues did not significantly alter the thermodynamic parameters or binding to soluble FcγRIIB.[28] These data would appear to be at variance with an NMR study[37] that concluded that in the absence of terminal galactose residues, the oligosaccharide moieties are mobile and do not interact with the protein. Although it has been proposed that glycans lacking galactose do not interact with the peptide chain resulting in the exposure of a previously covered region of the inner surface of C_H2,[38] studies using murine IgG2b indicated that even in the agalactosyl form, the glycans are buried in the protein.[39]

The oligosaccharide core can be further modified by the addition of fucose to the GlcNAc attached to Asn^{297} and by a bisecting GlcNAc. The presence of bisecting GlcNAc leads to an increase in affinity for FcγRIII with an accompanying increase in ADCC.[40,41] Although the presence of fucose did not influence binding to human FcγRI, antibodies lacking fucose showed a 50-fold enhanced binding to human FcγRIIIA accompanied by an increase in ADCC.[42]

B. COMPLEMENT

There are three different pathways through which the complement system can be activated: the classical, the lectin, and the alternative. The classical pathway, activated by the plasma protein C1, requires at least two C1q binding sites in close proximity, a feature of IgG antibody–antigen complexes and single molecules of polymeric IgM. The lectin pathway is triggered by mannose-binding lectin or protein (MBL or MBP), which recognizes N-acetylglucosamine > mannose > fucose > glucose (reviewed in Reference 43). MBL and C1 resemble each other in structure, and the classical pathway and the lectin pathway are identical downstream of the activating molecules. Binding of C1 or MBL leads to cleavage of the intermediate components C2, C3, C4, and C5, liberating fluid-phase products with inflammatory effects (C3a, C4a, and C5a) and the opsonins C3b and C4b, which are deposited on the target membrane surface. The alternative pathway is triggered by direct recognition of certain microbial surface structures. The three pathways intersect at the cleavage of C3. Complement receptors (CRs) on macrophages and neutrophils allow phagocytosis of particles coated with C4b or C3b and its degradation products. In addition, C4b and C3b bind to CR1 on erythrocytes, which carry soluble antigen–antibody complexes to the reticuloendothelial system, where they are eliminated. The complement cascade culminates in the formation of the membrane-attack complex, which forms a pore on the target cell surface, resulting in lysis. Organisms with a cell wall, such as fungi and Gram-positive bacteria, are resistant to lytic pore formation.

C1q binds to C_H2, although the precise site it recognizes may depend on the isotype of the antibody.[44,45] It is therefore not surprising that the carbohydrate attached at residue 297 impacts C1q binding. Aglycosylated IgG showed reduced[23,46] or undetectable[24,34] ability to activate complement. Replacement of residues Phe[241], Val[264], or Asp[265] (contact residues for carbohydrate) with Ala results in reduced recognition by guinea-pig complement and human C1q.[47] Although antibodies lacking terminal sialic acid or galactose showed varying reactivity with a monoclonal antibody (mAb) specific for C_H2, suggesting that the conformation of these proteins was altered by the different carbohydrate structures, these antibodies were not deficient in their ability to carry out complement-mediated hemolysis.[36] In fact, IgG1 lacking terminal galactose residues was more effective in binding C1q and assembling the terminal membrane attack complex in vitro.[36] However, the $Man_3GlcNAc_2Fuc$ and $ManGlcNAc_2Fuc$ glycoforms required approximately ten-fold higher levels of sensitization to achieve the same level of lysis as the native protein.[34] Murine IgG2b with either hybrid or high-mannose carbohydrate was proficient in complement-dependent cellular cytotoxicity (CDC) and ADCC.[48,49] Although mouse-human chimeric IgG1 produced in Lec 1 cells (with a carbohydrate of $GlcNAc_2,Man_5$) was deficient in activation of the classical pathway, it had a superior capacity to activate the alternative pathway.[35]

Rheumatoid arthritis is associated with a marked increase in IgG with carbohydrate lacking galactose and hence terminating in N-acetylglucosamine, which is recognized by MBL. Studies have shown that these terminal GlcNAc residues do indeed become accessible for MBL binding and that MBL binding results in the activation of complement.[38] IgG1 produced in Lec 8 cells has glycans lacking terminal galactose and was more effective in binding MBL in vitro.

The degree of sialylation and the linkage of the attached sialic acids can also influence the ability of IgG to activate complement. Human IgG1 is incompletely sialylated with NeuAc linked $\alpha2\rightarrow6$ to Gal. Wild-type chimeric IgG3 expressed in CHO-K1 cells contains little sialic acid. However, when Phe[243] was replaced by Ala,

53% of the glycans were $\alpha2{\rightarrow}3$-sialylated and the protein showed decreased ability to carry out complement-mediated lysis. Following transfection of a rat $\alpha2{\rightarrow}$ 6-sialytransferase (not expressed by CHO cells), 60% of the glycans were sialylated with approximately equal distribution of the two linkages and the protein showed enhanced ability to effect complement-mediated lysis.[50] Replacement of the carbohydrate contact residues Phe[241], Val[264], or Asp[265] by Ala in chimeric IgG3 produced in J558L resulted in increased galactosylation and sialylation relative to the wild-type oligosaccharide chains, and was accompanied by reduced recognition by guinea-pig complement and human C1q.[47]

A study has suggested that glycans of similar structure can assume different orientations and that these orientations can influence complement activation.[51] A pair of dinitrophenyl-specific murine monoclonal IgG2a antibodies with similar monosaccharide content differed in their binding to lectins and susceptibility to digestion with peptide N-glycosidase F, with the relative accessibility of the carbohydrate inversely related to the capacity of the antibodies to activate the classical pathway. Consistent with the hypothesis that the orientation of the more accessible N-glycan might inhibit C1q binding, enzymatic cleavage of the more accessible N-glycan resulted in enhanced Clq, C4b, and C3b deposition.

There are conflicting observations regarding the ability of IgA to fix complement. Solid-phase-deposited IgA was found to activate the alternative complement pathway and bind C3b.[52] Treatment with neuraminidase or N-glycosidase increased the C3b-fixing properties of serum IgA1 and IgA2 under these conditions, but it should be noted that deposition of the IgA on the solid phase may alter its conformation. Using antigen-coated plates, recombinant IgA1 with or without the N-linked glycosylation sites expressed in murine myeloma cells bound C3, although mutants lacking C_H3 carbohydrate showed reduced binding.[53] No activation of the alternative pathway as shown by Factor B or terminal complex binding was seen. IgA1, expressed in Sf9 cells, bound to plates through an interaction with antigen, and bound C3 when incubated with normal human serum.[54] However, the carbohydrate attached by Sf9 cells is oligomannose and activation may be through the lectin pathway. Using chimeric IgA2, it was found that immune complexes formed at equivalence would consume complement and activate the alternative pathway, as shown by C3 binding and the formation of the terminal complement complex, with aglycosylated IgA2 less effective than native IgA.[55]

C. HALF-LIFE

The rate at which a therapeutic agent is catabolized can play an important role in determining efficacy. The serum half-life of IgG is significantly longer than that of other antibody classes. Both human and murine IgG persist from 60 to 250 h in mice; human IgG persists still longer in humans. The half-life of IgG–antigen complexes is much shorter (approximately 6 h). The sites of IgG catabolism are widely distributed in visceral as well as peripheral tissue[56] with at least half of total IgG catabolism occurring in muscle and skin.

Recently, FcRn has been proposed to play a role in the catabolism of IgG. The expression of FcRn is widespread, including the capillary endothelium, the site at which catabolism may take place.[57,58] In mice that fail to express FcRn, the half-life of IgG is much shorter.[59] The concept has evolved that FcRn binds IgG that has been taken up in the fluid phase in the acidified endosome and diverts it from trafficking to the lysosome.[60] In this model, the cells involved in salvaging IgG are also responsible for IgG breakdown, and the model postulates a saturable receptor in the salvage pathway consistent with the observation that the half-life of IgG decreases with increasing serum concentration. Immune complexes may clear by different mechanisms than noncomplexed IgG, since raising the serum antibody concentration fails to alter the clearance of soluble immune complexes in humans.[61]

FcRn is a type I membrane glycoprotein similar in structure to class I MHC molecules consisting of a H chain of 50 kDa (α-chain) and β2-microglobulin. However, the x-ray crystallographic structure of the extracellular domains of rat and human FcRn[62] shows that the groove occupied by peptide in MHC class I molecules is occluded in FcRn.[63] Instead, Fc is bound at the interface between the C_H2 and C_H3 domains at a site located on the side of FcRn.[63,64] Given the site of FcRn binding, it is perhaps not surprising that many alterations in glycosylation do not influence the half-life of IgG. Aglycosylated chimeric IgG3, but not IgG1, showed a reduced half-life in mice; it is possible that the reduced half-life of IgG3 reflected its increased sensitivity to proteolysis.[24] Similarly, an aglycosylated chimeric IgG1 specific for TAG-72 was not substantially different from the glycosylated version in its plasma clearance in mice or primates.[26] However, the aglycosylated chimeric did show higher tumor/liver ratios. Chimeric IgG1, lacking terminal sialic acid or galactose, also showed no alterations in clearance even though IgG1 lacking galactose is preferentially bound by MBL.[36] In contrast, IgG1 terminating in oligomannose produced in Lec 1 cells was rapidly cleared, and that rapid clearance could be inhibited by injecting yeast mannan.[35] Surprisingly, the presence of variable-region carbohydrate also influences clearance rates, in some cases increasing the half-life (see below and Reference 65).

Growth in the presence of tunicamycin, an inhibitor of N-glycosylation, is frequently used to prepare proteins lacking N-linked carbohydrate. However, while aglycosylation of a murine IgG2b by site-directed mutagenesis had no effect on binding to FcRn *in vitro* or transport from the gut to blood *in vivo,* the treatment of cells producing either wild-type or aglycosyl mutant immunoglobulins with tunicamycin resulted in immunoglobulin that was less stable, that cleared more rapidly, and was transported slightly less efficiently.[66] Therefore, tunicamycin treatment may cause alterations in the immunoglobulin molecule in addition to those seen with the absence of N-linked carbohydrate.

The asialoglycoprotein receptor (ASGPR) binds oligosaccharides with terminal β-linked N-acetylgalactosamine (GalNAc) or galactose and can mediate the rapid clearance of glycoproteins bearing these terminal sugars. Indeed, it was found that recombinant chimeric IgA1 and IgA2 differ in their pharmacokinetic properties.[7] All three allotypes of IgA2 were rapidly cleared from the serum by the ASGPR present in the liver. In contrast, much less IgA1 was rapidly cleared from circulation by this route. IgA1 and IgA2 not eliminated by the ASGPR are both removed through an undefined ASGPR-independent pathway with half-lives of 4 and 10 h, respectively. The rapid clearance of IgA2 but not IgA1 may explain, in part, why the serum levels of IgA1 are greater than those of IgA2. Increasing the amount of sialic acid present on the carbohydrate decreases the interaction of IgA1 and IgA2 with the ASGPR.[67]

IgA bound to a human hepatoma cell line (HepG2) by the ASGPR[68] was endocytosed and rapidly catabolized. Although the liver accounted for more catabolism of monomeric murine IgA than all other tissue combined, IgG was catabolized equally in skin, muscle, and liver.[69,70] In the rat, ASGPRs are present on the surface of both hepatocytes and peritoneal macrophages. The hepatic receptor, but not the macrophage receptor, binds oligosaccharides with terminal GalNAc residues more tightly than ligands with terminal galactose residues.[71] Recent studies have shown that the hepatic receptor binds oligosaccharides with terminal Sia $\alpha2{\rightarrow}6$ GalNAc $\beta1{\rightarrow}4$ GlcNAc $\beta1{\rightarrow}2$ Man in addition to Gal and GalNAc.[72]

D. IMMUNOGENICITY

In several studies, the presence of carbohydrate on an antibody has been associated with increased immunogenicity. When the immunogenicity of a humanized IgG1 anti-CD3 mAb was investigated in transgenic mice expressing the target antigen, Fab-mediated cell-binding activity was a major determinant of immunogenicity.[73] However, at low dose levels the aglycosyl antibody consistently produced less of an antiglobulin response than the glycosylated antibody. In this case, it was proposed that the greater immunogenicity of the glycosylated antibody was a result of *in vivo* T cell activation as a consequence of crosslinking T cells to FcγR-bearing cells. The lower immunogenicity of the aglycosyl IgG1 CD3 mAb also correlated with a longer *in vivo* half-life and an improved capacity to block the target CD3 antigen. However, the aglycosyl mAb did not bring about T cell depletion from the peripheral circulation. The presence of variable-region carbohydrate in antidextran antibodies with constant-region carbohydrate was also found to increase immunogenicity (see below and Reference 65).

E. SECRETION

Following the cotranslational transfer of $Glc_3Man_9GlcNAc_2$ to the growing polypeptide chain, glucosidases I and II remove the two outermost glucose residues and the monoglucosylated oligosaccharide is recognized by the chaperones calriticulin and calnexin.[74] It is thought that the removal of the last glucose by glucosidase II results in the dissociation of the complex with chaperones, which will reform if the single glucose is added back by UDP-glucose:glycoprotein glucosyl-transferase, an enzyme that glucosylates unfolded but not native glycoproteins. The cycle is proposed to continue until a native conformation is achieved. Interestingly and in apparent conflict with this model, the IgY H chain has one high-mannose and one complex oligosaccharide, contains 27% monoglucosylated oligosaccharides, and can be used *in vitro* as a substrate for calreticulin binding.[75]

The degradation of unassembled IgM (μ) H chains and J chains, two glycoproteins with five and one N-linked glycans respectively, by cytosolic proteasomes can be prevented by inhibitors of mannosidase I.[76] Unassembled μ and J chains preferentially associate with the chaperone Bip. These results suggest that, for these proteins, removal of the terminal mannose from the central branch acts to initiate degradation. However, this pathway is not followed by all unassembled proteins, and multiple pathways appear to be operative in mammalian cells to eliminate misfolded proteins from the ER. ER mannosidases and proteaosome activity, but not glucose trimming, were also found to be essential for the degradation of mono- and di-glycosylated immunoglobulin L chains derived by mutagenesis from the nonsecreted L chain of NS1.[77]

The presence or absence of a carbohydrate structure can determine whether an immunoglobulin is assembled and secreted. Removal of either one or both of the N-linked addition sites from murine IgA resulted in inhibition of secretion and intra-cellular IgA (α) H chain degradation.[78] In contrast, removal of either or both of the two N-linked glycosylation sites from human IgA1 did not have a negative impact on its assembly and secretion.[53] Similarly, truncation of the O-linked carbohydrate in the hinge of IgA1 did not impair secretion.[79] Human IgD has three N-linked car-bohydrates. The one at position 354 near the hinge, but not the ones at either posi-tion 445 or 496, was required for secretion. IgD lacking the glycan at 354 did not transit into the Golgi, but remained in the ER.[79] Addition of a carbohydrate site to the heavy chain of an IgG antidansyl antibody by grafting in the CDR2 of a glyco-sylated antibody led to a failure of the chain to transit to the Golgi, resulting in intra-cellular degradation.[80] The growth of cells in the presence of tunicamycin showed that it was the presence of the glycan and not the altered amino acid sequence that was responsible for the failure of assembly and secretion.

IV. VARIABLE REGION GLYCOSYLATION

Serum IgG has 2.8 carbohydrates attached to each molecule. Two are from the con-served C_H2 carbohydrate with the remainder present in the variable region. Variable-region carbohydrate addition sites can be both generated and lost as a consequence of somatic hypermutation.[81] While variable-region carbohydrate is found on anti-bodies of many different specificities, in some B-cell malignancies, variable-region glycosylation appears to be an important aspect of the malignant phenotype.[82,83] A study has also shown increased ConA binding for tumor-reactive IgG in ovarian can-cer compared with control populations, but the structural alterations leading to this increased reactivity are not clear.[84]

As a general rule, the variable-region carbohydrate present on a molecule is more extensively processed than the constant-region carbohydrate of the same mol-ecule. Analysis of normal serum IgG showed two major glycan structures present on Fab (fucosylated digalacto-bianntenary with and without bisecting GlcNAc); the Fab contained relatively more galactose, bisecting GlcNAc, and sialic acid than the Fc.[37,85] The Fab N-glycans on pooled monomeric serum IgA were also found to be more sialylated than the Fc oligosaccharides, and the Fab region contained the majority of the triantennary oligosccharides.[8]

The presence of variable region carbohydrate has been shown to have different effects depending on the antibody. In several different polyspecific and anticarbohy-drate antibodies, variable-region carbohydrate improved the binding affinity. These include a polyspecifc human IgG1with λ L chains from a normal donor produced by a human–human–mouse heterohybridoma,[86,87] a murine IgM with κ L chains that recognized dextran and bovine serum albumin (BSA),[88] as well as the well-characterized antidextran myeloma proteins.[89,90] The carbohydrate present in the variable region did not affect the affinity of a humanized mouse mAb specific for IL-6[91] or CD22.[92] In contrast, the carbohydrate present in the V_H of an antibody specific for CD33 impaired binding.[93]

In some cases, there is asymmetric glycosylation of the variable region, and carbohydrate is present on only one of the Fabs. In the guinea pig, about 20% of the IgG1 and 10% of the IgG2 were asymmetrically glycosylated in the variable region.[94] A minor population with asymmetrically glycosylated heavy chains was observed in a murine IgG1 antigranulocyte antibody.[95] Four antidinitrophenol (DNP) asymmetrically glycosylated monoclonal murine IgG3 antibodies were capable of precipitating BSA-DNP only after the F(ab′)$_2$ fragments were treated with N-glycanase.[96]

Antidextran antibodies with glycosylation of variable regions of heavy and light chain have been extensively characterized. In initial studies, it was observed that the antibody produced by an antidextran hybridoma had a glycosylation site at Asn[58] in CDR2 of the V_H that had been introduced as a consequence of somatic hypermutation, and that glycosylation at this position increased its affinity.[89] Site-directed mutagenesis was then used to add sites at residue 54 or 60, and the resulting variable regions were expressed as chimeric antibodies with human constant regions.[90] Addition of carbohydrate at V_H60 increased the affinity for antigen, whereas the presence of carbohydrate at V_H54 interfered with the ability to bind antigen. Analysis of the N-linked sugar chains attached to the antidextrans showed that structures attached at Asn[54] and Asn[58] were complex-type but more highly sialylated than the Fc-associated sugars present on the same heavy chain.[97] Moreover, unlike the Fc-associated sugars, a significant population of Fab-associated sugars contained a Gal $\alpha1 \rightarrow 3$ Gal residue as a nonreducing terminus. Surprisingly, the carbohydrate attached at Asn[60] remained a high-mannose structure even though it was exposed on the surface of the antibody.[90]

When this study was extended by adding glycosylation sites to the other two CDRs of the H chain and the three CDRs of the L chain, only the site introduced at position 91 of the variable region of the L chain appeared to be completely used.[65] None of the attached carbohydrate was high mannose. Although most of the glycosylated antidextrans cleared with the expected half-life of 7 to 8 days, antibodies with carbohydrate at position 58 of the H chain and position 28 of the L chain cleared more rapidly (4 days), while antibodies with carbohydrate at either position 50 or 91 of the L chain cleared more slowly (10 days). While all of the antibodies with glycosylation in V_H elicited an immune response by 200 h, no immune response was observed to antibodies lacking variable region carbohydrate or glycosylated only in the L chain. The position of the variable-region carbohydrate also influenced the *in vivo* distribution of the antibody.

V. EXPRESSION SYSTEMS FOR THE PRODUCTION OF RECOMBINANT ANTIBODIES

One of the challenges of producing a recombinant antibody is the identification of the appropriate expression system. Two important considerations are the cost and functionality of the resulting antibody. Bacteria are often an attractive expression system for the production of recombinant proteins because of the low cost of the media in which they grow. However, *Escherichia coli,* the most commonly used expression system, do not add N-linked carbohydrate and hence cannot make antibodies with the full complement of functional properties. In bacteria, there is also the additional problem of the assembly of a four-chain protein with multiple intrachain and interchain disulfide bonds. Bacteria have been successfully used to make antibody fragments.

A. YEAST

Yeast is an attractive expression system. However, in *Saccharomyces cerevisiae,* hypermannosylated N-glycan structures with >100 mannose residues are often seen. Antibody produced in *S. cerevisiae* exhibited ADCC but not CDC.[98] However, in *Pichia pastoris,* hypermannosylation occurs less frequently and to a lower extent. Nonhuman glycosylation has been eliminated in Pichia by deleting the $\alpha 1 \rightarrow 6$ mannosyltransferase gene. When these yeast were transformed with libraries encoding catalytic domains of $\alpha 1 \rightarrow 2$ mannosidases and $\beta 1 \rightarrow 2$-N-acetylglucosaminyltransferas, a strain was isolated that produced the hybrid glycan $GlcNAc(Man)_5(GlcNAc)_2$ as the primary glycan.[99] Further introduction of mannosidase II and N-acetylglucosaminyl transferase II activity has now resulted in a yeast strain that produces glycoproteins uniformly glycosylated with $GlcNAc_2Man_3GlcNAc_2$.[100] These new yeast strains represent a significant advance and it will be interesting to explore their use as expressions systems for antibody production.

B. INSECT CELLS

Insect cells have also been used for antibody production. However, insect cells do not appear to contain all of the enzymes required to produce sialylated complex carbohydrates, although there appear to be differences among different insect lines. When the major virion envelope glycoprotein of the baculovirus *Autographa californica* (gp64) was produced in insect cells lines derived from *Spodopteria frugiperida* (Sf9), *Trichoplusia ni* (High 5), or *Estigmena acrea* (Ea), the carbohydrate contained mannose, fucose, and probably N-acetylglucosamine but lacked outer-chain galactose and sialic acid.[101] IFN-γ was differentially glycosylated depending on the insect cell line used for its expression and core $\alpha 1 \rightarrow 6$ fucosylation of oligomannose ($Man_3GlcNAc_2$), unknown in mammalian glycoproteins, was observed on recombinant IFN-γ produced by both Sf9 and Ea4 insect cells.[102]

A murine-human chimeric antibody expressed in Sf9 cells was N-glycosylated and had the same affinity and C1q binding activity as the same antibody produced in myeloma cells.[103] IgA1 expressed in Sf9 cells is assembled, secreted, and forms polymers if coexpressed with J chain.[54] However, in contrast to what was found for IgA produced in CHO cells,[8] IgA1 lacking carbohydrate in C_H2 was not bound by human cells expressing FcαR, possibly reflecting structural differences that occur because of the altered structure of either the N-linked carbohydrate at Asn^{459} or the O-linked glycans in the hinge.

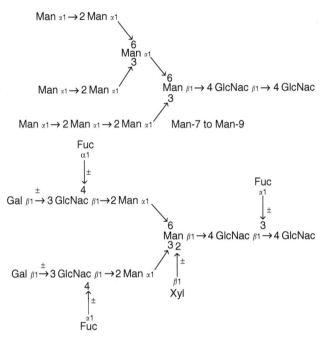

FIGURE 20.2 Structure of carbohydrates added to recombinant antibodies produced in tobacco. (From Bakker, H. et al., Proc. Natl. Acad. Sci. U.S.A., 98, 2899–2904, 2001. With permission.)

C. PLANTS

Plants are able to produce proteins with both high mannose and complex N-linked glycans having a core with two GlcNAc. However, this core is substituted by $\beta 1 \rightarrow$ 2-linked xylose and an $\alpha 1 \rightarrow 3$-linked core fucose instead of the $\alpha 1 \rightarrow 6$-linked core fucose found in mammals (Figure 20.2). Plants also lack $\beta 1 \rightarrow 4$ galactosyltransferase and hence do not add galactose to their complex N-glycans; sialic acid is also absent. When human $\beta 1 \rightarrow 4$ galactosyltransferase was expressed in tobacco plants, galactose was then added to the complex carbohydrate but there was no detectable effect on the occurrence of xylosylation and fucosylation of the N-glycans.[104]

The glycans present on several different antibodies produced in plants have been characterized. When a murine IgG1 mAb, specific for a cell-surface protein of *Streptococcus mutans*, was produced in transgenic tobacco plants, glycosylation occurred at both N-glycosylation sites located on the H chain (one at Asn[74], and the other at Asn[297]). Forty percent of the carbohydrate was high-mannose-type N-glycans, while 60% were processed with $\beta 1 \rightarrow 2$ xylose and $\alpha 1 \rightarrow 3$ fucose residues linked to the core Man$_3$GlcNAc$_2$, which contained zero, one, or two GlcNAc $\beta 1 \rightarrow 2$ residues.[105] The precise distribution of the different glycoforms on the Fab and Fc was not determined. In contrast, an antirabies virus expressed in transgenic tobacco contained mainly oligomannose N-glycans (90%) and had no potentially antigenic

$\alpha(1\rightarrow3)$-linked fucose residues.[106] The mAb was effective in virus neutralization but had a shorter half-life in BALB/c mice. The structure of the N-glycans on a mouse IgG (MGR48) was found to depend on the age of the leaf from which the antibodies were purified.[107] Antibodies isolated from young leaves had a relatively high amount of high-mannose glycans compared with antibodies from older leaves, which contain more terminal N-acetylglucosamine. Even when a KDEL ER anchoring domain was added to the antibody, seven different oligosaccharide structures were found representing oligosaccharides from high-mannose to complex-type N-glycans containing $\beta1\rightarrow2$ xylose and $\alpha1\rightarrow3$ fucose.[108]

A significant concern is whether the novel glycans attached by plants will be immunogenic. Studies have shown a role for $\beta1\rightarrow2$ xylose $\alpha1\rightarrow3$ fucose in recognition by IgE specific for plant glycoallergens.[109] However, when mice were immunized subcutaneously with a recombinant mouse monoclonal (IgG1, κ) produced in tobacco plants, only two of six immunized animals had weakly detectable antibody titers; however, the assay used detected only IgG2a, IgG2b, and IgG3 and not other isotypes including IgE.[110]

D. TRANSGENIC ANIMALS

Transgenic animals are attractive expression systems for the large-scale production of antibodies. However, there is species-specific variation in sialic acid structure, addition of bisecting GlcNAc, core fucosylation, and extent of galactose addition to IgG molecules.[111] Human and chicken IgG contain oligosaccharides with N-acetylneuraminic acid (NANA), whereas rhesus monkey, cow, sheep, goat, horse, and mouse IgGs contain oligosaccharides with N-glycolylneuraminic acid (NGNA), and IgGs from dog, guinea pig, rat, and rabbit contain both NANA and NGNA. Rat, horse, and dog IgGs contain very few galactosylated oligosaccharides, whereas about 90% of the sheep IgG oligosaccharides are galactosylated. The oligosaccharides from human IgG contain mostly core-fucosylated oligosaccharides, whereas those from rabbit IgG contain mostly nonfucosylated oligosaccharides. With the exception of chicken IgG, the predominant structure is biantennary complex. In chicken IgG, there is monoglucosylated $Man_9GlcNAc_2$ and an appreciable amount of high-mannose oligosaccharides, although complex biantennary structures with or without core fucose, terminal galactose, and bisecting GlcNAc are also present. For all species, there is an absence of hybrid and complex tri- and tetra-antennary structures suggesting that the degree of glycosylation is restricted within the Fc.

E. MAMMALIAN CELL LINES

Murine myeloma cells and CHO cells have been the most commonly used mammalian expression systems to date. Murine carbohydrates differ from those present in humans. An important modification to the core structure is the addition of a $\beta1\rightarrow4$-linked N-acetylglucosamine to the mannose $\beta1\rightarrow4$ in the trimannosyl core of N-linked sugars by β-D-mannoside β-1$\rightarrow$4-N-acetylglucosaminltransferase (GnT-III).[112] This modification is absent when proteins are produced in murine myelomas. For recombinant antibodies specific for CD20, neuroblastoma, and

CDw52 (CAMPATH-1H), antibody with a bisecting GlcNAc was consistently more active in ADCC (see References 40, 41, and 113).

It has been reported that a significant amount of circulating IgG in man specifically interacts with Gal $\alpha1\rightarrow3$ Gal $\beta1\rightarrow4$ GlcNAc, an epitope abundant on glycoconjugates of nonprimate mammals, prosimians, and New World monkeys, but absent from Old World monkeys, apes, and man.[114] Therefore, the presence of this carbohydrate structure on therapeutic proteins would have significant consequences; some of the carbohydrates introduced in murine expressions systems bear the Gal $\alpha1\rightarrow3$ Gal structure although its presence appears to be quite variable. H chains produced in J558L had oligosaccharides characteristic of mouse Ig including Gal $\alpha1\rightarrow3$ Gal and a molar ratio of 2.6:1 N-glycolylneuraminic acid:N-acetylneuraminic acid.[27,115] However, a recombinant humanized IgG1 produced by GS-NS0 cells contained predominantly variably galactosylated, fucosylated, biantennary structures with no bisecting N-acetylglucosamine or Gal $\alpha1\rightarrow3$ Gal residues.[116] The position of the carbohydrate also influences its structure with Gal $\alpha1\rightarrow3$ Gal residues present on the variable-region carbohydrate but absent from the constant region of the same antibody produced in mouse cells.[97]

CHO cells appear to lack the glycosyltransferases necessary for generating both the bisecting GlcNAc and $\alpha2\rightarrow6$-linked sialic acid.[117] Consistent with this, a humanized IgG antibody specific for lysozyme expressed in CHO-DUKX cells contained neutral, core-fucosylated biantennary oligosaccharides lacking sialic acid and bisecting GlcNAc (117). The degree of terminal agalactosylation was similar to normal serum with oligosaccharides, having only one terminal Gal exclusively galactosylated on the GlcNAc $\beta1\rightarrow2$ Man $\alpha1\rightarrow6$ Man $\beta1\rightarrow4$ antenna. The three major glycoforms present on CAMPATH-1H expressed in CHO cells are fucosylated, biantennary structures, containing zero, one, or two galactose residues. Although wild-type CHO cells produce sugars ending in NeuAc $\alpha2\rightarrow3$ Gal, CHO cells can be transfected to express an exogenous sialyltransferase that will attach NeuAc $\alpha2\rightarrow6$ to Gal.[118]

VI. CONCLUSIONS

Tremendous progress has been made in understanding the contribution of both the presence and the structure of carbohydrate to antibody function. Progress should continue at an ever accelerating rate as new tools for carbohydrate analysis are developed. Significant gaps still remain in our understanding of the factors that direct the addition and processing of O- and N-linked carbohydrates. Greater understanding of the role of carbohydrate coupled with improved, well-characterized expression systems should facilitate our ability to produced recombinant antibodies with the desired constellation of functional attributes.

ACKNOWLEDGMENTS

Work in the laboratory was supported by Grants AI29370, AI39187, AI51415, and CA87990 from the National Institutes of Health. I thank David Beenhouwer, Esther Yoo, and Letitia Wims for critically reading the manuscript.

REFERENCES

1. Kasturi, L., Chen, H., and Shakin-Eshleman, S.H., Regulation of *N*-linked core glycosylation: use of a site-directed mutagenesis approach to identify Asn-Xaa-Ser/Thr sequons that are poor oligosaccharide acceptors, *Biochem. J.*, 323, 415–419, 1997.

2. Gupta, R. and Brunak, S., Prediction of glycosylation across the human proteome and the correlation to protein function, *Pac. Symp. Biocomput.*, 310–322, 2002.

3. Rademacher, T.W., Parekh, R.B., Dwek, R.A., Isenberg, D., Rook, G., Axford, J.S., and Roitt, I., The role of IgG glycoforms in the pathogenesis of rheumatoid arthritis, *Springer Semin. Immunopathol.*, 10, 231–249, 1988.

4. Farooq, M., Takahashi, N., Arrol, H., Drayson, M., and Jefferis, R., Glycosylation of polyclonal and paraprotein IgG in multiple myeloma, *Glycoconj. J.*, 14, 489–492, 1988.

5. Chintalacharuvu, K.R., Raines, M., and Morrison, S.L., Divergence of human α-chain constant region gene sequences. A novel recombinant alpha 2 gene, *J. Immunol.*, 152, 5299–5304, 1994.

6. Torano, A., Tsuzukida, Y., Liu, Y.S., and Putnam, F.W., Location and structural significance of the oligosaccharides in human Ig-A1 and IgA2 immunoglobulins, *Proc. Natl. Acad. Sci. U.S.A.*, 74, 2301–2305, 1977.

7. Rifai, A., Fadden, K., Morrison, S.L., and Chintalacharuvu, K.R., The *N*-glycans determine the differential blood clearance and hepatic uptake of human immunoglobulin (Ig)A1 and IgA2 isotypes, *J. Exp. Med.*, 191, 2171–2182, 2000.

8. Mattu, T.S., Pleass, R.J., Willis, A.C., Kilian, M., Wormald, M.R., Lellouch, A.C., Rudd, P.M., Woof, JM and Dwek, R.A., The glycosylation and structure of human serum IgA1, Fab, and Fc regions and the role of *N*-glycosylation on Fc alpha receptor interactions, *J. Biol. Chem.*, 273, 2260–2272, 1998.

9. Chapman, A. and Kornfeld, R., Structure of the high mannose oligosaccharides of a human IgM myeloma protein. I. The major oligosaccharides of the two high mannose glycopeptides, *J. Biol. Chem.*, 254, 816–823, 1979.

10. Chapman, A. and Kornfeld, R., Structure of the high mannose oligosaccharides of a human IgM myeloma protein. II. The minor oligosaccharides of high mannose glycopeptide, *J. Biol. Chem.*, 254, 824–828, 1979.

11. Hickman, S., Kornfeld, R., Osterland, C.K., and Kornfeld, S., The structure of the glycopeptides of a human M-immunoglobulin. *J. Biol. Chem.*, 247, 2156–2163, 1972.

12. Fukuta, K., Abe, R., Yokomatsu, T., Kono, N., Nagatomi, Y., Asanagi, M., Shimazaki, Y., and Makino, T., Comparative study of the *N*-glycans of human monoclonal immunoglobulins M produced by hybridoma and parental cells, *Arch. Biochem. Biophys.*, 378, 142–150. 2000.

13. Mellis, S.J. and Baenziger, J.U., Structures of the oligosaccharides present at the three asparagine-linked glycosylation sites of human IgD, *J. Biol. Chem.*, 258, 11546–11556, 1983.

14. Mellis, S.J. and Baenziger, J.U., Structures of the O-glycosidically linked oligosaccharides of human IgD, *J. Biol. Chem.*, 258, 11557–11563, 1983.

15. Baenziger, J., Kornfeld, S., and Kochwa, S., Structure of the carbohydrate units of IgE immunoglobulin. I. Over-all composition, glycopeptide isolation, and structure of the high mannose oligosaccharide unit, *J. Biol. Chem.*, 249, 1889–1896. 1974.

16. Baenziger, J., Kornfeld, S., and Kochwa, S., Structure of the carbohydrate units of IgE immunoglobulin. II. Sequence of the sialic acid-containing glycopeptides, *J. Biol. Chem.*, 249, 1897–1903, 1974.

17.   Fridriksson, E.K., Beavil, A., Holowka, D., Gould, H.J., Baird, B., and McLafferty, F.W., Heterogeneous glycosylation of immunoglobulin E constructs characterized by top-down high-resolution 2-D mass spectrometry, *Biochemistry*, 39, 3369–3376, 2000.

18.   Wing, M.G., Moreau, T., Greenwood, J., Smith, R.M., Hale, G., Isaacs, J., Waldmann, H., Lachmann, P.J., and Compston, A., Mechanism of first-dose cytokine-release syndrome by CAMPATH 1-H: involvement of CD16 (FcgammaRIII) and CD11a/CD18 (LFA-1) on NK cells, *J. Clin. Invest.*, 98, 2819–2826, 1996.

19.   Sondermann, P., Huber, R., Oosthuizen, V., and Jacob, U., The 3.2-A crystal structure of the human IgG1 Fc fragment-Fc gammaRIII complex, *Nature*, 406, 267–273, 2000.

20.   Radaev, S., Motyka, S., Fridman, W.H., Sautes-Fridman, C., and Sun, P.D., The structure of a human type III Fcgamma receptor in complex with Fc, *J. Biol. Chem.*, 276, 16469–16477, 2001.

21.   Kato, K., Saut-Fridman, C., Yamada, W., Kobayashi, K., Uchiyama, S., Kim, H., Enokizono, J., Galinha, A., Kobayashi, Y., Fridman, W.H., Arata, Y., and Shimada, I., Structural basis of the interaction between IgG and Fcgamma receptors, *J. Mol. Biol.*, 295, 213–224, 2000.

22.   Radaev, S. and Sun, P., Recognition of immunoglobulins by Fcgamma receptors, *Mol. Immunol.*, 38, 1073–1083, 2002.

23.   Leatherbarrow, R.J., Rademacher, T.W., Dwek, R.A., Woof, J.M., Clark, A., Burton, D.R., Richardson, N., and Feinstein, A., Effector functions of a monoclonal aglyco-sylated mouse IgG2a: binding and activation of complement component C1 and inter-action with human monocyte Fc receptor, *Mol. Immunol.*, 22, 407–415, 1985.

24.   Tao, M.H. and Morrison, S.L., Studies of aglycosylated chimeric mouse-human IgG. Role of carbohydrate in the structure and effector functions mediated by the human IgG constant region, *J. Immunol.*, 143, 2595–2601, 1989.

25.   Leader, K.A., Kumpel, B.M., Hadley, A.G., and Bradley, B.A., Functional interactions of aglycosylated monoclonal anti-D with Fc gamma RI+ and Fc gamma RIII+ cells, *Immunology*, 72, 481–485, 1991.

26.   Hand, P.H., Calvo, B., Milenic, D., Yokota, T., Finch, M., Snoy, P., Garmestani, K., Gansow, O., Schlom, J., and Kashmiri, S.V., Comparative biological properties of a recombinant chimeric anti-carcinoma mAb and a recombinant aglycosylated variant, *Cancer Immunol. Immunother.*, 35, 165–174, 1992.

27.   Lund, J., Tanaka, T., Takahashi, N., Sarmay, G., Arata, Y., and Jefferis, R., A pro-tein structural change in aglycosylated IgG3 correlates with loss of huFc gamma R1 and huFc gamma R111 binding and/or activation, *Mol. Immunol.*, 27, 1145–1153, 1990.

28.   Mimura, Y., Sondermann, P., Ghirlando, R., Lund, J., Young, S.P., Goodall, M., and Jefferis, R., Role of oligosaccharide residues of IgG1-Fc in Fc gamma RIIb binding, *J. Biol. Chem.*, 276, 45539–45547, 2001.

29.   Ghirlando, R., Lund, J., Goodall, M., and Jefferis, R., Glycosylation of human IgG-Fc. influences on structure revealed by differential scanning micro-calorimetry, *Immunol. Lett.*, 68, 47–52, 1999.

30.   Krapp, S., Mimura, Y., Jefferis, R., Huber, R., and Sondermann, P., Structural analy-sis of human IgG-Fc glycoforms reveals a correlation between glycosylation and structural integrity, *J. Mol. Biol.*, 325, 979–989, 2003.

31.   Radaev, S. and Sun, P.D., Recognition of IgG by Fcgamma receptor. The role of Fc glycosylation and the binding of peptide inhibitors, *J. Biol. Chem.*, 276, 16478–16483, 2001.

32. Wilson, D.S., Wu, J., Peluso, P., and Nock, S., Improved method for pepsinolysis of mouse IgG(1) molecules to F(ab′)(2) fragments, *J. Immunol. Methods.*, 260, 29–36, 2002.

33. Jefferis, R., Lund, J., and Goodall, M., Recognition sites on human IgG for Fc gamma receptors: the role of glycosylation, *Immunol. Lett.*, 44, 111–117, 1995.

34. Mimura, Y., Church, S., Ghirlando, R., Ashton, P.R., Dong, S., Goodall, M., Lund, J., and Jefferis, R., The influence of glycosylation on the thermal stability and effector function expression of human IgG1-Fc: properties of a series of truncated glycoforms, *Mol. Immunol.*, 37, 697–706, 2000.

35. Wright, A. and Morrison, S.L., Effect of altered CH2-associated carbohydrate structure on the functional properties and *in vivo* fate of chimeric mouse-human immunoglobulin G1, *J. Exp. Med.*, 180, 1087–1096, 1994.

36. Wright, A. and Morrison, S.L., Effect of C2-associated carbohydrate structure on Ig effector function: studies with chimeric mouse-human IgG1 antibodies in glycosylation mutants of Chinese hamster ovary cells, *J. Immunol.*, 160, 3393–3402, 1998.

37. Wormald, M.R., Rudd, P.M., Harvey, D.J., Chang, S.C., Scragg, I.G., and Dwek, R.A., Variations in oligosaccharide-protein interactions in immunoglobulin G determine the site-specific glycosylation profiles and modulate the dynamic motion of the Fc oligosaccharides, *Biochemistry*, 36, 1370-1380, 1997.

38. Malhotra, R., Wormald, M.R., Rudd, P.M., Fischer, P.B., Dwek, R.A., and Sim, R.B., Glycosylation changes of IgG associated with rheumatoid arthritis can activate complement via the mannose-binding protein, *Nat. Med.*, 1, 237–243, 1995.

39. Yamaguchi, Y., Kato, K., Shindo, M., Aoki, S., Furusho, K., Koga, K., Takahashi, N., Arata, Y., and Shimada, I., Dynamics of the carbohydrate chains attached to the Fc portion of immunoglobulin G as studied by NMR spectroscopy assisted by selective 13C labeling of the glycans, *J. Biomol. NMR*, 12, 385–394, 1998.

40. Davies, J., Jiang, L., Pan, L.Z., LaBarre, M.J., Anderson, D., and Reff, M., Expression of GnTIII in a recombinant anti-CD20 CHO production cell line: Expression of antibodies with altered glycoforms leads to an increase in ADCC through higher affinity for FC gamma RIII, *Biotechnol. Bioeng.*, 74, 288–294, 2001.

41. Umana, P., Jean-Mairet, J., Moudry, R., Amstutz, H., and Bailey, J.E., Engineered glycoforms of an antineuroblastoma IgG1 with optimized antibody-dependent cellular cytotoxic activity, *Nat. Biotechnol.*, 17, 176–180, 1999.

42. Shields, R.L., Lai, J., Keck, R., O'Connell, L.Y., Hong, K., Meng, Y.G., Weikert, S.H., and Presta, L.G., Lack of fucose on human IgG1 N-linked oligosaccharide improves binding to human Fcgamma RIII and antibody-dependent cellular toxicity, *J. Biol. Chem.*, 277, 26733–26740, 2002.

43. Turner, M.W., Mannose-binding lectin: the pluripotent molecule of the innate immune system, *Immunol. Today*, 17, 532–539, 1996.

44. Duncan, A.R. and Winter, G., The binding site for C1q on IgG, *Nature*, 332, 738–740, 1988.

45. Idusogie, E.E., Presta, L.G., Gazzano-Santoro, H., Totpal, K., Wong, P.Y., Ultsch, M., Meng, Y.G., and Mulkerrin, M.G., Mapping of the C1q binding site on rituxan, a chimeric antibody with a human IgG1 Fc, *J. Immunol.*, 164, 4178–4184, 2000.

46. Dorai, H., Mueller, B.M., Reisfeld, R.A., and Gillies, S.D., Aglycosylated chimeric mouse/human IgG1 antibody retains some effector function, *Hybridoma*, 10, 211–217, 1991.

47. Lund, J., Takahashi, N., Pound, J.D., Goodall, M., and Jefferis, R., Multiple interactions of IgG with its core oligosaccharide can modulate recognition by complement and human Fc gamma receptor I and influence the synthesis of its oligosaccharide chains, *J. Immunol.*, 157, 4963–4969,1996.

48.   Nose, M. and Heyman, B., Inhibition of processing of asparagine-linked carbohydrate chains on IgG2a by using swainsonine has no influence upon antibody effector functions in vitro, *J. Immunol.*, 145, 910–914, 1990.

49.   Awwad, M., Strome, P.G., Gilman, S.C., and Axelrod, H.R., Modification of monoclonal antibody carbohydrates by oxidation, conjugation, or deoxymannojirimycin does not interfere with antibody effector functions, *Cancer Immunol. Immunother.*, 38, 23–30, 1994.

50.   Jassal, R., Jenkins, N., Charlwood, J., Camilleri, P., Jefferis, R., and Lund, J., Sialylation of human IgG-Fc carbohydrate by transfected rat α-2,6-sialyltransferase, *Biochem. Biophys. Res. Commun.*, 286, 243–249, 1994.

51.   White, K.D., Cummings, R.D., and Waxman, F.J., Ig *N*-glycan orientation can influence interactions with the complement system, *J. Immunol.*, 158, 426–435, 1997.

52.   Nikolova, E.B., Tomana, M., and Russell, M.W., The role of the carbohydrate chains in complement (C3) fixation by solid-phase-bound human IgA, *Immunology*, 82, 321–327, 1994.

53.   Chuang, P.D. and Morrison, S.L., Elimination of *N*-linked glycosylation sites from the human IgA1 constant region: effects on structure and function, *J. Immunol.*, 158, 724–732, 1997.

54.   Carayannopoulos, L., Max, E.E., and Capra, J.D., Recombinant human IgA expressed in insect cells, *Proc. Natl. Acad. Sci. U.S.A.*, 91, 8348–8352, 1994.

55.   Zhang, W. and Lachmann, P.J., Glycosylation of IgA is required for optimal activation of the alternative complement pathway by immune complexes, *Immunology*, 81, 137–141, 1994.

56.   Henderson, L.A., Baynes, J.W., and Thorpe, S.R., Identification of the sites of IgG catabolism in the rat, *Arch. Biochem. Biophys.*, 215, 1–11, 1982.

57.   Ghetie, V. and Ward, E.S., FcRn: the MHC class I-related receptor that is more than an IgG transporter, *Immunol. Today*, 18, 592–598, 1997.

58.   Ghetie, V. and Ward, E.S., Multiple roles for the major histocompatibility complex class I-related receptor FcRn, *Annu. Rev. Immunol.*, 18, 739–766, 2000.

59.   Roopenian, D.C., Christianson, G.J., Sproule, T.J., Brown, A.C., Akilesh, S., Jung, N., Petkova, S., Avanessian, L., Choi, E.Y., Shaffer, D.J., Eden, P.A., and Anderson, C.L., The MHC class I-like IgG receptor controls perinatal IgG transport, IgG homeostasis, and fate of IgG-Fc-coupled drugs, *J. Immunol.*, 170, 3528–3533, 2003.

60.   Junghans, R.P., Finally! The Brambell receptor (FcRB). Mediator of transmission of immunity and protection from catabolism for IgG, *Immunol. Res.*, 16, 29–57, 1997.

61.   Halma, C., Daha, M.R., van der Meer, J.W., Cohen, A., van Furth, R., Breedveld, F.C., Evers-Schouten, J.H., Pauwels, E.K., and van Es, L.A., Effect of monomeric immunoglobulin G (IgG) on the clearance of soluble aggregates of IgG in man, *J. Clin. Lab. Immunol.*, 35, 9–15, 1991.

62.   West, A.P., Jr. and Bjorkman, P.J., Crystal structure and immunoglobulin G binding properties of the human major histocompatibility complex-related Fc receptor, *Biochem.*, 39, 9698–9708, 2000.

63.   Burmeister, W.P., Gastinel, L.N., Simister, N.E., Blum, M.L., and Bjorkman, P.J., Crystal structure at 2.2 Å resolution of the MHC-related neonatal Fc receptor, *Nature*, 372, 336–343, 1994.

64.   Popov, S., Hubbard, J.G., Kim, J., Ober, B., Ghetie, V., and Ward, E.S., The stoichiometry and affinity of the interaction of murine Fc fragments with the MHC class I-related receptor, FcRn, *Mol. Immunol.*, 33, 521–530, 1996.

65. Coloma, M.J., Trinh, R.K., Martinez, A.R., and Morrison, S.L., Position effects of variable region carbohydrate on the affinity and *in vivo* behavior of an anti-(1→6) dextran antibody, *J. Immunol.*, 162, 2162–2170, 1999.

66. Hobbs, S.M., Jackson, L.E., and Hoadley, J., Interaction of aglycosyl immunoglobulins with the IgG Fc transport receptor from neonatal rat gut: comparison of deglycosylation by tunicamycin treatment and genetic engineering, *Mol. Immunol.*, 29, 949–956, 1992.

67. Basset, C., Devauchelle, V., Durand, V., Jamin, C., Pennec, Y.L., Youinou, P., and Dueymes, M., Glycosylation of immunoglobulin A influences its receptor binding, *Scand. J. Immunol.*, 50, 572–579, 1999.

68. Tomana, M., Kulhavy, R., and Mestecky, J., Receptor-mediated binding and uptake of immunoglobulin A by human liver, *Gastroenterology*, 94, 762–770, 1988.

69. Moldoveanu, Z., Epps, J.M., Thorpe, S.R., and Mestecky, J., The sites of catabolism of murine monomeric IgA, *J. Immunol.*, 141, 208–213, 1988.

70. Mestecky, J., Moldoveanu, Z., Tomana, M., Epps, J.M., Thorpe, S.R., Phillips, J.O., and Kulhavy, R., The role of the liver in catabolism of mouse and human IgA, *Immunol. Invest.*, 18, 313–324, 1989.

71. Iobst, S.T. and Drickamer, K., Selective sugar binding to the carbohydrate recognition domains of the rat hepatic and macrophage asialoglycoprotein receptors, *J. Biol. Chem.*, 271, 6686–6693, 1996.

72. Park, E.I., Manzella, S.M., and Baenziger, J.U., Rapid clearance of sialylated glycoproteins by the asialoglycoprotein receptor, *J. Biol. Chem.*, 278, 4597–4602, 2003.

73. Routledge, E.G., Falconer, M.E., Pope, H., Lloyd, I.S., and Waldmann, H., The effect of aglycosylation on the immunogenicity of a humanized therapeutic CD3 monoclonal antibody, *Transplantation*, 60, 847–853, 1995.

74. Patil, A.R., Thomas, C.J., and Surolia, A., Kinetics and the mechanism of interaction of the endoplasmic reticulum chaperone, calreticulin, with monoglucosylated (Glc1Man9GlcNAc2) substrate, *J. Biol. Chem.*, 275, 24348–24356, 2000.

75. Saito, Y., Ihara, Y., Leach, M.R., Cohen-Doyle, M.F., and Williams, D.B., Calreticulin functions in vitro as a molecular chaperone for both glycosylated and non-glycosylated proteins, *Embo. J.*, 18, 6718–6729, 1999.

76. Fagioli, C., and Sitia, R., Glycoprotein quality control in the endoplasmic reticulum. Mannose trimming by endoplasmic reticulum mannosidase I times the proteasomal degradation of unassembled immunoglobulin subunits, *J. Biol. Chem.*, 276, 12885–12892, 2001.

77. Chillaron, J., Adan, C., and Haas, I.G., Mannosidase action, independent of glucose trimming, is essential for proteasome-mediated degradation of unassembled glycosylated Ig light chains, *Biol. Chem.*, 381, 1155–1164, 2000.

78. Taylor, A.K. and Wall, R., Selective removal of alpha heavy-chain glycosylation sites causes immunoglobulin A degradation and reduced secretion, *Mol. Cell. Biol.*, 8, 4197–4203, 1988.

79. Gala, F.A. and Morrison, S.L., The role of constant region carbohydrate in the assembly and secretion of human IgD and IgA1, *J. Biol. Chem.*, 277, 29005–29011, 2002.

80. Gala, F.A. and Morrison, S.L., Variable region carbohydrate and antibody expression, *J. Immunol.*, 172, 5489–5494, 2004.

81. Dunn-Walters, D., Boursier, L., and Spencer, J., Effect of somatic hypermutation on potential *N*-glycosylation sites in human immunoglobulin heavy chain variable regions, *Mol. Immunol.*, 37, 107–113, 2000.

82. Zhu, D., McCarthy, H., Ottensmeier, C.H., Johnson, P., Hamblin, T.J., and Stevenson, F.K., Acquisition of potential *N*-glycosylation sites in the immunoglobulin variable region by somatic mutation is a distinctive feature of follicular lymphoma, *Blood*, 99, 2562–2568, 2002.

83. Zhu, D., Ottensmeier, C.H., Du, M.Q., McCarthy, H., and Stevenson, F.K., Incidence of potential glycosylation sites in immunoglobulin variable regions distinguishes between subsets of Burkitt's lymphoma and mucosa-associated lymphoid tissue lymphoma, *Br. J. Haematol.*, 120, 217–222, 2003.

84. Gercel-Taylor, C., Bazzett, L.B., and Taylor, D.D., Presence of aberrant tumor-reactive immunoglobulins in the circulation of patients with ovarian cancer, *Gynecol. Oncol.*, 81, 71–76, 2001.

85. Youings, A., Chang, S.C., Dwek, R.A., and Scragg, I.G., Site-specific glycosylation of human immunoglobulin G is altered in four rheumatoid arthritis patients, *Biochem. J.*, 314, 621–630, 1996.

86. Leibiger, H., Hansen, A., Schoenherr, G., Seifert, M., Wustner, D., Stigler, R., and Marx, U., Glycosylation analysis of a polyreactive human monoclonal IgG antibody derived from a human-mouse heterohybridoma, *Mol. Immunol.*, 32, 595–602, 1995.

87. Leibiger, H., Wustner, D., Stigler, R.D., and Marx, U., Variable domain-linked oligosaccharides of a human monoclonal IgG: structure and influence on antigen binding, *Biochem. J.*, 338, 529–538, 1999.

88. Fernandez, C., Alarcon-Riquelme, M.E., Abedi-Valugerdi, M., Sverremark, E., and Cortes, V., Polyreactive binding of antibodies generated by polyclonal B cell activation. I. Polyreactivity could be caused by differential glycosylation of immunoglobulins, *Scand. J. Immunol.*, 45, 231–239, 1997.

89. Wallick, S.C., Kabat, E.A., and Morrison, S.L., Glycosylation of a VH residue of a monoclonal antibody against α (1,6) dextran increases its affinity for antigen, *J. Exp. Med.*, 168, 1099–1109, 1988.

90. Wright, A., Tao, M.H., Kabat, E.A., and Morrison, S.L., Antibody variable region glycosylation: position effects on antigen binding and carbohydrate structure, *Embo. J.*, 10, 2717–2723, 1991.

91. Sato, K., Ohtomo, T., Hirata, Y., Saito, H., Matsuura, T., Akimoto, T., Akamatsu, K., Koishihara, Y., Ohsugi, Y., and Tsuchiya, M., Humanization of an anti-human IL-6 mouse monoclonal antibody glycosylated in its heavy chain variable region, *Hum. Antibodies Hybridomas*, 7, 175–183, 1996.

92. Leung, S.O., Goldenberg, D.M., Dion, A.S., Pellegrini, M.C., Shevitz, J., Shih, L.B., and Hansen, H.J., Construction and characterization of a humanized, internalizing, B-cell (CD22)-specific, leukemia/lymphoma antibody, LL2, *Mol. Immunol.*, 32:1413–1427, 1995.

93. Co, M.S., Scheinberg, D.A., Avdalovic, N.M., McGraw, K., Vasquez, M., Caron, P.C., and Queen, C., Genetically engineered deglycosylation of the variable domain increases the affinity of an anti-CD33 monoclonal antibody, *Mol. Immunol.*, 30, 1361–1367, 1993.

94. Malan Borel, I., Gentile, T., Angelucci, J., Margni, R.A., and Binaghi, R.A., Asymmetric Fab glycosylation in guinea-pig IgG1 and IgG2, *Immunology*, 70, 281–283, 1990.

95. Grebenau, R.C., Goldenberg, D.M., Chang, C.H., Koch, G.A., Gold, D.V., Kunz, A., and Hansen, H.J., Microheterogeneity of a purified IgG1 due to asymmetric Fab glycosylation, *Mol. Immunol.*, 29, 751–758, 1992.

96. Mathov, I., Plotkin, L.I., Squiquera, L., Fossati, C.A., Margni, R.A., and Leoni, J., *N*-glycanase treatment of F(ab')2 derived from asymmetric murine IgG3 mAb determines the acquisition of precipitating activity, Mol. Immunol., 32, 1123–1130, 1995.

97.   Endo, T., Wright, A., Morrison, S.L., and Kobata, A., Glycosylation of the variable region of immunoglobulin G—site specific maturation of the sugar chains, *Mol. Immunol.*, 32, 931–940, 1995.

98.   Horwitz, A.H., Chang, C.P., Better, M., Hellstrom, K.E., and Robinson, R.R., Secretion of Functional antibody and Fab fragment from yeast cells, *Proc. Natl. Acad. Sci. U.S.A.*, 85, 8678–8682, 1988.

99.   Choi, B.K., Bobrowicz, P., Davidson, R.C., Hamilton, S.R., Kung, D.H., Li, H., Miele, R.G., Nett, J.H., Wildt, S., and Gerngross, T.U., Use of combinatorial genetic Libraries to Humanize *N*-linked glycosylation in the Yeast *Pichia pastoris*, *Proc. Natl. Acad. Sci. U.S.A.*, 100, 5022–5027, 2003.

100. Hamilton, S.R., Bobrowicz, P., Bobrowicz, B., Davidson, R.C., Li, H., Mitchell, T., Nett, J.H., Rausch, S., Stadheim, T.A., Wischnewski, H., Wildt, S., and Gerngross, T.U., Production of complex human glycoproteins in yeast, *Science*, 301, 1244–1246, 2003.

101. Jarvis, D.L. and Finn, E.E., Biochemical analysis of the *N*-glycosylation pathway in baculovirus-infected lepidopteran insect cells, *Virology*, 212, 500–511, 1995.

102. Ogonah, O.W., Freedman, R.B., Jenkins, N., and Rooney, B.C., Analysis of human interferon-gamma glycoforms produced in baculovirus infected insect cells by matrix assisted laser desorption spectrometry, *Biochem. Soc. Trans.*, 23, 100S, 1995.

103. Jin, B.R., Ryu, C.J., Kang, S.K., Han, M.H., and Hong, H.J., Characterization of a murine-human chimeric antibody with specificity for the pre-S2 surface antigen of hepatitis B virus expressed in baculovirus-infected insect cells, *Virus Res.*, 38, 269–277, 1995.

104. Bakker, H., Bardor, M., Molthoff, J.W., Gomord, V., Elbers, I., Stevens, L.H., Jordi, W., Lommen, A., Faye, L., Lerouge, P., and Bosch, D., Galactose-extended glycans of antibodies produced by transgenic plants, *Proc. Natl. Acad. Sci. U.S.A.*, 98, 2899–2904, 2001.

105. Cabanes-Macheteau, M., Fitchette-Laine, A.C., Loutelier-Bourhis, C., Lange, C., Vine, N.D., Ma, J.K., Lerouge, P., and Faye, L., *N*-Glycosylation of a mouse IgG expressed in transgenic tobacco plants, *Glycobiology*, 9, 365–372, 1999.

106. Ko, K., Tekoah, Y., Rudd, P.M., Harvey, D.J., Dwek, R.A., Spitsin, S., Hanlon, C.A., Rupprecht, C., Dietzschold, B., Golovkin, M., and Koprowski, H., Function and glycosylation of plant-derived antiviral monoclonal antibody, *Proc. Natl. Acad. Sci., U.S.A.*, 100, 8013–8018, 2003.

107. Elbers, I.J., Stoopen, G.M., Bakker, H., Stevens, L.H., Bardor, M., Molthoff, J.W., Jordi, W.J., Bosch, D., and Lommen, A., Influence of growth conditions and developmental stage on *N*-glycan heterogeneity of transgenic immunoglobulin G and endogenous proteins in tobacco leaves, *Plant. Physiol.*, 126, 1314–1322, 2001.

108. Ramirez, N., Rodriguez, M., Ayala, M., Cremata, J., Perez, M., Martinez, A., Linares, M., Hevia, Y., Paez, R., Valdes, R., Gavilondo, J.V., and Selman-Housein, G., Expression and characterization of an anti-Hepatitis B surface antigen glycosylated mouse antibody in transgenic tobacco plants, and its use in the immunopurification of its target antigen, *Biotechnol. Appl. Biochem.*, 38, 223–230, 2003.

109. van Ree, R., Cabanes-Macheteau, M., Akkerdaas, J., Milazzo, J.P., Loutelier-Bourhis, C., Rayon, C., Villalba, M., Koppelman, S., Aalberse, R., Rodriguez, R., Faye, L., and Lerouge, P., β-(1,2)-xylose and α-(1,3)-fucose residues have a strong contribution in IgE binding to plant glycoallergens, *J. Biol. Chem.*, 275, 11451–11458, 2000.

110. Chargelegue, D., Vine, N.D., van Dolleweerd, C.J., Drake, P.M., and Ma, J.K., A murine monoclonal antibody produced in transgenic plants with plant-specific glycans is not immunogenic in mice, *Transgenic Res.*, 9, 187–194, 2000.

111. Raju, T.S., Briggs, J.B., Borge, S.M., and Jones, A.J., Species-specific variation in glycosylation of IgG: evidence for the species-specific sialylation and branch-specific galactosylation and importance for engineering recombinant glycoprotein therapeutics, *Glycobiology*, 10, 477–486, 2000.

112. Fukuta, K., Abe, R., Yokomatsu, T., Omae, F., Asanagi, M., and Makino, T., Control of bisecting GlcNAc addition to N-linked sugar chains, *J. Biol. Chem.*, 275, 23456–23461, 2000.

113. Lifely, M.R., Hale, C., Boyce, S., Keen, M.J., and Phillips, J., Glycosylation and biological activity of CAMPATH-1H expressed in different cell lines and grown under different culture conditions, *Glycobiology*, 5, 813–822, 1995.

114. Galili, U., Abnormal expression of α-galactosyl epitopes in man. A trigger for autoimmune processes? *Lancet*, 2, 358–361, 1989.

115. Lund, J., Takahashi, N., Hindley, S., Tyler, R., Goodall, M., and Jefferis, R., Glycosylation of human IgG subclass and mouse IgG2b heavy chains secreted by mouse J558L transfectoma cell lines as chimeric antibodies, *Hum. Antibodies Hybridomas*, 4, 20–25, 1993.

116. Hills, A.E., Patel, A., Boyd, P., and James, D.C., Metabolic control of recombinant monoclonal antibody N-glycosylation in GS-NS0 cells, *Biotechnol. Bioeng.*, 75, 239–251, 2001.

117. Routier, F.H., Davies, M.J., Bergemann, K., and Hounsell, E.F., The glycosylation pattern of humanized IgGI antibody (D1.3) expressed in CHO cells, *Glycoconj. J.*, 14, 201–207, 1997.

118. Lee, E.U., Roth, J., and Paulson, J.C., Alteration of terminal glycosylation sequences on N-linkedoligosaccharides of Chinese hamster ovary cells by expression of β-galactoside α-2,6-sialyltransferase, *J. Biol. Chem.*, 264, 13848–13855, 1989.

21 Recent Developments in the Synthesis of Oligosaccharides by Hyperthermophilic Glycosidases

Marco Moracci, Beatrice Cobucci-Ponzano, Giuseppe Perugino, Assunta Giordano, Antonio Trincone, and Mosé Rossi

CONTENTS

I. INTRODUCTION

Carbohydrates serve as the structural components and energy sources of the cell. Oligosaccharides, moreover, are involved in a variety of molecular recognition processes in intercellular communication.[1,2] The fact that these biomolecules play a key informative role in biology is becoming more and more evident, confirming their potential as therapeutic agents.[3] Excellent progress has been made in the past few years in the synthesis of oligosaccharides, making the potential of these molecules as therapeutic agents more realistic. In particular, chemical syntheses on polymer-supported carbohydrates or the use of whole microorganisms as cell factories are significant advances in this field.[4–6] However, this potential is limited because the complex structure of the oligosaccharides hampers the classical chemical synthesis. In particular, the demanding protecting group manipulations that are needed to control the stereospecificity and regiospecificity of the products are limiting steps for the efficient production of oligosaccharides, which is needed for biological testing.[7]

Enzymatic synthesis has emerged as the method of choice for the preparation of oligosaccharides on a large scale because the regio- and stereospecificity of the reaction can be controlled.[7] The approaches available so far are based on two major classes of enzymes: glycosyl transferases and glycosidases. The former approach suffers from the scarcity of the enzymes and from the relatively high costs of the nucleotide phospho-sugar donor substrates. However, the amount of available glycosyl transferase from the genomes of several organisms is sharply increasing,[8] and new methods for the preparation of nucleotide-activated sugars by enzymatic or biological methods have been recently reported;[9] thus, these problems are likely to be overcome gradually.

An alternative approach to sugar synthesis exploits glycoside hydrolases. These enzymes can be classified as either "retaining" or "inverting" depending on whether they retain or invert the configuration of the bond being broken during hydrolysis in the products. (Figure 21.1).[10] The active site of both classes of enzymes contains two carboxylic acid residues; in inverting glycosidases, one residue acts as an acid catalyst and the other as a base catalyst. The reaction takes place by following a single-displacement mechanism.[11] On the other hand, in the retaining reaction mechanism, which involves a nucleophile group and an acid/base catalyst, the reaction proceeds by a double-displacement mechanism in which a covalent glycosyl–enzyme intermediate is formed (glycosylation step) and hydrolyzed (deglycosylation step) in a general acid/base-catalyzed process. Both mechanisms involve oxocarbenium ion-like transition states.[12]

The residue acting as the catalytic nucleophile in retaining α- and β-glycosidases (Figure 21.1) can be identified using a variety of methods that are described in detail in several excellent reviews.[13,14] Once the active-site residues and the reaction mechanism of a particular enzyme have been experimentally determined, they can be easily extended to all the homologous enzymes by following the classification in families. So far, more than 2500 glycoside hydrolases are known and have been classified in to about 80 families on the basis of their amino acid sequence (http://afmb.cnrs-mrs.fr/CAZY/index.html).[15,16]

Glycosidases are exploited in synthetic reactions by reverse hydrolysis (equilibrium-controlled synthesis) or by transglycosylation (kinetically controlled process). The former method offers modest yields of oligosaccharide products,[17] whereas

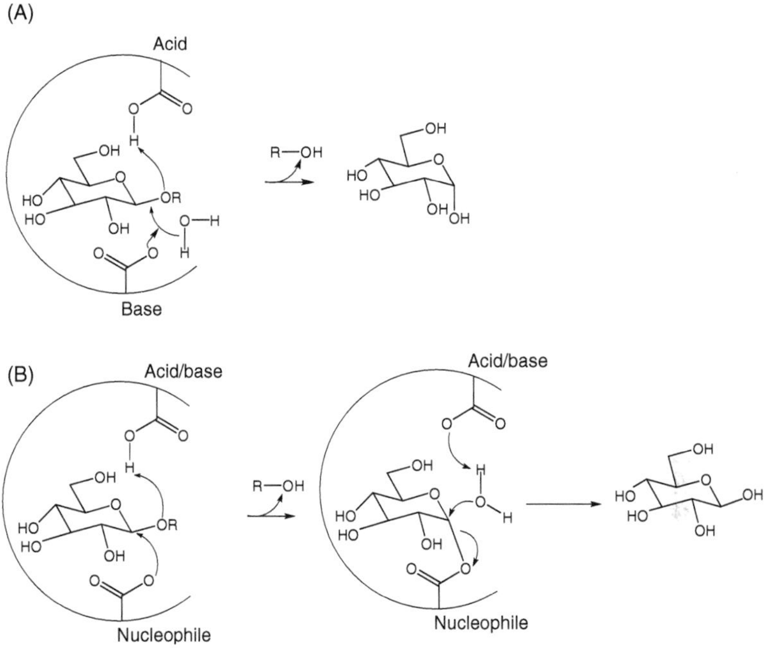

FIGURE 21.1 Reaction mechanism of (A) inverting and (B) retaining β-glycosidases.

kinetically controlled synthesis, which requires a retaining glycosidase (Figure 21.1), provides better yields (10 to 40%), but is not generally economical for large-scale synthesis. In fact, since the product of the reaction is a new substrate for the enzyme it can be hydrolyzed to reduce the final yields of the reaction. Thus, in order to maintain high yields, the reaction conditions have to be strictly controlled. A recent variation on this approach, in which the enzyme is modified by site-directed mutagenesis, eliminates the hydrolytic reaction and leads to the production of a novel class of enzymatic activities named glycosynthases.[18]

In this context, continuous efforts are being made in several laboratories to find new ways to synthesize oligosaccharides with increased yields and efficiency. In this regard, enzymes extracted from hyperthermophilic organisms, which optimally grow at temperatures of 80°C and above,[19] are often considered a useful alternative to conventional enzymes. In fact, their biodiversity and their intrinsic stability have allowed their exploitation in particular applications, as in the case of thermophilic DNA replicases in polymerase chain reactions. However, the use of enzymes from thermophilic sources is still limited because reactions under harsh conditions are restricted to special cases of demonstrated utility.

Biotransformations exploiting hyperthermophilic glycosidases have been reported in recent years,[20,21] suggesting that these enzymes can be useful tools for carbohydrate synthesis. In this chapter, we report on the most recent developments in the exploitation of hyperthermophilic glycosidases for the synthesis of oligosaccharides.

II. SYNTHESIS OF OLIGOSACCHARIDES BY THE β-GLYCOSIDASE FROM *SULFOLOBUS SOLFATARICUS*

A. GLYCOSIDES

The ability of the β-glycosidase from the hyperthermophilic archaeon *S. solfataricus* (Ssβ-gly) to synthesize glycosidic bonds has been exploited to prepare glycosides of different nature (β-D-glucosides, β-D-galactosides, and β-D-fucosides) using transglycosylation reactions with aryl donors and different acceptors. Various preparations of Ssβ-gly, such as the crude homogenate of *S. solfataricus* cells (the immobilized whole microorganism or lyophilized cells)[22] and the homogeneous enzyme purified from the archaeon or in recombinant form from *E. coli*,[23] were used as catalysts for the kinetically controlled glycosylation of various aglycones. The body of the stereochemical study indicates that the primary hydroxyl groups are always favored when compared with the secondary hydroxyl groups, although the overall regioselectivity of the reaction depends upon the complexity of the molecule. The yield depends upon chain length and OH position, but a marked improvement can generally be obtained by increasing the molar excess of the acceptor. In this respect, the stability of the thermophilic biocatalyst Ssβ-gly in organic solvents allowed the use of alcohol acceptors and organic cosolvent concentrations up to 95 to 97%. For instance, lyophilized *S. solfataricus* cells performed the synthesis in high yields (ca. 80%) of HEMA-β-D-galactoside (2-β-D-galactopyranosyl-oxyethyl methacrylate),[24] an anomerically pure β-galactosylated monomer, which is the starting point for the preparation of polymers with increased biological compatibility.[25] The yield for the enzymatic reaction was higher than that obtained by chemically coupling methyl α-D-galactopyranoside with a large excess of HEMA in the presence of 12-molybdophosphoric acid at 112°C.[26]

The knowledge on the selectivity of Ssβ-gly toward different hydroxyl groups allowed the planning of chemoenzymatic syntheses of useful target glycosides. Interesting compounds containing a double bond can be formed by the use of high concentrations of denaturing acceptors.[27] Examples of the preparation of naturally occurring glycosides are the synthesis and stereochemical determination of aleppotrioloside, a triol molecule possessing primary, secondary, and tertiary hydroxyl groups,[28] and the use of aromatic compounds bearing phenolic and aliphatic hydroxyl groups.[29,30]

The selectivity toward the pyranosidic ring has been studied in the glycosylation of salicin, the glucoside of 2-hydroxybenzyl alcohol; the primary hydroxyl group preference ($>60\%$) is again observed in the transglycosylation reaction to the pyranose O6 ring. The disaccharide for the glycosylation of the free hydroxymethylene on the aglycon of the salicin was not formed, but it was actively glucosylated when using 2-hydroxybenzyl alcohol as acceptor,[23] thus indicating the role of the OH groups of pyranose in positioning the acceptor on the active site of the enzyme for C6 functionalization.

B. 2-DEOXYGLYCOSIDES

The stereochemical study of the reactions of glycosidases with 1,2-unsaturated enol ether derivatives of sugars provided insights into the specificity and functioning of this class of biocatalysts.[31,32] These substrates operate both as glycosidase inhibitors and as substrates in the enzymatic synthesis of 2-deoxyglucosides. However, the glycal-based enzymatic inhibition is due to its functioning as a substrate; these enol ethers are hydrated by the enzyme with the formation and accumulation of a 2-deoxyglycosyl enzyme intermediate. Slow hydrolysis of the latter can lead to the formation of 2-deoxysugar or 2-deoxyglycosides of other acceptors present in the reaction mixture (Figure 21.2).

Owing to the occurrence of β-2-deoxyglycosyl moieties in many antibiotics and pharmaceutical compounds, interest was focused toward these derivatives as useful probes for the study of important enzymes and in therapeutics field. The chemical synthesis of these analogs presents a problem that has been resolved by stereodirecting auxiliary groups equatorially disposed at C2 position, whose removal in later steps often lowered the reaction yields.[33] In this respect, the anomerically selective enzymatic synthesis of β-2-deoxyglycosides represents a very interesting approach.

The study of the stereochemistry of the reaction of glucal with Ssβ-gly allowed one to define that the protonation of the double bond of glucal resulted in the equatorially disposed proton in the 2-deoxyglucoside formed. Moreover, different syntheses of 2-deoxyglycosides starting from glucal or galactal were reported.[34,35] However, when assayed in the presence of 4-nitrophenyl-β-D-glucoside (4-NP-β-D-Glc), glucal also functions as a competitive inhibitor showing a K_i of 8.44 mM. This value is 24-fold higher than the K_M of 4-NP-β-D-Glc (0.34 mM) and is similar to the K_M value of 6.88 mM, recently calculated for 4-NP-β-2-deoxy-glucoside (4-NP-β-2dGlc). These data indicate that Ssβ-gly has comparably low affinity for both glucal and 4-NP-β-2dGlc that form the same covalent 2-deoxy-glucosyl enzyme intermediate; presumably, the absence of the hydroxyl group in the C2 position of the intermediate determines the high K_M and K_i values obtained. These data show that glucal is a poor inhibitor of Ssβ-gly and explain why it is possible to use high concentrations of this compound in transglucosylation reactions with respect to those used in the catalysis with mesophilic enzyme.[34]

Different alkyl and pyranosidic acceptors were used for the synthesis of 2-deoxyglycosides as indicated in Table 21.1. Satisfactory yields (ca. 60%) were obtained with n-butanol and with other more useful alkyl acceptors (entries 4 and 5 in Table 21.1). The reaction of glucal and 4-penten-1-ol is a good example for a

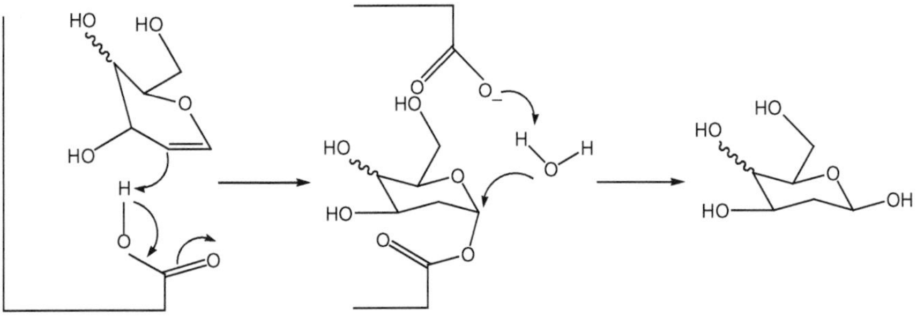

FIGURE 21.2 Mechanism of action of β-glycoside hydrolases on 1,2-enol ether derivatives of sugars.

TABLE 21.1
2-Deoxyglycosides Synthesized Using Ssβ-Gly

Entry	Acceptor	Yields (%)	Products[a] R = 2-deoxy-β-D-Glc	Molar excess
1	Methanol	50	CH$_3$-O-R	13
2	n-Butanol	65	CH$_3$- CH$_2$- CH$_2$- CH$_2$-O-R	16
3	2-Pentanol	20	CH$_3$- CH$_2$- CH$_2$- CH-(O-R)- CH$_3$	16
4	4-Penten-1-ol	80	CH$_2$ = CH$_2$-CH$_2$-CH$_2$-CH$_2$-O-R	20
5	3,4-Dimethoxybenzyl alcohol	19		3
6	Methyl-α-D-glucopyranoside	15		2
7	Phenyl-β-D-thioglucopyranoside	20		1.5
8	4-Nitrophenyl-α-D-glucopyranoside	32	and trisaccharides	1.1
9	2-Nitrophenyl-2-deoxy-N-acetyl-α-D-glucosamine	16	and trisaccharides	1.1
10	4-Nitrophenyl-2-deoxy-N-acetyl-β-D-glucosamine	23	and trisaccharides	1.1

[a] Arrows indicated the position of attack of 2-deoxy-β-D-Glc on pyranose ring.

simple one-pot enzymatic synthesis of 2-deoxygluco-*n*-pentenyl derivative useful in glycosylation chemistry by the Fraser–Reid methodology.[36] Ssβ-gly also promoted the synthesis of the 2-β-D-deoxyglucoside of 3,4-dimethoxybenzyl alcohol, which can be usefully applied in a new glycosylation procedure based on 2,3-dichloro-5, 6-dicyano-*p*-benzoquinone fragmentation.[37] The molar excess of acceptor used here was not higher than 3.3 at ca. 1.2 *M* glucal concentration. Under this condition an acceptable yield of 19% pure β-anomer was observed. The possibility of using high molar excess of inexpensive alcohols as acceptor, a high concentration of glucal, and of exploiting the residual activity of the thermophilic enzyme at 50°C are good characteristics for the practical application of the thermophilic biocatalyst in this field.

When probing the pyranosidic acceptors, a β-(1-6)-2-deoxygluco- derivative of thiophenyl β-D-glucopyranoside was synthesized as pure regioisomer with 20% yield in a reaction using synthetically useful molar excess of acceptor (entry 7, Table 21.1). This deoxythiodisaccharide is a useful starting material for oligosaccharide synthesis, where it can be used as a donor of the (2'-deoxy)-β-gentobiosyl building block by a simple chemical coupling methodology based on *N*-bromosuccinimide activation.[38] Other pyranosidic structures used in these reactions are reported in Table 21.1 along with the indication of di- and trisaccharidic products formed. The yields observed with these acceptors, used in a ~ 1:1 molar ratio, are in the range 20 to 40% depending on the donor, the reaction conditions, and the strict monitoring of the reaction. These values are of interest if one considers that 2-deoxyglycosides synthesis is not a trivial goal since: (i) different routes of chemical protection–deprotection are necessary for the synthesis of these compounds as for all glycosides and (ii) generally, a further step for removing the stereodirecting auxiliary groups at C2 is required.[33] The use of different pyranosidic acceptors permits the isolation of different regioisomers of deoxygluco- and galacto- derivatives based on the nitrophenyl leaving group. These compounds can be conveniently used for the kinetic study for the evaluation of C2 interactions in the active site with wild-type and mutant enzymes.

III. SYNTHESIS OF OLIGOSACCHARIDES BY THERMOPHILIC GLYCOSYNTHASES

The kinetically controlled synthesis of oligosaccharides requiring a retaining glycoside hydrolase (transglycosylation) provides good to medium yields, but it can be a demanding approach for large-scale synthesis. The major drawback of this approach is that the reaction products are themselves the target of the hydrolytic activity of the glycoside hydrolase; hence, reaction conditions must be carefully controlled, and the reaction must be closely monitored to increase final yields. The recent advent of *glycosynthases*, inactive retaining glycoside hydrolases specifically mutated in the catalytic nucleophile, represents a promising solution to these problems.[18]

When the active-site nucleophile is replaced with a non-nucleophile amino acid of reduced side chain, the resulting mutant is completely inactive since it cannot form the glycosyl–enzyme intermediate (Figure 21.1B). The mutant in the nucleophile residue can be reactivated in the presence of a glucosyl fluoride showing the opposite anomeric configuration (α-D) to that of the normal substrate (β-D) (Figure 21.3A). The enzyme transfers the donor to a suitable acceptor and the β-D-product cannot be hydrolyzed by the mutant and accumulate in the reaction. The mutant Glu358Ala of the β-glycosidase

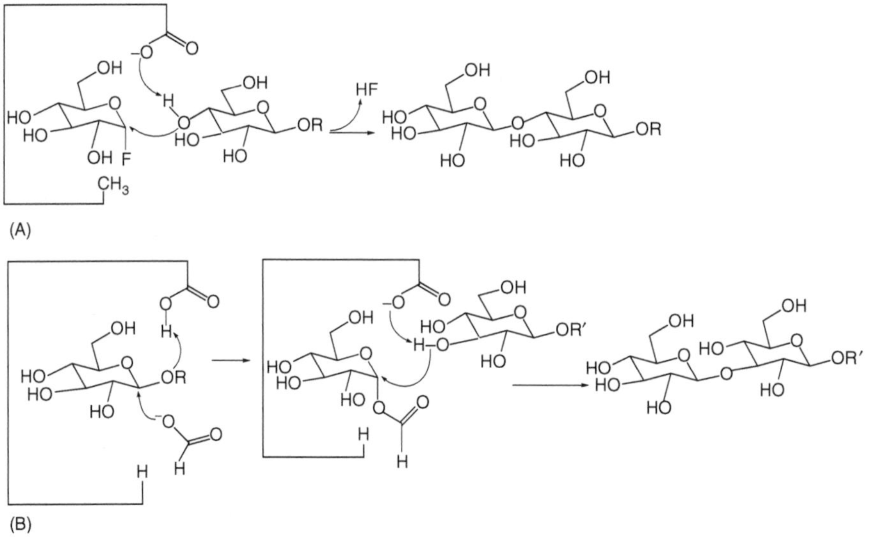

(A)

(B)

FIGURE 21.3   Reaction mechanisms of (A) inverting and (B) retaining glycosynthases.

from *Agrobacterium* was the first enzyme that was used in such an approach and was named glycosynthase.[18] The anomeric configuration of the products differs from that of the substrates; thus, the mutant behaves as an inverting synthase.[39]

In a complementary approach, Ssβ-gly mutated in the nucleophile was reactivated in the presence of external nucleophiles by using β-D-glucoside donors as substrate.[40] Owing to the stability of this enzyme in the presence of external nucleophiles, it could be used for the synthesis of oligosaccharides.

A.   HYPERTHERMOPHILIC β-GLYCOSYNTHASE FROM *S. SOLFATARICUS*

A crucial improvement in the synthetic activity of Ssβ-gly was obtained by exploiting the Glu387Gly mutant as an efficient thermophilic glycosynthase. By following an approach complementary to that described for mesophilic glycosynthases, Ssβ-gly was the first thermophilic glycosidase that reactivated in the presence of sodium formate as external nucleophile, and promoted the synthesis of branched oligosaccharides from 2-NP-β-glycosides.[40,41] As shown in figure 21.3B, in the first step of the reaction, the external nucleophile finds room in the active site of the enzyme depleted of the catalytic nucleophile and attacks the anomeric center of the substrate. The acid/base catalyst promotes the departure of the aglycon group of the substrate, whereas sodium formate and the glyconic group form a metastable intermediate. In the second step, the attack of the intermediate by an acceptor molecule, along with the action of the carboxylate group as a general base, completes the reaction allowing the formation of the product. The oligosaccharides produced by this enzyme accumulate in the reaction mixture because they are not activated donors and are not hydrolyzed by the glycosynthase. Since the products maintain the same anomeric configuration of the substrate, the enzyme was named retaining glycosynthase.[39]

The kinetic constants for hydrolysis at 65°C shown by wild-type and mutant Ssβ-gly for 2,4-dinitrophenyl- and 2-NP-β-D-Glc substrates are reported in Table 21.2. The maximal reactivation was observed with the 2,4-dinitrophenyl-β-D-Glc, whose specific activity was 15%, when compared with the wild-type enzyme without external nucleophile. The affinity to the substrate appeared unchanged upon mutation, whereas the specificity constant was 20%, compared with the wild-type enzyme. Peculiarly, significant reactivation of the Glu387Gly mutant on 2-NP-β-D-Glc, which has a relatively weak leaving group, was observed. Also in this case, the affinity to the substrate did not change upon mutation, but the catalytic efficiency was 8.4%, compared with the wild-type without an external nucleophile.[40]

Remarkably, the Ssβ-gly mutant is able to work either as an inverting or a retaining thermophilic glycosynthase by following both the approaches mentioned above (Figure 21.3). In the presence of α-glycosyl-fluoride as substrate, Ssβ-glyGlu387Gly synthesizes 2-NP-β-D-laminaribioside in 90% yields by following the mechanism shown in Figure 21.3A.[41] On the other hand, in the presence of 2 M sodium formate as an external nucleophile and activated β-glycoside donors, such as 2,4-dinitrophenyl- or 2-NP-β-D-Glc and 2-NP-β-D-fucoside, Ssβ-glyGlu387Gly follows the reaction mechanism shown in Figure 21.3B, producing branched oligoglucosides (85% total efficiency, with 50% disaccharides, 40% trisaccharides, and 10% tetrasaccharides by using 2-NP-β-D-Glc substrate).[41] The branching functionalization represents a unique characteristic of Ssβ-gly glycosynthase; the compounds produced can be of interest in the pharmaceutical and food fields[42] or can be used as new substrates or inhibitors for glycosidases.

TABLE 21.2
Steady-State Kinetic Constants of the Wild-Type and Glu387Gly Mutant Ssβ-gly at 65°C[a]

Enzyme	Substrate	Reaction Condition	K_M (mM)	k_{cat} (s^{-1})	k_{cat}/K_M (s^{-1}mM^{-1})	% Reactivation[b]
Ssβ-gly	2,4-DNP-Glc[c]	-2 M formate	0.17 ± 0.04	275.0 ± 16.0	1617.0	
	2,4-DNP-Glc	$+2$ M formate	0.61 ± 0.06	367.0 ± 12.0	602.0	
Ssβ-gly Glu387Gly	2,4-DNP-Glc	-2 M formate	ND[d]	ND	ND	
	2,4-DNP-Glc	$+2$ M formate	0.13 ± 0.02	42.0 ± 1.2	323.0	19.97
Ssβ-gly	2-NP-Glc	-2 M formate	1.01 ± 0.24	538.0 ± 11.0	533.0	
	2-NP-Glc	$+2$ M formate	0.50 ± 0.07	425.0 ± 11.0	850.0	
Ssβ-gly Glu387Gly	2-NP-Glc	-2 M formate	ND	ND	ND	
	2-NP-Glc	$+2$ M formate	1.17 ± 0.12	53.0 ± 1.2	45.0	8.44

[a] Kinetic constants were measured in 50 mM sodium phosphate buffer (pH 6.5).

[b] % of reactivation is calculated by taking as 100% the k_{cat}/K_M of the Ssβ-gly wild-type assayed at standard conditions.

[c] 2,4-DNP-Glc, 2,4-dinitrophenyl-glucoside; 2-NP-Glc, 2-nitrophenyl-glucoside.

[d] ND (not detectable) means that, using concentrations of enzyme of 10 μg/mL in the assay, the rates of change in absorbance did not vary in the experimental conditions and were approximately the same as the control without enzyme.

Although several retaining glycoside hydrolases mutated in the nucleophile are reactivated in the presence of sodium formate,[43–46] only the Ssβ-glyGlu387Gly mutant acts as retaining glycosynthase producing oligosaccharides.[39,41,47] Presumably, the intermediate formyl-glycoside, which has been identified in one case,[48] does not react efficiently with sugar acceptors in mesophilic enzymes. Interestingly, Gly mutations produced more efficient synthases when compared with Ala and Ser mutations in both hyperthermophilic and conventional enzymes,[40,49,50] although they follow different mechanisms of reactivation. This result suggests that the size of the substituting residue can play a relevant role in catalysis.

B. IMPROVING THE PERFORMANCE OF HYPERTHERMOPHILIC GLYCOSYNTHASE

Glycosynthase-catalyzed reactions produce excellent yields because the manipulation of the enzymes precludes the hydrolysis of the products; however, the reactions are sometimes slow and often require substantial quantities of mutant enzyme and extended incubation times. This led to a continuous search for more appropriated nucleophile amino acidic substitution and for reaction conditions that improve the performance of the synthase.[49,51] Recently, we reported that the synthetic activity of retaining glycosynthases could be greatly enhanced under acidic conditions.[47]

The reaction mechanism of retaining glycosynthases involves a glycosylation step in which the general acid/base catalyst and the formate ion cooperate (Figure 21.3B). The removal of the catalytic nucleophile in retaining glycosidases causes a downward shift in the pK_a of the acid/base catalyst.[52] Consequently, this group, at neutral pH, is converted into the ionized form, performing the first step of the reaction less efficiently. To convert the acid/base group in the protonated catalytically efficient form, glycosynthetic reactions were performed in sodium formate buffer (pH 3.0 to 6.0). Under these conditions, the Ssβ-glyGlu387Gly enzyme and two other retaining glycosynthases from hyperthermophilic Archaea showed improved efficiency in the synthetic reaction and enhanced synthetic repertoire.[47]

As shown in Table 21.3, the rate of hydrolysis of 2-NP-β-D-Glc substrate by the glycosynthases from *S. solfataricus* and *T. aggregans* (Taβ-gly) in sodium formate 50 mM (pH 3.0) was similar to that of its relative wild-type enzymes assayed at optimal conditions, resulting in the formation of most active glycosynthases known.[47] By contrast, the Glu372Ala mutant of the β-glycosidase from *P. furiosus* (CelB) was less active, confirming the superiority of Gly vs. Ala mutants. The same reaction, performed in 4 M sodium formate in phosphate buffer (pH 6.5), requires twice the amount of time and glycosynthase enzyme to obtain similar product yields.[41]

The different regioselectivity of these three hyperthermophilic glycosynthases is clearly shown in Table 21.4. *T. aggregans* glycosynthase produced mostly di- and trisaccharides, with a 14-fold higher amount of the β-1-4 regioisomer when compared with Ssβ-gly synthase. On the other hand, Ssβ-glyGlu387Gly showed a striking preference for the formation of the β-1-3 bond, as reported previously.[41] By contrast, the CelB synthase produced only the 2-NP-β-D-laminaribioside, which, presumably for the low activity of this glycosynthase, is not present in a sufficient

TABLE 21.3

Steady-State Kinetic Constants of Hyperthermophilic Glycoside Hydrolases and Relative Glycosynthases for 2-Nitrophenyl-β-D-glucoside at 65°C

Enzyme		Reaction Conditions			Kinetic Constants		
		Buffer	Formate	pH	K_M (mM)	k_{cat} (s^{-1})	k_{cat}/K_M (s^{-1} mM^{-1})
$Ss\beta$-gly							
	Wild type	Phosphate	—	6.5	1.01 ±0.24	538.0 ± 11.0	533.0
	Glu387Gly	Phosphate	—	6.5	ND[a]	ND	ND
		Phosphate	2 M	6.5	1.17 ±0.12	53.0 ± 1.2	45.0
		Formate	50 mM	3.0	16.4 ± 1.6	901.4 ± 32.9	55.0
		Formate	50 mM	4.0	4.9 ± 0.5	321.9 ± 12.0	65.9
		Formate	50 mM	5.0	2.2 ± 0.5	52.9 ± 3.1	23.8
		Formate	50 mM	6.0	1.1 ± 0.4	20.7 ± 1.6	18.5
$Ta\beta$-gly							
	Wild type	Phosphate	—	6.5	0.20 ±0.06	309.0 ± 18.5	1575.7
	Glu386Gly	Phosphate	—	6.5	ND	ND	ND
		Phosphate	2 M	6.5	1.01 ± 0.10	72.9 ± 2.4	72.9
		Formate	50 mM	3.0	6.6 ± 0.6	970.0 ± 39.9	147.4
		Formate	50 mM	4.0	3.3 ± 0.5	305.3 ± 13.6	91.0
		Formate	50 mM	5.0	3.1 ± 0.4	53.9 ± 2.0	17.2
		Formate	50 mM	6.0	2.8 ± 0.5	11.6 ± 0.6	4.2
CelB							
	Wild type	Phosphate	—	6.5	0.28 ± 0.06	1796.9 ±90.2	6480.0
	Glu372Ala	Phosphate	—	6.5	ND	ND	ND
		Phosphate	2 M	6.5	1.49 ± 0.12	3.9 ± 0.1	2.6
		Formate	50 mM	3.0	4.3 ± 0.2	47.6 ± 0.9	10.9
		Formate	50 mM	4.0	2.1 ± 0.4	6.7 ± 0.4	3.1
		Formate	50 mM	5.0	1.3 ± 0.1	0.45 + 0.01	0.3
		Formate	50 mM	6.0	ND	ND	ND

[a] See Table 21.2 for definition of ND.

amount to act as acceptor. However, the different product patterns observed could reflect the diversity in the active site of the three enzymes.

These studies indicate that the activity of glycosynthases, rescued by adding sodium formate, is largely improved at acidic conditions in which the acid/base catalyst is protonated, and that they can perform the first step of the reaction better (Figure 21.3B). This indication is supported by the observation that at pH 3.0, at which the three glycosynthases work optimally (Table 21.3), formate ($pK_a = 3.75$) is present at a concentration lower than that found at pH 4.0. Therefore, the activation observed is not correlated to the concentration of the external nucleophile but rather to the acidity of the reaction mixture. Moreover, the high efficiency observed in the glycosynthetic reaction indicates that the conditions used did not affect the second step of the reaction, in which the acid/base carboxylate should have been

TABLE 21.4

Final Products and Relative Ratios Obtained Using *Ssβ*-gly Glu387Gly, *Taβ*-gly Glu386Gly, and CelB Glu372Ala

Products[a]	*Ssβ*-gly Glu387Gly	*Taβ*-gly Glu386Gly	CelB Glu372Ala
Disaccharides	50%	86%	100%
Laminaribioside derivative (β-1,3)	80%	59%	100%
Cellobioside derivative (β-1,4)	2%	28%	ND[b]
Gentobioside derivative (β-1,6)	18%	12%	ND
Trisaccharides	40%	8%	ND
	70%	NC[c]	ND
	14%	NC	ND
Tetrasaccharides	10%	5%	ND
	54%	Tr[d]	ND

[a] Reaction performed at 65°C in 50 mM sodium phosphate buffer pH 4.0 after 30, 60 min and 7 h, respectively.

[b] ND = not detectable.

[c] NC = not characterized, the trisaccharides portion is formed by a mixture of products comigrating that were not further characterized.

[d] Tr = product detected only in traces.

ionized for the best performance. Acidic conditions could have been also beneficial in weakening the stability of the glycosyl-formate ester intermediate.[47]

More recently, the new reactivation conditions allowed the preparation of chromophoric oligosaccharides and the synthesis of the disaccharidic unit of xyloglucan oligosaccharides by using the Taβ-gly and the Ssβ-gly mutants.[53] By using 2-NP-β-D-Glc and 4-methylumbelliferyl-β-D-Glc (4-MU-Glc) as donor and acceptor, respectively, 4-MU-laminaribioside and –cellobioside disaccharides were obtained at pH 4.0 (Table 21.5). Almost complete conversion (93%) of the substrate was observed after 9 h, and the yields were remarkably high considering the low, 2-fold molar excess of the acceptor used.[53] Moreover, using 4-MU-Lam as acceptor, two trisaccharides were obtained in a reaction conducted at equimolar donor/acceptor ratio. It is worth noting that at these reaction conditions the two synthases also transfer the 2-NP-β-D-Gal to the xylosides of 4-penten-1-ol (Table 21.5), producing xyloglucan disaccharides that possess several biological activities.[54]

The success of this approach for three different enzymes indicates that it is of general applicability for glycosynthases from hyperthermophiles that, with their intrinsic stability can resist critical reaction conditions such as high temperature, low pH, and high concentration of organics, and allows their exploitation as synthetic tools.

IV. α-XYLOSIDASE FROM *S. SOLFATARICUS*

Recently, the genome of the hyperthermophilc Archaeon *S. solfataricus* has been completely sequenced[55] and has revealed genes encoding for 22 putative glycoside

TABLE 21.5
Syntheses Conducted with *Taβ*-gly Glu386Gly and *Ssβ*-Gly Glu387Gly

Enzyme	Donor	Acceptor	Products	Yield (%)
Taβ-gly Glu386Gly				
	2-NP-Glc	4-MU-Glc	4-MU-Lam, 4-MU-Cell	60–65
	2-NP-Glc	4-NP-Lac	—	—
	2-NP-Glc	P-S-Cell	—	Poor
	2-NP-Glc	4-MU-Lam	β-Glc-(1,4)-β-Glc-(1,3)-β-Glc-MU, β-Glc-β-Glc-β-Glc-MU	16
	2-NP-Gal	β-Xyl-4P	β-Gal-(1,3)-β-Xyl-4P	34
Ssβ-gly Glu387Gly				
	2-NP-Gal	α-Xyl-4P	β-Gal-(1,3)-α-Xyl-4P, β-Gal-(1,4)-α-Xyl-4P	33
	2-NP-Gal	β-Xyl-4P	β-Gal-(1,3)-β-Xyl-4P	40

2-NP-Glc = 2-nitrophenyl-β-D-glucoside; 2-NP-Gal = 2-nitrophenyl-β-D-galactoside; 4-MU-Glc = 4-methylumbelliferyl-β-D-glucoside; 4-NP-Lac = 4-nitrophenyl-β-D-lactoside; *P-S*-Cell, phenyl-thio-cellobioside; 4-MU-Lam = 4-methylumbelliferyl-β-D-laminaribioside; β-Xyl-4P = 4-penten-1-xyl-β-D-xylopyranoside; α-Xyl-4P = 4-penten-1-xyl-α-D-xylopyranoside.

hydrolases, the highest number among the sequenced Archaea, and 33 putative gly-cosyl transferases. Remarkably, 36% of the glycosidases map in a region of the genome of about 70 kb, and are likely to be involved in the degradation of sugars for energy metabolism.

One of the putative glycoside hydrolases present in this microorganism, belong-ing to the GH31 family of glycoside hydrolases that includes α-glucosidases and α-xylosidases, was efficiently expressed in *E. coli*.[56] The pure enzyme was com-pletely inactive on isomaltose, trehalose, and sucrose, and revealed low activity on maltose and different malto-oligosaccharides (Table 21.6). In contrast, the enzyme showed clear selectivity for xylose-containing substrates such as 4-NP-α-D-xylo-side (4-NP-α-Xyl), disaccharides isoprimeverose (α-D-xylopyranosyl-(1,6)-D-glu-copyranose), and 4NP-β-isoprimeveroside, which are the disaccharidic units of the hemicellulose xyloglucan. For clear specificity toward xylosides the enzyme was named XylS.[56] Qualitative analysis showed that the enzyme is also active on xyloglucan oligosaccharides, from which it produces xylose. These substrates were hydrolyzed from their nonreducing end, showing that XylS is an exo-glycosidase.[56] As expected for proteins from hyperthermophiles, the recombinant enzyme dis-plays maximal activity at 90°C and high stability to heat (half-life of 38 h at 90°C). It is worth noting that the extreme specificity of the exo-xylosidase activity of XylS has been exploited to characterize complex oligosaccharides. In fact, XylS, coupled to a β-glucosidase, was used to determine unequivocally the structure of a xyloglu-can oligosaccharide synthesized by a novel α-xylosyltransferase from *Arabidopsis*.[57]

Xyloglucan, the principal hemicellulose component in the primary cell wall and one of the most abundant storage polysaccharides in seeds, is widely distributed in plants. This polymer is composed of a β-(1,4)-glucan backbone, with α-(1,6)-D-xylose groups linked to about 75% of the glucosyl residues. Thus, the disaccharide isoprimeverose (α-D-xylopyranosyl-(1,6)-D-glucopyranose) represents the building block of xyloglucan. Additional ramifications of β-D-galactosyl-(1,2)-α-xylosyl and

TABLE 21.6
Substrate Specificity of the α-Xylosidase from *S. solfataricus* at 65°C

Substrates[a]	Specific activity (units/mg)[b]	Maximal Activity (%)
Isoprimeverose (100 mM)	16.0	100
4-NP-β-isoprimeveroside (100 mM)	8.15	50.9
4-NP-α-xyloside (37 mM)	2.33	14.6
4-NP-α-glucoside (14 mM)	0.03	0.2
Maltose (100 mM)	0.97	6.1
Maltose (10 mM)	0.47	2.9
Maltotriose (10 mM)	0.56	3.5
Maltotetraose (10 mM)	0.12	0.75
Maltopentaose (10 mM)	0.08	0.5

[a] No activity was observed on isomaltose, trehalose, and sucrose.

[b] Enzyme activity is given in μmoles of glycosyl bonds hydrolyzed per minute.

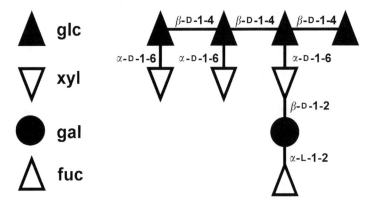

FIGURE 21.4 Schematic structure of a xyloglucan nonasaccharide.

TABLE 21.7

Oligosaccharide Synthesis by the α-Xylosidase from *S. solfataricus*

Donor	Acceptor	Products	Yields (%)
4-NP-α-Xyl	4-NP-β-Glc	R₁=α-Xyl	11
α-Xylosyl-F	Glucose		10–12
α-Xylosyl-F	4-NP-β-cellobioside	1. R₁=α-Xyl; R₂=R₃=H 2. R₁=R₂; R₃=α-Xyl 3. R₁=H; R₂=α-Xyl; R₃=H	15

α-L-fucosyl-(1,2)-β-D-galactosyl-(1,2)-α-xylosyl chains are α-(1,6)-linked at a lesser extent to the main backbone (Figure 21.4).[58] The solfataric fields, in which *S. solfataricus* grows under aerobic conditions, are rich in plant debris containing hemicellulosic material; XylS *in vivo* could recognize isoprimeverose units at the nonreducing end of xyloglucan fragments and promote the release of the xylose that can be used as an energy source.

XylS is able to transfer xylose in transxylosylation reactions. In fact, the enzyme, which follows the retaining reaction mechanism like the other members of GH31, forms 4-NP-β-isoprimeveroside by transxylosylation reaction from 4-NP-α-Xyl and 4-NP-β-Glc as donor and acceptor, respectively (Table 21.7).[56] In addition, different regioisomers, in which xylose is transferred to different glucose positions, possibly the OH in C3 and C4, were observed in trace amounts.[59] Interestingly, in preparative reactions, free isoprimeverose was obtained by using glucose and α-xylosyl fluoride as acceptor and donor, respectively (Table 21.7).[59] With the same α-xylosyl fluoride donor and 4-NP-β-cellobioside as acceptor, XylS synthesizes the trisaccharidic unit of xyloglucan XG in which xylose is attached at the C6 of the external glucose of the acceptor (Table 21.7). Because of the seven free OH groups of the acceptor, this synthesis is challenging since it may lead to multiple compounds that are difficult to purify.

These findings demonstrate that the typical exo-acting hydrolysis of this enzyme is also operative in a synthetic mode.[59] These results show that XylS could be used for the synthesis of the building blocks of xyloglucan oligosaccharides, which, acting as growth regulators of the plant cells, could find application in agro-food technology.[60]

V. α-L-FUCOSIDASE FROM *S. SOLFATARICUS*

α-L-fucosidases (3.2.1.51) are exo-glycosidases capable of cleaving α-linked L-fucose residues from glycoconjugates, in which the most common linkages are α-(1-2) to galactose and α-(1-3), α-(1-4), and α-(1-6) to *N*-acetylglucosamine residues. These compounds are involved in a variety of biological events as growth regulators and as the glucidic part of receptors in signal transduction, cell–cell interactions, and antigenic response.[61] The central role of fucose derivatives in biological processes explains the interest in α-L-fucosidase and fucosyl-transferase activities.

α-L-fucosidases in higher plants and in mammals are associated with different mechanisms of cell growth and regulation, since they are involved in the modification of fucosylated glucans.[62] In plants, α-L-fucosylated oligosaccharides derived from xyloglucan have been shown to regulate auxin- and acid pH-induced growth.[63] In mammals, oligosaccharides containing fucose have been found, for instance, in human milk and in blood group substances,[64] and are reported to play important roles in fertilization[65] and in adhesion processes of viruses, bacteria, and other parasites.[66] Changes in fucosylation patterns have been observed in several physiological events including pregnancy,[67] programed cell death of different cell types,[68] and in a variety of pathological events including diabetes[69] and colon and liver carcinomas.[70,71] In addition, the determination of α-fucosidase activity can be used to predict the development of colorectal, ovarian, and hepatocellular carcinomas,[72–74] whereas the deficiency in this enzyme causes fucosidosis, a well-known lysosomal storage disorder.[75] Family 29 of glycoside hydrolases classification (GH29) groups α-L-fucosidases from plants, vertebrates, and pathogenic microbes of plants and humans.[15]

Despite the importance of fucosylated sugars, at present, there are only limited data on α–fucosidases from eukaryal and bacterial microorganisms. The unusual high stability of enzymes from hyperthermophilic microorganisms stimulated to the search for genes encoding putative α-L-fucosidases in the sequenced genome of Archaea, which can be used as tools for the synthesis of target oligosaccharides.

A. IDENTIFICATION AND CHARACTERIZATION OF THE FIRST ARCHAEAL α-L-FUCOSIDASE

The first archaeal α-L-fucosidase, Ssα-fuc, has been identified and characterized only recently.[76] The analysis of the genome of *S. solfataricus*[55] revealed the presence of two open reading frames (ORFs), SSO11867 and SSO3060, encoding for 81 and 426 amino acid polypeptides that are homologous to the *N*- and the *C*-terminal parts, respectively, of full-length bacterial and eukaryal GH29 fucosidases.[15] The two ORFs are separated by a -1 frameshift and lead to a truncated product, suggesting that the gene is expressed *in vivo* by a mechanism called *recoding* consisting of a localized deviation from the standard translation rules.[76] To produce a single polypeptide, a single base was inserted, by site-directed mutagenesis, in the region of overlap between SSO11867 and SSO3060, restoring a single reading frame between the ORFs. The single ORF obtained was used to express the enzyme in *E. coli* as a fusion protein with glutathione-*S*-transferase (GST), which was subsequently removed. The native recombinant enzyme is a nonamer of 57 kDa molecular mass subunits (corresponding to the predicted full-length polypeptide of 495 amino acids) and it shows high specificity for the 4-NP-α-L-fucoside (4-NP-α-L-Fuc) substrate at 65°C (Table 21.8). Moreover, Ssα-fuc is thermoactive and thermostable, as expected for an enzyme from a hyperthermophilic microorganism. The optimal temperature of the enzyme is 95°C (Figure 21.5A) and it displayed high stability at 75°C, showing even 40% activation after 30 min of incubation and maintaining 60% residual activity after 2 h at 80°C (Figure 21.5B).[76]

It is worth noting that the mutation inserted to obtain the recombinant Ssα-fuc was designed on the basis of the programmed -1 frameshifting mechanism,[77] therefore, the functionality of the full-length enzyme gave support to the hypothesis that a translational *recoding* event, known so far only in Eukarya and Bacteria,[78] could be used to regulate the expression of this gene in *S. solfataricus*.

TABLE 21.8
Steady-State Kinetic Constants of Ssα-fuc and of the Mutant Asp242Gly at 65C°

	K_M (mM)	k_{cat} (s^{-1})	k_{cat}/K_M (s^{-1}mM^{-1})
Wild type	0.028 ± 0.004	287 ± 11	10,250
Asp242Gly	ND	ND	—
+ sodium azide	0.19 ± 0.02	9.66 ± 0.28	51.55
+ sodium formate	1.03 ± 0.11	5.91 ± 0.20	5.76

ND=not detectable.

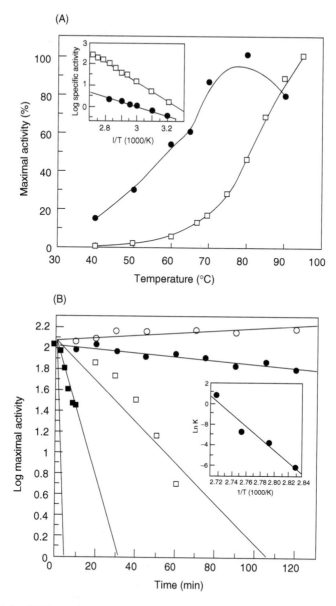

FIGURE 21.5 (A) Dependence on temperature of the wild-type Ssα-fuc (□) and Asp242Gly mutant in the presence of 2 *M* sodium azide (●). (B) Thermal stability of wild-type Ssα-fuc at 75°C (○), 80°C (●), 85°C (□), 90°C (■) and 95°C (–).

B.   TRANSFUCOSYLATION REACTIONS

Ssα-Fuc is capable of functioning in transglycosylation mode as reported for several mesophilic α-fucosidases.[79,80] The synthetic ability of the thermophilic enzyme was demonstrated by using, in transfucosylation reactions, 4-NP-α-L-Fuc and 4-NP-α-D-Glc as donor and acceptor, respectively. The fucosylated products, with 14% total yield with respect to 4-NP-α-L-Fuc, were disaccharides of the acceptor in which the α-L-fucose moiety of the donor is attached at positions 2 and 3 of Glc (α-L-Fuc$p$-(1-2)-α-D-Glc-O-4-NP and α-L-Fuc$p$-(1-3)-α-D-Glc-O-4-NP).[76] The α-anomeric configuration of the interglycosidic linkages in the products demonstrated, for the first time, by following transglycosylation reaction, that GH29 α–fucosidases follow a *retaining* reaction mechanism.[76]

The hydrolytic activity of Ssα-fuc on the disaccharide α-L-Fuc-(1-3)-α-L-Fuc-O-4-NP revealed that the enzyme is an exo-glycosyl hydrolase that attacks the substrates from their nonreducing end.[76] The disaccharide α-L-Fuc$p$-(1-3)-α-D-Glc-O-4-NP was previously synthesized, in 34% yield, by using the mesophilic α-L-fucosidase from *Penicillium multicolor*[80] with the fucosyl fluoride as donor and 4-NP-glucose as acceptor. Work is in progress to optimize the reaction conditions for Ssα-fuc by keeping as low as possible the acceptor/donor equivalent ratio useful for synthetic application with rare acceptors.

C. IDENTIFICATION OF THE NUCLEOPHILE

α-L-Fucosidases follow a retaining reaction mechanism for the hydrolysis of the substrate in which one carboxylic acid in the active site acts as the catalytic nucleophile, leading to the formation of the β-L-fucosyl intermediate, while the other plays the role of the general acid catalyst in the first step and the general base catalyst in the second step of the reaction (Figure 21.6). The nucleophile of GH29 α-L-fucosidases was identified, for the first time, by reactivation with sodium azide of the Ssα-fuc Asp242Gly inactivated mutant by analyzing the anomeric configuration of the fucosyl-azide product.[81]

The Asp242Gly mutant showed a turnover number (k_{cat}) of 0.24 s^{-1}, on 4-NP-α-L-Fuc, which is 1.0×10^{-3} times that of the wild-type activity (245 s^{-1}). In the presence of 2 M sodium azide the mutant revealed a k_{cat} value of 11 s^{-1}, indicating a 46-fold activation by azide. This activation falls in the range of chemically rescued activities of β-glycoside hydrolases (10-, 10^7-fold).[43,44] The specific activity of the mutant increases with temperature up to 80°C, indicating that, despite the mutation, the enzyme maintains its thermophilicity when reactivated in the presence of azide

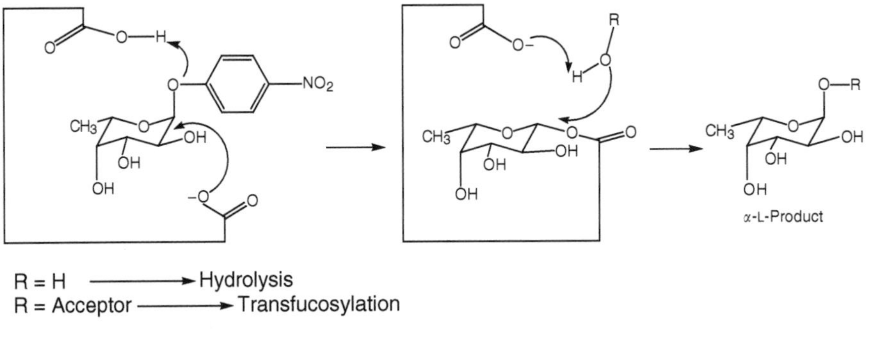

R = H ⟶ Hydrolysis
R = Acceptor ⟶ Transfucosylation

FIGURE 21.6 Retaining reaction mechanism of Ssα-fuc.

(Figure 21.5A). The reactivation experiment with sodium azide indicated that the Asp242Gly mutation affected a residue involved in catalysis in Ssα-fuc.

One of the methods used to identify the acid/base or the nucleophile of the gly-cosidases is to characterize the stereochemistry of the products of mutants reactivated by azide.[13] In the case of a fucosidase, reactivated mutants in the acid/base catalyst would produce glycosyl azide with the same anomeric configuration of the substrate (Figure 21.7A), whereas the isolation of glycosyl-azide products with an anomeric configuration opposite to that of the substrate would allow the identification of the cat-alytic nucleophile of the reaction (Figure 21.7B). The fucosyl-azide product obtained by the Asp242Gly mutant was found in the inverted (β-L) configuration compared with the substrate (Figure 21.7B). This finding allowed, for the first time, the unambiguous assignment of Asp242 and its homologous residues as the nucleophilic catalytic residues of GH29 α-L-fucosidases.[81] Later, by following a similar approach, the corre-sponding residue was also identified in the α-fucosidase from *Thermotoga maritima*.[82]

The azide rescue method has been used so far only for retaining β-D-glycosidases, whereas it has never been used for retaining α-(D/L)-glycosidases, and is generally based on substrates that are very reactive (showing leaving groups with pK_a<5) to facilitate the first step of the reaction.[46] Therefore, it is worth noting that Asp242Gly was efficiently reactivated on 4-NP-α-L-Fuc, which shows leaving group with pK_a = 7.18. By comparison, β-glycosidases from GH1 and GH52, mutated in the nucle-ophile, could not be reactivated by azide on 4-NP-glycosides, but require substrates with excellent leaving groups for an efficient chemical rescue of their activity.[40,46] The

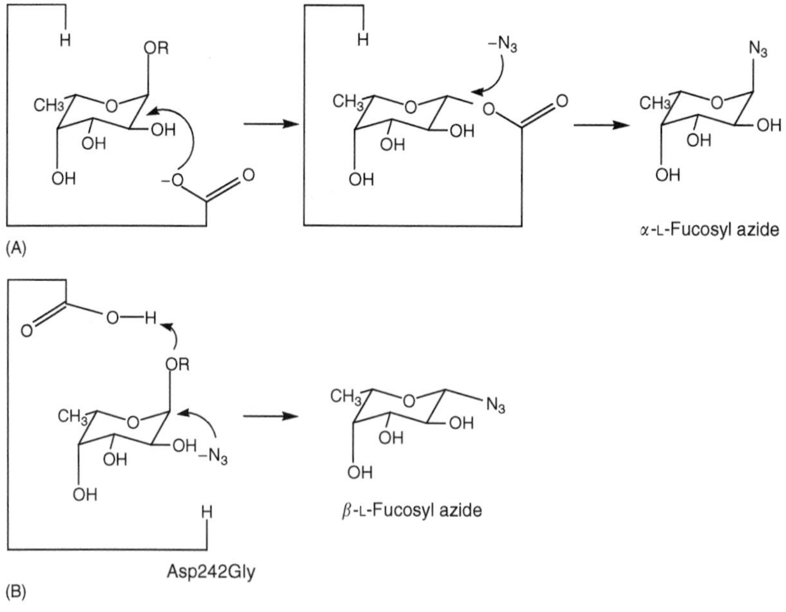

(A)

(B)   Asp242Gly

FIGURE 21.7 Chemical rescue of mutants of α-fucosidase in (A) the acid/base residue and (B) the nucleophile of the reaction.

identification, for the first time, of the nucleophile of a α-L-glycosidase by following this approach showed that it can be of general applicability for retaining enzymes.

The activity of the mutant Asp242Gly was also rescued on 4-NP-α-L-Fuc in the presence of sodium formate. The steady-state kinetic constants of Asp242Gly, determined in the presence of external nucleophiles, revealed that sodium azide and sodium formate produced about 0.5 and 0.056% of reactivation of the mutant, respectively, which was calculated by taking as 100% the k_{cat}/K_M of the wild type (Table 21.8).[81] The higher nucleophilicity of sodium azide, when compared with formate, explains the higher reactivation produced by the former.

As described in the third paragraph of this chapter, the hydrolytic activity of β-glycosynthases from hyperthermophilic Archaea can be rescued at levels comparable with the corresponding wild-type enzymes under acidic conditions.[47] A 16-fold reactivation of the Asp242Gly mutant was obtained at pH 4.0, strongly indicating that, in this case as well, the acidic conditions restored the protonated form of the acid/base residue.[81] The common hyperthermophilic nature of these archaeal enzymes, which resist the reaction conditions at low pH, allowed a similar chemical rescue of the activity, thus confirming that this approach can be of general applicability to α-L and β-D-glycosidases from hyperthermophiles.

The Asp242Gly mutant reactivated in sodium formate (pH 4.0) did not produce oligosaccharides, suggesting that the enzyme did not act as an α-L-glycosynthase.[81] This lack of activity is not surprising: hyperthermophilic β-glycosynthases efficiently produced oligosaccharides by using the substrate 2-NP-β-D-Glc, in which the 2-nitrophenol, although showing a pK_a similar to 4-nitrophenol (pK_a = 7.22 and 7.18 respectively), can form a chelate ring by H-bonding, increasing the leaving ability upon protonation.[47] The absence of aryl-α-L-fucosides with groups having better leaving ability has hampered the analysis of the reactivation of the Asp242Gly mutant with more reactive substrates. Work is in progress to test the α-fucosynthetic potential of this mutant on suitable substrates.

VI. CONCLUSIONS

Traditionally, the enzymes from thermophilic sources find most of their applications in biotransformations performed under harsh conditions. However, these applications, especially in the field of carbohydrate modifications, are of real interest in only a limited number of cases. We have shown here that glycosidases from hyperthermophiles can be used in oligosaccharide synthesis not only because of their intrinsic stability to heat, high molar excess of organics and nucleophilic compounds, and acid conditions, but also because of their intrinsic catalytic properties, such as efficiency in performing transglycosylation reactions, and their regio-, and stereospecificity. The exploration of the biodiversity of glycosidases from hyperthermophiles can widen their applicative perspective.

ACKNOWLEDGMENTS

This work was supported by Agenzia Spaziale Italiana project "Extremophilic Archaea as model systems to study origin and evolution of early organisms: molecular mechanisms of adaptation to extreme physical-chemical conditions" Contract no. I/R/365/02 and by MIUR project "Folding di proteine: l'altra metà del codice genetico."

REFERENCES

1. Varki, A., Biological roles of oligosaccharides: all of the theories are correct, *Glycobiology,* 2, 97–130, 1993.

2. Sears, P. and Wong, C.H., Intervention of carbohydrate recognition by proteins and nucleic acids, *Proc. Natl. Acad. Sci. USA,* 93, 12086–12093, 1996.

3. Zopf, D. and Roth, S., Oligosaccharide anti-infective agents, *Lancet,* 347, 1017–21, 1996.

4. Roberge, J.Y., Beebe, X., and Danishefsky, S.J., A strategy for a convergent synthesis of *N*-linked glycopeptides on a solid support, *Science,* 269, 202–204, 1995.

5. Endo, T. and Koizumi, S., Large-scale production of oligosaccharides using engineered bacteria, *Curr. Opin. Struct. Biol.,* 10, 536–541, 2000.

6. Seeberger, P.H., Solid phase oligosaccharide synthesis, in *Glycochemistry Principles, Synthesis, and Applications,* Wang, P.G., and Bertozzi, C.R., Eds., Marcel Dekker, New York, pp. 1–32, 2001.

7. Crout, D.H. and Vic, G., Glycosidases and glycosyl transferases in glycoside and oligosaccharide synthesis, *Curr. Opin. Chem. Biol.,* 2, 98–111, 1998.

8. Henrissat, B. and Davies, G.J., Glycoside hydrolases and glycosyltransferases, Families, modules, and implications for genomics, *Plant Physiol.,* 124, 1515–1519, 2000.

9. Sears, P. and Wong, C.H., Toward automated synthesis of oligosaccharides and glycoproteins, *Science,* 291, 2344–2350, 2001.

10. Koshland, D.E., Jr., Stereochemistry and the mechanism of enzymatic reactions, *Biol. Rev.,* 28, 416–436, 1953.

11. McCarter, J.D. and Withers, S.G., Mechanisms of enzymatic glycoside hydrolysis, *Curr. Opin. Struct. Biol.,* 4, 885–892, 1994.

12. Vasella, A., Davies, G.J., and Bohm, M., Glycosidase mechanisms, *Curr. Opin. Chem. Biol.,* 5, 619–629, 2002.

13. Ly, H.D. and Withers, S.G., Mutagenesis of glycosidases, *Annu. Rev. Biochem.* 68, 487–522, 1999.

14. Williams, S.J. and Withers, S.G., Glycosyl fluorides in enzymatic reactions, *Carbohydr. Res.,* 327, 27–46, 2000.

15. Henrissat, B., A classification of glycosyl hydrolases based on amino acid sequence similarities, *Biochem. J.,* 280, 309–316, 1991.

16. Henrissat, B. and Bairoch, A., New families in the classification of glycosyl hydrolases based on amino acid sequence similarities, *Biochem. J.,* 293, 781–788, 1993.

17. Withers, S.G., Mechanisms of glycosyl transferases and hydrolases, *Carb. Polymers.,* 44, 325–337, 2001.

18. Mackenzie, L.F., Wang, Q.P., Warren, R.A.J., and Withers, S.G., Glycosynthases: mutant glycosidases for oligosaccharide synthesis, *J. Am. Chem. Soc.,* 120, 5583–5584, 1998.

19. Vieille, C. and Zeikus, G.J., Hyperthermophilic enzymes: sources, uses, and molecular mechanisms for thermostability, *Microbiol. Mol. Biol. Rev.,* 65, 1–43, 2001.

20. Splechtna, B., Petzelbauer, I., Baminger, U., Haltrich, D., Kulbe, K.D., and Nidetzky, B., Production of a lactose-free galacto-oligosaccharide mixture by using selective enzymatic oxidation of lactose into lactobionic acid, *Enzyme Microb. Tech.,* 29, 434–440, 2001.

21. Moracci, M., Trincone, A., Cobucci-Ponzano, B., Perugino, G., Ciaramella, M., and Rossi, M., Enzymatic synthesis of oligosaccharides by two glycosyl hydrolases of *Sulfolobus solfataricus, Extremophiles,* 5, 145–152, 2001.

22. Trincone, A., Nicolaus, B., Lama, L., Morzillo, P., De Rosa, M., and Gambacorta, A., Enzyme-catalyzed synthesis of alkyl β-D-glycosides with crude homogenate of *Sulfolobus solfataricus, Biotechnol. Lett.,* 13, 235–240, 1991.

23. Trincone, A., Improta, R., Nucci, R., Rossi, M., and Gambacorta, A., Enzymatic synthesis of carbohydrate derivatives using β-glycosidase of *Sulfolobus solfataricus, Biocatalysis,* 10, 195–210, 1994.

24. Santin, M., Rosso, F., Sada, A., Peluso, G., Improta, R., and Trincone, A., Enzymatic synthesis of 2-β-D-galactopyranosyloxethyl methacrylate (GAlEMA) by the thermophilic archaeon *Sulfolobus solfataricus, Biotechnol. Bioeng.,* 49, 217–222, 1996.

25. Dordick, J.S., Enzymatic and chemo-enzymatic approaches to polymer synthesis, *TIBTECH,* 10, 287–293, 1992.

26. Kinomura, K., Kitazawa, S., Okumura, M., and Sakakibara, T., Potential application of some synthetic glycosides to food modification, *ACS Symp. Ser.,* 528, 221–228, 1993.

27. Trincone, A. and Pagnotta, E., Facile chemo-enzymatic access to monoglucosyl derivatives of 2,3-oxirane dimethanol, *Tetrahedron:Asymmetry,* 7, 2773–2774, 1996.

28. Trincone, A., Pagnotta, E., and Sodano, G., Chemo-enzymatic synthesis and stereochemistry of aleppotrioloside, a naturally occurring glucoside, *Tetrahedron Lett.,* 35, 1415–1416, 1994.

29. Trincone, A. and Pagnotta, E., Efficient chemoselective synthesis of 3′,4′-dihydroxypropiophenone 3-O-β-D-glucoside by thermophilic β-glycosidase from *Sulfolobus solfataricus, Biotechnol. Lett.,* 17, 45–48, 1995.

30. Trincone, A., Pagnotta, E., Fantin, G., and Fogagnolo, M., Enzymatic routes for the synthesis of rhododendrin and epi-rododendrin, *Biocat. Biotrans.,* 13, 245–253, 1996.

31. Sinnot, M.L., Catalytic mechanism of enzymic glycosyl transfer, *Chem. Rev.,* 90, 1171–1202, 1990.

32. Lai, E.C.K., Morris, S.A., Street, I.P., and Withers, S.G., Substituted glycals as probes of glycosidase mechanisms, *Bioorg. Med. Chem.,* 4, 1929–1937, 1996.

33. Marzabaldi, C.H. and Franck, R.W., The synthesis of 2-deoxyglycosides: 1988–1999, *Tetrahedron,* 56, 43, 8385–8417, 2000.

34. Trincone, A., Pagnotta, E., Rossi, M., Mazzone, M., and Moracci, M., Enzymatic synthesis of 2-deoxy-β-glucosides and stereochemistry of β-glycosidase from Sulfolobus solfataricus on glucal, *Tetraedron: Asymmetry,* 12, 2783–2787, 2001.

35. Trincone, A., Pagnotta, E., Giordano, A., Perugino, G., Rossi, M., and Moracci, M., Enzymatic synthesis of 2-deoxyglycosides using the β-glycosidase of the archeon *Sulfolobus solfataricus, Biocat. Biotrans.,* 21, 17–24, 2003.

36. Mootoo, D.R., Konradsson, P., Udodong, U., and Fraser-Reid, B., Armed and disarmed n-pentenyl glycosides in saccharide couplings leading to oligosaccharides, *J. Am. Chem. Soc.,* 110, 5583–5584, 1988.

37. Inanaga, J., Yokohma, Y., and Hanamoto, T., Utility of 3,4-dimethoxybenzyl (DMPM) glycosides. A new glycosylation triggered by 2,3-dichloro-5,6-dicyano-p-benzoquinone, *Chem. Lett.,* 85–88, 1993.

38. Nicolau, K.C., Seitz, S.P., and Papahatjis, D.P., A mild and general method for the synthesis of *O*-glycosides, *J. Am. Chem. Soc.,* 105, 2430–2434, 1983.

39. Moracci, M., Trincone, A., and Rossi, M., Glycosynthases: new enzymes for oligosaccharide synthesis, *J. Mol. Cat. B: Enzymatic,* 11, 155–163, 2001.

40. Moracci, M., Trincone, A., Perugino, G., Ciaramella, M., and Rossi, M., Restoration of the activity of active-site mutants of the hyperthermophilic β-glycosidase from *Sulfolobus solfataricus*: dependence of the mechanism on the action of external nucleophiles, *Biochemistry,* 37, 17262–17270, 1998.

41. Trincone, A., Perugino, G., Rossi, M., and Moracci, M., A novel thermophilic glycosynthase that effects branching glycosylation, *Bioorg. Med. Chem. Lett.,* 4, 365–368, 2000.

42. Kiho, T., Katsuragawa, M., Nagai, K., Ugai, S., and Haga, M., Structure and antitumor activity of a branched (1-3)-β-D-glucan from the alkaline extract of Amanita muscaria, *Carbohydr. Res.,* 224, 237–243, 1992.

43. Wang, Q., Graham, R.Q., Timbur, D., Warren, R.A.J., and Withers, S.G., Changing enzymatic reaction mechanism by mutagenesis: conversion of a retaining glucosidase to an inverting enzyme, *J. Am. Chem. Soc.,* 116, 11594–11595, 1994.

44. Viladot, J.L., de Ramon, E., Durany, O., and Planas, A., Probing the mechanism of Bacillus 1,3-1,4-β-D-glucan 4-glucanohydrolases by chemical rescue of inactive mutants at catalytically essential residues, *Biochemistry,* 37, 11332–11342, 1998.

45. MacLeod, A.M., Tull, D., Rupitz, K., Warren, R.A.J., and Withers, S.G., Mechanistic consequences of mutation of active site carboxylates in a retaining β-1,4-glycanase from *Cellulomonas fimi, Biochemistry,* 35, 13165–13172, 1996.

46. Shallom, D., Belakhov, V., Solomon, D., Shoham, G., Baasov, T., and Shoham, Y., Detailed kinetic analysis and identification of the nucleophile in α-L-arabinofuranosidase from *Geobacillus stearothermophilus* T-6, a family 51 glycoside hydrolase, *J. Biol. Chem.,* 277, 43667–43673, 2002.

47. Perugino, G., Trincone, A., Giordano, A., van der Oost, J., Kaper, T., Rossi, M., and Moracci, M., Activity of hyperthermophilic glycosynthases is significantly enhanced at acidic pH, *Biochemistry,* 42, 8484–8493, 2003.

48. Viladot, J.L., Canals, F., Batllori, X., and Planas, A., Long-lived glycosyl-enzyme intermediate mimic produced by formate re-activation of a mutant endoglucanase lacking its catalytic nucleophile, *Biochem. J.,* 355, 79–86, 2001.

49. Mayer, C., Jakeman, D.L., Mah, M., Karjala, G., Gal, L., Warren, R.A., and Withers, S.G., Directed evolution of new glycosynthases from Agrobacterium β-glucosidase: a general screen to detect enzymes for oligosaccharide synthesis, *Chem. Biol.,* 8, 437–443, 2001.

50. Hrmova, M., Imai, T., Rutten, S.J., Fairweather, J.K., Pelosi, L., Bulone, V., Driguez, H., and Fincher, G.B., Mutated varley (1,3)-β-D-glucan endohydrolases synthesize crystalline (1,3)-β-D-glucans, *J. Biol. Chem.,* 277, 30102–30111, 2002.

51. Mayer, C., Zechel, D.L., Reid, S.P., Warren, R.A., and Withers, S.G., The E358S mutant of Agrobacterium sp. β-glucosidase is a greatly improved glycosynthase. *FEBS Lett.,* 466, 40–44, 2000.

52. McIntosh, L.P., Hand, G., Johnson, P.E., Joshi, M.D., Korner, M., Plesniak, L.A., Ziser, L., Wakarchuk, W.W., and Withers, S.G., The pKa of the general acid/base carboxyl group of a glycosidase cycles during catalysis: a 13C-NMR study of *Bacillus circulans* xylanase, *Biochemistry,* 35, 9958–9966, 1996.

53. Trincone, A., Giordano, A., Perugino, G., Rossi, M., and Moracci, M., Glycosynthase-catalysed syntheses at pH below neutrality, *Bioorg. Med. Chem. Lett.,* 42, 9525–9531, 2003.

54. Fry, S.C., Oligosaccharin mutants, *Trends Plant Sci.,* 1, 326–328.

55. She, Q., Singh, R.K., Confalonieri, F., Zivanovic, Y., Allard, G., Awayez, M.J., Chan-Weiher, C.C., Clausen, I.G., Curtis, B.A., De Moors, A., Erauso, G., Fletcher, C., Gordon, P.M., Heikamp-de Jong, I., Jeffries, A.C., Kozera, C.J., Medina, N., Peng, X., Thi-Ngoc, H.P., Redder, P., Schenk, M.E., Theriault, C., Tolstrup, N., Charlebois, R.L., Doolittle, W.F., Duguet, M., Gaasterland, T., Garrett, R.A., Ragan, M.A., Sensen, C.W., and Van der Oost, J., The complete genome of the crenarchaeon *Sulfolobus solfataricus* P2, *Proc. Natl. Acad. Sci. USA,* 98, 7835–7840, 2001.

56. Moracci, M., Cobucci-Ponzano, B., Trincone, A., Fusco, S., De Rosa, M., van de Oost, J., Sensen, C.W., Charlebois, R.L., and Rossi, M., Identification and molecular characterization of the first α-xylosidase from an Archaeon, *J. Biol. Chem.,* 275, 22082–22089, 2000.

57. Faik, A., Price, N.J., Raikhel, N.V., and Keegstra, K., An Arabidopsis gene encoding an α-xylosyltransferase involved in xyloglucan biosynthesis, *Proc. Natl. Acad. Sci. USA,* 99, 7797–7802, 2002.

58. Crombie, H.J., Chengappa, S., Hellyer, A., and Reid, J.S.G., A xyloglucan oligosaccharide-active, transglycosylating beta-D-glucosidase from the cotyledons of nasturtium (*Tropaeolum majus* L) seedlings — purification, properties and characterization of a cDNA clone, *Plant J.* 15, 27–38, 1998

59. Trincone, A., Cobucci-Ponzano, B., Di Lauro, B., Rossi, M., Mitsuishi, Y., and Moracci, M., Enzymatic synthesis and hydrolysis of xylogluco-oligosaccharides using the first archaeal α-xylosidase from *Sulfolobus solfataricus, Extremophiles,* 5, 277–282, 2001.

60. Cutillas-Iturralde, A. and Lorences, E.P., Effect of Xyloglucan Oligosaccharides on Growth, Viscoelastic Properties, and Long-Term Extension of Pea Shoots, *Plant Physiol.,* 113, 103–109, 1997.

61. Vanhooren, P.T. and Vandamme, E.J., L-Fucose: occurrence, physiological role, chemical, enzymatic and microbial synthesis, *J. Chem. Technol. Biotechnol.,* 74, 479–497, 1999.

62. Staudacher, E., Altmann, F., Wilson, I.B., and Marz, L., Fucose in N-glycans: from plant to man, *Biochim. Biophys. Acta,* 1473, 216–236, 1999.

63. de La Torre, F., Sampedro, J., Zarra, I., and Revilla, G., AtFXG1, an Arabidopsis gene encoding alpha-L-fucosidase active against fucosylated xyloglucan oligosaccharides, *Plant Physiol.,* 128, 247–255, 2002.

64. Stahl, B., Thurl, S., Henker, J., Siegel, M., Finke, B., and Sawatzki, G., Detection of four human milk groups with respect to Lewis-blood-group-dependent oligosaccharides by serologic and chromatographic analysis, *Adv. Exp. Med. Biol.,* 501, 299–306, 2001.

65. Mori, E., Hedrick, J.L., Wardrip, N.J., Mori, T., and Takasaki, S., Occurrence of reducing terminal N-acetylglucosamine 3-sulfate and fucosylated outer chains in acidic N-glycans of porcine zona pellucida glycoproteins, *Glycoconj. J.,* 15, 447–456, 1998.

66. Listinsky, J.J., Siegal, G.P., Listinsky, C.M., Alpha-L-fucose: a potentially critical molecule in pathologic processes including neoplasia, *Am. J. Clin. Pathol.,* 110, 425–440, 1998.

67. Xiang, J. and Bernstein, I.A., Differentiative changes in fucosyltransferase activity in newborn rat epidermal cells, *Biochem. Biophys. Res. Commun.,* 189, 27–32, 1992.

68. Russell, L., Waring, P., and Beaver, J.P., Increased cell surface exposure of fucose residues is a late event in apoptosis, *Biochem. Biophys. Res. Commun.,* 250, 449–453, 1998.

69. Wiese, T.J., Dunlap, J.A., and Yorek, M.A., Effect of L-fucose and D-glucose concentration on L-fucoprotein metabolism in human Hep G2 cells and changes in fucosyltransferase and alpha-L-fucosidase activity in liver of diabetic rats, *Biochim. Biophys. Acta,* 1335, 61–72, 1997.

70. Rapoport, E. and Pendu, J.L., Glycosylation alterations of cells in late phase apoptosis from colon carcinomas, *Glycobiology,* 9, 1337–1345, 1999.
71. Noda, K., Miyoshi, E., Uozumi, N., Gao, C.X., Suzuki, K., Hayashi, N., Hori, M., and Taniguchi, N., High expression of alpha-1-6 fucosyltransferase during rat hepatocarcinogenesis, *Int. J. Cancer.* 75, 444–450, 1998.
72. Fernandez-Rodriguez, J., Ayude, D., de la Cadena, M.P., Martinez-Zorzano, VS., de Carlos, A., Caride-Castro, A., de Castro, G., and Rodriguez-Berrocal, F.J., Alpha-L-fucosidase enzyme in the prediction of colorectal cancer patients at high risk of tumor recurrence, *Cancer Detect. Prev.,* 24, 143–149, 2000.
73. Abdel-Aleem, H., Ahmed, A., Sabra, A.M., Zakhari, M., Soliman, M., and Hamed, H., Serum alpha L-fucosidase enzyme activity in ovarian and other female genital tract tumors, *Int. J. Gynaecol. Obstet.,* 55, 273–279, 1996.
74. Ishizuka, H., Nakayama, T., Matsuoka, S., Gotoh, I., Ogawa, M., Suzuki, K., Tanaka, N., Tsubaki, K., Ohkubo, H., Arakawa, Y., and Okano, T., Prediction of the development of hepato-cellular-carcinoma in patients with liver cirrhosis by the serial determinations of serum alpha-L-fucosidase activity, *Intern. Med.,* 38, 927–931, 1999.
75. Michalski, J.C. and Klein, A., Glycoprotein lysosomal storage disorders: alpha- and beta-mannosidosis, fucosidosis and alpha-*N*-acetylgalactosaminidase deficiency, *Biochim. Biophys. Acta,* 1455, 69–84, 1999.
76. Cobucci-Ponzano, B., Trincone, A., Giordano, A., Rossi, M., and Moracci, M., Identification of an archaeal alpha-L-fucosidase encoded by an interrupted gene. Production of a functional enzyme by mutations mimicking programmed −1 frameshifting, *J. Biol. Chem.,* 278, 14622–14631, 2003.
77. Farabaugh, P.J., Programmed translational frameshifting, *Annu. Rev. Genet.,* 30, 507–528, 1996.
78. Baranov, P.V., Gurvich, O.L., Fayet, O., Prere, M.F., Miller, W.A., Gesteland, R.F., Atkins, J.F., and Giddings, M.C., RECODE: a database of frameshifting, bypassing and codon redefinition utilized for gene expression, *Nucleic Acids Res.,* 29, 264–267, 2001.
79. Murata, T., Morimoto, S., Zeng, X., Watanabe, S., and Usui, T., Enzymatic synthesis of α-L-fucosyl-*N*-acetyllactosamines and 3'-*O*-α-L-fucosyllactose utilizing α-L-fucosidases, *Carbohydr. Res.,* 320, 192–199, 1999.
80. Farkas, E., Thiem, J., and Ajisaka, K., Enzymatic synthesis of fucose-containing disaccharides employing the partially purified alpha-L-fucosidase from *Penicillium multicolor, Carbohydr. Res.,* 328, 293–299, 2000.
81. Cobucci-Ponzano, B., Trincone, A., Giordano, A., Rossi, M., and Moracci, M., Identification of the catalytic nucleophile of the family 29 alpha-L-fucosidase from *Sulfolobus solfataricus* via chemical rescue of an inactive mutant, *Biochemistry,* 42, 9525–9531, 2003.
82. Tarling, C.A., He, S., Sulzenbacher, G., Bignon, C., Bourne, Y., Henrissat, B., and Withers, S.G., Identification of the catalytic nucleophile of the family 29 alpha-L-fucosidase from thermotoga maritima through trapping of a covalent glycosyl-enzyme intermediate and mutagenesis, *J. Biol. Chem.,* 279, 13119–13128, 2004.

22 Modification of Plant Cell Wall Polysaccharides Using Enzymes from *Aspergillus*

Ronald P. de Vries, Maureen C. McCann, and Jaap Visser

CONTENTS

I.   INTRODUCTION

Plant cell wall polysaccharides are the most abundant organic compounds found in nature and are used as raw material for many industrial processes. They are also the most important carbon source for a wide range of microorganisms, including both prokaryotes and lower eukaryotes. Some of these organisms are plant pathogens infecting the living plant, whereas others are saprophytes growing on dead plant material. In order to use plant cell wall polysaccharides as carbon source, these microorganisms produce a wide spectrum of enzymes, capable of breaking specific linkages in the polysaccharide. These enzymes have been purified and characterized from many different organisms, but several species from the fungal genus *Aspergillus* are particularly well studied with respect to the production of cell wall-degrading enzymes. Many of these enzymes are being used in industrial applications such as biobleaching of wood pulp[1,2] and enzymatic debarking,[3] dough improvement,[4–6] increasing the feed conversion efficiency of animal feed,[7] and clarifying juices.[8,9]

The genus *Aspergillus* consists of a large number of species, which include opportunistic pathogens, such as *A. fumigatus*. Most important for the production and industrial application of enzymes are some members of the group of black aspergilli (e.g., *A. niger and A. tubingensis*). Products from several black *Aspergillus* species have obtained a generally regarded as safe (GRAS) status, allowing them to be used in the food and feed industries. The main reasons for their success as industrial organisms are their good fermentation characteristics and their high level of protein secretion.

In this chapter, we will discuss the degradation and modification of plant cell wall polysaccharides using *Aspergillus* enzymes. First, we will review the structures of cell wall polysaccharides and discuss architectural features of the cell wall that may impact the accessibility of enzymes to their substrates. Then, we summarize the enzymes from *Aspergillus* involved in polysaccharide degradation and discuss the enzymatic modifications that are possible using these enzymes. A detailed description of the characteristics of individual enzymes will not be given as these have recently been reviewed.[10] These enzymes have been classified by Henrissat and co-workers[11–13] into families based on their amino acid sequence (see http://afmb.cnrs-mrs.fr/CAZY/index.html). This classification will not be discussed in detail, but the family numbers will be given. Finally, the regulatory systems involved in the production of these enzymes by *Aspergillus* will be discussed.

II. STRUCTURE OF PLANT CELL WALL POLYSACCHARIDES

A. CELLULOSE

Cellulose microfibrils form the main scaffolding within all plant cell walls. Cellulose is a linear polymer of glucose residues linked into $(1 \rightarrow 4)\beta$-D-glucan chains with a flat-ribbon conformation. Microfibrils are paracrystalline assemblies of several dozen such chains, hydrogen-bonded to one another along their length.[14] Microfibrils of angiosperms have been measured to be between 5 and 12 nm wide using an electron microscope.[15] Each $(1 \rightarrow 4)$ β-D-glucan chain is several thousand glucosyl units (~2–3 μm long), but individual chains begin and end at different places within the microfibril to allow a microfibril to reach lengths of hundreds of micrometers and contain thousands of individual glucan chains. By electron diffraction, the $(1 \rightarrow 4)$ β-D-glucan chains of cellulose are arranged parallel to one another, that is, all of the reducing ends of the chains point in the same direction.[16] The interior chains of the microfibril are highly crystalline, but chains on the surface of the microfibril may adopt a different conformation: the surface chains are more susceptible to chemical substitutions than would be predicted if they were hydrogen-bonded in a crystalline fashion.[17,18] Also, nuclear magnetic resonance (NMR) spectra of cell walls contain signals attributed to both amorphous and crystalline forms of cellulose.[19,20] The surface chains may be more susceptible to enzymatic hydrolysis by fungal attack.

B. XYLAN

Xylan is a linear polymer of $(1 \rightarrow 4)$ β-D-linked xylose residues with various side-chain substitutions of $(1 \rightarrow 2)$ β-4-*O*-methyl-α-D-glucuronic acid, acetyl, and arabinosyl residues.[21,22] Unbranched xylans can hydrogen-bond to cellulose or to each other. The attachment of the α-L-Ara and α-D-GlcA side groups to the xylan backbone of glucuronoarabinoxylans prevents the formation of hydrogen bonds, and, therefore, blocks cross-linking between two branched chains or to cellulose.[23] Xylans are particularly abundant in the thickened secondary walls of particular cell types, such as xylem.[24] However, all angiosperms contain at least small amounts of xylans in their primary cell walls. Xylan structure varies considerably with respect to the degree of substitution and position of attachment of α-L-Ara residues, from xylans with nearly every xylosyl unit branched to those with only 10% or less of xylosyl residues bearing side groups. In cell walls of dicotyledonous species, the

α-L-Ara units are more commonly found at the O-2 position. However, in the commelinoid monocots, where glucuronoarabinoxylans are major cell wall components, the Ara units are invariably at the O-3 position. In all glucuronoarabinoxylans, the α-D-GlcA units are attached to the O-2 position.[25]

The primary walls of the commelinoid orders of monocots contain significant amounts of aromatic substances in their unlignified cell walls — a feature that makes them fluorescent under UV light. A large fraction of plant aromatics consists of hydroxycinnamic acids, such as ferulic and *p*-coumaric acids. In grasses, these hydroxycinnamates are attached as carboxyl esters to the O-5 position of a few Ara units of glucuronoarabinoxylans.

C. GALACTO(GLUCO)MANNAN

Other less abundant noncellulosic polysaccharides, such as glucomannans, galactoglucomannans, and galactomannans, potentially bind to cellulose in some primary walls. These mannans are found in virtually all angiosperms. As their name suggests, galactomannans have a $(1\rightarrow 4)$ β-D-linked backbone of mannose residues with galactose residues attached at the O-6 position.[26] In contrast, glucomannan is a polymer with randomly distributed $(1\rightarrow 4)$ β-D-linked glucose residues and $(1\rightarrow 4)$ β-D-linked mannose residues.[27] *In vitro* synthesis of this polymer with GDP-mannose and GDP-glucose indicates that the synthase may have a single active center for which the two substrates compete to produce the heteropolysaccharide.[28] The glucomannan can be galactosylated at the O-6 position of the mannosyl residues[29] or at the O-6 of the glucosyl residues.[30]

D. XYLOGLUCAN

In dicotyledonous species, xyloglucans consist of linear chains of $(1\rightarrow 4)$ β-D-glucan with numerous α-D-xylose units linked at regular sites to the O-6 position of the glucose units. Some of the xylose residues are substituted further with α-L-Ara or β-D-Gal, depending on species, and sometimes the galactose is substituted further with α-L-fucose.[31,23] Xyloglucans are constructed in block-like unit structures containing 6 to 11 sugars, the proportions of which vary among tissues and species.[33] The glucan backbone of xyloglucan can hydrogen-bond to the surface of cellulose microfibrils, and some domains of the molecule may become entrapped within the microfibril.[34] With an average length of ~200 nm, xyloglucans are long enough to span the distance between two microfibrils and bind to each of them.[35] Thus, different domains of xyloglucan may be differentially accessible to hydrolysis.[34]

E. PECTIN

Pectins comprise a mixture of heterogeneous, branched, and highly hydrated polysaccharides rich in D-galacturonic acid. There are two classes of pectic polymer, galacturonans and rhamnogalacturonan I (RG I), distinguished by their backbones.[36] Galacturonans comprise homogalacturonan (HG), xylogalacturonan, apiogalacturonan, and rhamnogalacturonan (RG II) [23].[37–39] HG is a polymer of $(1\rightarrow 4)$-linked α-D-GalA residues, modified by either methyl esterification or substitution with acetyl groups and xylose or apiose. RG II is actually a modified HG bearing complex branched side chains of 11 sugars including apiose, aceric acid (3-C'-carboxy-5-deoxy-L-xylose), 2-O-methyl fucose, 2-O-methyl xylose, Kdo (3-deoxy-D-*manno*-2-octulosonic acid), and Dha (3-deoxy-D-*lyxo*-2-heptulosaric acid).[40,41] The RG II molecule is highly resistant to digestion by common hydrolytic enzymes, a fact that has aided its purification. It is not very abundant in cell walls, at about 3% of the dry mass, but its structural conservation across many different plant species indicates a distinct role in wall integrity for this constituent of pectin.

In contrast, RG I is composed of a repeating disaccharide unit [→2)-α-L-Rha$p$-(1→4)-α-D-Gal$p$A-(1→]$_n$, where n can be higher than 100.[42,43] The galacturonosyl residues can carry acetyl groups on O-2 and O-3 but are not methyl-esterified. The rhamnosyl residues can be substituted at O-4 with neutral sugars. The proportion of branched rhamnosyl residues generally varies from ~20 to ~80%, depending on the source of the polysaccharide, although essentially unbranched RG I molecules have been reported in seed mucilages.[44] The side chains can be single unit [β-D-Gal$p$-(1→4)], but also polymeric such as arabinogalactan I and arabinan (50 glycosyl residues or more). Arabinogalactan I is composed of a (1→4)-linked β-D-Gal$p$ backbone with α-L-Ara$f$ residues attached to the O-3 of the galactosyl residues.[37,38] The arabinans consist of a (1→5)-linked α-L-Ara$f$ backbone, which can be substituted with α-L-Ara$f$-(1→2)-,α-L-Ara$f$-(1→3)-, and α-L-Ara$f$-(1→3)-α-L-Ara$f$-(1→3) side chains.[45] These neutral polymers are pinned at one end to the RG I backbone but extend into, and are highly mobile in, the wall pores.[46] The precise structures of these side chains and their distribution along the backbone of RG I are not known.

III. MOLECULAR ARCHITECTURE OF THE CELL WALL

Cell walls are not homogeneous structures.[47] The primary wall is defined as the structure that participates in irreversible expansion of the cell. The middle lamella is a pectin-rich interface between the primary walls of neighboring cells, from which cellulose may be excluded. Some cells elaborate within the primary wall a secondary cell wall, enriched in cellulose, xylan, and lignin.[24] The molecular composition and arrangements of the wall polymers differ among species, among tissues of a single species, among individual cells, and even among regions of the wall around a single protoplast. Many cell wall polymers, particularly pectins, are developmentally regulated, appearing at specific times during the life of the cell. [48–51] Within a single wall there are zones of different architecture; the middle lamella, plasmodesmata, thickenings, channels, pit fields, and the cell corners, and there are also domains across the primary wall, in which the degree of pectin esterification and the abundance of RG I side chains differ.[39,47,52,53] A successful pathogen must be able to digest through these different cell wall domains with their own architectures to gain access to the plasma membrane. The hydrolysis of polymers from these domains depends on local cross-linking mechanisms.

The primary cell wall is made up of two, sometimes three, structurally independent but interacting networks.[14] The scaffolding framework of cellulose and cross-linking glycans lies embedded in a second network of matrix pectic polysaccharides. A third network consists of the structural proteins or a phenylpropanoid network. With the exception of cellulose, the polysaccharides, structural proteins, and a broad spectrum of enzymes are coordinately secreted in Golgi-derived vesicles and targeted to the cell wall. Many polymers are modified by esterification, acetylation, or arabinosylation for solubility during transport. Later, extracellular enzymes deesterify, de-acetylate or dearabinosylate to free sites along the polymers for cross-linking into the cell wall. A range of cross-linking possibilities exists, including hydrogen bonding, ionic bonding with Ca^{2+} ions, covalent ester linkages, ether linkages, and van der Waals' interactions. Cellulose is synthesized directly on the plasma membrane by cellulose synthase complexes.

Cellulose microfibrils are formed by large numbers of hydrogen bonds between the long glucan chains, but other neutral polysaccharides, xylans, mannans, and xyloglucans, may also become hydrogen-bonded to cellulose or to themselves. In grass species, a small proportion of the ferulic acid units of neighboring glucuronoarabinoxylans may cross-link by phenyl–phenyl or phenyl–ether linkages to interconnect these molecules into a large network.[54]

Diverse cross-links are implicated in creating a network of pectic polysaccharides within the cell wall. The amount and the distribution of methyl groups on the HG backbone are important in determining the extent of Ca^{2+} cross-linking within the wall. HG is secreted as highly methyl esterified polymers, and plant pectin methyl esterases located in the cell wall cleave some of the methyl groups to initiate binding of the carboxylate ions to Ca^{2+}. HG chains can condense by cross-linking with Ca^{2+} to form "junction zones," linking two parallel or two antiparallel chains.[55,56] Maximally strong junctions occur between two chains of at least seven unesterified GalA units each.[57] A "cable structure" has been proposed based on conformational analysis of galacturonans by solid-state ^{13}C-NMR spectroscopy.[58] This structure contains two levels of aggregation. First, single chains join in places to give the dimeric egg-box junction zones on which the egg-box model was originally based.[56] Secondly, in concentrated gels, the egg-box dimers also function as interjunction segments between junction zones with four or more chains in the 2_1 (180° rotation between GalA residues) and 3_1 (right-handed helix) conformations to form cables.[58] Hydrophobic interactions between methoxyl groups, and hydrogen bonds between undissociated carboxyl and secondary alcohol groups, may also be involved in holding pectic polysaccharides within the plant cell wall. Extracted HG with a high degree of methyl esterification forms gels held together by such mechanisms.[59, 60]

Although extraction with calcium chelating agents releases significant amounts of pectin from cell walls, a fuller extraction of pectins requires cold alkali, indicating the presence of covalent bonds that anchor pectin in the wall.[61] In onion cell walls, HG is primarily anchored within the wall by calcium bridges, whereas RG I is mostly covalently cross-linked.[62] Two molecules of RG II can complex with boron, forming a borate-diol ester.[63,64] Only the apiofuranosyl residues of the 2-O-methyl-D-xylose-containing side chains in each of the subunits of the dimer participate in the cross-linking.[64] Because RG II is covalently linked to HG or RG I,[65] borate-diol esters can cross-link two pectin chains.

There is some evidence for the natural occurrence of O-D-galacturonosyl ester cross-links, that is, esters that involve the carboxyl group of a galacturonosyl residue and the hydroxyl group of some other wall component, such as a polysaccharide.[66 67]

In rose suspension culture cells, about one third of the xyloglucan is covalently linked to RG I.[68] The arabinan side chains of RG I may be feruloylated, providing sites for oxidative cross-linking.[69] Feruloyl groups have been found ester-linked to the arabinosyl and galactosyl residues of extracted sugar-beet pectin.[69–71] Such residues are found in highly exposed pectin domains, and therefore accessible to wall peroxidases or laccases, which could catalyze oxidative coupling to create diferulate linkages.

The most obvious distinguishing feature of secondary walls is the incorporation of lignins, complex networks of aromatic compounds called phenylpropanoids. The phenylpropanoids, hydroxycinnamoyl alcohols or "monolignols," *p*-coumaryl, coniferyl, and sinapyl alcohol comprise most of lignin networks.[72] The monolignols are linked by ester, ether, or carbon–carbon bonds and form a waterproof structure impervious to most pathogens. Although the structural framework of the cell wall is largely carbohydrate, structural proteins may also form networks in the wall. There are three major classes of structural proteins named for their uniquely enriched amino acid: the hydroxyproline-rich glycoproteins (HRGPs), the proline-rich proteins (PRPs), and the glycine-rich proteins (GRPs), some members of which are expressed abundantly in specific cell types.[73] As in the case of phenylpropanoid network, the presence of large amount of proteins, cross-linked or not, may alter enzyme accessibility in specific cell types.

IV. ENZYMATIC DEGRADATION AND MODIFICATION OF PLANT CELL WALL POLYSACCHARIDES

A. CELLULOSE

1. Enzymes Involved in Cellulose Degradation

Three classes of enzymes are produced by *Aspergillus* for the biodegradation of cellulose: endo-β-1,4-glucanases (endoglucanases) (EC 3.2.1.4, glycoside hydrolase families 5, 6, 12, 74), cellobiohydrolases (exoglucanases) (EC 3.2.1.91, glycoside hydrolase family 6), and β-1,4-glucosidases (β-glucosidases) (EC 3.2.1.21, glycoside hydrolase families 1, 2).

In general, endoglucanases are small proteins with molecular mass between 20 and 40 kDa,[10] although some larger enzymes have also been reported.[74] They are mainly active on the amorphous regions of cellulose and hydrolyze cellulose to short oligosaccharides.

Cellobiohydrolases are exo-acting enzymes that release cellobiose from the terminal ends of cellulose. Most Aspergilli produce two cellobiohydrolases that are active only at the nonreducing end of cellulose,[75–77] although a subdivision of cellobiohydrolase that act either on the nonreducing or the reducing end of cellulose has been demonstrated for other fungi.[78]

Release of D-glucose from cellobiose and cellulose is achieved by the action of β-glucosidases, to produce a carbon source that is easily metabolized by the fungal cell. Large variations in molecular mass have been reported for *Aspergillus* β-glucosidases[10] ranging from 14 [79] to 137 kDa.[80] β-Glucosidases also play a role in the degradation of xyloglucan and galactoglucomannan (see below).

2.   Enzymatic Modification of Cellulose

In industrial processes, because of the linear structure of cellulose and limited types of enzymes involved in cellulose degradation, only a limited number of enzymatic modifications can be performed. Total hydrolysis of cellulose to D-glucose can be achieved by using a combination of all of the enzymes described above. Using pure endoglucanases, cellulooligosaccharides of varying length can be produced. The product profiles of oligosaccharides using different endoglucanases from *Aspergillus* have not been systematically compared, but it is likely that differences will exist. Production of cellobiose can be achieved by using β-glucosidase-free cellobiohydrolases alone or in combination with endoglucanases, depending on the subsite characteristics of the endoglucanase used.

B.   XYLAN

1.   Enzymes Involved in Xylan Degradation

In contrast to cellulose, biodegradation of xylans requires a wide range of enzymes as their structures are more heterogeneous. These enzymes can be divided into enzymes that act on the xylan backbone and those releasing side groups from the backbone. Two classes of enzymes, endo-β-1,4-xylanases (EC 3.2.1.8, glycoside hydrolase families 10, 11), commonly called endoxylanases, and β-1,4-D-xylosidases (EC 3.2.1.37, glycoside hydrolase family 2) act on the xylan backbone.

Endoxylanases are endo-acting enzymes that hydrolyze the β-1,4-linkages in the xylan backbone resulting in xylooligosaccharides of varying length and side groups. This group of enzymes has been extensively studied in several aspergilli[10] and in most species at least three different enzymes have been identified,[81–83] often with different properties. One of these properties is the activity on polysaccharides as was described for 5 endoxylanases purified from a commercial enzyme preparation.[84] Different specificities of these enzymes were observed for branched or debranched xylan and for polysaccharide or short oligosaccharides as substrate.

The oligosaccharides originating from the action of the endoxylanases on polymeric xylan can be further degraded to D-xylose by β-xylosidase. Only a single β-xylosidase activity has been reported during growth of *Aspergillus* on xylan. This exo-acting enzyme prefers small xylooligosaccharides as a substrate and is inhibited by the presence of other residues attached to the D-xylose residues.[85]

Several classes of enzymes remove side groups from the xylan backbone. Two classes of enzymes are involved in removing the α-1,2- and α-1,3-linked L-arabinose residues from xylan, α-L-arabinofuranosidases (EC 3.2.1.55, glycoside hydrolase families 2, 51, 54) and arabinoxylan arabinofuranohydrolases (glycoside hydrolase family 62). The former also releases L-arabinose from xyloglucan and pectin (see below), while the latter is specific for arabinosylated xylan. Arabinofuranosidases release both terminal L-arabinose residues and short arabino-oligosaccharides from xylan.[86] Two isozymes of arabinofuranosidase have been identified in many aspergilli,[87–90] with distinct physical properties.

A few reports mention the presence of D-galactose residues on the xylan backbone.[91,92] The removal of these side groups requires the activities of α- and β-1, 4-D-galactosidases (EC 3.2.1.22, glycoside hydrolase families 27, 36 and EC 3.2.1.23, glycoside hydrolase families 1, 35, respectively). The α-1,4-D-galactosidases also use galacto(gluco)mannan as substrate, while β-1,4-D-galactosidases are involved in degradation of xylan, xyloglucan, galacto(gluco)mannan, and pectin (see below). At least three different α-galactosidases have been identified from *A. niger*[93–96] that differ in physical properties. So far, only a single β-1,4-D-galactosidase has been identified, suggesting that this enzyme may have a broad substrate specificity.

α-Glucuronidases (EC. 3.2.1.139, glycoside hydrolase family 67) release (4-*O*-methyl-)glucuronic acid from xylan. The *Aspergillus* enzyme is mainly active on short oligosaccharides[97,98] and differs in this respect from the α-glucuronidase from the basidiomycete fungus *Schizophyllum commune* that is highly active on polymeric xylan.[99]

Acetyl residues attached to the xylan backbone are removed by acetyl xylan esterases (EC 3.1.1.72, carbohydrate esterase family 1). Unlike the other side-group-removing enzymes, acetyl xylan esterases are highly active on polymeric xylan.

Feruloyl esterases release ferulic acid and other aromatic acids linked to C5 of terminal L-arabinose residues. These enzymes depend on the degradation of the xylan backbone by endoxylanases for efficient liberation of ferulic acid. Several feruloyl esterases have been identified in *Aspergillus*,[100–108] of which FaeA (also named FaeIII) and FaeB (also named CinnAE) have been studied in most detail. They prefer a different substitution of the aromatic ring,[109] and have different substrate specificities with respect to the linkage of the aromatic moiety to the polysaccharide.[110] The enzymes are active on both xylan and pectin, although FaeA has a higher activity on xylan, while FaeB has a higher activity on pectin.

2. Enzymatic Modification of Xylan

Complete degradation of xylan to monomers can be achieved by using a cocktail of all of the enzymes involved in xylan degradation.[111] The xylanolytic system of *Aspergillus* is especially efficient due to large synergistic effects. Endoxylanase, β-xylosidase, arabinoxylan arabinofuranohydrolase, and acetyl xylan esterase all enhance each other's activity.[112,113] Similarly, the release of ferulic acid from xylan by feruloyl esterases is enhanced when endoxylanase is used in combination with feruloyl esterase,[101,114] while addition of endoxylanase and β-xylosidase increases the release of 4-*O*-methyl glucuronic acid by α-glucuronidase.[97]

Acetyl xylan esterase is especially important for the action of endoxylanases. Using endoxylanases I, II, and III, the release of xylose and short xylooligosaccharides increased by factors 1.9 to 4.4, 6.9 to 14.7, and 2.5 to 16.3, respectively (depending on the incubation time), when acetyl xylan esterase was added to the incubations.[115]

Studies of synergistic effects between all of the xylan-modifying enzymes[111] showed that xylose release by endoxylanase and β-xylosidase was affected by all accessory enzymes. Synergistic effects were also observed for L-arabinose release by α-arabinofuranosidase and arabinoxylan arabinofuranohydrolase, ferulic acid release by feruloyl esterase, and 4-*O*-methyl glucuronic acid release by α-glucuronidase.[111] Two L-arabinose-releasing enzymes, α-arabinofuranosidase B and arabinoxylan arabinofuranohydrolase, have complementary activities, indicating that these two enzymes act on different L-arabinose residues attached to the xylan backbone.[111] This study also demonstrated that sequential incubations (endoxylanase treatment, enzyme inactivation, second enzyme treatment) were less efficient for liberation of monomers than simultaneous incubations.[111]

As side chains limit the action of endoxylanases, incubations with purified endoxylanases result in fairly long xylooligosaccharides. Combined incubations of endoxylanases with specific accessory enzymes will allow the production of oligosaccharides with a more homogeneous length and specific substitutions. To design efficient strategies for the production of these oligosaccharides, additional research is required to determine the substrate specificity of the individual enzymes and the structure of the products obtained after combined and sequential incubations with xylan.

C.   GALACTO(GLUCO)MANNAN

1.   Enzymes Involved in Galacto(Gluco)Mannan Degradation

Six classes of enzymes are involved in the degradation of galacto(gluco)mannan: endo-β-1,4-D-mannanases (EC 3.2.1.78, glycoside hydrolase family 5), β-1, 4-D-mannosidases (EC 3.2.1.25, glycoside hydrolase family 1), α-1,4-D-galactosidases (EC 3.2.1.22, glycoside hydrolase families 27, 36), β-1,4-galactosidases (EC 3.2.1.23, glycoside hydrolase families 1, 35), β-1,4-glucosidases (EC 3.2.1.21, glycoside hydrolase families 1,2), and galactomannan acetyl esterases.

Endomannanases cleave the backbone of galacto(gluco)mannan to release manno-oligosaccharides (mainly mannobiose and mannotriose)[116–119] and their activity is affected by the side groups and the D-glucose to D-galactose ratio of the backbone.[120] Endomannanases prefer galacto(gluco)mannans with a low D-galactose substitution on the backbone.[117–121] However, the inhibitory effect of the D-galactose substitutions is reduced if all of the galactosyl residues are situated on the same side of the backbone,[122] suggesting that steric hindrance is the main cause of this effect.

β-Mannosidases release D-mannose from the nonreducing end of galacto(gluco)mannan and from manno-oligosaccharides produced by the activity of endomannanase. The activity of the enzyme is highest when one or more unsubstituted D-mannose residues are present adjacent to the terminal D-mannose residue.[123] So far, only a single β-mannosidase activity has been purified from several aspergilli.[123–126]

The α-linked D-galactose residues are removed from the galacto(gluco)mannan backbone by the action of α-galactosidases. Although these enzymes also play a role in xylan degradation, they have been mainly studied for their role in galacto(gluco)mannan degradation. The different α-galactosidases described for *Aspergillus* have different physical properties and substrate specificities.[93,95,117,127–132]

After endomannanases hydrolyze the linkage between D-mannose and D-glucose in the galactoglucomannan backbone, the resulting terminal D-glucose residues can be removed by β-glucosidase. Although several β-glucosidases have been purified from *Aspergillus* (see above), their specificity for galactoglucomannan has not been reported.

The acetyl residues attached to the D-mannose residues in the backbone of galacto(gluco)mannan are removed by galactomannan acetyl esterases. Two *Aspergillus* galactomannan acetyl esterases have been reported.[133,134] Both enzymes are active on polymeric galactomannan and mannooligosaccharides, and significantly enhance the activity of endomannanases.

2.   Enzymatic Modification of Galacto(Gluco)Mannan

Choosing the right endomannanase is a critical factor in obtaining the desired products from crude galacto(gluco)mannans. Different endomannanases generate distinct patterns of hydrolysis products from crude galacto(gluco)mannans, as described for ivory nut mannan.[135] A comparison of an *A. niger* endomannanase to other microbial endomannanases demonstrated that the *A. niger* endomannanase can cleave the linkage between an unsubstituted and a substituted D-mannose residue, while other endomannanases required at least two unsubstituted residues for efficient hydrolysis.[122] Thus, the *A. niger* endomannanase is very suitable for the production of short branched mannooligosaccharides, while the other endomannanases produce longer oligosaccharides, due to the frequency of two unsubstituted adjacent D-mannose residues in the polymer.

Mannooligosaccharides with D-galactose linked to the terminal mannosyl residue at the nonreducing end can be produced by the sequential action of endomannanase and β-mannosidase. As *A. niger* β-mannosidase is capable of releasing D-mannose residues up to, but not beyond, a substituted D-mannose residue,[123] this enzyme will trim the oligosaccharides resulting from endomannanase activity until a terminal substituted D-mannose residue is produced.

Selective removal of D-galactose residues from the polysaccharide is possible by choosing a specific α-galactosidase. It has been reported that α-galactosidase II (AglB) releases D-galactose residues from both terminal and internal D-mannose residues, while α-galactosidase I (AglC) can only remove D-galactose from terminal D-mannose residues.[93] This pattern of activity suggests that treatment of galacto(gluco)mannan with AglB will result in nearly unbranched polysaccharides, whereas the action of AglC will result in branched polysaccharides with unbranched terminal ends. AglB acts synergistically with endomannanase,[95] therefore, the addition of AglB to an incubation of galacto(gluco)mannan with endomannanase may produce shorter less-branched oligosaccharides. Similarly, treating a population of mannooligosaccharides obtained after endomannanase treatment with AglB or AglC will produce unbranched or branched oligosaccharides, respectively.

The activity of a galacto(gluco)mannan acetyl esterase from *A. oryzae* was not affected by the presence of other galacto(gluco)mannan-degrading enzymes,[136] indicating that this enzyme could be used to produce nonacetylated polymeric galacto(gluco)mannans. The action of the esterase did enhance the activity of endomannanase and α-galactosidase,[136] and could therefore be used in the production of short, nonbranched mannooligosaccharides.

D. XYLOGLUCAN

1. Enzymes Involved in Xyloglucan Degradation

The backbone of xyloglucan is identical to cellulose, and degradation of this backbone thus requires two cellulolytic enzyme classes: endo-β-1,4-glucanases (EC 3.2.1.4, glycoside hydrolase family 12) and β-1,4-glucosidases (EC 3.2.1.21, glycoside hydrolase families 1, 2). Six additional activities, are required for the degradation of the xyloglucan side chains: α-1,6-D-xylosidases, α-L-arabinofuranosidases (EC 3.2.1.55, glycoside hydrolase families 2, 51, 54), α-1,2-L-fucosidases, α-1,2-L-galactosidases, β-D-galactosidases (3.2.1.23, glycoside hydrolase families 1, 35), and xyloglucan acetyl esterases.

Endoglucanases that are active on cellulose will also be able to cleave the xyloglucan backbone when the side chains are removed. In addition, several xyloglucan-specific endoglucanases have been identified in *Aspergillus*[137,138] that are not active on cellulose. There are no reports on the specificity of β-glucosidases or arabinofuranosidases for substituted xyloglucan.

α-Xylosidase and α-fucosidase are required to degrade xyloglucan, but have not been extensively studied in *Aspergillus*. Two α-xylosidases were purified from *A. niger* that were active on xyloglucan-derived oligosaccharides,[139,140] while only one of the two enzymes purified from *A. flavus* had activity on these substrates.[141,142] An α-fucosidase purified from *A. niger* was reported to be active on fucose–galactose disaccharides,[143] but the enzyme was not tested for activity on xyloglucan.

2. Enzymatic Modification of Xyloglucan

A study comparing the hydrolysis products of two different *Trichoderma viride* endoglucanases demonstrated that these two enzymes have different specificities toward xyloglucan.[144] It is probable, although not yet demonstrated, that such differences will also exist for *Aspergillus* xyloglucan-active endoglucanases and specific enzymes could therefore be used to generate specific oligosaccharides.

The activity of *A. niger* β-galactosidase on xyloglucan oligosaccharides obtained after endoglucanase treatment was reported.[144] Endoglucanases may be used in various combinations with β-galactosidase, α-xylosidase, α-fucosidase, and α-arabinofuranosidase to generate specific sets of oligosaccharides.

E. PECTIN

1. Enzymes Involved in Pectin Degradation

Aspergillus produces a very large group of enzymes involved in pectin degradation.[10] A large number of enzymes are needed in part due to the complex structure of pectin, but also to the production of multiple isozymes. Nine classes of pectin backbone-degrading enzymes have been identified, including endopolygalacturonase (EC 3.2.1.15, glycoside hydrolase family 28), exopolygalacturonase (EC3.2.1.67, glycoside hydrolase family 28), endorhamnogalacturonan hydrolase (glycoside hydrolase family 28), rhamnogalacturonan rhamnohydrolase, rhamnogalacturonan galacturonohydrolase, α-rhamnosidase (EC 3.2.1.40, glycoside hydrolase family 78), endoxylogalacturonan hydrolase (glycoside hydrolase family 28), pectin lyase (EC 4.2.2.10, polysaccharide lyase family 1)), pectate lyase (EC 4.2.2.2, polysaccharide lyase family 1), and rhamnogalacturonan lyase (polysaccharide lyase family 4). In addition, 11 classes of side-chain-active enzymes are involved in pectin degradation: pectin acetyl esterase, pectin methyl esterase (EC 3.1.1.11, carbohydrate esterase family 8), rhamnogalacturonan acetyl esterase (carbohydrate esterase family 12), α-L-arabinofuranosidase (EC 3.2.1.55, glycoside hydrolase families 1, 51, 54), endo-α-1,5-arabinanase (EC 3.2.1.99, glycoside hydrolase family 43), β-1, 4-D-galactosidase (EC 3.2.1.23, glycoside hydrolase family 1, 35), β-1,3-endogalactanase, β-1,4-endogalactanase (EC 3.2.1.89, glycoside hydrolase family 53), β-1,6-endogalactanase, β-1,3-exogalactanase (EC 3.2.1.145, glycoside hydrolase family 55), and feruloyl esterase.

Endopolygalacturonases act on HG but can have different modes of action (processive vs. nonprocessive) and different substrate specificity depending on the degree of methylation.[145–150] Exopolygalacturonases also use HG as substrate as well as xylogalacturonan, but attack from the nonreducing end.[151–154] This enzyme releases both D-galacturonic acid as well as a D-xylose-D-galacturonic acid disaccharide, indicating that its activity is not hindered by the presence of D-xylose residues attached to the main chain.[155] One endoxylogalacturonan hydrolase from *Aspergillus* has been described.[156] This enzyme does not hydrolyze unsubstituted galacturonan, but releases xylogalacturonan oligosaccharides of varying length.[156] Pectin lyases and pectate lyases are endoenzymes that degrade the HG of pectin by a β-elimination reaction. Pectin lyases have a preference for pectins with a high degree of methylesterification, whereas pectate lyases prefer pectins with a low degree of esterification. Although multiple pectin lyases have been identified in *Aspergillus*, only one pectate lyase has been reported.[157–159]

Rhamnogalacturonan hydrolases and rhamnogalacturonan lyases cleave the RG I backbone.[160–163] The activity of rhamnogalacturonan hydrolase is inhibited by the presence of acetyl residues on the RG I backbone and is therefore dependent on the activity of rhamnogalacturonan acetyl esterase.[111] The activity of rhamnogalacturonan lyase is reduced by both acetylation and the presence of L-arabinose chains attached to the main chain.[161] However, the presence of D-galactose side chains enhances the activity of this enzyme.[161] Two exo-acting enzymes are involved in the degradation of the RG I backbone, rhamnogalacturonan rhamnohydrolase[164] and rhamnogalacturonan galacturonohydrolase.[165] The first enzyme is specific for terminal rhamnose residues, whereas the latter enzyme is specific for terminal rhamnosegalacturonic acid residues. The purification of two α-L-rhamnosidases from *A. niger* has also been reported,[166] but the activity of these enzymes has not been tested on RG I or RG I oligosaccharides.

The arabinan moiety of the pectic side chains is cleaved by α-L-arabinofuranosidases and endoarabinanases. Some arabinofuranosidases are specific for terminal α-1,3-linked L-arabinose residues,[167] whereas others can release terminal α-1,2- α-1,5-linked L-arabinose residues.[86] Endoarabinanases hydrolyze the internal α-1,5-linkages of arabinan,[168] and are important for its efficient hydrolysis.

Endogalactanases, exogalactanases, and β-galactosidases degrade the galactan side chains of pectin. Three types of linkages are found in galactan, β-1,3, β-1,4, and β-1,6, suggesting that all types of endo- and exogalactanases may be required for the complete degradation of galactan. To date, only endo-β-1,4-,[169–173] endo-β-1,6-,[174] exo-β-1,3-,[175] and exo-β-1,4-galactanases[176] have been reported.

Acetyl residues are removed from HG and RG I by pectin acetyl esterases and rhamnogalacturonan acetyl esterases, respectively.[111,177] Pectin methyl esterases hydrolyze methyl esters from HG.[178,179] While pectin methyl esterases expose the substrate for the activities of endopolygalacturonases and pectate lyases, rhamnogalacturonan acetyl esterase activity is essential for efficient hydrolysis of the RG I backbone by endorhamnogalacturonan hydrolase.[111,180]

Feruloyl esterases release the ferulic acid residues attached to terminal L-arabinose and D-galactose residues in the pectic side chains of some species, particularly Chenopodiaceae.[101,102,106,111] A comparison of the activities of two *A. niger* feruloyl esterases revealed that FaeA was active on ferulic acid residues linked to D-galactose, whereas CinnAE (FaeB) was able to release ferulic acid linked to both L-arabinose and D-galactose, with a preference for L-arabinose-linked ferulic acid.[110]

2. Enzymatic Modification of Pectin

A large number of enzymes from *Aspergillus* are involved in pectin degradation. Degradation of HG by the backbone-hydrolyzing enzymes is potentiated by the activities of pectin acetyl and methyl esterases. Thus, pectin methyl esterase strongly enhances the activity of endopolygalacturonases with a preference to pectins with a low degree of esterification.[178] It is therefore expected that they will have a negative effect on endopolygalacturonases with a preference for pectins with a higher degree of methyl-esterification,[181] possibly resulting in mixtures of different oligosaccharides. Synergy between pectin acetyl esterase, pectin methyl esterase, and pectin lyase has also been reported.[177] Production of specific sets of HG oligosaccharides will therefore depend on the choice of the endo-acting enzyme and can be influenced by the addition of pectin methyl or acetyl esterases.

Degradation of the RG I backbone by rhamnogalacturonan hydrolase and rhamnogalacturonan lyase is largely dependent on the action of rhamnogalacturonan acetyl esterase.[111,180] In the absence of the esterase, large oligosaccharides are produced from the backbone. In the presence of this enzyme, most of the backbone is hydrolyzed, resulting in arabinogalactan chains with only a small number of L-rhamnose and D-galacturonic acid residues attached to them. The inhibition of rhamnogalacturonan acetyl esterase in the presence of arabinogalactan side chains has been reported.[180] However, in another study, addition of arabinofuranosidase to incubations with the esterase did not increase the release of acetyl groups.[182]

The release of ferulic acid by a feruloyl esterase (FaeA) from *A. niger* was enhanced by the action of pectin lyase, but only to a small extent.[101] This finding was not surprising since pectin lyase is active on HG, while ferulic acid is attached to RG I. The increase in activity may be a consequence of rendering RG I more accessible to attack by other enzymes by hydrolysis of HG. The action of a second feruloyl esterase from *A. niger* (CinnAE or FaeB) was enhanced by endoarabinanase and arabinofuranosidase.[102,183] The activity of FaeA is enhanced in the presence of galactan-degrading enzymes (β-galactosidase and endogalactanase).[102] These results are consistent with the substrate specificities of FaeA (FAE-III) and FaeB (CinnAE) for oligosaccharides derived from sugar-beet pectin.[110] FaeA is active on galactose-linked ferulic acid residues, while FaeB can release both arabinose- and galactose-linked residues, but has a preference for L-arabinose-linked ferulic acid.[110]

Nearly all of the enzyme activities that degrade the side chains of pectin enhance the activity of each other.[111] However, release of L-arabinose by arabinofuranosidase B (AbfB) from *A. niger* did not depend on the action of other accessory enzymes and was in fact reduced by the degradation of the RG I backbone.[111]

It is clear that specific modifications of pectic polysaccharides require a careful selection of enzymes. The large number of enzymes involved in pectin degradation and their limitations in hydrolyzing pectin when acting alone, create possibilities for the production of specific sets of oligosaccharides with a low level of heterogeneity, and also complicate the choice of enzyme.

V.   REGULATION OF THE PRODUCTION OF PLANT CELL WALL POLYSACCHARIDE-DEGRADING ENZYMES IN *ASPERGILLUS*

A.   PRODUCTION OF PLANT CELL WALL POLYSACCHARIDE-DEGRADING ENZYMES ON CRUDE AND DEFINED CARBON SOURCES

The production of plant cell wall polysaccharide-degrading enzymes by *Aspergillus* has been reported on a wide variety of carbon sources, commonly crude plant compounds such as sugar-beet pulp or wheat bran. Although these crude carbon sources consist of predominantly one polysaccharide (arabinoxylan in wheat bran and pectin in sugar-beet pulp), they also contain other polysaccharides as well as oligosaccharides and monosaccharides, and therefore are not very informative with respect to the regulatory systems of the fungus involved in the production of these enzymes.

Aspergillus is not able to import polysaccharides, but can import certain oligosaccharides (lactose, sucrose).[184,185] It is therefore likely that the signals that activate the production of polysaccharide-degrading enzymes are monomeric or dimeric compounds that are derived from the polysaccharides. In many cases, it has been reported that the product of the enzyme reaction induces production of the enzyme. This regulatory mechanism suggests that small amounts of some enzymes are already present before induction. A small set of genes encoding enzymes may be constitutively expressed at a low level such that, in the presence of appropriate substrate, small amounts of the compounds are released that can induce higher levels of production of the required enzymes. Constitutive levels of expression have been reported for two polygalacturonases from *A. niger*,[149] and several genes encoding polysaccharide-degrading enzymes were identified in EST clones obtained from *A. niger* grown in minimal and complete medium using D-glucose as a carbon source (A. Tsang and R. Storms, only published in gene database).

Several regulatory systems have been identified in *Aspergillus* that affect the production of plant cell wall-degrading enzymes. These systems activate or repress the expression of the genes encoding these enzymes in response to signals from the environment.

B.   SPECIFIC SYSTEMS INVOLVED IN THE REGULATION OF PLANT CELL WALL POLYSACCHARIDES

1.   XlnR-Mediated Regulation of Xylanolytic and Cellulolytic Genes, Induced by D-Xylose

Regulation of xylanolytic and cellulolytic genes in *A. niger* is under the control of a transcriptional activator, XlnR, a DNA-binding protein of the GAL4-like family.[186] In the presence of D-xylose, this protein binds to specific sites in the promoters of its target genes. Initially, these promoter sites were defined as GGCTAAA,[186] but later modified to GGCTAA.[187] A recent study identified a functional GGCTAG binding site in the promoter of the gene encoding α-glucuronidase, modifying the consensus site to GGCTAR.[188] XlnR has also been characterized in *A. oryzae*, where the consensus of the binding site was determined as GGCTRA.[189] XlnR regulates genes encoding endoxylanases, β-xylosidase, α-glucuronidase, acetyl xylan esterase, arabinoxylan arabinofuranohydrolase, feruloyl esterase A, endoglucanases, cellobiohydrolases, and α- and β-galactosidases.[75,94,137,187] In addition, XlnR regulates the expression of the gene encoding D-xylose reductase,[190] the first step in D-xylose catabolism, but not the expression of the gene encoding D-xylulose kinase.[191]

2.   Gene Expression Induced by L-Arabinose/L-Arabitol

Expression of genes encoding L-arabinose-releasing enzymes (arabinofuranosidases, endoarabinanases, and arabinoxylan arabinofuranohydrolases) from *Aspergillus* was observed when L-arabinose and, more strongly, when L-arabitol is present in the medium.[192–195] The role of L-arabitol as the actual inducer of this system was further supported in a study using an *A. nidulans* L-arabitol dehydrogenase mutant that accumulated high levels of intracellular L-arabitol.[196] Coregulation of genes encoding L-arabinose-releasing enzymes and L-arabinose catabolic enzymes was reported for *A. niger*[195] and was recently confirmed by the characterization of two regulatory mutants that showed reduced growth on L-arabinose.[197] Not only are genes encoding L-arabinose-releasing enzymes induced by L-arabinose and L-arabitol, but also genes encoding D-galactose releasing enzymes (β-galactosidase and endogalactanase),[94,170,198] suggesting that this system may regulate the suite of enzymes required to degrade the side chains of the hairy regions of pectin as well as the catabolism of L-arabinose. To date, the transcription factor for this system has not been cloned and characterized.

3.   Gene Expression Induced by D-Mannose and D-Galactose

D-mannose and D-galactose are the main components of galacto(gluco)mannan. Expression of two α-galactosidase encoding genes (*aglA* and *aglB*) was detected in the presence of D-galactose and D-galactose-containing oligo- and polysaccharides[94,199] as was the expression of a β-galactosidase encoding gene.[94] A later study showed that a third α-galactosidase encoding gene (*aglC*) was induced by D-galactose.[200] All three genes encoding α-galactosidases as well as a gene encoding a β-mannosidase (*mndA*) are also induced by D-mannose.[200] However, no *mndA* expression was detected on D-galactose, suggesting that D-mannose and D-galactose induce two different regulatory systems.

4. Gene Expression Induced by D-Galacturonic Acid, D-Glucuronic Acid, and L-Rhamnose

The expression of a large number of pectinolytic genes from *A. niger* is induced in the presence of D-galacturonic acid.[198] This suggests the existence of a general regulatory system for genes involved in pectin degradation as was identified for xylan degradation (XlnR, see above). However, while all genes of the xylanolytic system appear to be expressed simultaneously,[187] the pectinolytic genes have large temporal variations in their expression profiles.[198] Based on these differences, and on expression levels of various carbon source combinations, several subsets of genes were identified.[198] A general pectinolytic regulatory system may exist in *A. niger* induced by D-galacturonic acid, but the expression of the individual genes may also be influenced by other factors. Some of the pectinolytic genes are also expressed in the presence of D-glucuronic acid, as is the gene encoding α-glucuronidase,[188] while expression of the gene encoding rhamnogalacturonan acetyl esterase is detected in the presence of L-rhamnose.[198] This finding suggests the presence of additional regulatory systems involved in pectin degradation.

5. Gene Expression Induced by Ferulic Acid and Other Aromatic Compounds

Xylan and pectin contain ferulic acid, *p*-coumaric acid, and other aromatic acids (see above) that can be released by feruloyl esterases. Two genes encoding feruloyl esterases from *A. niger* (*faeA* and *faeB*) are both expressed in the presence of ferulic acid.[102,201] However, the range of aromatic compounds that induces expression was different for each gene, suggesting the presence of multiple regulatory systems for different aromatic compounds.[102] In addition, *faeA* is regulated by XlnR (see above),[187,201] while *faeB* is not.[102]

C. WIDE DOMAIN REGULATORY SYSTEMS

1. Carbon Catabolite Repression

To use their energy efficiently, fungi only produce the enzymes they need for carbon source consumption in their present environment and prevent production of enzymes that are not necessary. The regulatory system involved in this process is known as carbon catabolite repression and is mediated by CreA, a zinc-finger protein that binds to specific sites (SYGGRG, with S = C or G, Y = C or T, and R = A or G) in the promoter of a large number of target genes,[202–204] thus abolishing or reducing their expression. CreA-mediated repression of genes encoding plant cell wall polysaccharide-degrading enzymes is a common phenomenon and has been described for genes involved in xylan, pectin, cellulose, and galactomannan degradation.[75,153,159,180,193,200,201,205–212]

Although CreA repression is predominantly studied in the presence of D-glucose, other monomeric carbon sources also result in repression of the target genes. The level of repression appears to depend on both the nature and concentration of the compound. Comparing the expression of four xylanolytic genes of *A. niger* in a wild-type and CreA-derepressed mutant at different D-xylose concentrations, demonstrated that D-xylose is not only the inducer of the xylanolytic system (see below) but at higher concentrations also represses, via CreA, the expression of the xylanolytic genes.[213] Similarly, a decrease in cellulase production was observed in *A. terreus* when the concentrations of D-glucose, D-xylose, and cellobiose in the growth medium were increased.[214] Recently, it was shown that the ferulic acid-induced expression of two feruloyl esterase encoding genes of *A. niger* was higher in a CreA-derepressed mutant than in the wild type in the presence of a variety of monosaccharides,[102] indicating that all these carbon sources can cause CreA repression.

2. pH Regulation

pH regulation is a common phenomenon in *Aspergillus* and is mediated by the DNA-binding protein PacC.[215,216] At alkaline pH, this protein activates alkaline-induced genes and represses acid-induced genes. At acidic pH, the protein is not active, resulting in the expression of acid-specific genes. PacC has been shown to affect the expression of genes encoding arabinofuranosidase and endoxylanases.[194,217] In addition, a role for pH regulation has been suggested in the production of poly-galacturonases and cellulases.[218,219]

VI. CONCLUDING REMARKS

The complex structure of the plant cell wall and the heterogeneity of the individual polysaccharides require a wide variety of enzymes to enable microorganisms to use these compounds as carbon sources. These microbial enzyme systems have evolved to complex, synergistic mixtures that not only include a wide range of activities, but also isoenzymes with distinct substrate specificities to be able to degrade specific parts of plant cell wall polysaccharides. This range of enzymes offers a high potential for the biotechnological production of monosaccharides, oligosaccharides, and modified polysaccharides. However, using microbial enzymes for poly- and oligosaccharide engineering still offers many challenges and requires a more detailed characterization of the reaction products of individual enzymes and enzyme cocktails than is currently available. These data will be essential before a reproducible production of the product of choice can even be envisioned. For each intended product, the options with respect to the choice of enzymes and whether to use these enzymes simultaneously or sequentially will need to be evaluated.

Many genes encoding plant cell wall polysaccharide-degrading enzymes have already been described allowing the production of large amounts of the corresponding enzymes in suitable expression hosts. Now that more and more microbial genome sequences are appearing, the number of genes encoding a specific enzyme will increase exponentially. Easy strategies to determine the substrate specificities of the corresponding enzymes will enable a fast selection of the right enzyme for a specific application. One of these strategies could be the use of pure poly- and oligosaccharides with a defined structure. Comparison of the product profiles of different enzymes using these substrates gives insight into the substrate specificities of the enzymes. The availability of pure poly- and oligosaccharides with a defined structure is still the limiting factor for these strategies, although the range of substrates in the market is increasing. As more enzymes are characterized and applied for the production of specific oligosaccharides and modified polysaccharides, more of these compounds will become available as substrates to test the specificity of other enzymes. In turn, these may then be useful for the production of other oligosaccharides and modified polysaccharides.

REFERENCES

1. Christov, L.P., Szakacs, G., and Balakrishnan, H.,. Production, partial characterization and use of fungal cellulase-free xylanases in pulp bleaching, *Process Biochem.*, 34, 511–517, 1999.
2. Viikari, L., Kantelinen, A., Sundquist, J., and Linko, M., Xylanases in bleaching — from an idea to the industry, *FEMS Microbiol. Rev.,*13, 335–350, 1994.
3. Rättö, M. and Viikari, L., Pectinases in wood debarking, in *Pectins and Pectinases,* Visser, J., and Voragen, A.G.J., Eds., Elsevier Science, Amsterdam, 1996, pp. 979–982.
4. Maat, J., Roza, M., Verbakel, J., Stam, H., Santos da Silva, M., Borrel, M., Egmond, M.R., Hagemans, M.L.D., van Gorcom, R.F.M., Hessing, J.G.M., van den Hondel, C.A.M.J.J., and van Rotterdam, C., Xylanases and their applications in bakery, in *Xylans and Xylanases,* Visser, J., Beldman, G., Kusters-van Someren, M.A., and Voragen, A.G.J., Eds, Elsevier Science, Amsterdam, 1992, pp. 349–360.
5. Petit-Benvegnen, M.-D., Saulnier, L., and Rouau, X., Solubilization of arabinoxylans from isolated water-unextractable pentosans and wheat flour doughs by cell wall degrading enzymes, *Cereal Chem.,* 75, 551–556, 1998.
6. Poutanen, K-., Enzymes. An important tool in the improvement of the quality of cereal foods, *Trends Food Sci. Technol.,* 8, 300–306, 1997.
7. Bedford, M.R. and Classen, H.L., The influence of dietary xylanase on intestinal viscosity and molecular weight distribution of carbohydrates in rye-fed broiler chicks, in *Xylans and Xylanases,* Visser, J., Beldman, G., Kusters-van Someren, M.A., and Voragen, A.G.J., Eds, Elsevier, Amsterdam, 1992, pp. 361–370.
8. Zeikus, J.G., Lee, C., Lee, Y.E., and Saha, B.C., Thermostable saccharidases. New sources, uses and biodesign, *ACS Symp. Ser.,* 460, 36–51, 1991.
9. Grassin, C. and Fauquembergue, P., Applications of pectinases in beverages, in *Pectin and Pectinases*, Visser, J. and Voragen, A.G.J., Eds. Elsevier Science, Amsterdam, 1996, pp. 453–462.
10. de Vries, R.P. and Visser, J., *Aspergillus* enzymes involved in degradation of plant cell wall polysaccharides, *Microb. Mol. Biol. Rev.,* 65, 497–522, 2001.
11. Henrissat, B., A classification of glycosidases based on amino-acid sequence similarities, *Biochem. J.,* 280, 309–316, 1991.
12. Henrissat, B. and Bairoch, A., New families in the classification of glycosidases based on amino acid sequence similarities, *Biochem. J.,* 293, 781–788, 1993.
13. Henrissat, B. and Bairoch, A., Updating the sequence based classification of glycosidases, *Biochem. J.,* 316, 695–696, 1996.

14. Carpita, N.C. and McCann, M.C., The cell wall, in *Biochemistry and Molecular Biology of Plants,* Buchanan, B.B., Gruissem, W., and Jones, R., Eds., American Society Plant Physiologists, Rockville, MD, 2000, pp. 52–109.

15. McCann, M.C., Wells, B., and Roberts, K., Direct visualisation of cross-links in the primary plant cell wall, *J. Cell Sci.,* 96, 323–334, 1990.

16. Koyama, M., Helbert, W., Imai, T., Sugiyama, J., and Henrissat, B., Parallel-up structure evidences the molecular directionality during biosynthesis of bacterial cellulose, *Proc. Nat. Acad. Sci. USA,* 94, 9091–9095, 1997.

17. Rowland, S.P. and Howley, P.S., Hydrogen bonding on accessible surfaces of cellulose from various sources and relationship to order within crystalline regions, *J. Polym. Sci. A,* 26, 1769–1778, 1988.

18. Verlhac, C., Dedier, J., and Chanzy, H., Availability of surface hydroxyl groups in Valonia and bacterial cellulose, *J. Polym. Sci. A,* 26, 1171–1177, 1990.

19. Ha, M.A., Apperley, D.C., Evans, B.W., Huxham, I.M., Jardine, W.G., Viĕtor, R.J., Reis, D., Vian, B., and Jarvis, M.C., Fine structure in cellulose microfibrils: NMR evidence from onion and quince, *Plant J.,* 16, 183–190, 1998.

20. Newman, R.H., Evidence for assignment of 13C NMR signals to cellulose crystallite surfaces in wood, pulp and isolated celluloses, *Holzforschung,* 52, 157–159, 1998.

21. Darvill, J.E., McNeil, M., Darvill, A.G., and Albersheim, P., Structure of plant cell walls. 11. Glucuronoarabinoxylan, a second hemicellulose in the primary cell walls of suspension-cultured sycamore cells, *Plant Physiol.,* 66, 1135–1139, 1980.

22. Carpita, N.C. and Whittern, D., A highly substituted glucuronoarabinoxylan from developing maize coleoptiles, *Carbohydr. Res.,* 146, 129–140, 1986.

23. Carpita, N.C. and Gibeaut, D.M., Structural models of primary cell walls in flowering plants: consistency of molecular structure with the physical properties of the walls during growth, *Plant J.,* 3, 1–30, 1993.

24. Milioni, D., Sado, P.E., Stacey, N.J., Domingo, C., Roberts, K., and McCann, M.C., Differential expression of cell-wall-related genes during the formation of tracheary elements in the *Zinnia* mesophyll cell system, *Plant Mol. Biol.,* 47, 221–238, 2001.

25. Carpita, N.C., Structure and biogenesis of the cell walls of grasses, *Annu. Rev. Plant Physiol. Plant. Mol. Biol.,* 47, 445–476, 1996.

26. Joshi, H. and Kapoor, V.P., *Cassia grandis* Linn f. seed galactomannan: structural and crystallographic studies, *Carbohydr. Res.,* 338, 1907–1912, 2003.

27. Goldberg, R., Gillou, L., Prat, R., Dupenhoat, C.H., and Michon, V., Structural features of the cell wall polysaccharides of *Asparagus officinalis* seeds, *Carbohydr. Res.,* 210, 263–276, 1991.

28. Piro, G., Zuppa, A., Dalessandro, G., and Northcote, D.H., Glucomannan synthesis in pea epicotyls — the mannose and glucose transferases, *Planta,* 190, 206–220, 1993.

29. Sims, I., Craik, D.J., and Bacic, A., Structural characterisation of galactoglucomannan secreted by suspension-cultured cells of *Nicotiana plumbaginifolia*, *Carbohydr. Res.,* 303, 79–92, 1997.

30. Katsuraya, K., Okuyama, K., Hatanaka, K., Oshima, R., Sato, T., and Matsuzaki K., Constitution of konjac glucomannan: chemical analysis and C-13 NMR spectroscopy, *Carbohydr. Polym.,* 53, 183–189, 2003.

31. York, W.S., van Halbeek, H., Darvill, A.G., and Albersheim, P., The structure of plant cell walls. 29. Structural analysis of xyloglucan oligosaccharides by H-1-NMR spectroscopy and fast atom bombardment mass spectrometry, *Carbohydr. Res.,* 200, 9–31, 1990.

32. Hisamatsu, M., Impallomeni, G., York, W.S., Albersheim, P., and Darvill, A.G., The structure of plant cell walls. 31. A new undecasaccharide subunit of xyloglucans with 2 alpha-L-fucosyl residues, *Carbohydr. Res.,* 211, 117–129, 1991.

33. Sims, I., Munro, S.L.A., Currie, G., Craik, D., and Bacic, A., Structural characterisation of xyloglucan secreted by suspension-cultured cells of *Nicotiana plumbaginifolia, Carbohydr. Res.,* 293, 147–172, 1996.
34. Pauly, M., Albersheim, P., Darvill, A., and York, W.S., Molecular domains of the cellulose/xyloglucan network in the cell walls of higher plants, *Plant J.,* 20, 629–639, 1999.
35. McCann, M.C. and Roberts, K., Architecture of the primary cell wall, in *The Cytoskeletal Basis of Plant Growth and Form*, Lloyd, C.W., Ed., Academic Press, New York, 1991, pp. 109–129.
36. Vincken, J.P., Schols, H.A., Oomen, R.J.F.J., McCann, M.C., Ulvskov, P., Voragen, A.G.J., and Visser, R.G.F., If homogalacturonan were a side chain of rhamnogalacturonan I. Implications for cell wall architecture, *Plant Physiol.,* 132, 1781–1789, 2003.
37. Mohnen, D., Biosynthesis of pectins and galactomannans in *Comprehensive Natural Products Chemistry,* Barton, D., Nakanishi, K., and Meth-Cohn, O., Eds., Elsevier, Dordrecht, The Netherlands, 1999, pp. 497–527.
38. Ridley, B.L., O'Neill, M.A., and Mohnen, D., Pectins: structure, biosynthesis, and oligogalacturonide-related signaling, Phytochemistry, 57, 929–967, 2001.
39. Willats, W.G.T., McCartney, L., MacKie, W., and Knox, J.P., Pectin: cell biology and prospects for functional analysis, *Plant Mol. Biol.,* 47, 9–27, 2001.
40. O'Neill, M.A., Warrenfeltz, D., Kates, K., Pellerin, P., Doci, T., Darvill, A.G., and Albersheim, P., Rhamnogalacturonan-II, a pectic polysaccharide in the walls of growing plant cell, forms a dimer that is covalently-linked by a borate ester, J. Biol. Chem., 271, 22923–22930, 1996.
41. Vidal, S., Doco, T., Williams, P., Pellerin, P., York, W.S., O'Neill, M.A., Glushka, J., Darvill, A.G., and Albersheim, P., Structural characterization of the pectic polysaccharide rhamnogalacturonan II: evidence for the backbone location of the aceric acid-containing oligoglycosyl side chain, *Carbohydr. Res.,* 326, 227–294, 2000.
42. McNeil, M., Darvill, A.G., and Albersheim, P., Structure of plant cell walls. X. Rhamnogalacturonan I, a structurally complex pectic polysaccharide in the walls of suspension-cultured sycamore cells, *Plant Physiol.,* 66, 1128–1134, 1980.
43. Albersheim, P., Darvill, A.G., O'Neill, M.A., Schols, H.A., and Voragen, A.G.J., An hypothesis: the same six polysaccharides are components of the primary cell walls of all higher plants, in Pectins and Pectinases,Visser, J. and Voragen, A.G.J., Eds., Elsevier Science, Amsterdam, 1996, pp. 47–55.
44. Penfield, S., Meissner, R.C., Shoue, D.A., Carpita, N.C., and Bevan, M.W., MYB61 is required for mucilage deposition and extrusion in the *Arabidopsis* seed coat, *Plant Cell,* 13, 2777–2791, 2001.
45. Schols, H.A. and Voragen, A.G.J., Complex pectins: structure elucidation using enzymes, in Pectin and Pectinases, Visser, J. and Voragen, A.G.J., Eds., Elsevier Science, Amsterdam, 1996, pp. 793–798.
46. Foster, T.J., Ablett, S., McCann, M.C., and Gidley, M.J., Mobility-resolved 13C-NMR spectroscopy of primary plant cell walls, *Biopolymers,* 39, 51–66, 1996.
47. McCann, M.C., Bush, M., Milioni, D., Sado, P., Stacey, N.J., Catchpole, G., Defernez, M., Carpita, N.C., Hofte, H., Wilson, R.H., and Roberts K., Approaches to understanding the functional architecture of the plant cell wall, Phytochemistry, 57, 811–821, 2001.
48. Bush, M.S., Marry, M., Huxham, I.M., Jarvis, M.C., and McCann, M.C., Developmental regulation of pectic epitopes during potato tuberisation, Planta, 213, 869–880, 2001.

49. Ermel, F.F., Follet-Gueye, M.L., Cibert, C., Vian, B., Morvan, C., Catesson, A.N., and Goldberg, R., Differential localization of arabinan and galactan side chains of rhamnogalacturonan 1 in cambial derivatives, *Planta,* 210, 732–740, 2000.

50. McCartney, L., Ormerod, A.P., Gidley, M.J., and Knox J.P., Temporal and spatial regulation of pectic $(1\rightarrow4)$-β-D-galactan in cells of developing pea cotyledons: implications and mechanical properties, *Plant J.,* 22, 105–113, 2000.

51. McCartney, L., Steele-King, C.G., Jordan, E., and Knox, J.P., Cell wall pectic $(1\rightarrow4)$-β-D-galactan marks the acceleration of cell elongation in the *Arabidopsis* seedling root meristem, *Plant J.,* 33, 447–454, 2003.

52. Willats, W.G.T., Orfila, C., Limberg, G., Buchholt, H.C., van Alebeek, G-J.W.M., Voragen, A.G.J., Marcus, S.E., Christensen, T.M.I.E., Mikkelson, J.D., Murray, B.S., and Knox, J.P., Modulation of the degree and pattern of methyl-esterification of pectic homogalacturonan in plant cell walls: implications for pectin methyl esterase action, matrix properties, and cell adhesion, *J. Biol. Chem.,* 276, 19404–19413, 2001.

53. Bush, M.S. and McCann, M.C., Pectic epitopes are differentially distributed in the cell walls of potato (*Solanum tuberosum*) tubers, *Physiol. Plant.,* 107, 201–213, 1999.

54. Schooneveld-Bergmans, M.E.F., Hopman, A.M.C.P., Beldman, G., and Voragen, A.G.J., Extraction and partial characterization of feruloylated glucuronoarabinoxylans from wheat bran, *Carbohydr. Polym.,* 35, 39–47, 1998.

55. Powell, D.A., Morris, E.R., Gidley, M.J., and Rees, D.A., Conformation and interactions of pectins: influence of residue sequence on chain association in calcium pectate gels, *J. Mol. Biol.,* 155, 517–531, 1982.

56. Jarvis M.C., Structure and properties of pectin gels in plant-cell walls. *Plant Cell Environ.,* 7, 153–164, 1984.

57. Daas, P.J.H., Boxma, B., Hopman, A.M.C.P., Voragen, A.G.J., and Schols, H.A., Nonesterified galacturonic acid sequence homology of pectins, *Biopolymers,* 58, 1–8, 2001.

58. Goldberg, R., Voragen, A.G.J., and Jarvis, M.C., Methyl-esterification, de-esterification and gelation of pectins in the primary cell wall, in *Pectins and Pectinases, Progress in Biotechnology,* vol. 14, Visser, J. and Voragen, A.G.J., Eds., Elsevier Press, Amsterdam, 1996, pp. 151–172.

59. Morris, E.R., Powell, D.A., Gidley, M.J., and Rees, D.A., Conformations and interactions of pectins: polymorphism between gel and solid states of calcium polygalacturonate, *J. Mol. Biol.,* 155, 507–516, 1982.

60. Oakenfull, D. and Scott A., Hydrophobic interaction in the gelation of high methoxyl pectins, *J. Food Sci.,* 49, 1093–1098, 1984.

61. Jarvis, M.C., The proportion of calcium-bound pectin in plant cell walls. *Planta,* 344, 344–346, 1982.

62. Redgwell, R.J. and Selvendran, R.R., Structural features of cell wall polysaccharides of onion *Allium cepa, Carbohydr. Res.,* 157, 183–199, 1986.

63. Kobayashi, M., Matoh, T., and Azuma J., Two chains of rhamnogalacturonan II are cross-linked by borate-diol ester bonds in higher plant cell walls, *Plant Physiol.,* 110, 1017–1020, 1996.

64. Ishii, T., Matsunaga, T., Pellerin, P., O'Neill, M.A., Darvill, A., and Albersheim, P., The plant cell wall polysaccharide rhamnogalacturonan II self-assembles into a covalently cross-linked dimer. *J. Biol. Chem.,* 274, 13098–13104, 1999.

65. Ishii, T. and Matsunaga, T., Pectic polysaccharide rhamnogalacturonan II is covalently linked to homogalacturonan, *Phytochemistry,* 57, 969–974, 2001.

66. Kim, J.B. and Carpita, N.C., Changes in esterification of the uronic-acid groups of cell wall polysaccharides during elongation of maize coleoptiles, *Plant Physiol.*, 98, 646–653, 1992.

67. Brown, J.A. and Fry, Novel, S.C., *O*-D-galacturonosyl esters in the pectic polysaccharides of suspension-cultured plant cells, *Plant Physiol.*, 103, 993–999, 1993.

68. Thompson, J.E. and Fry, S.C., Evidence for covalent linkage between xyloglucan and acidic pectins in suspension-cultured rose cells, *Planta*, 211, 275–286, 2000.

69. Ishii, T., Structure and functions of feruloylated polysaccharides, *Plant Sci.*, 127, 111–127, 1997.

70. Colquhoun, I.J., Ralet, M-C., Thibault, J-F., Faulds, C.B., and Williamson, G., Structure identification of feruloylated oligosaccharides from sugar beet pulp by NMR spectroscopy, *Carbohydr. Res.*, 263, 243–256, 1994.

71. Guillon, F. and Thibault, J.-F., Oxidative cross-linking of chemically and enzymically modified sugar-beet pectin, *Carbohydr. Polym.*, 12, 353–374, 1990.

72. Hatfield, R.D., Ralph, J., and Grabber, J.H., Molecular basis for improving forage digestibilities, *Crop Sci.*, 39, 27–37, 1999.

73. Cassab, G.I., Plant cell wall proteins, *Annu. Rev. Plant Physiol. Plant Mol. Biol.*, 49, 281–309, 1998.

74. Takada, G., Kawaguchi, T., Yoneda, T., Kawasaki, M., Sumitani, J-I., and Arai, M., Molecular cloning and expression of the cellulolytic system from *Aspergillus aculeatus*, in *Genetics, Biochemistry and Ecology of Cellulose Degradation*, Ohmiya, K., Karita, S., Hayashi, H., Kobayashi, Y., and Kimura, T., Eds., University Publishers Co., Tokyo, Ltd., 1999, pp. 364–373.

75. Gielkens, M.M.C., Dekkers, E., Visser, J., and de Graaff, L.H., Two cellobiohydrolase-encoding genes from *Aspergillus niger* require D-xylose and the xylanolytic transcriptional activator XlnR for their expression, *Appl. Environ. Microbiol.*, 65, 4340–4345, 1999.

76. Ivanova, G.S., Beletskaya, O.P., Okunev, O.N., Golovlev, E.L., and Kulaev, I.S., Fractionation of cellulase complex of the fungus *Aspergillus terreus*, *Appl. Biochem. Microbiol.*, 19, 275–281, 1983.

77. Hayashida, S., Mo, K., and Hosoda, A., Production and characteristics of avicel-digesting and non-avicel-digesting cellobiohydrolases from *Aspergillus ficuum*, *Appl. Environ. Microbiol.*, 54, 1523–1529, 1988.

78. Teeri, T.T., Crystalline cellulose degradation: new insight into the function of cellobiohydrolases, *Trends Biotechnol.*, 15, 160–167, 1997.

79. Bagga, P.S., Sandhu, D.K., and Sharma S., Purification and characterization of cellulolytic enzymes produced by *Aspergillus nidulans*, *J. Appl. Bacteriol.*, 68, 61–68, 1990.

80. Unno, T., Ide, K., Yazaki, T., Tanaka, Y., Nakakuki, T., and Okada, G., High recovery purification and some properties of a β-glucosidase from *Aspergillus niger*, *Biosci. Biotech. Biochem.*, 57, 2172–2173, 1993.

81. Fujimoto, H., Ooi, T., Wang, S.-L., Takiwaza, T., Hidaka, H., Murao, S., and Arai, M., Purification and properties of three xylanases from *Aspergillus aculeatus*, *Biosci. Biotech. Biochem.*, 59, 538–540, 1995.

82. Ito, K., Ogasawara, H., Sugimoto, T., and Ishikawa, T., Purification and properties of acid stable xylanases from *Aspergillus kawachii*, *Biosci. Biotech. Biochem.*, 56, 547–550, 1992.

83. Kormelink, F.J.M., Searle-van Leeuwen, M.J.F., Wood, T.M., and Voragen, A.G.J., Purification and characterization of three endo-(1,4)-β-xylanases and one β-xylosidase from *Aspergillus awamori*, *J. Biotechnol.*, 27, 249–265, 1993.

84.   Frederick, M.M., Frederick, J.R., Frayzke, A.R., and Reilly, P.J., Purification and characterization of a xylobiose- and xylose-producing endo-xylanase from *Aspergillus niger*, *Carbohydr. Res.*, 97, 87–103, 1981.

85.   Kormelink, F.J.M., Gruppen, H., Vietor, R.J., and Voragen, A.G.J., Mode of action of the xylan-degrading enzymes from *Aspergillus awamori* on alkali-extractable cereal arabinoxylans, *Carbohydr. Res.*, 249, 355–367, 1993.

86.   Beldman, G., Searle-van Leeuwen, M.J.F., de Ruiter, G.A., Siliha, H.A., and Voragen, A.G.J., Degradation of arabinans by arabinases from *Aspergillus aculeatus* and *Aspergillus niger*, *Carbohydr. Polym.*, 20, 159–168, 1993.

87.   Luonteri, E., Siika-aho, M., Tenkanen, M., and Viikari, L., Purification and characterization of three alpha-arabinofuranosidases from *Aspergillus terreus*, *J. Biotechnol.*, 38, 279–291, 1995.

88.   Ramon, D., van der Veen, P., and Visser, J., Arabinan degrading enzymes from *Aspergillus nidulans*: induction and purification, *FEMS Microbiol. Lett.*, 113, 15–22, 1993.

89.   Rombouts, F.M., Voragen, A.G.J., Searle-van Leeuwen, M.F., Geraeds, C.C.J.M., Schols, H.A., and Pilnik, W., The arabinases of *Aspergillus niger*: purification and characterization of two α-L-arabinofuranosidases and an endo-1,5-α-L-arabinase, *Carbohydr. Polym.*, 9, 25–47, 1988.

90.   van der Veen, P., Flipphi, M.J.A., Voragen, A.G.J., and Visser J., Induction, purification and characterisation of arabinases produced by *Aspergillus niger*, *Arch. Microbiol.*, 157, 23–28, 1991.

91.   Wilkie, K.C.B., and Woo, S.-L., A heteroxylan and hemicellulosic materials from bamboo leaves, and a reconsideration of the general nature of commonly occurring xylans and other hemicelluloses, *Carbohydr. Res.*, 57, 145–162, 1977.

92.   Ebringerová, A., Hromádková, Z., Petráková, E., and Hricovíni, M., Structural features of a water-soluble L-arabinoxylan from rye bran, *Carbohydr. Res.*, 198, 57–66, 1990.

93.   Ademark, P., Larsson, M., Tjerneld, F., and Stålbrand, H., Multiple α-galactosidases from *Aspergillus niger*: purification, characterization, and substrate specificities, *Enzyme Microbiol. Technol.*, 29, 441–448, 2001.

94.   de Vries, R.P., van den Broeck, H.C., Dekkers, E., Manzanares, P., de Graaff, L.H., Visser, J., Differential expression of three α-galactosidase genes and a single β-galactosidase gene from *Apergillus niger*, *Appl. Environ. Microbiol.*, 65, 2453–2460, 1999.

95.   Manzanares, P., de Graaff, L.H., and Visser, J., Characterization of galactosidases from *Aspergillus niger*: purification of a novel α-galactosidase activity, *Enzyme Microbiol. Technol.*, 22, 383–390, 1998.

96.   Scigelova, M. and Crout, D.H.G., Purification of α-galactosidase from *Aspergillus niger* for application in the synthesis of complex oligosaccharides, *J. Mol. Cat.*, 8, 175–181, 2000.

97.   de Vries, R.P., Poulsen, C.H., Madrid, S., and Visser, J., *aguA*, the gene encoding an extracellular α-glucuronidase from *Aspergillus tubingensis*, is specifically induced on xylose and not on glucuronic acid, *J. Bacteriol.*, 180, 243–249, 1998.

98.   Biely, P., de Vries, R.P., Vranská, M., and Visser, J., Inverting character of α-glucuronidase A from *Aspergillus tubingensis*, *Biochim. Biophys. Acta*, 1474, 360–364, 2000.

99.   Tenkanen, M. and Siika aho, M., An alpha-glucuronidase of *Schizophyllum commune* acting on polymeric xylan, *J. Biotechnol.*, 78, 149–161, 2000.

100. Barbe, C. and Dubourdieu, D., Characterisation and purification of a cinnamate esterase from *Aspergillus niger* industrial pectinase preparation, *J. Sci. Food Agric.*, 78, 471–478, 1998.

101. de Vries, R.P., Michelsen, B., Poulsen, C.H., Kroon, P.A., van den Heuvel, R.H.H., Faulds, C.B., Williamson, G., van den Hombergh, J.P.T.W., and Visser, J., The *faeA* genes from *Aspergillus niger* and *Aspergillus tubigensis* encode ferulic acid esterases involved in the degradation of complex cell wall polysaccharides, *Appl. Environ. Microbiol.*, 63, 4638–4644, 1997.

102. de Vries, R.P., Kester, H.C.M., vanKuyk, P.A., and Visser, J., The *Aspergillus niger faeB* gene encodes a second feruloyl esterase involved in pectin and xylan degradation, and is specifically induced on aromatic compounds, *Biochem. J.*, 363, 377–386, 2002.

103. Faulds, C.B. and Williamson, G., Ferulic acid esterase from *Aspergillus niger*: purification and partial characterization of two forms from a commercial source of pectinase, *Biotechnol. Appl. Biochem.*, 17, 349–359, 1993.

104. Faulds, C.B. and Williamson, G., Purification and characterization of a ferulic acid esterase (FAE-III) from *Aspergillus niger*: specificity for the phenolic moiety and binding to microcrystalline cellulose, *Microbiology*, 140, 779–787, 1994.

105. Koseki, T., Furuse, S., Iwano, K., and Matsuzawa, H., Purification and characterization of a feruloylesterase from *Aspergillus awamori*, *Biosci. Biotechnol. Biochem.*, 62, 2032–2034, 1998.

106. Kroon, P.A., Faulds, C.B., and Williamson, G., Purification and characterisation of a novel esterase induced by growth of *Aspergillus niger* on sugar-beet pulp, *Biotechnol. Appl. Biochem.*, 23, 255–262, 1996.

107. McCrae, S.I., Leith, K.M., Gordon, A.H., and Wood, T.M., Xylan degrading enzyme system produced by the fungus *Aspergillus awamori*: isolation and characterization of a feruloyl esterase and a *p*-coumaroyl esterase, *Enzyme Microbiol. Technol.*, 16, 826–834, 1994.

108. Tenkanen, M., Schuseil, J., Puls, J., and Poutanen, K., Production, purification and characterisation of an esterase liberating phenolic acids from lignocellulosics, *J. Biotechnol.*, 18, 69–84, 1991.

109. Kroon, P.A., Faulds, C.B., Brezillon, C., and Williamson, G., Methyl phenylalkanoates as substrates to probe the active sites of esterases, *Eur. J. Biochem.*, 248, 245–251, 1997.

110. Ralet, M-C., Faulds, C.B., Williamson, G., and Thibault, J-F., Degradation of feruloylated oligosaccharides from sugar beet pulp and wheat bran by ferulic acid esterases from *Aspergillus niger*, *Carbohydr. Res.*, 263, 257–269, 1994.

111. de Vries, R.P., Kester, H.C.M., Poulsen, C.H., Benen, J.A.E., and Visser, J., Synergy between accessory enzymes from *Aspergillus* in the degradation of plant cell wall polysaccharides, *Carbohydr. Res.*, 327, 401–410, 2000.

112. Kormelink, F.J.M. and Voragen, A.G.J., Degradation of different [(glucurono)arabino] xylans by combination of purified xylan-degrading enzymes, *Appl. Microbiol. Biotechnol.*, 38, 688–695, 1993.

113. Verbruggen, M.A., Beldman, G., and Voragen, A.G.J., Structures of enzymically derived oligosaccharides from sorghum glucuronoarabinoxylan, *Carbohydr. Res.*, 306, 265–274, 1998.

114. Bartolome, B., Faulds, C.B., Tuohy, M., Hazlewood, G.P., Gilbert, H.J., and Williamson, G., Influence of different xylanases on the activity of ferulic acid esterase on wheat bran, *Biotechnol. Appl. Biochem.*, 22, 65–73, 1995.

115. Kormelink, F.J.M., Lefebvre, B., Strozyk, F., and Voragen, A.G.J., The purification and characterisation of an acetyl xylan esterase from *Aspergillus niger*, *J. Biotechnol.*, 27, 267–282, 1993.

116. Ademark, P., Varga, A., Medve, J., Harjunpaa, V., Drakenberg, T., Tjerneld, F., and Stålbrand, H., Softwood hemicellulose-degrading enzymes from *Aspergillus niger*: purification and properties of a β-mannanase, *J. Biotechnol.*, 63, 199–200, 1998.

117.   Civas, A., Eberhard, R., le Dizet, P., and Petek, F., Glycosidases induced in
       Aspergillus tamarii Secreted α-D-galactosidase and β-D-mannanase, *Biochem. J.*,
       219, 857–863, 1984.
118.   Eriksson, K.-W. and Winell, M., Purification and characterization of a fungal β-man-
       nanase, *Acta Chem. Scand.*, 22, 1924–1934, 1968.
119.   Reese, E.T. and Shibata, Y., β-Mannanases of fungi, *Can. J. Microbiol.*, 11, 167–183, 1965.
120.   McCleary, B.V., Comparison of endolytic hydrolases that depolymerize 1,4-β-D-
       mannan, 1,5-α-L-arabinan, and 1,4-β-D-galactan, in *Enzymes in Biomass
       Conversion*, Leatham, G.F. and Himmel, M.E., Eds., American Chemical Society,
       Washington, 1991, pp. 437–449.
121.   McCleary, B.V. and Mathesen, N.K., Action patterns and substrate-binding require-
       ments of β-D-mannanase with mannosaccharides and mannan-type polysaccharides,
       Carbohydr. Res., 119, 191–219, 1983.
122.   McCleary, B.V., Modes of action of β-mannanase enzymes of diverse origin on
       legume seed galactomannans, *Phytochemistry*, 18, 757–763, 1979.
123.   Ademark, P., Lundqvist, J., Hagglund, P., Tenkanen, M., Torto, N., Tjerneld, F., and
       Stålbrand, H., Hydrolytic properties of a β-mannosidase purified from *Aspergillus
       niger*, *J. Biotechnol.*, 75, 281–289, 1999.
124.   Bouquelet, S., Spik, G., and Montreuil, J., Properties of a beta-D-mannosidase from
       Aspergillus niger, *Biochim. Biophys. Acta*, 522, 521–530, 1978.
125.   Elbein, A.D., Adya, S., and Lee, Y.C., Purification and properties of a β-mannosidase
       from *Aspergillus niger*, *J. Biol. Chem.*, 252, 2026–2031, 1977.
126.   Neustroev, K.N., Krylov, A.S., Firsov, L.M., Abroskina, O.L., and Khorlin, A.Y.,
       Isolation and properties of β-mannosidase from *Aspergillus awamori*, *Biokhimiya*,
       56, 1406–1412, 1991.
127.   Adya, S. and Elbein, A.D., Glycoprotein enzymes secreted by *Aspergillus niger*:
       Purification and properties of a α-galactosidase, *J. Bacteriol.*, 129, 850–856,
       1977.
128.   Civas, A., Eberhard, R., le Dizet, P., and Petek, F., Glycosidases induced in
       Aspergillus tamarii. Mycelial α-D-galactosidases, *Biochem. J.*, 219, 849–855, 1984.
129.   Knap, I.H., Carsten, M., Halkier, T., and Kofod, L.V., An α-Galactosidase Enzyme.
       International Patent WO 94/230221994, 1994.
130.   Rios, S., Pedregosa, A.M., Fernandez Monistrol, I., and Laborda, F., Purification and
       molecular properties of an α-galactosidase synthesized and secreted by *Aspergillus
       nidulans*, *FEMS Microbiol. Lett.*, 112, 35–42, 1993.
131.   Somiari, R.I. and Balogh, E., Properties of an extracellular glycosidase of *Aspergillus
       niger* suitable for removal of oligosaccharides from cowpea meal, *Enzyme Microb.
       Technol.*, 17, 311–316, 1995.
132.   Zapater, I.G., Ullah, A.H.J., and Wodzinski, R.J., Extracellular α-galactosidase (EC
       3.2.1.22) from *Aspergillus ficuum* NRRL 3135: purification and characterisation,
       Prep. Biochem., 20, 263–296, 1990.
133.   Puls, J., Schorn, B., and Schuseil, J., Acetylmannanesterase: a new component in the
       arsenal of wood mannan degrading enzymes, in *Biotechnology in Pulp and Paper
       Industry*, Kuwahara, M. and Shimada, M., Eds., University Publishers Co., Ltd.,
       Tokyo, 1992, pp. 357–363.
134.   Tenkanen, M., Thornton, J., and Viikari, L., An acetylglucomannan esterase of
       Aspergillus oryzae; purification, characterization and role in the hydrolysis of
       O-acetyl-galactoglucomannan, *J. Biotechnol.*, 42, 197–206, 1995.
135.   Gübitz, G.M., Haltrich, D., Latal, B., and Steiner, W., Mode of depolymerisation of
       hemicellulose by various mannanases and xylanases in relation to their ability to
       bleach softwood pulp, *Appl. Microbiol. Biotechnol.*, 47, 658–662, 1997.

136. Tenkanen, M., Puls, J., Ratto, M., and Viikari, L., Enzymatic deacetylation of galactoglucomannans, *Appl. Microbiol. Biotechnol.*, 39, 159–165, 1993.

137. Hasper, A.A., Dekkers, E., van Mil, M., van de Vondervoort, P.J.I., and de Graaff, L.H., EglC, a new endoglucanase from *Aspergillus niger* with major activity towards xyloglucan, *Appl. Environ. Microbiol.*, 68, 1556–1560, 2002.

138. Pauly, M., Andersen, L.N., Kaupinnen, S., Kofod, L.V., York, W.S., Albersheim, P., and Darvill, A., A xyloglucan-specific endo-β-1,4-glucanase from *Aspergillus aculeatus*: expression cloning in yeast, purification and characterization of the recombinant enzyme, *Glycobiology*, 9, 93–100, 1999.

139. Matsushita, J., Kato, Y., and Matsuda, K., Purification and properties of an α-D-xylosidase from *Aspergillus niger*, *J. Biochem.*, 98, 825–832, 1985.

140. Matsushita, J., Kato, Y., and Matsuda, K., Characterization of α-D-xylosidase II from *Aspergillus niger*, *Agric. Biol. Chem.*, 51, 2015–2016, 1987.

141. Yoshikawa, K., Yamamoto, K., and Okada, S., Isolation of *Aspergillus flavus* MO-5 producing two types of intracellular α-D-xylosidases: purification and characterization of α-D-xylosidase I., *Biosci. Biotech. Biochem.*, 57, 1275–1280, 1993.

142. Yoshikawa, K., Yamamoto, K., and Okada, S., Purification and characterization of an intracellular α-D-xylosidase II from *Aspergillus flavus* MO-5, *Biosci. Biotech. Biochem.*, 57, 1281–1285, 1993.

143. Bahl, O.P., Glycosidases of *Aspergillus niger*. II. Purification and general properties of 1,2-α-L-fucosidase. *J. Biol. Chem.*, 245, 299–304, 1970.

144. Vincken, J.P., Wijsman, A.J.M., Beldman, G., Niessen, W.M.A., and Voragen, A.G.J., Potato xyloglucan is built from XXGG-type subunits, *Carbohydr. Res.*, 288, 219–232, 1996.

145. Anjana Devi, N. and Appu Rao, A.G., Fractionation, purification, and preliminary characterization of polygalacturonases produced by *Aspergillus carbonarius*, *Enzyme Microb. Technol.*, 18, 59–65, 1996.

146. Benen, J., Kester, H., and Visser, J., Kinetic characterization of *Aspergillus niger* N400 endopolygalacturonases I, II, and C, *Eur. J. Biochem.*, 259, 577–585, 1999.

147. Kester, H.C.M., Magaud, D., Roy, C., Anker, D., Doutheau, A., Shevchik, V., Hugouvieux-Cotte-Pattat, N., Benen, J.A.E., and Visser, J., Performance of selected microbial pectinases on synthetic monomethyl-esterified di- and trigalacturonates, *J. Biol. Chem.*, 274, 37053–37059, 1999.

148. Kester, H.C.M. and Visser, J., Purification and characterization of polygalacturonases produced by the hyphal fungus *Aspergillus niger*, *Biotechnol. Appl. Biochem.*, 12, 150–160, 1990.

149. Parenicová, L., Benen, J.A.E., Kester, H.C.M., and Visser, J., *pga*A and *pga*B encode two constitutively expressed endopolygalacturonases of *Aspergillus niger*, *Biochem. J.*, 345, 637–644, 2000.

150. Parenicová, L., Kester, H.C.M., Benen, J.A.E., and Visser, J., Characterization of a novel endopolygalacturonase from *Aspergillus niger* with unique kinetic properties, *FEBS Lett.*, 467, 333–336, 2000.

151. Beldman, G., van den Broek, L.A.M., Schols, H.A., Searle-van Leeuwen, M.J.F., van Laere, K.M.J., and Voragen, A.G.J., An exogalacturonase from *Aspergillus aculeatus* able to degrade xylogalacturonan, *Biotechnol. Lett.*, 18, 707–712, 1996.

152. Hara, T., Lim, J.Y., Fujio, Y., and Ueda, S., Purification and some properties of exo-polygalacturonase from *Aspergillus niger* cultured in the medium containing Satsuma mandarin peel, *Nippon. Shokuhin. Kogyo. Gakkaishi.*, 31, 581–586, 1984.

153. Kester, H.C.M., Kusters-van Someren, M.A., Muller, Y., and Visser, J., Primary structure and characterization of an exopolygalacturonase from *Aspergillus tubingensis*, *Eur. J. Biochem.*, 240, 738–746, 1996.

154. Mikhailova, R.V., Sapunova, L.I., and Lobanok, A.G., Three polygalacturonases constitutively synthesized by *Aspergillus alliaceus*, *World J. Microbiol. Biotechnol.*, 11, 330–332, 1995.

155. Kester, H.C.M., Benen, J.A.E., and Visser, J., The exopolygalacturonase from *Aspergillus tubingensis* is also active on xylogalacturonan, *Biotechnol. Appl. Biochem.*, 30, 53–57, 1999.

156. van der Vlugt-Bergmans, C.J.B., Meeuwsen, P.J.A., Voragen, A.G.J., and van Ooyen, A.J.J., Endo-xylogalacturonan hydrolase, a novel pectinolytic enzyme, *Appl. Environ. Microbiol.*, 66, 36–41, 2000.

157. Benen, J., Parenicová, L., Kusters-van Someren, M., Kester, H., and Visser, J., Molecular genetic and biochemical aspects of pectin degradation in *Aspergillus*, in and Pectins and Pectinases, Visser, J. and Voragen, A.G.J., Eds., Elsevier Science, Amsterdam,1996, pp. 331–346.

158. Benen, J.A.E., Kester, H.C.M., Parenicova, L., and Visser, J., Characterization of *Aspergillus niger* pectate lyase A, *Biochemistry*, 39, 15563–15569, 2000.

159. Dean, R.A. and Timberlake, W.A., Regulation of the *Aspergillus nidulans* pectate lyase gene (*pelA*), *The Plant Cell*, 1, 275–284, 1989.

160. Kofod, L.V., Kauppinen, S., Christgau, S., Andersen, L.N., Heldt-Hansen, H.P., Dorreich, K., and Dalboge, H., Cloning and characterization of two structurally and functionally divergent rhamnogalacturonases from *Aspergillus aculeatus*, *J. Biol. Chem.*, 269, 29182–29189, 1994.

161. Mutter, M., Colquhoun, I.J., Schols, H.A., Beldman, G., and Voragen, A.G.J., Rhamnogalacturonase B from *Aspergillus aculeatus* is a rhamnogalacturonan α-L-rhamnopyranosyl-(1,4)-α-D-galactopyranosyluronide lyase, *Plant Physiol.*, 110, 73–77, 1996.

162. Schols, H.A., Geraeds, C.J.M., Searle-van Leeuwen, M.F., Kormelink, F.J.M., and Voragen, A.G.J., Rhamnogalacturonase: a novel enzyme that degrades the hairy regions of pectins, *Carbohydr. Res.*, 206, 104–115, 1990.

163. Suykerbuyk, M.E.G., Kester, H.C.M., Schaap, P.J., Stam, H., Musters, W., and Visser, J., Cloning and characterization of two rhamnogalacturonan hydrolase genes from *Aspergillus niger*, *Appl. Environ. Microbiol.*, 63, 2507–2515, 1997.

164. Mutter, M., Beldman, G., Schols, H.A., and Voragen, A.G.J., Rhamnogalacturonan α-L-rhamnopyranosylhydrolase. A novel enzyme specific for the terminal nonreducing rhamnosyl unit in rhamnogalacturonan regions of pectin, *Plant Physiol.*, 106, 241–250, 1994.

165. Mutter, M., Beldman, G., Pitson, S.M., Schols, H.A., and Voragen, A.G.J., Rhamnogalacturonan α-D-galactopyranosyluronohydrolase. An enzyme that specifically removes the terminal nonreducing galacturonosyl residue in rhamnogalacturonan regions of pectin, *Plant Physiol.*, 117, 153–163, 1994.

166. Manzanares, P., van den Broeck, H.C., de Graaff, L.H., and Visser, J., Purification and characterization of two different alpha-L-rhamnosidases, RhaA and RhaB, from *Aspergillus aculeatus*, *Appl. Environ. Microbiol.*, 67, 2230–2234, 2001.

167. Kaneko, S., Shimasaki, T., and Kusakabe, I., Purification and some properties of intracellular α-L-arabinofuranosidase from *Aspergillus niger* 5–16, *Biosci. Biotech. Biochem.*, 57, 1161–1165, 1993.

168. Dunkel, M.P.H., and Amado, R., Analysis of endo-(1–5)-α-L-arabinanase degradation patterns of linear (1–5)-α-L-arabino-oligosaccharides by high-performance anion-exchange chromatography with pulsed amperometric detection, *Carbohydr. Res.*, 268, 151–158, 1995.

169. Christgau, S., Sandal, T., Kofod, L.V., and Dalboge, H., Expression cloning, purification and characterization of a β-1,4-galactanase from *Aspergillus aculeatus*, *Curr. Genet.*, 27, 135–141, 1995.

170. de Vries, R.P., Parenicová, L., Hinz, S., Kester, H., Benen, J.A.E., Beldman, G., and Visser, J., Endogalactanase A from *Aspergillus niger* is specifically induced on L-arabinose and galacturonic acid and acts synergistically with *Aspergillus* rhamnogalacturonases, *Eur. J. Biochem.*, 269, 4985–4993, 2002.

171. Kimura, I., Yoshioka, N., and Tajima, S., Purification and characterization of an endo-1,4-β-D-galactanase from *Aspergillus sojae*, *J. Ferment. Bioeng.*, 85, 48–52, 1998.

172. Muzakhar, K., Hayashi, H., Kawaguchi, T., Sumitani, J., and Arai, M., Purification and properties of α-L-arabinofuranosidase and endo-β-D-1,4-galactanase from *Aspergillus niger* v. Tieghem KF-267 which liquified the Okara, in *Genetics, Biochemistry and Ecology of Cellulose Degradation,Ohmiya*, K., Sakka, K., Karita, S., Hayashi, H., Kobayashi, Y., and Kimura, T., Eds., University Publishers Co., Ltd., Tokyo, 1999, pp. 134–143.

173. van Casteren, W.H.M., Eimermann, M., van den Broek, L.A.M., Vincken, J.-P., Schols, H.A., and Voragen, A.G.J., Purification and characterisation of a beta-galactosidase from *Aspergillus aculeatus* with activity towards (modified) exopolysaccharides from *Lactococcus lactis* subsp. *cremoris* B39 and B891, *Carbohydr. Res.*, 329, 75–85, 2000.

174. Brillouet, J.-M., Williams, P., and Moutounet, M., Purification and some properties of a novel endo-β-(1-6)-D-galactanase from *Aspergillus niger*, *Agric. Biol. Chem.*, 55, 1565–1571, 1991.

175. Pellerin, P. and Brillouet, J.-M., Purification and properties of an exo-(1→3)-β-D-galactanase from *Aspergillus niger, Carbohydr. Res.*, 264,281–291, 1994.

176. Bonnin, E., Lahaye, M., Vigoureux, J., and Thibault, J.-F., Preliminary characterization of a new exo-β-(1,4)-galactanase with transferase activity, *Int. J. Biol. Macromol.*, 17, 3454, 1995.

177. Searle-van Leeuwen, M.J.F., Vincken, J-P., Schipper, D., Voragen, A.G.J., and Beldman, G., Acetyl esterases of *Aspergillus niger*: purification and mode of action on pectins, in Pectins and Pectinases, Visser, J. and Voragen, A.G.J., Eds., Elsevier Science, Amsterdam, 1996, pp. 793–798.

178. Christgau, S., Kofod, L.V., Halkier, T., Anderson, L.N., Hockauf, M., Dorreich, K., Dalboge, H., and Kauppinen, S., Pectin methyl esterase from *Aspergillus aculeatus*: expression cloning in yeast and characterization of the recombinant enzyme, *Biochem. J.*, 319, 705–712, 1996.

179. Khanh, N.Q., Ruttkowski, E.K., Leidinger, A H., and Gottschalk, M., Characterisation and expression of a genomic pectin methyl esterase-encoding gene in *Aspergillus niger*, *Gene*, 106, 71–77, 1991.

180. Kauppinen, S., Christgau, S., Kofod, L.V., Halkier, T., Dorreich, K., and Dalboge, H., Molecular cloning and characterization of a rhamnogalacturonan acetylesterase from *Aspergillus aculeatus*. Synergism between rhamnogalacturonan degrading enzymes, *J. Biol. Chem.*, 270, 27172–27178, 1995.

181. de Vries, R.P., Benen, J.A.E., de Graaff, L.H., and Visser, J., Plant cell wall degrading enzymes produced by *Aspergillus,* in *The Mycota X: Industrial Applications*, Osiewacz, H.D., Ed., Springer-Verlag, Heidelberg, 2001, pp. 263–280.

182. Searle-van Leeuwen, M.J.F., van den Broek, L.A.M., Schols, H.A., Beldman, G., V.A. G.J., Rhamnogalacturonan acetylesterase: a novel enzyme from *Aspergillus aculeatus*, specific for the deacetylation of hairy (ramified) regions of pectins, *Appl. Microbiol. Biotechnol.*, 38, 347–349, 1992.

183.   Kroon, P.A. and Williamson, G., Release of ferulic acid from sugar-beet pulp by using arabinase, arabinofuranosidase and an esterase from *Aspergillus* niger, *Biotechnol. Appl. Biochem.*, 23, 263–267, 1996.

184.   Liebs, P., On the localization of saccharose uptake in submerged mycelium of *Aspergillus niger, Arch. Mikrobiol.*, 61, 103–111, 1968.

185.   Fekete, E., Karaffa, L., Sandor, E., Seiboth, B., Biro, S., Szentirmai, A., and Kubicek, C.P., Regulation of formation of the intracellular beta-galactosidase activity of *Aspergillus nidulans, Arch. Microbiol.*, 179, 7–14, 2002.

186.   van Peij, N.N.M.E., Visser, J., and de Graaff, L.H., Isolation and analysis of *xlnR*, encoding a transcriptional activator coordinating xylanolytic expression in *Aspergillus niger, Mol. Microbiol.*, 27, 131–142, 1998.

187.   van Peij, N., Gielkens, M.M.C., de Vries, R.P., Visser, J., and de Graaff, L.H., The transcriptional activator XlnR regulates both xylanolytic and endoglucanase gene expression in *Aspergillus niger, Appl. Environ. Microbiol.*, 64, 3615–3619, 1998.

188.   de Vries, R.P., van de Vondervoort, P.J.I., Hendriks, L., van de Belt, M., and Visser, J., Regulation of the a-glucuronidase encoding gene (*aguA*) from *Aspergillus niger, Mol. Gen. Genet.*, 268, 96–102, 2002.

189.   Marui, J., Tanaka, A., Mimura, S., de Graaff, L.H., Visser, J., Kitamoto, N., Kato, M., Kobayashi, T., and Tsukagoshi, N., A transcriptional activator, AoXlnR, controls the expression of genes encoding xylanolytic enzymes in *Aspergillus oryzae, Fung. Genet. Biol.*, 35, 157–169, 2002.

190.   Hasper, A.A., Visser, J., and de Graaff, L.H., The *Aspergillus niger* transcriptional activator XlnR, which is involved in the degradation of the polysaccharides xylan and cellulose, also regulates D-xylose reductase gene expression, *Mol. Microbiol.*, 36, 193–200, 2000.

191.   vanKuyk, P.A., de Groot, M.J.L., Ruijter, G.J.G., de Vries, R.P., and Visser, J., The *Aspergillus niger* D-xylulose kinase gene is co-expressed with genes encoding arabinan degrading enzymes and is essential for growth on arabinose and xylose, *Eur. J. Biochem.*, 268, 5414–5423, 2001.

192.   Flipphi, M.J.A., Visser, J., van der Veen, P., and de Graaff, L.H., Arabinase gene expression in *Aspergillus niger*: indications for co-ordinated gene expression, *Microbiology*, 140, 2673–2682, 1994.

193.   Gielkens, M.M.C., Visser, J., and de Graaff, L.H., Arabinoxylan degradation by fungi: characterisation of the arabinoxylan arabinofuranohydrolase encoding genes from *Aspergillus niger* and *Aspergillus tubingensis, Curr. Genet.*, 31, 22–29, 1997.

194.   Gielkens, M.M.C., Gonzales-Candelas, L., Sanchez-Torres, P., van de Vondervoort, P.J.I., de Graaf, L.H., and Visser, J., The *abfB* gene encoding the major α-L-arabino-furanosidase of *Aspergillus nidulans*: nucleotide sequence, regulation and construction of a disrupted strain, *Microbiology* 145, 735–741, 1999.

195.   van der Veen, P., Flipphi, M.J.A., Voragen, A.G.J., and Visser, J., Induction of extra-cellular arabinases on monomeric substrates in *Aspergillus niger, Arch. Microbiol.*, 159, 66–71, 1993.

196.   de Vries, R.P., Flipphi, M.J.A., Witteveen, C.F.B., and Visser, J., Characterisation of an *Aspergillus nidulans* L-arabitol dehydrogenase mutant, *FEMS Microbiol. Lett.*, 123, 83–90, 1994.

197.   de Groot, M.J.L., van de Vondervoort, P.J.I., de Vries, R.P., vanKuyk, P.A., Ruijter, G.J.G., and Visser, J., Isolation and characterization of two specific regulatory *Aspergillus niger* mutants shows antagonistic regulation of arabinan and xylan metabolism, *Microbiology*, 149, 1183–1191, 2003.

198. de Vries, R.P., Jansen, J., Aguilar, G., Parenicová, L., Benen, J.A.E., Joosten, V., Wulfert, F., and Visser, J., Expression profiling of pectinolytic genes from *Aspergillus niger*, *FEBS Lett.*, 530, 41–47, 2002.

199. den Herder, I.F., Mateo Rosell, A.M., van Zuilen, C.M., Punt, P.J., and van den Hondel, C.A.M.J.J., Cloning and expression of a member of the *Aspergillus niger* gene family encoding α-galactosidase, *Mol. Gen. Genet.*, 233, 404–410, 1992.

200. Ademark, P., de Vries, R.P., Stålbrand, H., and Visser, J., Cloning and characterisation of genes encoding a β-mannosidase and an α-galactosidase from *Aspergillus niger* involved in galactomannan degradation, *Eur. J. Biochem.*, 268, 2982–2990, 2001.

201. de Vries, R.P. and Visser, J., Regulation of the feruloyl esterase (*faeA*) gene from *Aspergillus niger*, *Appl. Environ. Microbiol.*, 65, 5500–5503, 1999.

202. Dowzer, C.E.A. and Kelly, J.M., Analysis of the *creA* gene, a regulator of carbon catabolite repression in *Aspergillus nidulans*, *Mol. Cell. Biol.*, 11, 5701–5709, 1991.

203. Kulmburg, P., Mathieu, M., Dowzer, C., Kelly, J., and Felenbok, B., Specific binding sites in the *alcR* and *alcA* promoters of the ethanol regulon for the CreA repressor mediating carbon catabolite repression in *Aspergillus nidulans*, *Mol. Microbiol.*, 7, 847–857, 1993.

204. Ruijter, G.J.G. and Visser, J., Carbon repression in *Aspergilli*, *FEMS Microbiol. Lett.*, 151, 103–114, 1997.

205. Bussink, H.J.D., Buxton, F.P., and Visser, J., Expression and sequence comparison of the *Aspergillus niger* and *Aspergillus tubingensis* genes encoding polygalacturonase II, *Curr. Genet.*, 19, 467–474, 1991.

206. Fernandez-Espinar, M., Pinaga, F., de Graaff, L., Visser, J., Ramon, D., and Valles, S., Purification, characterization, and regulation of the synthesis of an *Aspergillus nidulans* acidic xylanase, *Appl. Microbiol. Biotechnol.*, 42, 555–562, 1994.

207. Kumar, S. and Ramon, D., Purification and regulation of the synthesis of a β-xylosidase from *Aspergillus nidulans*, *FEMS Microbiol. Lett.*, 135, 287–293, 1996.

208. Maldonado, M.C., Strasser de Saad, A.M., and Callieri, D., Catabolite repression of the synthesis of inducible polygalacturonase and pectinesterase by *Aspergillus* sp., *Curr. Microbiol.*, 18, 303–306, 1989.

209. Perez-Gonzalez, J.A., van Peij, N.N.M.E., Bezoen, A., MacCabe, A.P., Ramon, D., and de Graaff, L.H., Molecular cloning and transcriptional regulation of the *Aspergillus nidulans xlnD* gene encoding β-xylosidase, *Appl. Environ. Microbiol.*, 64, 1412–1419, 1998.

210. Pinaga, F., Fernandez-Espinar, M.T., Valles, S., and Ramon, D., Xylanase production in *Aspergillus nidulans*: induction and carbon catabolite repression, *FEMS Microbiol. Lett.*, 115, 319–324, 1994.

211. Ruijter, G.J.G., Vanhanen, S.I., Gielkins, M.M.C., van de Vondervoort, P.J.I., and Visser, J., Isolation of *Aspergillus niger creA* mutants: Effects on expression of arabinases and L-arabinose catabolic enzymes, *Microbiology*, 143, 2991–2998, 1997.

212. Solis-Pereira, S., Favela-Torres, E., Viniegra-Gonzalez, G., and Gutierrez-Rojas, M., Effects of different carbon sources on the synthesis of pectinase by *Aspergillus niger* in submerged and solid state fermentations, *Appl. Microbiol. Biotechnol.*, 39, 36–41, 1993.

213. de Vries, R.P., Visser, J., and de Graaff, L.H., CreA modulates the XlnR induced expression on xylose of *Aspergillus niger* genes involved in xylan degradation, *Res. Microbiol.*, 150, 281–285, 1999.

214. Ali, S. and Sayed, A., Regulation of cellulase biosynthesis in *Aspergillus terreus*, *World J. Microbiol. Biotechnol.*, 8, 73–75, 1992.

215. MacCabe, A.P., van den Hombergh, J.P.T.W., Tilburn, J., Arst, H.N.J., and Visser, J., Identification, cloning and analysis of the *Aspergillus niger* gene *pacC*, a wide domain regulatory gene responsive to ambient pH, *Mol. Gen. Genet.*, 250, 367–374, 1996.

216.   Tilburn, J., Sarkar, S., Widdick, D.A., Espeso, E.A., Orejas, M., Mungroo, J., Penalva, M.A., and Arst Jr., H.A., The *Aspergillus* PacC zinc finger transcription factor mediates regulation of both acidic and alkaline expressed genes by ambient pH, *EMBO J.*, 14, 779–790, 1995.

217.   MacCabe, A.P., Orejas, M., Perez-Gonzalez, J.A., and Ramon, D., Opposite patterns of expression of two *Aspergillus nidulans* xylanase genes with respect to ambient pH, *J. Bacteriol.*, 180, 1331–1333, 1998.

218.   Kojima, Y., Sakamoto, T., Kishida, M., Sakai, T., and Kawasaki, H., Acidic condition-inducible polygalacturonase of *Aspergillus kawachii*, *J. Mol. Catalysis.*, B, 6, 351–357, 1999.

219.   Stewart, J.C. and Parry, J.B., Factors influencing the production of cellulase by *Aspergillus fumigatus* (Fresenius), *J. Gen. Microbiol.*, 125, 33–39, 1981.

23 β-Glucosidases from Filamentous Fungi: Properties, Structure, and Applications

Jaime Eyzaguirre, Mauricio Hidalgo, Andrés Leschot

CONTENTS

I. INTRODUCTION

The recurrent crisis in the Persian Gulf illustrates the overdependence of the world on readily accessible and inexpensive sources of oil, and also shows the importance of developing alternative renewable sources of liquid fuels. Lignocellulosic materials, being inexpensive and abundant, represent one of the most promising raw materials for the production of such fuels. However, despite of numerous research efforts in different parts of the world, the biodegradation of lignocellulose (the main component of which is cellulose) is still not economically competitive, especially because of the high costs of the enzymes involved.[1]

Cellulose hydrolysis is performed in nature by the concerted action of a set of enzymes collectively called "cellulases." These enzymes are the endoglucanases (EC 3.2.1.4), which split internal β-glycosidic bonds, the exoglucanases (EC 3.2.1.91) that hydrolyze cellobiose residues from the nonreducing end of the cellulose fibers, and the β-glucosidases. Strictly speaking, β-glucosidase is not a cellulase, because it does not attack cellulose directly. However, it is considered important for the cellulolytic process, because it hydrolyzes soluble cellodextrins and cellobiose to glucose.

β-Glucosidase (or β-D-glucoside glucohydrolase, EC 3.2.1.21) hydrolyzes glycosidic bonds between two glucose residues or between a glucose residue and an aglycon, that can be an alkyl or aryl residue (Figure 23.1). As will be seen in more detail below, the specificity of β-glucosidase toward different substrates varies depending on the enzyme source. This enzyme is widely distributed in nature and can be found in animals,[2] plants,[3] bacteria,[4] fungi,[5] and yeasts.[6]

As indicated above, β-glucosidases, particularly those from microorganisms, are important in cellulose saccharification. Microorganisms that are poor producers of β-glucosidase do not degrade cellulose efficiently. Cellobiose, which is generated by the action of the endo- and exoglucanase, acts as inhibitor of those enzymes and has to be removed by β-glucosidase.[7] Due to its potential biotechnological importance, the β-glucosidase from numerous sources, particularly bacteria and fungi, have been purified and characterized. The work on β-glucosidases, until 1982, has been reviewed by Woodward and Wiseman;[8] later, Leclerc and co-authors[9] have surveyed the enzymes from yeast, and Stutzenberger[10] has reviewed thermostable β-glucosidases from mesophilic and thermophilic fungi. More recently, Bhatia and colleagues[11] have discussed microbial β-glucosidases particularly cloning, mode of action, and applications.

This review will focus on the fungal enzymes; special emphasis will be on the properties and structure of the enzymes and their possible biotechnological applications.

II. METHODS FOR ASSAY OF β-GLUCOSIDASE

Several methods, both sensitive and easy to use, have been developed for the determination of β-glucosidase activity. The most common are those using alkyl- or arylglucosides as substrates, which upon hydrolysis, release either colored or fluorescent products. The most utilized substrate is p-nitrophenyl-β-D-glucopyranoside (PNPG), which releases p-nitrophenol.[12] Some authors replace this substrate with the *ortho* isomer; however, it has been found that several fungal enzymes, such as the

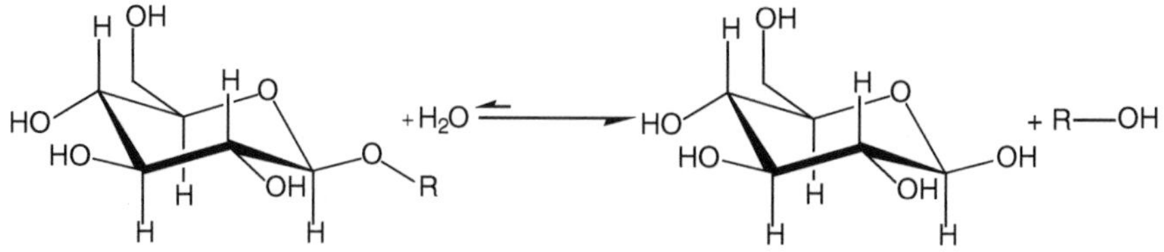

FIGURE 23.1 Reaction catalyzed by β-glucosidase. R can be glucose or an alkyl or aryl residue.

β-glucosidases from *Trichoderma koningii*[13] and *T. reesei*,[14] hydrolyze this isomer more slowly. Other substrates used are methyl-β-D-glucopyranoside, the natural glycosides salicin, esculin, and amygdalin, and the disaccharide cellobiose[12] (Figure 23.2).

If the criterion for assay is activity in cellulose biodegradation, cellobiose should be the substrate of choice.[15] Activity toward cellobiose is measured determining free glucose by the glucose oxidase–peroxidase method.[16] Unfortunately, this method can be influenced by inhibitory products from wood decomposition present in

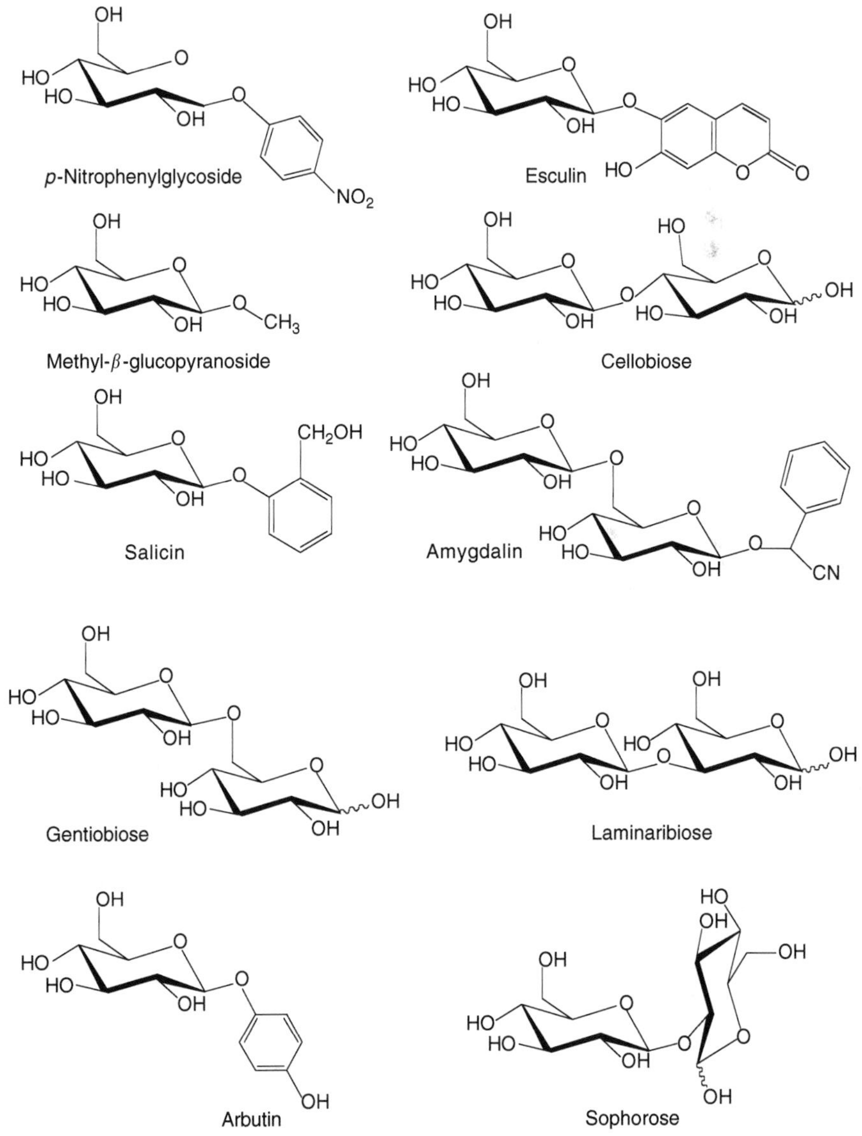

FIGURE 23.2 Substrates of β-glucosidase.

culture filtrates.[17] Glucose detection can also be accomplished by means of the coupled hexokinase/glucose-6- phosphate dehydrogenase assay.[18]

β-Glucosidase products can also be analyzed by high-performance liquid chromatography (HPLC);[19] the use of pulse amperometric detection is particularly sensitive.[20] This analysis is not affected by wood decomposition products.[21] For routine work, thin-layer chromatography (TLC)[22] and, more recently, high-performance thin-layer chromatography (HPTLC) have been used.[23] These techniques allow also the identification of transglycosylation products. Transglycosylating activity is common to many fungal β-glucosidases.[24–26]

Jackson[27] has described a procedure of high resolution for the separation of reducing saccharides, previously derivatized with the fluorophore 8 aminonaphtalene-1,3,6 trisulfonic acid (ANTS), by means of polyacrylamide gel electrophoresis. Amounts as small as 0.2 pmol of ANTS-labeled saccharides can be detected. This method should be useful in the analysis of β-glucosidase-generated products.

Fungal β-glucosidases are often produced in multiple enzyme forms that differ in some physicochemical properties.[14–15,26,28–37] Zymogram techniques are particularly useful in isozyme detection; they are based on gel electrophoresis or isoelectrofocusing, followed by visualization of the activities *in situ* in the gel. Two frequently used substrates are 6-bromo-2-naphtyl-β-D-glucoside and 4-methylumbelliferyl-β-D-glucoside;[38] 5-bromo-4-chloro-3-indolyl β-D-glucopyranoside, an alternative substrate, yields insoluble indigo upon hydrolysis.[39] PNPG and ONPG have also been used, although their reaction products diffuse quickly, and are less sensitive.

III. SUBSTRATE SPECIFICITY

Since fungal β-glucosidases are responsible for the hydrolysis of cellobiose in the saccharification of cellulose, these enzymes have also been called cellobiases. This name, however, is misleading, since it suggests a high substrate specificity for cellobiose. This is not the case, since β-glucosidases are also capable of hydrolyzing other β-1, 4 oligosaccharides with chain lengths of up to eight glucose units (cellooctaose).[19,40–42] β-Glucosidases, however, can be distinguished from the exoglucohydrolases (EC 3.2.1.74); the latter act more rapidly on longer cellooligosaccharides, while the speed of hydrolysis by β-glucosidases slows down with the increasing degree of polymerization (DP) of the substrate.[40,43] Besides, D-glucono-1,5-δ-lactone, a potent inhibitor of β-glucosidases,[43,44] does not inhibit the exoglucohydrolases.[43] Endo- and exoglucanases are also not significantly inhibited by this compound.[44]

β-Glucosidases show a great diversity in substrate specificity, and no definite patterns have been recognized so that every β-glucosidase has to be studied separately. However, the fungal enzymes can be broadly classified in two groups, based on their relative activities toward cellobiose and P(O)NPG:

(1) *Aryl-β-glucosidases*: these enzymes show higher relative activities toward P(O)NPG than cellobiose, or even show no activity toward this last substrate

(2) *Cellobiases*: enzymes that show higher activity toward cellobiose.

Penicillium herquei possesses two isoforms of β-glucosidase called G1 and G2;[28] these should be considered aryl-β-glucosidases, since their relative activities toward cellobiose are 91.3 and 40.4% as compared with PNPG. Similarly, the relative activity of the enzyme from *P. oxalicum* toward cellobiose is about 81%.[45] The *β*-glucosidase 3 from *Aspergillus aculeatus* shows only a relative activity toward cellobiose of 6.7%, an extreme case among fungal *β*-glucosidases with cellobiase activity.[46] Interestingly, the other two isoenzymes from this fungus (*β*-glucosidases 1 and 2) show relative activities toward cellobiose of 400 to 500%, so they should be considered as cellobiases.[46] However, these enzymes are also very active toward cellooligosaccharides with a degree of polymerization (DP) of 3 to 6, and their relative activities toward these substrates is higher than for cellobiose. These enzymes even show activity toward an insoluble cellooligosaccharide of DP=20 which equals that toward PNPG; for this reason they are also called cellooligoglucosidases.[46] On the other hand, *β*-glucosidase A purified from *A. niger*[47] and a *β*-glucosidase from *Stereum sanguinolentum*[48] are good examples of the cellobiase class.

With respect to substrate affinities (K_M), *β*-glucosidases also show great differences. Again, using their affinities toward P(O)NPG and cellobiose as references, they can be classified operationally into three groups:

(1) *β*-Glucosidases with affinities (K_M) similar for both substrates
(2) *β*-Glucosidases which show higher affinity (lower K_M) for cellobiose
(3) *β*-Glucosidases with higher affinities for P(O)NPG

Table 23.1 shows the K_M values for a set of fungal *β*-glucosidases for P(O)NPG and cellobiose. Values for P(O)NPG range from 0.055 mM (*Stachibotrys atra*[116]) to 34 mM (*β*-glucosidase II from *Phytophtora infestans*[107]) and for cellobiose from 0.031 (*Neocallimastix frontalis*[20]) to 340 mM (*β*-glucosidase II from *P. infestans*).[107] The majority of the enzymes listed can be classified in the first group. Few enzymes show significantly higher affinities for cellobiase, such as the enzyme from *N. frontalis*[97] and *Piromyces* sp.[107] More common is the finding of lower K_M values for P(O)NPG; examples are the *β*-glucosidases from *A. oryzae,*[68] *Monila* sp.[92] and *Talaromyces emersonii β*-glucosidase III.[25,116]

Besides hydrolyzing cellooligosaccharides of different lengths, *β*-glucosidases most often show activity toward both natural (plant) or synthetic aryl-glucosides, with a variety of aglycons. The relative activities toward these glycosides vary from enzyme to enzyme. A *β*-glucosidase purified from *A. niger* shows relative activities (cellobiose = 100%) toward the disaccharides laminaribiose (*β*1-3), gentiobiose (*β*1-6), sophorose (*β*1-2), and salicin (salicyl-glucose) of 92, 63, 48, 22, and 7%, respectively.[56] The G1 enzyme from *P. herquei* shows relative activities (PNPG =100%) of 82.7 and 70.3% toward gentiobiose and salicin, while those of the G2 isoenzyme are 8.7 and 54.5%, respectively.[28] This evidence indicates that differences are not only found between enzymes from different species but also between isoenzymes from the same microorganism.

Although in general *β*-glucosidases show a stronger activity toward oligosaccharides with (*β*1→4) linkages, several examples are found in the literature, where a higher activity is described toward glucans with $\beta(1\rightarrow2)$ and $\beta(1\rightarrow3)$ linkages. Enzymes with such properties are found in *A. fumigatus*[32] and *T. koningii,*[13] which are more active toward laminaribiose and sophorose than cellobiose. Such enzymes

TABLE 23.1

Physical Properties, Carbohydrate Content and K_M Values for Fungal β-Glucosidases

Organism	M_W ($\times 10^3$)		pI	Carbohydrate	K_M (mM)		Ref.
	Native	SDS–PAGE		Content (%)	PNPG (ONPG)	Cello-biose	
Acremonium persicinum	140	128	4.3	1.5	0.3	0.91	49
Aspergillus aculeatus							35
β-Glucosidase 1		133	4.7	23			
β-Glucosidase 2		132	4.3	22			
β-Glucosidase 3		136	3.6	15			
Aspergillus fumigatus							
β-Glucosidase$_{small}$	40,8		6.3	Glycoprotein			32
β-Glucosidase$_{large}$	340	170		Glycoprotein	0.88	0.84	33
Aspergillus japonicus	>240				(1.17)	5.7	50
Aspergillus kawashii							51
EX1		145		Glycoprotein			
EX2		130		Glycoprotein			
CB-1		120		Glycoprotein			
Aspergillus nidulans							
P-I	125	125	4.4		0.842	0.997	30
P-II	50	50	4.0		0.465	0.796	
β-Glucosidase I	240				0.53 (0.55)	1.17	52
β-Glucosidase II	78				0.22 (0.32)	0.89	
Aspergillus niger							
form I	117.4	121.9	4.2	17	0.47		29
form II	117.4	121.9	4.2	7.7	0.36		
β-Glucosidase A		118			0.43	0.50	47
β-Glucosidase B		109			0.11	0.27	
	325				0.82	1.33	53
β-Glucosidase I	96		4.6	17.5	0.616	0.446	37
β-Glucosidase II	96		3.8	17		0.201	
	230				0.75 (2.18)	1.23	54
					0.67	2.04	55
	137		3.8	12.5		0.85	42,56
Cellobiase A		88		8.8		0.90	57
Cellobiase B		80		9.4		1.63	
Cellobiase C		71		7.2		1.0	
	330	110			1.11		58
Novozym 188			5.0		1.03	5.63	59
					0.5		60
					0.63		61
β-Glucosidase I	105	49	3.2		21.7		62
β-Glucosidase II	360	120	4.0		2.2	15.4	63
	240	120	4.0		0.9	2.3	146

(*Continued*)

TABLE 23.1 (Continued)

Organism	M_W (×10³) Native	SDS–PAGE	pI	Carbohydrate Content (%)	K_M (mM) PNPG (ONPG)	Cello-biose	Ref.
Aspergillus ornatus					0.76		64
Aspergillus oryzae	40	43	4.2		0.55	7	65
Aspergillus phoenicis						0.75	66
Aspergillus roseus							
β-Glucosidase A	110	96		Glycoprotein	1.33	4.7	67
β-Glucosidase B	68	78		Glycoprotein	0.5	2	
Aspergillus sojae	250	118	3.8	23.8	0.14		68
Aspergillus terreus				65	0.78	0.40	69
Aspergillus tubingensis							70
I		131	4.2		0.76		
II		126	3.9		0.35		
III		54	3.7		3.2		
IV		54	3.6		6.2		
Aspergillus wentii							
A3	170		3.8				71
B2	145						72
					1.9	2.7	73
Botryodiplodia theobromae							
	331.6				0.33	1.0	74
Botrytis cinerea	350	88		Glycoprotein	0.06 (0.16)		75
	380	170					76,77
Chaetomium thermophilum var. *coprophilum*							
(deglycosylated)	40	43	7.7	73	0.76	3.13	78
Chaeomium trilaterale	240	118	4.86		5.1	1.9	79
Chalara paradoxa	170	167			0.52	0.58	80
Cladosporium resinae	98	98	4.2		0.07	2.3	81
Curvularia sp.					0.2		82
Evernia prunastri	311	60, 70	3.12		0.635	0.244	83
Fungal strain Y94	240	120	5.02			2.3	84
Fusarium oxysporum	110	110	3.8		0.093	1.07	41
Fusarium solani	400		3.15				85
Humicola grisea	156	82			0.316		86
	55	55		35	0.12	3.24	87
Humicola insolens		250	4.23		0.53	1.52	88
	45	45	8.1	10			89
Humicola lanuginosa	135	110		9	0.5	0.44	90
Lenzites trabea	320				0.41	1.4	91

(Continued)

TABLE 23.1 (Continued)

Organism	M_W ($\times 10^3$)		pI	Carbohydrate	K_M (mM)		Ref.
	Native	SDS–PAGE		Content (%)	PNPG (ONPG)	Cello-biose	
Macrophomina phaseolina							
β-glucosidase 1	323.6				(1.25)	1.6	34
β-glucosidase 2	213				(0.125)	0.9	
Monilia sp.	46.5	46.7	8.87	Glycoprotein	0.075	5.7	92
Mucor miehei							
G-a	250		4.3	23.4	0.58	1.04	93
G-b	220		4.3	13.0	0.58	1.04	
Myceliophtora thermophila					4.5		94
Nectria catalinensis					0.25		95
Neocallimastix frontalis		125	7.1		2.5		96
		85	6.95		0.67	0.053	97
	153	165	3.7			0.031	20
Orpinomyces sp.	87	85.4	3.95	8.5	0.35	0.25	98
Penicillium funiculosum					0.4	2.1	99
β-Glucosidase 1	230	120	4.55	10.7	0.23	0.2	100
β-Glucosidase 2	18				1.08		
Penicillium herquei							
G1		125	5.02	12.7	0.83	0.6	28
G2		122	5.24	16.1	0.63	1.3	
Penicillium oxalicum	133.5	110	4.0	Glycoprotein	0.37	10	45
Penicillium purpurogenum							
	90	90	4.2;6.0		0.085	0.8	101
	185	185		8	0.57	1.56	102
Phanerochaete chrysosporium							
Intracellular	410				0.11		103
Extracellular	90				0.16	0.53	
		46		4.64	5.3		104
		114			0.096	2.3	105
Phytophtora infectans							106
β-Glucosidase I	160–230			3.3	1.1	280	
β-Glucosidase II	32			4.7	34	340	
Piromyces sp.	46	44	4.15		1.53	0.05	107
	75.8				1.0		108
Pisolithus tinctorius	450	150	3.8		0.87		109

(*Continued*)

TABLE 23.1 (Continued)

Organism	M_W (×10³) Native	SDS–PAGE	pI	Carbohydrate Content (%)	K_M (mM) PNPG (ONPG)	Cello-biose	Ref.
Pleurotus ostreatus							110
F1		35	7.5				
F2		50	7.3				
F3		66	8.5				
Pyricularia oryzae	240	240	4.15		(1.43)	0.91	111
Schizophilum commune							
β-Glucosidase I		110	3.3	Glycoprotein	0.37	2.1	31
β-Glucosidase II		94	3.3	Glycoprotein	0.49	2.1	
Sclerotium rolfsii							40,112
β-Glucosidase 1	90	95.5	4.1		1.07	3.65	
β-Glucosidase 2	90	95.5	4.55		1.38	3.07	
β-Glucosidase 3	107	106	5.10		0.89	5.84	
β-Glucosidase 4	92	95.5	5.55		0.79	4.15	
Scytalidium lignicola							113
I		74			0.2	2.85	
II		69			0.13	5	
Sporotrichum pulverulentum							5
A1		165	4.8				
A2		172	4.52		0.15	4.5	
B1		165	5.15				
B2		172	4.87		0.21	3.7	
B3		182	4.56				
Sporotrichum thermophile							
β-Glucosidase A	440				(0.5)	Inactive	26
β-Glucosidase B	40				(0.18)	0.28	
	240	110			0.29	0.83	114
Stachibotrys atra							
Constitutive	500				0.070		115
Inducible	69		4.8	14.4	0.055		
Talaromyces emersonii							
β-Glucosidase I	138	138	3.4–4.1	51	0.14	0.58	25,116
β-Glucosidase III	40.7	40.7	3.60	26	1.03	23.7	
β-Glucosidase IV	52.5	52.5	4.4–4.5	12	0.81	1.47	
Termitomyces clypeatus	450	110	4.5	24	0.5	1.25	117
Thermoascus aurantiacus	85	89		33	0.52		118
	325	98		15.1			119
Gl-2		175			1.17		120
Gl-3		157			1.38		
	350	120			0.1137 (0.25)	0.637	121

(Continued)

TABLE 23.1   (Continued)

Organism	M_W (×10³)		pI	Carbohydrate	K_M (mM)		Ref.
	Native	SDS–PAGE		Content (%)	PNPG (ONPG)	Cello-biose	
Thermomonospora curvata							
Intracellular	66				5.6	30.3	122
Extracellular	66				1	0.7	
Thermomyces lanuginosus	200	105			0.075		123
Trametes gibbosa	640				1.48		124
Trichoderma sp.	178	109	4.8	75	0.096	0.8	125
Trichoderma harzianum	76	75	8.7		0.20 (0.80)		126
Trichoderma koningii							
β-Glucosidase 1	39.8		5.53	No carbohydrate	0.37	1.18	13
β-Glucosidase 2	39.8		5.85	2	0.85	0.86	
Trichoderma reesei	74.6		8.5	0.7	0.1	1.25	127
		78.5	8.3–8.4				39
		70	8.4	35	0.3	0.5	128
β-Glucosidase 1		71	8.7	0.1	0.18	2.1	14
β-Glucosidase 2		114	4.8	9	0.135	11.1	
	98		4.4			3.3	129
					3.5	1.9	130
Trichoderma viride	47	47	5.74	No carbohydrate	0.28	1.5	131
						5.6	132
					0.5	2.5	133
Xylaria regalis		85			1.72		134

may not be involved in cellulose saccharification and may have other biological functions. One can envision their participation, as part of the glucanase complex, in such processes as the utilization of endogenous carbohydrate reserves during conidiogenesis and conidial germination as well as in morphogenesis.[141]

Despite the large variability in relative activities toward cellooligosaccharides, β-1,2/1,3 β-glucans and aryl-glucosides, these enzymes are specific for the β-anomeric configuration. An exception may be the β-glucosidase from *Thermomyces lanuginosus,* which shows significant α-glucosidase activity.[123]

IV. EFFECT OF TEMPERATURE AND pH ON ACTIVITY AND STABILITY

The majority of the fungal β-glucosidases show optimum pH over the range 4.0 to 5.5 (Table 23.2). However, enzymes with optimum pH from as low as 2.5 up to 8.0 have been found; this allows a choice of enzymes for a variety of applications.

TABLE 23.2

Optimum pH and Temperature of Fungal β- Glucosidases

Organism	Optimum pH	Optimal Temperature (°C)	Ref.
Acremonium persicinum	5.5		49
Aspergillus aculeatus			46
β-Glucosidase 1	4–4.5	55	
β-Glucosidase 2	4–4.5	60	
β-Glucosidase 3	3	65	
Aspergillus fumigatus			
β-Glucosidase$_{small}$	5.0		32
β-Gglucosidase$_{large}$	4.0–5.0		33
Aspergillus japonicus	5.0	65	50
Aspergillus kawashii			51
EX1, EX2, EX3	4.5–5	60	
Aspergillus nidulans			
P-1	5.0	50	30
P-2	5.5	60	
β-Glucosidase I	5.0	55–60	52
β-Glucosidase II	5.5	55–60	
Aspergillus niger	4.6	60–65	53
β-Glucosidase A	4	60	47
β-Glucosidase B	3	70	
β-Glucosidase I	5.1		37
β-Glucosidase II	4.1		
	4.6	65	54
	5.0–5.5	60	55
Cellobiase A	4.5	55	57
Cellobiase B	4.5	55	
Cellobiase C	4.5	60	
	4.6–5.3	70	58
Novozym 188	4.5	60–70	59
	4		60
	4.5	60	61
β-Glucosidase I	5.0	55	62
β-Glucosidase II	4.5	60	63
	4.5	55	146
Aspergillus ornatus	4.6	60	64
Aspergillus oryzae	5.0	50	65
Aspergillus phoenicis	4.3		66
Aspergillus roseus			
β-Glucosidase A	4–4.5		67
β-Glucosidase B	4.8–5.2		

(Continued)

TABLE 23.2   (Continued)

Organism	Optimum pH	Optimal Temperature (°C)	Ref.
Aspergillus sojae	5.0	60	
Aspergillus terreus	4.8		69
Aspergillus tubingensis			70
I	4.6	65	
II	4.0	65	
III	5.0	60	
IV	5.0	60	
Aspergillus wentii	3.5–5.5	55–65	73
Aureobasidium	4	80	136
Botryodiplodia theobromae	5.0		74
Botrytis cinerea	6.5	50	75
	3.0	60	76,77
Chaetomium thermophilum var. coprophilum	5.5	60	78
Chaeomium trilaterale	4.2		79
Chalara paradoxa	4.0–5.0	45	80
Cladosporium resinae	4.5	50	81
Curvularia sp.	4	70	82
Evernia prunastri	4	50–60	83
Fungal strain Y94	4.8	60	84
Fusarium oxysporum	5-6	60	41
Humicola grisea	4–4.5	60	86
	6.0	50–60	87
Humicola insolens	5.0	50	88
			89
Humicola lanuginosa	4.5	60	90
Lenzites trabea	4.5	75	91
Macrophomina phaseolina			
β-Glucosidase 1	5.5	55	34
β-Glucosidase 2	4.8–6	65	
Monilia sp.	4–5	50	92
Mucor miehei			
G-a	8.0	60	93
G-b			
Myceliophtora thermophila	4.8	60	94
Nectria catalinensis	5.3	45	95
Neocallimastix frontalis	5.5–7.2	45	96
	6	50	20

(*Continued*)

TABLE 23.2 (Continued)

Organism	Optimum pH	Optimal Temperature (°C)	Ref.
Orpinomyces sp.	6.2	50	98
Penicillium funiculosum			
β-Glucosidase 1	4	70	100
Penicillium herquei			
G1	4–4.5	60	28
G2	4–4.5	60	
Penicillium oxalicum	5.5	55	45
Penicillium purpurogenum	3.5	60	101
	4.5	60–65	102
Phanerochaete chrysosporium			
Intracellular	7	45	103
Extracellular	5.5	45	
	5	60	104
	4–5.2		105
Phytophtora infestans			106
β-Glucosidase I	5.5	48	
β-Glucosidase II	5.25	30	
Piromyces sp.	6	55	107
	6	47	108
Pisolithus tinctorius	4	65	109
Pyricularia oryzae	5.5	55	111
Schizophilum commune			
β-Glucosidase 1	5.8		31
β-Glucosidase 2	5.1		
	5.4	52	135
Sclerotium rolfsii			112
β-Glucosidase 1	4.2–4.5	65–68	
β-Glucosidase 2	4.2–4.5	65–68	
β-Glucosidase 3	4.2–4.5	65–68	
β-Glucosidase 4	4.2–4.5	65–68	
Scytalidium lignicola			113
I	5.0	55	
II	4.8	55	
Sporotrichium thermophile			26
β-Glucosidase A	5.6	50	
β-Glucosidase B	6.2–6.3	50	
	5.2–5.4	65	114
Stachibotrys atra			115
Constitutive	4.7–5		
Inducible	6.7		

(Continued)

TABLE 23.2   (Continued)

Organism	Optimum pH	Optimal Temperature (°C)	Ref.
Talaromyces emersonii			
β-Glucosidase 1	4.1	70	116
β-Glucosidase 3	2,5	70	
β-Glucosidase 4	5.7	35	
Termitomyces clypeatus	5.0	65	117
Thermoascus aurantiacus	4.5–5	70	118
	4.2	71	119
Gl-2	4.5	75–80	120
Gl-3	4	75–80	
	4.5	80	121
Thermomyces lanuginosus	6	65	123
Trametes gibbosa	3.5	40	124
Trichoderma sp.	4.0	75	125
Trichoderma harzianum	5.0	45	126
Trichoderma reesei	4.5–5		127
	4.5	70	128
β-Glucosidase 1	4.6	65–70	14
β-Glucosidase 2	4	60	
	6.5		129
Trichoderma viride	4.0–4.5	65	133
Xylaria regalis	5.0	50	134

The optimum temperature ranges from 35 to 80°C. As can be seen from Table 23.2, most of the enzymes have an optimum at or above 55°C. The exception corresponds to the intracellular enzyme from *T. emersonii*[116] with optimal activity at 35°C.

The extracellular fungal β-glucosidases from mesophilic fungi are thermostable up to 60°C, while those of thermophilic fungi are stable to even higher temperatures.[10] A purified β-glucosidase from the mesophile *T. reesei* QM 9414[128] shows a high stability at temperatures of 50 to 55°C: after incubation for 7 h at pH 4.5 and 50°C no significant activity loss (less than 3%) was detected. A ($t_{1/2}$) of 5h was measured for this enzyme at 60°C. Similarly, the β-glucosidase purified from *P. oxalicum* has half-lives of 225 and 45 min at 50 and 60°C, respectively.[45] However, a very different situation is encountered with the intracellular enzyme from *T. reesei* QM 9123: 1 h at 40°C produces an inactivation of 95%, and total inactivation is observed when incubated at 60°C for the same period of time.[129] On the other hand, the thermophilic fungus *Thermoascus aurantiacus* possesses a β-glucosidase that is totally stable for 1 h at 70°C and retains 70% of its activity after 1h at 75°.[118] Curiously, the thermophilic fungus *Humicola lanuginosa* possesses a highly thermolabile β-glucosidase, much more labile than many enzymes from mesophilic origin.[90] This enzyme is stable at

30°C for 1 h, but at 60 and 70°C the enzyme is totally inactivated after incubation for a similar time period.

An important condition to achieve maximal stability is to keep the enzymes in a medium of appropriate pH. This statement can be illustrated with the enzymes from the fungi *A. japonicus* and *Macrophomina phaseolina*.[50,34] The former produces a *β*-glucosidase that shows no appreciable loss of activity after 40 min at 55°C, either at pH 5.8 or 7.0. This situation, however, changes drastically at 65°C. When the enzyme is incubated for 40 min at 65°C and pH 5.0 there is no appreciable loss of activity, while at the same temperature but at pH 7.0, a loss of activity of nearly 80% is observed. A similar phenomenon occurs with *β*-glucosidase I from *M. phaseolina* at 60°C; it is stable at pH 5.0 but loses 70% of its activity after 20 min at pH 7.0.

Cellulases, including *β*-glucosidases, are mostly glycoproteins. It has been shown that the carbohydrate moiety may significantly affect the thermostability of these enzymes. In the *β*-glucosidases of *Mucor miehei*[93] and *β*-glucosidases I, III, and IV from *T. emersonii*,[116] the enzymes with a higher carbohydrate content are more stable. The half-lives of the latter enzymes at 70°C and pH 5.0 are 410, 175, and 2 min, respectively. On the other hand, it has been observed in *A. aculeatus* F 50 and *P. herquei*[64,28] that the isoenzyme with lower carbohydrate content presents higher stability. For this reason, the role of the carbohydrates in thermostability is not conclusive. A comparison of the enzyme stability between the native, glycosylated enzyme, and a nonglycosylated form produced by genetic engineering or glycosidase treatment may help clarify this issue.

V. EFFECT OF INHIBITORS

Some cellulolytic fungi produce oxidoreductases, such as glucose oxidase, which are capable of oxidizing glucose, cellobiose, and other cellodextrins to the corresponding lactones.[137,138] Glucono-*δ*-lactone is produced in large amounts by grape-attacking fungi and can reach concentrations up to 2 g/L in wine.[65] This compound (Figure 23.3) is a potent competitive inhibitor of many *β*-glucosidases; values of K_i ranging from 0.0083 *μM* to 12.5 m*M* have been reported (see Table 23.3). However, some *β*-glucosidases show mixed inhibition by gluconolactone.[51,53,54,64,101] Inhibition by this compound has been explained by steric similarities between the lactone and the enzyme-bound substrate.[51,52] It may be important in the regulation of cellulose hydrolysis *in vivo*; therefore, this topic merits further investigation.

Some *β*-glucosidases are inhibited by high substrate concentration.[54,55] Besides, glucose, the product of the enzymic reaction, is a competitive inhibitor of many fungal *β*-glucosidases; K_i values for glucose between 17 *μM* and 1360 m*M* have been reported (Table 23.3). However, some *β*-glucosidases are noncompetitively inhibited,[57] thus, like gluconolactone, glucose can inhibit by different mechanisms. The degree of glucose inhibition may be an important factor in choosing a *β*-glucosidase for the industrial saccharification of cellulose.

When P(O)NPG is used as substrate, cellobiose may act as competitive inhibitor (see Reference 117 and unpublished observations by the authors). This result indicates that at least some *β*-glucosidases use the same active site for these two substrates.

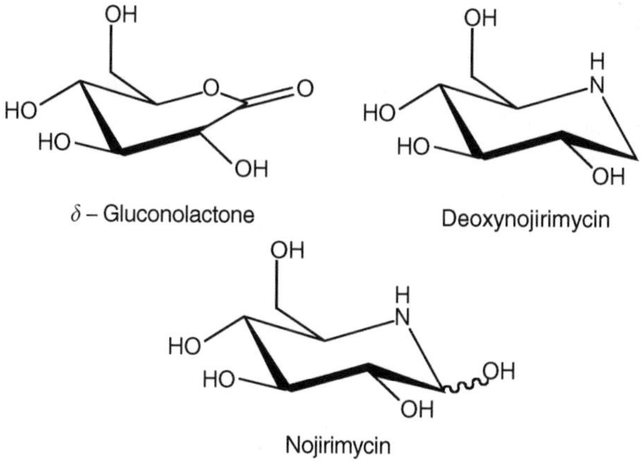

FIGURE 23.3   Inhibitors of β-glucosidase.

TABLE 23.3
Inhibition of β-Glucosidase

Organism	Substrate	Inhibitor	Inhibition type	$K_i(K_i')$ (μM)	Ref.
Acremonium persicinum	PNPG	Glucose	Competitive	350	49
	PNPG	Glulact	Competitive	5	
Aspergillus japonicus	ONPG	Glulact	Competitive	40	50
	ONPG	Nojirim	Competitive	1.8	
Aspergillus nidulans					
P-1	PNPG	Glulact	Not specified	3,480	30
	PNPG	Glucose	Not specified	9,170	
P-2	PNPG	Glulact	Not specified	850	
	PNPG	Glucose	Not specified	5,480	
β-Glucosidase I	Cellobiose	Glucose	Competitive	630	51
	Cellobiose	Glulact	Mixed	100 (950)	
β-Glucosidase II	PNPG	Glucose	Competitive	2,340	
	PNPG	Glulact	Mixed	110 (4860)	
Aspergillus niger					
β-Glucosidase A	PNPG	Glucose	Competitive	2,500	47
	PNPG	Gluclact	Competitive	150	
	PNPG	Denojir	Competitive	2	
β-Glucosidase B	PNPG	Glucose	Competitive	400	
	PNPG	Gluclact	Competitive	5	
	PNPG	Denojir	Competitive	1	
	PNPG	Glucose	Competitive	2,890	53
	PNPG	Gluclact	Mixed	700 (14,200)	
	PNPG	Glucose	Competitive	3,220	54
	PNPG	Gluclact	Mixed	240	

(Continued)

TABLE 23.3 (Continued)

Organism	Substrate	Inhibitor	Inhibition type	$K_i(K_i')$ (μM)	Ref.
Novozym 188	PNPG	Glucose	Part competitive	3,000	59
β-Glucosidase I	PNPG	Glucose	Competitive	543,000	62
β-Glucosidase II	PNPG	Glucose	Competitive	5,700	63
Aspergillus ornatus	PNPG	Glucose	Competitive	5,130	64
	PNPG	Gluclact	Mixed	240 (13 700)	
Aspergillus oryzae					
	PNPG	Glucose	Competitive	$1,360 \times 10^3$	65
	PNPG	Gluclact	Competitive	12,500	
BGI	PNPG	Glucose	Competitive	3,500	139
BGII	PNPG	Glucose	Competitive	953,000	
Aspergillus phoenicis	Cellobiose	Nojirim	Not specified	0.64	66
Aspergillus terreus	PNPG	Glucose	Competitive	3,500	69
Aspergillus tubingensis					70
I	PNPG	Glucose	Not specified	5,800	
II	PNPG	Glucose	Not specified	1,300	
III	PNPG	Glucose	Not specified	470,000	
IV	PNPG	Glucose	Not specified	600,000	
Aspergillus wentii	Cellobiose	Glucose	Competitive	14,000	73
Botrytis cinerea	PNPG	Glucose	Competitive	5,500	75
	PNPG	Glulact	Competitive	20	
Chaetomium trilaterale	PNPG	Glulact	Competitive	534	140
	PNPG	Nojirim	Competitive	19.5	
Chalara paradoxa	PNPG	Glucose	Competitive	11,020	80
Cladosporium resinae	PNPG	Glucose	Competitive	22,000	81
Evernia prunasti	PNPG	Glucose	Competitive	1,260	83
Fungal strain Y 94	cellobiose	Glucose	Competitive	1,350	84
Lenzites trabea	PNPG	Glucose	Noncompetitive	2,700	91
Macrophomina phaseolina					
βG1	ONPG	Glulact	Competitve	17	34
	ONPG	Nojirim	Competitve	0.5	
βG2	ONPG	Glulact	Competitive	30	
	ONPG	Nojirim	Competitve	1.6	
Monilia sp.	PNPG	Glucose	Competitive	670	92
Neocallimastix frontalis	Mumbgl	Glulact	Not specified	5.57	2
	Mumbgl	Glucose	Not specified	3,600	
	PNPG	Glucose	Competitive	10,500	96
	PNPG	Glulact	Competitive	240	
	PNPG	Glucose	Competitive	4,800	97
	PNPG	Glulact	Competitive	62	

(Continued)

TABLE 23.3 (Continued)

Organism	Substrate	Inhibitor	Inhibition type	$K_i(K_i')$ (μM)	Ref.
Orpinomyces sp.	PNPG	Glucose	Competitive	8 750	98
	PNPG	Glulact	Competitive	16.8	
	Cellobiose	Glulact	Competitive	2,570	
Penicillium funiculosum					
βG1	PNPG	Glucose	Competitive	920	100
	PNPG	Glulact	Competitive	67	
	Cellobiose	Glulact	Not specified	32	
βG2	PNPG	Glulact	Not specified	70	
	PNPG	Glucose	Competitive	1,700	99
	PNPG	Glulact	Competitive	8	
	Cellobiose	Glucose	Competitive	1,000	
Penicillium oxalicum	PNPG	Glulact	Not indicated	300	45
	PNPG	Glucose	Noninhibited up to 10 mM		
Penicillium purpurogenum	PNPG	Glucose	Competitive	1400	101
	PNPG	Glulact	Mixed	67	
Phanerochaete chrysosporium	PNPG	Glucose	Competitve	500	103
	PNPG	Glulact	Competitive	35	104
	PNPG	Denojir	Competitive	68.7	
	PNPG	Glucose	Competitive	270	105
	PNPG	Glulact	Competitive	4	
Piromyces sp.	PNPG	Glulact	Competitive	30	107
	PNPG	Glulact	Competitive	22	108
Pyricularia orizae	Salicin	Glulact	Competitive	7.4	111
Schizophyllum commune					
βG I	Cellobiose	Glucose	Competitive	1,000	31
	PNPG	Glucose	Competitive	3,700	
	Cellobiose	Glulact	Competitive	4	
βG II	Cellobiose	Glucose	Competitive	9,100	
	PNPG	Glucose	Competitive	7,700	
	Cellobiose	Glulact	Competitive	4.5	
	PNPG	Glucose	Competitive	17	135
Sclerotium rolfsii βG3	Cellobiose	Glucose	Competitive	550	112
	Cellobiose	Glulact	Competitive	10	
	Cellobiose	nojirim	Competitive	1	
Scytalidium lignicola					113
I	PNPG	Glucose	Noncompetitive	5,000	
II	PNPG	Glucose	Noncompetitive	7,500	
Sporotrichum pulverulentum					5
A	PNPG	Glulact	Competitive	0.35	
B	PNPG	Glulact	Competitive	1.5	
Stachybotrys atra	PNPG	Glulac	50% at 45 μM		115
	PNPG	Glucose	50% at 13 μM		

(Continued)

TABLE 23.3 (Continued)

Organism	Substrate	Inhibitor	Inhibition type	$K_i(K_i')$ (μM)	Ref.
Talaromyces emersonii					
β-glucosidase 1	PNPG	Glucose	Competitive	710	116
β-glucosidase 4	PNPG	Glucose	Competitive	52,000	
Termitomyces clypeatus	PNPG	Glucose	Competitive	1700	117
Thermoascus aurantiacus	PNPG	Glucose	Competitive	290	121
	PNPG	Glulact	Competitive	0.0083	
Tramites gibbosa	PNPG	Glucose	Competitive	5,200	124
Trichoderma sp.	PNPG	Glucose	Competitive	650	125
	PNPG	Glulact	Competitive	90	
Trichoderma harzianum	PNPG	Glucose	Competitive	1,400	126
	PNPG	Glulact	Competitive	1.8	
Trichoderma koningii					
*β*G1	ONPG	Glulact	Competitive	1.8	13
	ONPG	Glucose	Competitive	1050	
*β*G2	ONPG	Glulact	Competitive	1,17	
	ONPG	Glucose	Competitive	660	
Trichoderma reesei	Mumblg	Glucose	Competitive	700	127
*β*G1	PNPG	Glucose	Competitive	624	14
	PNPG	Glulact	Competitive	3,13	
	PNPG	Denojir	Competitive	1.26	
	Cellobiose	Glulact	Competitive	99.5	
	Cellobiose	Denojir	Competitive	3.93	
*β*G2	PNPG	Glucose	Competitive	189	
	PNPG	Glulact	Competitive	4.1	
	PNPG	Denojir	Competitive	1.19	
	Cellobiose	Glulact	Competitive	40	
	Cellobiose	Denojir	Competitive	1.72	
	Cellobiose	Glucose	Competitive	500	130
	PNPG	Glucose	Competitive	8,700	
Trichoderma viride	Cellobiose	Glucose	Noncompetitive	22,600 (28,300)	132
	PNPG	Glucose	Competitive	530	133
	Cellobiose	Glucose	Competitive	390	

Abbreviations: mumblg=methylumbelliferyl glucoside; glulact=gluconolactone; nojirim=nojirimycin; denojir=deoxynojirimycin

Other powerful inhibitors of *β*-glucosidase are nojirimycin and deoxynojir-imycin[44] (Figure 23.3), antibiotics produced by strains of *Streptomyces*. These compounds, as expected from their structure similarity to glucose, are competitive inhibitors and have K_i values in the range from 0.64 to 68.7 μM (Table 23.3).

Other *β*-glucosidase inhibitors include heavy metals such as Hg^{2+}, Cu^{2+}, Pb^{2+} and Co^{2+}, and *p*-chloromercuribenzoate. The following examples can be cited: 0.63 mM

CuSO$_4$ and 0.37 mM HgCl$_2$ inhibit *N. frontalis* β-glucosidase by 90%,[96] while 8 mM Pb(NO$_3$)$_2$ and 70 μM CuCl$_2$ inhibit β-glucosidase B from *Sporotrichum thermophile* by 50%.[26] Ethanol has been reported to act as inhibitor or activator, depending on the concentrations used: 30% ethanol activated while 80% inhibited the β-glucosidase from *T. aurantiacus*,[121] 10% ethanol activated by 30% while 40% inhibited by similar amount the β-glucosidase I from *A. niger*.[62] This dual effect may reflect different effects of ethanol on enzyme conformation. Some β-glucosidases may accept alcohols as acceptors in transglycosylation reactions.[80] Ethanol inhibition is of particular concern in certain industrial applications such as simultaneous cellulose saccharification and fermentation, because β-glucosidases may be exposed to substantial concentrations of this alcohol.

VI. ENZYME PURIFICATION

β-Glucosidases have been found in fungal culture supernatants, bound to the cell wall and cell membrane[51,142] and in the cytoplasm.[129] Most commonly, these enzymes are secreted to the medium. When purified from the culture supernatant, they are first concentrated by ultrafiltration or ammonium sulfate fractionation.[117] Less frequently used fractionation methods utilize ethanol precipitation[50,61] or bentonite adsorbtion.[61]

To obtain enzymes in their native form, not modified during the purification process, the number of purification steps should be kept to a minimum to avoid proteolytic cleavage among other unwanted modifications. To facilitate the purification, it is advantageous to choose culture conditions where the amount of impurities is minimized.

Purification procedures are under continuous development and several steps are usually necessary, based on different separation principles such as gel filtration, ion-exchange and hydrophobic interaction chromatographies. In addition, some investigators have employed adsorbtion and desorbtion from hydroxylapatite.[143] Affinity chromatography has also yielded good results. Rogalski and co-workers[144] have utilized controlled porosity glass activated by aminopropyltriethoxysilane and oxiranes and linked to salicin or cellobiose. Kmínkova and Kucera[145] have utilized concanavalin A coupled to Sepharose to purify a cellobiase from *T. viride*, taking advantage of its glycoprotein nature. Watanabe and colleagues[146] have used cellobiamine linked to epoxy-activated Eupergit C-30N. Others have successfully utilized isoelectrofocusing[5,101,128] and chromatofocusing.[37,146] These last two methods have allowed the isolation of β-glucosidase isoenzymes with different pI values, which are not easily separated by more conventional techniques.

Another strategy employed in the purification of β-glucosidases is the use of polyacrylamide gel electrophoresis under nondenaturing conditions, obtaining the enzyme by elution from the gel.[30,39]

The degree of purification attained is most commonly determined by SDS polyacrylamide gel electrophoresis. This technique is also used in molecular weight determination (see below). Isoelectrofocusing in polyacrylamide has also been used for purity determination, giving at the same time the pI values (see below).

VII.   MOLECULAR MASS AND ISOELECTRIC POINT

Native fungal β-glucosidases show molecular weight (M_w) in the broad range from 40,800 to 640,000 (Table 23.1). SDS gel electrophoresis gives single polypeptide chains from 35,000 through 250,000. Quaternary structures from monomers up to tetramers have been found (Table 23.1).

The molecular weight of the native enzymes has been mostly determined by gel filtration, although on occasions other procedures have been used. The molecular weight of the enzyme from A. *fumigatus* (48,800) was measured by low speed sedimentation equilibrium.[32] Nondenaturing polyacrylamide gel electrophoresis at different concentrations, with proteins of known molecular weight as standards, has also been employed.[147] Using this method, a M_w of 325,000 has been measured for a β-glucosidase from A. *niger*.[53] Caution must be exercised in the interpretation of molecular weight data of glycoproteins, as is the case of most β-glucosidases, since carbohydrate content may affect chromatographic or electrophoretic behavior. The presence of proteases in the culture medium may also cause artefactual results in molecular weight determinations.

Sequencing of β-glucosidases or of their genes or cDNAs allow a very precise determination of molecular mass, and comparison with the values obtained by physical means allow estimation of the degree of posttranslational modification (glycosylation).

Although the majority of the enzymes show pI values in the range of 4 to 6 (Table 23.1), some β-glucosidases are very acidic (pI of 3.1) or basic (pI of 8.87), suggesting significant differences in their amino acid composition.

VIII.   CARBOHYDRATE CONTENT

Fungal β-glucosidases are mostly glycoproteins; in some cases the carbohydrate portion of the molecule ranges up to 75% (w/w) (Table 23.1). The most commonly used method for carbohydrate quantification is the phenol–sulfuric acid method using glucose as standard.[148] Qualitatively, the glycoproteins can be visualized in polyacrylamide gels using the periodate–fuchsin stain.[32,149] Another frequently used option, although indirect, is to adsorb the protein to a concanavalin A–agarose column.[45,67,92] This method is based on the presence of α-D-mannopyranosyl and α-glucopyranosyl units in the carbohydrate portion.

The finding of carbohydrates in purified samples of β-glucosidases does not assure that these carbohydrates are covalently linked to the protein. For example, Witte and Wartenberg[37] have observed that two purified β-glucosidases from A.*niger* possess two types of carbohydrates. One type is weakly bound to the enzymes, and is lost during the purification, so that the carbohydrate content of the enzymes drops from 27 to 17%. Schmid and Wandrey[128] have found that a β-glucosidase from T. *reesei* QM 9414 is not a glycoprotein according to the gel staining method, although the quantitation of carbohydrates shows a percentage of 35%. These results suggest that the carbohydrate is not covalently bound to the enzyme. On the other hand, Messner and co-workers[150] have described the purification of a fungal cell-wall polysaccharide from T. *reesei*, which operates as an anchor glycan for the β-glucosidases from this organism. This polysaccharide is capable of reassociation with the extracellular enzyme and the enzyme released from the fungal cell wall. An interesting observation is the ability of this polysaccharide to increase the activity of the enzyme toward PNPG *in vitro*.

The importance of the carbohydrate moiety in β-glucosidase function is unknown. Some evidences point toward a relationship between degree of glycosylation and enzyme stability: a thermostable β-glucosidase from *M. miehei*, with a higher carbohydrate content yields a less thermostable form when subjected to digestion by a β-glucosaminidase preparation.[93] But, as has been pointed out before, the cases studied show contradictory results.

IX. ISOENZYME FORMS

A number of fungi have been found that produce multiple forms of β-glucosidase[5,14,25,28-37,70,116] (see Table 23.1). In many cases, the origin of this multiplicity is unclear, although some explanations have been postulated. β-Glucosidase heterogeneity in *S. pulverulentum* has been tentatively attributed to proteolysis.[5] Other possible posttranslational mechanism is glycosylation heterogeneity.[25] Protein–protein or protein–carbohydrate interactions are other possibilities that cannot be ruled out.[151] Two closely related forms of β-glucosidase (M_w 95,700 and 93,800) are secreted by *Schizophyllum commune;* they are postulated to arise from heterogeneity of the mRNA but from a single gene.[152] Alternatively, different forms of an enzyme may be the product of separate genes. Hypotheses concerning the multiplicity of β-glucosidase are put forward in numerous papers, but conclusive results are lacking.

Techniques such as isoelectrofocusing and chromatofocusing have often been successful in separating enzyme forms that could not be resolved by other methods. Witte and Wartenberg[37] have separated from *A. niger*, by means of chromatofocusing, two isoenzymes of similar M_w (96,000) but different pI values (3.8 and 4.6). Hidalgo and co-workers[101] have found, by both chromato- and isoelectrofocusing, two forms of β-glucosidase from *P. purpurogenum* with pI values of 4.2 and 6.0; again, both enzymes show the same molecular weight (90,000). Besides the pI values, many isoenzymes differ from each other in specific activity and molecular weight.

The isoenzymes may also have a different location in the cell, and this location may vary depending on the condition and age of the cultures and nutritional conditions. So, the localization of the β-glucosidase activity depends upon the carbon source as demonstrated in *S. pulverulentum*:[5] with cellobiose as the sole carbon source, only cell-wall-bound enzyme is produced, while the presence of cellulose may be necessary for secretion of the enzyme.

Several forms of β-glucosidases have been found in *T. reesei*, either in culture supernatants or bound to the cell wall and membrane, with pI values ranging from 4.4 to 8.7 (Table 23.1). A well-known fact is that *T. reesei* releases low amounts of β-glucosidase to the culture medium;[66] this has been explained by the observation that a large part of the enzyme remains bound to the cell wall during fungal growth.[141] Kubicek[153] has postulated that the membrane-bound β-glucosidase participates in the formation of sophorose, a potent inducer of cellulases. It has already been mentioned that the β-glucosidases from *T. reesei* may be involved in cell-wall metabolism during conidiogenesis, and thus may not be a true component of the cellulolytic enzyme system.[141] Inglin and colleagues[129] have isolated an intracellular β-glucosidase, which could have two possible functions: (1) in metabolic control of cellulase induction; and (2) as a proenzyme before transport through the cell membrane to the external milieu.

Analytical isoelectrofocusing of culture supernatants of *T. reesei* QM9414 shows that isoenzymes appear very early in the cultivation, suggesting that they are an inherent property of the fungus.[154] The pattern of isoenzymes may also vary with the age of the culture, as has been found in the wood-staining fungus *Botryodiplodia theobromae*, which produces a series of β-glucosidase forms in the culture filtrates. Enzymes of high M_w (350,000 to 380,000 Da) predominate in young cultures, while smaller forms (45,000 to 47,000) are more abundant in older filtrates.[155]

X. SEQUENCE AND STRUCTURE

Henrissat156 has developed a classification of glycosyl hydrolases based on amino acid sequence similarity. The enzymes are grouped in families, and as of January 2005, 99 families had been recognized. A website, which can be accessed at http://afmb.cnrs-mrs.fr/CAZY and is frequently updated, lists all the families and the enzymes assigned to them and gives direct access to the nucleotide and amino acid sequence databases.

β-Glucosidases are found in families 1, 3, and 9, and enzymes from filamentous fungi are present in families 1 and 3. Both family 1 and 3 glycosyl hydrolases have a catalytic mechanism with retention of the β-anomeric configuration,[157] and in the catalytic process participate a nucleophile/base, and a proton donor (acid–base catalyst).[157] The reaction mechanism and the role of different amino acids in catalysis and substrate binding have recently been discussed in detail by Bhatia and co-workers.[11]

Family 1 includes sequences of β-glucosidases from the filamentous fungi *A. niger, H. grisea, Orpinomyces* sp., *Piromyces* sp. (two sequences: Bgl1A and Cel1C), *T. emersonii, T. reesei* and *T. viride.* These sequences have been aligned in the bottom part of Figure 23.4. The legend to the figure indicates the GenBank accession number and references to publications where available. The conserved motif NEP (residues 430 to 433) includes the glutamate residue implicated as acid–base catalyst,[158] while the I(Y,V)(V,I)TENG motif (residues 893 to 899) contains the glutamate, which acts as nucleophile.[159] Significant similarity can also be observed in other regions of the sequence.

No three-dimensional structure of family 1 fungal β-glucosidase has been described so far. However, the structures of several bacterial and plant enzymes have been published (see References 160 and 161 and references therein). The structure found in all cases is that of a $(\alpha/\beta)_8$ barrel.

The family 3 fungal β-glucosidases include sequences from *Agaricus bisporus, A. aculeatus, A. kawachii, Botryotinia fuckelania, Coccidioides posadasii* (Bgl1 and Bgl2), *Phaeosphaeria avenaria, Phanerochaete chrysosporium* K3 and OGC101, *Piromyces* sp., *Septoria lycopersici, T. emersonii, T. reesei* (Cel3b and Cel3A/Bgl1), and *Volvariella volvacea.* These sequences have been aligned in the top part of

Figure 23.4. The alignment of this family of enzymes has been separated into two groups, which show high internal sequence similarity.

The first group includes the β-glucosidases from *A. bisporus* and *V. volvacea*, both basidiomycetes; these sequences have been found to be very similar to that of the enzyme from the slime mold *Dictyostelium discoideum*, which has also been included in this alignment for comparison purposes. No biochemical data are available for the two fungal enzymes; active-site residues for the *D. discoideum* protein have been recognized in the conserved sequence (T,S)DW (residues 411 to 413), as determined from a tryptic peptide of *A. wentii* β-glucosidase (162, 163) containing this sequence.

The remaining fungal enzymes are aligned in the central part of Figure 23.4. Several conserved motifs can be recognized. Among them are the sequences (T,S)DW and KH (F,Y,L) (residues 320 to 322), which seem to be common to all family 3 β-glucosidases; both sequences have been implicated as contributing to the active-site structure and catalysis. There is evidence supporting a role for the D in the (T,S)DW motif as the catalytic nucleophile.,[164] and it has been suggested that the proton donor may be the H in KH (F,Y,L).[11]

The only structure for a family 3 glycosyl hydrolase described so far is the barley β-glucan exohydrolase.[165] The amino-terminal segment constitutes an $(\alpha/\beta)_8$ barrel domain (such as found in family 1 enzymes), with a second domain arranged in a six-stranded β-sandwich. Two acidic amino acids (Asp^{285} and Glu^{491}) are probable catalytic residues. The first is in the (T,S)DW conserved sequence. Although the highly conserved sequence KH (F,Y,L) is also found in the barley enzyme, no specific function is assigned to the H residue; thus the nature of the proton donor in family 3 enzymes remains controversial.

FIGURE 23.4 Alignment of the sequences of β-glucosidases from filamentous fungi. The alignment has been performed using CLUSTAL W (www.ebi.ac.uk/clustalw) and further analyzed by means of AMAS (http://barton.ebi.ac.uk/servers/amas_server.html). The upper part of the alignment includes family 3 enzymes, two from the fungi *Agaricus bisporus* (AJ 293760) (AGABI) and *Volvariella volvacea* (AF 32973) (VOLVO), and a β-glucosidase from the slime mold *Dictyostelium discoideum* (L21014[166]) (DICDI). The central part of the alignment refers to family 3 enzymes from *Aspergillus aculeatus* (D64088[167]) (ASPAC), *Aspergillus kawachii* (AB003470[168]) (ASPAW), *Botryotinia fuckelania* (AJ0130890) (BOTFU), *Coccidioides posadasii* (U87805) (COCPO), *Coccidioides posadasii* (AF022893[169]) (COCPOBgl2), *Phaeosphaeria avenaria* (AJ276675[170]) (PHAAV), *Phanerochaete chrysosporium* K-3 (AB081121) (PHACHK-3), *Phanerochaete chrysosporium* OGC101 (AF036872[171]) (PHACHOGC101), *Piromyces sp.* (AY72977[164]) (PIRSP), *Septoria lycopersici* (SLU24701[172]) (SEPLY), *Talaromyces emersonii* (AY072918) (TALEM), *Trichoderma reesei* QM6a Cel3B (AY281374[173]) (TRIRECel3B), *Trichoderma reesei* QM9414 Cel3A/Bgl1 (U09580[174]) (TRIREBgl1). The lower part includes family 1 β-glucosidases from *Aspergillus niger* (AF 268911) (ASPNI), *Humicola grisea* (AB 0003109[175]) (HUMGR), *Orpinomyces* sp (AF016864) (ORPSP), *Piromyces* sp. Bgl1a (AJ276438[176]) (PIRSPBgl1A), *Piromyces* sp. Cel1C (AF500784[177]) (PIRSPCel1C), *Talaromyces emersonii* (AF439322) (TALEM_F1), *Trichoderma reesei* Cel1B (AY281377[173]) (TRIRE_F1), *Trichoderma viride* (AY 343988) (TRIVI). White letters on black background represent sequence identities and the gray boxes indicate sequence similarity. Numbers on top indicate the numbering of the amino acid sequence.

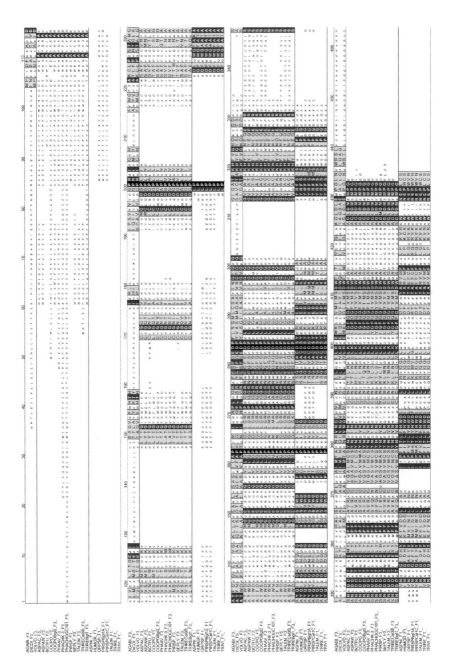

FIGURE 23.4

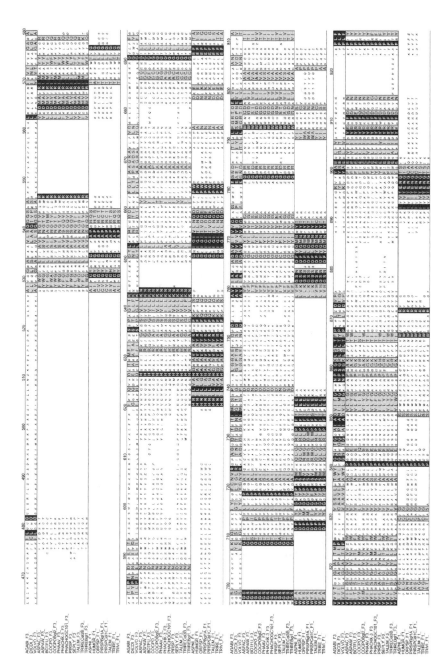

FIGURE 23.4

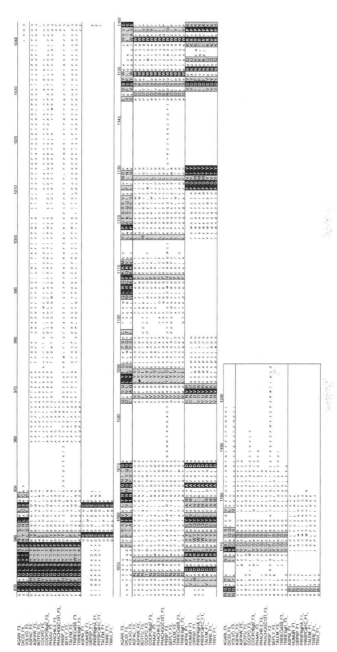

FIGURE 23.4

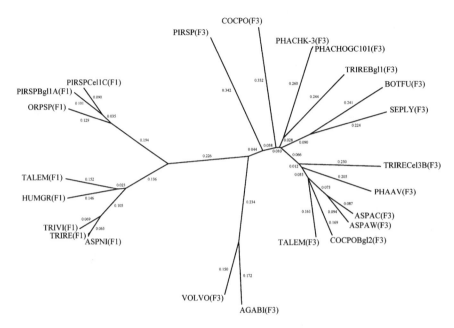

FIGURE 23.5   Phylogenetic tree of the β-glucosidases from filamentous fungi. The tree was constructed using CLUSTAL W and PHYLIP.[178] The numbers indicate the evolutionary distance between the enzymes analyzed in the multiple alignment of Figure 23.4. Those enzymes with the lowest similarity are at an arbitrary distance of 1. The abbreviations used have been defined in Figure 23.4.

Figure 23.5 presents a phylogenetic tree for the β-glucosidases from filamentous fungi. Family 1 enzymes form a clearly separate cluster. Although the enzymes from family 3 are closely related, those from *A. bisporus* and *V. volvacea* form a subgroup that can be differentiated, as evident from the sequence alignment.

XI.   BIOTECHNOLOGICAL APPLICATIONS

β-Glucosidases are produced at an industrial scale from cultures of *A. niger*, which is a GRAS (generally regarded as safe) microorganism, that is, its enzymes can be safely used in processing of foods and beverages for human consumption.[60] Spagna and co-workers[61] have developed a simple and inexpensive method for the purification of β-glucosidases from *A. niger,* which is applicable at the industrial level.

An important application of β-glucosidases that is receiving increasing attention is its use in flavor enhancement of fruit juices and wines by liberating flavor compounds from glucosidic precursors. Some fruits (such as apples and grapes) as well as their juices and fermentation products such as wine, in addition to a free fraction of volatile odorous terpenols, contain nonodorous and nonvolatile aroma precursors (i.e., terpenylglycosides). The terpenic residues (such as linalool, nerol, citronellol, and geraniol) (Figure 23.6) are bound to glucose by β-linkages; they are not efficiently liberated by the β-glucosidases from grapes and yeast present during

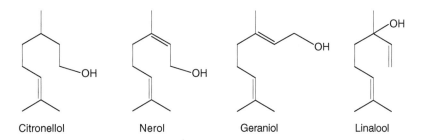

FIGURE 23.6 Terpenols present in some fruits (such as apples and grapes) as well as their juices and fermentation products such as wine.

winemaking,[65] and can be released by the addition of exogenous β-glucosidases.[179] Since the amount of bound monoterpenes is relatively high, there is a great potential in the use of this enzyme for flavor enhancement. Particularly valuable for this purpose are β-glucosidases which are active and stable at pH values 3-4 (the pH of wine and fruit juices) in the presence of ethanol, and that show a broad substrate specificity toward terpenyl glycosides.[179] Enzymes that fulfill these conditions are β-glucosidase HGT from *A. oryzae*[65] and β-glucosidase I from *A. niger.*[62] The flavor of the Japanese alcoholic beverage shochu, obtained by fermentation of sweet potatoes, is improved by β-glucosidase from *A. kawachii*, which is used to liberate nerol and geraniol from their respective glucosides.[180]

β-Glucosidases can also induce a loss of color in red fruit juices and wines by hydrolyzing antocyanins, which are responsible for this color. Antocyanins are mainly β-glucosides of antocyanidins (such as malvidin, peonidin, and cyanidin),[181] which upon hydrolysis yield colorless compounds.[182] It may be important in the processing of red wines to have enzymes available, which can be used for flavor enhancement but which show low decolorizing effect. Le Traon-Masson and Pellerin[47] have isolated two β-glucosidases from an *A. niger* preparation; one of them (β-glucosidase A) is highly active on cellobiose and geranyl β-D-glucoside but degrades malvidin–3-glucoside very slowly. Thus, this enzyme shows good potential for flavor enhancement without degradation of antocyanins. On the other hand, β-glucosidase B is a specific anthocyanase, with low activity on cellobiose and terpenyl glucosides; it may be useful for the specific decolorization of products derived from red fruits.

Citrus fruits contain glucosidic compounds such as prunin and naringin (4′,5, 7-trihydroxyflavanone-7-rhamnoglucoside) which impart a bitter taste to their juices. Debittering of citrus juices can be achieved through successive hydrolysis by α-rhamnosidase and β-glucosidase.[183]

As was pointed out at the beginning of this chapter, an important potential application is the use of β-glucosidase in the saccharification of cellulose for the production of glucose and eventual fermentation to obtain ethanol.[1] This technology, however, is not as yet commercially feasible, mainly due to the high cost of the enzymes involved. Endoglucanases and exoglucanases by attacking cellulose generate cellobiose, which is an inhibitor of cellulases,[8] and which is not directly fermentable by yeasts. β-Glucosidase facilitates the action of these enzymes by producing easily fermentable glucose. This compound, in turn, inhibits β-glucosidase (see discussion on inhibitors). An important development has been the production of glucose-tolerant β-glucosidases. Yan and Lin,[62] Günata and co-workers,[65,139] and Decker and co-workers[70] have obtained β-glucosidases with high K_i values for glucose (see Table 23.3) from different strains of *Aspergillus*. The cellulase system of *T. reesei* has been the most studied for cellulose hydrolysis; however, it has insufficient β-glucosidase for this purpose. The addition of external enzyme significantly reduces the time required for saccharification.[66] Current efforts are being made to reduce enzyme costs by a factor of ten in a program directed by the U.S. Department of Energy in conjunction with biotechnology companies such as Genencor and Novozyme. In the framework of this project, a study of the transcriptional regulation of the biomass-degrading enzymes from *T. reesei* has been published.[173]

Enzyme immobilization for the development of continuous systems, and more efficient use of catalysts and other purposes has received a growing interest in the last decade. This is also true for immobilization of β-glucosidase. Roy and co-workers[94] have evaluated the usefulness of immobilizing β-glucosidase from *Myceliophtora thermophila* for its eventual use in the saccharification of cellulose. The enzyme was immobilized using cross-linked polyacrylamide and cyanogen bromide-activated Sepharose gel beads, and it showed marked improvement in operational and storage stability. Aguado and colleagues[184,185] have immobilized a commercial cellulase from *P. funiculosum* containing β-glucosidase on nylon powder. A high-activity retention (67%) was obtained and the immobilization resulted in a remarkable increase in the thermal stability of the enzyme. Martino and co-workers[186,187] have immobilized β-glucosidase purified from a commercial preparation (CYTOLASE. PCL5, obtained from a selected GRAS strain of *A. niger*), obtaining best results when immobilization was carried out by cross-linking on chitosan with glutaraldehyde. The properties of the immobilized enzyme were analyzed to consider its use in improving the aromatic potential of wines[188] and the immobilized preparation was tested in laboratory-scale reactors.[189] Ortega and co-workers[190] have studied the optimization of the entrapment of β-glucosidase from *A. niger* in alginate and polyacrylamide gels and have compared the properties of the bound and free β-glucosidase. Changes in their kinetic parameters and pH–activity profile were observed on binding; therefore, in order to use immobilized enzymes effectively, their properties must be carefully analyzed. The results presented are examples of numerous studies performed at the laboratory level with immobilized enzymes. These results, however, await application at an industrial scale.

Some β-glucosidases, besides hydrolysis, catalyze transglycosylation reactions.[41,108]. This activity can be very useful in the synthesis of oligosaccharides and glycoconjugates. Several fungal enzymes have been studied for this purpose, such as β-glucosidase II from *A. niger*, used for the synthesis of alkyl β-glucosides[191] and cellooligosaccharides[192] from cellobiose. Applications of this property of β-glucosidases have been discussed recently in detail by Bhatia and colleagues.[11]

XII. CONCLUSIONS

β-Glucosidases have been studied from a large variety of fungal sources. They show ample differences in properties such as size, substrate specificity, resistance to inhibitors, and stability and activity at different pH and temperatures. Thus, these enzymes form an ample toolbox that can be effectively used in numerous and growing biotechnological applications.

ACKNOWLEDGMENTS

This work has been supported by DIPUC, by FONDECYT Grants 90-822, 1930673, 1960241, 8990004, and by UNIDO Grant 91-065 .

REFERENCES

1. Lee, J., Biological conversion of lignocellulosic biomass to ethanol, *J. Biotechnol.*, 56, 1–24, 1997.
2. McMahon, L.G., Nakano, H., Levy, M.D., and Gregory, J.F., Cytosolic pyridoxine-β-D-glucoside hydrolase from porcine jejunal mucosa. Purification, properties, and comparison with broad specificity β-glucosidase, *J. Biol. Chem.*, 272, 32025–32033, 1997.
3. Heyworth, R. and Walker, P.G., Almond emulsin β-D-glucosidase and β-D-galactosidase, *Biochem. J.*, 93, 331–335, 1962.
4. Han, Y.W. and Srinivasan, V.R., Purification and characterization of β-glucosidase of *Alcaligenes faecalis, J. Bacteriol.*, 100, 1355–1363, 1969.
5. Desphande, V., Eriksson, K.E., and Pettersson, B., Production, purification and partial characterization of 1,4 β-glucosidase enzymes from *Sporotrichum pulverulentum, Eur. J. Biochem.*, 90, 191–198, 1978.
6. Fleming, L.W. and Duerksen, J.D., Purification and characterization of yeast β-glucosidase, *J. Bacteriol.*, 93, 135–141, 1967.
7. Mandels, M., Cellulases, *Annu. Rep. Fermen. Process.*, 5, 35–78, 1982.
8. Woodward, J. and Wiseman, A., Fungal and other β-D-glucosidases—their properties and applications, *Enzyme Microb. Technol.*, 4, 73–79, 1982.
9. Leclerc, M., Arnaud, A., Ratomahenina, R., and Galzy, P., Yeast β-glucosidases, *Biotechnol. Genet. Eng. Rev.*, 5, 269–295, 1987.
10. Stutzenberger, F., Thermostable fungal β-glucosidases, *Lett. Appl. Microbiol.*, 11, 173–178, 1990.
11. Bhatia, Y., Mishra, S., and Bisaria, V.S., Microbial β-glucosidases: cloning, properties, and applications, *Crit. Rev. Biotechnol.*, 22, 375–407, 2002.
12. Wood, T.M. and Bhat, K.M., Methods for measuring cellulase activities, *Methods Enzymol.*, 160, 87–112, 1988.
13. Wood, T.M. and McCrae, S.I., Purification and some properties of the extracellular β-D-glucosidase of the cellulolytic fungus *Trichoderma koningii, J. Gen. Microbiol.*, 128, 2973–2982, 1982.
14. Chen, H., Hayn, M., and Esterbauer, H., Purification and characterization of two extracellular β-glucosidases from *Trichoderma reesei, Biochem. Biophys. Acta*, 1121, 54–60, 1992.
15. Khan, A.W., Meek, E., and Henschel, J.R., β-D-glucosidase: multiplicity of activities and significance to enzymic saccharification of cellulose, *Enzyme Microb. Technol.*, 7, 465–467, 1985.
16. Day, D.F. and Workman, W.E., A kinetic assay for cellulases, *Anal. Biochem.*, 126, 205–207, 1982.

17. Breuil, C. and Saddler, J.N., Limitations of using the D-glucose oxidase peroxidase method for measuring glucose derived from lignocellulosic substrates, *Biotechnol. Lett.*, 7, 191–196, 1985.

18. Bergmeyer, H.U., Bernt, E., Schmidt, F., and Stork, H., D-Glucose, Determination with hexokinase and glucose-6-phosphate dehydrogenase, in *Methods in Enzymatic Analysis*, Bergmeyer, H.U., Ed., Verlag Chemie, 1974, pp. 1196–1201.

19. Hounsell, E.F., Carbohydrates, in *HPLC of Small Molecules*, Lim, C.K., Ed., IRL Press, Oxford, 1986, pp. 49–68.

20. Wilson, C.A., McCrae, S.I., and Wood, T.M., Characterisation of a β-D-glucosidase from the anaerobic rumen fungus *Neocallimastix frontalis* with particular reference to attack on cello-oligosaccharides, *J. Biotechnol.*, 37, 217–227, 1994.

21. Schwald, W., Chan, M., Breuil, C., and Saddler, J.N., Comparison of HPLC and colorimetric methods for measuring cellulolytic activity, *Appl. Microbiol. Biotechnol.*, 28, 398–403, 1988.

22. Klaus, R. and Fischer, W., Quantitative thin-layer chromatography of sugars, sugar acids, and polyalcohols, *Methods Enzymol.*, 160, 159–175, 1988.

23. Doner, L.W., High-performance thin-layer chromatography of starch, cellulose, xylan, and chitin hydrolyzates, *Methods Enzymol.*, 160, 176–180, 1988.

24. Umezurike, G., Kinetic analysis of the mechanism of action of β-glucosidase from *Botryodiplodia theobromae* Pat, *Biochim. Biophys. Acta*, 397, 164–178, 1975.

25. Mc Hale, A. and Coughlan, M.P., Properties of the β-glucosidases of *Talaromyces emersonii*, *J. Gen. Microbiol.*, 128, 2327–2331, 1982.

26. Meyer, H.-P. and Canevascini, G., Separation and some properties of two intracellular β-glucosidases of *Sporotrichum (Chrysosporium) thermophile*, *Appl. Environ. Microbiol.*, 41, 924–931, 1981.

27. Jackson, P., The use of polyacrylamide-gel electrophoresis for the high-resolution separation of reducing saccharides labelled with the fluorophore 8-aminonaphtalene 1,3,6-trisulphonic acid, *Biochem. J.*, 270, 705–713, 1990.

28. Funaguma, T. and Hara, A., Purification and properties of two β-glucosidases from *Penicillium herquei* Banier and Sartory, *Agric. Biol. Chem.*, 52, 749–755, 1988.

29. Himmel, M.E., Adney, W.S., Fox, J.W., Mitchell, D.J., and Baker, J.O., Isolation and characterization of two forms of β-D-glucosidase from *Aspergillus niger*, *Appl. Biochem. Biotech.*, 39/40, 213–225, 1993.

30. Kwon, K.S., Kang, H.G., and Hah, Y.C., Purification and characterization of two extracellular β-glucosidases form *Aspergillus nidulans*, *FEMS Microbiol. Lett.*, 97, 149–153, 1992.

31. Lo, A.C., Barbier, J.-R., and Willick, G.E., Kinetics and specificities of two closely related β-glucosidases secreted by *Schizophyllum commune*, *Europ. J. Biochem.*, 192, 175–181, 1990.

32. Rudick, M.J. and Elbein, A.D., Glycoprotein enzymes secreted by *Aspergillus fumigatus*: Purification and properties of β-glucosidase, *J. Biol. Chem.*, 248, 6506–6513, 1973.

33. Rudick, M.J. and Elbein, A.D,. Glycoprotein enzymes secreted by *Aspergillus fumigatus*: Purification and properties of a second β-glucosidase, *J. Bacteriol.*, 124, 534–541, 1975.

34. Saha, S.C., Sanyal, A., Kundu, R.K., Dube, S., and Dube, D.K., Purification and characterization of two forms of extracellular β-glucosidase from jute pathogenic fungus *Macrophomina phaseolina*, *Biochem. Biophys. Acta*, 662, 22–29, 1981.

35. Sakamoto, R., Kanamoto, J., Arai, M., and Murao, S., Purification and physicochemical properties of three β-glucosidases from *Aspergillus aculeatus* No. F-50, *Agric. Biol. Chem.*, 49, 1275–1281, 1985.

36. Todorovic, R., Matavulj, M., Kandrak, J., and Grujic, S., Multiplicity of extracellular β-glucosidase from *Gliocladium virens* C2R1, *Microbios.*, 75, 217–226, 1993.

37. Witte, K. and Wartenberg, A., Purification and properties of two β-glucosidases isolated from *Aspergillus niger, Acta Biotechnol.*, 9, 179–190, 1989.

38. van Tilbeurgh, H., Loontiens, F.G., De Bruyne, C.K., and Claeyssens, M., Fluorogenic and chromogenic glycosides as substrates and ligands of carbohydrases, *Methods Enzymol.*, 160, 45–59, 1988.

39. Jackson, M.A. and Talburt, D.E., Purification and partial characterization of an extracellular β-glucosidase of *Trichoderma reesei* using cathodic run, polyacrylamide gel electrophoresis, *Biotechnol. Bioeng.*, 32, 903–909, 1988.

40. Sadana, J.C., Shewale, J.G., and Patil, R.V., β-D-glucosidase of *Sclerotium rolfsii*, Substrate specificity and mode of action, *Carbohydr. Res.*, 118, 205–214, 1983.

41. Christakopoulos, P., Goodenough, P.W., Kekos, D., Macris, B.J., Claeyssens, M., and Bhat, M.K., Purification and characterization of an extracellular β-glucosidase with transglycosylation and exo-glucosidase activities from *Fusarium oxysporum, Euro. J. Biochem.*, 224, 379–385, 1994.

42. Yazaki, T., Ohmishi, M., Rokushika, S., and Okada, G., Subsite structure of the β-glucosidase from *Aspergillus niger*, evaluated by steady-state kinetics with cello-oligosaccharides as substrates, *Carbohydr. Res.*, 298, 51–57, 1997.

43. Reese, E.T., Maguire, A.H., and Parrish, F.W., Glucosidases and exo-glucanases, *Can. J. Biochem.*, 46, 25–34, 1968.

44. Reese, E.T., Parrish, F.W., and Ettlinger, M., Nojirimycin, and D-glucono-1,5-lactone as inhibitors of carbohydrases, *Carbohydr. Res.*, 18, 381–388, 1971.

45. Copa-Patiño, J.L., Rodríguez, J., and Pérez-Leblic, M.I., Purification and properties of a β-glucosidase from *Penicillium oxalicum* autolysates, *FEMS Microbiol. Lett.*, 67, 191–196, 1990.

46. Sakamoto, R., Arai, M., and Murao, S., Enzymic properties of three β-glucosidases from *Aspergillus aculeatus* No. F-50, *Agric. Biol. Chem.*, 49, 1283–1290, 1985.

47. Le Traon-Masson, M.-P. and Pellerin, P., Purification and characterization of two β-D-glucosidases from an *Aspergillus niger* enzyme preparation: affinity and specificity toward glucosylated compounds characteristic of the processing of fruits, *Enzyme Microb. Technol.*, 22, 374–382, 1998.

48. Bucht, B. and Eriksson, K.E., Extracellular enzyme system utilized by the rot fungus *Stereum sanguinolentum* for the breakdown of cellulose. IV. Separation of cellobiase and aryl beta-glucosidase activities, *Arch. Biochem. Biophys.*, 129, 416–420, 1969.

49. Pitson, S.M., Seviour, R.J., and McDougall, B.M., Purification and characterization of an extracellular β-glucosidase from the filamentous fungus *Acremonium persicinum* and its probable role in beta-glucan degradation, *Enzyme Microb. Technol.*, 21, 182–190, 1997.

50. Sanyal, A., Kundu, R.K., and Dube DK Dube, S., Extracellular cellulolytic enzyme system of *Aspergillus japonicus* 2. Purification and characterization of an inducible extracellular β-glucosidase, *Enzyme. Microb. Technol.*, 10, 91–99, 1988.

51. Iwashita, K., Todoroki, K., Kimura, H., Shimoi, H., and Ito, K., Purification and characterization of extracellular and cell wall bound β-glucosidases from *Aspergillus kawachii, Biosci. Biotechnol. Biochem.*, 62, 1938–1946, 1998.

52. Hoh, Y.K., Yeoh, H.-H., and Tan, T.K., Isolation and characterization of β-glucosidases from *Aspergillus nidulans* mutant USDB 1183, *World J. Microbiol. Biotechnol.*, 9, 555–558, 1992.

53. Yeoh, H.H., Tan, T.K., Chua, S.L., and Lim, G., Properties of β-glucosidase purified from *Aspergillus niger, MIRCEN J.*, 4, 425–430, 1988.

54.   Hoh, Y.K., Yeoh, H.-H., and Tan, T.K., Properties of β-glucosidase purified from *Aspergillus niger* mutants USDB 0827 and USDB 0828, *Appl. Microbiol. Biotechnol.*, 37, 590–593, 1992.

55.   Singh, A., Agrawal, A.K., Abidi, A.B., and Darmwal, N.S., Properties of cellobiase from *Aspergillus niger, Appl. Microbiol. Biotechnol.*, 34, 356–358, 1990.

56.   Unno, T., Ide, K., Yazaki, T., Tanaka, Y., Nakakuki, T., and Okada, G., High recovery purification and some properties of a β-glucosidase from *Aspergillus niger, Biosci. Biotechnol. Biochem.*, 57, 2172–2173 1993.

57.   Abdel-Naby, M.A., Osman, M.Y., and Abdel-Fattah, A.F., Purification and properties of three cellobiases from *Aspergillus niger* A20, *Appl. Biochem. Biotechnol.*, 76, 33–44, 1999.

58.   Rashid, M.H. and Siddiqui, K.S., Purification and characterization of a β-glucosidase from *Aspergillus niger, Folia. Microbiol.*, 42, 544–550, 1997.

59.   Dekker, R.F.H., Kinetic, inhibition, and stability properties of a commercial β-D-glucosidase (cellobiase) preparation from *Aspergillus niger* and its suitability in the hydrolysis of lignocellulose, *Biotechnol. Bioeng.*, 28, 1436–1442, 1986.

60.   Martino, A., Pifferi, P.G., and Spagna, G., Production of β-glucosidase by *Aspergillus niger* using carbon sources derived from agricultural wastes, *J. Chem. Tech. Biotechnol.*, 60, 247–252, 1994.

61.   Spagna, G., Romagnoli, D., Angela, M., Bianchi, G., and Pifferi, P.G., A simple method for purifying glycosidases: α-L- arabinofuranosidase and β-D-glucopyranosidase form *Aspergillus niger* to increase the aroma of wine, Part I, *Enzyme Microb. Technol.*, 22, 298–304, 1998.

62.   Yan, T.-R. and Lin, C.-L., Purification and characterization of a glucose-tolerant β-glucosidase from *Aspergillus niger* CCRC 31494, *Biosci. Biotechnol. Biochem.*, 61, 965–970, 1997.

63.   Yan, T.-R., Lin, Y.-H., and Lin, C.-L., Purification and characterization of an extracellular β-glucosidase II with high hydrolysis and transglucosylation activities from *Aspergillus niger, J. Agric. Food. Chem.*, 46, 431–437, 1998.

64.   Yeoh, H.H., Tan, T.K., and Koh, S.K., Kinetic properties of β-glucosidase from *Aspergillus ornatus, Appl. Microbiol. Biotechnol.*, 25, 25–28, 1986.

65.   Riou, C., Salmon, J.-M., Vallier, M.-J., Günata, Z., and Barre, P., Purification, characterization, and substrate specificity of a novel highly glucose-tolerant β-glucosidase from *Aspergillus oryzae, Appl. Environ. Microbiol.*, 64, 3607–3614, 1998.

66.   Sternberg, D., Vijayakumar, P., and Reese, E.T., β-Glucosidase: Microbial production and effect on enzymatic hydrolysis of cellulose, *Can. J. Microbiol.*, 23, 139–147, 1977.

67.   Vodjani, G., Le Dizet, P., and Petek, F., Purification et propriétés de deux $(1{\rightarrow}4)$ β-D-glucosidases d' *Aspergillus roseus, Carbohydr. Res.*, 236, 267–279, 1992.

68.   Kimura, I., Yoshioka, N., and Tajima, S., Purification and characterization of a β-glucosidase with β-xylosidase activity from *Aspergillus sojae, J. Biosc. Bioeng.*, 87, 538–541, 1999.

69.   Workman, W.E. and Day, D.F., Purification and properties of β-glucosidase from *Aspergillus terreus, Appl. Environ. Microbiol.*, 44, 1289–1295, 1982.

70.   Decker, C.H., Visser, J., and Schreier, P., β-Glucosidase multiplicity from *Aspergillus tubingensis* CBS 843.92: purification and characterization of four β-glucosidases and their differentiation with respect to substrate specificity, glucose inhibition and acid tolerance, *Appl. Microbiol. Biotechnol.*, 55, 157–163, 2001.

71.   Legler, G., von Radloff, M., and Kempfle, M., Composition, *N*-terminal amino acids, and chain length of a β-glucosidase from *Aspergillus wentii, Biochem. Biophys. Acta.*, 257, 40–48, 1972.

72. Legler, G., Untersuchungen zum Wirkungsmechanismus glykosidspaltender Enzyme, II, *Hoppe-Seyler's Z Physiol Chem.,* 348, 1359–1366, 1967.
73. Srivastava, S.K., Gopalkrishnan, K.S., and Ramachandran, K.B., Kinetic characterization of a crude *β*-D-glucosidase from *Aspergillus wentii* Pt 2804, *Enzyme Microb. Technol.,* 6, 508–512, 1984.
74. Umezurike, G.M., The purification and properties of extracellular *β*-glucosidase from *Botryodiplodia theobromae Pat, Biochim. Biophys. Acta,* 227, 419–428, 1971.
75. Gueguen, Y., Chemardin, P., Arnaud, A., and Galzy, P., Purification and characterization of an intracellular *β*-glucosidase from *Botrytis cinerea, Enzyme Microb. Technol.,* 17, 900–906, 1995.
76. Sasaki, I. and Nagayama, H., *β*-Glucosidase from *Botrytis cinerea*: Its relation to the pathogenicity of this fungus, *Biosci. Biotech. Biochem.,* 58, 616–620, 1994.
77. Sasaki, I. and Nagayama, H., Purification and characterization of *β*-glucosidase from *Botrytis cinerea, Biosci. Biotech. Biochem.* 59, 100–101, 1995.
78. Venturi, L.L., de L Polizeli, M., Terenzi, H.F., Furriel, R.P.M., and Jorge, J.A., Extracellular *β*-glucosidase from *Chaetomium thermophilum* var coprophilum: production, purification and some biochemical properties, *J.Basic Microbiol.,* 42, 55–66, 2002.
79. Uziie, M., Matsuo, M., and Yasui, T., Purification and some properties of *Chaetomium trilaterale β*-xylosidase, *Agric. Biol. Chem.,* 49, 1159–1166, 1985.
80. Lucas, R., Robles, A., Alvarez de Cienfuegos, G., and Gálvez, A., *β*-Glucosidase from *Chalara paradoxa* CH32: Purification and properties, *J. Agr. Food. Chem.,* 48, 3698–3703, 2000.
81. Oh, K.-B., Hamada, K., Saito, M., Lee, H.-J., and Matsuoka, H., Isolation and properties of an extracellular *β*-glucosidase from a filamentous fungus, *Cladosporium resinae,* isolated from kerosene, *Biosci. Biotechnol. Biochem.,* 63, 281–287, 1999.
82. Banerjee, U.C., Production of *β*-glucosidase (cellobiase) by *Curvularia sp., Lett. Applied. Microbiol.,* 10, 197–199, 1990.
83. Yagüe, E. and Estévez, P., Purification and characterization of a *β*-glucosidase from *Evernia prunastri, Eur. J. Biochem.,* 175, 627–632, 1988.
84. Yamabone, T. and Mitsuishi, Y., Purification and properties of a *β*-glucosidase from fungal strain Y-94, *Agric. Biol. Chem.,* 53, 3359–3360, 1989.
85. Wood, T.M., The cellulase of *Fusarium solani.* Purification and specificity of the *β*-(1->4) glucanase and the *β*-D-glucosidase components, *Biochem. J.,* 121, 353–362, 1971.
86. Ferreira, E.X., Purification and characterization of a *β*-glucosidase from solid-state cultures of *Humicola grisea* var. thermoidea, *Can. J. Microbiol.,* 42, 1–5, 1996.
87. Peralta, R.M., Kadowaki, M.K., Terenzi, H.F., and Jorge, J.A., A highly thermostable *β*-glucosidase activity from the thermophilic fungus *Humicola grisea* var. thermoidea: purification and biochemical characterization, *FEMS Microbiol. Lett.,* 146, 291–295, 1997.
88. Yoshioka, H. and Hayashida, S., Purification and properties of *β*-glucosidase from *Humicola insolens* YH-8, *Agric. Biol. Chem.,* 44, 1729–1735, 1980.
89. Rao, U.S. and Murthy, S.K., Purification and characterization of a *β*-glucosidase and endocellulase from *Humicola insolens, Indian J. Biochem. Biophys.,* 25, 687–694, 1988.
90. Anand, L. and Vithayathil, J., Purification and properties of *β*-glucosidase from a thermophilic fungus *Humicola lanuginosa* (Griffon and Maublanc) Bunce, *J. Ferment. Bioeng.,* 67, 380–386, 1989.
91. Herr, D., Baumer, F., and Dellweg, H., Purification and properties of an extracellular *β*-glucosidase from *Lenzites trabea, Eur. J. Appl. Microbiol. Biotechnol.,* 5, 29–36, 1978.

92. Berry, R.K. and Dekker, R.F.H., Fractionation of the cellulolytic enzymes produced by a species of Monilia: purification and properties of an extracellular β-D-glucosidase, *Carbohydr. Res.*, 157, 1–12, 1986.

93. Yoshioka, H. and Hayashida, S., Relationship between carbohydrate moiety and thermostability of β-glucosidase from *Mucor miehei* YH-10, *Agric. Biol. Chem.*, 45, 571–577, 1981.

94. Roy, S.K., Raha, S.K., Dey, S.K., and Chakrabarty, S.L., Immobilization of β-glucosidase from *Myceliophthora thermophila, Enzyme Microb. Technol.*, 11, 431–435, 1989.

95. Pardo, A.G. and Forchiassin, F., Factors influencing β-glucosidase production, activity and stability in *Nectria catalinensis, Folia. Microbiol.*, 44, 71–76, 1999.

96. Li, X. and Calza, R.E., Purification and characterization of an extracellular β-glucosidase from the rumen fungus *Neocallimastix frontalis* EB188, *Enzyme Microb. Technol.*, 13, 622–628, 1991.

97. Li, X., and Calza, R.E., Kinetic study of a cellobiase purified from *Neocallimastix frontalis* EB 188, *Biochem. Biophys. Acta*, 1080, 148–154, 1991.

98. Chen H., Li, X., and Ljungdahl, L.G., Isolation and properties of an extracellular β-glucosidase from the polycentric rumen fungus *Orpinomyces* sp. strain PC-2, *Appl. Environ. Microbiol.*, 60, 64–70, 1994.

99. Parr, S.R., Some kinetic properties of the β-D-glucosidase (cellobiase) in a commercial cellulase product from *Penicillium funiculosum* and its relevance in the hydrolysis of cellulose, *Enzyme Microb. Technol.*, 5, 457–462, 1983.

100. Kantham, L. and Jagannathan, V., β-Glucosidase of *Penicillium funiculosum*. II. Properties and mycelial binding, *Biotechnol. Bioeng.*, 27, 786–791, 1985.

101. Hidalgo, M., Steiner, J., and Eyzaguirre, J., β-Glucosidase from *Penicillium purpurogenum*: purification and properties, *Biotechnol. Appl. Biochem.*, 15, 185–191, 1992.

102. Kamagata, Y., Sasaki, H., and Takao, S., Fractionation of extracellular cellulase produced by *Penicillium purpurogenum* and purification of β-glucosidase, *Microb. Utiliz. Renew. Resour.* 4, 124–133, 1984.

103. Smith, M.H. and Gold, M.H., *Phanerochaete chrysosporium* β-glucosidase: induction, cellular localization, and physical characterization, *Appl. Environ. Microbiol.*, 37, 938–942, 1979.

104. Copa-Patiño, J.L. and Broda, P., A *Phanerochaete chrysosporium* β-D-glucosidase/ β-D xylosidase with specificity for $(1\rightarrow3)$-β-D-glucan linkages, *Carbohydr. Res.*, 253, 265–275, 1994.

105. Lymar, E.S., Li, B., and Renganathan, V., Purification and characterization of a cellulose-binding β-glucosidase from cellulose-degrading cultures of *Phanerochaete chrysosporium, Appl. Environ. Microbiol.*, 61, 2976–2980, 1995.

106. Bodenmann, J., Heiniger, U., and Hohl, H.R., Extracellular enzymes of *Phytophtora infestans*: endo-cellulase, β-glucosidases and 1,3-β-glucanases. *Can. J. Microbiol.*, 31, 75–82, 1985.

107. Teunissen, M.J., Lahaye, D.H.T.P., Huis in't Veld, J.H.J., and Vogels, G.D., Purification and characterization of an extracellular β-glucosidase from the anaerobic fungus *Piromyces sp.* strain E2, *Arch. Microbiol.*, 158, 276–281, 1992.

108. Harhangi, H.R., Steenbakkers, P.J.M., Akhamanova, A., Jetten, M.S.M., van der Drift, C., and Op den Camp, H.J.M., A highly expressed family 1 β-glucosidase with transglycosylation capacity from the anaerobic fungus *Piromyces sp.* E2, *Biochem. Biophys. Acta*, 1574, 293–303, 2002.

109. Cao, W. and Crawford, D.L., Purification and some properties of β-glucosidase from the ectomycorrhizal fungus *Pisolithus tinctorius* strain SMF, *Can. J. Microbiol.*, 39, 125–129, 1993.

110. Morais, H., Ramos, C., Matos, N., Forgács, E., Cserháti, T., Almeida, V., Oliveira, J., Darwish, Y., and Illés, Z., Liquid chromatographic and electrophoretic characterisation of extracellular β-glucosidase of *Pleurotus ostreatus* grown in organic waste, *J. Chromatogr.*, B 770, 111–119, 2002.

111. Hirayama, T., Horie, S., Nagayama, H., and Matsuda, K., Studies on cellulases of a phytopathogenic fungus *Pyricularia oryzae* Cavara. II. Purification and properties of a β-glucosidase, *J. Biochem.*, 84, 27–37, 1978.

112. Shewale, J. and Sadana, J., Purification, characterization, and properties of β-glucosidase enzymes from *Sclerotium rolfsii, Arch. Biochem. Biophys.*, 207, 185–196, 1981.

113. Desai, J.D., Ray, R.M., Patel, N.P., Purification and properties of extracellular β-glucosidase from *Scytalidium lignicola, Biotechnol. Bioeng.*, 25, 307–313, 1983.

114. Bhat, K.M., Gaikwad, J.S., and Maheshwari, R., Purification and characterization of an extracellular β-glucosidase from the thermophilic fungus *Sporotrichum thermophile* and its influence on cellulase activity, *J. Gen. Microbiol.*, 139, 2825–2832, 1993.

115. De Gussem, R.L., Aerts, G.M., Claeyssens, M., and De Bruyne, C.K., Purification and properties of an induced β-D-glucosidase from *Stachybotrys atra, Biochem. Biophys. Acta*, 525, 142–153, 1978.

116. McHale, A. and Coughlan, M.P., The cellulolytic system of *Talaromyces emersonii*: purification and characterization of the extracellular and intracellular β-glucosidases, *Biochem. Biophys. Acta*, 662, 152–159, 1981.

117. Sengupta, S., Ghosh, A.K., and Sengupta, S., Purification and characterisation of a β-glucosidase (cellobiase) from a mushroom *Termitomyces clypeatus, Biochem. Biophys. Acta*, 1076, 215–220, 1991.

118. Tong, C.C., Cole, A.L., and Shepherd, M.G., Purification and properties of the cellulases from the thermophilic fungus *Thermoascus aurantiacus, Biochem. J.*, 191, 83–94, 1980.

119. Khandke, K.M., Vithayathil, P.J., and Murthy, S.K., Purification of xylanase, β-glucosidase, endocellulase and exocellulase from a thermophilic fungus *Thermoascus aurantiacus, Arch. Biochem. Biophys.*, 274, 491–500, 1989.

120. de Palma-Fernandez, E.R., Gomes, E., and da Silva, R., Purification and characterization of two beta-glucosidases from the thermophilic fungus *Thermoascus aurantiacus, Folia. Microbiol.*, 47, 685–690, 2002.

121. Parry, N., Beever, D., Owen, E., Vandenberghe, I., VanBeeumen, J., and Bhat, M.K., Biochemical characterization and mechanism of action of a thermostable β-glucosidase purified from *Thermoascus aurantiacus, Biochem. J.*, 353, 117–127, 2001.

122. Bernier, R. and Stutzenberger, F., Extracellular and cell-associated forms of beta-glucosidase in *Thermomonospora curvata, Lett. Appl. Microbiol.*, 7, 103–107, 1988.

123. Lin, J., Pillay, B., and Singh, S., Purification and biochemical characteristics of β-D-glucosidase from a thermophilic fungus, *Thermomyces lanuginosus, Biotechnol. Appl. Biochem.*, 30, 81–87, 1999.

124. Bhattacharjee, B., Roy, A., and Majumder, A.L., β-Glucosidase of a white-rot fungus *Trametes gibbosa, Biochem. Int.*, 28, 783–793, 1992.

125. Fadda, M.B., Curreli, N., Pompei, R., Rescigno, A., Rinaldi, A., and Sanjust, E., A highly active fungal β-glucosidase; purification and properties, *Appl. Biochem. Biotechnol.*, 44, 263–270, 1994.

126. Yun, S.-I., Jeong, C.-S., Chung, D.-K., and Choi, H.S., Purification and some properties of a β-glucosidase from *Trichoderma harzianum* Type C.-4, *Biosci. Biotechnol. Biochem.*, 65, 2028–2032, 2001.

127. Chirico, W.J. and Brown, R.D., Jr., Purification and characterization of a beta-glucosidase from *Trichoderma reesei, Eur. J. Biochem.*, 165, 333–341, 1987.

128.   Schmid, G. and Wandrey, C., Purification and partial characterization of a cellodextrin glucohydrolase (β-glucosidase) from *Trichoderma reesei* strain QM9414, *Biotechnol. Bioeng.*, 30, 571–585, 1987.

129.   Inglin, M., Feinberg, B.A., and Loewenberg, J.R., Partial purification and characterization of a new intracellular β-glucosidase of *Trichoderma reesei*, *Biochem. J.*, 185, 515–519, 1980.

130.   Woodward, J. and Arnold, S.L., The inhibition of β-glucosidase activity in *Trichoderma reesei* C30 cellulase by derivatives and isomers of glucose, *Biotechnol. Bioeng.*, 23, 1553–1562, 1981.

131.   Berghem, L.E.R. and Pettersson, L.G., The mechanism of enzymatic cellulose degradation: isolation and some properties of a β-glucosidase from *Trichoderma viride*, *Eur. J. Biochem.*, 46, 295–305, 1974.

132.   Hong, J., Ladisch, M.R., Gong, C.-S., Wankat, P.C., and Tsao, G.T., Combined product and substrate inhibition equation for cellobiase, *Biotechnol. Bioeng.*, 23, 2779–2788, 1981.

133.   Montero, M.A. and Romeu, A., Kinetic study on the β-glucosidase-catalysed reaction of *Trichoderma viride* cellulase, *Appl. Microbiol. Biotechnol.*, 28, 350–353, 1992.

134.   Wei, D.-D., Kirimura, K., Usami, S., and Lin, T.-H., Purification and characterization of an extracellular β-glucosidase from the wood-grown fungus *Xylaria regalis, Cur. Microbiol.*, 33, 297–301, 1996.

135.   Wilson, R.W. and Niederpruem, D.J., Control of β-glucosidases in *Schizophyllum commune, Can. J. Microbiol.*, 13, 1009–1020, 1967.

136.   Hayashi, S., Matsumoto, K., Wada, Y., Takasaki, Y., and Imada, K., Stable β-glucosidase from *Aureobasidium, Lett. Appl. Microbiol.*, 17, 75–77, 1993.

137.   Eriksson, K.-E., Pettersson, B., and Westermark, U., Oxidation: an important enzyme reaction in fungal degradation of cellulose, *FEBS. Lett.*, 49, 282–285, 1974.

138.   Westermark, U. and Eriksson, K.-E., Purification and properties of cellobiose: quinone-oxidoreductase from *Sporotrichum pulverulentum, Acta Chem. Scand. B.*, 29, 419–424, 1975.

139.   Günata, Z. and Vallier, M.-J., Production of a highly glucose-tolerant extracellular β-glucosidase by three *Aspergillus* strains, *Biotechnol. Lett.*, 21, 219–223, 1999.

140.   Uziie, M., Matsuo, M., and Yasui, T., Possible identity of β-xylosidase and β-glucosidase of *Chaetomium trilaterale*, *Agric. Biol. Chem.*, 49, 1167–1173, 1985.

141.   Jackson, M.A. and Talburt, D.E., Mechanism for β-glucosidase release into cellulose-grown *Trichoderma reesei* culture supernatants, *Exper. Mycol.*, 12, 203–216, 198.

142.   Messner, R. and Kubicek, C.P., Evidence for a single, specific β-glucosidase in cell walls from *Trichoderma reesei* QM 9414, *Enzyme. Microb. Technol.*, 12, 685–690, 1990.

143.   Kusama, S., Kusakabe, I., and Murakami, K., Purification and some properties of β-glucosidase from *Streptomyces sp, Agric. Biol. Chem.*, 50, 2891–2898, 1986.

144.   Rogalski, J., Wojtas-Wasilewska, M., and Leonowicz, A., Affinity chromatography of 1,4-β-glucosidase from *Trichoderma reesei* QM 9414, *Acta, Biotechnol.*, 11, 485–494, 1991.

145.   Kmínková, M. and Kucera, J., Separation of cellobiase (E.C. 3.2.1.21) from the crude cellulase system (E.C. 3.2.1.4) of *Trichoderma viride* using affinity chromatography on concanavalin A bound to agarose, *J. Chromatogr.* 244, 166–168, 1982.

146.   Watanabe, T., Sato, T., Yoshioka, S., Koshijima, T., and Kuwahara, M., Purification and properties of *Aspergillus niger* β-glucosidase, *Eur. J. Biochem.*, 209, 651–660, 1992.

147.   Hedrick, J.L. and Smith, A.J., Size and charge isomer separation and estimation of molecular weights of protein by disc gel electrophoresis, *Arch. Biochem. Biophys.*, 126, 155–164, 1968.

148. Dubois, M., Gilles, K.A., Hamilton, J.K., Rebers, P.A., and Smith, F., Colorimetric method for determination of sugars and related substances, *Anal. Chem.,* 28, 350–356, 1956.

149. Zacharius, R.M., Zell, T.E., Morrison, J.H., and Woodlock, J.J., Glycoprotein staining following electrophoresis on acrylamide gels, *Anal. Biochem.,* 30, 148–52, 1969.

150. Messner, R., Hagspiel, K., and Kubicek, C.P., Isolation of a *β*-glucosidase binding and activating polysaccharide from cell walls of *Trichoderma reesei, Arch. Microbiol.,* 154, 150–155, 1990.

151. Umezurike, G.M., The subunit structure of *β*-glucosidase from *Botryodiplodia theobromae* Pat, *Biochem. J.,* 145, 361–368, 1975.

152. Willick, G.E. and Seligy, V.L., Multiplicity in cellulases of *Schizophyllum commune:* derivation partly from heterogeneity in transcription and glycosylation, *Eur. J. Biochem.,* 151, 89–96, 1985.

153. Kubicek, C.P., Involvement of a conidial endoglucanase and a plasma-membrane-bound *β*-glucosidase in the induction of endoglucanase synthesis by cellulose in *Trichoderma reesei, J. Gen. Microbiol.,* 133, 1481–1487, 1987.

154. Labudová, I. and Farkas, V., Multiple enzyme forms in the cellulase system of *Trichoderma reesei* during its growth on cellulose, *Biochem. Biophys. Acta,* 744, 135–140, 1983.

155. Umezurike, G.M., The cellulolytic enzymes of *Botryodiplodia theobromae* Pat: separation and characterization of cellulases and *β*-glucosidases, *Biochem. J.,* 177, 9–19, 1979.

156. Henrissat, B., A classification of glycosyl hydrolases based on amino-acid sequence similarities, *Biochem. J.,* 280, 309–316, 1991.

157. Withers, S.G., Mechanism of glycosyl transferases and hydrolases, *Carbohydr. Polym.,* 44, 325–337, 2001.

158. Keresztessy, Z., Kiss, L., and Hughes, M.A., Investigation of the active site of the cyanogenic *β*-D-glucosidase (linamarase) from Manihot esculenta Crantz (cassava). II. Identification of Glu-198 as an active site carboxylate group with acid catalytic function, *Arch. Biochem. Biophys.,* 315, 323–30, 1994.

159. Withers, S.G., Warren, R.J., Street, I.P., Rupitz, K., Kempton, J.B., and Aebersold, R., Unequivocal demonstration of the involvement of a glutamate residue as a nucleophile in the mechanism of a retaining glycosidase, *J. Am. Chem. Soc.,* 112, 5887–5889, 1990.

160. Chi, Y.I., Martinez-Cruz, L.A., Jancarik, J., Swanson, R.V., Robertson, D.E., and Kim, S.H., Crystal structure of the beta-glycosidase from the hyperthermophile *Thermosphaera aggregans*: insights into its activity and thermostability, *FEBS Lett.,* 445, 375–83, 1999.

161. Hakulinen, N., Paavilainen, S., Korpela, T., and Rouvinen, J., The crystal structure of *β*-glucosidase from *Bacillus circulans sp.* alkalophilus: ability to form long polymeric assemblies, *J. Struct. Biol.,* 129, 69–79, 2000.

162. Bush, J., Richardson, J., and Cardelli, J., Molecular cloning and characterization of the full-length cDNA encoding the developmentally regulated lysosomal enzyme *β*-glucosidase in *Dictyostelium discoideum, J. Biol. Chem.,* 269, 146–1476, 1994.

163. Bause, E. and Legler, G., Isolation and structure of a tryptic glycopeptide from the active site of *β*-glucosidase A3 from *Aspergillus wentii, Biochim. Biophys. Acta,* 626, 459–465, 1980.

164. Steenbakkers, P.J.M., Harhangi, H.R., Bosscher, M.W., Van der Hooft, M.M.C., Keltjens, J.T., Van der Drift, C., Vogels, G.D., and Op den Camp, H.J.M., *β*-Glucosidase in cellulosome of the anaerobic fungus *Piromyces sp.* strain E2 is a family 3 glycoside hydrolase, *Biochem. J.,* 370, 963–970, 2003.

165. Varghese, J.N., Hrmova, M., and Fincher, G.B., Three-dimensional structure of a barley β-D glucan exohydrolase, a family 3 glycosyl hydrolase, *Structure*, 7, 179–190, 1999.

166. Bush, J., Richardson, J., and Cardelli, J., Molecular cloning and characterization of the full-length cDNA encoding the developmentally regulated lysosomal enzyme β-glucosidase in *Dictyostelium discoideum*, *J. Biol. Chem.*, 269, 146–1476, 1994.

167. Kawaguchi, T., Enoki, T., Tsurumaki, S., Sumitani, J., Ueda, M., Ooi, T., and Arai, M., Cloning and sequencing of the cDNA encoding β-glucosidase 1 from *Aspergillus aculeatus*, *Gene*, 173, 287–288, 1996.

168. Iwashita, K., Nagahara, T., Kimura, H., Takano, M., Shimoi, H., and Ito, K., The bglA gene of *Aspergillus kawachii* encodes both extracellular and cell wall-bound β-glucosidases, *Appl. Environ. Microbiol.*, 65, 5546–5553, 1999.

169. Hung, C.-Y., Yu, J.-J., Lehmann, P. F., and Cole, G .T., Cloning and expression of the gene which encodes a tube precipitin antigen and wall-associated β-glucosidase of *Coccidioides immitis*, *Infect. Immun.*, 69, 2211–2222, 2001.

170. Morrissey, J.P., Wubben, J.P., and Osbourn, A.E., Stagonospora avenae secretes multiple enzymes that hydrolyze oat leaf saponins, *Mol. Plant-Microbe Interact.*, 13, 1041–1052, 2000.

171. Li, B. and Renganathan, V., Gene cloning and characterization of a novel cellulose-binding β-glucosidase from *Phanerochaete chrysosporium*, *Appl. Environ. Microbiol.*, 64, 2748–2754, 1998.

172. Osbourn, A., Bowyer, P., Lunness, P., Clarke, B., and Daniels, M., Fungal pathogens of oat roots and tomato leaves employ closely related enzymes to detoxify different host plant saponins, *Molec. Plant-Microbe Interact.*, 8, 971–978, 1995.

173. Foreman, P.K., Brown, D., Dankmeyer, L., Dean, R., Diener, S., Dunn-Coleman, N.S., Goedegebuur, F., Houfek, T.D., England, G.J., Kelley, A.S., Meerman, H.J., Mitchell, T., Mitchinson, C., Olivares, H.A., Teunissen, P.J., Yao, J., and Ward, M., Transcriptional regulation of biomass-degrading enzymes in the filamentous fungus *Trichoderma reesei*, *J. Biol. Chem.*, 278, 31988–31997, 2003.

174. Barnett, G.C., Berka, R.M., and Fowler, T., Cloning and amplification of the gene encoding an extracellular β-glucosidase from *Trichoderma reesei*: evidence for improved rates of saccharification of cellulosic substrates, *Biotechnol.*, 9, 562–567, 1991.

175. Takashima, S., Nakamura, A., Hidaka, M., Masaki, H., and Uozumi, T., Molecular cloning and expression of the novel fungal β-glucosidase genes from Humicola grisea and *Trichoderma reesei*, *J. Biochem.*, 125, 728–736, 1999.

176. Harhangi, H.R., Steenbakkers, P.J., Akhmanova, A., Jetten, M.S., van der Drift, C., and Op den Camp, H.J., A highly expressed family 1 β-glucosidase with transglycosylation capacity from the anaerobic fungus *Piromyces sp*. E2. *Biochim. Biophys. Acta.*, 1574, 293–303, 2002.

177. Harhangi, H.R., Akhmanova, A., Steenbakkers, P.J.M., Jetten, M.S.M., van der Drift, C., and Op den Camp, H.J.M., Genomic DNA analysis of genes encoding (hemi-) cellulolytic enzymes of the anaerobic fungus *Piromyces sp*. E2, *Gene*, 314, 73–80, 2003.

178. Felsenstein, J., An alternating least squares approach to inferring phylogenies from pairwise distances, *Syst. Biol.*, 46, 101–11, 1997.

179. Gueguen, Y., Chemardin, P., Janbon, G., Arnaud, A., and Galzy, P., Use of β-glucosidase in the development of flavor in wines and fruit juices, *Carbohydr. Biotechnol. Protocols*, 10, 323–330, 1999.

180. Ohta, T., Ikuya, R., Nakashima, M., Morimitsu, Y., Samuta, T., and Saiki, H., Characteristic flavor of Kansho-shochu (sweet potato spirit), *Agric. Biol. Chem.*, 54, 1353–1357, 1990.

181. Francis, F.J., Food colorants: Anthocyanins, *Crit. Rev. Food Sci. Nutrit.*, 18, 273–314, 1989.

182. Yang, H.Y. and Steele, W.B., Removal of excess antocyanin pigments by enzyme, *Food Technol.*, 12, 517–519, 1958.

183. Roitner, M., Schalkhammer, T., and Pittner, F., Characterization of naringinase from *Aspergillus niger. Monatsch. Chem.*, 115, 1255–1267, 1984.

184. Aguado, J., Romero, M.D., and Rodríguez, L., Immobilization of *β*-glucosidase from *Penicillium funiculosum* on nylon powder, *Biotechnol. Appl. Biochem.*, 17, 49–55, 1993.

185. Aguado, J., Romero, M.D., Rodríguez, L.M., Calles, J.A., Thermal deactivation of free and immobilized *β*-glucosidase from *Penicillium funiculosum, Biotechnol. Prog.*, 11, 104–106, 1995.

186. Martino, A., Durante, M., Pifferi, P.G., Spagna, G., and Bianchi, G., Immobilization of *β*-glucosidase from a commercial preparation. Part 1. A comparative study of natural supports, *Process Biochem.*, 31, 281–285, 1996.

187. Martino, A., Pifferi, P.G., and Spagna, G., Immobilization of *β*-glucosidase from a commercial preparation. Part 2. Optimization of the immobilization process on chitosan, *Process Biochem.*, 31, 287–293, 1996.

188. Gallifuoco, A., D'Ercole, L., Alfani, F., Cantarella, M., Spagna, G., and Pifferi, P.G., On the use of chitosan-immobilized *β*-glucosidase in wine-making: kinetics and enzyme inhibition, *Process Biochem.*, 33, 163–168, 1998.

189. Gallifuoco, A., Alfani, F., Cantarella, M., Spagna, G., and Pifferi, P.G., Immobilized *β*-glucosidase for the winemaking industry: study of biocatalyst operational stability in laboratory-scale continuous reactors, *Process. Biochem.*, 35, 179–185, 1999.

190. Ortega, N., Busto, M.D., and Pérez-Mateos, M., Optimization of *β*-glucosidase entrapment in alginate and polyacrylamide gels, *Bioresour. Technol.*, 64, 105–111, 1998.

191. Yan, T.-R. and Liau, J.-C., Synthesis of alkyl *β*-glucosides from cellobiose with *Aspergillus niger β*-glucosidase II, *Biotech, Lett.*, 20, 653–657, 1998.

192. Yan, T.-R. and Liau, J.-C., Synthesis of cello-oligosaccharides from cellobiose with *β*-glucosidase II from *Aspergillus niger, Biotechnol. Lett.*, 20, 591–594, 1998.

24 Sialic Acids and Sialylmimetics: Useful Chemical Probes of Sialic Acid-Recognizing Proteins

Jennifer C. Wilson, Milton J. Kiefel, and Mark von Itzstein

CONTENTS

I. INTRODUCTION

Sialic acids are a family of naturally occurring nine-carbon sugars that are generally found as α-ketosidically linked components of cell-surface glycoconjugates. The most commonly found derivatives are those derived from 5-acetamido-D-*glycero-*

D-*galacto*-2-nonulosonic acid (Neu5Ac, **1**), although the 5-glycolyl (Neu5Gc, **2**) and 5-hydroxy (KDN, **3**) derivatives are also found in biological systems.[1,2] Natural modifications of these sialic acids generally involve acylation, phosphorylation or sulfation of the hydroxyl groups, with over 50 structurally distinct derivatives having been found in nature.[3] The structural diversity of the sialic acids, together with their position within glycoconjugates, results in them being intimately involved in a number of important physiological phenomena and disease states. Located at the terminus of numerous cell-surface oligosaccharides, sialic acids are ideally positioned to participate in biological processes including cell-biomolecule-mediated recognition phenomena, acting as receptors for viruses and bacteria, cell–cell communication, and as markers in certain diseases.[2,4,5]

The realization of the importance of sialic acids, and the proteins and enzymes that utilize them, in biological processes has resulted in intensive research activity over the last couple of decades. These efforts have been directed toward the development of a better understanding of the biological significance of sialic acids in certain physiological processes, particularly with regard to disease states. Much of this research has focussed on the synthesis of structurally modified sialic acid derivatives as probes or potential inhibitors of specific sialic acid-recognizing proteins, and several comprehensive overviews have been published.[1,6–11,12] An underlying theme in all of these efforts directed toward the development of novel sialic acid derivatives as biological probes is the complexity of chemical manipulations associated with sialic acids.[6–8,10] By virtue of the complexity of sialic acid chemistry, the last few years have seen a trend toward the development of sialylmimetics, especially with regard to investigations into carbohydrate-mediated recognition events in biological systems. In the broadest sense of the term, sialylmimetics can be considered as molecules that contain only those sialic acid structural features that are essential for interaction with a given biomolecule. Sialylmimetics have a number of potential advantages over the parent structure upon which they are based, especially if they are intended to be utilized as therapeutic agents. They can be designed to be stable to endogenous degradative enzymes, have increased affinity for the biomolecule under investigation, and can have improved pharmacokinetic profiles. Perhaps most important, however, is the fact that sialylmimetics can be significantly easier to prepare by chemical methods.

(1) R = NHAc (Neu5Ac)
(2) R = NHGc (Neu5Gc)
(3) R = OH (KDN)

(4) R = NHAc

Some recent works provide excellent insight into the broader area of glycomimetics as well as highlighting aspects of sialylmimetics.[6,12–16] Given the comprehensive nature of many of these works, our intention herein is to focus on the general themes of sialylmimetics as probes for influenza virus sialidase, *Vibrio cholerae* (VC) sialidase, and the selectins, with a particular emphasis on the literature of the last 5 years.

For the purposes of this chapter, we will focus on two broad groups of compounds. The first of these contains analogs of the known sialidase inhibitor Neu5Ac2en (**4**), which itself is considered a mimic of the transition-state species formed during the action of sialidases on sialylglycoconjugates (see below). A subset of these Neu5Ac2en analogs are those compounds that have been designed as mimics of the sialidase transition state. These compounds retain many of the key functional groups found in Neu5Ac2en, but are generally based on noncarbohydrate scaffolds. The second broad group of compounds discussed herein are referred to as sialylmimetics, wherein the sialic acid portion of a molecule of interest has been replaced by only those functional groups thought to be involved in important interactions. For many sialylglycoconjugates of interest this involves replacement of the entire sialic acid residue with a simple charged functionality (e.g., a carboxylate group), giving rise to sialylmimetics that bear little or no gross structural resemblance to the sialic acid. As will become apparent, such major structural renovation has, in some instances, led to compounds with significantly improved affinity for a given sialic acid-recognizing protein.

II. SIALIC ACIDS AND SIALYLMIMETICS AS PROBES FOR SIALIDASES

The focus of much attention in the production of sialylmimetics has been on the generation of inhibitors, based on sialylmimetic structures that inhibit the action of the exo-glycohydrolases known as sialidases. Mechanistically, sialidases cleave α-ketosidically linked terminal sialic acid residues from sialoglycoconjugates with retention of anomeric configuration as represented in Scheme 1. Initially, the α-Neu5Ac glycoside binds in a distorted half-chair arrangement to the active site. Kinetic isotope and molecular design studies[17–19] have provided evidence that support the proposal that a sialosyl cation transition-state intermediate **5** forms during the catalytic reaction. Neu5Ac is released from the active site as it's α-anomer, and to date the overall retaining outcome of this reaction has been found to be common to all naturally

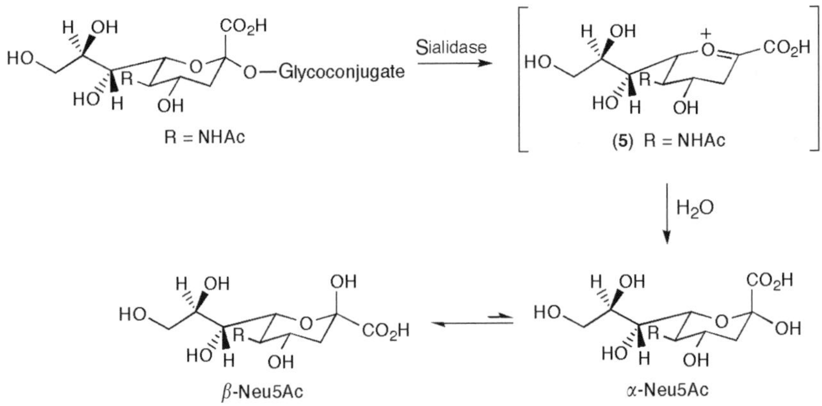

SCHEME 24.1

occurring sialidases.[20,21] The following discussion will concentrate on two sialidases, from influenza virus and VC, that play important functions in the lifecycle of the pathogenic microorganisms that cause influenza and cholera, respectively, and are therefore considered drug design targets.

A. INFLUENZA VIRUS SIALIDASE

The surface of the influenza virus is extensively decorated[22] with two major antigenic glycoproteins, a sialidase and a hemagglutinin, which play an integral role in the infection and spread of the virus that causes the highly contagious disease influenza. The replicative life cycle of the virus is assisted by these proteins: viral penetration is initiated when the hemagglutinin attaches to host cell-surface glycoconjugates bearing α-ketosidically linked terminal Neu5Ac residues. Fusion of the viral envelope and the host cell endosomal membrane[23–26] is mediated by the hemagglutinin, which in the low pH environment of the endosome[26,27] undergoes a dramatic conformational rearrangement. The sialidase assists in the life cycle and liberation of the viral progeny from the infected cell by a number of mechanisms.[28,29] Mucin contains N-acetylneuraminic acid residues and, therefore, the receptor-destroying properties of the sialidase assist the virus particles to penetrate respiratory secretions. The sialidase also inactivates hemagglutinin receptors on the cell membrane, allowing the newly formed virus particles to bud out from the cell surface. Another essential role for the sialidase is to prevent recognition of the N-acetylneuraminic acid residues attached to oligosaccharide chains of newly synthesized hemagglutinin and sialidases of neighboring virus particles. This sialidase function prevents viral aggregation by removal of N-acetylneuraminic acid.[28,29]

Extensive structural information has been provided by x-ray crystal structure analysis of these sialic acid-recognizing viral proteins,[30–54] and this structural information has been used to great advantage, in the case of influenza virus sialidase, to design therapeutics to treat influenza infection. In fact, this approach has been so successful that there are now two drugs commercially available, marketed under the trade names Relenza and Tamiflu that target the viral sialidase. Since the majority of sialic acid-based compounds produced to date target the sialidase, further discussion will be limited to this viral protein. Inhibition strategies directed toward influenza virus sialidase have previously been detailed in many excellent reviews in this field.[55–67] Strategies for the development of inhibitors to interfere with the functioning of the hemagglutinin have recently been discussed elsewhere.[67]

Influenza virus sialidase is a homo-tetramer, with one sialic acid binding site per 60 kDa monomeric unit that is viewed as a shallow depression in electron micrographs. Each monomeric unit is identical and has a fold typical of a sialidase — six, four-stranded antiparallel β-sheets arranged as if they were situated on the blades of a propeller. The tetrameric "head" is anchored to the viral membrane by a very long (100 Å), thin stalk. Critical to the successful design of the drugs to arrest influenza infection by targeting the sialidase was the observation that the amino acid residues important for catalytic functioning of the enzyme that line the active site and bind to N-acetylneuraminic acid, are invariant for both influenza A and B strains of the virus that infect human populations.[68]

The active site of influenza virus sialidase contains a number of well-formed adjoining pockets and is lined by a high proportion of charged and polar amino acid residues, although there are also some highly specific hydrophobic areas. The natural substrate for the enzyme, an α-ketosidically linked N-acetylneuraminic acid, resides in the active-site pocket in a distorted half-chair manner. Positioning of the substrate is achieved by way of three strategically arranged arginine residues (Arg118, Arg292, and Arg371), which form strong charge–charge associations with the negatively charged carboxylate group of Neu5Ac. Tyr406 further assists in the orientation of the Neu5Ac and is believed to stabilize the proposed sialosyl cation transition-state intermediate formed during the catalytic reaction. Other important interactions between the substrate, Neu5Ac, and the active site are observed close to the N-acetyl group and the glycerol side chain of Neu5Ac. The carbonyl group of the N-acetyl group forms a hydrogen bond with Arg152 and a buried water molecule. Important hydrophobic contacts are observed between the methyl of the N-acetyl group and conserved tryptophan and isoleucine residues (Trp178 and Ile222). This region of the catalytic site is referred to as the C-5 pocket. In the region of the catalytic site where the glycerol side chain is located (the C-6 pocket), the C-8 and C-9 hydroxyls of Neu5Ac form a bidentate hydrogen-bonding interaction with the acidic oxygens of Glu276. The C-4 hydroxyl of Neu5Ac is directed toward the carboxylate oxygen of Glu119 in the C-4 pocket. All the important, conserved interactions between Neu5Ac and the catalytic site amino acid residues are shown in Figure 24.1.[69]

In 1974, the first N-acetylneuraminic-based inhibitor of influenza virus sialidase, Neu5Ac2en (4), was described. This dehydrated, sialic acid derivative inhibited influenza virus at micromolar levels *in vitro*. It was not until much later when the x-ray structures of influenza virus sialidase were determined, that real progress was made in the hunt for a potent and selective inhibitor of the sialidase. In fact, in the early 1990s, determination of the crystal structures of N2, N9, and type B influenza virus sialidases in complex with Neu5Ac2en (4) provided an excellent opportunity for drug design.[48,70,71] Strikingly, these analyses revealed that critical amino acids within the catalytic pocket were conserved supporting the notion that a therapeutic agent designed to exploit these conserved features should provide an effective therapeutic against all strains of influenza virus. With the knowledge that Neu5Ac2en (4) was capable of inhibiting influenza virus sialidase already in hand, the crystal structure analysis was used to seek opportunities to further improve inhibitor binding. It was noted from these studies that a pocket situated within the catalytic site adjacent to the C-4 position of N-acetylneuraminic acid was lined with the acidic amino acid residues Glu119 and Glu227. Close inspection of this C-4 pocket revealed that basic substituents such as an amino or guanidinyl group replacing the C-4 hydroxyl of N-acetylneuraminic acid could be accommodated within this region of the active site and further, each of these substitutions could potentially interact with glutamic acid residues located in this region of the active-site pocket. Accordingly, these C-4-modified Neu5Ac2en derivatives were prepared through the oxazoline intermediate 6 by the synthetic protocol outlined in Scheme 2.[72,73] The oxazoline derivative 6 was prepared by treating peracetylated Neu5Ac2en1Me with $BF_3 \cdot Et_2O$. Exposure of 6 to azide afforded the fully protected 4-azido-4-deoxy-Neu5Ac2en derivative 7 which could be reduced, then deprotected to give the

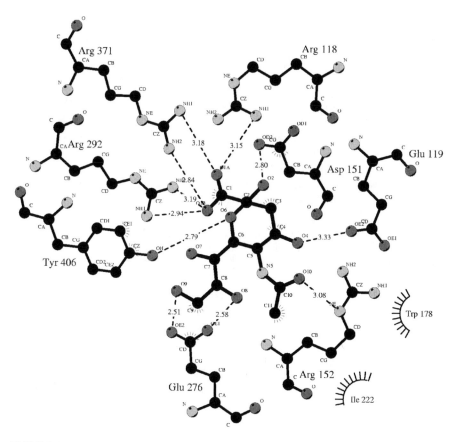

FIGURE 24.1 Interactions between Neu5Ac and the catalytic site residues of influenza virus sialidase.

SCHEME 24.2

desired target compound 4-amino-4-deoxy-Neu5Ac2en (**8**). 4-Deoxy-4-guanidino-Neu5Ac2en (**9**) was prepared by treating **8** with the guanidinylating reagent aminoiminomethanesulfonic acid in base. Modifications and improvements to the synthesis of **9** have been reported.[74,75] Both **8** and **9** were evaluated as sialidase inhibitors and found to be competitive inhibitors of influenza virus sialidase. Most significantly, both derivatives inhibited viral replication *in vitro*[76–79] and *in vivo*[77,80] for all strains of influenza A and B tested. The higher potency of 4-deoxy-4-guanidino-Neu5Ac2en (**9**), known generically as Zanamivir (and often in publications as GG167), led to its selection as a candidate for clinical trial evaluation for the treatment of influenza and its subsequent commercialization under the trade name Relenza (by the now Glaxo Smith Kline).

Zanamivir (**9**) acts as a slow-binding inhibitor. The slow-binding kinetics may arise from expulsion of a crystallographically observed water molecule from the C-4 pocket of the active site.[77,81] Moreover, Zanamivir appears to act in a selective manner for influenza virus sialidase since it displays only weak inhibition against sialidases from mammalian or bacterial sources.[57,76,82] X-ray crystallographic analyses of **8** and **9** in complex with influenza virus sialidase confirmed that indeed both derivatives were interacting with the sialidase catalytic site in the manner predicted by earlier molecular design studies.

Polymeric sialidase inhibitors bearing 4-deoxy-4-guanidino-Neu5Ac2en linked to a polymeric support via substitution at the C-7 hydroxyl, such as **10**, have been prepared.[83] This derivative was more active than the monomeric compound in plaque reduction assays. Polyvalent sialidase inhibitors bearing 4-deoxy-4-guanidino-Neu5Ac2en analogs on a polyglutamic acid backbone attached via a spacer of alkyl ether at C-7, such as **11** and **12**, have also been reported.[84] Although these multivalent derivatives showed less potent sialidase inhibitory activity against influenza A virus sialidase than Zanamivir, they were more active in plaque reduction assays against influenza A virus. Intranasal administration in a mouse model of one of the most potent polyvalent derivatives compared with Zanamivir showed increased survival rates.[84]

The success of Zanamivir as a potent and selective inhibitor of influenza virus sialidase that blocked viral replication resulted in extensive structure–activity relationship (SAR) studies being performed on the Neu5Ac2en (**4**) template as well as other noncarbohydrate templates. As will be demonstrated in the following sections, modifications to all positions around the Neu5Ac2en template have been explored in an attempt to optimize inhibitory activity, and to improve the physicochemical properties of the sialidase inhibitors.

1. C-4 Modification

Only a marginal decrease in inhibition compared with Neu5Ac2en was found on removal of the C-4 substituent to give 4-deoxy-Neu5Ac2en (**13**).[85] Interestingly, epimerization of the C-4 hydroxyl group has a much more detrimental effect on inhibition compared with complete removal of the C-4 substituent, with 4-*epi*-Neu5Ac2en (**14**) showing a 20-fold decrease in inhibition levels compared with Neu5Ac2en.[86] Substitution at C-4 with ether derivatives, such as compounds **15**, **16**, or **17**,[87] resulted in still poorer inhibition (**15**, **16**, and **17**: A/Memphis /1/71[H3N2]

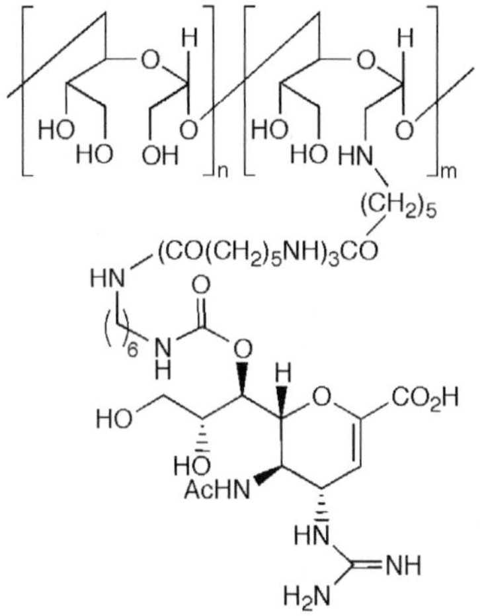

(10)

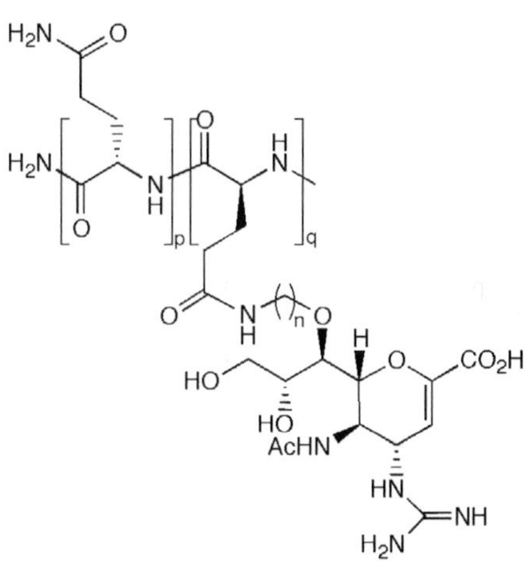

(11)

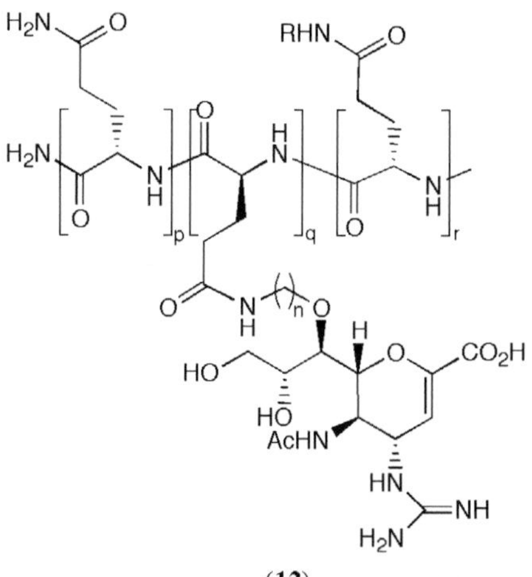

(12)

<30% at 1 mM compared with Neu5Ac2en>90%). The poor inhibition levels of **16** and **17** suggest that the increased chain length is not accommodated within the C-4 binding pocket, even when a basic functionality (e.g., the OCH$_2$C(=NH)NH$_2$ group in **17**) is provided to interact with the Glu119 and Glu227.

Tolerance to functionalization of the amino and guanidino groups of the potent influenza virus sialidase inhibitors 4-amino-4-deoxy- (**8**) and 4-deoxy-4-guanidino-Neu5Ac2en (**9**) has been explored. In all cases where the interaction with the glutamic acid residues was presumably disrupted, inhibition levels were observed to decrease dramatically.[72,74,76] Alkylation of the amine nitrogen of **8**, to give derivatives such as **18** and **19**, resulted in inhibition levels decreasing back to levels comparable with Neu5Ac2en.[72,76] Poorer inhibition (influenza A [N2], K_i=2×10^{-4} M) resulted when the nitrogen was N-acetylated, as in **20**. Functionalization of the N^3

(13) R = NHAc

(14) R = NHAc

(15) R^1 = NHAc; R^2 = OCH$_2$CN
(16) R^1 = NHAc; R^2 = OCH$_2$C(=O)NH$_2$
(17) R^1 = NHAc; R^2 = OCH$_2$C(=NH)NH$_2$

(18) R^1 = NHAc; R^2 = H, R^3 = CH$_2$CH=CH$_2$
(19) R^1 = NHAc; R^2 = R^3 = CH$_3$
(20) R^1 = NHAc; R^2 = H, R^3 = C(O)CH$_3$

nitrogen in the guanidinyl group of **9** with Me, C(O)OEt, NH$_2$, OH, and NO$_2$ groups led to detrimental effects on viral sialidase inhibition in all cases,[74] and again it was noted that inhibition dropped to micromolar levels. Inversion of configuration at C-4 for the amino[66] and azido[72] substituents resulted in approximately tenfold weaker inhibition than was observed for their equatorially arranged counterparts. Modeling studies for the 4-*epi*-amino-Neu5Ac2en showed that the enzyme could accommodate this modification without too much distortion within the active site.[66]

2. C-5 Modifications

It is clear from x-ray crystallographic studies that for sialidases that recognize *N*-acetamide-containing sialic acids, there exists a defined pocket within the active site that accommodates the *N*-acetyl group. Participation of key conserved amino acids in specific hydrogen bonding and hydrophobic interactions (described above) plays an important role in recognition and binding in this region of the active site. The importance of these interactions is highlighted when modifications to the *N*-acetyl group are considered. Replacement of the C-5 acetamido group of Neu5Ac2en with a di- or trifluoroacetamido group marginally improves its inhibitory capacity (fourfold).[88] Supporting the notion that hydrophobic contacts with Ile222 and Trp178 are important to binding, replacement of the *N*-acetamido group in 4-deoxy-4-amino-Neu5Ac2en (**8**) with the shortened *N*-formamido group, to give **21**, resulted in a 100-fold decrease in inhibition[88,89] while extension by a one carbon unit, to give **22**, resulted in only tenfold poorer inhibition. Greater than a 1000-fold drop in activity resulted from either methylation of the *N*-acetamido nitrogen of **8**, to give **23**, or introduction of a bulkier acylamido group (e.g., **24**) or the cyclic lactam derivative **25**.[89]

Complete removal of the *N*-acetamido group of Zanamivir (**9**), to give **26**, was reported to significantly quench its inhibitory activity (IC$_{50}$ influenza A [Aichi] **9**: 5×10^{-9} *M*, **26**: 130×10^{-6} *M*; influenza B [Victoria] **9**: 4×10^{-9} *M*, **26**: $>400 \times 10^{-6}$ *M*).[90] Modification of the *N*-acetamido group to the fluorinated amide (CF$_3$C(O)NH) or sulfonamide (CH$_3$SO$_2$NH) derivatives resulted in more modest (~tenfold) decreases in IC$_{50}$ values.[89]

(21) R = NHC(O)H
(22) R = NHC(O)CH$_2$CH$_3$
(23) R = NCH$_3$C(O)CH$_3$
(24) R = NHC(O)cyclopropyl

(25) R =

3. Glycerol Side Chain Modification

By far, the most extensive investigations on the Neu5Ac2en (**4**) template in recent years have focussed on modification or replacement of the stereochemically complex glycerol side chain.[91] A highly desirable feature of a strategy to produce derivatives where the glycerol side chain is replaced by simpler substituents, is the promise of relative ease of synthetic accessibility. It was also anticipated that this approach may provide access to a range of influenza virus sialidase inhibitors with improved physicochemical properties. In some cases, the glycerol side chain has been replaced with much simpler achiral substituents that can still participate in important interactions with conserved active-site residues within the C-6 pocket and therefore maintain inhibitory levels.[39,89,91–94] As previously noted, the C-8 and C-9 hydroxyl groups on the glycerol side chain form a bidentate interaction with Glu276, while the C-7 hydroxyl is essentially interaction-free. X-ray crystallographic and modeling analyses of some of these derivatives have provided some surprising results and highlighted structural differences between influenza A and B strains.[54,95]

Successful complete glycerol side-chain replacement has been exemplified by studies of the 4-amino and 4-guanidino-4*H*-pyran-6-carboxamides, represented by the general structures **27** and **28**, respectively. In these studies, the lipophilic glycerol side chain has been completely replaced in the majority of cases by hydrophobic groups. Access to derivatives such as **27** and **28** was provided through the C-6 carboxylate derivative **29** obtained by periodate oxidation of the glycerol side chain of **30** using amide coupling via the corresponding pentafluorophenyl ester **31** (91). Extensive SAR studies of these analogs have been performed to ascertain the factors affecting the inhibitory potency with respect to the carboxamide group[54,91,93] as well as modifications at C-4 and C-5.[96] Originally, it had been anticipated that carboxamide derivatives that could potentially mimic the bidentate interaction between the C-8/C-9 hydroxyls and Glu276 via either charged or hydrogen-bond donating groups would provide the most potent inhibition. Subsequently, it was demonstrated that derivatives with small hydrophobic substituents at R^1 and R^2 in compounds **27** and **28** were most potent. Some of these carboxamide derivatives (diethylamide, dipropylamide, and dimethylpyrrolidine) displayed inhibition levels against influenza A comparable with Zanamivir; however, this inhibition was selective because only micromolar levels of inhibition were observed for influenza B. Similar levels of inhibition were noted irrespective of whether the C-4 position contained an amino or guanidino moiety.[91] In this study, over 80 aliphatic amide derivatives were prepared and evaluated as sialidase inhibitors. In general, the secondary amides were weak inhibitors of influenza virus sialidases A and B, whereas the tertiary amide derivatives with one or two small alkyl groups showed potent inhibitory activity, particularly against influenza A sialidase. It was noted that the inhibition data correlated well with *in vitro* antiviral efficacy. Moreover, several of the most potent derivatives displayed useful antiviral activity *in vivo* when administered intranasally and evaluated in a mouse model of influenza virus A infection.[91]

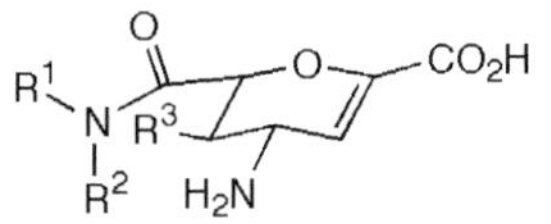

(27) R^3 = NHAc

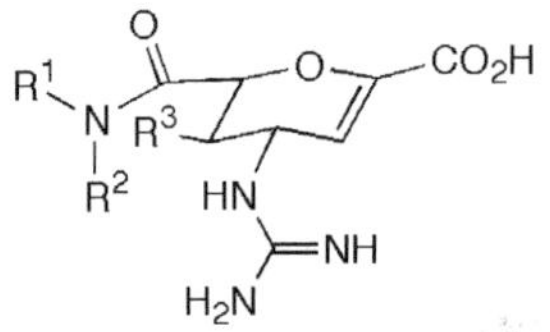

(28) R^3 = NHAc

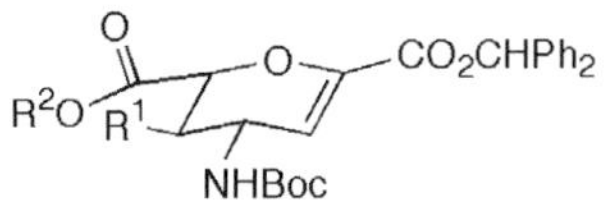

(29) R^1 = NHAc, R^2 = H
(31) R^1 = NHAc, R^2 = C_6F_5

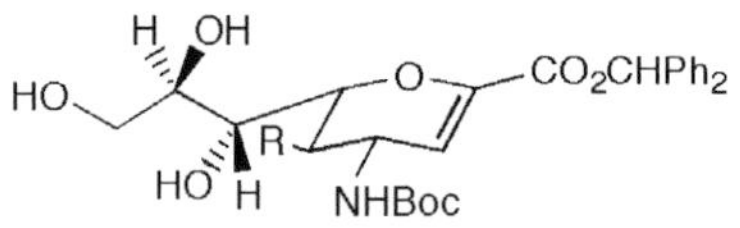

(30) R = NHAc

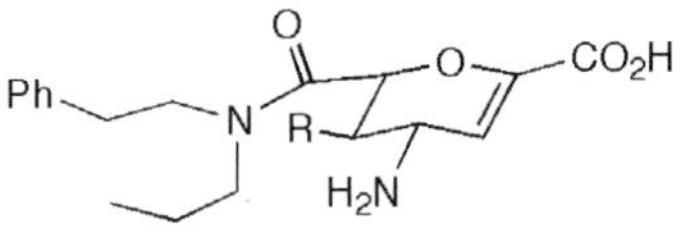

(32) R = NHAc

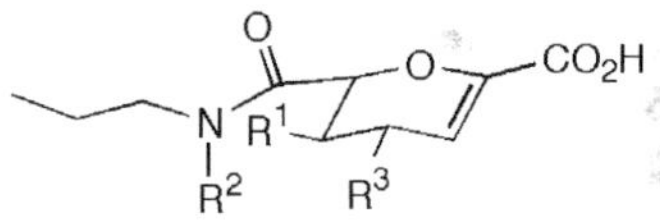

(33) R^1 = NHAc; R^2 = propyl, R^3 = NH_2
(34) R^1 = NHAc; R^2 = $PhCH_2CH_2$, R^3 = NH_2, OH, H

In a fascinating study, the structural basis for the selectivity of the C-6 carboxamide series for influenza A over influenza B was revealed by x-ray crystallographic analysis. Structural analysis of the most potent derivative, **32**, complexed with both influenza A and B was undertaken. For the influenza A virus sialidase complex it was revealed that binding of the tertiary carboxamide group in **32** was accompanied by the formation of an intramolecular planar salt bridge between two amino acids in the glycerol side-chain binding pocket. The formation of this intramolecular salt bridge can only take place if the side chain of the conserved residue Glu276 rearranges to form the salt bridge with Arg224 and if Glu276 also establishes hydrogenbonding interactions with His274 and a buried water molecule. The rearrangement and establishment of these new interactions for Glu276 does not appear to disturb surrounding amino acid residues for influenza A. In effect, this rearrangement changes the hydrophobicity profile of the C-6 binding pocket, and the increased affinity of the inhibitor is attributed to burial of hydrophobic surface area and formation of a salt bridge in an area of low dielectric constant. The scenario is different however for influenza B virus sialidase, where formation of this salt bridge causes major distortions in the amino acid backbone positioning of surrounding amino acids. It was proposed that the major distortions seen for influenza B were the result of steric crowding caused by a tryptophan residue (Trp405) and Arg224 in influenza B (the analogous residue is a glycine for influenza A).[39]

Modifications at C-4 were in general well tolerated in the carboxamide series represented by **33** and **34**.[96] Functionalization at C-4 in **34** with an amino group, a hydroxyl group, or complete removal of the amino group, results in compounds with comparable nanomolar activity. However, alkylation of the amino group in **33**, with either methyl or ethyl groups, resulted in a tenfold loss of activity compared with the corresponding C-4 amino derivative.

In an alternative strategy, the glycerol side-chain of Zanamivir (**9**) is retained but functionalized at C-7 with lipophilic substituents.[97,98] This approach was based on the notion that it is important to retain the bidentate interaction between the C-8 and C-9 hydroxyl groups of the glycerol side chain and the acidic oxygens of Glu276 for inhibitory potency. Moreover, by introducing modifications at C-7, which appears not to interact with sialidase active site residues by x-ray analysis, the physico-chemical properties of the inhibitors can be favorably altered. Using a chemoenzymatic approach, it was found that replacement of C-7 hydroxyl by small lipophilic groups, to give **35a** (R^2=F, N_3, OMe, OEt, NHAc), led to retention of nanomolar levels of sialidase inhibitory activity.[98] To investigate further the structure–activity relationship of this series, an extensive range of C-7 alkyl ether derivatives (e.g., **35b**) was prepared by direct alkylation of the C-7 hydroxyl, and their ability to inhibit influenza virus sialidase was evaluated.[97] It was found that unbranched alkyl ether derivatives at C-7 up to 12 carbons in length showed similar activity when compared with Zanamivir, but better efficacy in plaque reduction assays. Activity of the C-7-modified compounds **35b** was not affected greatly by the addition of terminal OH, N_3, NH_2, or NHAc functionality to the C-7 ether. Replacement of the C-8,C-9 diol with an ethyl group, such as in the series represented by **36**,[97] did not seem to be refractory to sialidase activity but resulted in poorer efficacy in plaque reduction assays for longer alkyl chains compared with series **35**.

(**35a**) R^1 = NHAc; R^2 = See text
(**35b**) R^1 = NHAc; R^2 = O(CH$_2$)$_n$X

(**36**) R^1 = NHAc; R^2 = O(CH$_2$)$_n$CH$_3$

C-7 substitution using a carbamate linkage to introduce a range of hydrophobic *N*-alkyl groups on the carbamate group, represented by structures **37** and **38**, rather than an ester linkage has also been described.[83,99] An advantage of this approach is that introduction of a carbamate linkage at C-7 avoids the known complication of acetate migration along the glycerol side chain of sialic acids.[100,101] Of this series of compounds, a number of *N*-monoalkylated C-7 carbamate derivatives of 4-deoxy-4-guanidino-Neu5Ac2en had inhibitory potency comparable to Zanamivir against influenza A virus sialidase and inhibition of viral replication in plaque reduction assays.[99] Dialkyl carbamates were generally 10 to 100 times less potent than 4-deoxy-4-amino-Neu5Ac2en or 4-deoxy-4-guanidino-Neu5Ac2en.[99]

(37) R = NHAc   (38) R = NHAc

The introduction of cyclic ether moieties at C-7 in 4-deoxy-4-guanidino-Neu5Ac2en represents a further extension of the work investigating C-7 modifications.[102] Representative structures such as **39**, **40**, and **41** have the C-7 hydroxyl group incorporated into either di-hydroxylated tetrahydro-furan-2-yl (**39**) or tetrahydro-pyran-2-yl (**40** and **41**) rings that totally replace the glycerol side chain. Of this series, the most potent was **40**, which had comparable activity to Zanamivir against influenza A virus sialidase and in plaque reduction assays. Of most significance was the finding that compound **40**, in mouse models of influenza infection, showed comparable *in vivo* efficacy to Oseltamivir (described later) an orally administered influenza therapeutic.

(39) R = NHAc   (40) R = NHAc   (41) R = NHAc

4. Cyclohexene Scaffold

Zanamivir is administered as a topical therapy, as an inhaled spray to the upper respiratory tract and lungs, where influenza virus infects human populations.[103,104] The polar nature of Zanamivir precludes the use of oral or intraperitoneal administration routes. Moreover, Zanamivir is rapidly eliminated from circulation by renal excretion.[103,104] Although Zanamivir was approved as a pharmaceutical, the opportunity to develop an orally active influenza drug still existed. Capitalizing on the extensive body of knowledge gained in the development of Zanamivir, new strategies to design influenza drugs were exploited with the most intense efforts concentrating on designing new therapies with improved physicochemical properties based on noncarbohydrate templates. Of particular interest were considerations such as the necessary balance between water solubility and lipophilicity for the optimal absorption of a drug in the gastrointestinal tract. Broadly speaking, major research efforts have focussed on replacement of the carbohydrate pyranose ring with noncarbohydrate core templates based on cyclohexene,[24,105–108] cyclopentane,[109–111] aromatic,[28,112–119] or pyrrolidine rings.[120–121] One attractive aspect of designing inhibitors based on simple chemical templates such as these is that the complexity of synthetic manipulations compared with using carbohydrate templates is significantly reduced. Particular success in this regard has culminated in the development of Oseltamivir (**42**) by Gilead/Hoffmann-LaRoche (also known in publications as GS4104 or commercially as Tamiflu) as a noncarbohydrate mimetic of 4-amino-4-deoxy-Neu5Ac2en.

(**42**)

    A number of factors highlighted the cyclohexene template as a judicious choice. First, it was proposed that the double bond and its position in the cyclohexene ring would mimic to some extent the putative sialosyl cation intermediate formed during the influenza virus sialidase catalyzed reaction. It was also noted that the versatile cyclohexene ring would be amenable to rapid and relatively facile chemical modification allowing many derivatives to be synthesized and tested within a short period of time and, further, that the cyclohexene template should be robust enough to withstand enzymatic degradation within the body. Cyclohexene analogs of Neu5Ac2en, **43** and **44** were prepared,[122] with a view to examining the effect of the positioning of the double bond on the inhibition of viral and bacterial sialidases. These derivatives showed micromolar levels of inhibition against influenza A, with better inhibition for isomer **44**. A similar trend and levels of inhibition were observed for derivatives **45** ($IC_{50} = 6.3 \ \mu M$) and **46** ($IC_{50} > 200 \ \mu M$) featuring a replacement of the glycerol side chain of Neu5Ac2en with an hydroxyl group and the C-4 hydroxyl with an amino group.[105]

    Early SAR studies based on cyclohexene scaffolds[105,106,108,123] provided the first insights that the replacement of the glycerol side chain with lipophilic substituents was not detrimental to influenza virus sialidase inhibition. The combined effect of glycerol side-chain replacement at C-6 with a 3-pentyloxy group and variations to other positions around the cyclohexene ring such as the C-4 or C-5 position[108] were investigated. Modification at C-4 with basic nitrogen-containing functional groups (C-5=N-acetyl; C6=pentyloxy), such as amine, amidine, guanidine, and N-methylguanidine revealed that the amidine derivative ($IC_{50} = 140$ nM, influenza A) showed similar levels of inhibition to the amine ($IC_{50} = 130$ nM, influenza A), but the guanidino derivative proved to be a much more potent inhibitor ($IC_{50} = 1.8$ nM, influenza A). C-5 replacement of

(43) R = NHAc

(44)

(45)

(46)

the N-acetyl group with CH_3CO^-, CF_3CO^-, CH_3CH_2CO, CH_3SO_2 with either azide or amine substituents at C-4 was considered. The N-acetyl and N-trifluoro acetyl (C-4=NH_2) showed similar levels of inhibiton, IC_{50}=130 and 100 nM (influenza A), respectively. Taken together this study suggested that the cyclohexene scaffold, combined with a lipophilic group at C-6, and retaining substituents at C-1 (-COOH), C-4 (NH_2 or guanidine) and C-5 would provide potent influenza virus sialidase inhibition.

Further comprehensive structure–activity studies exploring structures represented by 47, with intense focus on the various lipophilic alkyl side chains were undertaken.[105,106] As noted above, it was the intention that derivatives of this type would act as oral therapies against influenza virus infection, since it was rationalized that the oral bioavailability (compared for instance to Zanamivir) could be improved by replacement of the glycerol side chain with lipophilic groups to ameliorate some of the hydrophilic nature of the drugs. Accordingly, the size and geometry (straight-chain, branched, and cyclic) of the alkyl side chains were systematically modified to provide a profile of the factors that affected the sialidase inhibitory activity for this series of compounds. Replacement of R=H in 47 (IC_{50}=6300 nM, influenza A) with a methyl (IC_{50}=3700 nM) or ethyl (IC_{50}=2000 nM) to an n-propyl (IC_{50}=180 nM) produces a marked increase in inhibitory potency,[105] which may be a reflection of the ability of the n-propyl terminal methyl being close enough to participate in hydrophobic interactions with active-site residues. Interestingly, similar levels of inhibition were also noted even with n-alkyl chains as long as n-nonyl. Potent inhibition (IC_{50}<10 nM) was noted for branched derivatives such as 47a–c.[106] Differences in inhibitory potency for influenza A and B virus sialidase for cyclic analogs such as R=cyclohexane (IC_{50}=60 nM influenza A; IC_{50}=120 nM influenza B) and most strikingly for 47d (IC_{50}=16 nM influenza A; IC_{50}=6500 nM influenza B) which were not as dramatic for the straight chain or branched derivatives.

X-ray crystallographic analysis of some of the most potent compounds in this series, bound to influenza A virus sialidase, revealed that the lipophilic groups of some of these derivatives occupied the hydrophobic region of the glycerol binding pocket formed by residues Ile222, Arg224, and Arg226[105] and also the new hydrophobic pocket formed in this region of the active site when Glu276 rearranges its positioning to participate in a salt bridge with Arg224. This rearrangement increases the size of the glycerol side-chain binding pocket and reduces its hydrophilic nature.

This comprehensive study of the effect of replacing the glycerol side chain in Neu5Ac2en with simple hydrophobic ether groups led to the development of Oseltamivir (**42**, GS4071). Oseltamivir is a potent inhibitor ($K_i > 1$ nM) of influenza virus sialidase, showing slow-binding kinetics and similar levels of efficacy against influenza A and B (IC$_{50}$ 1 nM A/Pr/8/34 and 3 nM B/Lee/40).[25] The development of Oseltamivir has previously been reviewed.[61,107,124] Clinically, Oseltamivir is administered as its orally bioavailable ethyl ester prodrug (**48**, GS4104).[125] Following absorption in the gastrointestinal tract, **48** is enzymatically cleaved by esterases in blood and tissue[126,127] to liberate the active parent drug, **47**. Guanidinyl analogs, **49** and **50**, were also prepared and evaluated as influenza virus sialidase inhibitors. The oral bioavailabilities[24] of all the derivatives were evaluated and, interestingly, while the prodrug form of the amine derivative showed excellent oral bioavailability, those of the guanidino analog, **49** and its prodrug **50**, were poor.

X-ray crystal structure analysis of the complex of **42** with influenza virus sialidase reveals that the carboxylate, amine, and *N*-acetyl groups all interact with the sialidase active-site residues in the same way as the analogous functional groups of 4-amino-4-deoxy-Neu5Ac2en. The pendant arms of the hydrophobic 3-pentyl group of **42** occupy the hydrophobic regions formed when the side chain of Glu276 rearranges to form a charge–charge interaction with Arg224.

Similar investigations were also performed with a series of aza-substituted carbocyclic compounds represented by the general structure **51**.[128] For series **51**, various secondary and tertiary amines were prepared, and the three most potent analogs were the following **51a**: R_1=H, R_2=CH(CH$_2$CH$_3$)$_2$ (IC$_{50}$=11 nM A/Pr/8/34 and 100 nM B/Lee/40); **51b**: R_1=CH$_3$, R_2=CH(CH$_2$CH$_3$)$_2$ (IC$_{50}$=6 nM A/Pr/8/34 and 60=nM B/Lee/40); and **51c**: R_1=R_2=CH$_2$CH$_2$CH$_3$ (IC$_{50}$=12 nM A/Pr/8/34 and 60=nM B/Lee/40). Compound **51a** is the equivalent aza-analog to Oseltamivir and shows comparable activity. Confirmation that the pentyl group of **51b** binds in the hydrophobic pocket formed by the hydrocarbon side chains of Glu276, Ala246, Arg224, and Ile222, similar to Oseltamivir, was provided by x-ray analysis. Substituted and unsubstituted cycloalkylamines, such as **52**, were also investigated and although some showed nanomolar levels of inhibition against influenza A and B, none were as potent as Oseltamivir.[129]

(51)   (52)   (53)

Recently, the carbocyclic derivative 53 prepared from quinic acid as a starting material has been reported.[130] Conversion of the quinic acid derivative into the shikimic acid derivative, followed by several hydroxyl group functionalizations gave the target compound (3R,4S,5R)-4-acetamido-3-guanidino-5-hydroxycyclohex-1-ene-1-carboxylic acid (53). It has been proposed that the guanidino moiety is correctly positioned to occupy the region of the active site in which the glycerol side chain of the natural substrate would reside. Biological evaluation of this derivative has not been reported to date.

Alternative templates have explored the incorporation of one (e.g., 54)[131] or two (e.g., 55)[132] nitrogen atoms within a six-membered ring system. 6-Acetamido-5-amino- and -5-guanidino-3,4-dehydro-N-(2-ethylbutyryl)-3-piperidinecarboxylic acids 54a and 54b, respectively, were prepared from siastatin B (a natural bacterial sialidase inhibitor isolated from *Streptomyces* culture) but found to be inactive against influenza A. Evaluation of template 55 for viral sialidase inhibition, where R=–(C=O)CH(CH$_2$CH$_3$)$_2$ irrespective of whether R is in a *cis* or *trans* arrangement, gave micromolar inhibition against influenza A and 100 times poorer inhibition against influenza B. Marginally better inhibition (~ 10 fold) was observed for the corresponding guanindino derivative of 55. The poor inhibitory potency of these derivatives was revealed by NMR and x-ray analysis of these derivatives in complex with influenza A sialidase.

In solution the amine and N-acetyl groups of 55 occupy pseudo-axial positions rather than the preferred pseudo-equatorial positions that are adopted on binding to influenza sialidase. Further, x-ray analysis revealed that the ethyl groups both crowd into only one of the hydrophobic pockets, presumably due to the extra constraint imposed by the planar nature of the amide bond.

As part of the continuing search for novel scaffolds that mimic the geometry of the Neu5Ac transition state in the sialidase-catalyzed hydrolysis reaction, tetra-substituted bicyclo[3.2.1]octene 56[133] derivatives were prepared. Additionally, a bicyclo [2.2.2]octane 57[134] derivative based on a cyclohexane scaffold was investigated. The natural substituents at equivalent positions to Neu5Ac were retained for both scaffolds, but the lipophilic groups replacing the glycerol side chain resulted in only modest levels of inhibition (>μM) for influenza A and B.

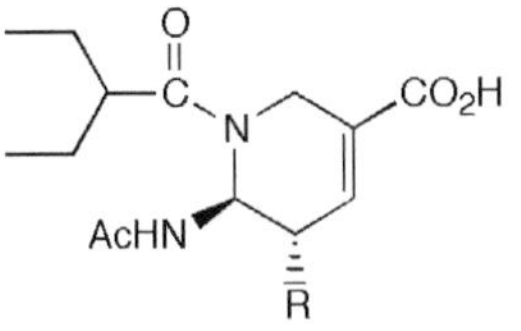

(54a) R = NH₂
(54b) R = NHC(=NH)NH₂

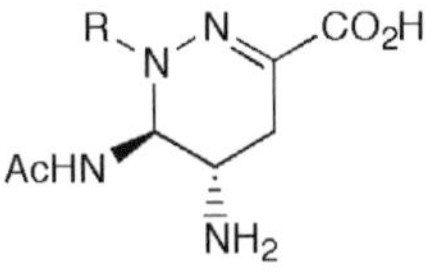

(55)

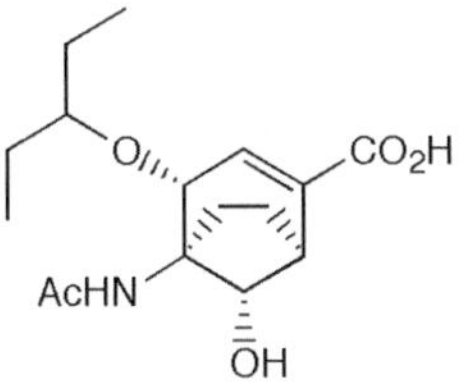

(56)

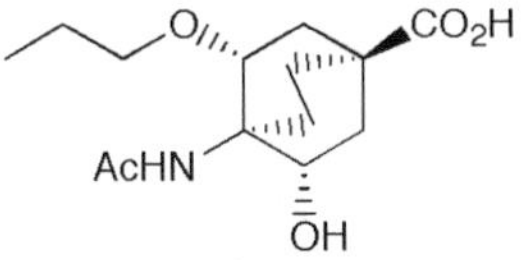

(57)

5. Cyclopentane Template

In the early 1990s, it was reported[135] that α-/β-6-acetylamino-3,6-dideoxy-D-*glycero-altro*-2-nonulofuranosonic acid (**58**) inhibited influenza virus sialidase at similar levels to Neu5Ac2en. Structurally, **58** retains some identifiable remnants of the important functional groups of Zanamivir that interact with the influenza virus sialidase active site. Notably, a carboxylate group interacts with the three arginine residues of the triarginyl cluster, a diol (mimicking the C-8 and C-9 hydroxyl groups of the glycerol side chain that participate in an important bidentate interaction with Glu276), and an *N*-acetyl group. Babu and coworkers[109] cocrystallized **58** with influenza virus sialidase (N9) and realized from this analysis that even though the cyclopentane ring was displaced from where the pyranose ring of Neu5Ac2en resides, the relative orientation of the important functional groups was roughly correct to allow interaction with active-site residues. In the same study, the trisubstituted cyclopentane derivative **59** was prepared, with a guanidino group replacing the amine and the triol side chain removed. Compound **59** was found to inhibit influenza virus sialidase with activity comparable to Neu5Ac2en. As predicted, x-ray crystallographic analysis of **59** in complex with influenza virus sialidase revealed that the guanidino group interacts with Asp151, Glu119, and Glu227 in a similar manner to the guanidino group of Zanamivir.[109] Continuing this study, the racemer **60** was prepared, which incorporated an *n*-butyl group to exploit the previously described hydrophobic pocket. Enantiomeric selection of the active isomer of **60** was achieved by a crystallization screening method, where the racemic mixture was soaked with influenza virus sialidase and data were then collected at 2 to 2.5 Å resolution. From these analyses, the active isomer was identified as containing the carboxylic acid and 1-(acetylamino)pentyl group in a *trans* orientation, while the carboxylic acid and the guanidino group were oriented in a *cis* arrangement with respect to each other.[109] Most unexpectedly, the guanidino group of **60** is found in a different orientation than that of the guanidino group of either **59** or Zanamivir. As noted for others classes of influenza sialidase inhibitors.[39,105] where the glycerol side chain was replaced by hydrophobic substituents, the hydrophobic groups interact with influenza A sialidase in a different way than that of influenza B.

The body of knowledge obtained from these SAR studies with **59** and **60** led to the development of **61**, known as BCX-1812, an orally active, potent inhibitor of influenza A and B (109). The active isomer of **61** was identified from crystal soaking experiments as (1S,2S,3R,4R,1′S)-3-(1′-acetamido-2′-ethyl)butyl-4-(amino-imino)-methylamino-2-hydroxycyclopentane-1-carboxylic acid, which was then purposely synthesized. BCX-1812 is orally active in mouse models of influenza,[111] inhibits influenza virus growth *in vitro*,[136] and shows nanomolar inhibition of influenza A and B.[109] Moreover, BCX-1812 is specific for influenza virus sialidase and is four-orders of magnitude less potent against sialidases from either mammalian or bacterial sources.[109] X-ray crystallographic analysis of BCX-1812 (**61**) in complex with influenza virus sialidase revealed that, similar to **60**, the guanidino group interacts with the active site of influenza virus sialidase in a different way when compared with either Zanamivir or even **59** from the same cyclopentane series. In adopting a different conformation the guanidino group displaces a water molecule in the active-site pocket normally occupied by the C-4 hydroxyl of Neu5Ac2en. The implications of the differences in the manner that the guanidino group interacts with the active site of influenza virus sialidase were found to have important implications for cross-reactivity of influenza virus resistant to Zanamivir. In fact, BCX-1812 retains its potent activity against a Zanamivir-resistant Glu119Gly variant with influenza A. BCX-1812 (**61**) was evaluated in Phase III clinical trials with more than 1200 patients, but was found to provide no statistically significant reduction in the duration of influenza symptoms and has subsequently been withdrawn from development.

(58)   (59)   (60)

(61)   (62)   (62a) R =   (62b) R =   (63)

Another approach to sialidase inhibition based on five-membered rings has explored tri- and tetra-substituted pyrrolidine analogs of the general type **62** and **63**. For this study, the "traditional medicinal chemistry approach", using x-ray crystallography, organic synthesis, and computational chemistry to explore scaffolds was combined with combinatorial chemistry methods such as high-throughput parallel synthesis.[137] A staggering number of derivatives (>600) of **62** were prepared, with variations at the pyrrole nitrogen including amides, carbamates, sulfoamides, and ureas. The vast majority were found to be inactive as influenza virus sialidase inhibitors. Micromolar levels of inhibition were observed, however, with some derivatives containing α-branched amides derived from aliphatic secondary amines such as **62a** and **62b**. For the tetra-substituted series, represented by **63**, only the acetamido ($R=COCH_3$) or trifluoroacetamido ($R=COCF_3$) derivatives showed low micromolar levels of activity against the sialidase from influenza A virus.

6. Aromatic Template

The development of sialidase inhibitors based on an aromatic template using benzoic acid has been extensively explored.[28,112–114,116–119,138] The rationale behind this approach is that the template structure only provides a scaffold to orient correctly the attached functional groups in the sialidase active site and is not itself directly involved in interactions. X-ray crystallographic examination of Zanamivir in complex with influenza virus sialidase provides support for this notion since the pyranose ring is not directly involved in any interactions with the active site of influenza virus sialidase.

Of the many derivatives based on the chemically simple benzoic acid template (called BANA) prepared and evaluated as influenza virus sialidase inhibitors,[113] 4-N-acetamido-3-guanidinobenzoic acid (**64**, named BANA113 or BCX-140) had the most potent inhibition with an IC_{50} of 2.5 μM. Interestingly, the guanidino moiety of **64** was observed to occupy the glycerol side-chain pocket and interact with Glu275, when examined crystallographically in complex with influenza B virus sialidase. Poorer inhibition ($IC_{50}=70$ μM) was observed with 4-N-acetamido-3, 5-diguanidinobenzoic acid **65**.[117]

Some examples of the types of structures investigated in this series include replacement of the guanindino group of **64** with amino imidazole structures, represented by **66**, which interestingly did not appear to affect the inhibitory activity significantly.[116] The 4-N-acetamido group of **64** has also been replaced by substituted pyrrolidines to give structures, such as **67**, but again only micromolar levels of inhibition were achieved with these derivatives.[118] More successful was the replacement of the guanidino group with the hydrophobic 3-pentylamino group to give **68**, which shows nanomolar levels of inhibition against influenza A sialidase.[118] It is perhaps not surprising that replacement of the guanidino group of **64** with a hydrophobic group leads to more potent inhibition because x-ray analysis showed that the guanidino group resided in the glycerol side-chain pocket of influenza virus sialidase. Indeed, when the x-ray crystal structure of **68** in complex with influenza A sialidase was examined it was found that the 3-pentylamino group occupies the hydrophobic pocket that is created in the vicinity of the glycerol side-chain pocket when Glu[275] positions itself to form a charge–charge interaction with Arg224.[114] Presumably,

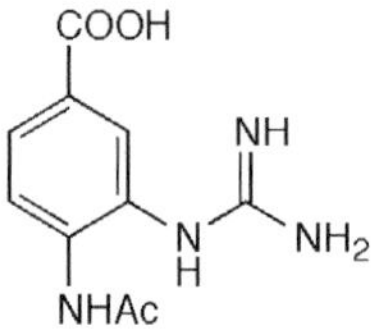

(64)

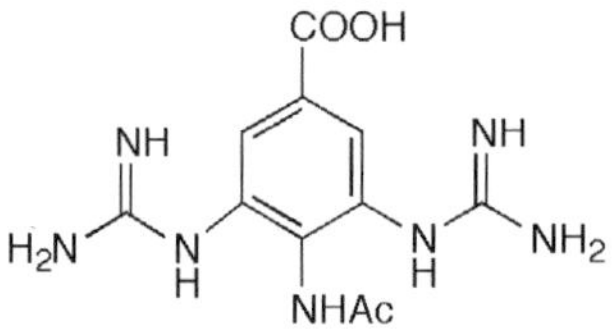

(65)

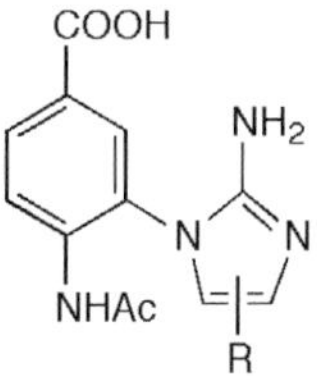

(66)

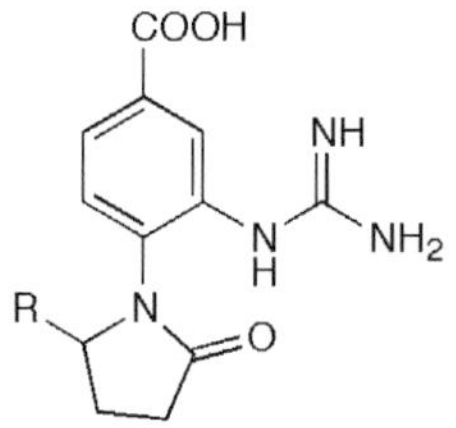

(67)

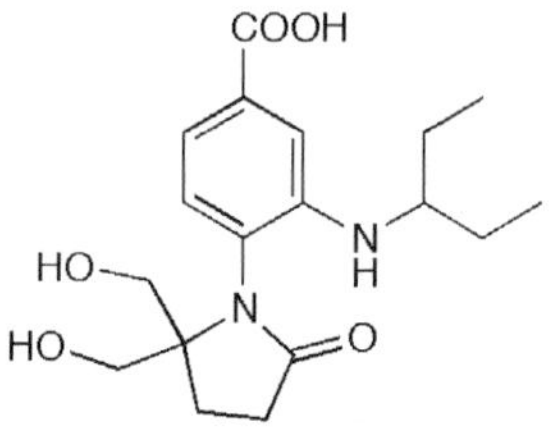

(68)

then, the significant entropy gains that result from the increased hydrophobic interactions lead to more potent activity.

B.   *VIBRIO CHOLERAE* SIALIDASE

VC is the bacterium responsible for cholera, a debilitating enteric disease causing widespread illness and mortality in low socioeconomic regions of the world where hygiene and living standards are suboptimal. The VC sialidase enzyme performs two functions in the life cycle and infectious process of the bacterium that assists progression of the disease by the microorganism. First, as part of a multienzyme mucinase complex,[139] the enzyme degrades gastrointestinal mucin, and in the process reduces its viscosity, facilitating access of the bacterium to the host epithelial cells. Secondly, the sialidase cleaves sialic acid residues from higher order gangliosides to unmask GM_1, the receptor for cholera toxin.[140,141] Given the recognized important role that sialidase performs in the pathogenicity of the VC microorganism, the sialidase has received some attention as a drug design target.[76,94,142–158]

The x-ray crystal structure of VC sialidase has been determined at 2.3 Å resolution,[82] and has revealed that the sialidase catalytic domain is flanked by two lectin domains. The catalytic domain is similar in many respects to that of other sialidases from bacterial and viral sources, which is not surprising given that the amino acid residues found within the active site known to be essential for catalytic activity are strictly conserved. One notable difference, however, is the requirement for a Ca^{2+} that plays an essential structural role. Other differences are also noted for the bacterial sialidase in comparison to the influenza virus sialidase. In particular, and most importantly for drug design efforts, the topography of the C-4 pocket differs significantly from the viral sialidase. As mentioned, influenza virus sialidase contains a large pocket in the vicinity of C-4, large enough to accommodate a guanidinyl moiety. This C-4 pocket is lined with two conserved amino acids, namely Glu119 and Glu227, conveniently exploited in the design of Zanamivir (**9**) to interact with the terminal nitrogen atoms of the guanidinyl group. For the bacterial sialidase, this pocket is much smaller because the Arg245 and Asp232 residues that interact with the C-4 hydroxyl of Neu5Ac occupy it. The notion that a guanidino group would be too sterically demanding to fit in the C-4 pocket of the bacterial sialidase was supported by inhibition data[76] for 4-deoxy-4-guanidino-Neu5Ac2en (**9**) against VC sialidase ($K_i = 6.0 \times 10^{-5}\,M$) as compared with the viral sialidase ($K_i = 1.0 \times 10^{-9}\,M$). For these reasons, the therapeutics currently available that target the viral enzyme are not as effective against this bacterial sialidase. Other differences for the bacterial sialidase are observed in the vicinity of the C-6 glycerol side chain of Neu5Ac.

Although the structure of VC sialidase has been available now for almost a decade, the significant advances in inhibitor design for influenza virus sialidase that occurred as a result of the availability of crystal structure coordinates, have not been paralleled for VC sialidase inhibitors to date. In fact, the astonishing capacity of the enzymes to accommodate a diverse range of modifications to the natural substrate have stymied most of the drug design strategies attempted so far, and only limited progress has been achieved. For a comprehensive account of this area prior to 1990, readers are directed to the excellent earlier review by Zbiral.[7]

Neu5Ac2en (**4**) inhibits VC sialidase with a K_i in the micromolar range ($K_i = 10^{-5}$ to $10^{-6}\,M$),[159] as first described in the late 1960s. Of Neu5Ac2en (**4**) and the various 5-acyl derivatives of Neu5Ac2en described by Tuppy and co-workers,[88,160] the most potent inhibition of VC sialidase was achieved by replacement of the *N*-acetyl group with a *N*-trifluoroacetyl moiety, resulting in a K_i of $1.8 \times 10^{-6}\,M$.[88,160] (The corresponding phosphonic acid analog, i.e., C-2 replacement of the carboxylate group for a phosphonate, showed inhibition levels of $7.5 \times 10^{-5}\,M$.[161]) Replacement of the *N*-acetyl group by amino, hydroxyl (KDN) and azido groups were all refractory to inhibition.[151] A decrease in K_i values by three-orders of magnitude on going from Neu5Ac2en to Neu2en (**69**) was suggested by the authors to highlight the requirement of a carboxamido functionality, and a 100-fold decrease in K_i values observed for the azido derivative suggested that the π-system of the azido group is not an adequate substitute for the *N*-acetyl group.[151] Earlier studies had revealed that the naturally occurring 2-deoxy-2,3-didehydro-*N*-glycolylneuraminic acid (**70**) showed slightly greater inhibition (50% at 10 μM) of VC sialidase than the *N*-acetyl analog (50% at 15 μM).[162]

(4) R = NHAc
(69) R = H

(70) R = NHC(O)CH$_2$OH

(71) R = NHAc

(72) R = NHAc

In light of the crystal structure analysis,[82] a molecular modeling study, using the program GRID and molecular dynamics calculations was undertaken to try to rationalize the structural requirements of inhibitor binding.[156] GRID is a molecular design tool that can be used to predict *in silico* steric and electronic interactions between various functional group probes and the amino acids that line the active site of VC sialidase. Alterations to the Neu5Ac2en framework that could optimize energetically favorable interactions with important active-site residues were considered as a result of this analysis. From the outset it was evident that Neu5Ac2en derivatives, which exploited the very large C-5 pocket, with extensive cavities adjacent to and below where the C-5 N-acetyl of Neu5Ac2en would reside, may provide access to potent inhibitors of the sialidase. Close inspection of the C5-binding domain led to the conclusion that an interaction between the active-site residue Asn318 and the C5-acylamino carbonyl is likely, supporting the previously reported view[151,152,163,164] that the carbonyl group is required for recognition by the sialidase.

Favorable interactions were predicted between the sialidase and Neu2en derivatives with hydroxyl or halogen-substituted acyl groups, at C-5. Neu5Ac2en derivatives modified at C-5 (replacement of N-acetyl with N-glycolyl, N-chloro, or the aliphatic derivatives, N-propanoyl, N-butanoyl, N-acryloyl) were modeled *in silico*. Conformational search protocols suggested that the N-glycolyl and N-chloro derivatives would be comfortably accommodated within the C-5 binding pocket. With regard to the aliphatic derivatives, however, while the N-glycolyl and N-acryloyl derivatives could be accommodated, alterations such as chain extension to the N-butanoyl or replacement of the methyl of N-acetyl group to a phenyl ring giving the N-benzyloyl-Neu2en were not accommodated for steric reasons. Although for VC sialidase the C-5 binding domain is a large cavity (extending adjacent to and below the C-5 N-acetyl group, as mentioned above), it does not extend much further directly from the C-5 N-acetyl methyl and therefore there is little scope for further C-5 chain extension. Given guidance by the molecular design studies, a range of C-5-modified Neu2en derivatives were prepared and their inhibition data against VC sialidase obtained (Table 24.1).[156] Although inhibition data for some of these derivatives have previously been reported,[88,165] it was felt necessary for completeness to reinvestigate all derivatives using the same assay protocols for a true comparison of their inhibitory activities. Most interestingly, the inhibition data appeared to correlate well with the predicted outcomes of the molecular modeling studies, with all compounds well accommodated within the enyzme's active-site architecture.

TABLE 24.1
Inhibition Data[156] against VC Sialidase

Inhibiter	K_i (μM)
N-acetyl-Neu2en	3.56
N-glycolyl-Neu2en	2.61
N-chloroacetyl-Neu2en	3.45
N-trichloroacetyl-Neu2en	250
N-acryloyl-Neu2en	2.55

In earlier investigations, functionalization at C-4 was investigated,[152] given that it had previously been noted that acetylation of the C-4 of Neu5Ac renders sialoglycoconjugates impervious to hydrolysis by bacterial and mammalian sialidases.[166,167] Although no significant improvement in inhibition, compared with Neu5Ac2en, was noted for the C-4-modified derivatives, comparable inhibition was observed with the 4-azido (**71**) ($K_i=7.5\times10^{-5}$ M) and 4-*epi*-4-azido (**72**) ($K_i=7.7\times10^{-5}$ M) derivatives.[152] More modest levels of inhibition were observed for the 4-amino, 4-*N*-formyl, 4-*N*-acetyl, and 4-*epi*-Neu5Ac2en derivatives all of which showed K_i values in the millimolar range.

With regard to the glycerol side chain of Neu5Ac2en, extensive exploration of possible modifications have been reported.[143,144] C-7, C-8, and C-7/C-8 glycerol side-chain epimers of Neu5Ac2en were found to be inhibitors of the sialidase. Deoxy-analogs of Neu5Ac2en were also explored,[144] with the 7-, 8-, and 9-deoxy derivatives showing comparable inhibition to Neu5Ac2en. 4,7-Dideoxy-Neu5Ac2en showed 100-fold poorer inhibition ($K_i=6.7$ mM), much weaker than the 4-deoxy- or 7-deoxy-Neu5Ac2en analogs ($K_i=0.5$ and 0.09 mM, respectively). Complete replacement of the glycerol side chain has also been trialed. A branched substitution at C-6 has been explored,[150] via the generation of 6-*C*-methyl (**73**) (nonpolar) and 6-*C*-(hydroxymethyl) (**74**) (polar) Neu5Ac2en analogs. These branched Neu5Ac2en derivatives were weak inhibitors (K_i value in the high millimolar range) of VC sialidase. Given that these modifications were tolerated by the bacterial sialidase, more recently[94] the synthesis of C-6-modified Neu5Ac2en and KDN2en sialylmimetics with an *iso*-propyl ether group replacing the glycerol side chain were prepared. These derivatives were of interest to determine if simple hydrophobic side chains would be tolerated by other sialidases, since it was well known that influenza virus sialidase was susceptible to strong inhibition when hydrophobic groups were substituted at this position of Neu5Ac2en mimetics.[39] Inhibition data for the C-6 *O-iso*-propyl ether Neu5Ac2en derivative **75** showed a 100-fold decrease in inhibition ($K_i=1.2\times10^{-4}$ M) against the bacterial enzyme when compared with Neu5Ac2en (**4**) ($K_i=3.4\times10^{-6}$ M) while for the viral enzyme, Neu5Ac2en (**4**) and **75** showed similar inhibitory levels. The evidence provided by the crystal-structure analysis and molecular modeling studies that significant differences in the active-site arrangement[39] exist between the viral and bacterial sialidases in the vicinity of the C-6 glycerol side chain provides a reasonable explanation for these observations.[82,156] A rearrangement of the polypeptide backbone, similar to that observed for influenza virus sialidase, does not appear as readily accommodated by the bacterial sialidase. In order to investigate further the interaction between this derivative and the bacterial enzyme, an NMR study using saturation transfer difference (STD) experiments and computational methodology, for example Dock, was undertaken.[168] The results of the STD studies suggested that the association between the bacterial sialidase and the mimetic **75** in the vicinity of the *O-iso*-propyl ether moiety is minimal. It appeared from the Dock experiments that sterically, the *O-iso*-propyl group should be accommodated within this region of the active site. Another possible explanation for the poorer inhibition data was suggested to be the extent of burial of the side chains of the inhibitor. The *N*-acetyl group is buried deep within a pocket of the active site while examination of the *O-iso*-propyl group shows that it is more exposed to the surface of the enzyme.

(73) R = NHAc   (74) R = NHAc   (75) R = NHAc

In strategies that moved away from utilizing the Neu5Ac2en template as a starting point for inhibitor design, Vasella and co-workers[122] investigated sialylmimetics based on pyrrolidine and piperidinose analogs, where the endocyclic oxygen of Neu5Ac is replaced by nitrogen. These compounds were found to be inhibitors of VC sialidase. This approach was supported by the notion that many naturally occurring α- and/or β-glycosidase inhibitors possess a basic nitrogen atom in a five- or six-membered ring (see Reference 145 and references cited therein). 6-Amino-2, 6-dideoxysialic acid analogs were prepared through a series of synthetic manipulations beginning with a Mitsunobu reaction of the nitroglycal **76**, to provide the formate **77** with inversion of configuration at C-3.[145] In their design strategy, the important C-4 OH functionality for sialidase activity[1,167] was retained and the known inhibitory ability of 4-*epi*-Neu5Ac2en analogs considered. Accordingly, derivatives **78**, **79**, and **80** were prepared. It was established that the 6-amino *N*-acetylneuraminic acid derivative **79**, with an axially disposed carboxylate functionality was inactive.[145] Interestingly, the authors note that pipecolinic acid **81**, where the axial carboxylic acid group in the preferred $^2C_5$ conformation is in the same orientation as in the naturally occurring α-D-glycoside of Neu5Ac, inhibited VC sialidase, albeit weakly. The 4-epimeric derivatives **78** and **80** with equatorially arranged carboxylate groups were found to be competitive inhibitors of the bacterial sialidase (K_i=0.12 and 0.19 m*M*, respectively). 2-Hydroxy methylation, such as in **82** and **83**, in either an axial or equatorial position led to attenuation of the sialidase inhibition.[147]

(76)   (77)   (78) R = NHAc

(79) R = NHAc   (80) R = NHAc   (81)

Elaboration of this approach was provided in a further study[153] of the piperidine template, retaining analogous substituents in the preferred natural Neu5Ac arrangement at C-2 (–COOH), C-4 (OH) and C-5 (*N*-acetyl) positions, but replacing or totally removing the glycerol side chain. These piperidinose-based mimetics of Neu5Ac were synthesized from *N*-acetyl-D-glucosamine via the azidoalkene **84**.

The piperidine derivative lacking the glycerol side chain, **85**, retains modest sialidase inhibition ($K_i = 5 \times 10^{-2}$ M) and exists in solution as a 2:1 mix of $^2C_5$ and $^5C_2$ conformers. Alkylation of the endocyclic nitrogen of **85**, to give **86**, was refractory to inhibition. Addition of functionalities, such as CH_2OH, CH_3F, or CH_3 at C-6, furnished piperidines with very weak inhibition, and for the hydroxymethyl derivative, further addition of a phenyl group at the ring nitrogen totally abolished activity. Interestingly, the hydroxymethyl epimeric derivative **87** proved a millimolar inhibitor of the bacterial sialidase.[153] This was rationalized on the basis that the C-7 OH group may participate in a hydrogen bond within the sialidase binding site that would, under natural circumstances, interact with the C-8 OH of Neu5Ac. Taken together these studies of C-6 and C-7 piperidine analogs supported the notion that an intact glycerol side chain is a requirement for sialidase inhibition for this series of analogs.

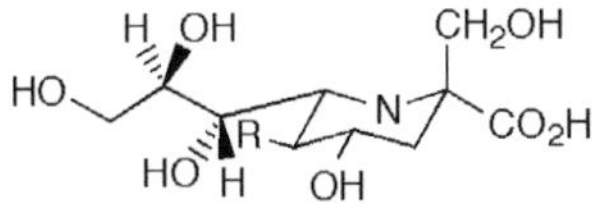

(82) R = NHAc 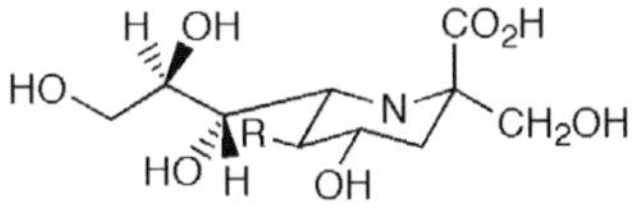   (83) R = NHAc 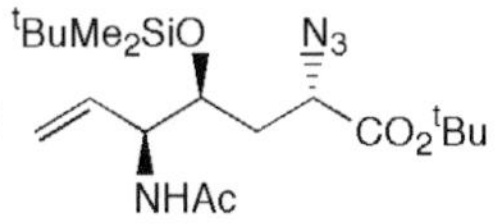   (84)

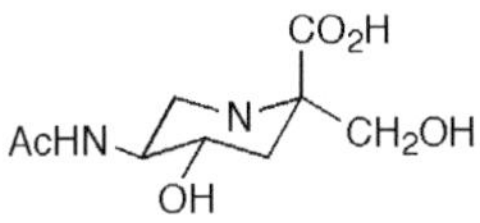

(85) 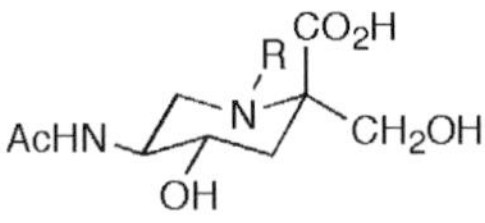   (86) R = CH_2CH_2OH 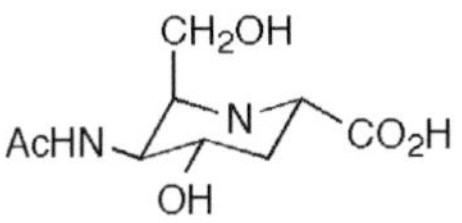   (87)

A series of pyrrolidine-based mimetics of Neu5Ac[148] were prepared from the methyl ester of Neu5Ac2en via a pyrrolidine-borane adduct **88**. Access to derivatives branched at the pyrrole nitrogen (**89**, **90**, and **91**) was achieved by N-acylation or N-methylation of derivative **92**. The phosphonate derivative **91** showed similar levels of inhibition to Neu5Ac2en, while **89** and **90** were very poor inhibitors ($>mM$ levels of inhibition).

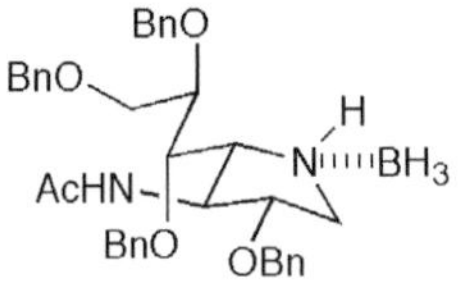

(88)

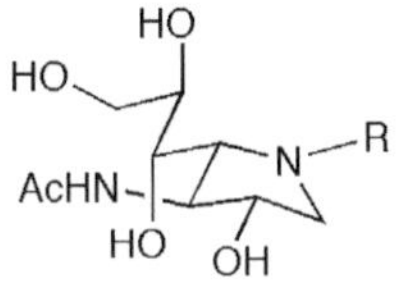

(89) R = CH_2CO_2H
(90) R = $COCO_2H$
(91) R = $CH_2PO_3H_2$
(92) R = H

Very early reports that a number of N-aryl and N-heteroaryloxamic acids were inhibitors of influenza virus sialidase[169] and that N-(4-nitrophenyl)oxamic acid was a competitive inhibitor of VC sialidase,[142] prompted further investigation of aromatic templates, represented by **93**.[155] Despite the extensive literature precedence reporting the importance of the glycerol side chain for activity against VC sialidase,[143,153,170] compounds of the type represented by **93** were found to be competitive inhibitors of VC sialidase. In addition to a number of *para* di-substituted compounds (e.g., **93**), several 1,2,4-tri-substituted compounds (e.g., **94** and **95**) were also reported.[155] The *N*-acyl-4-nitroanilines all exhibited poor water solubility. Noncompetitive inhibition of VC sialidase was seen for the acids (**93a** and **94c**) and for **95b** 4-nitro-*N*-(trifluoroacetyl)aniline.

(**93a**) R = COCO$_2$H
(**93b**) R = COCO$_2$Et
(**93c**) R = COCF$_3$
(**93d**) R = COCH$_2$I

(**94a**) R = H
(**94b**) R = COCO$_2$Et
(**94c**) R = COCO$_2$H

(**95a**) R = H
(**95b**) R = COCF$_3$

III. SIALYLMIMETICS AS PROBES FOR SELECTINS

The selectins are a family of carbohydrate-recognizing proteins that bind to fucosylated and sialylated glycoprotein ligands. Three selectins (E-, P-, and L-) have been identified, with the letter prefix indicating the cell type (Endothelial, Platelet, and Leukocyte, respectively) with which they are commonly associated. The selectins are involved in the trafficking of cells of the innate immune system, T-lymphocytes and platelets. The selectins are responsible for the early adhesion events in the recruitment of leukocytes to sites of inflammation, by promoting the initial tethering and rolling of leukocytes, eventually leading to the infiltration of leukocytes into the damaged surrounding endothelial tissue.[13,171–174] The overrecruitment of leukocytes can have damaging effects, including acute inflammatory diseases such as stroke and reperfusion injury during surgery, and chronic inflammatory diseases such as rheumatoid arthritis and asthma.[13,171–173] E- and P-selectin recognize the cell-surface sialyloligosaccharide sialyl Lewis x (sLe^x, **96**) as well as a P-selectin glycoprotein ligand PSGL-1, which displays sLe^x and a sulfated tyrosine, although this latter group is only required by P-selectin. L-selectin recognizes sialyl Lewis a (sLe^a, **97**), and sulfated oligosaccharides, such as heparin and sulfated sLe^x, although the precise pattern of sulfation required is not fully understood.[13,16,171–173,175–178] Despite the apparent structural diversity in the natural ligands for the selectins, all three selectins show important interactions with only a handful of functional groups within these ligands. These important interactions between the selectins and sLe^x (or its sulfated derivative) are summarized in Figure 24.2 and are based, in part, upon the NMR structure of sLe^x bound to E- and P-selectins,[179] crystallographic data,[180] and structure–activity studies.[13]

As can be seen (indicated by the emboldened atoms), only five hydroxyl groups and the carboxylate group of Neu5Ac are considered as participating in important

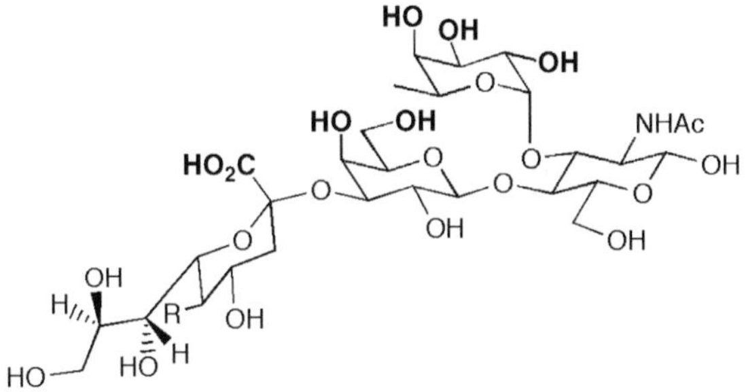

(96) R = NHAc

FIGURE 24.2 Important interactions (indicated by emboldened atoms) between sLe^x and the selectins.

interactions.[13,16] The relatively limited number of functional groups within sLe^x (**96**), considered to be important for interactions with selectins, makes this a perfect target for the development of smaller and simpler molecules that have the potential to exhibit affinity for the selectins comparable to or better than sLe^x itself. Indeed, the modest ($\sim 10^{-3}$ *M*) affinity of sLe^x (**96**) toward the selectins, together with the complexity of the chemistry associated with trying to make sLe^x derivatives, are the major reasons underlying the research into the development of mimetics of sLe^x. Some previous excellent literature describes in detail the synthesis of smaller and simpler molecules that have affinity for the selectins comparable to sLe^x itself.[6,12,13,16,172,178] In light of the comprehensive nature of some of these previous publications, this section will focus primarily on those mimetics of sLe^x that show high affinity for the selectins. Furthermore, for completeness, we will attempt to cover the majority of the different structural variants that have emerged as useful mimetics of sLe^x.

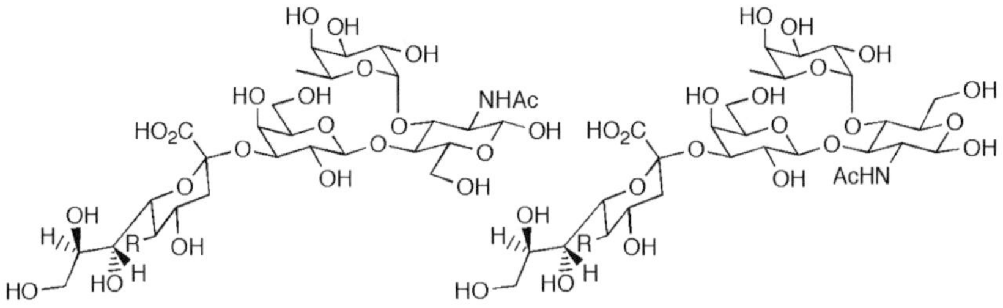

(96) R = NHAc

(97) R = NHAc

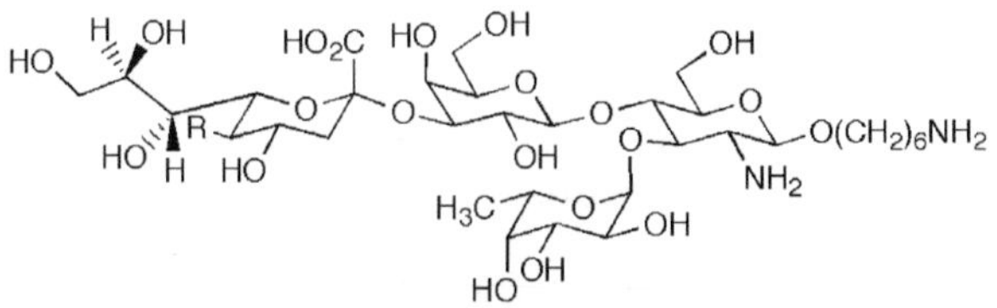

(98) R = NH$_2$

Given the relative structural complexity of sLe^x (**96**), some of the reported mimetics are comparatively simple molecules. It is perhaps appropriate, however, to begin with some examples of structurally modified sLe^x derivatives that have emerged recently that shed additional light on the structure–activity relationships between the selectins and their natural ligands. Following from the reported crystal structures of both E- and P-selectin bound to sLe^x,[181] molecular orbital calculations have shown that the neutral (i.e. protonated) form of sLe^x possesses the necessary structural characteristics for binding to E-selectin.[182] It was particularly noted that the LUMO level of the neutral sLe^x molecule is localized on both the carboxylic acid group of Neu5Ac and the glycosidic linkage between Neu5Ac and the penultimate galactose residue. This observation is interesting in that it is normally considered that the carboxylic acid group of Neu5Ac is charged (i.e., an anion) at physiological pH, and may have implications in the future design of mimetics for sLe^x. The reported[183] high affinity of the *bis*-de-*N*-acylated tetrasaccharide **98** (IC$_{50}$=160 μM for P-selectin; 270 μM for E-selectin) supports the notion that additional ionic interactions are important for binding.

Multivalency has long been believed to impart stronger binding interactions between specific carbohydrate sequences and the proteins that recognize them.[184] The use of multivalent sLe^x derivatives has thus far generally only shown modest improvement in affinity for the selectins,[16] although the use of four sLe^x epitopes on a branched oligosaccharide backbone has resulted in compounds with nanomolar affinity for L-selectin.[185] A report showing that bivalent sLe^x derivatives (e.g., **99**) had slightly improved affinity for E-selectin over monomeric sLe^x (IC$_{50}$=70 μM for **99** vs. 393 μM for sLe^x)[186] has led to the design of bivalent sLe^x derivatives such as **100**, which have restricted flexibility between the two sLe^x domains.[187] Although no biological data are available yet, it is anticipated that the extra rigidity in **100** will provide enhanced cooperative effects leading to molecules with significantly improved affinity for the selectins. Some polymeric examples of selectin inhibitors have also been described. It has been shown that presentation of both sLe^x and *O*-sulfotyrosine on the same polyacrylamide template resulted in a synergistic inhibitory effect on P-selectin.[188] More recently, the same authors have removed the sLe^x from the same polymeric template and found that the resultant *O*-sulfotyrosine polyacrylamide template is the most potent *in vitro* inhibitor of P-selectin reported to date.[189,190]

Interestingly, the monomeric sLe^x carboxylic acid derivative **101**, formed as a by-product during the cross-linking reaction to prepare compounds such as **99**[186] had an affinity for P-selectin around 40 times higher than sLe^x (**96**) itself, presumably due to the extrasulfated tyrosine-binding pocket in P-selectin. The presence of additional charged groups in sLe^x derivatives as biological probes, especially for L-selectin that is known to recognize sulfated sLe^x, led to an elegant synthesis of the mono-sulfated sLe^x analogs **102** and **103**.[191] Similarly, the 6-sulfo-de-*N*-acetylsialyl Lewis x derivative **104** was prepared as a ligand for L-selectin.[192] Compound **104** was found to be superior to all other molecules tested for their binding affinity toward L-selectin, including the corresponding analog that contained the *N*-acetyl group in the Neu5Ac ring. These authors also reported that structures with a sulfate group at C-6 in the Gal ring of sLe^x (such as **103**) have little or no binding affinity for L-selectin.[192]

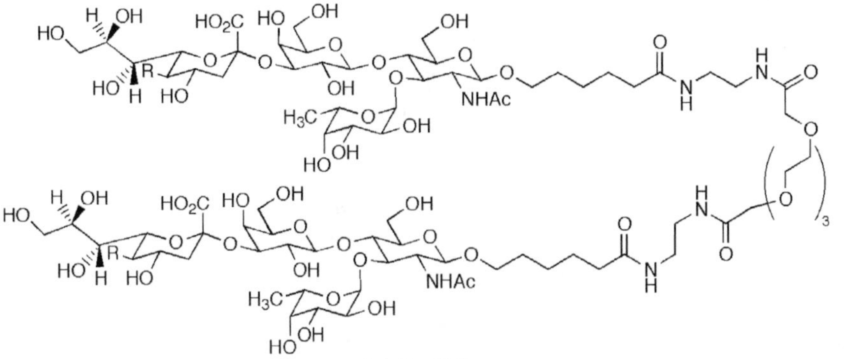

(99) R = NHAc

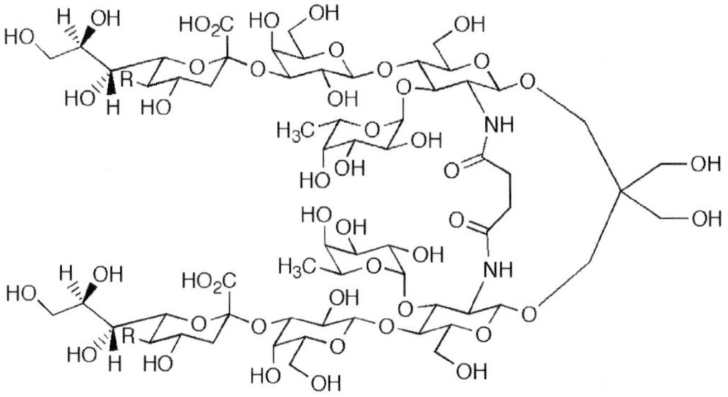

(100) R = NHAc

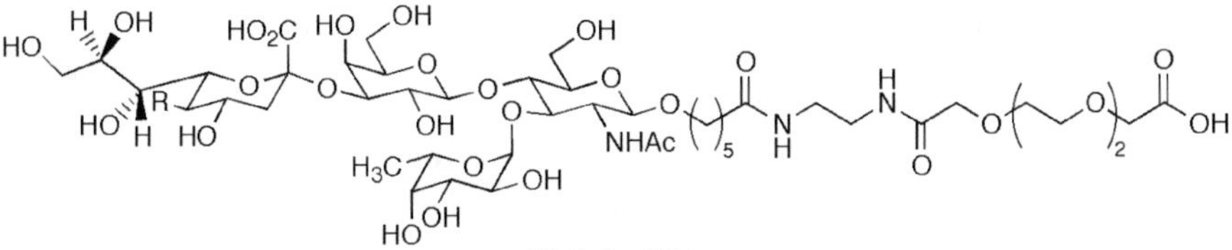

(101) R = NHAc

Moving away from structurally modified sLe^x derivatives as probes for the selectins, the majority of mimetics of sLe^x reported to date have the entire Neu5Ac portion replaced with simple charged groups (CO$_2^-$, SO$_3^-$, etc) while retaining some degree of carbohydrate backbone. Other researchers have moved completely away from carbohydrate-based scaffolds, and have reported peptide-based derivatives or aromatic derivatives that show affinity for the selectins. Of these sLe^x mimetics that contain a carbohydrate scaffold, derivatives have been prepared that are based upon penta-, tetra-, tri-, di-, and monosaccharide templates. For ease of discussion, this chapter will discuss the various types of sLe^x mimetics based upon the degree of carbohydrate character that remains in the final molecule.

(102) R₁ = SO₃Na; R₂ = H; R₃ = NHAc
(103) R₁ = H; R₂ = SO₃Na; R₃ = NHAc

(104) R = NH₂

A. PENTA- AND TETRASACCHARIDE-BASED sLeˣ MIMETICS

Some of the early reports of mimetics of sLeˣ were those molecules where the entire Neu5Ac portion of sLeˣ was replaced by a simple charged functionality. One such example is the penta-saccharide-based sLeˣ mimetic 105, which was prepared from the selectively protected trisaccharide 106 via glycosidation with a protected lactoside followed by the introduction of the sulfate group at C-3 of the Gal ring.[193] At that time, compound 105 was the most potent E-selectin ligand known. Although the tetrasaccharide-based analog 107,[194] which is the same as 105 except that it lacks the glucose unit at the reducing end, exhibited similar inhibitory activity to 105, the binding of E-selectin-expressing SC2 cells to the immobilized lipid-linked form of 105 was stronger than that of 107.[195] The trisaccharide analog 108 of 105, which was obtained from the advanced synthetic precursor to 106, was found to be less active than its pentasaccharide analog.[193]

Two mimetics of sLeˣ containing different charged groups replacing the Neu5Ac moiety were prepared from a single advanced precursor. Thus, selective deprotection of the levulinoyl group in 109 and subsequent sulfation or phosphorylation gave the sLeˣ mimetics 110 and 111, respectively.[196] In the same paper, these authors also reported the synthesis of the carboxylate functionalized tetrasaccharide sLeˣ mimetic 112. Compounds 110, 111, and 112 were evaluated for their affinity toward the selectins and were all found to show competitive inhibition compared with sLeˣ itself.[197]

A series of core 2 branched structures of the general formula 113 have provided some interesting insights into the different selectivities of the three selectins. While the doubly charged derivative 114 (R=Neu5Ac, R'=SO₃) showed improved affinity for E-selectin it lost affinity for L- and P-selectin compared with the asialo derivative 115 (R=H, R'=SO₃).[198] Interestingly, the inclusion of Neu5Ac on the other arm of the core 2 like structure, to give 116 (R=H, R'=Neu5Ac) gave significant inhibition of L- and P-selectin over E-selectin. The disulfated pentasaccharide derivative

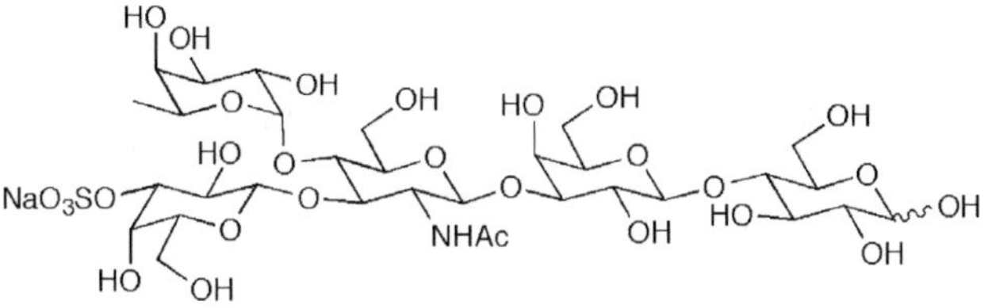

(105)

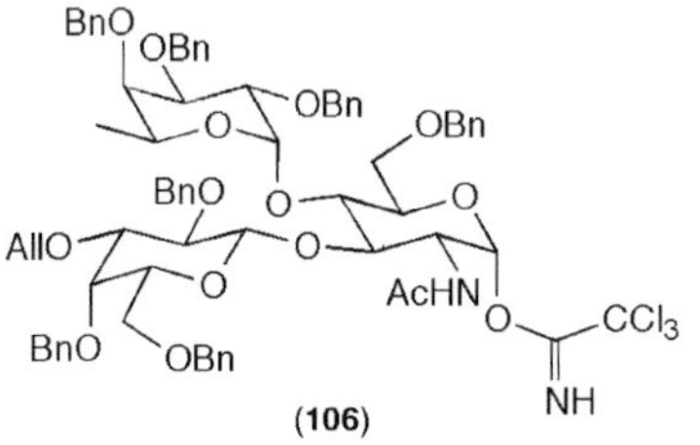

(106)

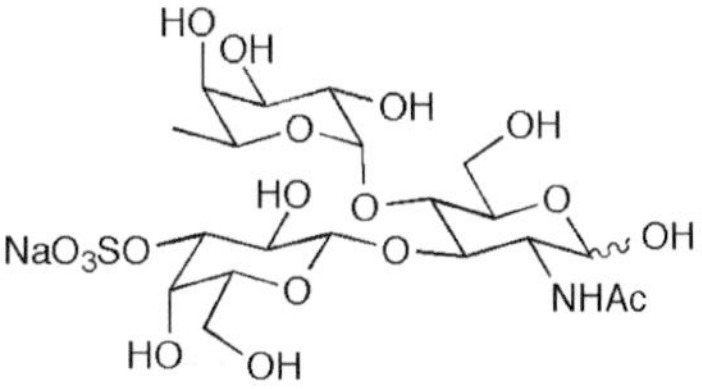

(108)

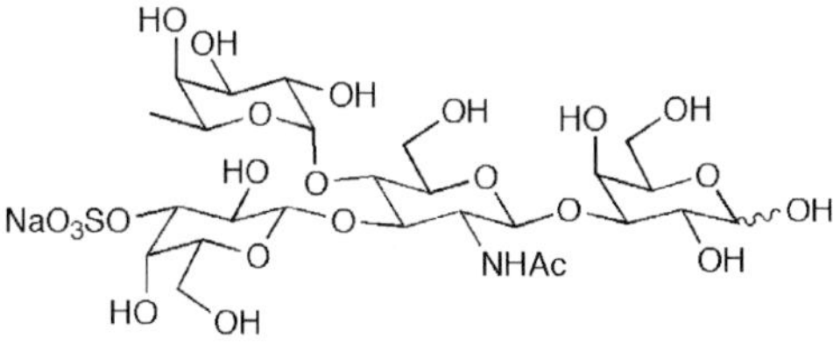

(107)

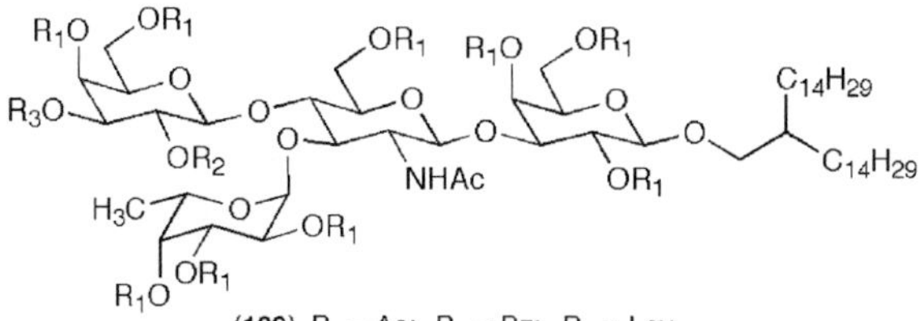

(109) R₁ = Ac; R₂ = Bz; R₃ = Lev
(110) R₁ = R₂ = H; R₃ = SO₃Na
(111) R₁ = R₂ = H; R₃ = P(O)(ONa)₂

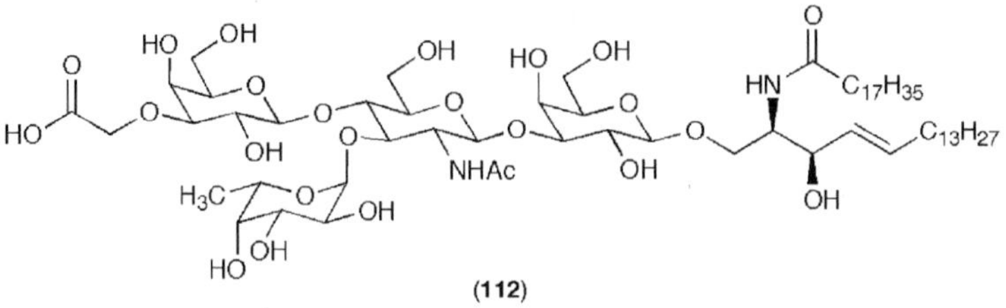

(112)

117 was prepared by chemoenzymic methods, with a key feature being the dibutyl-stannylene-mediated introduction of the second sulfate group into the derivative **118** that contains 11 free hydroxyl groups.[199] Other multiply sulfated sLeˣ mimetics as probes for the selectins include the derivatives **119** and **120**.[200] Interestingly, and in support of earlier observations,[199] it was noted that single sulfation at the 6-position of Gal in Le[x], to give the mimetic **121**, resulted in a loss of affinity toward L-selectin, whereas sulfation at the 6-position in the GlcNAc residue (e.g., as in compound **120**) led to enhanced affinity, suggesting that the 6-sulfate group in the GlcNAc residue is an important recognition motif for L-selectin.[200]

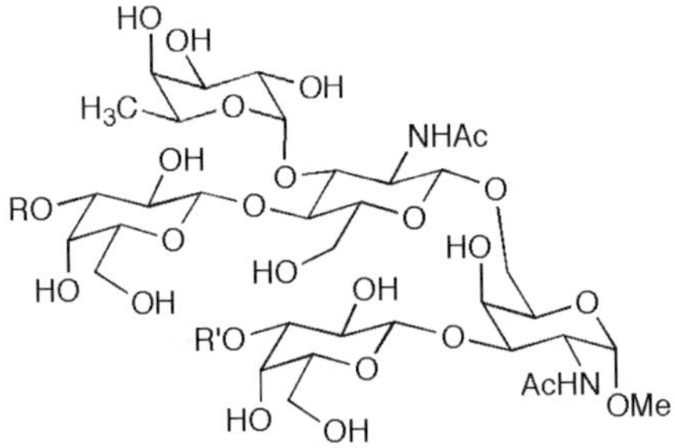

(113) R = H; R' = H
(114) R = Neu5Ac; R' = SO3
(115) R = H; R' = SO3
(116) R = H; R' = Neu5Ac

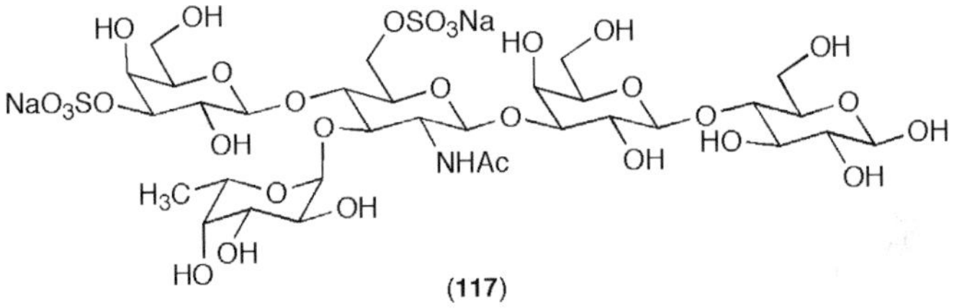

(117)

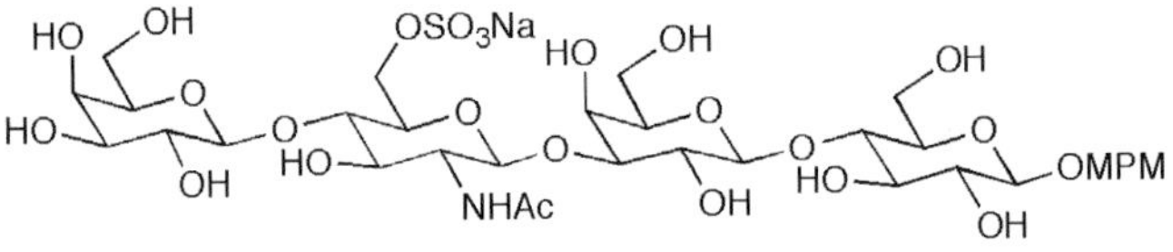

(118)

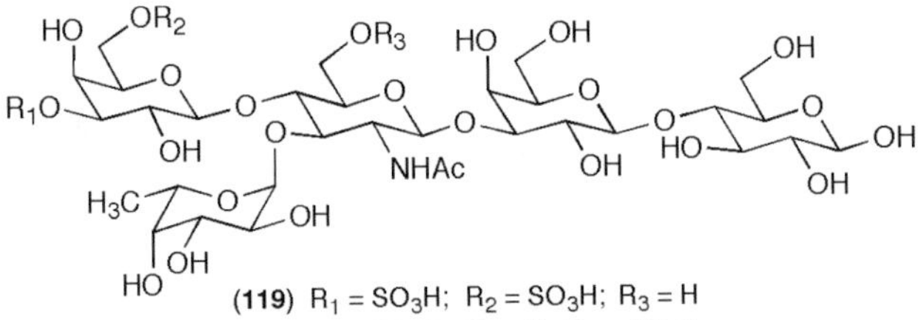

(119) R1 = SO3H; R2 = SO3H; R3 = H
(120) R1 = SO3H; R2 = H; R3 = SO3H
(121) R1 = H; R2 = SO3H; R3 = H

B. TRISACCHARIDE-BASED sLE^x MIMETICS

The realization that the entire N-acetylneuraminic acid residue of sLe^x (**96**) could be replaced by a simple charged group while retaining or even improving affinity for the selectins led researchers to consider less complex carbohydrate templates. Based upon their earlier observations that the sLe^x mimetics **122** and **123**, containing S-cyclohexyllactic acid as a replacement for Neu5Ac, were potent sLe^x mimetics,[201,202] these authors incorporated a modified glucosamine residue into their mimetics to give the trisaccharide-based sLe^x mimetic **124**.[203] Compound **124** was found to be 30-fold more potent than sLe^x in a static, cell-free ligand-binding assay that measures E-selectin inhibition under equilibrium conditions. Significantly, compound **124** also showed inhibition (IC$_{50}$ ~10 μM) of E-selectin in a dynamic *in vitro* assay that mimics *in vivo* conditions, unlike sLe^x (**96**), which failed to show any inhibition up to 1000 μM in this assay.[203] The potent E-selectin activity of the sLe^x mimetic **123** (IC$_{50}$ = 36 μM)[201] prompted the report of an alternative synthesis.[204] The key features of this alternative approach toward **123** involved the use of 2-thiopyridyl carbonate activating groups for the two glycosidation reactions needed to construct the trisaccharide template, as well as a Rh^{II}-catalyzed etherification reaction to introduce the C-3 substituent on the Gal ring.

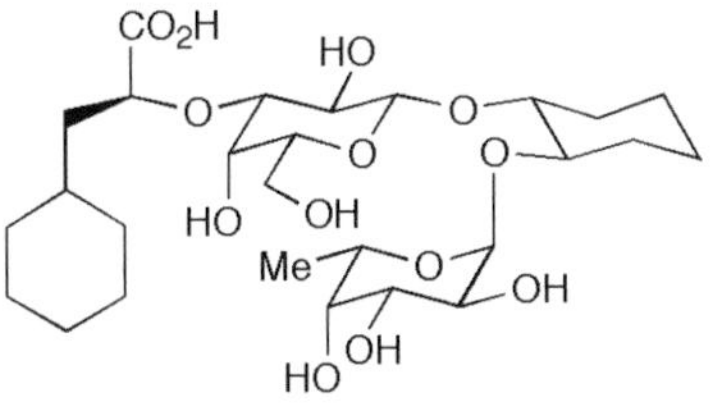

(122) 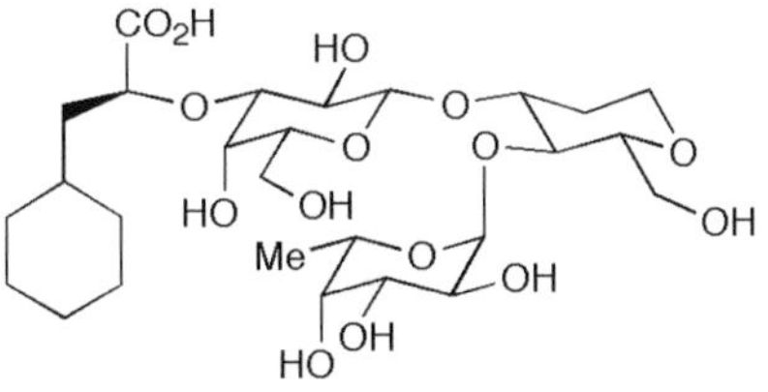   (123)

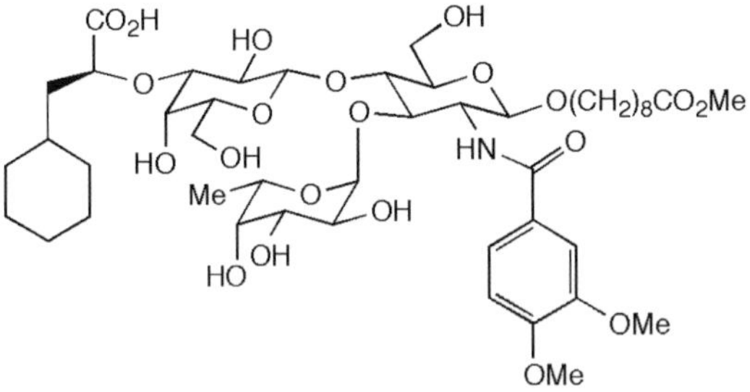

(124)

The development of polymerized liposomes of the general structure **125** containing the sLe^x mimetics **126** or **127** showed interesting activity against the three selectins.[205] Incorporation of additional charged functionalities into the lipid portion of **125** showed that, against L-selectin, strength of binding increased with increasing acidity of the lipid when the sLe^x mimetic **126** was attached. However, with the sulfated mimetic **127**, the nature of the lipid portion appeared to have little effect upon the binding affinity for L-selectin. A similar trend was observed for P-selectin, and the authors suggest that this evidence may support the notion that L-selectin, similar to P-selectin, has a second binding pocket for the recognition of anionic functionalities.[205] For E-selectin, only sulfate groups in the lipid portion together with the mimetic **126** showed any activity.

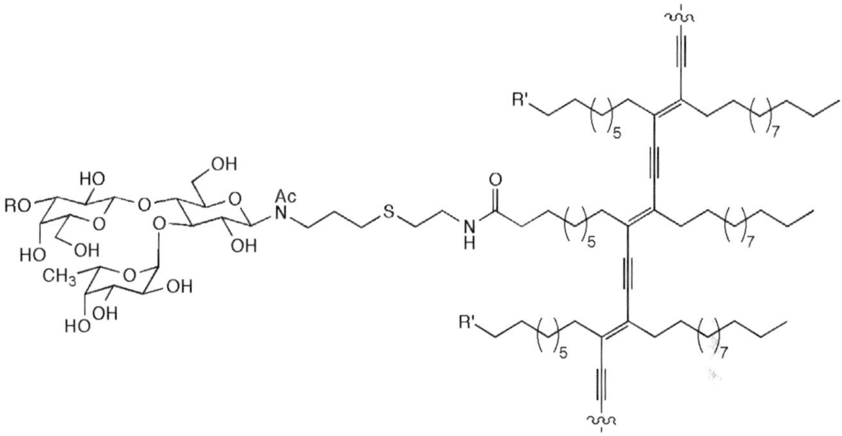

(125) R' = Anionic, cationic, or neutral
(126) R = CH$_2$CO$_2$H; R' = Anionic, cationic, or neutral
(127) R = SO$_3$H; R' = Anionic, cationic, or neutral

Commencing from lactose, the sLe^x mimetic **128** was prepared in 14 steps, with a key transformation being the introduction of the Gal C-3 substituent via dibutylstannylene-mediated alkylation of **129**.[206] Introduction of the fucose residue was achieved using *N*-iodosuccinimide-promoted glycosylation between the fucoside **130** and the lactoside **131** with excellent α-selectivity (α:β=10:1). Interestingly, the sLe^x mimetic **128** exhibited no activity when tested for its ability to inhibit the binding of leukocytes to immobilized E-selectin under shear stress conditions.

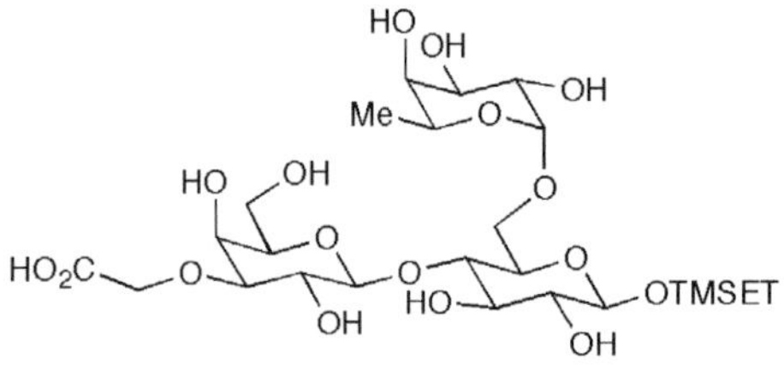

(128)

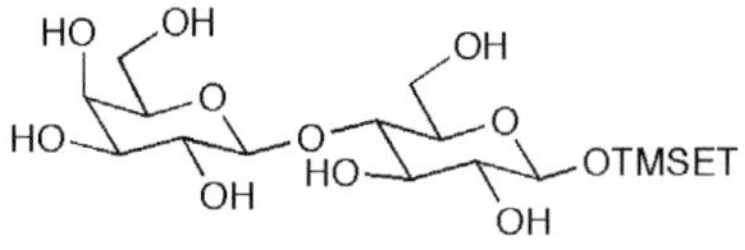

(129)

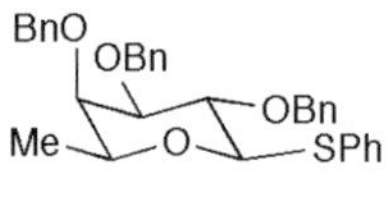

(130)

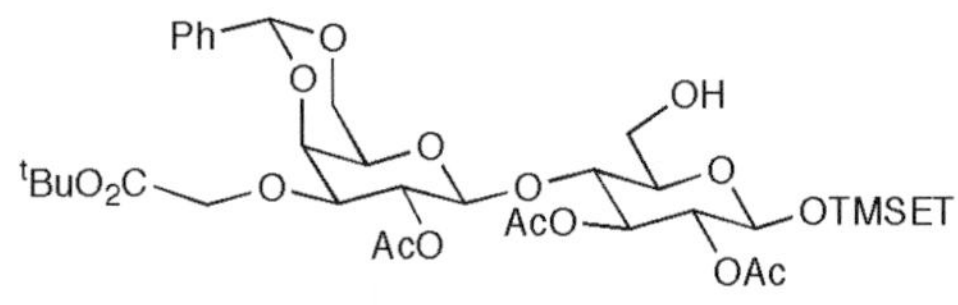

(131)

C.   DISACCHARIDE-BASED sLeˣ MIMETICS

The potent E-selectin activity of the trisaccharide-based sLeˣ mimetic **124** led these same authors to prepare a series of disaccharide-based sLeˣ mimetics with either rigid (e.g., **122**, **123**) or flexible (e.g., **132**) spacers between the two sugar units.[202] Interestingly, the flexible sLeˣ mimetic **132** exhibited no E-selectin activity. The introduction of a charged group at the C-6 position of the Gal unit in sLeˣ mimetics of the general structure **133** also resulted in complete elimination of E-selectin activity, indicating the importance of this group for recognition by E-selectin.[207]

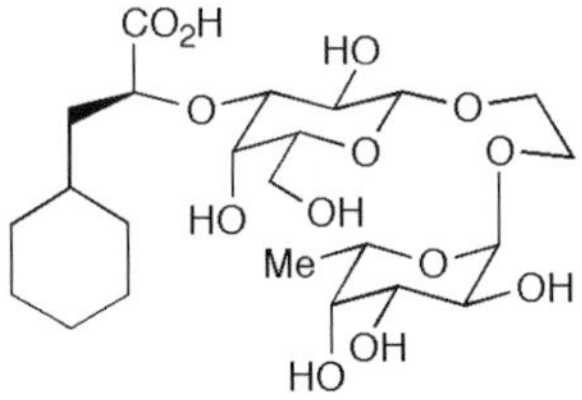

(132)

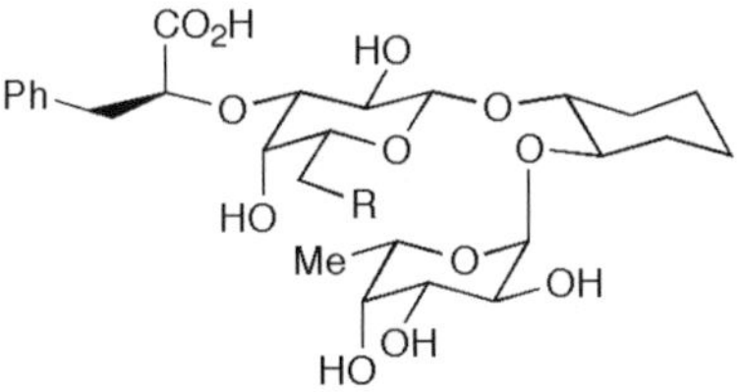

(133) R = NHC(O)R'; OSO₃H; CO₂H

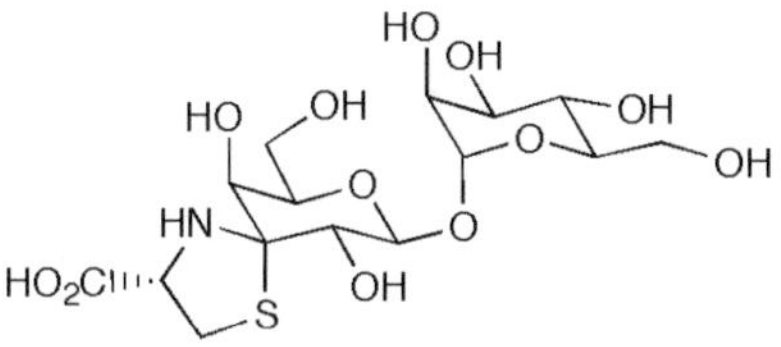

(135)

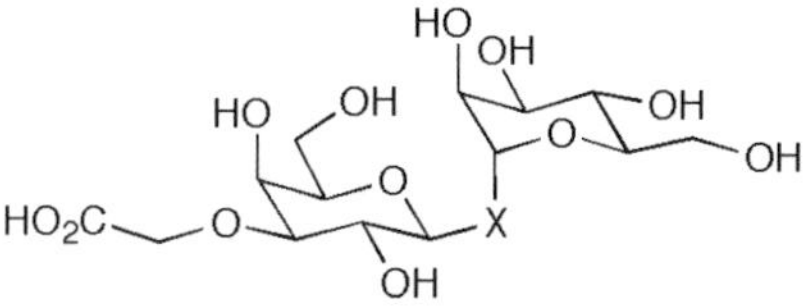

(134) X = O
(136) X = CH₂

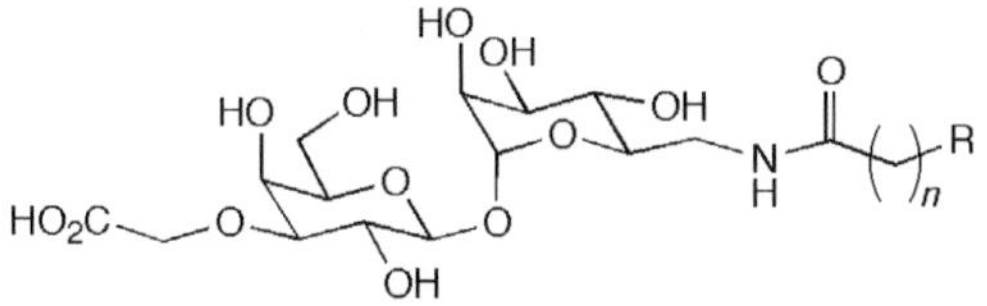

(137) n = 4, 9; R = Me, CH₂CO₂H, CH₂NH₂, CH₂OSO₃H

Several 1,1-linked disaccharides have been developed as the template for sLeˣ mimetics.[6,172] The earlier report of the Gal-(1,1)-Man-based sLeˣ mimetic **134**[208] prompted the development of the more rigid sLeˣ mimetic **135** that has improved potency toward P-selectin.[209] The improved potency of **135** ($IC_{50} = 19$ μM vs. 193 μM for **134**) is believed to be due to the increased hydrophobic character of **135** compared with **134** rather than the orientation of the carboxylate group. The excellent activity of **134** and **135** also led to the development of the *C*-disaccharide **136**[210] as well as the mannose C-6 amide functionalized derivatives **137**.[211] The latter compounds were prepared to try to capture additional hydrophobic interactions as well as to determine the effect of additional charged functionalities.

Some relatively simple allyl β-lactoside derivatives (e.g., **138** and **139**) have been reported as inhibitors of L-selectin.[212,213] Interestingly, the 6',6-disulfate derivative **139** was found to be twofold more potent than the 3',6'-disulfate analog **138** against L-selectin. Additionally, the sLe^x mimetic **139** was more active than sLe^x itself, and also more active than the 3'- or 6'-mono-sulfated lactoside analogs. A series of semisynthetic glucan sulfates (e.g., **140**) as well as heparin derivatives have recently been evaluated for their ability to block selectin-mediated rolling.[214] While it was found that none of these derivatives had an effect on E-selectin-mediated rolling, the specifically 2,4-bis-sulfated derivatives **140** (MW=14 or =170 kDa; either 1.1 or 1.6 sulfates per glucose unit) blocked the rolling induced by P-selectin with significantly higher potency than the heparin derivatives. It was also observed that the position of the sulfate groups and the molecular weight of the glucan-based compounds such as **140** were important for activity, although the position of sulfation was the overriding factor for potency.[214]

(138)   (139)

(140) SO$_3$H at the 2 and 4 positions, and
between 1.1 and 1.6 sulfates per Glc

D. MONOSACCHARIDE-BASED sLe^x MIMETICS

The desire to prepare selectin inhibitors with increased binding affinity decreased structural complexity, and having an improved feasibility of large-scale synthesis has led to a number of sLe^x mimetics that are based upon a monosaccharide template.[172] Much of the interest in this area is based upon the observations that it is principally only the carboxylate group of Neu5Ac, the 4- and 6-hydroxyl groups of the Gal unit and the hydroxyls of the fucose residue of sLe^x (**96**), that are involved in important interactions with the selectins. Based upon these observations, a number of groups have used simple monosaccharides as the template for constructing sLe^x mimetics. A series of glycopeptide derivatives with either fucose (e.g., **141**), or mannose (e.g. **142**) as the carbohydrate component have been described.[215] Of the several compounds reported, the fucose derivatives **141** exhibited no activity toward E-selectin, but were potent inhibitors of P- and L-selectin. Conversely, the mannose-based sLe^x mimetic **143** showed inhibition against all three selectins. A glycopeptide-based library of sLe^x

mimetics of the general structure **144** provided some useful structure–activity information, including the fact that a hydrophobic group at the 6-position of mannose in **144** results in a loss of activity against E-selectin.[216] However, none of the sLe[x] mimetics represented by **144** had a higher affinity than sLe[x] itself. Following from their synthesis of the *C*-mannoside derivatives **145** and **146** as E-selectin inhibitors,[217,218] Kaila and co-workers[219] have described the construction of some β-*C*-mannoside libraries of the general structure **147**. It was observed that the presence of a hydrophobic substituent at C-6 provided compounds with the greatest activity, due to the interaction of these groups with a hydrophobic region surrounding the top of the binding pocket in P- and E-selectin. However, none of these derivatives had a higher affinity for P-selectin than its natural substrate PSGL-1.

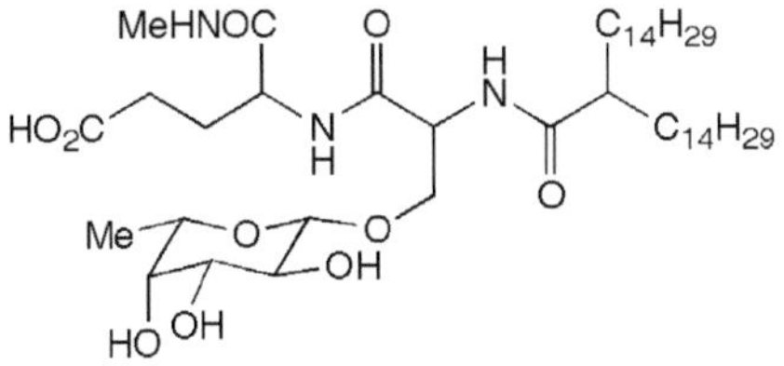

(141)

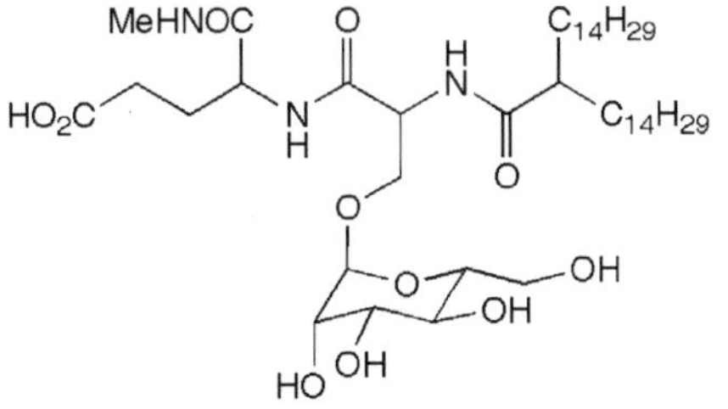

(142)

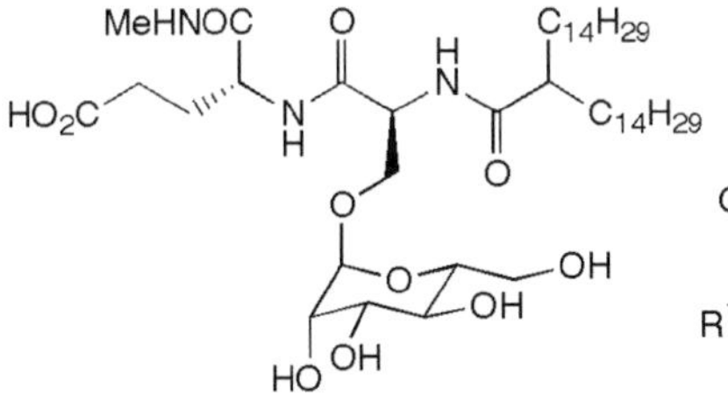

(143)

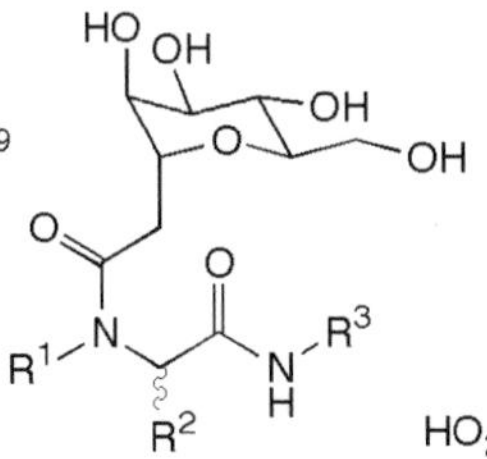

(144)

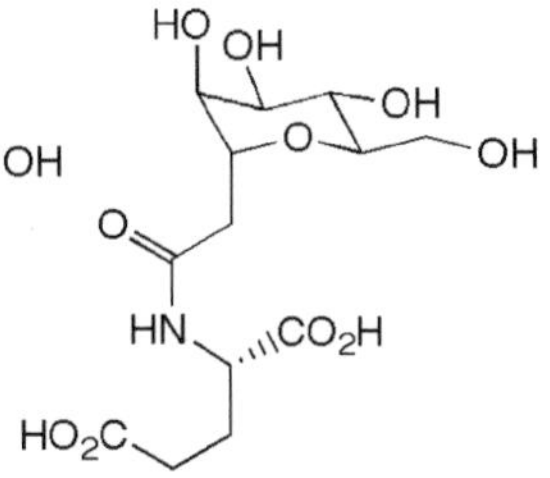

(145)

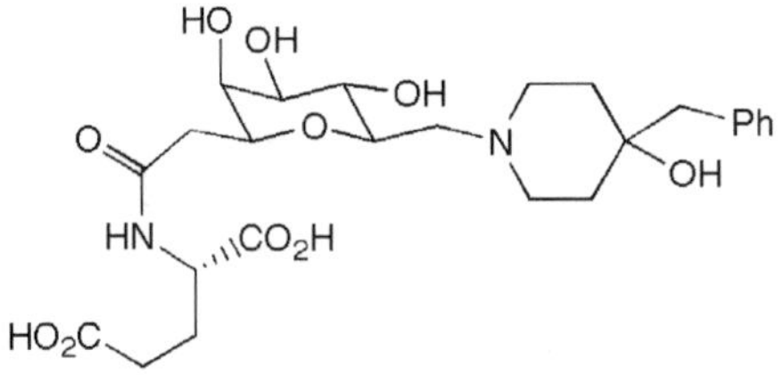

(146)

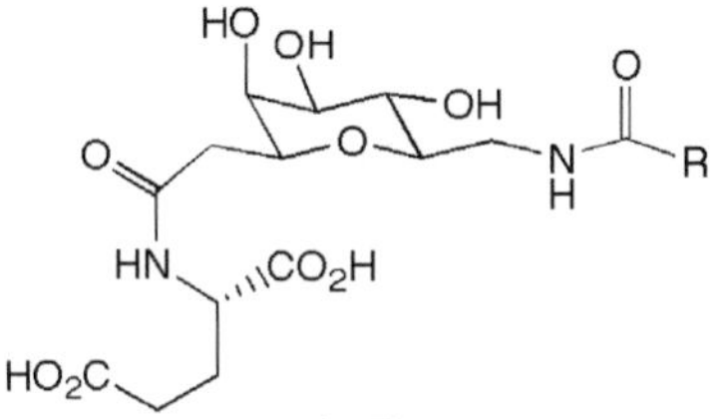

(147)

The mono- and bicyclic lactam derivatives **148** and **149**, together with some related derivatives, were prepared as β-turn surrogates and evaluated for their inhibition of E- and P-selectin.[220] Interestingly, while all of the lactam derivatives reported had no affinity for E-selectin, the sLe[x] mimetic **150** exhibited low micromolar inhibition of P-selectin. Several mannose-based sLe[x] mimetics with phosphate or phosphonate groups have also been described.[221] These compounds (e.g., **151** and **152**) were conveniently prepared using a chemoenzymic approach, and show low micromolar inhibition of P-selectin.

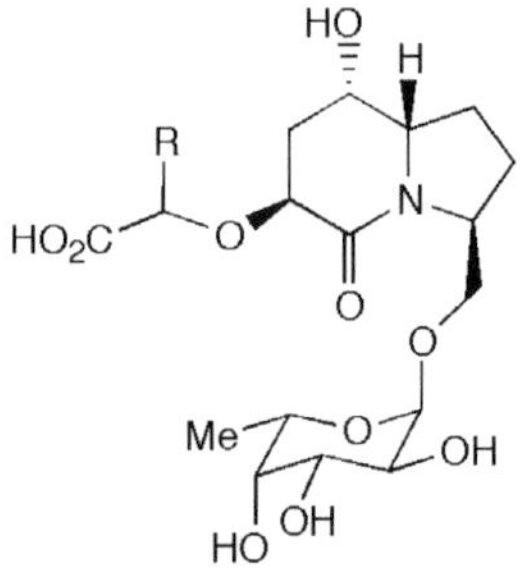

(**148**) R = Ph
(**150**) R = H

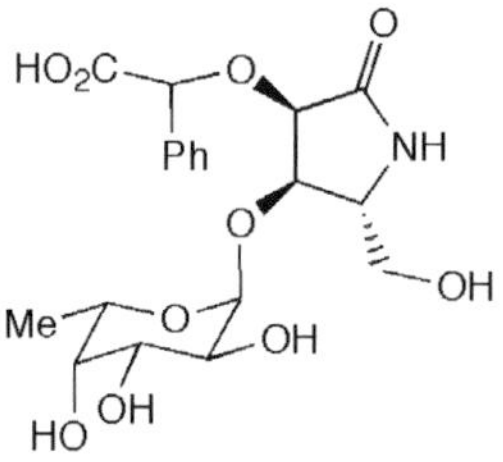

(**149**)

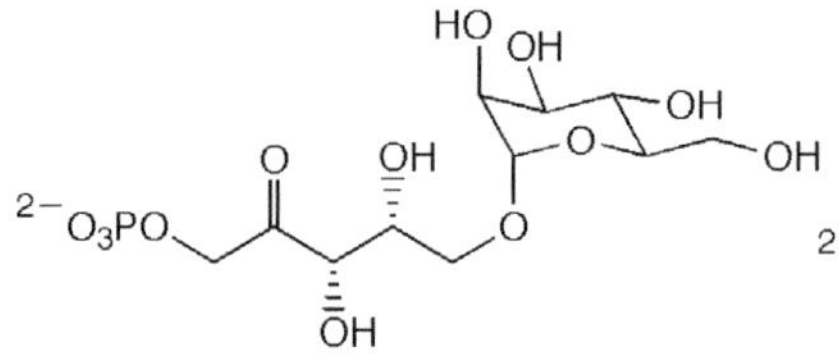

(**151**)

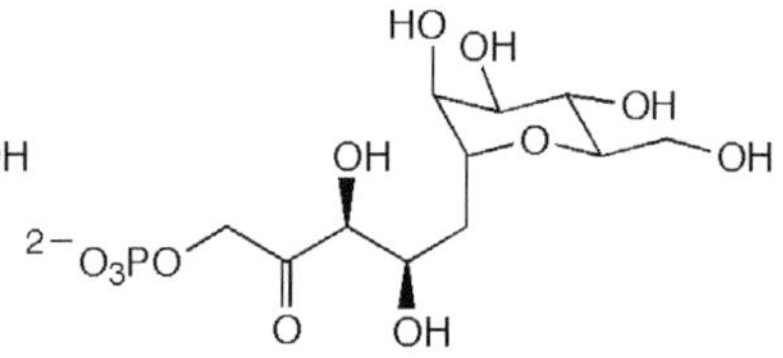

(**152**)

Two examples of multivalent mannosides as sLe^x mimetics are noteworthy. The first of these is referred to as TBC-1269 (**153**), which shows good activity against all three selectins (IC$_{50}$=0.5 mM for E-, 0.07 mM for P-, and 0.56 mM for L-selectin),[222] and is currently in phase-II clinical trials for the treatment of asthma, although preliminary results in human asthmatics (Avila, P.C., et al., *Clin. Exp. Allergy,* 34, 77–84, 2004) are not encouraging. Based upon the selectin inhibitory activity of the monomeric form of **153**,[223] the amino analog **154** of **153** has been prepared and coupled to diethylenetriaminepentaacetic acid (DTPA).[224] The complexation of **154** to DTPA through a flexible alkyl spacer was intended to provide selectin inhibitors that contained a contrasting agent for use in medical imaging.

Other carbohydrates have been used as the template for monosaccharide-based sLe^x mimetics. The 3-sulfated galactose "sulfatides" (e.g., **155**), which are known to bind to both P- and L-selectin,[225,226] have prompted other investigations into galactose-based sLe^x mimetics. The 3,6-disulfated galactose neoglycopolymer **156** was found to cause a dose-dependent loss of L-selectin from the cell surface of human neutrophils.[227] Interestingly, the monomeric disulfated precursor to **156** failed to exhibit any activity, supporting the notion that multivalency is a potentially beneficial feature of these low-molecular-weight sLe^x mimetics. In a substantial body of work,[228] a series of galactocerebrosides bearing malonate side chains (e.g., **157** and **158**) have been described. In the mimetics **157** and **158**, the malonate functionality

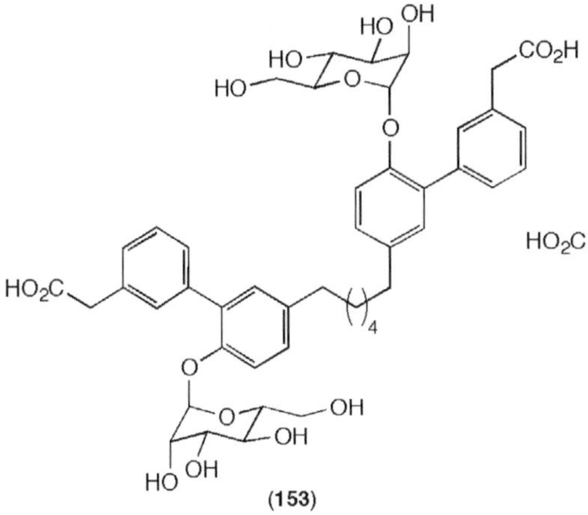

(153)

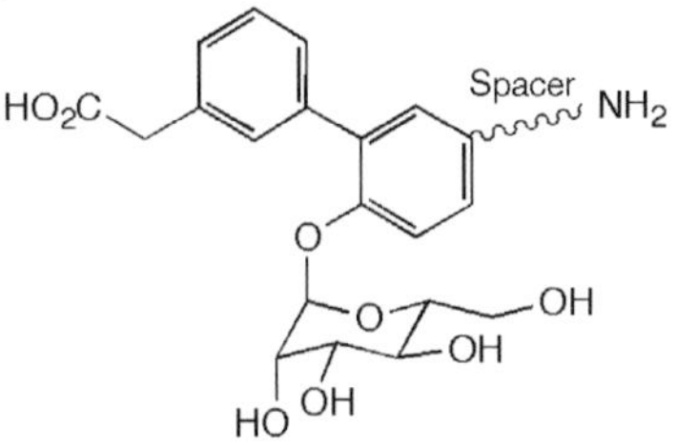

(154)

was designed to interact with two lysine residues in the active site of P-selectin, while the C-4 and C-6 hydroxyl groups on the Gal ring are arranged to chelate the calcium ion in the P-selectin-binding site. Of all the compounds described, derivative **159** showed the best *in vivo* activity against P-selectin. Furthermore, the structure–activity relationships stemming from this study, including the observation of a beneficial effect of having a *O*-benzoate at C-3 with the malonate at C-2 as well as the slightly improved activity of the α-sphingosine over the β-glycoside,[228] will undoubtedly provide the basis for further modifications of this template.

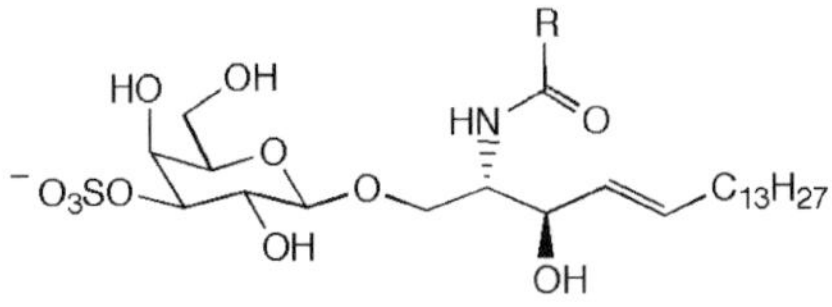

(155)

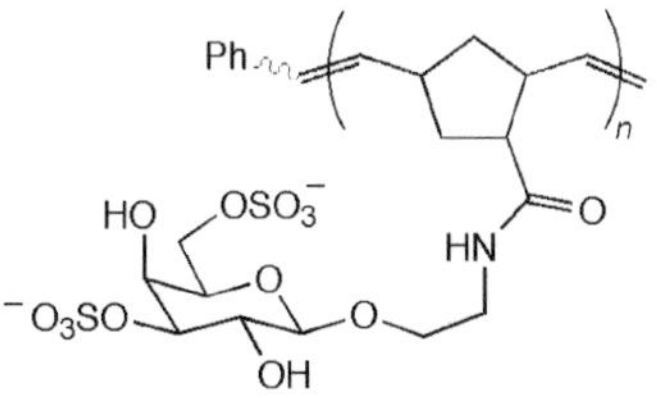

(156)

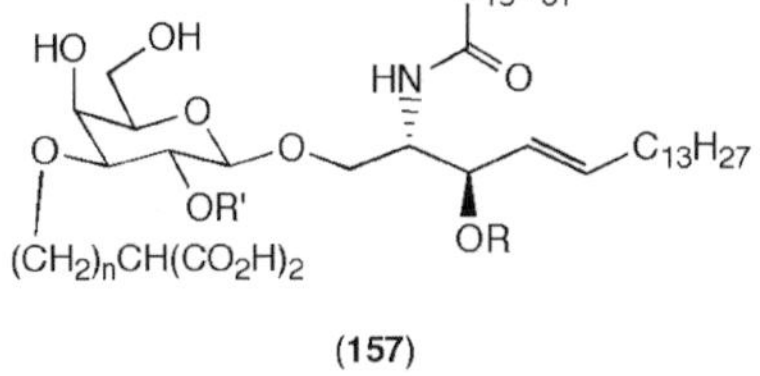

(157)

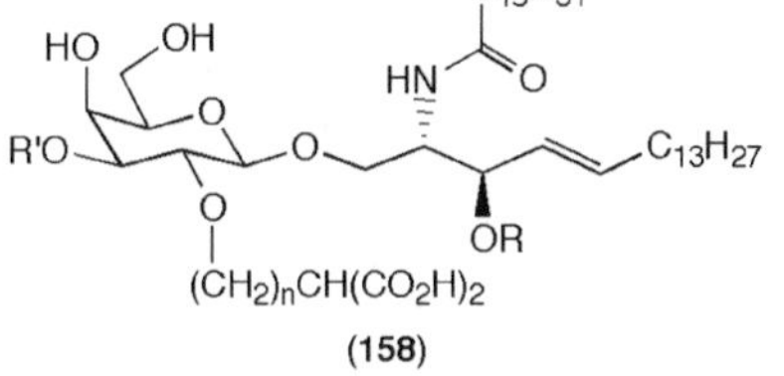

(158)

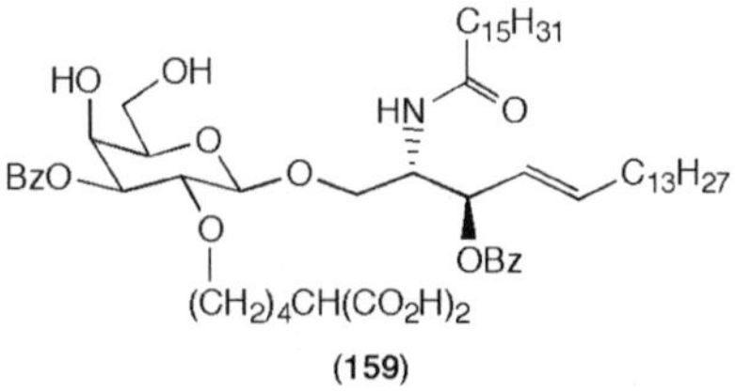

(159)

The fucose-based benzoic acid derivative **160** was found to have comparable affinity for E- (IC_{50}=2.4 m*M*) and P- (IC_{50}=1.6 m*M*) selectin when compared with sLe^x, and was the best inhibitor from a series of differentially substituted benzoic acid analogs of **160** reported.[229] Another benzoic acid derivative, the thioglycosidically linked fucoside **161**, was prepared as a sLe^x mimetic since it contains the three essential groups for P-selectin binding (viz, the fucose ring, a carboxylate group, and a hydrophobic portion).[230] Interestingly, while **161** showed potent inhibition of P-selectin (IC_{50}=10.6 μM), the substituted analog **162** exhibited no activity at all, suggesting that the distance between the aromatic ring (hydrophobic group) and the carboxylate group is important for favorable binding.

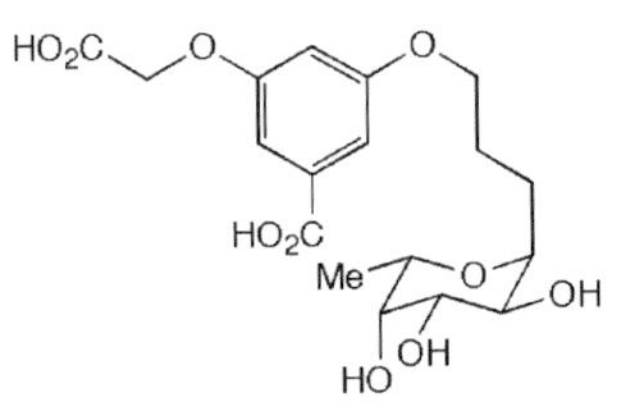

(160)

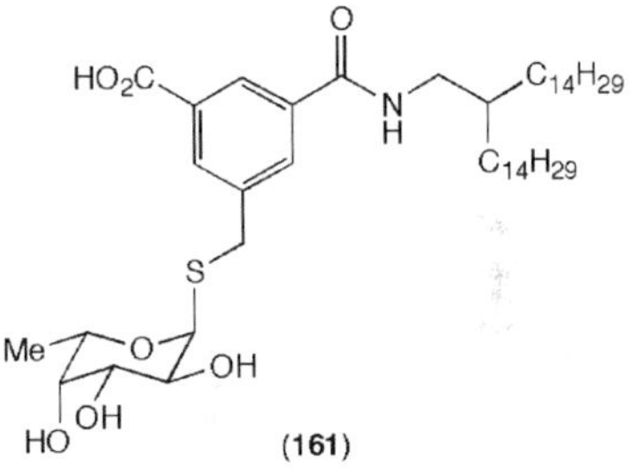

(161)

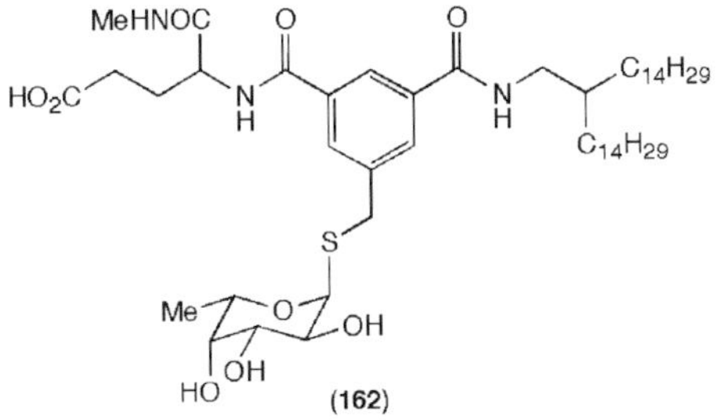

(162)

E. NONCARBOHYDRATE-BASED sLe^x MIMETICS

A series of bis-benzoic acid-based compounds (e.g., **163**) have been described for their inhibitory activity toward E-selectin.[231] The incorporation of varied hydrophobic chains and modified bis-benzoic acid cores resulted in the synthesis of a number of specifically substituted analogs of **163**.[232] The structure–activity relationships stemming from this work clearly show that the affinity of these non–carbohydrate-based sLe^x mimetics for the selectins is dramatically influenced by subtle structural changes. This work will no doubt assist in the development of highly specific inhibitors of individual selectins. The synthesis of several analogs of the imidazole-based sLe^x mimetic has led to the development of the potent (IC_{50}=0.3 μM) P-selectin inhibitor **164**.[233] The P-selectin inhibitor **165** (IC_{50}=0.2 μM) was initially obtained as a minor component from a mixture that was undergoing a random screen for P-selectin activity, although the structure was confirmed by synthesis.[234] It is interesting to note that **165** does not contain a carboxylate group and is a predominantly hydrophobic molecule.

One of the simplest sLe^x mimetics described to date that shows affinity for E-selectin is the *cis*-decalin derivative **166**. Compound **166** was designed using molecular modeling based upon the key interactions between sLe^x (**96**) and E-selectin and was prepared in nine steps from 4-benzyloxy-cyclohexanone.[235] Interestingly and somewhat surprisingly, based on the molecular modeling studies described, both the *cis*- and *trans*-decalinic enantiomers of **166** exhibited similar inhibition of E- (IC$_{50}$ ~5 mM) and P- (IC$_{50}$ ~5 mM) selectin when compared with sLe^x itself.

IV. SIALYLMIMETICS AS PROBES FOR OTHER SIALIC ACID-RECOGNIZING PROTEINS

Although the work toward the development of mimetics of sLe^x as inhibitors of the selectins has certainly received considerable interest, as exemplified in the section above, other sialic acid-recognizing proteins have also received attention with respect to the possibility of using sialylmimetics as potential probes or inhibitors. One such example is the sialyltransferase enzyme, which is involved in the biosynthesis of sialyl-glycoconjugates, and is hence crucial in diseases such as inflammation and cancer where the biosynthesis of sialylglycoconjugates is upregulated.[2,236,237] A recent review has described the latest advances in the development of sialyltransferase inhibitors,[237] and hence the topic will only be treated briefly here. Since sialyltransferases utilize CMP-Neu5Ac (**167**) as their natural substrate, it is perhaps not surprising that mimetics of CMP-Neu5Ac have been developed.[237,238] The CMP-quinic acid-based mimetics represented by the general structure **168** are some of the more interesting examples.[239]

The derivative **169** has an almost identical affinity for $\alpha(2,6)$-sialyltransferase ($K_i = 44$ μM) when compared with CMP-Neu5Ac itself, ($K_m = 46$ μM)[240] while the CMP-quinic acid derivative **170**[239] exhibited slightly improved inhibition of $\alpha(2,6)$-sialyltransferase with a $K_i = 20$ μM. These data suggest that the nature of the sugar portion is less important than the presence of cytidine for binding by the sialyltransferases.[240] Despite this, some examples of modified sugar analogs of CMP-Neu5Ac have been described as inhibitors of sialyltransferases.[237] Of the several other examples of "donor-based" mimetics for sialyltransferases, the sulfur-linked analog **171**[241] is stable but has less affinity for $\alpha(2,3)$-sialyltransferases, while the phosphonates **172** and **173** showed poorer affinity.[242]

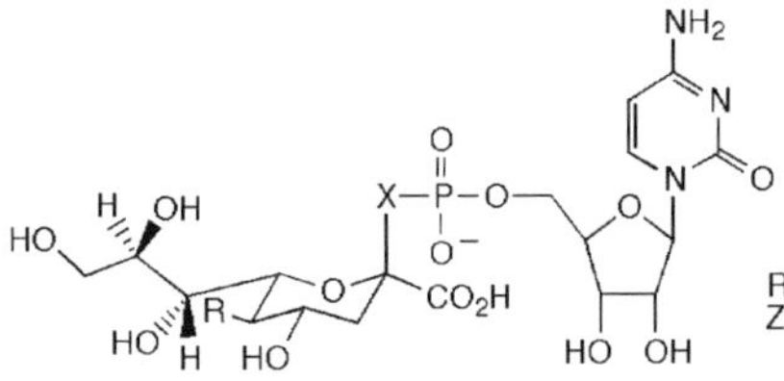

(167) R = NHAc; X = O
(171) R = NHAc; X = S

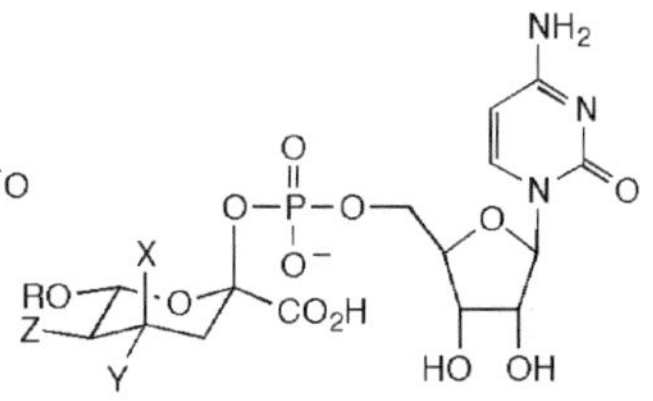

(168)

(169) R = H; X = OH; Y = H; Z = OH
(170) R = CH$_2$CH$_2$OH; X = OH; Y = H; Z = OH

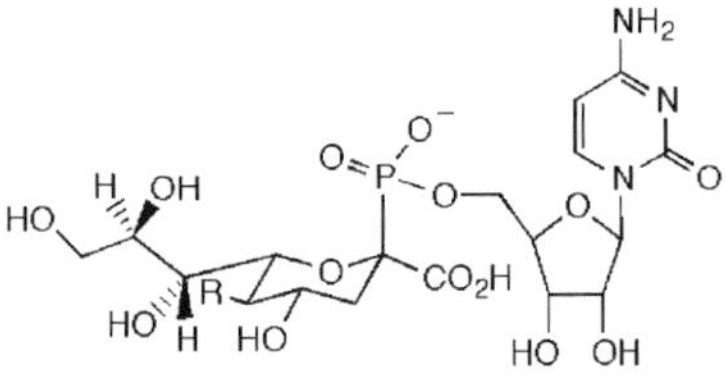

(172)

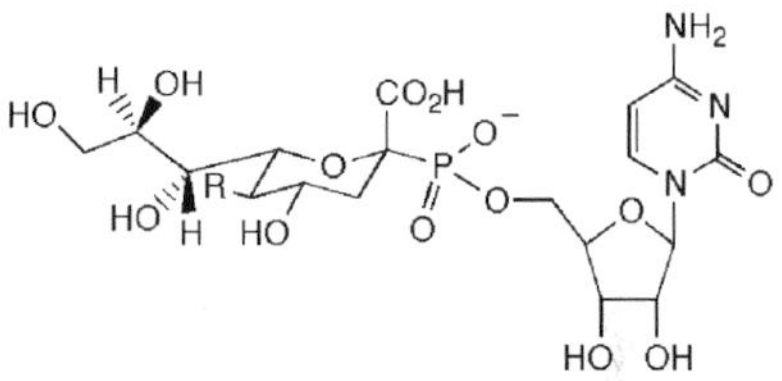

(173)

It is believed that the transfer of *N*-acetylneuraminic acid from CMP-Neu5Ac by sialyltransferases proceeds through a transition state where the Neu5Ac portion is distorted into a half-chair conformation with a developing oxocarbenium ion-like intermediate. Additionally, interpretation of kinetic data suggests that the transition-state species is bound around 10^{10} times tighter than the substrate, and has led a number of groups to investigate transition-state mimetics as potential inhibitors of sialyltransferases. Much of the initial work in this area was carried out by Schmidt and co-workers, and led to the development of the Neu5Ac2en-based derivative **174** which has a $K_i = 0.35$ μM against $\alpha(2,6)$-sialyltransferases.[243] The important structural feature of **174** for activity against $\alpha(2,6)$-sialyltransferase appears to be the additional charged group, since the analogous compound without the additional phosphate is essentially inactive. The discovery of **174** led to a series of compounds of the general structures **175** being developed, with the best compounds having either phenyl, furyl, or cyclohexenyl (e.g., **176**) substituents replacing the Neu5Ac2en residue in **174**.[240,244] The subsequent development of the $\alpha(2,6)$-sialyltransferase inhibitors such as compounds **177** ($K_i = 29$ nM) and **178** ($K_i = 25$ μM)[245] emphasizes the importance of the additional charged group for interaction with sialyltransferase, since compound **178** is derived from **177** by the formal loss of phosphate. The most recent example of sialylmimetics as inhibitors of sialyltransferases is the phosphoramidate analog **179**, which has $K_i = 68$ μM against $\alpha(2,6)$-sialyltransferase.[246] Compound **177** is the most potent inhibitor of sialyltransferases to date, and shows that both planarity about the anomeric carbon and the nature of the Neu5Ac side-chain replacement are important considerations in any future attempts at the development of these types of sialylmimetics as inhibitors of sialyltransferases.

Our own interest in the development of sialylmimetics has focussed both on the use of transition-state mimetics of Neu5Ac2en as inhibitors of influenza virus sialidase and VC sialidase (*vide infra*) as well on the development of substrate-based sialylmimetics as rotavirus inhibitors. Rotaviruses are recognized as the single most important cause of infantile gastroenteritis and are responsible for around 800,000 deaths annually, primarily in developing countries.[247–250] It is now recognized that rotavirus uses a carbohydrate-recognizing protein (called VP8*) during the initial adhesion events leading to infection of the host cell.[251] Although some debate remains as to what precise sequence of cell-surface carbohydrates is recognized by VP8*, it is generally accepted that the presence of sialic acids is important for binding.[251] Following from our initial studies on the use of thiosialosides (e.g., **180**) as inhibitors of rotaviral infection,[252] we have more recently reported that the use of lactose-based sialylmimetics of the general structure **181** provided molecules with modest inhibition of both animal and human strains of rotavirus.[253]

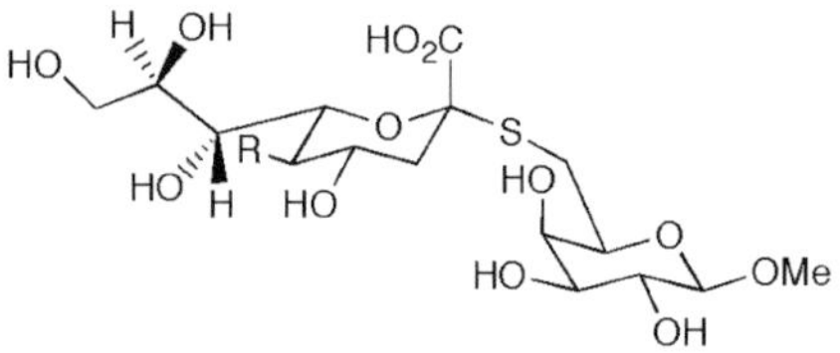

(180) R = NHAc 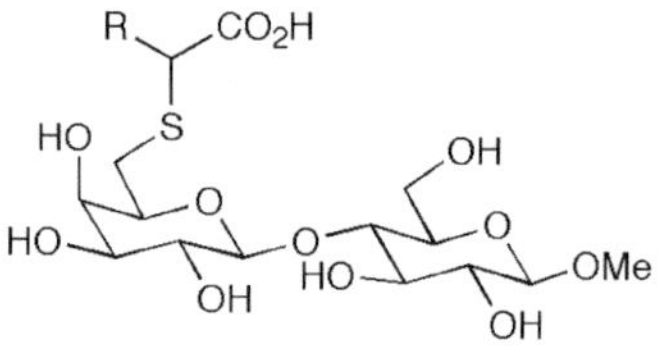   (181) R = alkyl, aryl

The use of lactose-based derivatives has also been reported for the inhibition of the cholera toxin B subunit, which is known to bind to the ganglioside GM$_1$ (**182**).[254] Following from their earlier report of the lactose derivative **183** as an inhibitor of the

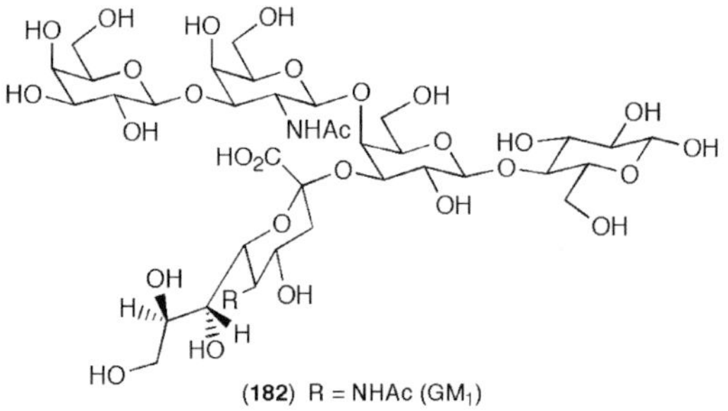

(182) R = NHAc (GM$_1$)

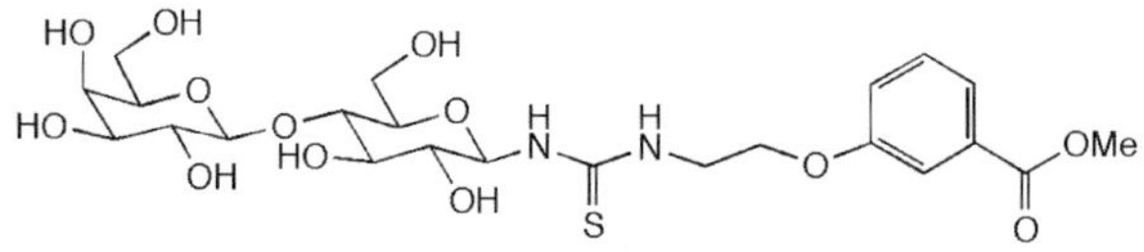

(183)

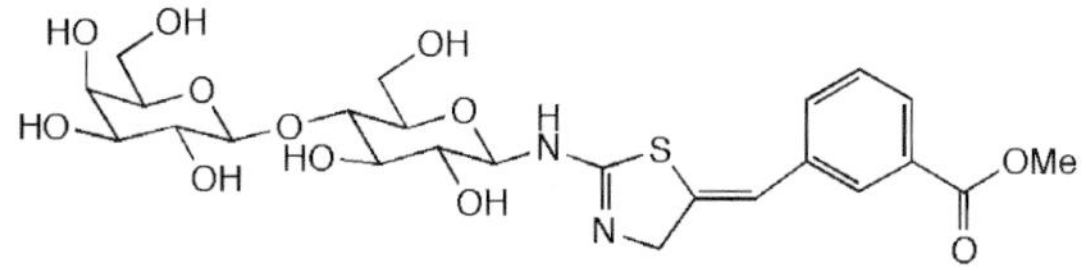

(184)

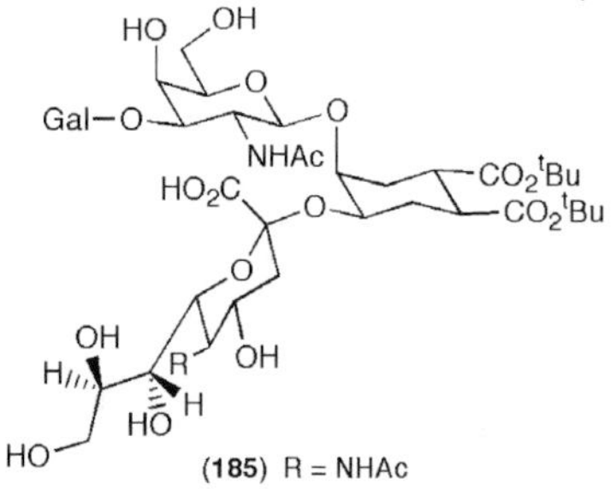

(185) R = NHAc

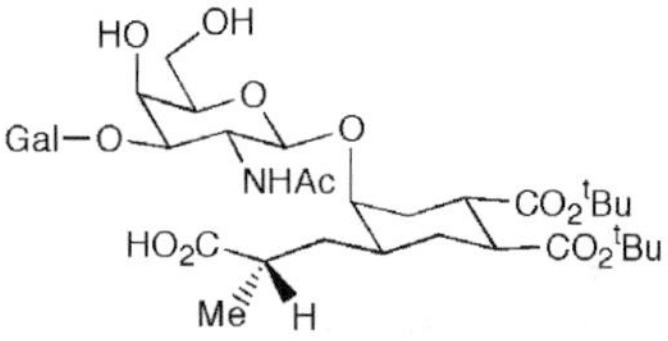

(186)

B subunit of cholera toxin (K_d=248 μM),[255] these same researcher have reported that
the more rigid lactoside derivative **184** has a K_d=23 μM against cholera toxin.[256]
Multivalent derivatives of **183** have also been described, with the best derivative
being the octavalent analog of **183** with a K_d=33 μM.[255] The finding that the pseu-
dosugar **185** had comparable affinity for cholera toxin when compared with GM_1[257]
prompted these same researchers to replace the Neu5Ac unit in **185** with simple
α-hydroxyacids.[258] Of the sialylmimetics prepared, the (R)-lactic acid derivative **186**
was found to have the highest affinity for cholera toxin, with a K_d=190 μM.[258]

V. CONCLUDING REMARKS

The excitement surrounding the chemistry and biochemistry of sialic acids and their
relevance to significant diseases has heightened over the last 20 years. The development of the anti-influenza drug Relenza and the subsequent discovery of the
Neu5Ac2en mimetic Oseltamivir has spawned great interest in the development of
carbohydrate-based compounds or mimetics of these compounds as potential pharmaceutical agents. As described above, there have been several examples of the benefits of using sialic acid analogs or sialylmimetics as biological probes. The excellent
progress in the development of sialylmimetics over recent years will undoubtedly
form the basis of many more studies into the use of such compounds as probes for
sialic acid-recognizing proteins over the next decade and beyond. While a great deal
is still to be achieved, based on the developments thus far, it is probable that we will
see many more examples of sialylmimetic-based pharmaceutical agents over the next
few years.

REFERENCES

1. Schauer, R., Chemistry, metabolism, and biological functions of sialic acids, *Adv. Carbohydr. Chem. Biochem.,* 40, 131–234, 1982.
2. Schauer, R. and Kamerling, J.P., Chemistry, biochemistry and biology of sialic acids, in *Glycoproteins II*, Montreuil, J., Vliegenthart, J.F.G., and Schachter, H., Eds., Elsevier Science, Amsterdam, 1997, pp. 243–402.
3. Angata, T. and Varki, A., Chemical diversity in the sialic acids and related α-keto acids. An evolutionary perspective, *Chem. Rev.,* 102, 439–469, 2002.
4. Varki, A., Biological roles of oligosaccharides: all of the theories are correct, *Glycobiology,* 3, 97–130, 1993.
5. Dwek, R.A., Glycobiology: toward understanding the function of sugars, *Chem. Rev.,* 96, 683–720, 1996.
6. Kiefel, M.J. and von Itzstein, M., Recent advances in the synthesis of sialic acid derivatives and sialylmimetics as biological probes, *Chem. Rev.,* 102, 471–490, 2002.
7. Zbiral, E., Synthesis of sialic acid analogs and their behaviour towards the enzymes of sialic acid metabolism and hemagglutinin X-31 of Influenza A-virus, In *Carbohydrates-Synthetic Methods and Applications in Medicinal Chemistry,* Ogura, H., Hasegawa, A., and Suams, T., Eds., VCH, Weinheim, 1992, pp. 304–339.
8. von Itzstein, M. and Kiefel, M.J., Sialic acid analogues as potential antimicrobial agents, in *Carbohydrates in Drug Design,* Witczak, Z.J. and Nieforth, K.A., Eds., Marcel Dekker, New York, 1997, 39–82.

9.   Brossmer, R. and Gross, H.J., Sialic acid analogs and application for preparation of neoglycoconjugates, *Methods Enzymol.,* 247, 153–176, 1994.

10.   Boons, G.-J. and Demchenko, A.V., Recent advances in *O*-sialylation, *Chem. Rev.,* 100, 4539–4565, 2000.

11.   Danishefsky, S.J. and Allen, J.R., From the laboratory to the clinic: a retrospective on fully synthetic carbohydrate-based anticancer vaccines, *Angew. Chem. Int. Ed.,* 39, 836–863, 2000.

12.   Roy, R., Sialoside mimetics as anti-inflammatory agents and inhibitors of flu virus infections, in *Carbohydrates in Drug Design*, Witczak, Z.J. and Nieforth, K.A., Ed., Marcel Dekker, New York, 1997, pp. 83–135.

13.   Sears, P. and Wong, C.-H., Carbohydrate mimetics: a new strategy for tackling the problem of carbohydrate-mediated biological recognition, *Angew. Chem. Int. Ed.,* 38, 2301–2324, 1999.

14.   Barchi, J.J., Jr., Emerging roles of carbohydrates and glycomimetics in anticancer drug design, *Curr. Pharm. Design,* 6, 485–501, 2000.

15.   Bols, M., Lillelund, V.H., Jensen, H.H., and Liang, X., Recent developments of transition-state analogue glycosidase inhibitors of non-natural product origin, *Chem. Rev.,* 102, 515–554, 2002.

16.   Simanek, E.E., McGarvey, G.J., Jablonowski, J.A., and Wong, C.-H., Selectin–carbohydrate interactions: from natural ligands to designed mimics, *Chem. Rev.,* 98, 833–862, 1998.

17.   Chong, A.K.J., Pegg, M.S., Taylor, N.R., and von Itzstein, M., Evidence for a sialyl cation transition state complex in the reaction of sialidase from influenza, *Eur. J. Biochem.,* 207, 335–343, 1992.

18.   Taylor, N.R. and von Itzstein, M., Molecular modeling studies on ligand binding to sialidase from influenza virus and the mechanism of catalysis, *J. Med. Chem.,* 37, 616–624, 1994.

19.   Guo, X., Laver, G.W., Vimr, E.R., and Sinnott, M.L., Catalysis by two sialidases with the same protein fold but different stereochemical courses: a mechanistic comparison of the enzymes from Influenza A virus and *Salmonella typhimurium, J. Am. Chem. Soc.,* 116, 5572–5578, 1994.

20.   Friebolin, H., Baumann, W., Brossmer, R., Keilich, G., Supp, M., Ziegler, D., and von Nicolai, H., Proton NMR spectroscopic evidence for the release of *N*-acetyl-α-D neuraminic acid as the first product of neuraminidase action, *Biochem. Int.,* 3, 321–326, 1981.

21.   Friebolin, H., Baumann, W., Keilich, G., Ziegler, D., Brossmer, R., and von Nicolai, H., ¹H NMR spectroscopy: a potent method for the determination of the substrate specificity of sialidases, *Hoppe Seyler's Z. Physiol. Chem.,* 362, 1455–1463, 1981.

22.   White, D.O., Influenza viral proteins: identification and synthesis, *Curr. Top. Microbiol. Immunol.,* 63, 1–48, 1974.

23.   Staschke, K.A., Hatch, S.D., Tang, J.C., Hornback, W.J., Munroe, J.E., Colacino, J.M., and Muesing, M.A., Inhibition of influenza virus hemagglutinin-mediated membrane fusion by a compound related to podocarpic acid, *Virology,* 248, 264–274, 1998.

24.   Li, W., Escarpe, P.A., Eisenberg, E.J., Cundy, K.C., Sweet, C., Jakeman, K.J., Merson, J., Lew, W., Williams, M., Zhang, L., Kim, C.U., Bischofberger, N., Chen, M.S., and Mendel, D.B., Identification of GS 4104 as an orally bioavailable prodrug of the influenza virus neuraminidase inhibitor GS 4071, *Antimicrob. Agents Chemother.,* 42, 647–653, 1998.

25.   Kati, W.M., Saldivar, A.S., Mohamadi, F., Sham, H.L., Laver, W.G., and Kohlbrenner, W.E., GS4071 is a slow-binding inhibitor of influenza neuraminidase from both A and B strains, *Biochem. Biophys. Res. Commun.,* 244, 408–413, 1998.

26. Dunn, C.J. and Goa, K.L., Zanamivir: a review of its use in influenza, *Drugs,* 58, 761–784, 1999.

27. Stamboulian, D., Bonvehi, P.E., Nacinovich, F.M., and Cox, N., Influenza, *Infect. Dis. Clin. North Am.,* 14, 141–166, 2000.

28. Singh, S., Jedrzejas, M.J., Air, G.M., Luo, M., Laver, W.G., and Brouillette, W.J., Structure-based inhibitors of influenza virus sialidase. A benzoic acid lead with novel interaction, *J. Med. Chem.,* 38, 3217–3225, 1995.

29. Nagai, T., Nishibe, Y., Makino, Y., Tomimori, T., and Yamada, H., Enhancement of *in vivo* anti-influenza virus activity of 5,7,4′-trihydroxy-8-methoxyflavone by drug delivery system using hydroxypropyl cellulose, *Biol. Pharm. Bull.,* 20, 1082–1085, 1997.

30. Weiss, W., Brown, J.H., Cusack, S., Paulson, J.C., Skehel, J.J., and Wiley, D.C., Structure of the influenza virus hemagglutinin complexed with its receptor, sialic acid, *Nature,* 333, 426–431, 1988.

31. Eisen, M.B., Sabesan, S., Skehel, J.J., and Wiley, D.C., Binding of the influenza A virus to cell-surface receptors: structures of five hemagglutinin-sialyloligosaccharide complexes determined by X-ray crystallography, *Virology,* 232, 19–31, 1997.

32. Sauter, N.K., Glick, G.K., Crowther, R.L., Park, S.-J., Eisen, M.B., Skehel, J.J., Knowles, J.R., and Wiley, D.C., Crystallographic detection of a second ligand binding site in influenza virus hemagglutinin, *Proc. Natl. Acad. Sci. USA,* 89, 324–328, 1992.

33. Sauter, N.K., Hanson, J.E., Glick, G.K., Brown, J.H., Crowther, R.L., Park, S.-J., Skehel, J.J., and Wiley, D.C., Binding of influenza virus hemagglutinin to analogs of its cell-surface receptor sialic acid: analysis by proton magnetic resonance spectroscopy and X-ray crystallography, *Biochemistry,* 31, 9609–9621, 1992.

34. Watowich, S.J., Skehel, J.J., and Wiley, D.C., Crystal structures of influenza virus hemagglutinin in complex with high affinity receptor analogs, *Structure,* 2, 719–731, 1994.

35. Weiss, W., Brunger, A.T., Skehel, J.J., and Wiley, D.C., Refinement of the influenza virus hemagglutinin by simulated annealing, *J. Mol. Biol.,* 212, 737–761, 1990.

36. Wiley, D.C. and Skehel, J.J., The structure and function of the hemagglutinin membrane glycoprotein of influenza virus, *Ann. Rev. Biochem.,* 56, 365–394, 1987.

37. Wilson, I.A., Skehel, J.J., and Wiley, D.C., Structure of the haemagglutinin membrane glycoprotein of influenza virus at 3 Å resolution, *Nature,* 289, 366–373, 1981.

38. Baker, A.T., Varghese, J.N., Laver, W.G., Air, G.M., and Colman, P.M., Three-dimensional structure of neuraminidase of subtype N9 from an avian influenza virus, *Proteins* 2, 111–117, 1987.

39. Taylor, N.R., Cleasby, A., Singh, O., Skarzynski, T., Wonacott, A.J., Smith, P.W., Sollis, S.L., Howes, P.D., Cherry, P.C., Bethell, R., Colman, P., and Varghese, J., Dihydropyrancarboxamides related to Zanamivir: a new series of inhibitors of Influenza virus sialidases. 2. Crystallographic and molecular modeling study of complexes of 4-amino-4H-pyran-6-carboxamides and sialidase from Influenza virus Types A and B, *J. Med. Chem.,* 41, 798–807, 1998.

40. Tulip, W.R., Varghese, J.N., Baker, A.T., van Donkelaar, A., Laver, W.G., Webster, R.G., and Colman, P.M., Refined atomic structures of N9 subtype influenza virus neuraminidase and escape mutants, *J. Mol. Biol.,* 221, 487–497, 1991.

41. Tulip, W.R., Varghese, J.N., Laver, W.G., Webster, R.G., and Colman, P.M., Refined crystal structure of the influenza virus N9 neuraminidase-NC41 Fab complex, *J. Mol. Biol.,* 227, 122–148, 1992.

42. Tulip, W.R., Varghese, J.N., Webster, R.G., Air, G.M., Laver, W.G., and Colman, P.M., Crystal structures of neuraminidase-antibody complexes, *Cold Spring Harbor Symp. Quant. Biol.* 54, 257–263, 1989.

43. Tulip, W.R., Varghese, J.N., Webster, R.G., Laver, W.G., and Colman, P.M., Crystal structures of two mutant neuraminidase-antibody complexes with amino acid substitutions in the interface, *J. Mol. Biol.,* 227, 149–159, 1992.

44. Varghese, J.N. and Colman, P.M., Three-dimensional structure of the neuraminidase of influenza virus A/Tokyo/3/67 at 2.2 Å resolution, *J. Mol. Biol.,* 221, 473–486, 1991.

45. Varghese, J.N., Colman, P.M., van Donkelaar, A., Blick, T.J., Sahasrabudhe, A., and McKimm-Breschkin, J.L., Structural evidence for a second sialic acid binding site in avian influenza virus neuraminidases, *Proc. Natl. Acad. Sci. USA,* 94, 11808–11812, 1997.

46. Varghese, J.N., Epa, V.C., and Colman, P.M., Three-dimensional structure of the complex of 4-guanidino-Neu5Ac2en and influenza virus neuraminidase, *Protein Sci.,* 4, 1081–1087, 1995.

47. Varghese, J.N., Laver, W.G., and Colman, P.M., Structure of the influenza virus glycoprotein antigen neuraminidase at 2.9 Å resolution, *Nature,* 303, 35–40, 1993.

48. Varghese, J.N., McKimm-Breschkin, J.L., Caldwell, J.B., Kortt, A.A., and Colman, P.M., The structure of the complex between influenza virus neuraminidase and sialic acid, the viral receptor, *Proteins,* 14, 327–332, 1992.

49. Varghese, J.N., Webster, R.G., Laver, W.G., and Colman, P.M., Structure of an escape mutant of glycoprotein N2 neuraminidase of influenza virus A/Tokyo/3/67 at 3 Å, *J. Mol. Biol.,* 200, 201–203, 1988.

50. Blick, T.J., Tiong, T., Sahasrabudhe, A., Varghese, J.N., Colman, P.M., Hart, G.J., Bethell, R.C., and McKimm-Breschkin, J.L., Generation and characterization of an influenza virus neuraminidase variant with decreased sensitivity to the neuraminidase-specific inhibitor 4-guanidino-Neu5Ac2en, *Virology,* 214, 475–484, 1995.

51. Colman, P.M., Laver, W.G., Varghese, J.N., Baker, A.T., Tulloch, P.A., Air, G.M., and Webster, R.G., Three-dimensional structure of a complex of antibody with influenza virus neuraminidase, *Nature,* 326, 358–363, 1987.

52. Colman, P.M., Tulip, W.R., Varghese, J.N., Tulloch, P.A., Baker, A.T., Laver, W.G., Air, G.M., and Webster, R.G., Three-dimensional structures of influenza virus neuraminidase-antibody complexes, *Philos. Trans. R. Soc. Lond. B,* 323, 511–518, 1989.

53. Colman, P.M., Varghese, J.N., and Laver, W.G., Structure of the catalytic and antigenic sites in influenza virus neuraminidase, *Nature,* 303, 41–44, 1983.

54. Smith, P.W., Sollis, S.L., Howes, P.D., Cherry, P.C., Cobley, K.N., Taylor, H., Whittington, H.R., Scicinski, J., Bethell, R., Taylor, N., Skarzynski, T., Cleasby, A., Singh, O., Varghese, J., and Colman, P., Novel inhibitors of influenza sialidases related to GG167 structure-activity, crystallographic and molecular dynamics studies with 4H-pyran-2-carboxylic acid 6-carboxamides, *Bioorg. Med. Chem. Lett.,* 6, 2931–2936, 1996.

55. Varghese, J.N., Development of neuraminidase inhibitors as anti-influenza virus drugs, *Drug Dev. Res.,* 46, 176–196, 1999.

56. Colacino, J.M., Staschke, K.A., and Laver, W.G., Approaches and strategies for the treatment of influenza virus infections, *Antivir. Chem. Chemother.,* 10, 155–185, 1999.

57. Taylor, G., Sialidases: structures, biological significance and therapeutic potential, *Curr. Opin. Struct. Biol.,* 6, 830–837, 1996.

58. Colman, P.M., A novel approach to antiviral therapy for influenza, *J. Antimicrob. Chemother.,* 44 (Suppl B), 17–22, 1999.

59. Wutzler, P. and Vogel, G., Neuraminidase inhibitors in the treatment of influenza A and B overview and case reports, *Infection,* 28, 261–266, 2000.

60. Varghese, J.N., Smith, P.W., Sollis, S.L., Blick, T.J., Sahasrabudhe, A., McKimm-Breschin, J.L., and Colman, P.M., Drug design against a shifting target: a structural basis for resistance to inhibitors in a variant of influenza virus neuraminidase, *Structure,* 6, 735–746, 1998.

61. Kim, C.U., Chen, X., and Mendel, D.B., Neuraminidase inhibitors as anti-influenza virus agents, *Antivir. Chem. Chemother.,* 10, 141–154, 1999.

62. Kiefel, M.J. and von Itzstein, M., Influenza virus sialidase: a target for drug discovery, *Prog. Med. Chem.,* 36, 1–28, 1999.

63. Air, G.M., Ghate, A.A., and Stray, S.J., Influenza neuraminidase as target for antivirals, *Adv. Virus Res.,* 54, 375–402, 1999.

64. Dowle, M.D. and Howes, P.D., Recent advances in sialidase inhibitors, *Expert Opin. Ther. Patents,* 8, 1461–1478, 1998.

65. Shigeta, S., Recent progress in anti-influenza chemotherapy. *Drugs R.D.,* 2, 153–164, 1999.

66. von Itzstein, M., Dyason, J.C., Oliver, S.W., White, H.F., Wu, W.Y., Kok, G.B., and Pegg, M.S., A study of the active site of influenza virus sialidase: an approach to the rational design of novel anti-influenza drugs, *J. Med. Chem.,* 39, 388–391, 1996.

67. Wilson, J.C. and von Itzstein, M., Recent strategies in the search for new anti-influenza therapies, *Curr. Drug Targets,* 4, 389–408, 2003.

68. Colman, P.M. and Ward, C.W., Structure and diversity of influenza virus neuraminidase, *Curr. Top. Microbiol. Immunol.,* 114, 177–255, 1985.

69. Wallace, A.C., Laskowski, R.A., and Thornton, J.M., LIGPLOT: a program to generate schematic diagrams of protein–ligand interactions, *Protein Eng.,* 8, 127–134, 1995.

70. Bossart-Whitaker, P., Carson, M., Babu, Y.S., Smith, C.D., Laver, W.G., and Air, G.M., Three-dimensional structure of influenza A N9 neuraminidase and its complex with the inhibitor 2-deoxy-2,3-dehydro-*N*-acetyl neuraminic acid, *J. Mol. Biol.,* 232, 1069–1083, 1993.

71. Burmeister, W.P., Henrissat, B., Bosso, C., Cusack, S., and Ruigrok, R.W.H., Influenza B virus neuraminidase can synthesise its own inhibitor, *Structure,* 1, 19–26, 1993.

72. von Itzstein, M., Wu, W.Y., Phan, T., Danylec, B., and Jin B., Preparation of derivatives and analogs of 2-deoxy-2,3-didehydro-*N*-acetylneuraminic acid as antiviral agents, WO9116320, CA117:49151y, 1991.

73. von Itzstein, M., Wu, W.Y., and Jin, B., The synthesis of 2,3-didehydro-2,4-dideoxy-4-guanidinyl-*N*-acetylneuraminic acid: a potent influenza virus sialidase inhibitor, *Carbohydr. Res.,* 259, 301–305, 1994.

74. Chandler, M., Bamford, M.J., Conroy, M., Lamont, B., Patel, B., Patel, V.K., Steeples, I.P., Storer, R., Weir, N.G., Wright, M., and Williamson, C., Synthesis of the potent influenza neuraminidase inhibitor 4-guanidino Neu5Ac2en-X-ray molecular structure of 5-acetamido-4-amino-2,6-anhydro-3,4,5-trideoxy-D-*erythro*-L-*gluco*-nononic acid, *J. Chem. Soc. Perkin Trans. 1,* 1173–1180, 1995.

75. Scheigetz, J., Zamboni, R., Bernstein, M.A., and Roy, B., A synthesis of 4-guanidino-2-deoxy-2,3-didehydro-*N*-acetylneuraminic acid, *Org. Prep. Proc. Int.,* 27, 637–644, 1995.

76. Holzer, C.T., von Itzstein, M., Jin, B., Pegg, M.S., Stewart, W.P., and Wu, W.Y., Inhibition of sialidases from viral, bacterial and mammalian sources by analogs of 2-deoxy-2,3-didehydro-*N*-acetylneuraminic acid modified at the C-4 position, *Glycoconjugate J.,* 10, 40–44, 1993.

77. von Itzstein, M., Wu, W.Y., Kok, G.B., Pegg, M.S., Dyason, J.C., Jin, B., Phan, T.V., Smythe, M.L., White, H.F., and Oliver, S.W., Rational design of potent sialidase-based inhibitors of influenza virus replication, *Nature,* 363, 418–423, 1993.

78. Pegg, M.S. and von Itzstein, M., Slow-binding inhibition of sialidase from influenza virus, *Biochem. Mol. Biol. Int.,* 36, 851–858, 1994.
79. Woods, J.M., Bethell, R., Coates, J.A., Healy, N., Hiscox, S.A., Pearson, B.A., Ryan, D.M., Ticehurst, J., Tilling, J., and Walcott, S.M., 4-Guanidino-2,4-dideoxy-2,3-dehydro-*N*-acetylneuraminic acid is a highly effective inhibitor of both the sialidase (neuraminidase) and of growth of a wide range of influenza A and B viruses *in vitro, Antimicrob. Agents Chemother.,* 37, 1473–1479, 1993.
80. Hayden, F.G., Treanor, J.J., Betts, R.F., Lobo, M., Esinhart, J.D., and Hussey, E.K., Safety and efficacy of the neuraminidase inhibitor GG167 in experimental human influenza, *JAMA,* 275, 295–299, 1996.
81. Hart, G.J. and Bethell, R.C., 2,3-didehydro-2,4-dideoxy-4-guanidino-*N*-acetyl-D-neuraminic acid (4-guanidino-Neu5Ac2en) is a slow-binding inhibitor of sialidase from both influenza A virus and influenza B virus, *Biochem. Mol. Biol. Int.,* 36, 695–703, 1995.
82. Crennell, S., Garman, E., Laver, G., Vimr, E., and Taylor, G., Crystal structure of *Vibrio cholerae* neuraminidase reveals dual lectin-like domains in addition to the catalytic domain, *Structure,* 2, 535–544, 1994.
83. Reece, P.A., Watson, K.G., Wu, W.-Y., Jin, B., and Krippner, G.Y., Preparation of poly(neuraminic acids) as influenza virus neuraminidase inhibitors, in *PCT International Applications,* (Biota Scientific Management Pty. Ltd., Australia), WO9821243, 1998.
84. Honda, T., Yoshida, S., Arai, M., Masuda, T., and Yamashita, M., Synthesis and anti-influenza evaluation of polyvalent sialidase inhibitors bearing 4-guanidino-Neu5Ac2en derivatives, *Bioorg. Med. Chem. Lett.,* 12, 1929–1932, 2002.
85. Driguez, P.A., Barrere, B., Quash, G., and Doutheau, A., Synthesis of transition-state analogues as potential inhibitors of sialidase from Influenza virus, *Carbohydr. Res.,* 262, 297–310, 1994.
86. Flashner, M., Kessler, J., and Tannenbaum, S.W., The interaction of substrate-related ketals with bacterial and viral neuraminidases, *Arch. Biochem. Biophys.,* 262, 297–310, 1983.
87. Ikeda, K., Sano, K., Ito, M., Saito, M., Hidari, K., Suzuki, T., Suzuki, Y., and Tanaka, K., Synthesis of 2-deoxy-2,3-didehydro-*N*-acetylneuraminic acid analogues modified at the C-4 and C-9 positions and their behavior towards sialidase from influenza virus and pig liver membrane, *Carbohydr. Res.,* 330, 31–41, 2001.
88. Meindl, P., Bodo, G., Palese, P., Schulman, J., and Tuppy, H., Inhibition of neuraminidase activity by derivatives of 2-deoxy-2,3-dehydro-*N*-acetylneuraminic acid, *Virology,* 58, 457–463, 1974.
89. Smith, B.J., Starkey, I.D., Howes, P.D., Sollis, S.L., Keeling, S.P., Cherry, P.C., von Itzstein, M., Wu, W.Y., and Jin, B., Synthesis and influenza virus sialidase inhibitory activity of analogues of 4-guanidino-Neu5Ac2en (GG167) with modified 5-substituents, *Eur. J. Med. Chem.,* 31, 143–150, 1996.
90. Starkey, I.D., Mahmoudian, M., Noble, D., Smith, P.W., Cherry, P.C., Howes, P.D., and Sollis, S.L., Synthesis and influenza virus sialidase inhibitory activity of the 5-deacetamido analogue of 2,3-didehydro-2,4-dideoxy-4-guanidinyl-*N*-acetylneuraminic acid, *Tetrahedron Lett.,* 36, 299–302, 1995.
91. Smith, P.W., Sollis, S.L., Howes, P.D., Cherry, P.C., Starkey, I.D., Cobley, K.N., Weston, H., Scicinski, J., Merritt, A., Whittington, A., Wyatt, P., Taylor, N., Green, D., Bethell, R., Madar, S., Fenton, R.J., Morley, P.J., Pateman, T., and Beresford, A., Dihydropyrancarboxamides related to Zanamivir: a new series of inhibitors of Influenza virus sialidases. 1. Discovery, synthesis, biological activity, and structure–activity relationships of 4-guanidino- and 4-amino-4H-pyran-6-carboxamides, *J. Med. Chem.,* 41, 787–797, 1998.

92.   Smith, P.W., Robinson, J.E., Evans, D.N., Sollis, S.L., Howes, P.D., Trivedi, N., and
      Bethell, R.C., Sialidase inhibitors related to Zanamivir: synthesis and biological
      evaluation of 4H-pyran 6-ether and ketone, *Bioorg. Med. Chem. Lett.*, 9, 601–604,
      1999.
93.   Sollis, S.L., Smith, P.W., Howes, P.D., Cherry, P.C., and Bethell, R., Novel inhibitors of
      influenza sialidase related to GG167, *Bioorg. Med. Chem.* Lett., 6, 1805–1808, 1996.
94.   Florio, P., Thomson, R.J., Alafaci, A., Abo, S., and von Itzstein, M., Synthesis of
      Δ4-β-D-glucopyranosiduronic acids as mimetics of 2,3-unsaturated sialic acids for
      sialidase inhibition, *Bioorg. Med. Chem. Lett.*, 9, 2065–2068, 1999.
95.   Taylor, N.R., Cleasby, A., Singh, O., Skarzynski, T., Wonacott, A.J., Smith, P.W.,
      Sollis, S.L., Howes, P.D., Cherry, P.C., Bethell, R., Colman, P., and Varghese, J.,
      Dihydropyrancarboxamides related to Zanamivir: a new series of inhibitors of
      influenza virus sialidases. 2. Crystallographic and molecular modeling study of com-
      plexes of 4-amino-4H-pyran-6-carboxamides and sialidase from influenza virus types
      A and B, *J. Med. Chem.*, 41, 798–807, 1998.
96.   Wyatt, P.G., Coomber, B.A., Evans, D.N., Jack, T.I., Fulton, H.E., Wonacott, A.J.,
      Colman, P., and Varghese, J., Sialidase inhibitors related to Zanamivir. Further SAR
      studies of 4-amino-4H-pyran-2-carboxylic acid-6-propylamides, *Bioorg. Med. Chem.*
      Lett., 11, 669–673, 2001.
97.   Honda, T., Masuda, T., Yoshida, S., Arai, M., Kaneko, S., and Yamashita, M.,
      Synthesis and anti-influenza virus activity of 7-*O*-alkylated derivatives related to
      Zanamivir, *Bioorg. Med. Chem. Lett.*, 12, 1925–1928, 2002.
98.   Honda, T., Masuda, T., Yoshida, S., Arai, M., Kobayashi, Y., and Yamashita, M.,
      Synthesis and anti-influenza virus activity of 4-guanidino-7-substituted Neu5Ac2en
      derivatives, *Bioorg. Med. Chem. Lett.*, 12, 1921–1924, 2002.
99.   Andrews, D.M., Cherry, P.C., Humber, D.C., Jones, P.S., Keeling, S.P., Martin, P.F.,
      Shaw, C.D., and Swanson, S., Synthesis and influenza virus sialidase inhibitory activity
      of analogues of 4-guanidino-Neu5Ac2en (Zanamivir) modified in the glycerol side-
      chain, *Eur. J. Med. Chem.*, 34, 563–574, 1999.
100. Kamerling, J.P., Schauer, R., Shukla, A.K., Stoll, S., and van Halbeek, H., Migration of
      O-acetyl groups in *N,O*-acetylneuraminic acids, *Eur. J. Biochem.*, 162, 601–607, 1987.
101. Reinhard, B. and Faillard H., Regioselective acetylations of sialic acid α-ketosides,
      Liebigs Ann. Chem., 193–203, 1994.
102. Masuda, T., Shibuya, S., Arai, M., Yoshida, S., Tomozawa, T., Ohno, A., Yamashita,
      M., and Honda, T., Synthesis and anti-influenza evaluation of orally active bicyclic
      ether derivatives related to Zanamivir, *Bioorg. Med. Chem. Lett.*, 13, 669–673,
      2003.
103. Ryan, D.M., Ticehurst, J., Dempsey, M., and Penn, C.R., Inhibition of influenza virus
      replication in mice by GG167 (4-guanidino-2,4-dideoxy-2,3-dehydro-*N*-acetylneu-
      raminic acid) is consistent with extracellular activity of viral neuraminidase (siali-
      dase), *Antimicrob. Agents Chemother.*, 10, 2270–2275, 1994.
104. Ryan, D.M., Ticehurst, J., and Dempsey, M.H., GG167 (4-guanidino-2,4-dideoxy-
      2,3-dehydro-*N*-acetylneuraminic acid) is a potent inhibitor of influenza virus in fer-
      rets, *Antimicrob. Agents Chemother.*, 39, 2583–2584, 1995.
105. Kim, C.U., Lew, W., Williams, H., Liu, L., Zhang, S., Swaminathan, S.,
      Bischofberger, N., Chen, M.S., Mendel, D.B., Tai, C.Y., Laver, W.G., and Stevens,
      R.C., Influenza neuraminidase inhibitors possessing a novel hydrophobic interaction
      in the enzyme active site: design, synthesis and structural analysis of carbocyclic
      sialic acid analogues with potent anti-influenza activity, *J. Am. Chem. Soc.*, 119,
      681–690, 1997.

106. Kim, C.U., Lew, W., Williams, M.A., Wu, H., Zhang, L., Chen, X., Escarpe, P.A., Mendel, D.B., Laver, W.G., and Stevens, R.C., Structure–activity relationship studies of novel carbocyclic influenza neuraminidase inhibitors, *J. Med. Chem.,* 41, 2451–2460, 1998.

107. Lew, W., Chen, X., and Kim, C.U., Discovery and development of GS 4104 (oseltamivir): an orally active influenza neuraminidase inhibitor, *Curr. Med. Chem.,* 7, 663–672, 2000.

108. Williams, M., Lew, W., Mendel, D.B., Tai, C.Y., Escarpe, P., Laver, G.W., Stevens, R.C., and Kim, C.U., Structure–activity relationships of carbocyclic influenza neuraminidase inhibitors, *Bioorg. Med. Chem. Lett.,* 7, 1837–1842, 1997.

109. Babu, Y.S., Chand, P., Bantia, S., Kotian, P., Dehghani, A., El Kattan, Y., Lin, T.H., Hutchison, T.L., Elliott, A.J., Parker, C.D., Ananth, S.L., Horn, L.L., Laver, G.W., and Montgomery, J.A., BCX-1812 (RWJ-270201): discovery of a novel, highly potent, orally active, and selective influenza neuraminidase inhibitor through structure-based drug design, *J. Med. Chem.,* 43, 3482–3486, 2000.

110. Sidwell, R. and Smee, D.F., Peramivir (BCX-1812, RWJ-270201) potential new therapy for influenza, *Expert Opin. Invest. Drugs,* 11, 859–869, 2002.

111. Sidwell, R.W., Smee, D.F., Huffman, J.H., Barnard, D.L., Bailey, K.W., Morrey, J.D., and Babu, Y.S., *In vivo* influenza virus-inhibitory effects of the cyclopentane neuraminidase inhibitor RWJ-270201, *Antimicrob. Agents Chemother.,* 45, 749–757, 2001.

112. Jedrzejas, M.J., Singh, S., Brouillette, W.J., Laver, W.G., Air, G.M., and Luo, M., Structures of aromatic inhibitors of influenza virus neuraminidase, *Biochemistry,* 34, 3144–3151, 1995.

113. Chand, P., Babu, Y.S., Bantia, S., Chu, N., Cole, L.B., Kotian, P.L., Laver, W.G., Montgomery, J.A., Pathak, V.P., Petty, S.L., Shrout, D.P., Walsh, D.A., and Walsh, G.M., Design and synthesis of benzoic acid derivatives as influenza neuraminidase inhibitors using structure-based drug design, *J. Med. Chem.,* 40, 4030–4052, 1997.

114. Finley, J.B., Atigadda, V.R., Duarte, F., Zhao, J.J., Brouillette, W.J., Air, G.M., and Luo, M., Novel aromatic inhibitors of influenza virus neuraminidase make selective interactions with conserved residues and water molecules in the active site, *J. Mol. Biol.,* 293, 1107–1119, 1999.

115. Brouillette, W.J., Ali, S.M., Finley, J., and Luo, M., New potent aromatic inhibitors of influenza virus sialidase. *Book of Abstracts,* 218th ACS National Meeting, New Orleans, Aug. 22–26, 1999, pp. MEDI-277.

116. Howes, P.D., Cleasby, A., Evans, D.N., Feilden, H., Smith, P.W., Sollis, S.L., Taylor, N., and Wonacott, A.J., 4-Acetylamino-3-(imidazol-1-yl)-benzoic acids as novel inhibitors of influenza sialidase, *Eur. J. Med. Chem.,* 34, 225–234, 1999.

117. Sudbeck, E.A., Jedrzejas, M.J., Singh, S., Brouillette, W.J., Air, G.M., Laver, W.G., Babu, Y.S., Bantia, S., Chand, P., Chu, N., Montgomery, J.A., Walsh, D.A., and Luo, M., Guanidinobenzoic acid inhibitors of influenza virus neuraminidase, *J. Mol. Biol.,* 267, 584–594, 1997.

118. Atigadda, V.R., Brouillette, W.J., Duarte, F., Ali, S.M., Babu, Y.S., Bantia, S., Chand, P., Chu, N., Montgomery, J.A., Walsh, D.A., Sudbeck, E.A., Finley, J., Luo, M., Air, G.M., and Laver, G.W., Potent inhibition of influenza sialidase by a benzoic acid containing a 2-pyrrolidinone substituent, *J. Med. Chem.,* 42, 2332–2343, 1999.

119. Atigadda, V.R., Brouillette, W.J., Duarte, F., Babu, Y.S., Bantia, S., Chand, P., Chu, N., Montgomery, J.A., Walsh, D.A., Sudbeck, E., Finley, J., Air, G.M., Luo, M., and Laver, G.W., Hydrophobic benzoic acids as inhibitors of influenza neuraminidase, *Bioorg. Med. Chem. Lett.,* 7, 2487–2497, 1999.

120. Kati, W.M., Montgomery, D., Carrick, R., Gubareva, L., Maring, C., McDaniel, K., Steffy, K., Molla, A., Hayden, F., Kempf, D., and Kohlbrenner, W., *In vitro* characterization of A-315675, a highly potent inhibitor of A and B strain influenza virus neuraminidases and influenza virus replication, *Antimicrob. Agents Chemother.*, 46, 1014–1021, 2002.

121. Degoey, D.A., Chen, H.-J., Flosi, W.J., Grampovnik, D.J., Yeung, M.C., Klein, L.L., and Kempf, D., Enantioselective synthesis of anti-influenza compound A-315675, *J. Org. Chem.*, 67, 5445–5453, 2002.

122. Vorwerk, S. and Vasella, A., Carbocyclic analogs of *N*-acetyl-2,3-didehydro-2-deoxy-D-neuraminic acid (Neu5Ac2en, DANA): synthesis and inhibition of viral and bacterial neuraminidases, *Angew. Chem. Int. Ed.*, 37, 1732–1734, 1998.

123. Kim, C.U., Lew, W., Williams, M., Zhang, S., Swaminathan, S., Bischofberger, N., Chen, D., Mendel, D.B., Li, W., Tai, L., Escarpe, P., Cundy, K.C., Eisenberg, E.J., Lacy, S.A., Sidwell, R., Stevens, R.C., and Laver, G.W., New potent, orally active neuraminidase inhibitors as anti-Influenza agents: *in-vivo* and *in vitro* activity of GS4071 and analogs, *36th Interscience Conference of Antimicrobial Agents and Chemotherapy,* American Society Microbiology, Washington, DC, 1996, p. 171.

124. Kim, C.U., Rational drug design of orally active influenza neuraminidase inhibitors: discovery and development of GS 4104, *Med. Chem. Res.,* 8, 392–399, 1998.

125. Mendel, D.B., Tai, C.Y., Escarpe, P.A., Li, W., Sidwell, R.W., Huffman, J.H., Sweet, C., Jakeman, K.J., Merson, J., Lacy, S.A., Lew, W., Williams, M.A., Zhang, L., Chen, M.S., Bischofberger, N., and Kim, C.U., Oral administration of a prodrug of the influenza virus neuraminidase inhibitor GS4071 protects mice and ferrets against influenza infection, *Antimicrob. Agents. Chemother.,* 42, 640–646, 1998.

126. Stella, V.J., Charman, W.N.A., and Naringrekar, V.H., Prodrugs: do they have advantages in clinical practice? *Drugs,* 29, 455–473, 1985.

127. Eisenberg, E.J., Bidgood, A., and Cundy, K.C., Penetration of GS4071, a novel influenza neuraminidase inhibitor, into rat bronchoalveolar lining fluid following oral administration of the prodrug GS4104, *Antimicrob. Agents. Chemother.,* 41, 1949–1952, 1997.

128. Lew, W., Wu, H., Mendel, D.B., Escarpe, P.A., Chen, X., Laver, W.G., Graves, B.J., and Kim, C.U., A new series of C3-aza carbocyclic influenza neuraminidase inhibitors: synthesis and inhibitory activity, *Bioorg. Med. Chem. Lett.,* 8, 3321–3324, 1998.

129. Lew, W., Wu, H., Chen, X., Graves, B.J., Escarpe, P.A., MacArthur, H.L., Mendel, D.B., and Kim, C.U., Carbocyclic influenza neuraminidase inhibitors possessing a C3-cyclic amine side chain: synthesis and inhibitory activity, *Bioorg. Med. Chem. Lett.,* 10, 1257–1260, 2000.

130. Bianco, A., Brufani, M., Manna, F., and Melchioni, C., Synthesis of a carbocyclic sialic acid analogue for the inhibition of influenza virus neuraminidase, *Carbohydr. Res.,* 332, 23–31, 2001.

131. Shitara, E., Nishimura, Y., Nerome, K., Hiramoto, Y., and Takeuchi, T., Synthesis of 6-acetamido-5-amino- and -5-guanidino-3,4-dehydro-*N*-(2-ethylbutyryl)-3-piperidinecarboxylic acids related to Zanamivir and Oseltamivir, inhibitors of influenza virus neuraminidases, *Org. Lett.,* 2, 3837–3840, 2000.

132. Zhang, L., Williams, M.A., Mendel, D.B., Escarpe, P.A., Chen, X., Wang, K.Y., Graves, B.J., Lawton, G., and Kim, C.U., Synthesis and evaluation of 1,4,5,6-tetrahydropyridazine derivatives as influenza neuraminidase inhibitors, *Bioorg. Med. Chem. Lett.,* 9, 1751–1756, 1999.

133. Jones, P.S., Smith, P.W., Hardy, G.W., Howes, P.D., Upton, R.J., and Bethell, R.C., Synthesis of tetrasubstituted bicyclo[3.2.1]octenes as potential inhibitors of influenza virus sialidase, *Bioorg. Med. Chem. Lett.,* 9, 605–610, 1999.

134. Smith, P.W., Trivedi, N., Howes, P.D., Sollis, S.L., Rahim, G., Bethell, R.C., and Lynn, S., Synthesis of a tetrasubstituted bicyclo[2.2.2]octane as a potential inhibitor of influenza virus sialidase, *Bioorg. Med. Chem. Lett.*, 9, 611–614, 1999.

135. Yamamoto, T., Kumazawa, H., Inami, K., Teshima, T., and Shiba, T., Synthesis of sialic acid isomers with inhibitory activity against neuraminidase, *Tetrahedron Lett.*, 33, 5791–5794, 1992.

136. Smee, D.F., Huffman, J.H., Morrison, A.C., Barnard, D.L., and Sidwell, R.W., Cyclopentane neuraminidase inhibitors with potent *in vitro* anti-influenza virus activities, *Antimicrob. Agents Chemother.*, 45, 743–748, 2001.

137. Wang, G.T., Chen, Y.W., Wang, S., Gentles, R., Sowin, T., Kati, W., Muchmore, S., Giranda, V., Stewart, K., Sham, H., Kempf, D., and Laver, W.G., Design, synthesis, and structural analysis of influenza neuraminidase inhibitors containing pyrrolidine cores, *J. Med. Chem.*, 44, 1192–1201, 2001.

138. Brouillette, W.J., Atigadda, V.R., Luo, M., Air, G.M., Babu, Y.S., and Bantia, S., Design of benzoic acid inhibitors of influenza neuraminidase containing a cyclic substitution for the *N*-acetyl grouping, *Bioorg. Med. Chem. Lett.*, 9, 1901–1906, 1999.

139. Stewart-Tull, D.E.S., Ollar, R.A., and Scobie, T.S., Studies on the *Vibrio cholerae* mucinase complex. I. enzymic activities associated with the complex, *J. Med. Microbiol.*, 22, 325–333, 1986.

140. Galen, J.E., Ketley, J.M., Fasano, A., Richardson, S.H., Wasserman, S.S., and Kaper, J.B., Role of *Vibrio cholerae* neuraminidase in the function of cholera toxin, *Infect. Immun.*, 60, 406–415, 1992.

141. Merritt, E.A., Sarfaty, S., van den Akker, F., L'Hoir, C., Martial, J.A., and Hol, W.G.J., Crystal structure of cholera toxin B-pentamer bound to receptor GM1 pentasaccharide, *Protein Sci.*, 3, 166–175, 1994.

142. Brossmer, R., Keilich, G., and Ziegler, D., Inhibition studies on *Vibrio cholerae* neuraminidase, *Hoppe-Seyler's Z. Physiol. Chem.*, 358, 391–396, 1977.

143. Zbiral, E., Brandstetter, H.H., Christian, R., and Schauer, R., Structural variations of *N*-acetylneuraminic acid. 7. Synthesis of the C-7-, C-8- and C-7, -8-side chain epimers of 2-deoxy-2,3-didehydro-*N*-acetylneuraminic acid and their behaviour towards sialidase from *Vibrio cholerae, Liebigs Ann. Chem.*, 781–786, 1987.

144. Zbiral, E., Schreiner, E., Christian, R., Kleineidam, R.G., and Schauer, R., Structural variations of *N*-acetylneuraminic acid. 10. Synthesis of 2,7-, 2,8-, and 2,9-dideoxy- and 2,4,7-trideoxy-2,3-didehydro-*N*-acetylneuraminic acids and their behavior towards sialidase from *Vibrio cholerae, Liebigs Ann. Chem.*, 159–165, 1989.

145. Baumberger, F., Vasella, A., and Schauer, R., Synthesis of new sialidase inhibitors, 6-amino-6-deoxysialic acids, *Helv. Chim. Acta.*, 71, 429-445, 1988.

146. Kijima-Suda, I., Ido, T., Ohrui, H., Itoh, M., and Tomita, K., Inhibition of sialidase activity by a newly synthesized derivative of sialic acid, *Sialic Acids Proceedings Japanese-German Symposium,* 1988, pp. 152–153.

147. Bernet, B., Murty, A.R.C.B., and Vasella, A., Analogues of sialic acids as potential sialidase inhibitors. Synthesis of 2-*C*-hydroxymethyl derivatives of *N*-acetyl-6-amino-2,6-dideoxyneuraminic acid, *Helv. Chim. Acta,* 73, 940–958, 1990.

148. Czollner, L., Kuszmann, J., and Vasella, A., Synthesis of pyrrolidine analogs of *N*-acetylneuraminic acid as potential sialidase inhibitors, *Helv. Chim. Acta,* 73, 1338–1358, 1990.

149. Wallimann, K. and Vasella, A., Phosphonic acid analogs of the *N*-acetyl-2-deoxyneuraminic acids: synthesis and inhibition of *Vibrio cholerae* sialidase, *Helv. Chim. Acta,* 73, 1359–1372, 1990.

150.   Vasella, A. and Wyler, R., Synthesis of the 6-C-methyl and 6-C-(hydroxymethyl) analogs of N-acetylneuraminic acid and of N-acetyl-2,3-didehydro-2-deoxyneuraminic acid, *Helv. Chim. Acta,* 73, 1742–1763, 1990.

151.   Schreiner, E., Zbiral, E., Kleineidam, R.G., and Schauer, R., Structural variations on N-acetylneuraminic acid. Part 21. 2,3-didehydro-2-deoxysialic acids structurally varied at C-5 and their behavior toward the sialidase from *Vibrio cholerae, Carbohydr. Res.,* 216, 61–66, 1991.

152.   Schreiner, E., Zbiral, E., Kleineidam, R.G., and Schauer, R., Structural variations on N-acetylneuraminic acid. 20. Synthesis of some 2,3-didehydro-2-deoxysialic acids structurally varied at C-4 and their behavior towards sialidase from *Vibrio cholerae, Liebigs Ann. Chem.,* 129–134, 1991.

153.   Gaenzer, B.I., Gyoergydeak, Z., Bernet, B., and Vasella, A., Analogs of sialic acids as potential sialidase inhibitors. Synthesis of C6 and C7 analogs of N-acetyl-6-amino-2, 6-dideoxyneuraminic acid, *Helv. Chim. Acta.,* 74, 343–369, 1991.

154.   Wallimann, K. and Vasella, A., C-glycosides of N-acetylneuraminic acid. Synthesis and study of their activity on *Vibrio cholerae* sialidase, *Helv. Chim. Acta.,* 74, 1520–1532, 1991.

155.   Engstler, M., Ferrero-Garcia, M.A., Parodi, A.J., Schauer, R., Storz-Eckerlin, T., Vasella, A., Witzig, C., and Zhu, X., N-(4-nitrophenyl)oxamic acid and related N-acylanilines are non-competitive inhibitors of *Vibrio cholerae* sialidase but do not inhibit *Trypanosoma cruzi* or *Trypanosoma brucei trans*-sialidases, *Helv. Chim. Acta,* 77, 1166–1174, 1994.

156.   Wilson, J.C., Thomson, R.J., Dyason, J.C., Florio, P., Quelch, K.J., Abo, S., and von Itzstein, M., The design, synthesis and biological evaluation of neuraminic acid-based probes of *Vibrio cholerae* sialidase, *Tetrahedron Asymm.,* 11, 53–73, 2000.

157.   Thobhani, S., Ember, B., Siriwardena, A., and Boons, G-J., Multivalency and the mode of action of bacterial sialidases, *J. Am. Chem. Soc.,* 125, 7154–7155, 2003.

158.   Chan, T.-H., Xin, Y.-C., and von Itzstein, M., Synthesis of phosphonic acid analogs of sialic acids (Neu5Ac and KDN) as potential sialidase inhibitors, *J. Org. Chem.,* 62, 3500–3504, 1997.

159.   Meindl, P. and Tuppy, H., 2-Deoxy-2,3-dehydrosialic acids, II. Competitive inhibition of *Vibrio cholerae* neuraminidase by 2-deoxy-2,3-dehydro-N-acylneuraminic acids, *Hoppe-Seyler's Z. Physiol. Chem.,* 350, 1088–1092, 1969.

160.   Meindl, P. and Tuppy H., 2-Deoxy-2,3-dehydrosialic acids. III. Synthesis and properties of 2-deoxy-2,3-dehydroneuraminic acid and of new N-acyl derivatives, *Montash Chem.,* 104, 402–414, 1973.

161.   Vasella, A. and Wyler, R., Synthesis of a phosphonic acid analog of N-acetyl-2, 3-didehydro-2-deoxyneuraminic acid, an inhibitor of *Vibrio cholerae* sialidase, *Helv. Chim. Acta,* 74, 451–463, 1991.

162.   Urlich, N., Ashok, S., Schroeder, C., Reuter, G., Schauer, R., Kamerling, J.P., and Vliegenthart, J.F., Structural parameters and natural occurrence of 2-deoxy-2,3-didehydro-N-glycoylneuraminic acid, *Eur. J. Biochem.,* 152, 459–463, 1985.

163.   Brossmer, R., Burk, G., Eschenfelder, V., Holmquist, L., Jackh, R., Neumann, B., and Rose, U., Recent aspects of the chemistry of N-acetyl-D-neuraminic acid, *Behring Inst. Mitt.,* 55, 119–123, 1974.

164.   Holmquist, L., Synthesis of N-acetylneuraminic acid derivatives and studies of their interaction with *Vibrio cholerae* neuraminidase, *FOA Report,* 9, 20, 1975.

165.   Noehle, U., Shukla, A.K., Schroeder, C., Reuter, G., Schauer, R., Kamerling, J.P., and Vliegenthart, J.F.G., Structural parameters and natural occurrence of 2-deoxy-2, 3-didehydro-N-glycoloylneuraminic acid, *Eur. J. Biochem.,* 152, 459–463, 1985.

166. Schauer, R. and Faillad, H., Action specificity of neuraminidase, action of bacterial neuraminidase on isomeric *N,O*-diacetylneuraminic acid glycosides in submaxillary mucin of horse and cow, *Hoppe-Seyler's Z.*, 349, 961–968, 1968.

167. Corfield, A.P., Sander-Wewer, M., Veh, R.W., Wember, M., and Schauer, R., The action of sialidases on substrates containing *O*-acetylsialic acids, *Hoppe-Seyler's Z.*, 367, 433–439, 1986.

168. Haselhorst, T., Wilson, J.C., Thomson, R.J., McAtamney, S., Menting, J.G., Coppel, R.L., and von Itzstein, M., Saturation transfer difference (STD) ¹H-NMR experiments and *in silico* docking experiments to probe the binding of *N*-acetylneuraminic acid and derivatives to *Vibrio cholerae* sialidase, *Proteins*, 54, 346–353, 2004.

169. Edmond, J.D., Johnston, R.G., Kidd, D., Rylance, H.J., and Sommerville, R.G., The inhibition of neuraminidase and antiviral action, *Br. J. Pharmacol. Chemother.*, 27, 415–426, 1966.

170. Clinch, K., Vasella, A., and Schauer, R., Synthesis of (2*R*,4*S*,5*S*)-5-acetamido-4-hydroxy-pipecolinic acid as a potential inhibitor of sialidases, *Tetrahedron Lett.*, 28, 6425–6428, 1987.

171. Ley, K., The role of selectins in inflammation and disease, *Trends Mol. Med.*, 9, 263–268, 2003.

172. Unger, F.M., The chemistry of oligosaccharide ligands of selectins:significance for the development of new immunomodulatory medicines, in: *Advances in Carbohydrate Chemistry and Biochemistry*, Horton, D., Ed., Academic Press, San Diego, 2001, pp. 207–435.

173. Lasky, L.A., Selectin–carbohydrate interactions and the initiation of the inflammatory response, *Ann. Rev. Biochem.*, 64, 113–139, 1995.

174. Vestwebber, D. and Blanks, J.E., Mechanisms that regulate the function of the selectins and their ligands, *Physiol. Rev.*, 79, 181–213, 1999.

175. Bowman, K.G., Cook, B.N., de Graffenried, C.L., and Bertozzi, C.R., Biosynthesis of L-selectin ligands: Sulfation of sialyl Lewis X-related oligosaccharides by a family of GlcNAc-6-sulfotransferases, *Biochemistry*, 40, 5382–5391, 2001.

176. Patel, A. and Lindhorst, T.K., A modular approach for the synthesis of oligosaccharide mimetics, *J. Org. Chem.*, 66, 2674–2680, 2001.

177. Feizi, T. and Galustian, C., Novel oligosaccharide ligands and ligand-processing pathways for the selectins, *Trends Biochem. Sci.*, 24, 369–372, 1999.

178. Kaila, N. and Thomas, B.E., Selectin inhibitors, *Expert Opin. Ther. Patents*, 13, 305–317, 2003.

179. Poppe, L., Brown, G.S., Philo, J.S., Nikrad, P.V., and Shah, B.H., Conformation of sLex tetrasaccharide, free in solution and bound to E-, P-, and L-selectin, *J. Am. Chem. Soc.*, 119, 1727–1736, 1997.

180. Graves, B.J., Crowther, R.L., Chandran, C., Rumberger, J.M., Li, S., Huang, K.S., Presky, D.H., Familletti, P.C., Wolitzky, B.A., and Burns, D.K., Insight into E-selectin/ligand interaction from the crystal structure and mutagenesis of the lec/EGF domains, *Nature*, 367, 532–538, 1994.

181. Somers, W.S., Tang, J.C., Shaw, G.D., and Camphausen, R.T., Insights into the molecular basis of leukocyte tethering and rolling revealed by structures of P- and E-selectin bound to sLex and PSGL-1, *Cell*, 103, 467–479, 2000.

182. Pichierri, F. and Matsuo, Y., Effect of protonation of the *N*-acetyl neuraminic acid residue of sialyl Lewis X. A molecular orbital study with insights into its binding properties toward the carbohydrate recognition domain of E-selectin, *Bioorg. Med. Chem.*, 10, 2751–2757, 2002.

183. Kuznik, G., Unverzagt, C., Horsch, B., and Kretzschmar, G., Chemical and enzymatic synthesis of modified sialyl Lewis X tetrasaccharides with high affinity for E and P-selectin, *J. Prakt. Chem.*, 342, 745–752, 2000.

184. Mammen, M., Choi, S.-K., and Whitesides, G.M., Polyvalent interactions in biological systems: implications for design and use of multivalent ligands and inhibitors, *Angew. Chem. Int. Ed.*, 37, 2754–2794, 1998.

185. Renkonen, O., Toppila, S., Penttila, L., Salminen, H., Helin, J., Maaheimo, H., Costello, C., Turunen, J., and Renkonen, R., Synthesis of a new nanomolar saccharide inhibitor of lymphocyte adhesion: different polylactosamine backbones present multiple sialyl Lewis x determinants to L-selectin in high-affinity mode, *Glycobiology*, 7, 453–461, 1997.

186. Wittmann, V., Takayama, S., Gong, K.W., Weitz-Schmidt, G., and Wong, C-H., Ligand recognition by E- and P-selectin: chemoenzymatic synthesis and inhibitory activity of bivalent sialyl Lewis X derivatives and sialyl Lewis X carboxylic acids, *J. Org. Chem.*, 63, 5137–5143, 1998.

187. Bintein, F., Auge, C., and Lubineau, A., Chemo-enzymatic synthesis of a divalent sialyl Lewis X ligand with restricted flexibility, *Carbohydr. Res.*, 338, 1163–1173, 2003.

188. Game, S.M., Rajapurohit, P.K., Clifford, M., Bird, M.I., Priest, R., Bovin, N.V., Nifant'ev, N.E., O'Beirne, G., and Cook, N.D., Scintillation proximity assay for E-, P-, and L-selectin utilizing polyacrylamide-based neoglycoconjugates as ligands, *Anal. Biochem.*, 258, 127–135, 1998.

189. Pochechueva, T.V., Ushakova, N.A., Preobrazhenskaya, M.E., Nifantiev, N.E., Tsvetkov, Y.E., Sablina, M.A., Tuzikov, A.B., Bird, M.I., Rieben, R., and Bovin, N.V., P-selectin blocking potency of multimeric tyrosine sulfates *in vitro* and *in vivo*, *Bioorg. Med. Chem. Lett.*, 13, 1709–1712, 2003.

190. Pochechueva, T.V., Galanina, O.E., Bird, M.I., Nifantiev, N.E., and Bovin, N.V., Assembly of P-selectin ligands on a polymeric template, *Chem. Biol.*, 9, 757–762, 2002.

191. Misra, A.K., Ding, Y., Lowe, J.B., and Hindsgaul, O., A concise synthesis of the 6-O- and 6'-O-sulfated analogues of the sialyl Lewis X tetrasaccharide, *Bioorg. Med. Chem. Lett.*, 10, 1505–1509, 2000.

192. Komba, S., Galustian, C., Ishida, H., Feizi, T., Kannagi, R., and Kiso, M., The first total synthesis of 6-sulfo-de-*N*-acetylsialyl Lewis X ganglioside: a superior ligand for human L-selectin, *Angew. Chem. Int. Ed.*, 38, 1131–1133, 1999.

193. Lubineau, A., Le Gallic, J., and Lemoine, R., First synthesis of the 3' - sulfated Lewis A pentasaccharide, the most potent human E-selectin ligand so far, *Bioorg. Med. Chem. Lett.*, 2, 1143–1151, 1994.

194. Nicolaou, K.C., Bockovich, N.J., and Carcanague, D.R., Total synthesis of sulfated LeX and LeA-type oligosaccharide selectin ligands, *J. Am. Chem. Soc.*, 115, 8843–8844, 1993.

195. Yuen, C-T., Bezouska, K., O'Brien, J., Stoll, M., Lemoine, R., Lubineau, A., Kiso, M., Hasegawa, A., Bockovich, N.J., Nicolaou, K.C., and Feizi, T., Sulfated blood group Lewis A, *J. Biol. Chem.*, 269, 1595–1598, 1994.

196. Yoshida, M., Kawakami, Y., Ishida, H., Kiso, M., and Hasegawa, A., Synthetic studies on sialoglycoconjugates 85: synthesis of sialyl Lewis X ganglioside analogs containing a variety of anionic substituents in place of sialic acid, *J. Carbohydr. Chem.*, 15, 399–418, 1996.

197. Wada, Y., Saito, T., Matsuda, N., Ohmoto, H., Yoshino, K., Ohashi, M., Kondo, H., Ishida, H., Kiso, M., and Hasegawa, A., Studies on selectin blockers. 2. Novel selectin blocker as potential therapeutics for inflammatory disorders, *J. Med. Chem.*, 39, 2055–2059, 1996.

198. Jain, R.K., Piskorz, C.F., Huang, B.G., Locke, R.D., Han, H.L., Koenig, A., Varki, A., and Matta, K.L., Inhibition of L- and P-selectin by a rationally synthesized novel core 2-like branched structure containing GalNAc-Lewis X and Neu5Acα(2-3)Galβ(1-3)GalNAc sequences, *Glycobiology*, 8, 707–717, 1998.

199. Lubineau, A., Augé, C., Le Goff, N., and Le Narvor, C., Chemoenzymatic synthesis of a 3IV, 6III-disulfated Lewis X pentasaccharide, a candidate ligand for human L-selectin, *Carbohydr. Res.*, 305, 501–509, 1998.

200. Galustian, C., Lubineau, A., Le Narvor, C., Kiso, M., Brown, G., and Feizi, T., L-selectin interactions with novel mono- and multisulfated Lewis X sequences in comparison with the potent ligand 3'-sulfated Lewis A, *J. Biol. Chem.*, 274, 18213–18217, 1999.

201. Thoma, G., Kinzy, W., Bruns, C., Patton, J.T., Magnani, J.L., and Bänteli, R., Synthesis and biological evaluation of a potent E-selectin antagonist, *J. Med. Chem.*, 42, 4909–4913, 1999.

202. Thoma, G., Magnani, J.L., Patton, J.T., Ernst, B., and Jahnke, W., Preorganization of the bioactive conformation of sialyl Lewis X analogues correlates with their affinity to E-selectin, *Angew. Chem. Int. Ed.*, 40, 1941–1945, 2001.

203. Thoma, G., Magnani, J.L., and Patton, J.T., Synthesis and biological evaluation of a sialyl Lewis X mimic with significantly improved E-selectin inhibition, *Bioorg. Med. Chem. Lett.*, 11, 923–925, 2001.

204. Hanessian, S., Mascitti, V., and Rogel, O., Synthesis of a potent antagonist of E-selectin, *J. Org. Chem.*, 67, 3346–3354, 2002.

205. Bruehl, R.E., Dasgupta, F., Katsumoto, T.R., Tan, J.H., Bertozzi, C.R., Spevak, W., Ahn, D.J., Rosen, S.D., and Nagy, J.O., Polymerized liposome assemblies: bifunctional macromolecular selectin inhibitors mimicking physiological selectin ligands, *Biochemistry*, 40, 5964–5974, 2001.

206. Chervin, S.M., Lowe, J.B., and Koreeda, M., Synthesis and biological evaluation of a new sialyl Lewis X mimetic derived from lactose, *J. Org. Chem.*, 67, 5654–5662, 2002.

207. Bänteli, R. and Ernst, B., Synthesis of sialyl Lewis X mimics. Modifications of the 6-position of galactose, *Bioorg. Med. Chem. Lett.*, 11, 459–462, 2001.

208. Hiruma, K., Kajimoto, T., Weitz-Schmidt, G., Ollmann, I., and Wong, C.-H., Rational design and synthesis of a 1,1-linked disaccharide that is 5 times as active as sialyl Lewis X in binding to E-selectin, *J. Am. Chem. Soc.*, 118, 9265–9270, 1996.

209. Shibata, K., Hiruma, K., Kanie, O., Wong, C.-H., Synthesis of 1,1-linked galactosyl mannosides carrying a thiazine ring as mimetics of sialyl Lewis X antigen: investigation of the effect of carboxyl group orientation on P-selectin inhibition, *J. Org. Chem.*, 65, 2393–2398, 2000.

210. Cheng, X., Khan, N., and Mootoo, D.R., Synthesis of the *C*-glycoside analogue of a novel sialyl Lewis X mimetic, *J. Org. Chem.*, 65, 2544–2547, 2000.

211. Hiruma, K., Kanie, O., and Wong, C.-H., Synthesis of analogs of 1,1-linked galactosyl mannoside as mimetics of sialyl Lewis X tetrasaccharide, *Tetrahedron*, 54, 15781–15792, 1998.

212. Glen, A., Leigh, D.A., Martin, R.P., Smart, J.P., and Truscello, A.M., The regioselective tert-butyldimethylsilylation of the 6'-hydroxyl group of lactose derivatives via their dibutylstannylene acetals, *Carbohydr. Res.*, 248, 365–369, 1993.

213. Bertozzi, C.R., Fukuda, S., and Rosen, S.D., Sulfated disaccharide with greater inhibitory potency for L-selectin than sialyl Lewis X, *Biochemistry*, 34, 14271–14278, 1995.

214. Hopfner, M., Alban, S., Schumacher, G., Rothe, U., and Bendas, G., Selectin-blocking semisynthetic sulfated polysaccharides as promising anti-inflammatory agents, *J. Pharm. Pharmacol.*, 55, 697–706, 2003.

215. Tsukida, T., Moriyama, H., Kurokawa, K., Achiha, T., Inoue, Y., and Kondo, H., Studies on selectin blockers. 7. Structure–activity relationships of sialyl Lewis X mimetics based on modified Ser-Glu dipeptides, *J. Med. Chem.*, 41, 4279–4287, 1998.

216. Tsai, C.-Y., Park, W.K.C., Weitz-Schmidt, G., Ernst, B., and Wong, C.-H., Synthesis of sialyl Lewis X mimetics using the Ugi four-component reaction, *Bioorg. Med. Chem. Lett.*, 8, 2333–2338, 1998.

217. Kaila, N., Yu, H.-A., and Xiang, Y., Design and synthesis of novel sialyl Lewis X mimics, *Tetrahedron Lett.*, 36, 5503–5506, 1995.

218. Kaila, N., Thomas, I., Bert, E., Thakker, P., Alvarez, J.C., Camphausen, R.T., and Crommie, D., Design and synthesis of sialyl Lewis X mimics as E-selectin inhibitors, *Bioorg. Med. Chem. Lett.*, 11, 151–155, 2001.

219. Kaila, N., Chen, L., Thomas, B.E., Tsao, D., Tam, S., Bedard, P.W., Camphausen, R.T., Alvarez, J.C., and Ullas, G., β-C-Mannosides as selectin inhibitors, *J. Med. Chem.*, 45, 1563–1566, 2002.

220. Hanessian, S., Huynh, H.K., Reddy, G.V., McNaughton-Smith, G., Ernst, B., Kolb, H.C., Magnani, J.L., and Sweeley, C., Exploration of β-turn scaffolding motifs as components of sialyl LeX mimetics and their relevance to P-selectin, *Bioorg. Med. Chem. Lett.*, 8, 2803–2808, 1998.

221. Lin, C.-C., Morís-Varas, F., Weitz-Schmidt, G., and Wong, C.-H., Synthesis of sialyl Lewis X mimetics as selectin inhibitors by enzymatic aldol condensation reactions, *Bioorg. Med. Chem.*, 7, 425–433, 1999.

222. Kogan, T.P., Dupre, B., Bui, H., McAbee, K.L., Kassir, J.M., Scott, I.L., Hu, X., Vanderslice, P., Beck, P.J., and Dixon, R.A.F., Novel synthetic inhibitors of selectin-mediated cell adhesion: synthesis of 1,6-*bis*[3-(3-carboxymethylphenyl)-4-(2-α-D-mannopyranosyloxy)phenyl]hexane (TBC1269), *J. Med. Chem.*, 41, 1099–1111, 1998.

223. Kogan, T.P., Dupre, B., Keller, K.M., Scott, I.L., Bui, H., Market, R.V., Beck, P.J., Voytus, J.A., Revelle, B.M., and Scott, D., Rational design and synthesis of small molecule, non-oligosaccharide selectin inhibitors: (α-D-mannopyranosyloxy)biphenyl-substituted carboxylic acids, *J. Med. Chem.*, 38, 4976–4984, 1995.

224. Fu, Y., Laurent, S., and Muller, R.N., Synthesis of a sialyl Lewis X mimetic conjugated with DTPA, Potential ligand of new contrast agents for medical imaging, *Eur. J. Org. Chem.*, 3966–3973, 2002.

225. Suzuki, Y., Toda, Y., Tamatani, T., Watanabe, T., Suzuki, T., Nakao, T., Murase, K., Kiso, M., Hasegawa, A., Tadano-Aritomi, K., Ishizuka, I., and Miyasaka, M., Sulfated glycolipids are ligands for a lymphocyte homing receptor, L-selectin (LECAM-1), binding epitope in sulfated sugar chain, *Biochem. Biophys. Res. Commun.*, 190, 426–434, 1993.

226. Aruffo, A., Kolanus, W., Walz, G., Fredman, P., and Seed, B., CD62/P-selectin recognition of myeloid and tumor cell sulfatides, *Cell*, 67, 35–44, 1991.

227. Gordon, E.J., Strong, L.E., and Kiessling, L.L., Glycoprotein-inspired materials promote the proteolytic release of cell surface L-selectin, *Bioorg. Med. Chem.*, 6, 1293–1299, 1998.

228. Marinier, A., Martel, A., Bachand, C., Plamondon, S., Turmel, B., Daris, J-P., Banville, J., Lapointe, P., Ouellet, C., Dextraze, P., Menard, M., Wright, J.J.K., Alford, J., Lee, D., Stanley, P., Nair, X., Todderud, G., and Tramposch, K.M., Novel mimics of sialyl Lewis X: design, synthesis and biological activity of a series of 2- and 3-malonate substituted galactoconjugates, *Bioorg. Med. Chem.*, 9, 1395–1427, 2001.

229. Kretzschmar, G., Synthesis of novel sialyl-Lewis X glycomimetics as selectin antagonists, *Tetrahedron*, 54, 3765–3780, 1998.

230.   Fukunaga, K., Tsukida, T., Moriyama, H., and Kondo, H., Drug design, synthesis, and evaluation of a non-sugar-based selectin antagonist, *Bioorg. Med. Chem. Lett.*, 11, 2365–2367, 2001.

231.   Hiramatsu, Y., Tsukida, T., Nakai, Y., Inoue, Y., and Kondo, H., Study on selectin blocker. 8. Lead discovery of a non-sugar antagonist using a 3D-pharmacophore model, *J. Med. Chem.*, 43, 1476–1483, 2000.

232.   Moriyama, H., Hiramatsu, Y., Kiyoi, T., Achiha, T., Inoue, Y., and Kondo, H., Studies on selectin blocker. 9. SARs of non-sugar selectin blocker against E-, P-, L-selectin bindings, *Bioorg. Med. Chem.*, 9, 1479–1491, 2001.

233.   Slee, D.H., Romano, S.J., Yu, J., Nguyen, T.N., John, J.K., Raheja, N.K., Axe, F.U., Jones, T.K., and Ripka, W.C., Development of potent non-carbohydrate imidazole-based small molecule selectin inhibitors with anti-inflammatory activity, *J. Med. Chem.*, 44, 2094–2107, 2001.

234.   Kaila, N., Xu, G-Y., Camphausen, R.T., and Xiang, Y., Identification and structural determination of a potent P-selectin inhibitor, *Bioorg. Med. Chem.*, 9, 801–806, 2001.

235.   De Vleeschauwer, M., Vaillancourt, M., Goudreau, N., Guindon, Y., and Gravel D., Design and synthesis of a new sialyl Lewis X mimetic: How selective are the selectin receptors? *Bioorg. Med. Chem. Lett.*, 11, 1109–1112, 2001.

236.   Burchell, J., Poulsom, R., Hanby, A., Whitehouse, C., Cooper, L., Clausen, H., Miles, D., and Taylor-Papadimitriou, J., An α-2,3 sialyltransferase (ST3Gal I) is elevated in primary breast carcinomas, *Glycobiology*, 9, 1307–1311, 1999.

237.   Wang, X., Zhang, L.-H., and Ye, XS., Recent developments in the design of sialyl-transferase inhibitors, *Med. Res. Rev.*, 23, 32–47, 2003.

238.   Compain, P. and Martin, O. R., Carbohydrate mimetics-based glycosyltransferase inhibitors, *Bioorg. Med. Chem.*, 9, 3077–3092, 2001.

239.   Schaub, C., Muller, B., and Schmidt, R.R., Sialyltransferase inhibitors based on CMP-quinic acid, *Eur. J. Org. Chem.*, 1745–1758, 2000.

240.   Schaub, C., Muller, B., and Schmidt, RR., New sialyltransferase inhibitors based on CMP-quinic acid: development of a new sialyltransferase assay, *Glycoconjugate J.*, 15, 345–354, 1998.

241.   Cohen, S.B. and Halcomb, R.L., Synthesis and characterisation of an anomeric sulfur analogue of CMP-sialic acid, *J. Org. Chem.*, 2000, 6145–6152, 2000.

242.   Muller, B., Martin, T.J., Schaub, C., and Schmidt, R.R., Synthesis of phosphonate analogues of CMP-Neu5Ac determination of α-(2-6)-sialyltransferase inhibition, *Tetrahedron Lett.*, 39, 509–512, 1998.

243.   Amann, F., Schaub, C., Muller, B., and Schmidt, R.R., New potent sialyltransferase inhibitors-synthesis of donor and of transition-state analogues of sialyl donor CMP-Neu5Ac, *Chem. Eur. J.*, 4, 1106–1115, 1998.

244.   Muller, B., Schaub, C., and Schmidt, R.R., Efficient sialyltransferase inhibitors based on transition-state analogues of the sialyl donor, Angew. *Chem. Int. Ed.*, 37, 2893–2897, 1998.

245.   Schworer, R. and Schmidt, R.R., Efficient sialyltransferase inhibitors based on glycosides of *N*-acetylglucosamine, *J. Am. Chem. Soc.*, 124, 1632–1637, 2002.

246.   Skropeta, D., Schworer, R., and Schmidt, R.R., Stereoselective synthesis of phosphoramidate α-(2-6)sialyltransferase transition state analogue inhibitors, *Bioorg. Med. Chem. Lett.*, 13, 3351–3354, 2003.

247.   Arias, C.F., Isa, P., Guerrero, C.A., Mendez, E., Zarate, S., Lopez, T., Espinos, R., Romero, P., and Lopez, S., Molecular biology of rotavirus cell entry, *Arch. Med. Res.*, 33, 356–361, 2002.

248. Lundgren, O. and Svensson, L., Pathogenesis of rotavirus diarrhea, *Microbes Infect.* 3, 1145–1156, 2001.

249. Parashar, U.D., Bresee, J.S., Gentsch, J.R., and Glass, R.I., Rotavirus, *Emerg. Infect. Dis.,* 4, 561–570, 1998.

250. Ciarlet, M. and Estes, M.K., Interactions between rotavirus and gastrointestinal cells, *Curr. Opin. Microbiol.,* 4, 435–441, 2001.

251. Kiefel, M.J. and von Itzstein, M., Carbohydrates as inhibitors of rotaviral infection, *Methods Enzymol.,* 363, 395–412, 2003.

252. Kiefel, M.J., Beisner, B., Bennett, S., Holmes, I.D., and von Itzstein, M., Synthesis and biological evaluation of *N*-acetylneuraminic acid-based rotavirus inhibitors, *J. Med. Chem.,* 39, 1314–1320, 1996.

253. Fazli, A., Bradley, S.J., Kiefel, M.J., Jolly, C., Holmes, I., and von Itzstein, M., Synthesis and biological evaluation of sialylmimetics as rotavirus inhibitors, *J. Med. Chem.,* 44, 3292–3301, 2001.

254. Tomasi, M., Battistini, A., Cardelli, M., Sonnino, S., and D'Agnolo, G., Interaction of cholera toxin with gangliosides:differential effects of the oligosaccharide of ganglioside GM1 and of micellar gangliosides, *Biochemistry,* 23, 2520–2526, 1984.

255. Vrasidas, I., de Mol, N.J., Liskamp, R.M.J., and Pieters, R.J., Synthesis of lactose dendrimers and multivalency effects in binding to the cholera toxin B subunit, *Eur. J. Org. Chem.,* 4685–4692, 2001.

256. Vrasidas, I., Kemmink, J., Liskamp, R.M.J., and Pieters, R.J., Synthesis and cholera toxin binding properties of a lactose-2-aminothiazoline conjugate, *Org. Lett.,* 4, 1807–1808, 2002.

257. Bernardi, A., Checchia, A., Brocca, P., Sonnino, S., and Zuccotto, F., Sugar mimics:an artificial receptor for cholera toxin, *J. Am. Chem. Soc.,* 121, 2032–2036, 1999.

258. Bernardi, A., Carrettoni, L., Grosso Ciponte, A., Monti, D., and Sonnino, S., Second generation mimics of ganglioside GM1 as artificial receptors for cholera toxin: replacement of the sialic acid moiety, *Bioorg. Med. Chem. Lett.,* 10, 2197–2200, 2000.

25 Engineering Carbohydrate Scaffolds into the Side Chains of Amino Acids and Use in Combinatorial Synthesis

Frank Schweizer

CONTENTS

I.   INTRODUCTION

Amino acids represent an important class of molecular building blocks used to generate diversity in nature. The ability of amino acids (α, β, γ, …) to form secondary structures in polyamides establishes the basis of three dimensional (3D) molecular architecture. Proteins and peptides, which are involved virtually in every biological process use a set of only 20 relatively simple α-amino acids to fulfill their tasks as enzymes, receptors, antibodies, lectins, and toxins. In the laboratory, amino acids are widely used as starting materials for natural product synthesis and increasingly as building blocks in combinatorial synthesis. For instance, amino acids have been used to prepare peptide libraries and small molecule libraries of heterocyclic pharmacophores such as benzodiazepines, β-lactams, pyrrolidines, and hydantoins.[1]

In order to extend the structural diversity of the naturally occurring amino acid building blocks and to modify the properties (stability, binding specificity, affinity, and pharmacokinetic behavior) of molecules that are amino acid-derived, such as proteins, peptides, small molecules, natural products, and combinatorial libraries, many unnatural amino acids have been prepared over the years. Usually the term unnatural refers to differences in the configuration (D instead of L), amino acid type (β-, γ-, …; instead of α-amino acid), nature of the side chain, additional substitution at the α-carbon, and N-substitution. Another approach for the design of unnatural amino acids can be found in the hybridization of amino acids with a polyfunctional sugar scaffold generating sugar-amino acids (SAAs), also known as glycosyl amino acids or glycosamino acids (GAA) and here generally referred to as sugar-amino acid hybrids (SAAHs) (Scheme 25.1).

In contrast to amino acids, SAAHs are (poly) hydroxylated amino acids where additional derivatization of the polyol can lead to increased diversity. SAAHs can be derivatized and oligomerized into compound libraries through well-established automated peptide protocols. This approach is particularly attractive for preparing glycomimetic libraries since oligosaccharide library synthesis has not yet reached the same level of automation as peptide synthesis. On the other hand, the engineering of an amino acid moiety into the sugar skeleton enables the SAAH to be incorporated into short peptide sequences, thus opening the door to novel peptidomimetics. Additional hydroxyl derivatization of the polyol could increase lipophilicity of the GAA and render them more likely to permeate cell membranes.

A.   NATURALLY OCCURRING SAAHS

SAAHs and their derivatives occur in nature in various forms (Scheme 25.1). For example, sialic acid, a sugar-δ-amino acid hybrid, plays an important role in inter- and intracellular molecular recognition events such as bacterial and viral infections, whereas muramic acid occurs in bacterial polysaccharides. O- and N-glycosylated amino acids are components of O- and N-linked glycoproteins and glycopeptides. Other naturally occurring SAAHs linked to proteins are the cysteine-linked

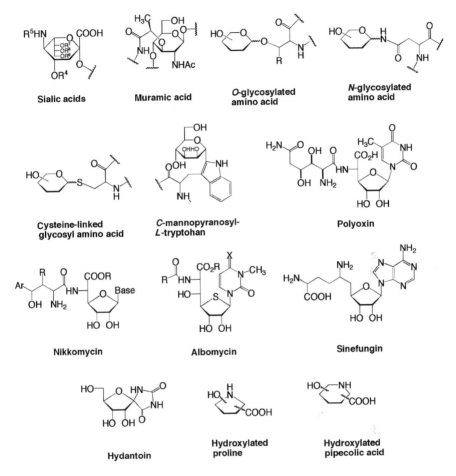

SCHEME 25.1   Examples of SAAHs found in nature.

glycoproteins[2,3] and *C*-linked mannopyanosyl-L-tryptophan.[4,5] The peptidyl nucleo-side antibiotics polyoxin, nikkomycin, albomycin, and sinefungin are SAAH-derived analogs.[6] In addition, the α,α-disubstituted SAAH analog hydantoin shows antiherbal activity with no toxicity to microorganisms and animals.[7] Replacement of the endocyclic oxygen atom in the ring with a nitrogen atom leads to azasugar-based SAAHs found in hydroxylated prolines and pipecolic acids.

II.   CLASSIFICATION AND PROPERTIES OF SAAH BUILDING BLOCKS

Incorporation of an amino acid into a monocyclic carbohydrate scaffold can occur in a variety of ways (Scheme 25.2). The amino acid moiety may either be incorporated directly into a sugar-derived five- or six-membered ring as in the SAAs (A_0 type), or tethered to the sugar ring, as in structures A_1 to A_4. Linking the amino acid moiety at a position adjacent to the endocyclic heteroatom provides the GAA

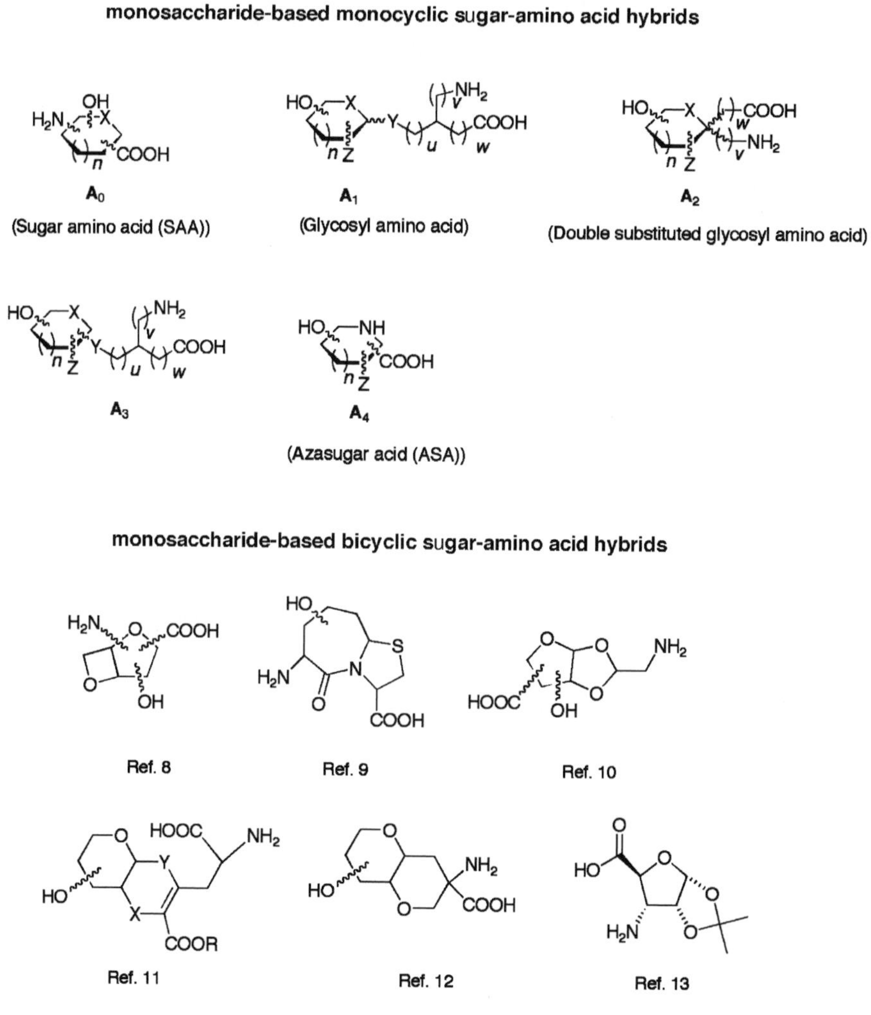

monosaccharide-based monocyclic sugar-amino acid hybrids

A₀

(Sugar amino acid (SAA))

A₁

(Glycosyl amino acid)

A₂

(Double substituted glycosyl amino acid)

A₃

A₄

(Azasugar acid (ASA))

monosaccharide-based bicyclic sugar-amino acid hybrids

Ref. 8

Ref. 9

Ref. 10

Ref. 11

Ref. 12

Ref. 13

SCHEME 25.2 Classification of SAAHs.

A_1 or the double-substituted GAA A_2. Linking to a position more distant from the endocyclic heteroatom affords the branched SAAHs A_3. Substitution of the endocyclic oxygen by nitrogen provides azasugar-based SAAHs, here referred to as azasugar acid (ASA) (Scheme 25.2). Substitutions of the endocyclic oxygen by other heteroatoms (S, P, Se, etc.) can also be envisaged. Recently, amino acids presented on more rigid bicyclic carbohydrate scaffolds have also appeared (Scheme 25.2).[8-13]

A. SUGAR AMINO ACID BUILDING BLOCKS

The incorporation of a carboxylic acid and an amino function into a regular cyclic carbohydrate skeleton provides conformationally restricted SAAs (Scheme 25.3).

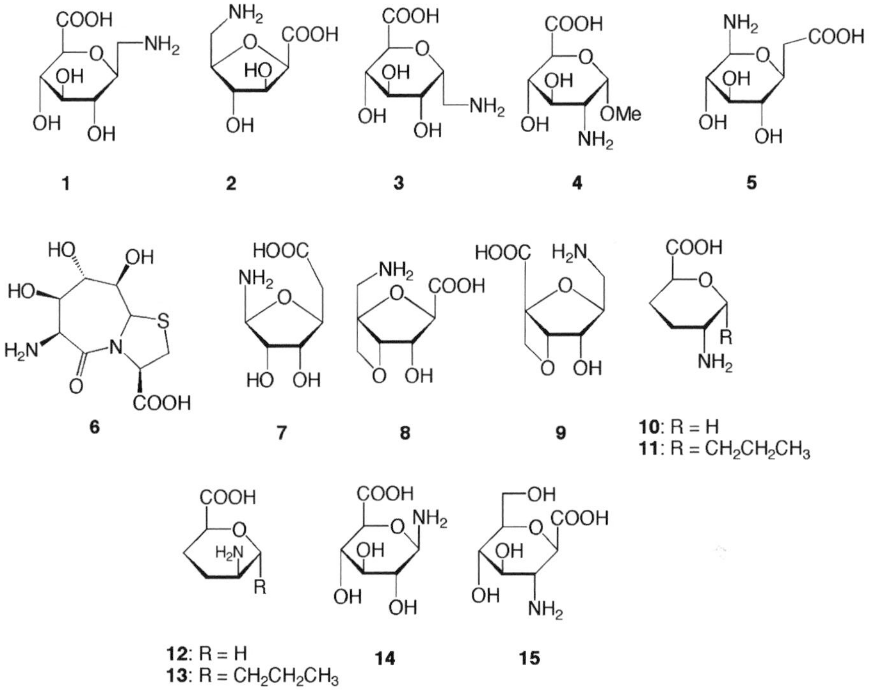

SCHEME 25.3 SAAHs (A_0 type) used in the synthesis of peptidomimetics.

The rigidity can be used to generate secondary structures provided that the carbohydrate–based amino acids are oligomerized or incorporated into short peptides.* For instance, the SAAHs **1** to **13** have been studied as dipeptide isosteres,[15–19] while **14** and **15** have been shown to mimic β- and γ-turns[15] (Scheme 25.3).

B. Glycosyl Amino Acids

GAAs (Type A_1) form another important class of amino acids involved in the modulation of protein folding, intra- and intercellular trafficking, receptor binding and signaling,[20†] enhancement of the thermal stability of proteins,[22] and protection against proteolytic degradation.[23,24] Attachment of carbohydrates to protein occurs via three major types of linkage: (a) N-glycosidic linkage between the reducing terminal sugar and the amide group of asparagine (Scheme 25.4), (b) O-glycosidic linkage between the sugar and a hydroxyl group of an amino acid, most commonly serine or threonine (Scheme 25.4), and also 5-hydroxylysine, 4-hydroxyproline, and tyrosine are known;[25] and (c) via an ethanolamine phosphate between the C-terminal residue of the protein and an oligosaccharide attached to phosphatidylinositol (GPI anchor). Other naturally

---

* For reviews on the synthesis of carbohydrate-based amino acids and use as peptidomimetics see Reference 14.

† For a review on the implications of carbohydrates in medicine and biology see Reference 21.

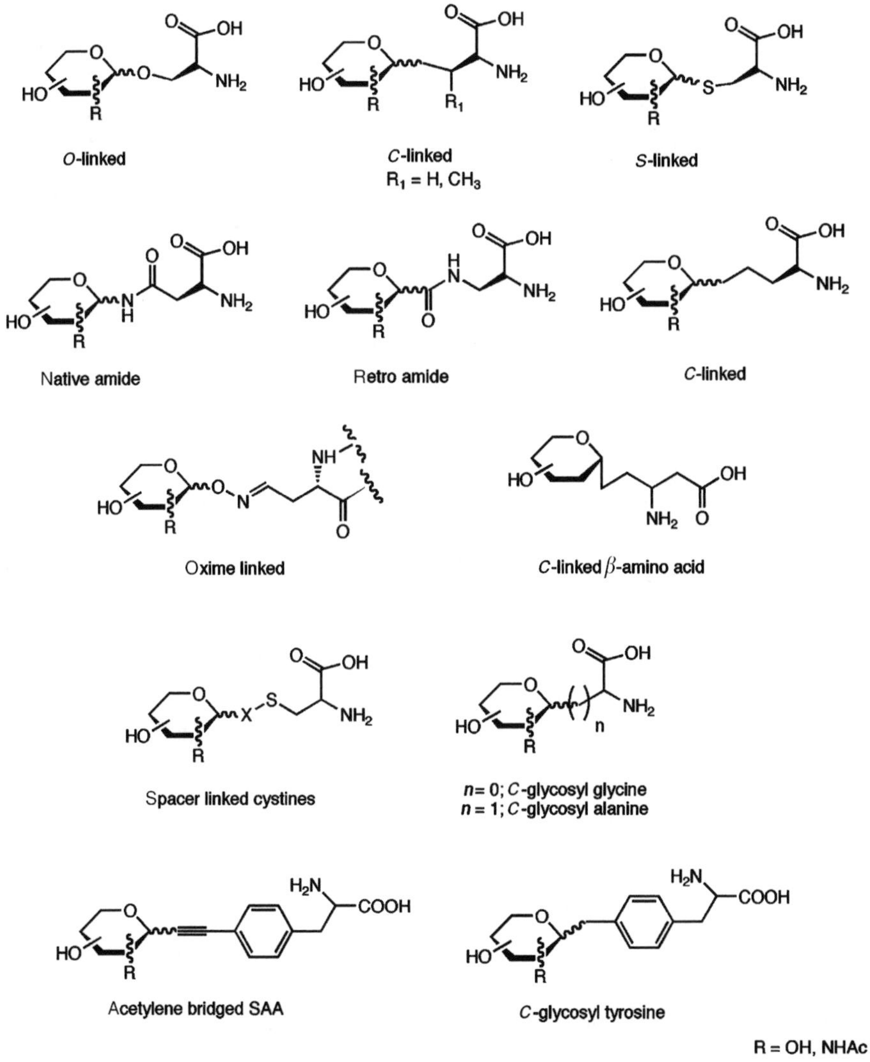

O-linked

C-linked
R₁ = H, CH₃

S-linked

Native amide

Retro amide

C-linked

Oxime linked

C-linked β-amino acid

Spacer linked cystines

n = 0; C-glycosyl glycine
n = 1; C-glycosyl alanine

Acetylene bridged SAA

C-glycosyl tyrosine

R = OH, NHAc

SCHEME 25.4 GAAs which can be used for the synthesis of natural and unnatural glycopeptides.

occurring linkages include the S-glycosidic linkage to cysteine[2,3] and C-glycosidic linkage to tryptophan[4,5] (Scheme 25.4). However, O- and N-linked glycopeptides are metabolically unstable towards glycosidases, an inherent limitation of these materials as potential drugs. Mimics of the naturally occurring glycopeptide linkages have been prepared in order to overcome these drawbacks[14a], (for a review see Reference 26) (Scheme 25.4). For instance, the exocyclic oxygen in the o-Ser (Thr) linkage has been replaced by a methylene unit[27-38] or by sulfur[39-45] (Scheme 25.4). On the other hand, N-linked glycopeptides may be mimicked by retro amides,[46,47] glycopeptoids,[48] and

C-linked glycopeptides in which the amide group has been replaced by an ethylene isostere[49–53] (Scheme 25.4). GAAs with an oxime linkage[54–56] between sugar and peptide have been prepared by chemoselective ligation[57] and various spacer-linked cysteines[58] have also appeared (Scheme 25.4). Other potential glycopeptide mimetics are the C-glycosyl glycines[59] and C-glycosyl alanines[60] (Scheme 25.4). Both types of SAAHs differ from the native O-linked SAA by a one- or two-carbon-shortened amino acid side chain. Acetylene-bridged GAAs[61] and C-glycosyl tyrosines[62] have also been synthesized and incorporated into C-glycopeptides. Recently, C-glycosyl-β-amino acids have also been prepared.[63]

C. DOUBLE-SUBSTITUTED GLYCOSYL AMINO ACIDS

Double-substituted A$_2$ SAAHs also show biological activity. For example, the naturally occurring spiroribofunranose hydantoin (Scheme 25.1) shows strong herbicidal activity, with no toxicity to microorganisms or animals,[7] whereas the disubstituted spirosugar α-amino acids 16,[64] 17,[65] and 18[66] (Scheme 25.5) are efficient specific inhibitors of muscle and liver glycogen phosphorylase, a major regulatory enzyme of blood sugar levels. In contrast, galactohydantoin 19 and compounds 20 and 21 showed no inhibitory effect against a variety of transferases and galactosidases.[67]

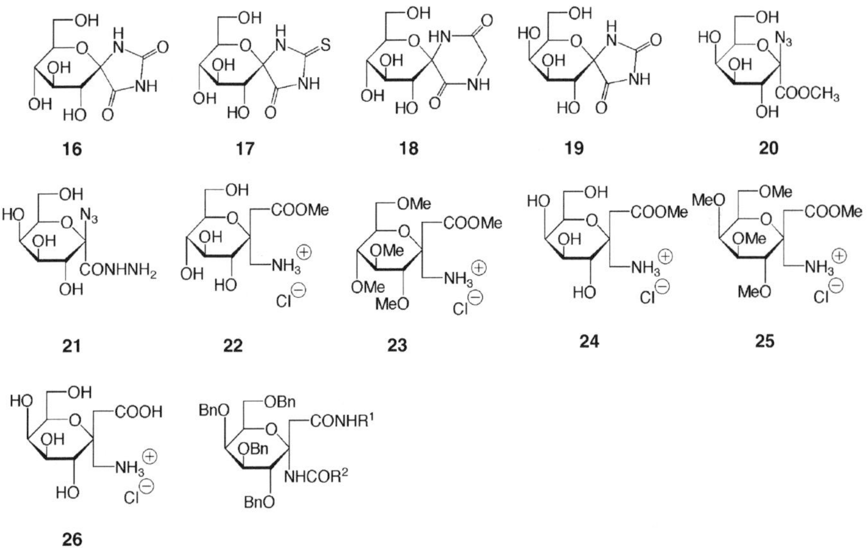

27: = R^1 =C$_7$H$_{13}$; R^2 = C$_6$H$_5$
28: = R^1 =CH$_2$COOMe; R^2 = CH$_3$
29: = R^1 =CH$_2$C$_6$H; R^2 = CH$_3$
30: = R^1 =CH$_2$C$_6$H; R^2 = CH$_3$
31: = R^1 =CH$_2$C$_6$H$_5$; R^2 = C$_7$H$_{13}$
32: = R^1 =CH$_2$C$_6$H; R$_2$ = CH$_2$C$_6$H$_5$

SCHEME 25.5 GAAs and derivatives which incorporate into the anomeric center of a carbohydrate GAA (A$_2$ type).

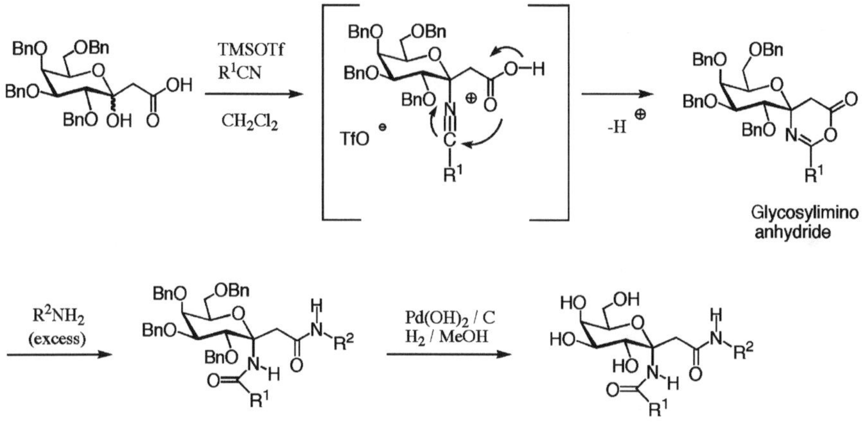

SCHEME 25.6 Synthesis of modified double-substituted glycosyl amino acids using a three-component Ritter reaction.

Recently, our research group has reported the synthesis of a variety of A_2 SAAHs such as the sugar-fused GABA analogs **22** to **26**[68] and sugar-β-amides **27–32**.[69] The sugar-β-amides were not accessible through acylation of the corresponding glycosylamines and were prepared through a three-component Ritter reaction (Scheme 25.6). Initially, a ketopyranoside reacts under Lewis acid-catalyzed conditions with a nitrile to form a glycosylimino anhydride intermediate which can be isolated. Exposure of this intermediate to simple primary amines produces novel sugar-β-peptides.

D. NONANOMERIC LINKING OF AN AMINO ACID TO A CARBOHYDRATE SCAFFOLD (A_3 SAAHs)

Nonanomeric linking of the amino acid moiety to the carbohydrate portion provides A_3 SAAHs which are found in a variety of antibiotics (Scheme 25.7). For instance, the C-glycofuranosyl α-amino acid **33**[70–74] is present in the polyoxin and nikkomycin antibiotics, whereas SAAH **34**[75–77] is a component of the nucleoside antibiotic ampurimycin. Analogs of the polyoxin-based SAAH such as **35** and **36** were initially described by Rosenthal,[78–80] but no biological data were presented. In addition, the glycopyranosyl α-amino acid **37** has been prepared and incorporated into a short peptide sequence.[72]

E. AZASUGAR ACIDS

Replacement of the endocyclic oxygen atom by a nitrogen atom leads to ASA (Scheme 25.8). Over the years, many azasugars have been found as potent inhibitors of glycosidases and glycosyltransferases. For instance, the ASAs **38** to **41** are potent inhibitors of glucuronidases,[81] while the mannose-based azasugar amides **42** and **43** are potent inhibitors of two β-N-acetylglucosaminidases (human placenta and bovine liver).[82] The L-rhamnopyranose-derived amides **44** and **45** are potent

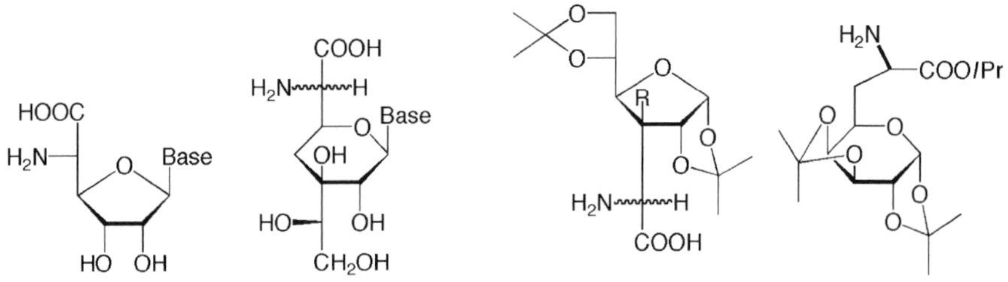

33 **34** **35** R = H, **36** R = OH **37**

SCHEME 25.7 SAAHs where the amino acid moiety forms the part of a branched carbohydrate structure GAA (**A**$_3$ type).

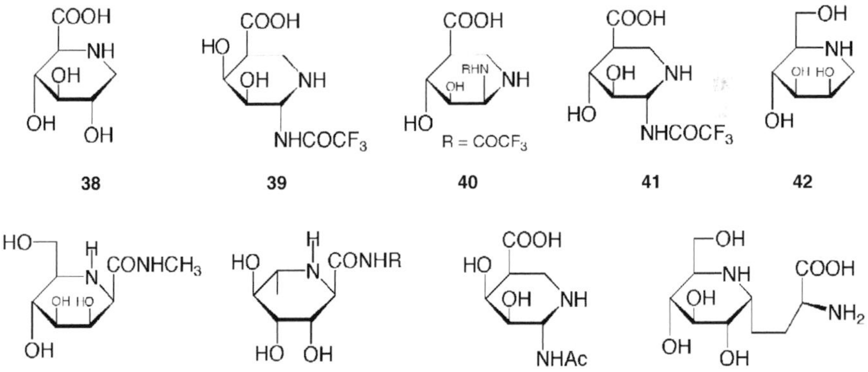

38 **39** **40** **41** **42**

43 **44** R = H **45** R = CH$_3$ **46** **47**

SCHEME 25.8 GAAs where the endocyclic oxygen has been substituted by an amino function termed aza sugar acid (ASA).

inhibitors of naringinase (L-rhamnosidase) and inhibit the biosynthesis of thymidine diphosphate-D-glucose.[83] Siastin B[84] (**46**) shows inhibitory effects against various neuraminidases, *β*-glucuronidases, and *N*-acetyl-*β*-D-glucosaminidase. Another interesting ASA is nojirimycinyl C-(l)-serine (**47**),[85] which combines a *C*-glycosyl amino acid and the glucosidase inhibitor nojirimycin. This compound, due to its expected metabolic stability and inhibitory activity against glycosidases, might be a promising building block to elucidate the biological function of protein glycosylation.

III. HOMOOLIGOMERIZATION OF SUGAR AMINO ACIDS — ACCESS TO OLIGOSACCHARIDE MIMETICS

SAAs, in which the amino, carboxylic acid, and polyol functions are directly incorporated into a sugar ring, have been used as oligosaccharide mimetics (Scheme 25.9).

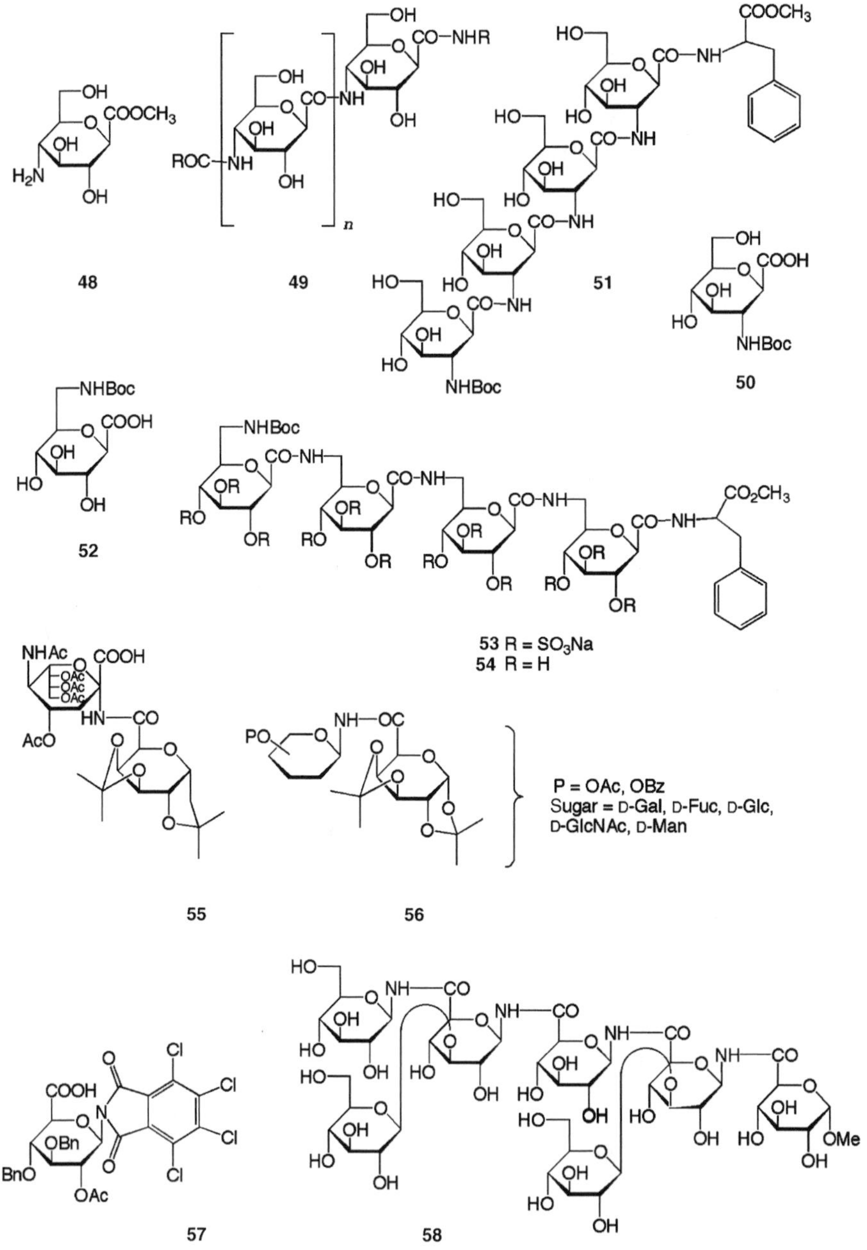

SCHEME 25.9 SAAHs (A_0 type) used in the synthesis of oligosaccharide mimetics.

Originally, the replacement of the glycosidic linkages by amide bonds was done to design polysaccharide analogs by Fuchs and Lehmann.[86] Exposing the amine **48** to methanolic sodium methoxide afforded polycondensed amide-linked oligomers **49**.[87]

The water-soluble condensation products were not further characterized but could be readily saponified with sodium hydroxide presumably due to the participation of the 7-OH of the sugar oligomer. Later, Nicolaou et al.[88] termed these analogs as carbopeptoids due to their hybrid character between carbohydrate and peptide. Particularly, the difficulties associated with the synthesis of oligosaccharide libraries, such as the sterocontrol of the newly formed glycosidic bond (α or β), efficiency of the glycosylation reaction and the susceptibility toward lysis by glycosidases, were expected to be overcome through the replacement of the glycosidic linkage by peptide bonds. The first characterized carbopeptoids were synthesized by Ichikawa and co-workers[89] using the central D-glucosamine-derived SAA building block **50**. Elongation of **50** afforded the tetramer **51**, which after O-sulfation, showed a strong inhibitory potency against HIV infection of CD4 cells.[89] Similarly, oligomerization of the Boc-protected SAA **52** followed by sulfation and deprotection gave the $\beta(1{\rightarrow}6)$ linked carbopeptoid **53** that showed a strong inhibitory potency of MT2 cells from HIV infection.[90] Interestingly, the unsulfated analog **54** did not show any measurable inhibitory effect in the same assay. Sabesan[91] prepared several amide-linked sugar dimers **55** and **56**, where the nitrogen of the peptide bond is connected to the anomeric carbon (Scheme 25.9). He suggested that this type of amide linkage might be sterically more compatible with that of a glycosidic oxygen but no biological data were provided. Similarly, the SAA **57** was used in the synthesis of the carbopeptoid **58**, an analog of the naturally occurring phytoalexin elicitor (Scheme 25.9). Unfortunately, heptamer **58** was devoid of biological activity.[92]

SAAHs, in analogy with amino acids, may induce secondary structures by oligomerization (Scheme 25.10).[93–101] Gervay and co-workers[95] synthesized dimeric through octameric (1→5)-linked sialo-oligomers **59** in solution, and on the solid phase,[96] and studied their behavior in water using circular dichroism (CD) and hydrogen/deuterium amide (NH/ND) exchange rates. The data obtained provided the first evidence that oligomers of constrained carbohydrate-derived amino acids form stable secondary structures in water. Fleet's group[93,94,97–102] investigated the behavior of several 5-aminomethyl-tetrahydrofuran-2-carboxylate oligomers for their ability to form secondary structures (Scheme 25.10). The tetrameric *trans*-SAAs **60** and **61** did not form strong intramolecular hydrogen bonds in chloroform, while octameric **62** showed a strong intramolecular hydrogen-bonding pattern reminiscent of a left-handed α-helix.[97] The tetrameric *cis*-SAAs **62** to **64** form secondary structures in solution (CDCl$_3$) more similar to a repeating β-turn. Tetramer **63**, which is the enantiomer of **62** (but with an isopropylidene rather than cyclohexylidene protecting group), had a very similar ^{1}H-NMR spectrum, indicating that the different ketal protecting group had very little effect on the solution conformation.[97,101] Recently, the synthesis of tetrameric **63** has been extended to the octamer and again a repeating β-turn was suggested as the solution structure in chloroform.[101] Interestingly, the unprotected tetramer **63**, when dissolved in methanol, exhibited secondary structures similar to the protected tetramer **63** in chloroform.[101] Oligomers having the structure **65**[100] and **66**[103] have also been described but do not form well-defined secondary structures (Scheme 25.10).

SAA oligomers, where the anomeric carbon is not part of the linkage region, have also been prepared (Scheme 25.11). Initial work on the synthesis of (2→6)

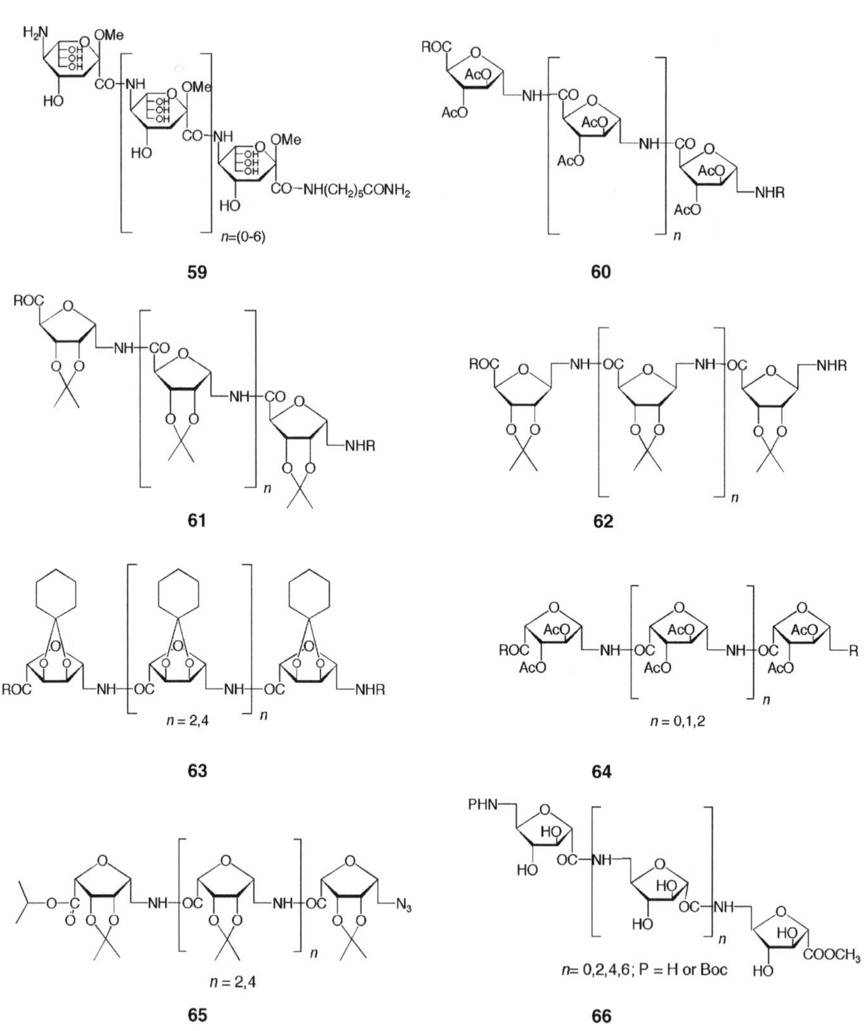

SCHEME 25.10 Homooligomerization of SAAHs (A_0 type). Some of the oligomers induce secondary structures in organic solvent.

amido-linked disaccharides **67** and **68** using 2-acetamido-2-deoxy-D-glucuronic and 2-acetamido-2-deoxy-D-manuronic acid, both components of the bacterial cell wall, was reported by Yoshimura and colleagues.[104] Wessel and co-workers[105] subsequently oligomerized the Fmoc-protected-2-amino-2-deoxy-D-glucuronic acid **69** to the tetramer **70** on solid phase. The same group also synthesized the tetramer **71** containing normuramic acid **72** as a building block.[106] Polymerization of 1-*O*-dodecyl-2-amino-2-deoxy-β-D-glucopyranosiduronic acid afforded poly (SAA) **73** that forms closely packed monolayers on spreading a dilute DMSO–CHCl$_3$ solution on a pure water surface.[107]

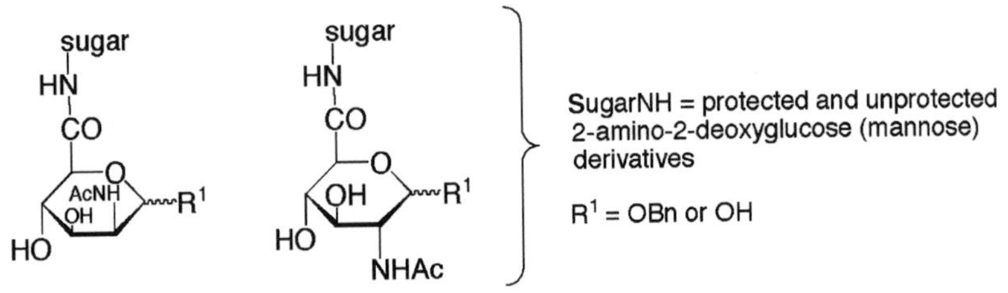

67 68

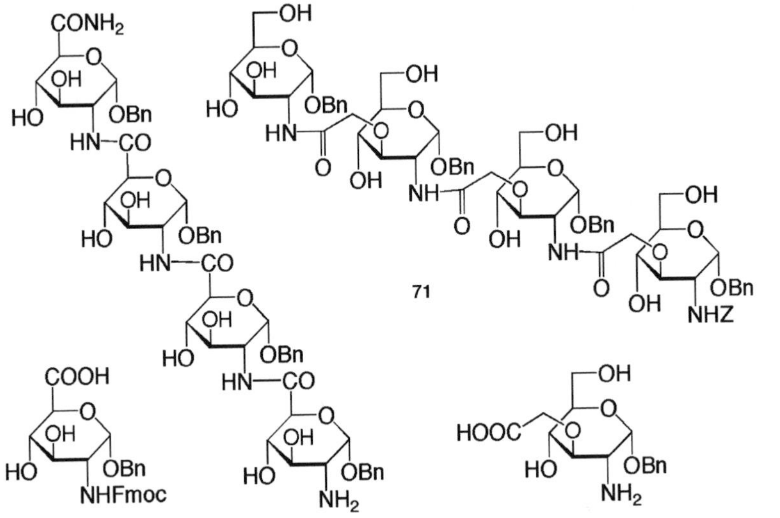

69 70 72

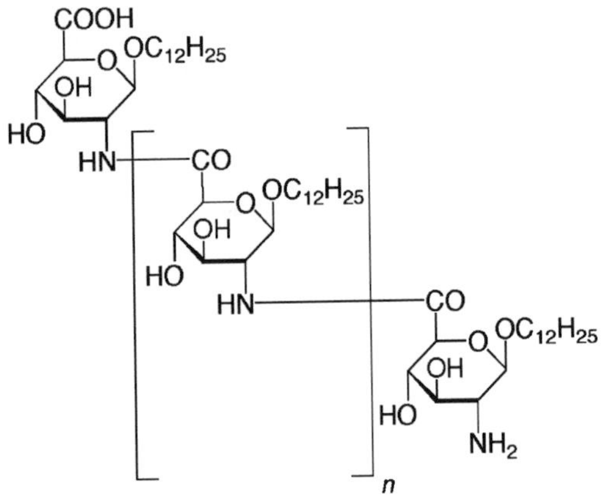

73

SCHEME 25.11 Amide-linked SAAH oligomers (**A**$_0$ type), which do not involve the anomeric (pseudo-anomeric) center.

IV.   INCORPORATION OF SAAHs INTO BIOACTIVE PEPTIDES

Incorporation of nonnatural SAAHs into native peptides may be a useful way to enhance resistance against enzymatic degradation and to induce conformational restraint in the peptide. This modification could improve the drug potential of peptides. In addition, polyhydroxylated SAAH-modified peptides may also improve the solubility in water and other organic solvents. The earliest example that illustrated the potential of SAA as peptidomimetics was provided by Kessler and co-workers.[15,16] Replacement of the Pro-Phe dipeptide sequence in the somatostatin analogue cyclo-(Pro-Phe-D-Trp-Lys-Thr-Phe) by the dipeptide isostere **1** resulted in a potent hexa-peptide mimetic **74** (Scheme 25.12). More recently, it was demonstrated that SAA **75** containing cyclic somatostatin analogs **76** to **79** exhibit strong antiprolifer-ative and apoptotic activity against multidrug-resistant hepatoma carcinoma[13] (Scheme 25.12). This is of particular interest, since resistance to chemotherapy has become a major problem in cancer therapy. Similarly, replacement of the D-Phe-Val sequence in the cyclic pentapeptide (-Arg-Gly-Asp-D-Phe-Val-) carrying the RGD (Arg-Gly-Asp) motif by the SAAHs **1**, **3** and the arabinose-derived bicyclic amino acid **83** afforded potent peptidomimetics **80** to **82** that retained their high affinity for the $\alpha_v \beta_3$ integrin receptor (Scheme 25.13). It is noteworthy that peptidomimetic **80**,

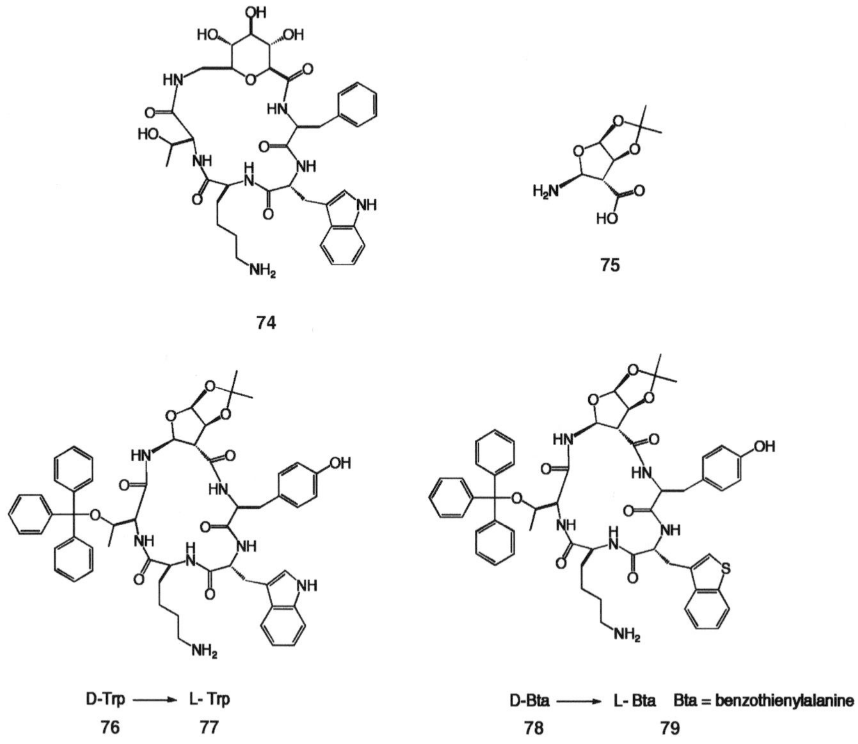

SCHEME 25.12   SAA-modified somatostatin analogs.

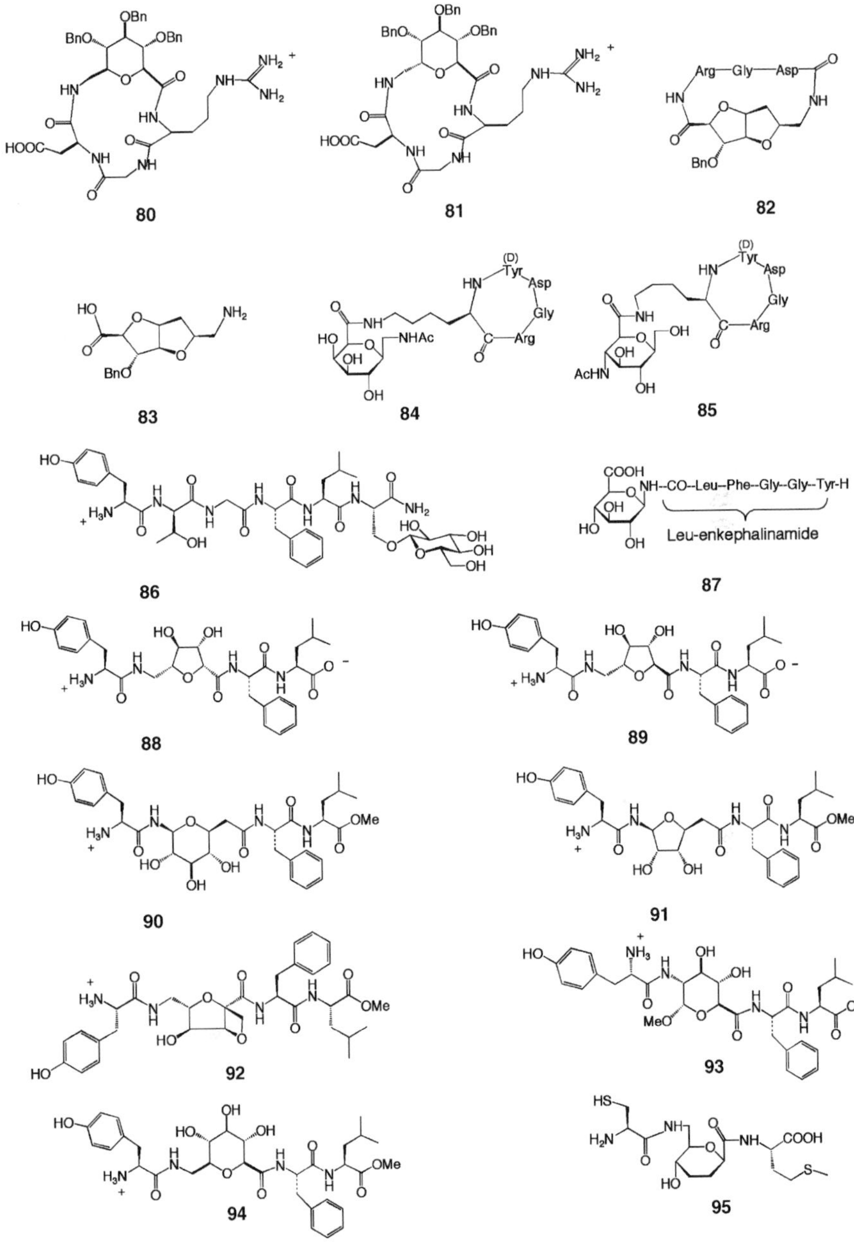

SCHEME 25.13 Incorporation of SAAs into bioactive peptides.

which contains a flexible *cis*-SAA, also exhibited a high affinity for the $\alpha_{IIb}\beta_3$ receptor, which is absent in peptide **81** incorporating the *trans*-SAA **3**.[18] Recently, Kessler and co-workers[18] demonstrated that the derivatization of the amino acid side chain in biologically active compounds with SAAHs influences activity and selectivity. For example, peptide **84**, which contains a SAAH-modified lysine side chain, showed the highest activity for the $\alpha_v\beta_5$ integrin receptor in a series of RGD peptides. Similarly RGD peptide **85**, which contains another lysine-modified SAAH, had a lower uptake in the liver than that of the nonglycosylated peptide. As a result, the initial blood concentrations were doubled.[18] These findings open the door to the development of novel receptor- and receptor-subtype-selective agents based on sugar platforms.[108–110] Besides, the somatostatin analog cyclo-(pro-Phe-D-Trp-Lys-Thr-Phe) and RGD peptides, and other medicinally interesting peptides such as opioid peptides have been modified by SAAHs. The initial interest in SAAH-modified opioid peptides derived from the fact that glycopeptide enkephalin analogs such as peptide **60** are capable of penetrating the blood–brain barrier and bind to targeted opioid receptors.[111–114] Originally, it was suggested that the GLUT-1 transporter provided the means for the transport of the glycosylated opioid peptides.[111] However, further testing had shown that the transport of these glycopeptides was not mediated by GLUT-1. Recent studies suggest that the incorporation of hydrophilic carbohydrate moieties into opioid peptides renders them amphipathic, promoting exchange between lipid and aqueous phases.[114] This property can enhance the ability of the resulting glycopeptide to insert reversibly into lipid phases, thus allowing for membrane-mediated transport across the endothelial layer (blood–brain barrier) via adsorptive endocytosis and subsequent excocytosis into the brain. Without the carbohydrate moiety, a lipophilic opioid peptide may remain within the lipid phase, inhibiting transport and exposing the peptide backbone to enkephalinases and other peptidases. The first example of an unnatural SAAH-modified opioid peptide was reported by Toth and co-workers.[115] For example, pharmacological evaluation showed that glycopeptide **87**, which contains the Leu-enkephalinamide sequence, was 40 times more potent than Leu-enkephalinamide itself.[115] Subsequently, Chakraborty and co-workers[17] substituted the Gly-Gly sequence in the Leu-enkephalins by the isostere *cis*-SAA **2**. The resulting peptide **88** exhibited biological activity similar to that of Leu-enkephalin. Conformational analysis and molecular dynamics revealed the presence of a nine-membered β-turn-like structure, absent in the inactive peptide **89**, which contains a *trans*-SAA.[17] The SAAH-modified Leu-enkephalin analogs **90** and **91** were synthesized by Overkleeft and co-workers,[19] but did not bind to the human μ-,δ-, and κ- opioid receptors using transfected hamster ovary cells. The same research groups also synthesized the Leu-enkephalin analog **92** containing a bicyclic furanoid SAAH.[8] In addition, isosteric substitution of Gly-Gly sequence in Leu-enkephalins by the SAAs **4** and **5** resulted in peptides **93** and **94**, which did not show any biological activity in the pig ileum assay.[15] SAAH-based protein farnesyltransferase (PFT) inhibitors have also been reported.[116] For example peptide **95**, which contains a dipeptide isosteric SAA, inhibits bovine PFT (IC_{50} = 214 μM)(Scheme 25.13). Recently, the SAAHs **10** to **13** have been used as dipeptide isosteres to replace the Ala-Tyr and Glu-Glu squences of the hsp65 p2-13 epitope of *Mycobacterium tuberculosis* and *Mycobacterium leprae* but so far no biological data have been provided for these neoglycopeptides.[116]

V. COMBINATORIAL SYNTHESIS WITH SAAHs

Monosaccharide building blocks, with incorporated amino and carboxylic acid functional groups (GAAs), have been proposed by Mc Devitt and Lansbury[117] as versatile building blocks for combinatorial synthesis. Particularly, the well-established chemistry of amide formation is a very attractive strategy to prepare compound libraries of glycosamino acid oligomers (glycotides). Synthetic glycotides were proposed as drug candidates since they would not be susceptible to glycosidases and may not be recognized by proteases due to the altered backbone relative to the natural substrate.[117] Functional group modifications of the polyol moiety could increase the lipophilicity of the molecules and render them more likely to permeate cell membranes.[117] Besides their use in peptide coupling reactions, GAAs may also be useful building blocks for other chemical transformations that are compatible with combinatorial strategies including reductive aminations[118] and multicomponent reactions.[119]

A. SUGAR-PLATFORM LIBRARIES ACCESSED BY SAA BUILDING BLOCKS

Carbohydrates have the potential to be used as highly functionalized rigid scaffolds.[108–110,120–123] In particular, the use of sugar-derived skeletons as nonpeptidomimetics of somatostatin (a cyclic tetradecapeptide)[109] demonstrated for the first time that sugars might be privileged platforms. Sofia and co-workers [124] elaborated on this idea and reported the synthesis of encoded trifunctionalized saccharide scaffolds termed "universal pharmacophore mapping libraries." Construction of the library employed the two-sugar building blocks **96** and **97** having a three-point attachment motif comprising a carboxylic acid moiety, a free hydroxyl group, and a protected amino group (Scheme 25.14). The free carboxylic acid was first reacted with eight amino acid-functionalized trityl-TentaGel resins, followed by carbamate formation at the free hydroxyl site with six isocyanates. Finally, the deprotected amino function was acylated with eight different carboxylic acids. Deacetylation (when necessary) and cleavage from the resin gave 16 × 48 sublibraries (Scheme 25.14). More recently, libraries have also been prepared using the scaffolds **98** and **99**.[125,126] Biological screening of platform libraries containing scaffold **99** led to the discovery of compounds exhibiting strong antibacterial activity.[126]

B. LIBRARY SYNTHESIS USING GAAs DERIVED FROM O- AND N-LINKED GLYCOPEPTIDES

GAAs are extremely useful building blocks for combinatorial synthesis of glycopeptides.[127] Similar to amino acids, GAAs can be oligomerized to generate glycopeptide libraries. One of the highlights in this field has been the synthesis of a 300,000-membered ladder-encoded glycopeptide library by Meldal and co-workers.[128] The pentafluorophenyl (Pfp) activated glycosyl amino esters **100** to **102**, and a variety of activated amino esters were used to generate a heptaglycopeptide library on the solid phase using a photocleavable linker. The synthesized library was screened "on bead" against the fluorescent-labelled lectin *Lathyrus odoratus*, which shows a weak specificity for the α-methyl glycosides of mannose, *N*-acetylglucosamine and glucose.

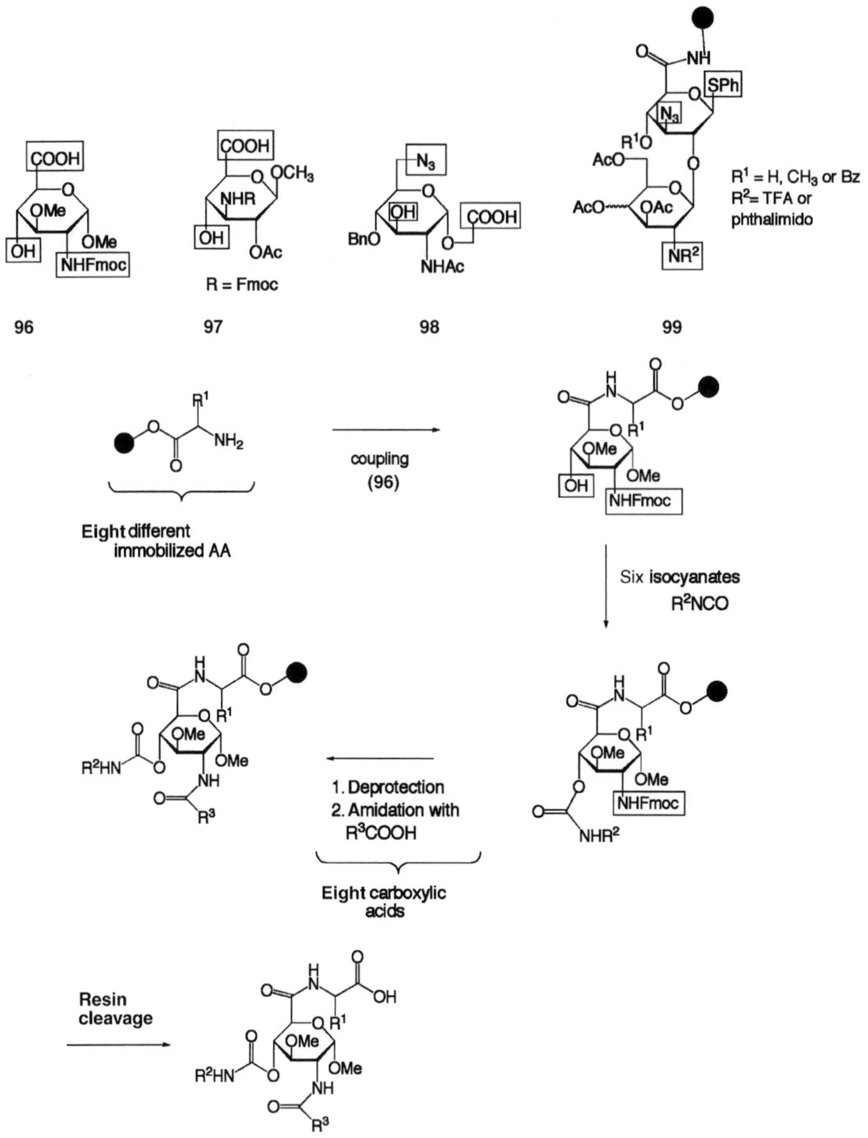

SCHEME 25.14 Synthesis of "universal pharmacophore libraries" based on a three-point attachment motif.

Interestingly, the most active compounds were glycopeptides containing only single mannose residues. Jobron and Hummel[129] used the sugar-unprotected N-glycopeptide building blocks **103** to **109** on a continuous surface for library synthesis termed SPOT synthesis. SPOT synthesis on cellulose is a highly effective method for the rapid preparation of spatially addressable peptides (Scheme 25.15). A library consisting of sialyl Lewis[X] mimetics was reported by Wong and co-workers (Scheme 25.15).[130]

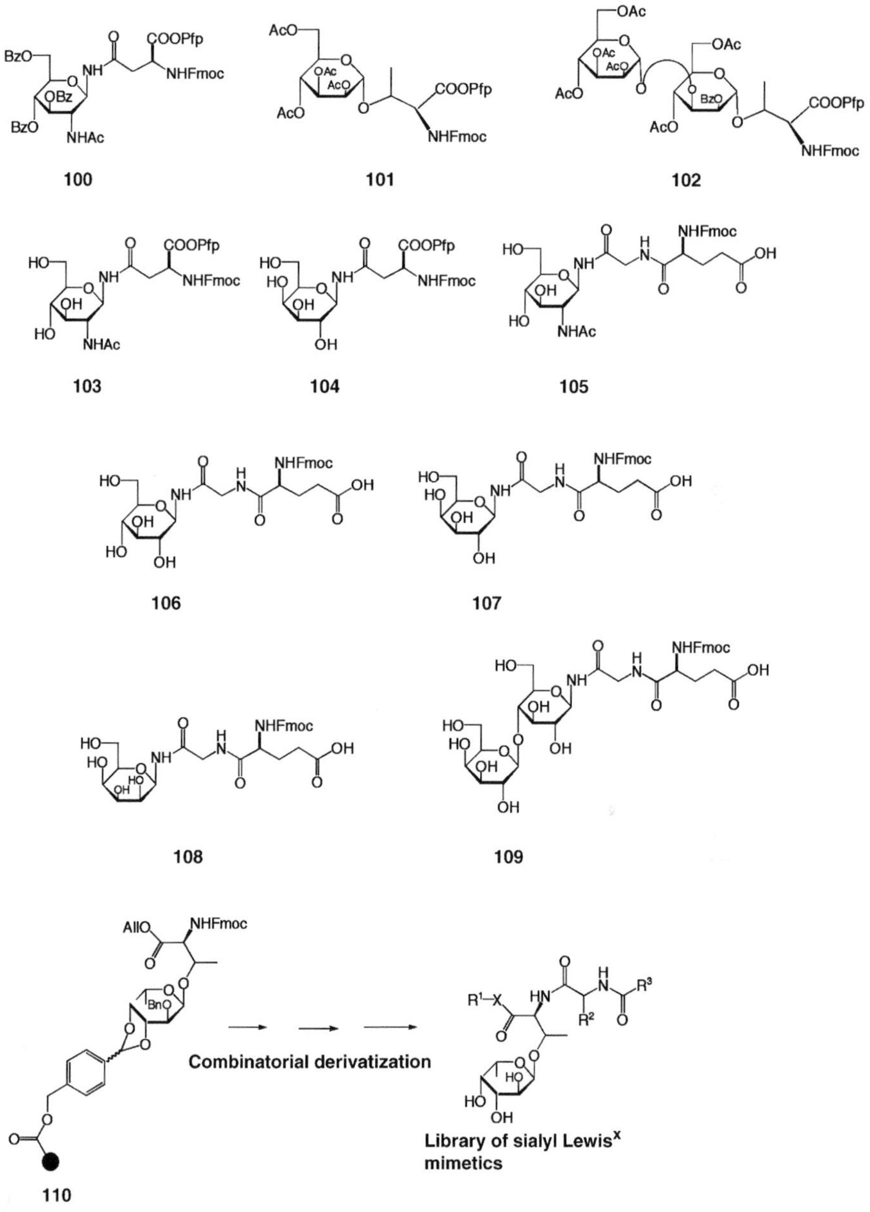

SCHEME 25.15 GAA building blocks used in combinatorial library synthesis.

Fucose, which contains the three hydroxyl groups required for recognition of sialyl Lewis^X by E-selectin, was retained as the only carbohydrate moiety, while the other three sugars were replaced with L-threonine and its derivatives. A fucosylated threonine amino acid **110** residue was immobilized on a carboxyl-functionalized resin via an acid-sensitive *cis*-1,2 diol protecting and anchoring group. The allyl group was first cleaved and the acid function was then derivatized as an ester or amide followed by a conventional peptide coupling cycle and capping (Scheme 25.15). The deprotected and purified compounds were tested against E- and P-selectins and showed moderate binding affinities. A comprehensive review on the combinatorial synthesis of glycopeptide libraries using GAA building blocks has recently been published.

C. LIBRARY SYNTHESIS USING GAAs WHICH ARE NOT DERIVED FROM *O*- AND *N*-LINKED GLYCOPEPTIDES

Among the mimics of the naturally occurring glycopeptides, the *S*-glycosyl amino acid family has attracted considerable interest over the years. The driving force behind the synthesis of *S*-glycosides has been the production of glycomimetics with enhanced stability toward chemical and enzymatic degradation. In addition, the synthesis of *S*-glycosides is less complicated than the synthesis of *C*-glycosides. The first combinatorial approach toward *S*-glycosamino acid libraries, termed carbohybrids, was reported by Hindsgaul and co-workers (Scheme 25.16).[118] A hydrophobically tagged 1-thiogalactose derivative **111** underwent Michael reactions with five different Michael acceptors to yield diastereomeric adducts. In the second step, the ketones present in the Michael adducts were reductively aminated with six different carboxyl-protected amino acids. After deprotection, 30 (5×6) thiosugaramino acids were obtained as mixtures of four diastereomers each. One of the library members was found to be competitive inhibitor of β-glactosidase from *Escherichia coli* with an inhibition constant of 1.7 μM.

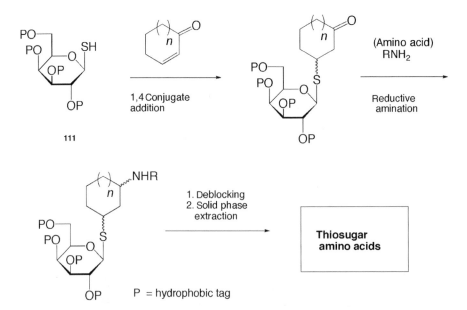

SCHEME 25.16 Synthesis of a glycotide-library containing thiosugar amino acids.

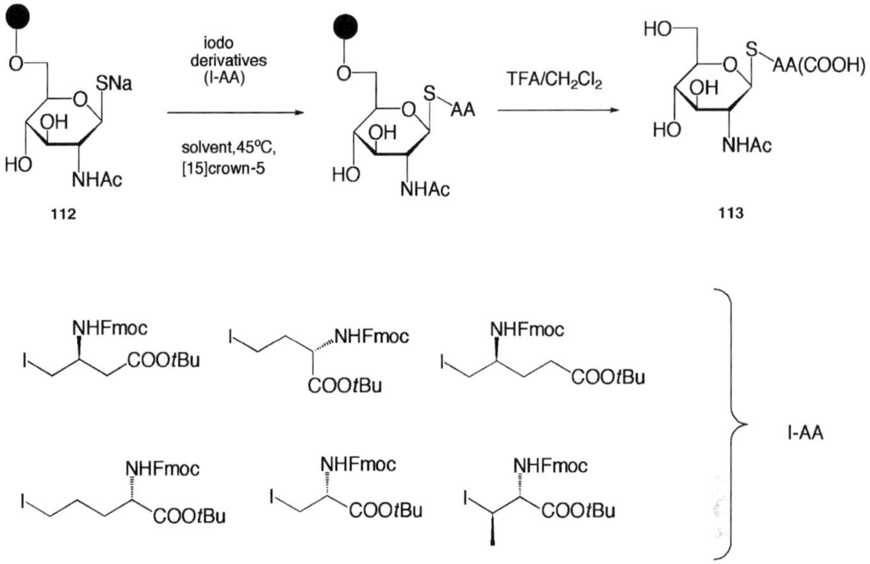

SCHEME 25.17 Synthesis of *S*-glycosamino acid building blocks on the solid phase.

Jobron and Hummel[43] described the synthesis of several unprotected *S*-glycosamino acid building blocks on the solid phase (Scheme 25.17). The key step in their synthesis is the reaction of an immobilized sodium thiolate **112** with amino acid iodo derivatives (I-AA) in the presence of[15] crown-5 as complexing agent. Resin cleavage with concomitant acid deprotection afforded various *S*-glycosamino acids **113** that can be used for solid phase and *S*-glycopeptide synthesis.

D. COMBINATORIAL APPROACHES TOWARD GLYCOPEPTIDES DERIVED FROM GAAS (A₂ TYPE)

Few reports of libraries containing A_2-based GAAs have appeared. One of the problems associated with sugar mimetics incorporating an α-amino acid moiety at the anomeric position is their tendency to equilibrate between two different stereoisomeric forms (Scheme 25.18).[67,131–134] Acylation of the amine usually prevents this equilibration,[67,134] while strongly basic conditions afford epimerized products.[132,134,135] The first syntheses of A2-based neoglycopeptides were described by Fleet and co-workers. Incorporation of the mannofuranose-based aminoester **114** and **115** into tetrapeptides afforded the glycopeptides **116** and **117** (Scheme 25.19).[134] In order to synthesize epimer-free glycopeptides with incorporated anomeric α-amino acids, Fleet et al.[137] treated the bicyclic L-rhamno lactone **118**, the D-gluco lactone **120**,[137] and the D-manno lactone **122**[136] with primary amines to yield the tripeptides **119**, **121**, and **123**, respectively, after deprotection (Scheme 25.19). Recently, our group has reported on the combinatorial synthesis of galacto-configured sugar-β-peptides using a multicomponent Ritter condensation.[138] Lewis acid promoted the reaction of α-D–galacto-2-deoxy-oct-3-ulopyranosonic acid **124** with nitriles afforded spiro dihydrooxazinone

SCHEME 25.18 Equilibration of amino acid incorporated into the anomeric centre of carbohydrates.

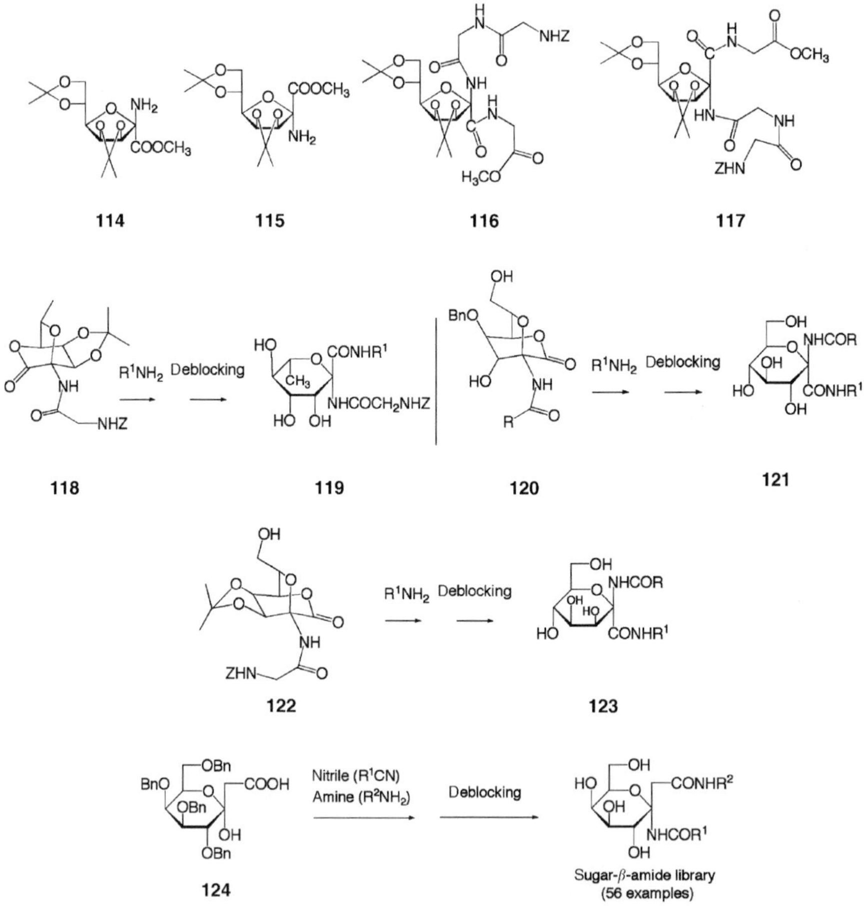

SCHEME 25.19 Lactone-strategy used for the synthesis of double-substituted glycosyl amino acids.

intermediates. Exposure of these intermediates to primary amines or amino acids produced a 56-member sugar-β-peptide library (Scheme 25.19). Interestingly, in all cases studied, the nitrile component occupies an axial position in the sugar scaffold. We are currently investigating the use of this highly functionalized α,α-disubstituted pyran scaffold as peptide mimetic and β-peptide synthesis.

E. COMBINATORIAL APPROACHES TOWARD BRANCHED SAAHS

Owing to the difficulties in their synthesis, branched SAAHs (A_3 type) have only rarely been described and appear to have only been incorporated into short peptide sequences.[72]

F. ACCESS TO AZA SUGAR ACID LIBRARIES

Among the ASAs the proline derivatives and pipecolic acids have attracted considerable interest over the years due to their potential to induce secondary structures in peptides and their potential as glycosidase inhibitor. Fleet's group[82,83,139,140] reported a lactone strategy that can be used to incorporate proline and pipecolic acids into aza sugar libraries (Scheme 25.20). Treatment of the lactones **125, 127,** and **129** with primary amines and deprotection (if necessary) provided the corresponding hydroxyproline derivatives **126,**[140] **128,**[140] and **130.**[139] Similarly, the pipecolic acids **131,**[83] **134,**[83] **136,**[102] and **137**[102] were accessible from the bicyclic lactones **131,133,** and **135** by simple reaction with amines followed by deprotection (Scheme 25.20). Interestingly, the ASA derivatives **136** and **137** showed strong competitive inhibition of β-N-acetylglucosaminidases from bovine liver and human placenta.[82]

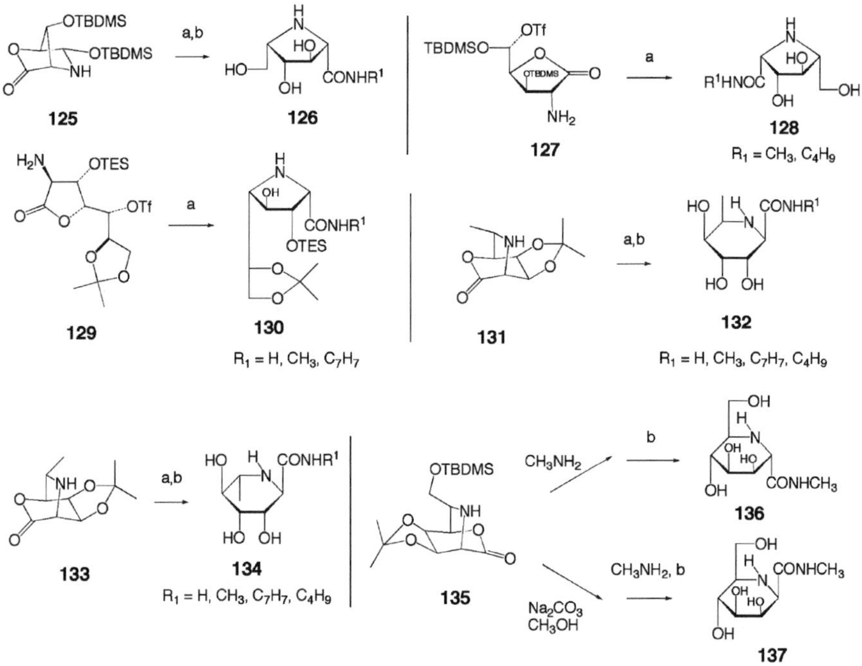

SCHEME 25.20 Lactone strategy to prepare aza sugar acid-based derivatives: (a) R_1NH_2, (b) deblocking.

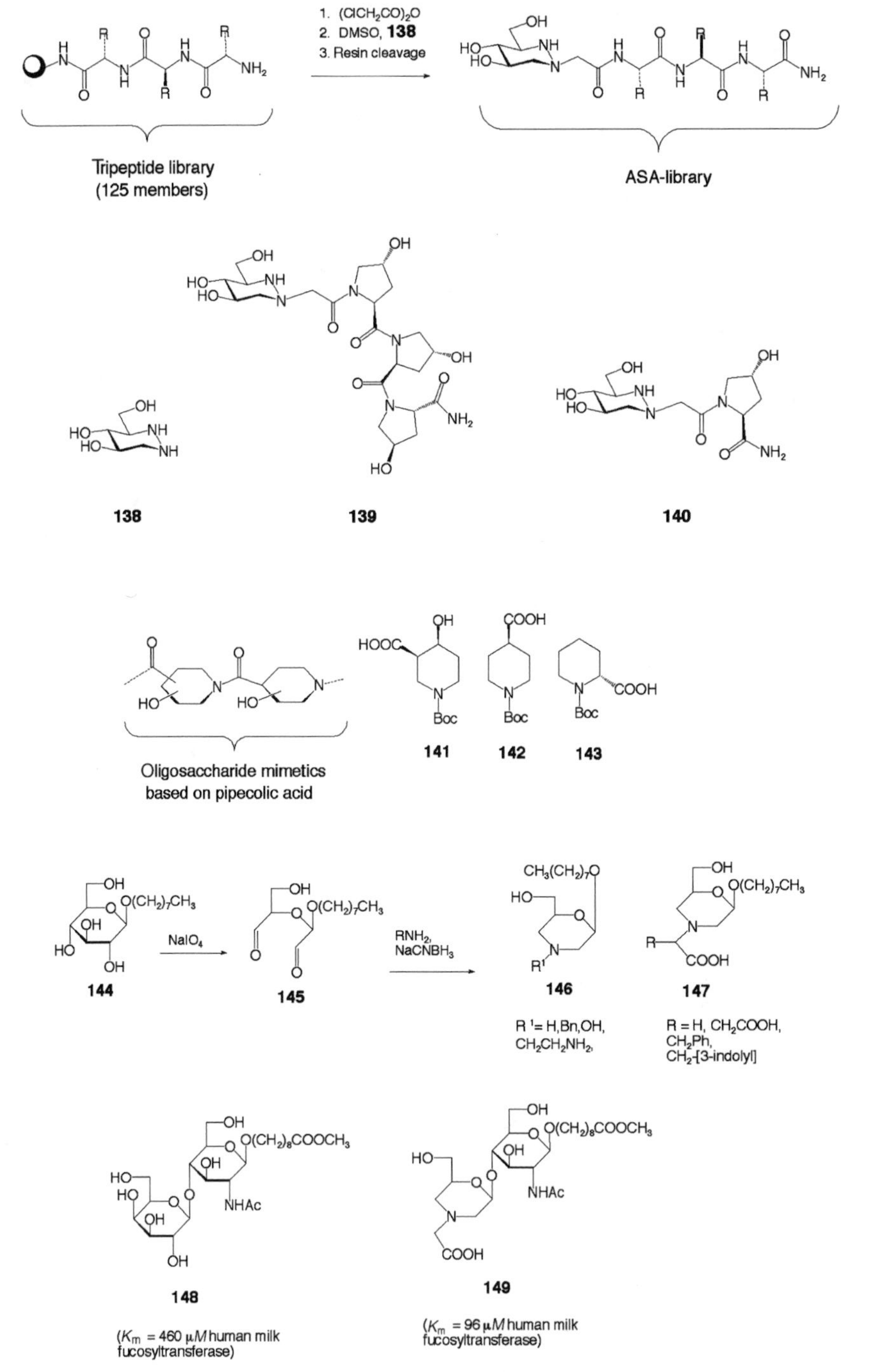

SCHEME 25.21 Combinatorial approaches toward aza sugar acid libraries.

A library of aza sugar glycosidase inhibitors has been prepared by Bol's group (Scheme 25.21).[141,142] Initially, a tripeptide library consisting of the L-amino acids of threonine, serine, 4-hydroxyproline, phenylalanine and alanine was synthesized on solid phase (4-methyl benzhydrylamine resin) using the split and mix method. Subsequently, the resulting five sublibraries of 25 peptides in each reactor were treated individually with chloroacetic anhydride followed by substitution with the aza sugar azafagomine **138** and cleavage from the resin. Deconvolution of the most active sublibraries yielded the potent almond β-glucosidase inhibitor **139** ($K_i = 40$ μM), bearing hydroxyprolines at all positions in the peptide. Interestingly, this compound was 40 times more active than compound **140** but still 60 times less active than the unmodified azafagomine, indicating that the triproline residue contributed to binding. Most of the synthesized compounds also inhibited glycogen phosphorylase.

A small 27-member amide library of piperidine-based carbohydrate mimics was synthesized on solid phase.[143] A split and mix synthesis using the three building blocks **141**, **142**, and **143** (Scheme 25.21) afforded 27 trimers that were subsequently tested for α-glucosidase (yeast), β-glucosidase (almonds), isomaltose (yeast), α-fucosidase (human placenta), β-mannosidase (snail) and β-galactosidase (E. coli). Unfortunately, none of the enzymes were inhibited significantly by the library compounds.

Du and Hindsgaul[144] developed a short pathway to morpholino-based SAAs (Scheme 25.21). Periodate oxidation of the octylglucoside **144** affords the dialde-hyde **145** that can be easily purified by simple adsorption onto reverse-phase C_{18} resins. Reductive amination of **145** with a variety of amines (benzylamine, hydrox-ylamine, and unprotected amino acids) affords morpholino-based ASAs **146** and **147**. The reaction was extended to the N-acetyllactosamine derivative **148** wherein only the terminal Gal residue contains vicinal diols. Periodate oxidation followed by reductive amination using glycine gave the 3'aza-disaccharide analog **149**. Interestingly, the disaccharide mimetic **149** was a better substrate than **148** for human milk fucosyltransferase (Scheme 25.21).

VI. CONCLUSIONS

SAAHs are versatile building blocks for combinatorial synthesis. Derivatization and oligomerization of the amino acid moiety can be applied to produce libraries with potential glycomimetic properties. Besides being sugar-like, SAAHs also have the potential to be used as peptidomimetics. Incorporation of SAAs into small peptides has resulted in biologically active peptides with improved biological and pharmaco-logical properties. Furthermore, SAAHs offer tremendous structural and functional diversity, which is largely unexplored and requires combinatorial strategies for efficient exploitation.

ACKNOWLEDGMENTS

Financial support from the Natural Sciences and Engineering Research Council of Canada (NSERC) and the Japanese Foundation for the Promotion of Science (JSPS) is gratefully acknowledged.

REFERENCES

1. Thompson, L.A. and Ellman, J., *Chem. Rev.*, 96, 555, 1996.
2. Weiss, J.B., Lote, C.J., and Bobinski, H., *Nature New Biol.*, 234, 25, 1971.
3. Lote, C.J. and Weiss, J.B., *FEBS Lett.*, 16, 81, 1971.
4. (a) Hofsteenge, J., Müller, D.R., Beer, T., Löffler, A., Richter, W.J., and Vliegenhart, J.F.G., *Biochemistry*, 33, 13524, 1994; (b) Beer, T., Vliegenhart, J.F.G., Löffler, A., and Hofsteenge, *Biochemistry*, 34, 11785, 1995; (c) Löffler, A., Doucey, M.-A., Jansson, A.M., Müller, D.R., Beer, T., Hess, D., Meldal, M., Richter, W.J., Vliegenhart, J.F.G., and Hofsteenge, J., *Biochemistry*, 35, 12005, 1996; (d) Kreig, J., Hartmann, S., Vicentini, A., Gläsner, W., Hess, D., Hofsteenge, J., *J. Mol. Biol. Cell*, 9, 301, 1998; (e) Vliegenhart, J.F.G. and Casset, F., *Curr. Opin. Struct. Biol.*, 8, 565, 1998; (f) Doucey, M.-A., Hess, D., Blommers, M.J.J., and Hofsteenge, J., *Glycobiology*, 9, 435, 1999.
5. (a) Manabe, S. and Ito, Y., *J. Am. Chem. Soc.*, 121, 9754, 1999; (b) Nishikawa, T., Ishikawa, M., and Isobe, M., *Synlett*, 123, 1999; (c) Manabe, S., Ito, Y., and Ogawa, T., *Chem. Lett.*, 919, 1998.
6. Isono, K., *J. Antibiot.*, 41, 1711, 1988 and references therein.
7. Nakajima, M., Itoi, K., Takamatsu, Y., Kinoshita, T., Okazaki, T., Kawakubo, K., Shindou, M., Hinma, T., Tohjigamori, M., and Haneishi, T., *J. Antibiotics*, 44, 293, 1991.
8. Van Well, R.M., Meijer, M.E.A., Overkleeft, H.S., Van Boom, J.H., van der Marel, G.A., and Overhand, M., *Tetrahedron*, 59, 2423, 2003.
9. (a) Geyer, A., Bockelmann, D., Weissenbach, K., and Fischer, H., *Tetrahedron Lett.*, 40, 477, 1999; (b) Geyer, A. and Moser, F., *Eur. J. Org. Chem.*, 1113, 2000; (c) Tremmel, P. and Geyer, A., *J. Am. Chem. Soc.*, 124, 8548, 2002.
10. (a) Peri, F., Cipolla, L., La Ferla, B., and Nicotra, F., *J. Chem. Soc. Chem. Commun.*, 2303, 2000; (b) Forni, E., Cipola, L., Caneva, E., La Ferla, B., Peri, F., and Nicotra, F., *Tetrahedron Lett.*, 43, 1355, 2002; (c) Peri, F., Bassetti, R., Caneva, E., De Gioia, L., La Ferla, F., Presta, M., Tanghetti, E., and Nicotra, F., *J. Chem. Soc. Perkin. Trans.*, 1, 638, 2002.
11. Bartolozzi, A., Li, B., and Franck, R.W., *Bioorg. Med. Chem.*, 11, 3021, 2003.
12. Palomo, C., Landa, A., Gonzalez-Regio, M.C., Garcia, J.M., Gonzalez, A., Odriozola, J.M., Martin-Pastor, M., and Linden, A ., *J. Am. Chem. Soc.*, 124, 8637, 2002.
13. Gruner, S.A.W., Keri, G., Schwab, R., Venetianer, A., and Kessler, H., *Org. Lett.*, 3, 3723, 2001.
14. (a) Dondoni, A. and Marra, A., *Chem. Rev.*, 100, 4395, 2000; (b) Gruner, S.A.W., Locardi, E., Lohof, E., and Kessler, H., *Chem. Rev.*, 102, 491, 2002; (c) Chakraborty, T.K., Ghosh, S., and Jayaprakash, J., *Curr. Med. Chem.*, 9, 421, 2002; (d) Peri, F., Cipolla, L., Forni, E., and Nicotra, F., *Monatsch. für Chemie*, 133, 369, 2002.
15. Graf von Roedern, E., Lohof, E., Hessler, G., Hoffmann, M., and Kessler, H., *J. Am. Chem. Soc.*, 118, 10156, 1996.
16. Graf von Roedern, E. and Kessler, H., *Angew. Chem. Int. Ed. Engl.*, 33, 687, 1994.
17. Chakraborty, T.K., Jayaprakash, S., Diwan, P.V., Nagaraj, R., Jampani, S.R.B., and Kunwar, A.C., *J. Am. Chem. Soc.*, 120, 12962, 1998.
18. Lohof, E., Planker, E., Mang, C., Burkhart, F., Dechantsreiter, M.A., Haubner, R., Wester, H.-J., Schwaiger, M., Hölzemann, G., Goodman, S.L., and Kessler, H., *Angew. Chem. Int. Ed.*, 39, 2761, 2000.
19. Kriek, N.M.A.J., van der Hout, E., Kelly, P., van Meijgaarden, K.E., Geluk, A., Ottenhoff, T.H.M., van der Marel, G.A., Overhand, M., van Boom, J.H., Valentijn, A.R.P.M., and Overkleeft, H.S., *Eur. J. Org. Chem.*, 2418, 2003.

20. Dwek, R.A., *Chem. Rev.*, 96, 683, 1996.
21. (a) Varki, A., *Glycobiology*, 3, 97, 1993; (b) Lee, Y.C. and Lee, R.T., *Acc. Chem. Res.*, 28, 322, 1995; (c) McAuliffe, J.C. and Hindsgaul, O., *Carbohydrates in medicine,* in *Molecular and Cellular Glycobiology, Frontiers in Molecular Biology*, Vol. 10, Fukuda, M. and Hindsgaul, O., Eds., Oxford University Press, Oxford, pp. 249–285, 2000.
22. Grochee, F.C., Gramer, M.J., Andersen, D.C., Bahr, J.B., and Rasmusssen, J.R., in *Frontier in Bioprocessing II,* Todd, C.P., Sikdar, S.K., and Bier, M., Eds., *American Chemical Society*, Washington DC, p. 199, 1992.
23. Fisher, J.F., Harrison, A.W., Bundy, G.L., Wilkinson, K.F., Rush, B.D., and Ruwart, M.J., *J. Med. Chem.*, 34, 3140,1991.
24. Mehta, S., Meldal, M., Duus, J.O., and Bock, K., *J. Chem. Soc. Perkin. Trans.*, 1, 1445, 1999 and references therein.
25. (a) Lis, H. and Sharon, N., *Eur. J. Biochem.*, 218, 1, 1993; (b) Hase, S., Nishimura, H., Kawabata, S., Iwanaga, S., and Kenaka, T., *J. Biol. Chem.*, 265, 1858, 1990; (c) Bock, K., Schuster-Kolbe, J., Altman, E., Allmaier, G., Stahl, B., Christian, R., Sletyr, U.B., and Messner, P., *J. Biol. Chem.*, 269, 7137, 1994.
26. Marcaurelle, L.A. and Bertozzi, C.R., *Chem. Eur. J.*, 5, 1384, 1999.
27. Bertozzi, C.R., Cook, D.G., Kobertz, W.R., Gonzalez-Sxarano, F., and Bednarski, M.D., *J. Am. Chem. Soc.*, 114, 10639, 1992.
28. Urban, D., Skrydstrup, T., and Beau, J.-M., *Chem. Commun.*, 955, 1998.
29. Campbell, A.D., Paterson, D.E., Raynham, T.M., and Taylor, R.J.K., *Chem. Commun.*, 1599, 1999.
30. Ben, R.N., Orellana, A., and Ayra, P., *J. Org. Chem.*, 63, 4817, 1998.
31. Dorgan, B.J. and Jackson, F.W., *Synlett*, 859, 1996.
32. Dondoni, A., Marra, A., and Massi, A., *Tetrahedron*, 54, 2827, 1998.
33. Dondoni, A., Marra, A., and Massi, A., *Chem. Commun.*, 1741, 1998.
34. Bertozzi, C.R., Hoeprich, J.P.D., and Bednarski, M.D., *J. Org. Chem.*, 57, 6092, 1992.
35. Fuchs, T. and Schmidt, R.R., *Synthesis*, 753, 1998.
36. Debenham, S.D., Debenham, J.S., Burk, M.J., and Toone, E.J., *J. Am. Chem. Soc.*, 119, 9897, 1997.
37. Tedebark, U., Meldal, M., Panza, L., and Bock, K., *Tetrahedron Lett.*, 39, 1815, 1998.
38. Gustafsson, T., Saxtin, M., and Kihlberg, J., *J. Org. Chem.*, 68, 2506, 2003.
39. Käsebeck, L. and Kessler, H., *Liebigs Ann./Recueil*, 1, 165, 1997.
40. Gerz, M., Matter, H., and Kessler, H., *Angew. Chem. Int. Ed. Engl.*, 32, 269, 1993.
41. (a) Elofsson, M., Walse, B., and Kihlberg, J., *Tetrahedron Lett.*, 32, 7613, 1991; (b) Monsigney, M.L.P., Delay, D., and Vaculik, M., *Carbohydr. Res.*, 59, 589, 1977.
42. Nicolaou, K.C., Chucholowski, A., Dolle, R.E., and Randall, J.L., *J. Chem. Soc. Chem. Commun.*, 1155, 1984.
43. Jobron, L. and Hummel, G., *Org. Lett.*, 2, 2265, 2000.
44. Knapp, S. and Myers, D.S., *J. Org. Chem.*, 67, 2995, 2002.
45. Cohen, S.B. and Halcomb, R.L., *J. Am. Chem. Soc.*, 124, 2534, 2002.
46. Hoffmann, M., Burkhart, F., Hessler, G., and Kessler, H., *Helv. Chim. Acta*, 79, 1519, 1996.
47. Frey, O., Hoffmann, M., and Kessler, H., *Angew. Chem. Int. Ed. Engl.*, 34, 2026, 1995.
48. Saha, U.K. and Roy, R., *Tetrahedron Lett.*, 36, 3635, 1995.
49. Dondoni, A., Marra, A., and Massi, A., *J. Org. Chem.*, 64, 933, 1999.
50. Burkhart, F., Hoffmann, M., and Kessler, H., *Angew. Chem. Int. Ed. Engl.*, 36, 1191, 1997.
51. Lay, L., Meldal, M., Nicotra, F., Panza, L., and Russo, G., *Chem. Commun.*, 1469, 1997.
52. Dondoni, A., Mariotti, G., and Marra, A., *Tetrahedron Lett.*, 41, 3483, 2000.

53. Dondoni, A., Mariotti, G., and Marra, A., *J. Org. Chem.*, 67, 4475, 2002.
54. Marcaurelle, L.A., Rodriguez, E.C., and Bertozzi, C.R., *Tetrahedron Lett.*, 39, 8417, 1998.
55. Peri, F., Cipolla, L., Rescigno, M., La Ferla, B., and Nicotra, F., *Bioconjugate Chem.*, 12, 325, 2001.
56. Cipolla, L., Rescigno, M., Leone, A., Peri, F., La Ferla, B., and Nicotra, F., *Bioorg. Med. Chem. Lett.*, 10, 1639, 2002.
57. (a) Marcaurelle, L.A. and Bertozzi, C.R., *Tetrahedron Lett.*, 39, 7279, 1998; (b) Werner, R.M., Shokek, O., and Davis, J.T., *J. Org. Chem.*, 62, 8243, 1997.
58. (a) Davis, N.J. and Flitsch, S.L., *Tetrahedron Lett.*, 32, 6793, 1991; (b) Wong, S.Y.C., Guile, G., Dwek, R., and Arsequell, G., *Biochem. J.*, 300, 843, 1994; (c) Bengtsson, M., Broddefalk, J., Dahmen, J., Henriksson, K., Kihlberg, J., Lohn, H., Srinivasa, B.R., and Stenvall, K., *Glycoconj. J.*, 15, 223, 1998.
59. (a) Schweizer, F. and Inazu, T., *Org. Lett.*, 3, 4115, 2001; (b) Dondoni, A., Junquera, F., Merchan, F.L., Merino, P., Schermann, M.-C., and Tejero, T., *J. Org. Chem.*, 62, 5484, 1997; (c) Simchen, G. and Pürkner, E., *Synthesis*, 525, 1990; (d) Colombo, L., Casiraghi, G., and Pittalis, A.J., *J. Org. Chem.*, 56, 3897, 1990; (e) Rosenthal, A. and Brink, A.J., *J. Carbohydr. Nucleosides Nucleotides*, 2, 343, 1975.
60. Vincent, S.P., Schleyer, A., and Wong, C.-H., *J. Org. Chem.*, 65, 4440, 2000.
61. Lowary, T., Meldal, M., Helmboldt, A., Vasella, A., and Bock, K., *J. Org. Chem.*, 63, 9657, 1998.
62. Pearce, A.J., Ramaya, S., Thorn, S.N., Bloomberg, G.B., Walter, D.S., and Gallagher, T., *J. Org. Chem.*, 64, 5453, 1999.
63. Palomo, C., Landa, A., Gonzalez-Regio, M.C., Garcia, J.M., Gonzalez, A., Odriozola, J.M., Martin-Pastor, M., and Linden, A., *J. Am. Chem. Soc.*, 124, 8637, 2002.
64. Bichard, J.C.F., Mitchell, E.P., Wormald, M.R., Watson, K.A., Johnson, L.N., Zographos, A.E., Koutra, D.D., Oikanomakos, N.G., and Fleet, G.W.J., *Tetrahedron Lett.*, 36, 2145, 1995.
65. (a) Somasak, L., Nagy, V., Docsa, T., Toth, B., and Gergely, P., *Tetrahedron Asymmetry*, 11, 405; (b) Osz, E., Somsak, L., Szilagyi, L., Kovacs, L., Docsa, D., Toth, B., and Gergely, P., *Bioorg. Med. Chem. Lett.*, 9, 1385, 1999.
66. Krülle, T.M., Watson, K.A., Gregoriou, M., Johnson, L.N., Crook, N.S., Watkin, D.J., Griffiths, R.C., Nash, R.J., Tsitanou, K.E., Zographos, S.E., Oikonomakos, G., and Fleet, G.W.J., *Tetrahedron Lett.*, 36, 8291, 1995.
67. Brandstetter, T.W., Wormald, M.R., Dwek, R.A., Butters, T.D., Platt, F.M., Tsitanou, K.E., Zographos, S.E., Oikonomakos, N.G., and Fleet, G.W.J., *Tetrahedron Asymmetry*, 7, 157, 1996.
68. Schweizer, F. and Hindsgaul, O., *Synlett*, 1743, 2001.
69. Schweizer, F., Lohse, A., Otter, A., and Hindsgaul, O., *Synlett*, 1434, 2001.
70. Ohrui, H., Kuzuhara, H., and Emoto, M., *Tetrahedron Lett.*, 45, 4267, 1971.
71. Neka, T., Hashizume, T., and Nishimura, M., *Tetrahedron Lett.*, 45, 95, 1971.
72. Coutrot, P., Grison, C., and Coutrot, F., *Synlett*, 393, 1998.
73. Vasella, A. and Voeffray, R., *Helv. Chim. Acta*, 65, 1134, 1982.
74. (a) Garner, P. and Park, J.M., *Tetrahedron Lett.*, 30, 5065, 1989; (b) Evina, C.M. and Guillerm, G., *Tetrahedron Lett.*, 37, 163, 1996; (c) Garner, P. and Park, J.M., *J. Org. Chem.*, 55, 3772, 1990.
75. Casiraghi, G., Colombo, L., Rassu, G., and Spanu, P., *J. Org. Chem.*, 56, 6523, 1991.
76. Hadrami, M.E., Lavergne, J.-P., Viallefont, P.H., Chiaroni, A., Riche, C., and Hasnaoui, A., *Synth. Commun.*, 23, 157, 1993.
77. Czernecki, S., Dieulesaint, A., and Valery, J.-M., *J. Carbohydr. Chem.*, 5, 469, 1986.

78. Rosenthal, A. and Richards, C.M., *Carbohydr. Res.*, 31, 331, 1973.
79. Rosenthal, A. and Shudo, K., *J. Org. Chem.*, 26, 4391, 1972.
80. Rosenthal, A., Richards, C.M., and Shudo, K., *Carbohydr. Res.*, 27, 353, 1973.
81. Nishimura, Y., Shitara, E., Adachi, H., Toyoshima, M., Nakajima, M., Okami, M., and Takeuchi, Y., *J. Org. Chem.*, 65, 2, 2000.
82. Shilvock, J.P., Nash, R.J., Lloyd, J.D., Winters, A.L., Asano, N., and Fleet, G.W.J., *Tetrahedron Asymmetry*, 9, 3505, 1998.
83. Shilvock, J.P., Wheatley, J.R., Nash, R.J., Watson, A.A., Griffiths, R.C., Butters, R.C., Müller, M., Watkin, D.J., Winkler, D.A., and Fleet, G.W.J., *J. Chem. Soc. Perkin. Trans.*, 1, 2735, 1999.
84. (a) Umezawa, H., Aoyagi, T., Komiyama, T., Morishima, H., Hamada, M., and Takeuchi, T., *J. Antibiot.*, 27, 963, 1974; (b) Nishimura, Y., Kudo, T., Kondo, S., and Takeuchi, T., *J Antibiot.*, 45, 963, 1992; (c) Nishimura, Y., Wang, W., Kondo, S., Aoyagi, T., and Umezawa, H., *J. Am. Chem. Soc.*, 110, 7249, 1988.
85. Fuchss, T. and Schmidt, R.R., *Synthesis*, 259, 2000.
86. (a) Fuchs, E.F. and Lehmann, J., *J. Chem. Ber.*, 108, 2254, 1975; (b) Fuchs, E.F. and Lehmann, J., *J. Carbohydr. Res.*, 45, 135, 1975.
87. Fuchs, E.F. and Lehmann, J., *J. Carbohydr. Res.*, 49, 267, 1975.
88. Nicolaou, K.C., Flörke, H., Egan, M.G., Barth, T., and Estevez, V.A., *Tetrahedron Lett.*, 36, 1775, 1995.
89. Suhara, Y., Ichikawa, M., Hildreth, J.E.K., and Ichikawa, Y., *Tetrahedron Lett.*, 37, 2549, 1996.
90. Suhara, Y., Hildreth, J.E.K., and Ichikawa, Y., *Tetrahedron Lett.*, 37, 1575, 1996.
91. Sabesan, S., *Tetrahedron Lett.*, 38, 3127, 1998.
92. Timmers, C.M., Turner, J.J., Ward, C.M., van der Marel, G.A., Kouwijzer, M.L.C.E., Grootenhuis, P.D.J., and van Boom, J.H., *Chem. Eur. J.*, 3, 920, 1997.
93. Smith, M.D., Long, D.D., Martin, A., Marquess, D.D., Claridge, T.D.W., and Fleet, G.W.J., *Tetrahedron Lett.*, 40, 2191, 1999.
94. Long, D.D., Hungerford, N.L., Smith, M.D., Brittain, D.E.A., Marquess, D.G., Claridge, T.S.W., and Fleet, G.W.J., *Tetrahedron Lett.*, 40, 2195, 1999.
95. Gervay, J., Flaherty, T.M., and Nguyen, C., *Tetrahedron Lett.*, 38, 1493, 1997.
96. Szabo, L., Smith, B.L., McReynolds, K.D., Parrill, A.L., Morris, E.R., and Gervay, J., *J. Org. Chem.*, 63, 1074, 1998.
97. Claridge, T.D.W., Long, D.D., Hungerford, N.L., Aplin, R.T., Smith, M.D., Marquess, D.G., and Fleet, G.W.J., *Tetrahedron Lett.*, 40, 2199, 1999.
98. Smith, M.D., Claridge, T.D.W., Tranter, G.E., Sansom, M.S.P., and Fleet, G.W.J., *Chem. Commun.*, 2041, 1998.
99. Long, D.D., Smith, M.M., Marquess, D.G., Claridge, T.D.W., and Fleet, G.W.J., *Tetrahedron Lett.*, 39, 9293, 1998.
100. Brittain, D.E.A., Watterson, M.P., Claridge, T.D.W., Smith, M.S., and Fleet, G.W.J., *J. Chem. Soc. Perkin. Trans.*, 1, 3655, 2000.
101. Hungerford, N.L., Claridge, T.D.W., Watterson, M.P., Aplin, R.T., Moreno, A., and Fleet, G.W.J., *J. Chem. Soc. Perkin. Trans.*, 1, 3666, 2000.
102. Long, D.D., Setz, R.J.E., Nash, R.J., Marquess, D.D., Lloyd, J.D., Winters, A.L., Asano, N., and Fleet, G.W.J., *J. Chem. Soc. Perkin. Trans.*, 1, 901, 1999.
103. Chakraborty, T.K., Jayaprakash, S., Srinivasu, P., Chary, M.G., Diwan, P.V., Nagaraj, R., Sankar, A.R., and Kunwar, A.C., *Tetrahedron Lett.*, 41, 8167, 2000.
104. Yoshimura, J., Ando, H., Sato, T., Tsuchida, S., and Hashimoto, H., *Bull. Chem. Soc. Jpn.*, 49, 2511, 1976.
105. Müller, C., Kitas, E., and Wessel, H.P., *J. Chem. Soc. Chem. Commun.*, 2425, 1995.

106. Wessel, H.P., Mitchell, C.M., Loboto, C.M., and Schmid, G., *Angew. Chem. Int. Ed. Engl.*, 32, 2712.
107. Nishimura, S.-I., Shinnosuke, N., and Kuriko, Y., *Chem. Commun.*, 617, 1998.
108. Nicolaou, K.C., Trujillo, J.L., and Chibale, K., *Tetrahedron*, 53, 8751, 1997.
109. (a) Hirschmann, R., Nicolaou, K.C., Pietranico, S., Leahy, E.M., Salvino, J., Arison, B., Cichy, M.A., Spoors, P.G., Shakespeare, W.C., Sprengeler, P.A., et al., *J. Am. Chem. Soc.*, 115, 12550, 1993; (b) Hirschmann, R., Hynes, J.J., Cichy-Knight, M.A., van Rijn, R.D., Sprengeler, P.A., Spoors, P.G., Shakespeare, W.C., Pietranico-Cole, S., Barbosa, J., Liu, J., et al., *J. Med. Chem.*, 41, 1382, 1998.
110. Smith III, A.B., Sasho, S., Barwis, B.A., Sprengeler, P., Barbosa, J., Hirschmann, R., and Cooperman, B.S., *Bioorg. Med. Chem. Lett.*, 8, 3133, 1998.
111. Polt, R., Porreca, F., Szabo, L.Z., Bilsky, E.J., Davis, P., Abbruscato, T.J., Davis, T.P., Horvath, R., Yamamura, H.I., and Hruby, V.J., *Proc. Natl. Acad. Sci. USA*, 91, 7114, 1994.
112. Bilsky, E.J., Egleton, R.D., Mitchell, S.A., Palian, M.M., Davis, P., Huber, J.D., Jones, H., Yamamura, H.I., Janders, J., Davis, T.P., Porreca, F., Hruby, V.J., and Polt, R., *J. Med. Chem.*, 43, 2586, 2000.
113. Polt, R. and Palian, M.M., *Drugs of the Future,* 26, 561, 2001.
114. Palian, M.M., Bogouslavsky, V.I., O'Brien, D.F., and Polt, R., *J. Am. Chem. Soc.*, 125, 5823, 2003.
115. Drouillat, B., Kellam, B., Dekany, G., Starr, M.S., and Toth, I., *Bioorg. Med. Chem. Lett.*, 7, 2247, 1997.
116. Overkleeft, H.S., Verhelst, S.H.L., Pieterman, E., Meeuwenoord, N.J., Overhand, M., Cohen, L.H., van der Marel, G.A., and van Boom, J.H., *Tetrahedron Lett.*, 40, 4103, 1999.
117. McDevitt, J.P. and Lansbury, P.T., *J. Am. Chem. Soc.*, 118, 3818, 1996.
118. Nilsson, U.J., Fournier, E.J.-L., and Hindsgaul, O., *Bioorg. Med. Chem.*, 6, 1563, 1998.
119. (a) Dömling, A. and Ugi, I., *Angew. Chem. Int. Ed. Engl.*, 39, 3168, 2000; (b) Ugi, I., *Angew. Chem. Int. Ed. Engl.*, 21, 810, 1982.
120. Wong, C.-H., Ye, X.-S., and Zhang, Z., *J. Am. Chem. Soc.*, 120, 7137, 1998.
121. (a) Wunberg, T., Kallus, C., Opatz, T., Henke, S., Schmidt, W., and Kunz, H., *Angew. Chem. Int. Ed.*, 37, 2503, 1998; (b) Kallus, C., Opatz, T., Wunberg, W., Schmidt, W., Henke, S., and Kunz. H., *Tetrahedron Lett.*, 40, 7783, 1999; (c) Opatz, T., Kallus, C., Wunberg, T., Schmidt, W., Henke, S., and Kunz, H., *Eur. J. Org. Chem.*, 8, 1527, 2003; (d) Opatz, T., Kallus, C., Wunberg, W., Schmidt, W., Henke, S., and Kunz, H., *Carbohydr. Res.*, 337, 2089, 2002.
122. Dinh, T.Q., Smith, C.D., Du, X., and Armstrong, R.W., *J. Med. Chem.*, 41, 981, 1998.
123. Streicher, B. and Wünsch, B., *Eur. J. Org. Chem.*, 6, 115, 2001.
124. Sofia, M.J., Hunter, R., Chan, T.Y., Vaughan, A., Dulina, R., Wang, H., and Gange, D., *J. Org. Chem.*, 63, 2802, 1998.
125. Ghosh, M., Dulina, R.R., Kakarla, R., and Sofia, M.J., *J. Org. Chem.*, 65, 8387, 2000.
126. Sofia, M.J., Allanson, N., Hatzenbuhler, N.T., Jain, R., Kakarla, R., Kogan, N., Liang, R., Liu, D., Silva, D.J., Wang, H., et al., *J. Med. Chem.*, 42, 3193, 1999.
127. For reviews on glycopeptide and glycoconjugate libraries: (a) Barkley, A. and Arya, P., *Chem. Eur. J.*, 7, 555, 2001; (b) Hilaire, P.M.S.T. and Meldal, M., *Angew. Chem. Int. Ed.*, 39, 1162, 2000; (c) Sofia, M.J., *Mol. Divers*, 3, 75, 1998.
128. St. Hilaire, P.M., Lowary, T.L., Meldal, M., and Bock, K., *J. Am. Chem. Soc.*, 120, 13312, 1998.
129. Jobron, L. and Hummel, G., *Angew. Chem. Int. Ed.*, 39, 1621, 2000.
130. Lampe, T.F.J., Weitz-Schmidt, G., and Wong, C.-H., *Angew. Chem. Int. Ed.*, 37, 1707, 1998.

131. Burton, J.W., Son, J.C., Fairbanks, A.J., Choi, S.S., Taylor, H., Watkin, D.J., Winchester, B.G., and Fleet, G.W.J., *Tetrahedron Lett.*, 34, 6119, 1993.

132. Brandstetter, T.W., Kim, Y.-H., Son, J.C., Taylor, H.M., Lilley, P.M., Watkin, D.J., Johnson, L.N., Oikonomakos, N.G., and Fleet, G.W.J., *Tetrahedron Lett.*, 36, 2149, 1995.

133. Estevez, J.C., Smith, M.D., Lane, A.L., Crook, S., Watkin, D.J., Besra, G.S., Brennan, P.J., Nash, R.J., and Fleet, G.W.J., *Tetrahedron Asymmetry*, 7, 387, 1996.

134. Estevez, J.C., Estevez, R.J., Ardon, H., Wormald, M.R., Brown, D., and Fleet, G.W.J., *Tetrahedron Lett.*, 35, 8885, 1994.

135. Brandstetter, T.W., de la Fuente, C., Kim, Y.-H., Johnson, L.N., Crook, S., deLilley, P.M., Watkin, D.J., Tsitanou, K.E., Zographos, S.E., Chrysina, E.D., Oikonomakos, N.G., and Fleet, G.W.J., *Tetrahedron*, 52, 10721, 1996.

136. Estevez, J.C., Long, D.D., Wormald, M.R., Dwek, R.A., and Fleet, G.W.J., *Tetrahedron Lett.*, 36, 8287,1995.

137. Fleet, G.W.J., Estevez, J.C., Smith, M.D., Blerot, Y., de la Fuente, C., Krülle, T.M., Besra, G.S., Brennan, P.J., Nash, R.J., Johnson, L.N., Oikonomakos, N.G., and Stalmans, W., *Pure Appl. Chem.*, 70, 279, 1998.

138. Lohse, A., Schweizer, F., and Hindsgaul, O., Combinatorial Chemistry & High Throughput Screening, 5, 389, 2002.

139. Lee, R.E., Smith, M.D., Pickering, L., and Fleet, G.W.J., *Tetrahedron Lett.*, 40, 8689, 1999.

140. Long, D.D., Frederiksen, S.M., Marquess, D.G., Lane, A.L., Watkin, D.J., Winkler, D.A., and Fleet, G.W.J., *Tetrahedron Lett.*, 39, 6091, 1998.

141. Lohse, A., Jensen, K.B., Lundgren, K., and Bols, M., *Bioorg. Med. Chem.*, 7, 1965, 1999.

142. Lohse, A., Jensen, K.B., and Bols, M., *Tetrahedron Lett.*, 40, 3033, 1999.

143. Brygesen, E., Nielsen, J., Willert, M., and Bols, M., *Tetrahedron Lett.*, 38, 5697, 1997.

144. Du, M. and Hindsgaul, O., *Synlett*, 395, 1997.

26 Effects of Short- and Long-Chain Fructans on Large Intestinal Physiology and Development of Preneoplastic Lesions in Rats

Morten Poulsen

CONTENTS

I. INTRODUCTION

Inulin-type fructans are a class of nondigestible carbohydrates naturally occurring in edible plants such as chicory, Jerusalem artichoke, and onions, and serve as the plants, main storage carbohydrate.[1] All the fructans of dietary interest are, as far as is known, of the inulin-type and will be the fructans of focus in this chapter. Fructans encompass both the short-chain fructooligosaccharides (FOS) and the longer chain inulin. The estimated average daily intake per capita of inulin-type fructans from natural sources ranges between 3 and 11 g in Western Europe. The corresponding figures for North America are 1 to 4 g.[2] In the past few years, a large number of food products containing inulin-type fructans have been introduced in the global food market.

The potential for the use of fructans in the food industry is obvious. Fructans have specific technological as well as nutritional assets being soluble upon heating and bland in taste. They can be incorporated into a wide variety of food products, preserving their intrinsic flavor without altering their texture or appearance. Owing to these qualities, fructans can be used as a fiber-like food ingredient. Furthermore, fructans help to provide body, good taste and appearance, making them useful as a fat replacer. Owing to their water-binding capacity, fructans also display gelling and thickening properties.[3] Short-chain fructans are much more soluble than longer chain fructans. They are even more soluble than sucrose, making them suitable ingredients to replace sucrose and at the same time decrease the energy content of the end product (i.e., by providing bulk with fewer calories). Their taste is reported to be very clean and without aftertaste.[4] These various technological advantages make fructans useful in many types of foods, such as sauces, desserts, dairy products, and soft drinks and could lead to a potential intake that is considerably higher than their intake through natural sources.[5]

Physiologically, though, fructans behave very much like soluble fibers and are promoted as such by the food industry, including statements regarding their potential health benefits. The incitements for bringing fructans to the market, therefore, seem to be, on the one hand, the technological uses and, on the other, "add value" to ordinary products.[5]

A variety of positive health effects have been associated with the ingestion of inulin-type fructans.[6,7] The extensive fermentation of inulin-type fructans in the large intestine is of special interest as it results in the formation of short-chain fatty acids (SCFA) such as acetate, propionate, and butyrate, a number of gases, and lactate.

These properties of fructans and their effects on a number of other gut-related parameters will be dealt with in this chapter using data from the literature and a recently published experimental animal study.[8]

II. DESCRIPTION AND NATURAL OCCURRENCE

Approximately 45,000 higher plant species, representing about ten families, use fructans (polyfructose molecules) as their main storage carbohydrate in leaves, stems, and roots.[5,9–12] Fructans can also be produced by algae, fungi, and bacteria.[13]

There is a substantial variation in both linkage types and the number of carbohydrate monomers making up the fructans. Differences in fructan length not only result from taxonomic variation but are also subject to environmental influence, the life cycle, and the time and conditions of storage.[5] The degree of polymerization (DP) of plant fructans varies from 3 to about 200 fructose units.[12]

Chemically, fructan molecules may be described in a simple way by the formula GF_n where G represents glucosyl unit, F the fructosyl unit, and n the number of fructosyl units linked ($n \geq 2$). The glucose moiety is linked to the end of the chain by an $\alpha(1–2)$ bond as in sucrose (Figure 26.1). The fructan group also contains minor amounts of F_m ($m \geq 2$) fructans, in which the glucosyl endunit is not present, and different kinds of branched molecules.[5]

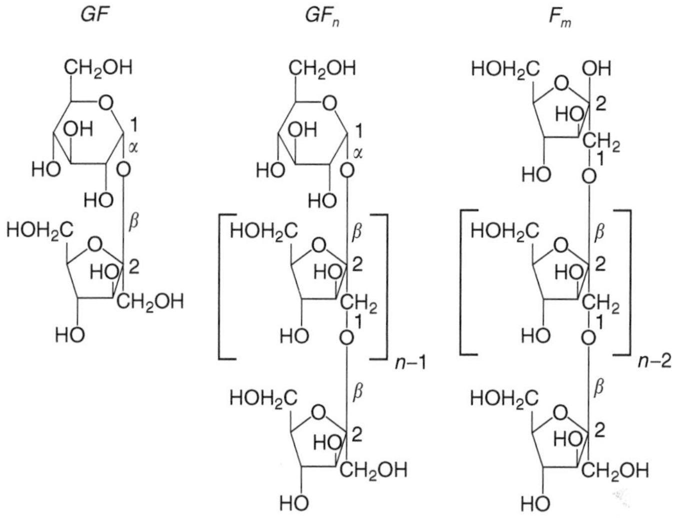

FIGURE 26.1 Chemical structures of sucrose (*GF*) and inulin-type fructans (*GF_n* and *F_m*). (Adapted from Nordic Working Group on Food Toxiology and Risk Evaluation, Safety evaluation of fructans. Nordic Council of Ministers, 1–117, 2000.)

Among food plants, chicory (*Cichorium intybus*) and Jerusalem artichoke (*Helianthis tuberosus*, topinambur) are rich sources of inulin and are well characterized as potential crops for commercial inulin production.[14–16]

Isolated inulin can be transformed either chemically or microbially. Transformations may proceed with or without previous hydrolysis to fructose. In the production of FOS, only partial hydrolysis is allowed. Many other conversions are characterized by an inherent initial hydrolysis of inulin to fructose. However, only some of these conversion products have the potential to be used in food applications.[5,14]

The main commercially available types of fructans are inulins (mainly DP 3–60) and FOS (mainly DP 3–10). Synonyms for FOS are fructosugars or oligofructose.

FOS are commercially obtained from inulin through enzymatic hydrolysis,[4] by enzymatic synthesis from sucrose and fructose or by bacterial enzymatic synthesis.[17,18] The enzyme source of FOS bacterial synthesis is derived either from plants, for example, Jerusalem artichoke or from bacteria and fungi such as *Aspergillus* sp. Depending on the enzyme source, the FOS products have different linkages and DPs.[5]

III. EPIDEMIOLOGY AND ANIMAL STUDIES

Fructans are defined as nonstarch polysaccharides. In general, the intake of nonstarch polysaccharides, such as cellulose, pectins, β-glucans, gums, and so on, has been shown to offer protection against colon and rectal cancer by most epidemiological studies;[19–22] however two recent prospective studies with 88,757 and 61,463 women, respectively, showed that the intake of nonstarch polysaccharides had no protective effect on the development of colorectal cancer.[23,24]

In experimental animal studies, nonstarch polysaccharides have been shown to protect, enhance, or have no effect on chemical-induced colon cancer. A protective effect is seen in the majority of the studies. However, the variability in data for the same compound could be due to differences in experimental design, diet formulation, and the exact composition and definition of the nonstarch polysaccharide.[25,26] It seems that one important property of nonstarch polysaccharides with respect to their effect on the large intestine and the development of colon cancer is their fermentability and solubility. A link is seen between solubility and fermentability as soluble carbohydrates are fermentable in most cases. Highly fermentable nonstarch polysaccharides, such as guar gum, pectin, carrageenan, and psyllium husk, are less protective, even promoting colon cancer, compared with less fermentable carbohydrates such as cellulose, wheat bran, barley bran, and so on.[26–29]

Owing to their novelty as food ingredients, the effect of fructans on the development of colorectal cancer has not been evaluated in any epidemiological study. Fructans are highly fermentable nonstarch polysaccharides. Despite the tendency of fermentable nonstarch polysaccharides to have no effect on or even promote colon cancer, a few animal studies have shown a decrease in the number of aberrant crypt foci (ACF) in the rat colon after the intake of fructans.[30,31] ACF is believed to be a preneoplastic lesion and indicative of a potential carcinogenic effect in the large bowel.[32–36] In the paper by Reddy and co-workers,[30] the long-chain fructan was more efficient than the short-chain one in reducing the number of ACF, but ACF was the only parameter addressed and the study was not designed to reveal the effect of dose level and duration of feeding.

In order to reveal further the observed positive effects of fructans, an animal study with both short- and long-chain fructans given at two dietary levels for 5 or 10 weeks was conducted at the Danish Institute of Food Safety and Nutrition.[8]

IV. EXPERIMENTAL DESIGN

The study by Poulsen and co-workers published in Nutrition and Cancer[8] was designed to examine the effect of chain length, dietary level, and duration of feeding of fructans on their potential chemopreventive properties against colon cancer, using ACF as an endpoint. As the difference in efficacy in reducing the number of ACF between the short- and long-chain fructan, observed earlier by Reddy and co-workers,[30] could be due to their difference in resistance to fermentation, the influence of chain length on gastrointestinal microflora, SCFA, pH, and cell proliferation, all parameters reflecting fermentation, were further investigated. In order to study how the putative fructan-induced changes in the microflora and the environment of the large intestine could affect the outcome of the parameters investigated, a 3-week-pretreatment period with two inulin-type fructans was included. The rats were initiated with 1,2-dimethylhydrazine dihydrochloride (DMH), a known colon carcinogen, given orally instead of subcutaneously, as this better simulated the dietary introduction of a carcinogen.

In the study, two commercially available fructans, Raftiline HP and Raftilose P95, were used. Raftiline HP is a long-chain inulin-type fructan and consists of chicory inulin from which the smaller molecules have been removed, resulting in an average chain length ≥23. Raftilose P95 is a short-chain inulin-type fructan and is a partial enzymatic hydrolysis product of chicory inulin with a chain length ranging between 2 and 8.

The animals, male Fischer 344 rats, were stratified and randomized to eight experimental groups of 20 each. Before the start of dosing with DMH, two groups were pretreated with 15% Raftilose or Raftiline for 3 weeks, while the remaining groups received a control diet. All groups were then dosed with DMH (20 mg/kg bw) by gavage once a week from week 0 to 3 (Figure 26.2), except the control group, which received 0.9% NaCl as vehicle. During and after the DMH dosing, the groups were fed a balanced purified diet containing, 0, 5, or 15% Raftilose or Raftiline (Table 26.1). Ten

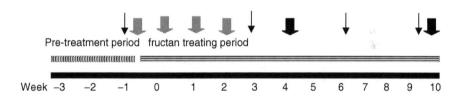

Pre-treatment period	fructan treating period	

Week   −3   −2   −1   0   1   2   3   4   5   6   7   8   9   10

DMH-initiation, 20 mg/kg bw by gavage, except the control group.

Sacrifice of 10 animals/group, 5 and 10 weeks after the first DMH-dose.

Collection of fresh feces for bacteriological quantification.

FIGURE 26.2   Experimental design of the study. (From Poulsen, M. et al. *Nutr. Cancer*, 42, 194–205, 2002. With permission.)

TABLE 26.1
Feeding Scheme

Group	N	Pretreatment Period Week −3 to −1	DMH Initiation Week 0–3	Treatment Period Week 0–4 or 0–9
Control	20	Control diet	−	Control diet
DMH control	20	Control diet	+	Control diet
5% Raftilose	20	Control diet	+	5% Raftilose
15% Raftilose	20	Control diet	+	15% Raftilose
15% Raftilose + pretreatment	20	15% Raftilose	+	15% Raftilose
5% Raftiline	20	Control diet	+	5% Raftiline
15% Raftiline	20	Control diet	+	15% Raftiline
15% Raftiline + pretreatment	20	15% Raftiline	+	15% Raftiline

Source: From Poulsen, M. et al. *Nutr. Cancer*, 42, 194–205, 2002. With permission.

animals from each group, selected at the initial randomization, were sacrificed 5 weeks after the first DMH dose, and the remaining after 10 weeks (Figure 26.2).

During the experimental period all animals were inspected twice daily. Body weight and food and water consumption were recorded weekly. During the experimental period, fresh fecal samples were taken for microbial analysis before DMH dosing, 1 week after the last DMH dose and 7 and 10 weeks after the first DMH dose (Figure 26.2). The fecal concentrations of bifidobacteria, enterococci, *Escherichia coli,* and *Clostridium perfringens* were measured. In addition, total counts of aerobic and anaerobic bacteria were made. At sacrifice, similar measurements of bacteria in the cecal content were made, and further cecal SCFA concentration, pH of cecum content, and weight of colorectum with and without content were determined. Animals sacrificed after 10 weeks were injected intraperitoneally with 5'-bromo-2'-deoxyuridine, one hour before sacrifice, for determination of cell proliferation. The organs of the abdominal and thoraxic cavity were macroscopically examined. The colorectum was cut longitudinally, rinsed in 0.9% NaCl, weighed, and divided into two pieces of equal length. The two pieces, proximal and distal segment, were pinned on a cork slab and fixed in buffered formaldehyde for approximately 24 h. The ACF of the entire colorectum were visualized by Giemsa staining. The staining of the intestinal mucosa with Giemsa is comparable with the methylene blue staining and is used to facilitate the recognition of the morphological aberrant crypts using a stereomicroscope. The ACF were grouped into small (1–3 crypts/focus), medium (4–6 crypts/focus), and large foci (≥7 crypts/focus).

V. EFFECT ON MICROFLORA

The composition of the bacterial flora of the large intestine is a valuable parameter when investigating compounds, such as fructans, that primarily exert their effect in the colon. However, the influence of the microflora on intestinal physiology and the development of colorectal cancer are far from being understood. The principal substrate for colonic bacterial growth is dietary carbohydrates that have escaped digestion in the small intestine, although there are also contributions from proteins, amino acids, and endogenously produced carbohydrates and glycoproteins.[37] When a large bulk of indigestible carbohydrate such as resistant starch or nonstarch polysaccharides such as fructans is fed to rats, it is possible to achieve changes in the composition and metabolic pattern of the bacteria flora resident in the large intestine. However, it seems that these changes are rarely consistent, and relatively large and continuous doses are needed to achieve any consistent effect because of the competition from the flora already colonized.

As the microflora seems to play a crucial role in the effect caused by the fructans, the concentration of a number of bacteria groups supposed to be affected by the ingestion of fructans was measured in the experimental study to see whether it was differently affected by different chain length and dose level. The fermentation of the inulin-type fructans is restricted to a limited number of bacteria genera colonizing the large intestine, such as bifidobacteria, which has the metabolic capacity to cleave the β(2-1)-glycosidic bonds.[38] Several studies in animal and man have shown a transient increase in the cecal and fecal number of bifidobacteria after the intake of short- and long-chain inulin-type fructans.[39–44] *In vitro* studies have shown that most enterobacteria, the *Bacteroides fragilis* group, peptostreptococci, and klebsiellae can ferment and grow on inulin-type fructans, although to a lesser extent.[45,46]

In the animal study currently under consideration,[8] it was difficult to interpret the microbial data due to large variations between the animals, the natural fluctuation during time, and the large buffer capacity of the large intestine, but some trends were deducted (data not shown). Overall, the short- and long-chain fructans seemed to affect the microflora quite similarly. A relationship between the intake of inulin-type fructans and an increase in bifidobacteria was only partly confirmed in our study, but was demonstrated earlier in rat and human studies.[39–42,44,47] A decrease in the enterococci present in both the cecal and fecal samples of the groups fed high levels of fructans is interesting. The results resemble findings in which a lower number of enterococci has been associated with a vegetarian-like diet rather than a western-type diet.[48]

A slight, but not significant, increase in *E. coli* seen in both fecal and cecal samples after the intake of fructans can be explained by their ability to ferment and grow on inulin-type fructans, as observed by Hartemink and Rombouts.[46] The general increase in cecal and fecal anaerobic bacteria and contemporaneous decrease in fecal aerobic bacteria observed in the groups given high amounts of fructans suggest a shift toward anaerobic species in the large intestine and are in accordance with previous observations.[42] In conclusion, fructans affect the microflora of the rats, an effect that seems to be independent of chain length.

VI. FERMENTATION, SCFA, AND pH

Owing to the configuration of their β-osidic bond, fructans resist hydrolysis by endogenous digestive enzymes (disaccharidases and α-amylase) in humans and animals, but these linkages are hydrolyzed to some extent under mild acidic conditions. The overall degradation of fructans in the small intestine is, however, relatively low.[5]

In an *in vitro* rat pancreatic or intestinal mucosa homogenates it was shown that the hydrolyzing activity toward a short-chain fructan (DP=3) in jejunal mucosa was only about 1/200 of that toward sucrose and about 1/900 of that toward maltose. The hydrolysis of a fructan with DP=4 was much less than that for the DP=3 fructan. Thus, fructans apparently become less digestible in the small intestine, the more fructose units the molecule contains.[49]

Nonstarch polysaccharides like fructans, reaching the cecum and colon, are fermentation substrates for the anaerobic microflora, which thereby generate SCFA and lactate. High SCFA concentrations are, therefore, seen after the administration of diets high in nonstarch polysaccharides.[42,50] The total amount and the profile of the SCFA can be altered by the consumption of different amounts and types of carbohydrates.[51–54] Of the SCFA generated, acetate, propionate, and butyrate account for approximately 90 to 95%.[52,55,56]

The production and presence of SCFA and lactate in the cecum and colon are supposed to be beneficial as they lower the pH, thereby creating a bacteriocidal environment for putative enteropathogens such as *E. coli* and *C. perfringens*.[57] In addition, SCFA are important substrates and intermediates in large intestinal cell metabolism.[58,59] Moreover, butyrate inhibits the growth of human colorectal tumor lines as well as other tumor cell lines *in vitro* as reviewed by Johnson.[60]

In the experimental animal study,[8] lower cecal pools and lower concentrations of propionate, butyrate (Figure 26.3), and total SCFA were seen in rats fed with 15% Raftiline compared to rats fed with 15% Raftilose. These levels reflect a lower degree of cecal fermentation of Raftiline and indicate that Raftilose is fermented more proximally. The observed increase in cecal pH of the groups given high amounts of fructans (data not shown) could be a consequence of the decreased concentration of SCFA. The increased relative colon wall weight seen in animals fed with Raftiline also indicated that Raftiline is fermented more distally than Raftilose. Further, the difference in fermentation pattern between Raftiline and Raftilose is reflected by the relative cecum wall weight, which tended to be higher in rats fed with 15% Raftilose compared to rats fed with 15% Raftiline (Figure 26.4). However, when evaluating the effect of fructans on the cecum and colon wall weight it must be taken into consideration that the wall weight consists mostly of muscles, and that the fructans are supposed to primarily affect the epithelium. The dose-dependent increase in cecal weight with content after the intake of Raftilose and Raftiline (Figure 26.4) probably reflects an increased amount of bacteria, undigested matter, and an increased water content due to osmotic activity. The observed increase in cecal and colon wall weight after the intake of fructans could be caused by a trophic effect of SCFA on the epithelial layer,[58,60,62] as the cecal pool of acetate, propionate, and butyrate in rats fed with fructans was markedly increased in a dose-dependent manner. However, as the measured number of cells per crypt profile is unchanged in colon, it is more likely that the increase in the colon wall weight reflects a general increase in the total area of the colon.

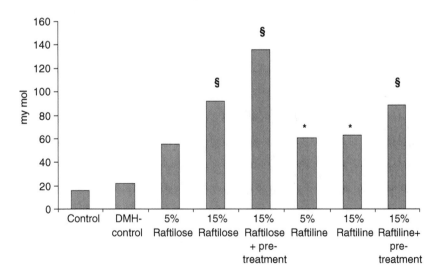

FIGURE 26.3 Total amount of butyrate in rat cecal content. Values are means calculated as weight of cecal content × cecal butyrate concentration. Statistical significance is as follows: *$P < 0.05$ and §$P < 0.001$ vs. DMH control. SE bars are not shown for clarity. (Adapted from Poulsen, M. et al. *Nutr. Cancer*, 42, 194–205, 2002. With permission.)

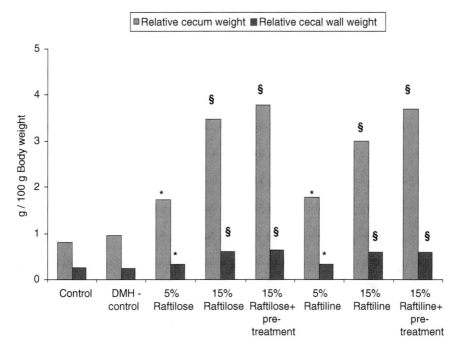

FIGURE 26.4 Relative cecum weight. Values are means. Statistical significance for each series is as follows: $^{*}P < 0.05$ vs. DMH control, $^{§}P < 0.01$ vs. *SE bars are not shown for clarity. (Adapted from Poulsen, M. et al. *Nutr. Cancer*, 42, 194–205, 2002. With permission.)

In conclusion, differences in chain length caused significant changes with regard to fermentation, SCFA, and pH.

VII. CELL PROLIFERATION

Proliferative epithelial cells occupy the basal regions of the crypts in colonic mucosa. Enhanced cell proliferation is generally supposed to increase the risk of colonic cancer development.[62–66] Measurement of cell proliferation at distinct locations in the colon is therefore a valuable parameter when evaluating the effect of dietary carbohydrates on the colonic environment. Known stimulators are the SCFA,[58,60,62] but certain luminal compounds such as calcium and deoxycholic acid may also influence colonic cell proliferation.[67] Nonstarch polysaccharides like fructans will most likely affect colonic cell proliferation as they provide substrates for colon that consequently lead to an increase in the production of SCFA.

In the animal study,[8] high levels of Raftilose or Raftiline for 10 weeks statistically and significantly decreased the proximal colon cell proliferation of the bottom and middle third of the crypt as well as the entire crypt when compared with control (data not shown). In contrast, the cell proliferation of the distal colon was not affected by fructan feeding. The total number of cells/crypt of neither the proximal nor the distal colon was affected (data not shown). The combination of decreased cell proliferation and the unchanged number of cells/crypt profile in the proximal colon observed in the study indicates a lower cell turnover and probably, therefore, a lower degree of apoptosis. However, Hughes and Rowland[68] showed a significantly higher number of apoptotic cells per colonic crypt after feeding rats with Raftiline and Raftilose diets compared with a control diet. To conclude, short- and long-chain fructans in the experimental animal study affect the colonic cell proliferation quite similarly. However, it must be emphasized that further studies including differentiation and apoptosis are needed to elucidate the fructan-induced changes in cell dynamics.

VIII. PRENEOPLASTIC CHANGES (ACF)

The most well-known and specific surrogate marker of colonic cancer used in short-term studies is the ACF.[69,70] ACF are morphological lesions in colonic mucosa and are recognized by their dilated irregular luminal openings, thicker epithelial lining, and pericryptal zone.[70] The frequency and multiplicity of ACF are increased by known colon carcinogens.[71] It is possible to measure ACF within weeks from the time of exposure, which makes it a good and feasible candidate as a biological biomarker of colorectal cancer. In general, ACF are believed to be true preneoplastic lesions; however, a clear relationship to the development of colon tumors does not always exist.[72,73] It is proposed that ACF with multiple crypts correlates best with tumor incidence.[74–76]

In experimental animals, inulin-type fructans have been found to inhibit the formation of chemically induced ACF, preneoplastic lesions in colon.[30,31,47,61] However, in the studies by Koo and Rao[61] and Gallaher and co-workers,[47] the reduction in ACF was observed only when the animals were given a short-chain fructan in combination with bifidobacteria. Rowland and co-workers[31] demonstrated a combination effect of a long-chain inulin-type fructan and bifidobacteria, but in addition showed that long-chain inulin-type fructan per se reduced the number of small ACF. Reddy and co-workers,[30] included both short- and long-chain inulin-type fructans in their study and found that both were capable of reducing the number of small ACF, with the longer chain ones being most efficient.

In the present rat study,[8] the long-chain fructan Raftiline was shown to be more efficient than the short-chain fructan Raftilose in reducing the total number of ACF as well as medium and small ACF (data not shown). The short-chain Raftilose, on the other hand, increased the number of medium and large ACF (Figure 26.5). The underlying mechanism could be that the longer chain Raftiline is fermented more distally, leading to a higher concentration of fermentation products where the ACF develop, and is thereby more efficient in modulating the ACF outcome. This hypothesis is supported by the observed differences in fermentation pattern, for example, cecal pool of SCFA, cecal pH, and colon and cecal wall weight. It is not clear which mechanisms are involved in promoting the effect of the short-chain fructan on the number of large ACF.

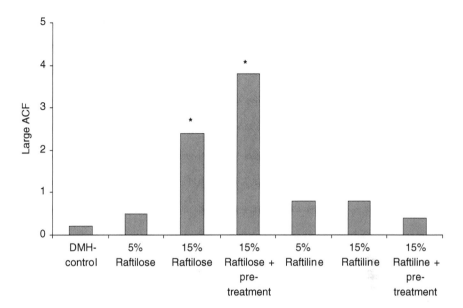

FIGURE 26.5   Number of large ACF in rats. Values are means. Large ACF > 7 crypts/focus. SE bars are not shown for clarity. Statistical significance is as follows: *P < 0.05 vs. DMH control. (Adapted from Poulsen, M. et al. *Nutr. Cancer*, 42, 194–205, 2002. With permission.)

IX. OTHER CLINICAL EFFECTS

In the experimental study,[8] animals given 15% Raftilose suffered from soft stool/diarrhea 3 to 5 weeks after the start of feeding, whereas feeding of 15% Raftiline caused soft stool/diarrhea for 1 to 2 weeks. No other diet-related clinical effects were seen in the study period.

The more severe and longer lasting diarrhea seen in the two groups given high levels of Raftilose correlates well with the higher osmotic value of the short-chain Raftilose. In fact, it was expected that Raftilose caused more diarrhea than Raftiline inasmuch as smaller molecules have a higher osmotic pressure and as slower fermented compounds are more easily tolerated than faster ones.[77] Similarly, Rumessen and Gudmand-Høyer[78] concluded, based on results from a human clinical study, that with the ingestion of fructans, abdominal symptoms increased with increase in dose and decrease in chain length. Soft stool and diarrhea have also been observed in other studies with fructans.[77,79]

The body weight of the rats in the experimental study[8] was negatively affected by the intake of fructans in a dose-dependent manner (Figure 26.6), which could be partly explained by the lower caloric density of fructans compared with sugar and starch. A caloric value of 4.2 to 9.5 kJ/g has been proposed.[7,80–82] Theoretically, the short-chain Raftilose and the long-chain Raftiline should have similar caloric values as both were found to be completely fermented in the large intestine of rats.[83] Rats fed with Raftilose, however, in general had a lower body weight gain than those fed with Raftiline, which

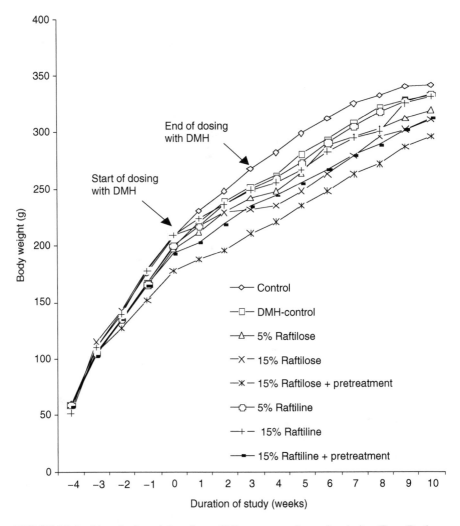

FIGURE 26.6 Mean body weights of rats. SE bars are not shown for clarity. (From Poulsen, M. et al. *Nutr. Cancer*, 42, 194–205, 2002. With permission.)

may imply that the severity of diarrhea negatively affected the body weight gain. Supporting this idea, the suppressing effect on body weight gain was most prominent during the period of diarrhea. The higher relative water intake seen in the two groups given high amounts of Raftilose could also be a consequence of the diarrhea.

X. CONCLUSIONS

The capability of nondigestible carbohydrates, such as the fructans, to give selective growth to an indigenous bacteria has given name to the prebiotic concept defined by Gibson and Roberfroid[84] as: "A prebiotic is a nondigestible food ingredient that beneficially affects the host by selectively stimulating the growth and/or activity of one or a limited number of bacteria in the colon, that can improve the host health." Fructans are in general believed to be prebiotics as most studies showed an increase in the number of bifidobacteria, which are supposed to support health. The microbiological data from the study by Poulsen and colleagues,[8] however, could not fully support the definition of fructans as prebiotics.

The purpose of this chapter was not only to focus on the effects of fructans on the microflora, but also to look at other fructan-mediated effects on the large intestine. Furthermore, the chapter concentrated on a biomarker, the ACF, that is indicative of a potential carcinogenic effect in the large intestine.

The results from the experimental animal study[8] showed that Raftiline was more efficient than Raftilose in inhibiting ACF development. In addition, with feeding Raftilose for 10 weeks increased the number of medium and large colonic ACF, whereas feeding with Raftiline either had no statistically significant effect or tended to decrease the number. The significance of ACF as a neoplastic marker is debatable, but it is generally accepted that large foci are more predictive of tumor development than smaller foci.[73–76]

The mechanism behind the ACF-progressing effect of Raftilose is not known, but as inulin-type fructans are incorporated in an increasing number of foodstuffs where it does not naturally occur, the effect especially of the shortchain Raftilose should be studied further. Data on SCFA and weight of cecum and colon indicate a difference in the site of fermentation between Raftilose and Raftiline, which most probably explains the opposite effect of the two fructans on ACF outcome. In conclusion, these observations underline the importance of chain length of otherwise similar compounds with regard to biological response toward the large intestine and overall health.

REFERENCES

1. Van Loo, J., Coussement, P., de Leenheer, L., Hoebregs, H., and Smits, G., On the presence of inulin and oligofructose as natural ingredients in the Western diet, *Crit. Rev. Food Sci. Nutr.*, 35, 525–552, 1995.

2. Moshfegh, A.J., Friday, J.E., Goldman, J.P., and Ahuja, J.K.C., Presence of inulin and oligofructose in the diets of Americans, *J. Nutr.*, 129,1407S–1411S, 1999.

3. Dysseler, P. and Hoffem, D., Inulin, an alternative dietary fibre. properties and quantitative analysis, *Eur. J. Clin. Nutr.*, 49, S145–S152, 1995.

4. ORAFTI, (1997) Product book of Raftilose® & Raftiline®, version 2.

5. Nordic Working group on Food Toxiology and Risk Evaluation, Safety evaluation of fructans. Nordic Council of Ministers, Copenhagen, 2000, TemaNord 2000:523, pp. 1–117.

6. Roberfroid, M. and Delzenne, N.M., Dietary fructans, *Annu. Rev. Nutr.*, 18, 117–143, 1998.

7. Roberfroid, M., Dietary fiber, inulin, and oligofructose: a review comparing their physiological effects, *Crit. Rev. Food Sci. Nutr.*, 33,103–148, 1993.

8. Poulsen, M., Mølck, A.M., and Jacobsen, B.L., Different effects of short- and long-chained fructans on large intestinal physiology and carcinogen-induced aberrant crypt foci in rats, *Nutr. Cancer*, 42, 194–205, 2002.

9. Frehner, M., Keller, F., and Wiemken, A., Localization of fructan metabolism in the vacuoles isolated from protoplasts of Jerusalem Artichoke tubers (*Heleanthus tuberosus* L.), *J. Plant Physiol.*, 116, 197–208, 1984.

10. Bancal, P. and Gaudillere, J.P., Oligofructan separation and quantification by high performance liquid chromatography, application to *Asparagus officinalis* and *Triticum aestivum*, *Plant Physiol. Biochem.*, 27, 745–750, 1989.

11. Hendry, G.A.F., Evolutionary origins and natural functions of fructans — a climatological, biogeographic and mechanistic appraisal, *New Phytol.*, 123, 3–14, 1993.

12. Pilon-Smits, E.A.H., Ebskamp, M.J.M., Paul, M.J., Jeuken, M.J.W., Weisbeek, P.J., and Smeekens, S.C.M., Improved performance of transgenic fructan–accumulating tobacco under drought stress, *Plant Physiol.*, 107, 125–130, 1995.

13. Nelson, C.J. and Smith, D., Fructans: their nature and occurrence, *Curr. Top. Plant. Biochem. Physiol.*, 5, 1–16, 1986.

14. Fuchs, A., Current and potential food and non-food applications of fructans, *Biochem. Soc. Trans.*, 19, 555–560, 1991.

15. de Leenheer, L. and Hoebregs, H., Progress in the elucidation of the composition of chicory inulin, *Starch*, 46, 193–196, 1994.

16. Pilon-Smits, E.A.H., Ebskamp, M.J.M., Jeuken, M.J.W., van der Meer, I.M., Visser, R.G.F., Weisbeek, P.J., and Smeekens, S.C.M., Microbial fructan production in transgenic potato plants and tubers, *Ind. Crops Prod.*, 5, 35–46, 1996.

17. Bornet, F.R.J., Undigestible sugars in food products, *Am. J. Clin. Nutr.*, 59S, 763S–769S, 1994.

18. Playne, M.J. and Crittenden, R., Commercially available oligosaccharides, Bull.IDF 313, 10–22, 1996.

19. Kaaks, R. and Riboli, E., Colorectal cancer and intake of dietary fibre, a summary of the epidemiological evidence, *Eu. J. Clin. Nutr.*, 49, S10–S17, 1995.

20. Kritchevsky, D., Epidemiology of fibre, resistant starch and colorectal cancer, *Eu. J. Cancer Prev.*, 4, 345–352, 1995.

21. Hill, M.J., Cereals, cereal fibre and colorectal cancer risk: a review of the epidemiological literature, *Eu. J. Cancer Prev.*, 6, 219–225, 1997.

22. Scheppach, W., Bingham, S., Boutron-Ruault, M.C., Gerhardsson de Verdier, M., Moreno, V., Nagengast, F.M., Reifen, R., Riboli, E., Seitz, H.K., and Wahrendorf, J., WHO consensus statement on the role of nutrition in colorectal cancer, *Eu. J. Cancer Prev.*, 8, 57–62, 1999.

23. Fuchs, C.S., Giovannucci, E.L., Golditz, G.A., Hunter, D.J., Stampfer, M.J., Rosner, B., Speizer, F.E., and Willett, W.C., Dietary fiber and the risk of colorectal cancer and adenoma in women, *J. Med.*, 340, 169–176, 1999.

24. Terry, P., Giovannucci, E., Michels, K.B., Bergkvist, L., Hansen, H., Holmberg, L., and Wolk, A., Fruit, vegetables, dietary fiber, and risk of colorectal cancer, *J. Natl. Cancer Inst.*, 93, 525–533, 2001.

25. Klurfield, D.M., Insoluble dietary fiber and experimental colon cancer — are we asking the proper questions? in *Dietary fiber: Chemistry, physiology, and health effects*, Kritchevsky, D., Bonfield, C., and Anderson, J.W., Eds., Plenum Press, New York, 1990, pp. 403–415.

26. Harris, P.J. and Ferguson, L.R., Dietary fibres may protect or enhance carcinogenesis, *Mutat. Res.*, 443, 95–110, 1999.

27. Jacobs, L.R., Influence of soluble fibers on experimental colon carcinogenesis, in *Dietary fiber: Chemistry, physiology, and health effects*, Kritchevsky, D., Bonfield, C., and Anderson, J.W., Eds., Plenum Press, New York, 1990, pp. 389–401.

28. Bingham, S.A., Epidemiology and mechanisms relating diet to risk of colorectal cancer, *Nutr. Res. Rev.*, 9, 197–239, 1996.

29. Lupton, J.R., Is fiber protective against colon cancer? where the research is leading us, *Nutrition*, 16, 558–561, 2000.
30. Reddy, B.S., Hamid, R., and Rao, C.V., Effect of dietary oligofructose and inulin on colonic preneoplastic aberrant crypt foci inhibition, *Carcinogenesis*, 18, 1371–1374, 1997.
31. Rowland, I.R., Rumney, C.J., Coutts, J.T., and Lievense, L.C., Effect of *Bifidobacterium longum* and inulin on gut bacterial metabolism and carcinogen-induced aberrant crypt foci in rats, *Carcinogenesis*, 19, 281–285, 1998.
32. Bird, R.P., McLellan, E.A., and Bruce, W.R., Aberrant crypts, putative precancerous lesions, in the study of the role of diet in the aetiology of colon cancer, *Cancer Surv.*, 8, 189–200, 1989.
33. Pretlow, T.P., Barrow, B.J., Ashton, W.S., O'Riordan, M.A., and Pretlow, T.G., Aberrant crypts: putative preneoplastic foci in human colonic mucosa, *Cancer Res.*, 51, 1564–1567, 1991.
34. Roncucci, L., Medici, M., Medline, A., Cullen, J.B., and Bruce, W.R., Identification and quantification of aberrant crypt foci and microadenomas in the human colon, *Hum. Pathol.*, 22, 287–294, 1991.
35. Roncucci, L., Medline, A., and Bruce, W.R., Classification of aberrant crypt foci and microadenomas in human colon, *Cancer Epidemiol. Biomarkers Prev.*, 1, 57–60, 1991.
36. Thorup, I., Histomorphological and immunohistochemical characterization of colonic aberrant crypt foci in rats: relationship to growth factor expression, *Carcinogenesis*, 18, 465–472, 1997.
37. Gibson, G.R., Willems, A., Reading, S., and Collins, M.D., Fermentation of non-digestible oligosaccharides by human colonic bacteria, *Proc. Nutr. Soc.*, 55, 899–912, 1996.
38. McKellar, R.C. and Modler, H.W., Metabolism of fructo-oligosaccharides by *Bifidobacterium* spp., *Appl. Microbiol. Biotechnol.*, 31, 537–541, 1989.
39. Gibson, G.R., Beatty, E.R., Wang, X., and Cummings, J.H., Selective stimulation of bifidobacteria in the human colon by oligofructose and inulin, *Gastroenterology*, 108, 975–982, 1995.
40. Bouhnik, Y.B., Flourié, M., Riottot, M., Bisetti, N., and Gailing, M.F., Effects of fructo-oligosaccharides ingestion on fecal bifidobacteria and selected metabolic indexes of colon carcinogenesis in healthy humans, *Nutr. Cancer*, 26, 21–29, 1996.
41. Buddington, R.K., Williams, C.H., Chen, S.C., and Witherly, S.A., Dietary supplement of neosugar alters the fecal flora and decreases activities of some reductive enzymes in human subjects, *Am. J. Clin. Nutr.*, 63, 709–716, 1996.
42. Campbell, J.M., Fahey, G.C., and Wolf, B.W., Selected indigestible oligosaccharides affect large bowel mass, cecal and fecal short-chain fatty acids, pH and microflora in rats, *J. Nutr.*, 127, 130–136, 1997.
43. Roberfroid, M., Van Loo, J., and Gibson, G.R., The bifidogenic nature of chicory inulin and its hydrolysis products, *J. Nutr.*, 128, 11–19, 1997.
44. Kruse, H.P., Kleessen, B., and Blaut, M., Effects of inulin on faecal bifidobacteria in human subjects, *Br. J. Nutr.*, 82, 375–382, 1999.
45. Mitsuoka, T., Hidaka, H., and Eida, T., Effect of fructo-oligosaccharides on intestinal microflora, *Die Nahr.*, 31, 427–436, 1987.
46. Hartemink, R. and Rombouts, F.M., Gas formation from oligosaccharides by the intestinal microflora, in *Non-digestible oligosaccharides: healthy food for the colon*, Hartemink, R., Ed., VLAG/WIAS, Wageningen, the Netherlands, 1997, pp. 57–66.
47. Gallaher, D.D., Stallings, W.H., Blessing, L.L., Busta, F.F., and Brady, L.J., Probiotics, cecal microflora, and aberrant crypts in the rat colon, *J. Nutr.*, 126, 1362–1371, 1996.

48.   Finegold, S.M., Sutter, V.L., Sugihara, P.T., Elder, H.A., Lehmann, S.M., and Phillips, R.L., Fecal microbial flora in Seventh Day Adventist populations and control subjects, *Am. J. Clin. Nutr.*, 30, 1781–1792, 1977.

49.   Oku, T., Tokunaga, T., and Hosoya, N., Nondigestibility of a new sweetener, "Neosugar", in the rat, *J. Nutr.*, 114, 1574–1581, 1984.

50.   Younes, H., Garleb, K., Behr, S., Rémésy, C., and Demigne, C., Fermentable fibers or oligosaccharides reduce urinary nitrogen excretion by increasing urea disposal in the rat cecum, *J. Nutr.*, 125, 1010–1016, 1995.

51.   Berggren, A.M., Björck, I.M.E., Nyman, E.M.G.L., and Eggum, B.O., Short-chain fatty acid content and pH in caecum of rats given various sources of carbohydrates, *J. Sci. Food Agric.*, 63, 397–406, 1993.

52.   Mortensen, P.B. and Nordgaard, I., Production of short-chain fatty acids from dietary fibre in the human large bowel, in *COST Action 92—Dietary fibre and fermentation in the colon*, Mälkki, Y. and Cummings, J.H., Eds., EC, Brussels, Belgium, 1996, pp. 130–139.

53.   Nyman, M., Berggren, A., Björck, I., and Eggum, B.O., Fermentation of dietary fibre and short-chain fatty acid production in rats, in *COST Action 92—Dietary fibre and fermentation in the colon*, Mälkki Y., and Cummings, J.H., Eds., EC, Brussels, Belgium, 1996, pp. 167–170.

54.   Salminen, S., Bouley, C., Boutron-Ruault, M.C., Cummings, J.H., Franck, A., Gibson, G.R., Isolauri, E., Moreau, M.C., Roberfroid, M., and Rowland, I., Functional food science and gastrointestinal physiology and function, *Br. J. Nutr.*, 80, S147–S171, 1998.

55.   Nyman, M. and Asp, N.G., Dietary fibre fermentation in the rat intestinal tract: effect of adaptation period, protein and fibre levels, and particle size, *Br. J. Nutr.*, 54, 635–643, 1985.

56.   Clausen, M.R., Butyrate and colorectal cancer in animals and in humans—mini–symposium: butyrate and colorectal cancer, *Eu. J. Cancer Prev.*, 4, 483–490, 1995.

57.   Gibson, G.R. and Wang, X., Regulatory effects of bifidobacteria on the growth of other colonic bacteria, *J. Appl. Bacteriol.*, 77, 412–420, 1994.

58.   Sakata, T., Stimulatory effect of short-chain fatty acids on epithelial cell proliferation in the rat intestine: a possible explanation for trophic effects of fermentable fibre, gut microbes and luminal trophic factors, *Br. J. Nutr.*, 58, 95–103, 1987.

59.   Sakata, T., Influence of short chain fatty acids on intestinal growth and functions, in Dietary fiber in health and disease, Kritchevsky, D. and Bonfeld, C., Eds., *Advances in experimental medicine and biology* 427, Plenum Press, New York, 1997, pp. 191–199.

60.   Johnson, I.T., Butyrate and markers of neoplastic change in the colon, *Eu. J. Cancer Prev.*, 4, 365–371, 1995.

61.   Koo, M. and Rao, A.V., Long-term effect of bifidobacteria and neosugar on precursor lesions of colonic cancer in CF1 mice, *Nutr. Cancer*, 16, 249–257, 1991.

62.   Zhang, J. and Lupton, J.R., Dietary fibers stimulate colonic cell proliferation by different mechanisms at different sites, *Nutr. Cancer*, 22, 267–276, 1994.

63.   Jacobs, L.R., Effect of dietary fiber on colonic cell proliferation and its relationship to colon carcinogenesis, *Prev. Med.*, 16, 566–571, 1987.

64.   Pretlow, T.P., Alterations associated with early neoplasia in the colon, in *Biochemical and Molecular Aspects of Selected Cancers*, Pretlow, T.G. and Pretlow, T.P., Eds., Academic Press, New York, 1994, pp. 93–141.

65.   Moore, M.A. and Tsuda, H., Chronically elevated proliferation as a risk factor for neoplasia, *Eu. J. Cancer Prev.*, 7, 353–385, 1998.

66. Akedo, I., Ishikawa, H., Ioka, T., Kaji, I., Narahara, H., Ishiguro, S., Suzuki, T., and Otani, T., Evaluation of epithelial cell proliferation rate in normal-appearing colonic mucosa as a high-risk marker for colorectal cancer, *Cancer Epidemiol. Biomarkers Prev.*, 10, 925–930, 2001.

67. Bartram, H.P., Scheppach, W., Schmid, H., Hofmann, A., Dusel, G., Richter, F., Richter, A., and Kasper, H., Proliferation of human colonic mucosa as an intermediate biomarker of carcinogenesis: effects of butyrate, deoxycholate, calcium, ammonia, and pH, *Cancer. Res.*, 53, 3283–3288, 1993.

68. Hughes, R. and Rowland, I.R., Stimulation of apoptosis by two prebiotic chicory fructans in the rat colon, *Carcinogenesis*, 22, 43–47, 2001.

69. Bird, R.P., Observation and quantification of aberrant crypts in the murine colon treated with a colon carcinogen: preliminary findings, *Cancer Lett.*, 37, 147–151, 1987.

70. Bird, R.P. and Good, C.K., The significance of aberrant crypt foci in understanding the pathogenesis of colon cancer, *Toxicol. Lett.*, 112–113, 395–402, 2000.

71. Kristiansen, E., The Application of a Short-Term Test for Detection of Modifying Effects of Dietary Factors in Rodent Colon Carcinogenesis, Ph.D. thesis, Danish Veterinary and Food Administration, Søborg, Denmark, 1997.

72. Kristiansen, E., Thorup, I., and Meyer, O., Influence of different diets on development of DMH-induced aberrant crypt foci and colon tumor incidence in Wistar rats, *Nutr. Cancer*, 23, 151–159, 1995.

73. Thorup, I., Meyer, O., and Kristiansen, E., Effect of potato starch, cornstarch and sucrose on aberrant crypt foci in rats exposed to azoxymethane, *Anticancer Res.*, 15, 2101–2106, 1995.

74. Pretlow, T.P., O'Riordan, M.A., Somich, G.A., Amini, S.B., and Pretlow, T.G., Aberrant crypts correlate with tumor incidence in F344 rats treated with azoxymethane and phytate, *Carcinogenesis*, 13, 1509–1512, 1992.

75. Zhang, X.M., Stamp, D., Minkin, S., Medline, A., and Corpet, D.E., Promotion of aberrant crypt foci and cancer in rat colon by thermolyzed protein, *J. Natl. Cancer Inst.*, 84, 1026–1030, 1992.

76. Magnuson, B.A., Carr, I., and Bird, R.P., Ability of aberrant crypt foci characteristics to predict colonic tumor incidence in rats fed cholic acid, *Cancer Res.*, 53, 4499–4504, 1993.

77. Carabin, I.G. and Flamm, W.G., Evaluation of safety of inulin and oligofructose as dietary fiber, *Regul. Toxicol. Pharm.*, 30, 268–282, 1999.

78. Rumessen, J.J. and Gudmand-Høyer, E., Fructans of chicory: intestinal transport and fermentation of different chain lengths and relation to fructose and sorbitol malabsorption, *Am. J. Clin. Nutr.*, 68, 357–364, 1998.

79. Tokunaga, T., Oku, T., and Hosoya, N., Influence of chronic intake of new sweetener fructooligosaccharide (Neosugar) on growth and gastrointestinal function of the rat, *J. Nutr. Sci. Vitaminol.*, 32, 111–121, 1986.

80. Hosoya, N., Dhorranintra, B., and Hidaka, H., Utilization of [U-^{14}C] fructooligosaccharides in man as energy resources, *J. Clin. Biochem. Nutr.*, 5, 67–74, 1988.

81. Roberfroid, M., Gibson, G.R., and Delzenne, N., The biochemistry of oligofructose, a nondigestible fiber: an approach to calculate its caloric value, *Nutr. Rev.*, 51, 137–146, 1993.

82. Molis, C., Flourié, B., Ouarne, F., Gailing, M.F., and Lartigue, S., Digestion, excretion, and energy value of fructooligosaccharides in healthy humans, *Am. J. Clin. Nutr.*, 64, 324–328, 1996.

83.   Nilsson, U. and Björck, I., Availability of cereal fructans and inulin in the rat intestinal tract, *J. Nutr.*, 118, 1482–1486, 1988.
84.   Gibson, G.R. and Roberfroid, M., Dietary modulation of the human colonic microbiota: introducing the concept of prebiotics, *J. Nutr.*, 125, 1401–1412, 1995.

27 Alginate as a Drug Delivery Carrier

Rajesh Pandey and G.K. Khuller

CONTENTS

I. INTRODUCTION

The use of drugs to alleviate human suffering is perhaps as old as disease itself. With an increase in the knowledge about the pathophysiology of various diseases, drug designing has flourished not only as a science, but also as an art. Drugs are almost never administered to a patient in an unformulated state. A "drug dosage formulation" consists of one or more active ingredients along with other molecules called excipients. Although often not realized, the use of excipients is as important as the drug itself. Excipients facilitate drug preparation and administration, enhance the

799

consistent release of the drug, and protect it from degradation. Thus, excipients, once considered to be inert substances, can potentially influence the rate and extent of drug absorption and thus determine the bioavailability of the drug. The term bioavailability implies the amount of drug available to the systemic circulation out of the total drug administered to a patient. Bioavailability is an important consideration in pharmaceutical dosage forms because it is only when the drug is in the systemic circulation that it can reach its target site to exert its therapeutic effect. The formulation of a drug so as to extract its maximum therapeutic benefit forms the concept of a drug delivery system (Figure 27.1).

There are four "D's" in a drug delivery system — drug, destination, disease, and delivery. The last is the only variable parameter. When a drug formulation is designed in such a way that the rate or the place of drug release is altered, it is called

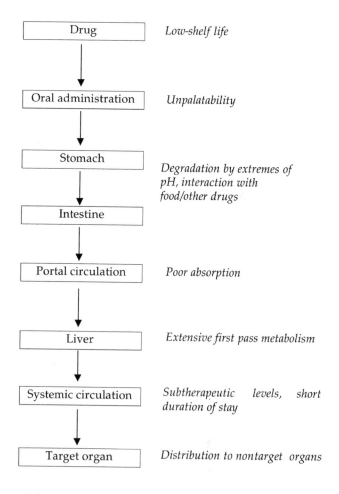

FIGURE 27.1   The variables (shown in italics) influencing the systemic bioavailability of a drug that are amenable to improvement using an appropriate delivery system.

a modified release system. Alternative terms include sustained release, controlled release, pulsatile release, slow release, extended release, prolonged release, and so on. Modified release is generally achieved by means of encapsulation. Encapsulation technology finds its application in several industries, including pharmaceuticals for controlled release of drugs. Polymers are extensively used, both as conventional excipients and more specifically as a tool in controlled drug delivery.[1] The drug release profile can be "tuned" depending on the choice of polymer (Table 27.1). One of the most versatile natural biopolymers is alginic acid.

II. ALGINIC ACID AND ALGINATE

The extraction of brown seaweed *(Phaeophyceae,* mainly *Laminaria)* with alkali and its subsequent treatment with mineral acids yields alginic acid, first reported by the British chemist E. C. Stanford at the end of 19th century. The large-scale production of alginate was introduced 50 years later. Alginic acid can be converted into various salts, for example, sodium alginate. A few bacteria, for example, *Azotobacter* and *Pseudomonas* species, are also capable of synthesizing alginate in an *O*-acetylated form. Alginate is the most abundant marine biopolymer and the second most abundant biopolymer (next to cellulose) in the world. The polymer can be obtained economically in an ultra-pure form and may be prepared in neutral or charged forms, which makes it compatible with a broad range of substances.[2,3]

Alginate is a family of unbranched polysaccharides composed of the epimers α-L-guluronic acid (G) and β-D-mannuronic acid (M) residues, arranged in homopolymeric blocks of each type (MM, GG) and also in heteropolymeric blocks (MG) (Figure 27.2). The M/G composition differs not only between one species of brown algae to another, but also between different parts of the same plant. The alginate

TABLE 27.1
Types of Polymers Used in Drug Delivery Systems

A.	Natural Polymers	B.	Synthetic Polymers
1.	Proteins and polypeptides	1.	Aliphatic polyesters and hydroxy acids
	Albumin		Polylactic acid[a]
	Fibrinogen/fibrin		Polyglycolic acid
	Collagen		Poly(lactide-co-glycolide)[a]
	Gelatin		Polyhydroxybutyric acid
	Casein		Polycaprolactone
2.	Polysaccharides		
	Alginic acid[a]	2.	Polyanhydrides
	Starch	3.	Polyorthoesters
	Dextran/dextrins	4.	Polyalkylcyanoacrylate
	Hyaluronic acid	5.	Polyamino acids
	Chitin	6.	Polyacrylamides
	Chitosan[a]	7.	Polyalkylcarbonates
3.	Cells and viruses		

[a] Most commonly used polymers.

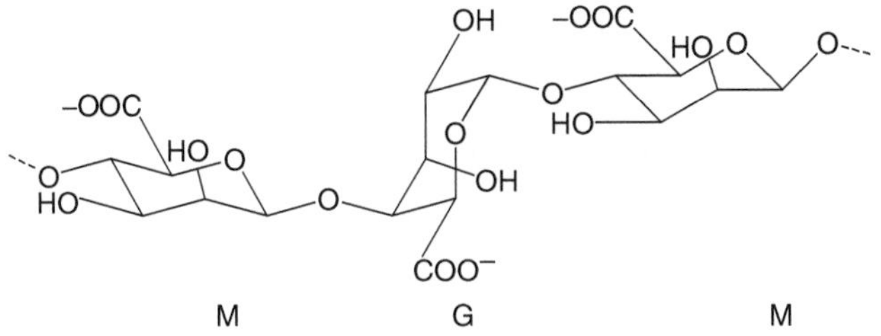

FIGURE 27.2 Molecular structure of alginate. M, β-D-mannuronic acid unit; G, α-L-guluronic acid unit.

monomer composition has a major impact on the drug release property.[4] Alginates with a high M content are best suited for thickening applications whereas those with a high G content are best for gelation. "Designer" alginates are currently being developed, involving the 5-epimerization of β-(1→4)–linked M residues to α-(1→4)-linked G residues in algal alginates using bacterial epimerases. With divalent cations (e.g., Ca^{2+}) alginates form strong gels characterized by strength and flexibility. The M/G ratio is an important consideration in the process — as the ratio decreases, the requirement of Ca^{2+} for cross-linking increases. The cations participate in the interchain binding between the G blocks to produce a three-dimensional network, often called an egg-box model (Figure 27.3). The gel network and homogeneity depend on the cation concentration. With an excess of $Ca,^{2+}$ a modified egg box having multiple alginate chains in the gelling zone may be formed with different physicochemical properties.[5]

Alginate gels with a high G content exhibit high porosity, low shrinkage during gelation, and do not swell after drying. When the M content increases, the gels become softer and more elastic, shrink, porosity reduces, and they swell after drying. The gelation occurs by the formation of egg-box junctions that associate the metal ions with the GG block of the alginate chain. The guluronic acid conformation leaves suitable distances between the COO^- and OH^- groups that allow a high degree of coordination with Ca^{2+} (Figure 27.4). A high G content ensures the formation of more rigid gels that are less prone to erosion so that the drug release is slower.[6]

As compared with neutral macromolecules, alginic acid possesses unique features because of its ability to form two types of gels depending upon the pH of the surrounding medium. At low pH, hydration of alginic acid leads to the formation of a high-viscosity acid gel due to intermolecular binding. The water molecules are trapped inside the alginate matrix but are free to migrate. This phenomenon finds its applications in cell immobilization and encapsulation. Divalent /multivalent cations, except Mg^{2+} form ionotropic hydrogels. The ability to form two types of gels leads to a large variation in physicochemical properties, so that alginate is suitable for granule formation, lyophilization, and direct compression into tablets. It is nontoxic, stable at room temperature, and sterilization may be performed by filtration or specific heat treatment.[7,8] The recently discovered alginate lyases, either mannuronate or guluronate lyases, catalyze the degradation of alginate. Thus, novel alginate polymers may be engineered for diverse applications.[9]

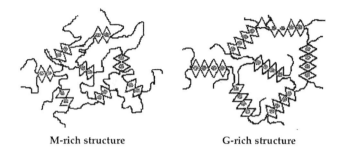

M-rich structure   G-rich structure

FIGURE 27.3   The egg-box model showing the binding zone between G blocks of alginate. Continuous lines represent M residues while boxes indicate cross-linked G-residues.

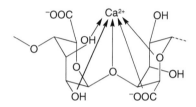

FIGURE 27.4   The mode of binding of calcium ions to guluronic acid units of alginate.

III. ALGINATE-BASED DRUG DELIVERY SYSTEMS

Alginate is already in clinical use for the supportive treatment of reflux esophagitis. It has found applications as a binding and disintegrating agent in tablets, a suspending and thickening agent in water-miscible gels/lotions/creams, and as a stabilizer for emulsions. Several attributes make alginate an ideal drug-delivery vehicle.[8,10] These features include a relatively high aqueous environment within the matrix, adhesive interactions with intestinal epithelium, a mild room temperature drug(s) encapsulation process free of organic solvents, a high gel porosity allowing high diffusion rates of macromolecules, the ability to control this porosity with simple coating procedures using polycations, such as chitosan or poly-L-lysine (PLL), and dissolution/biodegradation of the system under normal physiological conditions. Hence, it is not surprising that alginate has been used as a carrier for the controlled release of numerous molecules, for example, indomethacin,[11] sodium diclofenac,[12] nicardipine,[4] dicoumarol,[13] gentamicin,[14] vitamin C,[15] ketoconazole,[15] amoxycillin,[16] and antitubercular drugs.[17,18] There are two broad types of alginate-based drug delivery systems — the membrane system and the matrix system. In the membrane–reservoir system, the drug release from the inner reservoir core is controlled by the polymeric encapsulating membrane, which has a specific permeability. As the thickness of the coat/membrane increases, the release rate decreases. Moreover, the coencapsulation of certain nonpolar substances may further reduce the release rates. This property was advantageously used in the encapsulation and controlled release of indomethacin, where the sudden release of the drug is highly undesirable because it is well known that indomethacin is irritant to the gastrointestinal mucosa.[11] On the other hand, in the matrix system or more specifically, the swelling – dissolution – erosion system, the drug molecules are dispersed in a rate-controlled polymer matrix. The matrix swelling as well as dissolution/erosion occurring concomitantly at the matrix periphery are the factors that modulate drug release.[19]

IV. FACTORS INFLUENCING DRUG DELIVERY

Broadly speaking, factors related to the development of a particular formulation as well as factors encountered once the formulation is inside a living system can influence a drug delivery system to a great extent, as discussed below and summarized in Table 27.2.

A. THE pH OF SURROUNDING MEDIUM

The drug release mechanism from alginate beads or microspheres is strongly pH-dependent. For instance, in the stomach (acidic pH), the "swelling" phenomenon does not occur and the drugs are released by diffusion through the insoluble matrix. This process results in shrinkage of the alginate beads. In the intestine (neutral-alkaline pH), the swelling and erosion phenomena are markedly enhanced. Hence, alginate is undoubtedly an *environmentally responsive polymer*.

B. THE COMPOSITION OF ALGINATE

The drug release process is retarded when the alginates are rich in their G content, because the high degree of coordination between guluronic acid and Ca^{2+} forms more rigid gels that are less prone to swelling/erosion. The reverse is the case for alginates rich in their M content, where the gels are softer and more easily soluble. However, the situation is not the same if the drug molecule itself reacts with the mannuronic acid residues so that despite a high M content, the release of the drug is retarded, as occurs in the case of gentamicin sulphate.[14] The molecular weight and viscosity of alginate are also important variables affecting drug release. It has been shown that for low-molecular-weight alginate, the release rate of pindolol was minimum. In general, the higher the viscosity of alginate, the slower is the drug release rate.[8] The drug–polymer ratio is a critical consideration while designing an alginate-based drug delivery system with

TABLE 27.2
Factors Influencing Drug Delivery from Alginate-Based Systems

pH of the surrounding medium

The proportion of G and M residues in alginate

Molecular weight and viscosity of alginate

Drug-polymer ratio

Ionic nature of the drug

Nature and amount of cross-linker

Gelling time

Variation in microsphere size

Addition of regulatory molecules

predictable drug release kinetics. The release of nicardipine was shown to be slower from alginate particles having a 1:1 drug polymer ratio rather than from those with a 1:2 ratio.[4] In our studies with antitubercular drugs (ATDs), a 1:1 drug polymer ratio was as good as 1:2–5 ratios in terms of drug encapsulation and sustained release. In other words, increasing the amount of alginate did not confer any additional advantage to the formulation. The ionic nature of the encapsulated drug also determines the extent of drug loading as well as its release profile. Because of the overall negative charge of the alginate matrix, it is natural that cationic drugs (e.g., lidocaine hydrochloride) will be released more slowly as compared with anionic ones (e.g., sodium salicylate).

C. The Cross-Linker

Because of the important role of Ca^{2+} in the cross-linking process, a high Ca^{2+} content is desirable for the formation of stable beads.[4,20] The duration required for the formation of the appropriate number of cross-links to stabilize the alginate microspheres is called gelling/curing time, and it varies from drug to drug. In general, a long gelling time is favorable for bead maturation. Calcium alginate beads are more popular than formulations that contain Ba^{2+} or Sr^{2+} as the cross-linker because the former exhibit better sustained release kinetics. The gel strength decreases in the presence of metals in the following order: $Pb^{2+} > Cu^{2+} = Ba^{2+} > Sr^{2+} > Cd^{2+} > Ca^{2+} > Zn^{2+} > Co^{2+} > Ni^{2+}$. However, for the immobilization of living cells in alginate, toxicity is a critical factor and only Sr^{2+}, Ba^{2+} and Ca^{2+} are considered safe for these purposes.

D. Variation in Microsphere Size

If the microspheres of nonuniform size are mixed in a single formulation, a pulsatile drug release pattern may be obtained, as shown for dextran. This release pattern could be desirable if one intends to follow the biological circadian rhythm.[8]

E. Addition of Substances that Modulate the Properties of Alginate

The loading of the hydrophilic drug nimesulide increased in the presence of glutaraldehyde as an additional cross-linker. The presence of the copolymers ethyl cellulose and hydroxy propyl cellulose is also helpful in reducing drug release rates. However, the most successful of the copolymers used with alginate is chitosan. During the preparation of alginate beads, the latter are coated with chitosan by ionic interaction. The coating suppresses the gel matrix erosion and retards the release of encapsulated compounds, as observed for sodium diclofenac.[20] The drug release behavior of alginate – chitosan beads is again pH-dependent.[21] Chitosan is sparingly water soluble but protonation of the amine groups improves its solubility in the stomach (acidic pH). At alkaline pH of the intestine, the interpolymeric complex between alginate and chitosan swells and slowly disintegrates to release the drugs. The release rate is proportional to the degree of cross-linking between the two polymers. However, there is a certain limit up to which chitosan may be used with alginate while successfully maintaining the physicochemical characteristics of alginate. Our work has demonstrated that higher chitosan/alginate ratios drastically alter the morphology of the microspheres, and a 1:1 ratio is optimum as far as maintaining particle integrity *vis-à-vis* a satisfactory loading of ATD into the alginate–chitosan microspheres is concerned. In fact, increasing the amount of chitosan resulted in clumping of the microspheres, besides an unacceptable 20 to 25% reduction in drug loading (R. Pandey and G.K. Kuller, unpublished results).

Other nonpolymeric carbohydrates and lipids are also capable of modulating the release of drugs from alginate-based systems. The encapsulation of acetaminophen in alginate particles prepared along with lactose led to a faster dissolution of lactose (compared with alginate) in the intestine and site-specific colonic delivery of the drug.[8] A combination of alginate with liposomes (composed of a lipid shell surrounding an aqueous core) is an interesting "hybrid" drug delivery system that capitalizes on the advantages of the two systems, i.e., alginate and liposomes (Figure 27.5). The method involves suspending liposomes (encapsulating the drug of interest) in a solution of alginate and adding the mixture dropwise into a calcium chloride solution. The microspheres are recovered by filtration and may be coated with a polycation in order to achieve further stability. Our studies have shown that rifampicin, a highly hydrophobic drug, attains a high degree of encapsulation in liposomes and that, in turn, the liposomes may be coated with alginate. The formulation offers the possibility of oral administration of liposomes and, indeed, of maintaining sustained drug levels in the plasma for 3 to 5 days following a single oral administration to guinea pigs (R. Pandey and G.K. Kuller, unpublished results).

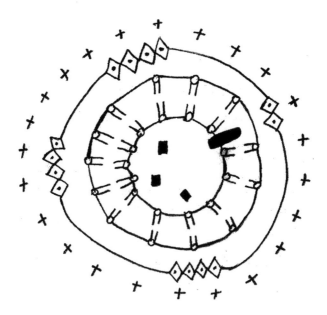

FIGURE 27.5 Schematic outline of a hybrid drug delivery system having three functional layers. *Innermost layer:* Phospholipid bilayer of a liposome encapsulating hydrophilic drug molecules (solid squares) in the aqueous core and hydrophobic drug (solid bar) in the lipid compartment. *Middle layer:* Alginate with G and M residues as depicted in Figure 27.3. *Outermost layer:* Poly-L-lysine or chitosan bearing multiple positive charges (+).

V. APPROACHES TO IMPROVE DRUG BIOAVAILABILITY

When a drug is administered through the intravenous (i.v.) route, all the drug molecules are instantly available to the systemic circulation. The i.v. route is an ideal therapeutic system where there are no chances of drug loss, compared with the oral route of drug administration, where several factors may be responsible for the loss of part of the administered dose, with the result that it never accesses the systemic circulation (Figure 27.1). This loss may be trivial or substantial. At best, the subcutaneous and intramuscular routes of drug administration are said to attain bioavailability profiles close to the i.v. route. Unfortunately, idealized systems are often not acceptable because of the pain and discomfort associated with parenteral routes of drug dosing and the requirement for medical expertise. Thus, the oral route has been, and will certainly continue to be, the route of choice for the administration of therapeutic agents. Drug delivery technologists are challenged with the awesome task of refining and improving on the fallacies of the conventional oral drug delivery system.

Enhancing the gastrointestinal residence time of drugs is an interesting approach toward improving the oral bioavailability of drugs. It can be achieved by three means — floating system, mucoadhesive system, and nanoparticulate system.

A. FLOATING OR GASTRORETENTIVE DRUG DELIVERY SYSTEM

This method entails the mixing of a drug (usually basic) with alginate and a pH-independent hydrocolloid, for example, hydroxypropyl methyl cellulose (HPMC, a gelling agent), all filled into a gelatin capsule. In the stomach, as water enters through the capsule, surface hydration of HPMC results in gelification and the trapping of air inside the powder bulk, thus accounting for the buoyant or floating nature of the capsule (Figure 27.6). At low pH, alginate exists as alginic acid, which further

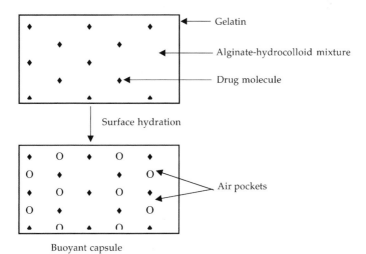

FIGURE 27.6 A schematic representation of the principle of a gastroretentive drug delivery system.

modifies the gel. The capsule is retained in the stomach for quite some time, and as gel erosion proceeds, the drug gradually diffuses out. Following the loss of buoyancy, the formulation enters the small intestine and encounters a higher pH.[22] Alginic acid turns into a more soluble alginate, the gel becomes more porous, and the drug readily diffuses out, compensating for its low dissolution at the higher pH. Floating liquid alginate preparations are already in clinical use to suppress gastroesophageal reflux and alleviate the symptoms of heartburn. The preparation consists of alginate and sodium bicarbonate, so that at acidic pH, alginic acid gel is formed and CO_2 is evolved. The gel becomes buoyant by trapping gas bubbles, and floats on gastric contents as a viscous layer, having a higher pH than the gastric contents.

B. MUCOADHESIVE OR BIOADHESIVE SYSTEM

The mucus lining of the digestive tract is rich in mucin, which contains an oligosaccharide chain with terminal sialic acid. Polyanions, especially polymers bearing carboxylic groups and a high charge density, serve as powerful "ligands" for mucin and are called mucoadhesive polymers, for example, alginate.[13] Bioadhesion offers several advantages: (a) longer gastrointestinal residence time improves drug absorption and bioavailability, (b) higher drug concentration at the site of adhesion–absorption creates a driving force for passive paracellular drug uptake, (c) fewer chances of drug dilution or degradation in luminal fluids, and (d) enhancement of topical action of certain drugs.

The cation-cross-linked alginate network can be degraded by the removal of Ca^{2+} ions by chelating agents, namely, lactate, citrate, or phosphate. Although these ions are practically nonexistent in human intestinal fluid, they can be introduced through the diet. Removal of Ca^{2+} reduces the cross-linking and destabilizes the gel, resulting in rapid drug release. Alginate can be made even more resistant in the presence of Ca^{2+} chelators by forming strong complexes with polycations, especially chitosan. Besides acting as a stabilizer, chitosan has other functions — it reduces the porosity of the alginate gel beads, and improves the bioavailability of the drugs because of its own mucoadhesive nature and its ability to modulate the intestinal tight junctions.[21] Alternatives to chitosan include PLL, polyarginine, polyornithine, spermine, and spermidine. The calcium alginate–PLL is the most promising encapsulation system yet developed. The better stabilizing properties of PLL over chitosan result from the numerous long-chain alkylamino groups extending from the polyamide backbone in various directions in PLL. These chains interact electrostatically with various alginate molecules, forming a highly cross-linked structure. Chitosan, on the other hand, has amino groups that are very close to the polysaccharide backbone. The interaction between positively charged amino groups of chitosan and negatively charged carboxylate groups of alginate is lessened due to steric repulsion between the two molecules. Nevertheless, the high cost of PLL prohibits its use on a large scale. Thus, the alginate-chitosan system is an economic and reliable microencapsulation system. Moreover, stabilizing molecules may be used to form more than one layer over alginate to achieve optimum drug release kinetics.[23]

C. ALGINATE NANOPARTICLES

Although it is well known that divalent cations induce the gelification of alginate, a critical adjustment in the relative proportions of Ca^{2+}/alginate allows the cation-induced rearrangement of the alginate molecules to form microdomains with high local concentrations of alginate instead of an infinite network of polymer. These microdomains are representative of a *pregel* state, that is, alginate nanoparticles, which can be recovered by high-speed centrifugation.[5] Particles in the nanosize range are advantageous over the more traditional microparticles/microspheres because nanoparticles are capable of being absorbed intact from the intestine via transcellular or paracellular pathways, thereby substantially improving the bioavailability of encapsulated drugs.[24] Nanoparticles achieve a higher drug encapsulation efficiency. Further, nanoparticles are suitable for i.v. administration, as has been shown for doxorubicin-loaded alginate nanoparticles. Alginate exhibits satisfactory hemocompatibility and being hydrophilic, avoids rapid clearance by the mononuclear-phagocyte system on i.v. administration. This imparts a long circulation half-life to alginate nanoparticles. The system is certainly advantageous over the more traditional neutral polymers or liposomes that require the additional incorporation of hydrophilic copolymers or polyethylene glycol fatty acid derivatives to enhance their hydrophilicity.[25] In our laboratory, we have prepared alginate nanoparticles stabilized with PLL/chitosan and encapsulating ATDs. The system is undergoing further evalution.

VI. ENCAPSULATION OF ANTITUBERCULAR DRUGS IN ALGINATE MICROSPHERES

Tuberculosis (TB) continues to be a leading cause of mortality attributable to any single infectious agent. The causative bacterium, *Mycobacterium tuberculosis*, causes two million deaths annually. The front line ATDs against TB include rifampicin, isoniazid, pyrazinamide, and ethambutol. Based on these drugs, an effective chemotherapeutic regimen is available for TB treatment; however, the fact that multiple ATDs need to be administered daily for a minimum period of 6 months makes TB treatment look perfect on the prescription but poor in practice. Patient noncompliance is a vexing problem that is responsible not only for treatment failure but also for the emergence of multidrug resistant cases. Dose-related side-effects, especially hepatitis, may also add to the worries of the physician. Ironically, except for long-acting rifamycins, no new ATD has been introduced of late. Therefore, formulating the already existing ATDs into a controlled release system is a desirable strategy to curtail the dose as well as dosing frequency and aid in improving patient compliance. The feasibility of such a system has been documented recently in our laboratory.[18] The method entails the encapsulation of ATDs during the cation-induced gelation of alginate, followed by the recovery of microspheres by filtration. The microspheres, on drying, measured 90 to 100 μm in diameter, encapsulating approximately 20 to 30% of isoniazid/pyrazinamide and a variable amount of rifampicin (40–70%). On oral administration to guinea pigs, a sustained release was observed in the plasma for 3 to 4 days based on which the chemotherapeutic efficacy was successfully evaluated in *M. tuberculosis*-infected guinea pigs, following once-a-week treatment schedule for 8 weeks. In order to improve further the drug encapsulation and pharmacokinetics of the formulation, a few critical modifications were introduced such as maintaining a constant drug/polymer ratio, addition of chitosan as a stabilizer, and reducing the gelling time. This process resulted in a reduction in microsphere size (65 to 75 μm in diameter), encapsulating approximately 85% of the initial amount of rifampicin and 70% of isoniazid/pyrazinamide.The microspheres, encapsulating ATDs at therapeutic doses, were first evaluated in laca mice by the oral route. The plasma profile, organ distribution, and key pharmacokinetic parameters including bioavailability (with respect to oral free drugs) were calculated. It was observed that all the drugs could be detected in the plasma for 7 days and in the organs (lung, liver, and spleen) for 9 days with a striking improvement in the pharmacokinetic parameters. In particular, the relative bioavailability was increased by 29 to 39-fold for all the three drugs. The study was repeated in other animal species to rule out the possibility of interspecies variation but similar results were obtained in rats, guinea pigs, and rabbits as well. Next, whether a reduction in drug doses would also be able to demonstrate the sustained release behavior or not was considered. The answer was a firm "yes"; indeed, drugs could be detected in the plasma for 5 days and in the organs for 7 days following a single oral dose of alginate-encapsulated ATDs at half the therapeutic dose. The key findings in guinea pigs are shown in Table 27.3 (unpublished results[26]). It was also critical to note that the formulation was nontoxic even at a very high dose tested in mice as opposed to an equivalent dose of free drugs, which proved to be lethal. It was observed that 150 times the therapeutic dose of drug-loaded microspheres did not produce any adverse effects, whereas the same dose of free drugs resulted in 100% mortality within 24 h.

Once the pharmacokinetics and safety of the microspheres were established, the chemotherapeutic efficacy studies were carried out in *M. tuberculosis*-infected guinea pigs, because TB progression in guinea pigs largely resembles human pathology. As predicted by the pharmacokinetic data, a schedule of microsphere administration every 10 days (therapeutic dose) as well as every 7 days (half therapeutic dose) was followed. Free drugs administered daily (as per the conventional chemotherapy) served as one of the control groups. As evident from Figure 27.7, five doses of microspheres (therapeutic dose) administered every 10 days or seven doses of microspheres (half therapeutic dose) administered every 7 days were as efficacious as 46 doses of free drugs (unpublished results[26]). The study clearly showed the feasibility of reducing the dosing frequency as well as the dose in TB chemotherapy. The implication of ATDs encapsulated in the alginate–chitosan microspheres is obvious — a substantial improvement in patient compliance would be possible besides simplifying the TB treatment schedule, pending the discovery of new and more potent drugs.

TABLE 27.3

In vivo **Biodistribution Profile of Alginate–Chitosan-Microsphere-Encapsulated Antitubercular Drugs in Guinea Pigs**

Parameter	Free Drugs (Therapeutic Dose)	Alginate–Chitosan Microspheres (Therapeutic Dose)	Alginate–Chitosan Microspheres (Half Therapeutic dose)
Duration of drug(s) stay in plasma (h)	12	168	120
Duration of drug(s) stay in tissues (h)	24	216	168
Relative bioavailability	1	17–19	20–30

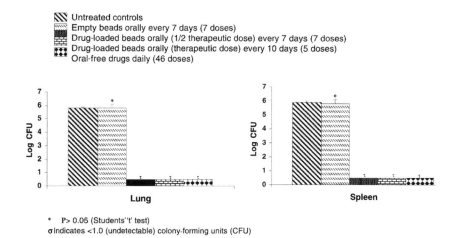

NN Untreated controls
Empty beads orally every 7 days (7 doses)
Drug-loaded beads orally (1/2 therapeutic dose) every 7 days (7 doses)
Drug-loaded beads orally (therapeutic dose) every 10 days (5 doses)
Oral-free drugs daily (46 doses)

*   P> 0.05 (Students' 't' test)
σ Indicates <1.0 (undetectable) colony-forming units (CFU)

FIGURE 27.7   Chemotherapeutic efficacy of alginate–chitosan-encapsulated antitubercular drugs against experimental tuberculosis in guinea pigs.

VII.   ENCAPSULATION OF PROTEINS, DNA, AND CELLS IN ALGINATE

Peptides and proteins are increasingly being recognized to be an important class of therapeutic agents. Although the parenteral route is commonly employed for their administration, several of its associated demerits such as thrombophlebitis, tissue necrosis, and poor patient compliance have stimulated the investigation of alternative nonparenteral routes, preferably oral. The latter route has its own problems, such as protein degradation owing to enzymatic proteolysis, extremes of pH, and an unfavorable partition coefficient to cross lipid bilayers. Natural polymer hydrogels such as alginate circumvent these problems because the proteins, once inside the shelter of the alginate matrix, are shielded from the harsh environment. Further, the encapsulation process itself being mild, avoiding exposure to elevated temperature and organic solvents along with the high inertness of the alginate toward proteins ensures that no loss of biological activity of the protein occurs.[27,28] However, cationic proteins may potentially compete with Ca^{2+} for carboxylic acid sites on alginate, resulting in a reduced diffusion rate or even protein inactivation. Incorporation of additives, for example, polyacrylic acid, protects the active ingredients from alginate. Formulations containing alginate have been used for the magnetically triggered delivery of insulin. The critical factors were the magnetic field and mechanical properties of the polymer matrix. Less rigid matrices showed higher release rates.[29]

Alginate gels have a remarkable capacity to retain low-molecular-weight oligonucleotides in the 24 to 36 kDa (40–60 bp) range. Alginate with a high G content impedes the release of oligonucleotides in this range, suggesting that a DNA–alginate interaction exists that hinders diffusion. However, the retention of <24 kDa DNA is also possible. The 5′ -tethering of DNA to alginate through a C-12 spacer effectively protects DNA during the attachment procedure, and the process may find application in identifying genes through target hybridization.[30]

The use of polymers for the microencapsulation of cells has been a challenge confronting biotechnologists. Cell encapsulation has distinct advantages over conventional suspension cultures, achieving higher cell densities, improved product recovery, and protection of cells from hydrodynamic shear forces resulting from agitation and aeration. The alginate/PLL gel encapsulation technique has been the most acceptable owing to its mild experimental conditions involving "cell-friendly" components at a pH, osmolarity, and temperature suitable for preserving cell viability.[31,32] The bioadhesive nature of alginate may make it useful as a delivery vehicle for biomolecules to nongastrointestinal mucosal tissue as well. Thus, the attachment of lectins to the surface of spermine-modified alginate beads allows for adhesion to the epithelium rather than the mucosa. This is important for developing oral/intranasal/respiratory vaccines. The encapsulation of proteins/DNA/cells in alginate aims at better management of chronic diseases such as diabetes mellitus, Parkinson's disease, cystic fibrosis, cancer, etc.[3,8] The examples in the following sections illustrate how alginate is being explored for encapsulating proteins of diverse clinical interest.

A. ENCAPSULATION OF VASCULAR ENDOTHELIAL GROWTH FACTOR

The formation of a blood vessel can occur by vasculogenesis or angiogenesis. Vasculogenesis is the formation of a primitive vascular plexus from embryonic precursor cells. The formation of additional vessels after vasculogenesis is called angiogenesis, and it is accomplished by sprouting from a preexisting vessel. Angiogenesis is the essential process in wound healing and tissue regeneration. Manipulation of angiogenesis using proangiogenic gene therapy has emerged as an alternative approach to replace the conventional surgical procedures in treating ischemic limb and myocardial diseases. Such applications have been tested in clinical trials to bypass the atherosclerotic vessels or to improve the cardiac and peripheral circulatory functions by establishing collateral arteries. Since all mammalian tissues require blood supply, angiogenesis is also a crucial element in the development of artificial organs, orthopedic scaffolds, and wound-dressing materials. Vascular endothelial growth factor (VEGF), a 45 kDa glycoprotein, is one of the most potent proangiogenic factors. Alginate is currently being evaluated for the encapsulation and controlled release of VEGF.[30]

B. ENCAPSULATION OF ANDROGEN-PRODUCING CELLS

Currently available androgen-replacement therapies include the oral administration of testosterone tablets/capsules, depot injections, sublingual treatment, and skin patches. However, side-effects such as the metabolic inactivation of testosterone, fluctuations in levels of the hormone, burning, rash, and skin necrosis may occur. These side-effects may be avoided through the application of encapsulated Leydig cells, which produce testosterone. Studies have shown that Leydig cells encapsulated in alginate/PLL/alginate microspheres are capable of secreting testosterone in culture and *in vivo*. Microencapsulated Leydig cells delivered intraperitoneally into castrated rats maintained a testosterone level of 0.51 ng/mL for >3 months. Similarly, the encapsulation of ovarian cells for the secretion of progesterone and estrogen is also being evaluated.[33]

C. ENCAPSULATION OF SCHWANN CELLS

Schwann cell (SC) transplantation has been proposed to encourage peripheral nerve regeneration, but an optimal SC-carrying matrix is urgently needed. The SC has been encapsulated in alginate and the addition of fibronectin (a protein known to help in nerve regeneration) to alginate hydrogel improved SC viability and growth profile. The nerve regeneration rate was enhanced and there was an additive effect when both SC and fibronectin were combined with alginate.[34]

VIII. SUMMARY

The field of drug designing entails not only the development of new and more potent drugs but also the formulation of already existing drugs into suitable delivery systems. In order to improve the bioavailability of drugs, various polymers have been researched as drug delivery vehicles, one of which is alginate. The attraction for this natural polymer stems from its easy availability in an economic fashion, its compatibility with neutral as well as charged molecules, the simplicity of the drug(s) encapsulation procedure, mucoadhesion, biodegradability under physiological conditions, lack of toxicity, and the ability to confer a sustained-release behavior. Besides the potential to formulate gastroretentive systems, nanoparticles as well as 'hybrid' systems are the additional virtues due to which alginate is fast becoming the polymer of choice in drug delivery (Table 27.4). The controlled-release phenomenon has been documented for a vast array of drugs ranging from vasodilators, antihypertensives, and bronchodilators, to antibiotics. In particular, the ability of alginate to co-encapsulate multiple ATDs and offer a controlled release profile is likely to have a major impact on enhancing patient compliance for a better management of tuberculosis. Further, the encapsulation of peptides/DNA/cells in alginate is also likely to eliminate some of the staggering problems that face biotechnologists. One can expect that in the years to come, alginate-based systems will play a major role in the practice of medicine.

TABLE 27.4
Alginate: The Polymer of Choice in Drug Delivery

A natural polymer
Large-scale production in an economic fashion
Compatible with a broad range of substances
Simple drug(s) encapsulation procedure
Mucoadhesive
Biodegradable and nontoxic
Formulation of different delivery systems
Sustained drug release
Enhanced bioavailability of drugs
Applications in biotechnology

REFERENCES

1. Dutt, M. and Khuller, G.K., Therapeutic efficacy of poly (DL-lactide-co-glycolide)-encapsulated antitubercular drugs against *M. tuberculosis* infection induced in mice, *Antimicrob. Agents Chemother.*, 45, 363–366, 2000.
2. Smidsrod, O. and Draget, K.I., Chemistry and physical properties of alginates, *Carbohydr. Eur.*, 14, 6–13, 1996.
3. Skjak-Braek, G. and Espevik, T., Application of alginate gels in biotechnology and biomedicine, *Carbohydr. Eur.*, 14, 19–25, 1996.
4. Takka, S. and Acarturk, F., Calcium alginate microparticles for oral administration: I: effect of sodium alginate type on drug release and drug entrapment efficiency, *J. Microencapsulation*, 16, 275–290, 1999.
5. Rajaonarivony, M., Vauthier, C., Couarraze, G., Puisieux, F., and Couvreur, P., Development of a new drug carrier made from alginate, *J. Pharm. Sci.*, 82, 912–917, 1993.
6. Ostberg, T., Vesterhus, L., and Graffner, C., Calcium alginate matrices for oral multiple unit administration: II. Effect of process and formulation factors on matrix properties, *Int. J. Pharm.*, 97, 183–193, 1993.
7. You, J.-O., Park, S.-B., Park, H.Y., Haam, S., Chung, C.H., and Kim, W.S., Preparation of regular sized Ca-alginate microspheres using membrane emulsification method, *J. Microencapsulation*, 18, 521–532, 2001.
8. Tonnesen, H.H. and Karlsen, J., Alginate in drug delivery systems, *Drug Dev. Ind. Pharm.*, 28, 621–630, 2002.
9. Wong, T.Y., Preston, L.A., and Schiller, N.L., Alginate lyase: review of major sources and enzyme characteristics, structure-function analysis, biological roles, and applications, *Annu. Rev. Microbiol.*, 54, 289–340, 2000.
10. Raj, N.K. and Sharma, C.P., Oral insulin — a perspective, *J. Biomater. Appl.*, 17, 183–196, 2003.
11. Joseph, I. and Venkataram, S., Indomethacin sustained release from alginate-gelatin or pectin-gelatin coacervates, *Int. J. Pharm.*, 126, 161–168, 1995.
12. Gonzalez-Rodriguez, M.L., Holgado, M.A., Sanchez-Lafuente, C., Rabasco, A.M., and Fini, A., Alginate-chitosan particulate systems for sodium diclofenac release, *Int. J. Pharm.*, 232, 225–234, 2002.
13. Chickering III, D.E., Jacob, J.S., Desai, T.A., Harrison, M., Harris, W.P., Morrell, C.N., Chaturvedi, P., and Mathiowitz, E., Bioadhesive microspheres:III. An *in vivo* transit and bioavailability study of drug loaded alginate and poly (fumaric-co-sebacic anhydride) microspheres, *J. Control Release*, 48, 35–46, 1997.
14. Lannuccelli, V., Coppi, G., and Cameroni, R., Biodegradable intraoperative system for bone infection treatment. I. The drug/polymer interaction, *Int. J. Pharm.*, 143, 195–201, 1996.
15. Cui, J.-H., Goh, J.-S., Park, S.-Y., Kim, P.-H., and Lee, B.-J., Preparation and physical characterization of alginate microparticles using air atomization method, *Drug Dev. Ind. Pharm.*, 27, 309–319, 2001.
16. Whitehead, L., Collett, J.H., and Fell, J.T., Amoxycillin release from a floating dosage form based on alginates, *Int. J. Pharm.*, 210, 45–49, 2000.
17. Lucinda-Silva, R.M. and Evangelista, R.C., Microspheres of alginate-chitosan containing isoniazid, *J. Microencapsulation*, 20, 145–152, 2003.
18. Ain, Q., Sharma, S., Khuller, G.K., and Garg, S.K., Alginate based oral drug delivery system for tuberculosis: pharmacokinetics and therapeutic effects, *J. Antimicrob. Chemother.*, 51, 931–938, 2003.

19. Takka, S., Ocak, O.H., and Acarturk, F., Formulation and investigation of nicardipine HCl-alginate gel beads with factorial design-based studies, *Eur. J. Pharm. Sci.*, 6, 241–246, 1998.

20. Fernandez-Hervas, M.J., Holgado, M.A., Fini, A., and Fell, J.T., *In vitro* evaluation of alginate beads of a diclofenac salt, *Int. J. Pharm.*, 163, 23–34, 1998.

21. Hejazi, R. and Amiji, M., Chitosan-based gastrointestinal delivery system, *J. Control Release*, 89, 151–165, 2003.

22. Singh, B.N. and Kim, K.H., Floating drug delivery systems: an approach to oral controlled drug delivery via gastric retention, *J. Control Release*, 63, 235–259, 2000.

23. Kumar, M.N.V.R., Nano and microparticles as controlled drug delivery devices, *J. Pharm. Sci.*, 3, 234–258, 2000.

24. Jiao, Y., Ubrich, N., Marchand-Arvier, M., Vigneron, C., Hoffman, M., Lecompte, T., and Maincent, P., *In vitro* and *in vivo* evaluation of oral heparin-loaded polymeric nanoparticles in rabbits, *Circulation*, 105, 230–235, 2002.

25. Labana, S., Pandey, R., Sharma, S., and Khuller, G.K., Chemotherapeutic activity against murine tuberculosis of once weekly administered drugs (isoniazid and rifampicin) encapsulated in liposomes, *Int. J. Antimicrob. Agents*, 20, 301–304, 2002.

26. Rajesh Pandey and Khuller, G.K., Chemotherapeutic potential of alginate chitosan microspheres as antitubercular drug carriers. *J. Antimicrob. Chemother.* 53, 635–640.

27. Gombotz, W.R., and Wee, S.F., Protein release from alginate matrices, *Adv. Drug Del. Rev.*, 31, 267–285, 1998.

28. Onal, S. and Zihnioglu, F., Encapsulation of insulin in chitosan-coated alginate beads: oral therapeutic peptide delivery, *Artif. Cells Blood Substit. Immobil. Biotechnol.*, 30, 229–237, 2002.

29. Saslawski, O., Weingarten, C., Benoit, J.P., and Couvreur, P., Magnetically responsive microspheres for the pulsed delivery of insulin, *Life Sci.*, 42, 1521–1528, 1988.

30. http://www.chemeng.queensu.ca/research/biochem/bioencapsulation/myweb4/ Research.htm.

31. Read, T.-A., Sorensen, O.R., Mahesparan, R., Enger, P.O., Timpl, R., Olsen, B.R., Hjelstuen, M.H.B., Haraldseth, O., and Sjerkvig, R., Local endostatin treatment of gliomas administered by microencapsulated producer cells, *Nat. Biotechnol.*, 19, 28–29, 2001.

32. Gugerli, R., Cantana, E., Heinzen, C., Stocker, U.V., and Marison, I.W., Quantitative study of the production and properties of alginate/poly-L-Lysine microcapsules, *J. Microencapsulation*, 19, 571–590, 2002.

33. Machluf, M., Orsola, A., and Atala, A., Controlled release of therapeutic agents: slow delivery and encapsulation, *World J. Urol.*, 18, 80–83, 2000.

34. Mosahebi, A., Wiberg, M., and Terenghi, G., Addition of fibronectin to alginate matrix improves peripheral nerve regeneration in tissue engineered conduits, *Tissue Eng.*, 9, 209–218, 2003.

28 Chitosan-Based Nonviral Vectors for Gene Delivery

Wen Guang Liu, William W. Lu, and Kang De Yao

CONTENTS

I. INTRODUCTION

Conceptually, gene therapy involves the introduction of an extraneous gene into a cell with the aim of tackling genetic diseases, slowing the progression of tumors, and fighting viral infections.[1] The success of gene therapy lies in achieving an efficient gene delivery into the target cells or tissue to a large extent without causing any vector-associated toxicity. To achieve this objective, various viral and nonviral vectors have been developed so far. Although viral vectors such as adenovirus, retroviral, and adeno-associated virus, which have been disabled of any pathogenic effects are relatively more efficient in gene transfection than nonviral counterparts, their oncogenic effects and immunogenicity remain an Achilles' heel.[2,3] Gelsinger's death from a gene therapy clinical trial in 1999, and the fact that two children developed leukemia after receiving gene therapy for several combined immunodeficiency disease prompted a detailed look at the safety issues of the viral vectors[4,5] and spurred a renewed interest in nonviral approaches.[6–9] Among the nonviral vectors currently being investigated, liposomes,[10,11] cationic polyelectrolytes such as polyethyleneimine,[12] poly(L-lysine),[13] dextran–spermine conjugate,[14] amphiphilic dendrimers,[15] poly(β-aminoester)s,[16]

N-isopropylacrylamide-ethyleneimine block copolymers,[17] gelatin,[18] carbohydrate-modified peptides,[19] have gained importance because polyelectrolyte complexes (PECs) can be formed in large quantities by simple electrostatic interactions between DNA and polycations. Additional advantages of this approach are that the entrapped DNA is shielded from contact with DNase and the loading capacity is not restricted by DNA size.[9] However, some intrinsic drawbacks with PECs, such as solubility, cytotoxicity, adsorption of serum proteins and blood cells, and low transfection efficiency have become a bottleneck to their applications *in vivo*.[9]

Many researchers have proposed methods to overcome the biological barriers related to the nonviral vector transgene systems; these barriers occur at both extracellular and intracellular levels as outlined by Brown and co-workers[20] and depicted in Figure 28.1. As shown in the diagram, to realize the ultimate gene delivery in nucleus, the exogenous gene must overcome multiple blockages. It is obvious that the ability of a vector to transport genes into targeted cells is the premise of improving gene transfer efficiency.

In the past decades, researchers found that chitosan excels in enhancing the transport of drugs across the mucosal epithelium although biocompatibility, biodegradability, and toxicity issues remained to be fully resolved.[21] Most recently, chitosan has been expanded to the field of gene delivery and many encouraging results have been published. This chapter is devoted to reviewing recent achievements in the study of the mechanisms of chitosan-mediated transcellular gene transport as well as considering the specific features of chitosan and its derivatives that provide this polysaccharide with favorable properties to serve as a nonviral vector for shuttling genes into cells.

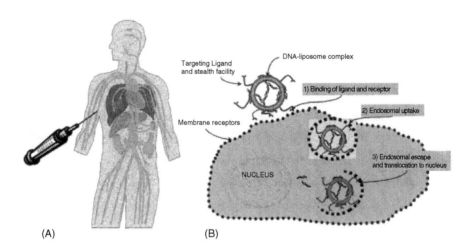

(A)   (B)

FIGURE 28.1 The barriers to nonviral vector-mediated gene delivery: (A) Extracellular Barriers: Degradation of DNA in plasma, Uptake of DNA by reticuloendothelial system, Inability to target DNA to specific organs, Largely ineffective via the oral route — except, for immunization, Transfection inhibited by mucus. (B) Intracellular Barriers: Endosomal escape of DNA, Lysosomal degradation of DNA, Cytoplasmic stability of DNA, Translocation of DNA to the nucleus.

II.   SOURCE OF CHITOSAN[22]

Chitosan[22] [α(1→4) 2-amino-2-deoxy β-D-glucan] is derived by the deacetylation of chitin, which is the second most universally abundant biopolymer existing in exoskeletons of crustaceans. Its chemical structure is illustrated in Figure 28.2. The chemical properties of chitosan are determined by its molecular weight and degree of deacetylation. Chitosan has an apparent pKa of 6.5. It aggregates in solutions at pH values>6. It is therefore only soluble in acidic solutions (pH=1 to 6), where most of the amino groups are protonated. Deacetylation of chitin is performed by boiling chitin from crab and shrimp shells in sodium hydroxide, and sometimes decolorized with oxides. Chitosan can be digested by lysozymes and chitinases secreted by intestinal microorganisms and also present in plant ingredients of food. The number of amino groups present and the molecular weight of chitosan both affect its biodegradability.[23] Because of its low production costs, biodegradability, biocompatibility, and recent FDA approval, recent years have witnessed a remarkable increase in the food and pharmaceutical applications of chitosan, which include use as a food preservative and for nasal, ocular, oral, parenteral, and transdermal drug delivery. Recently, its application in gene delivery has attracted increasing attention.[24–28]

III.   TRANSMEMBRANE CHARACTERISTICS

Interest in exploiting chitosan for gene delivery was sparked by observations that this compound could facilitate the transport of various drugs across cellular membranes. For instance, in one study chitosan was demonstrated to promote the nasal absorption of insulin in rats and sheep and to enhance the paracellular transport of peptides *in vitro* and *in vivo* by opening the tight junctions.[29] Holme and co-workers[30] used the transepithelial electrical resistance (TEER) to investigate the effect of chitosan with various molecular weights and degrees of deacetylation on the permeability of human intestinal epithelial cells (Caco-2). It was found that chitosan with a high degree of deacetylation and degree of polymerization (DP) >50 induced the greatest effect on the opening of tight junctions of cells. Chitosan has also been shown to bind mammalian and microbial cells by interacting with surface glycoproteins, and some studies have indicated that chitosan may actually be endocytosed into the cell. For example, chitosan microspheres were taken up by murine melanoma B16F10 cells

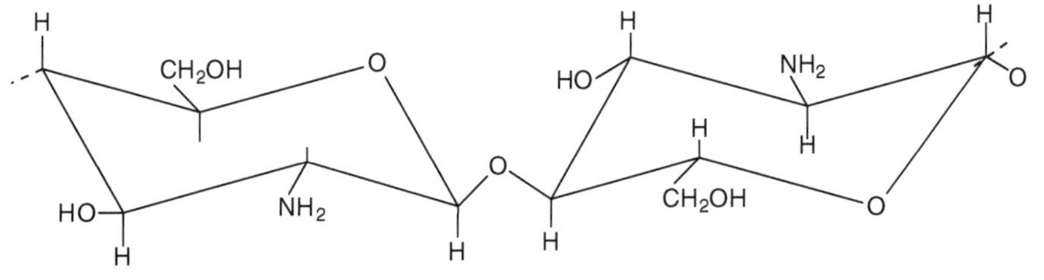

FIGURE 28.2   Chemical structure of chitosan.

via phagocytosis.[31] In repeated adhesion studies, chitosan has proved to have superior mucoadhesive properties in comparison with polycarbophiles.

Recent studies have shown that only the protonated form of soluble chitosan, i.e., the uncoiled configuration, can trigger the opening of the tight junctions and thereby facilitate the paracellular transport of hydrophilic compounds.[32] This property implies that chitosan would be effective as an absorption enhancer only in a limited area of the intestinal lumen, where the pH values are close to its pK_a (e.g., proximal duodenum). However, the results of an investigation on the interaction between low-molecular-weight (LMW) chitosan (M_w=4000, deacetylation degree=90%) with a lipid bilayer suggested that LMW chitosan could destabilize the lipid bilayer at neutral pH. Yang and colleagues[33] reckoned that the high solubility of LMW chitosan compensated for the relatively weak electrostatic and hydrophobic effect. Therefore, the transmembrane ability of chitosan is dependent upon the molecular weight, deacetylation degree and pH of the medium, and aqueous soluble LMW chitosans are preferentially selected as carriers in view of their nontoxicity and nonhemolysis. However, it should be pointed out that the standard with respect to molecular weight and deacetylation degree of chitosan for biomedical applications has not been fully established so far.

Leong's group[34] investigated the permeation mechanism of chitosan across dipalmitoyl-*sn*-glycerol-3-phosphocholine (DPPC) membrane bilayer. Differential scanning calorimetry (DSC) was used to elucidate the thermotropic behavior of DPPC–chitosan mixtures. Cross-polarization microscopy was applied to determine the structural features of multilamellar vesicles (MLV). The effect of hydrophobic driving force on the chitosan–DPPC interaction was also addressed. Their evidence showed that chitosan induced the fusion of multilamellar vesicles, and the attractive interchain and intermolecular forces of the hydrophobic core (acyl chains) in the DPPC bilayer were significantly reduced by chitosan–membrane interactions. The addition of chitosan also reduced the order in the two-dimensional packing of the acyl chains and increased the fluidity of DPPC bilayer. The study provided a valuable insight into the mechanism of chitosan-induced perturbation of a model membrane.

Berth and co-workers[35] have recently determined the radius of gyration of chitosan. Their study established the relationship between the (M_w) and radius of gyration (R_g) of chitosan in aqueous solution, and further revealed that chitosan behaved more like a Gaussian coil instead of the worm-like chain model found in common polyelectrolytes. At low pH, the primary amine along the backbone of chitosan is fully protonated. Therefore, the size of chitosan and pH are two important parameters that dictate its permeabilizing and perturbating effects on the cell membrane. In their subsequent work, Leong's group[36] investigated the effect of molecular weight and pH on the interactions of phospholipid bilayer with chitosan. Reduction of pH increased the number of protonated amines on the chitosan backbone and caused further disruption of the membrane organization. It was found that the cooperative unit of chitosan was significantly reduced with the increase in chitosan mole fraction. At a chitosan mole fraction of 0.04%, the increase in molecular weight from 113 to 213 kDa resulted in a dramatic reduction of cooperative unit from 155 to 43. Berth and co-workers[35] showed that R_g values of 113 and 213 kDa chitosan were 47 and 54 nm, respectively, in aqueous solution. Once the chitosan mole fraction was raised beyond 0.23%, the cooperative unit of 113 and 213 kDa chitosan reached a steady state of 52 and 24, respectively. The significant decrease in cooperative unit against chitosan molecular weight implied that chitosan swirled across the bilayer.

Chitosan becomes a polycation when its primary amines are protonated at a pH equal to its pK_a. At the same time, the presence of N-acetyl groups on the chitosan backbone imparts the polymer with hydrophobic properties. As a polyelectrolyte, chitosan tends to aggregate in aqueous solution. Recently, Philippova and co-workers[37] observed two types of hydrophobic aggregates in aqueous solutions of chitosan and its hydrophobically modified (HM) derivative. They proposed an aggregation model related to hydrophobic domains typical for different associating polymers with hydrophobic side chains and hydrophobic domains inherent to chitosan itself. It is noted that the substitution degree of chitosan used by Philippova is 88%, i.e., a large portion of the monosaccharide monomeric units of chitosan still exist with hydrophobic acetyl groups that can compete with electrostatic repulsion and lead to hydrophobic aggregates. Whereas our recent work revealed that chitosan with a degree of deacetylation of 99% did not aggregate in acetic solution,[38] we consider that for almost completely deacetylated chitosan, the hydrophobicity of the remaining acetyl groups can be neglected, and electrostatic repulsion weakens the hydrophobic intermolecular attraction to a certain degree thereby hindering the aggregation of chitosan itself.

In our previous publication,[39] we synthesized alkylated chitosans (ACSs) including butyl, octyl, and dodecyl chitosans and thermodynamically investigated the interactions of ACSs with DPPC membranes. The variations in enthalpies of the gel–crystalline transition in DPPC bilayer caused by the addition of ACSs demonstrated that DPPC membranes became more evidently perturbed upon elongating the side chains of chitosan derivatives in such a manner that increased the hydrophobicity of the ACSs. The same effects were also observed when ACSs were used as gene delivery vectors to improve transfection efficiency, which will be stated in the following section.

In terms of the accumulated knowledge of physiochemical properties of chitosan, it is reasonable to consider that a combined electrostatic–hydrophobic driving force from chitosan might induce the destabilization of cell membranes. For example, in chitosan-mediated transfection, Venkatesh and co-authors[40] argued that, in addition to ionic interactions, nonionic interactions between the carbohydrate backbone of chitosan and cell surface proteins might have an important role in the chitosan-mediated transfection of cells.

IV. CHITOSAN VECTOR IN GENE DELIVERY

A. MOLECULAR WEIGHT–TRANSFECTION RELATIONSHIP (MWTR)

Since Mumper et al.[41] pioneered the use of chitosan as a gene delivery system in 1995, many efforts have been made to explore the potential of this naturally occurring polysaccharide as a new nonviral vector. Researchers have evaluated a variety of chitosans of high-molecular weight (HMW) and (LMW) as well as chitosan oligomers.[42–44] However, in elucidating the effects of the molecular weight of chitosan on transfection efficiency, the results obtained by different authors seemed to be inconsistent. Mumper's group[44,45] investigated the influence of chitosan molecular weight, plasmid concentration, and charge ratio on complex diameter. The general trend was that the size of the complex particles increased as the molecular weight of the chitosan increased (Figure 28.3).

An analogous variation in particle size was observed with plasmid concentrations in experiments where, at selected plasmid concentrations, complexes constructed with a 1:6 $(-/+)$ ratio were larger in size than 1:2 $(-/+)$ complexes. Chitosan–plasmid complexes formulated with higher molecular weight chitosan were more stable to salt and serum challenge, and the complexes of a 1:2 $(-/+)$ charge ratio were shown to be most stable. In the absence of serum, the highest transfection level was obtained with the 102 kDa chitosan, but was still 250-fold less than that observed with the Lipofectamine™ positive control. In the presence of serum, the highest level of transfection was obtained with a 540 kDa chitosan complex and transfection efficiency showed a declining trend with decreasing molecular weight of chitosan. These results indicated that the molecular weight of chitosan had relatively limited influence on plasmid expression *in vitro*. Interestingly, after administration in the upper small intestine and colon of rabbits, chitosan formulations with lytic peptide modification performed better than lipid formulation although the absolute expression levels remained low. A speculation to account for low expression is that the uptake and decomplexation of chitoplexes, but not endosomal release, might be the critical rate-limiting steps in gene transfection. Other results also indicated that the molecular weights of the chitosans were not important for the transfection efficiency, at least not in the range 20 to 200 kDa.[46]

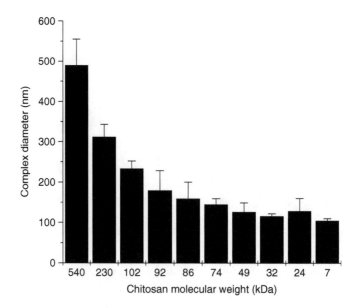

FIGURE 28.3 Effect of molecular weight of chitosan on the size of chitosan–plasmid complexes. Complexes were made at 1:6 $(-/+)$ with a plasmid concentration of 100 μg/mL.

The results presented above seem to support that HMW chitosan is superior to LMW in stability of complexes. However, the drawbacks associated with HMW chitosan such as high viscosity, complement activation, and insolubility at physiological pH need to be addressed prior to long-term applications *in vivo*.[47] Undoubtedly, chitosan polymers with relatively LMW cannot provide effective protection, for example from DNAses, for DNA. It is imaginable that if this fraction coexists with higher molecular weight ones, i.e., there is a high polydispersity of molecular weights, the LMW components fail to protect against DNAse attack. Moreover, coexisting different molecular weight fractions tend to form complexes with a wide range of structures and shapes, which affects the stability of the complexes and the resulting efficiency of gene transfection. Recently, Köping-Höggärd and co-workers[48] prepared monodisperse oligomers (6-, 8-, 10-, 12-, and 14-mer), 24-mer with very low polydispersity and ultrapure chitosan (UPC) whose chain is 40-fold longer than a 24-mer. Depending on the chain length of chitosan, its charge ratio, and buffer properties, chitosan–T4DNA complexes appeared in different physical shapes such as coils, soluble globules, soluble aggregates, precipitated globules, and precipitated aggregates. Chitosan–plasmid complexes followed the same pattern where, with the increase in chain length of the oligomers and charge ratios, the size distribution shifted from a dominance of extended coils to a dominance of soluble globules.

Agarose gel electrophoresis retardation assays demonstrated that only UPC and 24-mer chitosans could form stable complexes with DNA. These experiments clearly showed that the shape of complex influenced gene transfection. Complexes composed of coils and globules were inefficient in mediating transgene expression *in vitro*. In contrast, complexes with globular and aggregated shapes displayed a higher transfection level. Importantly, after pDNA was complexed with 24-mer or UPC and intratracheal administration to mouse lungs *in vivo* was done, efficient gene expression could be obtained. The results of kinetic analysis of gene expression indicated that the gene expression rate mediated by 24-mer was faster than for UPC. Köping-Höggärd and co-workers[48] supposed that 24-mer chitosan was more ideal than UPC and conventional HMW chitosans in mediating gene delivery, presumably due to rapid enzymatic degradation of oligomer chitosan and the subsequent rupture of the endolysosomal membranes.

B. APPLICATIONS OF CHITOSAN IN MESENCHYMAL STEM-CELL TRANSFECTION

To date, chitosan has been used to mediate gene delivery into various cell types such as human embryonic kidney cells (HEK293), human lung carcinoma cells (A549), B16 melanoma cells, COS-1, and HeLa cells.[24,42,44,49] The results obtained in these studies indicated that the transfection level was dependent on the cell type. So far, few researchers attempted to apply chitosan to the transfection of mesenchymal stem cells (MSCs). MSCs are prevalent in the bone marrow of adults. They can be isolated, expanded in culture, and stimulated to differentiate into bone, cartilage, muscle, marrow stroma, tendon, fat, and a variety of other connective tissues.[50] Because MSCs can be generated in culture, tissue-engineered constructs principally composed of these cells can be reintroduced into an *in vivo* setting. Furthermore, MSCs can be transfected with genetic vectors. These considerations render MSCs as potential candidates to deliver somatic gene therapies for local or systemic pathologies and, toward this objective, Corsi and co-workers[51] have recently begun to use chitosan to mediate MSCs transfection.

Chitosans of molecular weights 150, 400, and 600 kDa at a concentration of 0.02% were used for complexation with DNA. After extracting chitosan from complexes with chitosanase and lysozyme, the plasmid DNA released was shown to be intact. The physical shape of complexes appeared to be spherical with a mean size <100 nm (Figure 28.4), which was considered the probable route for pinocytosis. In addition, MSCs also did not seem to be affected from their incubation with complex nanoparticles which provided a dramatic benefit compared with a LipofectAMINE™ 2000 formulation that reduced cellular viability up to 40%. As seen in Figure 28.5, chitosan is effective in mediating MSCs transfection, although the transfection efficacy is not statistically significant relative to cells receiving solely naked nanoparticles. Furthermore, the transfection level is inferior to that seen with the LF formulation despite the toxicity of the LipofectAMINE™ 2000 formulation.

Vector-mediated MSCs transfection is of great importance in tissue engineering considering that the expression of growth factors by these cells can lead to bone formation. Lieberman et al.[52] reported that MSCs were transfected by an adenovirus carrying the gene for BMP-2 and then seeded on a demineralized bone matrix.[52] A complete healing was observed after the matrix was implanted into surgically produced

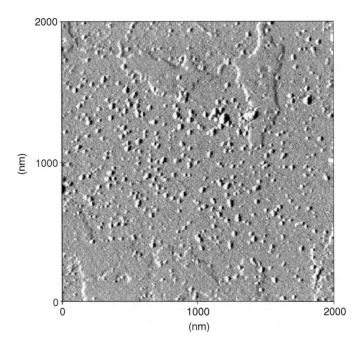

FIGURE 28.4 Atomic force microscopy analysis of complexes formed between chitosan and DNA.

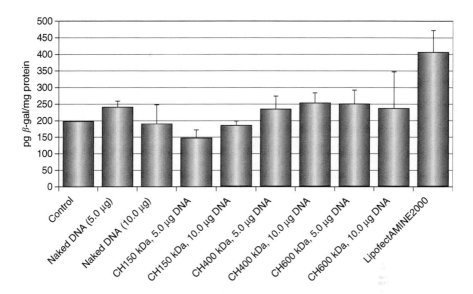

FIGURE 28.5 Transfection efficiency of chitosan–DNA nanoparticles incubated with MSCs in the presence of 10% serum. Cells received either 5.0 or 10 μg of naked DNA or nanoparticles composed of different molecular weights of 0.02% chitosan containing 5.0 or 10.0 μg of DNA.

gaps in the leg bone of rats but, as discussed earlier, the safety of viral vector remained concerned and an alternative approach may be to resort to nonviral vectors such as chitosan. The low transfection level of chitosan, however, is still a challenge. Also as has been noted, the order of magnitude of molecular weight of chitosan is in the range of hundred thousands. Such HMW material might lead to low transfection efficiency and selecting LMW chitosan might be a solution. Another approach to increase transfection efficiency is to couple a ligand specifically recognizing receptors on cellular membrane of MSCs to facilitate internalization of complexes into MSCs. Thus, identifying the receptors on MSC membrane is the premise of selecting targeting ligands.

C. ADMINISTRATION ROUTE OF CHITOPLEXES

Administration refers to the route of introduction of the chitosan-based composition into the body. The special delivery route of any selected vector construct will depend on the particular use for the nucleic acid associated with the chitosan-based composition. Administration includes, but is not limited to, intravenous, intramuscular, systemic, subcutaneous, subdermal, topical, nasal, or oral methods of delivery. Administration can be directly to a target tissue or through systemic delivery.[45] The ultimate objective of gene delivery is clinical application; hence, a proper administration route is one of the essential steps leading to clinical reality.

Food allergy is a common disease, among which peanut allergy is an increasingly important public health problem because of the ubiquitous use of peanut protein in a variety of food products. Intramuscular or intradermal administration of "naked" plasmid DNA encoding an antigen results in synthesis of the antigen protein by the cells and subsequent development of both cellular and humoral immune responses to the antigen. Despite the recent success of DNA-based immunization through intramuscular, intradermal, and subcutaneous routes, oral vaccination with "naked" DNA has been mostly ineffective.[53] Oral vaccination is useful not only because it generates mucosal immunity but also because of its high patient compliance, ease of administration and applicability for mass vaccination. Roy and co-workers[54] prepared nanoparticles by complexing HMW (390,000 kDa) with pCMVArah2 plasmid DNA encoding for the gene of a major peanut allergen. After the formulations were orally administered to AKR/J mice, a higher level of gene expression was observed in both the stomach and the small intestine. More specifically, mice fed with nanoparticles containing pCMVArah2 gene produced secretory IgA and serum IgG2a. Compared with control group (nonimmunized mice or mice treated with "naked" DNA), mice immunized with nanoparticles showed a remarkable reduction in anaphylaxis. These results indicate that oral allergen-gene immunization with chitosan–DNA nanoparticles hold a promise in prophylactically treating food allergy.

In their work,[55] Leong and co-workers attempted to treat mite allergy in mice by orally administering the chitosan–DNA nanoparticles. The results showed that oral immunization of nanoparticles synthesized from chitosan and plasmid DNA encoding house dust mite allergen, Der p1 could induce immune responses specific to both the left and right domains of Der p1, and successfully primed Th1-skewed immune responses against both domains of Der p1. It was proposed that chitosan–DNA nanoparticles were suitable for oral gene vaccination and moreover could be a general strategy for the development of vaccine regimes. In a slightly different approach to *in vivo* gene delivery, chitosan (102 kDa)-and chitosan oligomer (7 kDa)-based nanoparticles containing plasmid DNA were applied topically to the skin of mice. This method resulted in gene expression levels 300-fold higher than those of naked DNA, and furthermore, a significant antigen-specific IgG titer to expressed β-galactosidase at 28 day after the first application. [56]

The recent advances in gene therapy have opened a new avenue to the treatment of pulmonary diseases such as lung cancer, cystic fibrosis (CF), and allergen-induced airway hyperresponsiveness because direct access of a gene delivery system via the airway is possible. Nebulization is one of the practical systems for the administration of nonviral gene delivery systems. However, jet nebulization adversely affected the physical stability of lipid–DNA complexes. [57] Dry powder is another promising system for pulmonary gene delivery, while supercritical fluid technology offers the possibility to produce dry powder formulations suitable for inhalation or needle-free injection.

The powders of chitosan ($M_w = 3,000–30,000$) /pCMv-Luc plasmid complexes were prepared with supercritical carbon dioxide[58] and, as shown in Figure 28.6, had a rectangular shape and mean particle diameters of 12 to 13 μm. The chitosan–pDNA powders at various N/P ratios were administered to the mouse lung by using the apparatus shown in Figure 28.7, and luciferase activities were assayed after 9 h. The results revealed that the powders without chitosan increased the luciferase activity to 360% when compared with the control (mice were sacrificed 6 h after the administration of pCMV–Luc solution), whereas the powder chitosan–pDNAdp (N/P=5) showed the highest luciferase activities (1.1×10^6 RLU/mg of protein).

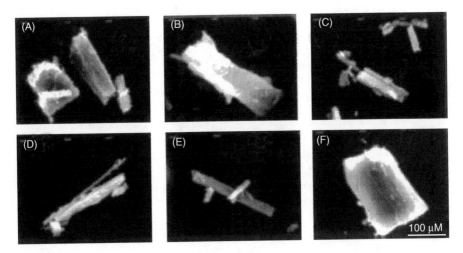

FIGURE 28.6 SEM of pDNA powders with mannitol as a carrier: (A) mannitol alone, (B) pDNAdp, (C) chitosan–pDNAdp (N/P=2.5), (D) chitosan–pDNAdp (N/P=5), (E) chitosan– pDNAdp (N/P=10), and (F) AcNa–pDNAdp.

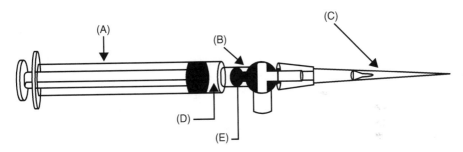

FIGURE 28.7 Apparatus for pulmonary administration of dry powder: (A) 1 mL syringe, (B) three-way stopcock; (C) disposable tip with dry powder; (D) compressed air; and (E) administration handle.

V. CHITOSAN DERIVATIVE VECTORS IN GENE DELIVERY

Although chitosan can mediate gene transfection effectively in some cases, the insolubility of HMW chitosan at neutral medium and the lack of specific recognition to its target cells currently limit its widespread application as DNA delivery vehicle. To overcome these limitations, researchers in individual cases have prepared a variety of chitosan derivatives by coupling conjugate or ligand with chitosan to improve its solubility or targeting. Herein, a ligand is a component of the delivery system or vehicle that binds to receptors, with an affinity for the ligand, on the surface or within compartments of a cell for the purpose of enhancing uptake or intracellular trafficking of the vector. It was reported that trimethylated chitosan with different degrees of quaternarization could increase the solubility of chitosan at neutral pH.[59] Deoxycholic acid-modified chitosan was developed as a colloidal carrier for

DNA.[60,61] A lactosylated chitosan was shown to ferry genes effectively into HeLa cells.[62] The above-mentioned work has been described in two reviews;[63,64] in the following section of this chapter the latest progress relevant to chitosan derivative-based transgene vectors is reviewed.

A. TRANSFERRIN–KNOB PROTEIN-CONJUGATED CHITOSAN VECTOR

Transferrin receptor responsible for iron import to the cells is found on many mammalian cells.[65] As a ligand, transferrin could efficiently transfer LMW drugs, non-bioactive macromolecules, and liposomes through a receptor-mediated endocytosis mechanism.[66] In the past decade, transferrin has been applied to deliver plasmid DNA and oligonucleotides.[67–69] To improve transfection efficiency, Mao and co-workers[70] explored two strategies to bind transferrin onto the surface of chitosan–DNA complex. The first was to insert aldehyde groups in transferrin (a glycoprotein) by periodate oxidation. The modified transferrin was then reacted with the amino groups of chitosan. This scheme ensured that the modification only occurred at the polysaccharide chains of transferrin. In addition, a long spacer arm provided by polysaccharide chain in transferrin minimized the loss of the cell-surface-binding activity of transferrin. The transfection experiment was performed on HEK293 using pRE-luciferase as a model plasmid. The transferrin-conjugated nanoparticles with 14.2% modification degree yielded levels twice as high as unmodified ones, whereas conjugation degrees of 5.3 and 32.7% generated levels similar to controls. A decrease in transfection ability at high levels of transferrin was attributed to self-conjugation.

In the second strategy, transferrin was linked to the nanoparticle surface through a disulfide bond. The transferrin-conjugated vector only resulted in a maximum of four-fold increase in transfection efficiency in HEK293 cells and only 50% increase in HeLa cells. To enhance transfection efficiency further, KNOB (C-terminal globular domain of the fiber protein) was conjugated to the chitosan by the disulfide linkages as well. The KNOB conjugation to the nanoparticles improved gene expression levels in HeLa cells by 130-fold.

B. ALKYLATED CHITOSAN VECTORS

Advantages of chitosan-based gene delivery vectors lie not only in their avoidance of cytotoxicity problems inherent with most synthetic polymeric vehicles but also in their unique capability for facilitating transcellular transport. One disadvantage, however, as shown with other polycations–DNA complexes is that chitosan–DNA complexes formed by electrostatic interactions between primary amino groups and phosphate groups are strong enough to resist DNA unpacking within cell to a certain degree. Okano et al.,[71] Sato et al.,[72] and Kabanov and Kabanov[73] have all reported that the incorporation of hydrophobic moieties designed to lessen these electrostatic interactions considerably increase transfection efficiency. In addition, by theoretical calculations, Kuhn et al.[74] found that for sufficiently hydrophobic amphiphiles, a charge neutralization or even a charge inversion of the DNA–amphiphile complexes could be achieved with rather low concentration of cationic amphiphile.

In our laboratory, chitosan was coupled with alkyl side chains of different lengths and the resulting alkyl chitosan derivatives were used as nonviral vectors for gene delivery for the first time.[39] The electrophoresis experiment demonstrated that the complex between CS and DNA started to form at a charge ratio $(+/-)$ of 1:1; ACS/DNA complexes started to form at a lower charge ratio $(+/-)$ of 1:4 (Figure 28.8). A small amount of alkylated chitosans played the same shielding role as chitosan in protecting DNA from DNase hydrolysis. Thermodynamic and morphological analyses demonstrated that ACSs exerted a larger and more evident perturbation effect on DPPC membrane compared with unmodified chitosan. CS and ACSs were used to transfer plasmid-encoding CAT into C_2C_{12} cell lines *in vitro*. Upon elongating the alkyl side chain, the transfection efficiency was increased and leveled off after the number of carbons in side chain exceeded eight (Figure 28.9). The higher transfection efficiency of ACS was attributed to the increasing entry into cells facilitated by hydrophobic interactions and easier unpacking of DNA from ACS carriers due to the weakening of electrostatic attractions between DNA and ACS.

C. GALACTOSYLATED CHITOSAN VECTORS

Experiments have established that asialoglycoprotein (ASGP) receptors, known as "hepatic lectin," are specifically recognized by ligands such as galactose, lactose, and apoprotein. The receptor is expressed exclusively by parenchymal hepatocytes which contain 100,000 to 500,000 binding sites per cell.[75] ASGP receptors have been utilized in gene delivery by coupling galactose-terminated oligosaccharide ligands to nonviral vectors resulting in markedly enhanced gene internalization into the hepatocytes. In their previous publications,[76,77] Park and co-workers developed galactosylated chitosan (GC)-*graft*-dextran and poly(ethylene glycol) (PEG) vectors. The chitosan derivatives can form stable complexes with DNA due to hydrophilic dextran or PEG protection. Also, while a small amount of the complexes were successfully transfected into the hepatocytes, the transfection efficiency mediated by the carrier was not satisfactory. The efficacy of the nonviral system was governed by an underlying delivery mechanism that was qualitatively studied by Park and co-workers by using confocal laser-scanning microscopy (CLSM) to observe the intracellular trafficking of GC-*graft*-PEG (GCP)–DNA complexes[78]. The results demonstrated that considerable ASGP-mediated endocytosis occured at 30 min after the addition of the

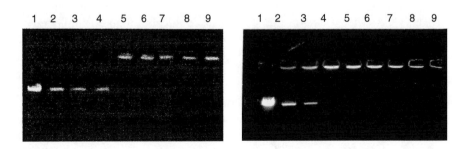

FIGURE 28.8 Electrophoresis of CS-DNA (a) and 12-CS-DNA (b) complexes on an agarose gel. lane 1, plasmid DNA; lane 2, +/− =1/10; lane 3, 1/1; lane 4, 1/4; lane 5, 1/1; lane 6, 2/1; lane 7, 4/1; lane 8, 8/1; lane 9, 10/1.

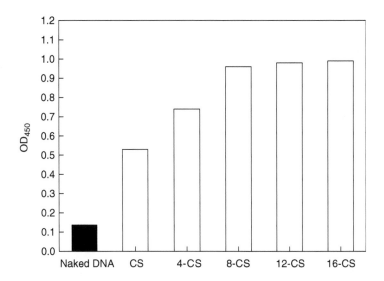

FIGURE 28.9 Transfection efficiency of pcDNA 3.1 plasmid encoding CAT mediated by CS and ACS. The prepared chitosan–DNA complexes were diluted with serum-free DMEM and added into the corresponding wells (5.0 μg of DNA/well). After 1.0 h incubation, the complexes were removed and the culture media were replaced by fresh serum-containing media and incubated for additional 48 h at 37°C. Then the cells were lysed with PBS solution containing 1.0% NP40 and 1 mmol/L PMSF and subjected to three cycles of deep freezing and thawing. The CAT concentration was determined with enzyme-linked immunosorbent assay (ELISA) method. The OD values were measured 450 nm by the plate-reader of Labsystem. The naked DNA was used as a control. See Ref. [39] for a detailed description.

complex to the cell and continued to accumulate, mainly in the cytoplasm, up to 3 h postincubation, suggesting that uptake of galactose-carrying complexes by ASGPR is both a high-affinity and high-capacity process. The complexes did not concentrate in multiple spots corresponding to intracellular vesicles but were diffused uniformly throughout the cytoplasm. This distribution pattern may partly result from rapid release of the complexes from endosomal compartments by the fusogenic property of PEG in the complex. Nevertheless, in some intracellular areas, plasmid was still bound to GCP vector. Therefore, the slow nuclear trafficking of plasmid followed by rapid escape from endosome was thought to be the rate-limiting step and may cause low transgene expression in GCP-mediated transfection.

Confocal microscopy can be applied to the study of intracellular trafficking of non-viral systems[79] by providing an ensemble-averaged technique that can provide a qualitative description of the transport process by determining the locations of complexes at discrete times. However, information associated with individual DNA carriers (the rates of individual particle movements, the mode of transport [e.g., random vs. directed or active], and the trajectory and directionality of the transport) remains a black box. Recently, Suh and co-workers[80] used multiple particle tracking (MPT) to investigate the individual motions of hundreds of polyethylenimine (PEI)–DNA nanocomplexes. Their results showed quantitatively that PEI–DNA nanocomplexes were efficiently

transported to the perinuclear region of the cell on microtubules, and PEI vectors fer-
ried genes by means of microtubule-based motor proteins thereby using the same
molecular mechanism as nature's most efficient DNA viruses. The results obtained by
MPT are contradictory to the common belief that the reduced efficiency of nonviral
gene carriers originates from a relatively slow random diffusion process as the vector
passes through the cell cytoplasm to the nucleus. Because vector-mediated gene delivery is a complex system, many problems remain to be resolved. Nonetheless, MPT
provides a new angle of view to improve the transfection efficiency of nonviral vectors
by overcoming barriers downstream of perinuclear accumulation.

In the following work,[81] Park and co-workers further modified galactosylated
chitosans of 10 and 50 kDa molecular weight with polyvinyl pyrrolidone (PVP)
considering better retainability of PVP in blood than PEG. The synthetic strategy is
depicted in Figure 28.10. The size of GC-graft-PVP (GCPVP)–DNA complexes was
in nanometer scale, and the secondary structure of DNA in GCP–DNA complexes
was intact. The results of gene transfection will be published soon.

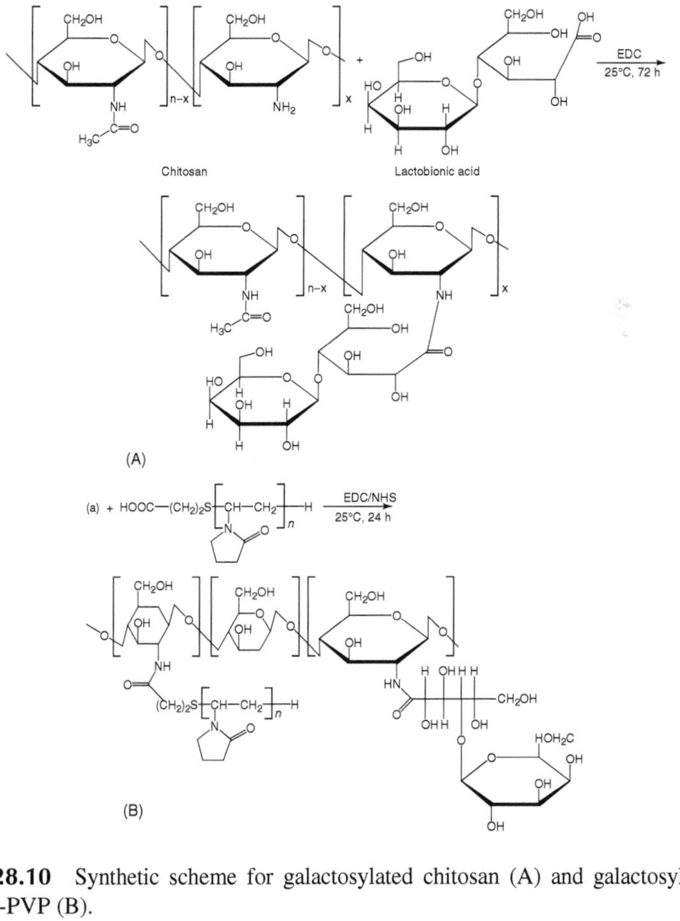

FIGURE 28.10 Synthetic scheme for galactosylated chitosan (A) and galactosylated chi-
tosan-graft-PVP (B).

REFERENCES

1. Verma, I.M. and Somia, N., Gene therapy — promises, problems and prospects, *Nature*, 389, 239–242, 1997.
2. Kay, M.A. and Nakai, H., Looking into the safety of AAV vectors, *Nature*, 424, 251, 2003.
3. Check, E., Harmful potential of viral vectors fuels doubts over gene therapy, *Nature*, 423, 573–574, 2003.
4. Ferber, D., Gene therapy: safer and virus-free?, *Science*, 294, 1638–1642, 2001.
5. Check, E., Regulators split on gene therapy as patient shows signs of cancer, *Nature*, 419, 545–546, 2002.
6. Cartier, R. and Reszka, R., Utilization of synthetic peptides containing nuclear localization signals for nonviral gene transfer systems, *Gene Ther.*, 9, 157–167, 2002.
7. Niidome, T. and Huang, L., Gene therapy progress and prospects: nonviral vectors, *Gene Ther.*, 9, 1647–1652, 2002.
8. Christopher, M., Wiethoff, C., and Russell, M., Barriers to nonviral gene delivery, *J. Pharm. Sci.*, 92, 203–217, 2003.
9. Schmidt-Wolf, G.D. and Schmidt-Wolf, I.G.H., Non-viral and hybrid vectors in human gene therapy: an update, *Trends Mol. Med.*, 9, 67–72, 2003.
10. Shi, G.F., Guo, W.J., Stephenson, S.M., and Lee, R.J., Efficient intracellular drug and gene delivery using folate receptor-targeted pH-sensitive liposomes composed of cationic/anionic lipid combinations, *J. Controlled Release*, 80, 309–319, 2002.
11. Torchilin, V.P., Levchenko, T.S., Rammohan, R., Volodina, N., Sternberg, B.P., and D'Souza, G.G.M., Cell transfection *in vitro* and *in vivo* with nontoxic TAT peptide–liposome–DNA complexes, *Proc. Natl. Acad. Sci.*, 100, 1972–1977, 2003.
12. Rudolph, C., Müller, R.H., and Rosenecker, J., Jet nebulization of PEI/DNA polyplexes: physical stability and *in vitro* gene delivery efficiency, *J. Gene. Med.*, 4, 66–74, 2002.
13. Midoux, P. and Monsigny, M., Efficient gene transfer by histidylated polylysine/pDNA complexes, *Bioconjug. Chem.*, 10, 406–411, 1999.
14. Azzam, T., Eliyahu, H., Makovitzki, A., and Domb, A.J., Dextran-spermine conjugate: an efficient vector for gene delivery, *Macromol. Symp.*, 195, 247–262, 2003.
15. Joester, D., Losson, M., Pugin, R., Heinzelmann, H., Walter, E., Merkle, H.P., and Diederich, F., Amphiphilic dendrimers: novel self-assembling vectors for efficient gene delivery, *Ang. Chem. Int. Ed.*, 42, 1486–1490, 2003.
16. Anderson, D.G., Lynn, D.M., and Langer, R., Semi-automated synthesis and screening of a large library of degradable cationic polymers for gene delivery, *Ang. Chem. Int. Ed.*, 42, 3153–3158, 2003.
17. Dinçer, S., Tuncel, A., and Pişkin, E., A potential gene delivery vector: *N*-isopropylacrylamide-ethyleneimine block copolymers, *Macromol. Chem. Phys.*, 203, 1460–1465, 2002.
18. Leong, K.W., Mao, H.Q., Truong-Le, V.L., Roy, K., Walsh, S.M., and August, J.T., DNA-polycation nanospheres as non-viral gene delivery vehicles, *J. Controlled Release*, 53, 183–193, 1998.
19. Niidome, T., Urakawa, M., Sato, H., Takahara, Y., Anai, T., Hatakayama, T., Wada, A., Hirayama, T., and Aoyagi, H., Gene transfer into hepatoma cells mediated by galactose-modified -helical peptides, *Biomaterials*, 21, 1811–1819, 2000.
20. Brown, M.D., Schätzlein, A.G., and Uchegbu, I.F., Gene delivery with synthetic (non viral) carriers, *Int. J. Pharm.*, 229, 1–21, 2001.
21. Dodane, V., Khan, M.A., and Merwin, J.R., Effect of chitosan on epithelial permeability and structure, *Int. J. Pharm.*, 182, 21–32, 1999.

22. van der Lubben, I.M., Verhoef, J.C., Borchard, G., and Junginger, H.E., Chitosan and its derivatives in mucosal drug and vaccine delivery, *Eur. J. Pharm. Sci.,* 14, 201–207, 2001.

23. Zhang, H. and Neau, S.H., *In vitro* degradation of chitosan by a commercial enzyme preparation: effect of molecular weight and degree of deacetylation, *Biomaterials,* 22, 1653–1658, 2001.

24. Ishii, T., Okahata, Y., and Sato, T., Mechanism of cell transfection with plasmid/chitosan complexes, *Biomembranes,* 1514, 51–64, 2001.

25. Cui, Z. and Mumper, R.J., Chitosan-based nanoparticles for topical genetic immunization, *J. Controlled Release,* 75, 409–419, 2001.

26. Leong, K.W., Mao, H.Q., Truong-Le, V.L., Roy, K., Walsh, S.M., and August, J.T., DNA-polycation nanospheres as non-viral gene delivery vehicles, *J. Controlled Release,* 53, 183–193, 1998.

27. Janes, K.A., Calvo, P., and Alonso, M.J., Polysaccharide colloidal particles as delivery systems for macromolecules, *Adv. Drug Deliv. Rev.,* 47, 83–97, 2001.

28. Ouji, Y., Terakura, A.Y., Hayashi, Y., Maeda, I., Kawase, M., Yamato, E., Miyazaki, J., and Yagi, K., Polyethyleneimine/chitosan hexamer-mediated gene transfection into intestinal epithelial cell cultured in serum-containing medium, *J. Biosci. Bioeng.* 94, 81–83, 2002.

29. Hejazi, R. and Amiji, M., Chitosan-based gastrointestinal delivery systems, *J. Controlled Release,* 89, 151–165, 2003.

30. Holme, H.K., Hagen, A., and Dornish, M., Influence of chitosans with various molecular weights and degrees of deacetylation on the permeability of human intestinal epithelial cells (Caco-2), in *Chitosan Per Os From Dietary Supplement To Drug Carrier,* R.A.A., Muzzarelli, Ed., Atec, Grottammare, Italy, 2000, pp. 127–136.

31. Carreno-Gomez, B. and Duncan, R., Evaluation of the biological properties of soluble chitosan and chitosan microspheres, *Int. J. Pharm.,* 148, 231–240, 1997.

32. Verhoef, J.C., Junginger, H.E., and Thanou, M., Chitosan and its derivatives as intestinal absorption enhancers, *Adv. Drug. Deliv. Rev.,* 50, S91–S101, 2001.

33. Yang, F., Cui, X.Q., and Yang, X.R., Interaction of low-molecular-weight chitosan with mimic membrane studied by electrochemical methods and surface plasmon resonance, *Biophys. Chem.,* 99, 99–106, 2002.

34. Chan, V., Mao, H.Q., and Leong, K.W., Chitosan-induced perturbation of dipalmitoyl-*sn*-glycero-3-phosphocholine membrane bilayer, *Langmuir,* 17, 3749–3756, 2001.

35. Berth, G., Dautzwnberg, H., and Peter, M.G., Physico-chemical characterization of chitosans varying in degree of acetylation, *Carbohydr. Polym.,* 36, 205–216, 1998.

36. Fang, N., Chan, V., Mao, H.Q., and Leong, K.W., Interactions of phospholipid bilayer with chitosan: effect of molecular weight and pH, *Biomacromol.,* 2, 1161–1168, 2001.

37. Philippova, O.E., Volkov, E.V., Sitnikova, N.L., and Khokhlov, A.R., Two types of hydrophobic aggregates in aqueous solutions of chitosan and its hydrophobic derivative, *Biomacromolecules,* 2, 483–490, 2001.

38. Liu, W.G., Sun, S.J., Zhang, X., and Yao, K.D., Self-aggregation behavior of alkylated chitosan and its effect on the release of a hydrophobic drug, *J. Biomater. Sci. Polym. Ed.,* 14, 851–859, 2003.

39. Liu, W.G., Zhang, X., Sun, S.J., Sun, G.J., and Yao, K.D., *N*-Alkylated chitosan as a potential nonviral vector for gene transfection, *Bioconjug. Chem.,* 14, 782–789, 2003.

40. Venkatesh, S. and Smith, T.J., Chitosan–membrane interactions and their probable role in chitosan-mediated transfection, *Biotechnol. Appl. Biochem.,* 27, 265–267, 1998.

41. Mumper, R.J., Wang, J., Claspell, J.M., and Rolland, A.P., Novel polymeric condensing carriers for gene delivery, *Proceedings of the International Symposium of Controlled Release of Bioactive Materials,* 22, 1995, pp. 178–179.

42. Sato, T., Ishii, T., and Okahata, Y., *In vitro* gene delivery mediated by chitosan. Effect of pH, serum, and molecular mass of chitosan on the transfection efficiency, *Biomaterials*, 22, 2075–2080, 2001.

43. Richardson, S.C.W., Kolbe, H.V.J., and Duncan, R., Potential of low molecular mass chitosan as a DNA delivery system: biocompatibility, body distribution and ability to complex and protect DNA, *Int. J. Pharm.*, 178, 231–243, 1999.

44. MacLaughlin, F.C., Mumper, R.J., Wang, J.J., Tagliaferri, J.M., Gill, I., Hinchcliffe, M., and Rolland, A.P., Chitosan and depolymerized chitosan oligomers as condensing carriers for *in vivo* plasmid delivery, *J. Controlled Release*, 56, 259–272, 1998.

45. Rolland, A. and Mumper, R.J., Chitosan Related Compositions and Methods for Delivery of Nucleic Acids and Oligonucleotides into a Cell, US Patent, 6,184,037, 2001.

46. Köping-Höggärd, M., Tubulekas, I., Guan, H., et al., Chitosan as a nonviral gene delivery system. structure–property relationships and characteristics compared with polyethyleneimine *in vitro* and after lung administration *in vivo*, *Gene Ther.*, 8, 1108–1121, 2001.

47. Fischer, D., Bieber, T., Li, Y., et al., A novel non-viral vector for DNA delivery based on low molecular weight, branched polyethyleneimine: effect of molecular weight on transfection efficiency and cytotoxicity, *Pharm. Res.*, 16, 1273–1279, 1999.

48. Köping-Höggärd, M., Mel'nikova, Y.S., Vårum, K.M., Lindman, B., and Artursson, P., Relationship between the physical shape and the efficiency of oligomeric chitosan as a gene delivery system *in vitro* and *in vivo*, *J. Gene. Med.*, 5, 130–141, 2003.

49. Mao, H.Q., Roy, K., Troung-Le, V.L., Janes, K.A., Lin, K.Y., Wang, Y., August, J.T., and Leong, K.W., Chitosan–DNA nanoparticles as gene carriers: synthesis, characterization and transfection efficiency, *J. Controlled Release*, 70, 399–421, 2001.

50. Caplan, A.I. and Bruder, S.P., Mesenchymal stem cells: building blocks for molecular medicine in the 21st century, *Trends Mol. Med.*, 7, 259–264, 2001.

51. Corsi, K., Chellat, F., Yahia, L., and Fernandes, J.C., Mesenchymal stem cells, MG63 and HEK293 transfection using chitosan–DNA nanoparticles, *Biomaterials*, 24, 1255–1264, 2003.

52. Lieberman, J.R., Daluiski, A., Stevenson, S., Wu, L., McAllister, P., Lee, Y.P., Kabo, J.M., Finerman, G.A., Berk, A.J., and Witte, O.N., The effect of regional gene therapy with bone morphogenetic protein-2-producing bone-marrow cells on the repair of segmental femoral defects in rats, *J. Bone. J. Surg. Am.*, 81, 905–917, 1999.

53. Etchart, N., Buckland, R., Liu, M.A., Wild, T.F., and Kaiserlian, D., Class I-restricted CTL induction by mucosal immunization with naked DNA encoding measles virus haemagglutinin, *J. Gen. Virol.*, 78, 1577–1580, 1997.

54. Roy, K., Mao, H.Q., Huang, S.K., and Leong, K.W., Oral gene delivery with chitosan–DNA nanoparticles generates immunologic protection in a murine model of peanut allergy, *Nat. Med.*, 5, 387–391, 1999.

55. Chew, J.L., Wolfowicz, C.B., Mao, H.Q., Leong, K.W., and Chua, K.Y., Chitosan nanoparticles containing plasmid DNA encoding house dust mite allergen, Der p 1 for oral vaccination in mice, *Vaccine*, 21, 2720–2729, 2003.

56. Cui, Z. and Mumper, R.J., Chitosan-based nanoparticles for topical genetic immunization, *J. Controlled Release*, 75, 409–419, 2001.

57. Eastman, S.J., Tousignant, J.D., Lukason, M.J., Murray, H., Siegel, C.S., Constantino, P., Harris, D.J., Cheng, S.H., and Scheule, R.K., Optimization of formulations and conditions for the aerosol delivery of functional cationic lipid:DNA complexes, *Hum. Gene Ther.*, 8, 313–322, 1997.

58. Okamoto, H., Nishida, S., Todo, H., Sakakura, Y., Iida, K., and Danjo, K., Pulmonary gene delivery by chitosan–pDNA complex powder prepared by a supercritical carbon dioxide process, *J. Pharm. Sci.* 92, 371–380, 2003.

59. Thanou, M., Florea, B.I., Geldof, M., Junginger, H.E., and Borchard, G., Quaternized chitosan oligomers as novel gene delivery vectors in epithelial cell lines, *Biomaterials*, 23, 153–159, 2002.

60. Lee, K.Y., Kwon, I.C., Kim, Y.H., Jo, W.H., and Jeong, S.Y., Preparation of chitosan self-aggregates as a gene delivery system, *J. Controlled Release*, 51, 213–220, 1998.

61. Kim, Y.H., Gihm, S.H., Park, C.R., Lee, K.Y., Kim, T.W., Kwon, I.C., Chung, H., and Jeong, S.Y., Structural characteristics of size-controlled self-aggregates of deoxycholic acid-modified chitosan and their application as a DNA delivery carrier, *Bioconjugate Chem.*, 12, 932–938, 2001.

62. Erbacher, P., Zou, S.M., Bettinger, T., Steffan, A.M., and Remy, J.S., Chitosan-based vector/DNA complexes for gene delivery: biophysical characteristics and transfection ability, *Pharm. Res.*, 15, 1332–1339, 1998.

63. Liu, W.G. and Yao, K.D., Chitosan and its derivatives — a promising non-viral vector for gene transfection, *J. Controlled Release*, 83, 1–11, 2002.

64. Borchard, G., Chitosans for gene delivery, *Adv. Drug Deliv. Rev.*, 52, 145–150, 2001.

65. Dautry-Varsat, A., Receptor-mediated endocytosis: the intracellular journey of transferrin and its receptor, *Biochimie*, 68, 375–381, 1986.

66. Deshpande, D., Velasquez, D.T., Wang, L.Y., et al., Receptor-mediated peptide delivery in pulmonary epithelial monolayers, *Pharm. Res.*, 11, 1121–1126, 1994.

67. Truong-Le, V.L., August, J.T., and Leong, K.W., Controlled gene delivery by DNA-gelatin nanospheres, *Hum. Gene Ther.*, 9, 1709–1717, 1998.

68. Truong-Le, V.L., Walsh, S.M., Schweibert, E., et al., Gene transfer by DNA-gelatin nanospheres, *Arch. Biochem. Biophys.*, 361, 47–56, 1999.

69. de Lima, M.C., Simoes, S., Pires, P., et al., Gene delivery mediated by cationic liposomes: from biophysical aspects to enhancement of transfection, *Mol. Membr. Biol.*, 6, 103–109, 1999.

70. Mao, H.Q., Roy, K., Troung-Le, V.L., Janes, K.A., Lin, K.Y., Wang, Y., August, J.T., and Leong, K.W., Chitosan–DNA nanoparticles as gene carriers: synthesis, characterization and transfection efficiency, *J. Controlled Release*, 70, 399–421, 2001.

71. Kurisawa, M., Yokoyama, M., and Okano, T., Transfection efficiency increases by incorporating hydrophobic monomer units into polymeric gene carriers, *J. Controlled Release*, 68, 1–8, 2000.

72. Sato, T., Kawakami, T., Shirakawa, N., and Okahata, Y., Preparation and characterization of DNA–lipoglutamate complexes, *Bull. Chem. Soc. Jpn.*, 68, 2709–2715, 1995.

73. Kabanov, A.V. and Kabanov, V.A., DNA complexes with polycations for the delivery of genetic materials into cells, *Bioconjug. Chem.*, 6, 7–20, 1995.

74. Kuhn, P.S., Levin, Y., and Barbosa, M.C., Charge inversion in DNA-amphiphile complexes: possible application to gene therapy, *Physica A*, 274, 8–18, 1999.

75. Spiess, M., The asialoglycoprotein receptor: a model for endocytic transport receptors, *Biochemistry*, 29, 10009–10018, 1990.

76. Park, Y.K., Park, Y.H., Shin, B.A., Choi, E.S., Park, Y.R., Akaike, T., and Cho, C.S., Galactosylated chitosan–graft-dextran as hepatocyte-targeting DNA carrier, *J. Controlled Release*, 69, 97–108, 2000.

77. Park, I.K., Kim, T.H., Park, Y.H., Shin, B.A., Choi, E.S., Chowdhury, E.H., Akaike, T., and Cho, C.S., Galactosylated chitosan-graft-poly(ethylene glycol) as hepatocyte-targeting DNA carrier, *J. Controlled Release*, 76, 349–362, 2001.

78. Park, I.K., Kim, T.H., Kim, S.I., Park, Y.H., Kim, W.J., Akaike, T., and Cho, C.S., Visualization of transfection of hepatocytes by galactosylated chitosan-graft-poly(ethylene glycol)/DNA complexes by confocal laser scanning microscopy, *Int. J. Pharm.*, 257, 103–110, 2003.

79.   Godbey, W.T., Wu, K.K., and Mikos, A.G., Tracking the intracellular path of poly(ethylenimine)/DNA complexes for gene delivery, *Proc. Natl. Acad. Sci. USA.*, 99, 7467–7471, 2002.

80.   Suh, J., Wirtz, D., and Hanes, J., Efficient active transport of gene nanocarriers to the cell nucleus, *Proc. Natl. Acad. Sci. USA*, 100, 3878–3882, 2003.

81.   Park, I.K., Ihm, J.E., Park, Y.H., Choi, Y.J., Kim, S.I., Kim, W.J., Akaike, T., and Cho, C.S., Galactosylated chitosan (GC)-graft-poly(vinyl pyrrolidone) (PVP) as hepato-cyte-targeting DNA carrier preparation and physicochemical characterization of GC-graft-PVP/DNA complex, *J. Controlled Release*, 86, 349–359, 2003.

29 Polysaccharides in Tissue Engineering Applications

Dietmar W. Hutmacher, David T.W. Leong, and Fulin Chen

CONTENTS

I.   INTRODUCTION

Originally, Tissue Engineering (TE) was defined from a very broad and general perspective as "the application of the principles and methods of engineering and life sciences toward the fundamental understanding of structure–function relationships in normal and pathological mammalian tissues and the development of biological substitutes to restore, maintain, or improve functions."[1] However, Vacanti and Langer's[2,3] group must be given the credit for laying down the ground work to make TE a discipline of its own in the field of biomedical sciences. Today, TE is driven by multidiscipline research thrusts (Figure 29.1). As a logical consequence, a large number of groups from different backgrounds study specific problems or hypotheses related to TE, for example, design and fabrication of scaffolds, cell isolation and characterization, cell proliferation, cell differentiation, bioreactors, etc. More recently, a smaller number of multidisciplinary groups have conceptualized their work in research thrusts that aim at moving a holistic TE platform step by step into clinical applications.[4]

Based on this background, it can be argued from a clinical point of view that downstream TE research aims at becoming one of the pillars in the area of Regenerative Medicine. In fact, most of the major universities and research institutes[5] inheriting regenerative medicine in their naming or mission statement cite TE and stem cell research as their main areas of interest.

At present, TE strategies can be classified into different categories, for example, as concepts that aim at tissue conduction, induction, and cell transplantation. In this case, the terms conduction and induction are borrowed from treatment concepts in oral and maxillofacial surgery, in which the concepts of osseointegration, osteoconduction, osteoinduction, and guided tissue regeneration (GTR aim at the regeneration of periodontal ligament), and guided bone regeneration (GBR) has been used for more than a decade.[6]

By reviewing the literature and creating a hierarchical framework, the authors of this chapter claim that TE can be subdivided into strategies or concepts as illustrated in Figure 29.1. The first strategy is purely cell-based and defined as cell transplantation; categorised into two broad types; a) cells in small volume of media (e.g., chondrocytes

───➤

FIGURE 29.1   The first generation of cell based TE concepts in the area of skin, cartilage, bone, and bone marrow regeneration were derived by isolation, expansion, and implantation of cells from the patients own tissue. Even so these concepts can claim a clinical success in selective treatment concepts, yet tissue engineers need to overcome major challenges to allow a widespread along with a clinically safe and predictable application. One challenge is to present the cells in a matrix to the implantation site to allow the cells to survive the wound-healing contraction forces, tissue remodeling and maturation. Hence, nowadays a great number of TE strategies are based on the development of a scaffold/cell construct. Based on this background it can be argued from a clinical point of view that downstream TE research aims at becoming one of the pillars in the area of regenerative medicine. In fact, most of the major universities and research institutes inheriting regenerative medicine in their naming or mission statement quote TE and stem cell research as their main area of interest.

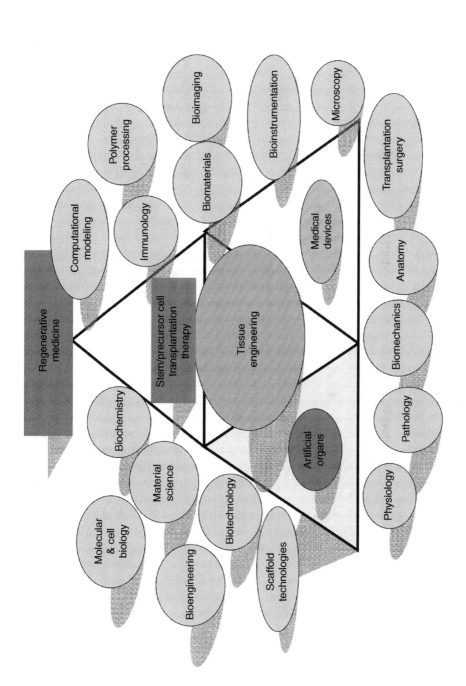

in combination with periosteal flap, injection of myocardiocytes into heart muscle, hemopoetic bone marrow cell transplantation) are injected by using minimal invasive concepts and b) cell sheets [cells + extra cellular matrix (EM), e.g., cell sheet formed by keratinocytes, cell sheet transplantation in cornea tissue engineering, etc] into the defect site. The other three methods are based on using different types of matrices (hydrogels, microspheres/beads, and three-dimensional scaffolds) individually or in combination with growth factors and cells. Hence, the strategies utilizing biochemical substances (growth factors, completely lyophilized cell fractions, peptides, etc) aim at delivering cues to the nonhomogenous cell population of the host tissue and blood clot. These cells are then supposed to migrate, proliferate, and finally differentiate into the defect site to repair or regenerate the tissue.

The fourth and most commonly applied strategy focuses on seeding and culturing specific cell types in a (3D) environment that aims at mimicking the natural ECM as closely as possible. Such 3D environments are achieved with sandwich cultures, hydrogels, and cells in specifically configured cellular solids also known as scaffolds.

The above discussion shows that TE is not only a complex and interdisciplinary area but also that the research team has multiple choices for developing strategies and technology platforms. The following sections provide a brief overview of the polymers used in TE, and the main part of the chapter is dedicated to TE research related to matrices and cell carriers made of polysaccharides.

II. POLYMER SCAFFOLDS USED IN TISSUE ENGINEERING

A. BASIC PRINCIPLES OF SCAFFOLD DESIGN AND FABRICATION

The design and fabrication of a scaffold should be governed by the physical and chemical properties as well as the biological requirements of a specific TE application.

Once the scaffold/cell construct is implanted, the regeneration of tissue relies not only on the scaffold's biological potential, but also on its physical properties at both the macroscopic and the microscopic level. Macroscopically, the scaffold must bear loads to provide stability to the tissues as they form and fulfill its volume maintenance function. At the same time, cell growth, differentiation, and ultimate tissue formation are dependent on the mechanical stimulation of the cells and the ECM. As a consequence, the scaffold must be able not only to withstand specific loads but also to transmit them in a controlled mode to the host tissues.

Envisioned mechanical performance of a scaffold/cell construct depends on the site of implantation (i.e., whether soft or hard tissue), and depending on the strategy, requires different properties such as elasticity, compressibility, viscoelastic behavior, compression, tensile, and shear strength.[7-10]

B. POROUS SOLIDS

According to Gibson and Ashby,[11,12] porous solids are classified into two general groups: honeycombs and foams. The ASTM terminology[13] divides pore morphologies into three groups: interconnecting (open pores), nonconnecting (closed pores), or a combination of both. When the pores are open, the foam material is usually drawn into struts forming the pore edges. A network of struts produces a low-density solid with pores connecting to each other through open faces. When the pores are closed, a network of interconnected plates produces a higher density solid. The virtually closed pores are sealed off from adjacent neighbors.

In the 1990s a number of different processing techniques were developed to fabricate 3D scaffolds based on the cellular solid design. These techniques include fiber bonding, solvent casting, particulate leaching, membrane lamination, melt molding, temperature-induced phase separation (TIPS), gas foaming, and 3D printing. A wide range of scaffold characteristics, such as porosity and pore size, have reported using such fabrication techniques.[14,15]

A review of the recent literature reflects a growing interest in the use of new techniques to design and fabricate scaffolds for TE applications. Advanced manufacturing technologies, also known as rapid prototyping (RP) or solid freeform fabrication (SFF) technologies, are being increasingly explored by tissue engineers.

RP is the process of creating a 3D object through repetitive deposition and processing of material layers using computer-controlled tools, based on 2D cross-sectional data obtained from slicing a computer-aided-design (CAD) model of the object. Several RP systems have been developed by various organizations including the stereo lithography apparatus (SLA) by 3D Systems, selective laser sintering (SLS) by DTM Corp., laminated object manufacturing (LOM) by Helisys, three-dimensional printing (3DP) by MIT, and fused deposition modeling (FDM) by Stratasys Inc. This rather new field has been already reviewed and the following papers have been recommended for further reading.[16–23]

C. TEXTILES

The first synthetic polymer scaffolds used for cell transplantation and tissue regeneration were mesh or felt-like materials based on polyglycolic acid (PGA) fibers, which were commercially available in the form of sutures.[2] Fibers provided a large surface area to volume ratio and were, from this point of view, suitable as scaffold matrix materials. PGA fibers in the form of tassels and felts were utilized as scaffolds in feasibility studies.[24,25] Textile processing technologies were also used to fabricate a series of nonwoven fabrics with fiber diameters at the micrometer scale, usually between 10 and 20 μm, by using different biodegradable polymers of synthetic and natural origin. A number of research groups developed research programs especially in bone,[26,27] cartilage,[28–30] and cardiac TE.[31,32]

D. HYDROGELS

A hydrogel is a 3D network composed of a polymer backbone, water (more than 30%), and a cross-linking agent to generate a complex network of high molecular weight. A plethora of hydrophilic polymer backbones have been explored for use as hydrogels. Cross-linking modalities range from covalent chemical cross-linking difunctional compounds with low molecular weight (e.g., glutaraldehyde or formaldehyde)[33,34] to radical polymerization via radical-forming polymer end-groups or difunctional compounds [i.e., oligo (PEG fumarate),[35] methacrylated dextran,[36] acrylated copolymers of poly (lactic acid), and PEG[37]]. Self cross-linking can also occur via inter- and intrapolymer chain condensation caused by high temperature, reduced pressures, and dehydration.[38] Because of their complex, 3D, hydrophilic structure, hydrogels are capable of absorbing large amounts of aqueous solution and undergo degradation at the same time either via erosion, hydrolysis, solubilization, or other biodegradation mechanisms. Hydrogels made of collagen, HA, and chitosan are degraded by enzymatic hydrolysis. Thus, hydrogels of this nature have been explored as drug delivery devices, wound-dressing materials, contact lenses, and cell carriers.[39]

Today, TE engineering and cell therapy approaches using injectable and *in situ* gel-forming systems are studied widely,[40] and this work is discussed in a number of excellent reviews.[41–45]

Selected injectable systems based on polysaccharides such as alginate, chitosan, and hyaluronan are evaluated in the context of the regeneration of bone, skin, liver, pancreas, etc. (see below). Injectable polymer formulations can gel *in vivo* in response to temperature change (thermal gelation), pH change, ionic cross-linking, or solvent exchange. The kinetics of gelation is directly affected by its mechanism. Injectable formulations might offer specific advantages over preformed scaffolds such as the possibility of a minimally invasive implantation, an ability to fill a desired shape, and the easy incorporation of various therapeutic agents.[46] Several factors need to be considered before an injectable gel can be selected as a candidate for TE applications. Apart from tissue-specific cell–matrix interactions, the following gel properties also need to be considered: gelation kinetics, matrix resorption rate, possible toxicity of degradation products and their elimination routes, and finally, the possible interference of the gel matrix with histogenesis. In addition, the accurate and instant space filling of a complex-shaped defect site is still a major challenge from a clinical point of view.

E. CELL ENCAPSULATION

Many diseases are closely tied to deficient or subnormal metabolic and secretory cell functions. Milder forms of these diseases can be managed by a variety of treatments. Currently, most enzyme and hormonal deficiencies are treated either by oral administration or by repeated injections of the missing substance. However, it is often extremely difficult or even impossible to imitate the metabolism and the complex roles of the hormone or enzyme that is not sufficiently produced by the body.

Hence, immunoisolated cell transplantation is one of the most promising approaches in regenerative medicine to overcome the limitations of current treatments. Cell encapsulation offers a method by which a substance can be released over long periods of time in a manner responsive to the metabolic needs of the individual patient. To date, a variety of primary cells including pancreatic islets, hepatocytes, and adrenal cortical cells have been encapsulated and studied using hydrogels.[47]

Another research direction utilizes encapsulated allogeneic and xenogeneic cells as well as genetically modified cell lines that release therapeutic substances out of immunoprotective microcapsules. Microcapsules technology aims at avoiding a lifetime of immunosuppressive therapy by excluding a clinically detectable immune response of the patient. Research in this direction has shown the feasibility of microcapsules based on hydrogels (particularly of alginate) for the transplantation of nonautologous cells and tissue fragments. Numerous technical accomplishments of the immunoisolation method have recently made the first successful long-term clinical applications (Phase I clinical trials) possible. However, full utilization of the potential of immunoisolated cell therapy requires several factors that have received limited attention in the past but that are important for the formulation of hydrogel-based immunoisolation systems that are highly versatile, potentially economical, and likely to win FDA approval.[48] This is because a number of barriers remain to be addressed, which include best source of functional cells, a stable, biocompatible membrane offering immune protection to the implant, a construct architecture ensuring cell viability and construct function, and the engineering of immune acceptance of the construct postimplantation.[49] The biomaterial-related issues are discussed in detail in the sections below.

III. SCAFFOLD MATERIALS

Today, the design and fabrication of scaffolds from a matrix point of view can be grouped into the following categories: (i) biodegradable and bioresorbable polymers that have been used for clinically established products such as collagen, fibrin glue, PEO, polyglycolide (PGA), polylactides (PLLA, PDLA), polycaprolactone (PCL), etc.; (ii) recent regulatory approved polymers such as polyorthoester (POE), polyanhydrides, polyhydroxyalkanoate (PHA), hyaluronic acid derivatives, and chitosan; and (iii) the entrepreneurial polymeric biomaterials, such as poly(lactic acid-co-lysine), which can selectively bond specific cell phenotypes and guide their differentiation and proliferation into the targeted functional premature or mature tissue.

A. SYNTHETIC POLYMERS

Synthetic degradable polyesters were adopted in surgery 30 years ago as materials for sutures, drug delivery, and bone fixation devices, and remain among the most widely used synthetic degradable polymers in medicine today. Degradable polyesters derived from three monomers lactide, glycolide, and caprolactone are in common clinical use and are characterized by degradation times ranging from days to years depending on composition and initial molecular weight. Hence, these types of polymers were most widely used to fabricate scaffolds in the early days of TE. However, over the last 10 years, a great number of new synthetic materials have been introduced to enhance the design, fabrication, and biological performance of scaffolds. A more comprehensive review on the work regarding synthetic polymer scaffolds is beyond the scope of this chapter. We recommend for further reading a selection of papers covering the topic in detail.[50–56]

B. NATURAL POLYMERS

Polymers of natural origin have frequently been used in TE applications because they are either components of or have macromolecular properties similar to the natural ECM. For example, collagens are the main proteins of the ECM and comprise 25% of the total protein mass of the body. Today, 20 different types of collagens have been identified.

Mechanical properties of collagens can be enhanced by using chemical cross-linkers (i.e., glutaraldehyde, formaldehyde, and carbodiimide), by cross-linking via physical treatments (i.e., UV irradiation, freeze-drying, and heating), and by blending them with other polymers (i.e., HA, PLA, PGA, poly(lactic-*co*-glycolic acid) (PLGA), chitosan, and PEO). Collagen is naturally degraded by metalloproteases (specifically, collagenase) and serine proteases, allowing for its degradation to be locally controlled by cells present in the engineered tissue. Several excellent reviews have summarized the advantages as well as the limitations of collagens in TE applications.[46,50,55,56]

Originally, fibrin glue was widely used as an adhesive in plastic and reconstructive surgery.[57] Fibrin glue has been shown to improve the percentage of skin graft take, especially when associated with difficult grafting sites or sites associated with unavoidable movement. Evidence also suggests improved hemostasis and a protective effect resulting in a higher immunity to bacterial infection. Fibrin, associated with fibronectin, has been shown to support keratinocyte and fibroblast growth both *in vitro* and *in vivo*, and may enhance cellular motility in the wound. When used as a delivery system for cultured keratinocytes and fibroblasts, fibrin glue may provide similar advantages to those proven with conventional skin grafts. Fibrin glue has also been shown to be a suitable delivery vehicle for exogenous growth factors that may in the future be used to accelerate wound healing.[58]

It can be argued that polysaccharides form a class of natural polymers that have been neglected in the earlier years of TE. Recently, recognition of their potential is growing in the field and is reflected by the number of papers published over the last 5 years. In particular, chitosan, alginate, and hyaloronic acid have been widely studied as cellular solid and hydrogel-based cell carriers.[59] Hence, the main part of this chapter will review the current knowledge and applications of polysaccharides as scaffold, matrix, and cell carrier in TE applications.

IV.   POLYSACCHARIDES

A.   Introduction

The growing interest in polysaccharide-based scaffold materials is a result of the following three factors: (A) the growing body of information pointing to the critical role of saccharide moieties in cell signaling pathways and in the area of immune recognition in particular; (B) the development of powerful and economical synthetic techniques that aim at automated synthesis of biologically active oligosaccharides. These techniques may allow us to finally decode and exploit the language of oligosaccharide signaling; and (C) the demand in TE research for new scaffold materials with specific, controllable biological activity in combination with different degradation and resorption kinetics.

B.   Agarose

Agarose is a seaweed-derived charged polysaccharide that forms a gel in the presence of supraphysiological concentrations of calcium ions. It exhibits a temperature-sensitive water solubility that can be utilized to entrap mammalian cells. A standard technique applied is that an agarose/cell suspension is transformed into liquid microbeads by extrusion or oil-in-water dispersion, which are then hardened by a reduction in temperature. Agarose entered TE as an experimental material for encapsulating endocrine cells, such as insulin-secreting pancreatic islets, for transplantation by injection. A homogeneous phase of agarose layer, without a permeoselective barrier, was shown to be sufficient for cell entrapment. A possible drawback, namely, the possibility of cellular protrusion through the agarose membrane, can be eliminated by coating the cells entrapped in agarose with an additional agarose layer.

Agarose gels seeded with chondrocytes have been utilized as vehicles for applying biological and mechanical stimuli to test factors that affect the *in vitro* and *in vivo* engineering of cartilage. Cell agarose constructs have been used to investigate components that affect chondrocyte metabolism, including various growth factors, cell morphology, synthesis of an extracellular matrix, *ex vivo* synthesis of cartilaginous material, cell phenotype, and the efficacy of cellular–agarose composites to promote repair of articular cartilage defects in animal models.[60,61]

Compressive strains applied to chondrocyte-seeded agarose matrices affect the synthesis of glycosaminoglycans (GAGs), DNA, and total protein.[62] Firm agarose gels support chondrocyte proliferation and are reported to sustain the differentiated phenotype. Even after chondrocytes propagated as a monolayer dedifferentiate as a consequence of serial passage, when propagated as a suspension culture in agarose gels they reexpress a differentiated phenotype as assessed by the reexpression of type II collagen.[63] Rabbit articular cartilage full-thickness defects implanted with agarose gels embedded with allograft chondrocytes exhibited significant healing when compared with controls. New subchondral bone formed and the newly synthesized tissue appeared to integrate with host articular cartilage.[64] However, at a very early stage, tissue engineers switched from agarose to alginates on account of biocompatibility-related issues.

C. ALGINATES

Alginates are copolymers made of β-D-mannuronic acid and α-L-guluronic acid and belong to a family of linear polysaccharides. Generally, they are produced by brown seaweeds and bacterial species belonging to the genera *Pseudomonas* and *Azotobacter*. They are composed of the two monomers, ManA and GulA linked to each other by 1–4 bonds. In contrast to most other polysaccharides, alginates do not contain repeating monomer sequence units along the polymer chains. Hence, there is a vast compositional heterogeneity among alginates isolated from different organisms and even among specimens obtained from different parts of the same organism.[65]

Bacterial alginates are acetylated to varying degrees, in contrast to those of seaweeds. The compositional variability is observed both at the level of fractional content of the two monomers, and also by the way they are sequentially distributed along the polymer chains. This variation in structural variability can only be adequately described in statistical terms.[66]

Alginate, currently commercially produced from the marine brown algae, can also be biologically produced by bacteria such as *Azotobacter vinelandii*, *A. chroococcum*, and several species of *Pseudomonas*. The ever-increasing applications of this polymer in the food and pharmaceutical sectors have led to the execution of major research programs, with objectives such as the better understanding of metabolic pathways, their physiological or biological functions, the regulation of their formation and composition, and the optimization of the microbial production process from a yield as well as quality point of view.[67]

The ability of alginate to form two types of gels dependent on pH, i.e., an acid gel and an ionotropic gel, gives the polymer unique properties compared with other natural macromolecules. The molecule can be tailor-made for a number of applications.[68] So far, more than 200 different alginate grades and a number of alginate salts are manufactured. The potential use of its various qualities as a pharmaceutical excipient has not been evaluated fully, but alginate is likely to make an important contribution in the development of polymeric drug delivery systems. However, as a prerequisite, it needs to be produced in an ultrapure form suitable for *in vivo* applications.

A number of purification procedures for alginates have been described[69] that do not interfere with the molecular composition of the alginate. These techniques are based on filtration, precipitation, and extraction steps. Purification substantially reduces the host tissue response, for example, when ultrapurified alginate is used for cell encapsulation, but the *in vivo* studies demonstrated two other significant effects in the clinical application. First, the duration of islets function is substantially prolonged but still limited to periods of 3 months to a year in spite of a virtual absence of host response.[70,71] Second, fibrous tissue overgrowth is always found in a small portion of the capsules. This latter observation shows that the biocompatibility of the microcapsules is influenced not only by matrix impurities, but also by the lack of reproducibility in the manufacturing process, and hence, batch-to-batch variations.

1. Wound Dressing

Dressings have been applied to open wounds for centuries. Traditionally, they have been absorbent, permeable materials, such as gauze, which could adhere to desiccated wound surfaces, inducing trauma on removal. With the advent of modern wound care products many dressings are now capable of absorbing large volumes of exudate, while still continuing to provide a moist wound healing environment. Equally important is their ability to lock exudate in the dressing (i.e., bacterial retention within the dressing matrix) such that upon removal from a wound surface, bacterial dispersion is minimized. Hydrocolloid dressings based on alginates are gaining increasing attention in this context.[72,73]

Hashimoto and co-workers[74] developed a new alginate dressing linked covalently with hybrid peptides. The hybrid peptides promoted attachment of NHDF, whereas neither Ac-KSIKVAV nor Ac-KVGVAPG promoted attachment. Although all the peptides we examined promoted the proliferation of NHDF to some extent, the hybrid peptide-coated plates showed strong NHDF proliferative activity compared with the other peptide. Our laboratory extended this work by creating alginate dressings linked with some of these peptides and examined their effectiveness in wound healing using a rabbit ear skin defect model *in vivo*. Ears with the alginate dressings linked with the hybrid peptides showed significantly greater epithelialization and a larger volume of regenerated tissue nine days after operation, compared with those treated with SIVAV-linked, VGVAPG-linked, and unlinked alginate dressings. These new alginate dressings linked with the hybrid peptides could be promising dressings especially for wounds with impaired healing.

2. Drug Delivery

The clinical requirements for extended and enhanced control of drug administration have increased the demand for custom-made polymers in the field of drug delivery. Based on this background, different formulations of alginates were developed.[75] The conventional use of alginate as a drug carrier in the various forms of matrices generally depends on its thickening, gel forming, and stabilizing properties. Mooney's group[76–78] did develop a major research thrust by using these highly versatile biopolymers of natural origin. They developed, characterized, and studied a number of alginate-based release systems. For example, vascular endothelial growth factors (VEGF) have been incorporated into ionically cross-linked alginate hydrogels.[79,80] *In vivo*, the growth factor is released from alginate matrices both by diffusion and by mechanical stimulation. Mooney's group concluded that the efficacy of this system has been demonstrated both *in vitro* and *in vivo* to enhance angiogenesis.

At the beginning of the 1990s, basic fibroblast growth factor (bFGF) was incorporated into heparin-alginate gels with the overall goal of enhancing angiogenesis.[81] Subsequently, the release system was studied in an animal model[82] and based on it, has proven efficient in a clinical pilot study.[83]

However, for some wound healing and TE applications it might be desirable to have an even slower release and also improve the alginate's biological activity after its incorporation into the hydrogel system.[84] *In vitro* experiments showed that a matrix made of heparin and alginate covalently cross-linked with ethylenediamine could significantly suppress the initial burst of bFGF. It was shown *in vitro* that bFGF was released over a period of 1 month under physiological conditions. *In vivo* cellular infiltration and angiogenesis were shown to occur in the drug delivery system, which was implanted subcutaneously in the dorsal area of rat with 1.0 mg of bFGF for up to 2 weeks.[85]

3. Cell Encapsulation

Alginate is also widely used to encapsulate cells of various derivations.[86,87] Human embryonic kidney 293 cells have been transfected with the gene for endostatin. These cells have been encapsulated in calcium cross-linked alginate gels and optimized for the secretion of endostatin. Alginate gel beads implanted into rat brain have shown only a moderate loss in cell viability and extended endostatin release for periods of up to 12 months.[88] After the transplantation of alginate microencapsulated porcine neonatal pancreatic cell clusters into the peritoneal cavity of streptozocin-induced diabetic B6AF1 mice, blood glucose levels were normalized in 81% of the animals that had received a transplant and remained normal until the termination of the experiments at 20 weeks. Hyperglycemic blood glucose levels after the explantation of the capsules confirmed the function of the encapsulated cells. The insulin content of the encapsulated cells was increased tenfold at 20 weeks after transplantation compared with pretransplantation levels.[89]

In a three-stage total hepatectomy rat model, survival time was longer in the rats given the microencapsulated hepatocytes than in the control rats. The blood ammonia concentrations increased soon after total hepatectomy but remained significantly lower in the rats with microencapsulated hepatocytes.[90] Alginate could also be modified by certain bioactive substances to improve its biocompatibility. Although it cannot completely eliminate the detrimental effects of poly-L-lysine on biocompatibility, the use of an epimerized coating on alginate–poly-L-lysine–alginate can improve the biocompatibility of such capsules.[91]

Dvir-Ginzberg and colleagues[92] reported that the hydrophilic nature of the alginate scaffold as well as its pore structure and interconnectivity enabled the efficient seeding of hepatocytes into the scaffolds, i.e., 70 to 90% of the initial cells depending on the seeding method. In the high-density cellular constructs, hepatocellular functions such as albumin and urea secretion and detoxification (cytochrome P-450 and phase II conjugating enzyme activities) remained high during the 7-day culture. The latest work on liver[93] and pancreas[94] TE was reviewed recently.

Based on the evidence that chondrocytes suspended in alginate gel have been shown to produce a substantial cell-associated matrix, Chia and co-workers[95] developed a new chondrocyte-expansion strategy. The objective of their study was to determine whether cartilage tissue could be generated using the alginate-recovered chondrocyte (ARC) method, in which chondrocytes are cultured in alginate as an intermediate step in tissue fabrication. Nasal septal chondrocytes from five patient donors were isolated by enzymatic digestion and then expanded in a monolayer culture. At confluency, a portion of these cells were seeded at high density onto a semipermeable membrane and cultured for 14, 21, or 28 days (monolayer group). The remaining cells were suspended in alginate and cultured until a cell-associated matrix was observed (10 to 17 days). The cells and their associated matrices were released from alginate (ARC group), seeded onto a semipermeable membrane, and cultured as already described. DNA (Hoechst 33258 Assay), glycosaminoglycan (GAG; dimethylmethylene blue assay), and collagen (hydroxyproline assay) were analyzed biochemically. Immunohistochemistry was performed to assess expressions of collagen type I and type II. Histochemistry was performed to localize cells accumulating sulfated GAG (Alcian Blue stain). The ARC constructs, in contrast to the monolayer constructs, had substantial structural stability and the histologic and gross appearance of cartilaginous tissue. ARC constructs demonstrated significantly greater GAG and collagen accumulation than monolayer constructs ($P<0.05$). Histologic analysis revealed substantial GAG and collagen type II production and only moderate collagen type I production. The composition of the matrix was thus similar to that of native human septal cartilage. The authors conclude that tissue-engineered human nasal septal cartilage using the ARC method has the histologic and gross appearance of native cartilage and a biochemical composition more like that of native cartilage than monolayer constructs.

4. Scaffold/Cell Constructs

Porous scaffolds made of RGD-modified alginate have been implanted into rats. The authors report that their data showed minimal immune response, modest tissue encapsulation, and good tissue ingrowth.[96]

Alginate-based hydrogels can be designed to have similar physical properties to cartilage, which is a highly hydrated tissue composed of chondrocytes embedded in type II collagen and GAGs. To date, numerous alginate systems mixed or seeded with chondrocytes have been designed and tested both *in vitro* (unpublished data) and *in vivo*. A large amount of information is available in the literature describing the use of alginate matrices of different types and forms in the field of articular and elastic cartilage TE (unpublished data; Figure 29.2A–F).

Alginate-based hydrogels have been used in many studies to deliver chondrocytes and are injected either subcutaneously or into different defect sites.[97–101] More recently, Bonassar's group[102] developed a concept to fabricate alginate/cell constructs by using a silicone-based molding technology. In a follow-up study, Vacanti's group[103] evaluated the novel technique of using a mold to create a complete, anatomically refined auricle in a large animal model. Mixtures of autogenous chondrocytes and biodegradable polymers were used inside a perforated, auricle-shaped hollow gold mold. Three biodegradable polymers (calcium alginate, pluronic F-127, and polyglycolic acid) were used to retain the seeded chondrocytes inside the mold. These molds, along with a control, were implanted subcutaneously in the abdominal area of ten animals (pigs and sheep). The constructs were removed after 8 to 20 weeks and were assessed by gross morphology and histology. All the gold implants were well tolerated by the animals. The implants using calcium alginate ($n=3$) generated constructs of the exact shape and size of a normal human ear; the histology demonstrated mostly normal cartilage with some persistent alginate. The implants with pluronic F-127 ($n=3$) resulted in cartilage with essentially normal histology, although leakage outside the molds and external cartilage generation was noted. Polyglycolic acid implants ($n=3$) produced no useful cartilage because of an inflammatory reaction with fibrosis. The control ($n=1$) demonstrated only a very small amount of fibrous tissue inside the mold.

From these animal studies it can be concluded that chondrocytes remain viable and produce ECM proteins consistent with cartilage as early as 4 weeks after implantation. However, the implanted constructs remain mechanically weak, achieving a compression modulus that is only about 15% that of the native cartilage. They also show a hydraulic permeability nearly 20 times that of the natural cartilage.

Liu and co-workers[104] confirmed earlier results from other groups[105] showing that chondrocytes seeded into alginate gels could maintain and regain their phenotype. Dedifferentiated chondrocytes expanded in monolayer cultures, regained their spherical shape, and actively proliferated. However, complete dedifferentiation took a period of several weeks, after which the synthesis of normal aggregating proteoglycans was maintained.

Alginate gel was used as a matrix to differentiate mesenchymal stem cells into chondrocytes.[106] Paige and colleagues[107] used alginate as a delivery system for isolated chondrocytes by means of injection to fabricate cartilage in nude mouse model. Pearly opalescent and firm specimens could be harvested as early as 6 weeks after injection. Yang and co-workers[108] postulated that the polymerization of alginate hydrogels can be controlled to allow injection of chondrocytes that produce new autologous cartilage at a subcutaneous dorsal site in rabbits.[108]

Hydrogel scaffolds are also being widely used in the area of nonload-bearing bone TE engineering because implantation can be performed in a minimally invasive

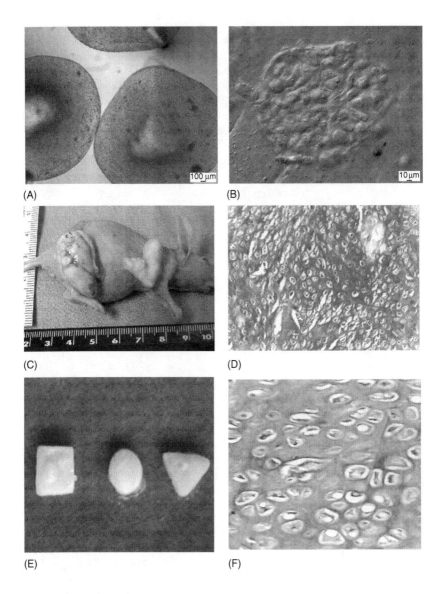

FIGURE 29.2 (A, B): Numerous alginate systems have been designed and tested as chondrocytes carrier both *in vitro* and *in vivo*. Chen et al. (unpublished data) studied proliferation and neotissue formation of chondrocytes embedded alginate *in vitro* (2A, B) and *in vivo* (2C–F). (C, D) Newly formed tissue in nude mouse model 2 months after injection of chondrocyte-alginate cultures (seeding density was 5×10^6 chondrocytes/mL). Safaranin-*O* staining of explanted specimen indicated the existence of significant amounts of GAG. (E, F) Gross morphology of cartilage grafts formed in predetermined shapes 2 months after implantation. Light micrographs of chondrocyte/alginate gel grafts fixed and stained with HE staining after 2 month implantation analysis showed cartilage-like tissue formation in specimen (F).

manner. In general, hydrogels do not possess the mechanical strength required in load-bearing applications; hence they are often used in combination with a scaffold that gives sufficient mechanical properties to the tissue-engineered graft. For example, Vacanti et al.[109] injected periosteal cells in combination with alginate into a coral-derived calcium phosphate scaffold to treat a part of the thumb of a patient via bone TE.

The degradable form of high-grade alginate, PAG, was mixed with primary rat calvarial osteoblasts, gelled, and injected into the backs of immunodeficient mice. In this system, mineralized bone was observed to form over a period of 9 weeks. However, due to diffusion limitations, this mineralization was limited to the outer one third of the implant.

Alginate scaffolds have also been designed that contain covalently bound adhesion and signaling peptides to specifically influence cell fate and bone formation. The presence of the adhesion peptide RGD significantly increased the amount of bone formation in an alginate scaffold when compared with gels without covalently bound peptide.[110] Another study indicated that a BMP-2-derived oligopeptide promoted ectopic bone formation when immobilized in alginate gels.[111] These studies indicated that cell interaction properties of hydrogels can be designed to enhance and possibly direct *in vivo* bone formation.

Alsberg and co-workers[112] studied the cotransplantation of chondrocytes and osteoblasts in alginate hydrogels modified with an RGD peptide sequence. New bone tissue was formed that grew in mass and cellularity by endochondral ossification in a manner similar to normal long-bone growth. The authors concluded from their histology that the transplanted cells organized into structures that morphologically and functionally resembled growth plates.

Steven and colleagues[113] developed a rapid-curing alginate gel system and demonstrated its utility as a scaffold for periosteum-derived chondrogenesis for articular cartilage TE applications. A homogeneous, mechanically stable gel was formulated by inducing gelation of a 2.0% (w/v) solution of a high G content alginate (65 to 75% G) with a 75 mM solution of $CaCl_2$. The gel exhibited near-elastic behavior at low levels of deformation (15%, $R^2 = 0.996$), Young's modulus of $0.17 +/- 0.01$ MPa, and rapid gelation kinetics (<1.0 min to completion). However, the *in vitro* cell culture of chondrocytes in the gel yielded alginate/cell constructs that lacked the continuous, interconnected collagen/proteoglycan network of hyaline cartilage.

D.   CHITIN/CHITOSAN

Chitosan is a semicrystalline polymer and the degree of crystallinity is a function of the degree of deacetylation. The degree of crystallinity is maximum for both chitin (i.e., 0% deacetylated) and fully deacetylated (i.e., 100%) chitosan. Minimum crystallinity is achieved at intermediate degrees of deacetylation. Owing to its stable, crystalline structure, chitosan is normally insoluble in aqueous solutions above pH 7. However, in dilute acids, the free amino groups are protonated and the molecule becomes fully soluble below ~pH 5. The pH-dependent solubility of chitosan provides a convenient mechanism for processing under mild conditions. Viscous solutions can be extruded and gelled in high pH solutions or baths of nonsolvents such as methanol. Such gel fibers can be subsequently drawn and dried to form high-strength fibers.[114] A strong reason for considering chitosan as a biomaterial is its abundance in nature. It is found in the shells of all shellfish such as crabs and shrimps and is second only to cellulose as the most abundant natural biopolymer.[115,116]

In the 1990s, chitosan was utilized in large amounts by the pharmaceutical industry.[117] The natural polymer is used in direct tablet compression, as a tablet disintegrant, for the production of controlled-release solid dosage forms or for the improvement of drug dissolution. Chitosan has been used for the production of controlled-release implant systems, for example, for the delivery of hormones. Recently, the transmucosal absorption-promoting characteristics of chitosan have been exploited for vaccine delivery,[118,119] and especially for nasal and oral delivery of polar drugs that could be combined with peptides and proteins.

Depending on the source and fabrication technique, the average molecular weight of chitosan's can range from 50 to 1000 kDa. Commercially available medical-grade materials have degrees of deacetylation ranging from 50 to 90%. Structurally, chitosan is a linear polysaccharide consisting of[120] linked D-glucosamine residues with a variable number of randomly located N-acetylglucosamine groups. It thus shares some characteristics with various GAGs and hyaluronic acid. GAGs have many specific interactions with growth factors, receptors, and adhesion proteins, and it is suggested that the molecular chains in chitosan may also have related bioactivities. In fact, chitosan oligosaccharides have been shown to have a stimulatory effect on macrophages, and the effect has been linked to the acetylated residues. Furthermore, chitosan and its parent molecule, chitin have been shown to exert chemotactic effects on neutrophils *in vitro* and *in vivo*.[120,121]

In general, these natural polymers have been found to evoke a mild foreign body reaction. In most cases, no major fibrous encapsulation has been observed after implantation. Formation of normal granulation tissue, often with accelerated angiogenesis, appears to be the typical course of healing. A significant accumulation of neutrophils in the vicinity of the implants is often seen, but this dissipates rapidly, and generally, a chronic inflammatory response does not develop. The stimulatory effects of chitosan and chitosan fragments on immune cells may play a role in inducing local cell proliferation and ultimately the integration of the implanted material with the host tissue.

In vivo, chitosan is degraded by enzymatic hydrolysis. The key players are lysozymes, which appear to target acetylated residues. However, there is some evidence that some proteolytic enzymes show low levels of activity with chitosan as well. The degradation products are chitosan oligosaccharides of variable length. The degradation kinetics appear to be inversely related to the degree of crystallinity, which is controlled mainly by the degree of deacetylation. Highly deacetylated forms (e.g., >85%) exhibit the lowest degradation rates and may last several months *in vivo*, whereas samples with lower levels of deacetylation degrade more rapidly. This issue has been addressed by derivatizing the molecule with side chains of various types. Such treatments alter molecular chain packing and increase the amorphous fraction, thus allowing more rapid degradation. They also inherently affect both the mechanical and solubility properties.[122]

Ikada and Tomihita[123] studied the degradation of chitin and chitosan films *in vitro* and *in vivo* in a rat model. The rate of *in vivo* degradation was high for chitin, reducing as the degree of deacetylation increased. Interestingly, these authors noted a mild foreign body reaction to chitosan. Onishi and Machida[124] investigated the *in vivo* biodegradation of water-soluble chitosan using a mouse model. They report that approximately 50% deacetylated material was found to be degraded within the first few days and cleared by the metabolization route of the animal. Tanaka and co-workers[125] found that when chitin and chitosan were administered orally or injected intraperitoneally by using a mice model, chitosan invoked a stronger reaction compared with chitin.

1. Wound Healing

The area of wound healing has been another major application of chitin- and chitosan-based matrices. Chitin's monomeric unit, *N*-acetylglucosamine, occurs in hyaluronic acid, an extracellular macromolecule that is important in wound repair. Therefore, chitin should possess the characteristics favorable for promoting rapid dermal regeneration and accelerated wound healing suitable for applications extending from simple wound coverings to sophisticated artificial skin matrices.[126]

Generally, chitosan has been proposed to enhance wound healing by retarding fibroblast proliferation and thereby inhibiting fibroplasia, and promoting organized tissue reconstruction. Chitosan applied to abdominal skin and subcutaneous incisions resulted in healing without fibroplasia or scarring.[128] Muzzarelli et al. used chitosan and chitosan ascorbate to replace dura mater in cats and reported complete polymer disappearance and normal tissue regeneration in 60 days.[129] Treating various dog tissues with chitosan solution resulted in the inhibition of fibroplasia (scarring) and enhanced tissue regeneration.

From a soft-tissue regeneration perspective, it is particularly interesting to note that chitosan-treated Goretex® vascular grafts[127] formed a unique blood clot layer that enhanced the proliferation of endothelial cells (progenitor cells from the lumen of an arterial graft) compared with the untreated Goretex graft. Unlike the typical clot with platelets and fibrin, the unique blood clot formed by chitosan consisted of a homogeneous layer of red blood cells that promoted the propagation of endothelium and smooth muscle over the clot layer. The chitosan-treated Goretex formed a clot layer that completely covered the surface of the graft, while the untreated Goretex graft material was not covered by any form of a clot. Examination at 1 to 4 months showed the chitosan-treated grafts to be encased in smooth muscle with a living endothelial cell lining. As is well known and expected, the untreated grafts were well tolerated but surrounded by fibrous connective tissue (i.e., scar). It is not determined whether the regeneration of tissues lining the graft was induced by the presence of chitosan or the layer of red blood cells aggregated by chitosan. In either case, chitosan appears to have prevented fibroplasia while enhancing the regeneration of normal tissues by progenitor cells present at the wound margin.

Okamoto and co-workers[130] reported that the migration of cells was affected by chitosan and glucosamine. Based on their study, the authors conclude that further work should be performed to elucidate the mechanisms that chitin and chitosan invoke during wound healing. Hence, in a second study, the group concluded that chitin, chitosan, and glucosamine affected the proliferation of mouse 3T6 fetal fibroblasts or HUVEC (human umbilical vascular endothelial cells).[131]

Several investigators, using various animal models, have evaluated the effect of chitosan on wound healing. The wound-healing acceleration properties of chitin and chitosan are well assessed in the literature. Minami and co-workers[131] noted accelerated regeneration of tissue with no visible scarring when treating various types of infected livestock wounds with chitin/polyester nonwoven dressings, chitin-cotton, and chitosan-cotton wound-filling materials. The degree of acceleration of the wound-healing process was determined in animal tests by measuring the tensile strength of the newly formed tissue of the wound.

Similar results were observed with domestic pets utilizing the same materials. In another study, 99% healing effectiveness in postoperative and traumatic wounds in livestock, new and old (nonhealing), and greater than 90% healing of infectious hoof inflammation using a topical chitosan spray were reported. Clinical results using the same product on human chronic leg ulcers showed increased granulation and epidermis formation and accelerated healing.[132]

Fujinaga and colleagues[133,134] studied the effect of chitosan on wound healing. They found that chitosan accelerated the infiltration of polymorphonuclear (PMN) cells at the wound site. PMN also stimulated the production of osteopontin that promotes cell attachment, essential in tissue reorganization at the wound site.

Ko and co-workers[135] reported the preparation of a wound-healing material based on water-soluble chitin (WSC) obtained by carefully deacetylating chitin to about 50% N-acetyl content. In a comparative study of chitin, chitosan, and WSC powders and WSC solution, wounded skin treated with WSC solution was found to have the highest tensile strength with the healing rate fastest for WSC solution, followed by WSC powder, chitin powder, and finally, chitosan powder. Similarly, Ueno and colleagues[136] demonstrated that chitosan in the form of chitosan–cotton was an accelerator of wound healing by playing a role in the infiltration of PMN cells at the wound site, an event essential in accelerated wound healing. Recently, Mizuno and co-workers[137] also concluded that the incorporation of bFGF into a chitosan matrix accelerated the rate of healing.[137]

The application of wound fluid-absorbing chitin beads has been proposed as a dressing material, too.[138] Alternatively, Ishihara and co-workers[139,140] have developed an azide-containing chitosan material, which, when activated by UV light, generates cross-links between the chitosan chains and an insoluble hydrogel. The dressing was studied in mice and showed accelerated wound closure and healing.

Mi and colleauges[141] developed an asymmetric chitosan membrane by immersion-precipitation phase-inversion method. This wound dressing consists of skin surface on top layer supported by a macroporous sponge-like sublayer. The thickness of the dense skin surface and the porosity of the sponge-like sublayer can be controlled by the modification of the phase-separation process using the per-evaporation method. The asymmetric chitosan membrane showed controlled evaporative water loss, excellent oxygen permeability, and promoted fluid drainage ability, and could also inhibit exogenous microorganism invasion due to the dense skin layer and inherent antimicrobial property of chitosan. The wound covered with the asymmetric chitosan membrane was hemostatic and healed quickly. Histological examination confirmed that the epithelialization rate was increased and the deposition of collagen in the dermis was well organized by covering the wound with this asymmetric chitosan membrane.

In an attempt to improve on the tensile properties of keratin films, Tanabe and colleagues included 10 to 30% chitosan to make a composite film.[142] Such a dressing exhibited improved resistance to bacterial growth and supported fibroblast attachment and proliferation. The direct use of *in situ* chitin with fungal mycelia from the fungus *Ganoderma tsugaue* to produce wound-healing sacchachitin membranes has also been demonstrated.[143] A nonwoven mesh obtained by first processing the mycelia to remove protein and pigment, followed by the isolation of fibers in the 10 to 50 μm diameter range, and final consolidation into a freeze-dried membrane under aseptic conditions was used in a wound model study. The initial cell culture (rat fibroblasts) and *in vivo* studies indicated good biocompatibility and immunogenicity.[144] The wound-healing properties of this fungal-based nonwoven mesh were further evaluated in two animal models.[145]

Chitin was combined with polyethylene glycol to form a partial gel-like wound dressing.[146] Results from animal studies showed improved wound healing in cases where such a gel concept was used. In another study, the preparation of a bilayered chitosan membrane by immersion–precipitation phase inversion proved effective for treating an infected wound site.[147] This membrane made use of a thin layer of chitosan as the antibacterial and moisture control barrier attached to a sponge layer that could absorb wound exudates. The membrane adhered well to the wound surface and promoted wound healing. This same bilayered chitosan membrane was subsequently loaded with silver sulfadiazine. The release of sulfadiazine showed a burst effect, while silver displayed a sustained longer term release. The combination release was found to be effective in controlling *Pseudomonas aeruginosa* and *Staphyloccocus aureus* populations in cell culture for up to 1 week.[148] A chlorhexidine containing chitosan-based wound dressing had antibacterial efficacy toward a number of bacteria at the wound site, namely *P. aeruginosa* and *S. aureus*.[149] The bilayered dressing was fabricated by combining two separately produced films, a carboxymethylated chitin hydrogel, which provided the exudate-absorbing component, interfaced with a chlorhexadine-loaded chitosan film.

Chitosan in combination with alginate as polyelectrolyte complex (PEC) films has also been prepared and evaluated as a wound-dressing materials.[150] These water-insoluble alginate polyelectrolyte (CS–AL PEC) membranes displayed greater stability to pH changes and are more effective as a controlled-release matrix than either the chitosan or alginate on its own.[151] Subsequently, it was found that separating the chitosan-alginate coacervates from the unreacted polymers prior to casting further enhanced film homogeneity.[152] The PEC membranes were found to promote accelerated healing of incisional wounds in a rat model.[153] Wounds closed at 14 days postoperatively and histological observations showed a mature epidermal architecture with a keratinized surface of normal thickness.

Hirano and co-workers[154] developed a matrix made of chitosan fibers, which were combined with hyaluronic acid, chondroitin sulfate, dermatan sulfate, and heparin. Filaments of approximately 25 cm were fabricated into cotton-like staple fibers. The fiber materials were found to contain between 5 and 33% of GAGs. The incorporation of the GAGs decreased the mechanical properties but accelerated the wound-healing process due to their biological activity.

Dung and co-authors[155] published the results of a clinical study in which chitin membranes (Vinachitin™) prepared from decrystallizing ricefield crabshells were used. In a 3-year clinical trial, the group treated more than 300 patients for deep burns, orthopedic trauma, and ulcers.

2. Drug Delivery

Chitosan possesses properties that make this polysaccharide of interest in sustained-release applications.[156] The work on N-succinylchitosan as a drug carrier, water-insoluble and water-soluble, was recently reviewed.[157,158] Lee and co-workers reported their work on the synthesis of poly(ethylene glycol) macromer and β-chitosan. The polymers were subsequently cross-linked via UV light to present semi-IPN hydrogels.[159] The gels were found to have high-equilibrium water content with reasonable mechanical strengths. Gupta and Ravikumar[160] developed a similar approach to devise a drug containing poly(ethylene glycol)/chitosan microspheres, using glutaraldehyde as the crosslinking agent.[160] The authors conclude that such a matrix allows up to 93% drug-loading capacity and that a near zero-order drug release profile can be obtained.

A novel inorganic–organic pH-sensitive membrane based on an interpenetrating network utilizing inorganic silicate and organic chitosan has been developed for drug delivery purposes.[161] The membrane was evaluated for its response to pH changes. The release of model compounds such as lidocaine–HCl, sodium salicylate, and 4-acetomidophenol was studied. The authors conclude that the matrix was sensitive to the external pH as well as the drugs' ionic interactions with chitosan. The membrane might also be sensitive to other factors such as temperature and light, which give new opportunities to design novel concepts for drug loading.

The *in situ* light-initiated polymerization of acrylic acid in the presence of chitosan was used to derive a novel membrane.[162] The interactions between the two polymers were determined to be based on hydrogen bonding. The strong adhesive property of this membrane rendered it suitable for transmucosal drug delivery applications. The loading and release study of triamcinolone acetonide (TAA), a drug used to reduce inflammation in the treatment of mouth ulcers, was subsequently reported.[163]

Another method of fabricating membranes for drug delivery is to make use of a combination of oxidized glucose dialdehyde and chitosan.[164] N-alkyl groups of varying chain lengths were used to modify the hydrophobicity of these chitosan membranes. Using vitamin B_2 as the drug model, the permeation and diffusion of B_2 decreased with an increase in pH and when the hydrophobicity of chitosan increased, i.e., as the alkyl chains increased in length. *In vitro* studies indicated no toxic side effects for this membrane system.

Another group[165] developed films made of chitosan-gelatin mixtures, which were fabricated without the need for crosslinking. A herbal extract was incorporated and implanted for delivery in the abdominal cavity of rats for up to 28 days. Degradation was accelerated when the gelatin content in the membrane increased.

The results indicated the potential of this drug delivery system in the treatment of anastomosis and muscle/tissue healing.

A number of groups developed microspheres and nanospheres made of chitosan. Mi and co-workers[166] produced microspheres that were cross-linked with ethylene glycol diglycidyl ether.[166] Subsequently, the group studied the influence of chemically altered carriers on the adsorption efficiency of the Newcastle disease virus antigen. A novel injectable chitosan-based delivery system with low cytotoxicity was also fabricated.[167] The chitosan microspheres with small particle size, low crystallinity, and good sphericity were prepared by a spray-drying method followed by treatment with a cross-linker. In this study, a naturally occurring cross-linking reagent (genipin), which has been used in herbal medicine and in the production of food dyes, was used to cross-link the chitosan microspheres. The group then developed chitin/PLGA- and chitin/PLA-based microspheres.[168] These biodegradable microspheres were prepared by polymer-blending and wet phase-inversion methods. The parameters such as selected nonsolvents, temperature of water, and ratio of polylactide/polyglycolide were adjusted to improve the thermodynamic compatibility of the individual polymers (chitin and PLGAs or chitin/PLA), which affects the hydration and degradation properties of the blend microspheres. A triphasic pattern of drug release is observed from the release of protein from the chitin/PLGA and chitin/PLA microspheres: the initially fast release (the first phase), the following slow release (the second phase), and the second burst release (the third phase). Formulations of the blends, which are based on the balance between the hydration rate of the chitin phase and the degradation of chitin/PLA and PLGA phase, can lead to a controllable release of bovine serum albumin (BSA).

Another group compared beads and microgranules made of chitosan and loaded with the antibiotic disodium diclofenac.[169] The authors concluded that the microgranules allowed better control when compared with the drug release profile of the beads. However, they also suggested that fine-tuning of the system was needed to get the desired zero-release curve. Onishi and colleagues[170] developed microspheres utilizing the complex co-acervation ability of Fe (III) ions with CM-chitin.[170] The addition of polyethylene glycol was found to be a condition *sine qua non* for the maintenance of the microspheres' spherical shape when they were subjected to a freeze-drying process. A mice study showed that 60% of the CM-chitin microspheres were cleared within 3 h after intravenous administration. However, the majority of the microspheres were retained by the liver and spleen. Hence, the authors propose that these microspheres could be used to target drug delivery to the liver.

Spray drying was applied to fabricate chitosan microspheres with small and spherical particle size.[171] The rationale behind this processing concept is that small size and the long *in vivo* half-life make this technique suitable for applications requiring injectable long-acting drug delivery systems. Chitosan microcores, in the range of 2 to 20 μm containing vitamin D_2, were fabricated via this process.[172] Attempts to coat the microcores with ethylcellulose produced nonspherical microparticles, attributed to the heterogeneous distribution of ethylcellulose on the chitosan microcore surface. Vitamin D_2 loading was determined to be above 80%. The release of vitamin D_2 was a function of the molecular weight of chitosan: the higher the molecular weight, the faster the release of vitamin D_2. The release was delayed if the microcores were coated with ethylcellulose.

Another group also worked on the concept to combine natural polymers with synthetic ones to increase the release rate.[173] An FDA-approved polylactic/glycolic acid (PLGA) copolymer was added to chitin solution to form a blend into which a drug could be added, and was subsequently made into microspheres. More rapid degradation was achieved when chitin was blended with PLGA, compared with PLGA alone. Similarly, drug release was more rapid when the chitin content was high. The group reports that their system is aimed at the delivery of cancer drugs.

Chitosan-based nanospheres have also been synthesized.[174] By means of the *in situ* polymerization of acrylic acid in the presence of chitosan, nanoparticles were subsequently obtained. The yield of nanoparticles was found to be a function of the molecular weight; the lower the molecular weight, the better the yield, with a maximum of around 70%. As a model drug, a silk peptide was incorporated into the nanoparticles. The release rate was defined as an initial burst followed by continued release for up to 10 days. The dependence of the release on pH was noticed, and the authors suggested that the nanospheres might find applications in the gastric cavity.

Takechi and colleagues[175] utilized chitosan as a major component of a local drug delivery system based on antiwashout fast-setting apatite cement. The antibiotic was incorporated during the paste formation process. The antibiotic displayed an initial burst followed by a decrease in the rate profile. Further histological evaluation on these materials clearly demonstrate the effectiveness of this antiwashout FSCPC-chitosan material that was attributed to the stronger initial mechanical properties achieved with this mixture.[176,177] In another variation, Zhang and Zhang[178] added calcium glass and β-TCP into β-chitosan solutions followed by freeze drying of the mixture to produce a scaffold.[178] The chitosan scaffold was soaked in gentamycin sulfate solution for 48 h to absorb the antibiotic. The release of antibiotic was found to be better mediated in the presence of the calcium additives. Cell culture results based on the MG63 osteoblast cell line indicated that the scaffold has the potential to be further studied *in vivo* for bone regeneration.

3. Gene Delivery

Originally, adenovirus, retrovirus, and adeno-associated virus have been used for gene delivery. Among them, adenovirus has several advantages over the others because of its broad host range, high viral titer, high gene transfer efficiency and its ability to infect nonproliferating cells. Thus, adenovirus is widely used for gene delivery to mammalian cells *in vitro* and *in vivo*. However, adenovirus also has serious drawbacks from a clinical point of view: transient gene expression following adenovirus infection because of the episomal location of adenovirus genome in the cells, cellular toxicity, immunogenicity, wild-type reversion and low infectivity to certain cells expressing low levels of the adenovirus receptor.[179]

Recent studies have shown the enhancement of adenovirus infectivity with several nonviral vectors such as cationic liposomes/polymers.[180] But although liposomes formed from cationic phospholipids offer several advantages over viral gene transfer, for example, low immunogenicity and ease of preparation, the success of the liposomal approach is limited. The toxicity of the cationic lipids and their relatively low transfection efficiency compared with viral gene delivery vectors are the main disadvantages.

DNA–polymer complexes involving cationic polymers, on the other hand, are more stable than cationic lipids. However, compared with viral vectors, the efficiency of gene delivery by synthetic cationic polymers is still relatively low. While it is possible to synthesize a large number of structurally different synthetic nonviral gene delivery vectors, by contrast, only a small number of polycations of natural origin are available. Yet, selected natural polymers may present a number of characteristics beneficial to gene delivery. Biocompatibility, low immunogenicity, and minimal cytotoxicity can render such polymers a good alternative to viral-or lipid-mediated transfections.[181]

Benchmarked against other natural polymers, chitosan is an inexpensive, biocompatible, biodegradable, and nontoxic cationic polymer that can form polyelectrolyte complexes with DNA. Hence, chitosan and chitosan derivatives have been studied as potentially safe and efficient cationic carriers for gene delivery. The transfection efficiency of chitosan as a gene delivery vehicle has been studied.[182] The molecular weight of chitosan, the charge ratio between the luciferase plasmid and chitosan, and the pH of the culture media were found to be determinants of the transfection efficiency *in vitro*. A more recent example is the use of chitosan for enhancing adenovirus infectivity in mammalian cells in gene therapy.[183] Lower concentrations of chitosan and lower molecular weight chitosans were better at enhancing adenovirus activity.

Erbacher and co-authors[184] concluded that plasmid DNA formulated with chitosan produced homogenous populations of complexes that were stable and had a diameter of approximately 50 to 100 nm. Discrete particles of condensed DNA had various geometries. Chitosan-DNA complexes efficiently transfected HeLa cells, independent of the presence of 10% serum, and did not require an added endosomolytic agent. In addition, gene expression gradually increased over time, from 24 to 96 h, whereas under the same conditions the efficacy of polyethylenimine-mediated transfection dropped by two orders of magnitude. At 96 h, chitosan was found to be ten times more efficient than PEI. However, chitosan-mediated transfection is highly dependent on the cell type.

Mao and colleagues[185] studied chitosan–DNA nanoparticles that were prepared using a complex coacervation process under defined conditions. The size of the particles was optimized to be in a narrow submicron range. The zeta potential of these nanoparticles was $+12$ to $+18$ mV at pH < 6, rapidly dropping to 0 at pH 7.2, which rendered the particles hydrophobic. This hydrophobic nature stabilized the nanoparticles without cross-linking and also protected the encapsulated plasmid DNA from nuclease degradation. The transfection efficiency of chitosan–DNA nanoparticles was cell-type-dependent. Higher gene expression levels were found in (HEK293) cells and IB-3-1 cells, compared with that in 9HTEo and HeLa cells. The presence of 10% fetal bovine serum did not interfere with their transfection ability.

To enhance the transfection efficiency, WSC was coupled with urocanic acid (UA) bearing imidazole ring, which can play a crucial role in endosomal rupture through the proton sponge mechanism. The urocanic acid-modified chitosan (UAC) was complexed with DNA, and UAC–DNA complexes were characterized. The sizes of UAC–DNA complexes under physiological conditions (109 to 342 nm) were almost same as those of chitosan–DNA complexes. UAC also showed good DNA-binding ability, high protection of DNA from nuclease attack, and low cytotoxicity. The transfection efficiency of chitosan into HEK293T cells was much enhanced after coupling with UA and increased with an increase in UA contents in the UAC.[186]

As this field is emerging very rapidly, a full review of this area is beyond the scope of this book. Readers who want to get an insight into this topic should refer to the cited reviews.[187–189]

4. Scaffold/Cell Constructs

Similar to many other polymers, chitosans can be processed into cellular solids for use in cell transplantation and tissue regeneration. Porous chitosan structures can be formed by a number of techniques.[190] A number of researchers have studied chitosan-based scaffolds in various cell culture studies and *in vivo* studies.[191–197]

The effects of chitin/chitosan and their oligomers/monomers on migrations of fibroblasts (3T6) and vascular endothelial cells (HUVEC) were evaluated *in vitro* by Okamato and co-workers.[198] In a direct migratory assay using the blind well chamber method, the migratory activity of 3T6 was seen to be reduced by chitin, chitosan, and the chitosan monomer (GlcN). The migratory activity of HUVECs was enhanced by chitin, chitosan, and the chitin monomer (GlcNAc), and was reduced by chitosan oligomers and GlcN. Supernatant of 3T6 preincubated with chitin or chitosan reduced the migratory activity of 3T6 cells. Supernatant of HUVECs preincubated with chitosan also reduced the migratory activity of HUVECs, but supernatant preincubated with chitin had no effect on them.

The effects of chitin and its partially deacetylated derivatives, chitosans, on the proliferation of human dermal fibroblasts and keratinocytes were examined *in vitro*[199] Chitosan with relatively high degrees of deacetylation strongly stimulated fibroblast proliferation while samples with lower levels of deacetylation showed less activity. Fraction CL313A, a shorter chain length 89% deacetylated chitosan chloride, was further evaluated using cultures of fibroblasts derived from a range of human donors. Some fibroblast cultures produced a positive mitogenic response to CL313A treatment, with proliferation rates increasing by approximately 50% over the control level at an initial concentration of 50 μg/mL, while others showed no stimulation of proliferation or even a slight inhibition (<10%). The stimulatory effect on fibroblast proliferation required the presence of serum in the culture medium, suggesting that the chitosan may be interacting with growth factors present in the serum and potentiating their effect. In contrast to the stimulatory effects on fibroblasts, fraction CL313A inhibited human keratinocyte mitogenesis, with up to 40% inhibition of proliferation being observed at 50 μg/mL. In general, highly deacetylated chitosans were more active than those with a lower degree of deacetylation.

Mori and co-workers[200] did confirmed these results by reporting that chitin scaffolds did not produce an adverse effect on L929 mouse fibroblasts proliferation. However, the authors conclude that chitosan appeared to exhibit an inhibitory effect, attributing this to the interaction of chitosan with growth factors expressed by the cells. This is in contrast to the report that found a high degree of deacetylation in chitosan more favorable for supporting cell growth, proliferation, and attachment.[201] The authors attributed this to the electropositive nature of the amino group permitting interactions between chitosan and cell surface elements.

Park and colleagues[202] studied the effect of carboxymethyl-chitosan (CM)-chitosan on the proliferation of normal human skin fibroblast and keloid fibroblast cultures.[202] They found that CM-chitosan did not restrict normal human skin fibroblasts but impeded keloid fibroblast by inhibiting type I collagen secretion, and suggested a role for wound healing in keloid control. VandeVord and co-workers[203] report a minimal immune response and little implant encapsulation to foam-like chitosan scaffolds.

Chitosan scaffolds have been evaluated extensively for cartilage and bone TE concepts. Zhang and co-authors[204, 205] reported characterizing the macroporous chitosan/calcium phosphate composite scaffolds (Figure 29.3A and B) by evaluating porosity and density measurements, mechanical testing, analysis of microstructure surface morphology biodegradation, and preliminary cell culture studies.

Khor and co-workers[206] made a series of porous chitin matrices by producing chitin gels from chitin solutions followed by lyophilization. Matrix pore sizes ranging from 100 to 500 μm could be obtained depending on the various pretreatment procedures of chitin gels prior to lyophilization.[206] Mouse and human fibroblast cell cultures exposed to these chitin matrices were found to be growing and proliferating, indicating the feasibility of using these porous chitin matrices for cell transplantation applications to regenerate tissues. Similarly, Wang and colleagues[207] have demonstrated the preparation of chitin-plasma-sprayed calcium HA matrices, while Ma and co-workers[208] also established the utility of chitosan scaffolds to support cell growth and proliferation.

In an attempt to extend the upper pore size limit of 500 μm for chitin matrices obtained by lyophilization, Chow and Khor[209] developed a novel method, defined as the internal bubbling process (IBP), for creating open and large-pored chitin matrices.

Chupa and colleagues[210] evaluated the potential of GAG–chitosan and dextran sulfate (DS)–chitosan materials for controlling the proliferation of vascular endothelial (EC) and smooth-muscle cells (SMC). GAG–chitosan complex membranes were generated seeded with human ECs or SMCs and cultured up to 9 days. In addition, scaffolds were implanted subcutaneously in rats to evaluate the *in vivo* response to these materials. The results indicated that while chitosan alone supported cell attachment and growth, GAG–chitosan materials inhibited spreading and proliferation of ECs and SMCs *in vitro*. In contrast, DS–chitosan surfaces supported the proliferation of both cell types. *In vivo*, heparin–chitosan and DS–chitosan scaffolds stimulated cell proliferation from the host tissue and the formation of a thick layer of dense granulation tissue. In the case of heparin scaffolds, the granulation layer was highly vascularized. These results indicate that the GAG–chitosan materials can be used to modulate the proliferation of vascular cells both *in vitro* and *in vivo*.

HA/chitosan–gelatin scaffolds have also been fabricated.[211] Essentially, each component was sequentially added and the resultant mixture was subjected to a lyophilization treatment to produce cellular solids. The scaffold was studied *in vitro* using

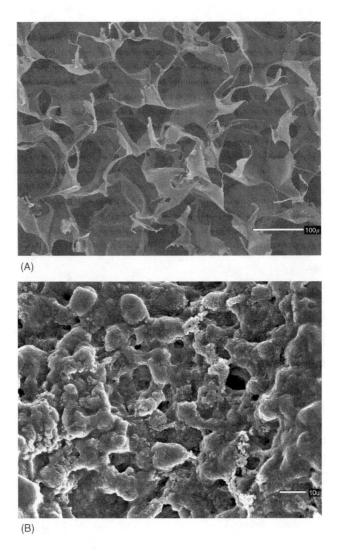

FIGURE 29.3 (A) SEM image of chitosan/hydroxyapatite composite scaffolds. (Courtesy of Dr. N. Zhang) The chitosan/hydroxyapatite composite scaffolds are macroporous and have an interconnected open-pore microstructure.[201,202] (B) SEM image of MG63 cells after 4 days of culture on the chitosan/hydroxyapatite composite scaffolds.[201,202] (Courtesy of Dr. N. Zhang)

osteoblast-like cells. Macroporous chitosan/calcium phosphate (β-tricalcium phosphate and calcium phosphate inverted glass) scaffolds have also been prepared. The composite scaffold was found to be stronger, bioactive, and biodegradable, the effect being dependent on the ratio of chitosan to the two types of calcium phosphates.[212]

Risbud and co-workers[213] have studied the interaction of chitosan–gelatin hydrogels with different cell lines.[213] Extending this approach, chitosan–polyvinyl pyrrolidone hydrogels have also been evaluated in endothelial cell culture studies.[214]

A membrane made of this material was used to demonstrate the possibility of supporting respiratory epithelial cells from the trachea. Chitosan has also been shown to exert a strong influence on nerve cell attachment and proliferation.[215] The chitosan system performed better when coated with extracellular matrix proteins such as laminin.

Cho and colleagues prepared chitosan/alginate scaffolds.[216] In their work, chitosan was modified with lactobionic acid to produce galactosylated chitosan that was added to a cross-linked alginate gel. Scaffolds were prepared by several freezing steps and a final lyophilization. The cellular solids exhibited the usual pore configurations, the size depending on the freezing pretreatments, the molecular weight of chitosan, and the amount of galactosylated chitosan. Hepatocytes attachment to the alginate-galactosylated chitosan scaffolds was demonstrated. Similarly, a photosensitive heterobifunctional cross-linking agent was attached to chitosan for coating onto polylactic acid (PLA) film surfaces.[217] Improved cell attachment was reported with this approach, while at the same time, chitosan could also be modified by interaction with heparin, which inhibited platelet adhesion and activation.

Zhu and co-workers utilized the reaction between the amino group on chitosan and the carboxylic acid group on amino acids with glutaraldehyde to attach various amino acids (lysine, arginine, aspartic acid, and phenyalanine) onto chitosan.[218] These amino acid functionalized chitosan moieties were subsequently entrapped onto PLA surfaces. The amino acid–chitosan–PLA membrane allowed chondrocyte attachment and proliferation. Yao and co-workers[219] reported similar results, using osteoblasts, for chitosan–PLA membranes to immobilize chitosan onto the PLA surface by the carbodiimide process.[219] Chung and co-workers recently reported their study on growing human endothelial cells on chitosans that had cell adhesive peptides photochemically grafted onto their surfaces.[220] Chitosan surfaces containing the grafted peptides were found to support the proliferation of human endothelial cells in contrast to chitosan only surfaces, where no adherence was observed.

Cui and colleagues[219] cultured islets in chitosan foams having pore size, of 200 to 500 μm. Twenty rat islets were seeded into the scaffolds and were cultured for up to 62 days. Insulin concentration both inside and outside was measured during the entire culture period. Changes in the morphology of islets were also observed. Freshly isolated islets had a loose appearance with an irregular border and most were seen as a single islet. Occasionally, a cluster consisting of 2 to 4 islets ranging mainly from 150 to 250 μm in diameter was observed. Islets cultured in different culture media retained their initial morphology, characterized by well-delineated smooth borders, for at least 53 days. The insulin release behavior of islets cultured in chitosan showed constant secretory capacities for 49 days, after, which a rapid and definitive decline took place. Until this stage, insulin concentration in chitosan was well maintained. The properties were dependent on the culture medium used and insulin diffusion released from the islets.[221]

Li and co-workers[222] prepared porous chitosan scaffolds by lyophilization of chitosan solution. The scaffolds were modified with water-soluble polyanionic species such as alginate and heparin. The pore structures of these scaffolds were viewed via light and scanning electron microscopy. The scaffolds prepared had a high porosity of approximately 90% with mean pore sizes ranging from 50 to 200 μm. They were used as substrates for hepatocytes culture. The cell attachment ratio was much higher than on the monolayer membrane, and hepatocytes exhibited a round cellular morphology with many microvilli evident on the surface of the cells. Metabolic activities of the cells were evaluated in terms of albumin secretion and urea synthesis.

Hutmacher's group designed and fabricated chitosan scaffolds via solid free form fabrication. Scaffolds made of chitosan and chitosan/TCP composites (Figure 29.4A–E) were designed and fabricated using a new RP robotic dispensing (RPBOD, Figure 29.5) system.[223] Characterizations of the macro-porosity, density, and surface area to volume ratio resulted in the fabrication of matrices with better mechanical properties, biocompatibility, and biodegradation. Those scaffolds were studied *in vitro* by using dermal fibroblasts and osteoblasts (Figure 29.3A–D).

E. DEXTRAN

Recently, dextran hydrogels have received increased attention on account of their potential utility in a number of biomedical applications. Due to their low tissue toxicity and high enzymatic degradability at different implantation sites, dextran hydrogels have been frequently considered as a potential matrix system for the delivery of bioactive agents. Several methods of preparing dextran hydrogels have been adopted.[224]

Hovgaard and co-workers[35,36] obtained hydrogels by cross-linking dextran with either 1,6-hexanediisocyanate or glutaraldehyde. Sequential reactions of dextran with glycidyl acrylate, followed by the polymerization of acrylated dextran, also led to the formation of the polymer network.[225] Recently, methacrylation of dextran has been conducted with full control of the degree of substitution (DS) by transesterification of glycidyl methacrylate (GMA) with dextran in DMSO. Hydrogels were obtained by the radical polymerization of methacrylated dextran (MA-dextran) in aqueous solution, using ammonium peroxydisulfate (APS) and N, N, N', N'-tetramethylethylenediamine (TMEDA) as initiator systems. The syntheses of methacrylated and acrylated

$\longrightarrow$

FIGURE 29.4   (A) In order to investigate the efficacy of the chitosan–TCP scaffolds for tissue culturing, pure chitosan, and chitosan–TCP scaffolds were fabricated using the microsyringing robotic technique described in Chapter 3. The scaffolds were cut into 8×8×5 mm specimens. The pure chitosan scaffolds were used as controls. Human osteoblasts 200,000 were seeded into each scaffold with fibrin glue to give the cells initial anchorage. The medium was changed every 3 days and the scaffolds were maintained in a self-sterilizable incubator at 37°C in 5% CO_2, 95% air and 99% relative humidity. Week 1: The cells were still embedded in fibrin glue. But, for the chitosan–TCP scaffolds, the cells began to reach out and attach themselves to the scaffold bars faster than the cells in the pure chitosan scaffolds, which tended to remain in the fibrin glue and attached at a slower rate. This phenomenon was generally observed in all the scaffolds. Hence, it can be said that the cells were attracted toward the TCP, which is a bone constituent. Week 2: The cells had attached to the scaffold bars for both the chitosan TCP and pure chitosan scaffolds. Toward the end of the second week the fibrin glue had degraded completely. This led to cells that were poorly attached to be washed away during media change. Week 3: By the third week, the cells in chitosan TCP scaffolds were well attached and had begun to stretch across the bars. Although the cells were attached to the bars in the pure chitosan scaffolds, they did not stretch across the bars as readily as can be observed. Week 4: The cells were attached and stretching in both types of scaffolds but the proliferation was slower in the pure chitosan scaffolds. The scale bar represents 100 μm.

dextrans have also been reported by reacting dextran with methacrylic anhydride, and with bromoacetyl bromide and sodium acrylate, respectively.[226,227] The swelling properties, enzymatic degradability, and release behavior of dextran hydrogels have also been extensively studied.[228–237]

Dextran hydrogels capable of undergoing volume change in response to pH variation also been reported.[238,239,240] Dextran hydrogels were then synthesized carried out by UV irradiation of MA-dextrans. Dextran hydrogels capable of undergoing pH-responsive swelling can be obtained by the activation of dextran with 4-nitrophenyl chloroformate, followed by conjugation with 4-aminobutyric acid and cross-linking with 1,10-diaminodecane.[240] Hydrogels were also obtained by the

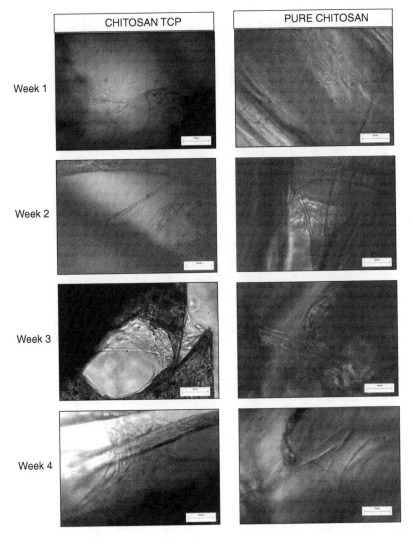

FIGURE 29.4 (A)

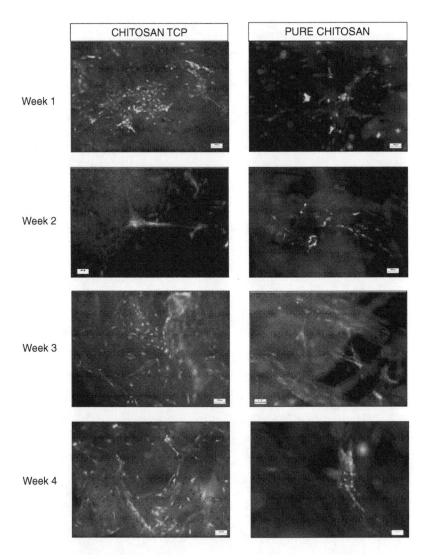

FIGURE 29.4 (B) Fluorescent microscopy after seeding with osteoblasts and coating with fibrin glue of chitosan-TCP scaffolds (left) and pure chitosan (right) scaffolds (8×8×5 mm). Viable cells show up with a green fluorescence while nonviable cells show up with as red fluorescence.

polymerization of maleic acid-carrying dextrans.[240] In our previous study, copolymerization of MA-dextran with AAc and *N-t*-butylacrylamide was performed, and the pH-dependent swelling properties were characterized.[239]

A dextranomer/HA co-polymer was employed as a bulking agent for vesicoureteral reflux. The authors report that only a single injection was necessary when compared with other treatment modalities. The blood clot formation (hematoma) and a clinically not detectable inflammatory response to the biodegradable material helped to stabilize the implant volume over the required period of time.[241] Cai and

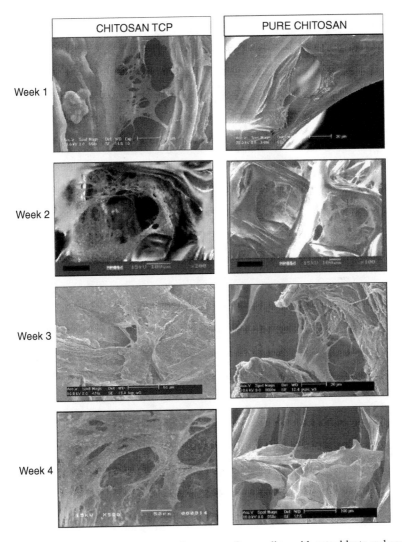

FIGURE 29.4 (C) Scanning electron microscopy after seeding with osteoblasts and coating with fibrin glue of chitosan-TCP scaffolds (left) and pure chitosan (right) ($8 \times 8 \times 5$ mm).

co-workers[211] developed a scaffold by blending PLA with natural biodegradable dextran, and a novel sponge-like scaffold made of it was fabricated thereof using solvent-casting and particle-leaching techniques.

The synthesis of polysaccharide-based sponges for use in TE was investigated by Ehrenfreund-Kleinman.[242] A comparative study of the branched polysaccharide arabinogalactan (AG) and the linear polysaccharide dextran in the formation of sponges by reaction with diamines or polyamines was conducted. Three AG-based sponges were synthesized from the cross-linking reaction with different amine molecules. The sponges obtained were highly porous, rapidly swelled in water, and were stable *in vitro*

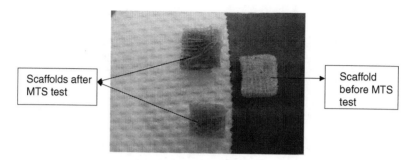

FIGURE 29.4   (D) Scaffolds before and after the MTS assay. The dye is still absorbed in the scaffold.

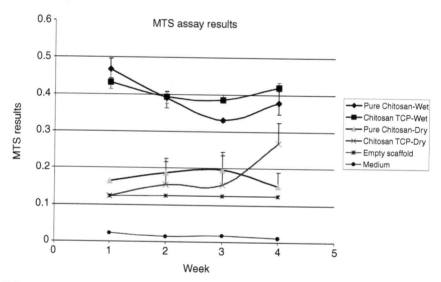

FIGURE 29.4   (E) Graph of MTS assay results.

for at least 11 weeks in aqueous media at 37° C. AG-chitosan sponges were chosen as most suitable to serve as scaffolds for cell growth in TE. The biocompatibility of these sponges *in vivo* was evaluated by histological staining and noninvasive MRI technique after implantation in BALB/c mice. Initially, the sponge-like material evoked an inflammatory response combined with neovascularization of the implant. The inflammatory reaction decreased with time, indicating a healing process.

Human parotid cells were cultured with two different types of commercially available microcarriers (Cytodex 3 and Cytopore 1) for up to 3 weeks *in vitro*. Positivity for amylase was detectable in 20 to 45% of cells growing on the microcarriers and especially on Cytodex 3. A decrease in amylase levels in the culture medium indicated functional deficiencies in the remaining acinar cells. TE of human salivary gland organoids on microcarriers is a new approach for the potential causative treatment of radiation-induced xerostomia. However, the authors concluded that before clinical

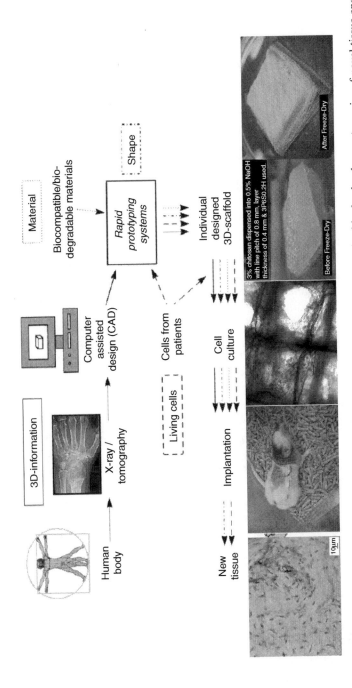

FIGURE 29.5 The utilization of computer-aided technologies in tissue engineering has evolved in the development a series of novel tissue engineering concepts. Flow diagram of a general rapid prototyping (RP) process. A computer-generated model based on CAD design or MRI or CT scan data in ".stl" format is imported into a RP's system software which allows the model to be "sliced" into thin horizontal layers, with the tool path specified for each layer. The "sliced" data is used to instruct the RP machine to build a scaffold layer by layer, based on the design of the morphology and shape of the computer model. Cells are then seeded on the scaffold and the construct is implanted directly or after a certain incubation period. The *in vivo* study will then show if the desired tissue is generated or regenerated.

application can be considered, significant improvements in the *in vitro* cultivation of salivary gland tissue and scaffold design have to be realized.[243]

Recently, dextran has been investigated as an alternative to PEG for low protein-binding, cell-resistant coatings on biomaterial surfaces. Although the antifouling properties of surface-grafted dextran and PEG are quite similar, the multivalent properties of dextran are advantageous when high-density surface immobilization of biologically active molecules to low protein-binding surface coatings is desired. Massia and colleagues[244] studied cell adhesion on untreated, aminated, and dextran-coated materials. Dextran coatings effectively limited cell adhesion and spreading on glass and PET surfaces in the presence of serum-borne cell adhesion proteins. With dextran-based surface coatings, it might be possible to develop well-defined surface modifications that promote specific cell interactions and perhaps better performance in long-term applications.[245–252]

F. GLYCOSAMINOGLYCANS

1. Introduction

GAGs are polysaccharides that occur ubiquitously within the ECM.[72] They are unbranched heteropolysaccharides, consisting of repeated disaccharide units. In their native form, several GAG chains are covalently linked to a central protein core, and the protein–polysaccharide complex is termed a proteoglycan. Proteoglycans play a major role in organizing and determining the properties and functionality of the ECM, and GAG chains are major factors in determining proteoglycan properties.

GAGs are synthesized in the Golgi system through a series of reactions that includes the C5-epimerization reaction along with extensive sulfation of the polymers. The single, Ca^{2+}-independent epimerase in heparin/heparan sulfate biosynthesis and the Ca^{2+}-dependent dermatan sulfate epimerase(s) also generate variable epimerization patterns, depending on other polymer-modification reactions.

There are six different types of GAGs: chondroitin sulfates, dermatan sulfate, keratin sulfate, heparin sulfate, heparin, and hyaluronic acid. The monosaccharides in GAGs are sulfated to varying degrees, with the exception of hyaluronic acid, which is not sulfated. Unlike the other GAGs, hyaluronate has a free, high-molecular weight molecule with no covalently attached protein.

GAGs used in scaffolds and matrixes are usually obtained as salts of sodium, potassium, or ammonia, and in this form they are all water soluble. The presence of strongly ionizing sulfate groups implies that the charge density of these molecules is much less pH-dependent than in the case of chitosan, and as a result, these molecules are soluble over a wide pH range. The disaccharide repeat structure and the resultant alternating glycosidic bond types prevent this family of materials from forming high-strength, crystalline structures in the solid state, and the dried or precipitated GAGs are essentially amorphous. The 3D configurations that form in solution tend to be flexible coil structures, extended as a result of the charge effects. For example, heparin is known to form single-chain helical structures in solution. The only GAG with significant gel-forming ability is hyaluronate. This high-molecular-weight molecule is the primary gelling agent in the vitreous humor of the eye and can form very high-viscosity solutions at concentrations as low as 0.3 wt%.

The potential of GAGs to be incorporated within the scaffold matrix stems not from their intrinsic physical properties but from their biological activity, which is a direct result of their numerous interactions with many proteins and growth factors of the body. This biological activity can be exploited by binding, complexing, or covalently linking GAG moieties to other polymers with superior structural or mechanical characteristics.

2. Hyaluronan

Hyaluronan (HA) or hyaluronic acid is a polysaccharide of the ECM. It is a main GAG having many structural, rheological, physiological, and biological functions in the body. It is a linear and monotonous anionic polymer, which is heterogeneously distributed in various soft tissues. Two modified sugars, glucuronic acid and *N*-acetyl glucosamine, form the disaccharide units. HA is a soluble molecule forming highly viscous solutions in water and interacts with binding proteins, proteoglycans, and growth factors, and also actively contributes to the regulation of the water balance acting on the osmotic pressure, low resistance, and selectively sieving the diffusion of plasma and matrix proteins. In the joints, HA acts as a lubricant, supporting the articular cartilage surfaces under shear stress. At a molecular level, HA acts as a scavenger molecule for free radicals.[253]

In the last decade, the use of medical-grade HA has been applied widely in joint surgery, corneal transplantation, treatment of cataract, intraocular lens implantation, and the treatment of vitroretinal diseases. It has been shown that HA can improve wound healing by means of HA degradation products, which induce endothelial cell proliferation and angiogenesis. This new class of hyaluronan-based biopolymers has good biocompatibility and controlled biodegradability.[254]

Despite its unique biological properities, water solubility, rapid resorption, and short residence time at the site of implantation led biomaterial scientists to modify their molecular structure to fabricate a scaffold material with sufficient physical properties for skin TE. Cross-linking and coupling reactions were the two ways considered for obtaining a matrix material with better mechanical properties. These chemical modifications were applied either to trap HA chains within a net of cross-linked proteins or to create covalent bonds between HA chains. The production of all these derivatives was driven by a concept similar to that which led to the production of cross-linked collagen. However, in a number of cases, concern has been expressed for the potential toxicity of some of the cross-linking agents utilized, such as glutaraldehyde, formaldehyde, and isocyanates.

Therefore, a new type of HA was obtained by creating cross-linking bonds by directly esterifying a certain percentage of the carboxyl groups of glucuronic acid along the polymeric chain with hydroxyl groups of the same or different hyaluronan molecules. Once the esterification of the polymer has taken place, the material can be easily processed into different scaffold types such as membranes, sponges, and microspheres via extrusion, lyophilization, or spray drying. Once wet, the benzylester loses part of its mechanical strength similar to many other natural polymers. However, under *in vitro* cell culture conditions, the material maintains its structural integrity for up to 3 weeks and does not shrink as fast as collagen-based materials.[256,257]

At present, HA-based scaffolds have been studied by a number of tissue engineers due to their excellent cell and tissue compatibility.[69,254,255,258-281]

Works on the various formulations of HA-based matrices, with respect to their applications in skin TE, has been discussed in a number of reviews.[282,283] It has been reported that an advantage of benzyl esters of HA (HYAFF-11, Fidia Advanced Biopolymers, Italy) is the good cell attachment and proliferation of fibroblast-like cells. The knowledge gained about processing large sheets of HYAFF-11 led to the development of a matrix for a tissue-engineered epidermal graft (Laserskin, Fidia Advanced Biopolymers, Italy). The membrane has microperforations (40 μm, 6000 perforations/cm^2) that allow the keratinocytes to communicate and grow toward the host tissue. Holes of 0.5 mm function as drainage for the wound exudate.[284,285] In contrast to the fibroblast-like cells, keratinocytes do not sufficiently attach and proliferate on the Laserskin matrix. Therefore, the manufacturer recommends culturing the keratinocytes with a fibroblast feeder layer.

Bakos and colleagues[286] developed a collagen/hyaluronan membrane to combine the material properties of a protein and polysaccharide. According to their results, interactions of these two polymers were very strong and resulted in improved degradation kinetics as well as mechanical strength. The physiochemical properties can be influenced from the point of view of material science by chemical cross-linking. The composites of collagen–hyaluronic acid have been used successfully in a number of clinical studies as epidermal matrix. Hutmacher's group (unpublished data) studied collagen/HA membrane for TE of the dermis (Figure 29.6). In recent years, the use of cell sheets for engineering functional 3D tissue constructs has proven to be an attractive option for tissue engineers. The resulting tissue constructs express mature ECM and resemble native tissues in terms of organization and function. However, it is a challenge to produce mechanically stable tissues by using only cell sheets alone, and forming tissues of specific shapes and sizes would be difficult. We therefore devised a novel technique of combining cell sheets with mechanically stable 3D matrices. To evaluate this technique, we cultured human dermal fibroblasts along with two hybrid matrices [poly(lactic-co-glycolic acid) mesh with lyophilized collagen (PLGA-c) and collagen–hyaluronic acid foam (CHA)] for the purpose of skin TE. To the best of our knowledge, this is the first study in which cell sheets are cultured together with 3D matrices for TE.

We found that over 2 weeks of in vitro culture, human dermal fibroblast formed a matured cell sheet that was strong enough to support the 3D hybrid matrices on its own. When the sheets were folded around the matrices, 3D dermal equivalents were formed. Abundant collagen was produced within the folded cell sheets but not within the cell aggregates between the cell sheet interfaces. PLGA-c supported a homogenous distribution of collagen-rich neotissue, but contracted over the culture period. CHA contracted marginally, and collagen-rich neotissue formation was only observed on the peripheral of the matrix and not within the matrix space.

These results, though preliminary, will serve as a basis for future explorations of culturing cell sheets in combinations with 3D matrices to engineer functional and mechanically stable skin, and possibly other tissues.

HA-based scaffolds promoted the attachment of human preadipocytes and differentiation along the adipogenic lineage.[287] In the treatment of at least 600 patients with

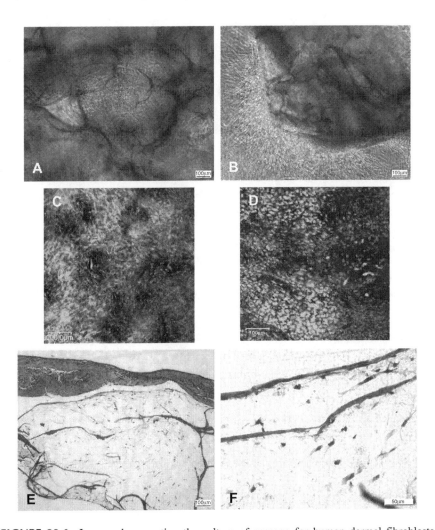

FIGURE 29.6 Images documenting the culture of passage for human dermal fibroblasts (HDFs) in collagen-hyaluronic acid (CHA) sponges. Phase-contrast light microscopy images showed HDFs proliferating at 3 weeks postseeding; (A) within the CHA and (B) outgrowing from the CHA to form a confluent cell sheet on the culture dish. (C,D) Confocal laser microscopy images indicated a high proportion of viable cells (green), 3 weeks postseeding, compared with nonviable cells (red) within CHA. Masson's Trichrome stained sections of cell sheet encapsulated CHA after 4 weeks in culture, showed that cell sheets formed a collagen-rich neotissue at the peripheral of CHA (E). However, only a loose cellular network with minimum ECM was formed within the CHA matrix space (F) Magnification — A,B,C,E, 100×; D, 200×; F, 400×.

acute knee cartilage lesions, autologous chondrocytes were grown onto HA-based scaffolds and then grafted into the knee. The authors conclude that the study yielded good clinical results.[288] A HYAFF-11 scaffold was seeded with bone marrow stromal cells (BMSCs) and implanted into a rat model with critical size defects, leading to defective mineralization and healing.[140] This response was enhanced via preincubation of BMSCs with bFGF. Little to no healing or mineralization was observed when the defect was left alone and when a blank scaffold was implanted.

3. Chondroitin Sulfates

In the 1970s chondroitin sulfates were incorporated into collagen-based hydrogels and scaffolds to enhance the biological performance of the natural polymer-based matrix. This type of matrix was extensively studied for skin TE. Histological as well as physical analysis of human skin samples led to the development of the "Integra™ Artificial Skin" by Yannas and co-workers,[289,290] in which dermal matrix is made of reconstituted bovine tendon type I collagen/chondroitin-6-sulfate. The design was based on the concept that the cellular elements and the supporting microvasculature of the host tissue are supposed to populate the collagen matrix and resemble normal connective tissue. The resulting commercial product Integra™ was recently also studied as a matrix for skin cell transplantation.[291] The work on this matrix was reviewed by one of its inventors.[292]

Subsequently, this type of matrix was evaluated for cartilage[293,294] and other TE applications.[295-299] However, the limited mechanical properties of such matrices and hydrogels are still issues that need to be resolved.[300] Based on this background, Collombel's group[301] developed a scaffold/neotissue construct composed of two compartments: (1) a dermal equivalent comprising an acellular dermal substrate populated by foreskin fibroblasts and (2) an epidermis tissue engineered from normal human keratinocytes seeded onto the dermal equivalent. The dermal substrate contains type I and III collagen and GAGs cross-linked by chitosan. Fibroblasts seeded into the porous structure of the dermal substrate provide a dermal equivalent suitable for supporting epidermal cells. Keratinocytes attach quickly, exhibit mitotic activity, and form a continuous and stratified epidermis. After 2 weeks of culture, histological sections show a basal layer with cuboidal cells attached to the dermal equivalent and several suprabasal cell layers including the stratum corneum. The authors concluded that this matrix concept based on our chitosan-cross-linked collagen — GAG matrix is morphologically equivalent to normal human skin and should thus provide a useful tool for *in vitro* toxicological studies as well as a suitable wound covering for the treatment of patients with severe burns.[302]

4. Heparin

Interactions with enzymes are the most extensively characterized of all GAG–protein interactions. In particular, the anticoagulant activity of heparin via activation of the protease inhibitor antithrombin III (ATIII) has been thoroughly characterized.[73,74] ATIII inactivates several proteases in the coagulation cascade, the most important of which is thrombin. The binding of heparin to ATIII increases the inactivation kinetics by as much as 2000-fold. ATIII binding and activation involves the highly specific recognition of a pentasaccharide sequence found in both heparin and heparin sulfate. Heparin and heparan sulfate are also known to bind and activate the enzyme lipoprotein particles in peripheral tissues. The fatty acids so formed are then available for uptake by adjacent cells. Lipoprotein lipase is normally found bound to endothelial cell surface heparan sulfate, where it is active. The soluble enzyme is inactive in the absence of GAG. This property has been used to reduce the levels of circulating lipoprotein by activating inactive lipase via injections of soluble GAG.

A wide variety of polypeptide growth factors bind avidly to heparin and heparin sulfate. Lower affinity binding also occurs with the other sulfated GAGs. Members of the fibroblast growth factor (FGF) family have been extensively studied, but heparin/GAG affinity has been demonstrated with epidermal growth factor (EGF), vascular endothelial growth factor (VEGF), and a variety of interleukins (e.g., IL-3) as well as hematopoietic factors such as granulocyte colony-stimulating factor (G-CSF), granulocyte macro-phase colony-stimulating factor (GM-CSF), and stem cell factor (SCF). GAG binding can have a variety of effects on growth factors depending on the ratios of the two molecular species and the binding environment. Growth factors can be bound and sequestered in the ECM by GAGS. In the bound state, the polypeptides are often protected from proteolytic attack, and can be subsequently released from the ECM under the action of GAG lyases, matrix metalloproteinases, or changes in the pH values. Growth factor binding to GAGs can also induce conformational changes that enhance receptor-growth factor binding. GAGs may also enhance growth factor activity indirectly by mediating the clustering of growth factors or receptor–ligand complexes on the cell surface.

The attachment of heparan sulfate to collagen matrices promotes the binding, modulation, and sustained release of signaling molecules. An example of this usage is the binding of bFGF to cross-linked collagenous matrices with HS. bFGF attachment increased threefold with HS compared with that without it. There was a more gradual and sustained release of bFGF *in vitro*. Subcutaneous implantation of these collagenous-HS/bFGF matrices in rats resulted in enhanced angiogenesis.[264] In another study, heparin integrated with chitosan hydrogels bound with EGF and FGF-1 and -2 were subcutaneously implanted into a mouse, and showed enhanced controlled release and vascularization effect.[303] These studies indicated the potential of HS as a good interface molecule between signaling molecules and matrices.

BD Matrigel™ Matrix is a solubulized basement membrane preparation extracted from EHS mouse sarcoma, a tumor rich in ECM proteins. Its major component is laminin, followed by collagen IV, heparan sulfate proteoglycans, and entactin. This hydrogel has been used in a vast number of *in vitro* and *in vivo* studies, but it is beyond the scope of this chapter to review them. References relevant to TE applications can be found on the referenced web page.[304]

Polystyrene foams, with pore sizes up to 100 μm, fabricated by phase separation from a homogeneous naphthalene solution, were derivatized with lactose and heparin, both of which are known to promote rat hepatocyte attachment and maintenance of their differentiated functions.[305]

V. NEW TRENDS AND DEVELOPMENTS

The transformation of polysaccharides from solution to gel takes place by chemical and physical processes. In TE applications where hydrogels are used to form structural and complex-shaped scaffolds, these chemical and physical processes are not feasible. These limitations can be overcome by applying vinylated biomacromolecule chemistry in combination with photofabrication technology, as illustrated with chitosan, heparin and hyaluronan. The polysaccharides are vinylated via a condensation reaction of their carboxyl or amino group with a vinyl monomer in the presence of a water-soluble condensation agent. The gel is then prepared by the polymerization of the vinylated polysaccharides. This polymerization takes places in the presence of water-soluble camphorquinone as a photo-cleavable radical producing agent and under visible light irradiation. Therefore, stereolithographically designed and fabricated microarchitectures using these vinylated photoreactive biomacromolecules can be used to fabricate complex scaffolds. Such applications include nerve tubular conduct packed with guide photocured fibres.[306]

The loss of cartilage phenotype by chondrocytes in 2D culture has been documented as early as 1977[307] and has remained an unresolved issue to date. Therefore, next to biomechanical stimulation,[308] the control of the redifferentiation of chondrocytes is likely to be the key to tissue engineering an adequately large mass of functional cartilage, and is currently regarded as the most challenging problem for cartilage regeneration via cell-based strategies.[309]

Current strategies of cartilage TE consist of two steps. First, the cells from a small biopsy of the patient's own tissue have to be multiplied. During this multiplication process, which takes place in a culture flask (2D environment), they lose their cartilage phenotype. In the second step, these cells have to be stimulated to reexpress their cartilage phenotype and produce cartilage matrix. However, most recent research has shown that chondrocytes maintain their phenotype when directly cultured after isolation in a spinner flask filled with microspheres made of or coated with polysaccharides and other natural polymers.[310,311] These data represent a paradigm shift with respect to the state of the art in monolayer culture techniques used in cartilage engineering.

Other desirable features are being incorporated into new gel systems to enhance their applicability in minimally invasive applications. Photopolymerization generally requires the introduction of a light source and is thus perceived as more invasive than a simple injection, but is highly suitable for applications where other "invasive" procedures such as angioplasty are employed. Gelation upon simple injection is highly desirable in applications such as the repositioning of the ureter. Although ionically induced gelation has been used for injection, the kinetics are rapid (~seconds) and difficult to control; further, inhomogenous gelation can result if the ion is not uniformly distributed throughout the initial macromer solution. Thermally induced gelation is one of the approach, provided a physiologically relevant temperature range for the sol–gel transition can be identified. A potentially versatile approach to controlling gelation kinetics over a wide range while allowing the incorporation of biologically active peptides is the use of enzymatic gelation,[62] where gelation kinetics can be controlled by the choice of substrate and the concentration of enzyme.

REFERENCES

1. Skalak, R. and Fox, C.F., Eds., *Tissue Engineering*, Liss, New York, USA, 1988.
2. Langer, R. and Vacanti. J.P., Tissue engineering, *Science,* 260, 920–926, 1993.
3. Vacanti, C.A. and Vacanti, J.P., The science of tissue engineering, *Orthop. Clin. North. Am.*, 31, 351–356, 2000.

4.   Pangarkar, N. and Hutmacher, D.W., Invention and business performance in the tissue engineering industry, *Tissue Eng.*, 9, 1313–1322, 2003.

5.   Hutmacher, D.W., Regenerative medicine and tissue engineering in Singapore, *Regen. Med.* (Japan), 2, 135–142, 2003.

6.   Hutmacher, D.W., Kirsch, A., Ackermann, K.L., and Huerzeler, M.B., Matrix and carrier materials for bone growth factors — state of the art and future perspectives, in *Biological Matrices and Tissue Reconstruction*, Stark, G. B., Horch, R., and Tancos, E., Eds., Springer, Heidelberg, Germany, 1998, pp. 197–206.

7.   Spector, M., Novel cell-scaffold interactions encountered in tissue engineering: contractile behavior of musculoskeletal connective tissue cells, *Tissue Eng.*, 8, 351–357, 2002.

8.   Hutmacher, D.W., Fu, X., Tan, B.K., and Schantz, J.T., Tissue engineering of elastic cartilage by using scaffold/cell constructs with different physical and chemical properties, in *Polymer Based Systems in Tissue Engineering, Replacement and Regeneration*, Reis, R. and Cohn, D., Eds., Kluwer Academic Publishers, The Netherlands, pp. 313–332, 2002.

9.   Fedewa, M.M., Oegema, T.R.J., Schwartz, M.H., MacLeod, A., and Lewis, J.L., Chondrocytes in culture produce a mechanically functional tissue, *J. Orthop. Res.*, 16, 227–236, 1998.

10. Cowin, S.C., How is a tissue built? *J. Biomech. Eng.*, 122, 553–569, 2000.

11. Gibson, L.J. and Ashby, M.F., *Proceedings of Royal Society (U.K.)*, 1982, p. 43.

12. Gibson, L.J. and Ashby, M.F., *Cellular Solids: Structure and Properties*, 2nd ed., Cambridge University Press, New York, 1997.

13. American Standard for Testing and Methods ASTM D883–99, Standard Terminology Relating to Plastics, 1999a.

14. Agrawal, C.M., Athanasiou, K.A., and Heckman, J.D., Biodegradable PLA–PGA polymers for tissue engineering in orthopedics, Materials Science Forum 1997, pp. 115–128.

15. Temenoff, J.S. and Mikos, A.G., Tissue engineering for regeneration of articular cartilage, *Biomaterials*, 21, 431–440, 2000.

16. Beaman, J.J., in *Solid Freeform Fabrication: A New Direction in Manufacturing*, Beamann, J.J., Barlow, J.W., Bourell, D.L., Crawford, R.H., Marcus, H.L., and McAlea, K.P., Eds., Kluwer Publishers, Boston, MA, 1997.

17. Taboas, J.M., Maddox, R.D., Krebsbach, P.H., and Hollister, S.J., Indirect solid free form fabrication of local and global porous, biomimetic and composite 3D polymer-ceramic scaffolds, *Biomaterials*, 24, 181–194, 2003.

18. Pham, D.T. and Gault, R.S., A comparison of rapid prototyping technologies, *Int. J. Mach. Tool. Manuf.*, 38, 1257–1287, 1998.

19. Sachs, E.M., Haggerty, J.S., Cima, M.J., and Williams, P.A., Three-Dimensional Printing Techniques, US Patent US5204055.

20. Landers, R. and Mülhaupt, R., Desktop manufacturing of complex objects, prototypes & biomedical scaffolds by means of computer-assisted design combined with computer-guided 3d plotting of polymers & reactive oligomers, *Macromol. Mater. Eng.*, 282, 17–21, 2000.

21. Zein, I., Hutmacher, D.W., Teoh, S.H., and Tan, K.C., Poly(ε-caprolactone) scaffolds designed and fabricated by fused deposition modeling, *Biomaterials*, 23, 1169–1185, 2002.

22. Ang, T.H., Sultana, F.S.A., Hutmacher, D.W., Wong, Y.S., Fuh, J.Y.H., Mo, X.M., Loh, H.T., Burdet, E., and Teoh, S.H., Fabrication of 3D chitosan-hydroxyapatite scaffolds using a robotic dispensing system, *Mater. Sci. Eng.*, 20, 35–42, 2002.

23.   Lam, C.X.F., Mo, X.M., Teoh, S.H., and Hutmacher, D.W., Scaffold development using 3D printing with a starch-based polymer, *Mater. Sci. Eng.*, 20, 49–56, 2002.

24.   Ma, P.X. and Langer, R., Degradation, structure and properties of fibrous nonwoven poly(glycolic acid) scaffolds for tissue engineering, in *Polymers in Medicine and Pharmacy*, Mikos, A.G., Ed., MRS, Pittsburgh, 1995, pp. 99–104.

25.   Rotter, N., Aigner, J., Naumann, A., Planck, H., Hammer, C., Burmester, G., and Sittinger, M., Cartilage reconstruction in head and neck surgery comparison of resorbable polymer scaffolds for tissue engineering of human septal cartilage, *J. Biomed. Mater. Res.*, 42, 347–356, 1998.

26.   Hutmacher, D.W. and Sittinger, M., Periosteal cells in bone tissue engineering, *Tissue Eng.*, 9(Suppl. 1), 45–63, 2003.

27.   Schmelzeisen, R., Schimming, R., and Sittinger, M., Making bone: implant insertion into tissue-engineered bone fro maxillary sinus floor augmentation — a preliminary report, *J. Craniomaxillfacial Surg.*, 31, 34–39, 2003.

28.   Vunjak-Novakovic, G., Searby, N., De Luis, J., and Freed, L.E., Microgravity studies of cells and tissues, *Ann. NY Acad. Sci.*, 974, 504–517, 2002.

29.   Moran, J.M., Pazzano, D., and Bonassar, L.J., Characterization of polylactic acid–polyglycolic acid composites for cartilage tissue engineering, *Tissue Eng.*, 9, 63–70, 2003.

30.   Mason, J.M., Breitbart., A.S., Barcia, M., Porti, D., Pergolizzi, R.G., and Grande, D.A., Cartilage and bone regeneration using gene-enhanced tissue engineering, *Clin. Orthop.*, 379(Suppl. 1), 171–178, 2000.

31.   Radisic, M., Euloth, M., Yang, L., Langer, R., Freed, L.E., and Vunjak-Novakovic, G., High-density seeding of myocyte cells for cardiac tissue engineering, *Biotechnol. Bioeng.*, 82, 403–414, 2003.

32.   Ozawa, T., Mickle, D.A., Weisel, R.D., Koyama, N., Ozawa, S., and Li, R.K., Optimal biomaterial for creation of autologous cardiac grafts, *Circulation*, 106(Suppl. 1), 176–82, 2002.

33.   Einerson, N.J., Stevens, K.R., and Kao, W.J., Synthesis and physicochemical analysis of gelatin-based hydrogels for drug carrier matrices, *Biomaterials*, 24, 509–523, 2003.

34.   Van Dijk-Wolthuis, W.N.E., Franssen, O., Talsma, H., Van Steenbergen, M.J., Kettenes-Van Den Bosch, J.J., and Hennink, W.E., Synthesis, characterization and polymerization of glycidyl methacrylate derivatized dextran, *Macromolecules*, 28, 6317–6322, 1995.

35.   Hovgaard, L. and Brondsted, H., Dextran hydrogels for colon-specific drug delivery, *J. Control. Release*, 36, 159–166, 1995.

36.   Brondsted, H., Anderson, C., and Hovgaard, L., Crosslinked dextran — a new capsule material for colon targeting of drugs, *J. Control. Release*, 53, 7–13, 1998.

37.   Van Dijk-Wolthuis, W.N.E., Kettenes-Van Den Bosch, J.J., Van Der Kerk-Van Hoof, A., and Hennink, W.E., Reaction of dextran with glycidyl methacrylate: an unexpected transesterification, Macromolecules, 30, 3411–3413, 1997.

38.   Van Dijk-Wolthuis, W.N.E., Hoogeboom, J.A.M., van Steenbergen, M.J., Tsang, S.K.Y., and Hennink, W.E., Degradation and release behavior of dextran-based hydrogels, *Macromolecules*, 30, 4639–4645, 1997.

39.   Cadée, J.A., Van Luyn, M.J.A., Van Wachem, P.B., Brouwer, L.A., De Groot, C.J., Den Otter, W., and Hennink, W.E., *In vivo* biocompatibility of dextran-based hydrogels, *J. Biomed. Mater. Res.*, 50, 397–404, 2000.

40.   Drury, J.L. and Mooney, D.J., Hydrogels for tissue engineering: scaffold design variables and applications, *Biomaterials*, 24, 4337–4351, 2003.

41. Hoffman, A.S., Hydrogels for biomedical applications, *Ann. NY Acad. Sci.*, 944, 62–73, 2001.

42. Hoffman, A.S., Hydrogels for biomedical applications, *Adv. Drug. Deliv. Rev.*, 17, 3–12, 2002.

43. Tirelli, N., Lutolf, M.P., Napoli, A., and Hubbell, J.A., Poly(ethylene glycol) block copolymers, *J. Biotechnol.*, 90, 3–15, 2002.

44. Nguyen, K.T. and West, J.L., Photopolymerizable hydrogels for tissue engineering applications, *Biomaterials*, 23, 4307–4314, 2002.

45. Byrne, M.E., Park, K., and Peppas, N.A., Molecular imprinting within hydrogels, *Adv. Drug Deliv. Rev.*, 54, 149–161, 2002.

46. Gutowska, A., Jeong, B., and Jasionowski, M., Injectable gels for tissue engineering, *Anat. Rec.*, 263, 342–349, 2001.

47. Lim, F., Microencapsulation of living cells and tissues — theory and practice, in *Biomedical Applications of Microencapsulation*, Lim, F., Ed., CRC Press, Boca Raton, FL, 1984, pp. 137–154.

48. Dvir-Ginzberg, M., Gamlieli-Bonshtein, I., Agbaria, R., and Cohen, S., Liver tissue engineering within alginate scaffolds: effects of cell-seeding density on hepatocyte viability, morphology, and function, *Tissue Eng.*, 9:757–766, 2003.

49. Sambanis, A., Engineering challenges in the development of an encapsulated cell system for treatment of type 1 diabetes, *Diabetes Technol. Ther.*, 2, 81–89, 2000.

50. Griffith, L.G., Acta Mater., 48, 263–277, 2000.

51. Hoffman, A.S., Synthetic polymeric biomaterials, in *Polymeric Materials and Artificial Organs*, Gebelein, C.G., Ed., American Chemical Society, 1984.

52. Widmer, M.S. and Mikos, A.G., Fabrication of biodegradable polymer scaffolds for tissue engineering, in *Frontiers in Tissue Engineering*, Patrick, C.W.J., Mikos, A.G., and McIntire, L.V., Eds., Elsevier Science Inc., New York, 1998, pp. 107–120.

53. Hutmacher, D.W., Huerzeler, M.B., and Schliephake, H., A review of material properties of biodegradable and bioresorbable polymers and devices for GTR and GBR applications, *Int. J. Oral. Maxillofac. Implants*, 11, 667–678, 1996.

54. Hutmacher, D.W., Polymeric scaffolds in tissue engineering bone and cartilage, *Biomaterials*, 21, 2529–2543, 2000.

55. Hutmacher, D.W., Scaffold design and fabrication technologies for engineering tissues — state of the art and future perspectives, *J. Biomat. Sci. Poly. Ed.*, 11, 107–124, 2001.

56. Chaignaud, B.E., Langer, R., and Vacanti, J.P., The history of tissue engineering using synthetic biodegradable polymer scaffolds and cells, in *Synthetic Biodegradable Scaffolds*, Atala, A. and Mooney D., Eds., Birkhauser, Boston, 1997, pp. 1–14.

57. Currie, L.J., Sharpe, J.R., and Martin, R., The use of fibrin glue in skin grafts and tissue-engineered skin replacements: a review, *Plast. Reconstr. Surg.*, 108, 1713–1726, 2001.

58. Hutmacher, D.W. and Vanscheidt, W., Matrices for tissue-engineered skin, *Drugs Today*, 38, 113–133, 2002.

59. Dornish, M., Kaplan, D., and Skaugrud, O., Standards and guidelines for biopolymers in tissue-engineered medical products ASTM alginate and chitosan standard guides, American Society for Testing and Materials, *Ann. NY Acad. Sci.*, 944, 388–397, 2001.

60. Minuth, W.W., Sittinger, M., and Kloth, S., Tissue engineering: generation of differentiated artificial tissues for biomedical applications, *Cell Tissue Res.*, 291, 1–11, 1998.

61. Lindenhayn, K., Perka, C., Spitzer, R., Heilmann, H., Pommerening, K., Mennicke, J., and Sittinger, M., Retention of hyaluronic acid in alginate beads: aspects for *in vitro* cartilage engineering, *J. Biomed. Mater. Res.*, 44, 149–155, 1999.

62. Benya, P.D. and Shaffer, J.D., Dedifferentiated chondrocytes reexpress the differentiated collagen phenotype when cultured in agarose gels, *Cell*, 30, 215–224, 1982.

63. Lee, D.A. and Bader, D.L., Compressive strains at physiological frequencies influence the metabolism of chondrocytes seeded in agarose, *J. Orthop. Res.*, 15, 181–188, 1997.

64. Rahfoth, B., Weisser, J., Sternkopf, F., Aigner, T., von der Mark, K., and Brauer, R., Transplantation of allograft chondrocytes embedded in agarose gel into cartilage defects of rabbits, *Osteoarthritis Cartilage*, 6, 50–65, 1998.

65. Sabra, W., Zeng, A.P., and Deckwer, W.D., Bacterial alginate: physiology, product quality and process aspects, *Appl. Microbiol. Biotechnol.*, 56, 315–325, 2001, Review.

66. Smidsrød, O. and Draget, K.I., Chemistry and physical properties of alginates, *Carbohydr. Eur.*, 14, 6–13, 1996.

67. Sabra, W., Zeng, A.P., and Deckwer, W.D., Bacterial alginate: physiology, product quality and process aspects, *Appl. Microbiol. Biotechnol.*, 56, 315–325, 2001.

68. Poncelet, D., Production of alginate beads by emulsification/internal gelation, *Ann. NY Acad. Sci.*, 944, 74–82, 2001.

69. Oerther, S., Le Gall, H., Payan, E., Lapicque, F., Presle, N., Hubert, P., Dexheimer, J., and Netter, P., Hyaluronate-alginate gel as a novel biomaterial: mechanical properties and formation mechanism, *Biotechnol. Bioeng.*, 63, 206–215, 1999.

70. Suzuki, K., Bonner-Weir, S., Trivedi, N., Yoon, K.H., Hollister-Lock, J., Colton, C.K., and Weir, G.C., Function and survival of macroencapsulated syngeneic islets transplanted into streptozocin-diabetic mice, *Transplantation*, 66, 21–28, 1998.

71. De Vos, P., De Haan, B.J., Wolters, G.H., Strubbe, J.H., and Van Schilfgaarde, R., Improved biocompatibility but limited graft survival after purification of alginate for microencapsulation of pancreatic islets, *Diabetologia*, 40, 262–270, 1997.

72. Thomas, S., Alginate dressings in surgery and wound management: part 3, *J. Wound Care*, 9, 163–166, 2000.

73. Hormbrey, E., Pandya, A., and Giele, H., Adhesive retention dressings are more comfortable than alginate dressings on split-skin-graft donor sites, *Br. J. Plast. Surg.*, 56, 498–503, 2003.

74. Hashimoto, T., Suzuki, Y., Tanihara, M., Kakimaru, Y., and Suzuki, K., Development of alginate wound dressings linked with hybrid peptides derived from laminin and elastin, *Biomaterials*, 25, 1407–1414, 2004.

75. Tonnesen, H.H. and Karlsen, J., Alginate in drug delivery systems, *Drug Dev. Ind. Pharm.*, 28, 621–630, 2002.

76. Chen, R.R. and Mooney, D.J., Polymeric growth factor delivery strategies for tissue engineering, *Pharm. Res.*, 20, 1103–1112, 2003.

77. Kong, H.J., Smith, M.K., and Mooney, D.J., Designing alginate hydrogels to maintain viability of immobilized cells, *Biomaterials*, 24, 4023–4029, 2003.

78. Lee, K.Y., Peters, M.C., Anderson, K.W., and Mooney, D.J., Controlled growth factor release from synthetic extracellular matrices, *Nature*, 408, 998–1000, 2000.

79. Peters, M.C., Isenberg, B.C., Rowley, J.A., and Mooney, D.J., Release from alginate enhances the biological activity of vascular endothelial growth factor, *J. Biomater. Sci. Polym.*, 12, 1267–1278, 1998.

80. Lee, K.Y., Peters, M.C., and Mooney, D.J., Controlled drug delivery from polymers by mechanical signals, *Adv. Mater.*, 13, 837–839, 2001.

81. Edelman, E.R., Mathiowitz, E., Langer, R., and Klagsbrun, M., Controlled and modulated release of basic fibroblast growth factor, *Biomaterials*, 12, 619–626, 1991.

82. Edelman, E.R., Nugent, M.A., Smith, L.T., and Karnovsky, M.J., Basic fibroblast growth factor enhances the coupling of intimal hyperplasia and proliferation of vasa vasorum in injured rat arteries, *J. Clin. Invest.*, 89, 465–473, 1992.

83. Sellke, F.W., Laham, R.J., Edelman, E.R., Pearlman, J.D., and Simons, M., Therapeutic angiogenesis with basic fibroblastic growth factor: technique and early results, *Ann. Thorac. Surg.*, 65, 1540–1544, 1998.

84. Elcin, Y.M., Dixit, V., and Gitnick, G., Extensive *in vivo* angiogenesis following controlled release of human vascular endothelial cell growth factor: implications for tissue engineering and wound healing, *Artif. Organs.*, 25, 558–565, 2001.

85. Tanihara, M., Suzuki, Y., Yamamoto, E., Noguchi, A., and Mizushima, Y., Sustained release of basic fibroblast growth factor and angiogenesis in a novel covalently crosslinked gel of heparin and alginate, *J. Biomed. Mater. Res.*, 56, 216–221, 2001.

86. Pangas, S.A., Saudye, H., Shea, L.D., and Woodruff, T.K., Novel approach for the three-dimensional culture of granulosa cell-oocyte complexes, *Tissue Eng,.* 9, 1013–1021, 2003.

87. Peirone, M., Ross, C.J.D., Horteland, G., Brash, J.L., and Change, P.L., Encapsulation of various recombinant mammalian cell types in different alginate microcapsules, *J. Biomed. Mater. Res.*, 42, 587–596, 1998.

88. Bjerkvig, R., Read, T.A., Vajkoczy, P., Aebischer, P., Pralong, W., Platt, S., Melvik, J.E., Hagen, A., and Dornish, M., Cell therapy using encapsulated cells producing endostatin, *Acta Neurochir. Suppl.*, 88, 137–141, 2003.

89. Omer, A., Duvivier-Kali, V.F., Trivedi, N., Wilmot, K., Bonner-Weir, S., and Weir, G.C., Survival and maturation of microencapsulated porcine neonatal pancreatic cell clusters transplanted into immunocompetent diabetic mice, *Diabetes*, 52, 69–75, 2003.

90. Umehara, Y., Hakamada, K., Seino, K., Aoki, K., Toyoki, Y., and Sasaki, M., Improved survival and ammonia metabolism by intraperitoneal transplantation of microencapsulated hepatocytes in totally hepatectomized rats, *Surgery*, 130, 513–520, 2003.

91. King, A., Strand, B., Rokstad, A.M., Kulseng, B., Andersson, A., Skjak-Braek, G., and Sandler, S., Improvement of the biocompatibility of alginate/poly-L-lysine/alginate microcapsules by the use of epimerized alginate as a coating, *J. Biomed. Mater. Res.*, 64, 533–539, 2003.

92. Dvir-Ginzberg, M., Gamlieli-Bonshtein, I., Agbaria, R., and Cohen, S., Liver tissue engineering within alginate scaffolds: effects of cell-seeding density on hepatocyte viability, morphology, and function, *Tissue Eng.*, 9, 757–766, 2003.

93. Allen, J.W. and Bhatia, S.N., Engineering liver therapies for the future, *Tissue Eng.*, 8, 725–737, 2002.

94. Giannoukakis, N., Pietropaolo, M., and Trucco, M., Genes and engineered cells as drugs for type I and type II diabetes mellitus therapy and prevention, *Curr. Opin. Investig. Drugs*, 3, 735–751, 2002.

95. Chia, S.H., Schumacher, B.L., Klein, T.J., Thonar, E.J., Masuda, K., Sah R.L., and Watson, D., Tissue-engineered human nasal septal cartilage using the alginate-recovered-chondrocyte method, *Laryngoscope*, 114, 38–45, 2004.

96. Loebsack, A., Greene, K., Wyatt, S., Culberson, C., Austin, C., Beiler, R., Roland, W., Eiselt, P., Rowley, J., Burg, K., Mooney, D., Holder, W., and Halberstadt, C., *In vivo* characterization of a porous hydrogel material for use as a tissue engineering bulking agent, *J. Biomed. Mater. Res.*, 57, 575–581, 2001.

97. Paige, K.T., Cima, L.G., Yaremchuk, M.J., Vacanti, J.P., and Vacanti, C.A., Injectable cartilage, *Plast. Reconst. Surg.*, 96, 1390–1398, 1995.

98. Bent, A.E., Tutrone, R.T., McLennan, M.T., Lloyd, L.K., Kennelly, M.J., and Badlani, G., Treatment of intrinsic sphincter deficiency using autologous ear chondrocytes as a bulking agent, *Neurourol. Urodynam.*, 20, 157–165, 2001.

99. Atala, A., Kim, W., Paige, K.T., Vacanti, C.A., and Retik, A.B., Endoscopic treatment of vesicoureteral reflux with a chondrocyte-alginate suspension, *J. Urol.*, 152, 641–643, 1994.

100. de Chalain, T., Phillips, J.H., and Hinek, A., Bioengineering of elastic cartilage with aggregated porcine and human auricular chondrocytes and hydrogels containing alginate, collagen, and κ-elastin, *J. Biomed. Mater. Res.*, 44, 280–288, 1999.

101. Lindenhayn, K., Perka, C., Spitzer, R.S., Heilmann, H.H., Pommerening, K., Mennicke, J., and Sittinger, M., Retention of hyaluronic acid in alginate beads: aspects for *in vitro* cartilage tissue engineering, *J. Biomed. Mater. Res.*, 44, 149–155, 1999.

102. Chang, S.C.N., Rowley, J.A., Tobias, G., Genes, N.G., Roy, A.K., Mooney, D.J., Vacanti, C.A., and Bonassar, L.J., Injection molding of chodrocyte/alginate constructs in the shape of facial implants, *J. Biomed. Mater. Res.*, 55, 503–511, 2001.

103. Kamil, S.H., Vacanti, M.P., Aminuddin, B.S., Jackson, M.J., Vacanti, C.A., and Eavey, R.D., Tissue engineering of a human sized and shaped auricle using a mold, *Laryngoscope*, 114, 867–870, 2004.

104. Liu, H., Lee, Y.W., and Dean, M.F., Re-expression of differentiated proteoglycan phenotype by dedifferentiated human chondrocytes during culture in alginate beads, *Biochim. Biophys. Acta*, 1425, 505–515, 1998.

105. van Susante, J.L., Buma, P., van Osch, G.J., Versleyen, D., van der Kraan, P.M., van der Berg, W.B., and Homminga, G.N., Culture of chondrocytes in alginate and collagen carrier gels, *Acta Orthop. Scand.*, 66, 549–556, 1995.

106. Kavalkovich, K.W., Boynton, R.E., Murphy, J.M., and Barry, F., Chondrogenic differentiation of human mesenchymal stem cells within an alginate layer culture system, *in Vitro Cell. Dev. Biol. Anim.*, 38, 457–466, 2002.

107. Paige, K.T., Cima, L.G., Yaremchuk, M.J., Vacanti, J.P., and Vacanti, C.A., Injectable cartilage, *Plast. Reconstr. Surg.*, 96, 1390–1398, 1995.

108. Yang, W.D., Chen, S.J., Mao, T.Q., Chen, F.L., Lei, D.L., Tao, K., Tang, L.H., and Xiao, M.G., A study of injectable tissue-engineered autologous cartilage, *Chin. J. Dent. Res.*, 3, 10–15, 2000.

109. Vacanti, C.A., Bonassar, L.J., Vacanti, M.P., and Shufflebarger, J., Replacement of an avulsed phalanx with tissue-engineered bone, *N. Engl. J. Med.*, 344, 1511–1154, 2001.

110. Alsberg, E., Anderson, K.W., Albeiruti, A., Franceschi, R.T., and Mooney, D.J., Cell-interactive alginate hydrogels for bone tissue engineering, *J. Dent. Res.*, 80, 2025–2029, 2001.

111. Suzuki, Y., Tanihara, M., Suzuki, K., Saitou, A., Sufan, W., and Nishimura, Y., Alginate hydrogel linked with synthetic oligiopeptide derived from BMP-2 allows ectopic osteoinduction *in vivo*, *J. Biomed. Mater. Res.*, 50, 405–409, 2000.

112. Alsberg, E., Anderson, K.W., Albeiruti, A., Rowley, J.A., and Mooney, D.J., Engineering growing tissues, *Proc. Natl. Acad. Sci.*, 99, 12025–12030, 2002.

113. Stevens, M.M., Qanadilo, H.F., Langer, R., and Prasad Shastri, V., A rapid-curing alginate gel system: utility in periosteum-derived cartilage tissue engineering, *Biomaterials*, 25, 887–894, 2004.

114. Hirano, S., Chitin biotechnology applications, *Biotechnol. Ann. Rev.*, 2, 237–258, 1996.

115. Majeti, N.V. and Kumar, R., A review of chitin and chitosan applications, *Reac. Funct. Polym.*, 46, 1–27, 2000.

116. Chandy, T. and Sharma, C.P., Chitosan — as a biomaterial, *Biomater. Art. Cells. Art. Org.*, 18, 1–24, 1990.

117. Dodane, V. and Vilivalam, V.D., Pharmaceutical applications of chitosan, *Pharm. Sci. Technol. Today*, 1, 246–253, 1998.

118. Bernkop-Schnurch, A., Chitosan and its derivatives: potential excipients for peroral peptide delivery systems, *Int. J. Pharm.*, 194, 1–13, 2000.

119. Felt, O., Buri, P., and Gurny, R., Chitosan: a unique polysaccharide for drug delivery, *Drug Dev. Ind. Pharm.*, 24, 979–993, 1998.

120. Shigemasa, Y. and Minami, S., Applications of chitin and chitosan for biomaterials, *Biotechnol. Genet. Eng. Rev.*, 13, 383–420, 1996.

121. Kas, H.S., Chitosan: properties, preparations and application to microparticulate systems, *J. Microencapsul.*, 14, 689–711, 1997.

122. Somashekar, D. and Joseph, R., Chitosanases — properties and applications: a review, *Bioresou. Technol.*, 55, 35–45, 1996.

123. Tomihata, K. and Ikada, Y., *In vitro* and *in vivo* degradation of films of chitin and its deacetylated derivatives, *Biomaterials*, 18, 567–575, 1997.

124. Onishi, H. and Machida, Y., Biodegradation and distribution of water-soluble chitosan in mice, *Biomaterials*, 20, 175–182, 1999.

125. Tanaka, Y., Tanioka, S.I., Tanaka, M., Tanigawa, T., Kitamura, Y., Minami, S., Okamoto, Y., Miyashita, M., and Nanno, M., Effects of chitin and chitosan particles on BALB/c mice by oral and parenteral administration, *Biomaterials*, 18, 591–595, 1997.

126. Khor, E. and Lee, Y.L., Implantable applications of chitin and chitosan, *Biomaterials*, 24, 2339–2349, 2003.

127. van der Lei, B. and Wildevuur, C.R., Improved healing of microvascular PTFE prostheses by induction of a clot layer: an experimental study in rats, *Plast. Reconstr. Surg.*, 84, 960–968, 1989.

128. Brandenberg, G., Leibrock, L.G., Shuman, R., Malette, W.G., and Quigley, H., Chitosan: a new topical hemostatic agent for diffuse capillary bleeding in brain tissue, *Neurosurgery*, 15, 9–13, 1984.

129. Muzzarelli, R., Baldassarre, V., Conti, F., Ferrara, P., Biagini, G., Gazzanelli, G., and Vasi, V., Biological activity of chitosan: ultrastructural study, *Biomaterials*, 9, 247–252, 1998.

130. Usami, Y., Okamoto, Y., Takayama, T., Shigemasa, Y., and Minami, S., Chitin and chitosan stimulate canine polymorphonuclear cells to release leukotriene B4 and prostaglandin E2, *J. Biomed. Mater. Res.*, 42, 517–522, 1998.

131. Okamoto, Y., Watanabe, M., Miyatake, K., Morimoto, M., Shigemasa, Y., and Minami, S., Effects of chitin/chitosan and their oligomers/monomers on migrations of fibroblasts and vascular endothelium, *Biomaterials*, 23, 1975–1979, 2002.

132. Minami, S., Okamoto, Y., Hamada, K., Fukumoto, Y., and Shigemasa, Y., Veterinary practice with chitin and chitosan, *EXS*, 87, 265–277, 1999.

133. Ueno, H., Nakamura, F., Murakami, M., Okumura, M., Kadosawa, T., and Fujinaga, T., Evaluation effects of chitosan for the extracellular matrix production by fibroblasts and the growth factors production by macrophages, *Biomaterials*, 22, 2125–2130, 2001.

134. Ueno, H., Murakami, M., Okumura, M., Kadosawa, T., Uede, T., and Fujinaga, T., Chitosan accelerates the production of osteopontin from polymorphonuclear leukocytes, *Biomaterials*, 22, 1667–1673, 2001.

135. Cho, Y.W., Cho, Y.N., Chung, S.H., Yoo, G., and Ko, S.W., Water-soluble chitin as a wound healing accelerator, *Biomaterials*, 20, 2139–2145, 1999.

136. Ueno, H., Yamada, H., Tanaka, I., Kaba, N., Matsuura, M., Okumura, M., Kadosawa, T., and Fujinaga, T., Accelerating effects of chitosan for healing at early phase of experimental open wound in dogs, *Biomaterials*, 20, 1407–1414, 1999.

137. Mizuno, K., Yamamura, K., Yano, K., Osada, T., Saeki, S., Takimoto, N., Sakura, T., and Nimura, Y., Effect of chitosan film containing basic fibroblast growth factor on wound healing in genetically diabetic mice, *J. Biomed. Mater. Res.*, 64, 177–181, 2003.

138. Yusof, N.L.M., Lim, L.Y., and Khor, E., Preparation and characterization of chitin beads as a wound dressing precursor, *J. Biomed. Mater. Res.*, 54, 59–68, 2001.

139. Ono, K., Saito, Y., Yura, H., Ishikawa, K., Kurita, A., Akaike, T., and Ishihara, M., Photocrosslinkable chitosan as a biological adhesive, *J. Biomed. Mater. Res.*, 49, 289–295, 2000.

140. Ishihara, M., Nakanishi, K., Ono, K., Sato, M., Kikuchi, M., Saito, Y., Yura, H., Matsui, T., Hattori, H., Uenoyama, M., and Kurita, A., Photocrosslinkable chitosan as a dressing for wound occlusion and accelerator in healing process, *Biomaterials*, 23, 833–840, 2002.

141. Mi, F.L., Shyu, S.S., Wu, Y.B., Lee, S.T., Shyong, J.Y., and Huang, R.N., Fabrication and characterization of a sponge-like asymmetric chitosan membrane as a wound dressing, *Biomaterials*, 22, 165–173, 2001.

142. Tanabe, T., Okitsu, N., Tachibana, A., and Yamauchi, K., Preparation and characterization of keratin–chitosan composite film, *Biomaterials*, 23, 817–825, 2002.

143. Su, C.H., Sun, C.S., Juan, S.W., Hu, C.H., Ke, W.T., and Sheu, M.T., Fungal mycelia as the source of chitin and polysaccharides and their applications as skin substitutes, *Biomaterials*, 18, 1169–1174, 1997.

144. Hung, W.S., Fang, C.L., Su, C.H., Lai, W.F.T., Chang, Y.C., and Tsai, Y.H., Cytotoxicity and immunogenicity of sacchachitin and its mechanism of action on skin wound healing, *J. Biomed. Mater. Res.*, 56, 93–100, 2001.

145. Su, C.-H., Sun, C.-S., Juan, S.-W., Ho, H.-O., Hu, C.-H., and Sheu, M.-T., Development of fungal mycelia as skin substitutes: effects on wound healing and fibroblast, *Biomaterials*, 20, 61–68, 1999.

146. Lee, Y.M., Kim, S.S., Park, M.H., Song, K.W., Sung, Y.K., and Kang, I.K., β-Chitin-based wound dressing containing sulfurdiazine, *J. Mater. Sci. Mater. Med.*, 11, 817–823, 2000.

147. Mi, F.L., Shyu, S.S., Wu, Y.B., Lee, S.T., Shyong, J.Y., and Huang, R.N., Fabrication and characterization of a sponge-like asymmetric chitosan membrane as a wound dressing, *Biomaterials*, 22, 165–173, 2001.

148. Mi, F.-W., Wu, Y.-B., Shyu, S.-S., Schoung, J.-Y., Huang, Y.-B., Tsai Y.-H., and Hao, J.-Y., Control of wound infections using a bilayer chitosan wound dressing with sustainable antibiotic delivery, *J. Biomed. Mater. Res.*, 59, 438–449, 2002.

149. Loke, W.K., Lau, S.K., Lim, L.Y., Khor, E., and Chow, K.S., Wound dressing with sustained anti-microbial capability, *J. Biomed. Mater. Res.*, 53, 8–17, 2000.

150. Yan, X.-L., Khor, E., and Lim, L.Y., Chitosan-alginate films prepared with chitosans of different molecular weights, *J. Biomed. Mater. Res.*, 58, 358–365, 2001.

151. Yan, X.L., Khor, E., and Lim, L.Y., PEC films prepared from chitosan-alginate coacervates, *Chem. Pharm. Bull.*, 48, 941–946, 2000.

152. Wang, L.S., Khor, E., and Lim, L.Y., Chitosan-alginate-$CaCl_2$ system for membrane coat application, *J. Pharm. Sci.*, 90, 1134–1142, 2001.

153. Wang, L.S., Khor, E., Wee, A., and Lim, L.Y., Chitosan-alginate PEC membrane as a wound dressing: assessment of incisional wound dressing, *J. Biomed. Mater. Res.*, 63, 610–618, 2002.

154. Hirano, S., Zhang, M., and Nakagawa, M., Release of glycosaminoglycans in physiological saline and water by wet-spun chitin-acid glycosaminoglycan fibers, *J. Biomed. Mater. Res.*, 56, 556–561, 2001.

155. Dung, P.L., Pham, T.D., Nguyen, T.M., Nguyen, K.T., Chu, D.K., Le, T.S., Trinh, B., Nguyen, T.B., Bach, H.A., and Cao, V.M., Vinachitin, an artificial skin for wound healing, in *Chitin and chitosan chitin and chitosan in life sciences*, Uragami, T., Kurita, K., and Fukamizo, T., Eds., Kodansha Scientific Ltd., Tokyo, 2001, pp. 27–230.

156. Skaugrud, O., Hagen, A., Borgersen, B., and Dornish, M., Biomedical and pharmaceutical applications of alginate and chitosan, *Biotechnol. Genet. Eng. Rev.*, 16, 23–40, 1999.

157. Kato, Y., Onishi, H., and Machida, Y., *N*-succinyl-chitosan as a drug carrier: water-insoluble and water-soluble conjugates, *Biomaterials*, 25, 907–915, 2004.

159. Lee, Y.M., Kim, S.S., and Kim, S.H., Synthesis and properties of poly(ethylene glycol) macromer/β-chitosan hydrogels, *J. Mater. Sci. Mater. Med.*, 8, 537–541, 1997.

160. Gupta, K.C. and Ravi Kumar, M.N.V., pH dependent hydrolysis and drug release behavior of chitosan/poly(ethylene glycol) polymer network microspheres, *J. Mater. Sci. Mater. Med.*, 12, 753–759, 2001.

161. Park, S.B., You, J.O., Park, H.Y., Haam, S.J., and Kim, W.S., A novel pH-sensitive membrane from chitosan: preparation and its drug permeation characteristics, *Biomaterials*, 22, 323–330, 2001.

162. Ahn, J.S., Choi, H.K., and Cho, C.S., A novel mucoadhesive polymer prepared by template polymerization of acrylic acid in the presence of chitosan, *Biomaterials*, 22, 923–928, 2001.

163. Ahn, J.S., Choi, H.K., Chun, M.K., Ryu, J.M., Jung, J.H., Kim, Y.U., and Cho, C.S., Release of triamcinolone acetonide from mucoadhesive polymer composed of chitosan and poly(acrylic acid) *in vitro*, *Biomaterials*, 23, 1411–1416, 2002.

164. Li, F., Liu, W.G., and Yao, K.D., Preparation of oxidized glucose-crosslinked *N*-alkylated chitosan membrane and *in vitro* studies of pH-sensitive drug delivery behavior, *Biomaterials*, 23, 343–347, 2002.

165. Zhao, H.R., Wang, K., Zhao, Y., and Pan, L.Q., Novel sustained-release implant of herb extract using chitosan, *Biomaterials*, 23, 4459–4462, 2002.

166. Mi, F.W., Shyu, S.S., Chen, C.T., and Schoung, J., Porous chitosan microsphere for controlling the antigen release of Newcastle disease vaccine: preparation of antigen-adsorbed microsphere and *in vitro* release, *Biomaterials*, 20, 1603–1612, 1999.

167. Mi, F.L., Tan, Y.C., Liang, H.F., and Sung, H.W., *In vivo* biocompatibility and degradability of a novel injectable-chitosan-based implant, *Biomaterials*, 23, 181–191, 2002.

168. Mi, F.L., Shyu, S.S., Lin, Y.M., Wu, Y.B., Peng, C.K., and Tsai, Y.H., Chitin/PLGA blend microspheres as a biodegradable drug delivery system: a new delivery system for protein, *Biomaterials*, 24, 5023–5036, 2003.

169. Gupta, K.C. and Ravi Kumar, M.N.V., Drug release behavior of beads and microgranules of chitosan, *Biomaterials*, 21, 1115–1119, 2000.

170. Hata, H., Onishi, H., and Machida, Y., Preparation of CM-chitin microspheres by complexation with iron(III) in w/o emulsion and their biodisposition characteristics in mice, *Biomaterials*, 21, 1779–1788, 2000.

171. Mi, F.L., Tan, Y.C., Liang, H.F., and Sung, H.W., *In vivo* biocompatibility and degradability of a novel injectable-chitosan-based implant, *Biomaterials*, 23, 181–191, 2002.

172. Shi, X.-Y. and Tan, T.-W., Preparation of chitosan/ethylcellulose complex microcapsule and its application in controlled release of Vitamin D_2, *Biomaterials*, 23, 4469–4473, 2002.

173. Mi, F.L., Lin, Y.M., Wu, Y.B., Shyu, S.S., and Tsai, Y.H., Chitin/PLGA blend microspheres as a biodegradable drug-delivery system: phase-separation, degradation and release behavior, *Biomaterials*, 23, 3257–3267, 2002.

174. Hu, Y., Jiang, X., Ding, Y., Ge, H., Yuan, Y., and Yang, C., Synthesis and characterization of chitosan-poly(acrylic acid) nanoparticles, *Biomaterials*, 23, 3193–3201, 2002.

175. Takechi, M., Miyamoto, Y., Momota, Y., Yuasa, T., Tatehara, S., Nagayama, M., Ishikawa, K., and Suzuki, K., The *in vitro* antibiotic release from anti-washout apatite cement using chitosan, *J. Mater. Sci. Mater. Med.*, 13, 973–978, 2002.

176. Takechi, M., Miyamoto, Y., Ishikawa, K., Toh, T., Yuasa, T., Nagayama, M., and Suzuki, K., Initial histological evaluation of anti-washout type fast-setting calcium phosphate cement following subcutaneous implantation, *Biomaterials*, 19, 2057–2063, 1998.

177. Takechi, M., Ishikawa, K., Miyamoto, Y., Nagayama, M., and Suzuki, K., Tissue responses to anti-washout apatite cement using chitosan when implanted in the rat tibia, *J. Mater. Sci. Mater. Med.*, 12, 597–602, 2001.

178. Zhang, Y. and Zhang, M., Calcium phosphate/chitosan composite scaffolds for controlled *in vitro* antibiotic drug release, *J. Biomed. Mater. Res.*, 62, 378–386, 2002.

179. Roth, J.A. and Cristiano, R.J., Gene therapy for cancer: what have we done and where are we going? *J. Nat. Cancer Inst.*, 89, 21–39, 1997.

180. Rolland, A.P., From genes to gene medicines: recent advances in nonviral gene delivery, *Crit. Rev. Ther. Drug Carrier Syst.*, 15, 143–198, 1998.

181. Garnet, M.C., Gene delivery systems using cationic polymers, *Crit. Rev. Ther. Drug Carrier Syst.*, 16, 147–207, 1999.

182. Sato, T., Ishii, T., and Okahata, Y., *In vitro* gene delivery mediated by chitosan. Effect of pH, serum, and molecular mass of chitosan on the transfection efficiency, *Biomaterials*, 22, 2075–2080, 2001.

183. Kawamata, Y., Nagayama, Y., Nakao, K., Mizuguchi, H., Hayakawa, T., Sato T., and Ishii, N., Receptor-independent augmentation of adenovirus-mediated gene transfer with chitosan *in vitro*, *Biomaterials*, 23, 4573–4579, 2002.

184. Erbacher, P., Zou, S., Bettinger, T., Steffan, A.M., and Remy, J.S., Chitosan-based vector/DNA complexes for gene delivery: biophysical characteristics and transfection ability, *Pharm. Res.*, 15, 1332–1339, 1998.

185. Mao, H.Q., Roy, K., Troung-Le, V.L., Janes, K.A., Lin, K.Y., Wang, Y., August, J.T., and Leong, K.W., Chitosan–DNA nanoparticles as gene carriers: synthesis, characterization and transfection efficiency, *J. Control. Release*, 70, 399–421, 2001.

186. Kim, T.H., Ihm, J.E., Choi, Y.J., Nah, J.W., and Cho, C.S., Efficient gene delivery by urocanic acid-modified chitosan, *J. Control. Release*, 93, 389–402, 2003.

187. Borchard, G., Chitosans for gene delivery, *Ad. Drug Delivery Rev.*, 52, 145–150, 2001.

188. Wen, G.L., Kang, D.Y., and Yao, D., Chitosan and its derivatives — a promising nonviral vector for gene transfection, *J. Control. Release*, 83, 1–11, 2002.

189. Pouton, C.W. and Seymour, L.W., Key issues in non-viral gene delivery, *Adv. Drug Del. Rev.*, 34, 3–19, 1998.

190. Suh, J.K.F. and Matthew, H.W.T., Application of chitosan-based polysaccharide biomaterials in cartilage tissue engineering: a review, *Biomaterials*, 21, 2589–2598, 2000.

191. Ma, J., Wang, H., He, B., and Chen, J., A preliminary *in vitro* study on the fabrication and tissue engineering applications of a novel chitosan bilayer material as a scaffold of human neofetal dermal fibroblasts, *Biomaterials*, 22, 331–336, 2001.

192. Onishi, H. and Machida, Y., Biodegradation and distribution of water-soluble chitosan in mice, *Biomaterials*, 20, 175–82, 1999.

193. Eser, A., Elcin, Y.M., and Pappas, G.D., Neural tissue engineering: adrenal chromaffin cell attachment and viability on chitosan scaffolds, *Neurol. Res.*, 20, 648–654, 1998.

194. Elcin, Y.M., Dixit, V., and Gitnick, G., Hepatocyte attachment on biodegradable modified chitosan membranes: *in vitro* evaluation for the development of liver organoids, *Artif. Organs*, 22, 837–846, 1998.

195. Madihally, S.V. and Matthew, H.W., Porous chitosan scaffolds for tissue engineering, *Biomaterials*, 20, 1133–1142, 1999.

196. Sechriest, V.F., Miao, Y.J., Niyibizi, C., Westerhausen-Larson, A., Matthew, H.W., Evans, C.H., Fu, F.H., and Suh, J.K., GAG-augmented polysaccharide hydrogel: a novel biocompatible and biodegradable material to support chondrogenesis, *J. Biomed. Mater. Res.*, 15, 534–541, 2000.

197. Chupa, J.M., Foster, A.M., Sumner, S.R., Madihally, S.V., and Matthew, H.W., Vascular cell responses to polysaccharide materials: *in vitro* and *in vivo* evaluations, *Biomaterials*, 21, 2315–2322, 2000.

198. Okamoto, Y., Watanabe, M., Miyatake, K., Morimoto, M., Shigemasa, Y., and Minami, S., Effects of chitin/chitosan and their oligomers/monomers on migrations of fibroblasts and vascular endothelium, *Biomaterials*, 23, 1975–1979, 2002.

199. Howling, G.I., Dettmar, P.W., Goddard, P.A., Hampson, F.C., Dornish, M., and Wood, E.J., The effect of chitin and chitosan on the proliferation of human skin fibroblasts and keratinocytes *in vitro*, *Biomaterials*, 22, 2959–2966, 2001.

200. Mori, T., Okumura, M., Matsura, M., Ueno, K., Tokura, S., Okamoto, Y., Minami, S., and Fujinaga, T., Effects of chitin and its derivatives on the proliferation and cytokine production of fibroblasts *in vitro*, *Biomaterials*, 18, 947–951, 1997.

201. Prasitsilp, M., Jenwithisuk, R., Kongsuwan, K., Damrongchai, N., and Watts, P., Cellular responses to chitosan *in vitro*: the importance of deacetylation, *J. Mater. Sci. Mater. Med.*, 11, 773–778, 2000.

202. Chen, X.G., Wang, Z., Liu, W.S., and Park, H.J., The effect of carboxymethyl-chitosan on proliferation and collagen secretion of normal and keloid skin fibroblasts, *Biomaterials*, 23, 4609–4614, 2002.

203. VandeVord, P.J., Matthew, H.W.T., DeSilva, S.P., Mayton, L., Wu, B., and Wooley, P.H., Evaluation of the biocompatibility of a chitosan scaffold in mice, *J. Biomed. Mater. Res.*, 59, 585–590, 2002.

204. Zhang, Y. and Zhang, M., Synthesis and characterization of macroporous chitosan/calcium phosphate composite scaffolds for tissue engineering, *J. Biomed. Mater. Res.*, 55, 304–312, 2001.

205. Zhang, Y., Ni, M., Zhang, M.Q., and Ratner, B., Calcium phosphate/chitosan composite scaffolds for bone tissue engineering, *Tissue Eng.*, 9, 337–345, 2003.

206. Chow, K.S., Khor, E., and Wan, A.C.A., Porous chitin matrices for tissue engineering: fabrication and *in-vitro* cytotoxic assessment, *J. Polym. Res.*, 8, 27–35, 2001.

207. Wang, M., Chen, L.J., Ni, J., Weng, J., and Yue, C.Y., Manufacture and evaluation of bioactive and biodegradable materials and scaffolds for tissue engineering, *J. Mater. Sci. Mater. Med.*, 12, 855–860, 2001.

208. Ma, J., Wang, H., He, B., and Chen, J., A preliminary *in vitro* study on the fabrication and tissue engineering applications of a novel chitosan bilayer material as a scaffold of human neofetal dermal fibroblasts, *Biomaterials*, 22, 331–336, 2001.

209. Chow, K.S. and Khor, E., Novel fabrication of open-pore chitin matrixes, *Biomacromolecules*, 1, 61–67, 2000.

210. Chupa, J.M., Foster, A.M., Sumner, S.R., Madihally, S.V., and Matthew, H.W., Vascular cell responses to polysaccharide materials: *in vitro* and *in vivo* evaluations, *Biomaterials*, 22, 2315–2322, 2000.

211. Zhao, F., Yin, Y., Lu, W.W., Leong, J.C., Zhang, W., Zhang, J., Zhang, M., and Yao, K., Preparation and histological evaluation of biomimetic three-dimensional hydroxyapatite/chitosan–gelatin network composite scaffolds, *Biomaterials*, 23, 3227–3234, 2002.

212. Zhang, Y. and Zhang, M., Synthesis and characterization of macroporous chitosan/calcium phosphate composite scaffolds for tissue engineering, *J. Biomed. Mater. Res.*, 55, 304–312, 2001.

213. Risbud, M., Endres, M., Ringe, J., Bhonde, R., and Sittinger, M., Biocompatible hydrogel supports the growth of respiratory epithelial cells: possibilities in tracheal tissue engineering, *J. Biomed. Mater. Res.*, 56, 120–127, 2001.

214. Risbud, M.V., Bhonde, M.R., and Bhonde, R.R., Effect of chitosan-polyvinyl pyrrolidone hydrogel on proliferation and cytokine expression of endothelial cells: implications in islet immunoisolation, *J. Biomed. Mater. Res.*, 57, 300–305, 2001.

215. Haipeng, G., Zhong, Y., Li, J., Gong, Y., Zhao, N., and Zhang, X., Studies on nerve cell affinity of chitosan-derived materials, *J. Biomed. Mater. Res.*, 52, 285–295, 2000.

216. Chung, T.W., Yang, J., Akaike, T., Cho, K.Y., Nah, J.W., Kim, S.I., and Cho, C.S., Preparation of alginate/galactosylated chitosan scaffold for hepatocyte attachment, *Biomaterials*, 23, 2827–2834, 2002.

217. Zhu, A., Zhang, M., Wu, J., and Shen, J., Covalent immobilization of chitosan/heparin complex with a photosensitive hetero-bifunctional crosslinking reagent on PLA surface, *Biomaterials*, 23, 4657–4665, 2002.

218. Zhu, H., Ji, J., Lin, R., Gao, C., Feng, L., and Shen, J., Surface engineering of poly(D,L-lactic acid) by entrapment of chitosan-based derivatives for the promotion of chondrogenesis, *J. Biomed. Mater. Res.*, 62, 532–539, 2002.

219. Cai, K., Yao, K., Cui, Y., Lin, S., Yang, Z., Li, X., Xie, H., Qing, T., and Luo, J., Surface modification of poly (D,L-lactic acid) with chitosan and its effects on the culture of osteoblasts *in vitro*, *J. Biomed. Mater. Res.*, 60, 398–404, 2002.

220. Chung, T.-W., Lu, Y.-F., Wang, S.-S., Lin, Y.-S., and Chu, S.-H., Growth of human endothelial cells on photochemically grafted Gly–Arg–Gly–Asp (GRGD) chitosans, *Biomaterials*, 23, 4803–4809, 2002.

221. Cui, W., Kim, D.H., Imamura, M., Hyon, S.H., and Inoue, K., Tissue-engineered pancreatic islets: culturing rat islets in the chitosan sponge, *Cell Transplant*, 10, 499–502, 2001.

222. Li, J., Pan, J., Zhang, L., Guo, X., and Yu, Y., Culture of primary rat hepatocytes within porous chitosan scaffolds, *J. Biomed. Mater. Res.*, 67A, 938–943, 2003.

223. Ang, T.H., Sultana, F.S.A., Hutmacher, D.W., Wong, Y.S., Fuh, J.Y.H., Mo, X.M., Loh, H.T., Burdet, E., and Teoh, S.H., Fabrication of 3D chitosan-hydroxyapatite scaffolds using a robotic dispensing system, *Mater. Sci. Eng.*, 20, 35–42, 2002.

224. Cascone, M.G., Barbani, N., Cristallini, C., Giusti, P., Ciardelli, G., and Lazzeri, L., *J. Biomater. Sci. Polym. Ed.*, 12, 267–281, 2001.

225. Chiu, H.C., Lin, Y.F., and Hsu, Y.H., Effects of acrylic acid on preparation and swelling properties of pH-sensitive dextran hydrogels, *Biomaterials*, 23, 1103–1112, 2002.

226. Kim, S.H., Won, C.Y., and Chu, C.C., Synthesis and characterization of dextran-based hydrogels prepared by photocrosslinking, *Carbohydr. Polym.*, 40, 183–190, 1999.

227. Kim, S.H. and Chu, C.C., Synthesis and characterization of dextran-methacrylate hydrogels and structural study by SEM, *J. Biomed. Mater. Res.*, 49, 517–527, 2000.

228. Franssen, O., Vos, O.P., and Hennink, W.E., Delayed release of a model protein from enzymatically-degrading dextran hydrogels, *J. Control. Release*, 44, 237–245, 1997.

229. Franssen, O., Van Ooijen, R.D., De Boer, D., Maes, R.A.A., Herron, J.N., and Hennink, W.E., Enzymatic degradation of methacrylated dextrans, *Macromolecule*, 30, 7408–7413, 1997.

230. Franssen, O., Vandervennet, L., Roders, P., and Hennink, W.E., Degradable dextran hydrogels: controlled release of a model protein from cylinders and microspheres, *J. Control. Release*, 60, 211–221, 1999.

231. Franssen, O., Van Ooijen, R.D., De Boer, D., Maes, R.A.A., and Hennink, W.E., Enzymatic degradation of cross-linked dextrans, *Macromolecules*, 32, 2896–2902, 1999.

232. De Smedt, S.C., Lauwers, A., Demeester, J., Van Steenbergen, M.J., Hennink, W.E., and Roefs, S.P.F.M., Characterization of the network structure of dextran glycidyl methacrylate hydrogels by studying the rheological and swelling behavior, *Macromolecules*, 28, 5082–5088, 1995.

233. De Smedt, S.C., Meyvis, T.K.L., Demeester, J., Van Oostveldt, P., Blonk, J.C.G., and Hennink, W.E., Diffusion of macromolecules in dextran methacrylate solutions and gels as studied by confocal scanning laser microscopy, *Macromolecules*, 30, 4863–4870, 1997.

234. Cai, Q., Yang, J., Bei, J., and Wang, S., A novel porous cells scaffold made of poly-lactide-dextran blend by combining phase-separation and particle-leaching techniques, *Biomaterials*, 23, 4483–4492, 2002.

235. Hennink, W.E., Talsma, H., Borchert, J.C.H., de Smedt, S.C., and Demeester, J., Controlled release of proteins from dextran hydrogels, *J. Control. Release*, 39, 47–55, 1996.

236. Hennink, W.E., Franssen, O., van Dijk-Wolthuis, W.N.E., and Talsma, H., Dextran hydrogels for the controlled release of proteins, *J. Control. Release*, 48, 107–114, 1997.

237. Stenekes, R.J.H. and Hennink, W.E., Equilibrium water content of microspheres based on cross-linked dextran, *Intern. J. Pharm.*, 189, 131–135, 1999.

238. Chiu, H.C., Hsiue, G.H., Lee, Y.P., and Huang, L.W., Synthesis and characterization of pH-sensitive dextran hydrogels as a potential colon-specific drug delivery system, *J. Biomater. Sci. Polym. Edn.*, 10, 591–608, 1999.

239. Chiu, H.C., Wu, A.T., and Lin, Y.F., Synthesis and characterization of acrylic acid-containing dextran hydrogels, *Polymer*, 42, 1471–1479, 2001.

240. Kim, S.H., Won, C.Y., and Chu, C.C., Synthesis and characterization of dextran-maleic acid based hydrogel, *J. Biomed. Mater. Res.*, 46, 160–170, 1999.

241. Läckgren, G., Wåhlin, N., Sköldenberg, E., and Stenberg, A., Long-term follow up of children treated with dextranomer/hyaluronic acid copolymer for vesicoureteral reflux, *J. Urol.*, 166, 1887–1892, 2001.

242. Ehrenfreund-Kleinman, T., Gazit, Z., Gazit, D., Azzam, T., Golenser, J., and Domb, A.J., Synthesis and biodegradation of arabinogalactan sponges prepared by reductive amination, *Biomaterials*, 23, 4621–4631, 2002.

243. Bucheler, M., Wirz, C., Schutz, A., and Bootz, F., Tissue engineering of human salivary gland organoids, *Acta Otolaryngol.*, 122, 541–545, 2002.

244. Massia, S.P., Stark, J., and Letbetter, D.S., Surface-immobilized dextran limits cell adhesion and spreading, *Biomaterials*, 21, 2253–2261, 2000.

245. Zhang, Y. and Chu, C.C., Biodegradable dextran-polylactide hydrogel networks: their swelling, morphology and the controlled release of indomethacin, *J. Biomed. Mater. Res.*, 59, 318–28, 2002.

246. Lebugle, A., Rodrigues, A., Bonnevialle, P., Voigt, J.J., Canal, P., and Rodriguez, F., Study of implantable calcium phosphate systems for the slow release of methotrexate, *Biomaterials*, 23, 3517–3522, 2002.

247. Trudel, J. and Massia, S.P., Assessment of the cytotoxicity of photocrosslinked dextran and hyaluronan-based hydrogels to vascular smooth muscle cells, *Biomaterials*, 23, 3299–3307, 2002.

248. Iskakov, R.M., Kikuchi, A., and Okano, T., Time-programmed pulsatile release of dextran from calcium-alginate gel beads coated with carboxy-n-propylacrylamide copolymers, *J. Control. Releases.*, 23, 57–68, 2002.

249. Chan, C., Thompson, I., Robinson, P., Wilson, J., and Hench, L., Evaluation of Bioglass/dextran composite as a bone graft substitute, *Int. J. Oral. Maxillofac. Surg.*, 31, 73–77, 2002.

250. Kamimura, W., Ooya, T., and Yui, N., Transience in polyion complexation between nicotinamide-modified dextran and carboxymethyl dextran during enzymatic degradation of dextran, *J. Biomater. Sci. Polym. Edn.*, 12, 1109–1122, 2001.

251. Letourneur, D., Machy, D., Pelle, A., Marcon-Bachari, E., D'Angelo, G., Vogel, M., Chaubet, F., and Michel, J.B., Heparin and non-heparin-like dextrans differentially modulate endothelial cell proliferation: *in vitro* evaluation with soluble and crosslinked polysaccharide matrices, *J. Biomed. Mater. Res.*, 60, 94–100, 2002.

252. Zhang, Y. and Chu, C.C., *In vitro* release behavior of insulin from biodegradable hybrid hydrogel networks of polysaccharide and synthetic biodegradable polyester, *J. Biomater. Appl.*, 16, 305–325, 2002.

253. Campoccia, D., Hunt, J.A., Doherty, P.J., Zhong, S.P., O'Regan, M., Benedetti, L., and Williams, D.F., Quantitative assessment of the tissue response to films of hyaluronan derivatives, *Biomaterials*, 17, 963–975, 1996.

254. Campoccia, D., Doherty, P., Radice, M., Brun, P., Abatangelo, G., and Williams, D.F., Semisynthetic resorbable materials from hyaluronan esterification, *Biomaterials*, 19, 2101–2127, 1998.

255. Madihally, S.V. and Matthew, H.W.T., Porous chitosan scaffolds for tissue engineering, *Biomaterials*, 20, 1133–1142, 1999.

256. Campoccia, D., Hunt, J.A., Doherty, P.J., Zhong, S.P., O'Regan, M., Benedetti, L., and Williams, D.F., Quantitative assessment of the tissue response to films of hyaluronic derivatives, *Biomaterials*, 17, 963–975, 1996.

257. Collier, J.H., Camp, J.P., Hudson, T.W., and Schmidt, C.E., Synthesis and characterization of polypyrrole-hyaluronic acid composite biomaterials for tissue engineering applications, *J. Biomed. Mater. Res.*, 50, 574–584, 2000.

258. Zacchi, V., Soranzo, C., Cortivo, R., Radice, M., Brun, P., and Abatangelo, G., *In vitro* engineering of human skin-like tissue, *J. Biomed. Mater. Res.*, 40, 187–194, 1998.

259. Robinson, D., Halperin, N., and Nevo, Z., Regenerating hyaline cartilage in articular defects of old chickens using implants of embryonal chick chondrocytes embedded in a new natural delivery substance, *Calcif. Tissue. Int.*, 46, 246–253, 1990.

260. Galassi, G., Brun, P., Radice, M., Cortivo, R., Zanon, G.F., Genovese, P., and Abatangelo, G., *In vitro* reconstructed dermis implanted in human wounds: degradation studies of the HA-based supporting scaffold, *Biomaterials*, 21, 2183–2191, 2000.

261. Harris, P.A., di Francesco, F., Barisoni, D., Leigh, I.M., and Navsaria, H.A., Use of hyaluronic acid and cultured autologous keratinocytes and fibroblasts in extensive burns, *Lancet*, 353, 35–36, 1999.

262. Tyrell, D.J., Kilfeather, S., and Page, C.P., Therapeutic uses of heparin beyond its traditional role as an anticoagulant, *Trends Pharmacol. Sci.*, 16, 198–204, 1995.

263. Sears, P., Tolbert, T., and Wong, C.H., Enzymatic approaches to glycoprotein synthesis, *Genet. Eng.*, 23, 45–68, 2001.

264. Pieper, J.S., Hafmans, T., van Wachem, P.B., van Luyn, M.J., Brouwer, L.A., Veerkamp, J.H., and van Kuppevelt, T.H., Loading of collagen-heparan sulfate matrices with bFGF promotes angiogenesis and tissue generation in rats, *J. Biomed. Mater. Res.*, 62, 185–194, 2002.

265. Acarturk, F. and Takka, S., Calcium alginate microparticles for oral administration: II. effect of formulation factors on drug release and drug entrapment efficiency, *J. Microencapsul.*, 16, 291–301, 1999.

266. Takka, S. and Acarturk, F., Calcium alginate microparticles for oral administration: I. effect of sodium alginate type on drug release and drug entrapment efficiency, *J. Microencapsul.*, 16, 275–290, 1999.

267. Sezer, A.D. and Akbuga, J., Release characteristics of chitosan treated alginate beads: I. sustained release of a macromolecular drug from chitosan treated alginate beads, *J. Microencapsul.*, 16, 195–203, 1999.

268. Sezer, A.D. and Akbuga, J., Release characteristics of chitosan treated alginate beads: II. sustained release of a low molecular drug from chitosan treated alginate beads, *J. Microencapsul.*, 16, 687–696, 1999.

269. Torre, M.L., Giunchedi, P., Maggi, L., Stefli, R., Machiste, E.O., and Conte, U., Formulation and characterization of calcium alginate beads containing ampicillin, *Pharm. Dev. Technol.*, 3, 193–198, 1998.

270. Draget, K.I., Skjak-Braek, G., and Smidsrod, O., Alginate based new materials, *Int. J. Biol. Macromol.*, 21, 47–55, 1997.

271. Shiraishi, S., Imai, T., and Otagiri, M., Controlled-release preparation of indomethacin using calcium alginate gel, *Biol. Pharm. Bull.*, 16, 1164–1168, 1993.

272. Becker, T.A., Kipke, D.R., and Brandon, T., Calcium alginate gel: a biocompatible and mechanically stable polymer for endovascular embolization, *J. Biomed. Mater. Res.*, 54, 76–86, 2001.

273. Tapia, C., Costa, E., Terraza, C., Munita, A.M., and Yazdani-Pedram, M., Study of the prolonged release of theopylline from polymeric matrices based on grafted chitosan with acrylamide, *Pharmazie*, 57, 744–749, 2002.

274. Tapia, C., Costa, E., Moris, M., Sapag-Hagar, J., Valenzuela, F., and Basualto, C., Study of the influence of the pH media dissolution, degree of polymerization, and degree of swelling of the polymers on the mechanism of release of diltiazem from matrices based on mixtures of chitosan/alginate, *Drug Dev. Ind. Pharm.*, 28, 217–224, 2002.

275. Lin, S.Y. and Ayres, J.W., Calcium alginate beads as core carriers of 5-aminosalicylic acid, *Pharm. Res.*, 9, 1128–31, 1992.

276. Becker, T.A., Kipke, D.R., Preul, M.C., Bichard, W.D., and McDougall, C.G., In vivo assessment of calcium alginate gel for endovascular embolization of a cerebral arteriovenous malformation model using the Swine rete mirabile, *Neurosurgery*, 51, 453–458; discussion 458–459, 2002.

277. Thomas, S., Alginate dressings in surgery and wound management:—Part 1, *J. Wound Care*, 9, 56–60, 2000.

278. Thomas, S., Alginate dressings in surgery and wound management: Part 2, *J. Wound Care*, 9, 115–119, 2000.

279. Thomas, S., Alginate dressings in surgery and wound management: Part 3, *J. Wound Care*, 9, 163–166, 2000.

280. Stabler, C.L., Sambanis, A., and Constantinidis, I., Effects of alginate composition on the growth and overall metabolic activity of beta TC3 cells, *Ann. NY Acad. Sci.*, 961, 130–133, 2002.

281. Murphy, C.L. and Sambanis, A., Effect of oxygen tension and alginate encapsulation on restoration of the differentiated phenotype of passaged chondrocytes, *Tissue Eng.*, 7, 791–803, 2001.

282. Rastrelli, A., Skin graft polymers, in Dumitriu, S., Ed., *Polymeric Biomaterials*, Marcel Dekker, New York, 1994, pp. 313–324.

283. Pomahac, B., Svensjo, T., Yao, F., Brown, H., and Eriksson, E., Tissue engineering of skin, *Crit. Rev. Oral Biol. Med.*, 9, 333–344, 1998.

284. Lam, P.K., Chan, E.S., To, E.W., Lau, C.H., Yen, S.C., and King, W.W., Development and evaluation of a new composite Laserskin graft, *J. Trauma*, 47, 918–22, 1999.

285. Zacchi, V., Soranzo, C., Cortivo, R., Radice, M., Brun, P., and Abatangelo, G., *In vitro* engineering of human skin-like tissue, *J. Biomed. Mater. Res.*, 40, 187–194, 1998.

286. Bakos, D., Jorge-Herrero, E., and Koller, J., Resorption and calcification of chemically modified collagen/hyaluronan hybrid membranes, *Polim. Med.*, 30, 57–64, 2000.

287. Halbleib, M., Skurk, T., de Luca, C., von Heimburg, D., and Hauner, H., Tissue engineering of white adipose tissue using hyaluronic acid-based scaffolds. I: *in vitro* differentiation of human adipocyte precursor cells on scaffolds, *Biomaterials*, 24, 3125–3132, 2003.

288. Pavesio, A., Abatangelo, G., Borrione, A., Brocchetta, D., Hollander, A.P., Kon, E., Torasso, F., Zanasi, S., and Marcacci, M., Hyaluronan-based scaffolds (Hyalograft C) in the treatment of knee cartilage defects: preliminary clinical findings, *Novartis Found. Symp.*, 249, 203–217; discussion 229–233, 234–238, 239–241, 2003.

289. Yannas, I.V., Burke, J.F., Huang, C., and Gordon, P.L., Correlation of *in vivo* collagen degradation rate with in vitro measurements, *J. Biomed. Mater. Res.*, 9, 623–628, 1975.

290. Yannas, I.V., Burke, J.F., Gordon, P.L., Huang, C., and Rubenstein, R.H., Design of an artificial skin. II. control of chemical composition, *J. Biomed. Mater. Res.*, 14, 107–132, 1980.

291. Sethi, K.K., Yannas, I.V., Mudera, V., Eastwood, M., McFarland, C., and Brown, R.A., Evidence for sequential utilization of fibronectin, vitronectin and collagen during fibroblast-mediated collagen contraction, *Wound Repair Regen.*, 10, 397–408, 2002.

292. Ellis, D.L. and Yannas, I.V., Recent advances in tissue synthesis *in vivo* by use of collagen-glycosaminoglycan copolymers, *Biomaterials*, 17, 291–299, 1996.

293. Nehrer, S., Breinan, H., Ashkar, S., Shortkroff, S., Minas, T., Sledge, C., Yannas, J., and Spector, M., Characteristics of articular chondrocytes seeded in collagen matrices *in vitro*, *Tissue Eng.*, 4, 175–183, 1998.

294. Lee, C.R., Grodzinsky, A.J., and Spector, M., The effects of cross-linking of collagen-glycosaminoglycan scaffolds on compressive stiffness, chondrocyte-mediated contraction, proliferation, and biosynthesis, *Biomaterials*, 22, 3145–3154, 2001.

295. Rong, Y., Sugumaran, G., Silbert, J.E., and Spector, M., Proteoglycans synthesized by canine intervertebral disc cells grown in a Type I collagen-glycosaminoglycan matrix, *Tissue Eng.*, 8, 1037–1047, 2002.

296. Marty-Roix, R., Bartlett, J.D., and Spector, M., Growth of porcine enamel-, dentin-, and cementum-derived cells in collagen-glycosaminoglycan matrices *in vitro*: expression of alpha-smooth muscle actin and contraction, *Tissue Eng.*, 9, 175–86, 2003.

297. Butler, C.E., Navarro, F.A., Park, C.S., and Orgill, D.P., Regeneration of neomucosa using cell-seeded collagen-GAG matrices in athymic mice, *Ann. Plast. Surg.*, 48, 298–304, 2002.

298. Hsu, W.C., Spilker, M.H., Yannas, I.V., and Rubin, P.A., Inhibition of conjunctival scarring and contraction by a porous collagen-glycosaminoglycan implant, *Invest. Ophthalmol. Vis. Sci.*, 41, 2404–2411, 2000.

299. Grzesiak, J.J., Pierschbacher, M.D., Amodeo, M.F., Malaney, T.I., and Glass, J.R., Enhancement of cell interactions with collagen/glycosaminoglycan matrices by RGD derivatization, *Biomaterials*, 18, 1625–1632, 1997.

300. Freyman, T.M., Yannas, I.V., Yokoo, R., and Gibson, L.J., Fibroblast contractile force is independent of the stiffness which resists the contraction, *Exp. Cell Res.*, 272, 153–162, 2002.

301. Shahabeddin, L., Berthod, F., Damour, O., and Collombel, C., Characterization of skin reconstructed on a chitosan-cross-linked collagen-glycosaminoglycan matrix, *Skin Pharmacol.*, 3, 107–114, 1990.

302. Augustin, C., Collombel, C., and Damour, O., Use of dermal equivalent and skin equivalent models for identifying phototoxic compounds *in vitro*, *Photodermatol. Photoimmunol. Photomed.*, 13, 27–36, 1997.

303. Ishihara, M., Obara, K., Ishizuka, T., Fujita, M., Sato, M., Masuoka, K., Saito, Y., Yura, H., Matsui, T., Hattori, H., Kikuchi, M., and Kurita, A., Controlled release of fibroblast growth factors and heparin from photocrosslinked chitosan hydrogels and subsequent effect on *in vivo* vascularization, *J. Biomed. Mater. Res.*, 64, 551–559, 2003.

304. http://www.bdbiosciences.com/discovery_labware/Products/cell_environments _and_ECMs/extracellular_matrix/basement_membrane_matrix/#ref.

305. Gutsche, A.T., Lo, H., Zurlo, J., Yager, J., and Leong, K.W., Engineering of a sugar-derivatized porous network for hepatocyte culture, *Biomaterials*, 17, 387–393, 1996.

306. Matsuda, T. and Magoshi, T., Preparation of vinylated polysaccharides and photofabrication of tubular scaffolds as potential use in tissue engineering, *Biomacromolecules*, 3, 942–950, 2002.

307. von der Mark, K., Gauss, V., von der Mark, H., and Muller, P., Relationship between cell shape and type of collagen synthesized as chondrocytes lose their cartilage phenotype in culture, *Nature*, 267, 531–532, 1977.

308. Davisson, T., Kunig, S., Chen, A., Sah, R., and Ratcliffe, A., Static and dynamic compression modulate matrix metabolism in tissue engineered cartilage, *J. Orthop. Res.*, 20, 842–848, 2002.

309. Hutmacher, D.W., Ng, K.W., Kaps, C., and Sittinger, M., Elastic cartilage engineering using novel scaffold architectures in combination with a biomimetic cell carrier, *Biomaterials*, 24, 4445–4458, 2003.

310. Werkmeister, J.A., Tebb, T.A., Tsai, W.B., Tsai, C.C., Glattauer, V., Chang, K.Y., Thissen, H., and Ramshaw, J.A.M., Control of chondrocyte phenotype for repair of articular cartilage, *12th Annual Conference of the Australian Society for Biomaterials*, Australian National University, Canberra, 2002.

311. Anderer, U. and Libera, J., *In vitro* engineering of human autogenous cartilage, *J. Bone. Miner. Res.*, 17, 1420–1429, 2002.

Index